Biology Now

with Physiology

THIRD EDITION

ANNE HOUTMAN
EARLHAM COLLEGE

MEGAN SCUDELLARI
SCIENCE JOURNALIST

CINDY MALONE
CALIFORNIA STATE UNIVERSITY, NORTHRIDGE

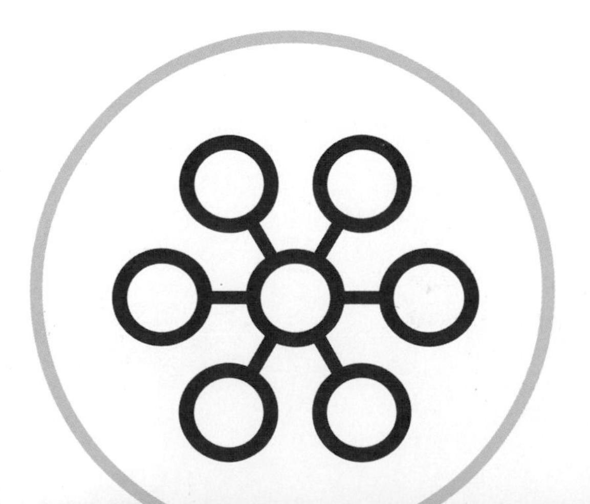

W. W. NORTON & COMPANY
Independent Publishers Since 1923

W. W. Norton & Company has been independent since its founding in 1923, when William Warder Norton and Mary D. Herter Norton first published lectures delivered at the People's Institute, the adult education division of New York City's Cooper Union. The firm soon expanded its program beyond the Institute, publishing books by celebrated academics from America and abroad. By midcentury, the two major pillars of Norton's publishing program—trade books and college texts—were firmly established. In the 1950s, the Norton family transferred control of the company to its employees, and today—with a staff of five hundred and hundreds of trade, college, and professional titles published each year—W. W. Norton & Company stands as the largest and oldest publishing house owned wholly by its employees.

Printed in Canada

Editor: Betsy Twitchell
Assistant Editor: Danny Vargo
Editorial Assistant: Maggie Stephens
Developmental Editor: Kurt Wildermuth
Project Editor: Layne Broadwater
Managing Editor, College: Marian Johnson
Managing Editor, College Digital Media: Kim Yi
Associate Director of Production, College: Benjamin Reynolds
Media Editor: Kate Brayton
Associate Media Editor: Jasmine Ribeaux
Media Editorial Assistant: Kara Zaborowsky
Media Project Editor: Jesse Newkirk
Marketing Manager, Biology: Ruth Bolster
Content Development Specialist: Todd Pearson
Design Director: Jillian Burr
Text Design: Anne DeMarinis
Photo Editor: Ted Szczepanski
Photo Researcher: Fay Torresyap
Permissions Manager: Megan Schindel
Permissions Consultant: Elizabeth Trammell
Composition by Graphic World / Project Manager: Sunil Kumar
Illustrations by Dragonfly Media Group
Manufacturing: Transcontinental—Beauceville QC

Permission to use copyrighted material is included in the backmatter of this book.

Library of Congress Cataloging-in-Publication Data

Names: Houtman, Anne (Anne Michelle), author. | Scudellari, Megan, author. | Malone, Cindy, author.
Title: Biology now with physiology / Anne Houtman, Earlham College, Megan Scudellari, Science Journalist, Cindy Malone, California State University, Northridge.
Description: Third edition. | New York : W.W. Norton & Company, [2021] | Includes bibliographical references and index.
Identifiers: LCCN 2020040773 | **ISBN 9780393422818** (paperback) | **ISBN 9780393533606 (epub)**
Subjects: LCSH: Biology. | Biology—Research—Anecdotes. | Biologists—Anecdotes.
Classification: LCC QH308.2 .H82 2021 | DDC 570.72—dc23
LC record available at https://lccn.loc.gov/2020040773

ISBN 978-0-393-42281-8

W. W. Norton & Company, Inc., 500 Fifth Avenue, New York, NY 10110-0017
wwnorton.com

W. W. Norton & Company Ltd., 15 Carlisle Street, London W1D 3BS

1 2 3 4 5 6 7 8 9 0

Brief Contents

Contents

INTRODUCTION

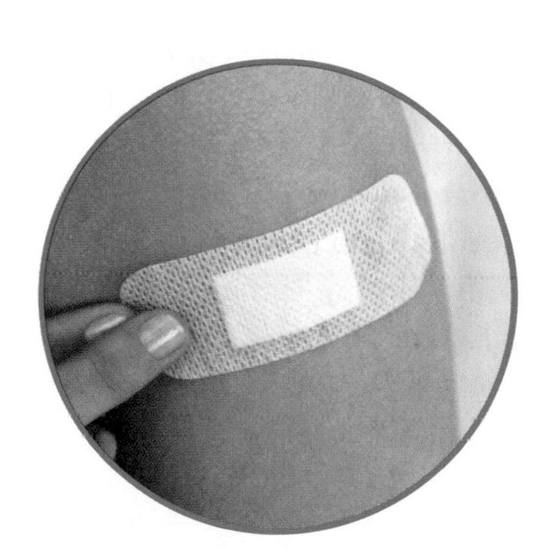

UNIT 1: CELLS

CHAPTER 3: Chemistry of Life 41

Breaking Good

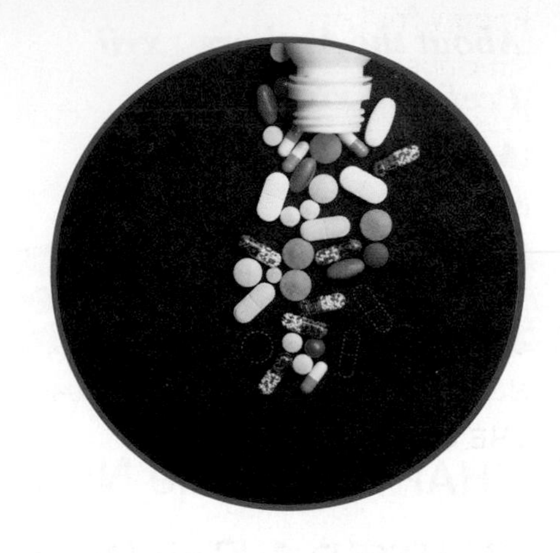

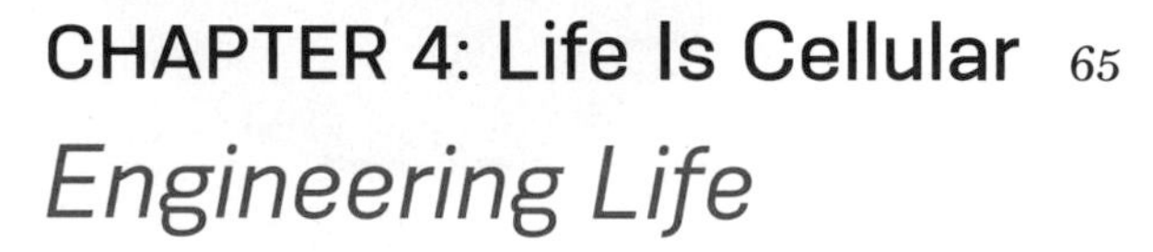

CHAPTER 4: Life Is Cellular 65

Engineering Life

UNIT 2: GENETICS

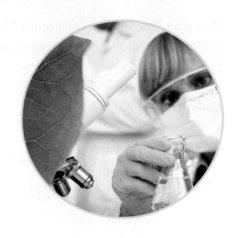

UNIT 3: EVOLUTION

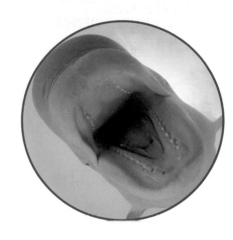

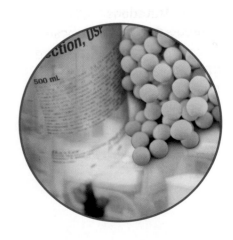

CHAPTER 13: Adaptation and Speciation 245

Penguins on Thin Ice

UNIT 4: BIODIVERSITY

CHAPTER 14: The History of Life 265

Dinosaur Hunters

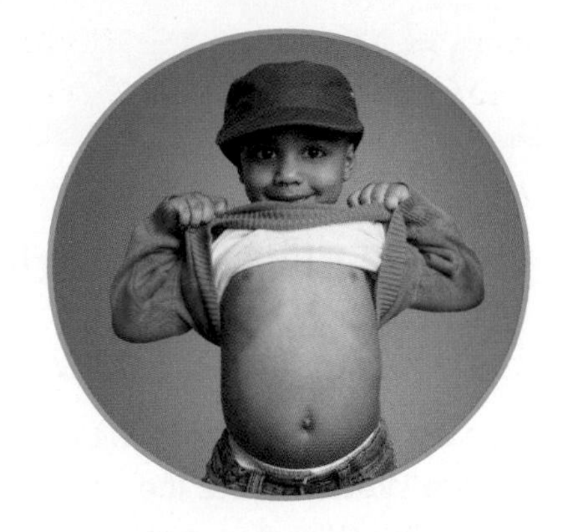

UNIT 5: ECOLOGY

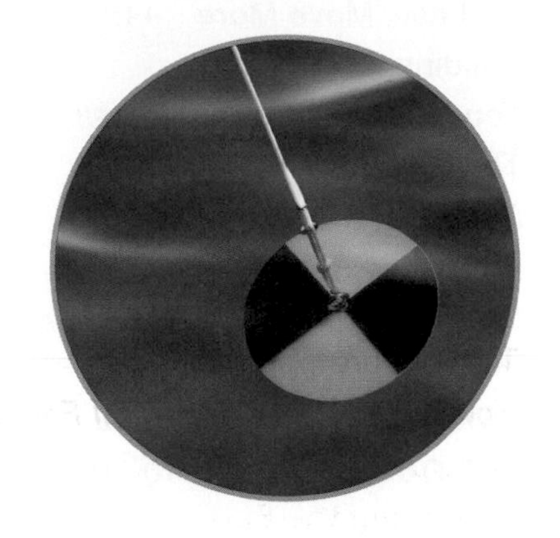

UNIT 6: PHYSIOLOGY

CHAPTER 23: Muscular, Skeletal, and Digestive Systems 443

CHAPTER 24: Circulatory, Respiratory, Urinary, and Nervous Systems 463

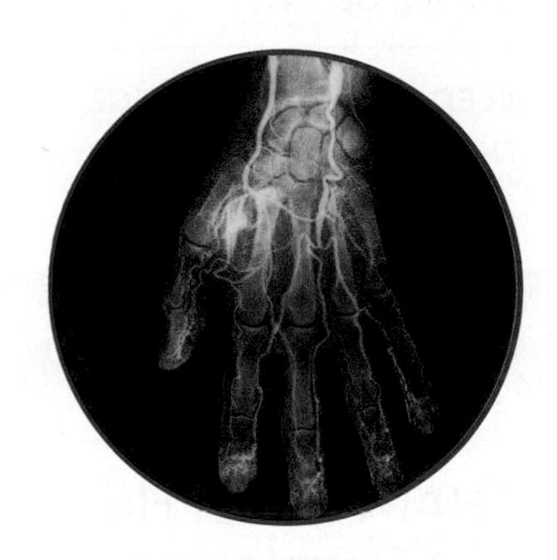

About the Authors

ANNE HOUTMAN is the president of Earlham College and professor of biology. She has over 25 years of experience teaching nonmajors biology at a variety of private and public institutions, which gives her a broad perspective of the education landscape. She is strongly committed to evidence-based, experiential education and has been an active participant in the national dialogue on STEM (science, technology, engineering, and math) education for over 25 years. Anne's research interests are in the ecology and evolution of hummingbirds. She grew up in Hawaii, received her doctorate in zoology from the University of Oxford, and conducted postdoctoral research at the University of Toronto.

MEGAN SCUDELLARI is an award-winning freelance science writer and journalist based in Boston, Massachusetts, specializing in the life sciences. She has contributed to *Newsweek*, *Scientific American*, *Discover*, *Nature*, and *Technology Review*, among others, and she was a health columnist for the *Boston Globe*. For 5 years she worked as a correspondent and later as a contributing editor for *The Scientist* magazine. In 2013, she was awarded the prestigious Evert Clark/Seth Payne Award in recognition of outstanding reporting and writing in science. She has also received accolades for investigative reporting on traumatic brain injury and a feature story on a new class of drugs targeting "undruggable" proteins. Megan received an MS from the Graduate Program in Science Writing at the Massachusetts Institute of Technology and worked as an educator at the Museum of Science, Boston.

CINDY MALONE began her scientific career wearing hip waders in a swamp behind her home in Illinois. She earned her BS in biology at Illinois State University and her PhD in microbiology and immunology at UCLA. She continued her postdoctoral work at UCLA in molecular genetics. She is currently a distinguished educator and a professor at California State University, Northridge, where she is the director of the CSUN-UCLA Stem Cell Scientist Training Program funded by the California Institute for Regenerative Medicine. Her research is aimed at training undergraduates and master's degree candidates to understand how genes are regulated through genetic and epigenetic mechanisms that alter gene expression. She has been teaching nonmajors biology for over 20 years and has won teaching, mentorship, and curriculum enhancement awards at CSUN.

Preface

Biology is a part of so many decisions that students will need to make as individuals and as members of society. It helps them see the value of vaccination, because they will understand what viruses are and how the immune system works. It helps them decide how to respond to the ongoing cleanup from hurricanes, floods, wildfires, and other natural disasters because they will understand how an ecosystem functions. And it helps them make more informed decisions about their own nutrition because they will understand the effects of fat, cholesterol, and vitamins and minerals on our health. The examples are endless. Making informed decisions on these real-world issues requires students to be comfortable with scientific concepts and the process of scientific discovery.

How do we instill that capability in students? The last decade has seen an explosion of research on how students learn best. In a nutshell, they learn best when they see the relevance of a subject to their lives, when they are actively engaged in their learning, and when they are given opportunities to practice critical thinking. Furthermore, they retain knowledge that is shared in the context of a story – which is exactly how *Biology Now* is structured.

Most faculty who teach nonmajors biology would agree that our goal is to introduce students to both the key concepts of biology (for example, cells, DNA, evolution) and the tools to think critically about biological issues. Many would add that they want their students to leave the class with an appreciation for the value of science to society and with an ability to distinguish between science and the nonscience or pseudoscience that bombards them on a daily basis.

How can a textbook help combine the ways students learn best with the goals of a nonmajors biology class? At the most basic level, if students don't read the textbook, they can't learn from it. When students do read them, traditional textbooks are adept at teaching key concepts, and they have recently begun to emphasize the relevance of biology to students' lives. But students may be intimidated by the length of chapters and the amount of difficult text, and they often cannot see the connections between the story and the science. More important, textbooks have not been successful at helping students become active learners and critical thinkers, and none emphasize the process of science or how to assess scientific claims. It was our goal to make *Biology Now* relevant and interactive and to be sure that it emphasized the process of science in short chapters that students *want* to read, while still covering the essential content found in other nonmajors biology textbooks.

Following the model of the First and Second Editions, each chapter in our book covers science concepts important for nonmajors to know, integrated into a current news story about people *doing* science, reported firsthand by Megan, an experienced journalist who specializes in reporting scientific findings in a compelling and accurate way. Anne and Cindy then flesh out the chapter with figures and end-of-chapter materials that support understanding of the content.

For this third edition, we wrote six new stories on pressing, timely events—such as the opioid epidemic and the melting of Greenland—that will help instructors keep their courses grounded in real-world events. We also added updated content requested by adopters of our first and second editions with additional data questions in the end-of-chapter materials, as instructors felt these were so helpful to students.

We sincerely hope you enjoy the fruits of our long labors.

Anne Houtman
Megan Scudellari
Cindy Malone

What's New in the Third Edition?

- New chapter stories on fresh, engaging topics such as HPV vaccination, opioid addiction, and glacial melting in Greenland. New stories include the following:

Chapter 2: Evaluating Scientific Claims—Sex, Cancer, and a Vaccine
Laura Brennan was only 25 when she was diagnosed with terminal cervical cancer, a disease that can be prevented through HPV vaccination. Her successful campaign to increase vaccination rates in Ireland and promote scientific literacy embodies the spirit of this book and the aims of this chapter.

Chapter 3: Chemistry of Life—Breaking Good
Edwin Chindongo, a recent college graduate with a new job, became addicted to opioids after receiving a doctor's prescription for oxycodone for nerve damage in his legs. Alongside Edwin's harrowing journey into addiction and out the other side, this chapter describes neuroscientist Dr. Laura Bohn's research to make opioids less addictive by altering their chemical structure.

Chapter 8: Chromosomes and Human Genetics—Curing the Incurable
Spinal muscular atrophy, sickle cell disease, and severe combined immunodeficiency are just three of many diseases that were untreatable before the rise of gene therapy. In this updated chapter, we tell three new stories of children with these devastating genetic diseases and the science behind their treatments.

Chapter 13: Adaptation and Speciation—Penguins on Thin Ice
When a team of scientists from New Zealand set out to map the family tree of penguins, they found a surprisingly strong correlation between when species diverged and new islands emerged, illustrating the concept of speciation. In this chapter, flamboyant and regal penguins guide us through the process of how organisms adapt to their environments.

Chapter 18: Introduction to Ecology—Climate Meltdown
As part of NASA's OMG project, glaciologist Josh Willis and a team survey the alarming speed of melting ice in Greenland. From the skies and mountains of Greenland, this chapter explores the biosphere, climate, water and carbon cycles, and global warming with a strong focus on the science of measuring and mitigating glacial melting.

Chapter 23: Muscular, Skeletal, and Digestive Systems—Made to Move
Just how bad is not exercising for your health? In this chapter, Cleveland Clinic cardiologist Wael Jaber investigates why the life expectancy of Americans has been declining. Using data from a 20-year volunteer study, we follow Jaber and others through an understanding of the benefits of exercise on the muscular, skeletal, and digestive systems.

- An expanded end-of-chapter review, "Need-to-Know Science," condenses essential information from each chapter in a digestible, useful format.
- New nutrition content, added to Chapters 3 and 23, emphasizes the science behind healthy eating—and why it matters.
- A new feature in every chapter called "Debunked!" presents common myths about science and concise explanations of the facts. This feature, supported by published scientific papers cited in the back of the book, reminds students to evaluate scientific claims that they encounter in everyday life.
- A new icon in the end-of-chapter section signposts questions that ask students to analyze, interpret, and make inferences using data.
- New online Evaluating Scientific Claims activities take students step-by-step through the process of evaluating scientific claims to help them develop the scientific literacy they need to make informed choices in their everyday lives.
- Updated Smartwork and InQuizitive courses include animation, interactive and visually based questions that promote critical thinking, interaction with data, and engagement with biology in the real world.
- New resources are available in *Biology Now*'s online Interactive Instructor's Guide, which provides instructors with the ability to easily search for active learning resources by chapter, phrase, topic, and type of resource.

The perfect balance of science and story

Every chapter leads the reader through an important, timely story about people doing science, stimulating students' curiosity about biological concepts and motivating them to read.

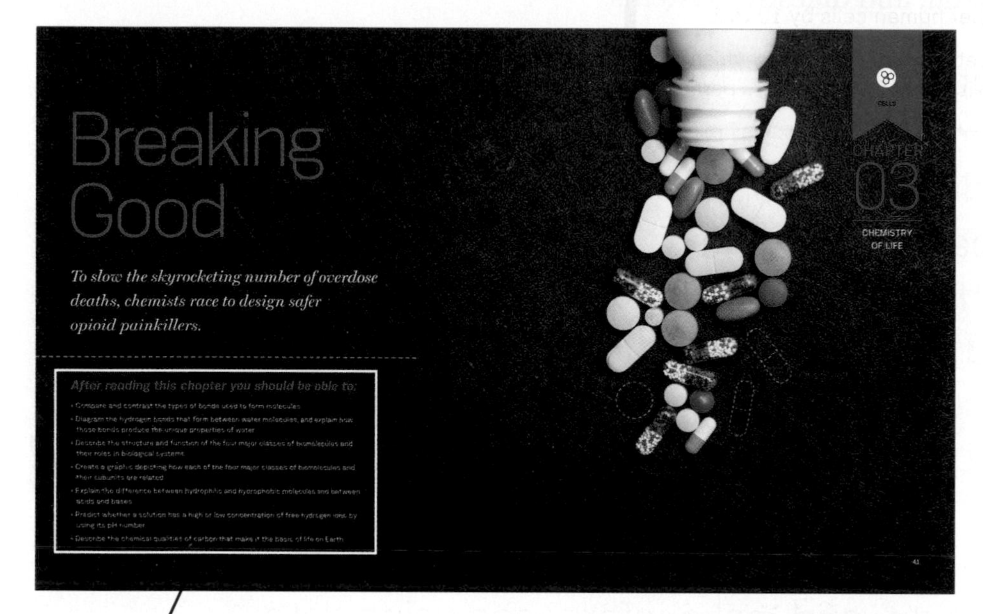

Dynamic chapter-opening spreads inspired by each chapter's story draw students into the material.

"After reading this chapter you should be able to" introduces learning outcomes that preview the concepts presented in each chapter.

DEBUNKED! presents common myths about science and concise explanations of the facts.

DEBUNKED!

MYTH: Adults can't grow new brain cells.

FACT: In 2013, researchers in Europe found—and others have since confirmed—that the adult human brain structure associated with the formation of memories (called the hippocampus) produces new neurons via mitosis throughout life, even in old age. In addition, exercise, sex, and stress relief seem to enhance the number of newly made neurons.

DEBUNKED!

MYTH: Bacteria cells in our bodies outnumber human cells by 10 to 1.

FACT: According to a 2016 study, the ratio between microbes and human cells is approximately 1 to 1, although it varies from person to person.

DEBUNKED!

MYTH: Five-second rule! Food dropped on the floor and left for less than 5 seconds is safe to eat because microbes need time to transfer.

FACT: Some bacteria transfer onto food instantaneously, in less than 1 second, so think twice about eating that cookie off the floor.

DEBUNKED!

MYTH: Blood in veins is blue.

FACT: Human blood is always red. It can look blue through our skin due to an optical illusion resulting from how tissues absorb light and how our eyes see color. Some animals do have blue blood, however, such as lobsters and squid.

DEBUNKED!

MYTH: Women can't get pregnant if they have sex while on their period.

FACT: It's unlikely, but if the timing is right, a woman can become pregnant while menstruating (see **Figure 22.8**). Because sperm can live inside the human body for up to 5 days, if a woman ovulates early, or more than once per cycle as happens in approximately 10 percent of women, sperm could find and fertilize an egg.

An inquiry-based approach that builds science skills—asking questions, thinking visually, and interpreting data

Most **figures** in the book are accompanied by three questions that promote understanding and encourage engagement with the visual content. Answers are provided at the back of the book, making the questions a useful self-study tool.

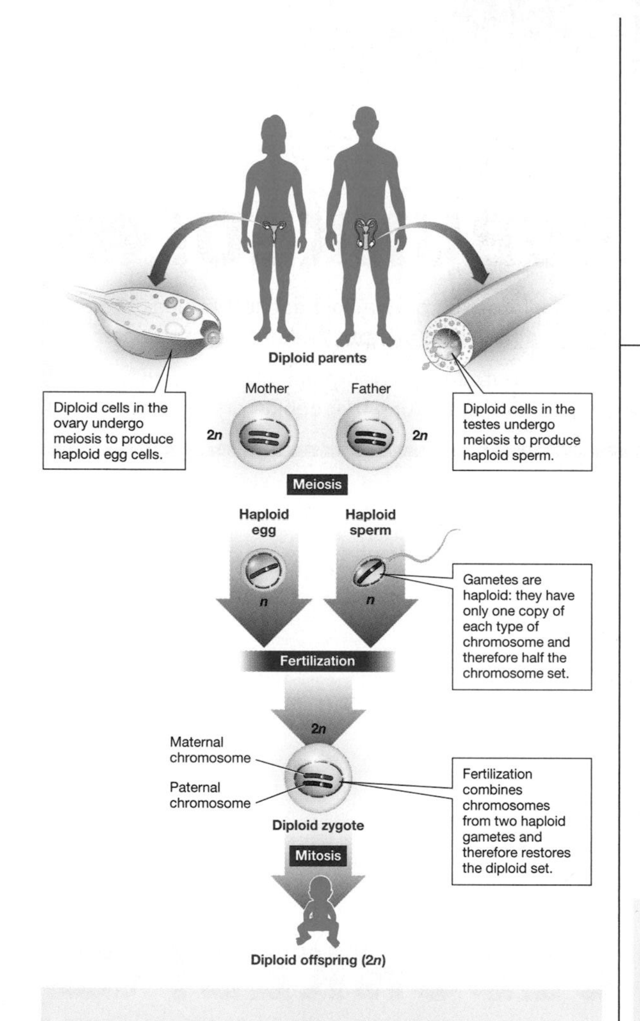

Q1: Is a zygote haploid or diploid?

Q2: Which cellular process creates a baby from a zygote?

Q3: How might long-term, significant exposure to BPA experienced by a mother or father prior to conceiving a child explain potential birth defects in the fetus?

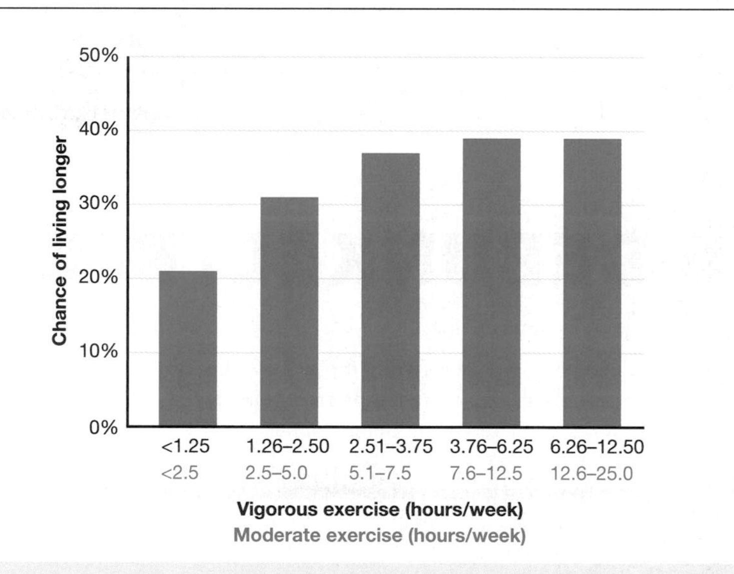

Q1: Participants who vigorously exercised 1.25–2.5 hours per week or moderately exercised 2.5–5.0 hours per week, as recommended by the guidelines, increased their chance of living longer by what percentage?

Q2: How much moderate exercise increased participants' chance of living longer by 20 percent? How much moderate exercise maxed out participants' increased chance of living longer? What was the maximum percentage chance?

Q3: Previous studies suggested that 12.5 hours of vigorous exercise per week, such as elite athletes commonly perform, can be harmful. Does this study support that conclusion? Why?

Engaging, data-driven **infographics** appear in every chapter. Topics range from opioid overdoses (Chapter 3) to the effects of climate change on extinction rates (Chapter 18) and many more. The infographics expose students to scientific data in an engaging way.

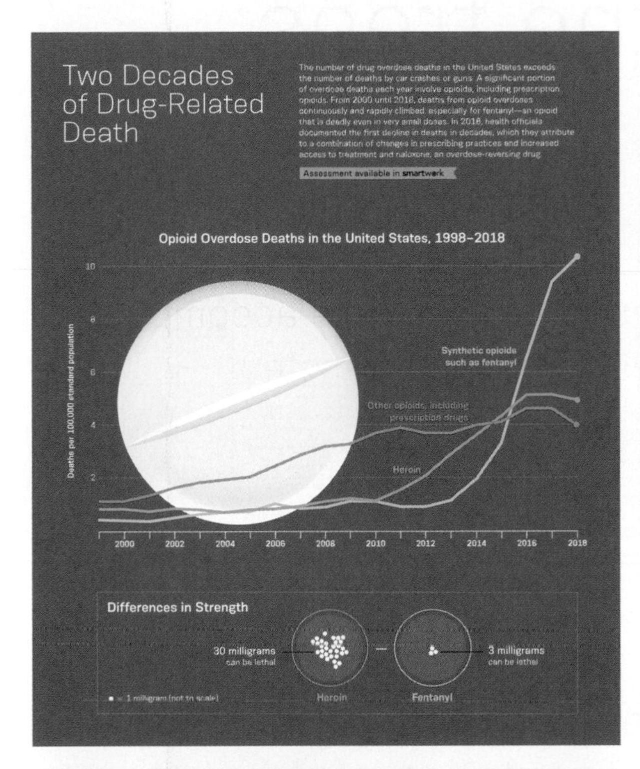

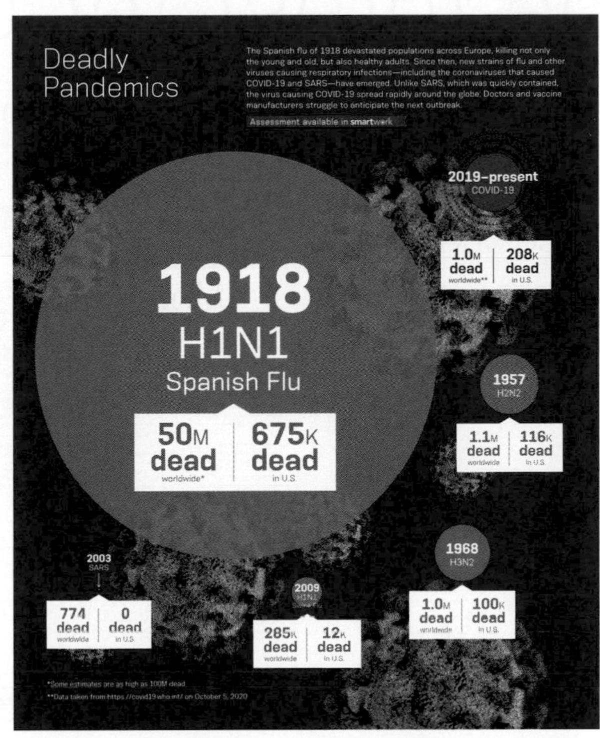

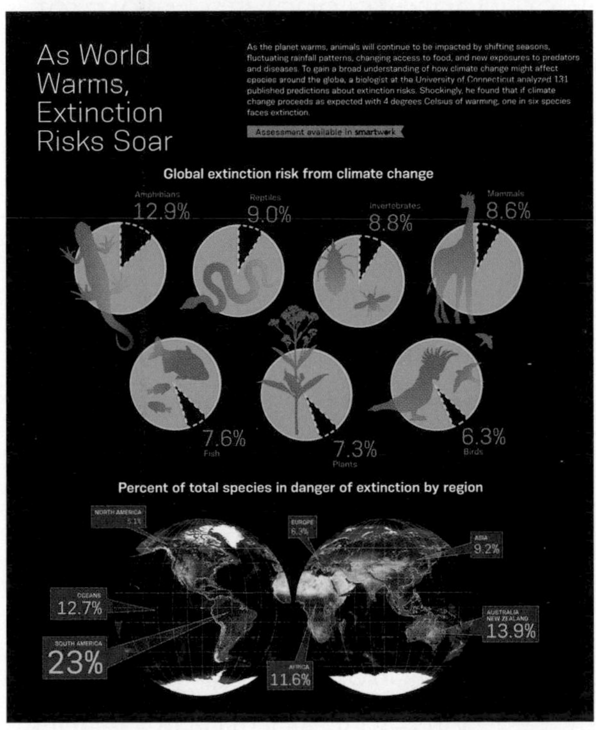

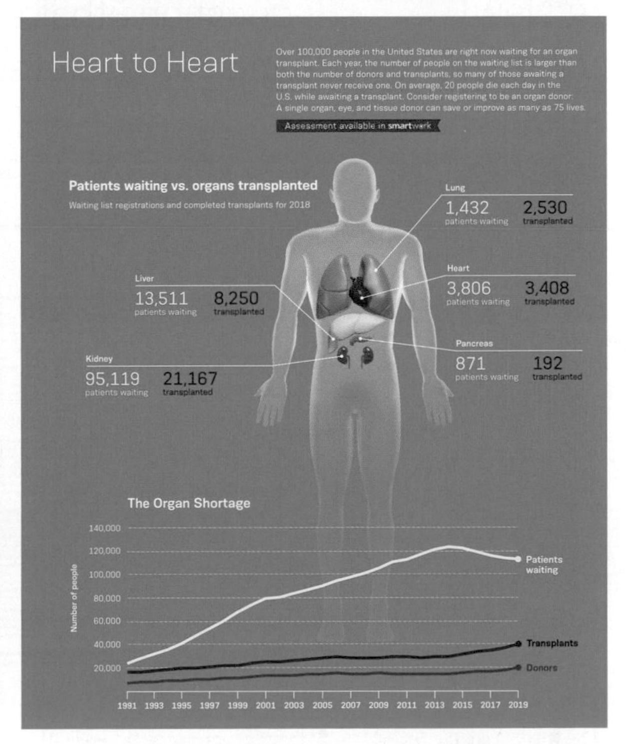

Extensive end-of-chapter review ensures that students see the forest for the trees.

Need-to-Know Science identifies each chapter's key terms and science concepts, providing students with a guide for studying.

NEED-TO-KNOW SCIENCE

- The physical world is composed of **matter** (page 42), which is anything that has mass and occupies space. One type of matter is an element, and 98 distinct **elements** (page 42) exist naturally on Earth. An **atom** (page 43) is the smallest unit of an element that maintains its unique properties. At its core, every atom has a **nucleus** (page 43), which contains positively charged **protons** (page 43) and uncharged **neutrons** (page 43), and is surrounded by negatively charged **electrons** (page 43).
- The number of protons in an atom's nucleus is called its **atomic number** (page 43) and is unique to each element. **Isotopes** (page 43) of an element—atoms of the same species—have the same number of protons but different numbers of neutrons. The sum of the number of protons and the number of neutrons is the **atomic mass number** (page 43) of an isotope. The atomic mass number reflects the amount of mass in an element.
- The chemical interactions that cause atoms to associate with each other are known as **chemical bonds** (page 45). When an atom loses or gains electrons, it becomes, respectively, a positively or negatively charged **ion** (page 46). Ions of opposite charge are held together by **ionic bonds** (page 46).
- **Covalent bonds** (page 45) are formed by the sharing of electrons between atoms. A **molecule** (page 43) contains at least two atoms that are held together by chemical bonds.
- **Hydrogen bonds** (page 46) are weak associations between two molecules such that a partially positive hydrogen atom within one molecule is attracted to a partially negative region of another molecule.
- **Chemical compounds** (page 43) are molecules that contain atoms from at least two different elements. Carbon atoms can link with each other and with other atoms to generate a great diversity of chemical compounds called **organic molecules** (page 43). Multiple organic molecules bound together are an **organic compound** (page 43).
- The four main types of large organic molecules, or **biomolecules** (page 47), include the following:
  - **Proteins** (page 47) are **polymers** (page 47) of the 20 amino acid **monomers** (page 47) in countless different combinations; they make possible the endless functions in the cell.
  - **Carbohydrates** (page 47) range in size from simple monosaccharides and disaccharides to complex polymers that may contain thousands of monomers.
  - **Nucleic acids** (page 47) are polymers of nucleotide monomers: DNA is composed of deoxyribonucleotides, and RNA is composed of ribonucleotides.
  - **Lipids** (page 49) are not polymers and are better known as fats, oils, and steroids.
- Proteins, carbohydrates, and lipids are important considerations in **nutrition** (page 52), the study of what we eat.
- Partial electrical charges result from the unequal sharing of electrons between atoms, giving rise to **polar molecules** (page 51). The polarity of individual water molecules and the hydrogen bonding across water molecules explain nearly all of the special properties of water.
- A compound that mixes completely with water is said to be **soluble** (page 51). A **solution** (page 51) is any combination of a dissolved substance, known as the **solute** (page 51), and a fluid into which the solute has dissolved, known as the **solvent** (page 51).
- Ions and polar molecules are **hydrophilic** (page 52); they readily dissolve in water. Nonpolar molecules cannot associate with water and are therefore **hydrophobic** (page 52).
- When liquid water heats, it transitions into gas. During this process, called **evaporation** (page 53), hydrogen bonds break, and molecules spread apart. During the opposite process, called **condensation** (page 53), water vapor cools, molecules slowly re-form hydrogen bonds, and the gas returns to liquid.
- In **chemical reactions** (page 56), bonds between atoms are formed or broken. **Reactants** (page 56), the participants in a chemical reaction, are modified to give rise to **products** (page 56), new ions or molecules.
- An **acid** (page 56) is a hydrophilic compound that loses one or more hydrogen ions (H^+) when dissolved in water. The chemical opposite of an acid is a **base** (page 56), a hydrophilic compound that gains one or more H^+ ions when it dissolves in water. The concentration of free H^+ ions in water is expressed by the **pH scale** (page 56) and reflects whether a solution is **acidic** (page 56), **basic** (alkaline; page 56), or neutral.

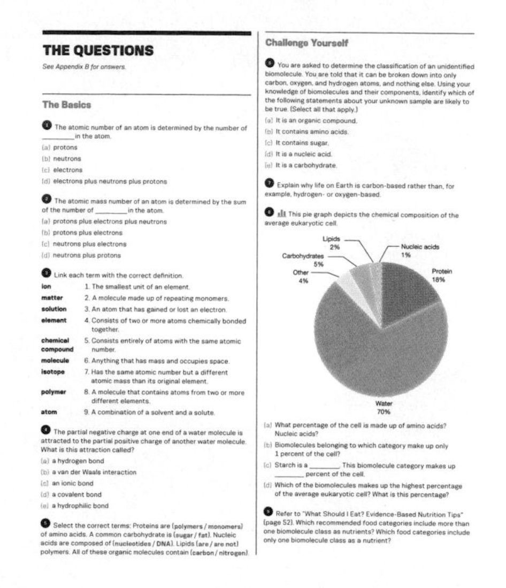

THE QUESTIONS

See Appendix B for answers.

The Basics

The atomic number of an atom is determined by the number of ________ in the atom.
(a) protons
(b) neutrons
(c) electrons
(d) electrons plus neutrons plus protons

The atomic mass number of an atom is determined by the sum of the number of ________ in the atom.
(a) protons plus electrons plus neutrons
(b) protons plus electrons
(c) neutrons plus electrons
(d) neutrons plus protons

Link each term with the correct definition.

ion	1. The smallest unit of an element.
matter	2. A molecule made up of repeating monomers.
solution	3. An atom that has gained or lost an electron.
element	4. Consists of two or more atoms chemically bonded together.
chemical compound	5. Consists entirely of atoms with the same atomic number.
molecule	6. Anything that has mass and occupies space.
isotope	7. Has the same atomic number but a different atomic mass than its original element.
polymer	8. A molecule that contains atoms from two or more different elements.
atom	9. A combination of a solvent and a solute.

The partial negative charge at one end of a water molecule is attracted to the partial positive charge of another water molecule. What is this attraction called?
(a) a hydrogen bond
(b) a van der Waals interaction
(c) an ionic bond
(d) a covalent bond
(e) a hydrophilic bond

Select the correct terms: Proteins are (**polymers / monomers**) of amino acids. A common carbohydrate is (**sugar / fat**). Nucleic acids are composed of (**nucleotides / DNA**). Lipids (**are / are not**) polymers. All of these organic molecules contain (**carbon / nitrogen**).

Challenge Yourself

You are asked to determine the classification of an unidentified biomolecule. You are told that it can be broken down into only carbon, oxygen, and hydrogen atoms, and nothing else. Using your knowledge of biomolecules and their components, identify which of the following statements about your unknown sample are likely to be true. (Select all that apply.)
(a) It is an organic compound.
(b) It contains amino acids.
(c) It contains sugar.
(d) It is a nucleic acid.
(e) It is a carbohydrate.

Explain why life on Earth is carbon-based rather than, for example, hydrogen- or oxygen-based.

This pie graph depicts the chemical composition of the average eukaryotic cell.

(a) What percentage of the cell is made up of amino acids? Nucleic acids?
(b) Biomolecules belonging to which category make up only 1 percent of the cell?
(c) Starch is a ________. This biomolecule category makes up ________ percent of the cell.
(d) Which of the biomolecules makes up the highest percentage of the average eukaryotic cell? What is this percentage?

Refer to "What Should I Eat? Evidence-Based Nutrition Tips" (page 52). Which recommended food categories include more than one biomolecule class as nutrients? Which food categories include only one biomolecule class as a nutrient?

End-of-chapter questions follow Bloom's taxonomy, moving from review (The Basics), to synthesis (Try Something New), to critical thinking (Challenge Yourself), to application (Leveling Up).

A new data interpretation icon signposts questions that ask students to analyze, interpret, and make inferences using data.

Leveling Up questions, based on questions the authors use in their classrooms, prompt students to relate biology concepts to their own lives. The questions focus on one of the following themes: "Doing science," "Is it science?," "Life choices," "Looking at data," "What do *you* think?," and "*Write Now* biology."

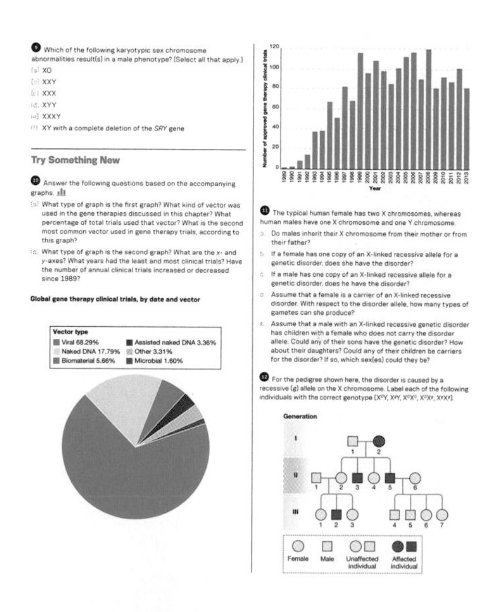

Which of the following karyotypic sex chromosome abnormalities result(s) in a male phenotype? (Select all that apply.)
(a) XO
(b) XXY
(c) XXX
(d) XYY
(e) XXXY
(f) XY with a complete deletion of the *SRY* gene

Try Something New

Answer the following questions based on the accompanying graphs.
(a) What type of graph is the first graph? What kind of vector was used in the gene therapies discussed in this chapter? What percentage of total trials used that vector? What is the second most common vector used in gene therapy trials, according to this graph?
(b) What type of graph is the second graph? What are the x- and y-axes? What years had the least and most clinical trials? Have the number of annual clinical trials increased or decreased since 1989?

Global gene therapy clinical trials, by date and vector

The typical human female has two X chromosomes, whereas human males have one X chromosome and one Y chromosome.
a. Do males inherit their X chromosome from their mother or from their father?
b. If a female has one copy of an X-linked recessive allele for a genetic disorder, does she have the disorder?
c. If a male has one copy of an X-linked recessive allele for a genetic disorder, does he have the disorder?
d. Assume that a female is a carrier of an X-linked recessive disorder. With respect to the disorder allele, how many types of gametes can she produce?
e. Assume that a male with an X-linked recessive genetic disorder has children with a female who does not carry the disorder allele. Could any of their sons have the genetic disorder? How about their daughters? Could any of their children be carriers for the disorder? If so, which sex(es) could they be?

For the pedigree shown here, the disorder is caused by a recessive (*g*) allele on the X chromosome. Label each of the following individuals with the correct genotype (X^GY, X^gY, X^GX^G, X^GX^g, X^gX^g).

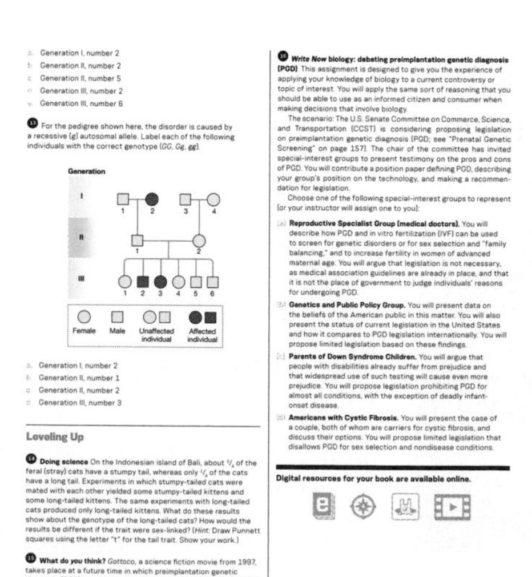

a. Generation I, number 2
b. Generation II, number 2
c. Generation II, number 5
d. Generation III, number 2
e. Generation III, number 6

For the pedigree shown here, the disorder is caused by a recessive (*g*) autosomal allele. Label each of the following individuals with the correct genotype (*GG*, *Gg*, *gg*).

a. Generation I, number 2
b. Generation II, number 1
c. Generation II, number 2
d. Generation III, number 3

Leveling Up

Doing science On the Indonesian island of Bali, about 3/4 of the feral (stray) cats have a stumpy tail, whereas only 1/4 of the cats have a long tail. Experiments in which stumpy-tailed cats were mated with each other yielded some stumpy-tailed kittens and some long-tailed kittens. The same experiments with long-tailed cats produced only long-tailed kittens. What do these results show about the genotype of the long-tailed cats? How would the results be different if the trait were sex-linked? (*Hint:* Draw Punnett squares using the letter "t" for the tail trait. Show your work.)

What do you think? *Gattaca*, a science fiction movie from 1997, takes place at a future time in which preimplantation genetic diagnosis (PGD) is used to create genetically superior children, and social status is determined mainly by one's DNA. Watch the movie and take careful notes as you observe how PGD has been misused by society. Do you think these outcomes are likely if we continue to develop PGD? Are the benefits worth this risk?

***Write Now* biology: debating preimplantation genetic diagnosis (PGD)** This assignment is designed to give you the experience of applying your knowledge of biology to a current controversy or topic of interest. You will apply the same sort of reasoning that you should be able to use as an informed citizen and consumer when making decisions that involve biology.

The scenario: The U.S. Senate Committee on Commerce, Science, and Transportation (CCST) is considering proposing legislation on preimplantation genetic diagnosis (PGD; see "Prenatal Genetic Screening" on page 157). The chair of the committee has invited special-interest groups to present testimony on the pros and cons of PGD. You will contribute a position paper defining PGD, describing your group's position on the technology, and making a recommendation for legislation.

Choose one of the following special-interest groups to represent (or your instructor will assign one to you):

(a) **Reproductive Specialist Group (medical doctors).** You will describe how PGD and in vitro fertilization (IVF) can be used to screen for genetic disorders or for sex selection and "family balancing," and to increase fertility in women of advanced maternal age. You will argue that legislation is not necessary, as medical association guidelines are already in place, and that it is not the place of government to judge individuals' reasons for undergoing PGD.
(b) **Genetics and Public Policy Group.** You will present data on the beliefs of the American public in this matter. You will also present the status of current legislation in the United States and how it compares to PGD legislation internationally. You will propose limited legislation based on these findings.
(c) **Parents of Down Syndrome Children.** You will argue that people with disabilities already suffer from prejudice and that widespread use of such testing will cause even more prejudice. You will propose legislation prohibiting PGD for almost all conditions, with the exception of deadly infant-onset disease.
(d) **Americans with Cystic Fibrosis.** You will present the case of a couple, both of whom are carriers for cystic fibrosis, and discuss their options. You will propose limited legislation that disallows PGD for sex selection and nondisease conditions.

Digital resources for your book are available online.

Powerful resources for teaching and assessment

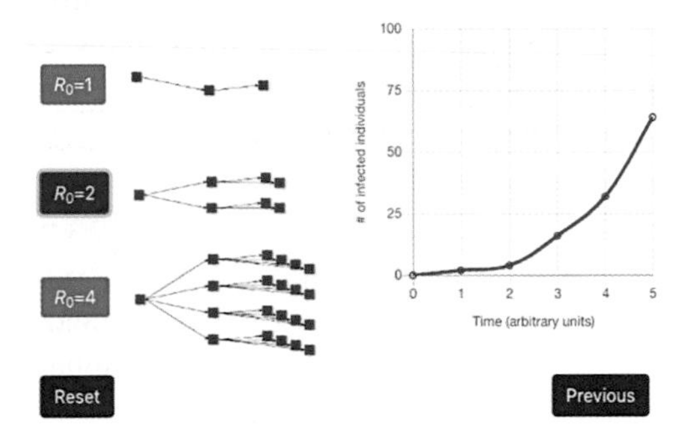

NEW online Evaluating Scientific Claims activities take students step-by-step through the process of evaluating scientific claims to help them develop the scientific literacy they need to make informed choices in their everyday lives. These activities are available in the ebook and can be assigned for a grade in InQuizitive and Smartwork.

The **Interactive Instructor's Guide** helps instructors bring *Biology Now*'s inquiry-based approach into the classroom by delivering a wealth of resources within a searchable database. Useful in a variety of classroom sizes and setups, including online learning, these resources include suggested online videos with discussion questions, activities with downloadable handouts, descriptions of animations with discussion questions, Leveling Up rubrics, lecture PowerPoints featuring clicker questions, and more.

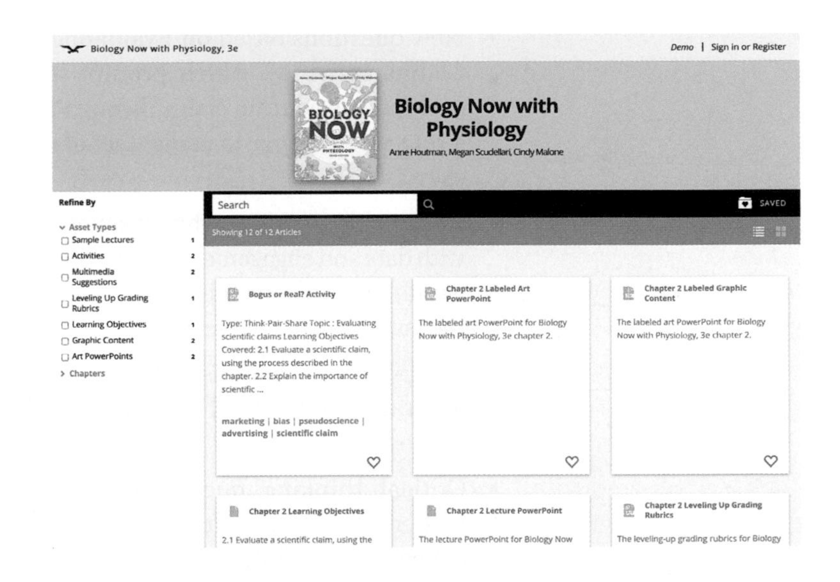

Other presentation tools for instructors

InQuizitive InQuizitive is Norton's easy-to-use, adaptive-learning, and quizzing tool that improves student understanding of important learning objectives. A variety of question types, including animation questions, story-based questions, and critical thinking questions, test students' knowledge in different ways across the learning objectives in each chapter. Students are motivated by the engaging game-like elements that allow them to set their confidence level on each question to reflect their knowledge, track their own progress, earn bonus points for high performance, and review learning objectives they might not have mastered. Instructors can assign InQuizitive activities out of the box or use simple tools to customize the learning objectives they want students to work on or how much time they want students to spend on each module.

Smartwork Smartwork delivers engaging, interactive online homework to students, helping instructors and students reach their teaching and learning goals. The Smartwork course for *Biology Now* includes the following:

- New questions based on Evaluating Scientific Claims activities, which prompt students to evaluate a scientific claim through the lens of a real-world scenario using a set of interactive exercises.
- Infographic questions, which promote interaction with data and engagement with biology in the real world, while making this popular visual feature of the text an assignable activity.
- Story-based questions, which help students learn and understand the science behind the stories in the text.
- Critical thinking questions, which prompt students to think critically about important concepts in biology.
- Animation questions, which engage students with the book-specific animations covering biology concepts.

Resources for your LMS You can easily add high-quality Norton digital resources to your online, hybrid, or lecture courses. Get started building your course with our easy-to-use integrated resources; all activities can be accessed within your existing learning management system.

Ebook Norton ebooks give students and instructors an enhanced reading experience at a fraction of the cost of a print textbook. Students are able to have an active reading experience and can take notes, bookmark, search, highlight, and read offline. Instructors can even add their own notes for students to see as they read the text. Norton ebooks can be viewed on—and synced among—all computers and mobile devices.

Animations Key concepts and processes are explained clearly through high-quality, ADA-compliant animations developed from the meticulously designed art in the book. These animations are available for lecture presentation in the Interactive Instructor's Guide, are embedded in InQuizitive and Smartwork questions, and are linked to the ebook.

Test Bank Each chapter's test bank includes 75 or more questions structured around the learning objectives from the textbook and conforms to Bloom's taxonomy. Questions are further classified by text section and difficulty and are provided in multiple-choice, fill-in-the-blank, and short-answer forms. Infographic questions in every chapter help test student interpretation of charts and graphs. Norton Testmaker brings Norton's high-quality testing materials online. Create assessments from anywhere with an Internet connection, without installing specialized software. Search and filter test bank questions by chapter, type, difficulty, learning objective, and other criteria. You can also customize questions to fit your course. Then, easily export your tests to Microsoft Word or Common Cartridge files for import into your LMS.

Squarecap Classroom Response Questions Squarecap is a next-generation classroom response system that promotes a high level of student engagement in the classroom through quality assessment in a variety of question types. Students can answer questions directly from their smartphones; the Squarecap system offers instructors in-depth analytics on student performance in real time. *Biology Now* can be packaged with review questions specific to each chapter.

Art Files All art and photos from the book are available, in presentation-ready resolution, as both JPEGs and PowerPoints for instructor use.

Lecture Slides Complete lecture PowerPoints thoroughly cover chapter concepts and include images and clicker questions to encourage student engagement.

Acknowledgments

We could not have created this textbook without the enthusiasm and hard work of many people. First and foremost, we'd like to thank our indefatigable editor, Betsy Twitchell, for her market-driven approach and terrific visual sense. Katie Callahan and Danny Vargo ensured we had feedback on each and every chapter and kept us moving forward through many challenges. Thanks to Kurt Wildermuth for his focused commitment to improving the readability of this textbook and to Andrew Sobel for his careful reading of the finished chapters.

Thank you to our talented project editor, Layne Broadwater, for tirelessly working to improve the layout of this book and for keeping our chapters moving. Thank you to our talented copy editor, Donna Mulder, for being so meticulous with our manuscript and so pleasant to work with.

We are grateful to photo researcher Fay Torresyap for her reliable and creative work and to Ted Szczepanski for managing the photo process. Production manager Ben Reynolds skillfully oversaw the translation of our raw material into the beautiful book you hold in your hands; he, too, has our thanks. Special thanks to design director Jillian Burr and cover designer Eric Nyquist for creating such an extraordinary and truly gorgeous book.

Media editor Kate Brayton, associate editor Jasmine Ribeaux, and media assistant Kara Zaborowsky worked so hard to create the instructor and student resources accompanying our book. Their determination, creativity, and positive attitude resulted in supplements of the highest quality that will truly make an impact on student learning. Jesse Newkirk's commitment to quality as media project editor ensured that every element of the resource package meets Norton's high standards.

We appreciate the analytical mind of content development specialist Todd Pearson and his efforts to improve the supplements for this book. Thanks also to director of marketing Steve Dunn and marketing manager Ruth Bolster for her creativity. We thank director of sales Erik Fahlgren and every single one of Norton's extraordinary salespeople for spreading the word about our book. Finally, we thank Marian Johnson, Julia Reidhead, Roby Harrington, Drake McFeely, Mike Wright, and everyone at Norton for believing in our book.

We would be remiss not to thank also all of our colleagues in the field who gave their time and expertise in reviewing, class testing, and contributing to *Biology Now* and its many supplements and resources. Thank you all.

Reviewers

Third Edition

Marilyn Baguinon, Kutztown University
Tiffany Bensen, University of Mississippi
Linda Berlin, Seminole State College of Florida
Jennifer Bess, Hillsborough Community College
Margie Beucher, Utah Valley University
Cal Borden, Saginaw Valley State University
Kelly Burke, College of the Canyons
Lisa Carloye, Washington State University
Michelle Cawthorn, Georgia Southern University
Alyson Center, Normandale Community College
Alexander Cheroske, Moorpark College
Kimberly Cline-Brown, University of Northern Iowa
Susan Dalterio, University of Texas at San Antonio

Jill Devito, University of Texas at Arlington
Joseph D'Silva, Norfolk State University
Kamal Dulai, University of California Merced
Steve Edwards, Central Oregon Community College
Eric Ekdale, San Diego State University
Sara Emerson, California State University, Stanislaus
Robert Ewy, SUNY Potsdam
Robert Fowler, San Jose State University
Michele Garrett, Guilford Technical Community College
Tracy Gaskin, Coastal Carolina University
Danielle Goodspeed, Sam Houston State University
William Gosnell, University of Cincinnati Clermont College
Tamar Goulet, University of Mississippi
Krista Hahn, California State University Fullerton
Rachael Hannah, University of Alaska
Janet Harouse, New York University
Brynn Heckel, California State University Dominguez Hills
Tina-Maria Hopper, Missouri State University
Virginia Irintcheva, Truckee Meadows Community College
Evelyn Jackson, University of Mississippi
Jamie Jensen, Brigham Young University
Carl Johansson, Fresno City College
Anthony Jones, Tallahassee Community College
Nikolay Kukushkin, New York University
Amy Kutay, Lock Haven University
Ann LeMaster, Ivy Tech Community College
Cynthia Littlejohn, University of Southern Mississippi
Krista Lucas, Pepperdine University
Michael Marlen, Southwestern Illinois College
Jill Maroo, University of Northern Iowa
Roy Mason, Mount San Jacinto College
Catarina Mata, Borough of Manhattan Community College
Loren Matthews, Georgia Southern University
Kathy McCarthy, Duxbury High School
Gabrielle McLemore, Morgan State University
Mindy Murray, San Jacinto College
Terina Nusinov, Seminole State College of Florida
Eva Nyutu, Saginaw Valley State University
Kevin Oley, Appalachian State University
Margaret Olney, Saint Martin's University
Wiline Pangle, Central Michigan University
Chris Parker, Texas Wesleyan University
Rachel Portinga, University of Wisconsin
Vanessa Quinn, Purdue University
Sami Raut, University of Alabama
Mario Raya, Bristol Community College
Lisa Regula Meyer, University of Akron
Debra Rinne, Seminole State College of Florida
Jennifer Ripley-Stueckle, West Virginia University
Anne Rizzacasa, Tahquitz High School
Sydha Salihu, West Virginia University
Pramila Sen, Houston Community College
Wendy Sera, Houston Community College
Erin Shanle, Longwood University
Viji Sitther, Morgan State University
Brad Smit, Saugatuck Public Schools
Jessica Snoberger, Cape Fear Community College
Anna Bess Sorin, University of Memphis
Sunni Taylor, Texas State University
Chad Thompson, SUNY Westchester Community College
Jeffrey Walck, Middle Tennessee State University
Jeff Wesner, University of South Dakota, Main Campus
Clay White, Lone Star College
Satya Witt, University of New Mexico
Julie Zeller, Birmingham Charter School
Joyce Zimmer, Greenville High School

Second Edition

Anne Artz, Preuss School, University of California San Diego
Allan Ayella, McPherson College
Erin Baumgartner, Western Oregon University
Joydeep Bhattacharjee, University of Louisiana, Monroe
Rebecca Brewer, Troy High School
Victoria Can, Columbia College Chicago
Lisa Carloye, Washington State University, Pullman
Michelle Cawthorn, Georgia Southern University
Craig Clifford, Northeastern State University

Beth Collins, Iowa Central Community College
Julie Constable, California State University, Fresno
Gregory A. Dahlem, Northern Kentucky University
Danielle M. DuCharme, Waubonsee Community College
Robert Ewy, SUNY Potsdam
Clayton Faivor, Ellsworth Community School
Michael Fleming, California State University, Stanislaus
Kathy Gallucci, Elon University
Kris Gates, Pikes Peak Community College
Heather Giebink, Pennsylvania State University
Candace Glendening, University of Redlands
Sherri D. Graves, Sacramento City College
Cathy Gunther, University of Missouri
Meshagae Hunte-Brown, Drexel University
Douglas P. Jensen, Converse College
Ragupathy Kannan, University of Arkansas–Fort Smith
Julia Khodor, Bridgewater State University
Jennifer Kloock, Garces Memorial High School
Karen L. Koster, University of South Dakota
Dana Robert Kurpius, Elgin Community College
Joanne Manaster, University of Illinois
Mark Manteuffel, St. Louis Community College
Jill Maroo, University of Northern Iowa
Tsitsi McPherson, SUNY Oneonta
Kiran Misra, Edinboro University of Pennsylvania
Jeanelle Morgan, University of North Georgia
Lori Nicholas, New York University
Fran Norflus, Clayton State University
Christopher J. Osovitz, University of South Florida
Christopher Parker, Texas Wesleyan University
Brian K. Paulson, California University of Pennsylvania
Carolina Perez-Heydrich, Meredith College
Thomas J. Peri, Notre Dame Preparatory School
Kelly Norton Pipes, Wilkes Early College High School
Gordon Plague, SUNY Potsdam
Benjamin Predmore, University of South Florida
Jodie Ramsey, Highland High School
Logan Randolph, Polk State College
Debra A. Rinne, Seminole State College of Florida
Michael L. Rutledge, Middle Tennessee State University
Celine Santiago Bass, Kaplan University
Steve Schwendemann, Iowa Central Community College
Sonja Stampfler, Kellogg Community College
Jennifer Sunderman Broo, Saint Ursula Academy
J. D. Swanson, Salve Regina University
Heidi Tarus, Colby Community College
Larchinee Turner, Central Carolina Technical College
Ron Vanderveer, Eastern Florida State College
Calli A. Versagli, Saint Mary's College
Mark E. Walvoord, University of Oklahoma
Lisa Weasel, Portland State University
Derek Weber, Raritan Valley Community College
Danielle Werts, Golden Valley High School
Elizabeth Wright, Athenian School
Steve Yuza, Neosho County Community College

First Edition

Joseph Ahlander, Northeastern State University
Stephen F. Baron, Bridgewater College
David Bass, University of Central Oklahoma
Erin Baumgartner, Western Oregon University
Cindy Bida, Henry Ford Community College
Charlotte Borgeson, University of Nevada, Reno
Bruno Borsari, Winona State University
Ben Brammell, Eastern Kentucky University
Christopher Butler, University of Central Oklahoma
Stella Capoccia, Montana Tech

Kelly Cartwright, College of Lake County
Emma Castro, Victor Valley College
Michelle Cawthorn, Georgia Southern University
Jeannie Chari, College of the Canyons
Jianguo Chen, Claflin University
Beth Collins, Iowa Central Community College
Angela Costanzo, Hawai'i Pacific University
James B. Courtright, Marquette University
Danielle DuCharme, Waubonsee Community College
Julie Ehresmann, Iowa Central Community College
Laurie L. Foster, Grand Rapids Community College
Teresa Golden, Southeastern Oklahoma State University
Sue Habeck, Tacoma Community College
Janet Harouse, New York University
Olivia Harriott, Fairfield University
Tonia Hermon, Norfolk State University
Glenda Hill, El Paso Community College
Vicki J. Huffman, Potomac State College, West Virginia University
Carl Johansson, Fresno City College
Victoria Johnson, San Jose State University
Anthony Jones, Tallahassee Community College
Hinrich Kaiser, Victor Valley College
Vedham Karpakakunjaram, Montgomery College
Dauna Koval, Bellevue College
Maria Kretzmann, Glendale Community College
MaryLynne LaMantia, Golden West College
Brenda Leady, University of Toledo
Lisa Maranto, Prince George's Community College
Roy B. Mason, Mt. San Jacinto College
Gabrielle L. McLemore, Morgan State University
Paige Mettler-Cherry, Lindenwood University
Rachel Mintell, Manchester Community College
Kiran Misra, Edinboro University of Pennsylvania
Lori Nicholas, New York University
Louise Mary Nolan, Middlesex Community College
Fran Norflus, Clayton State University
Brian Paulson, California University of Pennsylvania
Carolina Perez-Heydrich, Meredith College
Ashley Ramer, University of Akron
Nick Reeves, Mt. San Jacinto College
Tim Revell, Mt. San Antonio College
Eric Ribbens, Western Illinois University
Kathreen Ruckstuhl, University of Calgary
Michael L. Rutledge, Middle Tennessee State University
Brian Sato, University of California Irvine
Malcolm D. Schug, University of North Carolina at Greensboro
Craig M. Scott, Clarion University of Pennsylvania
J. Michael Sellers, University of Southern Mississippi
Marieken Shaner, University of New Mexico
David Sheldon, St. Clair County Community College
Jack Shurley, Idaho State University
Daniel Sigmon, Alamance Community College
Molly E. Smith, South Georgia State College, Waycross
Lisa Spring, Central Piedmont Community College
Steven R. Strain, Slippery Rock University of Pennsylvania
Jeffrey L. Travis, SUNY Albany
Suzanne Wakim, Butte College
Mark E. Walvoord, University of Oklahoma
Sherman Ward, Virginia State University
Lisa Weasel, Portland State University
Jennifer Wiatrowski, Pasco-Hernando State College
Rachel Wiechman, West Liberty University
Bethany Williams, California State University, Fullerton
Satya M. Witt, University of New Mexico
Donald A. Yee, University of Southern Mississippi

Focus Group Participants

Michelle Cawthorn, Georgia Southern University
Marc Dal Ponte, Lake Land College
Kathy Gallucci, Elon University
Tamar Goulet, University of Mississippi
Sharon Gusky, Northwestern Connecticut Community College
Krista Henderson, California State University, Fullerton
Tara Jo Holmberg, Northwestern Connecticut Community College
Brenda Hunzinger, Lake Land College
Jennifer Katcher, Pima Community College
Cynthia Kay-Nishiyama, California State University, Northridge
Kathleen Kresge, Northampton Community College
Sharon Lee-Bond, Northampton Community College
Suzanne Long, Monroe Community College
Boriana Marintcheva, Bridgewater State University
Roy B. Mason, Mt. San Jacinto College
Gwen Miller, Collin College
Kimo Morris, Santa Ana College
Fran Norflus, Clayton State University
Tiffany Randall, John Tyler Community College
Gail Rowe, La Roche College
J. Michael Sellers, University of Southern Mississippi
Uma Singh, Valencia College
Patti Smith, Valencia College
Bethany Stone, University of Missouri
Willetta Toole-Simms, Azusa Pacific University
Bethany Williams, California State University, Fullerton

Class Test Participants

Bruno Borsari, Winona State University
Jessica Brzyski, Georgia Southern University
Beth Collins, Iowa Central Community College
Christopher Collumb, College of Southern Nevada
Jennifer Cooper, University of Akron
Julie Ehresmann, Iowa Central Community College
Michael Fleming, California State University, Stanislaus
Susan Holecheck, Arizona State University
Dauna Koval, Bellevue College
Kiran Misra, Edinboro University of Pennsylvania
Marcelo Pires, Saddleback College
Michael L. Rutledge, Middle Tennessee State University
Jack Shurley, Idaho State University
Uma Singh, Valencia College
Paul Verrell, Washington State University
Daniel Wetzel, Georgia Southern University
Rachel Wiechman, West Liberty University

Instructor and Student Resource Contributors

Holly Ahern, SUNY Adirondack
Steven Christenson, Brigham Young University–Idaho
Beth Collins, Iowa Central Community College
Julie Ehresmann, Iowa Central Community College
Jenny Gernhart, Iowa Central Community College
Julie Harless, Lone Star College
Janet Harouse, New York University
Vedham Karpakakunjaram, Montgomery College
Dauna Koval, Bellevue College
Brenda Leady, University of Toledo
Boriana Marintcheva, Bridgewater State University
Paige Mettler-Cherry, Lindenwood University
Lori Nicholas, Northwest Vista College
Christopher Osovitz, University of South Florida
Tiffany Randall, John Tyler Community College
Lori Rose, Sam Houston State University
Suzanne Wakim, Butte College
Bethany Williams, California State University, Fullerton

Laura Manno Christian, Rensselaer Polytechnic Institute
James Rudnicky, Thomas Nelson Community College
Grace Mavodza, John Tyler Community College
Debra Rinne, Seminole State College of Florida
Erin Baumgartner, Western Oregon University
Mark Walvoord, University of Oklahoma
Chad Smith, Southwestern Illinois College
Paul Strode, Fairview High School
William Heyborne, Southern Utah University
Jill Maroo, University of Northern Iowa
Amy Dewald, Eureka College
Megan O'Brien, Pine Crest School
Jenna Scheub, Interlochen Center for the Arts
Constance Pope, University of Dayton
Karen Cruse Suder, Summit Country Day School
Lisa Carloye, Washington State University
Amy Stockman, Cameron University
Joel Piperberg, Millersville University
Matt Badtke, Jackson College
Matthew Abbott, Des Moines Area Community College
Margaret Olney, Saint Martin's University
William Gosnell, University of Cincinnati
Christopher Parker, Texas Wesleyan University
Suzanna Brauer, Appalachian State University

This book wouldn't have happened without Anne's husband, Will, who took care of every single other thing in her life so that she could write while still holding down a day job. His support and love and that of her children, Abi, Ben, and Al, are what keep her going every day. Megan thanks her husband, Ryan, and three children, May, Parker, and Tyler, who keep a much-loved, crayon-covered copy of this textbook in their bedroom bookshelf and ask to read chapters before bedtime. Cindy thanks her husband, Mike, and her children, Ben and Lily, along with their numerous pets for the chaotic lifestyle that inspired her to step up her game. Cindy also thanks her friends and students who laugh at her jokes and keep her grounded in reality.

Perhaps most of all, we are indebted to the many scientists and citizen-scientists who shared their time and stories for these chapters. To the women and men we interviewed for this book, we thank you. Your stories will inspire the next generation of biologists.

Biology Now

with Physiology

Caves of Death

Scientists scramble to identify a mysterious scourge decimating bat populations.

After reading this chapter you should be able to:

- Caption a diagram of the scientific method, identifying each step in the process.
- Develop a hypothesis from a given observation and suggest one or more predictions based on that hypothesis.
- Design an experiment using appropriate variables, treatments, and controls.
- Interpret data collected during a scientific study and identify appropriate conclusions.
- Give specific examples of a scientific fact and a scientific theory.
- Create a graphic showing the levels of biological organization.

SCIENCE

CHAPTER 01

THE NATURE OF SCIENCE

Every spring for 30 years, Alan Hicks laced up his hiking boots, packed his camera, and set out to count bats in caves in upstate New York. A scientist with the New York State Department of Environmental Conservation, Hicks led one of the few efforts in the country to collect annual data on bat populations. Since 1980, he had never missed the annual cave trip—until March 17, 2007.

"That day, of all days in my entire career, I stayed at my desk," recalls Hicks, who had remained behind to write a report for his supervisor. A couple of hours after his crew left to inspect some local caves, 15 miles from the Albany office, Hicks's cell phone rang.

"Hey, Al. Something weird is going on here," said a nervous voice. "We've got dead bats. Everywhere."

The line went quiet. "What are we talking here?" asked Hicks. "Hundreds of dead bats?"

"No," said the voice. "Thousands."

At first, Hicks conjectured that the bats had died in a flood, which had happened in that particular cave before. But the next day, a young volunteer who had been out with the team told Hicks to check his email. The volunteer had sent him a picture taken the day before of eight little brown bats (*Myotis lucifugus*) hanging from a cave outcropping. Each one had a fuzzy white nose. This was a surprise because little brown bats do not have white noses.

Hicks emailed the picture to every bat researcher he knew. The fuzzy white material looked like a fungus (plural "fungi")—a type of living organism that includes things like yeast and mushrooms. Yet there was no previous record of a fungus killing bats. As scientist after scientist looked at the picture, they all replied the same way: "What is that?" Hicks resolved to find out what was killing the bats and whether the white fuzz was involved.

DEBUNKED!

MYTH: Bats are blind.

FACT: Bats have eyes and can see quite well, especially in low-light conditions such as dusk and dawn. In fact, bat vision is better than human vision at those times of day. The phrase "blind as a bat" may have originated from the fact that bats fly in erratic patterns, as if they cannot see where they are going.

Why was Hicks so interested in saving the bats? Hicks was a biologist, a scientist who studies organisms. He knew that bats devour insects that would otherwise destroy agricultural crops and forests (see "Bug Zappers" on page 14). Bats also eat mosquitoes, the most deadly animal to humans; by transmitting malaria and other insect-borne diseases, mosquitoes kill hundreds of thousands of people each year.

As a biologist, Hicks took a scientific view of the world—relying on logic, striving for objectivity, and valuing evidence over other ways of discovering the truth. **Science** is a body of knowledge about the natural world, but it is much more than just a mountain of data. Science is an evidence-based process for acquiring that knowledge.

- Science deals with the natural world, which can be detected, observed, and measured.
- Science is based on evidence that can be demonstrated through observations and/or experiments.
- Science is subject to independent validation and peer review.
- Science is open to challenge by anyone at any time on the basis of evidence.
- Science is a self-correcting enterprise.

To gather knowledge, Hicks would apply the **scientific method** (**Figure 1.1**). The scientific method is not a set recipe that scientists follow rigidly. Instead, the term is meant to capture the core logic of how science works. Some people prefer to speak of the **process of science** rather than the scientific method. Whatever we call it, the practices that produce scientific knowledge can be applied across a broad range of disciplines—including bat biology.

Keep in mind that, as powerful as the scientific method is, it is restricted to seeking natural causes to explain the workings of our world. There are other areas of inquiry that science cannot address. The scientific method cannot tell us what is morally right or wrong. For example, science can inform us about the differences between humans and other animals, but it cannot identify the morally correct way to act on that information. Science also cannot address the existence of God or any other supernatural being. Nor can it tell us what is beautiful or ugly, which poems are most lyrical, or which paintings

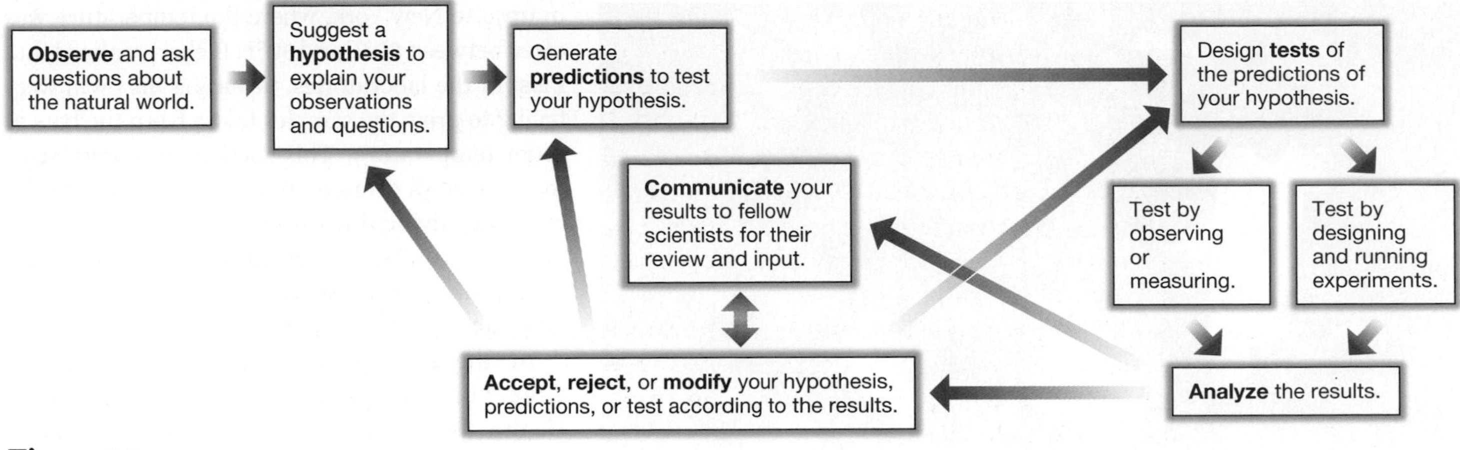

Figure 1.1

The scientific method

The scientific method is a logical process that helps us learn more about the natural world.

Q1: What were the original observation and question of the scientists studying the sick bats?

Q2: At what point in the scientific method would a scientist decide on the methods to use to test a hypothesis?

Q3: How might you explain the process of science to someone who complains that "scientists are always changing their minds; how can we trust what they say?"

See Appendix A for answers to the figure questions.

are most inspiring. So, although science exists comfortably alongside different belief systems—religious, political, and personal—it cannot answer all questions.

But science is the best way to answer questions about the natural world. The first two steps of the scientific method are to *gather observations* and *form a hypothesis*. Hicks didn't waste a moment of time before applying the scientific method to the question of the white fuzz. Bats were dying. "Bats are part of the planet and vital members of the ecosystem," says Hicks. "They play an important role in the environment in which we live."

Bat Crazy

On March 18, the day after the first dead bats were discovered, Hicks entered the cave to make observations—a key part of the scientific process. An **observation** is a description, measurement, or record of any object or phenomenon. Hicks's team observed that the sick bats had not only white noses but also depleted fat reserves, meaning that the bats did not have enough stored energy to get through the winter. The bats also had white fuzz on their wings with scarred and dying wing tissue, and they were behaving abnormally, waking up early from hibernation and leaving the cave when it was still too cold outside to hunt.

Hicks's team also observed that the illness cut across species—many different types of bats were getting sick—and the bats exhibited a high rate of death: in some locations, up to 97 percent of infected bats died. Hicks and others began to call the illness white-nose syndrome (WNS).

"For the first few years, we were just sleuthing," says Paul Cryan, a research biologist with the U.S. Geological Survey (USGS), and one of the scientists who received the original picture Hicks had sent by email. From that first picture, Cryan was involved in trying to pinpoint the cause. "We were trying to understand something that had never happened before in a group of animals that was poorly understood."

In the caves, Hicks began collecting dead bats and sending them to laboratories around the United States. In those labs, technicians scraped samples from the bats' noses and wings, rubbed the samples into petri dishes (shallow glass or plastic plates containing nutrients used

Figure 1.2

Preparing to enter the bat cave
Scientists suit up to collect more observations on the infected bats and the environmental conditions in the bats' roosting cave.

Q1: Which step(s) in the scientific method does this photograph illustrate?

Q2: What types of environmental data might the researchers have collected in the cave?

Q3: Why would the researchers in this photograph have worn protective gear?

See Appendix A for answers to the figure questions.

to grow microorganisms), and watched to see whether the white fuzz would grow. Time after time, many different types of bacteria and fungi grew on the dishes, speckling them with dots of different-colored colonies, but none of the samples were unusual. Nothing special or dangerous appeared to be present on the bats.

One researcher, a young microbiologist named David Blehert, decided to try something different. Blehert worked at the USGS National Wildlife Health Center in Madison, Wisconsin. In December 2007, Hicks called Blehert, who listened carefully as Hicks described how WNS was spreading. "He said we had a major problem on our hands," recalls Blehert. "It turns out he was 100 percent right."

Hicks described to Blehert the conditions under which the bats lived during hibernation in caves in upstate New York, where the temperature was often between 30°F and 50°F. Blehert realized that most of the laboratories, including his own, were trying to grow the samples taken from the bats at room temperature. This method was conducive to the growth of many types of fungi. But in the caves, any living thing would have to grow at cold temperatures. So Blehert and his technicians took samples from dead bats, put them on petri dishes, and placed the dishes in the fridge.

At the same time, Melissa Behr, then an animal disease specialist at the New York State Health Department, accompanied Hicks on a trip to a local cave (**Figure 1.2**). Behr swabbed a sample of the white fuzz directly from a bat in the cave, immediately spread it onto a glass slide, and looked at it under a microscope. A unique fungus was visible in little white fuzzy patches of cells. Up close, the individual spores of this single-celled organism appeared crescent-shaped—different from all the other "normal" microbes growing on the bats' skin and different from any fungus known to the researchers.

But Behr's single observation wasn't enough evidence to convince anyone that the strange-looking fungus was the cause of WNS. To be useful in science, an observation must be repeatable, preferably by multiple techniques. Independent observers should be able to see or detect the same object or phenomenon, at least some of the time.

In this case, Blehert was able to reproduce Behr's results by an independent technique. After letting his petri dishes sit in the fridge for a few weeks, Blehert removed them and observed white patches of the same strange, crescent-shaped fungal spores. "OK, we now have in laboratory culture what Melissa captured when she collected white material in the caves," thought Blehert. "We've got it."

Prove Me Wrong

In science, as in everyday life, observations lead to questions, and questions lead to potential explanations. For example, if you flip on a light switch but the light does not turn on, you wonder why, and then you look for an explanation: Is the lamp plugged in? Has the lightbulb burned out? You then identify one of these explanations as the most likely hypothesis for why the light did not turn on.

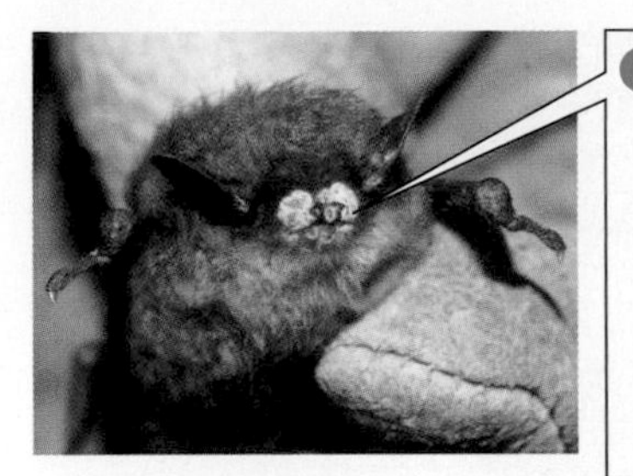

2 **Hypothesis:** Bats with white noses are infected with a fungus, and this fungus is causing death.

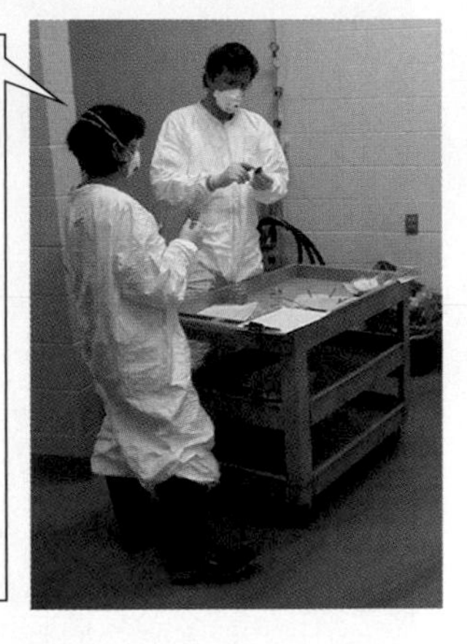

3 **Predictions:** *If* the white noses are caused by a transmissible fungus, *then* healthy bats that hibernate in contact with affected bats should develop the condition. *If* the white noses are caused by a deadly fungus, *then* healthy bats inoculated with the fungus should develop white noses and die at higher rates.

Figure 1.3

Experimental design

Researchers follow a path from observation to hypothesis to testable prediction.

A scientific **hypothesis** (plural "hypotheses") is an informed, logical, and plausible explanation for observations of the natural world. From the start, Hicks hypothesized that a new, cold-loving fungus was the primary cause of death in the bats. After observing the unique crescent-shaped fungal spores, Behr and Blehert agreed with this hypothesis. "It was the simplest solution," says Blehert. "We had bats with a white fungus that nobody had ever seen before growing on them, so that was the most likely thing that was doing it."

But other scientists disagreed. A fungus itself is rarely deadly to a mammal. More often, a fungus causes an annoying, but not lethal, skin infection, or is a secondary response after an animal gets sick from a viral or bacterial infection. So scientists proposed other hypotheses for the cause of WNS. Some suggested the fungus was a secondary effect of an underlying condition, such as a viral infection. Others hypothesized that an environmental contaminant, such as a pesticide, was the cause of death. "There were so many different hypotheses," says Cryan. "But that's what is beautiful about the scientific process. You observe as much as you can, and from those observations you can form multiple hypotheses. Science doesn't proceed by just landing on the right hypothesis the first time."

One of the joys, and challenges, of the scientific method is that after scientists suggest competing hypotheses, they then test their own hypotheses against those of others. A scientific hypothesis must be constructed in such a way that it is potentially **falsifiable**, or refutable. In other words, it must make predictions that can be clearly determined to be true or false (**Figure 1.3**). A well-constructed hypothesis is precise enough to make predictions that can be expressed as "if . . . then" statements.

For example, *if* WNS is caused by a transmissible fungus, *then* healthy bats that hibernate in contact with affected bats should develop the condition. *If* the fungus is secondary to an underlying condition, *then* the infection will occur in bats only after the primary underlying condition is present. *If* an environmental contaminant is the cause, *then* bats with WNS symptoms will have elevated levels of that contaminant in their blood or on their skin.

In each "if . . . then" case, it is possible to design tests able to demonstrate that a prediction is true or false. Although predictions can be shown to be true or false, the same is not true of hypotheses. Hypotheses can be *supported*, but no amount of testing can *prove* a hypothesis is correct with complete certainty (**Figure 1.4**).

The reason a hypothesis cannot be proved is that there might be another factor, unmeasured or unobserved, that explains why the prediction is true. For example, consider the first prediction stated in the previous paragraph—that healthy bats hibernating in contact with affected bats will develop WNS. If this is true, the reason might be that the healthy bats were infected by a fungus from their neighbor, supporting the hypothesis that the disease is caused by a transmissible fungus. Alternatively, related bats may tend to hibernate together in the same cave, and the disease, or at least vulnerability to it, might be genetically based. The hypothesis that the disease is fungal is *supported* but not *proved* by the correctness of this prediction.

Blehert set out to test the hypothesis that he, Behr, and Hicks had put forward: that a unique, cold-loving fungus was the primary cause of death in the bats. One can test a hypothesis through observational studies or experimental studies. Blehert's first study was observational. Observational studies

Figure 1.4

Hypotheses are supported or not supported but never proved

This vintage advertisement for cigarettes may seem ridiculously dated, but "science" is still used to sell products. Most Americans see thousands of advertisements every day, and many of these make "scientific" claims that are exaggerated or inaccurate.

Q1: According to this advertisement, what hypothesis was scientifically tested?

Q2: What prediction comes from this hypothesis? Is this prediction testable? Why or why not?

Q3: Why can't the hypothesis in this ad be "proved"?

See Appendix A for answers to the figure questions.

can be purely **descriptive**—reporting information (**data**) about what is found in nature. Observational studies can also be **analytical**—looking for (analyzing) patterns in the data and addressing how or why those patterns came to exist. **Statistics**—a branch of mathematics that can measure the reliability of data—helps scientists determine how well the patterns support the hypothesis. Observational studies usually rely on both descriptive and analytical methods to test predictions that follow from a hypothesis.

In 2009, Blehert, Behr, and Hicks published a scientific paper in which they described the results from inspecting 117 dead bats. They identified microscopic damage caused by a specific kind of fungus in 105 of the bats and isolated and identified the fungus from a subset of 10 of them. It was a type of cold-loving fungus belonging to a group of fungi called *Geomyces*. They named this new species *Geomyces destructans*.

Their observational study revealed a relationship between white fungus on the noses of bats and bat illness and death. Observational studies suggest possible causes for a phenomenon, but they do not establish a cause-and-effect relationship. To demonstrate that the fungus was actually causing the illness—and not just correlated with it—Blehert designed and conducted an experiment. Testing scientific hypotheses often involves both observational and experimental approaches (**Figure 1.5**).

Catching the Culprit

An **experiment** is a repeatable manipulation of one or more aspects of the natural world. Blehert's experiment was to take healthy bats into his laboratory and expose them to the fungus. Like analytical observational studies, experimental studies use statistics to determine whether the experimental results support or refute the hypothesis being tested.

In studying nature—whether through observations, experiments, or both—scientists focus on **variables**. Variables are characteristics of an object or an individual organism that can change. In a scientific experiment, a researcher typically manipulates a single variable, known as the **independent variable**. In this case, Blehert's independent variable was fungal exposure. Some bats were exposed; others were not. A **dependent variable** is any variable that responds, or could potentially respond, to changes in the independent variable. Blehert's dependent variable was any sign of WNS on the healthy bats.

How do these variables help establish a cause-and-effect relationship? If we think of the independent variable as the cause, then the dependent variable is the effect. In the most basic experimental design, a researcher manipulates a single

One research team

Three research approaches

Descriptive:
Scientist wearing night goggles examines bats for evidence of WNS.

Analytical:
Bats are weighed and measured to evaluate health impacts of WNS.

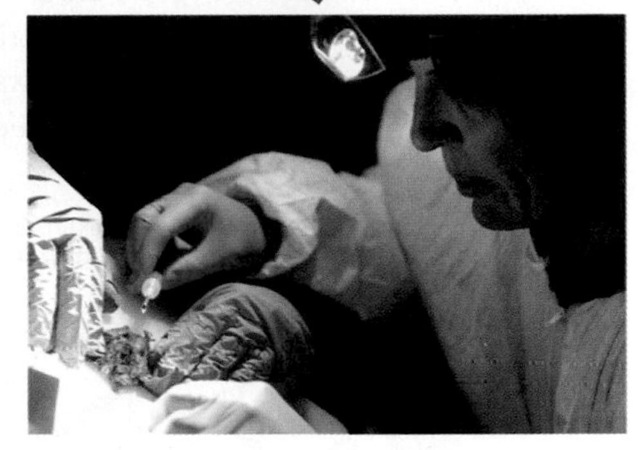

Experimental:
Scientists inject bats with fungicide to test whether it will protect bats from WNS.

Figure 1.5

Testing hypotheses using multiple approaches

Scientists set up an underground laboratory in Tennessee's New Mammoth Cave. To test hypotheses about white-nose syndrome (WNS), the researchers used descriptive, analytical, and experimental approaches.

Q1: State a possible hypothesis that could be tested with the analytical approach shown here.

Q2: State the hypothesis being tested with the experimental approach.

Q3: Why can't an observational study show a cause-and-effect relationship? Why is an experimental study the only research method that can show a cause-and-effect relationship?

See Appendix A for answers to the figure questions.

independent variable and tracks how that manipulation changes the value of a dependent variable. Blehert manipulated his independent variable (exposing some bats to the fungus but not others) and then tracked his dependent variable (whether the bats showed symptoms of WNS).

Blehert made sure his experiment was a controlled experiment. A **controlled experiment** measures the value of the dependent variable for two or more groups of subjects that are comparable in all respects, except that one group is exposed to a change in the independent variable and the other group is not. In this case, healthy bats were either exposed to the fungus or not exposed. Typically, a researcher obtains a sufficiently large sample of study subjects and assigns them randomly to two groups. Randomization helps ensure that the two groups are comparable at the start.

One group, the **control group**, is maintained under a standard set of conditions with no change in the independent variable. Blehert had 34 healthy bats in his control group. He kept these bats in the laboratory, under the same conditions as the other bats except that he did not expose them to *Geomyces destructans.*

The other group, known as the experimental or **treatment group**, is maintained under the same standard set of conditions as the control group, but the independent variable is manipulated. Blehert exposed 83 healthy bats to the fungus. Of these, 36 were exposed to the fungus through the air, 18 were put into close contact with naturally infected bats, and 29 had the fungus applied directly to their wings (**Figure 1.6**).

As noted earlier, when scientists test a prediction that follows from a hypothesis and find that prediction upheld, the hypothesis is said to be supported. Scientists can be relatively confident in a supported hypothesis, but they cannot say that the hypothesis has been proved true. Even well-established scientific ideas can be overturned if new evidence against the prevailing view comes to light. Albert Einstein is famously reported to have said, "No amount of experimentation can ever prove me right; a single experiment can prove me wrong."

After watching both the control group and the treatment groups, Blehert saw what he had predicted: physical exposure to the fungus caused white-nose syndrome, but exposure through the air did not. Healthy bats that had fungus applied directly to their wings or were caged with naturally infected bats had high rates of WNS by the experiment's end. It was the first direct evidence that the fungus was the primary cause of WNS. A follow-up study by Blehert and others showed that the fungus not only leads to

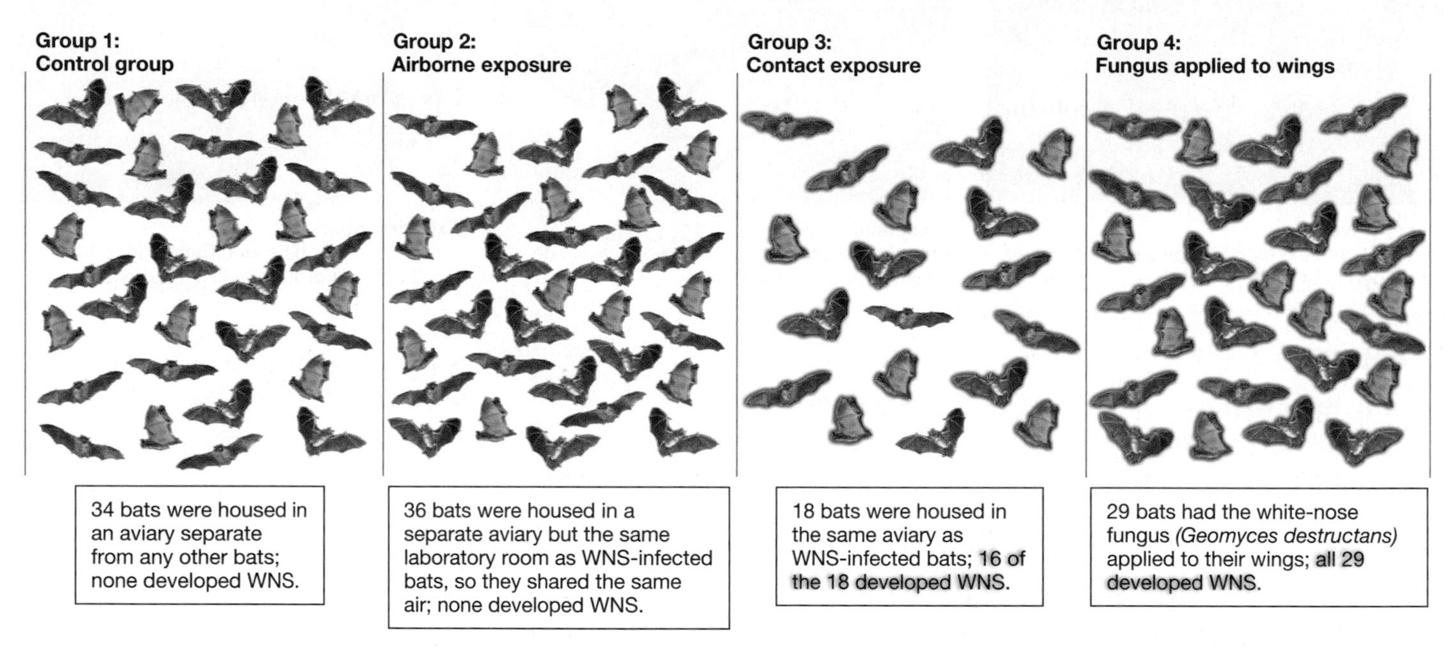
Group 1:
Control group
Group 2:
Airborne exposure
Group 3:
Contact exposure
Group 4:
Fungus applied to wings
34 bats were housed in an aviary separate from any other bats; none developed WNS.
36 bats were housed in a separate aviary but the same laboratory room as WNS-infected bats, so they shared the same air; none developed WNS.
18 bats were housed in the same aviary as WNS-infected bats; 16 of the 18 developed WNS.
29 bats had the white-nose fungus (Geomyces destructans) applied to their wings; all 29 developed WNS.

Figure 1.6

Blehert's experimental design

Blehert and his colleagues captured 117 healthy bats and brought them into the laboratory. They divided the bats into control and treatment groups and observed them for 102 days.

Q1: Explain how each treatment group differs from the control group. What is the independent variable for each group? What is the dependent variable for all groups?

Q2: What hypothesis was being tested in this experiment?

Q3: In one or two sentences, state the conclusions you can draw from the experiment. Did the results support the hypothesis? Why or why not?

See Appendix A for answers to the figure questions.

symptoms of WNS in bats but also is sufficient to cause death (**Figure 1.7**).

Over the years, other hypotheses about the cause of WNS have not been upheld. For example, scientists were not able to identify a single environmental contaminant at elevated levels in infected bats. When a prediction is not upheld, the hypothesis is reexamined and changed, or it is rejected.

One of the greatest strengths of science is that scientific knowledge is tentative and therefore open to challenge at any time by anyone. An absolute requirement of the scientific method is that evidence be based on observations, experiments, or both. Furthermore, the observations and experiments that furnish the evidence must be subject to testing by others; independent researchers should be able to make the same observations or obtain the same results if they use the same conditions. In addition, the evidence must be collected as objectively as possible—that is, as free of personal or group bias as possible. Blehert's controlled experiment fit all these conditions.

The main mechanism for policing personal or group bias and even outright fraud in science is **peer-reviewed publication**. Peer-reviewed publications are found in scientific journals that publish original research after it has passed the scrutiny of experts who have no direct involvement in the research under review. Before Blehert's research was published, it was reviewed by numerous scientists who had not participated in the experiment. If reviewers have concerns during the peer-review process, such as whether the evidence is strong enough to support a hypothesis, they can ask the paper's authors to address those concerns (for example, by gathering additional evidence) and to resubmit the paper. Blehert's paper passed the peer-review process and was published in the scientific journal *Nature* in 2011. At that point, says Blehert, the evidence that *G. destructans* causes WNS was strong enough that "I think we'd convinced most people." In 2013, scientists renamed *G. destructans* as *Pseudogymnoascus destructans* when the fungus's genus was reclassified.

But identifying the cause of WNS did not stop the disease from spreading. By March 2008, just a year after Hicks's team found the thousands of dead bats near Albany, more bats were found dead and dying in caves across Vermont, Massachusetts, and Connecticut.

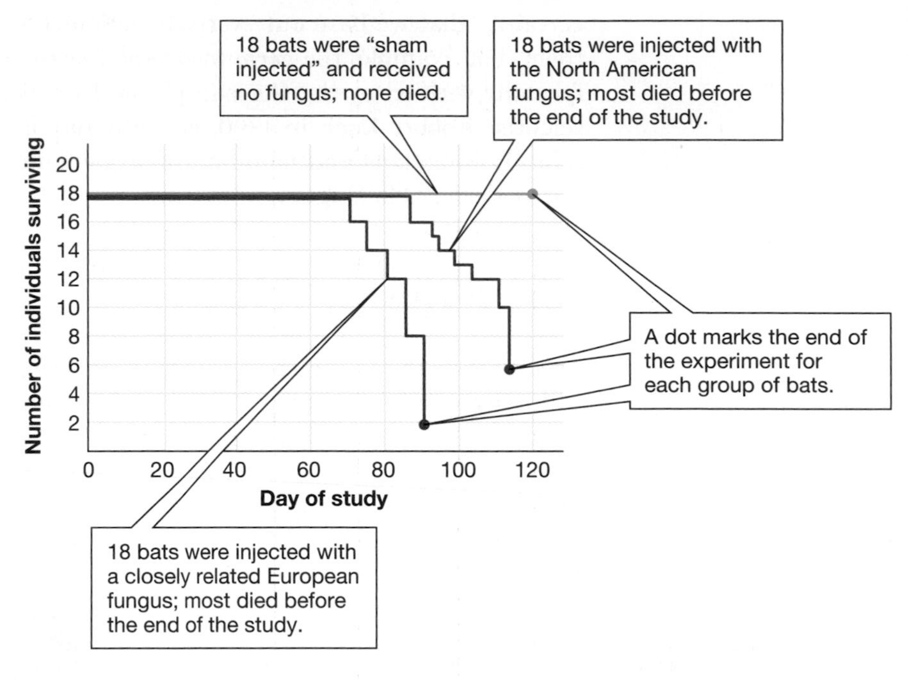

Figure 1.7

Follow-up experiment supports the hypothesis that the fungus causes WNS and higher mortality in bats

The experiment whose results are plotted here was conducted by some of the same researchers interviewed for this story.

Q1: What is the control group in this experiment, and what are the two treatment groups?

Q2: At day 40, approximately how many individuals were alive in each treatment group? At day 80? At day 100?

Q3: In one or two sentences, state the conclusions you can draw from the experiment. Did the results support the hypothesis? Why or why not?

See Appendix A for answers to the figure questions.

Within a year, the disease had spread as far as Tennessee and Missouri. The spread of WNS is a scientific fact: bats around the United States are dying from this disease. In casual conversation, we typically use the word "fact" to mean a thing that is known to be true. A scientific **fact** is a direct and repeatable observation of any aspect of the natural world.

Scientific fact should not be confused with scientific **theory**. Outside of science, people often use the word "theory" to mean an unproven explanation. In science, a theory is a hypothesis, or a group of related hypotheses, that has received substantial confirmation through diverse lines of investigation by independent researchers. Scientific theories have such a high level of certainty that we base our everyday actions on them. For example, the *germ theory of disease*, formally verified by the German physician and scientist Robert Koch in 1890, is the basis for treating infections and maintaining hygiene in the modern world (**Figure 1.8**).

No End in Sight

According to the U.S. Fish and Wildlife Service, white-nose syndrome has killed more than 7 million bats across the United States since 2007 and shows no signs of slowing. As of this writing, the disease has spread to 33 U.S. states and 7 Canadian provinces. Almost all species of bats that hibernate in these regions have been affected, including little brown bats and endangered Indiana bats, both of which have been particularly hard-hit.

The fungus appears to be related to a type of fungus common in caves in Europe. Most likely, a human traveler from Europe accidentally carried it across the Atlantic and into the Albany cave, where it infected its first bat in the United States. Researchers continue to explore

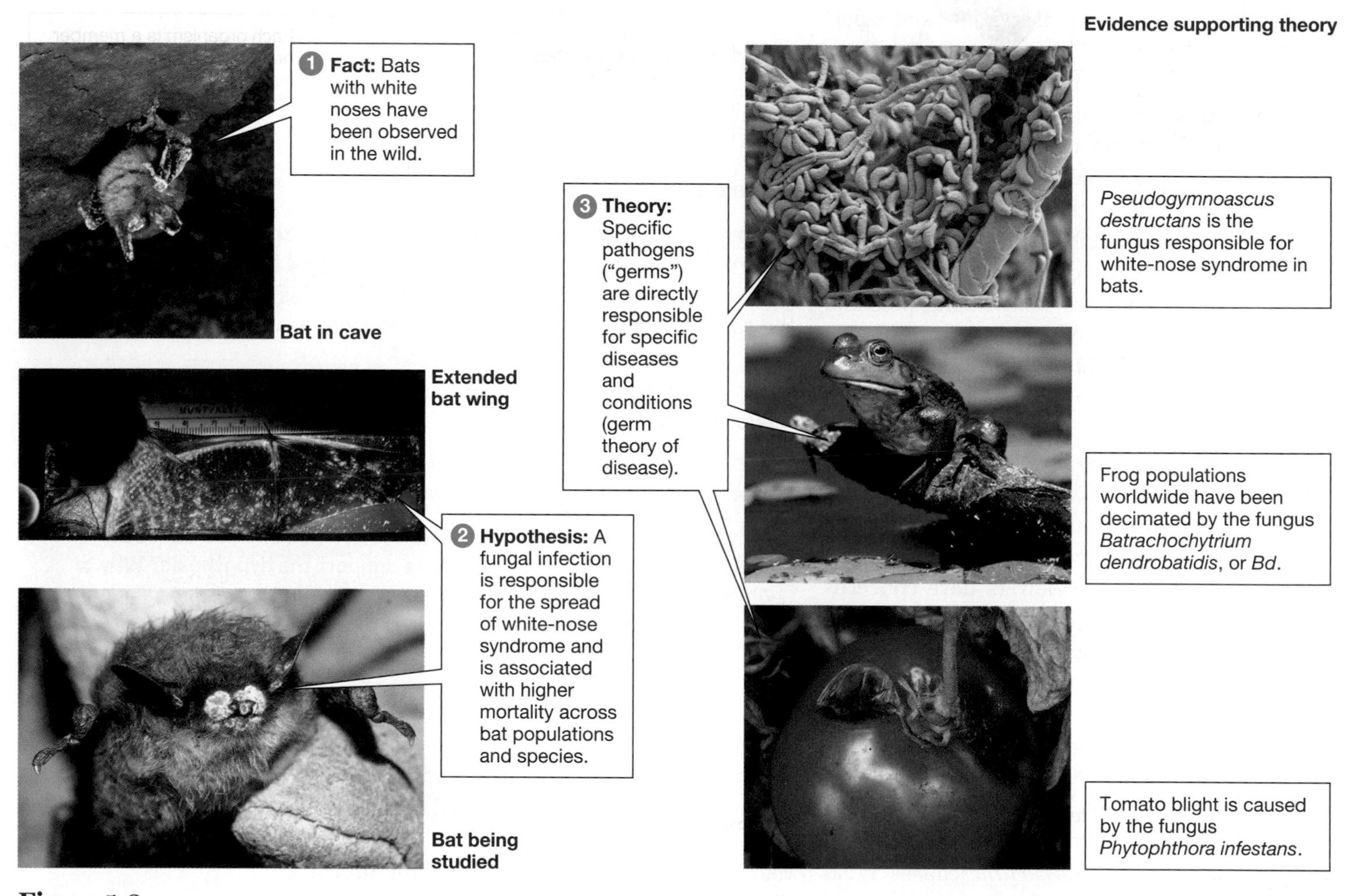

Figure 1.8

Facts, hypotheses, and theories

It is important to distinguish among facts, hypotheses, and theories when thinking and talking about science.

Q1: Give one *fact* about bats that you learned from this chapter.

Q2: What is another example of evidence for the *germ theory of disease*? (*Hint*: Think about human diseases.)

Q3: Explain in your own words the difference between a fact and a hypothesis and between a hypothesis and a theory.

See Appendix A for answers to the figure questions.

exactly how the fungus kills the bats. It appears to wake them from hibernation so many times during the winter that the bats use up their fat reserves too soon and do not survive the months of cold weather. The fungus also eats through the bats' delicate wings, which are important not only for flight but also for maintaining healthy levels of water, oxygen, and carbon dioxide in the bats' bodies. The impact of this fungus on bats is a powerful example of how a microorganism can affect many levels of life, from individual tissues and organs up to whole populations, communities, and even ecosystems. For scientists, the term **biological hierarchy** refers to the many levels at which life can be studied, from microscopic molecules and cells to whole communities and ecosystems (**Figure 1.9**).

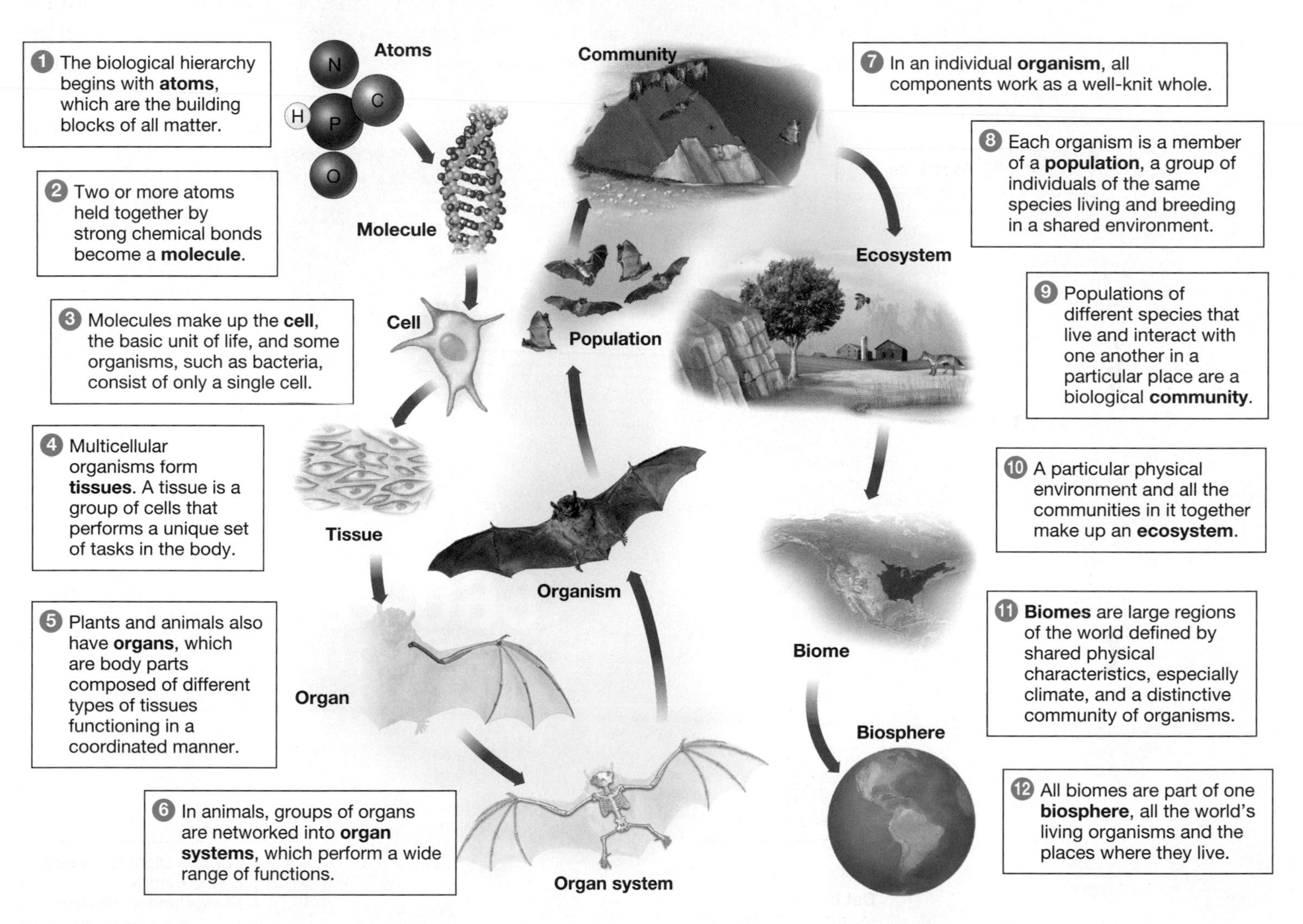

1 The biological hierarchy begins with **atoms**, which are the building blocks of all matter.
Atoms
N
C
H
P
O
2 Two or more atoms held together by strong chemical bonds become a **molecule**.
Molecule
3 Molecules make up the **cell**, the basic unit of life, and some organisms, such as bacteria, consist of only a single cell.
Cell
4 Multicellular organisms form **tissues**. A tissue is a group of cells that performs a unique set of tasks in the body.
Tissue
5 Plants and animals also have **organs**, which are body parts composed of different types of tissues functioning in a coordinated manner.
Organ
6 In animals, groups of organs are networked into **organ systems**, which perform a wide range of functions.
Organ system
Organism
Population
Community
7 In an individual **organism**, all components work as a well-knit whole.
8 Each organism is a member of a **population**, a group of individuals of the same species living and breeding in a shared environment.
Ecosystem
9 Populations of different species that live and interact with one another in a particular place are a biological **community**.
10 A particular physical environment and all the communities in it together make up an **ecosystem**.
Biome
11 **Biomes** are large regions of the world defined by shared physical characteristics, especially climate, and a distinctive community of organisms.
Biosphere
12 All biomes are part of one **biosphere**, all the world's living organisms and the places where they live.

Figure 1.9

Life can be studied at many levels

The **biological hierarchy** is a way to visualize the breadth and scope of life, from the smallest structures to the broadest interactions between living and nonliving systems.

Q1: Give examples of other kinds of organs that mammals such as bats have. (*Hint:* Think of the organs in your own body.)

Q2: Are bats in California part of the community of bats in upstate New York if they are of the same species? Why or why not?

Q3: Is the soil in a cave where bats live a part of the bats' population, community, or ecosystem? Explain your reasoning.

See Appendix A for answers to the figure questions.

Bug Zappers

Bats are skilled natural exterminators, consuming billions upon billions of insects each year, including crop pests and mosquitoes. For this reason, bats play a critical role in agriculture and potentially in human health with the onset of increased mosquito-borne viruses such as West Nile and Zika. The loss of bat populations to white-nose syndrome would have a significant impact on farms, forests, and people around the country.

Assessment available in smartwork

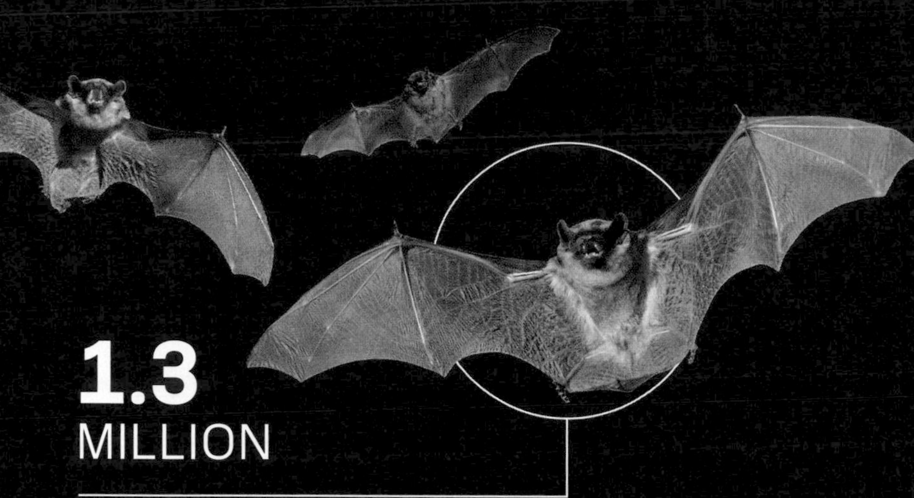

1.3
MILLION

A single colony of **150 big brown bats** *(Eptesicus fuscus)* in Indiana is estimated to eat nearly **1.3 million pest insects each year.**

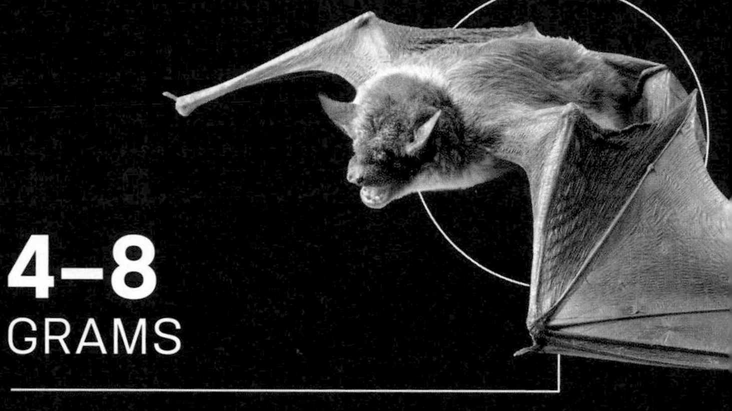

4–8
GRAMS

A single little brown bat *(Myotis lucifugus)* can consume **4–8 grams of insects each night** during the active season.

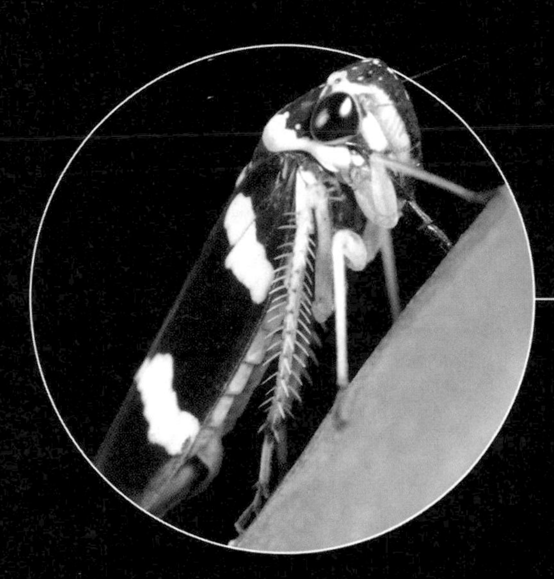

660–1,320
METRIC TONS

Extrapolating the diet of a single bat to 1 million bats estimated to have died from WNS, between **660 and 1,320 metric tons of insects** are no longer being consumed each year in WNS-affected areas.

$22.9
BILLION

If bats disappeared entirely from the United States, it would cost the agricultural industry roughly **$22,900,000,000 per year** to save crops by dealing with the insects no longer being eaten by bats.

According to the U.S. Fish and Wildlife Service, WNS is one of the worst wildlife diseases of modern times. The disease continues to spread, most recently into Kansas and central Texas, where officials called the fungus a "looming disaster."

As of yet, there is no cure for white-nose syndrome, though scientists continue to study how to slow or stop the disease. An international team of scientists identified populations of bats in Asia and the United States that survived the fungal infection, so it is possible that bats can evolve resistance to the deadly disease. Some research teams have also had luck treating infected bats with antifungal agents like ultraviolet light and nontoxic chemicals, but it has been difficult to deliver such treatments to large groups of bats.

Today, bats continue to die across America every winter. Hicks, now retired, worries that students, hikers, and tourists will visit caves in the United States, not see any bats, and think that's normal. "In 2006, in one big cave in the Adirondacks, we counted over 185,000 bats. Anywhere you shined a light, there was a bat," says Hicks. "You go in now, and there's not a bat in sight."

NEED-TO-KNOW SCIENCE

- **Science** (page 4) is both a body of knowledge about the natural world and an evidence-based process for generating that knowledge.
- The **scientific method** (page 4) represents the core logic of the process by which scientific knowledge is generated. The scientific method requires that we (1) make **observations** (page 5), (2) devise a **hypothesis** (page 7) to explain those observations, (3) generate predictions from that hypothesis, (4) test those predictions, and (5) share the results of the tests for **peer review** (page 11) by fellow scientists.
- A hypothesis cannot be proved true; it can only be supported or not supported. If the predictions of a hypothesis are not supported, the hypothesis is rejected or modified. If the predictions are upheld, the hypothesis is supported.
- We can test hypotheses by making further observations or by performing **experiments** (controlled, repeated manipulations of nature; page 8) that will either uphold the predictions or show them to be incorrect.
- In a scientific experiment, the **independent variable** (page 8) is manipulated by the investigator. Any variable that can potentially respond to the changes in the independent variable is called a **dependent variable** (page 8).
- A scientific **fact** (page 11) is a direct and repeatable observation of any aspect of the natural world. A scientific **theory** (page 11) is a major idea that has been supported by many observations and experiments.
- The term **biological hierarchy** (page 13) refers to the many levels at which life can be studied: atom, molecule, cell, tissue, organ, organ system, organism, population, community, ecosystem, biome, and biosphere.

THE QUESTIONS

See Appendix B for answers.

The Basics

1 Which of the following statements is true of science?

(a) It is a body of knowledge about the natural world.

(b) It is the process by which we generate this knowledge.

(c) It is an evidence-based process.

(d) all of the above

(e) none of the above

2 When scientists use the word "theory," they mean

(a) an educated guess.

(b) a fact proved by many experiments.

(c) wild speculation.

(d) an experimental prediction.

(e) a major idea that has been supported by many observations and experiments.

3 Select the correct terms: The process of science begins with a(n) (**prediction** / **observation**) about the natural world. A scientist then proposes a (**hypothesis** / **prediction**), which is the basis of one or more testable (**observations** / **predictions**).

4 Place the following steps of the scientific method in the correct order by numbering them from 1 to 7.

____ a. Make observations about the natural world.

____ b. Test the predictions by designing an experiment or collecting observational data.

____ c. Run the experiment or collect the data, then analyze the results.

____ d. Generate predictions to test the hypothesis.

____ e. Share the results with fellow scientists so that they can review and evaluate them.

____ f. Develop a hypothesis to explain the observations.

____ g. Accept, reject, or modify the hypothesis depending on the results.

5 Identify the level of biological organization for each of the following.

____ a. the kidney of a bat

____ b. an oak tree outside a cave in upstate New York

____ c. bats in a cave in upstate New York

____ d. the physical and biological components of a cave in upstate New York

____ e. the respiratory system (including the nose, mouth, lungs) of a bat

____ f. all the species living and interacting within a cave in upstate New York

Challenge Yourself

6 Describe one observation, one hypothesis, and one experiment from the white-nose syndrome research discussed in this chapter.

7 Which of the following statements is a scientific hypothesis (that is, makes testable predictions)? (Select only one.)

(a) The Atkins diet helps people lose more weight and keep it off than Weight Watchers does.

(b) Even though no one else can see him, the ghost of my dog lives in my backyard.

(c) People born under the sun sign Aquarius are kinder and cuter than those born under Scorpio.

(d) It is unethical to text while driving.

(e) none of the above

8 Consider an experiment in which subjects are given a pill to test its effectiveness in reducing the duration of a cold. Which of the following is the best way to treat the control group?

(a) Give the control group two pills instead of one.

(b) Do nothing with the control group.

(c) Give the control group a pill that looks like the test pill but does nothing.

(d) Let the control group choose whether or not to take any pills.

(e) Tell the members of the control group that they are the control group.

Try Something New

9 Mad cow disease, or bovine spongiform encephalopathy, appears to be caused by a novel infectious agent: a protein that replicates, or copies itself, by causing related proteins to modify their structure from a harmless shape to a dangerous one. These prions (short for "proteinaceous infectious particles") also appear to be the cause of several other spongiform encephalopathy diseases, such as scrapie in sheep, and kuru and Creutzfeldt-Jakob disease in humans. Which of the following observations or experiments would *not* support the hypothesis that a prion causes spongiform encephalopathy? (Select only one.)

(a) The brains of many sheep with scrapie contain prion proteins, but the brains of most sheep without scrapie do not.

(b) There is a high incidence of kuru in populations of people who consume brain tissue from prion-infected animals.

(c) When prions are fed to sheep, most of them subsequently develop scrapie, whereas sheep not fed prions do not develop scrapie.

(d) Coyotes that feed on cows with mad cow disease do not subsequently develop spongiform encephalopathy.

(e) When introduced into sheep brain cells in culture, prions cause the normal proteins to change shape into dangerous prion proteins.

10 Label each of the following statements as an observation, a hypothesis, an experiment, or a result.

_____ a. People who ingest a liquid contaminated with a rotavirus will subsequently experience acute diarrhea.

_____ b. Out of 20 students given rotavirus-contaminated chocolate bars, 17 students subsequently experienced acute diarrhea.

_____ c. While 10 students were given a liquid containing a rotavirus, 10 other students were given uncontaminated liquid.

_____ d. People exposed to rotaviruses often experience acute diarrhea.

_____ e. The latest outbreak of acute diarrhea on campus was due to a rotavirus-contaminated elevator button in the dorms.

11 Which levels of the biological hierarchy did the scientists in this chapter work within? Identify at least three levels, and explain your reasoning for each level.

12 Review the map showing the confirmed and suspected outbreaks of WNS (an updated map can be found at http://www.whitenosesyndrome.org/static-page/where-is-wns-now). Is each statement that follows a testable, properly formulated scientific hypothesis for the spread of WNS to bats in Washington State? Explain why or why not. (*Hint*: Think about the criteria for a scientific hypothesis.)

(a) Did bats infected with WNS arrive in Washington State via human smuggling for research purposes?

(b) The WNS fungus was transported from the Midwest to Washington State via a bat found in Flight 1701's cargo hold.

(c) Bats in Washington State were infected by contaminated clothing and equipment from visiting researchers checking for WNS in these colonies.

(d) How did bats end up with WNS in Washington State, so far away from all other sites of contamination?

(e) A mysterious cloud of fungal particles must have floated over Washington State before the infected bat colonies were discovered.

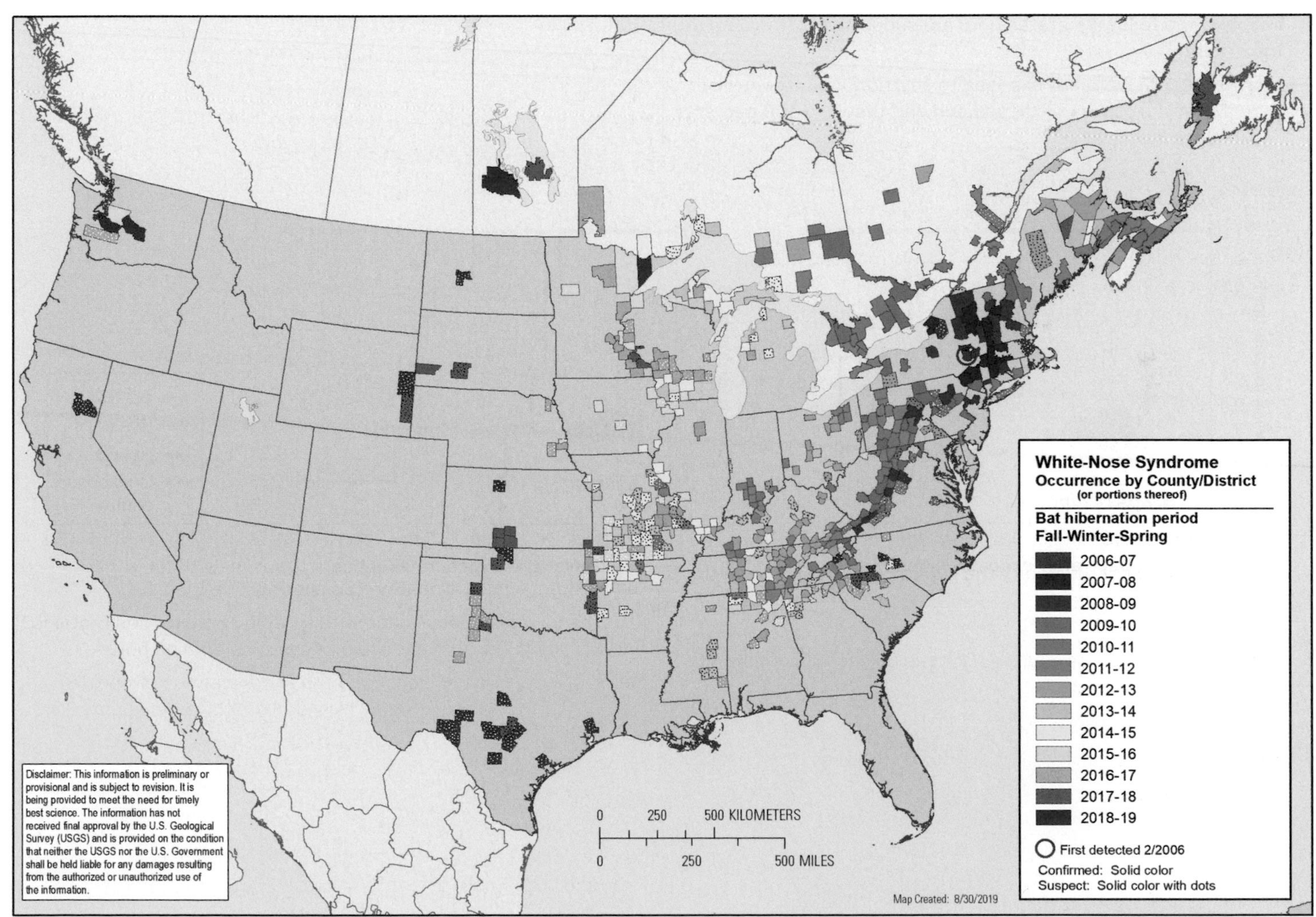

Citation: White-nose syndrome occurrence map - by year (2019). Data Last Updated: 8/30/2019. Available at: https://www.whitenosesyndrome.org/static-page/wns-spread-maps.

Leveling Up

13 **Doing science** Suppose you are a biologist with expertise in white-nose syndrome in bats. You have been asked to contribute your scientific expertise to understanding the fungal infection that is decimating frog populations worldwide: *Batrachochytrium dendrobatidis*, or *Bd*. To prepare for this involvement, read the March 2019 *Science* article on how *Bd* has been spread around the world: https://www.sciencemag.org/news/2019/03/fungus-has-wiped-out-more-species-any-other-disease

Now look at the accompanying graphs that are based on data from a scientific paper published in 2018. They show the results of an experiment involving two species of frogs: (a) *Brachycephalus ephippium* and (b) *Ischnocnema parva*. Each species was exposed to—and thus infected with—three kinds of chytrid fungus: *Bd*-GPL, *Bd*-Brazil, and a hybrid version. Each *x*-axis shows the time period of the experiment in days. Each *y*-axis shows the percentage of frogs that died over the time period (that is, after exposure/infection). The dotted line shows the survivorship of the control—frogs that were not exposed to the fungus but that otherwise experienced the same experimental conditions.

(a) Which of the following statements are supported by the data in the graph?
1. No frogs that were not exposed to chytrid fungus (controls) died over the course of the experiment; they had 100 percent survivorship.
2. The hybrid chytrid fungus killed all *B. ephippium* frogs that were exposed to it.
3. All *I. parva* frogs died within 20 days of exposure to the hybrid chytrid fungus.
4. The *Bd*-GPL fungus was equally deadly to both species of frogs.
5. *Bd*-Brazil appears to be the least deadly of the chytrid fungi for these two frog species.

(b) For each species of frog and each fungus, identify the approximate time in the experiment (day number) at which 50 percent of the exposed frogs had died.

(c) Propose a hypothesis to explain the results depicted in the graphs and then identify a testable prediction from that hypothesis. Design an experiment to test the prediction, identifying your independent and dependent variables, the control group, and the treatment conditions. Create a graph to show the results you expect to find (1) if your hypothesis is supported and (2) if your hypothesis is not supported.

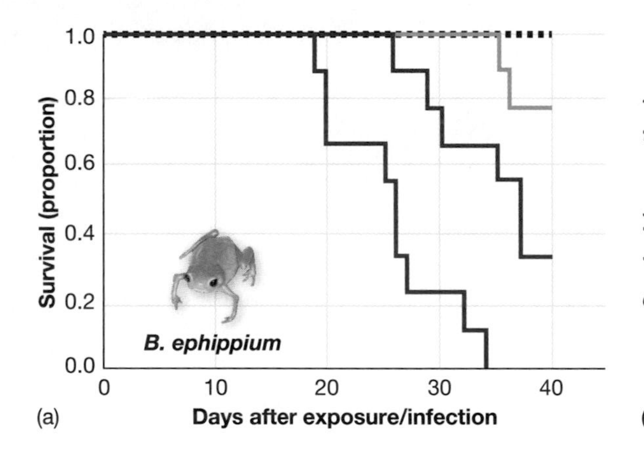

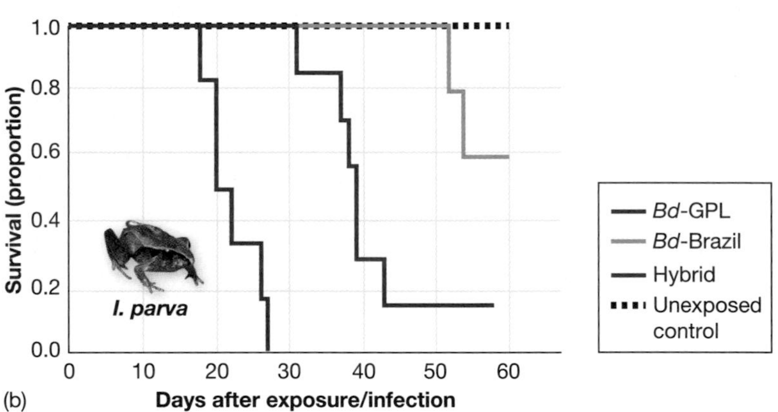

14 **What do *you* think?** Most bats are insectivores—that is, they eat insects. Such bats save the agriculture industry billions of dollars each year because they eat insects that damage crops (see "Bug Zappers" on page 14). Other species of bats pollinate crops. Based on this information, what relationship would you predict between farming and WNS? How does your prediction affect your level of concern about WNS? Explain your reasoning.

Digital resources for your book are available online.

Sex, Cancer, and a Vaccine

Human papillomavirus (HPV) causes cancer.
A widely available vaccine prevents HPV infection.
So why are so few young people getting the vaccine?

After reading this chapter you should be able to:

- Evaluate a scientific claim, using the process described in the chapter.
- Explain the importance of scientific literacy for making informed decisions.
- Distinguish between secondary and primary literature, and explain the role of peer review in the latter.
- Compare and contrast basic and applied research, and give an example of each.
- Distinguish between correlation and causation, and give an example of each.
- Determine whether a scientific claim is based on real science or pseudoscience.

SCIENCE

CHAPTER 02

EVALUATING SCIENTIFIC CLAIMS

Laura Brennan didn't have time to see a doctor. In the fall of 2016, the recent college graduate from Clare, Ireland, was constantly on the road in her new job as a sales manager, crisscrossing Ireland by car (**Figure 2.1**). But as the irregular bleeding continued—something more than a typical period—Laura made a quick stop at a clinic. The doctor suspected the bleeding was due to a bacterial infection.

But when Laura's primary care physician examined Laura's cervix, she saw signs of a large growth and urgently sent Laura to the hospital for an ultrasound. The ultrasound led to a magnetic resonance imaging (MRI) scan that led to a biopsy. When the head gynecologist walked into the room after the biopsy, Laura knew it was bad news. She asked if there was any chance it wasn't cancer. "I don't think so," he said.

The doctors found a tumor the size of a lime on Laura's cervix. A subsequent scan showed the cancer had spread to the lymph nodes in her pelvis. In December 2016, the 24-year-old was diagnosed with cervical cancer caused by HPV-16.

Figure 2.1

Laura Brennan and her dog Bailey

Laura Brennan was an eloquent, passionate, and tireless advocate for the HPV vaccine.

Virtually all cases of cervical cancer are caused by human papillomavirus (HPV), a virus passed on during sex, including vaginal, anal, and oral sex. HPV is the most common sexually transmitted infection in the United States; 80 percent of sexually active people will be infected with HPV at some point. In 9 out of 10 infected people, however, HPV clears from the body within 2 years and does not cause any health problems, so most individuals never even know they've had it.

But sometimes, for unknown reasons, the infection persists and can cause cells to become abnormal, especially in areas exposed to the virus during sex, including genital areas, the anus, the rectum, the mouth, and the throat. Those abnormal cell changes can lead to cancer in those areas of the body. Cervical cancer is the most common cancer caused by HPV in women, and cancer in the back of the throat, known as oropharyngeal cancer, is the most common in men.

Of the more than 150 types of HPV, 12 types are known to cause cancer. Two of those, HPV-16 and HPV-18, cause 70 percent of cervical cancers, according to the National Institutes of Health (NIH). Each year in the United States, an estimated 11,000 women are diagnosed with cervical cancer, and 4,000 women die of the disease. HPV-16 causes more than 90 percent of cases of head and neck cancers in men, as well as most anal and penile cancers. In fact, because cervical cancer rates are declining and rates of HPV-caused oropharyngeal (back of the throat) cancer are increasing so rapidly, the throat has replaced the cervix as the most common site for HPV-associated cancers.

But starting in 2006, humankind gained a weapon in the fight against HPV. In the United States, the Food and Drug Administration (FDA) approved two vaccines, Gardasil® and Cervarix®, which prevent infections of HPV-16 and HPV-18. Gardasil also prevents infections of HPV-6 and HPV-11 that cause genital warts, and the latest version of Gardasil also protects against another five types of the virus that cause cancer.

Vaccination is the injection of material—typically an inactivated and harmless infectious organism or parts of such an organism (for example, a single protein)—that stimulates the immune system to protect against future exposure to that pathogen (**Figure 2.2**). When the body's immune system is exposed to a vaccine, it

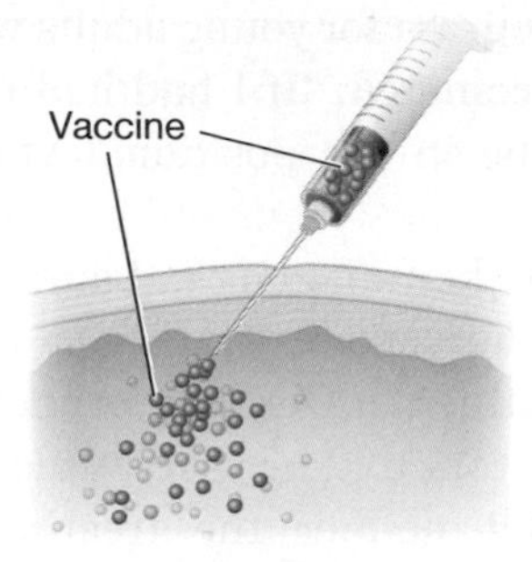

A vaccine with an inactive form or harmless amount of a virus (or other organism) is injected under the skin.

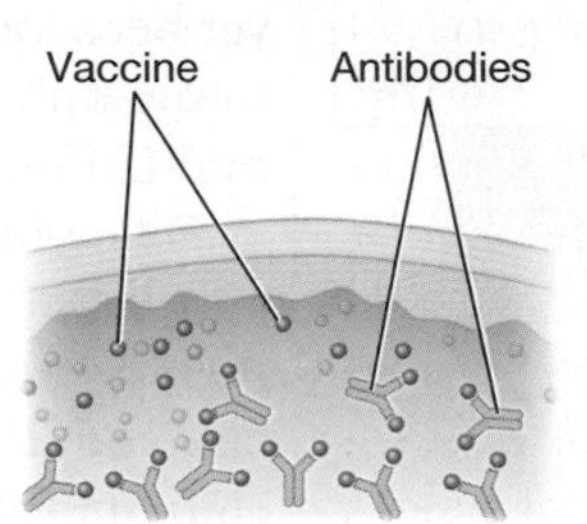

The vaccine stimulates the immune system to produce antibodies (in green) that recognize the virus.

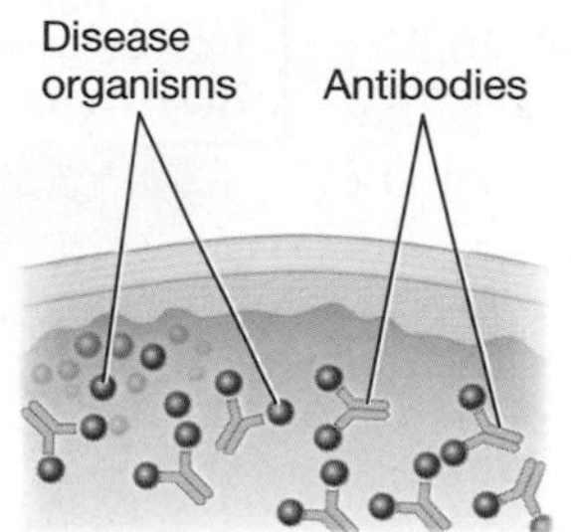

When the individual is exposed to the virus after vaccination, the new antibodies are primed to attack the invader.

Figure 2.2

How vaccines work

Each vaccine trains the body's immune system to fight a particular kind of infection.

Q1: How does a vaccine create immunity to a virus?

Q2: Why is it important that the virus in the vaccine is inactive or harmless?

Q3: Natural immunity occurs without a vaccine, just by exposure to a particular illness. For example, why will someone who has had chicken pox not get it a second time?

See Appendix A for answers to the figure questions.

recognizes the inactivated organism (or its parts) as an invader and mounts an attack against it. Upon later contact with the infectious organism, a vaccinated individual's immune system remembers the inactivated organism from the vaccine and is already armed and bristling.

Before vaccines, children died in large numbers from diphtheria, pertussis (whooping cough), smallpox, polio, tetanus (lockjaw), and many other diseases (**Figure 2.3**). Seven to eight million children died every year from measles alone. Over the past 200 years, scientists have developed vaccines to protect against dozens of pathogens, starting with the smallpox virus. Those infectious pathogens still exist in our environment but because vaccines protect people from infection, when most of the population is vaccinated, we rarely see the resulting disease.

Today, the Centers for Disease Control and Prevention (CDC), the public health branch of the U.S. government, recommends a series of vaccines for children between birth and 18 years, including the annual influenza ("flu") vaccine, also recommended for all adults (see "It's Time for Your Flu Shot" on page 23). The HPV vaccine is on that list, recommended in a two-dose series for boys and girls aged 11 to 12

It's Time for Your Flu Shot

Why do you need a flu shot every year? Why doesn't one shot protect you for life, as many other vaccines do? In fact, each year's flu vaccine *does* protect you for life from a particular strain of the flu virus. Unfortunately, the virus mutates rapidly, so every year's flu strain is likely to be different from the previous year. To the immune system, the new strain looks like a completely new virus. The pesky flu virus can even combine itself with other flu viruses, from animals such as birds and pigs, to create deadly flu pandemics. For example, in 2009 the H1N1 strain, or "swine flu," caused widespread illness around the globe.

Each flu season, the CDC uses current flu cases to predict what the next season's flu viruses will be. Then vaccine manufacturers produce a vaccine to protect us from the anticipated viruses (you will learn about the physical process of producing vaccines in Chapter 10). Sometimes the experts are extremely accurate, and the flu vaccine is highly effective against that season's viruses. Sometimes, however, the experts are not so accurate, and the vaccine is not as protective. It's a viral genetics puzzle and a bit of a guessing game. Still, even in years when the vaccine does not provide full coverage for that season, it is likely to protect you from future versions of the flu.

Here's the bottom line: The more yearly flu vaccines you receive, the more flu viruses your immune system will be ready to fight, and the less likely you will be to catch the seasonal flu. More important, you will be less likely to succumb to a deadly flu pandemic.

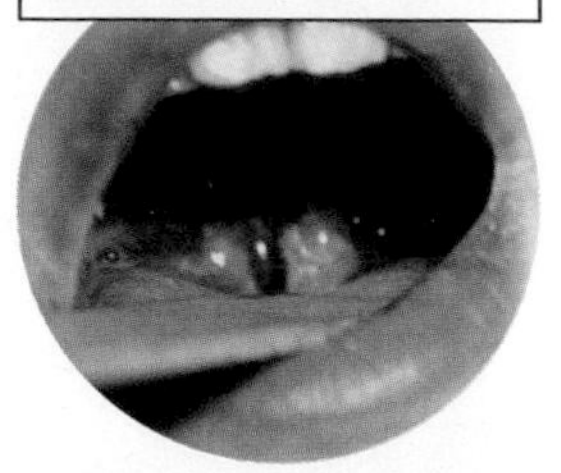

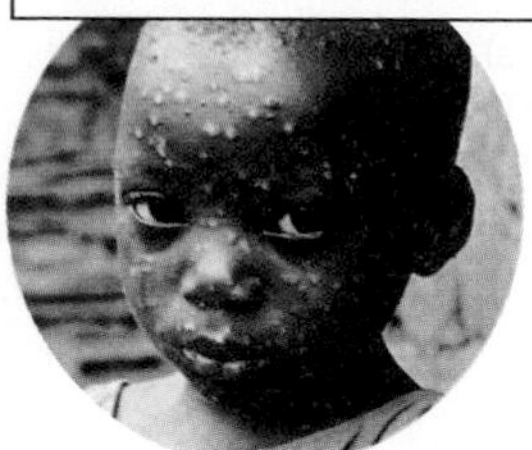

Pertussis
1922–25: 7,363 (5%)
2000–19: up to 20 (0.1%)

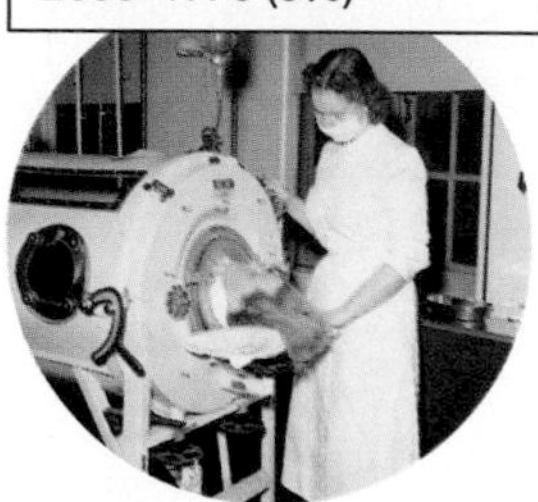

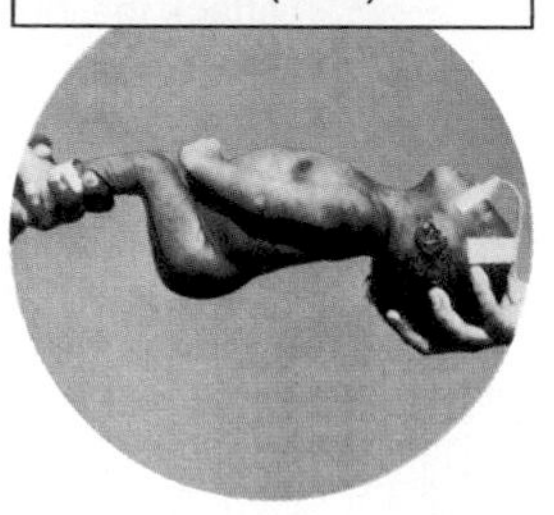

Figure 2.3

Vaccines save lives

Each box gives the particular disease's average annual number of deaths and, in parentheses, the percentage mortality rate (i.e., cases proving fatal) in the United States. The top line represents the prevaccination period (e.g., 1920–22 for diphtheria), and the bottom line represents the postvaccination period of the twenty-first century.

Q1: Before vaccinations, which diseases had the highest and lowest mortality rates?

Q2: After vaccinations, which diseases had the highest and lowest mortality rates?

Q3: When you look at all the numbers and percentages in this figure, what is your takeaway?

See Appendix A for answers to the figure questions.

and for teens and adults up to age 26 who were not previously vaccinated. The HPV vaccine is recommended at a young age because vaccination should be given before potential exposure to HPV through sexual activity. In addition, the vaccine produces the best immune response in the body at younger ages.

As of 2019, more than 100 countries had an HPV vaccine program. In Ireland, the vaccine is offered and administered at schools to every girl 11 and 12 years old and now is offered to boys as well. Tragically, Laura was too old to get the HPV vaccine when it first became available in Irish schools, and at the time there was no catch-up program for young adults who had not yet been vaccinated. "If I had had the vaccine, I wouldn't be in the position I am in today," said Laura.

Following her diagnosis, Laura underwent a series of "horrendous" radiation and chemotherapy treatments. Her long brunette hair fell out, and doctors told her she would never have children. But neither the treatment nor its effects dimmed Laura's outgoing personality and infectious smile. "It is curable," she assured her brother Ferg with a grin. "I got this."

Two months later, an MRI said she was all clear. But Laura didn't feel right. She went back in for another scan. This time, the lymph nodes in her chest lit up. The cancer had metastasized inside her body and was continuing to spread (see "Cancer: Uncontrolled Cell Division" on page 113). There was no longer hope of a cure.

Battling Rumors

Although the CDC has long had the goal of vaccinating at least 80 percent of the adolescents in the United States against HPV, since the vaccine became available, the first-dose vaccination rate has hovered around 60 percent with even smaller percentages of youth being fully vaccinated (**Figure 2.4**). Despite the availability of a vaccine that protects against two deadly forms of cancer, parents are not giving it to their children. Why?

Let's think for a moment about how people share information: in conversations, on the radio, television, the internet, social media, and so on. In 2011, Republican presidential candidate Representative Michele Bachmann from Minnesota told a story on television about a crying mother who said her daughter received the HPV vaccine and then suffered from intellectual disabilities "thereafter." The American Academy of Pediatrics (AAP), a reputable professional organization of 67,000 pediatricians, immediately refuted Bachmann's claim, asserting that "there is absolutely no scientific validity to this statement." But the story spread nonetheless, as did other claims about the HPV vaccine, suggesting that it wasn't necessary and that it would encourage young people to have sex.

Stories and anecdotes (short stories with a point) are powerful. They can change how

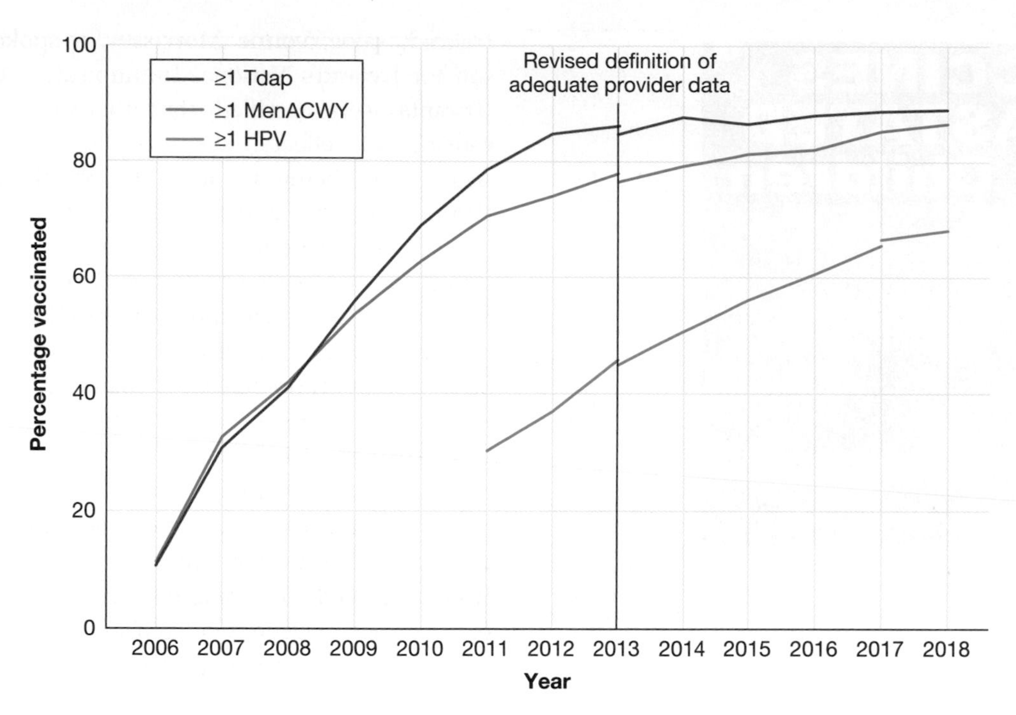

Figure 2.4

Estimated vaccine coverage of teenagers in the United States, 2006–18

The CDC conducts a national phone survey each year to estimate the number of teenagers (aged 13–17) who have received the recommended vaccinations for their age group. In 2006, the CDC's Advisory Committee on Immunization Practices (ACIP) began recommending Tdap vaccinations (protection from tetanus, diphtheria, and pertussis) and MenACWY vaccinations (protection from meningitis) for all teenagers. In 2011, the ACIP recommended HPV vaccinations for girls and in 2016 it recommended HPV vaccinations for boys.

Q1: Which vaccine has the highest uptake rate (percentage of population vaccinated), based on the most recent year? Which has the lowest? What are the approximate uptake rates in each case?

Q2: Why are there no data for HPV vaccine uptake before 2011?

Q3: What do you predict the uptake rate will be for the HPV vaccine in the United States 10 years from now? Explain your reasoning.

See Appendix A for answers to the figure questions.

we feel or think about a subject. But anecdotal evidence is not scientific evidence. Anecdotes are not representative of collected data or collected scientific observations; therefore, they cannot reliably give us the big picture of a subject or phenomenon. Scientific evidence can.

The statement that the HPV vaccine causes disability—or the opposite statement that the vaccine is safe—is a **scientific claim**, a statement about how the natural world works that can be tested using the scientific method (discussed in Chapter 1). We are exposed to scientific claims every day—dozens of them, in fact. They are not all true.

For example, editors at the magazine *Popular Science* asked one of their journalists to write down every scientific claim he heard in a day and evaluate each one. He recorded a whopping 106 claims and investigated each of them, from Cheerios' claim that it reduces cholesterol (supported by scientific evidence) to the claim that a face cream infused with vitamin A would revitalize skin (not supported). The majority of scientific claims come from advertisers. Though companies are legally bound to tell the truth, not all of them do. Of the 106 claims, most were bogus.

Scientific claims also come from special-interest groups and organizations that exist to

Figure 2.5

Anti-vaccine ads are professionally produced and misleading

The website stopmandatoryvaccination.com relies on GoFundMe donations to fund slickly produced advertising. This example implies that you must choose between trusting a medically trained professional and trusting parents who know their child best. But most parents—and essentially all medical doctors—agree that children need to be vaccinated for their safety and the safety of others.

advance certain causes, often for political or religious reasons. These claims include statements about global warming, evolution, and medical care. Bachmann's claim, for example, was likely politically motivated. Such claims may be presented in slick or professional ways, but a slick or professional appearance does not necessarily mean a claim is accurate (**Figure 2.5**). Therefore, it is important to question the "truth" of a claim when you hear it.

Just as in the United States, false scientific claims about the HPV vaccine spread across Ireland, says Yvonne Morrissey, a spokesperson for Ireland's National Immunisation Office. "Parents were terrified that the vaccine was causing side effects," she says. "Serious health issues were being reported on social media." Parents feared these anecdotal side effects, despite the fact that side effects of vaccines are typically minor and severe allergic reactions are extremely rare (**Figure 2.6**). The uptake of the HPV vaccine in Ireland was even lower than in the United States, with only 50 percent of eligible girls receiving it. "Unfortunately, sometimes rumors carry more strength than facts. It was a very difficult time to try and combat it," says Morrissey.

Just 3 days after finding out her cancer was terminal, Laura wrote to the Health Service Executive (HSE), Ireland's national health services organization, on Facebook and asked if her story might help encourage girls and young women to get the vaccine. "It was incredible," recalled Morrissey, who remembers Laura's message. "We stayed in contact, then decided to make a video with her."

In footage from the recording of her first video, Laura sits nervously in a metal chair in front of a white background, trying to remember her lines. "I got this," she says with a nervous laugh and a big smile. Then she tells her story. This recording became the first part of an extended campaign to promote scientific literacy about the HPV vaccine.

We, the public, including Laura, are not simply consumers of science and technology. We are participants. By voting on issues that have a scientific underpinning and spending money on products backed (or not backed) by the scientific method, we shape the course of science and influence which technologies are used, as well as where and how they're used. Although our personal values and political leanings are likely to influence how we vote, the underlying science should also be taken into consideration. **Scientific literacy**, an understanding of the basics of science and the scientific process, enables us to make informed decisions about the world around us and to communicate our knowledge to others. Our hope is that this book will help you become scientifically literate.

Scientific claims directly affect our lives because we make decisions based on them. Some of these are small decisions: Should I

DEBUNKED!

MYTH: Vaccines can cause the disease that the vaccines are supposed to prevent.

FACT: Vaccines may cause mild symptoms resembling the diseases they protect against, but these symptoms do not indicate infections. The safety of vaccines is highly regulated, and today's vaccines do not cause disease.

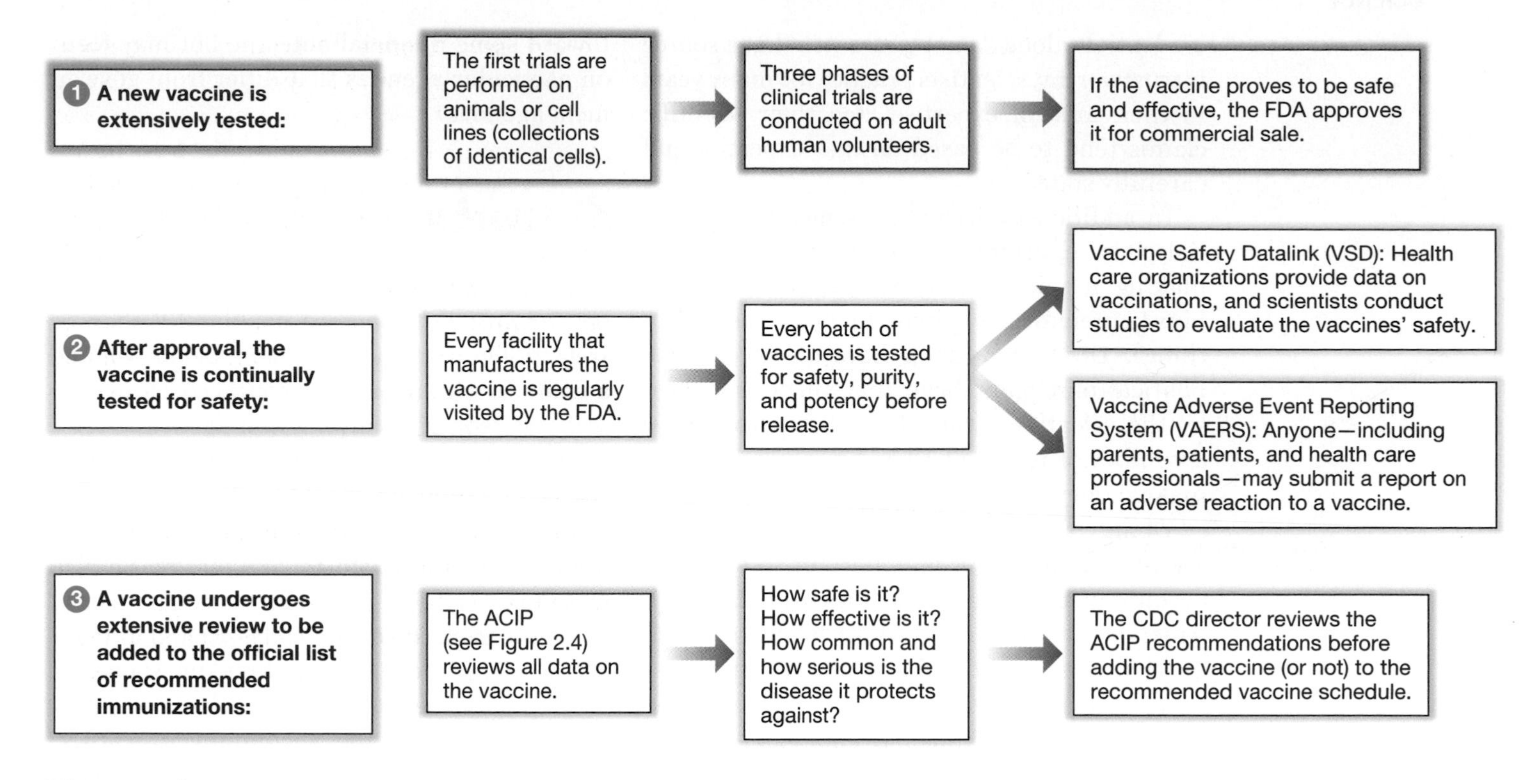

Figure 2.6

Evaluating vaccines: an ongoing process

Vaccines are continually tested and evaluated for effectiveness, safety, and side effects. A minor side effect might be a sore arm or a low-grade fever. A major side effect is very rare. A serious allergic reaction to the measles, mumps, and rubella (MMR) vaccine, for example, happens less than once in a million doses. The severity of the disease a vaccine works against, as well as how likely a person is to be infected with the disease, is also considered in determining whether to recommend the vaccine.

Q1: Why do the manufacturers of vaccines begin with tests on animals or cell lines before moving on to adult human subjects?

Q2: What types of ongoing testing and reporting are vaccines subjected to?

Q3: What do FDA, ACIP, and CDC stand for, and what is the role of each in evaluating vaccines?

See Appendix A for answers to the figure questions.

take a multivitamin every morning? How often should I exercise? Others are larger decisions: Do I support taxes meant to help the environment? Should I use my cell phone less, based on the latest science? Should I get vaccinated? The good news is that you can learn to evaluate scientific claims. You can be skeptical about claims and, using *critical thinking*, make scientifically literate decisions for yourself and the world.

Credentials, Please

Laura's campaign videos encouraged families to have their daughters get the HPV vaccine. One of her key messages was that parents and young people should get their health information from reputable sources and ask questions of health care professionals.

But how do you know when a source is reputable? Which professionals should you trust? There are many easy steps you can take to assess a scientific claim. First, find out if the person making the claim has the right **credentials**, the qualifications to be recognized as an authority on the subject. If the person has an advanced degree, such as a PhD (doctor of philosophy) or an MD (doctor of medicine), is the degree in the field related to the particular claim? PhDs in physics do not have training in germ theory, for example, and medical doctors do not have training in atmospheric science. Although good credentials alone do not guarantee that a source is trustworthy, scientists practice for many years in their area of expertise, and their scientific claims tend to be based on that expertise and carefully stated.

In addition to evaluating someone's credentials, it is important to assess whether the person making a scientific claim has an agenda or **bias** (a prejudice or opinion for or against something). Does the person have an ideological, political, or religious belief that will be supported by the scientific claim? Does the person have a conflict of interest? Does he or she stand to make money if others accept the claim?

Doing scientific research almost always requires money, so it is important to take into account where the money comes from. In North America, the vast majority of **basic research** in science is funded by the federal government—that is, by taxpayers. Basic research is intended to expand the fundamental knowledge base of science. In the United States, Capitol Hill appropriates more than $39 billion each year for basic and medical research in the life sciences, including biomedicine and agriculture. Researchers must compete vigorously for the limited funds, and this competition helps ensure that public money goes toward supporting high-quality science. Research funded by the government is normally not considered biased, as the funding comes from taxpayers (who do not, as a group, have direct financial stakes in the outcome). For example, studies that the CDC and the NIH sponsor about vaccines and vaccine safety are considered free of bias because they are funded by taxpayers rather than the pharmaceutical industry (which does have a financial stake in the outcome).

Industries and businesses also spend a great deal of money funding science. Often they support **applied research**, in which scientific knowledge is applied to human issues and often commercial applications. In some cases, researchers funded by industry may have a bias in favor of whatever that industry is selling. Funding from industry does not necessarily mean that a scientific claim is incorrect, but the claim should be looked at closely to rule out bias. Philanthropic organizations, such as the Bill & Melinda Gates Foundation, a nonprofit charity, also fund scientific research. Not-for-profit companies typically have less bias toward some financial outcome but may focus on particular agendas that differ from government agendas.

Reputable Sources

In 2017, Laura was invited to share her story on *The Late Late Show*, a popular TV show in Ireland. On camera with a long blond wig and a smile as usual, Laura encouraged parents to seek out reputable information about the HPV vaccine. "It's natural to be scared of things," she said. "I probably would have been scared of things if I didn't have the full information." She urged parents to go onto reputable websites to "get the facts."

As you seek to get the facts—that is, investigate a scientific claim—your first stop should be the library or the internet to get a basic overview of the topic from the **secondary literature**, which summarizes and synthesizes an area of research. Good secondary sources include textbooks, review articles, reputable news outlets (such as the *New York Times*), and popular science magazines (such as *Popular Science* and *Discover*).

For secondary literature on the internet, try to visit sites that are affiliated with the government (such as the CDC's website), a university, or a respected institution (such as a major hospital or museum). Wikipedia often has overview articles that link to science blogs and review articles in science journals. It's important to check the credentials of the person or people behind a resource, especially on the internet. Anonymous sources are not to be trusted.

When evaluating a scientific claim, you may need more detailed information than is available in the secondary literature, especially if you're dealing with a particularly important life decision or if the area of science involved is changing rapidly. In that case, you should next review the **primary literature**, where scientific research is first published (**Figure 2.7**). Primary sources include technical reports, conference proceedings, and dissertations, but the most important primary sources are peer-reviewed scientific journals, such as *JAMA* (*Journal of the American Medical Association*) and *Ecology* (published by the Ecological Society of America).

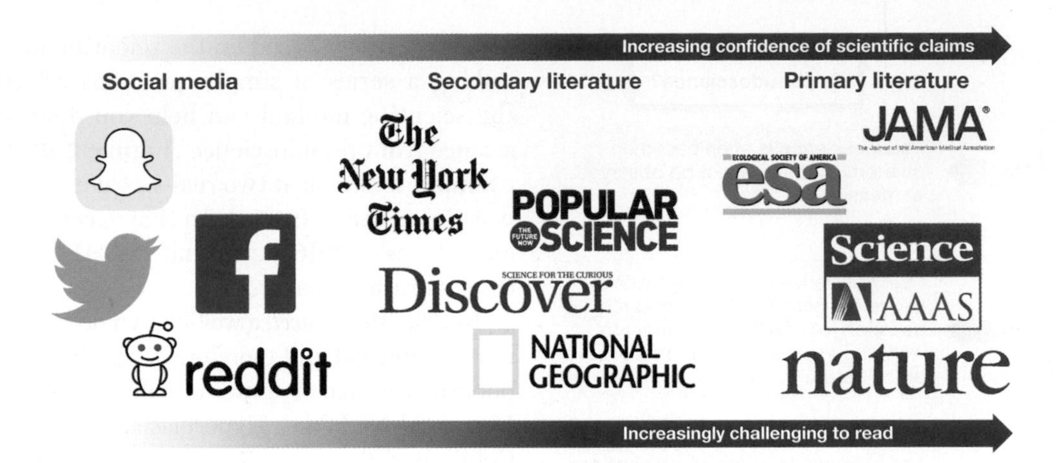

Figure 2.7

Scientific claims in the media and literature

It is easy to find and read scientific claims in social media. However, this is not a good source of scientific information. For help in making important life choices, it is important to go to the secondary literature or even the primary scientific literature for accurate and reliable information.

Q1: Why are we less confident of scientific claims made over social media?

Q2: Where would you place a blog in this figure? Would it matter whether the blog was written by a practicing scientist? Explain your reasoning.

Q3: Give an example of when you would rely on secondary literature to evaluate a scientific claim and an example of when you would go to the primary literature. What is the basis of that decision?

See Appendix A for answers to the figure questions.

Laura took her own advice and did continuous background research into the HPV vaccine. She could cite safety studies with the best of them, Morrissey recalls, and was always the first to share new primary literature about the vaccine. Her efforts continue to be available on her Twitter feed (**Figure 2.8**).

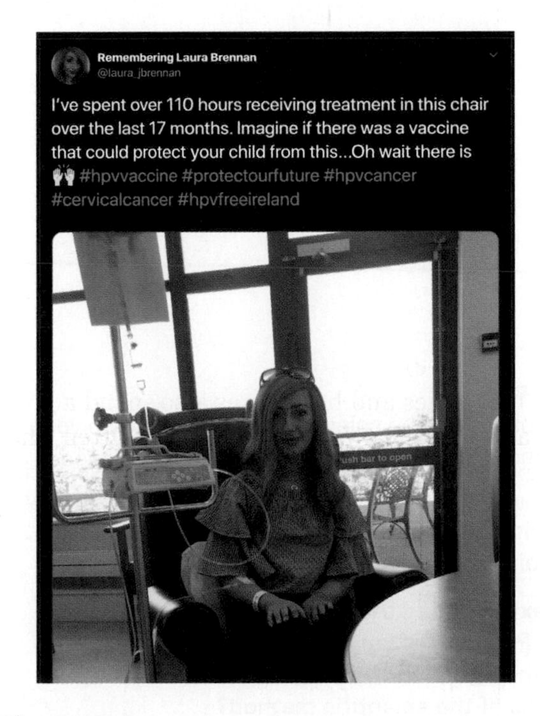

Figure 2.8

@laura_jbrennan

Laura Brennan used her Twitter feed to share the literature on HPV and cervical cancer.

Confronting Pseudoscience

After Laura's appearance on *The Late Late Show*, the World Health Organization (WHO) invited her to spread the message about the HPV vaccine across Europe to help combat misinformation being spread about the vaccine. Instead of turning to reputable sources, parents were being guided by pseudoscience.

A claim that superficially looks as though it is backed by science is called **pseudoscience**. Pseudoscience is characterized by scientific-sounding statements, beliefs, or practices that

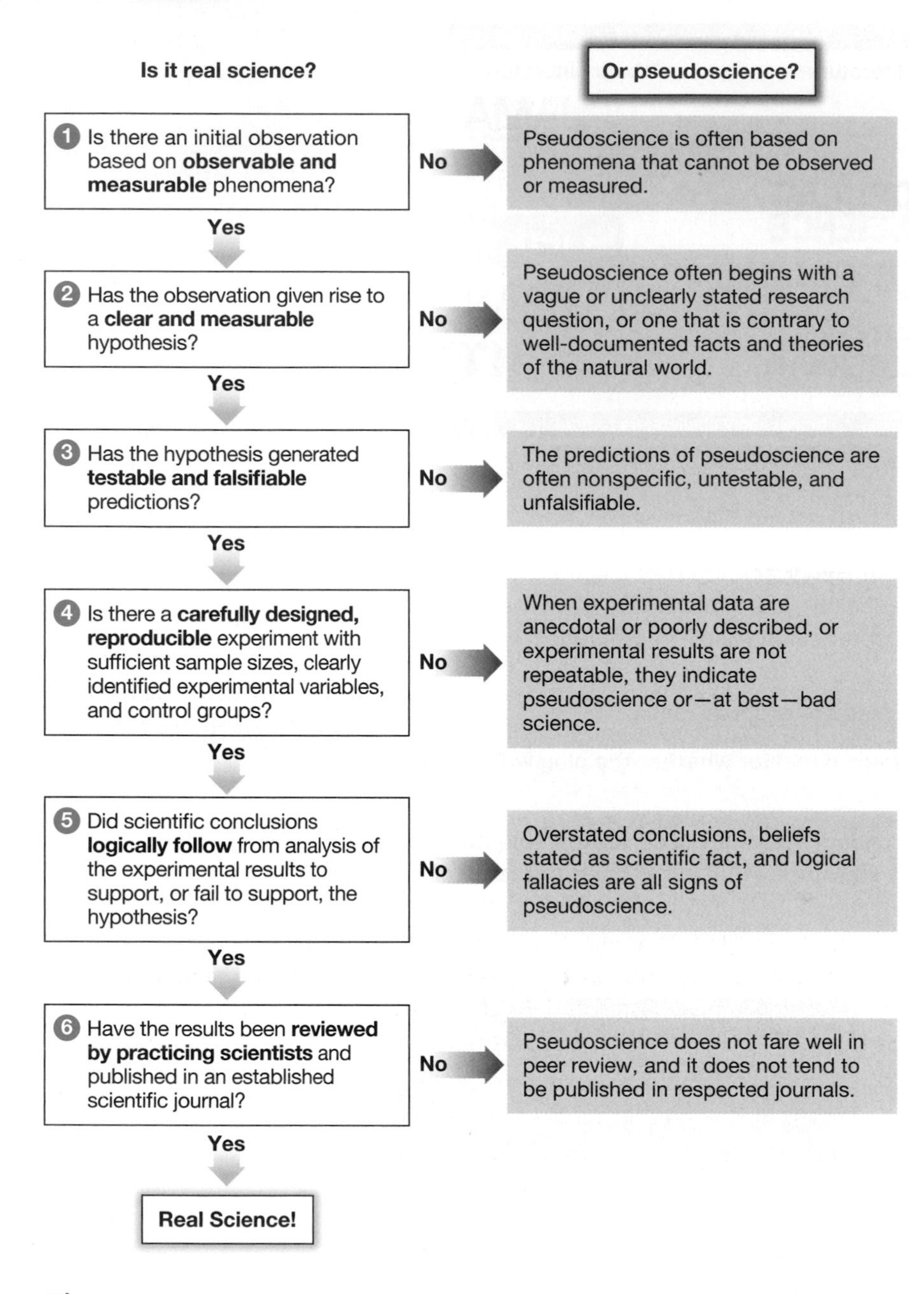

Is it real science?
Or pseudoscience?
1 Is there an initial observation based on observable and measurable phenomena?
No
Pseudoscience is often based on phenomena that cannot be observed or measured.
Yes
2 Has the observation given rise to a clear and measurable hypothesis?
No
Pseudoscience often begins with a vague or unclearly stated research question, or one that is contrary to well-documented facts and theories of the natural world.
Yes
3 Has the hypothesis generated testable and falsifiable predictions?
No
The predictions of pseudoscience are often nonspecific, untestable, and unfalsifiable.
Yes
4 Is there a carefully designed, reproducible experiment with sufficient sample sizes, clearly identified experimental variables, and control groups?
No
When experimental data are anecdotal or poorly described, or experimental results are not repeatable, they indicate pseudoscience or—at best—bad science.
Yes
5 Did scientific conclusions logically follow from analysis of the experimental results to support, or fail to support, the hypothesis?
No
Overstated conclusions, beliefs stated as scientific fact, and logical fallacies are all signs of pseudoscience.
Yes
6 Have the results been reviewed by practicing scientists and published in an established scientific journal?
No
Pseudoscience does not fare well in peer review, and it does not tend to be published in respected journals.
Yes
Real Science!

Figure 2.9

Science or pseudoscience?

A series of simple questions based on the scientific method can help you determine whether a "scientific" study is real science or pseudoscience.

Q1: Why do all the green arrows continue down to "real science," whereas each red arrow points directly to "pseudoscience"?

Q2: Return to the main text and read the rest of this section and the following section, "Real or Pseudo?" Then come back and answer this question: What part(s) of the figure show where Wakefield's study failed to meet the standards of the scientific method?

Q3: What is the scientific claim behind the vaccine-autism controversy? What is an alternative scientific claim?

See Appendix A for answers to the figure questions.

are not actually based on the scientific method. Asking a series of simple questions related to the scientific method can help you distinguish science from pseudoscience (**Figure 2.9**). As an example, let's look at two real-world examples of pseudoscience: a 1998 claim that vaccines cause autism and a 2016 claim that the HPV vaccine causes brain damage.

In 1998, the *Lancet*, a well-known peer-reviewed medical journal and therefore a reputable source, published a paper unremarkably titled "Ileal-Lymphoid-Nodular Hyperplasia, Non-specific Colitis, and Pervasive Developmental Disorder in Children." The paper presented a study of 12 children ranging in age from 3 to 10 who had experienced a loss of language skills—a symptom of autism spectrum disorders—as well as diarrhea and abdominal pain. Parents of 8 of the 12 children said the onset of symptoms occurred shortly after the children's immunization with the measles, mumps, and rubella (MMR) vaccine.

The authors of the paper, a team of 12 researchers, concluded that more research was needed, but in a press conference when the paper was published, one of the authors, a British doctor named Andrew Wakefield, made the claim that single vaccines, rather than the MMR triple vaccine, were likely to be "safer" for children. The study and press conference sparked widespread fear among parents that the MMR vaccine could cause autism. This fear has led directly to the anti-vaccination, or anti-vax, movement, which continues to generate controversy and headlines.

The Wakefield et al. study was published around the time that officials began documenting an increase in the rate of autism. The CDC, which funds basic research to follow trends of disease in the United States, said that the rise was likely due to heightened awareness of autism, more screening within schools, and a willingness to label the condition. But after Wakefield publicly suggested that the MMR vaccine might be causing autism, the press and other organizations began to report that the rise in autism cases was caused by the increased use of vaccines.

Linking the rising rate of autism with increased use of vaccines is a correlation. **Correlation** means that two or more aspects of the natural world behave in an interrelated manner. That is, if one aspect shows a particular value, we can predict a value for the other aspect(s). But correlation does not prove **causation**. In other words, just because two aspects are correlated does not mean that a change in one aspect *causes* a change in another. A correlation may suggest possible causes for a phenomenon or phenomena, but it does not establish a cause-and-effect relationship. For example, there is a correlation between organic food sales and the increase in autism. From 1998 to 2007, organic food sales increased hand in hand with the number of autism diagnoses. These changes are correlated, but there is no scientific evidence that eating organic food causes autism (**Figure 2.10**).

Another correlation that spurred fears about vaccines is that the onset of autism symptoms occurs at about the same age that children receive vaccinations. Most children receive the MMR vaccine at about 15 months old, which is shortly before the first symptoms of autism are often noticed. Parents of children with autism saw their children begin to exhibit symptoms of the illness after their vaccinations and therefore directly observed a correlation between the injection of the vaccine and the onset of autism spectrum disorder.

Yet, again, correlation does not prove causation. Only scientific experiments can demonstrate causation. So was Wakefield's claim about the link between MMR vaccine and autism based on good science or pseudoscience? Did the MMR vaccine cause autism?

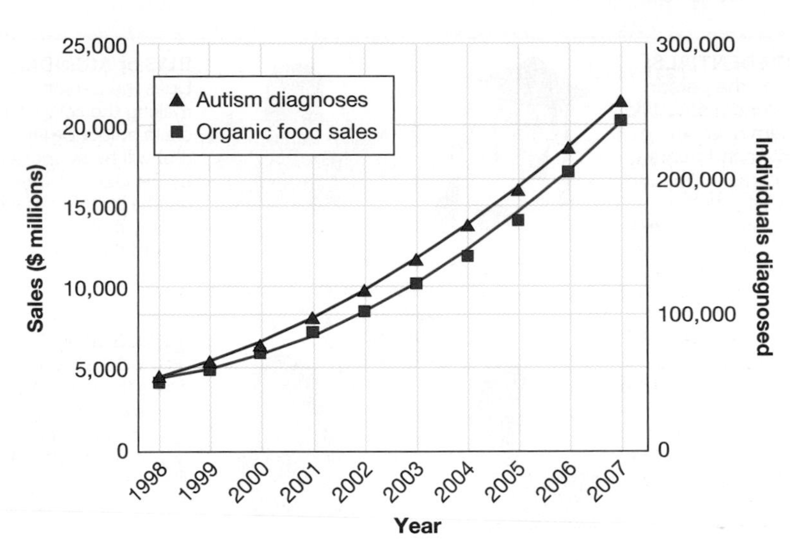

Figure 2.10

Correlation is not causation: organic food and autism

Reddit user Jasonp55 created a tongue-in-cheek demonstration of why it is important not to assume that correlation means causation. Jason used real data on organic food sales and the increase of autism from 1998 to 2007. As this graph shows, the two phenomena are highly correlated. However, the graph cannot show scientific evidence that one causes the other.

Q1: How much did organic food sales grow during the period covered in the graph? How much did the autism diagnoses grow?

Q2: Why might both organic food sales and autism diagnoses have increased during this time period? A Reddit user in the original discussion thread suggested that both might be affected by increasing wealth in the United States. How might an increase in wealth affect these variables?

Q3: How has the vaccine-autism debate been confused by people misinterpreting correlation as causation?

See Appendix A for answers to the figure questions.

Real or Pseudo?

It was quickly obvious that Wakefield's study did not live up to the standards of good scientific research (**Figure 2.11**). First, the study was small—only 12 children participated—yet the conclusions were grand: Wakefield suggested that all children should stop receiving the MMR vaccine. Sample size is extremely important in observational studies, where small samples may skew data to one extreme. Large sample sizes are more likely to represent a whole population, are less likely to be affected by outliers, and provide the power to draw more accurate conclusions.

A second problem with the Wakefield study was that it did not use a random sample of children. Instead, the participants were picked specifically for their symptoms. Third, the study did not have a control group, such as children who had been vaccinated but did not show signs of autism or children who showed signs of autism but had not been vaccinated. Finally, and most important, the finding could not be repeated. In the 20 years after the Wakefield paper was published, the proposed link between the MMR vaccine and autism was studied exhaustively, in over 100 published papers that included millions of children. There is no evidence supporting a link: vaccination does not cause autism, and the two are not even directly correlated. In fact, studies first published in March 2014 provide evidence that autism begins in the womb, before birth, and it appears likely that the causes of

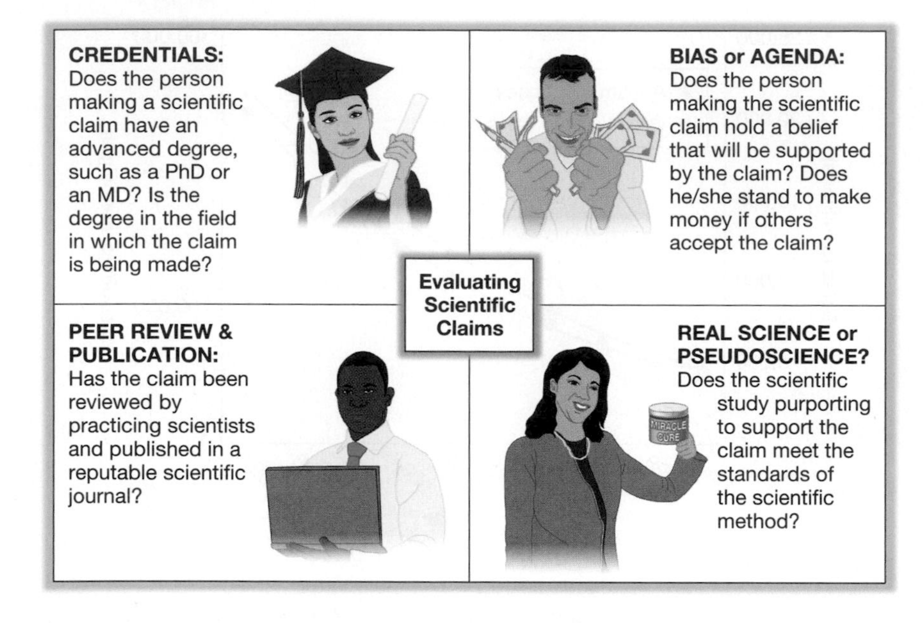

Figure 2.11

Evaluating scientific claims

We are constantly bombarded with scientific claims. A few simple questions will help you evaluate which of these claims are valid and which are not.

Q1: Why is it important to know the education and expertise of a person making a scientific claim?

Q2: List at least five biases that people making scientific claims might have.

Q3: Describe a situation in which you might *not* dismiss the scientific claim of a person who did not have appropriate credentials or who had a bias toward the claim.

See Appendix A for answers to the figure questions.

autism are many, including genetic and environmental factors but not vaccines.

The paper that started the vaccine-autism debate, Wakefield's 1998 *Lancet* paper, proposed a claim that is incorrect. All told, it's okay for science to be incorrect. There is no expectation that every single one of the thousands of studies published each year will be right. Newly published papers can and do present new ideas, and then the scientific community evaluates and tries to falsify those new ideas. Sometimes further research does falsify an idea—that is, the hypothesis is not supported. Sometimes further research reaches the same conclusions, and the hypothesis is supported.

Yet in the case of the *Lancet* paper, there was bias and wrongdoing. Years after publication of the paper, it came to light that Andrew Wakefield had received large amounts of money as a paid expert for lawyers who were suing vaccine manufacturers. Wakefield had also applied for a patent on a vaccine that would rival the most commonly used MMR vaccine.

Ultimately, Wakefield lost his medical license for "irresponsible and dishonest" conduct, the other authors of the paper disavowed the study and its conclusions, and in February 2010, the *Lancet* officially retracted the paper—a rare action for publishers (**Figure 2.12**). A peer-reviewed paper is retracted (withdrawn as untrue or inaccurate) by a publisher or author when its findings are no longer considered trustworthy because of error, plagiarism, a violation of ethical guidelines, or other scientific misconduct. Following his disgrace in the United Kingdom, Wakefield moved to Texas, where he continues to peddle his discredited claims.

Now for the second case of pseudoscience—a 2016 paper claiming the HPV vaccine causes brain damage. The paper, from a team led by Toshihiro Nakajima of Tokyo Medical University in Japan, was published in the reputable journal *Scientific Reports*. It described impaired movement and brain damage in mice given the HPV vaccine, but the experiment had been poorly conducted. As analyzed by two independent research groups, the original experiment involved giving mice enormous doses of the HPV vaccine—a thousand times greater than what is normally given to people, who are significantly larger than mice—along with a toxin that caused leaks in the barrier that protects the brain. In addition, according to other scientists, the conclusions were overstated and did not match the data. The paper was not scientifically sound.

In 2018, the journal retracted the paper, yet HPV researchers worry the pseudoscience was in circulation too long and will be used by the anti-vaccination lobby to stoke fears about the HPV vaccine.

Fears versus Facts

In the United States, HPV vaccination rates remain "critically low," according to a 2019 study in the *Journal of Infectious Disease*. Only 16 percent of U.S. adolescents had been fully vaccinated against HPV by the time they turned 13, the study authors found.

A recent **meta-analysis**—a research paper that combines results from different studies—suggests how to improve HPV vaccination rates. By analyzing the methods and conclusions of 62 studies, scientists found that strong physician recommendation for the vaccine had the greatest influence on parents' decision to vaccinate, followed by alleviating HPV safety concerns and promoting positive belief in vaccines.

The best way to counter the spread of anti-vaccine pseudoscience is with clear information, and the scientific evidence in support of the HPV vaccine continues to grow. In countries around the world, study after study finds that routine vaccination of girls aged 11 to 13 has led to a dramatic reduction in cervical precancers later in life. Data on the impact of the vaccine on boys and men are forthcoming, as the vaccine was introduced later for males than for females.

At the CDC, researchers called epidemiologists, who investigate patterns and causes of disease, have been tracking the number of cervical precancers in the years since the HPV vaccine became available. Precancers are lesions on the cervix indicating abnormal cell changes that occur much earlier than cervical cancer, which can take decades to develop. By tracking precancers, epidemiologists can get a quick indicator of how commonly cervical disease is occurring. "We know the vaccine was highly effective in clinical trials, but monitoring health outcomes is important to understand the real-world impact of vaccination," says Nancy McClung, an epidemiologist with the CDC's HPV Vaccine Impact Monitoring Project.

Using data from over 10,000 biopsy specimens collected between 2008 and 2014, McClung and colleagues found that in women aged 18 to 39 who got the HPV vaccine, the proportion of cervical precancers due to HPV types 16 and 18 decreased from 55 percent to 33 percent. "This is evidence the vaccine is working to prevent cervical disease in young women," says McClung. They published their results in a peer-reviewed paper in 2019.

The CDC team also found a decline, although a smaller one, in the number of precancers among unvaccinated women due to HPV-16/18. McClung says that decline is likely due to herd protection, which occurs when a significant proportion of a population has developed immunity to a disease though vaccination, making it less likely to spread. When a critical portion of a population is vaccinated, typically 80 percent to 95 percent, then the spread of disease is contained, and herd protection becomes **herd immunity** (**Figure 2.13**). In other words, vaccinating a large number of people keeps germs out of circulation and protects the vulnerable members of the community.

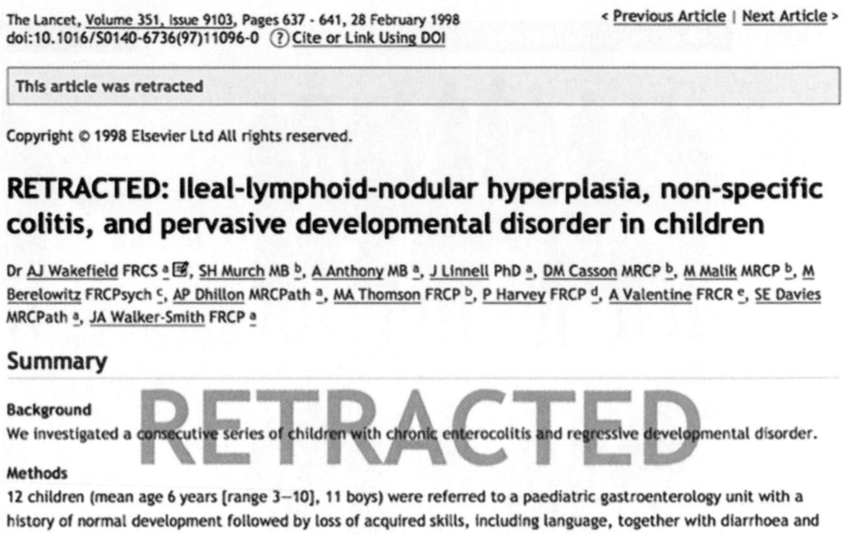

When parents opt not to vaccinate their children, herd immunity decreases or disappears. Rates of vaccination in the United States have been hovering around 75 percent because of misconceptions about the safety of vaccines and the spread of pseudoscience. This low level of immunization puts at risk not only the children who are not vaccinated but also those who *cannot* be vaccinated, such as infants too young for a vaccine or people who are genetically unable to respond to a vaccine.

Because of that loss of herd immunity, infectious diseases of the past, some thought to be eliminated, are roaring back. In 2000, measles—a highly contagious infection that spreads through coughing and sneezing and can

The Lancet, Volume 351, Issue 9103, Pages 637 - 641, 28 February 1998
doi:10.1016/S0140-6736(97)11096-0 (?) Cite or Link Using DOI

< Previous Article | Next Article >

This article was retracted

Copyright © 1998 Elsevier Ltd All rights reserved.

RETRACTED: Ileal-lymphoid-nodular hyperplasia, non-specific colitis, and pervasive developmental disorder in children

Dr AJ Wakefield FRCS [a], SH Murch MB [b], A Anthony MB [a], J Linnell PhD [a], DM Casson MRCP [b], M Malik MRCP [b], M Berelowitz FRCPsych [c], AP Dhillon MRCPath [a], MA Thomson FRCP [b], P Harvey FRCP [d], A Valentine FRCR [e], SE Davies MRCPath [a], JA Walker-Smith FRCP [a]

Summary

Background

We investigated a consecutive series of children with chronic enterocolitis and regressive developmental disorder.

RETRACTED

Methods

12 children (mean age 6 years [range 3–10], 11 boys) were referred to a paediatric gastroenterology unit with a history of normal development followed by loss of acquired skills, including language, together with diarrhoea and

Figure 2.12

Wakefield et al.

The paper that precipitated the vaccine-autism scare is debunked and retracted.

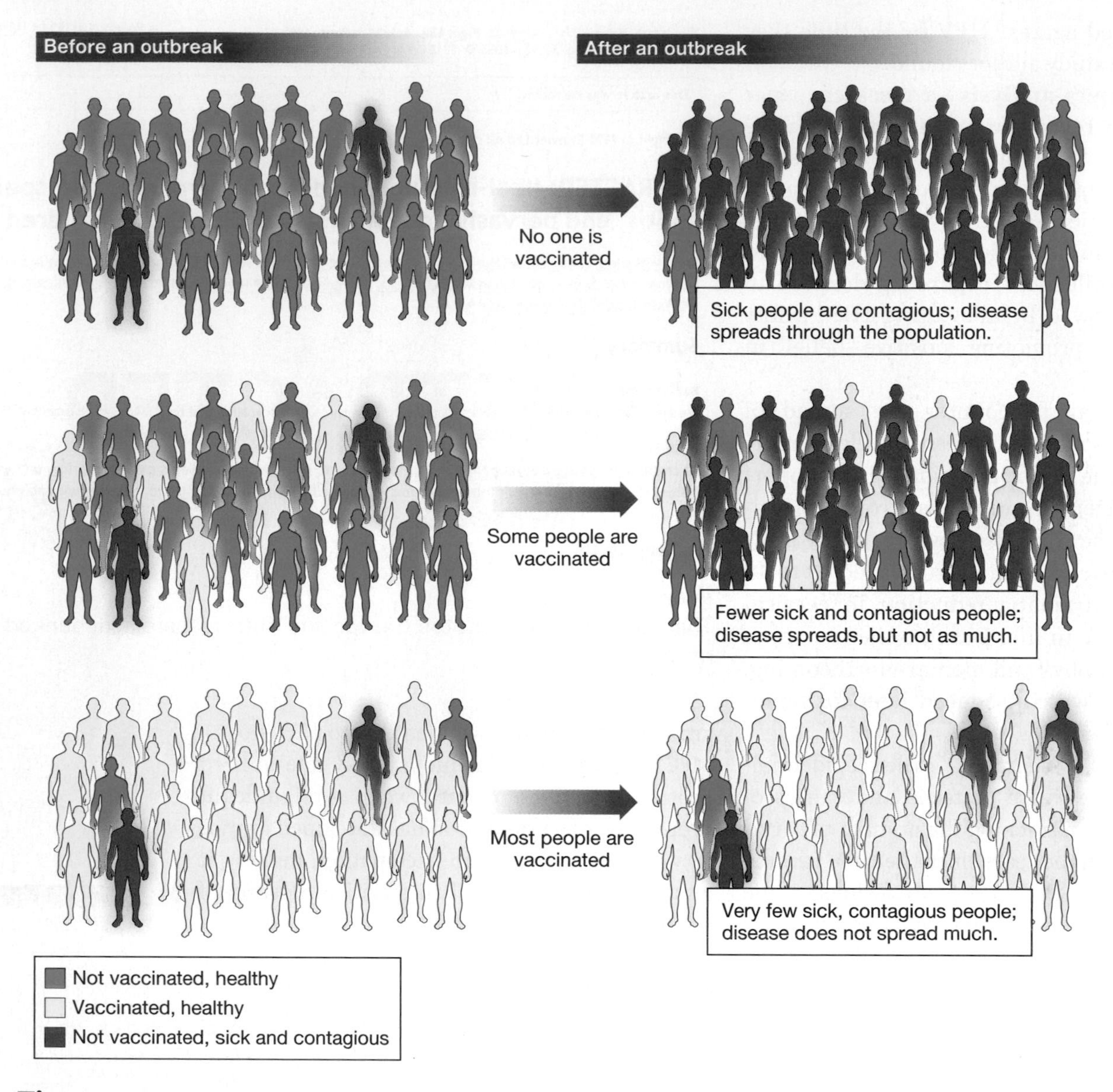
Before an outbreak
After an outbreak
No one is vaccinated
Sick people are contagious; disease spreads through the population.
Some people are vaccinated
Fewer sick and contagious people; disease spreads, but not as much.
Most people are vaccinated
Very few sick, contagious people; disease does not spread much.
Not vaccinated, healthy
Vaccinated, healthy
Not vaccinated, sick and contagious

Figure 2.13

Vaccine prevalence and herd immunity

When the majority of the population is not vaccinated, an outbreak of disease will spread further and may cause disease and death in members of the population who are too young to be vaccinated or have a compromised immune system. However, when most of the population is vaccinated against a contagious disease, the disease will spread little during an outbreak; this tighter containment of a disease is the result of what is known as herd immunity. Image modified from the National Institute of Allergy and Infectious Diseases (NIAID).

Q1: What happens to an immunized person when a disease spreads through a population? (*Hint:* In the graphic, follow an immunized individual before and after a disease spreads.)

Q2: How does vaccination of an individual help that person's community?

Q3: Explain why a disease is less likely to spread to vulnerable members of a population if most people are immunized.

See Appendix A for answers to the figure questions.

Safety in Numbers

In 1999, the United Kingdom began a meningitis vaccination program for children up to age 18. Meningitis rates rapidly fell off throughout the population—children and adults alike—thanks to both the effectiveness of the vaccine and the protective effects of herd immunity. Within a decade, rates were close to 0. Models show that without herd immunity, the number of meningitis cases would have rebounded. The U.S. CDC recommends meningitis vaccination up to age 23, as outbreaks have been reported on college campuses in recent years. Meningitis, an inflammation of the membranes (meninges) surrounding the brain and spinal cord, can be life-threatening.

Assessment available in smartwork

Meningitis cases in England and Wales

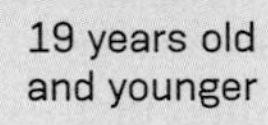

19 years old and younger

20 years old and older

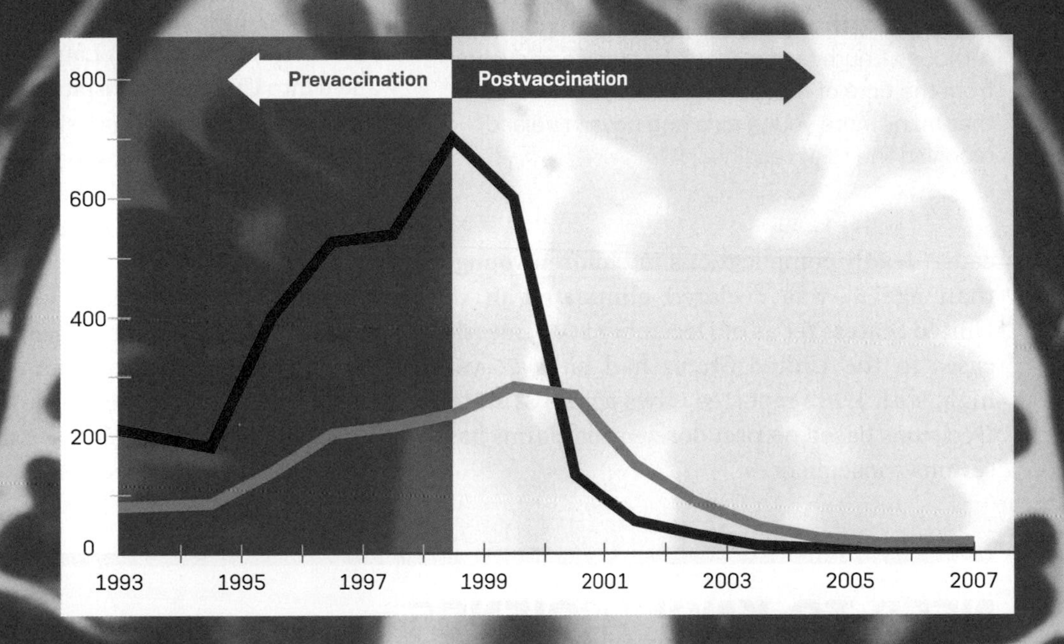

Observed cases

Predicted cases (with herd immunity)

Predicted cases (no herd immunity)

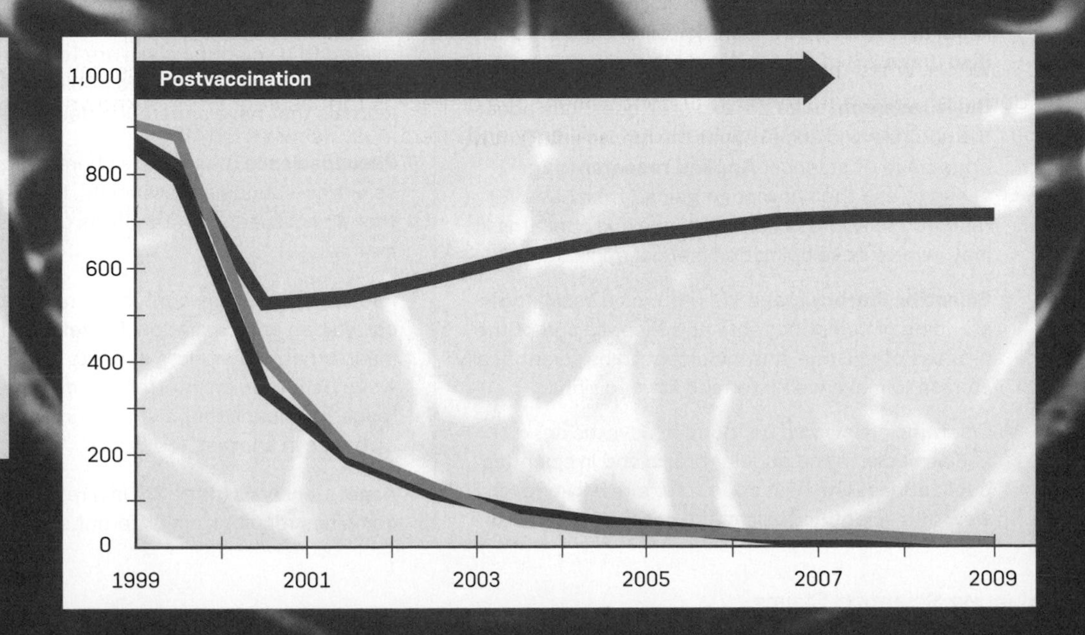

Figure 2.14

Laura Brennan, gone but not forgotten
Although Laura died in 2019, her tireless advocacy from the time of her diagnosis in 2016 ensured that many more young girls and boys in Ireland received the HPV vaccine.

cause deadly complications in children younger than age 5—was declared eliminated in the United States. Yet as of December 2019, measles cases in the United States had hit a 25-year high, with 1,282 reported cases across 31 states. Decisions based on pseudoscientific claims have serious consequences.

"This Vaccine Saves Lives. . . . It Can Save Yours"

Laura Brennan's campaign to promote the HPV vaccine worked. Since she first contacted the HSE in 2017, the uptake of the HPV vaccine in Ireland has increased dramatically, from 51 percent in 2017 to 70 percent in 2019. In March 2019, over 22,000 girls got the vaccine. At age 26, Laura was awarded the first ever Patient Advocacy Medal by the Royal College of Physicians in Ireland, whose spokesperson said, "What Laura has done is immeasurable."

On March 19, 2019, Laura and her family walked into University Hospital Limerick for what they expected to be a short stay. The next morning, Laura took a turn for the worse. She died just before noon, surrounded by her family (see tribute video at https://youtu.be/UeVdgSWv28s).

After Laura's death, people around Ireland and the world retweeted her words from a 2018 speech: "I'll be gone soon. Once I am gone there is nothing more I can do about it, so I hope you all listen to me while you have a chance. This vaccine saves lives. It could have saved mine, but it can save yours" (**Figure 2.14**).

NEED-TO-KNOW SCIENCE

- A **scientific claim** (page 25) is a statement about how the natural world works that can be tested using the scientific method. To evaluate scientific claims, it is important to look at the **credentials** (page 27) and **bias** (page 28) of those making the claim.
- **Basic research** (page 28) explores questions about the natural world and expands the fundamental knowledge base of science. **Applied research** (page 28) seeks to use the knowledge gained from basic research to address human issues and concerns. It may involve developing commercial applications.
- **Scientific literacy** (page 26) requires a basic understanding of scientific facts and theories and of the process of science. It is important to be scientifically literate to make well-informed life decisions.
- Scientific claims can be found in advertising, in social media, in the popular press, and in scientific publications. The best source for a review or overview of a scientific topic is the **secondary literature** (page 28), including the popular press, reputable websites, and review articles in scientific journals.
- The actual experimental or observational results related to a scientific claim are found in the **primary literature** (page 28), articles published in scientific journals that have undergone peer review.
- **Pseudoscience** (page 29) is characterized by scientific-sounding statements, beliefs, or practices that do not meet the standards of the scientific method.
- Pseudoscience may confuse or misrepresent correlation and causation. **Correlation** (page 30) means that two or more aspects of the natural world behave in an interrelated manner. **Causation** (page 31) means that a change in one aspect causes a change in another.
- A **meta-analysis** (page 33) is a research paper that combines results from different studies.

THE QUESTIONS

See Appendix B for answers.

The Basics

1 Which of the following items should receive the *least* consideration when evaluating a scientific claim?

(a) the scientific credentials of the person making the claim

(b) your personal beliefs and values

(c) whether a study supporting the claim has been published in a peer-reviewed scientific journal

(d) whether a study supporting the claim meets the standards of the scientific method

(e) any possible biases of the person making the claim

2 Scientific literacy means that you

(a) are able to easily read and understand a scientific journal article.

(b) have taken a university-level science course.

(c) understand the process of science and basic scientific facts and theories.

(d) enjoy reading current science news in newspapers and blogs.

(e) are a good critical thinker.

3 "Correlation does not prove causation" means that

(a) if two variables are correlated, one is likely to have caused the other.

(b) if changes in one variable cause changes in another, the variables are not correlated.

(c) only experimental research can answer questions about the natural world.

(d) although two variables are interrelated, changes in one do not necessarily cause changes in the other.

(e) none of the above

4 Link each term with the correct definition.

scientific literacy	1. Uses scientific knowledge to address human issues.
basic research	2. Helps in making informed life choices.
applied research	3. Consists of peer-reviewed scientific journal articles.
secondary literature	4. Gives an overview of scientific findings on a subject.
primary literature	5. Contributes to fundamental science knowledge.

5 Select the correct terms:

Evaluating a scientific claim begins with reviewing the (**credentials** / **fame**) of the person or people making the claim. It is also important to know whether those making a particular claim have a (**detachment** / **bias**), a vested interest in whether or not the claim is true. To gain an overview of scientific studies related to the claim, it is helpful to read the (**primary literature** / **secondary literature**).

6 You are trying to determine whether a scientific claim is based on real science or pseudoscience. Place the following questions you will address in the correct order by numbering them from 1 to 6.

_____ a. Is the initial observation based on observable and measurable phenomena?

_____ b. Has the study been reviewed by practicing scientists and published in an established scientific journal?

_____ c. Has the hypothesis generated specific, testable, and falsifiable predictions?

_____ d. Has the observation given rise to a clear and measurable hypothesis?

_____ e. Are the experimental design and analysis well described, well designed, reproducible, and conducted with a sufficient sample size?

_____ f. Are the scientific conclusions logical, based on evidence, and justified, given the study results?

Challenge Yourself

7 An example of basic research is

(a) studying how hummingbirds learn song.

(b) investigating how the melting of polar ice caps affects agriculture.

(c) looking at possible genetic contributions to autism spectrum disorder.

(d) designing more effective vaccines for dangerous infectious diseases.

(e) exploring how agricultural waste can be turned into fuel.

8 For each of the following, select the term that best describes the type of literature it represents: primary, secondary, or neither.

_____ a. A research study from Dr. Drake and colleagues on the blood sugar levels of diabetic rats that eat only kale, compared to control diets, is published in a peer-reviewed scientific journal.

_____ b. In the infomercial selling his own health care products, Dr. Horton states, "I believe my personal cure of daily meditation and yoga for diabetes is more effective than any drug I've ever prescribed in 20 years of practicing medicine."

_____ c. Dr. DeBellard insists in her blog that people will lose weight in a healthy manner if they acquire some of her personal parasitic tapeworm.

_____ d. In an article published in the *Annual Review of Nutrition*, a peer-reviewed journal, Dr. Pepper summarizes the last 10 years of basic research on diet and diabetes.

9 Which of the following situations has the greatest potential for biased or inaccurate results in an experimental process? Why?

(a) Ms. Ochoa-Bolton is an outside consultant who is conducting a health and wellness survey for a pharmaceutical company. She knows neither the name of the company nor the name of the drug being tested for the survey.

(b) Ms. Adamian is a research technician surveying a study group on responses to a new cold remedy. She knows only the email address of each participant and asks each one identical questions by computer.

(c) Dr. Wisidagama is evaluating cancer patients for their responses to a new therapeutic drug. She knows which patients are receiving the placebo and which are receiving the drug.

(d) Dr. Waters is analyzing biopsy samples from rats that have been given either a placebo or an experimental drug believed to reduce inflammation. Each sample is identified by a code number, so she cannot tell which treatment each rat received.

(e) Ms. Nuno is conducting a survey of weight loss regimens as part of her master's degree project. Her online survey is anonymous and asks each participant the same questions.

Try Something New

10 Determine whether each of the following statements is likely to represent real science or pseudoscience. In each case of likely pseudoscience, identify which scientific standard is not met.

(a) On his TV show, Dr. Oz says that green coffee beans burn fat, so they help you lose weight without dieting.

(b) The *Wall Street Journal* reports that climate scientists have conspired to exaggerate the effects of global climate change.

(c) Researchers report at a national scientific conference that they have found a genetic link to autism.

(d) Many astrologers agree that people born under the sun sign Aquarius are more intelligent than those born under Scorpio.

(e) A study published in the scientific journal *Diabetes* finds that sleeping in a cooler room may increase metabolic rate and insulin sensitivity.

11 The following graphs illustrate the incidence of pertussis (whooping cough) cases in the United States. The first graph organizes the data by year from 1922–2017, with the inset showing a zoomed-in view of 1990–2017. The second graph organizes the data by age group from 1990–2017. (The 2018 data are not complete in either graph.) DTP, Tdap, and DTaP are different formulations of the vaccine that covers tetanus, diphtheria, and pertussis.

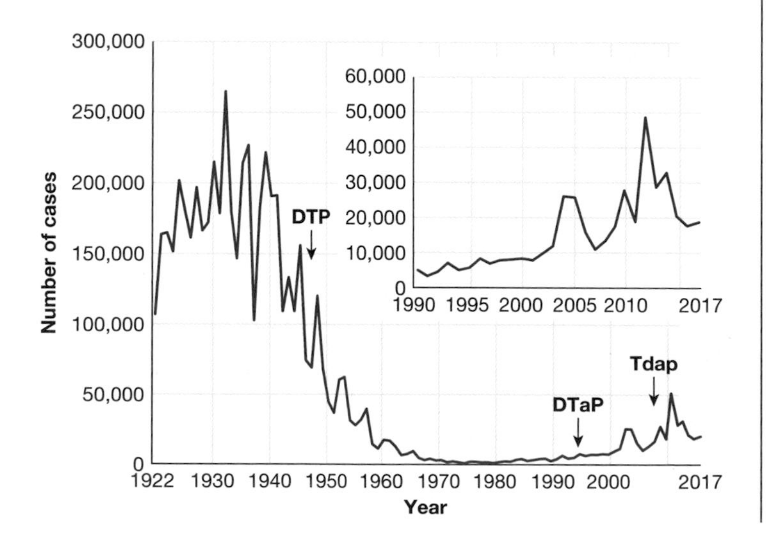

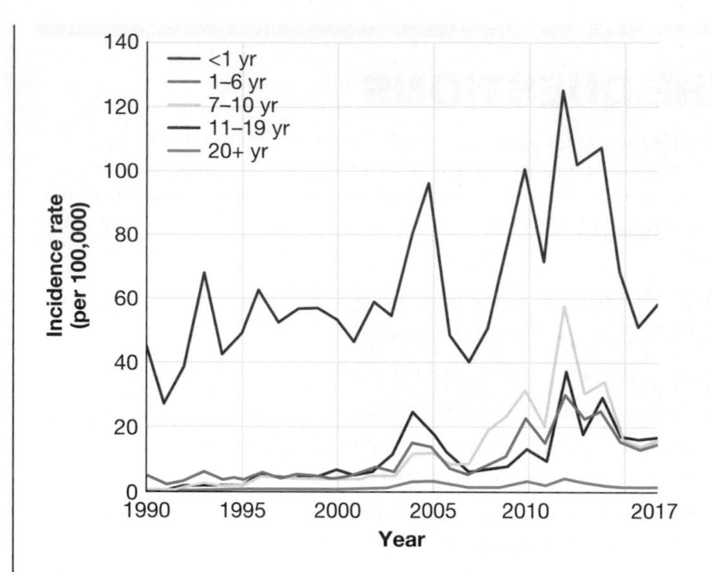

(a) Describe what the first graph shows. What do each of the axes represent? What does any point on the line show? What general trend is seen if all the data shown in the graph are considered?

(b) Why do you think there was an increase in cases of pertussis in the first decade of the twenty-first century?

(c) Describe what the second graph shows. What do each of the axes represent? What does any point on each of the different lines show? What general trend is seen if all the lines are considered? What differs among the lines?

(d) Compare the incidence of pertussis cases in children under 1 year old and people over 20 years old.

(e) Summarize your reflections on reviewing these graphs and the relative risks of pertussis and the pertussis vaccine. What would you recommend to someone trying to decide whether to vaccinate a child?

12 Vaccines don't stop when you are older! Besides the annual flu vaccine, the Advisory Committee on Immunization Practices (ACIP) and CDC recommend that adults over age 50 be vaccinated with Shingrix, which protects against shingles, a devastatingly painful condition caused by the same virus that causes chicken pox. What steps would you go through (when you are 50 or older) to decide whether to be vaccinated with the Shingrix vaccine? (*Hint*: Use the method for evaluating scientific claims, including assessing real science versus pseudoscience.)

13 Cognitive biases are systematic errors in thinking that all of us experience to some extent. Choose one of the following cognitive biases, read about it online (using reputable sources!) and explain how it relates to anti-vaccine beliefs: Dunning-Kruger effect, confirmation bias, neglect of probability, omission bias, or illusory correlation.

Leveling Up

14 **Life choices** Most American universities require students to have the following immunizations: MMR (measles, mumps, and rubella), varicella (chicken pox), and Tdap (tetanus, diphtheria, and pertussis/whooping cough). Most also recommend or require hepatitis A and B, meningococcal conjugate (meningitis), HPV (human papillomavirus), poliovirus, and the annual flu vaccine.

(a) Which of these vaccines have you received? If you have not received one or more of them, why not?

(b) At the time this book was written, there was not yet a vaccine available for SARS-CoV-2, the novel coronavirus that set off a global pandemic. If or when a vaccine is available, will you be vaccinated? Why or why not?

(c) Some school districts allow parents to opt out of vaccines because of their belief system. Do you think this option should be allowed for university students? Why or why not? What are the possible consequences of allowing people to opt out of vaccinations?

15 ***Write Now* biology: evaluating scientific claims** In a 2014 episode of *Last Week Tonight*, John Oliver discussed a poll finding that one in four Americans was skeptical of global climate change. According to a late 2018 poll, that number has climbed to three in four. Still, Oliver dismissed such polls and compared them to a poll asking, "Which number is bigger, 15 or 5?" or "Do owls exist?" Why does Oliver feel that what the American public believes about climate change is not relevant? (*Hint*: It could be argued that each of the people polled was making a scientific claim, either in support of or against the scientific consensus on climate change.)

16 **Is it science?** The 2011 movie *Contagion* depicts the spread of a fictional virus. Watch the movie and make a list of the scientific concepts it presents, noting whether you think each one is real science or pseudoscience. Then use this textbook and internet research to determine whether you were correct or incorrect in each case. If you evaluated any of the concepts incorrectly, reflect on how you came to your initial conclusions.

17 **What do *you* think?** Read *Small Steps: The Year I Got Polio*, Peg Kehret's 1996 memoir about contracting polio as a seventh-grader. Keep a reflective journal, documenting your emotional reactions to each chapter. Also note what you learn about the biology of the polio virus, the symptoms of polio, and the treatment options (or lack thereof) in the late 1940s. In your final entry, reflect on how this novel did or did not change your feelings toward vaccination.

Digital resources for your book are available online.

Breaking Good

To slow the skyrocketing number of overdose deaths, chemists race to design safer opioid painkillers.

After reading this chapter you should be able to:

- Compare and contrast the types of bonds used to form molecules.
- Diagram the hydrogen bonds that form between water molecules, and explain how those bonds produce the unique properties of water.
- Describe the structure and function of the four major classes of biomolecules and their roles in biological systems.
- Create a graphic depicting how each of the four major classes of biomolecules and their subunits are related.
- Explain the difference between hydrophilic and hydrophobic molecules and between acids and bases.
- Predict whether a solution has a high or low concentration of free hydrogen ions by using its pH number.
- Describe the chemical qualities of carbon that make it the basis of life on Earth.

CHEMISTRY OF LIFE

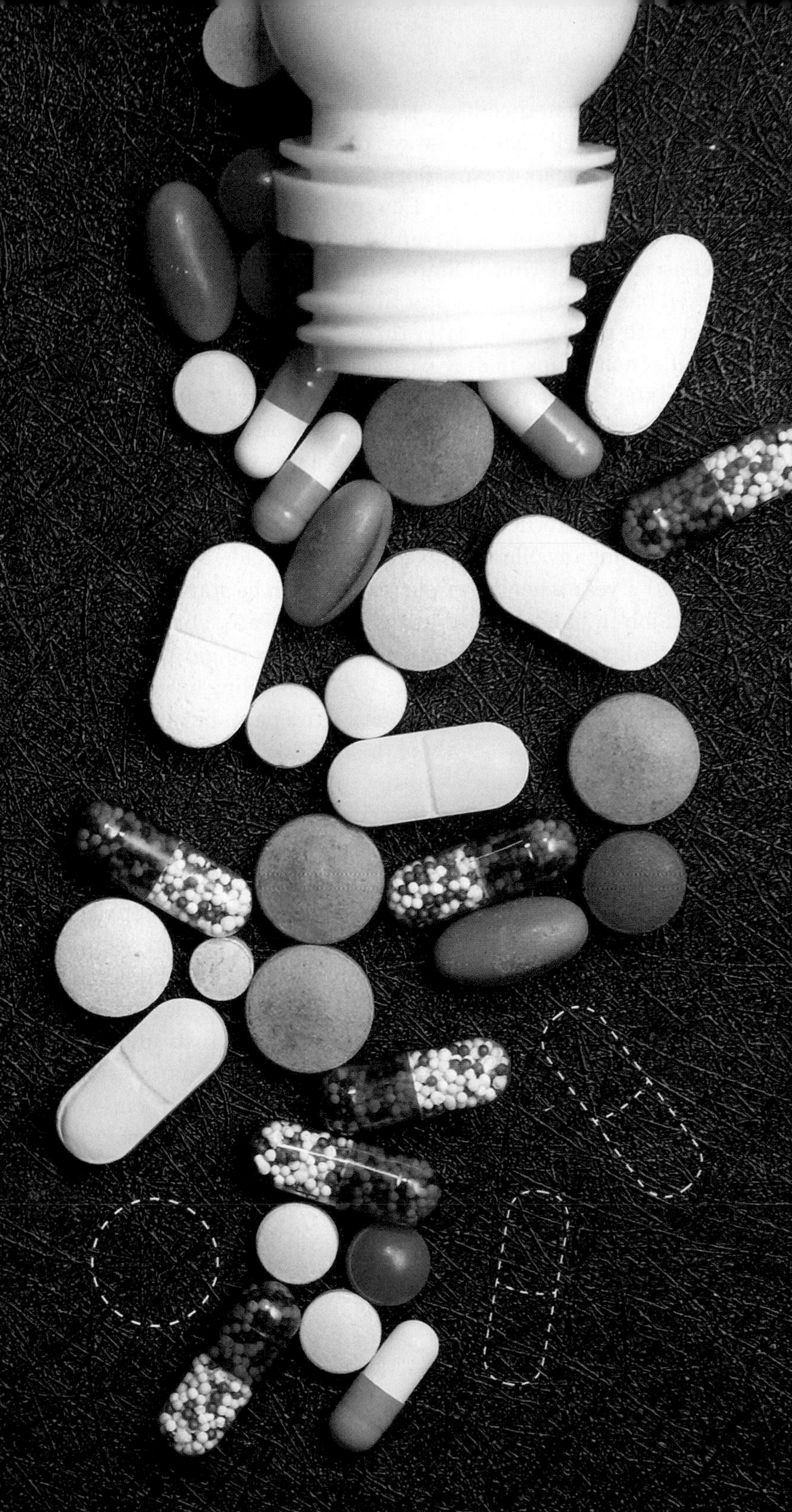

By the time he graduated college, Edwin Chindongo was well on his way to achieving the American dream. Born in Zimbabwe and orphaned at the age of 11, Edwin was adopted with his older sister by an aunt in Lowell, Massachusetts. In the United States, the young man worked hard and became a star high school linebacker with good grades. After turning down football scholarships to stay close to family, he attended the University of Massachusetts Lowell, where he played defensive back and studied criminal justice and psychology. Someday, he thought, he'd like to be a cop.

While in school, Edwin interned at Enterprise Rent-A-Car, and the company offered him a full-time job at $40,000 a year when he graduated. With a degree and job in hand, he also joined a local semi-pro football team (**Figure 3.1**). Life was good.

Then, one morning, Edwin awoke with searing pain in his legs. At the emergency room, his blood sugar was astronomically high—so high that the attending physician said the young man should be in a coma. Edwin was diagnosed with diabetes and put on a treatment plan.

For the ongoing leg pain, a result of nerve damage caused by the diabetes, a doctor prescribed Percocet. This name-brand drug contains acetaminophen (a nonopioid pain reliever) and oxycodone (an opioid pain reliever). Edwin was directed to take one pill three times a day.

Figure 3.1

Edwin Chindongo before the drug crisis

Edwin was a star athlete in high school and continued playing football through college and after. Opioid addiction can happen to anyone, including people who live healthy, active lifestyles.

At first, the doctor prescribed a pill containing 5 milligrams (mg) of oxycodone. He soon increased the dose to 10 mg, then 15 mg, then 30 mg per pill. Around the sixth month, Edwin began to take the pills more frequently than they were prescribed. When the prescription ran out after 7 months, Edwin asked around, and a friend of a friend who sold weed directed him to a guy who sold pills.

While on the promotion track at work, Edwin had purchased a condo, drove a Range Rover, and had saved close to $100,000 in the bank. Then he started buying pills on the street.

"I was spending $500 to $600 a day just trying to get high on these pills. At the time, I was telling myself that I was still taking them for the pain, but obviously I wasn't anymore," says Edwin. "Within a year, everything was gone."

A Growing Epidemic

Opioid abuse is epidemic in America. An estimated 2 million people around the country are abusing or are addicted to prescription opioids.

And it's killing them: According to the National Institute on Drug Abuse, more than 130 people die every day from an opioid overdose. Many of them are young; one recent study found that a shocking 1 in 5 deaths among Americans aged 25 to 34 were a result of opioid overdose, up from 1 in 100 in 2001.

How did we end up here? To understand this health crisis, it helps to look at its roots in history and in chemistry. The first opioids, synthetic or semisynthetic drugs that work like opiates, were derived from their natural counterpart, an opiate called morphine. Morphine was first created by a German chemist, Friedrich Wilhelm Sertürner, in 1804.

Like his fellow chemists throughout history, Sertürner studied the composition and properties of **matter**. Matter makes up the physical world. It is anything that has mass and occupies a volume of space. One type of matter is an **element**, a pure substance that has distinct physical and chemical properties and cannot be broken down into other substances by ordinary chemical methods. There are 98 natural elements known to us, and another 24 have been created in laboratories.

An **atom** is the smallest unit of an element that retains the element's distinctive properties. Atoms make up all common materials, including this book, the air, and you. Every atom has a dense core, a **nucleus** (plural "nuclei"), consisting of positively charged **protons** and electrically neutral **neutrons**. A cloud of negatively charged **electrons** surrounds the nucleus (**Figure 3.2**, bottom). Electrons have significantly less mass than protons and neutrons have. In other words, electrons have much less material inside them, and that material, the mass, determines their weight. So if an electron weighed as much as a 1-liter bottle of water, a proton or a neutron would be as heavy as a car.

The number of protons in an atom's nucleus is called its **atomic number** and is unique to that element (**Figure 3.2**, top). **Isotopes** of an element—atoms of the same species—have the same number of protons but different numbers of neutrons. The sum of the number of protons and the number of neutrons is the **atomic mass number** of an isotope. The atomic mass number is how much mass is in an element. For example, the most common isotope of carbon has 6 protons and 6 neutrons, giving it an atomic mass number of 12, and we call it carbon-12 (^{12}C). The isotope of carbon with 6 protons and 8 neutrons is carbon-14 (^{14}C).

Atoms linked by bonds form **molecules**. Molecules that contain atoms from at least two different elements are **chemical compounds**. Molecules that include at least one carbon atom are **organic molecules**, and multiple organic molecules bound together are molecules called **organic compounds**. Morphine, which Sertürner purified from the seeds of the poppy plant, is an organic compound. It is also one of the most powerful painkillers we know.

But it has side effects. Historians suspect Sertürner took morphine and experienced these side effects, including difficulty breathing, constipation, drug tolerance, withdrawal, and addiction. *Tolerance* occurs when a person no longer responds to a drug in the same way he or she used to and must take a larger dose to achieve the same responses. *Withdrawal* is the experience of physical or psychological symptoms when a person stops taking an addictive drug. *Addiction* is drug use that remains compulsive despite its negative consequences.

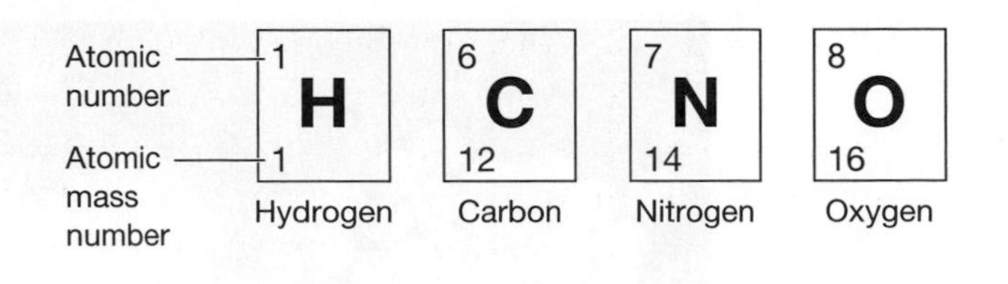

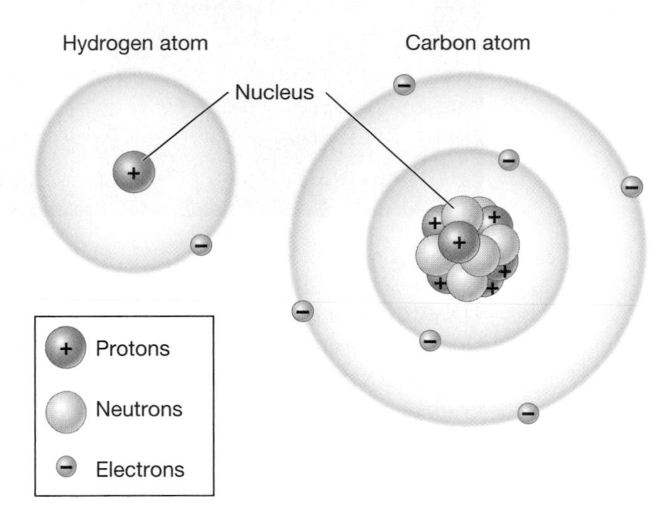

Figure 3.2

Atomic structure

The four boxes at the top of this figure are part of the periodic table (see Appendix C), which shows all the elements. Each entry in the table has the element symbol in the center, the atomic number in the upper left corner, and the atomic mass number in the lower left corner. The protons, neutrons, and electrons of the hydrogen and carbon atoms at the bottom of this figure are shown greatly enlarged in relation to the size of the whole atom.

Q1: How many protons, neutrons, and electrons does the hydrogen atom shown here have? What are the atomic number and the atomic mass number of the hydrogen atom?

Q2: What are the atomic number and the atomic mass number of the carbon isotope shown?

Q3: Nitrogen-11 is an isotope of nitrogen that has 7 protons and 4 neutrons. What are the atomic number and atomic mass number of nitrogen-11?

See Appendix A for answers to the figure questions.

The medical community quickly recognized that morphine was addictive, and chemists began trying to make synthetic versions of the drug, opioids, with similar painkilling effects but fewer side effects. Since the 1800s, numerous opioids have been synthesized, including heroin. In 1916, in an effort to create a nonaddictive opioid, two German scientists developed oxycodone, which is about one and a half times stronger than morphine. In 1995, the U.S. Food and

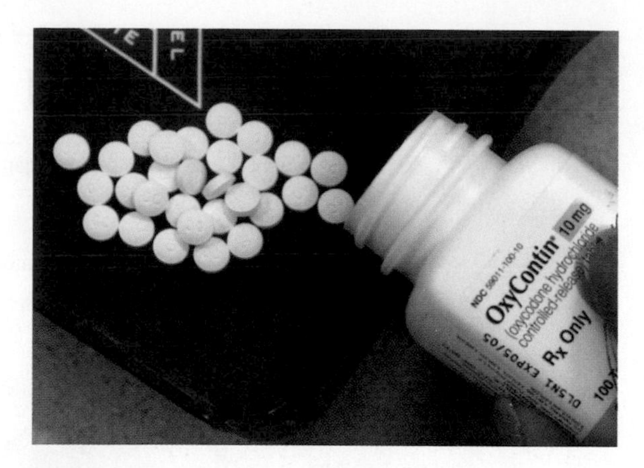

Figure 3.3

Oxycodone and OxyContin

Under the brand name OxyContin, the highly addictive semisynthetic opioid oxycodone was originally falsely marketed as a nonaddictive pain reliever.

Drug Administration (FDA) approved oxycodone, under the brand name OxyContin, for sale by the drugmaker Purdue Pharma (**Figure 3.3**).

Purdue aggressively marketed OxyContin, and doctors began widely prescribing the opioid. The company trained its sales representatives to share the message with doctors that patients taking the pills had a less than 1 percent chance of becoming addicted.

That, however, is absolutely not true.

To Hell and Back

Soon after Edwin started buying pills on the street, he missed mortgage payments on his condo. He moved in with a friend, and soon his life savings ran dry. He began stealing money from family. "At the time, the pills were controlling my life, so I didn't care," recalls Edwin. He even traveled with his dealer to pain clinics in Florida, where he learned how to get cheap pills in large quantities to sell back in Massachusetts.

Edwin's life took a sharp turn on the day his dealer ran out of pills. The dealer offered heroin instead, and Edwin hesitated—he'd never done heroin and didn't want to start. Although its effects are similar to those of oxycodone, heroin is more expensive with a well-earned reputation as a dangerous "street drug." As encouragement, the dealer offered Edwin the drug free of charge and showed him how to mix the powder with water and inject it.

Alone in his room, a needle loaded with heroin in his hand, Edwin prepared to shoot up. But he couldn't. He hesitated not only because of his fear of needles but also because of a feeling of being outside of himself. Edwin was aware of the line he was about to cross.

He called his sister, a nurse practitioner. She immediately drove over, picked him up, and took him to a detox program in Boston. The program was full, and Edwin had to wait for weeks in a homeless shelter until a bed in the rehab program became available. After that, Edwin entered an intensive post-detox program. "It was the worst couple months of my life," Edwin recalls. Not only did the program require commitment and long hours, but also his body was simultaneously going through withdrawal.

During withdrawal, a person experiences a lot of pain, agitation, and cravings. Bodily changes can include sweating, nausea, and muscle cramping. This is the brain's attempt at

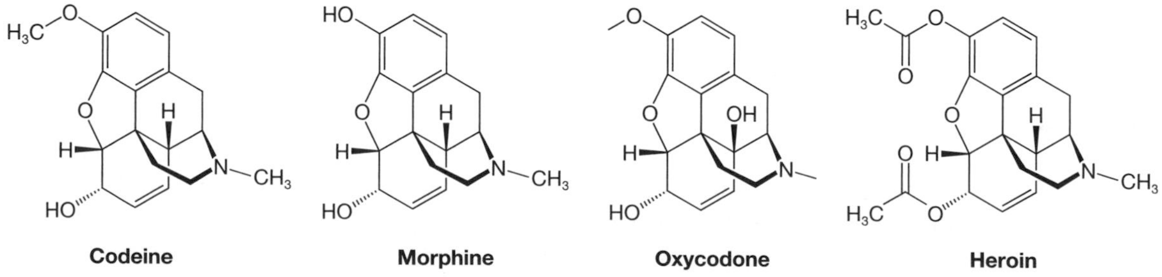

Codeine | Morphine | Oxycodone | Heroin

Figure 3.4

Opiates and opioids have similar chemical structures

Codeine and morphine are opiates. These drugs are natural substances that can be extracted from the seeds of opium poppy plants. Oxycodone and heroin are opioids, made by modifying the chemical structure of morphine.

adjusting to the lack of opioids and restoring a healthy, functioning system.

Forming Bonds

It's amazing that a simple organic compound can alter a person's behavior so dramatically, but opioids are very powerful. The drugs attach to various cells in the body, including brain cells, and send signals that block feelings of pain and promote feelings of euphoria.

Structurally, all opioid and opiate compounds are similar to each other. Just slight differences in structure can affect how long an opioid or opiate lasts, its effect on pain, and its potential for addiction (**Figure 3.4**). Those differences in structure result from **chemical bonds.** Three major types of chemical bonds attach atoms to one another: covalent bonds, ionic bonds, and hydrogen bonds.

Atoms interact with other atoms via electrons; they can donate electrons, accept electrons, and even share electrons. When two atoms share electrons, they form a **covalent bond** (**Figure 3.5**, top). Electrons moving around an atom's nucleus have different energy levels. We can think of the electrons as segregated into rings or shells, each

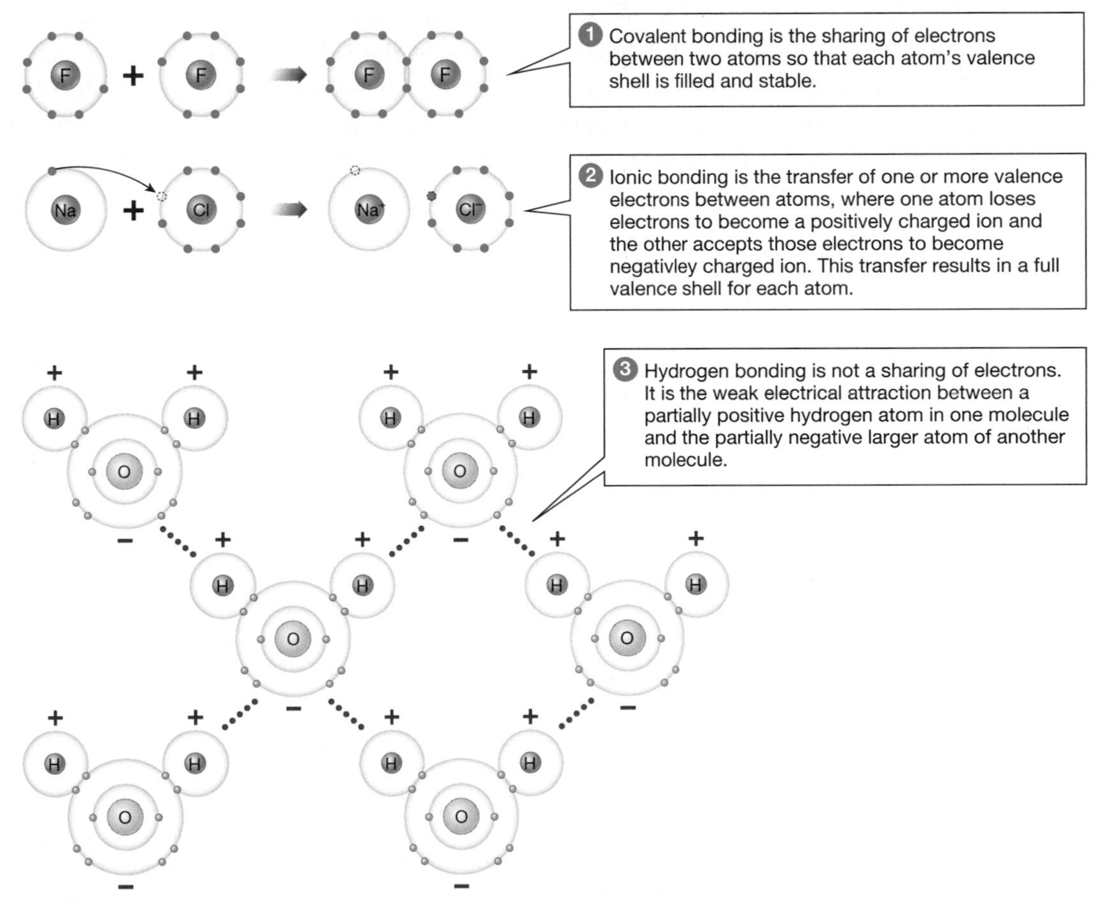

F
F
F
F
1 Covalent bonding is the sharing of electrons between two atoms so that each atom's valence shell is filled and stable.
Na
Cl
Na+
Cl−
2 Ionic bonding is the transfer of one or more valence electrons between atoms, where one atom loses electrons to become a positively charged ion and the other accepts those electrons to become negativley charged ion. This transfer results in a full valence shell for each atom.
H
O
3 Hydrogen bonding is not a sharing of electrons. It is the weak electrical attraction between a partially positive hydrogen atom in one molecule and the partially negative larger atom of another molecule.

Figure 3.5

Three major types of chemical bonds

Covalent, ionic, and hydrogen bonds allow atoms to interact with each other.

Q1: Which type of bond results in one positive ion and one negative ion?

Q2: Which type of bond is not a covalent bond but requires the atoms to be covalently bonded to another atom?

Q3: What is a common feature of the valence shells of both atoms involved in covalent and ionic bonds?

See Appendix A for answers to the figure questions.

of which can contain up to a fixed number of electrons. The innermost shell, closest to the nucleus, can hold up to two electrons. Moving outward, the next two shells can each hold up to eight electrons. A shell needs to be full to be stable, so in an effort to fill their shells, atoms may form covalent bonds. In a covalent bond, atoms share electrons in their outermost shell of electrons, the *valence shell*.

Atoms that have lost or gained valence electrons are called **ions**. Because electrons are negatively charged, an atom that has gained an electron is a negative ion, and an atom that has lost an electron is a positive ion. When a negatively charged ion and a positively charged ion are in the same vicinity, they will chemically attract each other and form an **ionic bond** (**Figure 3.5**, middle). Common table salt, sodium chloride (NaCl), is composed of sodium and chlorine held together by ionic bonds. Unlike a covalent bond, no electrons are shared in an ionic bond.

The third means of attaching atoms to one another is called a **hydrogen bond**. Atoms covalently bonded to hydrogen share the electrons in the bond unequally, resulting in partial electrical charges on each atom within the molecule. The hydrogens are partially positively charged, whereas their counterpart atoms in the covalent bond are partially negatively charged. The partial positive hydrogen of one molecule is attracted to the partial negative atom of another molecule, and a hydrogen bond forms when these atoms are in close proximity. This hydrogen bond is a weak electrical attraction between the partial positive charge and the partial negative charge—opposites attract! For example, molecules of water bind to each other through hydrogen bonds because the partially negatively charged oxygen end of one water molecule attracts one of the partially positively charged hydrogen ends of another water molecule nearby (**Figure 3.5**, bottom). A single hydrogen bond is about 20 times weaker than a covalent bond, hydrogen bonds make up for that lack of strength with sheer quantity. The collective cross-linking of many, many hydrogen bonds amounts to a potent force.

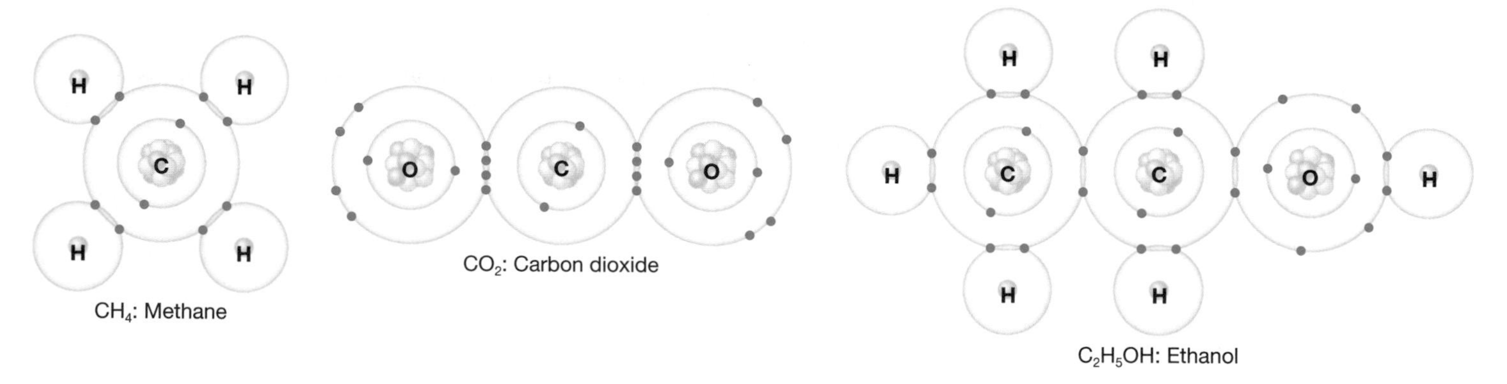

Figure 3.6

Versatile carbon

All organic compounds are built from carbon atoms interacting with atoms of other elements through strong covalent bonds. A carbon atom can bond with up to four other atoms because it has four valence electrons in its outer shell. Recall that atoms must fill their outer shells to be stable and that covalent bonds require the sharing of valence electrons. Count the electrons circling each of the atoms in the molecules shown; all the outer shells are full.

Q1: In methane gas, how many electrons is each hydrogen atom sharing? How many is the carbon atom sharing?

Q2: Carbon dioxide is not technically considered an organic compound, even though it contains a carbon atom. What essential atom is found in the organic compounds shown here that is not included in carbon dioxide?

Q3: Draw a molecule of formaldehyde (CH_2O). How many electrons is the oxygen atom sharing with the carbon atom? How many is the carbon atom sharing with the oxygen atom and with each hydrogen atom?

See Appendix A for answers to the figure questions.

The Building Blocks of Life

At the most basic level, our bodies function through the actions of large organic molecules, or **biomolecules** (sometimes called "macromolecules"). Four major classes of biomolecules critical for living cells are proteins, carbohydrates, nucleic acids, and lipids.

Each of these biologically important molecules is built on a framework of covalently bonded carbon atoms. Carbon is the predominant element in living systems, partly because it can form large molecules that contain thousands of atoms. A single carbon atom can form strong covalent bonds with up to four other atoms (**Figure 3.6**). Carbon atoms can also bond to other carbon atoms, forming long chains, branched molecules, and even rings. No other element is as versatile as carbon in the sheer diversity of complex molecules that can be assembled from it. In addition to carbon, hydrogen, nitrogen, oxygen, phosphorus, and sulfur round out the six most important elements that make up most biomolecules on Earth (**Figure 3.7**).

Of the biomolecules, **proteins** are the most numerous and versatile. Like carbohydrates and nucleic acids, they are **polymers**, long strands of repeating units of small molecules called **monomers**. The monomers making up proteins are amino acids. Different combinations of the 20 amino acid monomers allow for countless proteins that vary in size and shape and, therefore, vary as well in their function (**Figure 3.8**).

For example, enzymatic proteins, such as polymerases, enable DNA to be copied in cells (DNA replication is discussed in Chapter 9). Structural proteins give cells shape. Hormone and receptor proteins, such as insulin and its receptor, enable cells to take up sugars for use as energy. Membrane transport proteins help move substances into and out of cells. Antibodies are proteins that protect us from disease. Storage proteins such as LDL and HDL (low-density and high-density lipoproteins, respectively) carry cholesterol, venoms, and toxins. Opioid receptors, on the outside of cells, are also proteins.

Carbohydrates are the next-most-versatile biomolecules. They range in size from simple sugar monomers (monosaccharides) and two-monomer sugars (disaccharides) to complex polymers (polysaccharides) that may contain thousands of monomers (**Figure 3.9**). Simple sugars are the cell's direct fuel to make ATP (adenosine triphosphate), the molecular energy source essential for all cellular work (see Chapter 4 for more on ATP). Other carbohydrates, such as glycogen in animal muscle tissue and starch in plant tissue, are used for energy storage. (However, a body can't live on energy alone; see "What Should I Eat? Evidence-Based Nutrition Tips" on page 52.) Three additional complex carbohydrates provide structural support to cells: cellulose, also known as fiber, helps plants grow tall; chitin forms a hard outer covering to protect organisms without an internal skeleton, such as insects, spiders, and crustaceans; and peptidoglycan is a major component of bacterial cell walls.

The third major category of the **nucleic acids**—DNA (deoxyribonucleic acid) and RNA (ribonucleic acid)—forms the basis of life itself. These polymers consist of nucleotide monomers: DNA is composed of deoxyribonucleotides, and RNA is composed of ribonucleotides (**Figure 3.10**). DNA provides living organisms with long-term, stable genetic information storage in a form that is easily copied and passed on to future generations. Our genes are DNA.

That sounds important, but what about RNA? Without RNA, the information stored in our genes would be stuck there, like a blueprint in a foreign language that no one can decipher. RNA comes in many forms and plays many

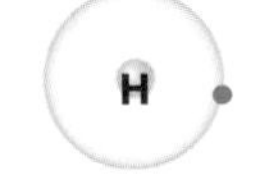

Hydrogen

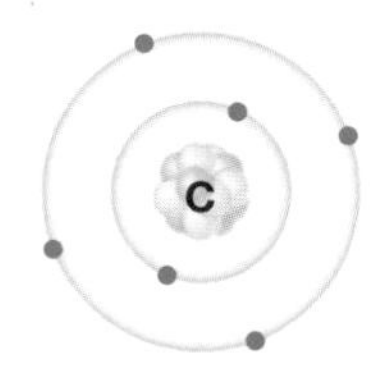

Carbon

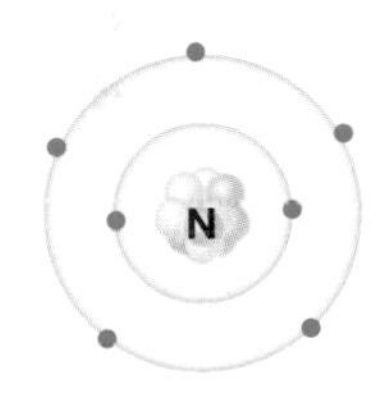

Nitrogen

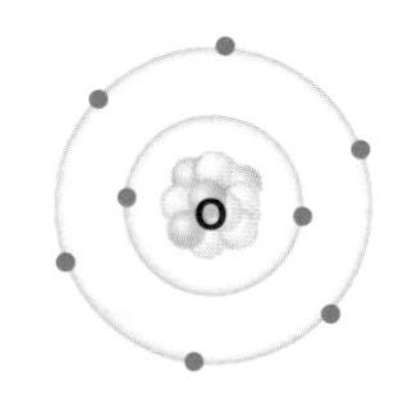

Oxygen

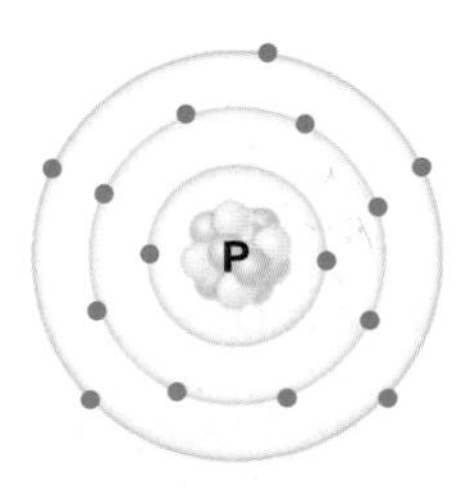

Phosphorus

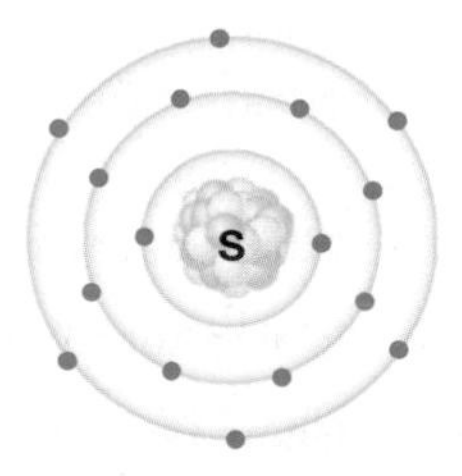

Sulfur

Figure 3.7

The big six elements of life

Although more than 25 types of elements can be found in biomolecules, the six elements shown here are the most common. In fact, just four of these—namely, hydrogen, carbon, nitrogen, and oxygen—make up 96 percent of the human body's mass.

Q1: Which element has the most electrons in its valence shell? The least?

Q2: Which elements have three electron shells? Which elements have two? Which element has only one?

Q3: To fill the valence shells of each element, how many electrons are needed?

See Appendix A for answers to the figure questions.

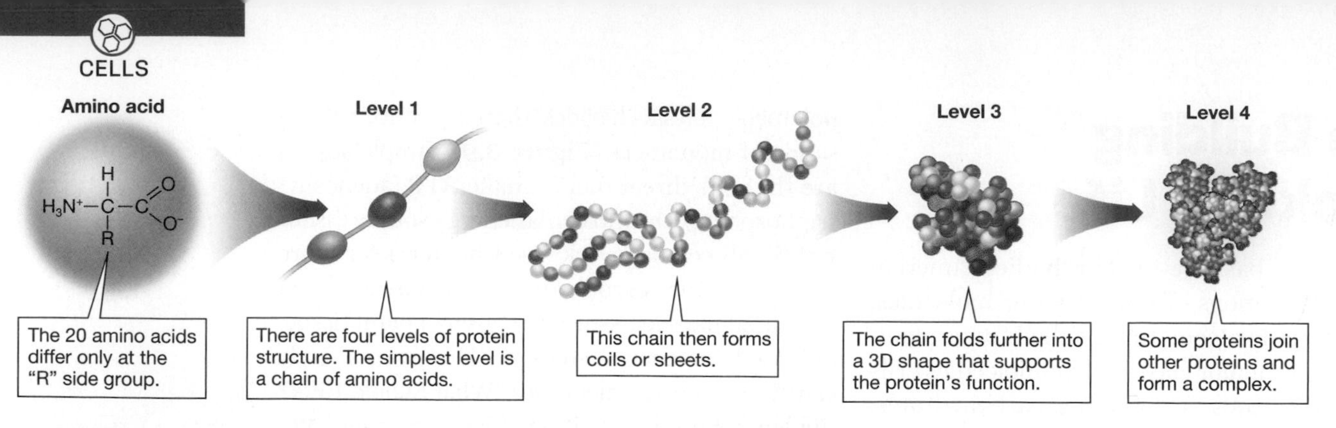

Amino acid
Level 1
Level 2
Level 3
Level 4
The 20 amino acids differ only at the "R" side group.
There are four levels of protein structure. The simplest level is a chain of amino acids.
This chain then forms coils or sheets.
The chain folds further into a 3D shape that supports the protein's function.
Some proteins join other proteins and form a complex.

Figure 3.8

Proteins make up the majority of biomolecules present in a cell

Proteins are polymers of amino acid monomers and the most diverse biomolecules in living organisms. They help maintain life by regulating where and when chemical reactions occur in cells, serving a structural function by providing internal and external support to protect and maintain cell shape, aiding in movement or motility, transmitting signals through cell membranes, transporting molecules into and out of cells, and protecting living organisms by identifying and flagging invaders for destruction.

Q1: Which monomers make up proteins, and how many monomers are there?

Q2: Describe the four levels of protein structure.

Q3: Which of the function(s) listed in the caption do you think opioid receptors in brain cell membranes perform?

See Appendix A for answers to the figure questions.

Monosaccharides pair up to form disaccharides (e.g., sucrose).

Long chains of monosaccharides create polysaccharides, such as starch, chitin, peptidoglycan, cellulose, and glycogen.

Tubers like potatoes store energy as starch.

CH_2OH Simple sugar

The "strings" in celery are cellulose fibers.

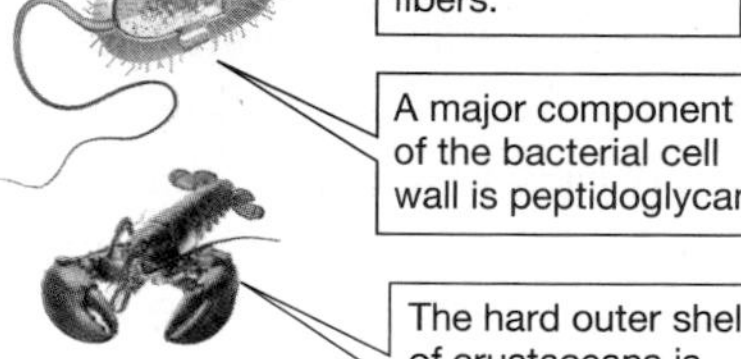

A major component of the bacterial cell wall is peptidoglycan.

Common sugar is the disaccharide sucrose.

Glycogen is stored in muscle tissue as a quick energy source.

The hard outer shell of crustaceans is made of chitin.

Fructose is the monosaccharide that gives honey its sweetness. Monosaccharides are simple sugars.

Figure 3.9

Carbohydrates are the most abundant biomolecule on Earth

Carbohydrates are polymers of simple sugars that are used for energy storage and structural support. Cells also attach carbohydrate molecules to proteins and lipids, modifying their structures to enhance functionality. It is estimated that more than 50 percent of the total carbon content of Earth takes the form of carbohydrate compounds. In plants and bacteria, carbohydrates are the main component of the cell wall.

Q1: Which of the big six elements on Earth are found in carbohydrates?

Q2: Which carbohydrate is used for energy storage in animal muscle tissue? Plant tissue?

Q3: Name two carbohydrates used as structural support.

See Appendix A for answers to the figure questions.

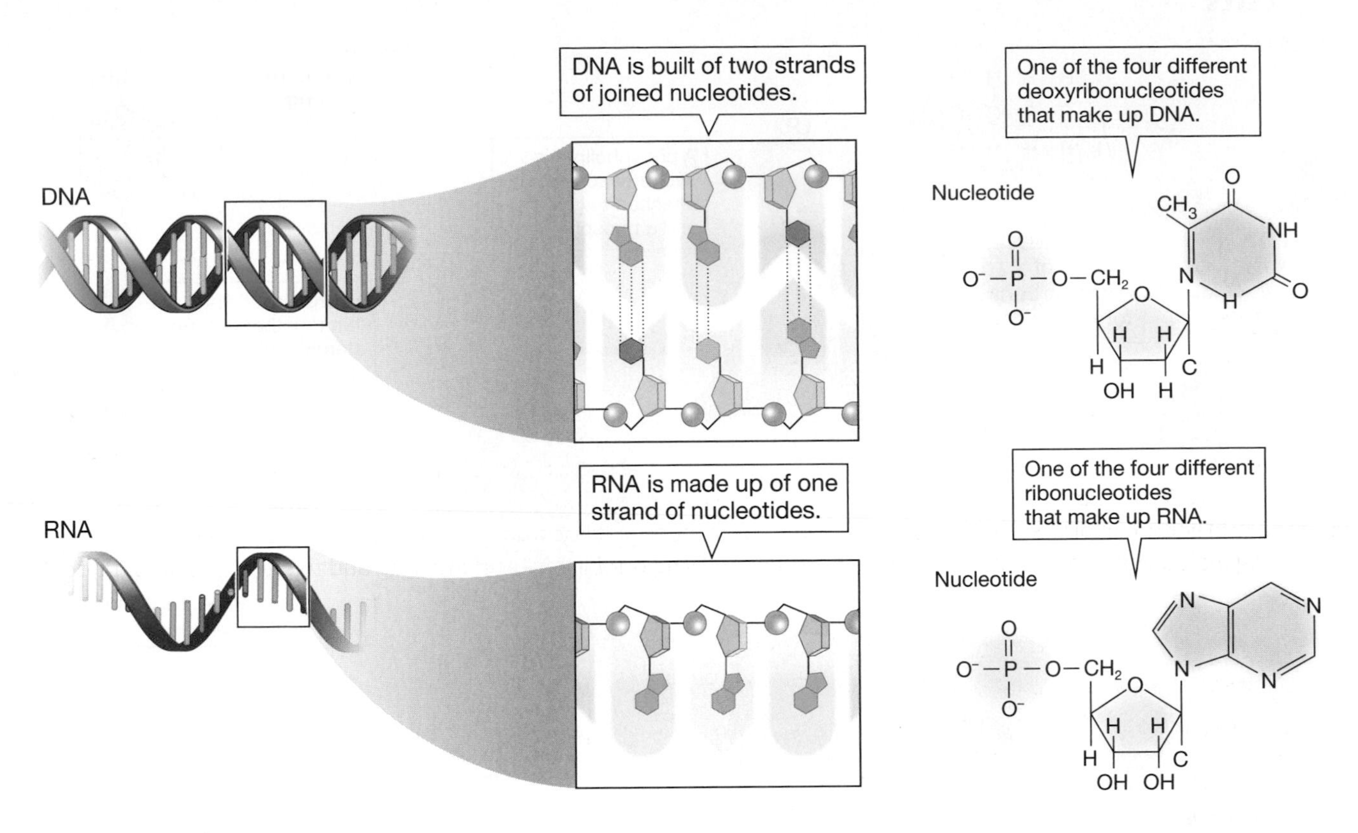

Figure 3.10

Nucleic acids are biomolecules that are essential to all known forms of life

The nucleic acids are polymers of nucleotides that function to store and transmit genetic information via DNA and RNA, respectively. Nucleotide monomers consist of three components: a five-carbon sugar, a phosphate group, and a nitrogenous base.

Q1: Which nucleic acid is built of two strands of joined nucleotides?

Q2: How many different types of ribonucleotide monomers make up RNA molecules?

Q3: Chromosomes are made up of which type of nucleotide monomers?

See Appendix A for answers to the figure questions.

roles, but its most important job is providing a readable genetic language that enables genes to be expressed as proteins.

A final valuable type of biomolecule is the **lipids**, which are better known as fats, oils, and steroids. Lipids are not polymers, because their structure is not composed of a chain of monomers (**Figure 3.11**). Nevertheless, the lipids form a diverse group of biomolecules that are made up of combinations of hydrocarbons (compounds of carbon and hydrogen molecules), fatty acids, and/or glycerol molecules. For example, triglycerides—composed of three fatty acid molecules linked to a glycerol molecule—provide long-term energy storage in both plants (as oils) and animals (as fats) and serve as insulation against the cold in animals. And as discussed in Chapter 4, phospholipids are crucial molecules, being the main component of all cellular membranes.

DEBUNKED!

MYTH: Opioid addiction occurs because of poor choices or weak morals.

FACT: Opioids are powerfully addictive. Even when used *exactly* as prescribed for the treatment of pain, opioids can cause physical dependence. Opioid abuse can and sometimes does begin with a legitimate prescription.

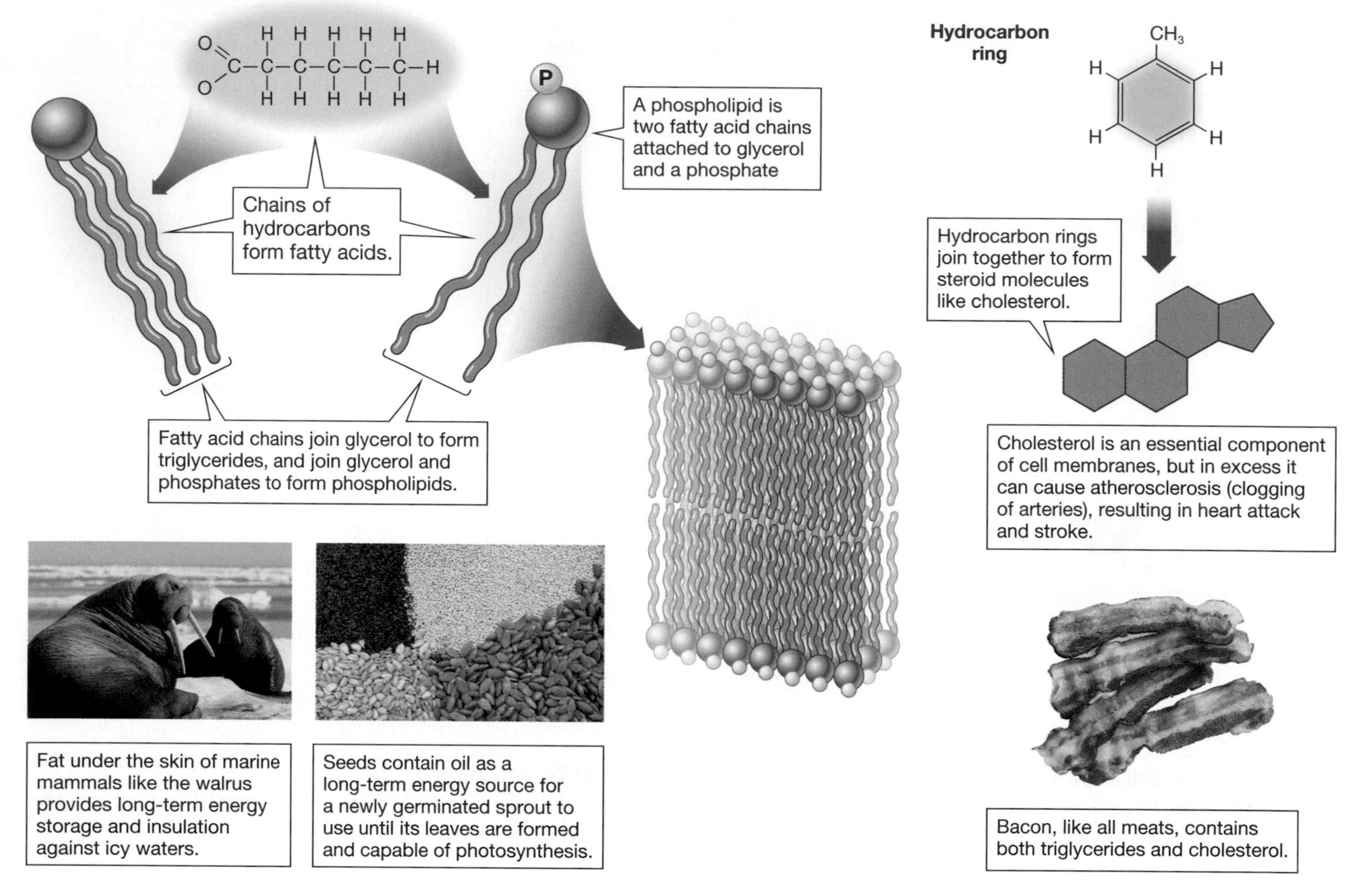

Chains of hydrocarbons form fatty acids.
A phospholipid is two fatty acid chains attached to glycerol and a phosphate
P
Fatty acid chains join glycerol to form triglycerides, and join glycerol and phosphates to form phospholipids.
Hydrocarbon ring
Hydrocarbon rings join together to form steroid molecules like cholesterol.
Cholesterol is an essential component of cell membranes, but in excess it can cause atherosclerosis (clogging of arteries), resulting in heart attack and stroke.
Fat under the skin of marine mammals like the walrus provides long-term energy storage and insulation against icy waters.
Seeds contain oil as a long-term energy source for a newly germinated sprout to use until its leaves are formed and capable of photosynthesis.
Bacon, like all meats, contains both triglycerides and cholesterol.

Figure 3.11

Lipids are a structurally diverse group of biomolecules

Unlike proteins, carbohydrates, and nucleic acids, lipids are not polymers. Instead, they are composed of hydrocarbons (hydrogen and carbon) and oxygen. They play vital roles in many cellular processes, including energy storage, insulation, structural support, protection, and communication.

Q1: **How many fatty acid chains attach to glycerol to form a triglyceride? How many are needed to form a phospholipid?**

Q2: **What type of lipid is formed from hydrocarbon rings?**

Q3: **What type of molecules do chains of hydrocarbons form?**

See Appendix A for answers to the figure questions.

Making It Hard to Get High

Opioids ease terrible pain but often at a terrible toll. In an effort to reduce that toll, researchers and pharmaceutical companies have sought to make safer painkillers. The goal is to provide pain relief for people like Edwin without the danger of addiction.

Consider that Purdue Pharma is now the defendant in close to 2,000 lawsuits for its alleged role in downplaying addiction risks and profiting from the opioid crisis. When the company first released OxyContin, it promoted this pinky-nail-sized pill as extending the release of oxycodone in the body over 12 hours. Users quickly learned to bypass that mechanism. "When it comes to an addict, they always find a way," says Edwin. Users

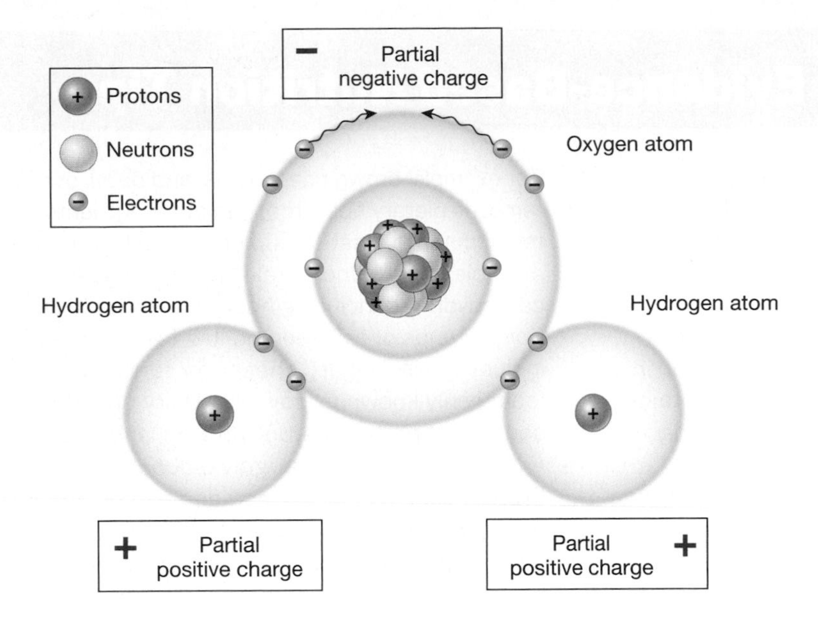

Figure 3.12

Water molecules are polar

The oxygen (O) in a water molecule (H_2O) pulls the negatively charged electrons closer to itself than to the two hydrogens (H) in the covalent bonds, creating an uneven distribution of charge within the molecule. The partial positive and partial negative charges are what make molecules polar.

Q1: Where are the covalent bonds in this figure?

Q2: This figure shows a water molecule. A hydrogen molecule (H_2) consists of two hydrogen nuclei that share two electrons. Draw a simple diagram of a hydrogen molecule indicating the positions of the two electrons.

Q3: When table salt (sodium chloride, NaCl) dissolves in water, it separates into a sodium ion (Na^+) and a chloride ion (Cl^-). Which portion of a water molecule would attract the Na^+ ion, and which portion would attract the Cl^- ion?

See Appendix A for answers to the figure questions.

could get high more quickly by crushing the pills and snorting the pill powder or dissolving the pills in water and injecting the resulting solution.

A compound that mixes completely with water is said to be **soluble**. Oxycodone and other compounds are soluble because of water's special properties. A water molecule is made up of two hydrogen atoms and one oxygen atom held together by shared electrons—that is, by covalent bonds. In a water molecule, the oxygen atom uses an electron from each of two hydrogen atoms to fill its outer shell, increasing its count of electrons from six to eight. The hydrogen atoms also benefit from the bond; they fill their outer (and only) shell with the two electrons needed to be stable.

However, the electrons in a water molecule are not shared equally; they spend more time near the oxygen atom than near the hydrogen atoms. Because electrons are negatively charged particles, the oxygen end of a water molecule takes on a slightly negative charge, and the hydrogen ends become slightly positively charged. This lopsided electron sharing means that water is a **polar molecule** (**Figure 3.12**), and as discussed earlier, molecules with partial positive and partial negative atoms form hydrogen bonds with each other.

It is polarity that makes water especially good at dissolving other substances into solutions. A **solution** is any combination of a **solute** (a dissolved substance, such as sugar) and a **solvent** (the fluid, such as water, into

What Should I Eat? Evidence-Based Nutrition Tips

You've been reading about drug addiction and biomolecules. Quick: What's the connection? If you answered "chemistry," you're correct. If you answered "health," you're also correct. To maintain good health, it is important to avoid addiction. In addition, you need to supply your body with the raw materials it needs to grow and function. Some of those raw materials come to you in the form of biomolecules, such as carbohydrates, proteins, and lipids. Yes, that protein shake you had this morning delivered protein biomolecules into your system.

The study of what we eat—a.k.a. our diet—is the science of **nutrition**. Nutrition is based on *nutrients*, the components of food that an organism needs to survive and thrive. Humans rely on six groups of essential micronutrients and macronutrients to stay healthy and cut the risk of disease related to diet: *vitamins* and *minerals* (micronutrients) and proteins, carbohydrates, fats, and water (macronutrients).

First, let's meet the micronutrients. Your body requires 13 essential vitamins: A, C, D, E, K, and the B vitamins, which are thiamine (B_1), riboflavin (B_2), niacin (B_3), pantothenic acid (B_5), pyroxidine (B_6), biotin (B_7), folate (B_9), and cobalamin (B_{12}). The minerals your body needs consist of seven micronutrients—namely, calcium, phosphorus, magnesium, sodium, potassium, chloride, and sulfur—and small amounts of trace minerals, including iron, manganese, copper, iodine, zinc, cobalt, fluoride, and selenium.

Countless fad diets focus on just one or two macronutrients, such as proteins or carbohydrates, and don't say much or anything about vitamins and minerals. According to nutrition experts, however, healthy eating includes all the essential micronutrients and macronutrients in moderation. Here's an example of a healthy daily eating pattern, as backed by science:

- *Vegetables*: 2 cups of a variety of vegetables from all five subgroups. The veggie subgroups are dark green, red, and orange legumes (beans and peas), starch, and others (for example, asparagus, cabbage, beets). These foods are healthy sources of carbohydrates, including fiber. Legumes are also healthy sources of protein.
- *Fruit*: 2 cups of fruits, which can include whole fruits in fresh, canned, frozen, or dried forms and 100 percent fruit juices (that is, not fruit-flavored drinks). Fruits are a healthy source of carbohydrates, including fiber.
- *Grains*: $^3/_4$ cup of grains, which can be either whole, containing the entire kernel including the bran and germ (for example, brown rice, quinoa, and oats), or refined, which removes dietary fiber, iron, and other nutrients. Doctors recommend at least $^1/_2$ cup of your daily grain intake consist of whole grains. Whole grains provide a healthy source of carbohydrates, including fiber, and also provide protein.
- *Dairy*: 3 cups of fat-free and low-fat (1 percent) dairy, including milk, yogurt, cheese, or fortified soy beverages (commonly known as soymilk). Dairy provides healthy sources of protein, carbohydrates, and lipids.
- *Protein Foods*: $^2/_3$ cup (roughly 5 ounces) per day of protein. Protein food subgroups include seafood, meats (for example, beef, pork, and lamb), poultry, eggs, nuts, seeds, and soy products. Legumes and dairy are also considered protein foods. These foods provide not only proteins but also carbohydrates and lipids.
- *Oils*: 5–7 teaspoons per day on average based on age, biological sex, and activity level. Oils, being lipids, are the major source of essential fatty acids in our diets; they are also a major source of vitamin E.

And, of course, there are a few things to avoid in your diet:

- *Saturated fats, trans fats, and cholesterol*: Saturated fats are found naturally in foods containing butter, palm and coconut oils, cheese, and red meat. Trans fats are manufactured to make liquid vegetable oils more solid and are listed as "partially hydrogenated oils." Cholesterol is found naturally in animal products and plays important roles in body function at low levels. However, too much cholesterol clogs arteries and leads to heart disease. There is no nutritional need for saturated fats, trans fats, or high levels of cholesterol in people over 2 years old, so these substances together should not be more than 10 percent of your calories per day. In addition, eating foods such as nuts and fatty fish helps reduce cholesterol.
- *Added sugars*: After checking lists of ingredients for sugar, limit these added sugars to 10 percent of your calorie intake. They are "empty calories" that contribute no essential nutrients to your diet.

To form healthy eating habits, make small shifts in your food choices. For example, whenever possible drink water instead of a sugary drink; grab a handful of almonds instead of chips; have strawberries instead of cake as a dessert. Over time, little changes in your nutrition habits can have big health benefits.

which the solute has dissolved). Water is called the "versatile solvent" because it successfully dissolves so many substances. The polar nature of water molecules, however, means that they will *not* interact with uncharged or nonpolar substances, such as fat or oil. Molecules that are soluble in water (such as salt or vinegar) are called **hydrophilic** ("water-loving"); molecules that don't dissolve well in water (such as fat or oil) are called **hydrophobic** ("water-fearing"). **Figure 3.13** shows these processes in action.

Water can change form. In fact, it can exist in all three states of matter: liquid, gas, and solid. When liquid water heats up, it transitions from the liquid state to the gas state in a phenomenon known as **evaporation**. During this process, hydrogen bonds break, and water molecules spread out away from each other. Think about boiling water on the stove. You can see steam coming up out of the pot, and then it disappears. Steam is actually tiny water droplets. As more and more hydrogen bonds break, individual water molecules move farther and farther apart—they can no longer be seen as tiny liquid droplets. The water is now water vapor, a gas. By the opposite reaction, as water vapor cools, molecules slowly re-form hydrogen bonds, and the gas returns to the liquid state—a process known as **condensation**.

Hydrogen bonds explain the physical properties of water in its three states. Water forms a liquid at room temperature because hydrogen bonds stick water molecules together, keeping them close. Those hydrogen bonds are constantly forming and breaking in water, creating a nonstop jostling that gives water its liquid form (**Figure 3.14**, left).

When water chills, its molecules cannot move about as vigorously (**Figure 3.14**, right). As water turns into ice at 0°C (32°F), a stable network of hydrogen bonds emerges. The molecules become spaced farther apart, locked into an orderly pattern known as a crystal lattice. That spacing is the reason ice occupies more space than liquid water. Normal ice is 9 percent less dense than liquid water (density is mass divided by volume), which explains why ice floats on water. This property of water is quite unusual—most substances are *more* dense in the solid state than in the liquid state—but it has helped shape life as we know it. If ice did not float on water, then each time a lake or river froze in the winter, the frozen top layer would sink to the bottom and the new, liquid top layer would freeze. This process would repeat until the entire lake or river was frozen solid, no matter the depth. All the aquatic creatures and plants would be killed in the process, and terrestrial plants and animals would no longer have access to freshwater.

With water in its liquid state and opioids as solid pills, opioid abusers looking to inject the drugs were able to draw on the chemical properties of both the water and the drugs. To prevent the pills from being crushed into powder and dissolved, Purdue Pharma and other pharmaceutical companies, encouraged by the FDA, began making tamper-resistant pills. Covered with special coatings, the pills either broke into chunks or turned into a gel when crushed. Officials hoped this measure would lower the likelihood that a person would inject the drug or snort it and thus lower rates of drug abuse.

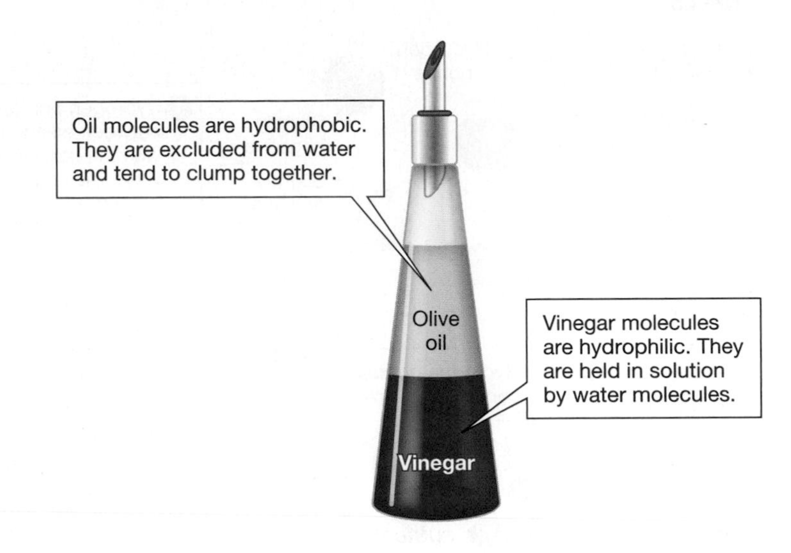

Oil molecules are hydrophobic. They are excluded from water and tend to clump together.
Olive oil
Vinegar molecules are hydrophilic. They are held in solution by water molecules.
Vinegar

Figure 3.13

Hydrophilic substances dissolve in water, but hydrophobic substances do not

Just try to shake, stir, or whisk oil and vinegar together. Eventually, they will always separate because oil and vinegar are made of very different types of molecules. Oils are lipids (see "What Should I Eat? Evidence-Based Nutrition Tips" on page 52) and are nonpolar molecules. Most of the atoms in oils share electrons evenly, so these molecules do not have partial charges. Nonpolar molecules always clump and reclump together, and they exclude polar molecules. As a result, they are hydrophobic. By contrast, vinegars are hydrophilic. Most of them are solutions of acetic acid and water, polar molecules that are held together in solution through hydrogen bonding.

Q1: Describe what will happen to the molecules of olive oil if you shake the bottle and then let it sit for an hour. What about the molecules of vinegar?

Q2: What will happen if you add another fat to the bottle, such as warm bacon grease, and shake it?

Q3: Given how sugar behaves when it is mixed into coffee or tea, would you predict that it is hydrophobic or hydrophilic?

See Appendix A for answers to the figure questions.

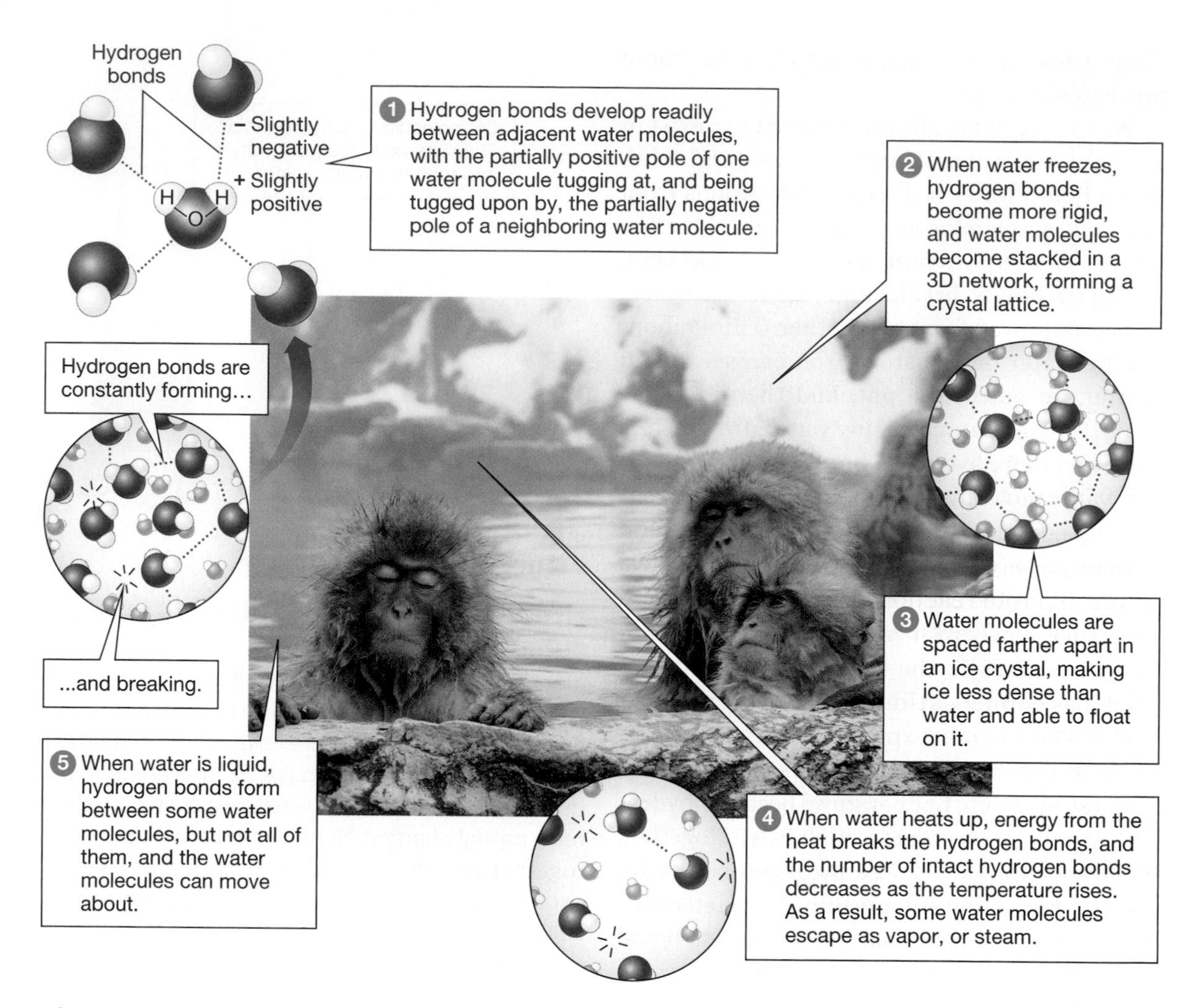

Hydrogen bonds
– Slightly negative
+ Slightly positive
H
O
H
1 Hydrogen bonds develop readily between adjacent water molecules, with the partially positive pole of one water molecule tugging at, and being tugged upon by, the partially negative pole of a neighboring water molecule.
2 When water freezes, hydrogen bonds become more rigid, and water molecules become stacked in a 3D network, forming a crystal lattice.
Hydrogen bonds are constantly forming...
...and breaking.
3 Water molecules are spaced farther apart in an ice crystal, making ice less dense than water and able to float on it.
5 When water is liquid, hydrogen bonds form between some water molecules, but not all of them, and the water molecules can move about.
4 When water heats up, energy from the heat breaks the hydrogen bonds, and the number of intact hydrogen bonds decreases as the temperature rises. As a result, some water molecules escape as vapor, or steam.

Figure 3.14

Water molecules change state as hydrogen bonds increase or decrease

Japanese snow macaques escape the cold with a daily dip in natural hot springs. Water can be seen here in its liquid, solid, and gas states.

Q1: Identify where in the picture water can be seen in its liquid, solid, and gas states.

Q2: In the gas state, water molecules move too rapidly and are too far apart to form hydrogen bonds. Compare the volumes occupied by an equal number of water molecules in the liquid, solid, and gas states.

Q3: Explain in your own words how ice floats on water.

See Appendix A for answers to the figure questions.

Unfortunately, the tamper-resistant pills often cost twice as much as conventional opioids. A 2018 study in Australia found the new pills made no difference in the number of opioid overdoses—people simply purchased the normal pills. As opioid abuse continues, chemists are now taking a different route to solving the problem—altering the molecular structures of the drugs.

This Is Your Brain on Drugs

As a college student, Laura Bohn liked biochemistry, but it was a class about nutrition in the brain that really captivated her. "I really liked the thought of how what we ingest . . . can affect mood and perception," recalls Bohn, now a

Figure 3.15

Dr. Laura Bohn
Dr. Bohn has made a career out of studying how opioids affect the brain. Her research focuses on how to make painkillers safer and nonaddictive.

neuroscientist at Scripps Research, a nonprofit biomedical research institute in La Jolla, California, and Jupiter, Florida (**Figure 3.15**).

Bohn eventually did research for her PhD degree on how the brain responds to opioids. Opioids and opiates must bind to proteins—one of the four major classes of biomolecules described earlier—to elicit a physiological response. Specifically, they bind to opioid receptor proteins that act like little doorbells to activate signals inside the cell. "So many things happen that are dependent on those receptors," says Bohn. "When a person swallows, snorts, or injects an opioid, the drug binds to opioid receptors all over the body. The entire body is affected," she adds. The person's mood changes, the pupils shrink, the cough reflex is suppressed, blood pressure drops, and digestion slows.

Most critically, when opioids bind to receptors in the brain, they activate different cascades of chemical activity, like a falling domino that triggers two parallel lines of dominoes (**Figure 3.16**). The signals sent by the opioid receptors through brain cell membranes activate the medically targeted pathway resulting in pain relief. These signals can also initiate a second pathway associated with restricted breathing. "If you give [a person] an opiate, you instantly see pain relief and you instantly see a drop in blood oxygen levels and breathing rates," Bohn explains. This second pathway is, in fact, the main reason people die from using

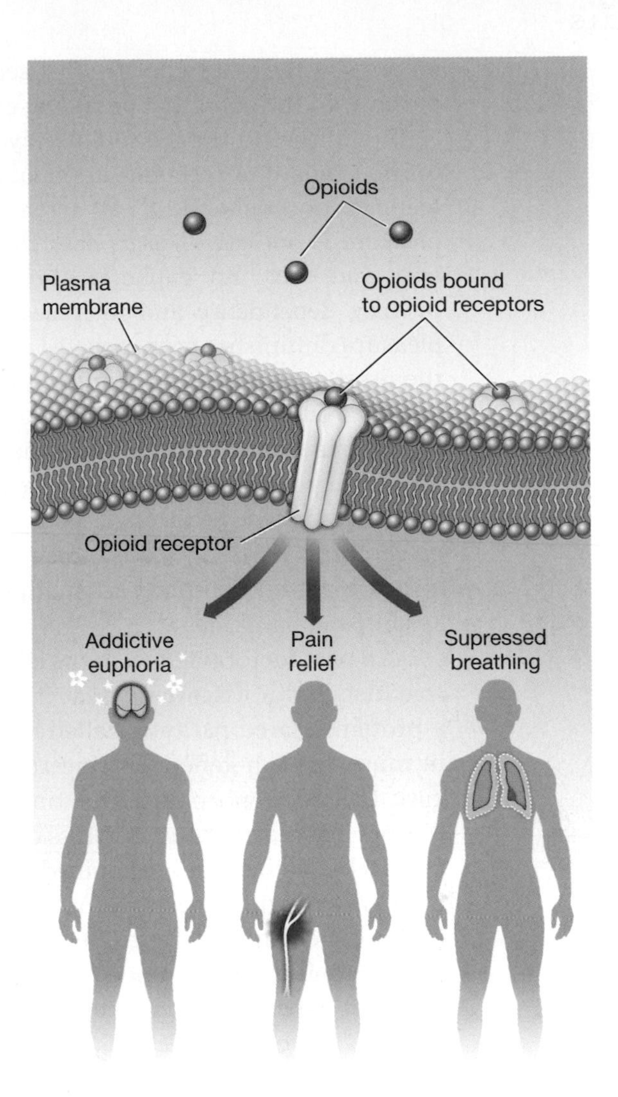

Figure 3.16

The three pathways for opioid responses
Opioid receptors are membrane proteins that, when bound by opioids, send signals into the cell. These signals can result in addictive euphoria, pain relief, and changes in body physiology including suppressed breathing.

Q1: Medically administered opioids—namely, prescription drugs—are meant to target which pathway(s)?

Q2: Most, if not all, opioid abusers were originally prescribed opioids for medical pain relief. These people were "hooked" into illicit drug abuse by which pathway(s)?

Q3: Opioid receptor proteins are lodged in cell membranes. Their centers interact with hydrophobic fatty acids of phospholipids, and their ends interact with the hydrophilic phosphate groups of phospholipids. Are opioid receptors hydrophilic or hydrophobic? Explain.

See Appendix A for answers to the figure questions.

an opioid—the drug activates that second brain pathway to the point that breathing can stop.

In addition to these two pathways, when an opioid binds to a receptor, it stimulates strong feelings of euphoria. That intense rush of pleasure is addictive—users seek the sensation over and over. Yet euphoria also contributes to drug dependence and tolerance. When the pleasure diminishes with repeated opioid abuse, the users take more and more of a drug to get the same effect.

What if a drug could activate the first pathway, pain relief, without activating the second one, suppressed breathing? Bohn and other researchers have dedicated their careers to answering this question, attempting to make safer opioid painkillers.

As a postdoctoral fellow in the lab of another scientist, Bohn discovered that if she blocked a protein-related pathway called beta-arrestin in mice through genetic engineering and then gave the mice morphine over time, the mice did not have breathing suppression and, with repeated dosing, they remained sensitive to the pain-relieving effects. In other words, they did not become tolerant. Because tolerance did not build in the mice, they most likely were not experiencing the euphoria-producing effects. "It would suggest if you take out this beta-arrestin component from the signaling pathways, maybe we can improve some therapeutics," says Bohn.

In her own lab, Bohn worked closely with a team of medicinal chemists—scientists who design and synthesize pharmaceutical drugs—to literally build new compounds from scratch through chemical reactions. The process of breaking existing chemical bonds and creating new ones is known as a **chemical reaction**. The **reactants**—in this case, substances in the lab—undergo a chemical change and form new ions or molecules, called **products**. Some chemical reactions require added energy to begin. Others release energy.

In 2017, Bohn's group tested six of 60 potential compounds that structurally looked like they should lean toward one of the two pathways. "Finding the initial structure was serendipity," says Bohn, "but this was followed by years of testing many compounds." Finally, the researchers put six of the compounds to the test.

In mice, all the compounds were as potent if not more potent than morphine at pain relief, and several of them were less likely to activate the beta-arrestin pathway and reduce oxygen levels. Some of these compounds did not create tolerance in mice. These findings suggest it may be possible to synthesize a drug that relieves pain but does not suppress breathing and is safer than current opioids.

In 2017, a team in Berlin, Germany, published details in the journal *Science* of a newly synthesized opioid that soothed pain in rats without addiction or breathing suppression. By modifying the chemical structure of fentanyl, scientists were able to make the drug NFEPP, a synthetic opioid similar to morphine but 50–100 times more potent (**Figure 3.17**). NFEPP appeared to bind to opioid receptors only under acidic conditions, as in inflamed and therefore painful tissue injuries, but not in the brain or elsewhere in the body, where it might cause negative side effects such as breathing suppression. An **acid** is a hydrophilic compound that dissolves in water and loses one or more hydrogen ions (H^+). By *donating* H^+ ions to water, acids *increase* the concentration of free H^+ ions in an aqueous (water-based) solution—creating an **acidic** solution. H^+ ions are extremely reactive and can disrupt or alter chemical reactions.

The chemical opposite of an acid is a **base**. Although bases are also hydrophilic, they differ from acids in *accepting* H^+ ions from aqueous surroundings. By removing H^+ ions, bases *reduce* the concentration of free H^+ ions in an aqueous solution—creating a **basic** solution. Strong bases, like strong acids, can be dangerous because they disrupt chemical reactions important to life. However, acids react with bases to have an overall neutralizing effect, reducing the concentration of reactive H^+ ions.

H^+ ion concentration is commonly expressed on a scale from 0 to 14, where 0 represents an extremely high concentration of free H^+ ions and 14 represents the lowest concentration. This scale is called the **pH scale**, but the original meaning of the "p" is unclear; the abbreviation "pH" may mean "potential of hydrogen" or "power of hydrogen." The system is logarithmic, with each pH unit indicating

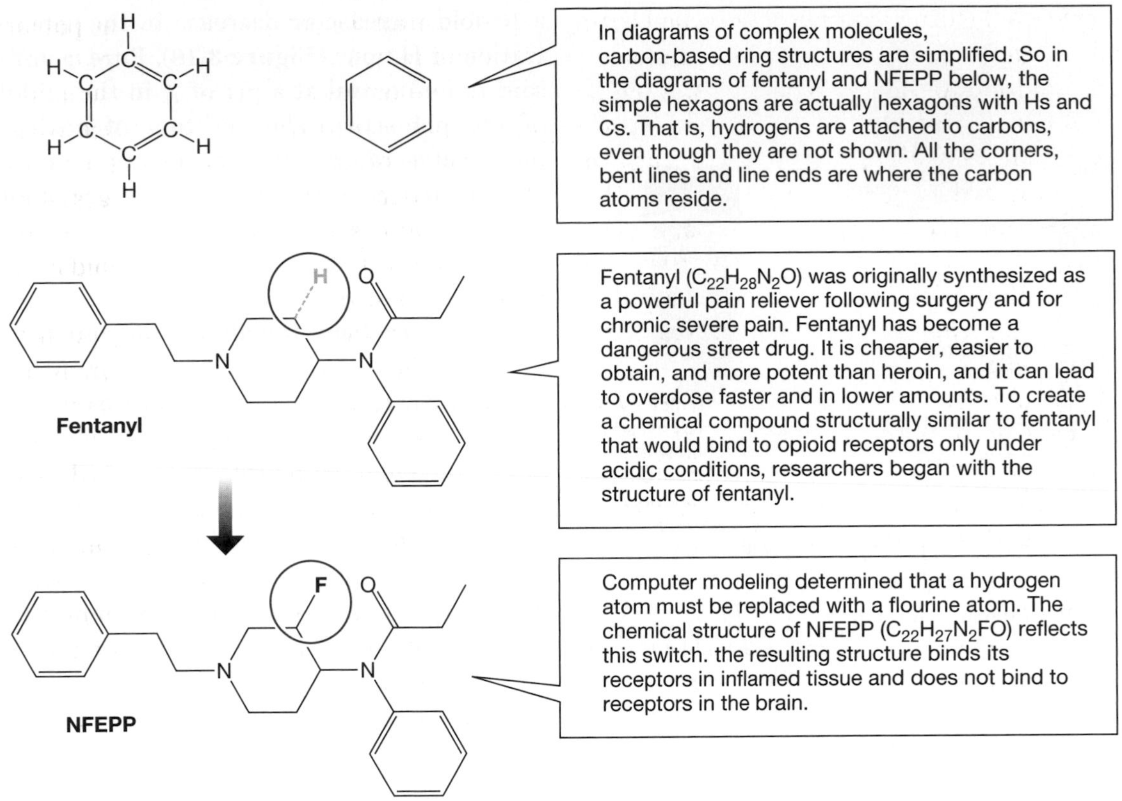

In diagrams of complex molecules, carbon-based ring structures are simplified. So in the diagrams of fentanyl and NFEPP below, the simple hexagons are actually hexagons with Hs and Cs. That is, hydrogens are attached to carbons, even though they are not shown. All the corners, bent lines and line ends are where the carbon atoms reside.

Fentanyl

Fentanyl ($C_{22}H_{28}N_2O$) was originally synthesized as a powerful pain reliever following surgery and for chronic severe pain. Fentanyl has become a dangerous street drug. It is cheaper, easier to obtain, and more potent than heroin, and it can lead to overdose faster and in lower amounts. To create a chemical compound structurally similar to fentanyl that would bind to opioid receptors only under acidic conditions, researchers began with the structure of fentanyl.

NFEPP

Computer modeling determined that a hydrogen atom must be replaced with a flourine atom. The chemical structure of NFEPP ($C_{22}H_{27}N_2FO$) reflects this switch. the resulting structure binds its receptors in inflamed tissue and does not bind to receptors in the brain.

Figure 3.17

Fentanyl and NFEPP differ structurally at only one atom

Fentanyl was originally synthesized as a powerful pain reliever for use after surgery and for chronic severe pain—such pain may be long-lasting or permanent. Unfortunately, as a street drug, fentanyl is cheaper and easier to buy and more potent than heroin. It is also likely to produce an overdose faster and in lower amounts than heroin. Researchers used this powerful synthetic opioid as the basis for a new synthetic opioid that would not present the same dangers. To create an analog of fentanyl that would bind to opioid receptors only under acidic conditions, computer modeling determined that a hydrogen atom must be replaced with a fluorine atom.

Q1: List one reason why carbon ring structures might be diagrammed without the hydrogens and carbons in the structures of complex molecules.

Q2: Single covalent bonds are represented by one line between atoms, whereas double covalent bonds between atoms are represented by two lines. How many double covalent bonds are in the chemical structure of fentanyl? How many are in the chemical structure of NFEPP?

Q3: If NFEPP is approved and marketed as a safe and potent pain reliever, do you think fentanyl drug addictions and overdoses will decrease, increase, or stay the same? Why do you think so? What about oxycodone (OxyContin) addictions and overdoses?

See Appendix A for answers to the figure questions.

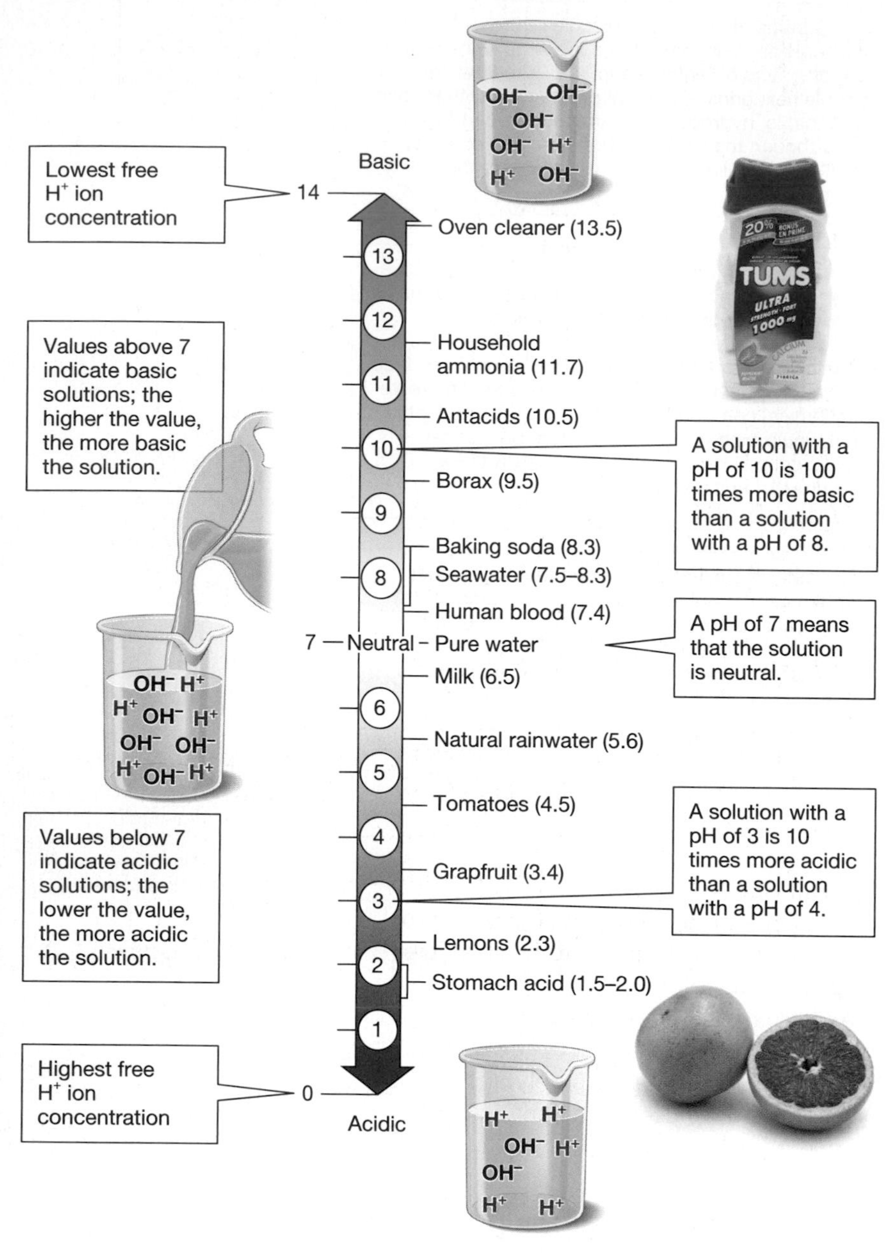

Figure 3.18

The pH scale indicates hydrogen ion concentration

A pH value indicates whether an aqueous solution is acidic, basic (alkaline), or neutral.

Q1: **Which has a higher concentration of free hydrogen (H^+) ions: vinegar with a pH of 2.8 or milk with a pH of 6.5?**

Q2: **What happens to the concentration of free H^+ ions in your stomach when you drink milk?**

Q3: **Black coffee has a pH of 5. Does adding coffee to water (pH of 7) increase or decrease the concentration of free H^+ ions in the liquid?**

See Appendix A for answers to the figure questions.

a 10-fold increase or decrease in the concentration of H^+ ions (**Figure 3.18**). Pure water is said to be neutral at a pH of 7, in the middle of the pH scale. The addition of acids to pure water raises the concentration of free H^+ ions, making the solution more acidic and pushing the pH below the neutral value of 7. Adding a base lowers the concentration of free H^+ ions in the solution, making the resulting solution more basic (alkaline) and raising the pH above 7.

NFEPP was designed to be sensitive to pH because, as noted previously, acidic conditions often surround inflamed tissue. Inflammation typically causes a drop in pH from 7.4 to 5.4. Because NFEPP is activated when pH is at that lower level, it works only at sites of pain and injury. The next step, the study authors say, is to test the drug in humans to find out if it works without the negative side effects of other synthetic opioids.

Talking about Addiction

When a doctor first prescribed oxycodone to Edwin in 2010, he did not say it was an addictive substance. "Even now, looking back, I get mad about it," says Edwin. "If someone had said, 'Your chances of getting addicted are very, very high. You might lose everything you own.' . . . I might have said, 'I'll stick to the Tylenol.'"

Edwin has now been sober for 3 years. He has his own apartment again and a job working as an outreach worker for the City of Boston. Pain still occasionally radiates through his legs, but Edwin avoids medication. Instead, he goes on long walks and plays with his 10-year-old son.

Someday, Edwin says, he will tell his son the whole story of his addiction, sparing no details. "For the longest time, it's been a taboo thing people didn't want to talk about," says Edwin. But with rising numbers of opioid abuse and deaths, that is no longer a luxury—it is something that society and families urgently need to address, he argues. "There needs to be more education about addiction. We must talk about it."

Two Decades of Drug-Related Death

The number of drug overdose deaths in the United States exceeds the number of deaths by car crashes or guns. A significant portion of overdose deaths each year involve opioids, including prescription opioids. From 2000 until 2018, deaths from opioid overdoses continuously and rapidly climbed, especially for fentanyl—an opioid that is deadly even in very small doses. In 2018, health officials documented the first decline in deaths in decades, which they attribute to a combination of changes in prescribing practices and increased access to treatment and naloxone, an overdose-reversing drug.

Assessment available in **smart**work

Opioid Overdose Deaths in the United States, 1998–2018

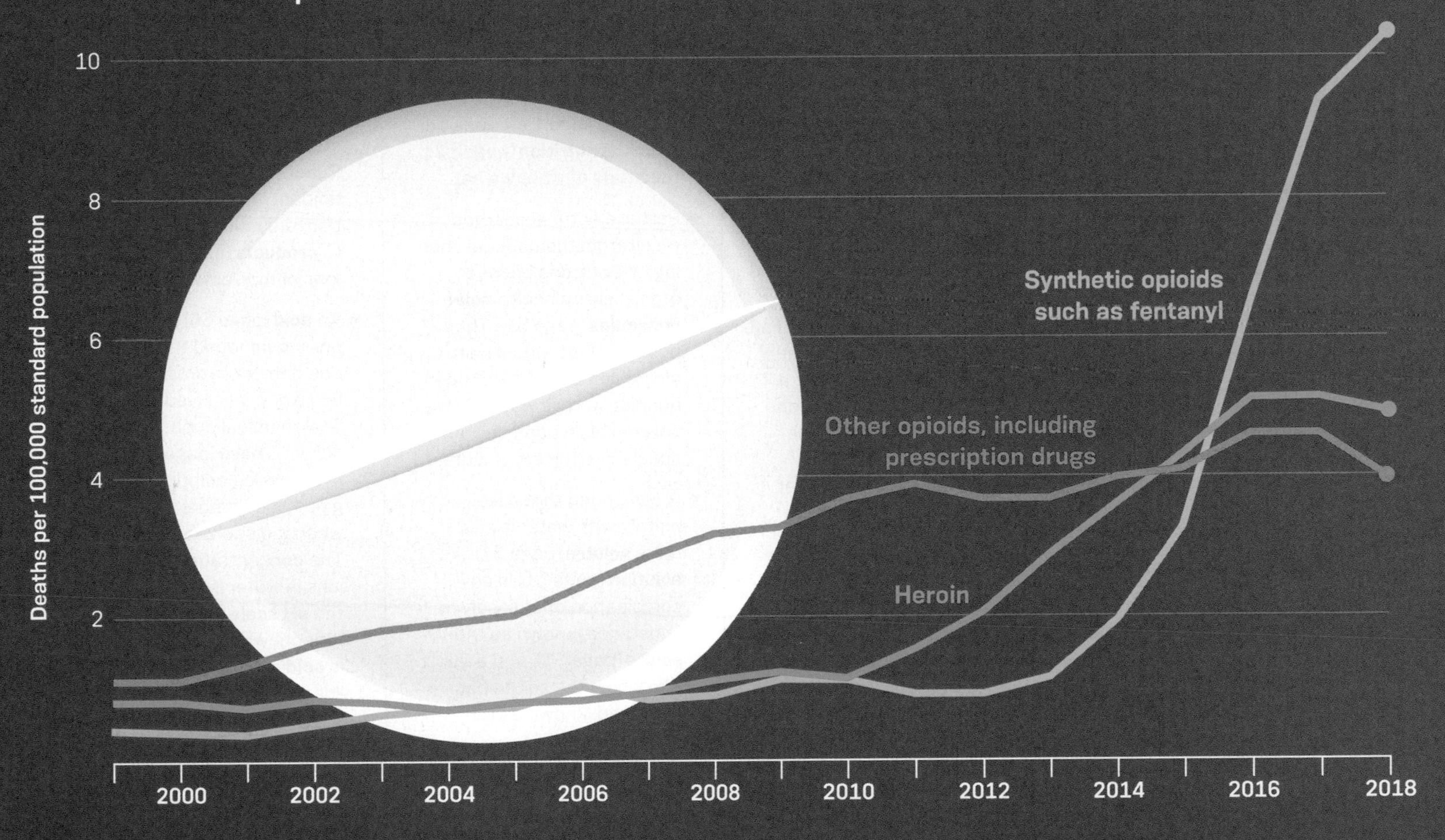

Differences in Strength

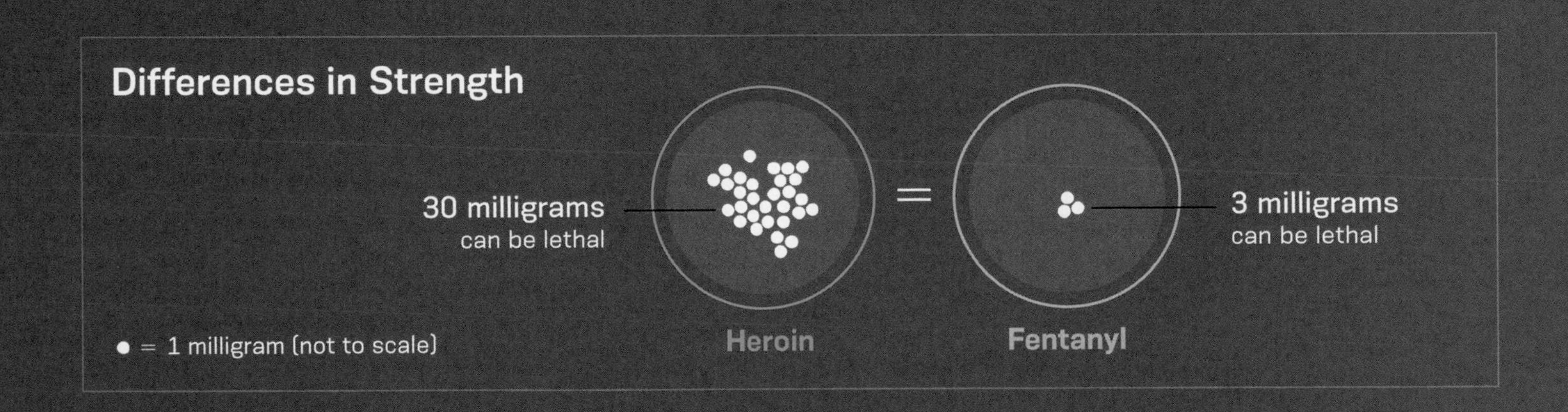

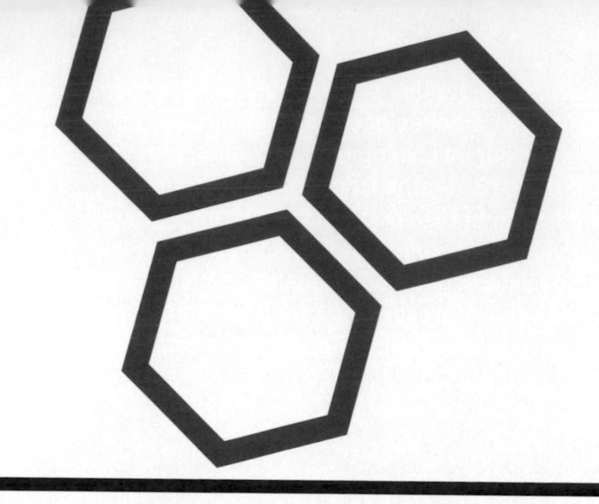

NEED-TO-KNOW SCIENCE

- The physical world is composed of **matter** (page 42), which is anything that has mass and occupies space. One type of matter is an element, and 98 distinct **elements** (page 42) exist naturally on Earth. An **atom** (page 43) is the smallest unit of an element that maintains its unique properties. At its core, every atom has a **nucleus** (page 43), which contains positively charged **protons** (page 43) and uncharged **neutrons** (page 43), and is surrounded by negatively charged **electrons** (page 43).
- The number of protons in an atom's nucleus is called its **atomic number** (page 43) and is unique to each element. **Isotopes** (page 43) of an element—atoms of the same species—have the same number of protons but different numbers of neutrons. The sum of the number of protons and the number of neutrons is the **atomic mass number** (page 43) of an isotope. The atomic mass number reflects the amount of mass in an element.
- The chemical interactions that cause atoms to associate with each other are known as **chemical bonds** (page 45). When an atom loses or gains electrons, it becomes, respectively, a positively or negatively charged **ion** (page 46). Ions of opposite charge are held together by **ionic bonds** (page 46).
- **Covalent bonds** (page 45) are formed by the sharing of electrons between atoms. A **molecule** (page 43) contains at least two atoms that are held together by chemical bonds.
- **Hydrogen bonds** (page 46) are weak associations between two molecules such that a partially positive hydrogen atom within one molecule is attracted to a partially negative region of another molecule.
- **Chemical compounds** (page 43) are molecules that contain atoms from at least two different elements. Carbon atoms can link with each other and with other atoms to generate a great diversity of chemical compounds called **organic molecules** (page 43). Multiple organic molecules bound together are an **organic compound** (page 43).
- The four main types of large organic molecules, or **biomolecules** (page 47), include the following:
  - **Proteins** (page 47) are **polymers** (page 47) of the 20 amino acid **monomers** (page 47) in countless different combinations; they make possible the endless functions in the cell.
  - **Carbohydrates** (page 47) range in size from simple monosaccharides and disaccharides to complex polymers that may contain thousands of monomers.
  - **Nucleic acids** (page 47) are polymers of nucleotide monomers: DNA is composed of deoxyribonucleotides, and RNA is composed of ribonucleotides.
  - **Lipids** (page 49) are not polymers and are better known as fats, oils, and steroids.
- Proteins, carbohydrates, and lipids are important considerations in **nutrition** (page 52), the study of what we eat.
- Partial electrical charges result from the unequal sharing of electrons between atoms, giving rise to **polar molecules** (page 51). The polarity of individual water molecules and the hydrogen bonding across water molecules explain nearly all of the special properties of water.
- A compound that mixes completely with water is said to be **soluble** (page 51). A **solution** (page 51) is any combination of a dissolved substance, known as the **solute** (page 51), and a fluid into which the solute has dissolved, known as the **solvent** (page 51).
- Ions and polar molecules are **hydrophilic** (page 52); they readily dissolve in water. Nonpolar molecules cannot associate with water and are therefore **hydrophobic** (page 52).
- When liquid water heats, it transitions into gas. During this process, called **evaporation** (page 53), hydrogen bonds break, and molecules spread apart. During the opposite process, called **condensation** (page 53), water vapor cools, molecules slowly re-form hydrogen bonds, and the gas returns to liquid.
- In **chemical reactions** (page 56), bonds between atoms are formed or broken. **Reactants** (page 56), the participants in a chemical reaction, are modified to give rise to **products** (page 56), new ions or molecules.
- An **acid** (page 56) is a hydrophilic compound that loses one or more hydrogen ions (H^+) when dissolved in water. The chemical opposite of an acid is a **base** (page 56), a hydrophilic compound that gains one or more H^+ ions when it dissolves in water. The concentration of free H^+ ions in water is expressed by the **pH scale** (page 56) and reflects whether a solution is **acidic** (page 56), **basic** (alkaline; page 56), or neutral.

THE QUESTIONS

See Appendix B for answers.

The Basics

1 The atomic number of an atom is determined by the number of _________ in the atom.

(a) protons

(b) neutrons

(c) electrons

(d) electrons plus neutrons plus protons

2 The atomic mass number of an atom is determined by the sum of the number of _________ in the atom.

(a) protons plus electrons plus neutrons

(b) protons plus electrons

(c) neutrons plus electrons

(d) neutrons plus protons

3 Link each term with the correct definition.

ion	1. The smallest unit of an element.
matter	2. A molecule made up of repeating monomers.
solution	3. An atom that has gained or lost an electron.
element	4. Consists of two or more atoms chemically bonded together.
chemical compound	5. Consists entirely of atoms with the same atomic number.
molecule	6. Anything that has mass and occupies space.
isotope	7. Has the same atomic number but a different atomic mass than its original element.
polymer	8. A molecule that contains atoms from two or more different elements.
atom	9. A combination of a solvent and a solute.

4 The partial negative charge at one end of a water molecule is attracted to the partial positive charge of another water molecule. What is this attraction called?

(a) a hydrogen bond

(b) a van der Waals interaction

(c) an ionic bond

(d) a covalent bond

(e) a hydrophilic bond

5 Select the correct terms: Proteins are (**polymers / monomers**) of amino acids. A common carbohydrate is (**sugar / fat**). Nucleic acids are composed of (**nucleotides / DNA**). Lipids (**are / are not**) polymers. All of these organic molecules contain (**carbon / nitrogen**).

Challenge Yourself

6 You are asked to determine the classification of an unidentified biomolecule. You are told that it can be broken down into only carbon, oxygen, and hydrogen atoms, and nothing else. Using your knowledge of biomolecules and their components, identify which of the following statements about your unknown sample are likely to be true. (Select all that apply.)

(a) It is an organic compound.

(b) It contains amino acids.

(c) It contains sugar.

(d) It is a nucleic acid.

(e) It is a carbohydrate.

7 Explain why life on Earth is carbon-based rather than, for example, hydrogen- or oxygen-based.

8 This pie graph depicts the chemical composition of the average eukaryotic cell.

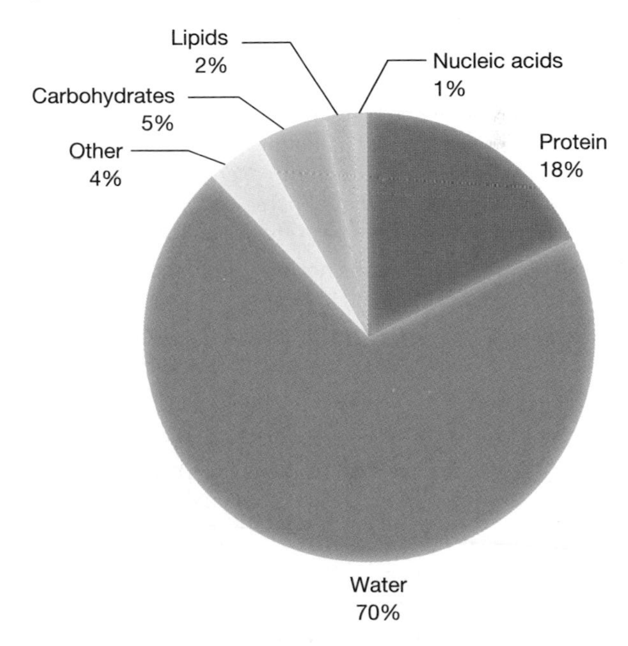

(a) What percentage of the cell is made up of amino acids? Nucleic acids?

(b) Biomolecules belonging to which category make up only 1 percent of the cell?

(c) Starch is a _________. This biomolecule category makes up _________ percent of the cell.

(d) Which of the biomolecules makes up the highest percentage of the average eukaryotic cell? What is this percentage?

9 Refer to "What Should I Eat? Evidence-Based Nutrition Tips" (page 52). Which recommended food categories include more than one biomolecule class as nutrients? Which food categories include only one biomolecule class as a nutrient?

Try Something New

10 In the accompanying figure, a carbon atom resides at each unlabeled corner of the hexagons and wherever a "C" is shown. Each "O" is an oxygen atom, and each "H" is a hydrogen atom. What is this structure?

(a) a fatty acid monomer

(b) two simple sugar molecules linked together

(c) a nucleotide polymer

(d) a hydrocarbon ring polymer

(e) an amino acid

11 Lipoproteins are relatively large, combined clumps of both protein and lipid molecules that circulate in the blood of mammals. They come in two forms, called HDL and LDL, and they act like suitcases to move cholesterol, fatty acid remnants, triglycerides, and phospholipids from one place to another through the bloodstream. (LDL recirculates lipids throughout the body [bad], whereas HDL takes lipids to the liver to excrete them in feces [good!].) Given that lipids are hydrophobic and proteins can be hydrophilic, which of the following statements is correct?

(a) The lipid portion of LDL does not dissolve in the bloodstream, whereas the lipid portion of HDL does.

(b) The protein portions of both LDL and HDL can dissolve or interact with the water molecules in the bloodstream.

(c) Neither the protein nor the lipid portions of LDL molecules can interact with water molecules in the bloodstream.

(d) Both the protein and the lipid portions of HDL molecules can interact with water molecules in the bloodstream.

(e) none of the above

12 Answer a–d using the following data from the United States Department of Agriculture.

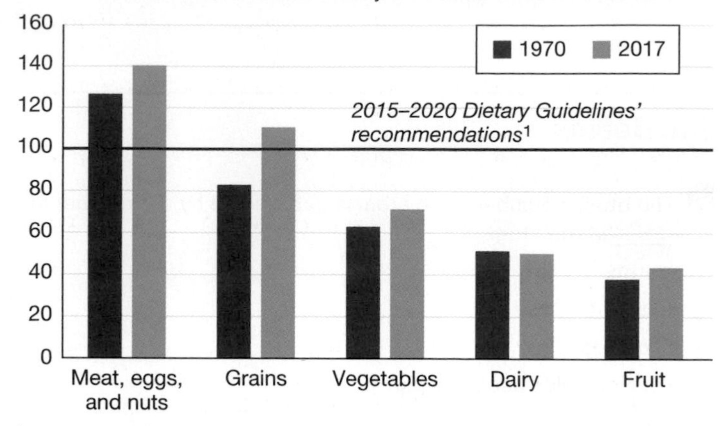

[1]Based on a 2,000-calorie-per-day diet.
Loss-adjusted food availability data are proxies for consumption. Rice availability data were discontinued and thus are not included in the grains group.
Source: USDA, Economic Research Service, Loss-Adjusted Food Availability Data and *2015–2020 Dietary Guidelines.*

(a) In which category or categories did consumption fail to meet guidelines in 1970 but exceed guidelines in 2017?

(b) In which category or categories did consumption fail to meet guidelines in 2017 but exceed guidelines in 1970?

(c) In which category or categories did consumption fail to meet guidelines in both 1970 and 2017?

(d) In which category or categories did consumption exceed guidelines in both 1970 and 2017?

13 You have found an unknown cleaning solution in your roommate's under-sink cabinet. You have several spring cleaning projects to take care of and would like to use this product if it has the correct pH for the particular job. You know you need an extremely basic solution to clean your oven but an acidic solution, such as vinegar, to clean your coffee maker. You borrow a few pH indicator strips from your biology lab TA and test the solution. For each of the following scenarios, select the best answer.

(a) The solution tested at a pH of 11; you will be able to clean your (**oven / coffee maker / both / neither**).

(b) The solution tested at a pH of 2; you will be able to clean your (**oven / coffee maker / both / neither**).

(c) The solution tested at a pH of 7; you will be able to clean your (**oven / coffee maker / both / neither**).

(d) The solution tested at a pH of 9; you will be able to clean your (**oven / coffee maker / both / neither**).

(e) The solution tested at a pH of 4; you will be able to clean your (**oven / coffee maker / both / neither**).

14 Laundry detergent molecules have a short polar end and a long nonpolar end, enabling them to bind to water (on the polar end) and oils (on the nonpolar end). Suppose you spill some oil-and-vinegar salad dressing on your shirt. Explain how washing your shirt with detergent will help remove the dressing.

Leveling Up

15 **What do *you* think?** There's no shortage of movies depicting drug abuse, addiction, and ruined lives. A few of the more disturbing films are *Beautiful Boy* (2018), *Requiem for a Dream* (2000), *Gia* (1998), *Trainspotting* (1996), and *The Panic in Needle Park* (1971). With all these tragic stories splayed across the big screen, how is it that so many people unsuspectingly get hooked on opioids? The Centers for Disease Control and Prevention (CDC) claims that new laws being enacted in many states to limit the numbers of opioids doctors can prescribe will decrease the potential for abuse, addiction, and overdose. Do you believe the changing legal landscape will have the desired effects? Use your internet research skills to find out what evidence exists for or against the CDC's claim.

16 **Is it science?** Life on Earth evolved as carbon-based. Silicon shares many of the chemical properties of carbon, yet it is not a building block of life on Earth. Watch the episode of the TV series *The X-Files* titled "Firewalker" (http://www.hulu.com/watch/158588) or read through its story line (https://en.wikipedia.org/wiki/Firewalker_%28The_X-Files%29). While doing so, list the scientific concepts presented, and note whether you think they represent real science or pseudoscience (for a refresher on the difference, see Chapter 2). Then use your textbook and internet research to determine whether you were correct or incorrect in each case. If you evaluated any of the concepts incorrectly, reflect on how you came to your initial conclusions.

Digital resources for your book are available online.

Engineering Life

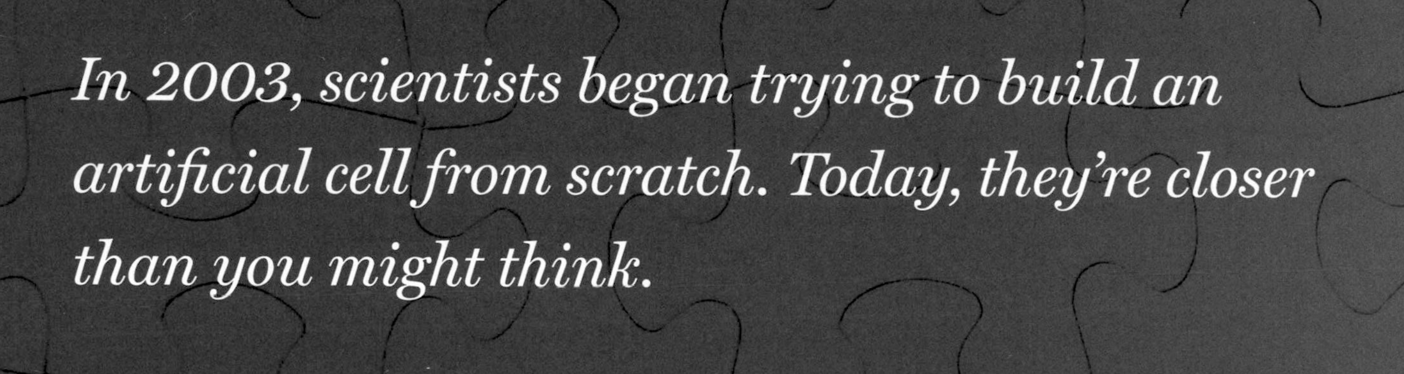

In 2003, scientists began trying to build an artificial cell from scratch. Today, they're closer than you might think.

After reading this chapter you should be able to:

- Explain cell theory and why it is central to the study of life.
- Determine whether something is living or nonliving based on the characteristics of living things.
- Describe the differences in the structures of viruses, prokaryotes, and eukaryotes.
- Diagram a plasma membrane, showing how the structure allows some substances in and keeps others out.
- Compare and contrast passive and active transport of materials into and out of cells.
- Differentiate between exocytosis and the three types of endocytosis.
- Describe the role of any given organelle in a eukaryotic cell.
- Identify the main differences between a plant cell and an animal cell.

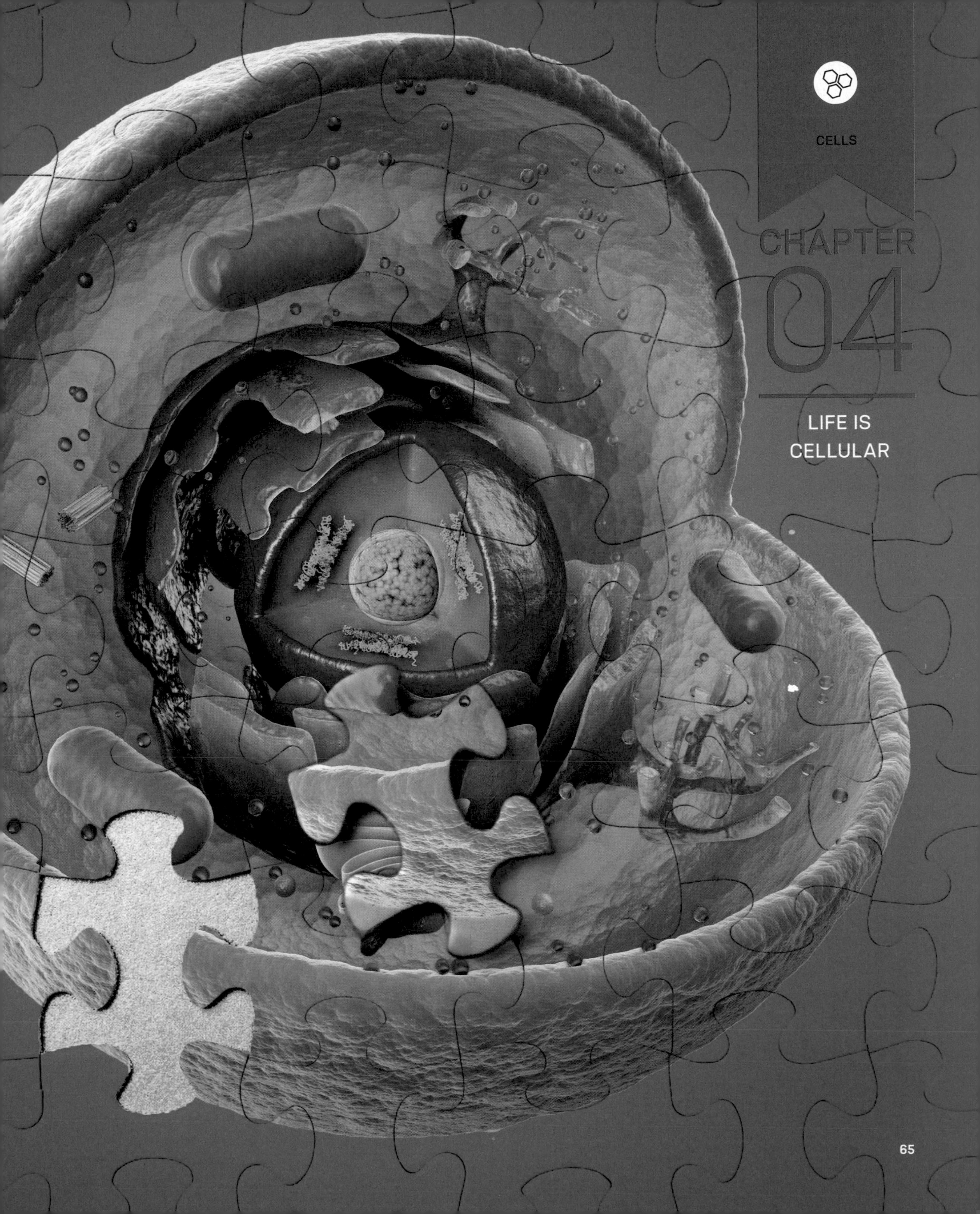

CHAPTER 04

LIFE IS CELLULAR

The sky was still dark when Daniel Gibson hurried into the J. Craig Venter Institute (JCVI) in La Jolla, California. At 5:00 a.m., his footsteps echoed through the empty halls of the building. He reached a laboratory door and slipped inside. There, Gibson peered into a warm incubator, his eyes scanning rows of palm-sized petri dishes. His stomach was in knots. For 3 months, the experiment had failed. Would this day—Monday, March 29, 2010—be any different?

Gibson is part of a team at the JCVI with a single, audacious goal: to create life. For more than a decade, this team of scientists and engineers has attempted to build a synthetic, or human-made, **cell**. Cells are the smallest and most basic unit of life—microscopic, self-contained units enclosed by a protective membrane (**Figure 4.1**). The human body is composed of approximately 100 trillion (10^{14}) cells. On that day in 2010, however, the JCVI was trying to synthesize just one cell—a bacterium, a type of single-celled microorganism.

Gibson's team had sequenced a bacterium's complete genetic information, its **genome**; built a synthetic version of that genome using basic laboratory chemicals; and, finally, replaced the natural DNA of another species of bacterium with the synthetic DNA. DNA (deoxyribonucleic acid) is a large and complex molecule that acts as a set of instructions for building an organism, like a blueprint. Almost every cell of every living organism contains DNA. Because DNA transfers information from parents to offspring, it is essential for reproduction. Life, no matter how simple or how complex, uses this inherited genetic code to direct the structure, function, and behavior of every cell. DNA is made up of many nucleotides held together in a structure called the double helix, a ladderlike assembly twisted along its length into a spiral (see Figure 3.10).

Gibson's boss, the famous geneticist J. Craig Venter, worked for more than 15 years and spent millions of dollars to construct a synthetic DNA helix from chemicals in the laboratory, but Gibson and the team had been unable to get that synthetic DNA to work inside a cell. Every Friday for 3 months, they transplanted the synthetic DNA into a bacterial cell whose own DNA had been removed. The synthetic DNA included a gene to make the cells turn bright blue, so every Monday Gibson hurried to the incubator and checked the petri dishes for a colony of blue cells. But Monday after Monday, the dishes were barren. "We did the genome transplantation again and again," he recalls, "but nothing was working."

Then, in mid-March, Gibson identified an error in a single gene in the synthetic DNA. A gene is a segment of DNA that codes for a distinct genetic characteristic, such as having O-type blood or a dimpled chin. In Gibson's bacterium, the gene with the error was responsible for DNA replication. When it wasn't working, the bacterium couldn't replicate its DNA, and it died. So in late March, Gibson fixed the DNA error, transplanted the genome yet again, and waited.

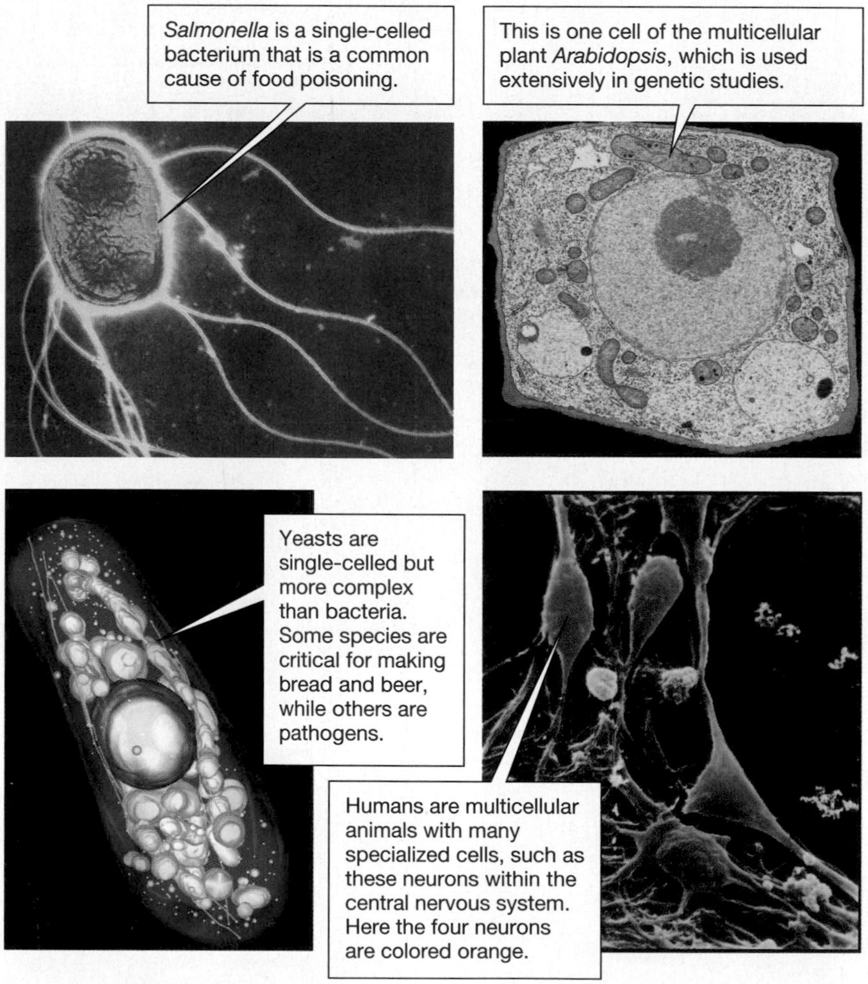

Salmonella is a single-celled bacterium that is a common cause of food poisoning.

This is one cell of the multicellular plant *Arabidopsis*, which is used extensively in genetic studies.

Yeasts are single-celled but more complex than bacteria. Some species are critical for making bread and beer, while others are pathogens.

Humans are multicellular animals with many specialized cells, such as these neurons within the central nervous system. Here the four neurons are colored orange.

Figure 4.1

An individual organism may consist of a single cell or many cells

These photos of cells were taken using electron microscopes, which have higher power than light microscopes, so they can reveal smaller structures. Color has been added to the images to differentiate structures within the cells. Note the membrane appearing near the outer edge of each cell.

Life, Rewritten

Scientists are pushing the boundaries of **cell theory**, one of the unifying principles of biology, in their efforts to put synthetic DNA into another cell devoid of its own genetic material. Cell theory has two main parts: every living organism is composed of one or more cells, and all cells living today came from a preexisting cell. By trying to engineer a cell in the laboratory, Venter, Gibson, and others are challenging the second part of the definition. "It's going to be a big challenge to create a totally synthetic cell," says James Collins, a synthetic biologist at Boston University, "but it's fundamentally intriguing to explore how life may have arisen on the planet."

The JCVI's first step toward a synthetic cell was a small one. In 2003, Venter's team flexed its scientific muscles by synthesizing the 11-gene, 5,386-base-pair genome of phiX174, a virus that infects bacteria. A **virus** is a small, infectious agent that can replicate only inside a living cell. Most viruses are little more than stripped-down genetic material wrapped in proteins, yet these pathogens attack and devastate organisms in every kingdom of life, from bacteria to plants and animals. Though the JCVI team successfully

The Characteristics of Living Organisms

All living things share certain features that characterize life.

1. *They are composed of one or more cells.* The cell is the smallest and most basic unit of life; all organisms are made of one or more cells. Larger organisms are made up of many different kinds of specialized cells and are known as *multicellular organisms*.
2. *They reproduce autonomously using DNA.* All living organisms are able to reproduce, that is, to make new individuals like themselves. DNA is the genetic material that transfers information from parents to offspring. A segment of DNA that codes for a distinct genetic characteristic is called a gene. Life, no matter how simple or complex, uses this inherited genetic code to direct the structure, function, and behavior of every cell.
3. *They obtain energy from the environment to support metabolism.* Because all organisms need energy to survive, they use a wide variety of methods to capture energy from their environment. The process by which living organisms capture, store, and use energy is known as *metabolism*.
4. *They sense the environment and respond to it.* Living organisms sense many aspects of their external environment, from the direction of sunlight to the presence of food and mates. All organisms gather information about the environment by sensing it and then respond appropriately.
5. *They maintain a constant internal environment.* Living organisms sense and respond to not only the external environment but also to their internal conditions. All organisms maintain constant internal conditions—a process known as *homeostasis*.
6. *They can evolve as groups.* Evolution is a change in the genetic characteristics of a group of organisms over generations. When a characteristic becomes more or less common across generations, evolution has occurred within the group.

	Rock	Virus	Fungus	Plant	Animal
Composed of one or more cells	✕	✕	✓	✓	✓
Autonomously reproduce themselves	✕	✕	✓	✓	✓
Obtain energy from their environment	✕	✕	✓	✓	✓
Sense their environment and respond to it	✕	✕	✓	✓	✓
Maintain a constant internal environment (homeostasis)	✕	✕	✓	✓	✓
Can evolve as groups	✕	✓	✓	✓	✓
Living	✕	?	✓	✓	✓

created a virus with a synthetic genome, it was not considered the first synthetic life-form because scientists debate whether viruses are alive. (For more on this debate, see "Viruses—Living or Not?" on page 69.)

Next, Venter and his colleagues moved on to a bacterium. In 2010, they sequenced and built the genome of *Mycoplasma mycoides* (*M. mycoides*), a bacterium that causes the mammary glands of goats to swell. They constructed the genome—a 1.1-million-base-pair DNA sequence—using the four necessary ingredients: adenine (A), thymine (T), guanine (G), and cytosine (C). These nucleotides are the building blocks of DNA. Organized in different combinations, A, T, G, and C carry all the instructions for everything a cell does.

The team used a machine to read the nucleotide sequence of the *M. mycoides* genome and then "print out" little bits of that code, creating synthetic strands of DNA about 50–80 bases long. They then strung these pieces together using living cells as factories, inserting the short segments into yeasts and *Escherichia coli* (*E. coli*)—single-celled microorganisms. These microorganisms interpreted the strands as broken pieces of DNA and stitched them together, creating longer and longer sequences of synthetic DNA. It was like building the Eiffel Tower from a massive box of Legos, constructing a single support beam at a time. The effort—with many mistakes along the way—took years. "It was very complex," said Venter. "It was a long, involved process." Ultimately, this process yielded a DNA sequence of 1.1 million base pairs, which other researchers expanded to 4.4 million in 2019.

Congratulations, It's a . . . Cell

Once the DNA sequence of *Mycoplasma mycoides* was complete and intact, the JCVI team attempted to transfer it into another species and make it work. This was the experiment that almost drove Gibson crazy. The researchers removed the DNA from a cell of a closely related bacterium, *Mycoplasma capricolum* (*M. capricolum*), and replaced it with the *M. mycoides* synthetic DNA.

After months of trying, on that Monday morning at 5:00 a.m., Gibson cautiously scanned the petri dishes. There, on a single dish, was a group of bright-blue cells—evidence that the *M. capricolum* cell had "booted up" the *M. mycoides* DNA and transformed itself into an *M. mycoides* cell (**Figure 4.2**). Gibson was ecstatic. Moments later, he sent a text message to Venter, waking him up. Within the hour, Venter was in the lab

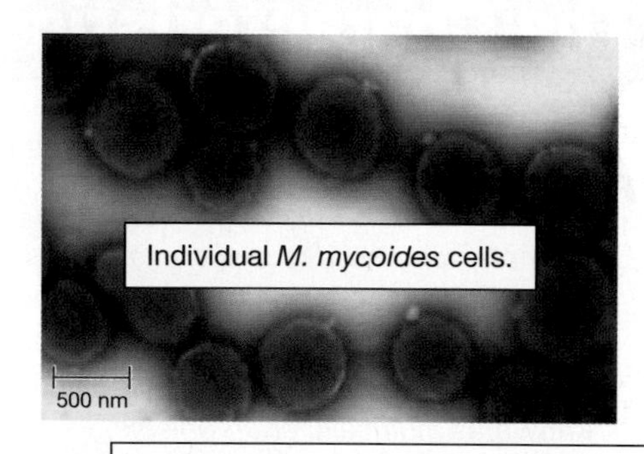

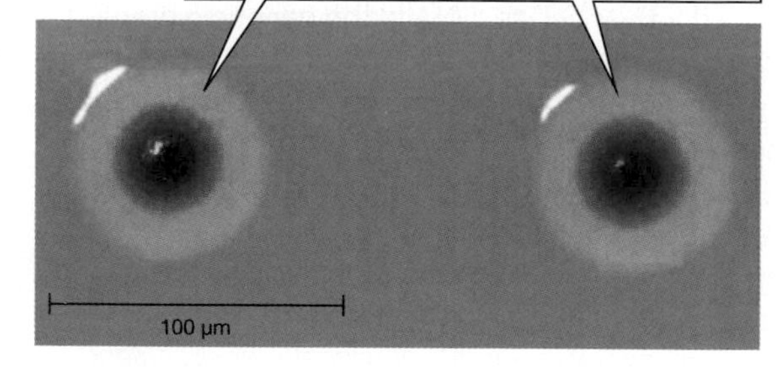

Figure 4.2

The first synthetic organism

The bacterium *M. capricolum* was transformed by the insertion of synthetic DNA of a closely related species of bacterium, *M. mycoides*. The synthetic DNA included a gene that codes for blue pigment.

Q1: Why did the researchers insert the gene that codes for blue pigment into the synthetic DNA?

Q2: What part of the transformed bacterium was synthetic?

Q3: Do you think this experiment created life?

See Appendix A for answers to the figure questions.

DEBUNKED!

MYTH: Bacteria cells in our bodies outnumber human cells by 10 to 1.

FACT: According to a 2016 study, the ratio between microbes and human cells is approximately 1 to 1, although it varies from person to person.

with a camera, taking pictures of the tiny blue dollop in the dish. "How does it feel to create life?" Venter asked Gibson. They opened champagne and toasted their success.

Over the following weeks, Gibson repeated the experiment hundreds of times to make sure the blue cells were not an accident or a fluke. Every time, the cells with a synthetic genome survived. The team had done it—created the first synthetic cell. "They are living cells," Venter told *The Scientist* magazine when the research was published 2 months later. "The only difference is that they have no natural history. Their parents were the computer."

Once the research was published, the reaction from the academic community, captured in *Nature* magazine, was swift and divided. Some called it a significant advance: "We now have an unprecedented opportunity to learn about life," said Mark Bedau, a professor of philosophy at Reed College in Oregon. Arthur Caplan, a bioethicist at the University of Pennsylvania, said, "Venter's achievement would seem to extinguish the argument that life requires a special force or power to exist. In my view, this makes it one of the most important scientific achievements in the history of mankind."

Others were more hesitant. "Has [Venter] created 'new life'?" asked George Church, a prominent geneticist at Harvard Medical School. "Not really. . . . Printing out a copy of an ancient text isn't the same as understanding the language." Gibson and Venter agree that they did not create life from scratch, but they argue that they did create new life from existing life.

Viruses—Living or Not?

You've heard their names: Ebola virus, Zika virus, Influenza virus, Coronavirus. Viruses—microscopic, noncellular infectious particles—are perhaps the smallest biological agents with the greatest impact on human health. Like living organisms, viruses reproduce and evolve, yet they lack some of the key characteristics of life—which is why most scientists today regard viruses as nonliving. For one thing, viruses are not made up of cells. A virus is much simpler than a cell, usually consisting of a small piece of genetic material (for example, DNA) that is wrapped in a protein coat. Some viruses also have an envelope, a lipid layer usually stolen from a cell's plasma membrane, enclosing the central core of genetic material and protein.

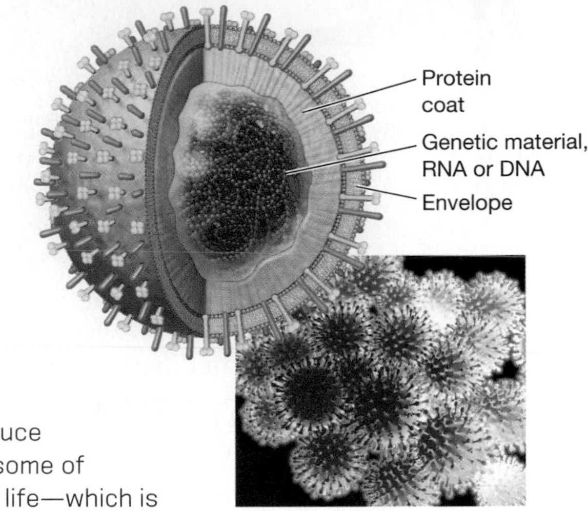

Another difference, compared to living organisms, is that viruses lack the many structures within cells that are necessary for critical cellular functions such as homeostasis, autonomous reproduction, and metabolism. To gain these functions, they become "body snatchers": Viruses make the cells of the organisms they infect do their work for them. They accomplish this feat by invading cells, releasing viral genetic material into the cell interior, and "hijacking" the host cell's machinery. Viruses multiply to huge numbers, and viral offspring escape from a host cell either by causing it to burst open or by budding off from the cell, wrapped in a layer of the host cell's plasma membrane.

Uniquely, unlike the case with living organisms, the genetic material that viruses pass from one generation to the next is not always DNA; sometimes it is RNA. Viruses are generally classified by the type of genetic material they possess (type of DNA or RNA molecule), their shape and structure, the type of organism (host) they infect, and the disease they produce. The variant forms of a particular type of virus are called **viral strains**, or serotypes. Viruses evolve new strains within a host so quickly that sometimes an antiviral drug or vaccine developed to fight an older strain becomes useless against a new strain.

A Different Approach

As Venter and Gibson were making headlines, other scientists were quietly pursuing a different approach to building a cell. One young scientist in California decided to start by building one of the simplest, yet most vital, components of a cell: the layer of molecules that surrounds it.

In the chemistry department at UC San Diego, assistant professor Neal Devaraj was fascinated by the idea of building life. "Most molecular biologists study what exists," says Devaraj, "but when you're a chemist and constantly make new compounds, you want to engineer something from scratch." So instead of taking a cell apart and determining how it works, Devaraj decided to try to build a cell artificially, using materials not typically found

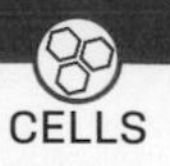

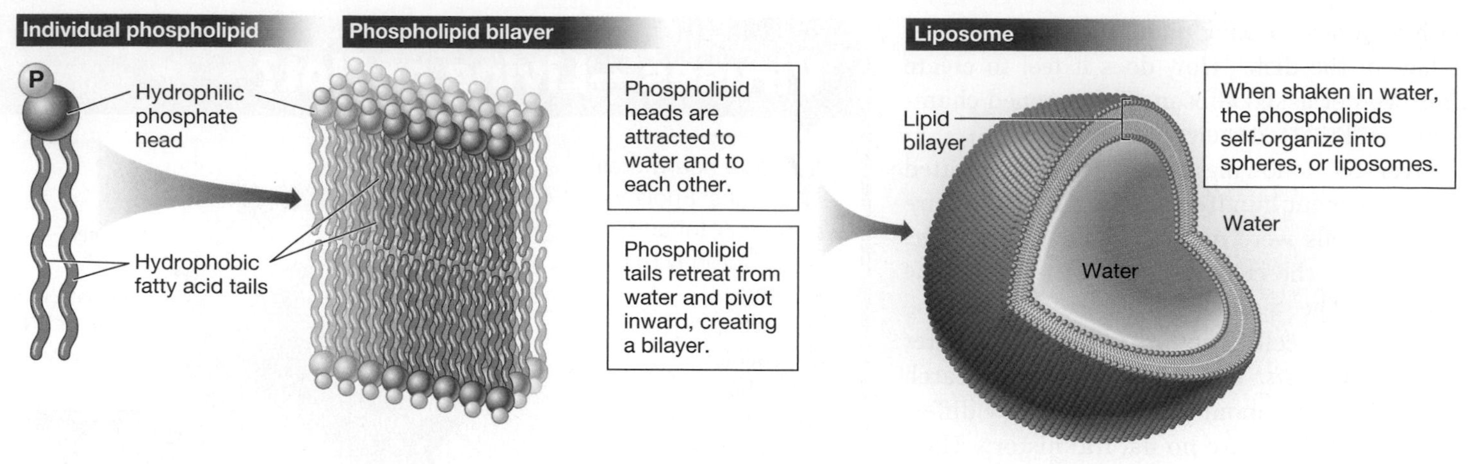
Individual phospholipid
P
Hydrophilic phosphate head
Hydrophobic fatty acid tails
Phospholipid bilayer
Phospholipid heads are attracted to water and to each other.
Phospholipid tails retreat from water and pivot inward, creating a bilayer.
Liposome
Lipid bilayer
When shaken in water, the phospholipids self-organize into spheres, or liposomes.
Water
Water

Figure 4.3

Liposomes form when phospholipids and water are shaken together

When you shake a mixture of phospholipids and water, the phospholipid bilayers bend and link together to form spheres called liposomes. This simple structure is remarkably similar to the basic structure of a cell.

Q1: Why is it important that the phosphate head of a phospholipid is hydrophilic?

Q2: What genetic component of most organic cells do liposomes lack? What does that omission mean for liposomes?

Q3: Could the tendency of phospholipid bilayers to spontaneously form spheres have played a role in the origin of life? (*Hint*: Refer to "The Characteristics of Living Organisms" on page 67.)

See Appendix A for answers to the figure questions.

in nature. "If you want to really understand the principles by which life operates and evolves, the best way to do so is to build a cell from the ground up," he says.

Scientists suspect that one of the first events at the beginning of life on Earth was the formation of a **plasma membrane**, a barrier separating a cell from its external environment. A plasma membrane is made of two layers of **phospholipids**, a type of organic molecule. Phospholipids have a phosphate head that is water-loving, or *hydrophilic*, and a lipid (fatty acid) tail that is water-fearing, or *hydrophobic*. In water, these molecules form a double layer with heads out and tails in. This barrier separates the contents of the cell from what lies outside the cell. Thus, the membrane is a **phospholipid bilayer**, a mostly impermeable barrier. When a phospholipid bilayer forms a sphere, or **liposome**, the fluid inside the liposome can have a different composition from the fluid outside (**Figure 4.3**). The ability to maintain an internal environment separate from the external environment is one of the most critical functions of the plasma membrane of a cell.

Given a container of phospholipids, anyone can make a membrane, says Devaraj. "Making membranes is almost a trivial thing," he explains. "You take natural or synthetic phospholipids, add water, and they form membranes." Yet researchers had been unable to form a membrane from scratch—without using preexisting phospholipids. In nature, new phospholipids are created by enzymes embedded in the cell.

Instead of trying to engineer new phospholipids, Devaraj wanted to start with something simpler. He worked with graduate student Itay Budin, who now works at UC Berkeley, to create a self-assembling membrane. They first mixed together oil and a detergent. Then they added copper, a metal ion, as a catalyst to spark a chemical reaction. With the addition of copper, sturdy membranes begin to bud off the oil; these were self-assembling structures. "There's no equivalent whatsoever in nature," says Devaraj. "Our goal was simply to mimic biology."

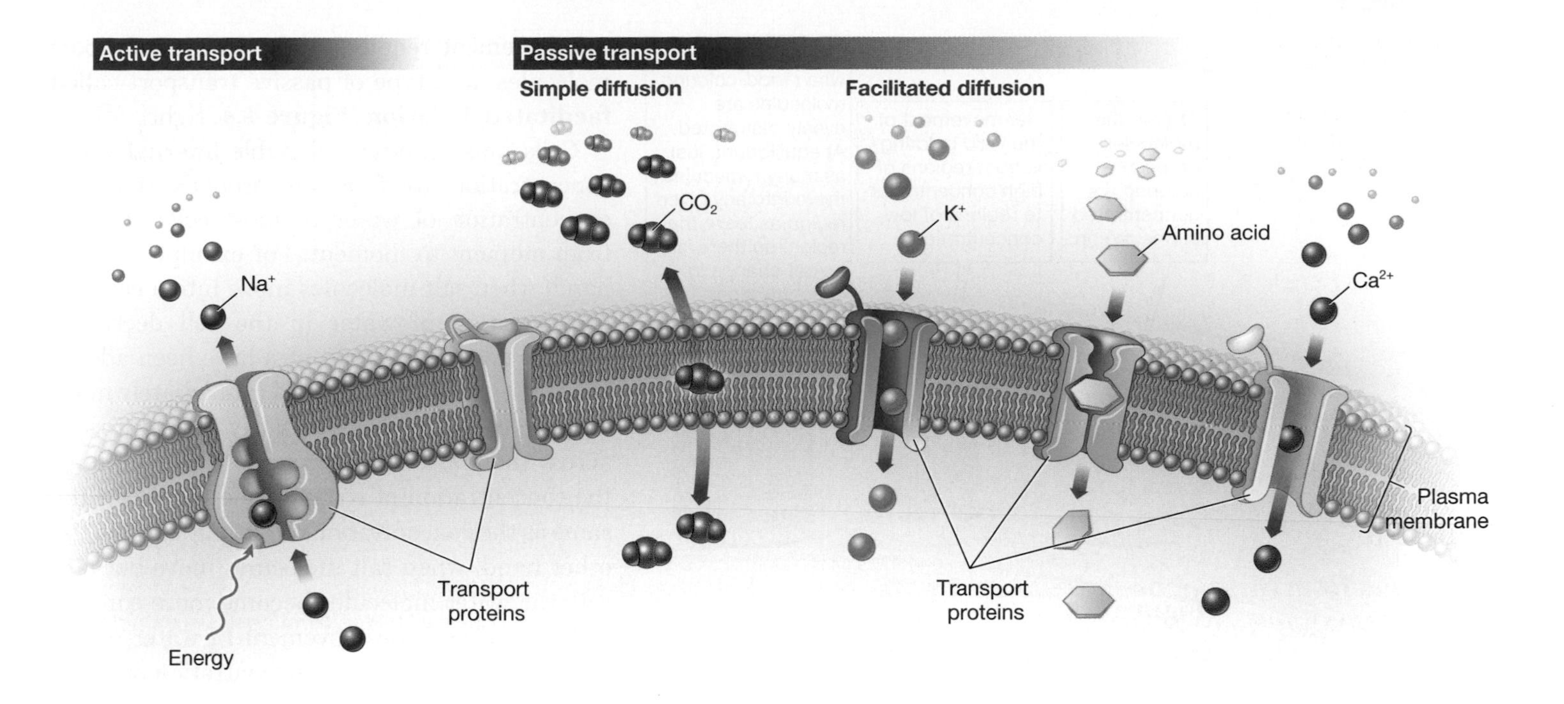

Figure 4.4

The plasma membrane is a barrier and a gatekeeper

The plasma membrane moves substances selectively. Whether molecules pass through the membrane is determined largely by the types of membrane proteins embedded in the phospholipid bilayer.

Q1: How is the plasma membrane a barrier? How is it a gatekeeper?

Q2: Why can't ions (such as Na^+) cross the plasma membrane without the help of a transport protein?

Q3: If no energy were available to the cell, what forms of transport would not be able to occur? What forms of transport could occur?

See Appendix A for answers to the figure questions.

Through the Barrier

Devaraj admits that his artificial membrane is far simpler than a real cell's plasma membrane, which is dotted with numerous proteins, including transport proteins. **Transport proteins** are the gates, channels, and pumps that allow molecules to move into and out of the cell, contributing to the selective permeability of the plasma membrane. **Selective permeability** means that some substances can cross the membrane on their own, others are excluded, and still others can pass through the membrane when aided by transport proteins (**Figure 4.4**).

All movement of substances through the plasma membrane occurs by either active or passive transport. Some transport proteins facilitate **active transport**, the movement of a substance that requires an input of energy (**Figure 4.4**, left). Molecules move across the plasma membrane by active transport when they need to move from a region of *lower* concentration to a region of *higher* concentration. In contrast, **passive transport** is the movement of a substance without the addition of energy (**Figure 4.4**, middle and right). Movement via passive transport is spontaneous.

A primary type of passive transport is **diffusion**, the movement of a substance from a region of *higher* concentration to a region of *lower* concentration, known as moving down a

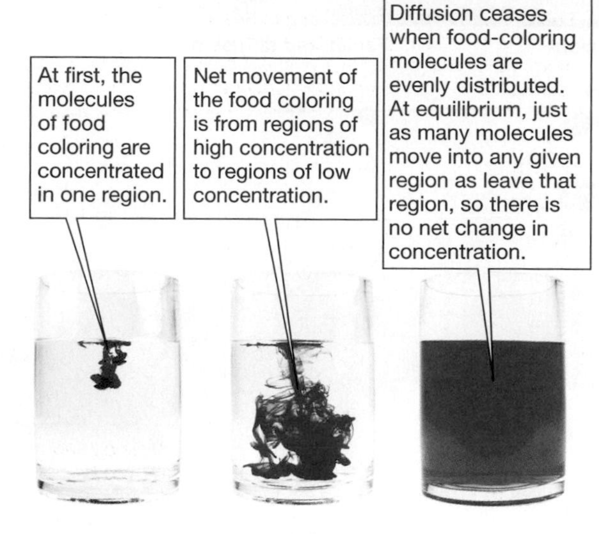

At first, the molecules of food coloring are concentrated in one region.
Net movement of the food coloring is from regions of high concentration to regions of low concentration.
Diffusion ceases when food-coloring molecules are evenly distributed. At equilibrium, just as many molecules move into any given region as leave that region, so there is no net change in concentration.

Figure 4.5

Food coloring in water illustrates diffusion

Q1: Is the dye at equilibrium in any of these glasses? Describe how the first glass will look when the dye is at equilibrium with the water.

Q2: Will diffusion mix the molecules of dye evenly through the water, or is it necessary to shake the container to get a uniform mixture?

Q3: Will diffusion mix the dye faster in hot water than in cold water? Why or why not? (*Hint:* Review the discussion of the behavior of water molecules at different temperatures in Chapter 3.)

See Appendix A for answers to the figure questions.

concentration gradient (**Figure 4.5**). Oxygen and carbon dioxide enter and leave cells by **simple diffusion**. These small, uncharged molecules slip between the large molecules in the phospholipid bilayer without much hindrance.

Water moves in and out of cells (and compartments inside cells) by **osmosis**. Osmosis is a form of diffusion because the water molecules are moving from areas of higher concentration to areas of lower concentration (**Figure 4.6**). Osmosis is critical for cellular processes because most cells are at least 70 percent water, and nearly all cellular processes take place in a water-rich environment. In some cells, water can simply flow through the plasma membrane. In other cells, the movement requires a channel of transport molecules, in a type of passive transport called **facilitated diffusion** (**Figure 4.4**, right).

Cells must maintain a stable internal water concentration to function properly, but the concentration of water in most cells changes from moment to moment. For example, on one hand, when salt molecules move into a cell, the concentration of water in the cell decreases because additional molecules have been added. In response, water molecules immediately move into the cell by osmosis—that is, they diffuse—across the plasma membrane into the cell until the concentration of water inside the cell is the same as the concentration on the outside. On the other hand, when salt molecules move out of a cell, the water molecules become more concentrated, and osmotic movement of water out of the cell then restores the concentration of water in the cell to its correct level.

You can imagine, then, that the concentration of solutes, the dissolved substances within a cell, in relation to the concentration *outside* the cell is of critical importance. When cells are surrounded by fluid with the same solute concentration as the cell interior, the extracellular and intracellular environments are said to be **isotonic** to each other (*iso*, "equal"). If the extracellular environment has a higher solute concentration, it is **hypertonic** to the cell interior (*hyper*, "more"). If it has a lower solute concentration than the cell's interior, it is said to be **hypotonic** to it (*hypo*, "less"). The take-home point here is that in a hypertonic solution, a cell loses water; in a hypotonic solution, a cell gains water; in an isotonic solution, there is no net movement of water, meaning a cell gains and loses water at the same rate.

The careful balance of concentrations is particularly vital in human blood, where red blood cells are typically in an isotonic solution. Because the solute concentrations inside and outside these cells are the same, dehydrated patients receive an IV drip of saline, a solution of salt in water. If a dehydrated patient received water instead of a saline solution, the water would dilute the patient's blood, making it hypotonic. Osmosis would then occur, causing water to rapidly diffuse into red blood cells to the point where the cells could burst and die.

Most hydrophobic molecules, even fairly large ones, can pass through the plasma membrane via simple diffusion because they mix readily with the hydrophobic tails that form the core of the phospholipid bilayer. But hydrophilic substances such as sodium ions (Na^+), hydrogen ions (H^+), and larger molecules, including sugars and amino acids, cannot cross the plasma membrane without assistance. These substances move across the plasma membrane by facilitated diffusion (see **Figure 4.4**, right).

Devaraj's artificial membrane does not contain any transport proteins, so large hydrophilic molecules cannot pass through it. But small molecules such as oxygen and carbon dioxide can pass through his membrane via simple diffusion.

The plasma membrane also contains **receptor proteins**, which are sites where molecules released by other cells can bind. The binding of a molecule to a receptor protein starts a chain of events inside the cell that causes the cell to do something. For example, the receptors in nerve cells receive molecular signals from other nerve cells that cause the cells to fire. Receptor proteins are key components of a cell's communication system, enabling it to respond appropriately to changes in its surroundings.

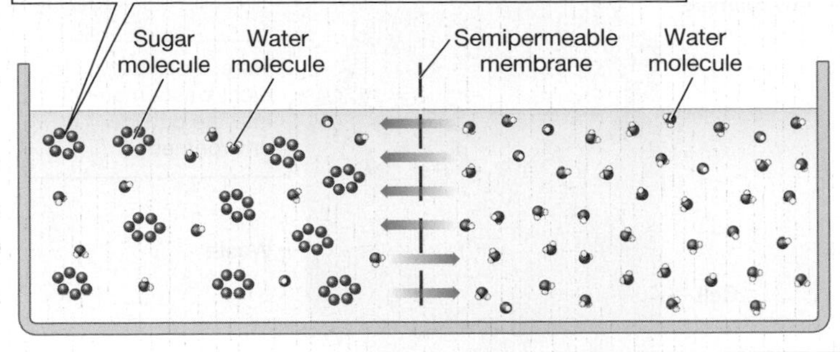

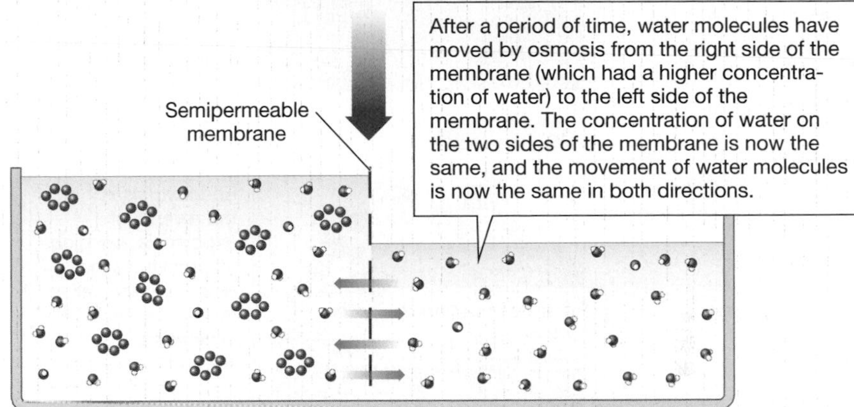

Semipermeable membrane
After a period of time, water molecules have moved by osmosis from the right side of the membrane (which had a higher concentration of water) to the left side of the membrane. The concentration of water on the two sides of the membrane is now the same, and the movement of water molecules is now the same in both directions.

Figure 4.6

In osmosis, water diffuses across a semipermeable membrane

Osmotic movement of water between a cell and the external environment is critical for maintaining a constant water concentration in the cell, which it needs to function properly.

Q1: What would the second diagram look like if the pores in the semipermeable membrane were too small to allow water molecules to pass through?

Q2: What would the second diagram look like if the pores were large enough to let both water molecules and sugar molecules pass through?

Q3: The fluid in an IV bag is isotonic to blood. What change would you see in the red blood cells of a patient if a bag of a hypertonic solution was used in error?

See Appendix A for answers to the figure questions.

Another Way Through

In addition to transport proteins, there is another way that molecules move into and out of a cell. Sections of the plasma membrane can bulge inward or outward to form packages called **vesicles**. Vesicles move molecules from place to place inside a cell but also transport substances into and out of the cell. Some of these processes require plasma membrane receptor proteins.

Cells expel materials in vesicles via **exocytosis** (**Figure 4.7**). The substance to be exported from the cell is packaged into a vesicle; as the vesicle approaches the plasma membrane, a portion of the vesicle's membrane fuses with the plasma membrane. The inside of the vesicle then opens to the exterior of the cell, discharging its contents.

Endocytosis (**Figure 4.7**) is the opposite of exocytosis. In this process, a section of plasma membrane bulges inward to form a pocket around extracellular fluid, molecules, or particles. The pocket deepens until the opening in the membrane pinches off and the membrane breaks free as a closed vesicle, now wholly contained within the cell. Endocytosis can be nonspecific or specific. In nonspecific endocytosis, all of the material in the immediate area is

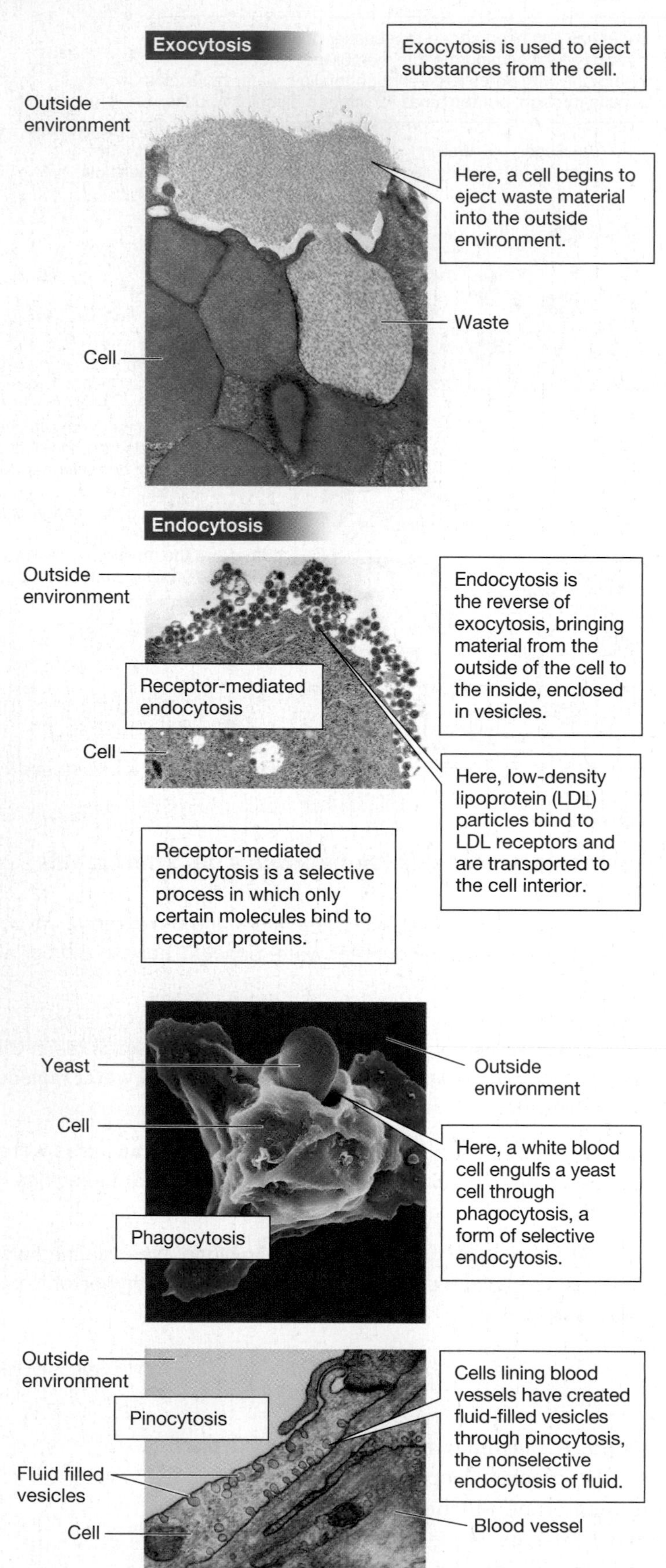

Exocytosis
Exocytosis is used to eject substances from the cell.
Outside environment
Here, a cell begins to eject waste material into the outside environment.
Waste
Cell
Endocytosis
Outside environment
Endocytosis is the reverse of exocytosis, bringing material from the outside of the cell to the inside, enclosed in vesicles.
Receptor-mediated endocytosis
Cell
Here, low-density lipoprotein (LDL) particles bind to LDL receptors and are transported to the cell interior.
Receptor-mediated endocytosis is a selective process in which only certain molecules bind to receptor proteins.
Yeast
Outside environment
Cell
Here, a white blood cell engulfs a yeast cell through phagocytosis, a form of selective endocytosis.
Phagocytosis
Outside environment
Pinocytosis
Cells lining blood vessels have created fluid-filled vesicles through pinocytosis, the nonselective endocytosis of fluid.
Fluid filled vesicles
Cell
Blood vessel

surrounded and enclosed; in specific endocytosis, one particular type of molecule is enveloped and imported.

There are three types of endocytosis. **Receptor-mediated endocytosis (Figure 4.7)** is a form of endocytosis in which receptor proteins embedded in the membrane recognize specific surface characteristics of substances to be incorporated into the cell. For example, human cells use receptor-mediated endocytosis to take up cholesterol-containing packages called low-density lipoprotein (LDL) particles. **Phagocytosis (Figure 4.7)**, or "cellular eating," is a large-scale version of endocytosis in which particles considerably larger than biomolecules are ingested. Specific cells in the immune system use phagocytosis to ingest an entire bacterium or virus. **Pinocytosis (Figure 4.7)** is a form of endocytosis that is often described as "cellular drinking" because it involves the capture of fluids. However, the cell does not attempt to collect particular solutions. Pinocytosis is nonselective: the vesicle budding into the cell contains whatever happened to be dissolved in the fluid when the cell "drank."

There is a long way to go before an artificial plasma membrane will perform processes like endocytosis and exocytosis, says Neal Devaraj. "But just because a research problem is difficult doesn't mean we shouldn't tackle it," he adds.

Figure 4.7

Endocytosis and exocytosis

Cells take substances in through endocytosis and move them out through exocytosis.

Q1: If endocytosis itself is nonspecific, how does receptor-mediated endocytosis bring only certain molecules into a cell?

Q2: What sorts of molecules could be moved by endocytosis or exocytosis but not by diffusion?

Q3: How does the fluid that enters a cell via pinocytosis differ from the fluid that enters by osmosis?

See Appendix A for answers to the figure questions.

Sizing Up Life

Cells range dramatically in size, and viruses are even smaller. The largest cell shown here, the *Amoeba proteus*, measures 500 μm (micrometers = millionth of a meter) and can only be seen through a light microscope. In order to observe a virus, like those shown here, one would have to use an electron microscope, which relies on a high-voltage beam of electrons to magnify an image.

Assessment available in **smart**work

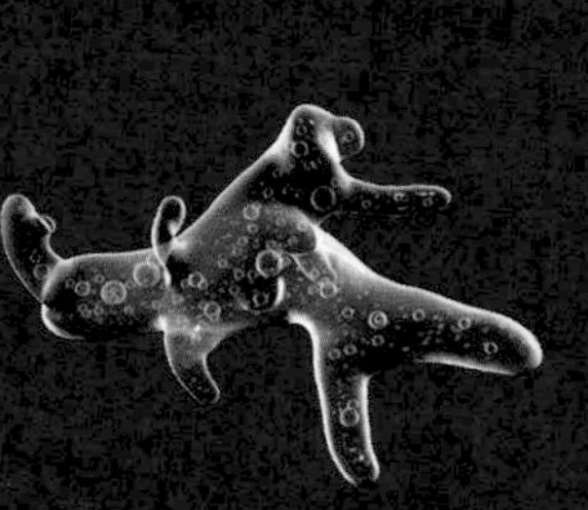

500 μm

Amoeba proteus

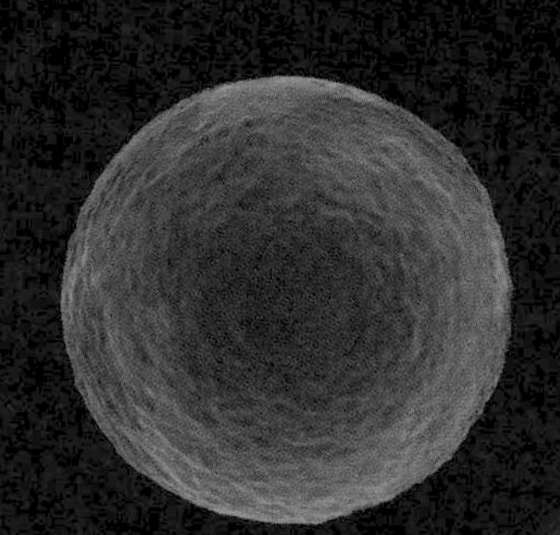

130 μm

Human egg cell

(For size comparison, Amoeba proteus *is shown in the background.)*

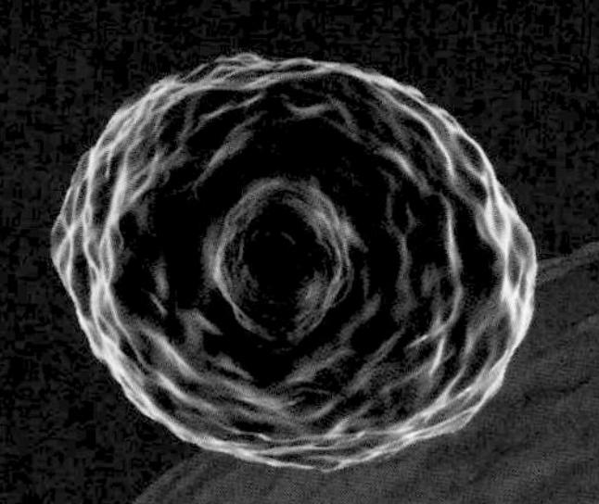

30 μm

Skin cell

(For size comparison, a human egg cell is shown in the background.)

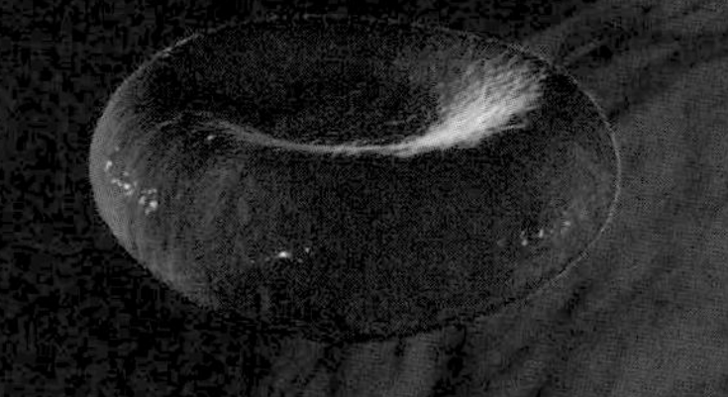

8 μm

Red blood cell

(For size comparison, a skin cell is shown in the background.)

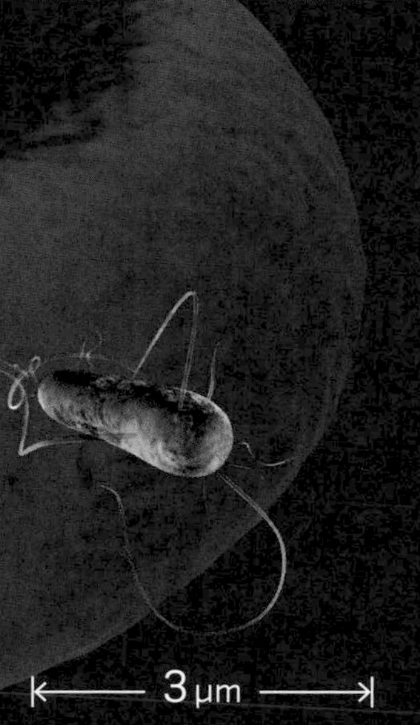

3 μm

***E. coli* bacterium**

(For size comparison, a red blood cell is shown in the background.)

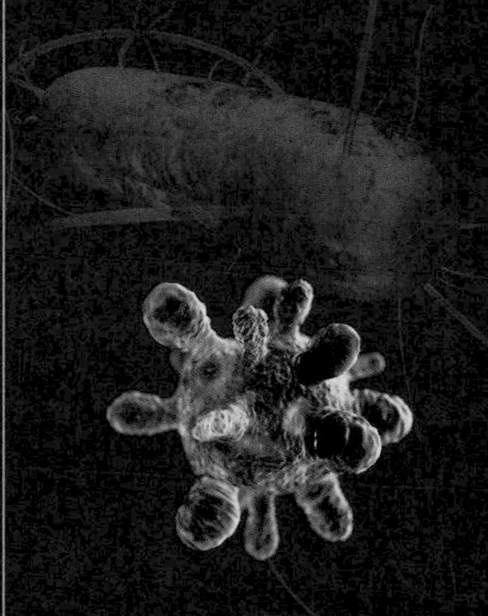

.22 μm

Measle virus

(For size comparison, E. coli bacterium *is shown in the background.)*

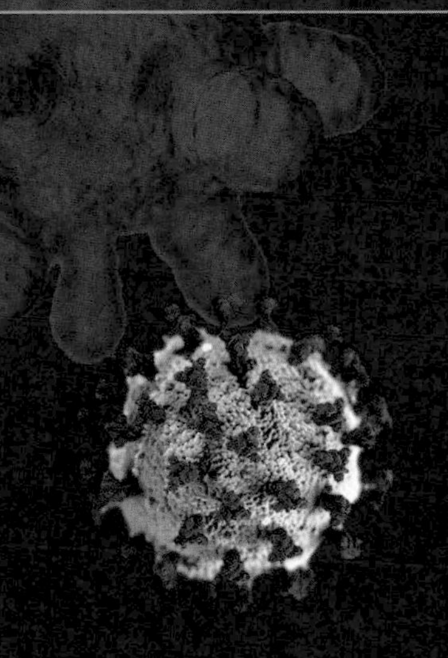

.13 μm

SARS-CoV-2 virus

(For size comparison, a measle virus is shown in the background.)

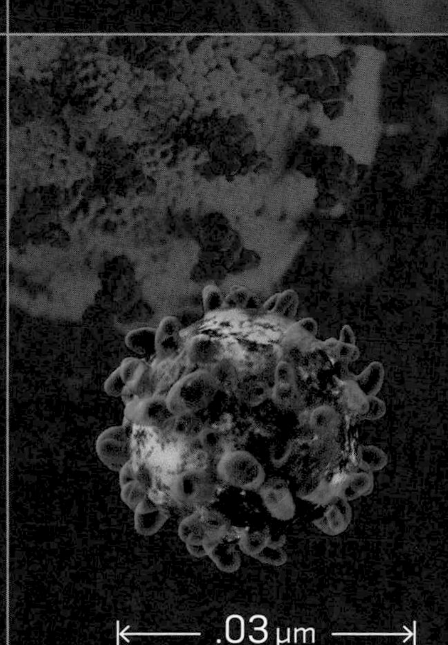

.03 μm

Rhinovirus

(For size comparison, a SARS-CoV-2 virus is shown in the background.)

Prokaryotes versus Eukaryotes

Devaraj's team continues to pursue that ideal. In 2015, the group succeeded in designing and synthesizing an artificial membrane that sustained continual growth, just like a living cell. In 2019, they built a simple system—consisting of a bacterial enzyme and phospholipid precursors—to make functional phospholipids. The new phospholipids then self-assembled into vesicle-forming membranes.

Membranes are important not only because they form the plasma membrane, and therefore the structure of all cells, but also because they compartmentalize and separate processes within eukaryotic cells. Depending on the fundamental structure of their cells, all living organisms can be sorted into one of two groups: **prokaryotes** or **eukaryotes**. *Mycoplasma mycoides* and all other bacteria are prokaryotes, but virtually all the organisms you see every day, including all plants, animals, fungi, and protists (single-celled organisms such as amoeba and algae) are eukaryotes. All cells—prokaryotes and eukaryotes, although very different—do share four common features: the plasma membrane; the cytoplasm, the fluid portion inside cells; DNA, the genetic material; and ribosomes, the molecular machines that produce proteins.

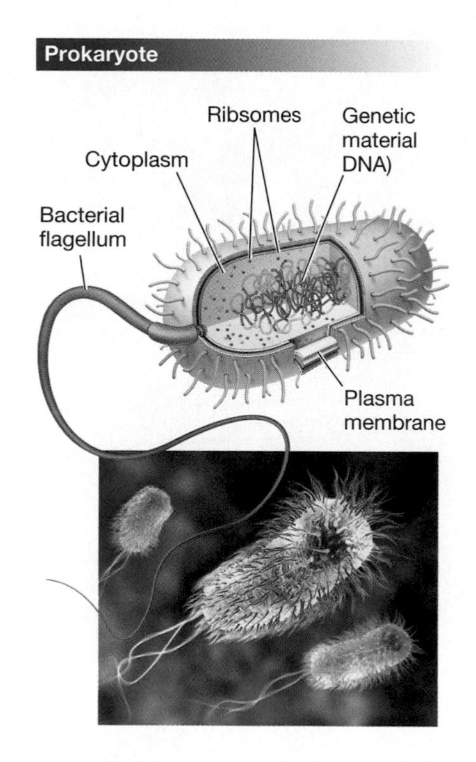

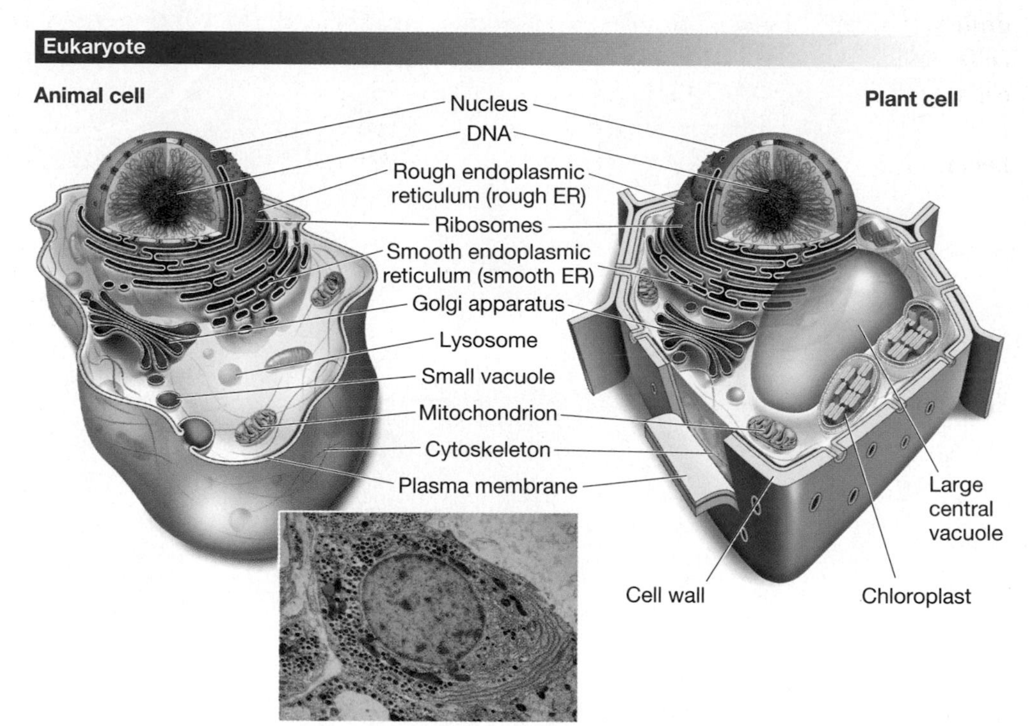

Figure 4.8

Prokaryotic and eukaryotic cells

Prokaryotic cells, like all cells, contain DNA, cytoplasm, ribosomes, and a plasma membrane. Plant cells and many prokaryotes have a cell wall that serves as a kind of exoskeleton for added support. To enable movement, some bacteria possess one or more flagella.

Q1: What structures do prokaryotic and eukaryotic cells have in common?

Q2: What structures are unique to prokaryotic cells? To eukaryotic cells?

Q3: Both plants and animals are eukaryotes, but there are differences in their cellular structure. What are those differences?

See Appendix A for answers to the figure questions.

Eukaryotic cells are larger and more complex than prokaryotic cells (**Figure 4.8**). They are roughly 10 times wider with a cell volume about a thousand times greater. Unlike prokaryotic cells, eukaryotic cells have a double-membrane-enclosed **nucleus** (plural "nuclei") that contains the organism's DNA, and they have a variety of membrane-enclosed subcellular compartments called **organelles**. Through specialization and division of labor, these organelles act like cubicles in a large office, allowing the cell to localize different processes in different places. In contrast, prokaryotic cells are like an open floor plan: they lack a cell nucleus or any membrane-encased organelles. How did organisms come to exist in such complex forms from such simple origins? According to the **endosymbiotic theory**, simple prokaryotic cells endocytosed other simple prokaryotic cells approximately 1.5 billion years ago, and over evolutionary time the endocytosed cells became organelles.

Though the first fully artificial cell will most likely be a simple prokaryotic cell, synthetic biologists aspire to build a eukaryotic cell. "We're not quite there yet, but it's interesting to think about making complex structures like organelles," says Devaraj.

What's in a Cell?

The nucleus is the control center of the eukaryotic cell. It contains most of the cell's genetic material, DNA, and may occupy up to 10 percent of the space inside the cell. Inside the nucleus, long strands of DNA are packaged with proteins into a remarkably small space.

The boundary of the nucleus, called the **nuclear envelope**, consists of two concentric phospholipid bilayers that make up the double membrane of this organelle. The nuclear envelope is speckled with thousands of small openings called **nuclear pores**. These pores allow chemical messages to enter and exit the nucleus (**Figure 4.9**).

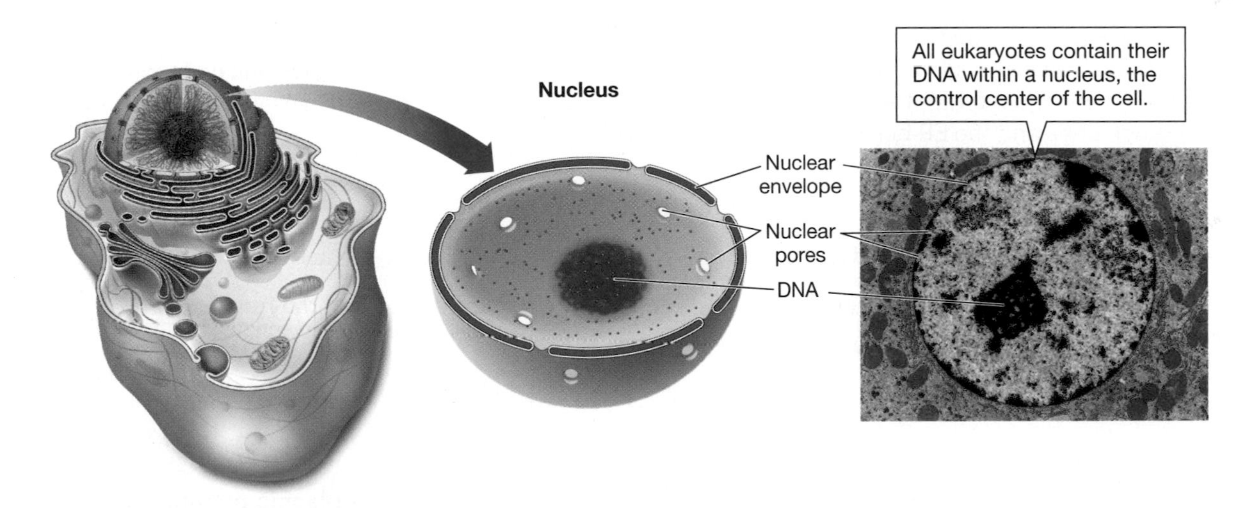
Nucleus
All eukaryotes contain their DNA within a nucleus, the control center of the cell.
Nuclear envelope
Nuclear pores
DNA

Figure 4.9

The nucleus

The double-membrane-bound nucleus is the hallmark organelle of eukaryotic cells.

Q1: **What is the main function of the nucleus?**

Q2: **What are nuclear pores, and what do they do?**

Q3: **If prokaryotic cells do not have a nucleus, where do they keep their genetic material?**

See Appendix A for answers to the figure questions.

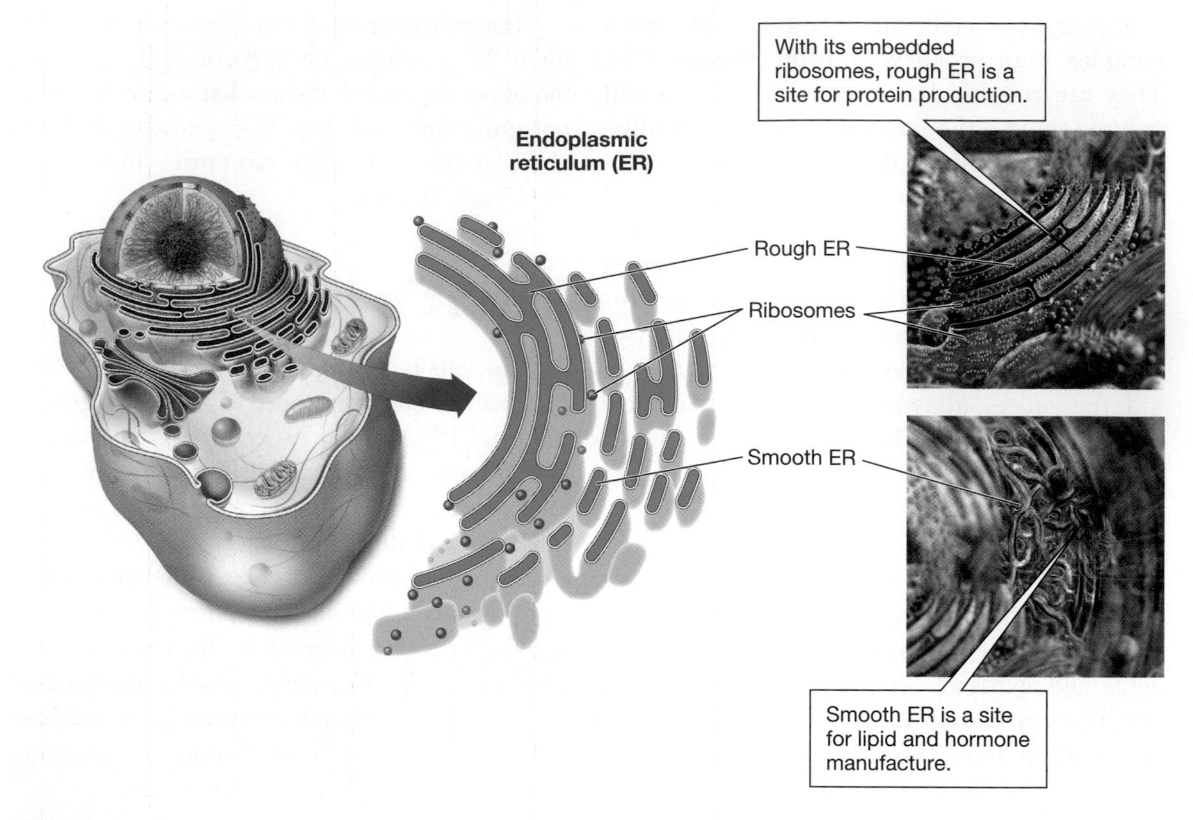

With its embedded ribosomes, rough ER is a site for protein production.
Endoplasmic reticulum (ER)
Rough ER
Ribosomes
Smooth ER
Smooth ER is a site for lipid and hormone manufacture.

Figure 4.10

Endoplasmic reticulum

The endoplasmic reticulum (ER) comes in rough and smooth forms, each with its own unique function. The ER is contiguous with the nuclear membrane: it is almost impossible to determine where the nuclear envelope ends and the ER begins at the transition area.

Q1: What are the functions of the smooth ER and the rough ER?

Q2: Where is the ER located in the cell?

Q3: If prokaryotic cells do not have an ER, where do you think they perform the functions mentioned in Q1?

See Appendix A for answers to the figure questions.

The **endoplasmic reticulum (ER)** is an extensive and interconnected network of sacs. It is made of a single membrane that is continuous with the outer membrane of the nuclear envelope (**Figure 4.10**). The membranes of the ER are classified into two types based on their appearance: smooth and rough. Enzymes associated with the surface of the **smooth ER** manufacture lipids and hormones. In some cell types, smooth-ER membranes also break down toxic compounds. **Ribosomes** embedded in the **rough ER** give it the knobby appearance from which it gets its name. Ribosomes on the rough ER assemble proteins that will be inserted into the cell's plasma membrane or its organelles.

Proteins and lipids produced by the ER are packaged into transport vesicles, which bud off from the ER membrane. The molecules are then delivered to and fused with the membranes of the **Golgi apparatus**, which resemble a pile of flattened balloons (**Figure 4.11**). The Golgi apparatus works like a post office, repackaging and directing the proteins and lipids to their final destinations either inside or outside the cell. The Golgi transport vesicles deliver their cargo to these destinations by fusing with the membranes of the target compartment

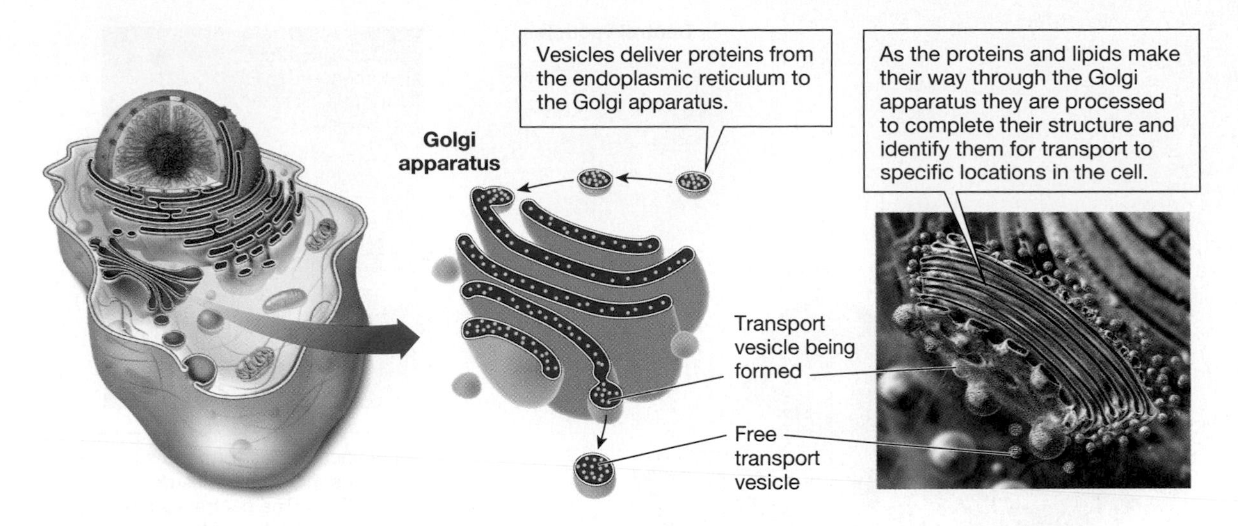
Vesicles deliver proteins from the endoplasmic reticulum to the Golgi apparatus.
As the proteins and lipids make their way through the Golgi apparatus they are processed to complete their structure and identify them for transport to specific locations in the cell.
Golgi apparatus
Transport vesicle being formed
Free transport vesicle

Figure 4.11

Golgi apparatus

Vesicles budding from the smooth and rough ER fuse with the Golgi apparatus. New vesicles, in turn, bud off from the Golgi apparatus for transport to other membranes of cellular organelles or to the plasma membrane of the cell.

Q1: What is the function of the Golgi apparatus?

Q2: Where is the Golgi apparatus located in the cell? What is the significance of that location?

Q3: Why do you think prokaryotic cells can survive without a Golgi apparatus? (*Hint*: Eukaryotic cells are very large, and prokaryotic cells are very small.)

See Appendix A for answers to the figure questions.

membranes. The Golgi apparatus targets these destinations by adding a specific chemical tag to each molecule it receives, like attaching a shipping label to a package.

Transport vesicles bring large molecules that will be discarded to **lysosomes**, organelles that act as garbage disposals and recycling centers in both plants and animals (**Figure 4.12**, top). Lysosomes contain a variety of enzymes that degrade biomolecules, and they release the breakdown products into the cell interior to be discarded or reused. In plant cells, lysosomes are called **vacuoles** and perform many functions, but the main function of the plant vacuole is water storage (**Figure 4.12**, bottom). Some plant vacuoles stockpile noxious compounds that can deter herbivores from eating plants.

In most eukaryotic cells, the main source of energy is the **mitochondrion** (plural "mitochondria"), a tiny power plant that fuels cellular activities (**Figure 4.13**, top). Mitochondria are made up of double membranes—a smooth external membrane and a folded internal membrane—that form a mazelike interior. Mitochondria use chemical reactions to transform the energy of food molecules into ATP (adenosine triphosphate), the universal cellular fuel, in a process called *cellular respiration*. Mitochondria provide ATP to all eukaryotic cells (plant, animal, fungi, and protist), but the cells of plants and some protists have additional organelles called **chloroplasts** that capture energy from sunlight and use it to manufacture food molecules via *photosynthesis* (**Figure 4.13**, bottom). We explore cellular respiration and photosynthesis in Chapter 5.

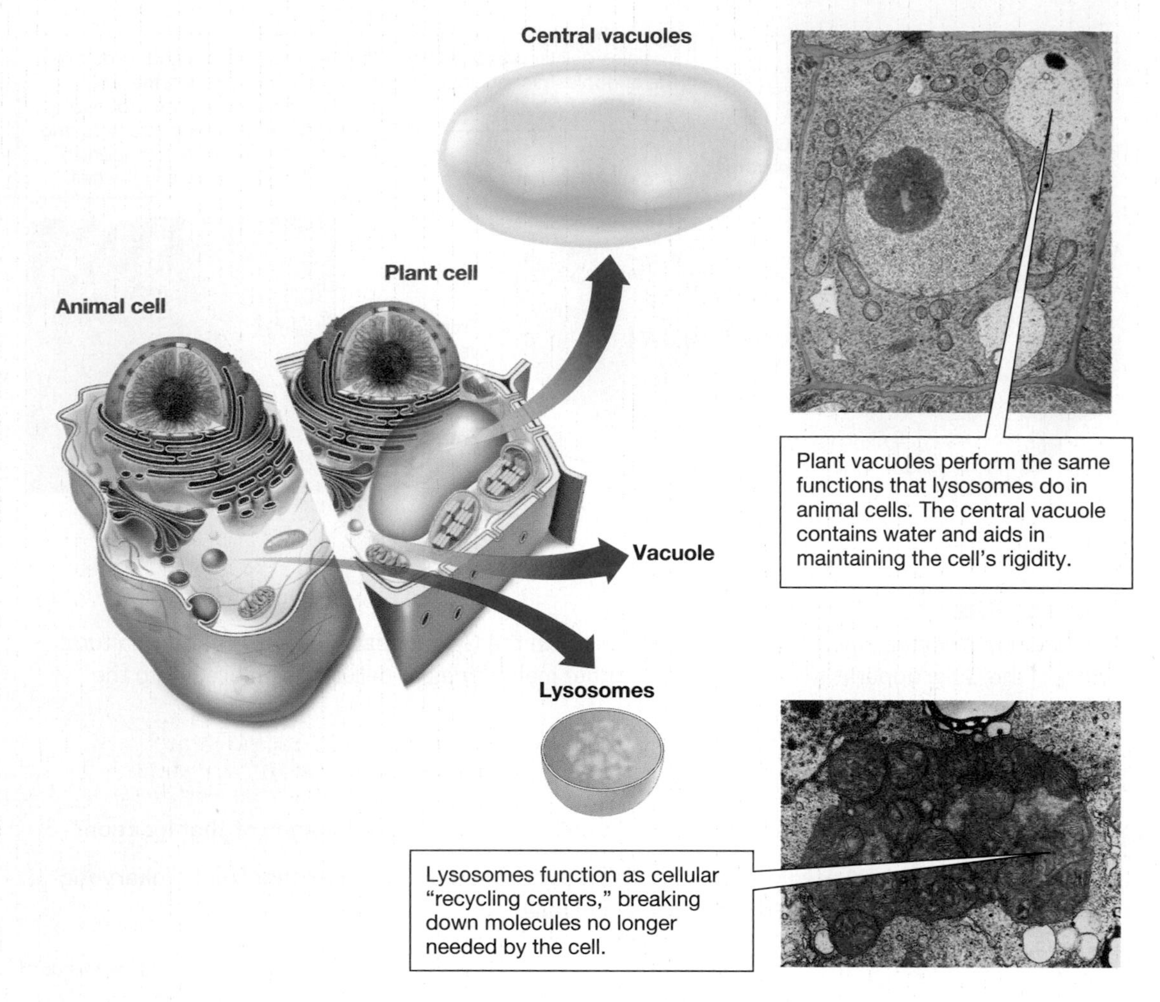

Central vacuoles
Plant cell
Animal cell
Plant vacuoles perform the same functions that lysosomes do in animal cells. The central vacuole contains water and aids in maintaining the cell's rigidity.
Vacuole
Lysosomes
Lysosomes function as cellular "recycling centers," breaking down molecules no longer needed by the cell.

Figure 4.12

Lysosomes and vacuoles

Both plant cells and animal cells have vesicles called lysosomes, which contain enzymes that catalyze the breakdown of the biomolecules including proteins, lipids, carbohydrates, and nucleic acids. Lysosomes also control the breakdown and recycling of the biomolecule components of old organelles. Plant cells usually contain one large central vacuole in charge of water storage and water balance in the cell. Vacuoles are also found in fungi, which are multicellular eukaryotes, and protists, which are single-celled eukaryotes. The numbers and functions of vacuoles found in these cells are too numerous to detail here, but many contribute to water balance or store toxic wastes.

Q1: What are the functions of lysosomes? What is the main function of vacuoles?

Q2: Which types of cells have lysosomes? Which types have vacuoles?

Q3: Lysosomes and vacuoles were once thought to have analogous functions in animal cells and plant cells, respectively. Now that you have learned about both of these types of organelles, explain why they are not analogous structures.

See Appendix A for answers to the figure questions.

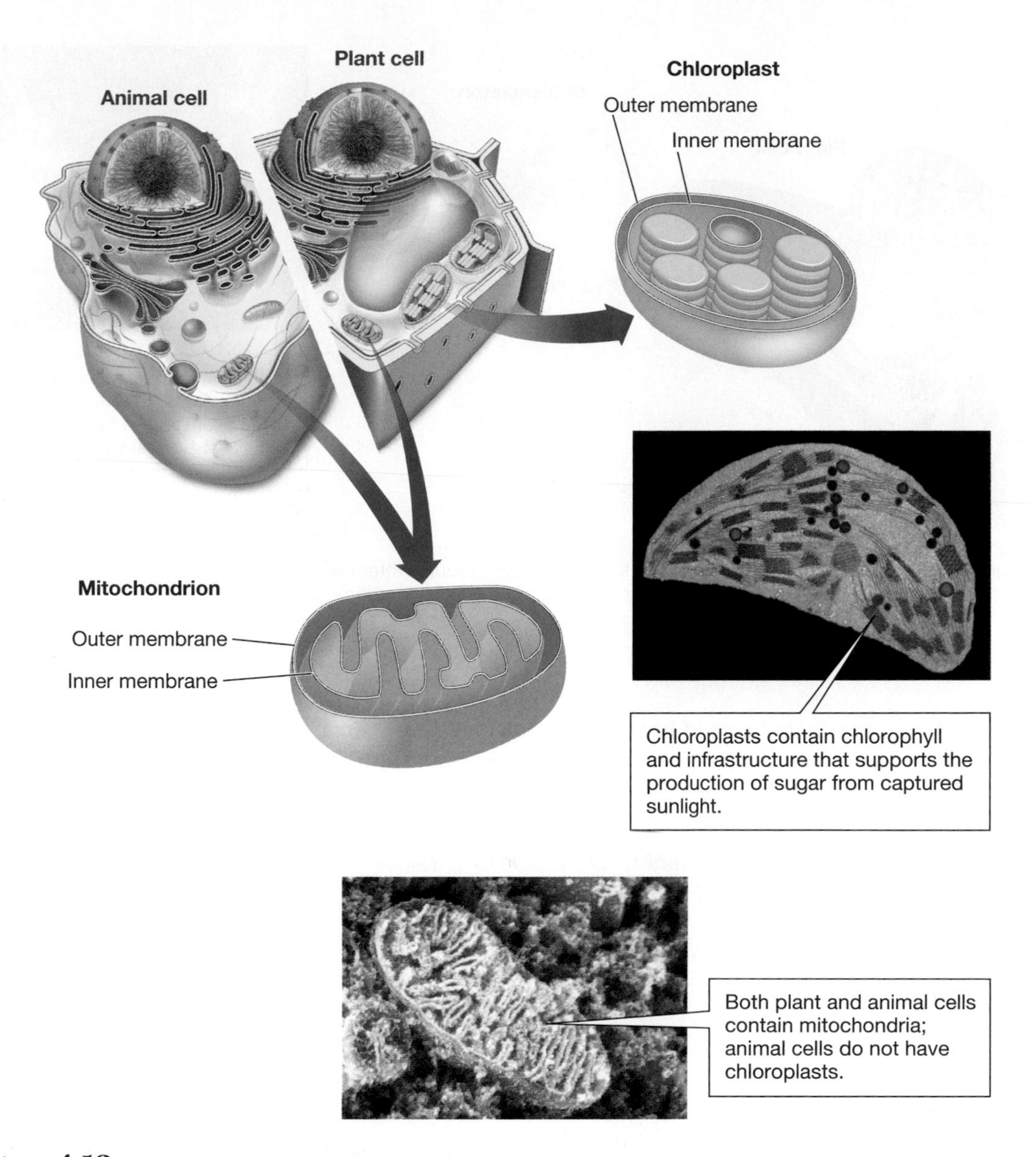
Plant cell
Animal cell
Chloroplast
Outer membrane
Inner membrane
Mitochondrion
Outer membrane
Inner membrane
Chloroplasts contain chlorophyll and infrastructure that supports the production of sugar from captured sunlight.
Both plant and animal cells contain mitochondria; animal cells do not have chloroplasts.

Figure 4.13

Mitochondria and chloroplasts

Mitochondria and chloroplasts have similar structure and function. Each has an outer membrane and an extensive inner membrane compartment, and each has its own DNA but depends on nuclear genes for some functions. Both are involved in the conversion of energy. Mitochondria break down sugar to produce cellular energy, known as ATP, and chloroplasts convert solar energy into the chemical energy of sugar.

Q1: What is the main function of mitochondria? Of chloroplasts?

Q2: Which types of cells have mitochondria? Which types have chloroplasts?

Q3: According to the endosymbiotic theory, mitochondria and chloroplasts were once free-living single-celled organisms that were absorbed by another single-celled organism. Over the millennia, they evolved into the organelles of today. Before they were absorbed, what processes must these single-celled organisms have been able to perform on their own?

See Appendix A for answers to the figure questions.

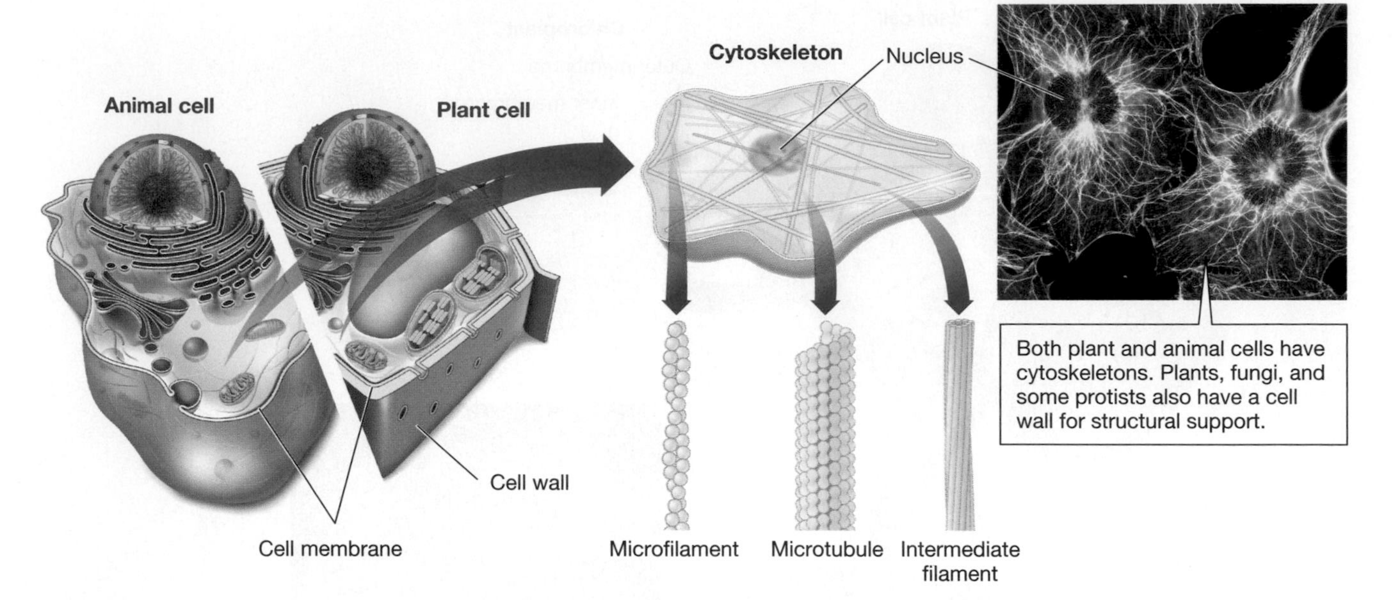
Animal cell
Plant cell
Cytoskeleton
Nucleus
Both plant and animal cells have cytoskeletons. Plants, fungi, and some protists also have a cell wall for structural support.
Cell wall
Cell membrane
Microfilament
Microtubule
Intermediate filament

Figure 4.14

Cytoskeleton

The cytoskeleton is made up of three main types of fibers: microfilaments, intermediate filaments, and microtubules. Interactions among these three fiber types help maintain cell structure, but they also permit changes in cell shape and allow cell movement.

Q1: What is the main function of the cytoskeleton?

Q2: Which types of cells have a cytoskeleton?

Q3: Postulate: The cytoskeleton of a cell is similar in function to the skeleton of a human. Give an example in support and one in rebuttal of this statement.

See Appendix A for answers to the figure questions.

The network of protein cylinders and filaments collectively known as the **cytoskeleton** forms the framework of a cell (**Figure 4.14**). The three main types of cytoskeletal fibers are microfilaments, intermediate filaments, and microtubules. Through these fibers, the cytoskeleton organizes the interior of a eukaryotic cell, supports the intracellular movement of organelles such as transport vesicles, and enables whole-cell movement in some cell types. It also gives shape to eukaryotic cells. Fungi and plant cells separately evolved cell walls for additional cell structure and support, but animal cells rely only on the cytoskeleton.

Life Goes On

From Venter's complex synthetic genome to Devaraj's simple self-assembling membrane, scientists debate whether the top-down or bottom-up approach will be more successful in the effort to build an artificial cell. But the two can be complementary, says James Collins. "I'm not sure one will win out. I think they bring different things to the table," he notes. Gibson agrees: "I would like to one day be able to combine all of a cell's parts from nonliving components, including the genome, and incubate them, and see if we can get

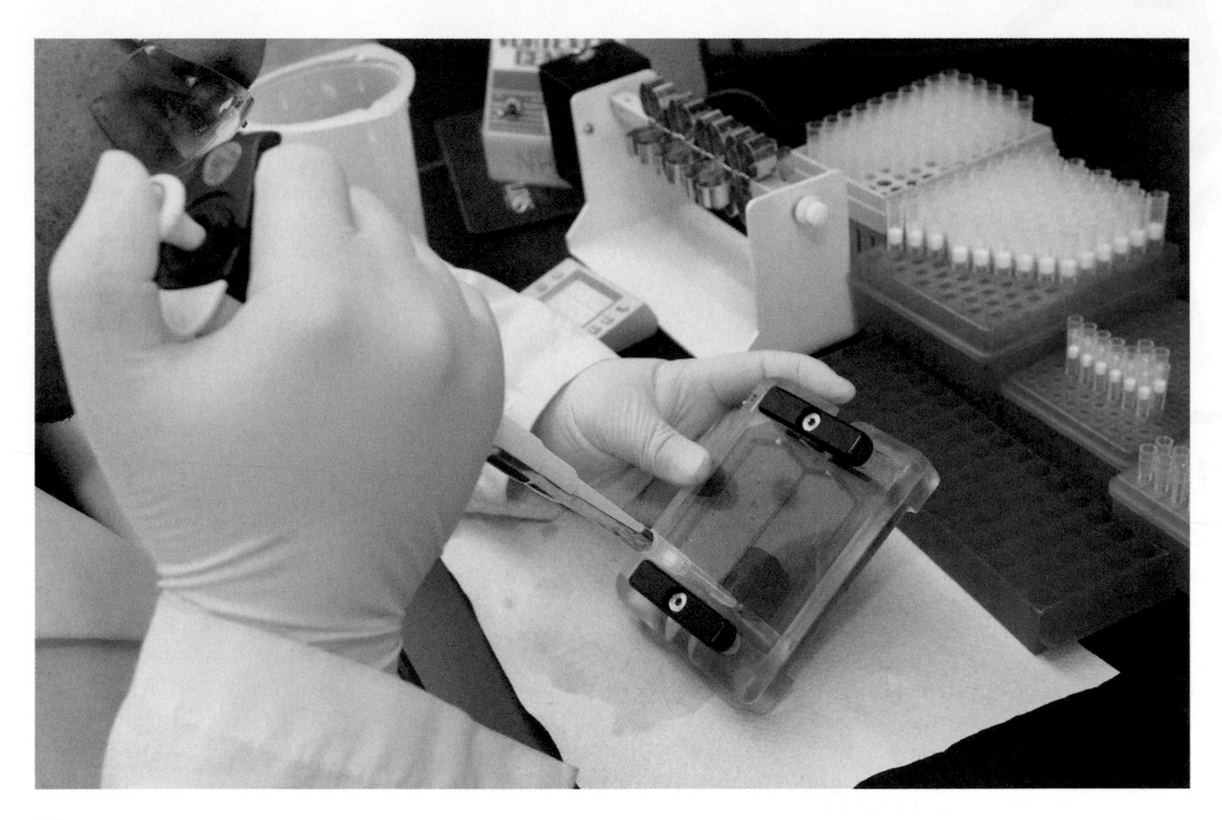

Figure 4.15

Scientist in action

Laboratory technician Javier Quinones sets up the procedure used to sequence the *M. mycoides* genome in the sequencing laboratory at the J. Craig Venter Institute located in Rockville, Maryland.

life out of those nonliving components," he says. "It would help us better understand how cells work."

Today, the work on both ends continues. While Devaraj devises new ways to build membranes, Gibson and his team recently simplified the *Mycoplasma mycoides* genome (**Figure 4.15**). With the goal of determining the smallest set of genes needed to maintain life, they broke the *M. mycoides* genome into eight DNA segments and mixed and matched them to see which combinations would produce viable cells. Eventually, they narrowed the genome down to just 473 genes capable of sustaining life. Amazingly, the team could not identify the function of 149 of the 473 genes. "We don't know about a third of essential life, and we're trying to sort that out now," Venter told *Nature* magazine in 2016.

Knowing both the identities and function of a core set of genes necessary for life will make it easier to build other synthetic cells with specific functions, says Gibson. "There's not a single organism in the world where we understand what every gene does," he notes, but reaching such an understanding with those 473 genes in the *M. mycoides* genome "would be the first example of that."

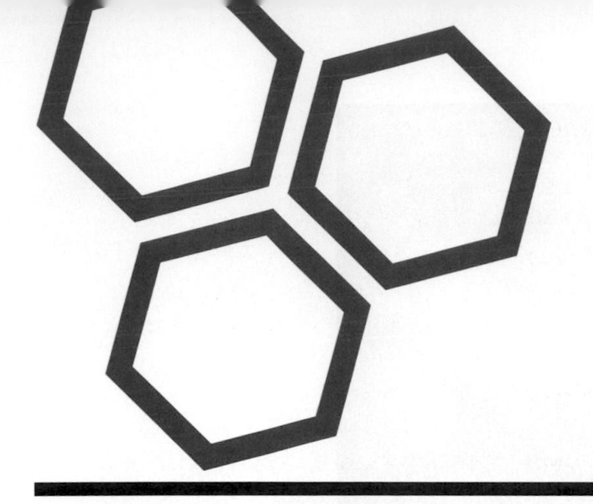

NEED-TO-KNOW SCIENCE

- **Cells** (page 66) are the basic units of all living organisms. **Cell theory** (page 67) states that all living things are composed of one or more cells and that all cells living today came from a preexisting cell. In addition, all living organisms reproduce, using DNA to pass genetic information from parent to offspring; take in energy from their environment; sense and respond to their environment; exhibit homeostasis, maintaining constant internal conditions; and can evolve as groups.
- The **genome** (page 66) is made up of DNA and is a cell's complete genetic information. The four necessary ingredients of the genome are adenine (A), thymine (T), guanine (G), and cytosine (C).
- A **virus** (page 67) is a small, infectious agent that can replicate only inside a living cell.
- Every cell is surrounded by a **plasma membrane** (page 70) that separates the chemical reactions inside the cell from the surrounding environment. The plasma membrane is formed by a **phospholipid bilayer** (page 70) embedded with proteins that perform a variety of functions. When the phospholipid bilayer forms a sphere, it is called a **liposome** (page 70).
- **Transport proteins** (page 71) help make the plasma membrane **selectively permeable** (page 71). In **passive transport** (page 71), substances move across the plasma membrane without the direct expenditure of energy. **Active transport** (page 71) by cells requires energy. **Diffusion** (page 71) is the passive transport of a substance from a region where it is at a higher concentration to a region where it is at a lower concentration. Water, oxygen, and carbon dioxide usually enter and leave cells by **simple diffusion** (page 72). **Facilitated diffusion** (page 72) requires transport proteins.
- **Osmosis** (page 72) is the diffusion of water across a selectively permeable membrane. In a **hypotonic** (page 72) solution, a cell gains water. In a **hypertonic** (page 72) solution, a cell loses water. In an **isotonic** (page 72) solution, there is no net movement of water. Cells can actively balance their water content through osmosis.
- Through packages called **vesicles** (page 73), cells export materials by **exocytosis** (page 73) and import materials by **endocytosis** (page 73).
- In **receptor-mediated endocytosis** (page 74), **receptor proteins** (page 73) in the plasma membrane recognize and bind the substance to be brought into the cell. **Phagocytosis** (page 74) is endocytosis in which large particles are ingested. **Pinocytosis** (page 74) is endocytosis involving the capture of fluids.
- **Prokaryotes** (page 76) are single-celled organisms that do not have a nucleus or complex internal compartments.
- **Eukaryotes** (page 76) may be single-celled or multicellular. By volume, eukaryotic cells can be a thousand times larger than prokaryotic cells. Their cells typically have many membrane-enclosed compartments, or **organelles** (page 77), that concentrate and organize cellular chemical reactions. The organelles include the nucleus, the smooth endoplasmic reticulum (ER), the rough ER, the Golgi apparatus, lysosomes, mitochondria, and chloroplasts.
- According to the **endosymbiotic theory** (page 77), eukaryotes evolved from simple prokaryotes, which incorporated other prokaryotic cells through endocytosis. The internal prokaryotic cells eventually developed into mitochondria, chloroplasts, and possibly other organelles.
- The **nucleus** (page 77) is the control center of a eukaryotic cell and contains DNA. It is bound by the **nuclear envelope** (page 77), which has **nuclear pores** (page 77) that allow communication between the nucleus and the cell interior.
- Lipids are made in the **smooth ER** (page 78). Some proteins are manufactured by **ribosomes** in the **rough ER** (page 78).
- The **Golgi apparatus** (page 78) receives proteins and lipids, sorts them, and directs them to their final destinations.
- **Lysosomes** (page 79) break down biomolecules such as proteins into simpler compounds that can be used by the cell. Plant **vacuoles** (page 79) store water and lend physical support to plant cells.
- **Mitochondria** (page 79) produce chemical energy for eukaryotic cells in the form of ATP.
- **Chloroplasts** (page 79) harness the energy of sunlight to make sugars through photosynthesis.
- Eukaryotic cells depend on the **cytoskeleton** (page 82) for structural support and for the ability to move and change shape. Plants, fungi, and some protists also have a cell wall that provides structural support.

THE QUESTIONS

See Appendix B for answers.

The Basics

1 Which of the following is a living organism? (Select all that apply.) If it is not living, which criterion, or criteria, does it not meet?

(a) an oak tree
(b) an influenza virus
(c) the fungus that causes white-nose syndrome in bats
(d) a diamond
(e) your teacher

2 To grow bacteria from cells of *Mycoplasma capricolum*, Daniel Gibson replaced some *M. capricolum* DNA with synthetic DNA of *Mycoplasma mycoides*. To recognize the colony that came from the synthetic DNA, Gibson added a ________ that coded for a blue pigment.

(a) cell
(b) chromosome
(c) gene
(d) bacterium

3 A phospholipid has a phosphate head and a lipid (fatty acid) tail. Which of the following statements correctly describes the nature of those two components?

(a) The phosphate head is hydrophobic, and the lipid tail is hydrophilic.
(b) Both the phosphate head and the lipid tail are hydrophilic.
(c) Both the phosphate head and the lipid tail are hydrophobic.
(d) The phosphate head is hydrophilic, and the lipid tail is hydrophobic.

4 Link each process with the correct definition.

receptor-mediated endocytosis	1. A cell ingests a large particle, such as a bacterial cell.
phagocytosis	2. Receptor proteins embedded in the membrane recognize specific surface characteristics of substances.
pinocytosis	3. A transport vesicle inside the cell approaches the plasma membrane of the cell, fuses with it, and releases its contents to the outside of the cell.
exocytosis	4. A vesicle containing whatever molecules are in solution outside the cell bulges inward, pinches off, and enters the cell.

5 Link each structure with the correct function.

chloroplast	1. Location of the cell's DNA.
Golgi apparatus	2. Site of protein synthesis.
lysosome	3. Site of lipid synthesis.
mitochondrion	4. Adds chemical tags to newly synthesized proteins to direct them to their correct location.
nucleus	5. Breaks down biomolecules by enzymatic action.
rough endoplasmic reticulum	6. Site of cellular respiration.
smooth endoplasmic reticulum	7. Site of photosynthesis.

6 In the following table, indicate whether the specified cellular component is found in each type of cell by placing an "X" in the relevant columns.

		Eukaryotes	
Component	**Prokaryotes**	**Animals**	**Plants**
Plasma membrane			
Cellulose cell wall			
Nucleus			
Endoplasmic reticulum			
Golgi apparatus			
Ribosomes			
Cytoskeleton			
Mitochondria			
Chloroplasts			

Challenge Yourself

7 How does the phospholipid bilayer of a liposome differ from the phospholipid bilayer of the plasma membrane of a cell?

(a) The phospholipid bilayer of a liposome contains only phospholipids without the proteins that are embedded in the plasma membrane of a cell.
(b) The phospholipid bilayer of a liposome contains two bilayers of phospholipid molecules, whereas the plasma membrane of a cell contains only one.
(c) The phospholipid bilayer of a liposome completely envelops the liposome, whereas the plasma membrane of a cell does not completely envelop the cell.
(d) The phospholipid molecules in the phospholipid bilayer of a liposome are oriented with the lipid ends on the outside of the bilayer and the phosphate groups on the inside.

❽ In the accompanying graph, label the purple curve and the green curve as representing either simple diffusion or facilitated diffusion. Explain how you determined which curve was which.

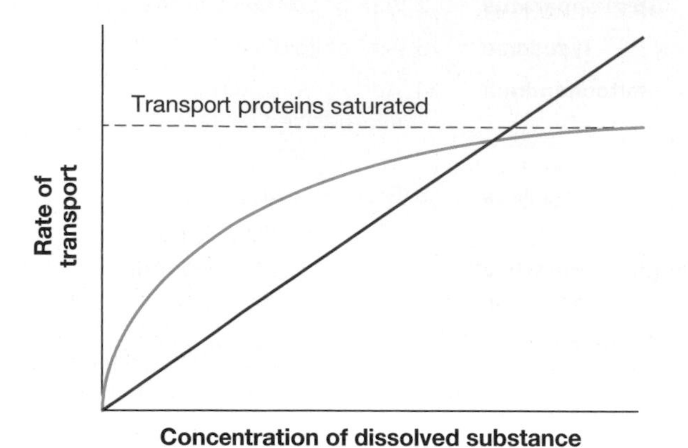

❾ Which of the following statements about transport through the plasma membrane of a cell is correct?

(a) Both passive and active transport require the input of energy.

(b) Passive transport does not require the input of energy, and active transport does.

(c) Passive transport does require the input of energy, and active transport does not.

(d) Neither active nor passive transport requires the input of energy.

Try Something New

❿ If a virus was discovered that had the ability to reproduce itself without the help of a cell and its machinery, several experiments could be performed to show that it should be categorized as "alive." For each experiment, specify which one of the six criteria for being classified as "alive" would be tested.

_____ a. Place a set number of viral particles in a nutrient broth and check for an increase in viral particles after several days.

_____ b. Place a set number of viral particles in a broth where all the nutrients are corralled to one side of the container and check for viral particle accumulation in the nutrient-rich area of the container.

_____ c. Place a set number of viral particles in a nutrient broth and check for waste products (broken-down nutrients) in the broth after several days.

_____ d. Place a set number of viral particles in a nutrient broth for several months, adding different nutrients daily. Randomly isolate viral particles and sequence their DNA to determine whether there have been changes in genetic characteristics of this population of viral particles compared to the DNA of the original population before culturing.

_____ e. Perform biochemical and microscopic analyses on a number of viral particles to determine whether they have the main components of a living cell, including a plasma membrane, ribosomes, and DNA.

⓫ The picture below is a beaker divided by a semipermeable membrane like the one in Figure 4.6. Imagine that you put 5 grams of sugar in the left side of the beaker and 10 grams of sugar in the right side, and then add water to bring both sides to the same depth. Select the correct terms in the following description of what happens next:

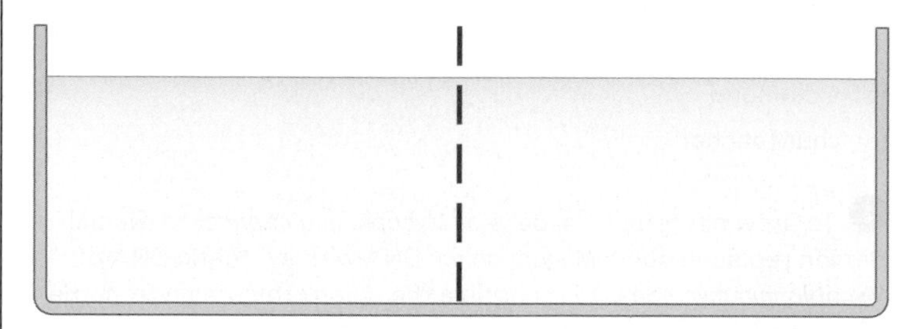

The water will move by osmosis to the (**left side / right side**) of the beaker because there are (**more / fewer**) sugar molecules and (**more / fewer**) water molecules on that side.

⓬ When the chemical oligomycin is added to cells, the production of ATP is blocked. Which of the following processes would be most affected by oligomycin? Why?

(a) simple diffusion

(b) facilitated diffusion

(c) active transport

(d) All of these processes would be equally affected by oligomycin.

(e) None of these processes would be affected by oligomycin.

⓭ Following are three images of red blood cells. Beneath each image are three statements related to it. For each image, interpret the conditions of the cell(s) to select the correct terms in the statements. Note: Refer back to the "Sizing Up Life" infographic (page 75) to see the normal shape of red blood cells.

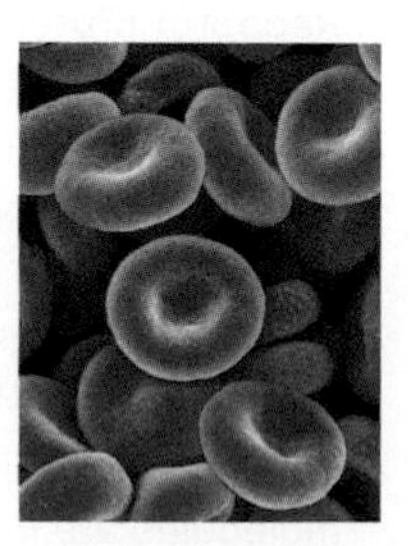

(a) These red blood cells are in a(n) (**hypertonic / hypotonic / isotonic**) solution.

(b) In this environment, normal red blood cells (**gain / lose / neither gain nor lose**) water.

(c) The total solute concentration outside the cell is (**lower than / higher than / equal to**) the total solute concentration inside it.

(d) This red blood cell is in a(n) (**hypertonic / hypotonic / isotonic**) solution.

(e) In this environment, normal red blood cells (**gain / lose / neither gain nor lose**) water.

(f) The total solute concentration outside the cell is (**lower than / higher than / equal to**) the total solute concentration inside it.

(g) This red blood cell is in a(n) (**hypertonic / hypotonic / isotonic**) solution.

(h) In this environment, normal red blood cells (**gain / lose / neither gain nor lose**) water.

(i) The total solute concentration outside the cell is (**lower than / higher than / equal to**) the total solute concentration inside it.

Leveling Up

14 **What do *you* think?** Viruses display many of the characteristics of living organisms (see "Viruses—Living or Not?" on page 69). In particular, they reproduce, creating new virus particles. During reproduction, viruses make copies of their genetic material, and some of the copies contain mutations that are beneficial to the virus. For example, HIV (human immunodeficiency virus), the virus that causes the disease AIDS (acquired immunodeficiency syndrome) mutates so often that its surface proteins change faster than we can develop antiviral drugs. New drug-resistant strains of HIV are appearing constantly. However, viruses can reproduce only after entering a living cell of an organism, because viruses hijack the cell's machinery and use it to produce new viruses. Where does that combination of characteristics place viruses on the scale of nonlife to life? Are viruses living organisms? Nonliving? If neither of those categories fits the properties of viruses, how should they be classified?

15 **What do *you* think?** Did Venter create "new life" by inserting synthetic DNA copied from the genome of *Mycoplasma mycoides* into a cell of *Mycoplasma capricolum* from which the DNA had been removed? Be prepared to support your opinion in a class discussion.

16 **Doing science** Go to the National Science Foundation website (http://www.nsf.gov), and search for and read "Biologists Replicate Key Evolutionary Step in Life on Earth" (News Release 12-009, January 16, 2012). State the hypothesis the biologists were testing, describe their experimental design, and explain their results. Was their hypothesis supported? What experiment do you think they should try next? State a hypothesis that you would test using their multicellular yeast samples.

Digital resources for your book are available online.

Rock Eaters

Unusual "electric" microbes could provide the energy storage solution we've been looking for.

After reading this chapter you should be able to:

- Explain how photosynthesis and cellular respiration support life on Earth.
- Differentiate between anabolic and catabolic metabolic processes.
- Define the importance of ATP in a cell.
- Label a diagram of the light-dependent reactions and the light-independent reactions of photosynthesis.
- Describe the role of enzymes in metabolic pathways.
- List the three stages of cellular respiration and explain the function of each.
- Compare and contrast photosynthesis and cellular respiration.

CELLS

CHAPTER 05

HOW CELLS WORK

Annie Rowe stands knee-deep in cold, clear salt water. Sweating, she pushes a shovel into the ground and lifts a mound of heavy mud. It's a sunny day on Santa Catalina Island, a rocky retreat off the coast of California, where Rowe has come to collect samples. Once the bucket is full, she sits on a rock and retrieves a mesh filter to sieve out any small invertebrates, such as snails and worms, from the sand. They are not part of the experiment.

A postdoctoral research fellow at the University of Southern California, Rowe pours the filtered sediment into one of her five 10-gallon aquariums. When each aquarium is about two-thirds full, she and a graduate student transport the tanks to a laboratory on the island. There, they hook each aquarium up to a steady stream of seawater from the harbor.

To the untrained eye, the aquariums appear sterile, empty. In reality, both the water and sand in the tanks are teeming with microscopic organisms, or *microbes*, and Rowe is hunting for a rare, strange one. Deep in the sand of each aquarium, she buries a metal electrode buzzing with electrical current to attract her quarry: a microbe that eats electricity (**Figure 5.1**).

"I went into microbiology because there's so much diversity in things that microbes can do," says Rowe. As a PhD student, she began studying unusual microbes with the hope of someday putting them to use toward one of the greatest human challenges of our time: developing sources of alternative energy.

Humans rely heavily on *fossil fuels* such as oil, coal, and petroleum to power our cars, trains, planes, and more. Hundreds of millions of years ago, single-celled organisms used carbon dioxide (CO_2) and sunlight to make cell materials, which then became buried and concentrated into oil and other substances in Earth's crust. Burning those substances releases the ancient carbons back into Earth's atmosphere as CO_2, and a consequence of this increase in carbon dioxide is that the surface of our planet has begun to warm. This *global warming* is changing *climates*, regions' long-term weather

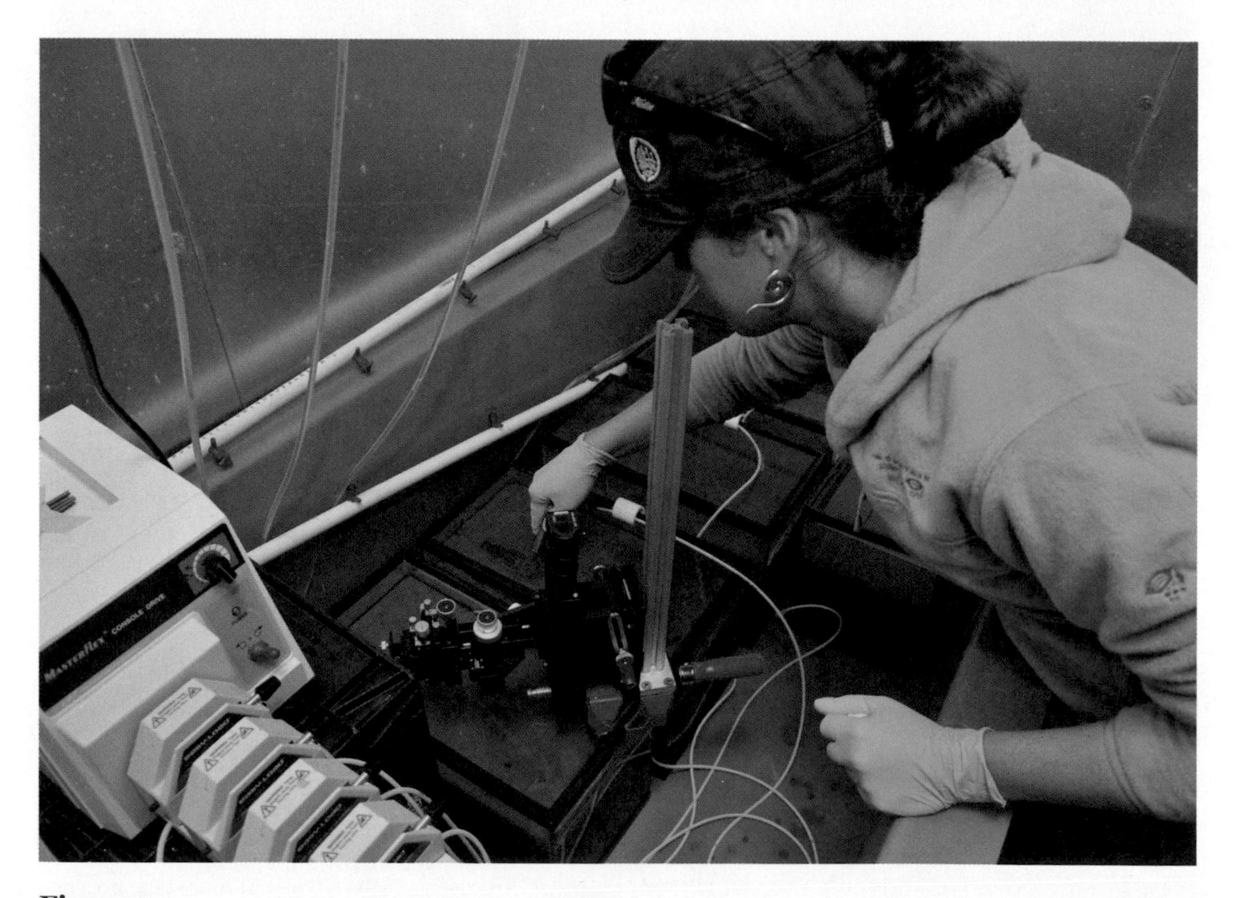

Figure 5.1

Researcher at work

Annie Rowe buries electrodes in her sediment aquariums.

patterns. To prevent further *climate change* and lessen our dependence on fossil fuels, scientists and engineers have developed clean, renewable ways to capture energy, such as solar, wind, and water power.

Yet to switch from a fossil fuel–based economy to one based on these cleaner technologies, we need efficient and inexpensive ways to store renewable energy. With the right chemical stabilizers, oil can be left in a barrel for decades, but a direct current of electricity from solar panels, wind turbines, or a dam needs to be used immediately or fed into the electrical grid. Today's batteries aren't a good storage option because they are bulky, expensive, and not terribly efficient.

Now imagine a tool that could siphon that solar, wind, or water electricity from the source and transform it into a liquid fuel. That's where electricity-eating microbes come in.

Energy for Life

All living cells require energy. Organisms use energy for growth, reproduction, and defense and to manufacture the many chemical compounds that make up living cells. They must obtain energy from the living or nonliving components of their environment, and the very core of making and storing energy is the transfer of *electrons*—subatomic particles with a negative charge. Electrons play an essential role in electricity, magnetism, and many other physical phenomena that shape the world we know. Much of our technology, including computers and cell phones, depends on shaping and controlling the flow of electrons. But long before electrons flowed through devices, they moved through cells. As the Nobel Prize–winning physiologist Albert Szent-Györgyi reportedly said, "Life is nothing but an electron looking for a place to rest."

According to the first law of thermodynamics, energy cannot be created or destroyed, but it can be changed from one form to another. In other words, cells cannot create energy from nothing. They must change one form of energy into another form. However, organisms cannot simply perform the biological equivalent of plugging into an electrical socket and sucking up electricity. That's because the plasma membrane is an electrically neutral zone that prevents charged particles from sneaking through, including electrons. So, to move electrons into and out of a cell, living organisms attach electrons to molecules. Plants and algae, for example, smuggle electrons into the cell via water molecules. Humans and other organisms obtain electrons from food, such as sugars, proteins, and fats.

Cells store and use energy by transferring electrons among molecules via chemical reactions. Thousands of different types of chemical reactions are required to sustain life in even the simplest cell. The term **metabolism** describes all the chemical reactions that occur inside living cells, including those that capture, store, or release energy. Most chemical reactions in a cell occur in chains of linked events known as **metabolic pathways**. Metabolic pathways produce key biological molecules in a cell, including important chemical building blocks such as amino acids and nucleotides.

Two metabolic pathways drive most of the life around us. The sun is the ultimate source of energy for most living organisms, and in the first process, known as **photosynthesis**, organisms capture energy from the sun and use it to create sugars from carbon dioxide and water (**Figure 5.2**, steps 1–2). In this way, photosynthetic organisms (plants, algae, and some bacteria) transform light energy into chemical energy stored in the covalent bonds of glucose, a sugar molecule. Glucose fuels the cell's activities and is used as the fuel to create all the biomolecules needed for life. Extra glucose is converted to lipids for long-term energy storage for future needs.

The second important process is **cellular respiration**, a process that complements photosynthesis. During cellular respiration, cells break down glucose into usable energy (**Figure 5.2**, step 3). Specifically, photosynthetic organisms, such as plants, algae, and some bacteria, get their energy by breaking down the glucose they

DEBUNKED!

MYTH: Respiration is the same thing as breathing.

FACT: Respiration is a biochemical process that involves the production of energy from glucose, and it occurs in every cell in the body. Breathing is the physical process of inhaling and exhaling air though the lungs.

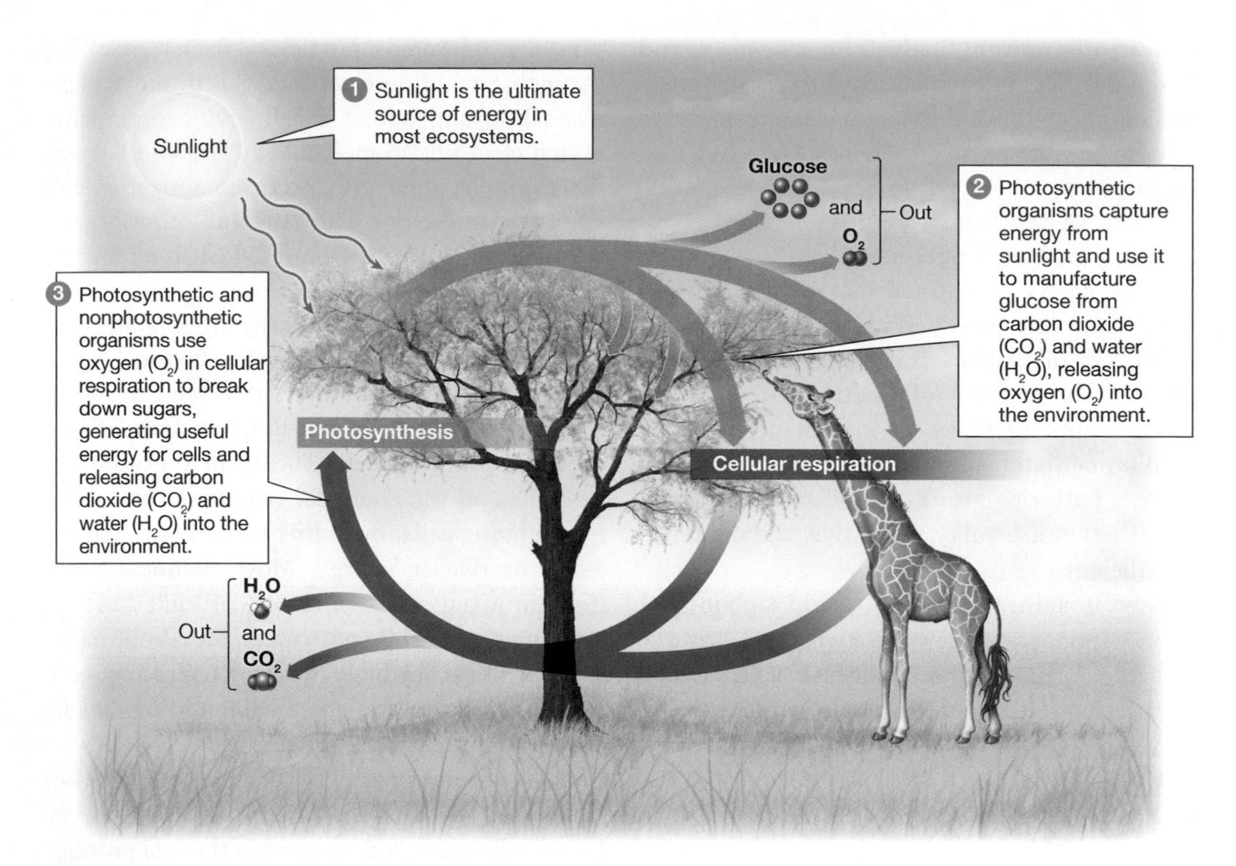

Figure 5.2

From sunlight to usable energy

Photosynthesis transforms the energy in sunlight into the energy in sugar molecules within the cell, with help from carbon dioxide and water, releasing oxygen as a by-product. Without photosynthesis, we would not have access to the sun's energy, and we would not have oxygen in our atmosphere to support life. In a complementary process, cellular respiration breaks down these sugar molecules, with help from oxygen, allowing organisms to access the energy stored in them.

Q1: Why is the pink line showing cellular respiration originating at both the giraffe and the tree?

Q2: How does animal life depend on photosynthesis?

Q3: Explain how photosynthesis and cellular respiration are "complementary" processes.

See Appendix A for answers to the figure questions.

made during photosynthesis. Nonphotosynthetic organisms, including animals, get their energy by eating plants or other animals and breaking down sugars and other nutrients in that food. Crazily enough, electrons are transferred back and forth during these two metabolic processes. Metabolism, at its core, is a feat of juggling electrons.

On Santa Catalina Island, Rowe's experiment focused on electrons. After setting up her tanks, Rowe let them sit for 3 months, untouched. During that time, her instruments began to detect evidence of life: the negative charge flowing through the electrodes in the sand began to steadily increase. Something was stealing the electrons, consuming more and more electricity.

An Unusual Diet

Scientists in the 1980s first discovered two types of bacteria that expel electrons onto metal. If nature produced microbes that spit out electrons, why could it not also produce microbes

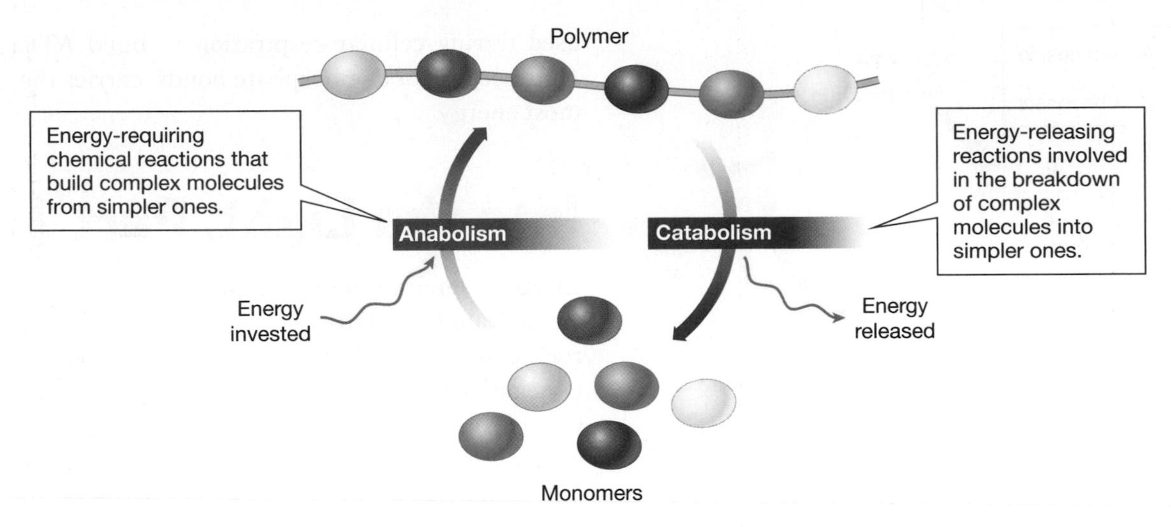

Polymer
Energy-requiring chemical reactions that build complex molecules from simpler ones.
Energy-releasing reactions involved in the breakdown of complex molecules into simpler ones.
Anabolism
Catabolism
Energy invested
Energy released
Monomers

Figure 5.3

Anabolism builds biomolecules; catabolism breaks them down

Metabolic reactions either build or break down molecules. Molecule-building reactions (anabolism) cost energy, and molecule-breakdown reactions (catabolism) release energy.

Q1: What source of energy would plants use for anabolic reactions? Would an animal use the same kind of energy?

Q2: What source of energy would plants release in catabolic reactions? Would an animal release the same kind of energy?

Q3: What would happen to a plant if catabolism stopped in all its cells? If catabolism stopped in all of an animal's cells, would the same result occur?

See Appendix A for answers to the figure questions.

that directly ingest electrons? These "rock eaters" are what Annie Rowe was looking for in her sediments.

All living cells have two main types of metabolism: anabolism and catabolism. **Anabolism** refers to metabolic pathways that create complex molecules from simpler compounds—a process used by all cells to make the basic building blocks of life. Because it takes energy to form chemical bonds, an anabolic reaction requires the input of energy. Photosynthesis is an example of anabolism: as plants sit in the noonday sun, the cells in their green leaves capture energy to make glucose from carbon dioxide; thus, a complex molecule is assembled from simpler ones. **Catabolism** refers to the opposite: metabolic pathways that release chemical energy in the process of breaking down complex molecules. Breaking chemical bonds releases energy, which is why catabolic reactions provide energy for other functions. Plant cells rely on a catabolic process, cellular respiration, to break down the glucose made by photosynthesis (**Figure 5.3**).

All of this metabolic activity requires energy, so cells need **energy carriers** to deliver energy where it is needed around the cell. Every living cell uses **ATP** (adenosine triphosphate)—a small organic molecule—to carry energy from one part of the cell to another. All living things create ATP during the catabolic reactions of cellular respiration. ATP powers almost all activities in the cell, such as moving molecules and ions in and out of the cell, sending nerve impulses, triggering muscle contractions, and moving organelles around inside the cell. In addition to powering a cell's activities, ATP fuels metabolic reactions and most other enzymatic reactions. If a cell exhausts its ATP supply, it will die.

As the name indicates, the ATP molecule consists of one adenosine and three phosphate groups. Much of the usable energy in ATP is held in its energy-rich phosphate bonds. To release

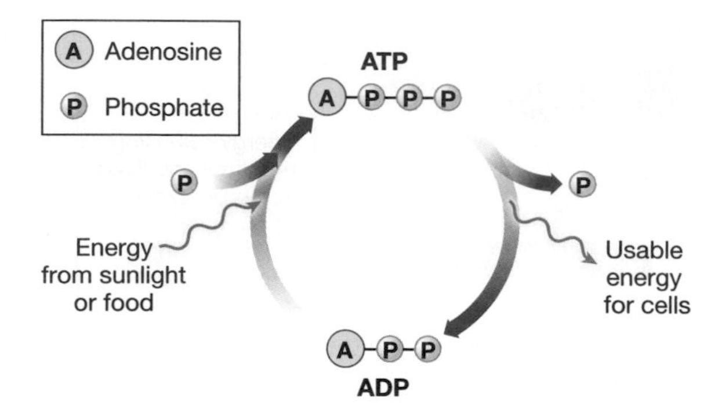

Figure 5.4

Cells store and deliver energy using ATP

Loading a high-energy phosphate (P) on ADP transforms the otherwise sedate molecule into the live wire that is ATP. But making ATP—the universal energy currency of cells—by creating a bond between ADP and a phosphate group requires metabolic energy, as does the creation of all chemical bonds. Breaking off a phosphate from ATP releases its energy for use by cells.

Q1: Define ATP in your own words.

Q2: When ATP is transformed to ADP, what two things are released in the process?

Q3: Arsenic disrupts the production of ATP from ADP. Why would this characteristic cause arsenic to be a potent poison?

See Appendix A for answers to the figure questions.

that energy, one of the bonds must be broken. When the phosphate on the end of an ATP molecule breaks off, energy is released along with the leftover molecule of **ADP** (adenosine diphosphate). As its name indicates, that ADP molecule consists of one adenosine and the remaining two phosphate groups. By contrast, to form a molecule of ATP, the opposite has to happen: metabolic energy is used to form a chemical bond between ADP and a free phosphate (**Figure 5.4**).

To perform photosynthesis and cellular respiration, cells rely on ATP and other energy carriers: NADPH, NADH, and $FADH_2$. In these molecules, the energy is stored in the chemical bonds of hydrogen rather than phosphate. Each carrier is a specialist in terms of the amount of energy it carries and the types of chemical reactions to which it supplies energy and from which it receives energy. NADPH is an energy carrier used during photosynthesis to build glucose. NADH and $FADH_2$ are energy carriers used during cellular respiration to build ATP. ATP, because of its phosphate bonds, carries the most energy.

Into the Light, Part 1

In 2009, environmental engineers at Pennsylvania State University identified the first elusive rock eaters by exposing a mix of microbes to a negatively charged electrode (**Figure 5.5**). One of the species, *Methanobacterium palustre*, survived on the electricity, taking in electrons and using them in a metabolic pathway to convert carbon dioxide to methane, a gaseous organic compound. One news article summed up the experiment with the headline BUG EATS ELECTRICITY, FARTS BIOGAS. The Penn State engineers proposed, but were unable to show, that the bacteria ("bugs") were ingesting naked electrons straight from the surface of the electrode.

A few years later, Alfred Spormann launched his own investigation into bacteria that seemed to be "eating" electrons. Spormann, a microbiologist at Stanford University, has long been interested in outliers, microbes with unusual metabolisms. Tall and thin with a throaty German accent, Spormann has spent more than 30 years studying such microbes, from those in the human gut that contribute to irritable bowel syndrome to bacteria that could be used to help decontaminate groundwater. "I'm really interested in exceptions to the rules

Figure 5.5

A negatively charged electrode

To grow rock eaters, researchers used an electrode, a conductor that can collect electrons and become negatively charged.

and in understanding the plasticity of microbial metabolism," says Spormann. Typically, the microbes he studies are **anaerobic**. That is, they grow in places with little or no oxygen because they don't need oxygen for metabolism. In fact, some of these cells are poisoned by oxygen.

But unlike anaerobic microbes, most animals require oxygen to breathe. That oxygen comes from photosynthesis. For this reason, and because photosynthesis is the way energy from the sun gets stored on Earth, many scientists consider photosynthesis to be the most important life process on our planet. In the cells of many eukaryotes—specifically, algae and plants—photosynthesis takes place inside chloroplasts. If viewed under a microscope, these organelles look like green grapes. Chloroplasts contain an extensive network of structures called *thylakoids*, piled up like stacks of pancakes, which contain complex proteins needed for photosynthesis. Also embedded in the membranes of the thylakoids is the green pigment **chlorophyll**, which is specialized for absorbing light energy (**Figure 5.6**). Photosynthetic prokaryotes—bacteria—do not have chloroplasts. Instead, chlorophyll is embedded directly in their plasma membranes.

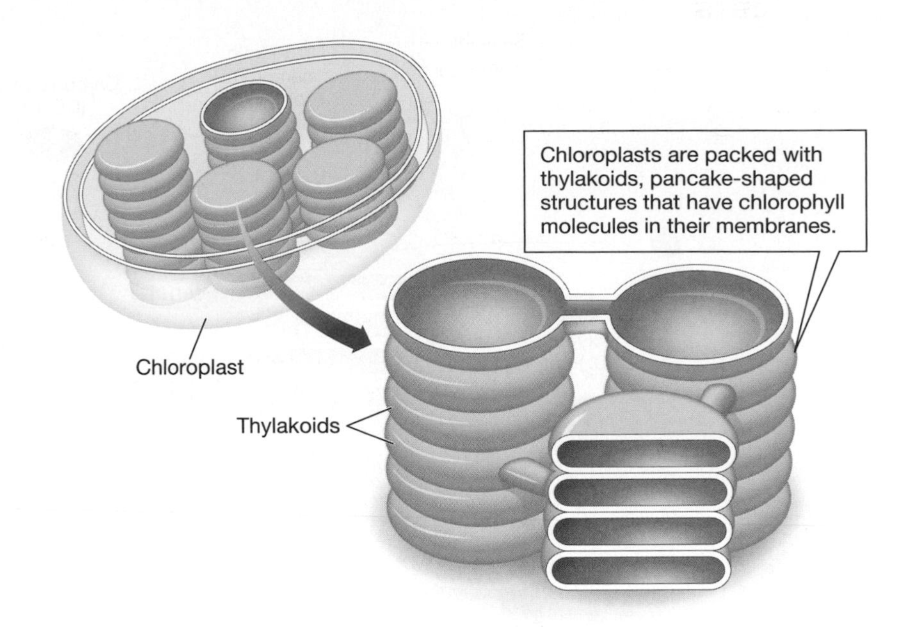

Chloroplasts are packed with thylakoids, pancake-shaped structures that have chlorophyll molecules in their membranes.
Chloroplast
Thylakoids

Figure 5.6

Chloroplasts and thylakoids

In many eukaryotes (algae and plants), photosynthesis occurs in chloroplasts. These structures include thylakoids, which contain chlorophyll.

Q1: How does embedding chlorophyll in the membranes of thylakoids versus the membrane of the chloroplast increase the amount of chlorophyll each chloroplast can accommodate?

Q2: Do all photosynthetic organisms have their chlorophyll embedded in the membranes of thylakoids?

Q3: What would cells with higher numbers of chloroplasts be better at absorbing?

See Appendix A for answers to the figure questions.

Photosynthesis takes place in two principal stages: the light-dependent reactions and the light-independent reactions, also called the *Calvin cycle*. During the **light-dependent reactions**, chlorophyll molecules absorb energy from sunlight and use that energy to break the bonds of water molecules. The splitting of water (H_2O) produces hydrogen (H^+) and oxygen gas (O_2)—the oxygen that we breathe—as a by-product released into the atmosphere.

As detailed in the lower left corner of **Figure 5.7**, the light-dependent reactions are an extremely complex cascade of chemical events. These events depend on protein complexes embedded in the thylakoid membranes, including photosystems II and I (PSII and PSI), additional proteins that complete the **electron transport chain**, and ATP synthase. The electrons and protons of the hydrogen (H^+) that result from the splitting of water are handed over to PSII and then PSI, using intervening proteins before and after PSI to complete the electron transport chain. During this process, PSI generates the energy carrier NADPH by adding a hydrogen to the electron carrier $NADP^+$. ATP synthase ultimately generates ATP by adding a phosphate group to ADP (a process detailed in **Figure 5.4**) by means of a proton gradient and proton pump. ATP and NADPH are used in the next stage of photosynthesis, the light-independent reactions.

The next stage consists of **light-independent reactions**, a series of chemical reactions to convert carbon dioxide (CO_2) into glucose. This process is also known as the **Calvin cycle** and **carbon fixation**. As shown in the lower right corner of **Figure 5.7**, the reactions use energy delivered by ATP (resulting in ADP) and electrons and hydrogen ions donated by NADPH (resulting in $NADP^+$). By capturing carbon atoms from CO_2 in the air and converting them into glucose or fixing them into glucose, the light-independent reactions make carbon from the nonliving world available to the photosynthetic organisms and eventually to other living

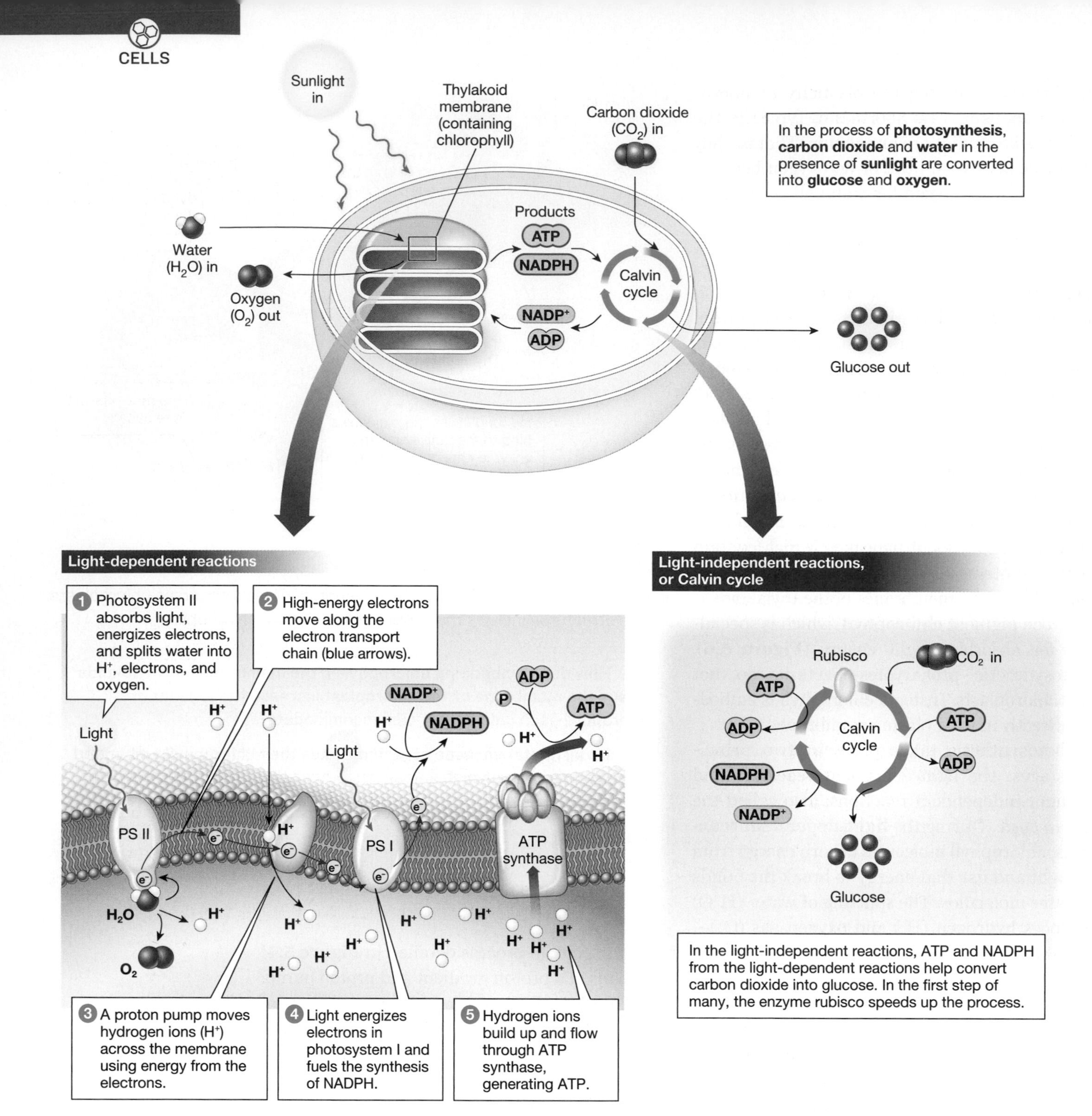

Figure 5.7

Photosynthesis occurs in two stages

In the first stage of photosynthesis, the light-dependent reactions generate energy carriers. In the second stage, the light-independent reactions create glucose.

Q1: What are the three reactants (types of input) needed for photosynthesis to occur?

Q2: Which products of the light-dependent reactions of photosynthesis are used by the light-independent reactions?

Q3: What are the two major overall products (types of output) of photosynthesis?

See Appendix A for answers to the figure questions.

organisms that eat them, including us. Enzymes catalyze these reactions at each step; the enzyme needed in the first step—and the most abundant enzyme on the planet—is **rubisco.**

Catalyzing Reactions

Around 2014, Spormann's team began studying an "electric" microbe that lives in water, *Methanococcus maripaludis*. They wanted to see whether this bacteria directly ingested electrons, but it didn't. The researchers found that *M. maripaludis* actually excreted an enzyme onto the surface of an electrode to do its dirty work.

Nearly all metabolic reactions are facilitated by enzymes. **Enzymes** are biological *catalysts*—molecules that speed up chemical reactions. Without the action of enzymes—most of which are proteins—metabolism would be extremely slow. In fact, life as we know it could not exist. Enzymes work by positioning **substrates**—molecules that will react to form new products—in an orientation that favors the making or breaking of chemical bonds (**Figure 5.8**).

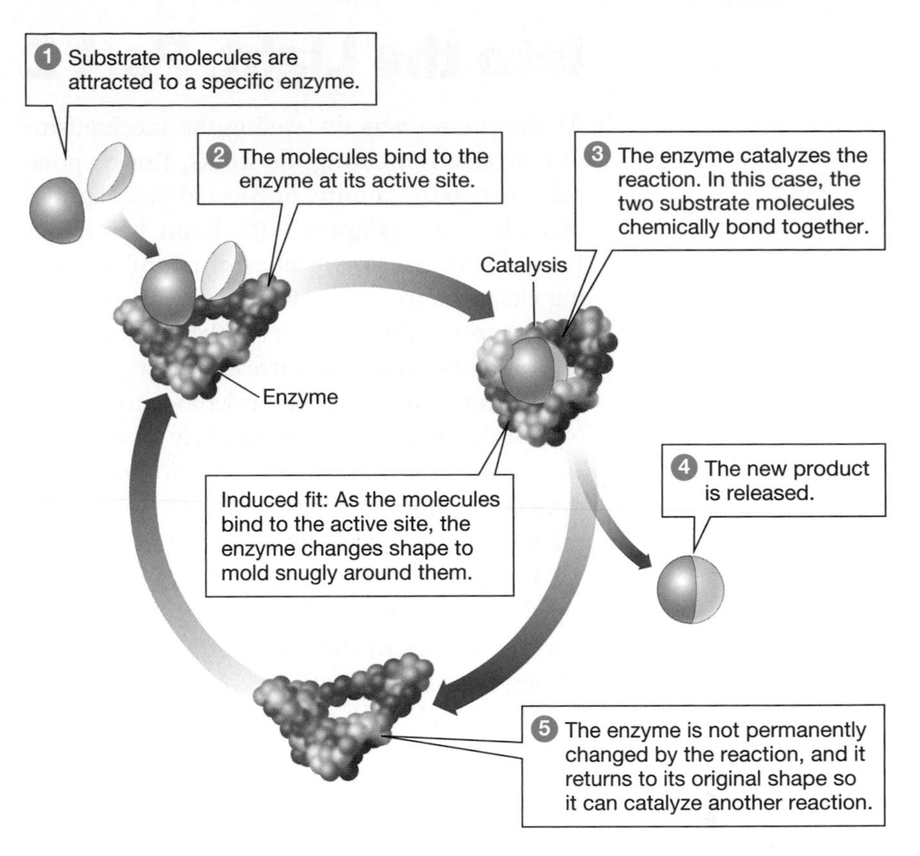

1 Substrate molecules are attracted to a specific enzyme.
2 The molecules bind to the enzyme at its active site.
3 The enzyme catalyzes the reaction. In this case, the two substrate molecules chemically bond together.
Catalysis
Enzyme
Induced fit: As the molecules bind to the active site, the enzyme changes shape to mold snugly around them.
4 The new product is released.
5 The enzyme is not permanently changed by the reaction, and it returns to its original shape so it can catalyze another reaction.

Figure 5.8

Enzymes are molecular matchmakers

Enzymes dramatically increase the rate of chemical reactions. They do this by positioning a substrate molecule in an orientation that favors the creation or destruction of a chemical bond. Here, a bond is formed between two molecules.

Q1: Why is it important that enzymes are not permanently altered when they bind with substrate molecules?

Q2: If the active site of an enzyme were blocked, how would the production of a product be affected?

Q3: If a cell could not produce more of a particular enzyme and a poison blocked this enzyme from resuming its original shape after an induced fit, what would happen to the production of that enzyme's product?

See Appendix A for answers to the figure questions.

Each enzyme binds only to a specific substrate, and it catalyzes a specific chemical reaction, such as rubisco catalyzing the first step of the light-independent reactions. An enzyme's function is based on its chemical characteristics and the three-dimensional shape of its **active site**—the location within the enzyme where substrates bind. When molecules bind to the active site, the enzyme changes shape—a process called **induced fit**. The enzyme's shape and, therefore, its activity can be affected by temperature, pH, and salt concentration. Because the enzyme's active site is not permanently changed as reactions are catalyzed, the enzyme can perform its role again and again. As a result, the enzyme does not have to be present in great numbers or need to be continually produced.

The enzyme that Spormann and his team identified was unusual because it was excreted to the exterior of the cell. Most enzymes work on the inside of cells. A multistep metabolic pathway can proceed rapidly and efficiently because the required enzymes are physically close together in the cell and the products of one enzyme-catalyzed chemical reaction serve as the basis for the next reaction in the series, as is especially true for metabolic pathway reactions.

In the case of *M. maripaludis*, the excreted enzyme's role was to grab an electron from the metal, pair it with a proton from water, and create a hydrogen atom—a familiar food for microbes that easily pass through the plasma membrane. The bacteria "found a way to produce a compound that is easily metabolized by the cells," says Spormann. And although this particular microbe wasn't eating naked electrons, Spormann has since isolated a microbe that does directly take up electrons. His team has yet to publish the details.

Into the Light, Part 2

As Spormann was unraveling the mechanisms of how electron-eating life works, Rowe's prospecting trip to Catalina uncovered more species of rock eaters (**Figure 5.9**). From her tanks, Rowe identified 30 varieties of microbes sucking electrons from the electrodes. She was even able to grow a few of them on plates in the laboratory, using minerals such as sulfur and iron as electron sources for the microbes to munch. Not all the species relied exclusively on electrons as a food source; that is, they didn't just live off mineral sediments. Like eukaryotic cells, most microbial cells use the traditional route of cellular respiration for obtaining energy, converting the energy in the bonds of sugar molecules into the energy in the bonds of ATP molecules through catabolic reactions.

Have a look back at **Figure 5.2**. As detailed there, photosynthesis produces glucose. During cellular respiration, the complementary process, carbon-carbon bonds in glucose molecules are broken. Each carbon atom is released into the environment in a molecule of carbon dioxide (CO_2), with water (H_2O) as a by-product. Thus, cellular respiration is the reciprocal process of

Figure 5.9

Rock eaters of Catalina

In the jar is sediment containing rock eaters collected off the coast of Santa Catalina Island.

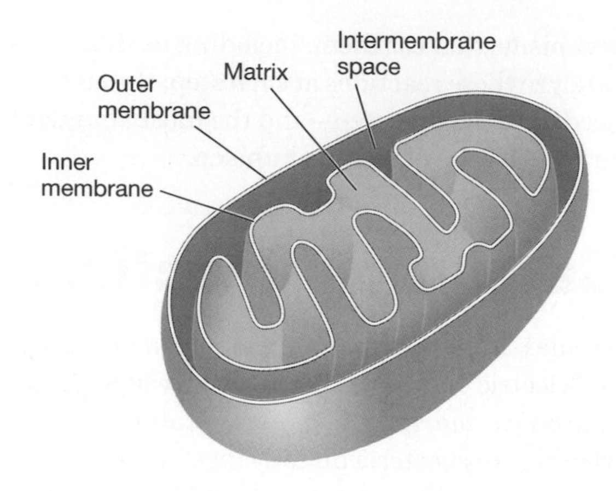

Figure 5.10

The parts of a mitochondrion

In eukaryotes, cellular respiration occurs within this simple but vital structure.

photosynthesis. In most eukaryotes, cellular respiration is a multistep process that occurs in the cell's mitochondria (singular "mitochondrion"; **Figure 5.10**). Cellular respiration is the main way that eukaryotic cells break down sugars to make usable cellular ATP energy. Photosynthetic organisms first make their own glucose during photosynthesis and then perform cellular respiration on that self-made glucose. Humans and other animals, by contrast, digest glucose directly from food sources, break it down into carbon dioxide, hydrogen, and electrons, and then process those captured electrons for energy.

Cellular respiration consists of three stages: glycolysis, the Krebs cycle, and oxidative phosphorylation. The first stage, **glycolysis**, takes place in the cytoplasm of the cell. During glycolysis, the carbon bonds in sugars (mainly glucose) are broken to make three-carbon compounds called pyruvate. This process releases energy and results in the creation of two useful molecules of ATP and two molecules of NADH for each glucose molecule that is split (**Figure 5.11**, top left: step 1). In other words, glycolysis converts some of the chemical energy of glucose into the chemical energy of NADH and ATP.

In terms of the evolution of life, glycolysis was probably the earliest means of producing ATP from food molecules. It is still the primary means of energy production in many prokaryotes, including some of Rowe's microbes. However, because glucose is only partially broken down through this process, the ATP energy yield from glycolysis is small. For most

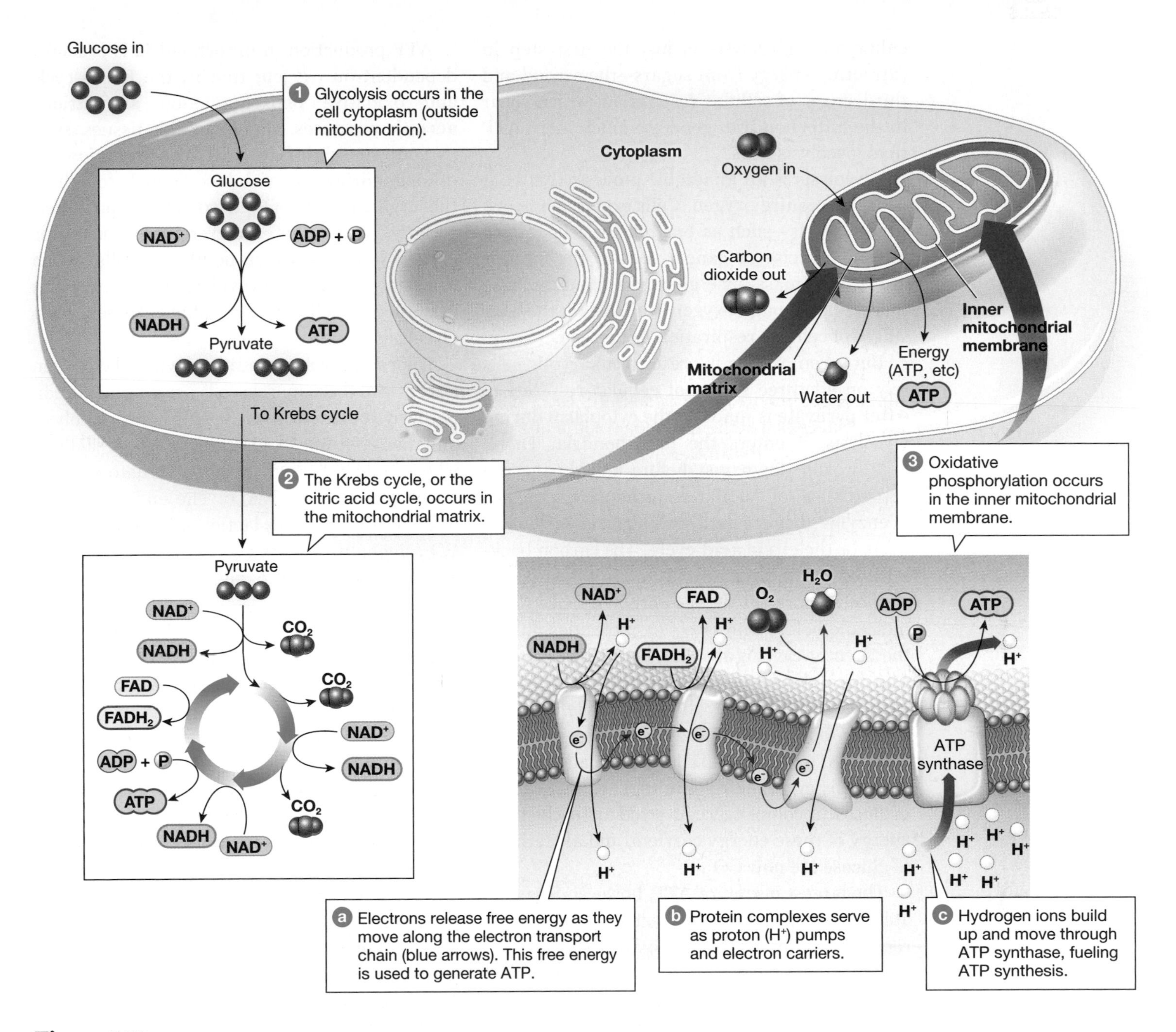

Figure 5.11

Three steps of cellular respiration in eukaryotes

In glycolysis, each six-carbon glucose molecule is converted into two three-carbon molecules of pyruvate, during which NADH and ATP are generated. In the next step, the Krebs cycle (or citric acid cycle), carbon dioxide is released and energy carriers are created, including ATP, NADH, and $FADH_2$. In the third and final step, oxidative phosphorylation, more ATP is produced than through any other metabolic pathway.

Q1: What are the end products of cellular respiration after all three steps are complete?

Q2: Considering the inputs and products of each process, why is cellular respiration considered the reciprocal process to photosynthesis?

Q3: Which of the three stages of cellular respiration—namely, glycolysis, the Krebs cycle, or oxidative phosphorylation—could organisms have used 4 billion years ago before photosynthesis by cyanobacteria released oxygen into the atmosphere?

See Appendix A for answers to the figure questions.

eukaryotes, glycolysis is just the first step in extracting energy from sugars—the second and third stages of cellular respiration, which occur in the mitochondria, generate much more ATP than glycolysis does.

Glycolysis is an anaerobic process; that is, it does not require oxygen. Cells experiencing low oxygen levels—such as our muscle cells during intense exercise or cancer cells in the interior of a tumor mass—use glycolysis alone because there is not enough oxygen available for the later stages of cellular respiration to proceed.

But when oxygen is present, eukaryotic cells rely on all three stages of cellular respiration. After pyruvate is made in the cytoplasm during glycolysis, it enters the mitochondria. There, it is broken down just before and during the second stage of cellular respiration: a sequence of enzyme-driven reactions known as the **Krebs cycle** or the **citric acid cycle**. The carbon backbone of the pyruvate molecule is taken apart, step by step, releasing one carbon dioxide (CO_2) molecule before and two CO_2 molecules during the Krebs cycle (**Figure 5.11**, bottom left: step 2). The breakdown and release of stored energy of the carbon bonds by the Krebs cycle produces a large bounty of energy carriers, including ATP, NADH, and $FADH_2$. Essentially, during the Krebs cycle the remaining chemical energy of glucose is completely converted to the chemical energy of these energy carriers, and all carbons of glucose are now CO_2.

The largest output of ATP, however, is generated during the third and last stage of cellular respiration: **oxidative phosphorylation**. In a nutshell, hydrogen (H^+) atoms are removed from NADH and $FADH_2$ and handed over to oxygen (O_2) through an electron transport chain (**Figure 5.11**, bottom right: step 3a). In this way, electrons from hydrogen atoms are dumped onto oxygen (step 3b), and hydrogen protons are used to generate a proton gradient and proton pump that in turn produce very large amounts of ATP through ATP synthase (**Figure 5.11**, detailed in bottom right: step 3c). Thus, during oxidative phosphorylation, the chemical energy of NADH and $FADH_2$ is converted into the chemical energy of ATP. In fact, oxidative phosphorylation can generate 15 times as much ATP as can glycolysis alone. However, oxidative phosphorylation absolutely needs oxygen to occur.

ATP production in mitochondria is crucially dependent on oxygen; that is, the Krebs cycle and oxidative phosphorylation are strictly **aerobic** processes. Highly aerobic tissues, such as your muscles, have high concentrations of mitochondria and a rich blood supply to deliver the large amounts of O_2 needed to support their activity. Even so, intense exercise exhausts this oxygen supply, at which point your cells switch to glycolysis only, as previously mentioned. But how do the cells extract sufficient energy via glycolysis alone?

Under anaerobic conditions when the oxygen level is low, these muscle cells use a fermentation pathway. **Fermentation** begins with glycolysis and then continues with a special set of reactions whose only role is to help perpetuate glycolysis by regenerating NAD^+. The end product of fermentation is either lactic acid or ethanol and CO_2, depending on the organism. This process enables organisms to generate ATP through glycolysis alone when aerobic ATP production (cellular respiration) is limited (**Figure 5.12**). If the low amount of ATP generated by glycolysis cannot be perpetuated, the cells will run out of ATP and die.

Many species of bacteria that live in places such as oxygen-deficient swamps, sewage, or deep layers of soil never use oxygen or are actually poisoned by oxygen. Many microbes can dump electrons using elements other than oxygen, such as iron or sulfur compounds, to generate large amounts of ATP through ATP synthase. When these other elements run out, these organisms use fermentation to regenerate NAD^+ needed to perpetuate glycolysis. These microbes generate ATP through ATP synthase and through glycolysis without ever using oxygen.

Most eukaryotic cells can perform both aerobic (cellular respiration) and anaerobic (glycolysis) ATP production, depending on the oxygen levels in their environment. Because the Krebs cycle and oxidative phosphorylation produce magnitudes more ATP than glycolysis and fermentation yield, these cells always use oxygen and aerobic cellular respiration when oxygen levels permit.

Together, photosynthesis and cellular respiration enable cells to store and use energy from the sun. These processes are two sides of the same coin; the products of one reaction are the ingredients for the other. Photosynthesis, an anabolic

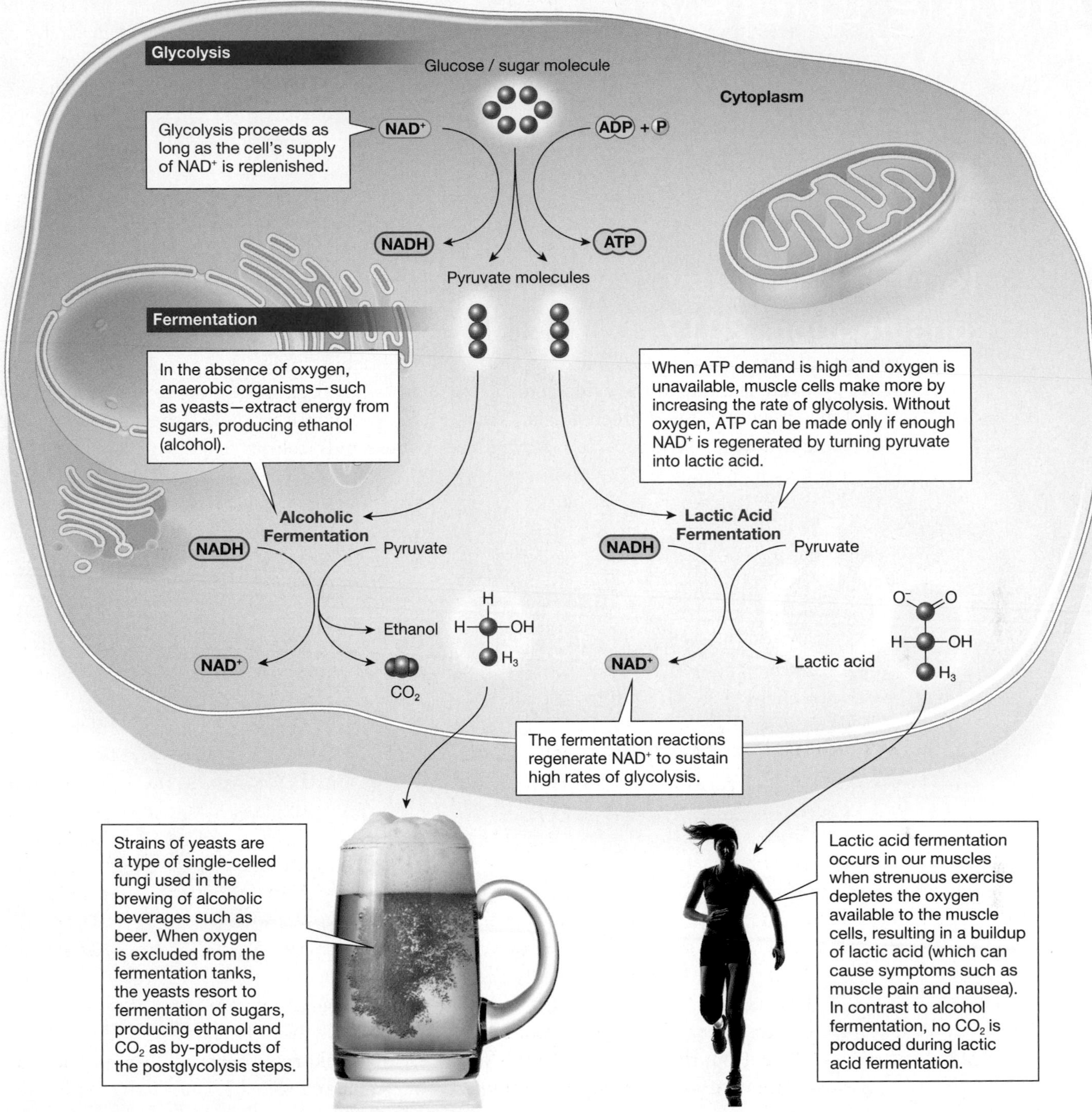

Figure 5.12

Fermentation produces energy in the absence of oxygen

When the oxygen supply is inadequate to support ATP production through cellular respiration, fermentation enables ATP production through glycolysis alone.

Q1: Which product released by fermentation accounts for the bubbles in beer?

Q2: Bread makers rely on yeast to cause their dough to rise before baking. What must be included in the dough for fermentation to occur?

Q3: Explain in your own words why lactic acid builds up in your muscles during strenuous physical activity.

See Appendix A for answers to the figure questions.

Sourcing Energy

Microbes that "eat" electricity may someday provide an efficient way to store renewable energy. In the meantime, renewable energy sources provide almost 20 percent of the world's power. Here's a breakdown of the latest numbers.

Assessment available in smartwork

Total world energy consumption, 2017

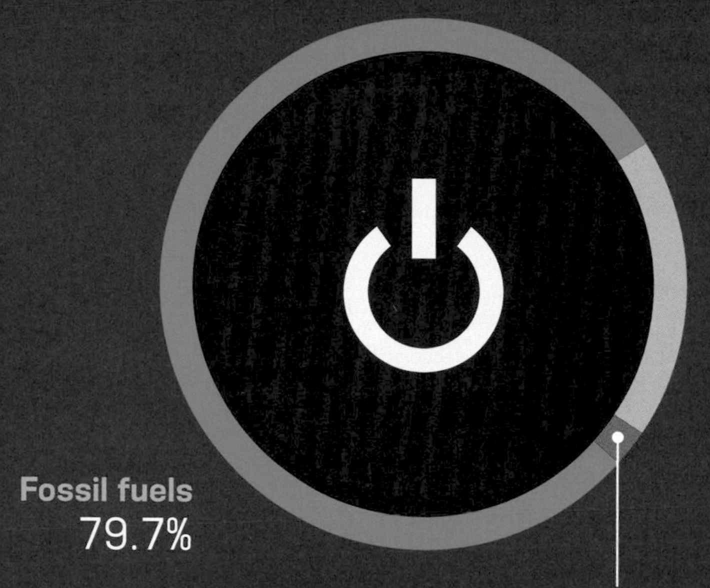

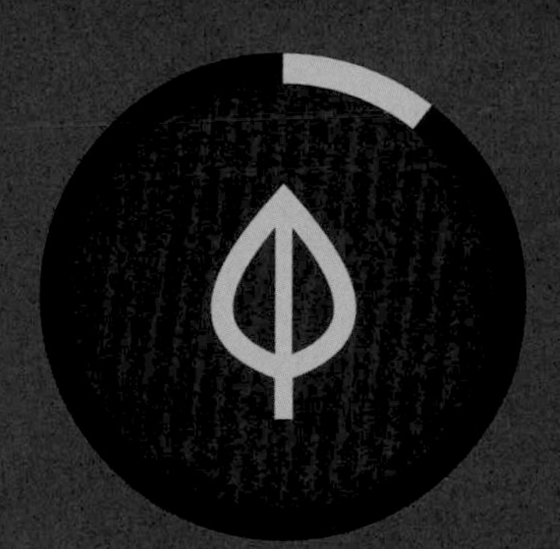

Renewable heat energy from biomass, geothermal, and solar technologies
4.2%

Hydropower
3.6%

Renewable power from wind, biomass, geothermal, solar, and ocean technologies
2.0%

Biofuels
1.0%

Totals may not add up due to rounding.

process, requires energy (sunlight) and CO_2 and releases O_2 and glucose. Cellular respiration, a catabolic process, requires O_2 and glucose and releases CO_2 and energy.

Unlike the situation with photosynthesis and cellular respiration—metabolic processes that have been well studied for decades—researchers don't yet know the details of how rock eaters store and use energy. "There isn't really any pathway worked out yet," says Rowe, who now runs her own lab at the University of Cincinnati. "Truthfully, it's amazing that it works." She's eager to find out how. Using the species Rowe isolated from the salty water of Santa Catalina Island, she is currently developing tools to identify which enzymes are involved in their metabolism. Meanwhile, the hunt for additional species of rock eaters continues.

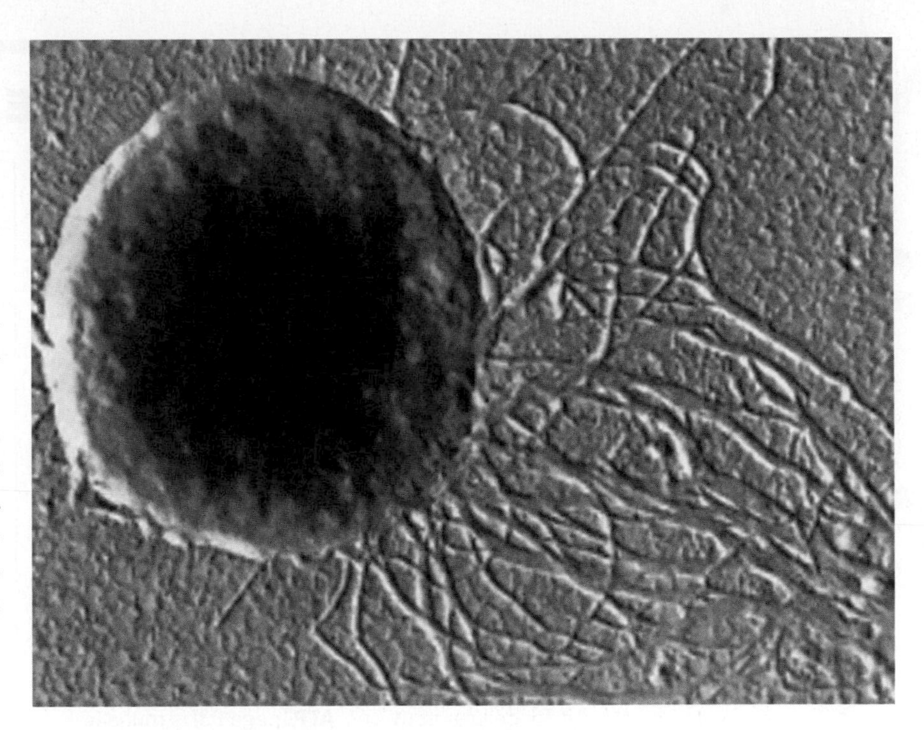

Figure 5.13

The rock eater *M. maripaludis*
The hairlike pili extending from this bacterium are electrically conductive.

Bacterial Batteries

In 2010, El-Naggar found that pili (singular "pilus"), the tiny hairlike structures on bacteria, are conductive (**Figure 5.13**)—just as conductive, in fact, as silicon, the basis of most electronics. One single nano-sized pilus, which researchers refer to as a bacterial "nanowire," can transport about 106 electrons per second—enough to sustain the respiration of a whole cell.

Today, Rowe, Spormann, and others are focusing on how to use these unique bacteria in "bacterial batteries." What if energy from a solar panel, wind turbine, or dam could be passed into a tank of pili-bearing bacteria, where the microbes would gobble up the electricity and store it in chemical bonds in the form of methane or another natural gas? In that case, the gas could be liquefied and stored until needed.

"If you could convert electrical energy, through microbial processes, to a liquid fuel, that could be a really valuable way of storing energy," says Rowe. This electricity-to-fuel scenario could also be run in reverse, using microbes to digest a fuel source such as raw sewage and convert it into electricity.

There's also the intriguing scenario that someday our electronic devices will be powered by microorganisms that manipulate the flow of electrons. In this rapidly evolving field, the possibilities are endless. "We don't know how many organisms do or don't do it," adds Spormann. "It's entirely possible that rock eating is more widespread than previously thought. We just don't know because it's so new."

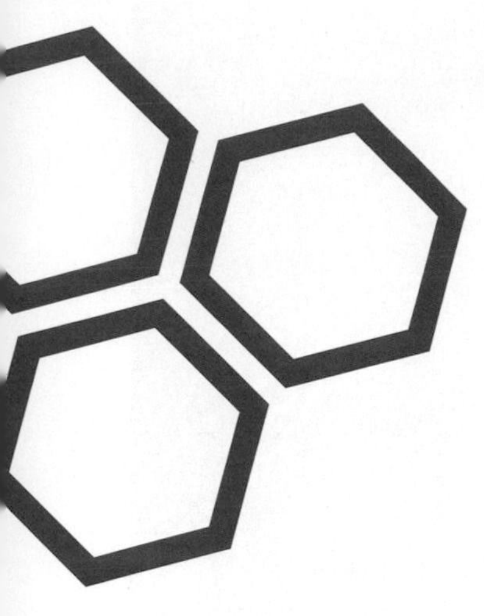

NEED-TO-KNOW SCIENCE

- A **metabolic pathway** (page 91) is a multistep sequence of chemical reactions.
- The sun is the source of energy fueling most living organisms. Plants, algae, and some bacteria gain energy from their environment through **photosynthesis** (page 91). Most organisms use **cellular respiration** (page 91) to extract usable energy from sugar molecules. In chemical terms, photosynthesis and cellular respiration are opposite processes.
- All the chemical reactions involved in the capture, storage, and use of energy by living organisms are collectively known as **metabolism** (page 91). Energy-requiring synthesis reactions, such as photosynthesis, are **anabolism** (page 93). Energy-releasing breakdown reactions, such as cellular respiration, are **catabolism** (page 93).
- **Energy carriers** (page 93) store energy and deliver it for cellular activities. **ATP** (page 93) is found in all cells and is the energy carrier used to do cellular work.
- Photosynthesis takes place in chloroplasts and occurs in two stages. In the **light-dependent reactions** (page 95), energy is absorbed using pigment molecules that include **chlorophyll** (page 95) as electrons flow along the **electron transport chain** (page 95). The light-dependent reactions split water molecules, releasing hydrogen and oxygen, and they create the energy carriers ATP and NADPH.
- During the **light-independent reactions** (page 95), also known as the **Calvin cycle** (page 95) and **carbon fixation** (page 95), the energy carriers produced during the light-dependent reactions are used to convert carbon dioxide into glucose. The first step of this process is catalyzed by the enzyme **rubisco** (page 97).
- **Enzymes** (page 97) are biological catalysts, which are usually proteins that speed up chemical reactions. An enzyme's function is based on its chemical characteristics and the three-dimensional shape of its **active site** (page 97).
- Cellular respiration occurs in three stages: (1) **glycolysis** (page 98), which yields small amounts of ATP and NADH; (2) the **Krebs cycle** (page 100), which releases carbon dioxide and produces NADH, $FADH_2$, and ATP; and (3) **oxidative phosphorylation** (page 100), which generates many molecules of ATP.
- In the absence of oxygen, **fermentation** (page 100) converts the products of glycolysis into alcohol or lactic acid, in order to regenerate the NAD^+ needed for glycolysis to continue in the absence of oxygen.

THE QUESTIONS

See Appendix B for answers.

The Basics

1 Metabolic pathways

(a) always break down large molecules into smaller units.

(b) only link smaller molecules together to create polymers.

(c) are often organized as a multistep sequence of reactions.

(d) occur only in mitochondria.

2 Enzymes

(a) speed up reactions that would otherwise occur much more slowly.

(b) spur reactions that would otherwise never occur.

(c) provide energy for anabolic but not catabolic pathways.

(d) are consumed during the reactions they catalyze.

3 The most common energy-carrying molecule in all organisms is

(a) carbon dioxide.

(b) water.

(c) ATP.

(d) rubisco.

4 The major product of photosynthesis is

(a) lipids.

(b) glucose.

(c) amino acids.

(d) nucleotides.

5. Which of the following statements is *not* true?

(a) Glycolysis is the first stage of cellular respiration.

(b) Glycolysis can proceed under low oxygen levels with the assistance of fermentation.

(c) Glycolysis produces less ATP than does either the Krebs cycle or oxidative phosphorylation.

(d) Glycolysis produces most of the ATP required by aerobic organisms, such as humans.

6. Select the correct terms:
Photosynthesis and cellular respiration are chemically (**identical** / **opposite**) processes. Cellular respiration is an example of (**anabolism** / **catabolism**), which (**produces** / **expends**) energy.

7. In the accompanying diagram of photosynthesis, fill in each blank with the appropriate term(s): Sunlight in, CO_2 in, O_2 out, H_2O in, ATP and NADPH, ADP and $NADP^+$, Sugar, Light-dependent reactions, Light-independent reactions.

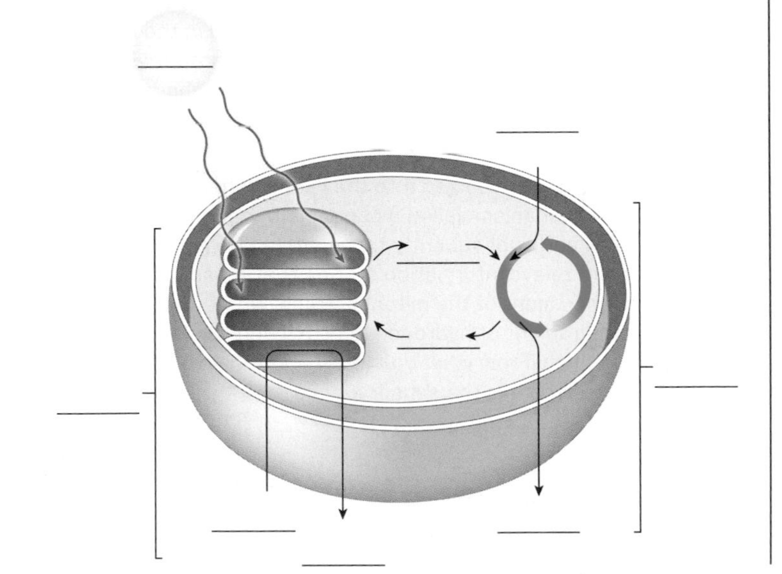

8. Place the following steps of cellular respiration in the correct order by numbering them from 1 to 4.

_____a. The Krebs cycle produces the energy carriers NADH, $FADH_2$, and ATP.

_____b. If oxygen levels are adequate, pyruvate is transported into the mitochondrion. If oxygen levels are very low, fermentation proceeds.

_____c. Glucose is broken down to produce ATP and NADH.

_____d. An electron transport chain produces ATP from ADP.

Challenge Yourself

9. Use the accompanying figure to answer the following questions. In Labrador retrievers, coat color is determined by an enzymatic pathway that converts reactant (input) pigment molecules into product (output) pigment molecules, using enzymes at each step. If enzymes 1, 2, and 3 are all functional in a Labrador, what color will its fur be? If only enzymes 1 and 2 are functional, what color will the dog's fur be? If only enzyme 1 is functional, what color will the dog's fur be?

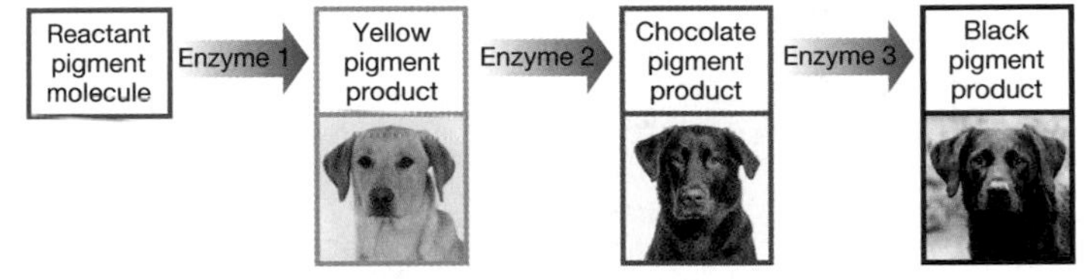

10 After strenuous exercise, you may notice that your muscles burn and feel sore the next day. Which statement best explains this phenomenon?

(a) Proteins in muscle cells are being digested to provide energy.
(b) Carbon dioxide is building up in muscle cells and changing their pH.
(c) Spontaneous combustion occurs during strenuous exercise, so avoid it at all costs!
(d) ATP is accumulating in muscle cells, causing a burning sensation.
(e) Without adequate oxygen, muscle cells are fermenting pyruvate into lactic acid.

11 The accompanying graph depicts the activation energy, or the amount of energy needed for a reaction to proceed, with and without an enzyme.

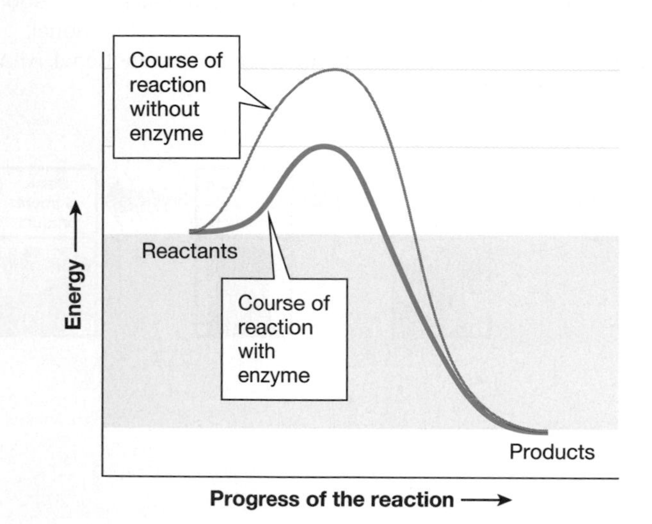

a. Which reaction requires more energy to proceed—the one with or without an enzyme? How do you know?
b. Is this reaction anabolic or catabolic? How do you know?

Try Something New

12 In 2012, an Illinois man was killed by cyanide poisoning after he won a million dollars in the lottery. Cyanide is a lethal poison because it interferes with the electron transport chain in mitochondria. What effect would cyanide have on cellular respiration?

(a) Glycolysis, the Krebs cycle, and oxidative phosphorylation would all be inhibited.
(b) The Krebs cycle would be inhibited, but oxidative phosphorylation would not.
(c) Oxidative phosphorylation would be inhibited.
(d) Glycolysis, the Krebs cycle, and oxidative phosphorylation would all be stimulated.

13 Plants in the genus *Ephedra* have been harvested for their active substance ephedrine for centuries. Used to reduce the symptoms of bronchitis and asthma, ephedrine is also taken as a stimulant and study aid as well as a dietary supplement (see question 16). It is also the main ingredient in the illegal production of methamphetamine. Since 2006, the sale of ephedrine and related substances has been limited and monitored in the United States. One effect of ingesting ephedrine is greatly increased metabolism, which has been known to kill users of ephedrine. How might an increased rate of metabolic processes cause death?

14 You have been transported into the future, where an extremely tiny mechanism called a nanosensor can be inserted into a living cell and subsequently travel into any organelle. The nanosensor relays information stating that it has lodged itself in a compartment of the mitochondrion, where there is a concentration gradient of hydrogen ions (H^+) and levels of ATP production are high. From what you know about cellular respiration in mitochondria, where is the sensor lodged?

(a) inner mitochondrial membrane
(b) mitochondrial matrix
(c) chloroplast stroma
(d) cell cytoplasm
(e) plasma membrane

Leveling Up

15 **What do *you* think?** What would happen if a virus destroyed all photosynthetic organisms on Earth?

16 ***Write Now* biology: calorie-burning fat** Your friend has emailed you a link (see below) to a *New York Times* article on "brown" fat. He has been trying to lose weight and wants to know whether you think it would be a good idea for him to spend more time in the cold rather than continue to exercise regularly. He is also interested in the article's mention of ephedrine's ability to stimulate brown fat, and he asks if you think he should begin to take ephedrine supplements. Compose a reply to your friend addressing the following points (using one or two paragraphs for each point). [*Note:* Your friend doesn't have to be male, and you may need to do further reading to answer (b) and (c).]

a. Explain in detail how brown fat burns calories when someone is chilled.

b. Explain how ephedrine affects metabolism and what its possible side effects are.

c. Contrast (a) and (b) with the effect of exercise on metabolism, both in the short term and in the longer term by increasing muscle mass.

d. In your final paragraph, advise your friend as to whether he should begin spending time in the cold to increase weight loss or take ephedrine supplements and whether he should continue to exercise regularly. Justify your opinion with data and logic.

To research your answer, begin with the following *New York Times* article, published on April 8, 2009: "Calorie-Burning Fat? Studies Say You Have It" (http://www.nytimes.com/2009/04/09/health/research/09fat.html). Consult and reference at least two additional sources in your answer.

17 **Is it science?** Wouldn't it be amazing if humans could manufacture their own food through photosynthesis like plants do? Use internet research to determine whether this idea is real science or science fiction. Is there evidence that animals can use the sun's energy directly and/or perform photosynthesis in nature? In the lab? If so, how would that capability affect life as we know it?

Digital resources for your book are available online.

Toxic Plastic

Two ruined experiments expose the health risks of chemicals in everyday products.

After reading this chapter you should be able to:

- Label a figure of the major stages of the cell cycle, and explain the processes that occur during each stage.
- Compare and contrast cell division by binary fission, mitosis, and meiosis.
- Distinguish between sister chromatids and homologous chromosomes.
- Diagram, using the appropriate terms, the steps in mitosis and in meiosis.
- Explain the importance of the checkpoints in the cell cycle and the consequences of bypassing those checkpoints.
- Identify the ways in which meiosis and fertilization together produce genetically diverse offspring.

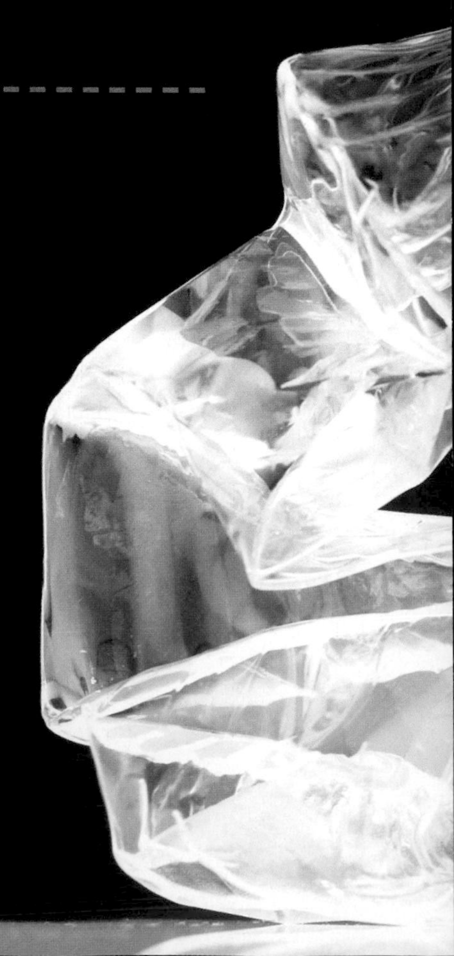

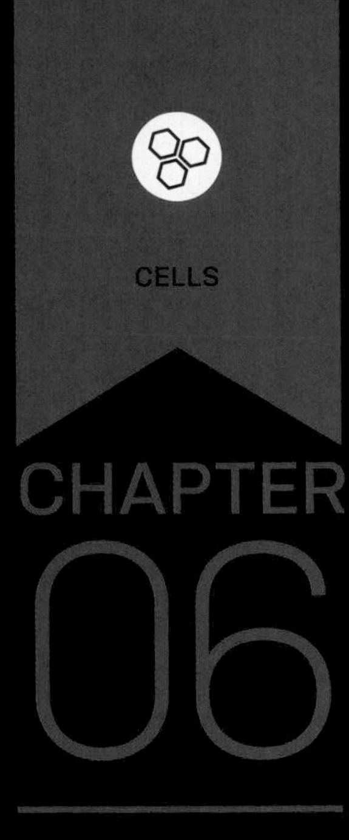

CHAPTER 06

CELL DIVISION

It began as a run-of-the-mill experiment. In 1989, biologists Ana Soto and Carlos Sonnenschein at Tufts University in Massachusetts were studying how the hormone estrogen regulates the growth of cells in the female reproductive system. For their research, the duo developed an experimental setup consisting of human breast cancer cells growing in plastic bottles called cell culture flasks (**Figure 6.1**, top). The flasks were filled with a liquid containing an ingredient to prevent the cells from multiplying in the absence of estrogen. When estrogen was added to the flasks, the cells grew. When estrogen was absent, they didn't.

One day, suddenly and surprisingly, cells in the flasks began growing even when estrogen hadn't been added. "What had worked for years didn't work anymore," says Sonnenschein. The two scientists immediately stopped their experiments and began searching for the cause. "It smacked of contamination," recalls Soto, as if estrogen had somehow gotten into the flasks. But after weeks of searching, Soto and Sonnenschein still couldn't identify a source of contamination. They became so paranoid that they suspected someone was entering the lab at night and secretly dripping estrogen into their flasks.

Almost 10 years later, in 1998, geneticist Patricia Hunt at Case Western Reserve University in Ohio stared dumbfounded at another experimental anomaly. Hunt was studying why older women are at increased risk of having children with abnormal chromosomes. An individual with Down syndrome, for example, has 47 chromosomes—the tiny, string-like structures in cells that contain genes— instead of the usual 46 (see Chapter 8 for more on chromosomes). Hunt hypothesized that hormone levels can increase the risk of such conditions. To test her hypothesis, Hunt and her research team raised groups of mice with varying levels of hormones and checked their egg cells for abnormal numbers of chromosomes (**Figure 6.1**, bottom).

The experiment was almost complete when Hunt went in to check on the control mice one last time. As discussed in Chapter 1, a control population is a necessary baseline for comparison against an experimental population. In this case, the control was a group of healthy mice whose hormone levels had not been altered. Using a light microscope, Hunt examined mouse oocytes—that is, immature egg cells. She was shocked. The cells were a mess, the chromosomes scrambled. A whopping 40 percent of the resulting eggs had chromosomal defects. "The controls were completely bonkers," says Hunt. "One week they were fine, the next week they weren't. That's when we knew something was going on."

Like Soto and Sonnenschein, Hunt scrutinized every method and every piece of lab equipment used in the experiment, looking for the culprit. But as weeks passed, she couldn't figure out what had ruined her experiment.

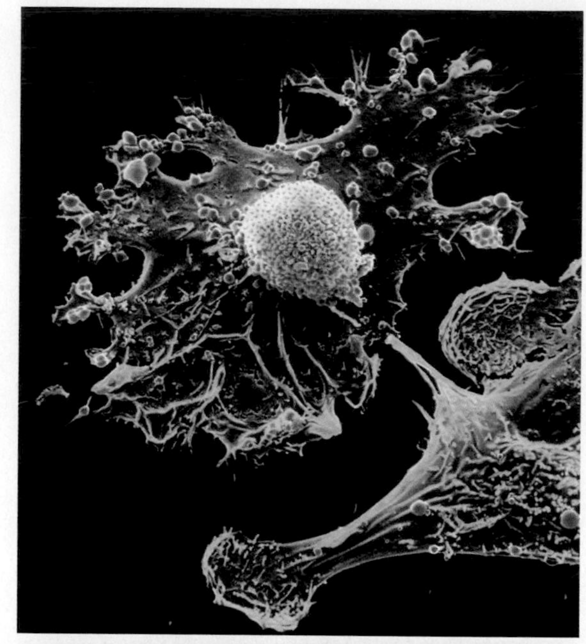

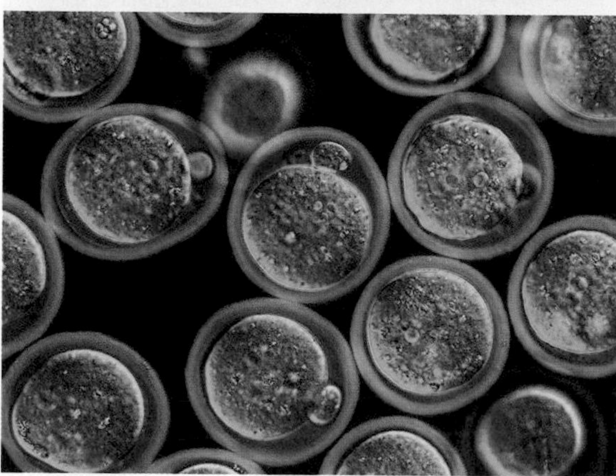

Figure 6.1

Two biological materials for study

To study the hormone estrogen, a research team grew human breast cancer cells (top). In a separate study, on hormone levels and chromosomes, a researcher examined immature mouse eggs (bottom).

Soto, Sonnenschein, and Hunt didn't know it at the time, but their botched experiments would change the course of their scientific careers forever. The three would spend the next decade identifying, tracking, and investigating a toxic chemical that pervades our environment.

Divide and Conquer

In Soto and Sonnenschein's experiment, the breast cancer cells were multiplying under the wrong circumstances. In Hunt's experiment, the mouse oocytes were not producing egg cells correctly. In both cases, something was interfering with the ability of the cells to divide—disrupting the cell cycle.

The **cell cycle** is a sequence of events that makes up the life of a typical eukaryotic cell, from the moment of its origin to the time it divides to produce two daughter cells. The time it takes to complete a cell cycle depends on the organism, the type of cell, and the life stage of the organism. Human cells, for example, typically have a 24-hour cell cycle, whereas mouse oocytes can take days to complete a cycle. Some fly embryos, by contrast, have cell cycles that are only 8 minutes long.

The two main stages in the cell cycle of eukaryotes are interphase and cell division. Each stage is marked by distinctive cell activities (**Figure 6.2**). **Interphase** is the longest stage of the cell cycle; most cells spend 90 percent or more of their life span in interphase. During this phase, a cell takes in nourishment, manufactures proteins and other substances, expands in size, and conducts special functions depending on the cell type. Neurons in the brain transmit electrical impulses, for example, whereas beta cells in the pancreas release the hormone insulin.

Interphase can be divided into three main intervals: G_1, S, and G_2. The **G_1 phase** (for "gap 1") is the first phase of interphase. In cells that are destined to divide, preparations for cell division begin during the next phase of interphase, **S phase** ("S" stands for "synthesis"). A critical event during the S phase is the copying, or *replication*, of all the cell's DNA molecules, which contain the organism's genetic information. The last phase of interphase, **G_2 phase** (for "gap 2"), occurs after the S phase but before the start of cell division. Cells make their final preparations for cell division during G_2.

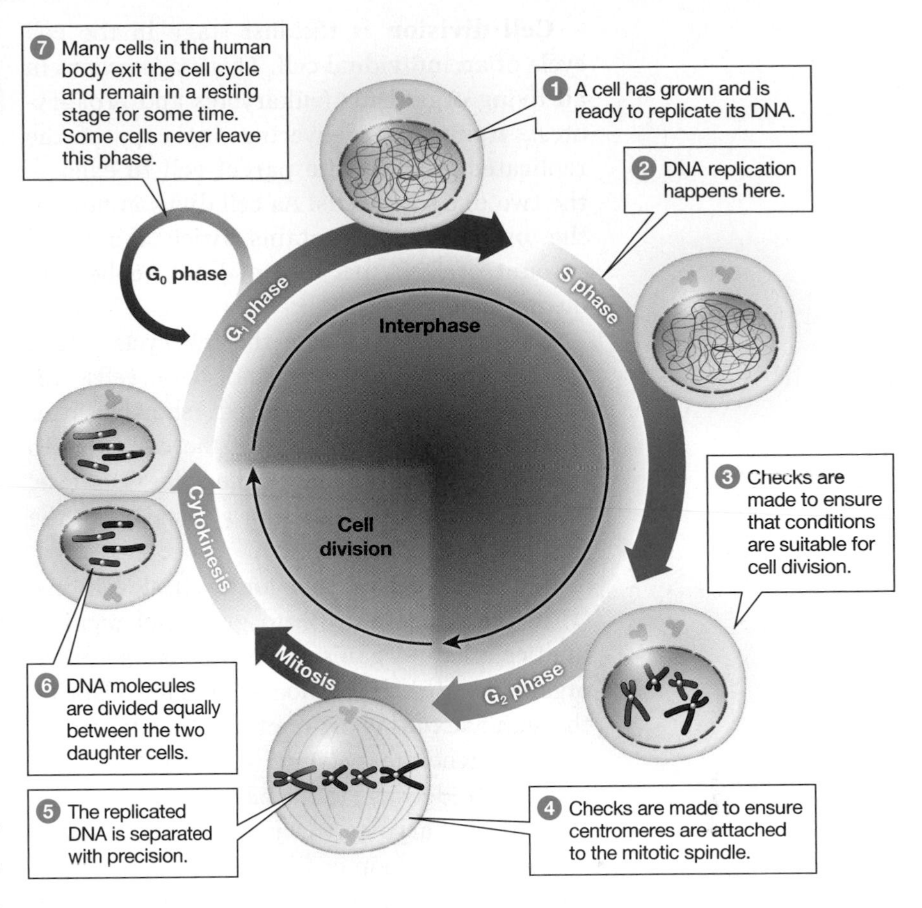

7 Many cells in the human body exit the cell cycle and remain in a resting stage for some time. Some cells never leave this phase.
1 A cell has grown and is ready to replicate its DNA.
2 DNA replication happens here.
G_0 phase
G_1 phase
S phase
Interphase
Cytokinesis
Cell division
3 Checks are made to ensure that conditions are suitable for cell division.
Mitosis
G_2 phase
6 DNA molecules are divided equally between the two daughter cells.
5 The replicated DNA is separated with precision.
4 Checks are made to ensure centromeres are attached to the mitotic spindle.

Figure 6.2

The cell cycle

The eukaryotic cell cycle consists of two major stages: interphase and cell division. DNA is in the chromatin state throughout interphase (G_1, S_1, and G_2) but is depicted as condensed chromosomes in G_2 to illustrate that the chromosomes have replicated during S phase.

Q1: When is DNA replicated during the cell cycle?

Q2: When in the cell cycle does the cell separate into two genetically identical daughter cells?

Q3: If a cell does not complete the cell cycle, what phase does it enter? Is this part of the cell cycle?

See Appendix A for answers to the figure questions.

Early cell biologists bestowed the term "gap" on the G_1 and G_2 phases because they believed them to be less significant than the S phase and cell division. We now know that the "gap" phases are often periods of growth during which both the size of the cell and its protein content increase. Furthermore, each "gap" phase serves as a checkpoint to prepare the cell for the phase immediately following it, ensuring that the cell cycle does not progress unless all conditions are suitable.

Cell division is the last stage in the cell cycle of an individual cell. This stage occurs in all living organisms (eukaryotes and prokaryotes). It involves delivering one copy of the replicated DNA in the parent cell to each of the two daughter cells. As cell division begins, the original cell contains twice the usual amount of DNA because of DNA replication during the S phase.

Not all cells complete the cell cycle. Many types of cells—neurons and beta cells, for example—become specialized shortly after entering G_1. They pull out of the cell cycle to enter a nondividing state called the G_0 phase. The **G_0 phase** can last from a few days to the lifetime of the organism.

Cells begin cell division for two basic reasons: (1) to reproduce and (2) to grow and repair a multicellular organism. Most single-celled organisms use cell division to produce offspring through **asexual reproduction.** Asexual reproduction generates *clones,* offspring that are genetically identical to the parent.

Most prokaryotes carry their genetic material in just one loop of DNA. These cells reproduce through **binary fission** (**Figure 6.3**). In this type of cell division, the prokaryotic cell simply copies the circular chromosome and each daughter cell receives one copy of the DNA loop—resulting in two cells that are genetically identical.

Soto, Sonnenschein, and Hunt, however, were studying eukaryotic cells, where cell division is more complicated than binary fission. Instead of being in one loop, eukaryotic DNA consists of multiple, distinct linear chromosomes wrapped around proteins and coiled into fibers that have to be unwound, replicated, and equally distributed between the two daughter cells. In addition to this complication, eukaryotic DNA lies inside a nucleus, enclosed by the double layer of membranes that makes up the nuclear envelope. In most eukaryotes, the nuclear envelope must be disassembled in the dividing cell and then reassembled in each of the daughter cells toward the end of cell division.

To make things even more complicated, eukaryotic cells actually undergo two types of division, depending on the cell type: asexual reproduction through *mitosis,* and sexual

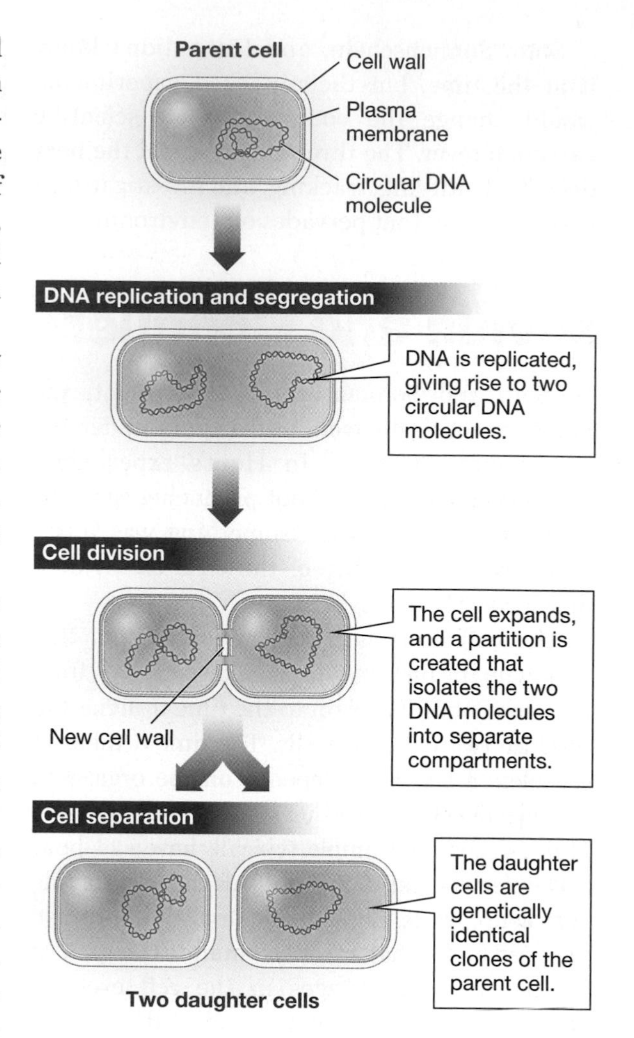

Figure 6.3

Cell division in a prokaryote

Many prokaryotes reproduce asexually in a type of cell division known as binary fission.

Q1: When single-celled organisms go through cell division, why is it called asexual reproduction?

Q2: What is asexual reproduction referred to specifically in prokaryotes?

Q3: Although prokaryotes do not have an "interphase" during cell division like eukaryotes do, they must perform one of the main processes of eukaryotic interphase. What is this process in prokaryotes that must occur before the cell physically divides in two?

See Appendix A for answers to the figure questions.

reproduction—the production of sperm and eggs—through *meiosis*.

Trade Secret

Back in Massachusetts, Soto and Sonnenschein spent 4 months trying to figure out why their experiment had stopped working—how unknown estrogen was getting into their cell culture flasks and causing the cells to divide. By trial and error, they determined that a compound seemed to be shedding from the walls of the plastic tubes in which they stored the liquid being added to the cell culture flasks.

They called the tube manufacturer to confirm that an ingredient was added to make the tubes more impact resistant. But the company refused to reveal the identity of the "trade secret" ingredient. Soto and Sonnenschein spent a year purifying the secret ingredient and finally identified a compound called nonylphenol, a chemical used to make detergents and hard plastics. The reason for all their problems became clear: nonylphenol mimics the action of estrogen.

Like estrogen, nonylphenol activates **mitotic division**, a type of cell division that generates two genetically identical daughter cells from a single parent cell in eukaryotes. Mitotic division consists of two steps that occur after interphase: mitosis and cytokinesis. The first step, **mitosis**, is the division of the copied chromosomes in the nucleus. Mitosis includes four main phases: *prophase*, *metaphase*, *anaphase*, and *telophase*. Each phase is defined by easily identifiable events (**Figure 6.4**).

A parent cell sets up for an upcoming mitotic division by replicating its DNA during the S phase of interphase, well before mitosis begins. During the G_1, S, and G_2 phases, the cell's DNA, in the nucleus, is not packed tightly. The double-stranded DNA remains loosely associated with proteins, and each is referred to as a chromatin fiber. The chromatin fibers must be accessible for conducting the business of the cell and for replication. When chromatin fibers are replicated—during the S phase of interphase—two identical chromatin fibers, now called **sister chromatids,** are produced. These sister chromatids are firmly attached at a precise region of the

Cancer: Uncontrolled Cell Division

Cancer accounts for over 600,000 deaths in the United States each year—nearly one in every four deaths. Only heart disease kills more people. Over the course of a lifetime, an American male has a nearly one in two chance of being diagnosed with cancer; American women fare slightly better, with a one in three chance of developing cancer. There are more than 200 different types of cancer, but the big four—lung, prostate, breast, and colon cancers—combine to account for about half of all cancers. More than 15 million Americans alive today have been diagnosed with cancer and are either in remission or undergoing treatment. The National Cancer Institute estimates that the collective price tag for treating the various forms of cancer is more than $145 billion per year.

Every cancer begins with a single cell that goes rogue and starts dividing without the checkpoints of a normal cell. This runaway cell division rapidly creates a cell mass known as a **tumor**. Tumors that remain confined to one site are **benign**. Because benign tumors can usually be surgically removed, they are generally not a threat to the patient's survival. However, an actively growing benign tumor is like a cancer-in-training. Because these tumor cells are not subject to the monitoring that occurs at checkpoints during the cell cycle of a normal cell, their descendants can become increasingly abnormal—changing shape, increasing in size, and ultimately ceasing normal cell functions. As tumor cells progress toward a cancerous state, they begin secreting substances that cause **angiogenesis**, the formation of new blood vessels. The resulting increase in blood supply to the tumor is important for delivering nutrition to it and whisking waste away from it, allowing the tumor to grow larger.

Most cells in the adult animal body are firmly anchored in one place and will stop dividing if they are detached from their surroundings. This phenomenon is known as **anchorage dependence**. But some tumor cells may acquire anchorage *in*dependence, the ability to divide even when released from their attachment sites. When tumor cells gain anchorage independence and start invading other tissues, they are transformed into **cancer cells**, also known as **malignant cells**. Cancer cells may break loose from their attachment sites and enter blood vessels or lymph vessels to emerge in distant locations throughout the body, where they form new tumors. The spread of a disease from one organ to another is known as **metastasis**. Metastasis typically occurs at later stages in cancer development. Once a cancer has metastasized to form tumors in multiple organs, it may be very difficult to fight.

Cancer cells multiply rapidly wherever they establish themselves, overrunning neighboring cells, monopolizing oxygen and nutrition, and starving normal cells in the vicinity. Without restraints on their growth and migration, cancer cells steadily destroy tissues, organs, and organ systems. The normal function of these organs is then seriously impaired, and cancer deaths are ultimately caused by the failure of vital organs.

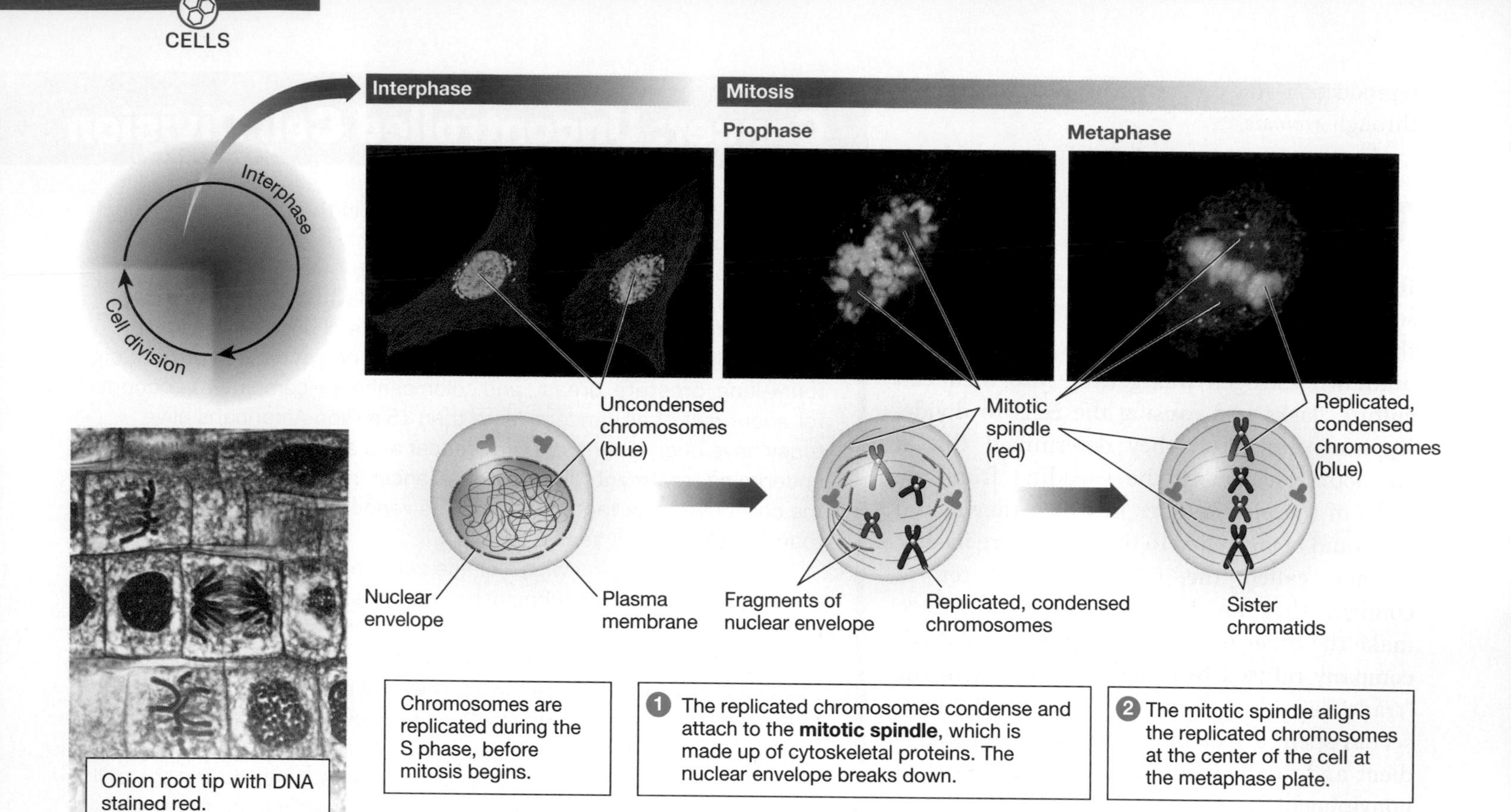
Interphase
Mitosis
Prophase
Metaphase
Interphase
Cell division
Uncondensed chromosomes (blue)
Mitotic spindle (red)
Replicated, condensed chromosomes (blue)
Nuclear envelope
Plasma membrane
Fragments of nuclear envelope
Replicated, condensed chromosomes
Sister chromatids
Chromosomes are replicated during the S phase, before mitosis begins.
1 The replicated chromosomes condense and attach to the **mitotic spindle**, which is made up of cytoskeletal proteins. The nuclear envelope breaks down.
2 The mitotic spindle aligns the replicated chromosomes at the center of the cell at the metaphase plate.
Onion root tip with DNA stained red.

Figure 6.4

Cell division in a eukaryote

Mitotic cell division is composed of two main stages: mitosis (with four substages) and cytokinesis.

Q1: The image at far left in the figure is a cross section of an onion root tip stained to display the chromosomes. From the arrangement of the chromosomes, it is clear that all of these cells are not in the same stage of the cell cycle. What stages do you see, and how do you know?

Q2: What happens between the end of interphase and early prophase that changes the appearance of the chromosomes?

Q3: Explain in your own words the role of the mitotic spindle in mitosis.

See Appendix A for answers to the figure questions.

chromatids called the **centromere** (**Figure 6.5**). As cell division begins—during prophase of mitosis—each sister chromatid with its tightly attached proteins pack into a chromosome for cell division. This packing, or condensing, is necessary because every chromatin fiber is enormously long, even in the simplest eukaryotic cell. Unpacked chromatin fibers would be a tangled mess, prone to breakage.

One of the main objectives of mitosis is to separate those sister chromatids. During anaphase, they are pulled apart at the centromere, and one of each is delivered to the opposite ends of the parent cell. Eukaryotic cells have evolved an elaborate choreography to minimize the risk of mistakes during the equal and symmetrical separation of the replicated genetic material. Normally, no daughter cell winds up lacking a chromosome, nor does it gain a duplicate. Unless an error occurs, each daughter cell inherits the same genetic information that the parent cell had in the G_1 phase of its life.

After the replicated DNA has been divided in two, half to each end of the parent cell, the cytoplasm is divided by a process called **cytokinesis** ("cell movement"). In animal cells, the plasma membrane pinches together at the midline to break into two separate cells. By contrast,

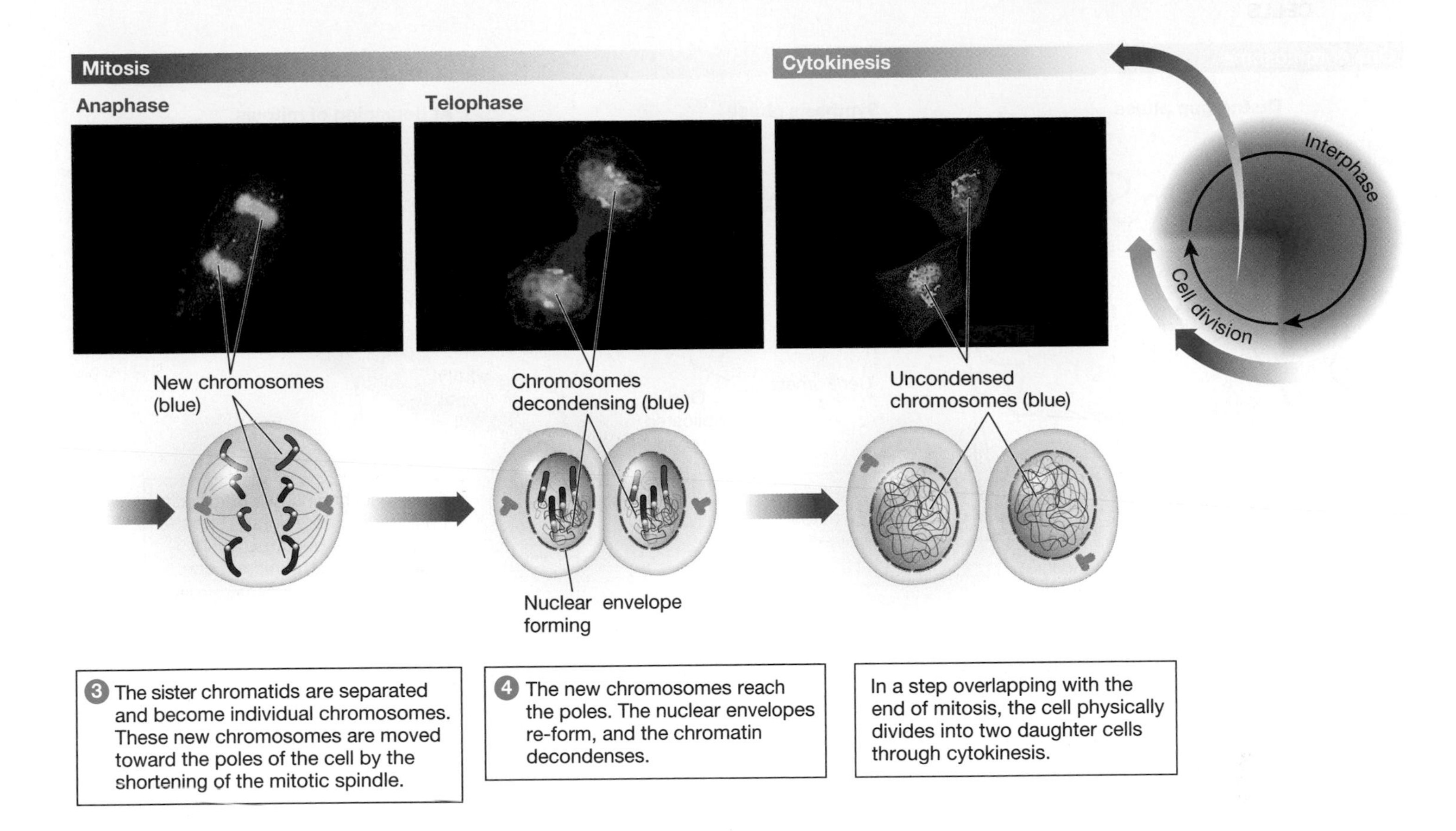

Mitosis
Anaphase
Telophase
Cytokinesis
Interphase
Cell division
New chromosomes (blue)
Chromosomes decondensing (blue)
Uncondensed chromosomes (blue)
Nuclear envelope forming
3 The sister chromatids are separated and become individual chromosomes. These new chromosomes are moved toward the poles of the cell by the shortening of the mitotic spindle.
4 The new chromosomes reach the poles. The nuclear envelopes re-form, and the chromatin decondenses.
In a step overlapping with the end of mitosis, the cell physically divides into two daughter cells through cytokinesis.

plant cells build new plasma membranes at the midline with cell walls in between. Both cytokineses give rise to two self-contained daughter cells that are clones of each other.

Mitotic division can serve both the eukaryotic organism's need to replace itself (to reproduce) and its need to add new cells to its body. Many multicellular eukaryotes use mitotic division to reproduce asexually, including seaweeds, fungi, and plants, and some animals, such as sponges and flatworms. All multicellular organisms also rely on mitotic division for the growth of tissues and organs and the body as a whole and for repairing injured tissue and replacing worn-out cells. Mitosis is why children grow taller and why skin closes over a cut.

Good Cells Gone Bad

Cell division is not always a good thing. Runaway cell division can create a tumor (see "Cancer: Uncontrolled Cell Division" on page 113). In a developing organism, it can also cause an organ such as the heart or liver to form incorrectly and not function properly. It is little wonder, then, that the cell cycle is carefully controlled in healthy individuals.

The commitment to proceed through the cell cycle and divide is made during the G_1 phase in response to internal and external signals. In humans, external signals that influence the commitment include hormones and proteins called growth factors. When some hormones and growth factors are present, they act like the gas pedal in a car and push a cell toward cell division. If these hormones and growth factors are not present, there is no cell cycle progression or cell division.

After a cell enters the cell cycle, special cell cycle regulatory proteins are activated. These

DEBUNKED!

MYTH: Adults can't grow new brain cells.

FACT: In 2013, researchers in Europe found—and others have since confirmed—that the adult human brain structure associated with the formation of memories (called the hippocampus) produces new neurons via mitosis throughout life, even in old age. In addition, exercise, sex, and stress relief seem to enhance the number of newly made neurons.

One chromosome
During gap phase
Synthesis phase
At beginning of mitosis
One replicated chromosome
Centromere
Centromere
Two sister chromatids
One replicated chromosome
Single chromatin fiber.
Each chromatin fiber is replicated during this phase, before mitosis.
Two sister chromatids
As mitosis begins, the chromatin packs even more tightly, and the chromosome becomes easily visible with a light microscope.

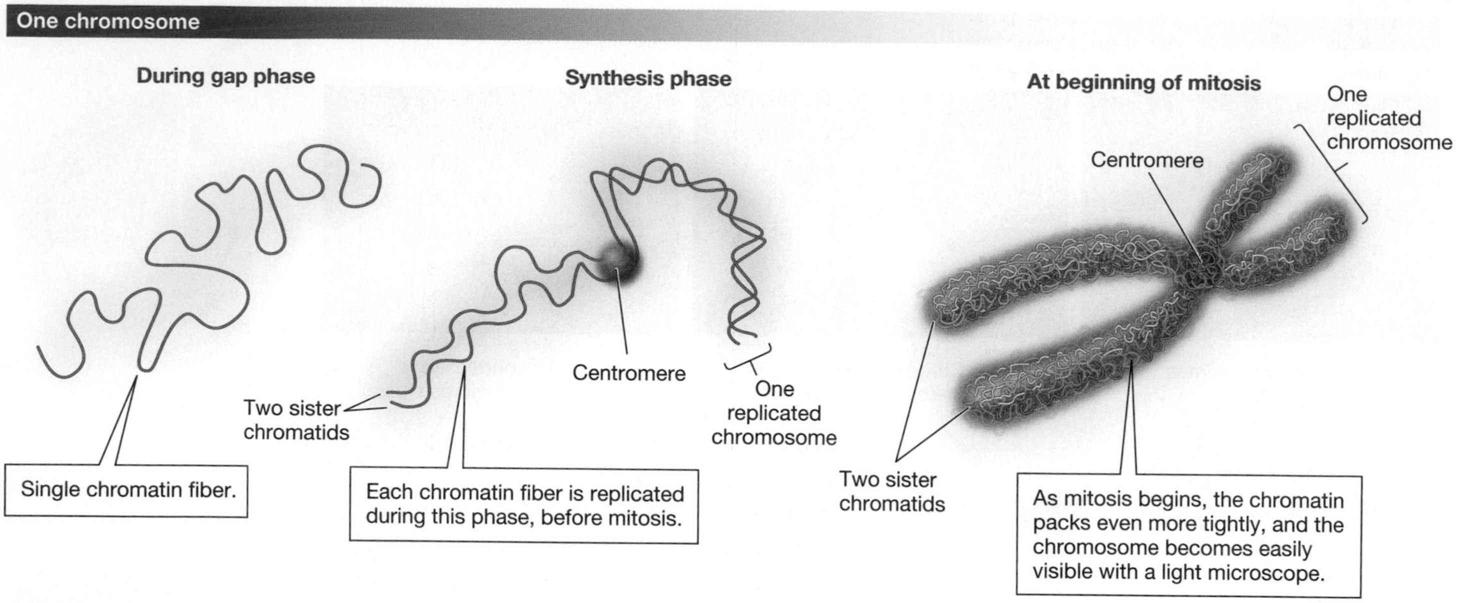

Figure 6.5

Chromosomes are copied and condensed in preparation for cell division

Chromosomes spend the majority of the cell cycle unpacked (left). Each of these chromosomes will be copied, or replicated, during synthesis (middle). They are tightly packed, or condensed, during early mitosis (right).

Q1: Why is it important for a chromosome to be copied before mitosis?

Q2: Are sister chromatids attached at the centromere considered to be one or two chromosomes?

Q3: Why is the chromosome's DNA tightly packed and condensed with proteins for mitosis and cytokinesis instead of remaining exceptionally long and loose?

See Appendix A for answers to the figure questions.

proteins "throw the switch" that enables the cell to pass through critical checkpoints and proceed from one phase of the cell cycle to the next (**Figure 6.6**). For example, upon receiving the appropriate signals, cell cycle regulatory proteins at the G_1 checkpoint advance a cell from the G_1 phase to the S phase by triggering chromosome replication and other processes associated with it.

Cell cycle regulatory proteins also respond to negative internal or external control signals. Internal signals will pause a cell at the G_1 checkpoint under any of the following conditions: the cell is too small, nutrition is inadequate, or the cell's DNA is damaged. The G_2 checkpoint pauses under the same circumstances, as well as when chromosome duplication in the S phase is incomplete for any reason.

Nonylphenol interferes with the G_1 checkpoint. In essence, it gives the cell a green light to enter the cell cycle at a time when the cell would not normally divide. So, when Soto and Sonnenschein realized that nonylphenol enables human breast cells to divide inappropriately, they became concerned. If nonylphenol was being used in everyday plastics, it was possible that healthy human cells were exposed to it on a regular basis.

In a 1991 paper detailing their discovery, Soto and Sonnenschein wrote that nonylphenol might be interfering with science experiments like theirs and that, even more important, it could be harmful to humans. "From the very beginning, we realized that this could be a health problem," says Sonnenschein.

Unequal Division

In Ohio, Patricia Hunt was having no success determining why her control mouse eggs divided abnormally, with either too many or too

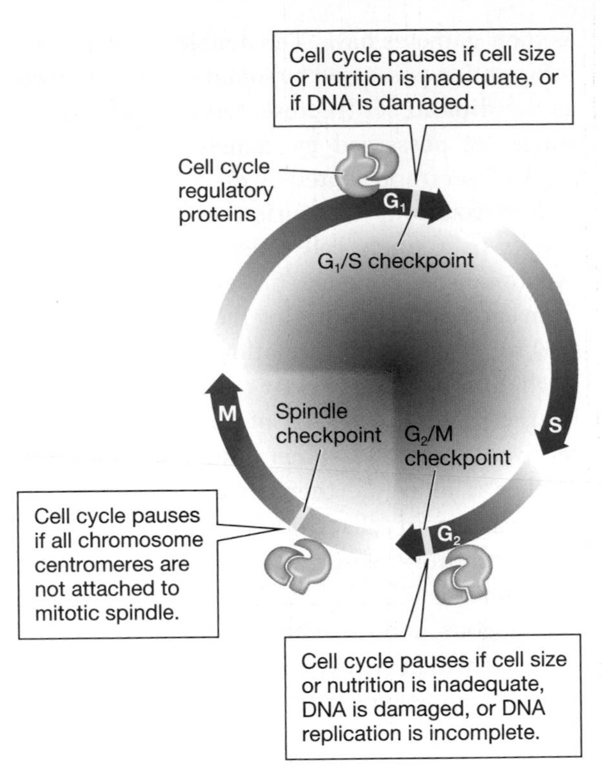

Figure 6.6

The cell cycle must pass through checkpoints to proceed

Three checkpoints operate during the cell cycle.

Q1: What could happen if the cell's checkpoints are disabled?

Q2: What is the advantage of stopping the cell cycle if the cell's DNA is damaged?

Q3: Which cell cycle checkpoint may have been influenced in Soto and Sonnenschein's breast cancer cell experiments?

See Appendix A for answers to the figure questions.

few chromosomes. After searching in vain for months, one day she noticed that something was wrong with the plastic mouse cages: the water bottles were leaking, and the plastic cage walls were hazy.

She asked around and found out that months earlier a substitute janitor had used detergent with a high pH, instead of the normal low-pH detergent, to clean the cages and bottles. A detergent with a higher pH—that is, a stronger base—may have more cleaning power but also be more corrosive. With some lab work, she identified a chemical oozing from the corroded plastic. "It took essentially one washing with the wrong detergent," says Hunt, "and once it was damaged, [the plastic] started to leach bisphenol A."

Bisphenol A, or BPA, was first synthesized in 1891. It is a synthetic hormone that, like nonylphenol, mimics estrogen. In the 1930s, BPA was tested by clinicians seeking a hormone replacement therapy for women who needed estrogen, but it wasn't as effective as other substitutes. In the 1940s and 1950s, the chemical industry found another use for BPA, as a chemical component of a clear, strong plastic. Manufacturers began to incorporate it into eyeglass lenses, water and baby bottles, the linings of food and beverage cans, and more.

Unfortunately, as Hunt found out, BPA doesn't necessarily stay in those products. Not all the BPA used to make plastics gets locked into chemical bonds, and what doesn't get locked into bonds can break free, or leach out—especially if the plastic is heated, such as when a baby bottle is sterilized, or if the plastic is exposed to a harsh chemical, as Hunt's mouse cages were. Because of its prevalence in products and its ability to leach out of them, BPA is one of the most common chemicals we are exposed to in everyday life.

To confirm the hypothesis that BPA caused the mouse egg abnormalities, Hunt's team recreated the original event. They intentionally damaged a set of new cages and put healthy female mice in them. They also had the mice drink from damaged water bottles. Later, when they examined the eggs of these mice, they saw the same toxic effects as before: 40 percent of the eggs had abnormal chromosomes. The eggs showed that errors in meiosis had occurred.

Meiosis is a specialized type of cell division. It kicks off **sexual reproduction**, the process by which genetic information from two individuals is combined to produce offspring. Sexual reproduction has two steps: cell division through meiosis and fertilization. BPA was affecting the first of these steps—meiosis.

As you learned earlier, mitosis produces daughter cells with the same number of chromosomes as the parent cell. These non–sex cells are called **somatic cells**. In contrast, meiosis produces **gametes**, daughter cells that have half the chromosome count of the parent cell. The differences between mitosis and meiosis are the reason why the somatic cells of plants and animals have twice as much genetic information

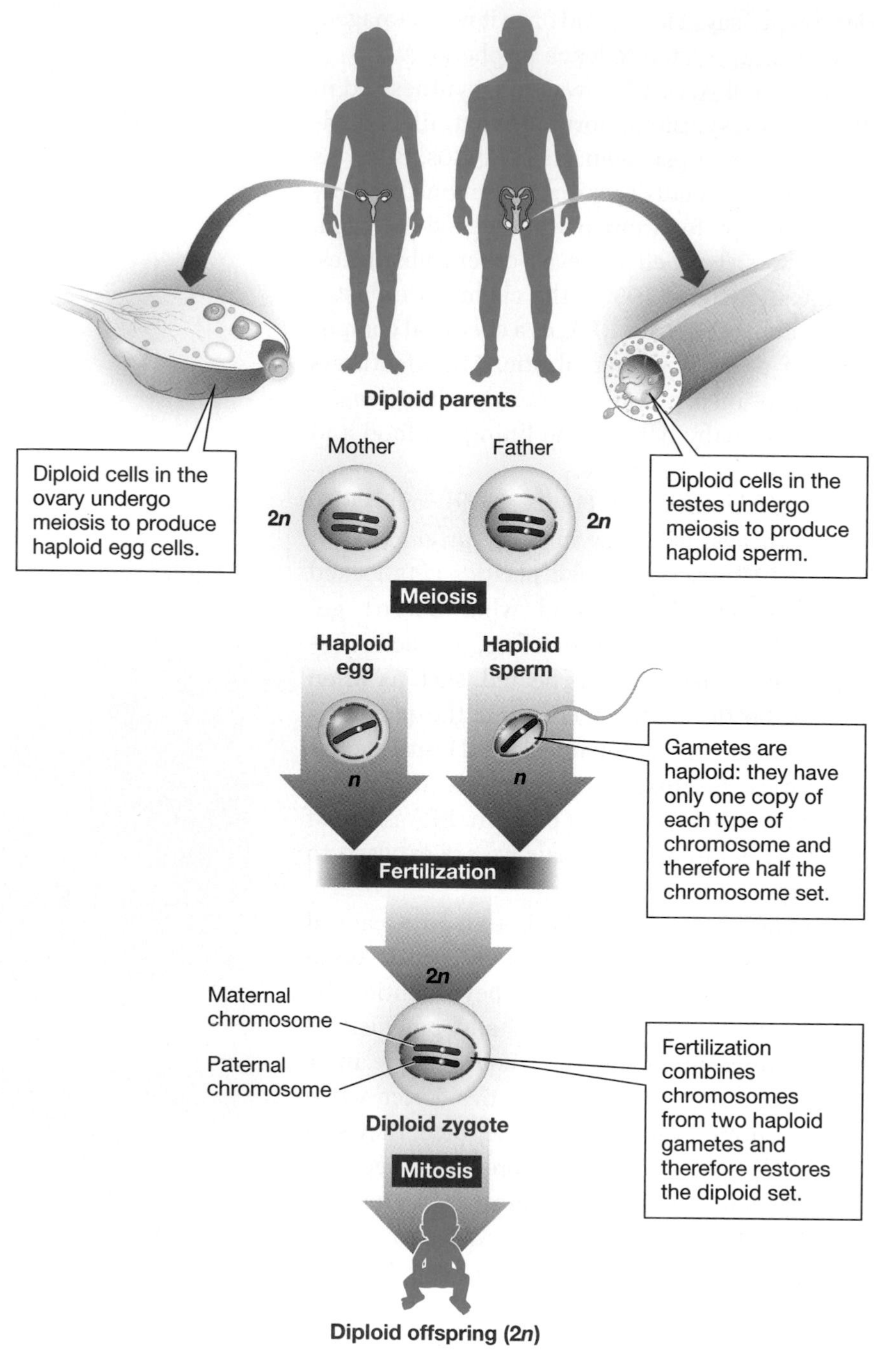
Diploid parents
Mother
Father
Diploid cells in the ovary undergo meiosis to produce haploid egg cells.
2n
2n
Diploid cells in the testes undergo meiosis to produce haploid sperm.
Meiosis
Haploid egg
Haploid sperm
n
n
Gametes are haploid: they have only one copy of each type of chromosome and therefore half the chromosome set.
Fertilization
2n
Maternal chromosome
Paternal chromosome
Fertilization combines chromosomes from two haploid gametes and therefore restores the diploid set.
Diploid zygote
Mitosis
Diploid offspring (2n)

Figure 6.7

Fertilization creates a zygote from the fusion of two gametes

In species with two sexes, female gametes are *eggs* and male gametes are *sperm*. This figure shows only one of the 23 homologous pairs found in human cells.

Q1: Is a zygote haploid or diploid?

Q2: Which cellular process creates a baby from a zygote?

Q3: How might long-term, significant exposure to BPA experienced by a mother or father prior to conceiving a child explain potential birth defects in the fetus?

See Appendix A for answers to the figure questions.

as their gametes have. The double set of genetic information possessed by somatic cells is known as the **diploid** set (represented by $2n$), and the single set possessed by gametes is called the **haploid** set (represented by n).

Fertilization, the fusion of two gametes, results in a single cell called the **zygote**. The zygote inherits a haploid (n) set of chromosomes from each gamete, restoring the complete diploid ($2n$) set of genetic information to the offspring. Each **homologous pair** of chromosomes in the zygote consists of one chromosome received from the father and one from the mother. The zygote then divides by mitosis to create a mass of cells that will eventually develop into a mature organism (**Figure 6.7**). All cells in a mature organism are diploid, containing homologous pairs of chromosomes, except for gametes, which are haploid and contain only one of each homologous pair.

The two stages of meiosis occur after interphase. Recall that during the S phase of interphase, the cell's DNA is replicated. **Meiosis I** reduces the chromosome number to haploid by separating one of each homologous pair into two different daughter cells. Each homologous chromosome lines up with its partner and then separates to the two ends of the cells, and cytokinesis occurs. These two cells now proceed through **meiosis II**; the remaining chromosomes separate and each sister chromatid moves into the two ends of the cells, cytokinesis occurs, and two different daughter cells result from each of the daughter cells of meiosis I. Unlike meiosis I, the phases of the division cycle of meiosis II are almost exactly like those of mitosis: sister chromatids separate to the two ends of the cells and cytokinesis occurs, leading to an equal segregation of chromatids into two new daughter cells (**Figure 6.8**).

In summary, meiosis I produces two haploid cells (n). Each of these cells includes one of each pair of duplicated chromosomes. In meiosis II, the sister chromatids separate in these two haploid cells. Then the cells divide, producing a total of four haploid cells (n) with unduplicated chromosomes. Each haploid gamete now has half of the chromosome set found in the original diploid cell ($2n$) that underwent meiosis.

BPA is toxic because it disrupts the process of meiosis; it hinders the ability of the chromosomes to separate into four haploid cells. Hunt realized that if BPA was disrupting meiosis in

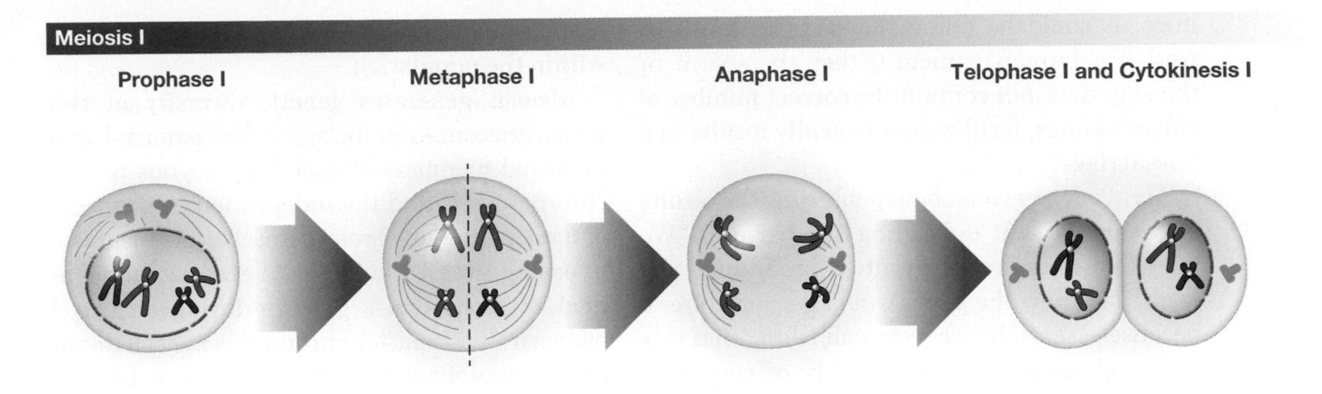

1. Each replicated chromosome pairs with its partner, or homologue.
2. Homologous chromosome pairs line up at the metaphase plate, aided by the spindle fibers.
3. The paternal and maternal homologous chromosomes separate through the shortening of the spindle fibers.
4. The first cytokinesis takes place, producing two haploid cells.

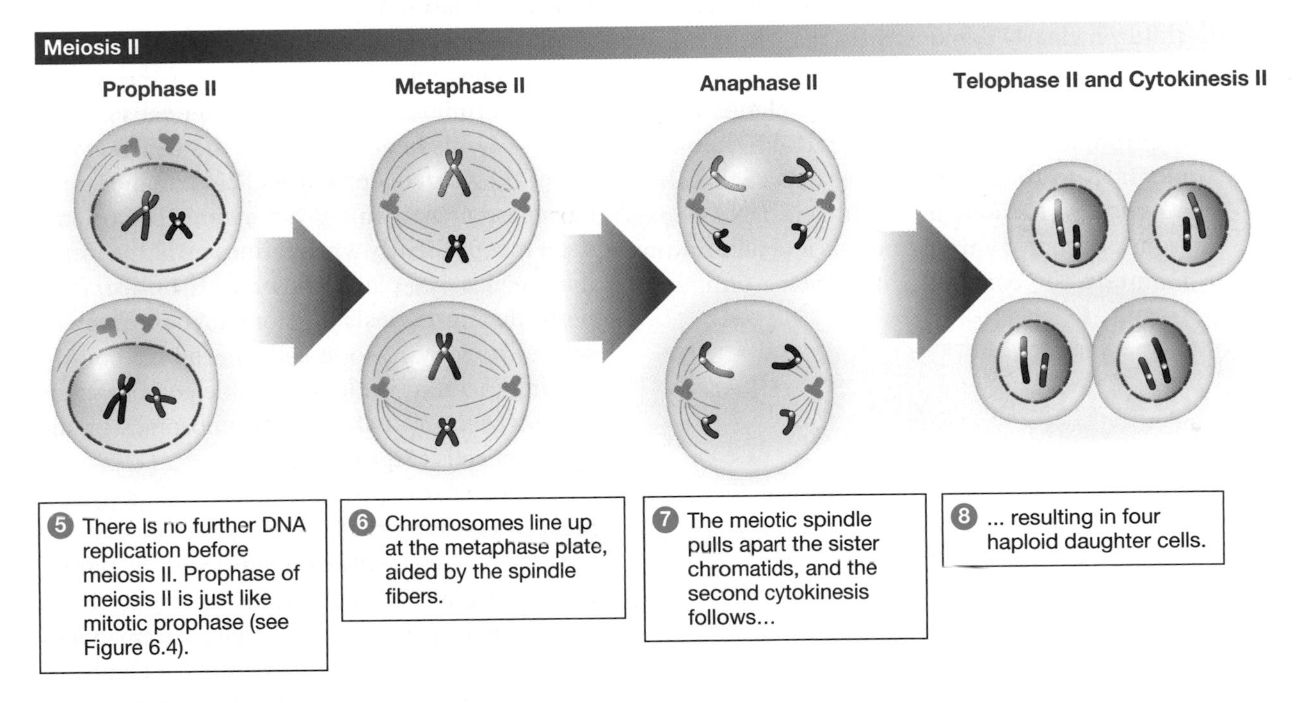

5. There is no further DNA replication before meiosis II. Prophase of meiosis II is just like mitotic prophase (see Figure 6.4).
6. Chromosomes line up at the metaphase plate, aided by the spindle fibers.
7. The meiotic spindle pulls apart the sister chromatids, and the second cytokinesis follows...
8. ... resulting in four haploid daughter cells.

Figure 6.8

Meiosis with cytokinesis creates haploid daughter cells

After interphase, meiosis I occurs: the homologous chromosomes of a diploid cell are paired and then separated into two haploid cells. During meiosis II, sister chromatids of these haploid cells are each separated into two haploid cells. Cytokinesis occurs at the end of meiosis I and at the end of meiosis II.

Q1: Is a daughter cell haploid or diploid after the first meiotic division? How about after the second meiotic division?

Q2: What is the difference between homologous chromosomes and sister chromatids?

Q3: If the skin cells of house cats contain 19 homologous pairs of chromosomes, how many chromosomes are present in the egg cells they produce?

See Appendix A for answers to the figure questions.

mice, it could be doing the same in humans. And if a human gamete (either the sperm or the egg) does not contain the correct number of chromosomes, fertilization typically results in a miscarriage.

Hunt was nervous about publishing the results of her experiment and calling BPA harmful. "We knew we were stepping onto a landmine," she says. "We knew the paper would get some press, because essentially we were publishing that this chemical—used in a wide variety of consumer products and that we are probably all exposed to—can cause an increased risk of miscarriage and babies with birth defects."

Shuffling the DNA

A glance at a pair of parents with their biological children clearly demonstrates that the offspring resulting from sexual reproduction are similar to their parents. But unlike the clones resulting from asexual reproduction, children are not identical to their parents. Because each half of a sexually reproducing organism's DNA comes from a different parent, meiosis and fertilization maintain the constant chromosome number of a species while allowing for genetic diversity within the population.

Meiosis generates genetic diversity in two ways: crossing-over between the paternal and maternal members of each homologous pair of chromosomes, and the independent assortment of these paired-up chromosomes during meiosis I. **Crossing-over** is the physical exchange of identical chromosomal segments during meiosis I between the nonsister chromatids in each duplicated homologous pair. These nonsister chromatids make physical contact at random sites along their length, and each exchange the exact same segments of DNA (**Figure 6.9**). The chromatids are said to be *recombined*, and the exchange of DNA segments is known as **genetic recombination**. Without crossing-over, every chromosome inherited by a gamete would be just the way it was in the parent cell.

The **independent assortment** of chromosomes—that is, the random distribution of the homologous chromosomes into daughter cells during meiosis I—also contributes to the genetic variety of the gametes produced. Each homologous chromosome pair in a given meiotic cell orients itself independently when it lines up at the imaginary equatorial plane known as the *metaphase plate* during meiosis I. Homologous chromosomes in every meiotic cell line up differently, leading to a multitude of possible combinations of maternal and paternal chromosomes in the daughter cells (**Figure 6.10**).

As with crossing-over, the independent assortment of chromosomes creates gametes that are likely to be genetically different from the parent and also from each other. Then, during fertilization, the fusion of two gametes adds a tremendous amount of genetic variation because it combines a one-in-a-million egg with a one-in-a-million sperm (actually, in terms of chromosomal variety in humans, it's 1 in 8.4 million for either gamete!). These three processes together—crossing-over, independent assortment, and fertilization—give each of us our genetic uniqueness.

What Can You Do?

There are things you can do to reduce your own risk of exposure to BPA. Today, the U.S. Food and Drug Administration recommends that individuals not put hot or boiling liquid intended for consumption in plastic containers made with BPA. (Some, but not all, plastics that are marked with the recycling code 3 or 7 may be made with BPA.) The organization also recommends discarding all bottles with scratches, which may harbor bacteria and, if the plastic contains BPA, may lead to greater release of the chemical.

"Get educated about your world," says Heather Patisaul, a BPA researcher at North Carolina State University. If you are concerned about BPA exposure, it is possible to cut it down by making lifestyle changes such as not eating canned food and not putting plastic in the microwave. "You can be empowered," says Patisaul. "Those types of things can effect great change."

Still, BPA and similar chemicals are ubiquitous in modern life, says Ana Soto. Ultimately, the best way to avoid them will be for government regulators to take a stand and outlaw the use of these chemicals in consumer products, she notes. She encourages individuals to contact their representatives and ask them to push legislation limiting the use of harmful chemicals, including BPA, in manufacturing.

Ten Years Later

Hunt and her team published their results in 2003. "We got a firestorm," she recalls with a grimace. The press reported the findings extensively, and many people and companies were

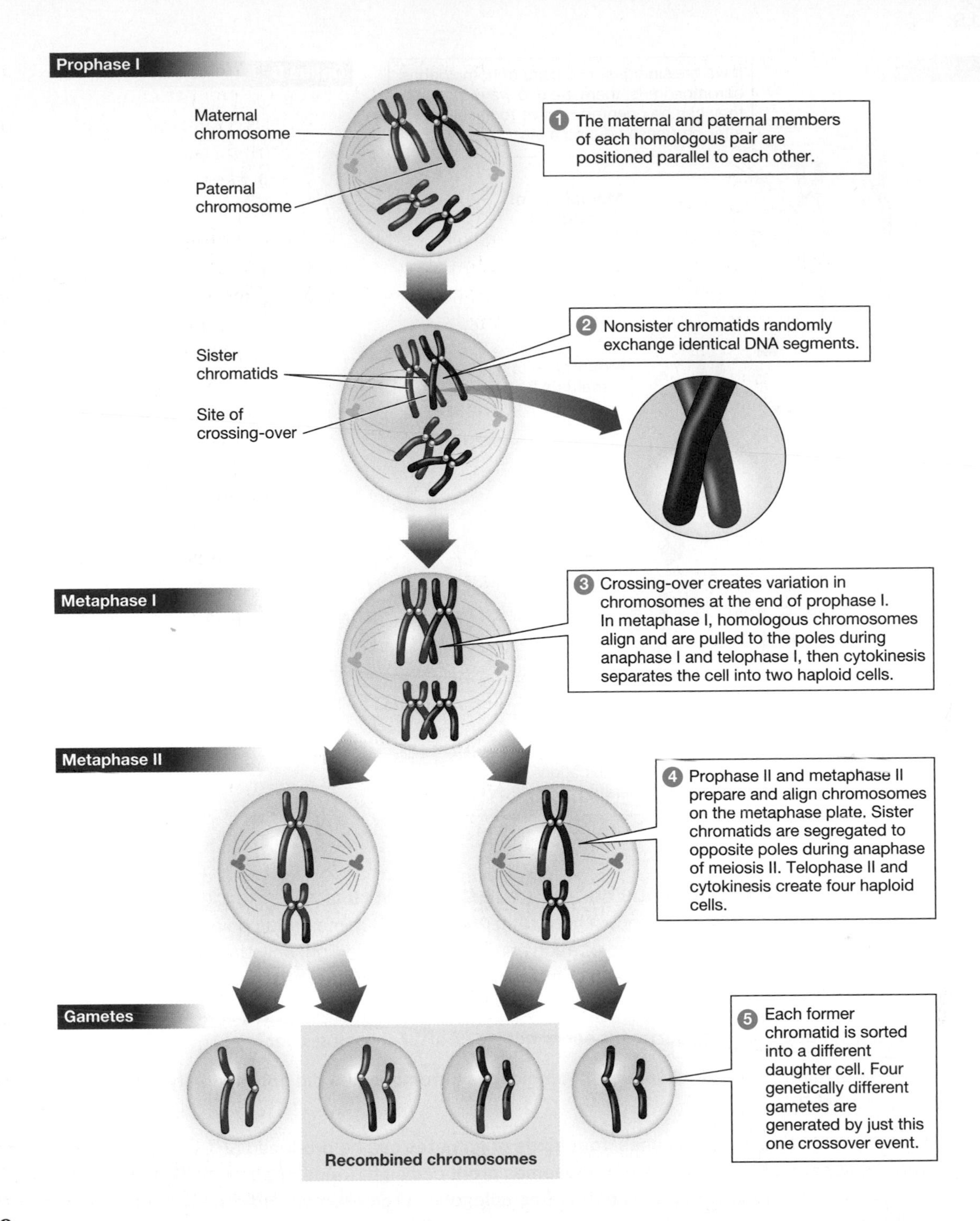

Prophase I
Maternal chromosome
Paternal chromosome
1 The maternal and paternal members of each homologous pair are positioned parallel to each other.
Sister chromatids
Site of crossing-over
2 Nonsister chromatids randomly exchange identical DNA segments.
Metaphase I
3 Crossing-over creates variation in chromosomes at the end of prophase I. In metaphase I, homologous chromosomes align and are pulled to the poles during anaphase I and telophase I, then cytokinesis separates the cell into two haploid cells.
Metaphase II
4 Prophase II and metaphase II prepare and align chromosomes on the metaphase plate. Sister chromatids are segregated to opposite poles during anaphase of meiosis II. Telophase II and cytokinesis create four haploid cells.
Gametes
Recombined chromosomes
5 Each former chromatid is sorted into a different daughter cell. Four genetically different gametes are generated by just this one crossover event.

Figure 6.9

Crossing-over produces chromosomes with new combinations of DNA

Only two maternal and two paternal chromosomes are depicted here, two pairs of homologous chromosomes, rather than the 23 pairs of homologous chromosomes found in humans. In addition, only the resulting cells of prophase I, metaphase I, metaphase II, and the resulting gametes of the process of meiosis are depicted here. The transitional steps are omitted.

Q1: Why is the term "crossing-over" appropriate for the exchange of DNA segments between homologous chromosomes?

Q2: At what stage of meiosis (I or II) does crossing-over occur?

Q3: What would be the effect of crossing-over between two sister chromatids?

See Appendix A for answers to the figure questions.

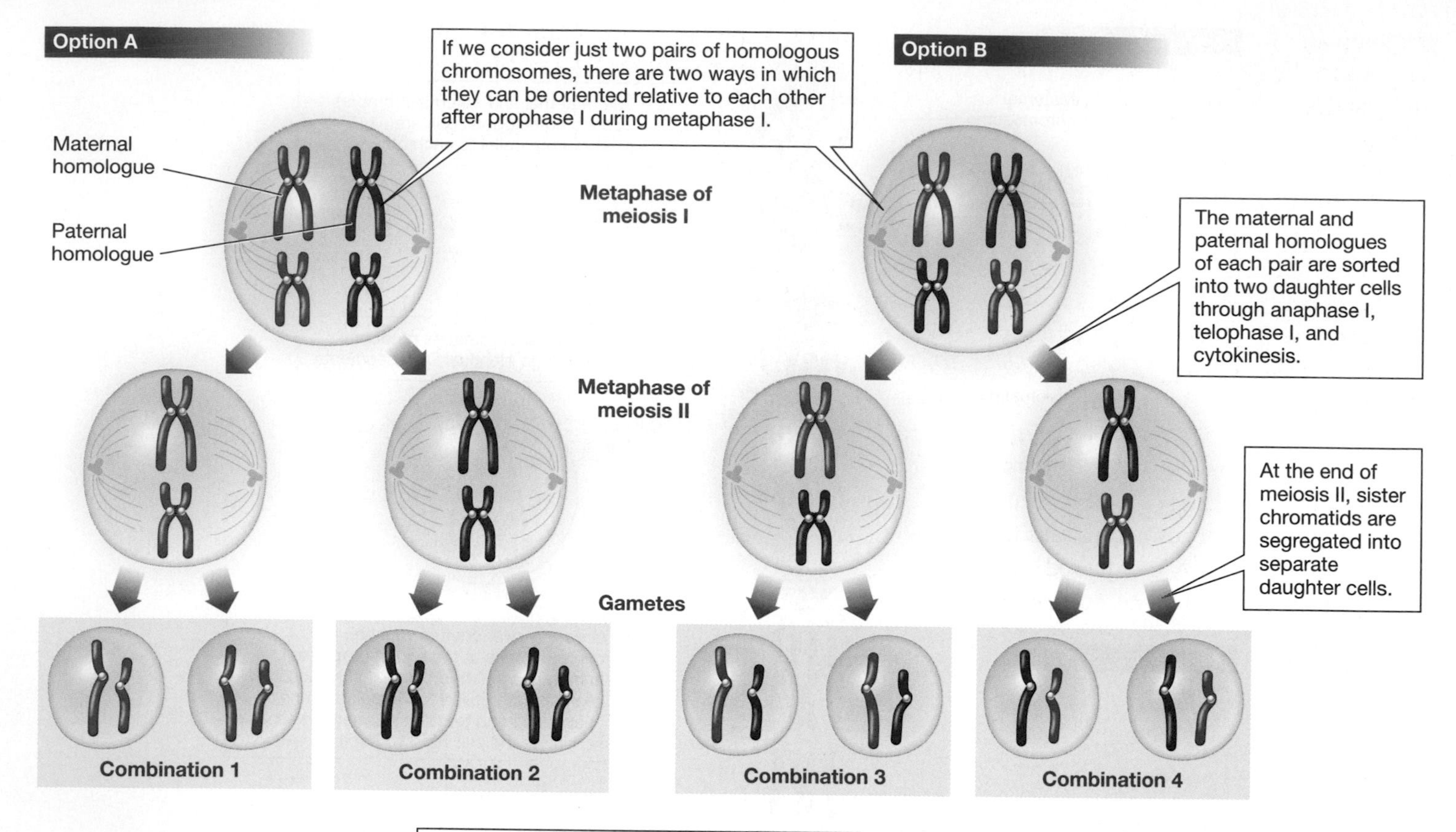
Option A
Option B
If we consider just two pairs of homologous chromosomes, there are two ways in which they can be oriented relative to each other after prophase I during metaphase I.
Maternal homologue
Paternal homologue
Metaphase of meiosis I
The maternal and paternal homologues of each pair are sorted into two daughter cells through anaphase I, telophase I, and cytokinesis.
Metaphase of meiosis II
At the end of meiosis II, sister chromatids are segregated into separate daughter cells.
Gametes
Combination 1
Combination 2
Combination 3
Combination 4
Four possible combinations of chromosomes in gametes generated by meiosis II.

Figure 6.10

The independent assortment of homologous chromosomes generates chromosomal diversity among gametes
Only two pairs of homologous chromosomes are shown here rather than the 23 homologous pairs in human cells. Each gamete will receive either a maternal or a paternal homologue of each chromosome. In addition, only the resulting cells of metaphase I, metaphase II, and the resulting gametes of the process of meiosis are depicted here. The transitional steps are omitted.

Q1: During meiosis, does independent assortment occur before or after crossing-over?

Q2: What would be the effect on genetic diversity if homologous chromosomes did not randomly separate into the daughter cells during meiosis?

Q3: With two pairs of homologous chromosomes, four kinds of gametes can be produced (2^2). In the same way, a mosquito (*Aedes aegypti*) with three pairs of homologous chromosomes produces eight kinds of gametes (2^3). How many kinds of gametes can a spinach plant produce with its six pairs of homologous chromosomes? How many kinds of gametes can be produced with the 23 homologous pairs of chromosomes in human cells?

See Appendix A for answers to the figure questions.

upset over the allegations that BPA was toxic. Members of the plastics industry who did not agree with the paper's conclusions criticized Hunt's work. But there was more supporting research to come. Soto and Sonnenschein had also turned their attention to BPA because in everyday products it is a far more common chemical than nonylphenol and is therefore of greater concern.

At the same time that Soto, Sonnenschein, and Hunt were doing their work, Frederick vom Saal at the University of Missouri found that male mice that had been exposed to BPA in utero—even at very low doses—had dramatically enlarged prostates in adulthood that were hypersensitive to hormones. This study suggests that men are also at risk of health effects from BPA.

In 2007, Hunt followed up her original work with a study that she says made the first paper look like "child's play." Her team exposed pregnant mice to BPA just as their female fetuses were producing a supply of eggs in their ovaries. When that second generation of females became adults, their eggs were also damaged, Hunt found, demonstrating that BPA exposure affects not just adult females but also two generations of their offspring.

Exposure aside, not everyone agrees that BPA is toxic. Numerous companies that manufacture plastics have conducted studies and gotten results that do not match the results of Hunt's and vom Saal's research. To reach a scientific consensus, on November 28, 2006, Soto, Sonnenschein, Hunt, vom Saal, and 34 other researchers from across the United States gathered at the University of North Carolina in Chapel Hill to summarize the research on BPA. The result of their two-day meeting was the "Chapel Hill Bisphenol A Consensus Statement," summarizing hundreds of studies done in vitro (taking place outside a living organism, such as in a test tube or Petri dish) and in vivo (taking place inside a living organism) over the previous 10 years. Completing this analysis, the group concluded firmly that BPA exposure at current levels in our environment presents a risk to human health (see "What Can You Do?" on page 120). "It was quite clear that there is a serious problem," says Soto.

Over time, many baby-bottle manufacturers took BPA out of their bottles, even as government regulators were slower to respond. "As scientists, our role is to call attention to what is wrong, but it is the role of the politicians to act on it and try to straighten it out," says Sonnenschein. In July 2012, the U.S. Food and Drug Administration (FDA) banned BPA from baby bottles and children's drinking cups, though the prohibition does not apply to the use of BPA in other types of containers (**Figure 6.11**). Many scientists and activists are still concerned, however, about the chemicals that have replaced BPA, some of which also mimic estrogen.

Scientists all over the world have now shown that BPA disrupts meiosis and mitosis and causes a plethora of health and behavioral problems in mice and rats, including breast and prostate cancer, miscarriage and birth defects, diabetes and obesity, and attention-deficit/hyperactivity disorder (ADHD). Whether BPA is causing similar diseases in humans remains unknown; such hypotheses are hard to test because most people already have BPA in their bodies, making experimental control groups difficult to set up. BPA has been found in human blood, urine, breast milk, and amniotic fluid. In 2016, Hunt and her colleagues found "near-universal exposure" to BPA in a group of pregnant women in the United States. A major source of that exposure, they found, was cash register receipts, which can be hard to avoid touching.

Hunt, Soto, and Sonnenschein continue to explore the effects of BPA; they are studying how exposure to low doses of BPA affects monkeys, a model animal that more accurately represents the human system. And Hunt is now analyzing chemicals being used as replacements for BPA: in 2018, she and her team published evidence that BPS, a common BPA alternative, also disrupts meiosis and causes genetic abnormalities in mice, suggesting it may not be a safe alternative. "We're slowly raising awareness," says Hunt, "and slowly changing things."

Figure 6.11

BPA-free bottles and cans are now widely available
If you are concerned about being exposed to BPA, check labels before you buy.

Top 10 cancer deaths by rate of incidence

Incidence rates per 100,000

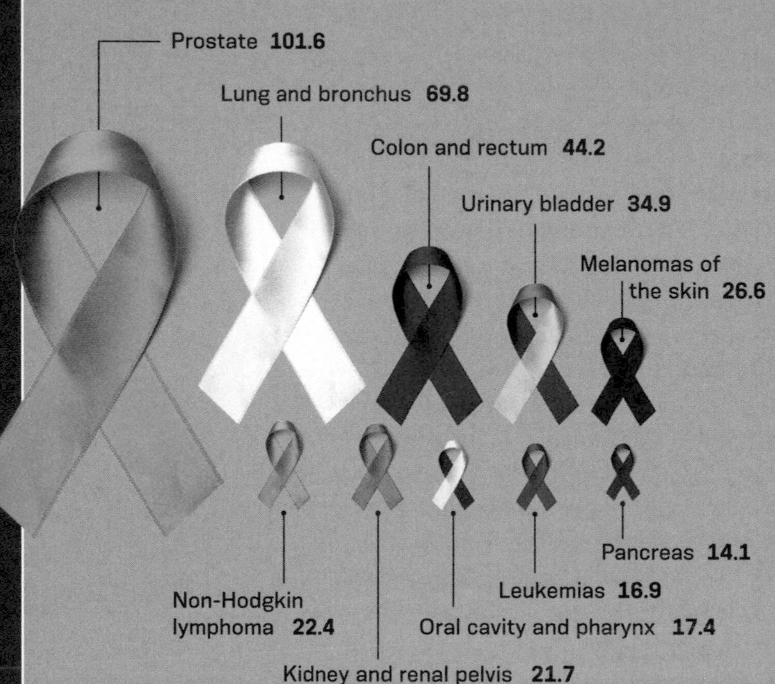

Male
Prostate 101.6
Lung and bronchus 69.8
Colon and rectum 44.2
Urinary bladder 34.9
Melanomas of the skin 26.6
Pancreas 14.1
Leukemias 16.9
Non-Hodgkin lymphoma 22.4
Oral cavity and pharynx 17.4
Kidney and renal pelvis 21.7

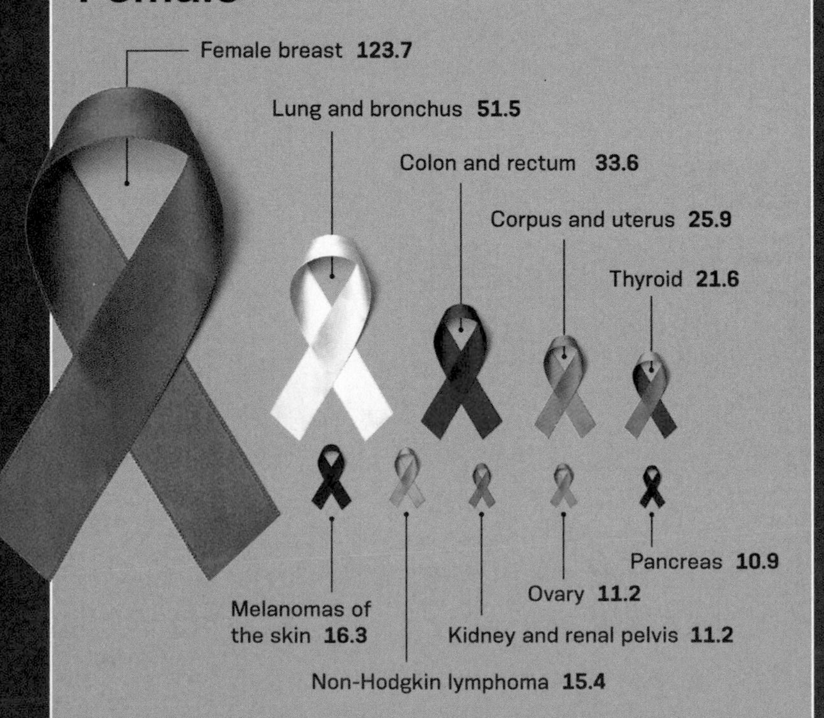

Female
Female breast 123.7
Lung and bronchus 51.5
Colon and rectum 33.6
Corpus and uterus 25.9
Thyroid 21.6
Pancreas 10.9
Ovary 11.2
Melanomas of the skin 16.3
Kidney and renal pelvis 11.2
Non-Hodgkin lymphoma 15.4

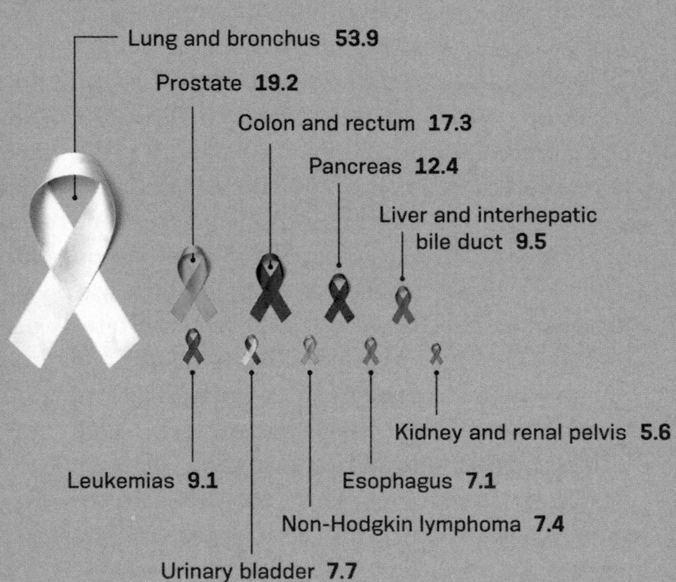

Male
Lung and bronchus 53.9
Prostate 19.2
Colon and rectum 17.3
Pancreas 12.4
Liver and interhepatic bile duct 9.5
Kidney and renal pelvis 5.6
Leukemias 9.1
Esophagus 7.1
Non-Hodgkin lymphoma 7.4
Urinary bladder 7.7

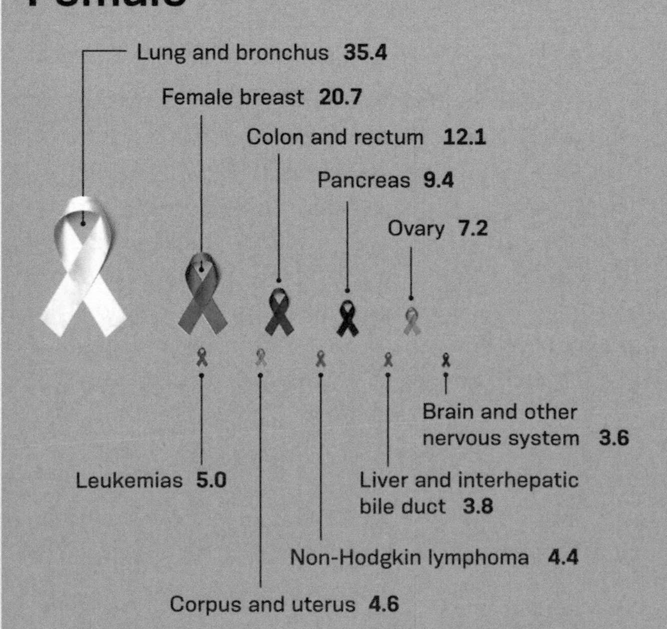

Female
Lung and bronchus 35.4
Female breast 20.7
Colon and rectum 12.1
Pancreas 9.4
Ovary 7.2
Brain and other nervous system 3.6
Leukemias 5.0
Liver and interhepatic bile duct 3.8
Non-Hodgkin lymphoma 4.4
Corpus and uterus 4.6

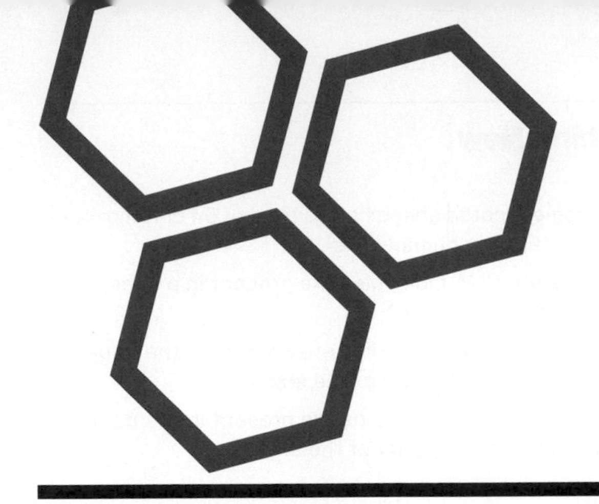

NEED-TO-KNOW SCIENCE

- The **cell cycle** (page 111) is the sequence of events over the life span of a eukaryotic cell that will divide. **Interphase** (page 111) and **cell division** (page 112) are the two main stages of the cell cycle. Interphase is longer and consists of the **G_1**, **S**, and **G_2 phases** (page 111). DNA is replicated in the S phase. Cell division is the last phase in the life of an individual cell. Cells that will not divide exit the cell cycle and enter the **G_0 phase** (page 112).
- Cell division is necessary for growth and repair in multicellular organisms and for **asexual reproduction** (page 112) and **sexual reproduction** (page 117) in all types of organisms. Many prokaryotes divide through **binary fission** (page 112), a form of asexual reproduction.
- Eukaryotes perform cell division through **mitosis** (page 113) followed by **cytokinesis** (page 114), producing daughter cells that are genetically identical to each other and to the parent cell. The four main phases of mitosis are prophase, metaphase, anaphase, and telophase. Through these phases, the chromosomes of a parent cell are condensed and positioned appropriately, and the sister chromatids are separated to opposite ends of the cell. During cytokinesis, the cytoplasm of the parent cell is physically divided to create two daughter cells.
- The **somatic cells** (page 117) of eukaryotes have two of each type of chromosome matched together in **homologous pairs** (page 118). One chromosome in each pair is inherited from the mother; the other is from the father. Chromosomal replication produces two identical **sister chromatids** (page 113) that are held together firmly at the **centromere** (page 114).
- The cell cycle is carefully regulated. Checkpoints ensure that the cycle does not proceed if conditions are not right. Unregulated cell growth and cancer can occur when checkpoints fail.
- **Meiosis** (page 117) is critical for sexual reproduction. In animals, the products of meiosis are sex cells, called **gametes** (page 117), that fuse during **fertilization** (page 118) to give rise to a **zygote** (page 118). Meiosis—consisting of two rounds of nuclear and cytoplasmic divisions—produces **haploid** (page 118) gametes containing only one chromosome from each homologous pair. When two gametes fuse during fertilization, a **diploid** (page 118) zygote is formed, having one of each homologous chromosome from both parents.
- During **meiosis I** (page 118), the maternal and paternal members of each homologous pair are sorted into two daughter cells. **Meiosis II** (page 118) is similar to mitosis in that sister chromatids are segregated into separate daughter cells at the end of cytokinesis.
- Meiosis produces genetically diverse gametes through **crossing-over** (page 120) of homologous chromosomes, leading to **genetic recombination** (page 120) and then the **independent assortment** (page 120) of homologous chromosomes. Meiosis and fertilization together introduce genetic variation into populations.

THE QUESTIONS

See Appendix B for answers.

The Basics

1 Homologous chromosomes are

(a) the same thing as sister chromatids.
(b) a pair of chromosomes of the same kind.
(c) identical copies of the same chromosome.
(d) always haploid.

2 Which of the following is *not* a contributor to genetic variation?

(a) binary fission
(b) crossing-over of homologous chromosomes
(c) independent assortment of homologous chromosomes
(d) fertilization

3 Link each cell phase to the event that occurs within it.

cytokinesis	1. Each of the chromosomes in a human cell contains two sister chromatids by the end of this phase.
S phase	2. Most cell growth occurs during this phase.
G_1 phase	3. Cells that will never replicate leave the cell cycle and enter this phase.
G_0 phase	4. Two separate daughter cells are produced at the end of this phase.

4 Select the correct terms in the following sentences:

(**Mitosis / Meiosis**) produces daughter cells with half the number of chromosomes that the parent cell has. Cell division in prokaryotes is called (**mitosis / binary fission**). Meiosis I separates (**sister chromatids / homologous chromosomes**); meiosis II separates (**sister chromatids / homologous chromosomes**) into separate daughter cells.

5 Place the following events of sexual reproduction in the correct order by numbering them from 1 to 5.

_____ a. Separation of homologous chromosomes
_____ b. Separation of sister chromatids
_____ c. Mitosis within the zygote, leading to a multicellular organism
_____ d. Cytokinesis, leading to four haploid daughter cells
_____ e. Fusion of two gametes

6 Loss of cell cycle control may lead to
(a) pregnancy.
(b) cancer.
(c) fertilization.
(d) crossing-over of homologous chromosomes.

Challenge Yourself

7 Suppose a scientist identifies a protein that prevents a cell from entering mitosis if there are any signs of DNA damage. This protein would be classified as a type of _______ protein.
(a) chromatid
(b) cell cycle checkpoint
(c) benign
(d) malignant
(e) angiogenesis

8 You've been reduced in size by a misfire from a "shrink-inator" gun! You realize that you are inside a cell during prophase I of meiosis. You see two linear molecules of DNA compacted and attached to one another by a centromere. What exactly are you looking at?
(a) a homologous pair of chromosomes
(b) cell cycle checkpoints
(c) the metaphase plate
(d) sister chromatids
(e) gametes

9 Scientists are able to isolate cultured cancer cells in various phases of the cell cycle in the laboratory. The accompanying graph shows the amount of DNA in a cell during the phases of the cell cycle. How do you explain the changes indicated by the curve?

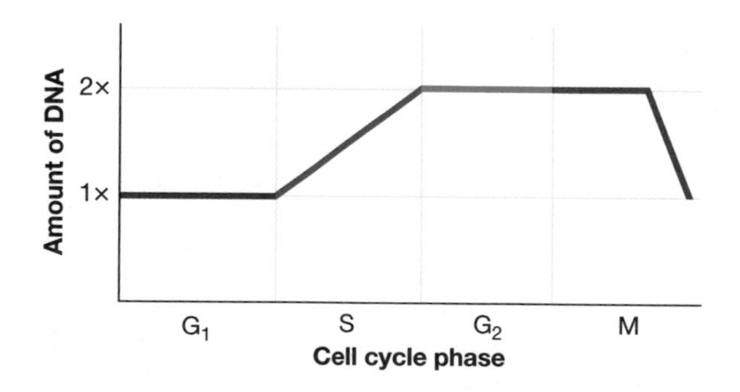

Try Something New

10 The cells of domesticated sheep have a total of 54 chromosomes (versus 46 chromosomes in human cells).
(a) How many separate DNA molecules are present in a sheep liver cell at the end of the G_1 phase?
(b) How many separate DNA molecules are present in the daughter cells after meiosis I in the ovary of the sheep?
(c) How many separate DNA molecules are present in the daughter cells after meiosis II in the ovary of the sheep?

11 Biopsies from aggressive cancers often have cells that contain several nuclei per cell when viewed through a microscope. Which scenario could explain how such a multinucleated cell might have come to be?
(a) The cell underwent repeated mitosis with simultaneous cytokinesis.
(b) The cell had multiple S phases before it entered mitosis.
(c) The cell underwent repeated mitosis, but cytokinesis did not occur.
(d) The cell underwent repeated cytokinesis but no mitosis.
(e) The cell actually went through meiosis and not mitosis.

12 Describe the likely consequences of bypassing the G_1 and G_2 checkpoints in the cell cycle. Why do chemicals such as nonylphenol lead to the multiplication of abnormal cells?

13 According to the accompanying graph, which normal tissue type has the highest percentage of dividing cells? What about for cancer? Which tissue type shows the greatest increase in dividing cells when comparing normal cells to cancer cells of that tissue?

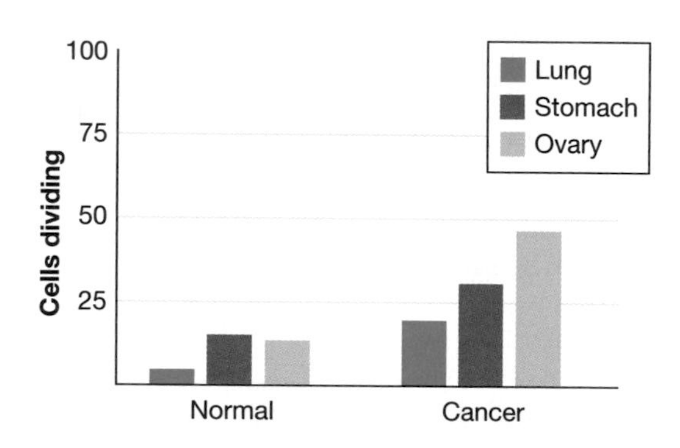

Leveling Up

14 **What do *you* think?** According to the latest research, only 5–10 percent of cancers are directly attributable to genetic causes. Many cancers might be prevented if people behaved differently, such as if they stopped smoking or never started.

A "sin tax" is a tax on a product or activity to offset negative effects of that product or activity. For example, a sin tax on tobacco might decrease the amount that people smoke (because of the increased cost) and could also partially fund the costs of medical care necessitated by increased rates of cancer and other diseases caused by smoking. Critics point out that sin taxes have historically triggered smuggling and black markets, have a disproportionate effect on poor people because the wealthy can more easily afford to pay the higher prices, and infringe on people's freedom to live as they choose.

Do you favor a sin tax on tobacco? Would fewer people smoke, or would they smoke less, if tobacco were more expensive? Would higher health care premiums for smokers, based on their higher risk for cancer, cause more people to stop smoking or never start? Do the potential benefits to society justify such measures, or should people be left alone to decide for themselves?

15 ***Write Now* biology: BPA effects** The studies of BPA described in this chapter used an inbred strain of mice that is known to be especially susceptible to estrogen and estrogen-like chemicals, such as BPA. The plastics industry maintains that the susceptibility of this strain of mice to estrogen renders these studies invalid as a basis for estimating the effects of BPA on humans. BPA researchers respond that the current situation, exposing millions of people to unknown levels of BPA, constitutes a massive uncontrolled experiment. They maintain that even a small risk of harm is too great to be allowed when so many people are exposed. Bills banning the use of BPA in food and beverage containers were introduced in Congress in 2009, 2011, 2013, 2014, 2015, and 2016. All failed to pass. Do a web search for "BPA research" and read up on both sides of this issue. Be sure to examine the sources of your information to determine whether they represent scientific organizations, health-advocacy groups, or branches of the chemical industry. Then sift through your findings to determine a position: does the latest science justify a ban, or are the warnings unnecessarily alarmist? Draft a letter urging your congressional representative to support or oppose the next bill banning the use of BPA in manufacturing.

Digital resources for your book are available online.

Dog Days of Science

Two canine-loving researchers unravel the genetic secrets of man's best friend.

After reading this chapter you should be able to:

- Distinguish between the genotype and phenotype of a given genetic trait.
- Describe the importance of Gregor Mendel's experiments to our understanding of inheritance.
- Illustrate Mendel's laws of segregation and independent assortment.
- Create a Punnett square to predict the phenotype of offspring from parents with a known genotype—both for single genes and for two independent genes.
- Give examples of Mendelian traits and of traits with complex inheritance.
- Explain how an individual's phenotype may be determined by multiple genes that interact with one another and with the environment.

GENETICS

CHAPTER 07

PATTERNS OF INHERITANCE

Gordon Lark's best friend was dying. Soft and shaggy, with tousled black hair, Georgie hadn't left Lark's side in 10 years, since his daughter had first purchased Georgie as a puppy from two kids by the side of the road. But as she aged, Georgie had become ill with Addison's disease. This disorder caused her body's immune system to attack and destroy her own tissues. Georgie passed away in 1996.

Lark was heartbroken. To ease his grief, he decided to adopt another dog of the same breed: a Portuguese water dog (PWD) (**Figure 7.1**). He contacted Karen Miller, a PWD breeder on a farm in rural New York. As part of the owner screening process, Miller asked Lark about his profession. He responded that he worked as a scientist at the University of Utah in Salt Lake City. "I said I was a soybean geneticist," Lark recalls, "but all she heard was 'blah, blah, genetics, blah, blah.' And she got really excited."

As a breeder, Miller was keenly interested in how dogs inherit characteristics from their parents, so Miller and Lark began talking by phone each week about genetics. When it came time to pick up his new puppy, Mopsa, Lark requested the bill. But Miller didn't want Lark's money. She had something else in mind: she gave Mopsa to him free of charge, in the hope that Lark would start researching dog genetics. "That's silly," Lark told her. He wasn't a dog researcher. Lark had spent a career studying the genetic traits of bacteria and soybeans.

A **genetic trait** is any inherited characteristic of an organism. Some genetic traits are **invariant**, meaning they are the same in all individuals of the species. All soybeans, for example, have pods that contain seeds. Other genetic traits are **variable**. For example, soybean seeds occur in various sizes and colors, including black, brown, and green.

Apart from his love for canines, Lark wasn't intimately familiar with dogs' physical and biochemical traits. **Physical traits**, such as the shape of a dog's face, are easy to observe. **Biochemical traits**, such as a dog's susceptibility to Addison's disease, are often more difficult to observe. It is easy to collect physical and biochemical information from a field of soybeans but far more difficult to collect it from domesticated animals in homes all over the country. And to study dogs, Lark would also need data on **behavioral traits**, such as shyness and extroversion—factors he didn't have to take into account with soybeans. All of these traits—physical, biochemical, and behavioral—are influenced by genes.

Figure 7.1

A Portuguese water dog

This breed is named for its history of helping Portuguese fishermen with their work.

Getting to the Genes

A **gene** is the basic unit of information affecting a genetic trait. At the molecular level, a gene consists of a stretch of DNA on a **chromosome**. A chromosome is a threadlike molecule of DNA found in the nucleus of a eukaryotic cell (**Figure 7.2**). A gene contains the information, or "code," for a particular genetic trait.

To study PWD traits, Lark would need dog DNA, which can be obtained from blood or saliva. Once he had that DNA, he could begin to search for **alleles**—different versions of a given gene—and link them to genetic traits. Alleles of a gene are caused by changes in the DNA that makes up the gene (this process, called mutation, is discussed further in Chapter 9). Genetic diversity in a species, whether that species is soybeans or dogs or humans, comes about because the species as a whole contains many different alleles of its genes.

Dogs are the champions of genetic variation. All dogs are the same species, yet a Pekingese weighs only a couple of pounds, whereas a Saint Bernard can weigh over 180 pounds. Dogs, in fact, are reported to have more variation in the size and shape of their species than any other living land mammal on Earth, with the possible exception of humans.

On a whim, Lark agreed to dabble in dog genetics, but to do so he would need to compare individual dog **genotypes** (their genetic makeup, controlled by their combinations of alleles) with their **phenotype** (the physical expression of their genetic makeup). The genotype of a given trait is the pair of alleles that codes for its phenotype. To identify genes responsible for dog traits, Lark would need both types of information for each of many dogs.

Miller was already on the case. Three months after Lark got Mopsa, Miller sent him 5,000 PWD pedigrees—detailed health and breeding records for individual dogs. Lark was astonished. It was the first of many times that the enthusiasm and generosity of dog owners would contribute to his research.

"That was literally how this started," says Lark. Lark and Miller's unlikely partnership blossomed into a national research project producing valuable knowledge about the genetic basis of health and disease in both man and man's best friend. What's more, their effort demonstrated how tiny genetic changes can create huge variation in a single species.

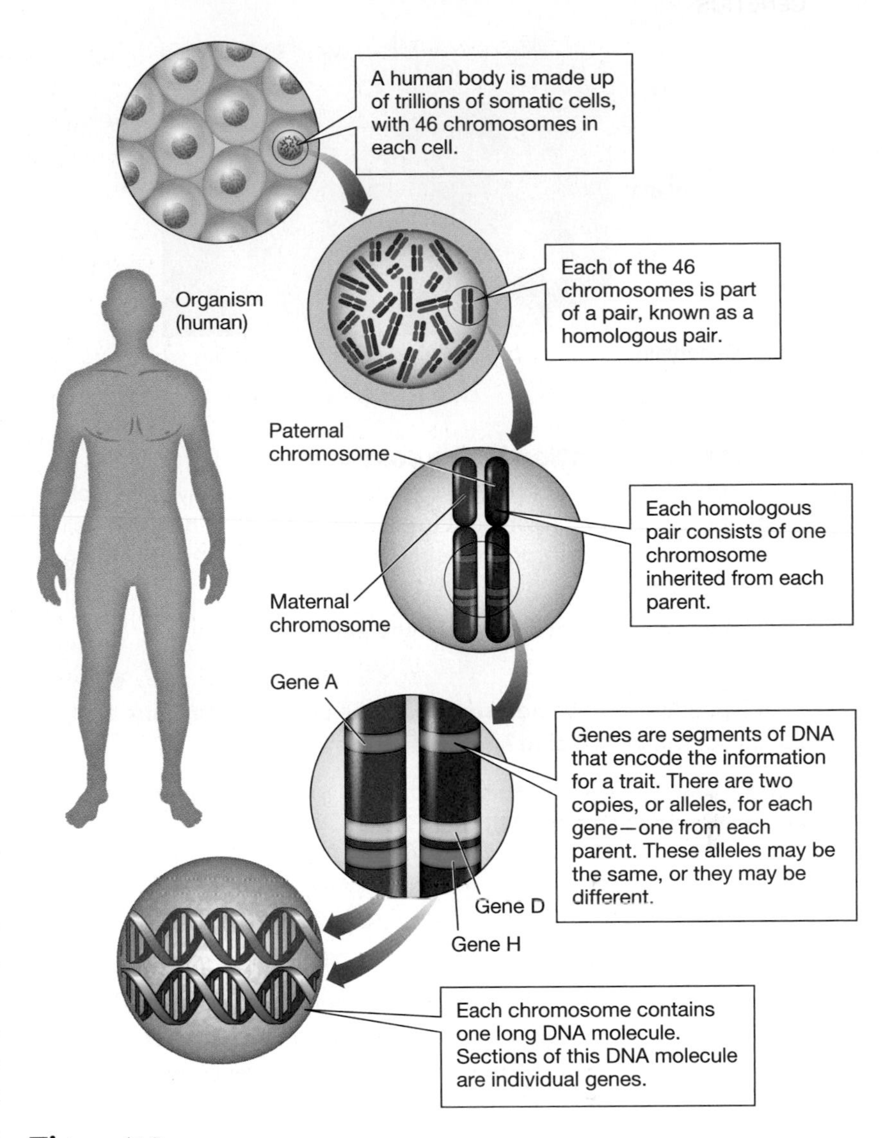

Figure 7.2

Genes are segments of DNA that confer genetic traits

Somatic cells (cells of the body) contain chromosomes. Chromosomes are made of DNA. Within the DNA there are two copies of most genes, known as alleles.

Q1: How are chromosomes related to DNA and genes?

Q2: How many copies of each gene are found in the diploid cells in a woman's body? (*Hint*: See Chapter 6 for a refresher on diploid versus haploid cells.)

Q3: There are 46 chromosomes in a human diploid cell; how many chromosomes come from the person's mother and how many from the father?

See Appendix A for answers to the figure questions.

Pet Project

The Georgie Project, as Lark fondly named it, officially began in 1996. Lark's first task was to collect genotypes and phenotypes from PWDs. To his pleasant surprise, PWD owners were enthusiastic and began flooding him with pedigrees, blood samples, and X-rays taken by their veterinarians. In short order, Lark had DNA from more than 1,000 dogs and detailed body measurements for over 500. Then the hard work began.

Using the dogs' genotypes and phenotypes, Lark set out to pinpoint the alleles for particular traits. Some genes have alleles that are **dominant** when paired with another allele. That is, one allele prevents a second allele from affecting the phenotype when the two alleles are paired together. The black-fur allele (*B*), for example, is dominant in dogs. An allele that has

Figure 7.3

Poodles illustrate variation in the coat color gene

These poodles, close cousins of the Portuguese water dog, may have a black coat (dominant allele *B*) or a brown coat (recessive allele *b*). Other coat colors, with different inheritance patterns, are found in poodles and other dog breeds.

Q1: Which might you observe directly: the genotype or the phenotype?

Q2: Which poodle could be heterozygous: the one with the black coat or the one with the brown coat?

Q3: Can you identify with certainty the genotype of a black poodle? A brown poodle? In each case, why?

See Appendix A for answers to the figure questions.

no effect on the phenotype when paired with a dominant allele is said to be **recessive**. In dogs, the brown-fur allele (*b*) is recessive. (When a gene has one dominant and one recessive allele, we generally use an uppercase letter for the dominant allele and a lowercase letter for the recessive allele.)

An individual that carries two copies of the same allele (such as *BB* or *bb*) is **homozygous** for that gene. An individual whose genotype consists of two different alleles (*Bb*) is **heterozygous** for that gene. Having one dominant allele and one recessive allele, a heterozygous individual will show the dominant phenotype; a dog that is heterozygous for the fur color gene (*Bb*), for example, will be black (**Figure 7.3**).

The first dog trait that Lark decided to investigate was size. What makes a Great Dane large and a Chihuahua small? To find out, Lark asked for help from the "mother of all dog projects," as Lark calls her—a researcher named Elaine Ostrander, whose entry into dog research was almost as strange as Lark's.

Crisscrossing Plants

In 1990, Ostrander was a young, enthusiastic researcher who had just completed her postdoctoral studies in molecular biology at Harvard University and was ready to start her own laboratory in California. But first she had to decide which organism to study. Typical choices included fruit flies, worms, or plants—organisms that are easy to grow and manipulate. Ostrander picked plants, just as Gregor Johann Mendel, an Austrian monk who later became known as the "father of modern genetics," had done in the mid-1800s.

Mendel famously bred pea plants in a garden at his monastery. Through his work with pea plants, Mendel discovered patterns of inheritance that today form the foundation of genetics for scientists like Ostrander. "Mendel's laws," as they are now called, describe how genes are passed from parents to offspring. These laws allow us to use parental genotypes to predict offspring genotypes and phenotypes.

Each time Mendel bred two pea plants together, he was performing a **genetic cross**, or just "cross" for short. A genetic cross is a controlled mating experiment performed to examine how a particular trait is inherited. In a series of genetic crosses, the organisms involved in the first cross are called the **P generation** ("P" for "parental").

For example, Mendel investigated the inheritance of flower color by crossing pea plants that had different flower colors (**Figure 7.4**). He had noticed that some plants always "bred true" for flower color; that is, the offspring always produced flowers that had the same color as the parents and were therefore homozygous. He performed a genetic cross with a P generation in which one parent bred true for purple flowers (*PP*) and the other bred true for white flowers (*pp*). The first generation of offspring of a genetic cross is called the **F_1 generation** ("F" for "filial," a word that refers to a son or daughter). When the individuals of the F_1 generation are crossed with each other, the resulting offspring are said to belong to the **F_2 generation**. Mendel allowed the F_1 generation pea plants to self-fertilize to produce the F_2 generation.

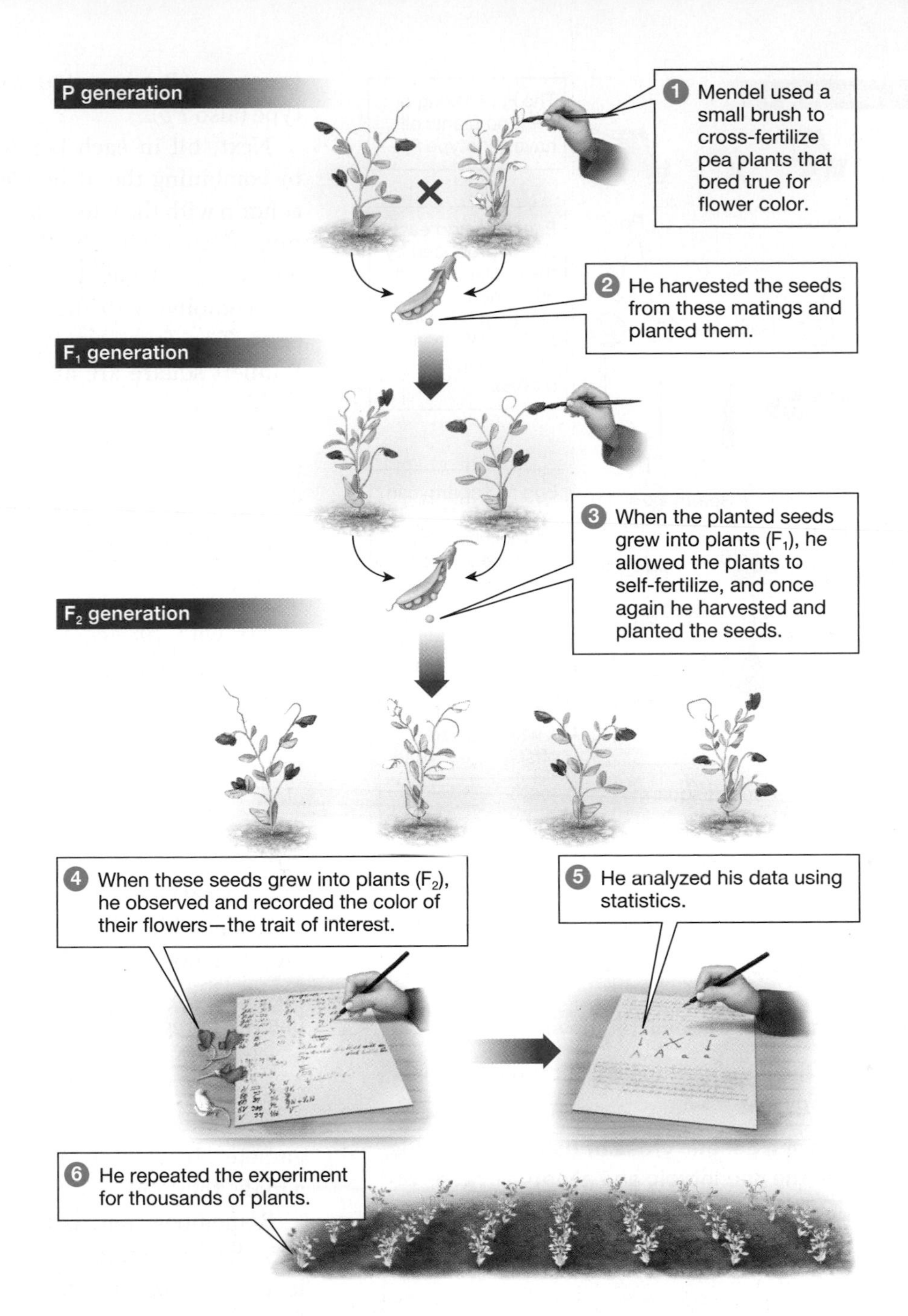

P generation
1 Mendel used a small brush to cross-fertilize pea plants that bred true for flower color.
2 He harvested the seeds from these matings and planted them.
F1 generation
3 When the planted seeds grew into plants (F1), he allowed the plants to self-fertilize, and once again he harvested and planted the seeds.
F2 generation
4 When these seeds grew into plants (F2), he observed and recorded the color of their flowers—the trait of interest.
5 He analyzed his data using statistics.
6 He repeated the experiment for thousands of plants.

Figure 7.4

Mendel's careful experiments

Mendel was meticulous in conducting his research and making observations, following a very careful protocol. Controlled breeding is possible in flowering plants because they have both male and female reproductive structures. Mendel manipulated his P generation plants by removing all the female structures from one plant and all the male structures from the other, thus preventing self-fertilization. He could then perform the initial cross by transferring pollen from the "male" plant to the flower of the "female" plant.

Q1: What would you predict about the color of the F_1 plants' flowers?

Q2: Why was it important for Mendel to begin with pea plants that he knew bred true for flower color? Why couldn't he simply cross a purple-flowered plant and a white-flowered plant?

Q3: Over the years, Mendel experimented with more than 30,000 pea plants. Why did he collect data on so many plants? Why didn't he study just one cross? (*Hint*: Read "What Are the Odds?" on page 136 before answering.)

See Appendix A for answers to the figure questions.

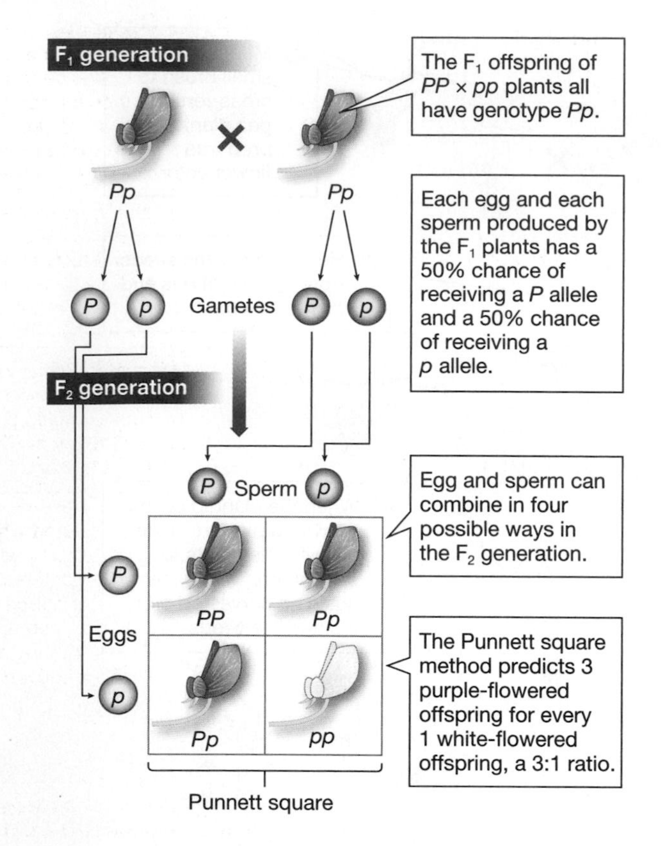

F_1 generation
The F_1 offspring of $PP \times pp$ plants all have genotype Pp.
Pp
Pp
Each egg and each sperm produced by the F_1 plants has a 50% chance of receiving a P allele and a 50% chance of receiving a p allele.
P
p
Gametes
P
p
F_2 generation
P
Sperm
p
Egg and sperm can combine in four possible ways in the F_2 generation.
P
Eggs
p
PP
Pp
Pp
pp
The Punnett square method predicts 3 purple-flowered offspring for every 1 white-flowered offspring, a 3:1 ratio.
Punnett square

Figure 7.5

Punnett squares

Punnett squares are a tool scientists use to predict the offspring of genetic crosses.

Q1: Why did Mendel's entire F_1 generation look the same?

Q2: The phenotypic ratio in the F_2 generation is 3:1 purple-to-white flowers. What is the genotypic ratio?

Q3: Draw a Punnett square for a genetic cross of two heterozygous, black-coated dogs. What are the phenotypic and genotypic ratios of their offspring?

See Appendix A for answers to the figure questions.

We can predict the results of a genetic cross by using a grid-like diagram called a **Punnett square** (**Figure 7.5**). A Punnett square shows all possible ways that two alleles can be brought together through fertilization. To create a Punnett square showing how a trait is inherited, list the alleles of the male genotype across the top of the grid, writing each unique allele just once. List the alleles of the female genotype along the left edge of the grid, again writing each unique allele only once. In the case of Mendel's cross of the F_1 generation, the male genotype (*Pp*) is mated with the female genotype (also *Pp*).

Next, fill in each box (or "cell") in the grid by combining the male allele at the top of each column with the female allele listed at the beginning of each row. The Punnett square shows all four ways in which the two alleles in the sperm can combine with the two alleles found in the eggs. The four genotypes shown within the Punnett square are all equally likely outcomes of this cross.

Using the Punnett square method, we can predict that ¼ of the F_2 generation is likely to have genotype *PP*, ½ to have genotype *Pp*, and ¼ to have genotype *pp*. Because the allele for purple flowers (*P*) is dominant, plants with *PP* or *Pp* genotypes have purple flowers, whereas plants with *pp* genotypes have white flowers. Therefore, we predict that ¾ (75 percent) of the F_2 generation will have purple flowers and ¼ (25 percent) will have white flowers—a 3:1 (¾:¼) ratio of phenotypes. This prediction is very close to the actual results that Mendel obtained. Of a total 929 F_2 plants that Mendel raised, 705 (76 percent) had purple flowers and 224 (24 percent) had white flowers.

Results like these supported Mendel's first law, the **law of segregation**. Mendel did not know about DNA, but in modern terms, the law of segregation states that the two alleles of a gene are separated during meiosis I, the specialized type of cell division during sexual reproduction, and they end up in different gametes. In a female, they end up in egg cells; in a male, they end up in sperm cells.

Remember that, before meiosis, one of the two alleles is found on one of the chromosomes in a homologous pair, and the other allele is found on the other chromosome in the pair (see **Figure 7.2**). Mendel's law of segregation is basically about how homologous chromosome pairs are divided into separate daughter cells during meiosis. This division enables us to predict how a single trait will be inherited from the particular genes of two parents.

You can try this on your own by making a Punnett square to predict the ratio of black and brown offspring that would result if two heterozygous (*Bb*) black-coated dogs were mated (see Q3 for **Figure 7.5**). It is important to understand that the predicted ratios simply give the *probability* that a particular offspring will have a certain phenotype or genotype; the actual ratio will vary (see "What Are the Odds?" on page 136).

Peas in a Pod

Mendel's research on peas led to his second law, the **law of independent assortment**. This law states that when gametes form, the two alleles of any given gene segregate during meiosis independently of any two alleles of other genes. For example, pea seeds can have a round or wrinkled shape, and they can be yellow or green. Two different genes control the two different traits: the *R* gene, with alleles *R* (round) and *r* (wrinkled), controls seed shape, whereas the *Y* gene, with alleles *Y* (yellow) and *y* (green), controls the color of the seed. Neither gene affects the inheritance of the other.

Mendel tested the idea of independent assortment in the set of experiments illustrated in **Figure 7.6**. This time, he tracked two traits instead of one: seed shape, the trait controlled by the *R* and *r* alleles, and seed color, controlled by the *Y* and *y* alleles. The test of his hypothesis came when he examined the phenotypes of the offspring produced by crossing the heterozygous F_1 plants (*RrYy*). As predicted by the hypothesis, two new phenotypic combinations were found among the F_2 offspring: plants with round, green seeds (*RRyy* or *Rryy*) and plants with wrinkled, yellow seeds (*rrYY* or *rrYy*). **Figure 7.6** summarizes the ratios of the two parental phenotypes and the two new, nonparental phenotypes.

It is important to note that most traits are controlled by many genes and influenced by the organism's environment. Traits that are controlled by a single gene and unaffected by environmental conditions are called **Mendelian traits**. But when Mendel described his laws of inheritance, he had no idea what genes were made of, where they were located within a cell, or how they segregated and independently assorted.

Today we know that genes are located on chromosomes and that these chromosomes are the basis for all inheritance. We call these assertions the **chromosome theory of inheritance**. This theory explains the mechanism underlying Mendel's laws: chromosomes are paired, each homologous chromosome in a pair has one allele for a gene, alleles are shuffled and recombined and then separated randomly into sperm and

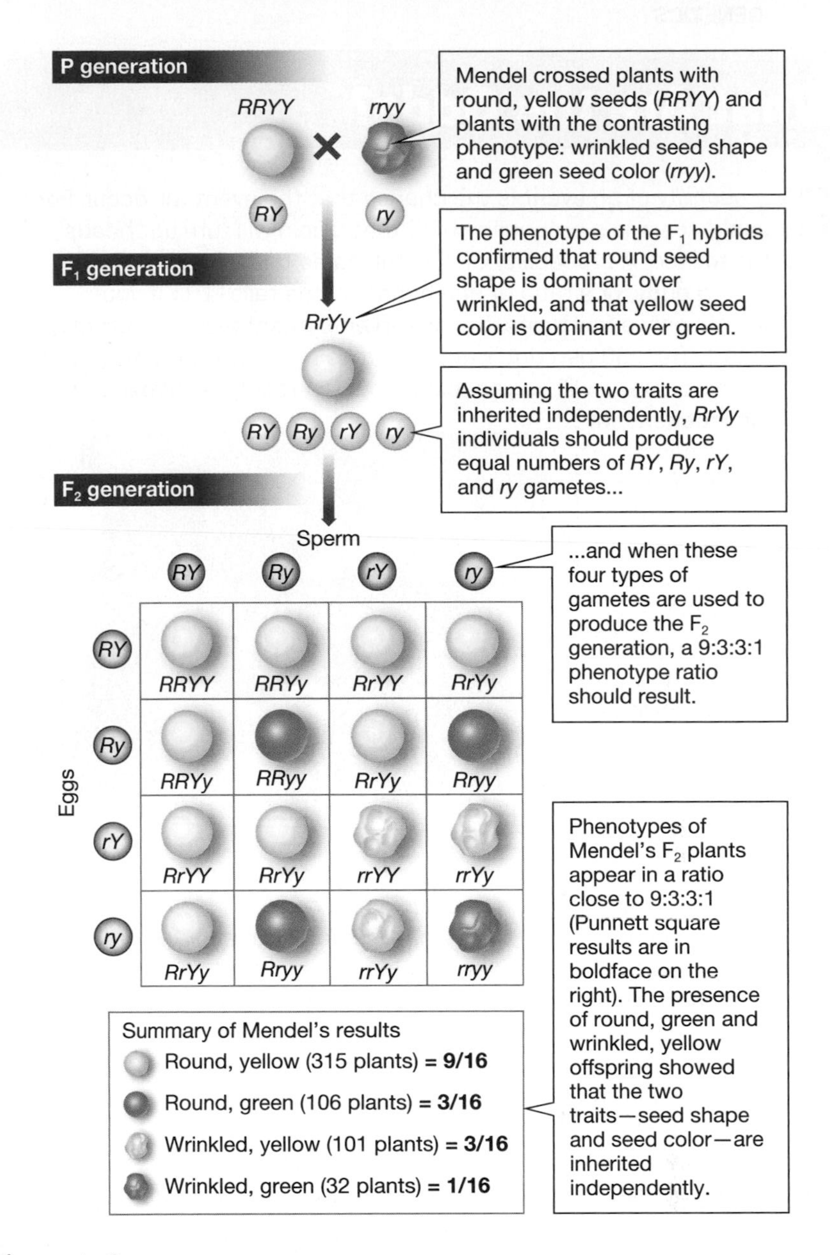

Figure 7.6

Independent assortment of pea color and shape

Mendel used two-trait breeding experiments, called **dihybrid crosses**, to test the hypothesis that the alleles of *two different genes* are inherited independently from each other.

Q1: List all the possible offspring phenotypes and genotypes.

Q2: What is the offspring phenotypic ratio?

Q3: Complete a Punnett square for a genetic cross of two true-breeding Portuguese water dogs—one with a black, wavy coat (homozygous dominant, *BBWW*) and one with a brown, curly coat (homozygous recessive, *bbww*). What is the phenotypic ratio of their offspring (F_1)? Now fill out another Punnett square, crossing two of the offspring. What is the phenotypic ratio of the F_2 generation?

See Appendix A for answers to the figure questions.

What Are the Odds?

The *probability* of an event is the chance that the event will occur. For example, there is a probability of .5 that a coin will turn up "heads" when it is tossed. A probability of .5 is the same thing as a 50 percent chance, or 1/2 odds, or a ratio of one heads to one tails (1:1). If you toss the coin only a few times, the observed percentage of heads may differ greatly from 50 percent. But if you toss it many, many times, that observed percentage will be very close to 50 percent. Each toss of a coin is an independent event, in the sense that the outcome of one toss does not affect the outcome of the next toss. The probability of getting two heads in a row is a product of the separate probabilities of each individual toss: .5 × .5, which is .25. In a genetic cross of two heterozygous (*Bb*) black-coated dogs, the probability of getting a brown puppy is 1/4, or .25. The probability of getting a black puppy is 3/4, or .75.

We cannot know with certainty what the actual phenotype or genotype of a particular offspring is going to be, except when true-breeding individuals are crossed. For example, two brown dogs, both of whom have a *bb* genotype, will have only *bb*-genotype, brown-phenotype offspring. Moreover, the probability that a particular offspring will display a specific phenotype is completely unaffected by how many offspring there are. The likelihood that we will see the 3:1 black-to-brown outcome, however, increases when we analyze a larger number of offspring, just as Mendel analyzed thousands of pea plants.

egg cells during meiosis (see Figures 6.7 and 6.8 in Chapter 6). Then, during fertilization, a one-in-a-million sperm fuses with a one-in-a-million egg to create a unique individual. That is how offspring can have genotypes and phenotypes that were not present in either parent, such as a brown puppy born to two black dogs.

DEBUNKED!

MYTH: Eye color is a simple genetic trait, and blue eyes are recessive.

FACT: At least 12 genes contribute to eye color, and blue eyes are not determined by a recessive allele at one gene. Two blue-eyed parents can, in fact, have a brown-eyed child.

Going to the Dogs

Like Mendel, Elaine Ostrander planned to study plants to unravel the secrets of genetics and inheritance. But when she arrived at UC Berkeley to open her lab, the space was not yet available. So she wandered down the hall and into the office of Jasper Rine, a geneticist who normally studied yeast but was looking for someone to start a mammalian genome research project. Ostrander volunteered.

But which mammal to study? "I was allergic to cats, and I didn't know enough about cows or pigs or horses," she recalls, so she picked dogs. Not only was Ostrander a dog lover, but also the American Kennel Club had just begun offering funding to researchers trying to identify genes associated with dog diseases.

In 1993, Ostrander began identifying all the genes in dogs—that is, making a map of the dog genome. Some colleagues expressed their concern, suggesting that no one would give her money to support the research. But Ostrander is nothing if not persistent, and she knew the potential value of the research: dogs and humans share many genes, and more than 360 genetic disorders occur in both humans and dogs, including cancer, epilepsy, heart disease, and Addison's disease, the illness that killed Lark's dog Georgie. The genetics of bladder cancer is difficult to study in humans, for example, but the disease is quite common in Scottish terriers and would be easier to study in a dog species. By cracking the genetic code of dogs, Ostrander hoped to uncover causes and potential treatments for human diseases.

In 2005, she published the first full dog genome sequence for a female boxer named Tasha. The achievement gained her scientific fame and raised awareness among scientists of the importance of dog genetics to human health. "Of the more than 5,500 mammals living today, dogs are arguably the most remarkable," Ostrander's coauthor, Eric Lander, a professor of biology at the Massachusetts Institute of Technology, said when the first dog genome sequence was published. "The incredible physical and behavioral diversity of dogs—from Chihuahuas to Great Danes—is encoded in their genomes. It can uniquely help us understand embryonic development, neurobiology, human disease, and the basis of evolution" (**Figure 7.7**).

Shadow, a standard poodle, was the first dog to have its genome partially (about 80%) sequenced.

Tasha, a boxer, was the first dog to have its complete genome sequenced. Boxers are vulnerable to hip, thyroid, and heart problems. Scientists identified a gene in boxers for cardiomyopathy, a heart disorder also found in humans.

Pembroke Welsh corgis may develop a fatal neurodegenerative condition similar to amyotrophic lateral sclerosis (ALS) in humans. The human gene mutation associated with ALS was also found in corgis with the condition.

Psychiatric disorders often have a genetic component. Doberman pinschers are susceptible to canine compulsive disorder, similar to obsessive-compulsive disorder in humans. The responsible gene in Dobermans has been linked to autism disorders in humans.

Golden retrievers are prone to cancers of the bone marrow. Ostrander's research group is analyzing the genomes of hundreds of goldens with and without cancer, hoping to identify the genes responsible.

Figure 7.7

Man's best friend

The Dog Genome Project has identified the genetic basis of several diseases and conditions in dogs, and in some cases it has been able to link the gene to a similar gene in humans.

Q1: Boxers are far more inbred than poodles. Why does that inbreeding make boxers a better target for genetic studies of disease than poodles are?

Q2: Explain why a geneticist interested in finding a gene linked to cancer would want to look at the DNA of senior golden retrievers with *and* without cancer.

Q3: Obsessive-compulsive disorder (OCD) in humans is characterized by obsessive thoughts and compulsive behavior, such as pacing. Canine compulsive disorder (CCD) is characterized by compulsive behavior such as "flank sucking," sometimes seen in Doberman pinschers. Would you predict that the medications given to humans with OCD would decrease compulsive behaviors in CCD dogs? Why or why not?

See Appendix A for answers to the figure questions.

But years before Ostrander completed the dog genome, she had begun a different pet project. In 2001, Ostrander received a call from a scientist in Utah who wanted to talk about dogs. It was Gordon Lark, who told her he was collecting genetic trait information about Portuguese water dogs (PWDs). "The day I met Gordon was the best day of my life," says Ostrander. "I knew it was golden."

In 2002, the duo published a paper pinpointing genes that control dog body shape, from the tall, lanky look of a greyhound to the short, stocky frame of a pit bull. In the acknowledgments of the paper, they thanked the PWD breeder Karen Miller and all the PWD owners who had contributed pedigree information.

In the spring of 2006, Lark and Ostrander began their second collaboration, this time to identify the genetic basis of dog size. Lark collected skeletal measurements of 92 PWDs and DNA samples from each dog. Ostrander used that genotype and phenotype information to identify a key gene for body size—*IGF1*, which controls the activity of a growth factor and is known to influence body size in mice and humans. This gene's two alleles are called *I* and *B*. Lark and Ostrander discovered that PWDs homozygous for allele *I* (*II*) were usually large dogs, and those homozygous for allele *B* (*BB*) were always small dogs. That single gene accounted for whether a PWD was large or small.

Interestingly, neither *IGF1* allele is dominant or recessive. Instead, heterozygous dogs, with an *IB* genotype, are medium-sized dogs. This is an example of a trait inherited by **incomplete dominance** in which neither allele is able to exert its full effect, so a heterozygote displays an intermediate phenotype. Dogs with an *IB* genotype aren't large or small but rather medium-sized (**Figure 7.8**).

Early in the twentieth century, geneticists identified yet another type of interaction among alleles—namely, codominance—that Mendel had not observed among his pea plants. A pair of alleles will show **codominance** when the effects of the two alleles are equally visible in the phenotype of the heterozygote. In dogs, gum color is codominant: a dog's gums can be pink, black, or pink with black spots. In the latter case, both alleles are fully on display, and neither is diminished nor diluted by the presence of the other allele (as in incomplete dominance) or suppressed by a dominant allele (as in the case of dominant and recessive alleles). In humans, the blood type AB is a codominant trait.

Genotype: *II*
Phenotype: Large

Genotype: *BB*
Phenotype: Small

Genotype: *IB*
Phenotype: Medium

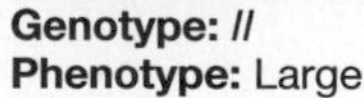

Figure 7.8

Incomplete dominance of body size alleles

Great Danes (left) and Chihuahuas (middle) illustrate the extreme size variation found in domestic dogs. Dogs heterozygous for the main body size gene show an intermediate size, such as that of the Cocker Spaniel pictured here (right).

Q1: What are the genotypes of a large dog and a small dog?

Q2: Is it possible to have a heterozygous large dog? Explain why or why not.

Q3: Crossing a Great Dane and a Chihuahua is likely to be unsuccessful, even though they are members of the same species (and thus have compatible sperm and egg). Why is that? What are some potential risks of such a cross?

See Appendix A for answers to the figure questions.

It's Complicated

Many of the traits people tend to be curious about—body weight, intelligence, athleticism, and musical talent, to name a few—are even more complicated. A **complex trait** is a genetic trait whose pattern of inheritance cannot be predicted by Mendel's laws of inheritance. Complex traits do not fit the straightforward single-gene, single-phenotype pattern discussed so far.

Sometimes a *single gene* influences a number of *different traits*. Such cases are examples of

Most Chronic Diseases Are Complex Traits

A disease is a condition that impairs health. It may be caused by an external factor, such as infection by a virus or bacterium, or by injury from a harmful chemical or high-energy radiation. A lack of nutrition can also lead to disease. Inadequate vitamin C consumption, for example, produces scurvy, once common among sailors and pirates. Disease may also be caused by the malfunction of one or more genes. Diseases caused exclusively by gene malfunction are described as genetic disorders, distinguishing them from infections and other types of diseases.

But many of the diseases that are most common in industrialized countries—heart disease, cancer, stroke, diabetes, asthma, and arthritis, for example—are caused by multiple genes interacting in complex ways with each other and with external factors. They are complex traits: malfunctions in key genes make a person susceptible to developing these diseases, but environmental factors affect whether the disease will actually appear and how severe the symptoms will be. A large percentage of the estimated risk of developing chronic diseases is preventable by lifestyle choices such as maintaining good nutrition, exercising regularly (see graph), and avoiding tobacco. ("Chronic" means unceasing—a reference to the fact that people who develop one of these diseases will have it for the rest of their lives.)

A major goal of modern genetics is to identify genes that contribute to human disease. Researchers have identified alleles associated with increased risk of a number of common ailments, including high blood pressure, heart disease, diabetes, Alzheimer disease, several types of cancer, and schizophrenia. The hope is that one day soon, genetic tests will tell us whether we are predisposed to a disease before we become ill with it. Then a person carrying a risky allele might take preventive measures to reduce the chance of developing the condition, and treatment could be customized to fit the particular allele involved. This tailored approach to treatment, called "personalized medicine," is already being used to treat breast cancer and other chronic diseases.

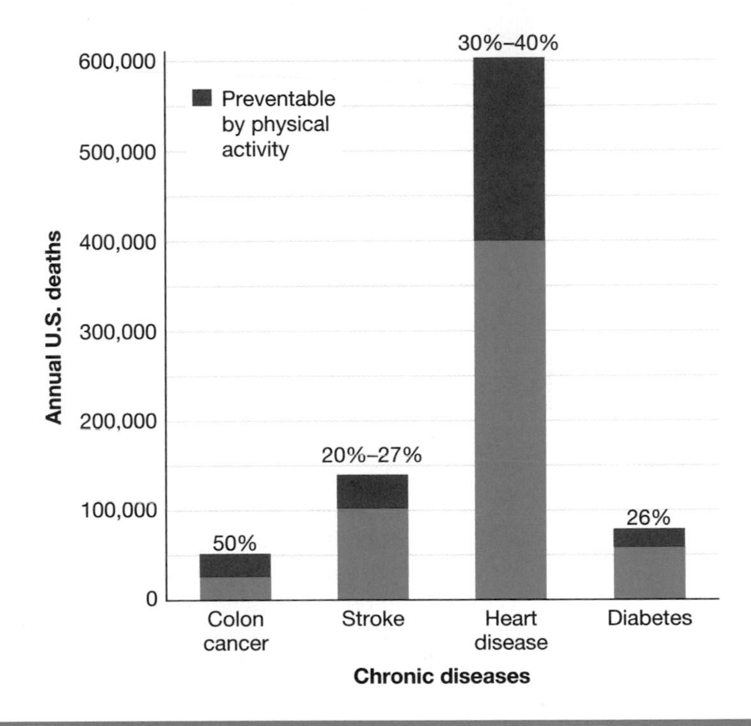

pleiotropy (*pleio*, "many"; *tropy*, "change"). In PWDs, Lark found that single genes can control multiple related skeletal traits. The shape of a dog's head and the shape of its limbs are controlled by a single gene. That connection makes sense, says Lark, as a fast dog benefits from having a small head and long legs, whereas a strong dog, such as a pit bull, uses both its massive jaw and short, thick legs for power.

Another good example of pleiotropy comes from a long-term breeding experiment to tame silver foxes. A researcher found that as foxes became tamer, their appearances and behaviors changed (see "The New Family Pet?" on page 141).

Patterns of inheritance can get even more complicated. *Single traits* governed by *more than one gene* are called **polygenic traits.** In humans, polygenic traits include eye color, skin color, running speed, blood pressure, and body size. Of the thousands of human genetic traits, governed by an estimated 24,000 genes, fewer than 4,000 are known or suspected to be controlled by a single gene with a dominant and a recessive allele. The rest are polygenic traits.

To make matters more complicated, a gene can have more than two alleles. Recall from earlier in this chapter that AB blood type in humans is an example of codominance. Blood type, in fact, is governed by a gene with three alleles. Throughout the human population, there are three alleles for blood type, but each individual has only two of those alleles.

Another twist on inheritance is **epistasis.** This is a phenomenon where two genes are inherited independently, yet the alleles of one gene affect

Figure 7.9

Epistasis in coat color

These Labrador retrievers show the complex inheritance of coat color. The yellow dog carries two alleles that interfere with the deposition of melanin in hair. Both the brown and the black dogs must carry at least one allele that allows melanin deposition. A dash indicates that one allele is unknown, based on the dog's phenotype.

Q1: What are the possible genotypes of the black dog? The brown dog? The yellow dog?

Q2: Draw a Punnett square showing possible matings between the black dog (assuming it is heterozygous at both genes) and the yellow dog (assuming it is heterozygous at the *B* gene). List all the possible phenotypes of their offspring. (See Figure 7.6 for an example of a Punnett square made with two traits.)

Q3: If you wanted the most variable litter possible, what colors of Labrador retrievers would you cross? Assume that your Labradors are true-breeding for color.

See Appendix A for answers to the figure questions.

the expression of the other gene's alleles. The coat color of Labrador retrievers, for example, is affected by epistasis (**Figure 7.9**). Dog fur has a dominant allele (*B*) that leads to black fur and a recessive allele (*b*) that produces brown fur, yet another gene, called the expression gene, can completely eliminate these effects depending on which allele (*E* or *e*) is present. When the dominant *E* allele is present (*EE* or *Ee*), the dog's fur includes a pigment called melanin that makes it possible for the fur color genotype (*BB*, *Bb*, or *bb*) to be expressed. But when a dog has the recessive *ee* genotype, its fur does not have melanin. As a result, the dog will be yellow, regardless of the genotype at the *B*/*b* gene.

If the environment affects the phenotype, it becomes nearly impossible to predict the phenotype when given only the genotype of an individual or its parents. The effects of many genes depend on internal and external environmental conditions, such as body temperature, carbon dioxide levels in the blood, external temperature, and amount of sunlight.

For example, cats have a gene that produces the enzyme tyrosinase, which is involved in melanin production. Siamese cats have a special allele of the gene that makes a tyrosinase that works well at colder temperatures but does not function at warmer temperatures, so the production of melanin depends on the temperature of the cat's surroundings (**Figure 7.10**). Because a cat's extremities—ears, nose, paws, and tail—tend to be colder than the rest of its body, those parts tend to be dark in Siamese cats because tyrosinase produces melanin there.

In fact, if a patch of light hair is shaved from the body of a Siamese cat and the skin is covered with an ice pack, the hair that grows back will be dark. Similarly, if dark hair is shaved from the tail and allowed to grow back under warm conditions, it will be light-colored.

Man's Best Friend

After describing the inheritance of size in Portuguese water dogs, Lark and Ostrander looked at the *IGF1* gene in over 350 dogs representing 14 small breeds and 9 giant breeds. The genotype *BB* was common in small dogs and virtually nonexistent in large dogs. "All small dog breeds had them. It didn't matter when they were bred or how; they all had the exact same pattern," says Ostrander.

"It was amazing," adds Lark. Breeders have, over time, been selecting for these alleles to create smaller and smaller dogs. "What mankind can do, without any genetic tools but just knowledge of heritability, is just extraordinary," he says.

After their success with dog size, Lark and Ostrander identified genes responsible for other traits, such as fur color, leg length, and skull shape. They also identified genes related to cancer and other complex traits that might tell us something

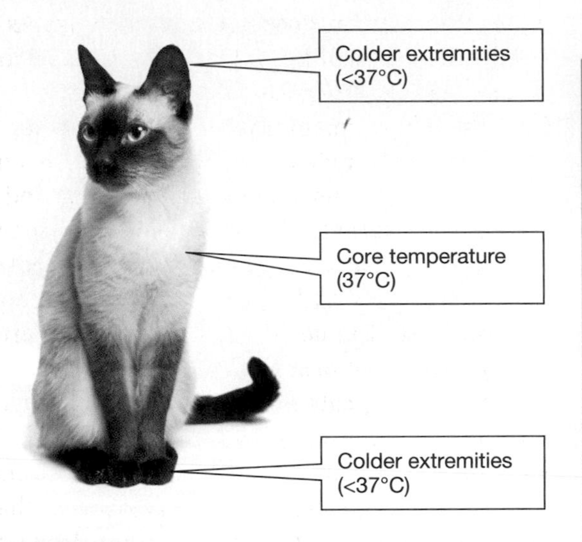

Figure 7.10

The environment can alter the effects of genes

Coat color in Siamese cats is controlled by a special allele. This allele produces an enzyme that does or does not produce melanin, depending on the temperature of a specific body part.

Q1: The gene that brings about the pale Siamese body fur is also partly responsible for the typical blue eyes of the species. What is the term for this type of inheritance?

Q2: Siamese kittens that weigh more tend to have darker fur on their bodies. Why might this be?

Q3: The Siamese cat pictured is called a "seal point" because it has seal-colored (dark brown) extremities. Some Siamese cats show the same color pattern, but the dark areas are of a lighter color or even a different shade—for example, lilac point, red point, blue point. What results would you predict if the tests described in the text (shaving the cat and then increasing or decreasing temperature) were performed on cats with these color patterns? Why?

See Appendix A for answers to the figure questions.

The New Family Pet?

The silver fox is the same species as the more familiar red fox: *Vulpes vulpes*. Because of its soft, silver coat, it has been bred in captivity for over 100 years to provide fur for coats, stoles, and hats.

In 1959, a Russian geneticist, Dmitry Belyaev, began to conduct breeding experiments on silver foxes he had purchased from a fur breeder, pairing only the tamest individuals of each generation. He determined how tame a fox was by observing its response when approached and offered food. As the foxes became tamer in each generation, they did not show a fear response until they were older—9 weeks instead of 6 weeks. (Domestic dogs develop a fear response at about 8–12 weeks.) In addition, in the tame foxes, the hormones associated with a fear response did not increase until later. These traits were all clearly influenced by the same gene or genes—an example of pleiotropy.

Only 50 years and 35 generations separate this tame silver fox from its wild relative

Another surprising result was that the foxes' appearance began to change along with their behavior. They developed shorter legs, curlier tails, wider faces, and floppier ears. These changes made the adult foxes look more puppy-like, and they are similar to the differences between domestic dogs and their ancestors, wolves. Scientists conjecture that in both cases, tameness and associated changes in development, physiology, and anatomy were brought about by breeding for juvenile features.

about human disease (see "Most Chronic Diseases Are Complex Traits" on page 139). In border collies, Ostrander's team identified a gene involved in an eye disease that causes blindness in both humans and dogs. Her lab also identified a gene involved in kidney cancer in dogs that causes a similar syndrome in humans.

Although the Georgie Project has ended, research in the area of dog genetics is flourishing. Ostrander is now Chief and NIH Distinguished Investigator of the Cancer Genetics and Comparative Genomics Branch at the National Institutes of Health, where her team is exploring the genetics of sport-hunting dogs, the evolution of modern Italian dogs, and hairlessness in domestic dog breeds, among many other topics. In 2019, Ostrander's group published a massive amount of data from the genomes of dogs, documenting over 91 million variations

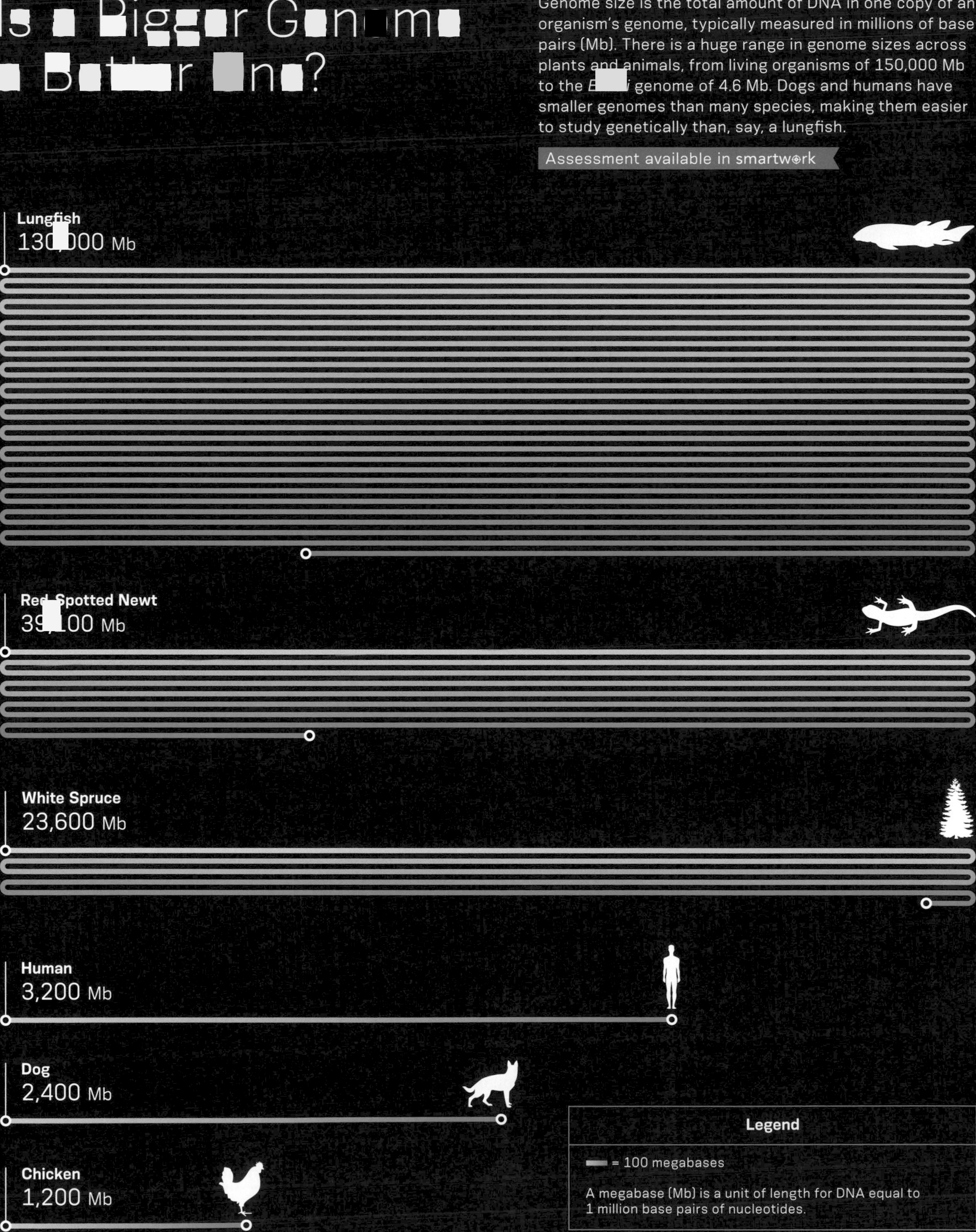

Genome size is the total amount of DNA in one copy of an organism's genome, typically measured in millions of base pairs (Mb). There is a huge range in genome sizes across plants and animals, from living organisms of 150,000 Mb to the E genome of 4.6 Mb. Dogs and humans have smaller genomes than many species, making them easier to study genetically than, say, a lungfish.
Assessment available in smartwork
Lungfish
130,000 Mb
Red-Spotted Newt
39,100 Mb
White Spruce
23,600 Mb
Human
3,200 Mb
Dog
2,400 Mb
Chicken
1,200 Mb
Legend
= 100 megabases
A megabase (Mb) is a unit of length for DNA equal to 1 million base pairs of nucleotides.

in dog genes across 144 modern breeds. "I'm really happy with the growth of the field," says Ostrander, "and there's a lot of research coming down the pike."

Mopsa, the puppy that Karen Miller gave Lark in 1996 in return for studying PWDs, died in April 2012, just a week short of her sixteenth birthday (**Figure 7.11**). But Lark has a new best friend, a PWD puppy he named Chou (pronounced "shoo"), for the French *petit chou*, meaning "little cabbage."

"We often make a mistake and call Chou 'Mopsa,' because Chou looks so similar," says Lark. After all, he adds, PWDs share similar genotypes and thus similar phenotypes. And these genetic traits make them the cuddly, devoted pets that they are.

Figure 7.11

The scientist and his dog
Gordon Lark holds Mopsa.

NEED-TO-KNOW SCIENCE

- A **genetic trait** (page 130) is any characteristic that is inherited and may be **physical** (page 130), **biochemical** (page 130), or **behavioral** (page 130). All of these types of traits are either **invariant** (the same in all members of a species, page 130) or **variable** (different in different members of a species, page 130).
- A **gene** (page 130) is a stretch of DNA that affects one or more genetic traits. Genes are found on **chromosomes** (page 130), threadlike molecules made of DNA and proteins.
- The **genotype** (page 131) is an individual's genetic makeup or, more specifically, the pair of different versions of a given gene (**alleles**, page 130) that determines a given trait. The **phenotype** (page 131) is the observable expression of an individual's genetic makeup or, more specifically, the expression of a version of the given trait.
- An allele is **dominant** (page 131) when it prevents a second allele from affecting the phenotype. This second allele is said to be **recessive** (page 132) because it has no effect on the phenotype when paired with a dominant allele.
- When a genotype consists of two copies of the same allele, it is **homozygous** (page 132) for that gene. A genotype that consists of two different alleles is **heterozygous** (page 132) for that gene.
- A grid-like diagram called a **Punnett square** (page 134) helps predict the probability of genotypes and phenotypes resulting from a **genetic cross** (page 132). In a series of genetic crosses, the first (parental) generation is the **P generation** (page 132). The first offspring of a genetic cross is the **F_1 generation** (page 132). Offspring of the F_1 generation are the **F_2 generation** (page 132).
- Mendel's experiments enabled him to deduce two laws of inheritance. The **law of segregation** (page 134) states that two alleles of a gene are separated during meiosis and end up in different gametes. The **law of independent assortment** (page 135) states that during meiosis, the two alleles of any given gene segregate independently of any two alleles of any other gene.
- **Mendelian traits** (page 135) are genetic traits controlled by a single gene and unaffected by environmental conditions.
- The **chromosome theory of inheritance** (page 135) explains the mechanism underlying Mendel's laws: genes occupy specific locations on chromosomes, and those chromosomes are randomly shuffled and recombined during meiosis.
- When a heterozygote displays an intermediate phenotype, neither allele is able to exert its full effect. This condition is called **incomplete dominance** (page 138). When the effects of the two alleles are equally visible in the phenotype of the heterozygote, the pair of alleles shows **codominanoo** (page 138).
- **Complex traits** (page 138) have patterns of inheritance that Mendel's laws cannot predict. In **pleiotropy** (page 139), for example, a single gene influences a number of different traits. **Polygenic traits** (page 139), by contrast, are governed by more than one gene. In **epistasis** (page 139), two genes are inherited independently, yet the alleles of one affect the expression of the other gene's alleles.

THE QUESTIONS

See Appendix B for answers.

The Basics

1 Link each term with the correct definition.

genotype	1. An individual that carries one copy each of two different alleles (for example, an *Aa* individual or an *IB* individual).
phenotype	2. An individual that carries two copies of the same allele (for example, an *AA*, *aa*, or *II* individual).
heterozygote	3. An allele that does not affect the phenotype when paired with a different allele in a heterozygote.
homozygote	4. The genetic makeup of an individual; more specifically, the two alleles of a given gene that affect a specific genetic trait in a given individual.
dominant	5. The specific version of a genetic trait that is displayed by a given individual.
recessive	6. An allele that controls the phenotype when paired with a different allele in a heterozygote.

2 Select the correct terms:

The (**gene** / **allele**) for coat color has two (**genes** / **alleles**)—one for brown coloring and one for black.

3 Select the correct terms:

Cells undergo (**mitosis** / **meiosis**) to become gametes. This process sorts the alleles of a gene into separate gametes, which is the basis for Mendel's law of (**segregation** / **independent assortment**). Genes on different chromosomes also sort into separate gametes during this process, which is the basis for Mendel's law of (**segregation** / **independent assortment**).

4 Identify whether each of the following traits is an example of Mendelian inheritance (M) or a more complex form of inheritance (C).

_____ a. brown versus black coat color in dogs

_____ b. body size in dogs

_____ c. coat color in Siamese cats

_____ d. skin color in humans

_____ e. flower color in pea plants

5 A single phenotype that results from a combination of two different genes in which one gene interferes with the expression of another gene is known as

(a) epistasis.

(b) complete dominance.

(c) incomplete dominance.

(d) codominance.

(e) pleiotropy.

6 Before Mendel conducted his experiments with pea plants, people believed that offspring were a "blend" of their parents and would show intermediate levels of their parents' traits. What would Mendel's F_1 pea flowers have looked like if this were true?

(a) white

(b) purple

(c) red

(d) yellow

(e) light purple

Challenge Yourself

7 A riddle: In my type of inheritance, the F_1 offspring of a true-breeding black parent and a true-breeding white parent are all gray. What type of inheritance am I?

8 Some time ago, you noticed that the sunflowers in your garden were either tall or short—nothing in between. You bred the tall sunflowers for many generations, until you felt confident that they "bred true," and you did the same for the short sunflowers. You then set up a parental cross (P), and all of the resulting F_1 offspring were short sunflowers. From this experiment, you conclude that the short phenotype is

(a) pleiotropic.

(b) recessive.

(c) true-breeding in the F_1 generation.

(d) dominant.

(e) incompletely dominant.

9 In chickens, a mutant gene called *frizzle* causes not only feathers that curl outward like a Labradoodle's fur but also an abnormal body temperature, an increased metabolism, and fewer eggs laid than by a normal chicken. From this information, you can conclude that the *frizzle* gene is _________.

Try Something New

10 The silver fox belongs to the same species as the red fox (see "The New Family Pet?" on page 141). Two silver foxes always breed true for silver offspring. A silver fox bred to a red fox will produce either all red offspring or, occasionally, half red and half silver offspring. Red foxes bred together usually produce all red offspring, but they occasionally produce silver offspring in the ratio of 3 red to 1 silver. (*Hint*: Make up your own genotype for silver/red fox fur and then draw Punnett squares showing these predicted results.) Which of the following statements is/are consistent with the information provided here about inheritance of coat color in these types of foxes? (Select all that apply.)

(a) Red foxes are all homozygous.

(b) Silver foxes are all homozygous.

(c) Red is dominant to silver.

(d) Some silver foxes are homozygous and some are heterozygous.

(e) Some red foxes are homozygous and some are heterozygous.

11 In your garden you grow Big Boy (round) and Roma (oval) tomatoes. You love the taste of Big Boys, but you think it's easier to slice Roma tomatoes. You decide to cross-pollinate a Big Boy and a Roma to see whether you can create a new strain of "Long Boys." In the first generation, all the tomatoes are round. How would you explain this result? What would your next cross be? Write out the cross in a Punnett square, using parental genotypes. What proportion of the next generation, if any, would be oval?

12 For several hundred years, goldfish have been selectively bred in China and Japan for body color and shape, tail shape, bulging eyes, and even fleshy head growths.

Wild goldfish | Pet-shop goldfish | Black moor goldfish

Imagine that you have a tank of pet-shop goldfish and have just added a couple of black moor goldfish, hoping they will breed. When the eggs laid by the black moor female (P generation) hatch and the young fish (F_1 generation) begin to develop, you are shocked to see that they are orange. How would you explain this result in terms of the inheritance of body color in goldfish? What breeding experiment could you conduct to test your hypothesis?

13 In 2009, a large team of researchers including Elaine Ostrander and Gordon Lark published the results of research on coat inheritance in dogs. The study began by focusing on dachshunds and Portuguese water dogs but then widened to more than 80 breeds. The scientists were able to explain 95 percent of the variation in dog coat types with just two alleles at each of three genes, each inherited independently of the other. These genes coded for hair length (L = short / l = long), wave or curl in the coat (W = wave / w = curl), and the presence of "furnishings" (F = furnishings / f = no furnishings), which are the moustache and eyebrows often seen in wire-haired dogs (see middle dog in the photo). Long-haired dogs carry two copies of the long-hair allele, which is recessive to the short-hair allele. Dogs with furnishings can be either homozygous or heterozygous for the furnishings allele; dogs without furnishings are homozygous for the no-furnishings allele.

(a) At the hair length and furnishings genes, what is the genotype of a long-haired dog without furnishings?

(b) At the hair length and furnishings genes, what are all the possible genotypes of a short-haired dog with furnishings?

(c) Create a Punnett square of two dogs heterozygous for hair length and furnishings. What is the offspring phenotype ratio for those two traits?

Leveling Up

14 **Doing science** Do you want to get involved in dog research? If you have a purebred as a pet, you can. Find out whether the Dog Genome Project at the National Institutes of Health is doing research on your pet's breed. If they are, you can send in a swab of your dog's saliva and contribute to science. Visit the NIH website (http://research.nhgri.nih.gov/dog_genome) for more information.

15 **Looking at data** Humans and dogs have lived together for at least 15,000 years. However, not all people are interested in sharing their life with a dog. In a May 2019 study, researchers investigated whether being a "dog person" has a genetic component. The accompanying figure shows the estimated genetic contribution (heritability) of having a dog, based on a study of more than 35,000 twins.

(a) What is the approximate genetic estimate at age 30? At age 60?

(b) At what age is the genetic estimate highest?

(c) Would you predict that cat ownership has a genetic component? Why or why not?

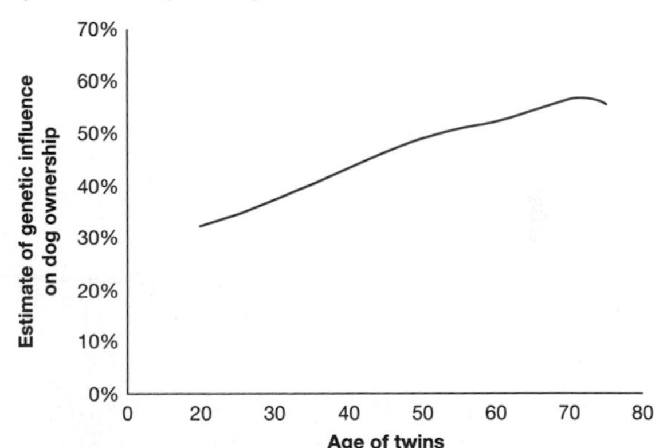

16 **What do *you* think?** Many people are critical of those who breed or purchase purebred dogs, arguing that there are many mixed-breed dogs waiting to be adopted from shelters. They also point out that mixed-breed dogs are less likely than purebred dogs to suffer from genetic diseases. Those who prefer a particular breed argue that there is a strong genetic influence on dog personality and behavior and that they don't want any surprises when they add a new member to their family. What do you think?

Digital resources for your book are available online.

Curing the Incurable

Gene therapies halt, even reverse, diseases once considered a death sentence. The medical revolution has begun.

After reading this chapter you should be able to:

- Review a human karyotype to identify the sex chromosomes and any abnormalities in chromosome number.
- Diagram a chromosome, identifying genes, alleles, and loci.
- Describe the genotype and phenotype of a "genetic carrier" and use this term appropriately.
- Explain how sex is genetically determined in humans, and how sex determination relates to the inheritance of sex-linked traits.
- Compare and contrast the inheritance of recessive, dominant, and sex-linked genetic disorders.
- Calculate the probability of inheriting a particular genetic disorder by using a Punnett square.
- Interpret a human pedigree to determine whether a given condition is recessive, dominant, or sex-linked.

GENETICS

CHAPTER 08

CHROMOSOMES AND HUMAN GENETICS

Three-month-old Arabella Smygov couldn't lift her head and kept her arms tucked tightly against her body in fetal position, never reaching for toys. Doctors soon identified the cause—an uncommon allele of a gene that produces a protein critical for muscle function.

As a teenager, Jennelle Stephenson struggled to walk upstairs. She expected to die young. Her genome harbored a gene that leads to abnormally shaped red blood cells, which clumped together to block her blood vessels and caused "bone-crushing" pain.

Eleven-month-old Samuel Evangelista developed a life-threatening infection after a doctor's visit. An abnormal gene in his genome meant Samuel didn't have a functioning immune system, so he had no way to ward off infections or respond to vaccines.

Arabella, Jennelle, and Samuel were born with severe genetic disorders (**Figure 8.1**). A **genetic disorder** is a disease caused by a defective gene or genes. It can be caused by an individual gene or an abnormality in chromosome number or structure. In all cases, these disorders result from genetic mutation. A mutation can be inherited—that is, passed down from parent to child—or it can occur spontaneously, with no family history of the illness.

Figure 8.1

Jennelle Stephenson

Before gene therapy, Jennelle suffered extreme pain in her bones and joints due to sickle cell disease.

Twenty years ago, a genetic disorder such as Arabella's, Jennelle's, or Samuel's disorder was considered incurable, set in stone. Because it was coded in the individual's DNA, the disorder was etched into the very blueprint of the body. But what if that faulty blueprint could be corrected?

Gene therapy is a technique for correcting abnormal genes responsible for disease development. By using genes as medicine, to replace or remove mutated DNA, scientists hope to stop or slow disease. Gene therapy is one form of **genetic engineering**, the permanent introduction of one or more new or altered genes into a cell, tissue, or organism.

When it was first proposed in the 1960s, gene therapy was a pipe dream. Early versions garnered publicity and excitement, but they didn't work, and the field was plagued with safety issues. Over time, new advances in medicine and technology led to some of the first gene therapy successes. "In the 2000s, trials started to show benefits," says Donald Kohn, a doctor and researcher at the University of California, Los Angeles, who has led numerous gene therapy trials since 1993.

In 2012, the first-ever gene therapy was approved for use: European health officials gave the green light to gene therapy for a genetic disorder called lipoprotein lipase deficiency (LPLD). Five years later, the U.S. government approved a gene therapy for a form of inherited childhood blindness caused by a mutation that prevented light receptors in the eye from working.

Both of those are rare diseases, occurring in a small percentage of the population. Today, gene therapy is moving mainstream, tackling some of the most common genetic diseases. These one-time, potentially curative treatments are especially successful in diseases involving blood, because doctors can easily access blood cells for use. "It's working now for at least a dozen diseases, and there's probably at least another dozen coming to the clinic," says Kohn.

Gene therapy is a twenty-first-century medical success story, but the field was almost abandoned at the start.

Putting Viruses to Work

In September 1999, an 18-year-old named Jesse Gelsinger took part in one of the earliest gene therapy clinical trials (**Figure 8.2**). As noted in Chapter 2, a clinical trial is a study in which a drug or treatment is tested for safety and efficacy in a small group of people. Jesse's genetic disorder was under control with medications and diet, but he wanted to help others overcome the same rare illness he had, a metabolic condition called ornithine transcarbamylase deficiency.

Jesse was the eighteenth person in the trial to receive the therapy. Unlike the other participants, within a day his body had a violent immune reaction to a virus used as part of the gene therapy. Four days after receiving the therapy, Jesse died.

Genes cannot be inserted directly into a cell. Because they need a vehicle to transport them inside, doctors use viruses as transport vehicles, or vectors, in gene therapy (**Figure 8.3**). Although nonviral vectors can carry DNA, inactivated viruses—that is, viruses that are engineered by scientists to not carry disease—are the most common vectors used in gene therapy because they are very good at sneaking into a cell's nucleus where they deliver the healthy copy of a gene to the genome.

Initially, the types of viral vectors used in gene therapy didn't work well, and some were unsafe. The one Jesse Gelsinger was exposed to, for example, caused his immune system to react as if he had a terrible infection. In another case, a viral vector that was used in gene therapy sparked leukemia, a form of blood cancer, in children.

But by the early 2000s, researchers had discovered better and safer viral vectors to deliver healthy genes to human cells. A series of clinical trials starting in 2007 showed that gene therapy could help patients with genetic disorders and produce almost no side effects.

Whether inherited or spontaneous, genetic disorders result from mutations: changes to specific DNA sequences, chromosome number, or chromosome structure. Every species has a characteristic number of chromosomes. For example, humans have 23 pairs of homologous (matching) chromosomes, for a total of 46, whereas mosquitoes have only 3 pairs, or 6 in total. Of the human chromosomes, 22 of the 23 pairs are **autosomes**. The final pair of human chromosomes are the **sex chromosomes**, which determine a person's biological sex.

Autosomes are homologous chromosomes exactly alike in terms of length, shape, and the genes they carry (**Figure 8.4**). Human autosomes are labeled with the numbers 1 through 22 from largest to smallest (for example,

Figure 8.2

Volunteer and early casualty

Jesse Gelsinger died as a result of his participation in an early gene therapy clinical trial.

DEBUNKED!

MYTH: There are only two biological sexes.

FACT: Although most people are either male or female, there are many people whose genes, chromosomes, hormones, and sex organs do not all fit easily into one category or another.

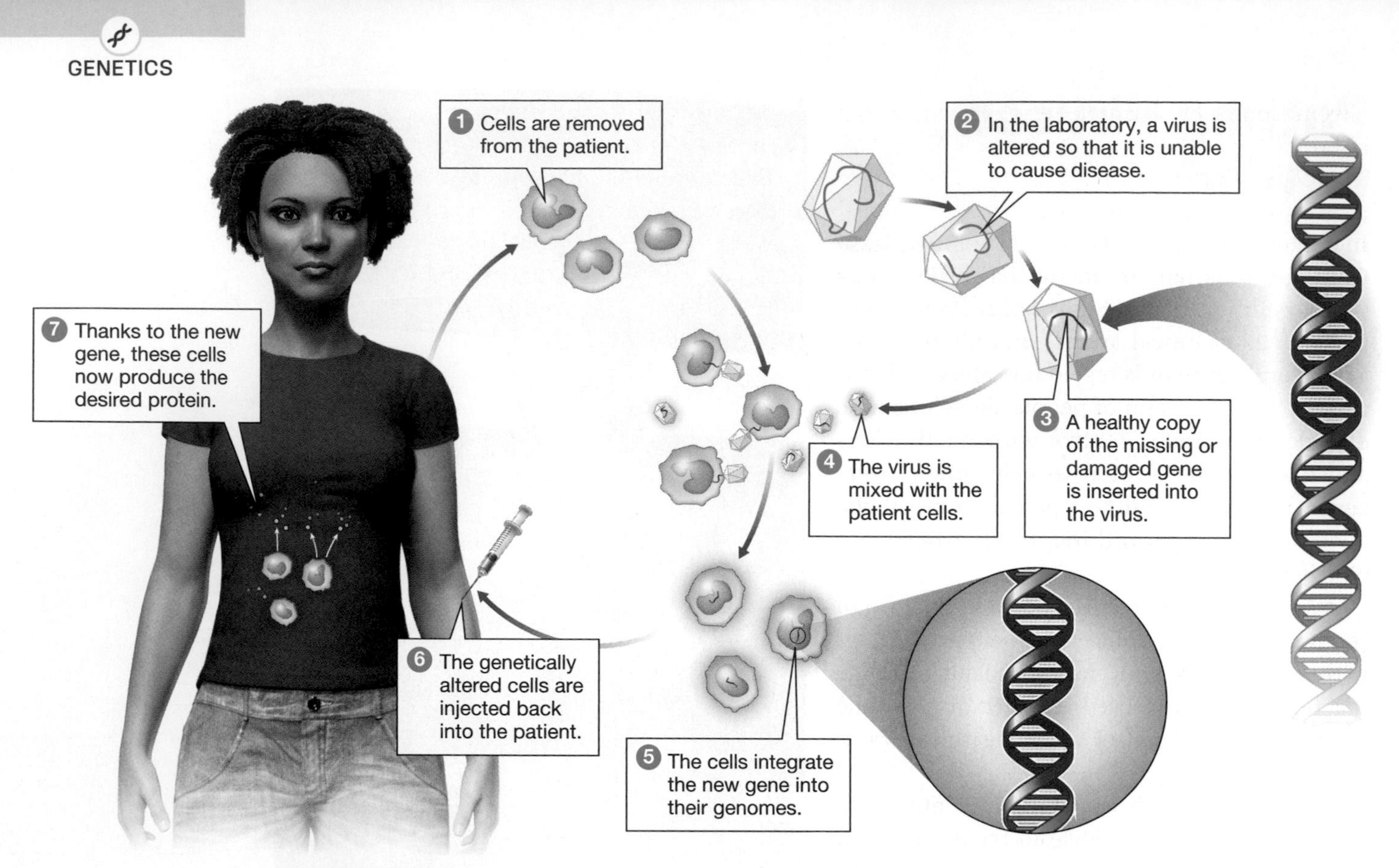

Figure 8.3

Gene therapy

In gene therapy, genetic information is transferred into cells to achieve a desired effect. Gene therapy may be used to compensate for an abnormal gene that causes the cell to malfunction.

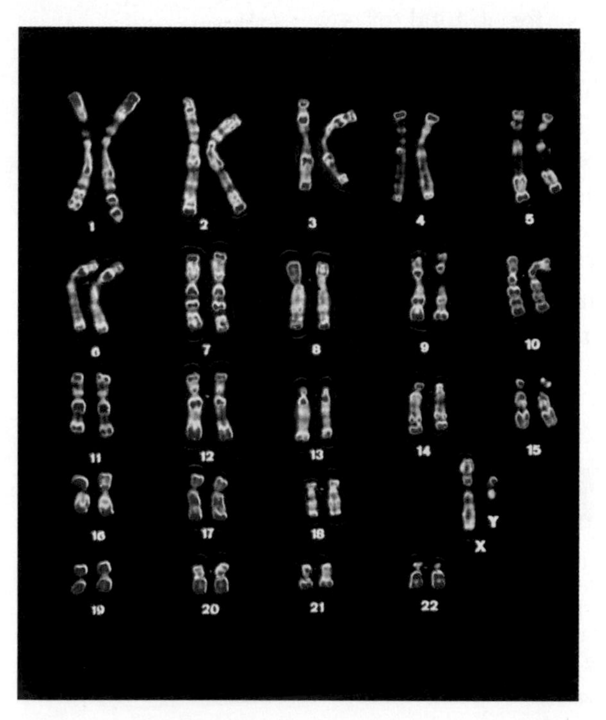

Figure 8.4

A human karyotype

To research chromosomal abnormalities, a scientist will photograph a cell's chromosomes during mitosis (see Chapter 6) and then pair up each set of homologous chromosomes to create a **karyotype**. In humans, the autosomes are numbered 1 through 22, and the sex chromosomes are designated X or Y.

Before answering the following questions, return to the text and read the rest of this section.

Q1: Is this illustration the karyotype of a male or a female?

Q2: How would the karyotype of a person with Down syndrome differ from this karyotype?

Q3: The size of a chromosome correlates roughly with the number of genes residing on it. Why are an extra copy of chromosome 21 and a missing Y chromosome two of the least damaging chromosomal abnormalities?

See Appendix A for answers to the figure questions.

chromosome 4). Jennelle was ill due to an abnormal gene on chromosome 11. Arabella's altered gene was on chromosome 5. Sex chromosomes, by contrast, are assigned letter names; in humans, males have one X chromosome and one Y chromosome, whereas females have two X chromosomes. The Y chromosome in humans is much smaller than the X chromosome. Samuel's mutation was on the X chromosome.

Each chromosome has a particular structure, with genes arranged on it in a precise sequence. Any change in the chromosome number or structure, compared to what is typical for a species, is considered a **chromosomal abnormality** (**Figure 8.5**).

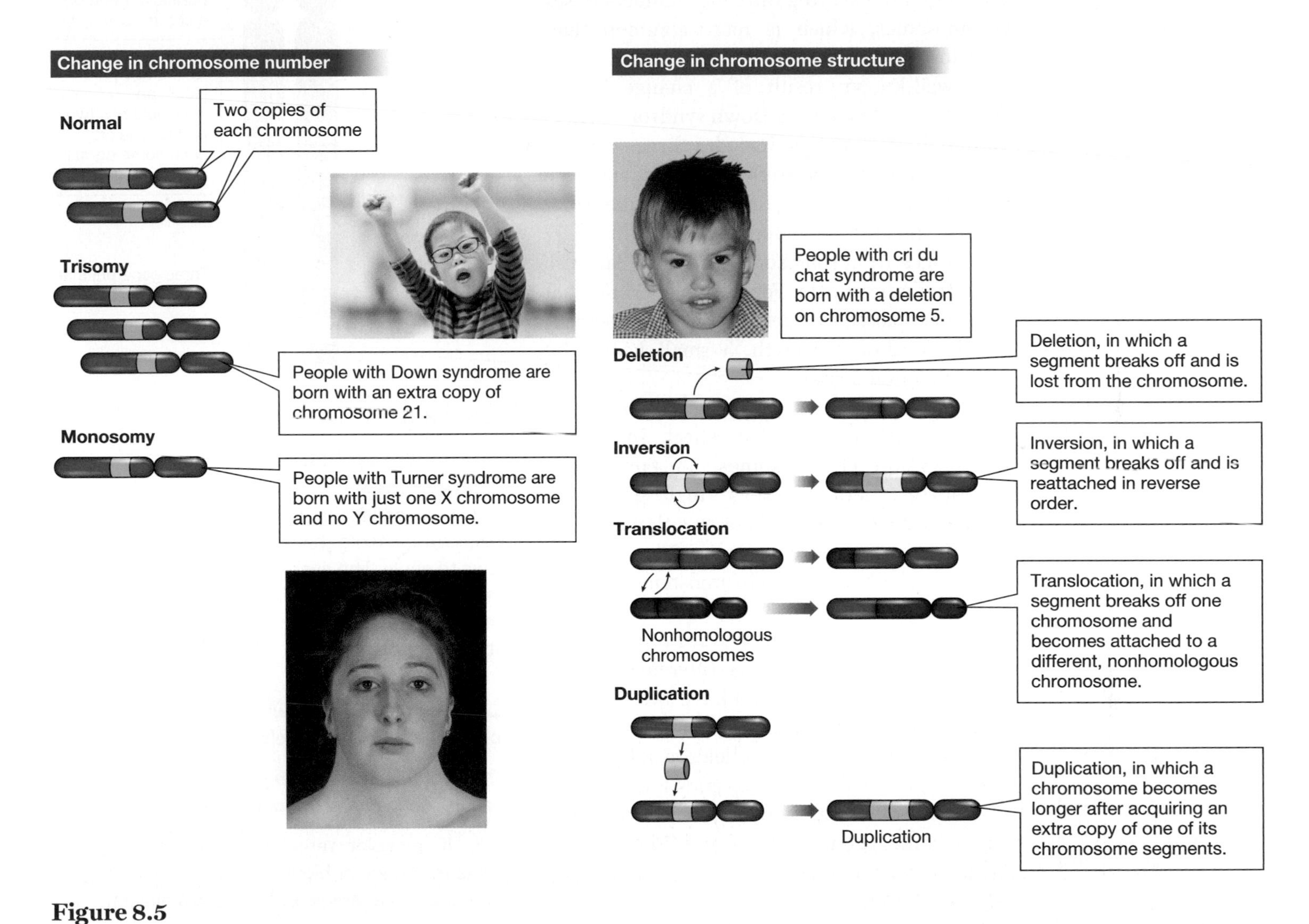

Figure 8.5

Chromosomal abnormalities can cause serious genetic disorders

Any increase or decrease in the number of chromosomes almost invariably results in spontaneous abortion of the fetus, which is estimated to occur in up to 20 percent of all pregnancies. Down syndrome and a missing or additional sex chromosome are exceptions. Changes in chromosome structure may have relatively minor or more severe effects, depending on the size and location of the change.

Q1: Why are changes in chromosome number almost always more severe than changes in chromosome structure?

Q2: In which part of meiosis would you predict that chromosomal abnormalities are produced? (Refer back to Chapter 6 if necessary.)

Q3: Create a mnemonic (a code) to help you remember the four kinds of structural changes to chromosomes.

See Appendix A for answers to the figure questions.

Changes in the number of chromosomes in humans are usually lethal because of the number of genes carried on each chromosome. An exception to this rule is Turner syndrome, in which females receive only one X chromosome. Individuals with Turner syndrome tend to live long, healthy lives with mild to moderate reproductive issues. Several other disorders are related to carrying different numbers of sex chromosomes, which is more common than carrying different numbers of autosomes.

A well-known result of a change in the number of autosomes is Down syndrome. Here, an individual's genome includes three copies of chromosome 21. A person with Down syndrome can live a relatively long life but may suffer from mild to moderate developmental disabilities.

Changes in chromosome structure can also have dramatic effects. For example, cri du chat syndrome, caused by a deletion on chromosome 5, results in slowed growth, a small head, and developmental delays.

Just as mutations can affect entire chromosomes, they can also affect individual genes within chromosomes. In fact, many genetic disorders involve a single gene, and these disorders are often not as lethal as those involving entire chromosomes. The physical location of a gene on a chromosome is called its **locus** (plural "loci"). The locus for Arabella's defect is a gene called *SMN1* on chromosome 5. Because a gene can occur in different versions, or alleles, a diploid cell can have two different alleles at a given locus on a pair of homologous chromosomes. If the two alleles at a locus are different, the cell is heterozygous for the gene. If the two alleles at a locus are identical, the cell is homozygous for the gene (**Figure 8.6**). Arabella was homozygous for an abnormal allele of *SMN1*.

If a person's cells are homozygous for a disease-causing mutation, gene therapy can introduce a new, healthy copy of that gene into the cell. In most cases of gene therapy, doctors start by removing stem cells from a patient's body. **Stem cells** are unique, unspecialized cells that can make identical copies of themselves for long periods of time. They are mixed in a dish with a virus vector carrying the healthy gene. Then the cells are frozen and repeatedly tested to make sure the gene was successfully delivered into the genome of the stem cells.

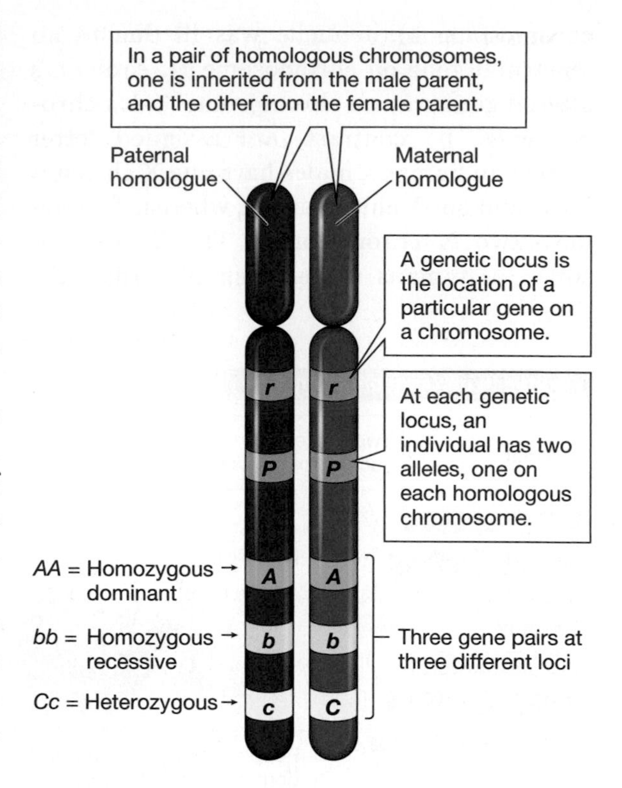
In a pair of homologous chromosomes, one is inherited from the male parent, and the other from the female parent.
Paternal homologue
Maternal homologue
A genetic locus is the location of a particular gene on a chromosome.
At each genetic locus, an individual has two alleles, one on each homologous chromosome.
r
r
P
P
A
A
b
b
c
C
AA = Homozygous dominant
bb = Homozygous recessive
Cc = Heterozygous
Three gene pairs at three different loci

Figure 8.6

Genetic loci on homologous chromosomes

The genes shown here take up a larger portion of the chromosome than they would if they were drawn to scale. The average human chromosome has more than a thousand different genes interspersed with large stretches of DNA that do not code for proteins.

Q1: How do we know whether two chromosomes are homologous?

Q2: In one sentence, explain how the terms "gene," "locus," and "chromosome" are related.

Q3: If hair color were determined by a single gene for brown or blonde hair (it is not), what would be an example of the gene's alleles?

See Appendix A for answers to the figure questions.

Once doctors are satisfied that the cells have integrated the new, healthy copy of the gene, the genetically modified cells are injected back into the patient, where they replicate, spread through the body, and begin to produce the desired protein. It's like a medication that, once taken, stays in the body and keeps working indefinitely.

The process to insert a healthy version of a gene does not take long, says Marina Cavazzana of Necker Enfants Malades Hospital, Paris, who has led numerous clinical trials for gene therapies. Her work recently led to a gene therapy for the blood disease beta thalassemia; the treatment was approved in Europe in June 2019. It takes about 4 days to grow the patient's cells in the presence of a virus vector followed by 1 month of testing the cells, according to Cavazzana. "We do a number of tests to be sure the level of correction we have obtained is enough to cure the patient."

Going Mainstream

The first set of gene therapies to become commercially available treated rare diseases. Now, Kohn, Cavazzana, and others are bringing gene therapy to mainstream medicine by treating illnesses that are more common.

One of these illnesses is sickle cell disease, the blood disorder that Jennelle Stephenson suffered from, and that approximately 100,000 people in the United States and millions of people around the world have inherited. In the United States, about 15 infants per 1,000 births are born with sickle cell disease. The disorder is more common among certain ethnic groups, including people of African, Mediterranean, or southwest Asian descent.

People with sickle cell disease have a genetic mutation that causes abnormal hemoglobin, a protein in red blood cells. Because of this abnormality, the red blood cells become rigid, sticky, and deformed. Their crescent moon shape resembles a farm tool used to cut wheat called a sickle, hence, the name sickle cell disease. These irregular cells get stuck in small blood vessels, blocking the flow of blood and oxygen to parts of the body and leading to extreme pain in bones and joints as well as bone deterioration, strokes, and organ failure.

Sickle cell disease is a **recessive genetic disorder** carried on autosomes. Both males and females have two copies of autosomal chromosomes and therefore identical odds of being homozygous or heterozygous for a disorder allele. Several thousand human genetic disorders are inherited as recessive traits on autosomes. These include Tay-Sachs disease and the most common fatal genetic disease in the United States, cystic fibrosis (**Figure 8.7**).

Recessive genetic disorders vary in severity. Some have relatively mild effects. Adult-onset lactose intolerance, for example, is caused by a single recessive allele that leads to a shutdown in the production of lactase, the enzyme that digests milk sugar. Others are lethal. Arabella Smygov, the infant who couldn't lift her head, was diagnosed with spinal muscular atrophy type 1 (SMA1). This severe recessive genetic disorder typically kills those with it by age 2.

The only individuals who have disorders caused by an autosomal recessive allele (*a*) are those who have two copies of that allele (*aa*). Usually, when a child inherits a recessive genetic disorder, both parents are heterozygous; that is, they both have the genotype *Aa* (**Figure 8.8**). It is also possible for one or both parents to have the genotype *aa* and thus the disease. Because the *A* allele is dominant and does not cause the disorder, heterozygous individuals (*Aa*) are **genetic carriers** of the disorder; they carry the disorder allele (*a*) but do not have the disease.

If two carriers of a recessive genetic disorder have children, the patterns of inheritance are the same as for any other recessive trait. Each child, male or female, has a 25 percent chance of not carrying the disorder allele (genotype *AA*), a 50 percent chance of being a carrier (genotype *Aa*), and a 25 percent chance of inheriting the disorder (genotype *aa*).

These percentages reveal one way in which lethal recessive disorders such as SMA1 can persist in the human population. Although homozygous recessive individuals (with genotype *aa*) often die before they are old enough to have children, carriers (with genotype *Aa*) are not harmed by the disorder. In a sense, the *a* alleles can hide in heterozygous carriers, and those carriers are likely to pass the disorder allele to half of their children. Recessive genetic disorders can also arise in the human population because new mutations can produce new copies of the recessive alleles.

In early 2019, Arabella received gene therapy in a clinical trial at Seattle Children's Hospital to replace her abnormal gene with a healthy one. One month after receiving an injection of a new, working copy of the gene carried by a viral vector, Arabella was sitting up in a swing with her head upright. Soon, she was able to roll over and sit on her own and hold out her hands

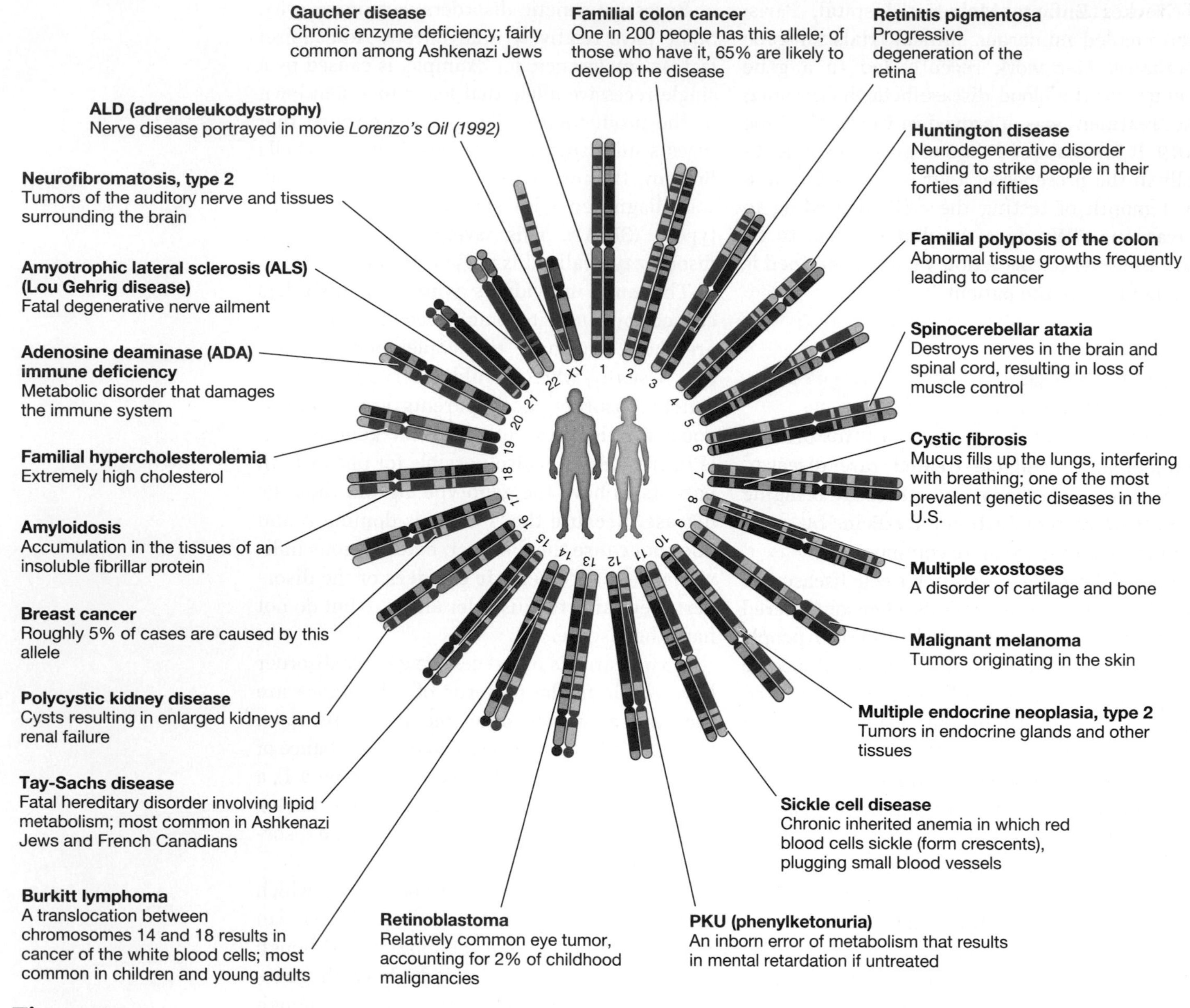
Gaucher disease
Chronic enzyme deficiency; fairly common among Ashkenazi Jews
Familial colon cancer
One in 200 people has this allele; of those who have it, 65% are likely to develop the disease
Retinitis pigmentosa
Progressive degeneration of the retina
ALD (adrenoleukodystrophy)
Nerve disease portrayed in movie *Lorenzo's Oil (1992)*
Huntington disease
Neurodegenerative disorder tending to strike people in their forties and fifties
Neurofibromatosis, type 2
Tumors of the auditory nerve and tissues surrounding the brain
Familial polyposis of the colon
Abnormal tissue growths frequently leading to cancer
Amyotrophic lateral sclerosis (ALS) (Lou Gehrig disease)
Fatal degenerative nerve ailment
Spinocerebellar ataxia
Destroys nerves in the brain and spinal cord, resulting in loss of muscle control
Adenosine deaminase (ADA) immune deficiency
Metabolic disorder that damages the immune system
Familial hypercholesterolemia
Extremely high cholesterol
Cystic fibrosis
Mucus fills up the lungs, interfering with breathing; one of the most prevalent genetic diseases in the U.S.
Amyloidosis
Accumulation in the tissues of an insoluble fibrillar protein
Multiple exostoses
A disorder of cartilage and bone
Breast cancer
Roughly 5% of cases are caused by this allele
Malignant melanoma
Tumors originating in the skin
Polycystic kidney disease
Cysts resulting in enlarged kidneys and renal failure
Multiple endocrine neoplasia, type 2
Tumors in endocrine glands and other tissues
Tay-Sachs disease
Fatal hereditary disorder involving lipid metabolism; most common in Ashkenazi Jews and French Canadians
Sickle cell disease
Chronic inherited anemia in which red blood cells sickle (form crescents), plugging small blood vessels
Burkitt lymphoma
A translocation between chromosomes 14 and 18 results in cancer of the white blood cells; most common in children and young adults
Retinoblastoma
Relatively common eye tumor, accounting for 2% of childhood malignancies
PKU (phenylketonuria)
An inborn error of metabolism that results in mental retardation if untreated
XY
1
2
3
4
5
6
7
8
9
10
11
12
13
14
15
16
17
18
19
20
21
22

Figure 8.7

Single-gene disorders

Mutations of single genes that lead to genetic disorders are found on the X chromosome and on each of the 22 autosomes in humans. In each of these mutations, the healthy allele at that locus codes for an important function; for example, the sickle cell allele is a mutation in the gene that codes for the hemoglobin protein, critical for carrying oxygen in the blood. For clarity, this figure shows only one such genetic disorder per chromosome.

Q1: Which chromosome contains the gene for cystic fibrosis? For Tay-Sachs disease? For sickle cell disease?

Q2: No known genetic disorders are encoded on the Y chromosome. Why do you think this is?

Q3: In your own words, explain why most single-gene disorders are recessive rather than dominant.

See Appendix A for answers to the figure questions.

for toys. The therapy that helped Arabella was approved in May 2019 by the U.S. government.

Cavazzana and her team in Paris published the results of the first successful gene therapy in a teenager with sickle cell disease in 2017, and they continue to treat patients in clinical trials.

In the United States, on the day after Christmas in 2017, Jennelle Stephenson received a gene therapy for sickle cell disease as part of a clinical trial at the National Institutes of Health. "Merry Christmas to me," said Jennelle with a smile, sitting on a hospital bed and holding a clear bag filled with 500 million of her own stem cells, genetically modified via a viral vector to contain a healthy version of the hemoglobin gene.

Nine months after that infusion, when doctors checked Jennelle's blood, there wasn't a sickle cell in sight. Although she once struggled to walk up a flight of stairs, Jennelle was now able to begin going to the gym, swimming, and taking martial arts classes. "I couldn't do anything, and now I've broken free," she later said in a news interview. "I'm in charge now. My cells are no longer in charge."

Deadly with One Allele

Sickle cell disease is an example of a recessive genetic disorder, in which a child inherits two recessive copies of a disorder allele. A rarer type of inherited disease is a **dominant genetic disorder**, caused by an autosomal dominant allele (*A*). In this case, the allele that causes a disorder cannot "hide" in the same way that a recessive allele can: *AA* and *Aa* individuals have the disorder; only *aa* individuals are symptom-free (**Figure 8.9**). Dominant genetic disorders are rarer than recessive disorders because a dominant disorder often produces serious negative effects immediately upon birth, and individuals with the *A* allele may not live long enough to reproduce. Hence, few people with a dominant genetic disorder pass the allele on to their children.

For this reason, most cases of a dominant genetic disorder are produced by a new, spontaneous mutation in a generation. For example, achondroplasia, a form of dwarfism, is caused by a mutation in a gene involved in bone growth.

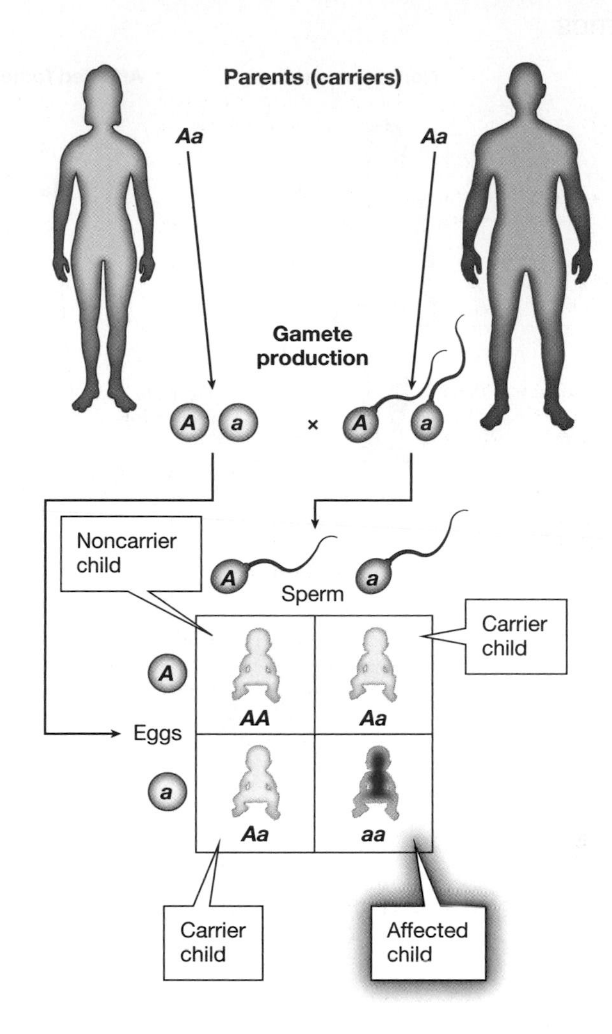

Figure 8.8

Inheritance of spinal muscular atrophy type 1 (SMA1), an autosomal recessive disorder

The patterns of inheritance for a human autosomal recessive genetic disorder are the same as for any recessive trait (compare this figure with the pattern shown by Mendel's pea plants in Figure 7.5). Recessive disorder alleles are denoted *a*. Dominant, normal alleles are denoted *A*. Here, the parents are a carrier female (genotype *Aa*) and a carrier male (genotype *Aa*).

Q1: Which of the children in this Punnett square represents Arabella? What is her genotype?

Q2: If Arabella's parents had another child, what is the probability that the child would have SMA1? What is the probability that the child would be a carrier of SMA1?

Q3: If Arabella is able to have a child of her own someday, and the other parent is not a carrier of SMA1 (he would likely be tested before they chose to have children; see "Prenatal Genetic Screening" on page 157), what is the probability that the child would have SMA1? What is the probability that the child would be a carrier? (Arabella's treatment does not change sex cells.)

See Appendix A for answers to the figure questions.

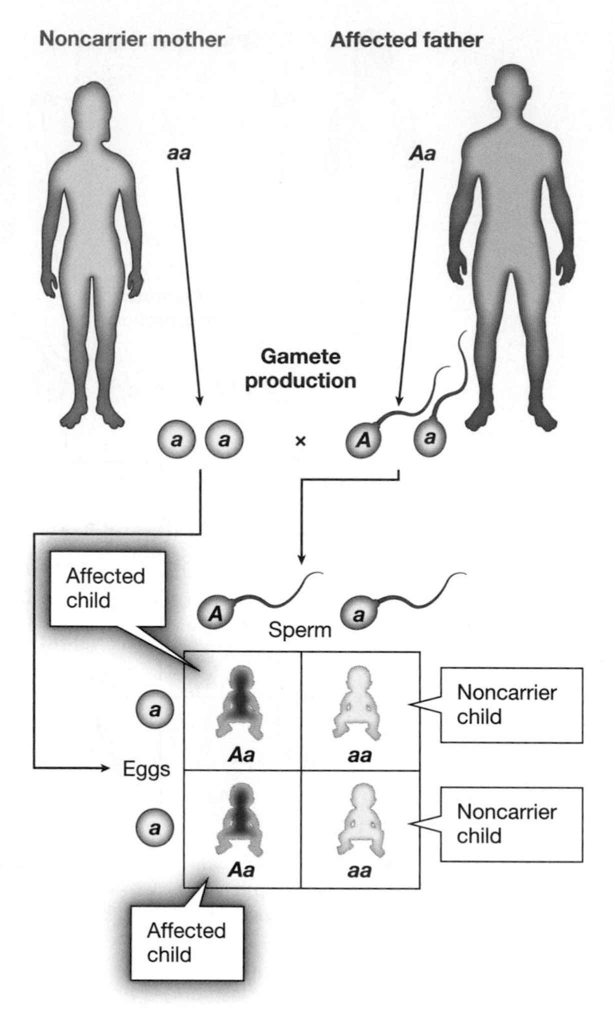

Figure 8.9

Inheritance of an autosomal dominant disorder

The pattern of inheritance for a human autosomal dominant genetic disorder is the same as for any other dominant trait. This Punnett square shows the possible children of a normal female (genotype *aa*) and an affected male (genotype *Aa*).

Q1: What is the probability that a child with one parent who has an autosomal dominant disorder will inherit the disease?

Q2: Why are there no carriers with a dominant genetic disorder?

Q3: Because dominant genetic disorders are rare, it is extremely rare for both parents to have the condition (genotype *Aa*). Draw a Punnett square with two *Aa* parents. What proportion of the offspring would have the disorder? What proportion would not have it?

See Appendix A for answers to the figure questions.

People with achondroplasia have a decreased life span, so few of them live long enough to pass the mutation on to their offspring. Instead, infants with achondroplasia are born to unaffected parents at a rate of between one in 10,000 and one in 100,000. Almost all of these births are to fathers older than 35 years, who produce this mutation during sperm production.

Huntington disease, another dominant genetic disorder, is an exception: it is often passed down from parent to child. Symptoms of Huntington disease, such as uncontrolled movements and loss of intellectual faculties caused by dying brain cells, arise in a person's forties, after the person carrying that allele has had the opportunity to reproduce. In this way, the allele is readily passed from one generation to the next. For this reason some couples whose families have a history of Huntington disease may choose to screen their developing fetus for the gene that causes the disorder (see "Prenatal Genetic Screening").

Trapped in a Bubble

Some genetic diseases are due to sex-linked inheritance—that is, an inherited mutation on the X or Y chromosome. Recall the case of Samuel Evangelista, the little boy without an immune system.

When Samuel was 3 months old, he was diagnosed with severe combined immunodeficiency (SCID). This illness is also known as *bubble boy disease*, because of a young boy in Texas, David Vetter, who had it. From birth until age 12 when he died, David lived in a plastic bubble to protect him from infection. When he was 6, David got a chance to walk and play outdoors thanks to NASA engineers, who built him a special spacesuit (**Figure 8.10**).

There are several types of SCID, each caused by a mutation of a gene that makes an immune protein. Any of these mutations results in a malfunctioning immune system. A person's lack of immunity then leads to frequent infections from bacteria, fungi, and viruses. Samuel's disease was caused by a mutation in a gene that makes a protein essential for normal immune function, the interleukin-2 receptor subunit gamma (*IL2RG*) gene, which is located on the X chromosome. Roughly 1,240

Prenatal Genetic Screening

How is the baby? This is one of the first questions we ask after a child is born. Usually everything is fine, but sometimes, as with Arabella, the answer can be devastating. Today, some parents choose to have prenatal genetic screening performed to check the health of the fetus before birth.

In one method, **amniocentesis**, a needle is inserted through the abdomen into the uterus to extract a small amount of amniotic fluid from the pregnancy sac that surrounds the fetus. This fluid contains fetal cells (often sloughed-off skin cells) that can be tested for genetic disorders. Another method is **chorionic** (kohr-ee-AH-nik) **villus sampling (CVS)**. Here, a physician uses ultrasound to guide a narrow, flexible tube through a woman's vagina and into her uterus, where the tip of the tube is placed next to the villi, a cluster of cells that attaches the pregnancy sac to the wall of the uterus. Cells are removed from the villi by gentle suction and then tested for genetic disorders.

Risks associated with amniocentesis and CVS—including vaginal cramping, miscarriage, and premature birth—have declined dramatically in recent years because of advances in technology and more extensive training. Recent studies suggest that the risk of miscarriage after CVS and amniocentesis is essentially the same: about 0.06 percent.

The tests are widely used by parents who know they face an increased chance of giving birth to a baby with a genetic disorder. Older parents, for example, might want to test for Down syndrome, as the risk of that condition increases with the age of the mother and perhaps the father. A couple in which one parent carries an allele for a dominant genetic disorder (such as Huntington disease), or both parents are carriers for a recessive genetic disorder (such as cystic fibrosis), might also choose prenatal genetic screening.

Couples that elect to have such tests performed have only two choices if their fears are confirmed: they can either abort the fetus or they can give birth to a child with a genetic disorder. Prior to conception, however, couples at risk of having a child with a genetic disorder can choose a different route, which involves two steps:

- If they are willing and can afford the procedure, a couple can choose to have a child by **in vitro fertilization (IVF)**, in which an egg is fertilized by a sperm in a petri dish, after which one or more embryos are implanted into the mother's uterus.
- During **preimplantation genetic diagnosis (PGD)**, one or two cells are removed from the developing embryo in the dish, usually 3 days after fertilization. The cell or cells removed from the embryo are tested for genetic disorders. Finally, one or more embryos that are free of disorders are implanted into the mother's uterus, and the rest of the embryos, including those with genetic disorders, are frozen. Typically, parents who opt for PGD either have a serious genetic disorder or carry alleles for one.

Like all other genetic screening methods, PGD raises ethical issues. People who support the use of PGD feel that amniocentesis and CVS provide parents with a bleak set of moral choices. In their view, PGD enables parents to determine their offspring's fate at the earliest possible opportunity (for example, as an embryo at the 4- to 12-cell stage rather than as a well-developed fetus). Those opposed to PGD argue that once fertilization has occurred, a new life has formed and it is immoral to end that life (even at the 4- to 12-cell stage). What do you think?

Figure 8.10

The original "bubble boy"
David Vetter plays outside wearing the protective suit made for him by NASA.

Genetic Diseases Affecting Americans

Spinal muscular atrophy, sickle cell disease, and severe combined immunodeficiency are just a few of the many genetic conditions seen in humans. Here are some of the most common genetic diseases in the United States, most of which can be identified in newborns through genetic testing.

Assessment available in smartwork

U.S. births per year: 4,000,000

• = 1 birth

6,037 births
Down Syndrome

1,140 births
Cystic Fibrosis

800 births
Marfan Syndrome

Male U.S. births per year: 2,050,000

• = 1 male birth

590 births
Duchenne Muscular Dystrophy

400 births
Hemophilia

400 births
Fragile X Syndrome

of the estimated 20,000 human genes are found on the X and Y chromosomes. Because of their location, these 1,240 genes are said to be **sex-linked**.

Approximately 1,180 of those 1,240 genes are located on the X chromosome, whereas only about 60 genes are located on the much smaller Y chromosome. Genes on the X chromosome are **X-linked**. *IL2RG* is found on the X chromosome, so Samuel's disease is called X-linked SCID, or XSCID. X chromosomes contain genes known to be involved in many human genetic traits and disorders, although the latter are rare compared to autosomal disorders.

Genes on the Y chromosome are **Y-linked**. Because human females have two copies of the X chromosome, all the gametes (eggs) they produce contain one X chromosome, passed on to their offspring. Males, however, have one X chromosome and one Y chromosome, so half of their gametes (sperm) will contain an X chromosome and half will contain a Y chromosome (**Figure 8.11**). There are no well-documented cases of disease-causing Y-linked genes. One critical gene found on the Y chromosome is *SRY*, the gene that kicks off male sex determination in humans. Mutations within *SRY* lead to sexual development disorders.

We can use a Punnett square to illustrate how the X-linked recessive mutation for SCID is inherited. We label the recessive mutated *IL2RG* allele *a*, and in the Punnett square we write this allele as X^a to emphasize the fact that it is on the X chromosome. Then we label the dominant, healthy allele *A* and write this allele as X^A in the Punnett square (**Figure 8.12**).

In this illustration, the mother is a carrier for XSCID, with the genotype X^AX^a. The father is a noncarrier, with the genotype X^AY. Each of their sons will have a 50 percent (1 in 2) chance of getting the disorder. Males of genotype X^aY suffer from the condition because the Y chromosome does not have a copy of that gene. In other words, because males cannot be heterozygous for any X-linked genes, the effects of an *a* allele cannot be masked. In general, males are more likely than females to have recessive X-linked disorders, because they need to inherit only a single copy of the disorder allele to exhibit the disorder. Females, by contrast, must inherit two copies to be affected. X-linked recessive inheritance explains why boys are more likely to get XSCID than girls are.

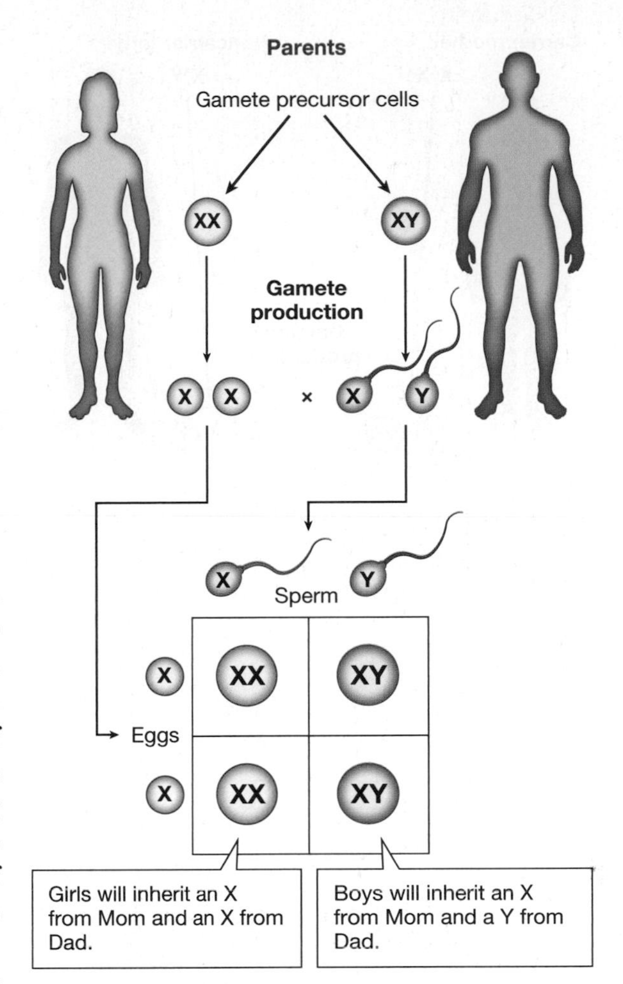

Figure 8.11

Sex chromosomes determine biological sex

A mother's gametes (eggs) always deliver an X chromosome to her offspring. A father's sperm may deliver an X chromosome or a Y chromosome.

Q1: What are the odds that a given egg cell will contain an X chromosome? A Y chromosome? What are those odds for a sperm cell?

Q2: If a couple has two daughters, does that mean their next two children are more likely to be sons? Explain your reasoning. (*Hint*: Refer back to "What Are the Odds?" on page 136.)

Q3: Sisters share the same X chromosome inherited from their father, but they may inherit different X chromosomes from their mother. What is the probability that brothers share the same Y chromosome? What is the probability that brothers share the same X chromosome?

See Appendix A for answers to the figure questions.

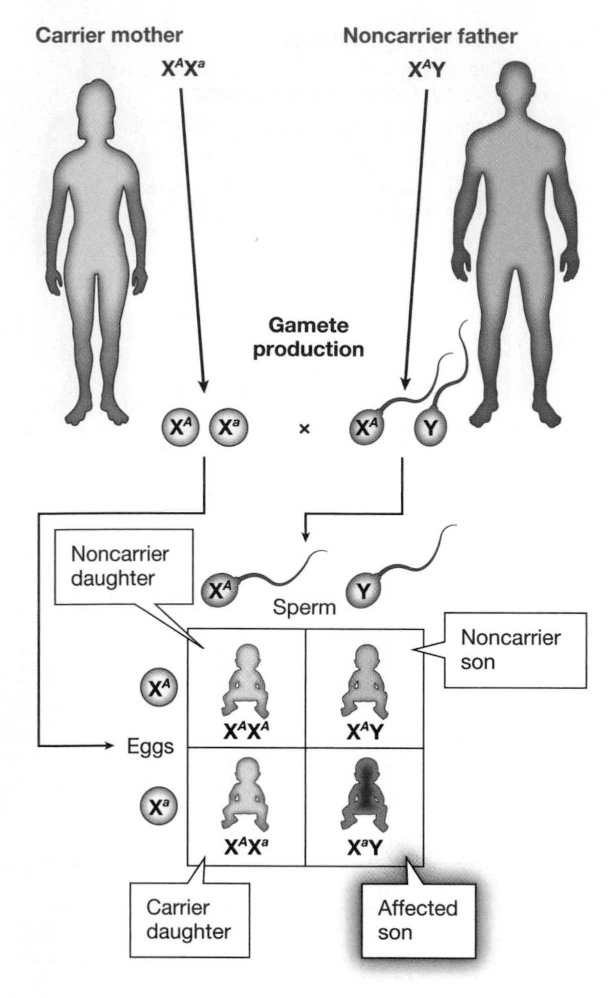

Figure 8.12

X-linked recessive conditions are more common in males

The recessive disorder allele (*a*) is located on the X chromosome and is denoted by X^a. The dominant normal allele (*A*), also on the X chromosome, is denoted by X^A.

Q1: Which of the children specified in this Punnett square represents Samuel? What is his genotype?

Q2: Explain why Samuel is neither homozygous nor heterozygous for the *XSCID* gene.

Q3: Create a Punnett square to illustrate the offspring that could result if Samuel had children with a noncarrier woman. What is the probability that a son would have XSCID? What is the probability that a daughter would be a carrier of XSCID? (Samuel's treatment does not change sex cells.)

See Appendix A for answers to the figure questions.

Other X-linked genetic disorders in humans include red-green color blindness, hemophilia, and Duchenne muscular dystrophy, a lethal disorder that causes muscles to waste away, often leading to death at a young age. All of these X-linked disorders are caused by recessive alleles—but how do geneticists know that? And how do they know that Arabella's SMA1 and Jennelle's sickle cell disease are coded by recessive, autosomal genes?

Geneticists use a family pedigree to identify the inheritance of these traits. Similar to a family tree, a **pedigree** is a chart that shows genetic relationships among family members over two or more generations of a family's medical history. The pedigree for a recessive disorder, such as SMA1 or sickle cell disease, has a different pattern (**Figure 8.13**) than that for an X-linked disorder, such as XSCID (**Figure 8.14**).

Samuel was one of eight infants who received a gene therapy for XSCID developed at St. Jude Children's Hospital in Tennessee. In 2019, the scientists who made that therapy announced that it was safe and effective in all eight children, and today those kids are growing normally with functioning immune systems. Samuel returned home after the therapy with no restrictions. Gone was a future of life inside a bubble—instead he is a healthy child, able to attend school, play with friends, and run outside to his heart's content.

Million-Dollar Cures

It was back in 1993, at UCLA, that Donald Kohn began developing a gene therapy for ADA-SCID, an autosomal recessive form of SCID. Today, 53 patients have been treated in a series of clinical trials for that therapy. In 2016, a similar therapy developed by an Italian team was approved for use in Europe. Now it's time to begin tracking the long-term effects of gene therapies, says Kohn, including how long gene therapy will last or if there are long-term safety risks to the treatment.

"If you get enough stem cells fixed and corrected, they should last forever," says Kohn. Patients who received gene therapy 14 years ago appear to still be producing healthy proteins today, he says, but the question remains: "Are they going to last seven decades? We won't know that for another six decades."

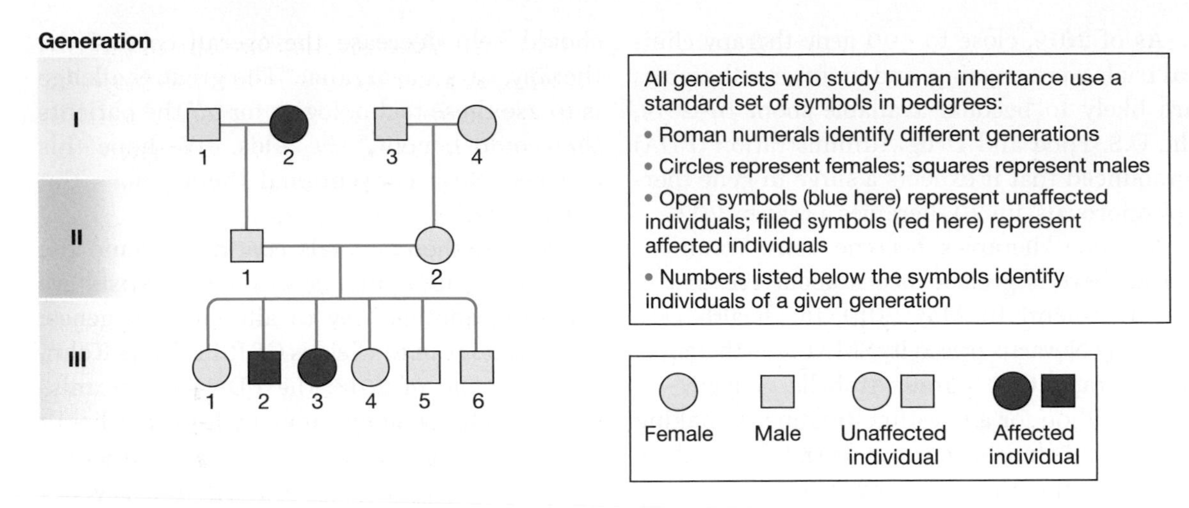
Generation
I
II
III
1
2
3
4
1
2
1
2
3
4
5
6
All geneticists who study human inheritance use a standard set of symbols in pedigrees:
• Roman numerals identify different generations
• Circles represent females; squares represent males
• Open symbols (blue here) represent unaffected individuals; filled symbols (red here) represent affected individuals
• Numbers listed below the symbols identify individuals of a given generation
Female
Male
Unaffected individual
Affected individual

Figure 8.13

Patterns of inheritance can be analyzed in family pedigrees

This sickle cell disease pedigree shows six children (generation III), two of whom are affected with the disease. An SMA1 pedigree would be quite similar. Note that individuals who join the family by marriage do not show a connecting line from the previous generation.

Q1: Which two children in this pedigree have sickle cell disease? How do you know?

Q2: Do either of their parents have sickle cell disease? If so, which one(s)? How do you know?

Q3: Do any of their grandparents have sickle cell disease? If so, which one(s)? How do you know?

See Appendix A for answers to the figure questions.

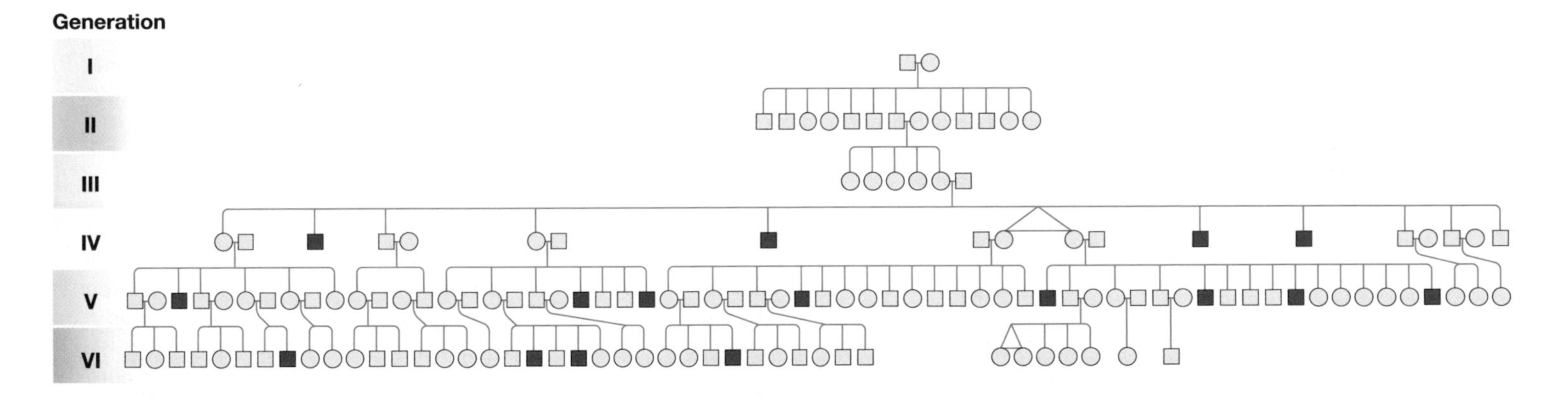
Generation
I
II
III
IV
V
VI

Figure 8.14

Pedigree of a family with a history of XSCID syndrome, an X-linked disorder

As in Figure 8.13, circles represent females and squares represent males in this pedigree. Individuals affected by XSCID are shaded red. Note that individuals who join the family by marriage do not show a connecting line from the previous generation.

Q1: How many male and how many female descendants (individuals who did not join the family by marriage) does generation IV of the pedigree include?

Q2: What proportions of the male and female descendants in generation IV were affected by the disorder?

Q3: Why would a geneticist hypothesize that the disease was X-linked?

See Appendix A for answers to the figure questions.

As of 2019, close to 400 gene therapy clinical trials were ongoing, and more gene therapies are likely to become available soon. In 2019, the U.S. Food and Drug Administration (FDA) announced that it expects a surge in gene therapy approvals, up 20 therapies a year by 2025.

As gene therapies become more commonplace, there is great concern about the cost of the treatment. In May 2019, the health care company Novartis priced its SMA1 gene therapy—the therapy that cured Arabella Smygov—at $2.13 million for a one-time treatment, making it the world's most expensive drug. In June 2019, Bluebird Bio, the company funding some of Marina Cavazzana's clinical trials, announced that the gene therapy for beta thalassemia, approved in Europe, would cost $1.78 million per person.

Pharmaceutical companies say these high prices are due to the investment required to develop the therapies and that the prices reflect the ability to avoid a lifetime of expensive treatments. Once companies are able to produce large quantities of viral vectors, that should help decrease the overall cost of the therapy, says Cavazzana. "The great challenge is to use these technologies for *all* the patients that could benefit," she adds. "We hope this can become a conventional therapy for a lot of patients."

As gene therapy trials continue around the world for genetic diseases, some scientists are exploring another way to alter human genes: "The big hope in the field is CRISPR," says Kohn.

At the end of 2018, the FDA gave permission to biotechnology company Editas to begin a human clinical trial using a virus to deliver not a healthy gene into eye cells but a gene-editing tool called CRISPR (clustered regularly interspaced short palindromic repeats), where it works like a pair of scissors to cut and correct a single mutation that causes inherited childhood blindness.

CRISPR has changed the way scientists are able to interact with and manipulate DNA. In the next chapter we explore how and why it is being used at one of the most prestigious universities in the world to edit the genomes of pigs.

NEED-TO-KNOW SCIENCE

- A **genetic disorder** (page 148) is a disease or condition caused by a genetic abnormality, either inherited or occurring spontaneously. Genetic disorders can be caused by mutations in individual genes or by **chromosomal abnormalities** (changes in chromosome number or structure, page 151).
- **Gene therapy** (page 148) is a **genetic engineering** (page 148) technique for correcting abnormal genes responsible for disease development.
- **Stem cells** (page 152) are unique, unspecialized cells that can make identical copies of themselves for long periods of time. In most cases of gene therapy, stem cells from the patient are mixed in a dish with a virus vector carrying the healthy gene.
- Humans have two sets of 23 chromosomes for a total of 46 chromosomes. Homologous **autosomes** (non–sex chromosomes, page 149) make up 22 of the 23 pairs. Every person has a set of **sex chromosomes** (page 149). Males have one X and one Y chromosome, and females have two X chromosomes.
- The physical location of a gene on a chromosome is called its **locus** (page 152). **Sex-linked** (page 159) genes are found solely on the X chromosome (**X-linked**, page 159) or Y chromosome (**Y-linked**, page 159).
- A **recessive genetic disorder** (page 153) is caused by two recessive alleles. Heterozygous individuals (*Aa*) are **genetic carriers** (page 153) of the disorder; they carry the disorder allele (*a*) but do not have the disease. Several thousand human genetic disorders are inherited as recessive traits on autosomes. A **dominant genetic disorder** (page 155) is much rarer and is caused by a dominant allele on an autosome.
- A Punnett square can be used to calculate the probability of inheriting a particular genetic trait.
- A family **pedigree** (page 160) can be used to determine whether a given condition is recessive, dominant, or sex-linked.

THE QUESTIONS

See Appendix B for answers.

The Basics

1 Use these terms correctly in the following sentence: **alleles, chromosomes, genes, loci**. Two homologous ________ contain the same ________, found at the same ________, but may have the same or different copies of ________.

2 A particular person is said to be a carrier of a genetic trait. What does this tell you about the person's phenotype?

(a) The person physically shows the trait.

(b) The person physically shows the trait more than a noncarrier would.

(c) The person physically shows a partial version of the trait.

(d) The person does not physically show the trait.

3 Match each of the following terms with the correct definition.

gene therapy	1. A procedure in which cells are gently suctioned from a pregnant woman's uterus to test for genetic disorders in the fetus.
in vitro fertilization	2. A procedure in which a small amount of fluid, including the fetal cells within it, is carefully extracted from a pregnant woman's uterus to test for genetic disorders in the fetus.
preimplantation genetic diagnosis (PGD)	3. A treatment approach that seeks to correct a genetic disorder by inserting healthy copies of the mutated genes responsible for the disorder.
chorionic villus sampling (CVS)	4. A procedure in which one or two cells are removed from a developing embryo and tested for genetic disorders; embryos that are free of genetic disorders may then be implanted into a woman's uterus.
amniocentesis	5. A procedure in which an egg is fertilized in a petri dish, after which one or more embryos are implanted into a woman's uterus.

4 In the karyotype shown here, identify the sex chromosomes. Is this individual a male or a female?

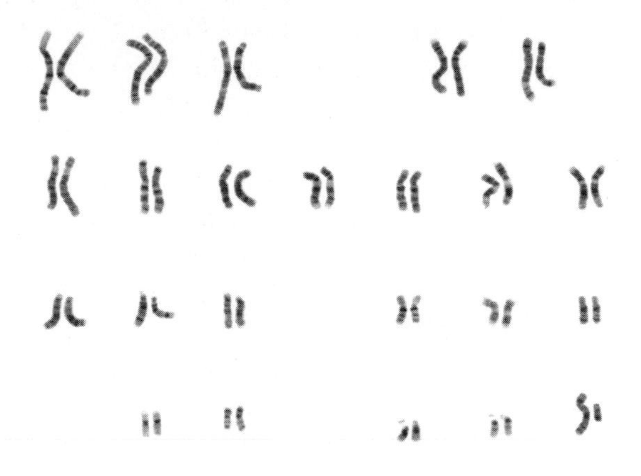

Challenge Yourself

5 Sickle cell disease is inherited as a recessive genetic disorder in humans; the normal hemoglobin allele (*H*) is dominant to the sickle cell allele (*h*). For two parents of genotype *Hh* (carriers), construct a Punnett square to show the possible genotypes of their children.

6 Which of the following correctly predicts the possible genotypes of the children from question 5?

(a) $1/4$ *HH*, $1/2$ *Hh*, $1/4$ *hh*

(b) $1/2$ *HH*, $1/2$ *Hh*

(c) $1/2$ *HH*, $1/2$ *hh*

(d) $1/4$ *HH*, $1/4$ *Hh*, $1/2$ *hh*

(e) $1/4$ *HH*, $3/4$ *Hh*

7 Which of the following correctly predicts the possible phenotypes of the children from question 5?

(a) $1/4$ normal, $3/4$ sickle cell disease

(b) all normal

(c) all sickle cell disease

(d) $1/2$ normal, $1/2$ sickle cell disease

(e) $3/4$ normal, $1/4$ sickle cell disease

8 Given that the normal hemoglobin allele (*H*) is dominant to the sickle cell allele (*h*), each time two *Hh* individuals have a child together, what is the chance that the child will have sickle cell disease?

(a) 0 percent

(b) 75 percent

(c) 25 percent

(d) 50 percent

(e) 100 percent

9 Which of the following karyotypic sex chromosome abnormalities result(s) in a male phenotype? (Select all that apply.)

(a) XO
(b) XXY
(c) XXX
(d) XYY
(e) XXXY
(f) XY with a complete deletion of the *SRY* gene

Try Something New

10 Answer the following questions based on the accompanying graphs.

(a) What type of graph is the first graph? What kind of vector was used in the gene therapies discussed in this chapter? What percentage of total trials used that vector? What is the second most common vector used in gene therapy trials, according to this graph?

(b) What type of graph is the second graph? What are the *x*- and *y*-axes? What years had the least and most clinical trials? Have the number of annual clinical trials increased or decreased since 1989?

Global gene therapy clinical trials, by date and vector

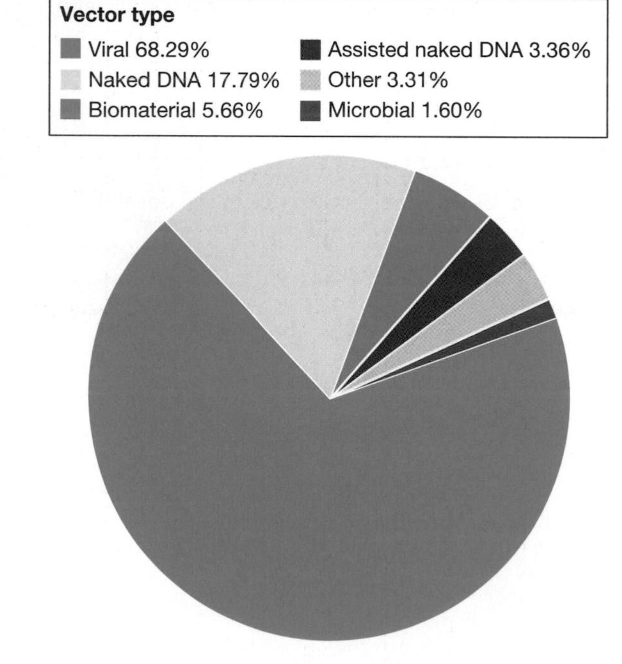

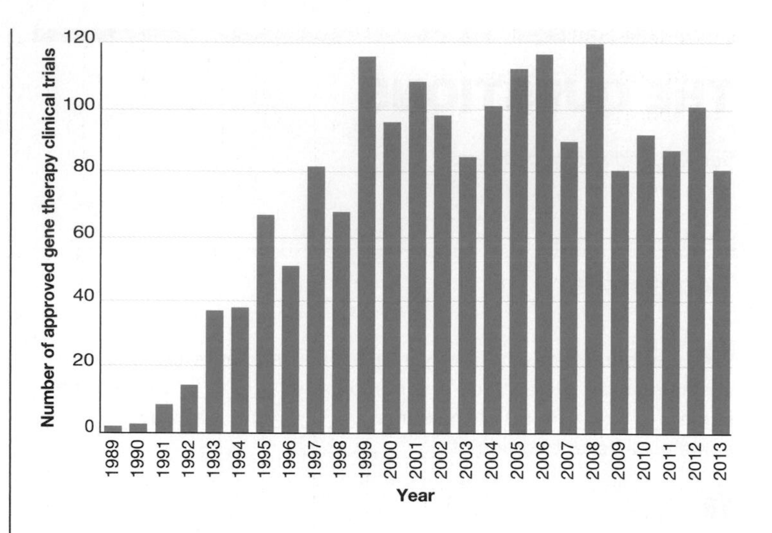

11 The typical human female has two X chromosomes, whereas human males have one X chromosome and one Y chromosome.

a. Do males inherit their X chromosome from their mother or from their father?
b. If a female has one copy of an X-linked recessive allele for a genetic disorder, does she have the disorder?
c. If a male has one copy of an X-linked recessive allele for a genetic disorder, does he have the disorder?
d. Assume that a female is a carrier of an X-linked recessive disorder. With respect to the disorder allele, how many types of gametes can she produce?
e. Assume that a male with an X-linked recessive genetic disorder has children with a female who does not carry the disorder allele. Could any of their sons have the genetic disorder? How about their daughters? Could any of their children be carriers for the disorder? If so, which sex(es) could they be?

12 For the pedigree shown here, the disorder is caused by a recessive (g) allele on the X chromosome. Label each of the following individuals with the correct genotype (X^GY, X^gY, X^GX^G, X^GX^g, X^gX^g).

Generation
I: 1 2
II: 1 2 3 4 5 6
III: 1 2 3 4 5 6 7

Female | Male | Unaffected individual | Affected individual

a. Generation I, number 2
b. Generation II, number 2
c. Generation II, number 5
d. Generation III, number 2
e. Generation III, number 6

13 For the pedigree shown here, the disorder is caused by a recessive (*g*) autosomal allele. Label each of the following individuals with the correct genotype (*GG*, *Gg*, *gg*).

Generation

I: 1, 2, 3, 4

II: 1, 2

III: 1, 2, 3, 4, 5, 6

Female | Male | Unaffected individual | Affected individual

a. Generation I, number 2
b. Generation II, number 1
c. Generation II, number 2
d. Generation III, number 3

Leveling Up

14 **Doing science** On the Indonesian island of Bali, about $^{3}/_{4}$ of the feral (stray) cats have a stumpy tail, whereas only $^{1}/_{4}$ of the cats have a long tail. Experiments in which stumpy-tailed cats were mated with each other yielded some stumpy-tailed kittens and some long-tailed kittens. The same experiments with long-tailed cats produced only long-tailed kittens. What do these results show about the genotype of the long-tailed cats? How would the results be different if the trait were sex-linked? (*Hint:* Draw Punnett squares using the letter "t" for the tail trait. Show your work.)

15 **What do *you* think?** *Gattaca*, a science fiction movie from 1997, takes place at a future time in which preimplantation genetic diagnosis (PGD) is used to create genetically superior children, and social status is determined mainly by one's DNA. Watch the movie and take careful notes as you observe how PGD has been misused by society. Do you think these outcomes are likely if we continue to develop PGD? Are the benefits worth this risk?

16 ***Write Now* biology: debating preimplantation genetic diagnosis (PGD)** This assignment is designed to give you the experience of applying your knowledge of biology to a current controversy or topic of interest. You will apply the same sort of reasoning that you should be able to use as an informed citizen and consumer when making decisions that involve biology.

The scenario: The U.S. Senate Committee on Commerce, Science, and Transportation (CCST) is considering proposing legislation on preimplantation genetic diagnosis (PGD; see "Prenatal Genetic Screening" on page 157). The chair of the committee has invited special-interest groups to present testimony on the pros and cons of PGD. You will contribute a position paper defining PGD, describing your group's position on the technology, and making a recommendation for legislation.

Choose one of the following special-interest groups to represent (or your instructor will assign one to you):

(a) **Reproductive Specialist Group (medical doctors).** You will describe how PGD and in vitro fertilization (IVF) can be used to screen for genetic disorders or for sex selection and "family balancing," and to increase fertility in women of advanced maternal age. You will argue that legislation is not necessary, as medical association guidelines are already in place, and that it is not the place of government to judge individuals' reasons for undergoing PGD.

(b) **Genetics and Public Policy Group.** You will present data on the beliefs of the American public in this matter. You will also present the status of current legislation in the United States and how it compares to PGD legislation internationally. You will propose limited legislation based on these findings.

(c) **Parents of Down Syndrome Children.** You will argue that people with disabilities already suffer from prejudice and that widespread use of such testing will cause even more prejudice. You will propose legislation prohibiting PGD for almost all conditions, with the exception of deadly infant-onset disease.

(d) **Americans with Cystic Fibrosis.** You will present the case of a couple, both of whom are carriers for cystic fibrosis, and discuss their options. You will propose limited legislation that disallows PGD for sex selection and nondisease conditions.

Digital resources for your book are available online.

Pigs to the Rescue

Using CRISPR, an unprecedented genome-editing tool, scientists hope to create a steady stream of transplant organs—from pigs.

After reading this chapter you should be able to:

- Describe the structure of DNA using appropriate terminology.
- Use base-pairing rules to determine a complementary strand of DNA based on a given template strand.
- Detail how the genome-editing tool CRISPR-Cas9 works.
- Label a diagram of DNA replication, identifying the location of each step in the process.
- Describe PCR, and explain how it relates to DNA replication.
- Explain the cause of DNA replication errors, and describe how they are repaired.
- Give an example of a mutation and its potential effects on an organism.

GENETICS

CHAPTER 09

WHAT GENES ARE

"Pigs," thought Marc Güell. "What if we could modify pigs?"

Güell, then a young biochemist at Harvard Medical School in Massachusetts, had been working with colleagues on techniques to manipulate **DNA (deoxyribonucleic acid)**, the genetic code of life. In 2012, Güell and his colleague Luhan Yang, along with their boss, the geneticist George Church, had begun adding and deleting genes from organisms, a process known as *genome editing*. They were using a new technological breakthrough called CRISPR-Cas9—CRISPR for short.

Before the discovery of CRISPR (pronounced "crisper"), editing DNA was difficult, time-consuming, and expensive. Since the 1970s, researchers have used bulky enzymes isolated from bacteria, called restriction enzymes, to cut DNA at specific locations. This genetic editing typically was done only in model organisms such as mice and fruit flies—species for which biologists had developed a solid tool kit for manipulating DNA. CRISPR, however, made genome editing inexpensive and available to all genetics labs, not just well-funded ones. "It was the biggest change in my career," says Güell.

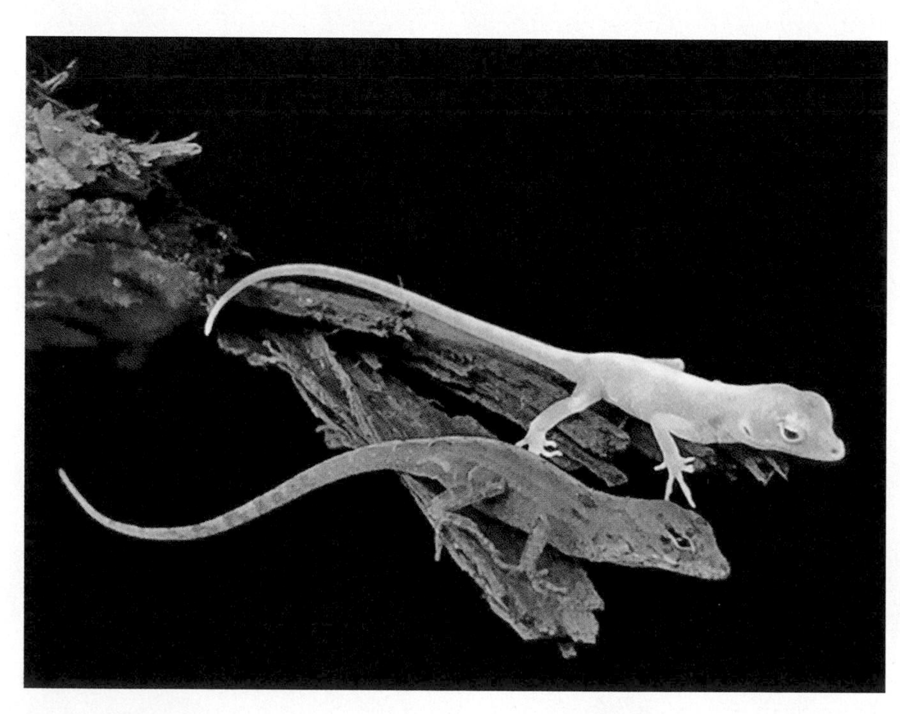

Figure 9.1

Gene editing in lizards

A single edited gene is the only genetic difference between a normally colored and an albino Anolis lizard.

CRISPR is a simple but creative combination of a protein and two single-stranded molecules of **RNA (ribonucleic acid)**, a macromolecule that stores genetic data. Like a "molecular scalpel," as one of the CRISPR inventors called it, CRISPR allows researchers to alter the DNA of nearly any organism—fungi, plants, mammals, even lizards. In 2019, researchers at the University of Georgia produced albino Anolis lizards by editing a single gene, which affects pigmentation, in the mother's egg cells (**Figure 9.1**).

Church's team decided to focus on a much larger, much pinker species: *Sus scrofa domesticus*, the domestic pig. "We started to think about what would be a good application to use this technology," says Güell, now a professor at Pompeu Fabra University in Spain. "One of the things that seemed like a very interesting, very difficult problem was the lack of organs for transplantation. We thought, what if we could modify pigs to make them compatible enough to be an unlimited source of organs?"

Deep in the DNA

Each day in the United States, on average 22 people die while waiting for an organ transplant. Over 113,000 men, women, and children are currently on the national transplant waiting list, each hoping their name is called before it's too late.

Researchers have explored many ways to grow and store organs for transplantation—from freezing them to building them from scratch—but one of the most promising, if you can look past the mud and flies, is pigs. Our porcine friends have long been considered an excellent potential source of organs because their organs—including the heart, liver, and kidneys—are relatively close in size to human organs and because pigs and humans have similar anatomies (**Figure 9.2**). In addition, pigs are an easier sell to the public: people tend to prefer the idea of transplants from pigs over transplants from mammals more closely related to us, such as baboons. If we were able to transplant organs from nonhuman animals into humans, a process called *xenotransplantation*, healthy organs could be available in essentially limitless supply.

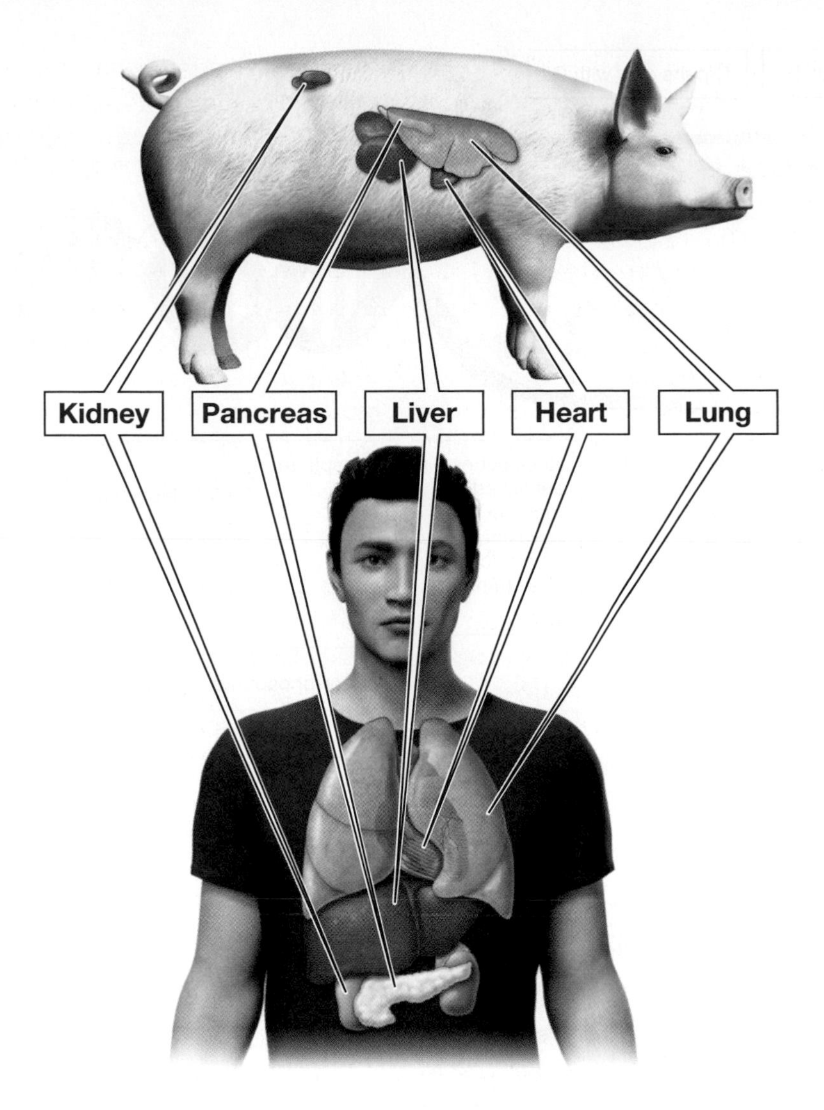

Figure 9.2

Pig organs and human organs are remarkably similar in size

For a nonhuman-to-human organ transplant to succeed, the nonhuman organ must fit into the space where the human organ was removed. Organs that are too big will not fit. Organs that are too small will not function at the level necessary to sustain life.

Yet there has been a barrier to harvesting pig organs for humans: the pig genome is dotted with DNA from a family of viruses called porcine endogenous retroviruses, or PERVs. Because of the presence of this viral DNA in the pig genome, pig cells produce and release PERVs—and two of the three subtypes of PERVs can infect human cells, making it risky to transplant pig organs into people for fear of making the recipients sick.

After pigs acquired the viruses, PERV DNA slipped easily into the pig genome because it has the same structure as pig DNA. In fact, all living things share the same DNA structure, and species often share and swap DNA with each other. As discussed in Chapters 3–4, DNA is built from two parallel strands of repeating units called **nucleotides**. Each nucleotide is composed of the sugar deoxyribose, a phosphate group, and one of four **bases**: adenine, cytosine, guanine, or thymine. We identify nucleotides by their bases, using "adenine nucleotide" as shorthand for "nucleotide with an adenine base."

The nucleotides of a single strand are connected by covalent bonds between the phosphate group of one nucleotide and the sugar of the next nucleotide. The two DNA strands are connected by hydrogen bonds linking the bases on one strand to the bases on the other, like the

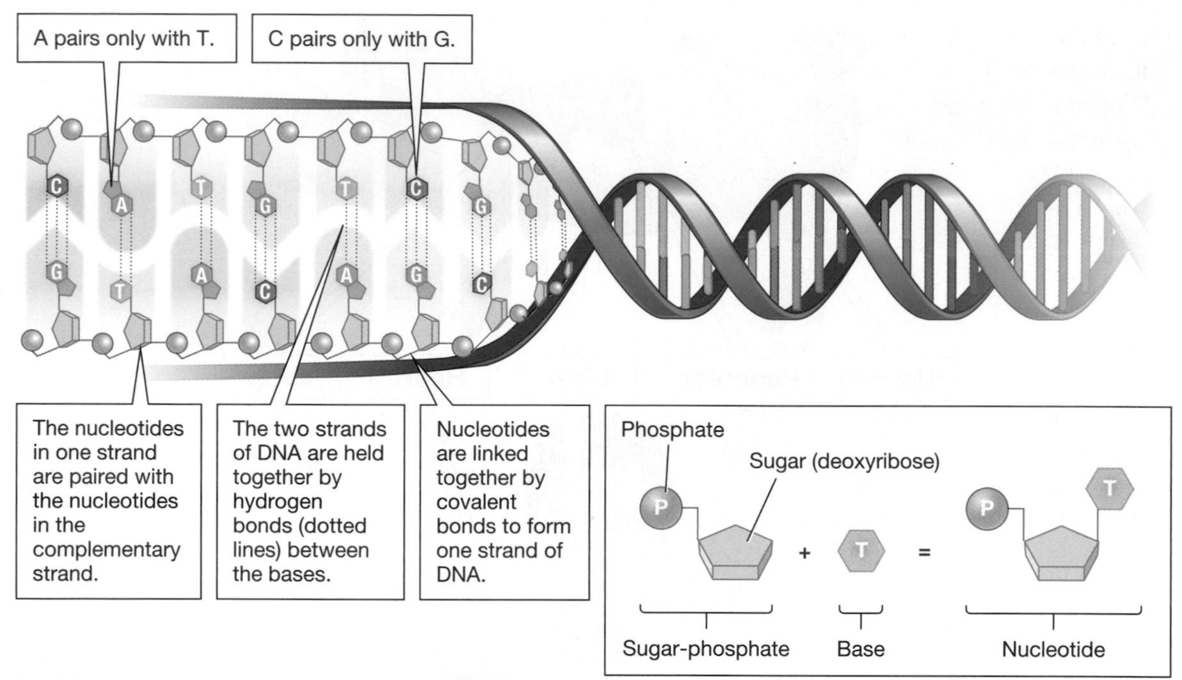

A pairs only with T.
C pairs only with G.
The nucleotides in one strand are paired with the nucleotides in the complementary strand.
The two strands of DNA are held together by hydrogen bonds (dotted lines) between the bases.
Nucleotides are linked together by covalent bonds to form one strand of DNA.
Phosphate
Sugar (deoxyribose)
Sugar-phosphate
Base
Nucleotide

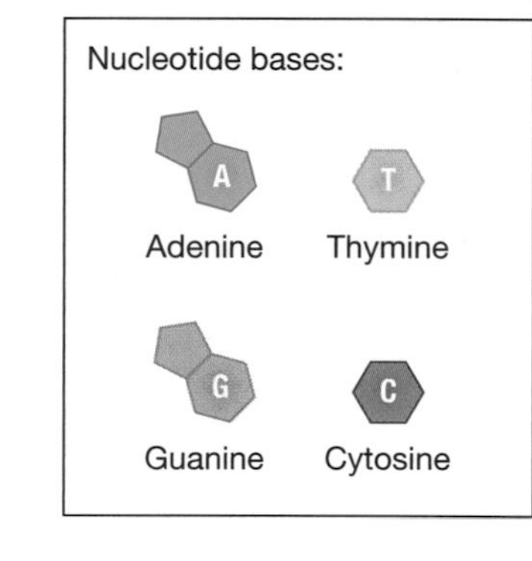

Nucleotide bases:
Adenine
Thymine
Guanine
Cytosine

Figure 9.3

The DNA double helix and its building blocks

A molecule of DNA consists of two complementary strands of nucleotides that are twisted into a spiral around an imaginary axis, rather like the winding of a spiral staircase.

Q1: Name two base pairs.

Q2: Why is the DNA structure referred to as a "ladder"? What part of the DNA represents the rungs of the ladder? What part represents the sides?

Q3: Is the hydrogen bond that holds the base pairs together a strong or weak chemical bond? Why is that important?

See Appendix A for answers to the figure questions.

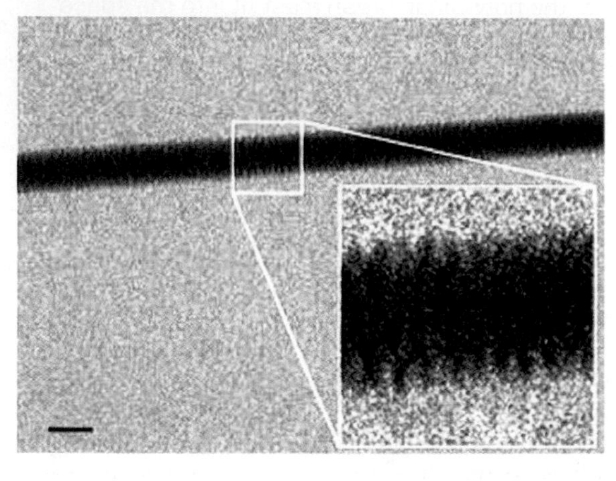

Figure 9.4

What DNA actually looks like

In November 2012, Italian researchers used an electron microscope to directly visualize DNA for the first time. This is the single thread of double-stranded DNA that they saw.

rungs that connect the two sides of a ladder (**Figure 9.3**). Covalent bonds are strong, which is important in maintaining the specific order of the nucleotides. The weaker hydrogen bonds allow the two DNA strands to be pulled apart for replication.

The term **base pair**, or nucleotide pair, refers to two nucleotides held together by bonds between their bases; that is, a base pair corresponds to one rung of the DNA ladder. The ladder twists into a spiral called a **double helix** (**Figure 9.4**). Within the long, winding double helix of the pig genome, short sections of DNA from PERVs are scattered about. These PERV sections are made up of the same four nucleotides as the rest of the DNA, but they encode information for viral proteins instead of pig proteins.

Nucleotides do not form base pairs willy-nilly. As shown in **Figure 9.3**, the adenine

(A) nucleotide on one strand can pair only with thymine (T) on the other strand; cytosine (C) on one strand can pair only with guanine (G) on the other strand. These **base-pairing rules**, which provide **complementary base-pairing** between two nucleic acid strands, have an important consequence: when the sequence of nucleotides on one strand of the DNA molecule is known, the sequence of nucleotides on the other, complementary strand of the molecule is automatically known as well. The fact that A can pair only with T and that C can pair only with G allows the original strands to serve as "template strands" on which new strands can be built through complementary base-pairing. (In Chapter 10, we delve more deeply into building new DNA strands, including how RNA can pair with DNA, which CRISPR takes advantage of.)

Still, the four nucleotides can be arranged in any order along a single strand of DNA, and each DNA strand is composed of millions of these nucleotides, so a tremendous amount of information can be stored in a DNA sequence and in a genome. The genome of the domestic pig, for example, has about 3 billion base pairs, the human genome has about 3.2 billion base pairs, a tomato has only about 900 million base pairs, and the bacterium *Escherichia coli* (better known as *E. coli*) has a measly 4.6 million. The sequence of nucleotides in DNA differs among species and among individuals within a species, and these differences in genotype can result in different phenotypes (**Figure 9.5**).

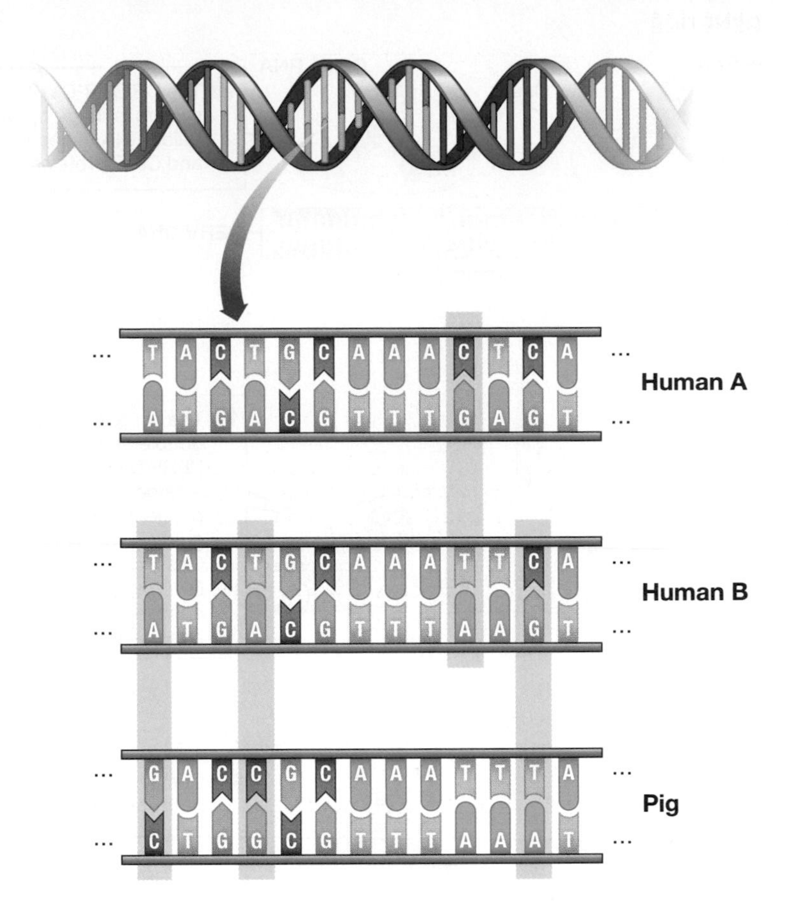

Figure 9.5

The sequence of bases in DNA differs among species and among individuals within a species

The sequence of bases in a hypothetical gene is compared for two humans (A and B) and a pig. Base pairs highlighted in blue are variant; that is, they differ between the genes of persons A and B and between the same genes in humans and pigs.

Q1: If all genes are composed of just four nucleotides, how can different genes carry different types of information?

Q2: Would you expect to see more variation in the sequence of DNA bases between two members of the same species (such as humans) or between two individuals of different species (for example, humans and pigs)? Explain your reasoning.

Q3: Do different alleles of a gene have the same DNA sequence or different DNA sequences?

See Appendix A for answers to the figure questions.

In the mid-1990s, scientists became very excited about the idea of using pig organs in humans, but testing stalled because of the fear that humans would become infected with PERVs. Just breeding pigs in sterile conditions can't get rid of the virus; it's integrated right there in the double helix. The Harvard Medical School team believed CRISPR might be able to solve that problem by inactivating the PERV DNA in pig cells once and for all.

Precise Cuts

The DNA sequences called **CRISPR**, a blessedly easy acronym for *clustered regularly interspaced short palindromic repeats*, did not start out as a gene-editing tool. They are actually part of a defense system used by bacteria. Bacteria are constantly bombarded by viruses that try to sneak into and take over their genomes, so bacteria evolved a set of defensive measures, including a tool to recognize and cut foreign, invading DNA. The CRISPR-Cas9 system—pioneered by microbiologists Jennifer Doudna

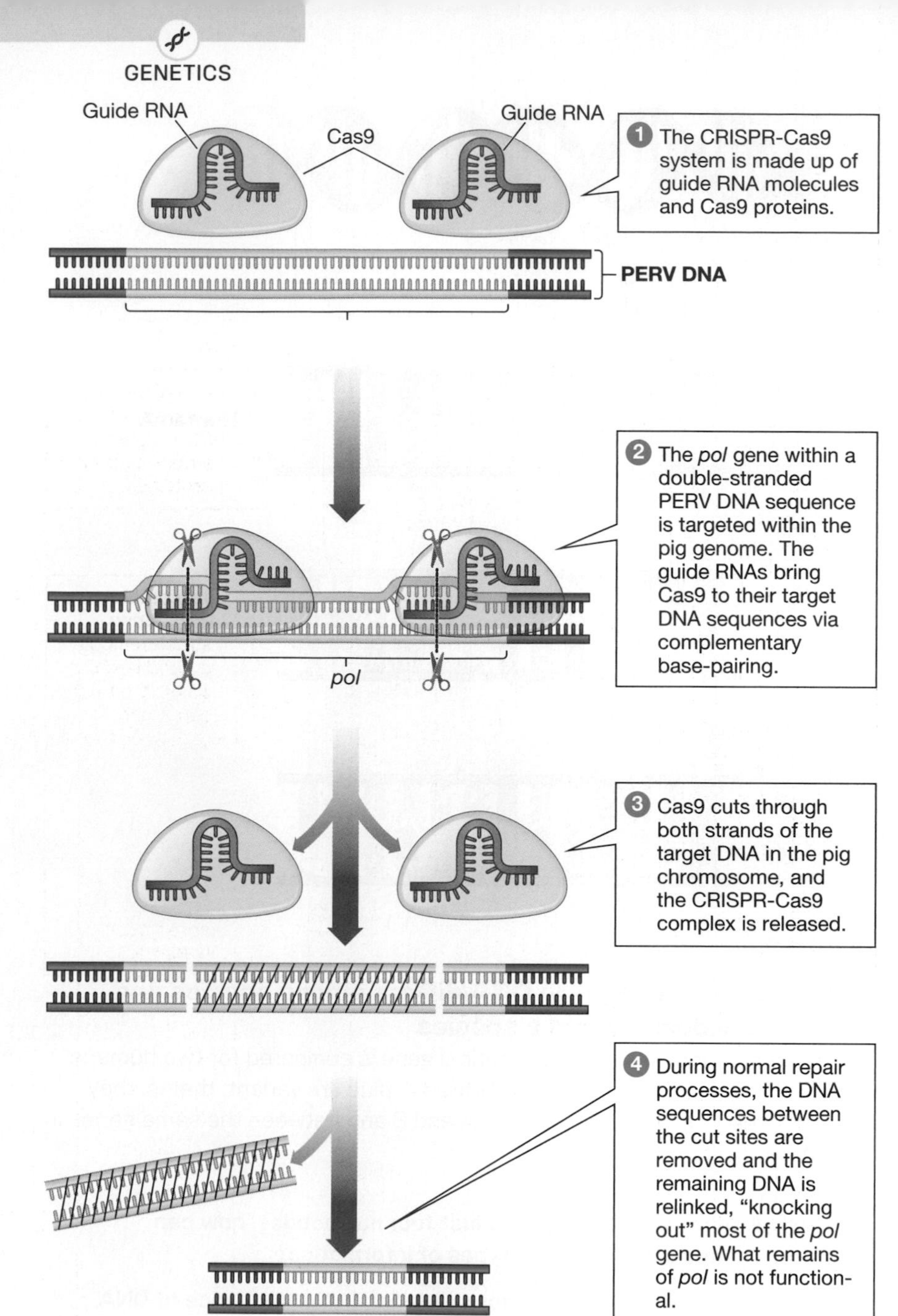

Guide RNA
Cas9
Guide RNA
1 The CRISPR-Cas9 system is made up of guide RNA molecules and Cas9 proteins.
PERV DNA
2 The pol gene within a double-stranded PERV DNA sequence is targeted within the pig genome. The guide RNAs bring Cas9 to their target DNA sequences via complementary base-pairing.
pol
3 Cas9 cuts through both strands of the target DNA in the pig chromosome, and the CRISPR-Cas9 complex is released.
4 During normal repair processes, the DNA sequences between the cut sites are removed and the remaining DNA is relinked, "knocking out" most of the pol gene. What remains of pol is not function-al.

Figure 9.6

Genome editing with CRISPR-Cas9, an efficient and cost-effective tool

The *pol* gene is involved in integrating viral DNA into a host genome. Here, CRISPR is used to remove that gene from pig PERV DNA.

Q1: What common mechanism is employed by the guide RNA to find its target DNA sequence?

Q2: How many strands of DNA must Cas9 cut to be effective?

Q3: Does Cas9 also cause the deletion of DNA from the genome?

See Appendix A for answers to the figure questions.

of UC Berkeley and Emmanuelle Charpentier at Umeå University in Sweden—is composed of two RNA molecules that guide Cas9 proteins to chosen, precise sites in a genome. There, Cas9 efficiently cuts both strands of the DNA (**Figure 9.6**).

The CRISPR system is like a "Delete" key or, in certain experiments, the cut-and-paste tool of a word processing program—except with nucleotides and genes instead of letters and words. Although most new laboratory tools can take months or years to become widely used, CRISPR was immediately a lab favorite. Scientists jumped right into editing the zebra fish genome to study gene function, correcting genetic disease mutations in adult mice, enhancing pest resistance in wheat, and more.

Shortly after Doudna and Charpentier's discovery, two laboratories—namely, Feng Zhang at the Broad Institute of MIT and Harvard and George Church's lab at Harvard Medical School— engineered CRISPR systems for genome editing in eukaryotic cells. With that innovation, the Church team was almost ready to edit the pig genome, but first they needed to identify how many copies of the PERV DNA the genome contained. Using laboratory tools to sequence the pig DNA, they identified 62 copies of PERVs scattered throughout the genome.

Although 62 may sound like a lot, it's a drop in the bucket compared to the number of genes in a mammalian genome. Pigs have over 1,600 genes for their sense of smell alone. An estimated 19,000 protein-coding genes are packed into the human genome, surrounded by a vast amount of DNA that does not code for proteins but may have other functions. Cells stuff an enormous amount of DNA into a small space. They use various packaging proteins to wind, fold, and compress the DNA double helix, going through several levels of packing to create the DNA-protein complex that we call a chromosome (**Figure 9.7**).

Short lengths of double-stranded DNA are wound around "spools" of proteins, known as **histone proteins**, to create a beads-on-a-string structure consisting of many histone beads, called **nucleosomes**, which are connected by strings of DNA. This beads-on-a-string structure is compressed and coiled into a more compact form known as the **chromatin fiber**. The chromatin fiber is looped back and forth to further coil and condense into a chromosome.

Double or Nothing

DNA from viruses is present in all mammalian genomes that we know of, but not all these viral instructions behave the same. Viral DNA has been part of the human genome for a long, long time, yet it is no longer active. Viral DNA wove into the pig genome much more recently and is still active.

Viral DNA remains in genomes because it is passed from generation to generation via **DNA replication**, the duplication of a DNA molecule. DNA replication is ongoing in our bodies. It occurs before a cell enters mitosis (see Figure 6.4), so that there is a copy of the DNA to pass along to the new cell. DNA replication occurs feverishly when a new embryo is being formed. And it occurs when viruses hijack our cell machinery to copy their own DNA.

DNA replication occurs in three steps:

1. The DNA molecule unwinds through special proteins that bind the DNA at sequences known as **origins of replication**. The proteins then break the hydrogen bonds connecting the two strands of DNA.
2. Each separated strand is then used as a template for the construction of a new strand of DNA. **DNA polymerase**—a key enzyme in the replication of DNA—builds a new strand of DNA by first connecting to a **primer**, a short chain of nucleotides near the origin of replication, and then adding one complementary nucleotide after another.
3. When construction is complete, there are two identical copies of the original DNA molecule. Each copy is composed of a template strand of DNA (from the original DNA molecule) and a newly synthesized strand of DNA.

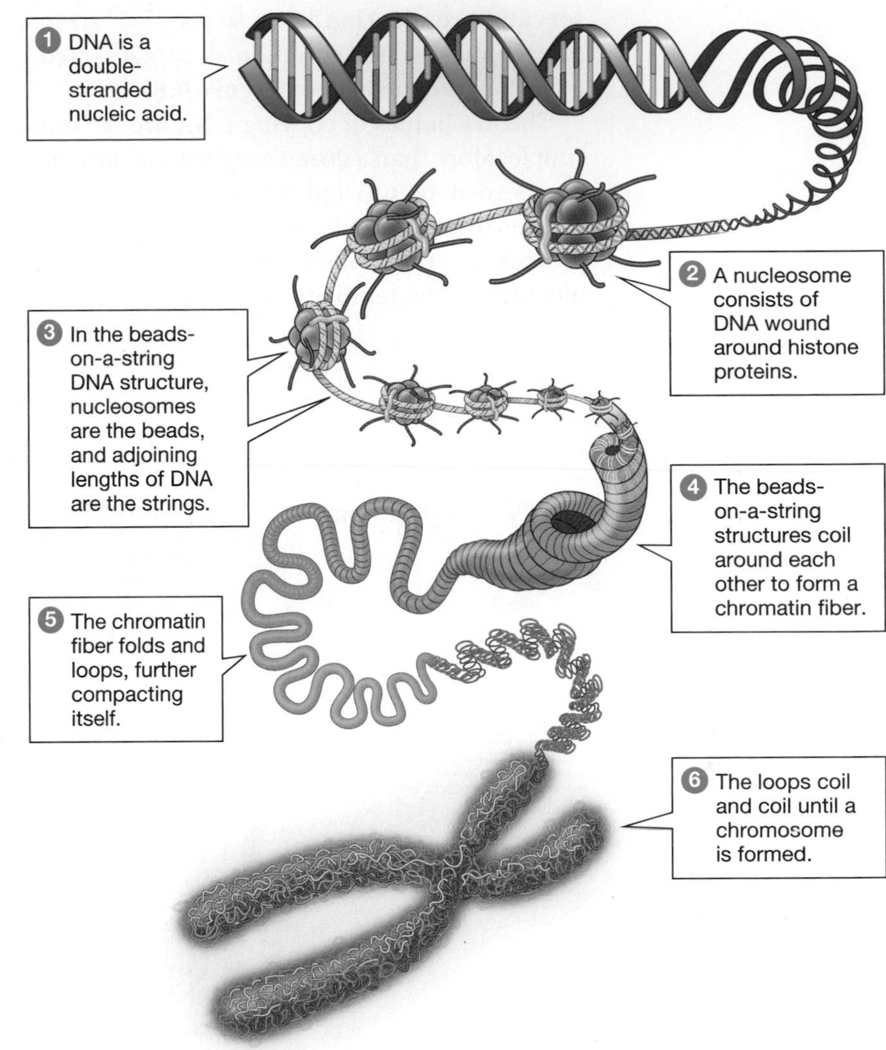

1 DNA is a double-stranded nucleic acid.
2 A nucleosome consists of DNA wound around histone proteins.
3 In the beads-on-a-string DNA structure, nucleosomes are the beads, and adjoining lengths of DNA are the strings.
4 The beads-on-a-string structures coil around each other to form a chromatin fiber.
5 The chromatin fiber folds and loops, further compacting itself.
6 The loops coil and coil until a chromosome is formed.

Figure 9.7

Chromosomes are meticulously organized DNA-protein complexes

The DNA double helix is continuously coiled and packaged around proteins until it is compacted into a chromosome.

Q1: What structures result from the first level of DNA coiling around proteins?

Q2: What makes up a "bead" and what makes up a "string" in the beads-on-a-string structure of DNA?

Q3: What structure is more compact than the beads-on-a-string structure but less compact than an actual chromosome?

See Appendix A for answers to the figure questions.

This mode of replication is known as **semiconservative replication** because one "old" strand (the template strand) is retained, or "conserved," in each new double helix (**Figure 9.8**).

The mechanics of copying DNA are far from simple. More than a dozen enzymes and proteins are needed to unwind the DNA, to stabilize the separated strands, to start the replication process, to attach nucleotides that are complementary to the template strand, to "proofread" the results, and to join partly replicated fragments of DNA to one another, ultimately forming a replicated chromosome.

Despite the complexity of this task, cells can copy DNA molecules containing billions of nucleotides in a matter of hours—about 8 hours in humans (over 100,000 nucleotides per second). This speed is achieved in part by starting the replication of each DNA molecule at thousands of different origins of replication simultaneously.

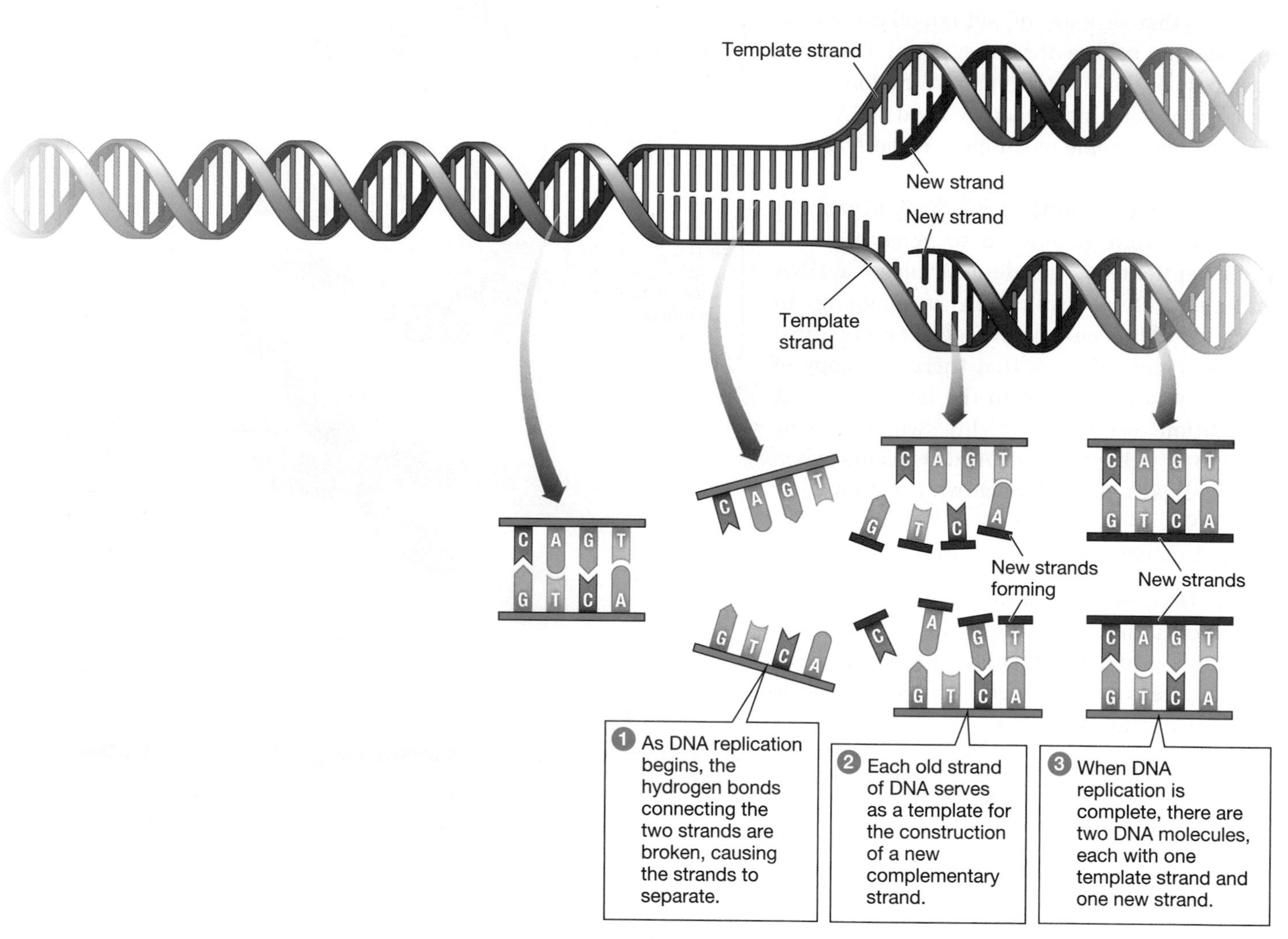
Template strand
New strand
New strand
Template strand
C A G T
G T C A
C A G T
G T C A
C A G T
G T C A
New strands forming
New strands
C A G T
G T C A
As DNA replication begins, the hydrogen bonds connecting the two strands are broken, causing the strands to separate.
Each old strand of DNA serves as a template for the construction of a new complementary strand.
When DNA replication is complete, there are two DNA molecules, each with one template strand and one new strand.

Figure 9.8

DNA replication is semiconservative

In this simplified overview of DNA replication, the template DNA strands are blue, and the newly synthesized strands are magenta.

Q1: Where do the DNA template strands come from? Why are they called "template" strands?

Q2: What must be broken before replication can begin?

Q3: In your own words, explain why replication is described as "semiconservative."

See Appendix A for answers to the figure questions.

Before CRISPR was developed, perhaps the most popular genetic technology was a tool for replicating DNA quickly in a test tube. In the 1980s, scientists developed this fast way to mimic natural DNA replication in a laboratory. The **polymerase chain reaction** (**PCR**) is a technique that makes it possible to produce millions of copies of a DNA sequence—to "amplify" the DNA—in less than an hour even with an extremely small initial amount of DNA (**Figure 9.9**).

PCR relies on heat, not special proteins, to cause DNA to unwind and its strands to separate. PCR also uses targeted primers to start the replication process at points determined by the scientist, rather than at DNA's natural origins of replication. The researcher creates a solution containing DNA polymerase, primers, and loose nucleotides, and the solution goes into a PCR machine. The machine cycles between heating, which separates the strands, and cooling, which enables the DNA polymerase to build two new strands of DNA through complementary base-pairing. These heating and cooling steps are repeated 25–40 times to replicate millions of exact copies of the targeted region of DNA.

PCR remains a central tool in most biology labs, as it is especially useful to turn a small

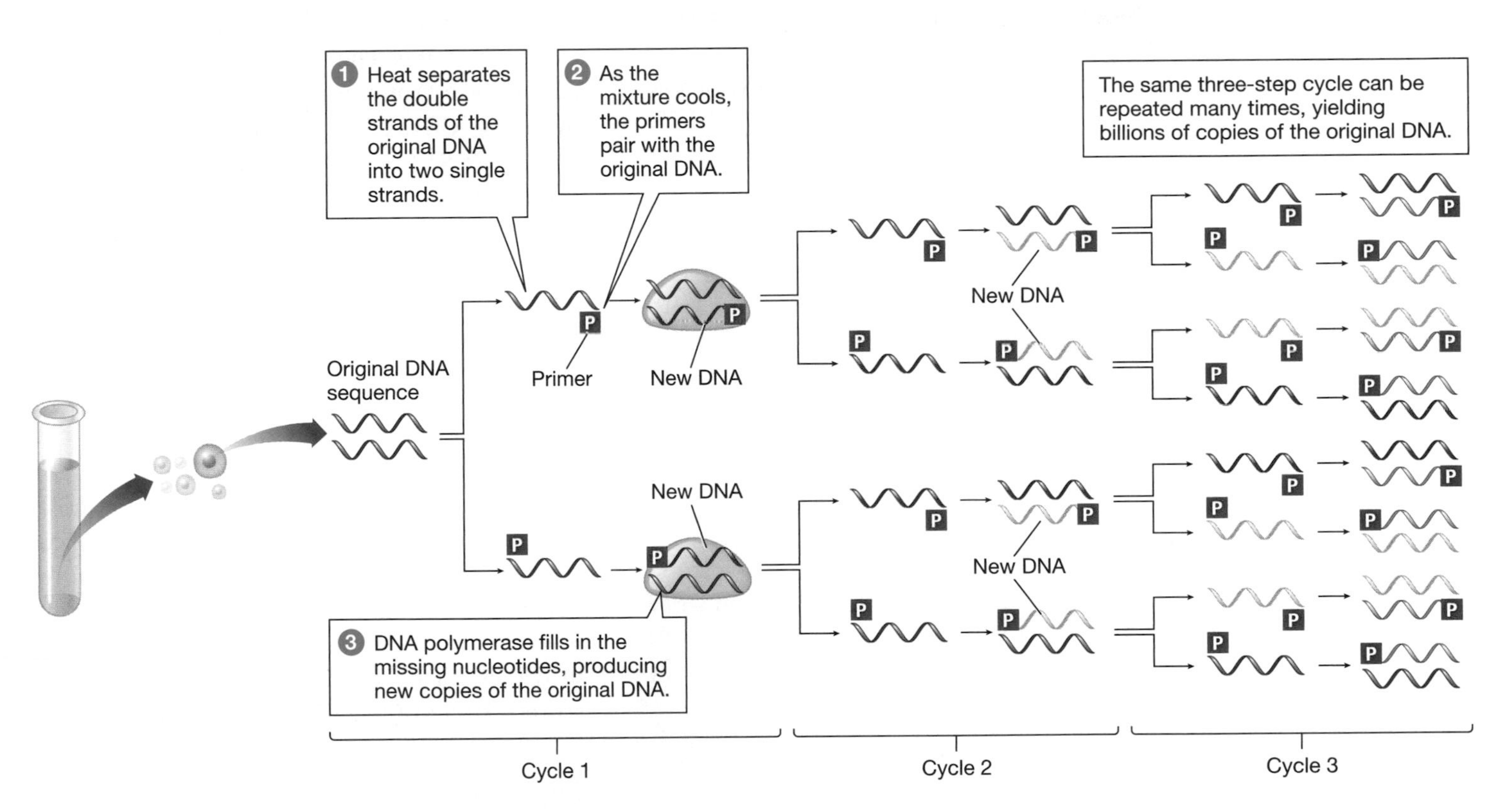

Figure 9.9

PCR can amplify small amounts of DNA more than a millionfold

In this laboratory technique, short primers made up of synthetic strings of nucleotides are mixed in a test tube with a sample of the target DNA, the enzyme DNA polymerase, and all four nucleotides (A, C, G, and T). The primers form base pairs with the two ends of the gene to be replicated. A machine then processes the mixture and repeatedly doubles the number of double-stranded versions of the template sequence.

Q1: PCR replicates DNA many times to increase the amount available for analysis. Why is this process called "amplification"?

Q2: During the PCR cycle, what causes the DNA strands to separate?

Q3: What is one major difference between PCR and DNA replication? (*Hint*: Think about what is being copied.)

See Appendix A for answers to the figure questions.

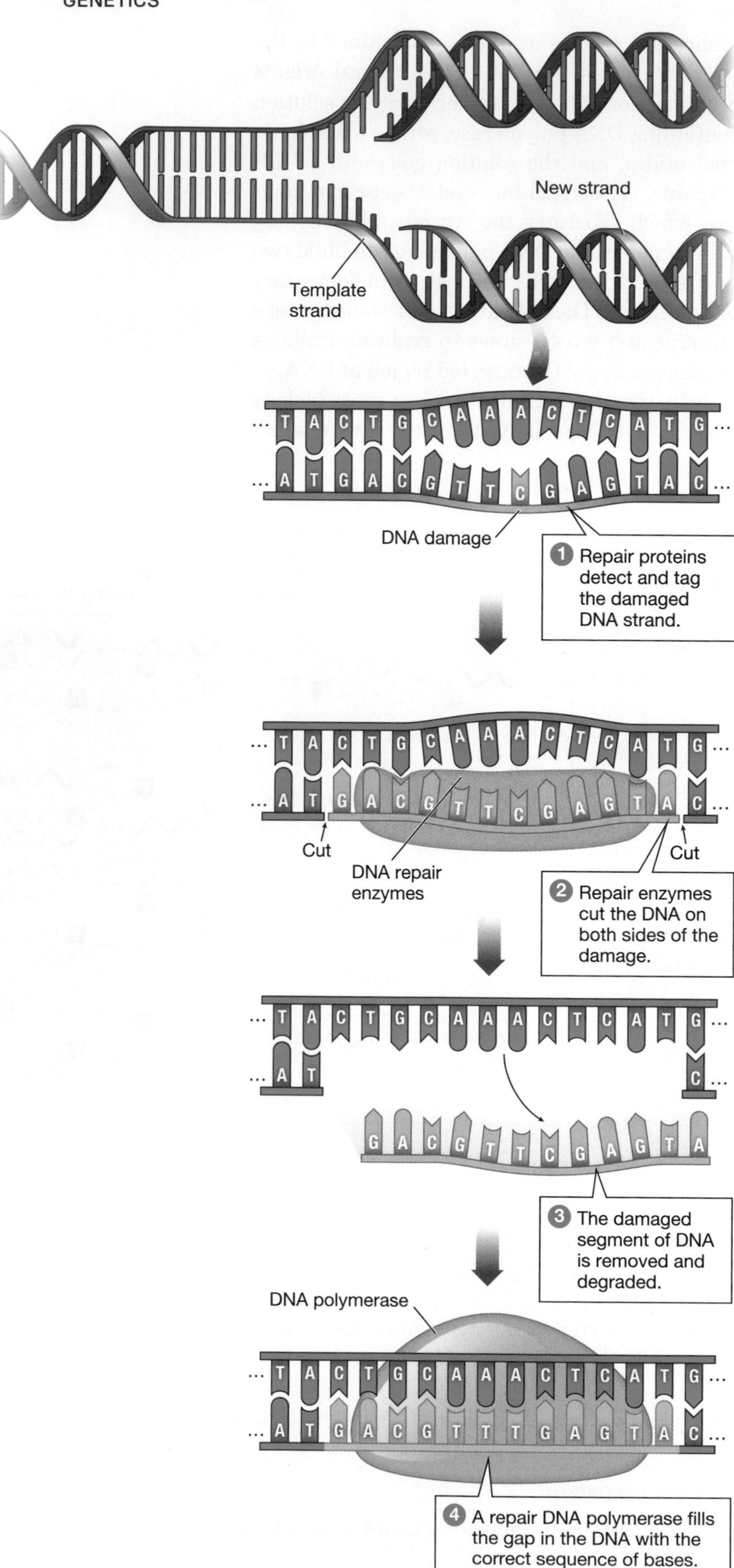

amount of DNA into an amount large enough to use in experiments. The Harvard Medical School team, for example, used PCR to replicate pig DNA and then sequenced that DNA. DNA sequencing involves machines that produce a data file listing the complete sequence of nucleotides in a DNA sample. Through PCR and then sequencing, the scientists were able to identify all the places in the pig genome containing PERV DNA.

Making Mutations

Church, Güell, and the team brainstormed a way to inactivate all 62 copies of PERV in the pig genome. The plan was to prevent DNA polymerase from replicating the PERV DNA by using CRISPR to mutate a small region of the PERV DNA: a gene named *pol* (see **Figure 9.6**). A change to the sequence of nucleotides in an organism's DNA is called a **mutation**. The extent of a mutation can range from the change of one nucleotide in a single base pair (known as a **point mutation**) to the addition or deletion of one or more whole chromosomes (a *chromosomal abnormality*, described in Chapter 8).

Güell was using CRISPR to artificially mutate genes, yet point mutations often occur naturally and randomly, especially during DNA replication. When DNA is copied right before mitosis, there are many opportunities for mistakes. Enzymes that copy DNA sometimes insert an incorrect nucleotide in the newly synthesized strand. In addition, DNA in cells is constantly

Figure 9.10

Repair proteins fix DNA damage

Large complexes of DNA repair proteins work together to fix damaged DNA.

Q1: Summarize how DNA repair works and why the repair mechanisms are essential for the normal functioning of cells and of whole organisms.

Q2: Is DNA repair 100 percent effective?

Q3: What would happen to an organism if its DNA repair became less effective?

See Appendix A for answers to the figure questions.

being damaged by chemical, physical, and biological agents, including energy from radiation or heat, collisions with other molecules in the cell, attacks by viruses (such as PERVs), and random chemical accidents (some of which are caused by environmental pollutants but most of which result from normal metabolic processes).

Replication errors and damage to DNA—especially to essential genes—disrupt normal cell functions. If not repaired, DNA damage leads to malfunctioning proteins, which can cause disease. The infamous *p53* gene, for example, is damaged or missing in most cancers. DNA damage can cause the death of cells and, ultimately, the death of an organism. Thankfully, cells have a way to recover: DNA polymerase immediately corrects almost all mistakes during DNA replication, "proofreading" complementary base pairs as they form.

Yet DNA polymerase is not infallible. When a point mutation occurs and it escapes proofreading by DNA polymerase, a mismatch error has occurred. This happens about once in every 10 million nucleotides. Luckily, cells have another backup safety program: repair proteins that correct 99 percent of mismatch errors, reducing the overall chance of an error to one mistake in every billion nucleotides (**Figure 9.10**).

On the rare occasion when a mismatch error is not corrected by repair proteins, the DNA sequence is changed, and the new sequence is reproduced the next time the DNA is replicated. If the mutation occurs within a gene, it will result in the formation of a new allele. Most new alleles are either neutral or harmful, but occasionally a mutation may be beneficial. In fact, all species, including humans, evolved thanks to the mutation of genes and the spread of new alleles through a population (we explore these processes in Chapter 12).

Three types of point mutations can alter a gene's DNA sequence: substitutions, insertions, and deletions. In a **substitution** point mutation, one nucleotide is substituted for another in the DNA sequence of the gene. An **insertion** or **deletion** point mutation occurs when a nucleotide is, respectively, inserted into or deleted from a DNA sequence. Sickle cell disease, a human genetic blood disorder, is caused by a substitution point mutation (**Figure 9.11**). Sometimes, however, changing a few nucleotides in a gene's DNA sequence has little or no effect. In such cases, a mutation is said to be "silent" because it produces no change in the function of the protein and, therefore, no change in the phenotype of the organism.

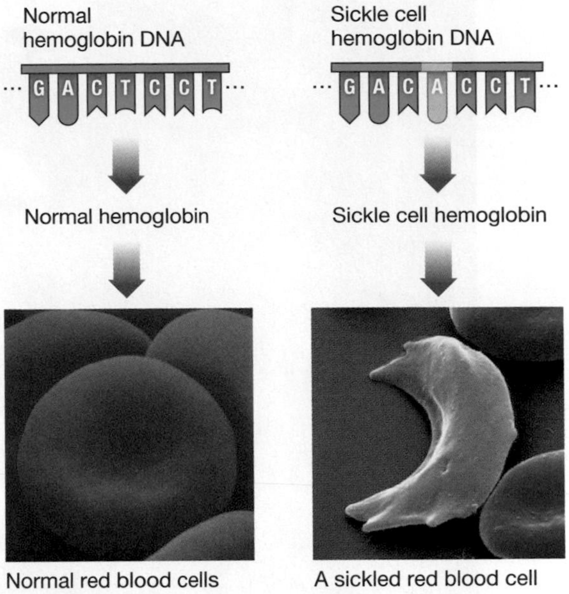

Insertions and deletions can be point mutations, but they can also involve more than one

Figure 9.11

A point mutation in the hemoglobin gene leads to sickle cell disease

In people with the genetic disorder sickle cell disease, a single base is altered in the gene that makes hemoglobin, an important protein involved in oxygen transport in red blood cells. The red blood cells of people with sickle cell disease can clog blood vessels, leading to serious effects such as heart and kidney failure.

Q1: What are the three types of point mutations? Which one of these causes sickle cell disease?

Q2: Sickle cell disease is an autosomal recessive genetic disorder. How many mutated hemoglobin alleles do people with sickle cell disease have?

Q3: In previous decades, sickle cell disease was often fatal in childhood. Because of improved treatments, individuals with sickle cell disease are now living into their forties, fifties, or longer. How might this extension of life span affect the prevalence of sickle cell disease in the population?

See Appendix A for answers to the figure questions.

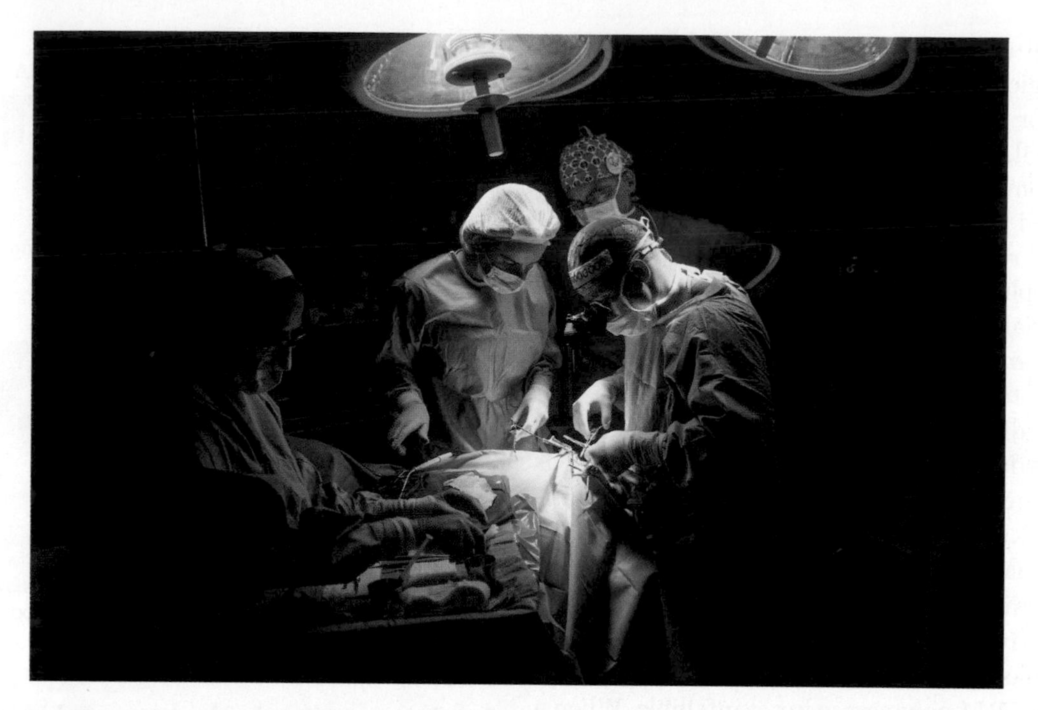

Figure 9.12

Experimental pig-to-baboon lung transplant

A lung from a pig engineered by CRISPR to prevent rejection is tested for safety and efficacy in primates by being transplanted into a baboon.

nucleotide; sometimes thousands may be added or deleted. Large insertions and deletions almost always result in the production of a protein that cannot function properly. Güell's plan to silence the PERVs in the pig genome was to use CRISPR to delete bits of DNA in the *pol* gene. The *pol* gene is essential for PERVs to replicate, so a large mutation should halt the formation of viral particles. In addition to being essential to the virus, *pol* is present in all 62 copies of PERVs but not elsewhere in the pig genome, so targeting CRISPR to mutate *pol* would inactivate the PERV DNA while leaving the pig DNA alone. "My biggest concern is always not to destroy the genome," says Güell. He and the team designed two CRISPR guides to target the "molecular scalpel" right to *pol* and to mutate the heck out of that gene.

After much trial and error, the team inserted its CRISPR tool into pig embryos, and the system began its work. The CRISPR system made 455 different cuts in the PERV *pol* genes throughout the genome, resulting in deletions ranging from 1 to 148 base pairs. About 80 percent of the engineered mutations were small deletions of fewer than nine base pairs—tiny, but enough to disrupt the gene.

Pigs Are People Too?

In late 2015, Güell, Church, and the team presented the results of their initial work: using CRISPR, they had simultaneously inactivated all 62 PERVs in pig embryos. In previous gene-editing experiments that didn't involve CRISPR, PERVs from the pig cells infected human cells when put in close proximity. This time, when the team placed the CRISPR-edited pig cells next to human cells, there was zero transmission of the virus. "The gene editing completely disrupted the ability of the virus DNA to replicate and form viral particles," says Güell.

DEBUNKED!

MYTH: Only certain people have "disease genes."

FACT: A gene's purpose is not to cause disease. Genetic diseases are caused by *mutations* in genes we all have. For example, everyone has *BCRA1* (the "breast cancer gene"). People with mutations in that gene are at higher risk for cancer.

Within a year of that breakthrough, Church, Yang, and Güell started a company, Massachusetts-based eGenesis, to further develop the technology. In August 2017, the company announced the birth of 37 PERV-free baby pigs. The first piglet was named Laika, after the first dog in space. The black-and-white miniature pig breed grows no bigger than 150 pounds, just the right size for organs that could be transplanted into adult humans (**Figure 9.12**).

Other researchers are pursuing a different path: growing actual human organs in pigs. Again, CRISPR is the workhorse for the experiment. In 2016, a team at UC Davis used CRISPR to remove from a pig embryo the genes that encode for the pig pancreas. The procedure created a void in the embryo, which they then filled with human stem cells. Stem cells have the potential to develop into almost all human cells and organs including, in this case, a fully functional human pancreas.

Some research groups are trying to use this method to grow other organs. **Figure 9.13** shows the process by which scientists could use CRISPR and human stem cells to grow a pig with human kidneys, which would then be used to donate a kidney to a patient in need of a transplant. No researchers have yet completed this process,

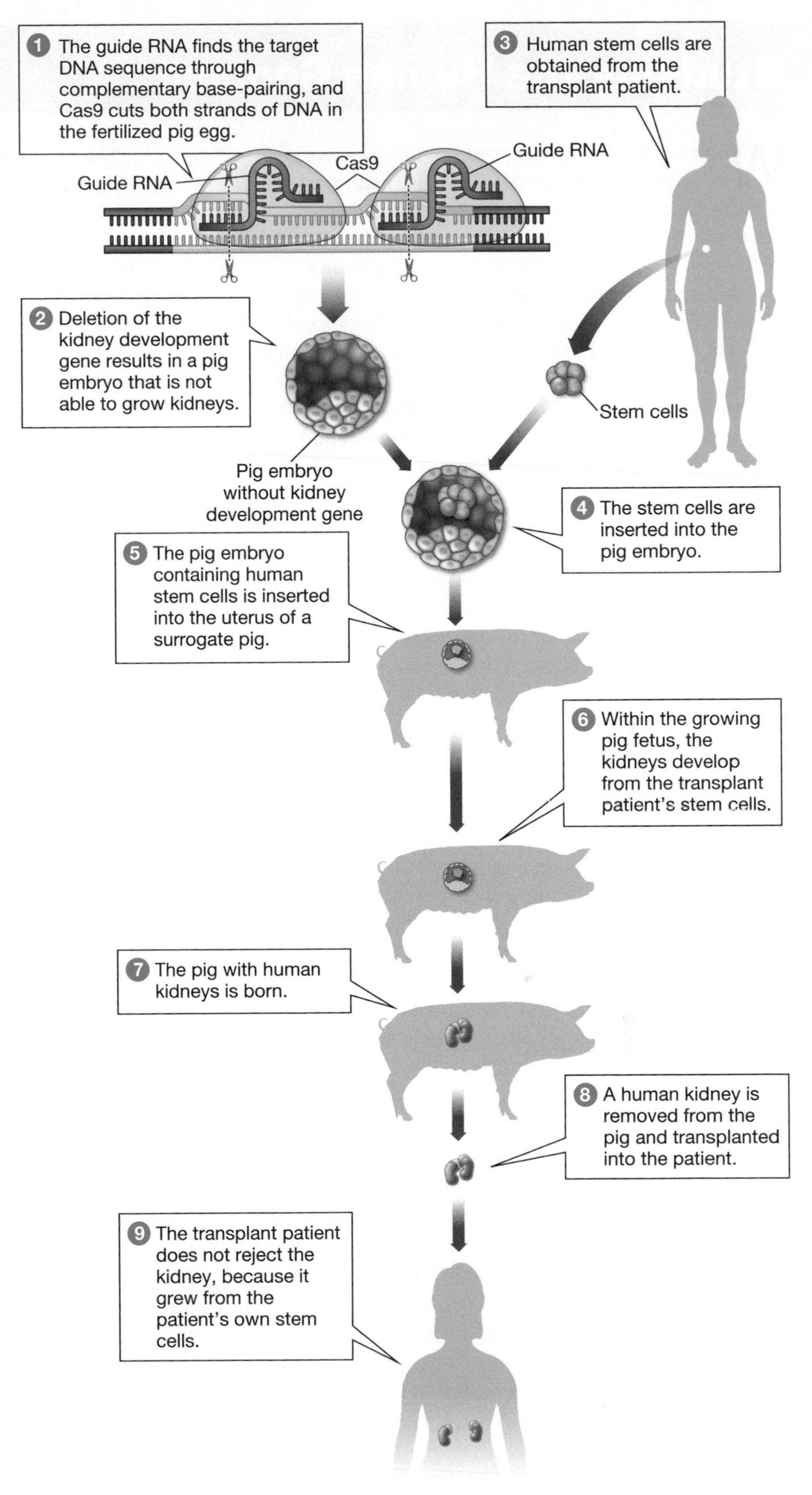

1 The guide RNA finds the target DNA sequence through complementary base-pairing, and Cas9 cuts both strands of DNA in the fertilized pig egg.
Guide RNA
Cas9
Guide RNA
3 Human stem cells are obtained from the transplant patient.
2 Deletion of the kidney development gene results in a pig embryo that is not able to grow kidneys.
Stem cells
Pig embryo without kidney development gene
4 The stem cells are inserted into the pig embryo.
5 The pig embryo containing human stem cells is inserted into the uterus of a surrogate pig.
6 Within the growing pig fetus, the kidneys develop from the transplant patient's stem cells.
7 The pig with human kidneys is born.
8 A human kidney is removed from the pig and transplanted into the patient.
9 The transplant patient does not reject the kidney, because it grew from the patient's own stem cells.

Figure 9.13

Human organs could be grown in pigs

Human organs for transplant could be grown in pigs modified by CRISPR to lack those organs. Here, a hypothetical process produces a kidney.

Q1: What step in this process is similar to one in the original CRISPR method that removed the PERVs from the pig genome? (*Hint*: Review Figure 9.6.)

Q2: What step in this process was not included in the original CRISPR method that removed the PERVs from the pig genome?

Q3: What parts of this process would scientists need to change in order to develop several different human organs in a single pig?

See Appendix A for answers to the figure questions.

Studying the Human Genome

When Gregor Mendel first proposed his laws of inheritance in the mid-1800s (as described in Chapter 7), the Austrian monk had no idea what genes were made of. Neither did anyone else until more than 100 years later, when the combined efforts of scientists led to the discovery of the DNA double helix. At King's College, London, Rosalind Franklin and Maurice Wilkins produced the first images of DNA's structure, using a technique called X-ray crystallography (see photo). Meanwhile, at nearby Cambridge University, Francis Crick and James Watson, inspired by a presentation by Wilkins, began making models of DNA, building off the work of many scientists before them. With critical information from Franklin's X-ray photographs (obtained without her permission), Watson and Crick identified the correct structure of DNA. All four scientists published their work in the same issue of *Nature* in 1953. In 1962, the three men received the Nobel Prize in Physiology or Medicine for their work. Franklin, who had passed away in 1958 from cancer, was not honored.

Once scientists knew the structure of DNA, it was only a matter of time before they identified all the estimated 20,000–25,000 human genes and sequenced all 3 billion bases in the genome. The international Human Genome Project (HGP) began in 1990 and lasted 13 years. During that period, the U.S. government spent an estimated $437 million on human genome sequencing and $3.8 billion on a wide range of projects related to genomics.

A government summary called the HGP "arguably the single most influential investment to have been made in modern science." The knowledge and techniques gained through the HGP have led to advances in agriculture, energy, the environment, and, perhaps most notably, human medicine.

The understanding that each person has a unique molecular and genetic profile led directly to the rise of personalized medicine, also called precision medicine, in which medical treatments are tailored to an individual based on his or her predicted risk of disease. One of the earliest examples of personalized medicine was a drug called trastuzumab. The drug reduces recurrence of breast cancer in women who make too much of a protein called HER2, which can be detected with a blood test. Similar examples, such as targeting a treatment to a specific group of people based on their unique disease susceptibility, are emerging across all areas of medicine and health.

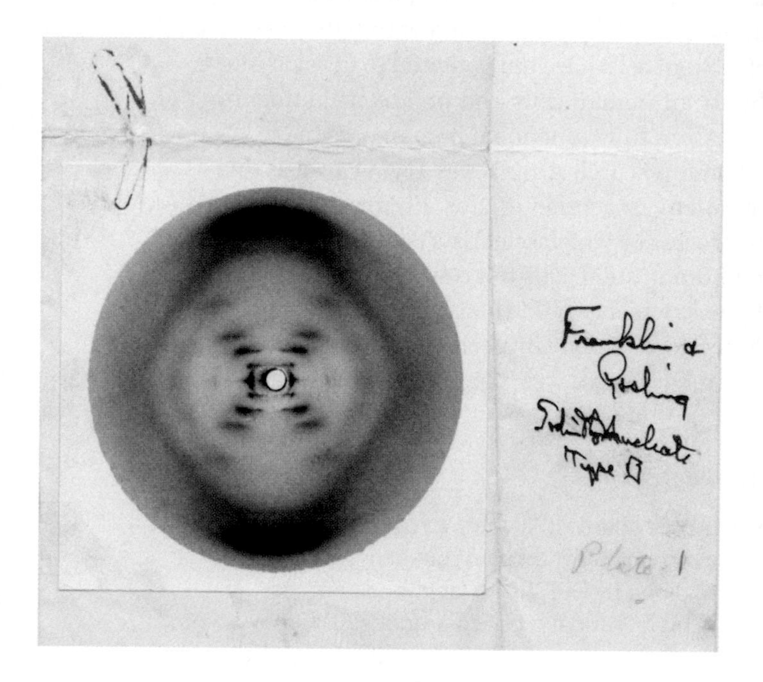

and the work remains controversial, especially because of concerns that the human cells might migrate to the developing pig's brain and make the pig, even in some small way, more human.

Overall, CRISPR has revitalized the idea of a safe, clean, limitless source of healthy organs. "I think this is a magical point in the field of [animal transplants]," William Westlin, eGenesis's executive vice president for research and development, told *The Guardian* in 2019. "It's no longer a question of if. It's just a question of when."

The Meteoric Rise of CRISPR

The gene editing tool CRISPR has taken molecular biology laboratories by storm over the past five years. It has been used to edit the genomes of crops and livestock to improve breeding and production, to control populations of disease-carrying insects, to silence genetic disorders in animal models, and more. Here are a few highlights from the short but shining history of CRISPR.

Assessment available in **smartwork**

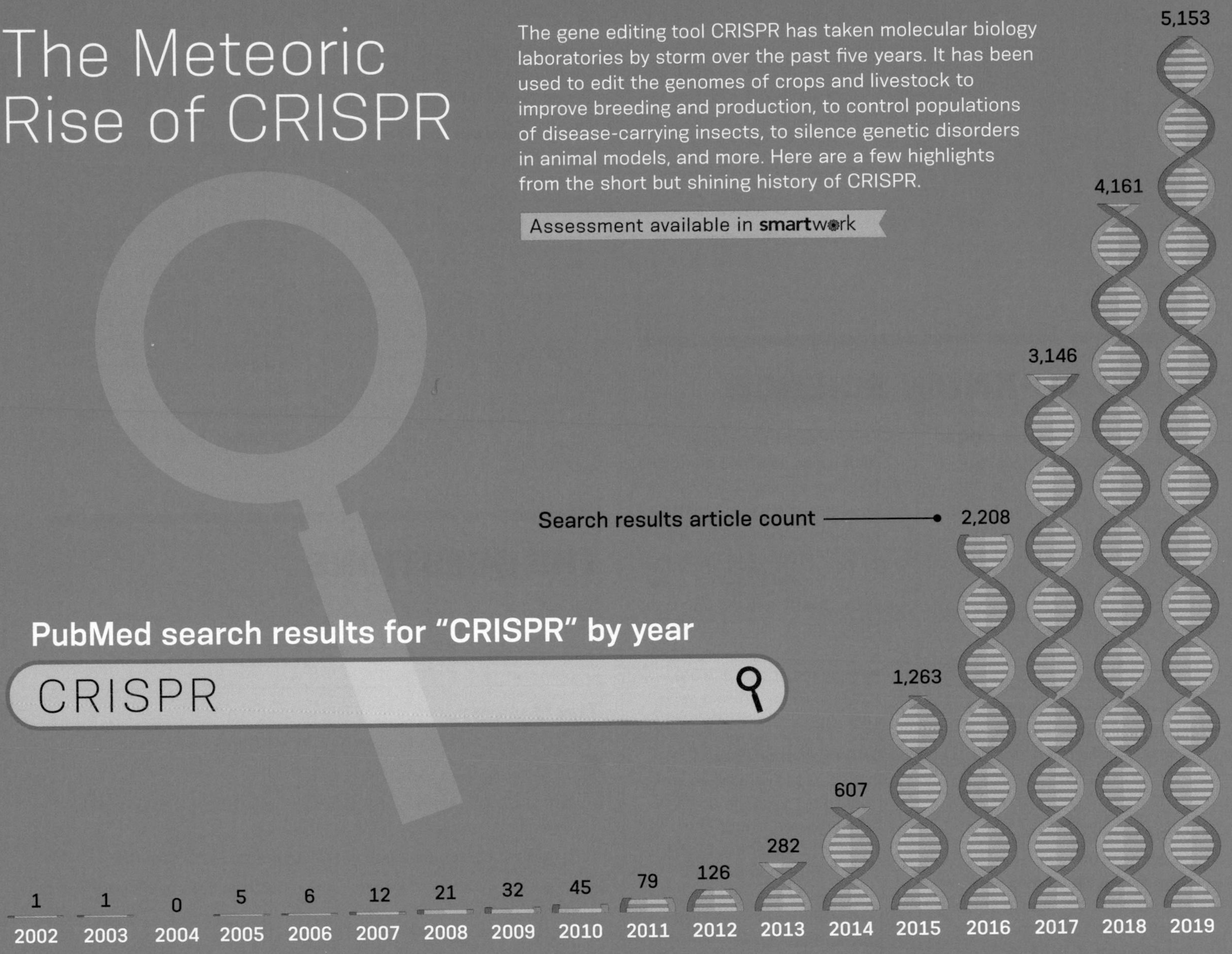

PubMed search results for "CRISPR" by year

CRISPR

Search results article count

2002	2003	2004	2005	2006	2007	2008	2009	2010	2011	2012	2013	2014	2015	2016	2017	2018	2019
1	1	0	5	6	12	21	32	45	79	126	282	607	1,263	2,208	3,146	4,161	5,153

1987
CRISPR repeats are first observed in bacterial genomes. Their significance is not yet known.

2002
The term CRISPR is coined by researchers in Spain and the Netherlands.

2006
Researchers propose that CRISPR functions in nature as part of a bacterial adaptive immune system.

2011
The final necessary piece for the genome-editing system is identified: a second small RNA needed to guide Cas9 to its targets.

2013
The CRISPR-Cas9 system is used to edit targeted genes in both human and mouse cells, and later plant cells.

2016
The first human trial to use CRISPR gene editing gets approval from the National Institutes of Health, in a cancer therapy trial to edit a patient's own immune system cells.

2018
A Chinese scientist announces the birth of the first gene-edited babies—twin girls with a gene altered to reduce the risk of contracting HIV. Scientists around the world react with outrage because there is an international ban on editing the human genome, and the scientist's work is suspended.

2020
Immune cells whose genomes have been altered with CRISPR are well-tolerated by three people with cancer in a University of Pennsylvania clinical trial.

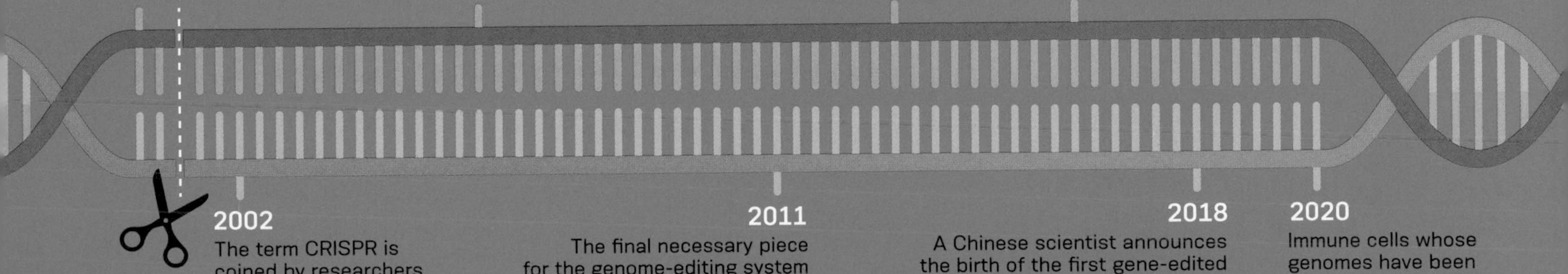

NEED-TO-KNOW SCIENCE

- Genes are composed of **DNA** (page 168), which consists of two parallel strands of repeating units called **nucleotides** (page 169) twisted into a **double helix** (page 170).
- The four nucleotides of DNA contain the **bases** (page 169) adenine (A), cytosine (C), guanine (G), and thymine (T). The nucleotides exhibit **complementary base-pairing** (page 171) according to **base-pairing rules** (page 171): base A can pair only with T, and C can pair only with G.
- DNA is wrapped around **histone proteins** (page 172), forming **nucleosomes** (page 172). The nucleosome structures can further compact the DNA by coiling around themselves to form a **chromatin fiber** (page 173). Chromatin fibers further coil around themselves to form chromosomes.
- The **CRISPR**-Cas9 editing system (page 171) is composed of two pieces of **RNA** (page 168) designed to form base pairs at precise locations in a gene. This DNA-RNA interaction guides the CRISPR proteins to the sites where they efficiently cut the DNA, resulting in a gene deletion after normal repair processes take place. Additional genetic manipulations are required to generate a gene insertion.
- **DNA replication** (page 173) occurs in all living organisms prior to mitosis. The double helix unwinds, and the two strands break apart. Each strand of DNA serves as a template from which a new strand is copied. **DNA polymerase** (page 173) builds each new strand of DNA using **primers** (page 173) located near the **origins of replication** (page 173). This mode of replication is known as **semiconservative replication** (page 174) because one "old" strand (the template strand) is retained, or "conserved," in each new double helix.
- The **polymerase chain reaction** (**PCR**, page 175) is a laboratory technique to amplify the DNA from a small initial amount to millions of copies. Amplified DNA can then be sequenced to examine specific genes or mutations.
- DNA is subject to damage by physical, chemical, and biological agents, and errors in DNA replication are common. DNA polymerase "proofreads" the DNA during replication and corrects most mistakes. Repair proteins are a backup repair mechanism and correct any errors that DNA polymerase misses.
- A change to the sequence of bases in an organism's DNA is called a **mutation** (page 176). Three types of mismatch mutations can alter a gene's DNA sequence: **substitutions**, **insertions**, and **deletions** (page 177). If only a single base is altered, it is a **point mutation** (page 176).

THE QUESTIONS

See Appendix B for answers.

The Basics

1 DNA replication results in

(a) two DNA molecules—one with two old strands, and one with two new strands.

(b) two DNA molecules, each of which has two new strands.

(c) two DNA molecules, each of which has one old strand and one new strand.

(d) none of the above

2 The DNA of cells is damaged

(a) constantly.

(b) by collisions with other molecules, chemical accidents, and radiation.

(c) not very often and only by radiation.

(d) both a and b

3 The DNA of different species differs in the

(a) sequence of bases.

(b) complementary base-pairing.

(c) number of nucleotide strands.

(d) location of the sugar-phosphate portion of the DNA molecule.

4 Mutation

(a) can produce new alleles.

(b) can be harmful, beneficial, or neutral.

(c) is a change in an organism's DNA sequence.

(d) all of the above

5 Link each term with the correct definition.

nucleotide	1. Two complementary bases joined by hydrogen bonds.
base pair	2. One component of a nucleotide; there are four variants of this component.
DNA molecule	3. A strand of nucleotides linked together by covalent bonds between a sugar and a phosphate; two strands are linked by hydrogen bonds between complementary bases.
base	4. A unit consisting of the sugar deoxyribose, a phosphate group, and a base (adenine, cytosine, guanine, or thymine).

6 In the diagram of replication shown here, fill in the blanks with the appropriate terms: (a) base pair, (b) base, (c) nucleotide, (d) template strand, (e) newly synthesized strand, (f) separating strands.

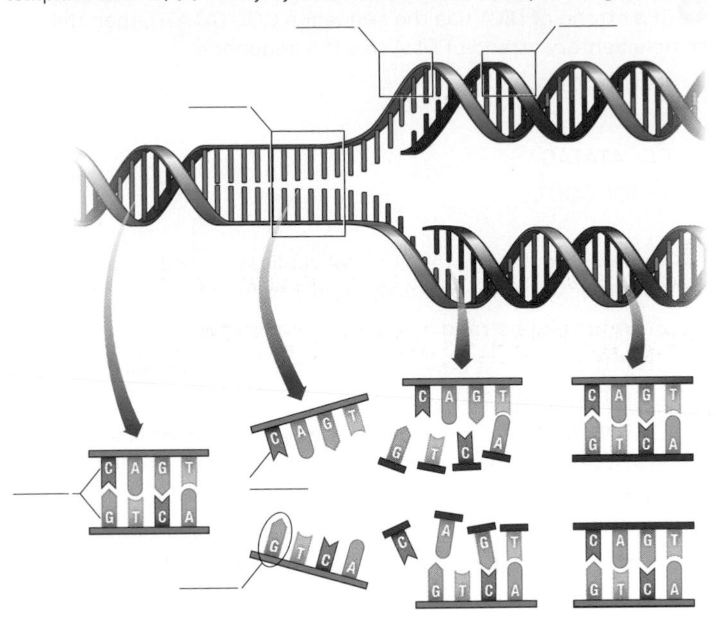

7 Select the correct terms:

To work on removing PERV DNA from the pig genome, researchers first replicate the DNA many times, using (**PCR** / **CRISPR**). This process increases the amount of DNA so that they are able to run experiments to edit the genome, using (**PCR** / **CRISPR**).

Challenge Yourself

8 If a strand of DNA has the sequence CGGTATATC, then the complementary strand of DNA has the sequence

(a) ATTCGCGCA.

(b) GCCCGCGCT.

(c) GCCATATAG.

(d) TAACGCGCT.

9 Place the following steps of DNA replication and repair in the correct order by numbering them from 1 to 5.

____ a. A template strand begins to be replicated.

____ b. If the incorrect base is not identified and replaced, it remains as a point mutation in the DNA.

____ c. DNA polymerase identifies and replaces most incorrect bases with the correct base, complementary to the base on the template strand.

____ d. An incorrect base is added to the growing strand of DNA.

____ e. Proteins identify and replace any incorrect bases missed by DNA polymerase.

10 Given the template DNA sequence GCAGCATGTT, identify each of the following mutations to the complementary strand as an insertion, a deletion, or a substitution.

____ a. CGTCGTACA

____ b. CGTGGTACAA

____ c. CGTCGTACTAA

11 If the target sequence for PCR on one strand of DNA is ATGCAAATCCTGG, what is the sequence of each of the two strands of the double-stranded DNA produced?

Try Something New

12 Using base-pairing rules to guide you, for a DNA double helix that contains 20 percent adenine (A), specify the percentage of thymine (T), guanine (G), and cytosine (C).

13 The following data are from a 2018 study on American opinions about gene editing using technologies such as CRISPR. The Pew Research Center for Science and Society is a nonpartisan nonprofit organization that conducts public opinion polls and other kinds of social science research. Below are data from a Pew study on opinions about the use of gene-editing technologies such as CRISPR:

(a) What percentage of people questioned believed that it was appropriate to use gene editing to treat a serious condition a child would have at birth?

(b) What percentage of people believed that it was inappropriate to use gene editing to reduce the risk of a disease that could occur over a lifetime?

(c) Of the options presented, what did people feel was the least appropriate use of gene editing?

(d) For each option, do you feel that using gene editing would be appropriate or inappropriate?

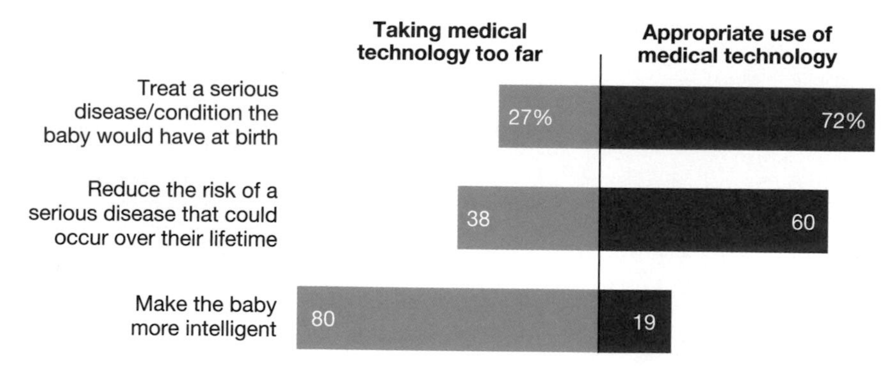

14 Scientists estimate that genes encoding proteins make up less than 1.5 percent of the human genome. Other genes in our cells encode different types of nonprotein RNA molecules. The rest of our genome consists of various types of *noncoding DNA*, which does not code for any kind of functional RNA. Some of the remaining DNA has regulatory functions—for example, controlling gene expression. Some of it has architectural functions, such as giving structure to chromosomes or positioning them at precise locations within the nucleus. Given that these sections of DNA are noncoding, does it matter if they contain replication errors? Explain your reasoning.

Leveling Up

15 **What do *you* think?** Read the article "Genome-Edited Baby Claim Provokes International Outcry," about the scientist in China who used CRISPR to edit the genes of human embryos so that the resulting children (now twin preschool girls) would be resistant to some strains of HIV (https://www.nature.com/articles/d41586-018-07545-0). Given your responses to question 13, what is your opinion of the research that the scientist He Jiankui conducted? Was this an appropriate or inappropriate use of gene-editing technology? Explain your reasoning.

16 ***Write Now* biology: next-generation genetically modified organisms (GMO)** Read the article "CRISPR Brings an Early Harvest" (http://www.genengnews.com/gen-news-highlights/crispr-brings-an-early-harvest/81253507), which describes a new way to produce GMO tomatoes without using standard gene-transfer techniques. Write a letter to your state or federal legislator expressing your opinion either for or against this new technology for creating GMO foods. Cover the following points in your letter.

(a) Explain how CRISPR works, comparing and contrasting it to traditional GMO technology.

(b) Describe how the CRISPR system was applied to the plants in the article.

(c) In your final paragraph, try to persuade your government representative to either endorse or condemn this technology for improving crop yield and quality, using specific examples.

Digital resources for your book are available online.

Tobacco's New Leaf

One company's quest to produce tomorrow's drugs in today's plants.

After reading this chapter you should be able to:

- Explain how gene expression creates a phenotype from a genotype.
- Label a diagram of transcription, identifying each step in the process and the relevant molecules.
- Label a diagram of translation, identifying each step in the process and the role of each type of RNA.
- Determine the correct amino acid from a given codon.
- Justify how the genetic code fits the definitions of "redundant" and "universal."
- Define gene expression, and discuss its importance for an organism.

GENETICS

CHAPTER 10

HOW GENES WORK

The greenhouse is vast, the size of half a football field. A loud, steady thrum reverberates in the room as massive metal fans push air through the hot, humid space. Mike Wanner, a tall, serious man, walks through rows of leafy plants, each a foot and a half tall. He stoops to rub a leaf between his fingers, then raises his hand to his nose and sniffs. "That's what tobacco smells like," Wanner shouts, straining to be heard over the noise of the fans.

Wanner is the North Carolina site manager for operations at Medicago, a Canadian biotechnology company growing tobacco at a facility outside Durham, once home to Lucky Strike cigarettes, and in Canada. But these plants are not being grown to smoke, chew, or dip. Instead, they serve a dramatically different purpose: this tobacco will be used to make influenza vaccines (**Figure 10.1**).

Flu vaccines are normally grown in chicken eggs, and this lengthy process takes months. As an alternative, companies and researchers have begun experimenting with "biopharming": manufacturing vaccine proteins in plants.

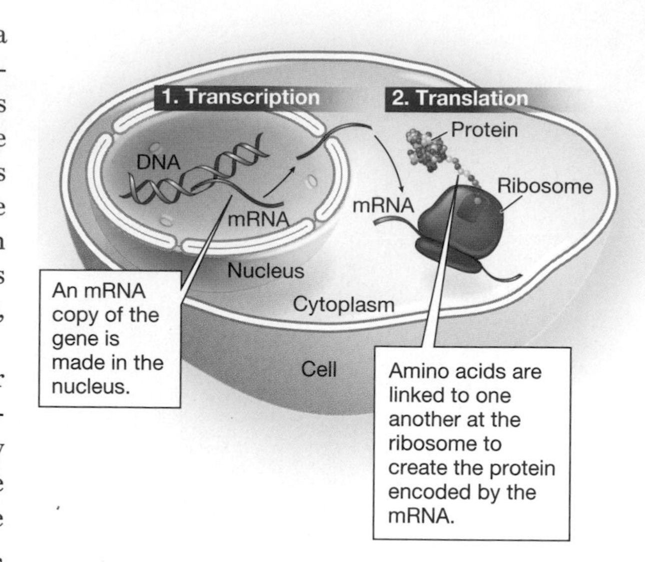

Figure 10.2

An overview of gene expression

In gene expression, genetic information flows from DNA to RNA to protein. This process consists of two steps: (1) The transcription of a protein-coding gene produces an mRNA molecule, which is then transported to the cytoplasm. (2) There translation occurs as the protein is made with the help of ribosomes.

Figure 10.1

Growing tobacco to treat the flu

Tobacco plants grow in a Medicago greenhouse in Quebec City, Canada.

Scientists insert a gene that codes for the protein they want to create into the cells of a plant, which will then produce that protein. This production of a protein from a gene occurs via **gene expression**—the process by which genes are transcribed into RNA and then translated to make proteins (**Figure 10.2**).

Proteins participate in virtually every process inside and between cells. Because proteins are so important, gene expression is the fundamental way that genes influence the structure and function of a cell or organism. It is through this process that an organism's genotype gives rise to its phenotype (discussed in Chapter 7). Gene expression occurs in all prokaryotes, eukaryotes, and viruses.

In addition to being involved in every process in a cell, proteins are the key components of many vaccines, not just the flu vaccine. Injected pieces of viral proteins activate the human immune system to defend against that virus in the future (see Figure 2.2). To make these drugs, scientists must produce large quantities of viral proteins. The traditional method involves injecting chicken eggs with a virus, letting the virus multiply in the chicken cells, then extracting the

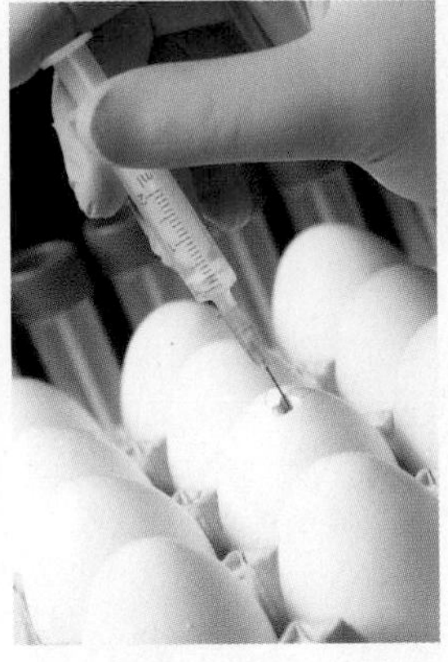

	Tobacco	Egg
Speed from outbreak to vaccine production:	1 month	6 months
Cost to produce 50 million flu vaccines:	$36 million	$400 million
Allergy risks:	Minimal	Individuals with egg allergies cannot receive the vaccine
Availability:	Not yet approved by the Food and Drug Administration (FDA)	Approved and currently the most widely used technology to produce flu vaccines

Figure 10.3

Tobacco or egg?

The use of tobacco plants to produce proteins for influenza vaccines has several advantages over the traditional approach that uses chicken eggs.

Q1: Why is it faster to produce vaccines through biopharming with plants than by injecting viruses in eggs? Why is speed important?

Q2: How much cheaper is biopharming with plants than creating vaccines in eggs? Why is cost important?

Q3: The FDA is responsible for ensuring the safety of drugs and food. Why must tobacco-derived vaccines, like any new medications, be approved by the FDA?

See Appendix A for answers to the figure questions.

virus, removing its genetic material, and preparing a vaccine from the leftover viral proteins.

Unlike chicken eggs, however, plants can be grown in vast quantities, and they grow rapidly—often in just days or weeks. "The big advantage of plant systems is that they can produce massive amounts of proteins very inexpensively," says James Roth, director of the Center for Food Security and Public Health at Iowa State University, who studies biopharming in plants and animals (and is not associated with Medicago). In the event of a pandemic flu outbreak, Medicago could produce vaccines 6 times faster and 12 times more cheaply than traditional egg manufacturing could, the company claims (**Figure 10.3**).

In April 2012, Medicago put its tobacco plants to the test, running the Durham facility at full tilt for 30 days. The U.S. Department of Defense had given the company millions of dollars to test whether it could quickly produce enough pandemic flu vaccine from the tobacco to stem an outbreak. The pressure is on, says Wanner. Standing inside the facility, he looks out over his crop of mini protein factories. "We'll see what happens."

Fighting the Flu with Tobacco

In 1997, Louis-Philippe Vézina, then a research scientist at Agriculture and Agri-Food Canada and an associate professor at Université Laval in Quebec City, Canada, decided to start a company. A plant biotechnologist, Vézina wanted to explore the possibility of manufacturing proteins in alfalfa plants. For this reason, he called his company Medicago, after the genus (group of species) to which alfalfa belongs. Later, as the company grew, Vézina and his team discovered that tobacco produces higher yields of proteins in a shorter time frame than alfalfa does, so the company switched plants (and caught the attention of the cigarette manufacturer Phillip Morris, who was seeking ways to diversify beyond the cigarette market and quickly invested in Medicago).

By 2005, Medicago had begun to receive calls asking about its product and when it would begin clinical trials, the first step toward getting a drug approved by the U.S. Food and Drug Administration (FDA). "I was surprised. Even

Figure 10.4

Swine flu

In 2009, people around the world donned masks as a precaution against influenza virus H1N1, nicknamed "swine flu" because it contained DNA from human, bird, and swine flu viruses. The virus caused a pandemic, killing an estimated 284,500 people.

the FDA was asking, 'When are we going to see your vaccine?'" recalls Vézina. "They were keen on technologies like this because they know how much vaccines cost, and they're interested in anything that can decrease that cost."

One of the first vaccines that Medicago produced was a vaccine for influenza virus H1N1, or swine flu, the most common cause of the flu in 2009 (**Figure 10.4**). In the previous H1N1 pandemic, it had taken months for vaccines grown in chicken eggs to reach the market. By contrast, Medicago produced its vaccine, ready for testing, in just 19 days.

To make a flu vaccine in plants, Vézina and his team use genetic engineering in much the same way that it is used to create gene therapies, as described in Chapter 8. They identify and synthesize a single viral gene that codes for hemagglutinin, a protein found on the surface of flu viruses (**Figure 10.5**). To make large quantities of that viral protein, the scientists at Medicago first insert the hemagglutinin gene into small, rod-shaped bacteria called agrobacteria, which infect plants. Next, a robotic arm lifts a tray containing 5-week-old tobacco plants secured to the surface, flips the tray upside down, and dips the plants into a liquid solution swimming with the gene-infused agrobacteria. Once the plants are immersed in the liquid, the technicians turn on a vacuum, which sucks air out of the leaves. As the air is pulled out, agrobacteria are absorbed in. This process is like dipping a sponge into water, squeezing it, and then releasing it to soak up the liquid.

When the plants are returned to the greenhouse, their leaves are floppy and almost translucent, like wet tissue paper. Now things are cooking. Once the agrobacteria are inside the leaves, the bacterial cells release the viral hemagglutinin gene into the plant cells, where it is transported into plant cell nuclei to begin the process of gene expression: making a protein from DNA via the two steps of transcription and translation.

"When you get the *Agrobacterium* solution into the leaf, it will invade the cells and there will be a burst of viral gene expression in the plant cell," said Vézina. "The plant takes over and uses its machinery to produce the protein."

Two-Step Dance, Transcription: DNA to RNA

The first step in gene expression—in this case, for a flu gene in a plant cell—is **transcription**. Transcription creates a segment of RNA based on a DNA template. In the nucleus, an enzyme called **RNA polymerase** binds to a segment of DNA, near the beginning of the gene, called a **promoter**. The promoter contains a specific sequence of DNA bases that the RNA polymerase recognizes and binds to (**Figure 10.6**). At Medicago, scientists attach specific promoters to the hemagglutinin gene so that the plant cell's RNA polymerase can identify the gene and actively transcribe it, maximizing the rate of transcription, says Vézina.

Once bound to the promoter, the RNA polymerase unzips the DNA double helix at the beginning of the gene, separating a short portion of the two strands. Only one of the two DNA strands is used as a template, and thus it is called the **template strand**. The RNA

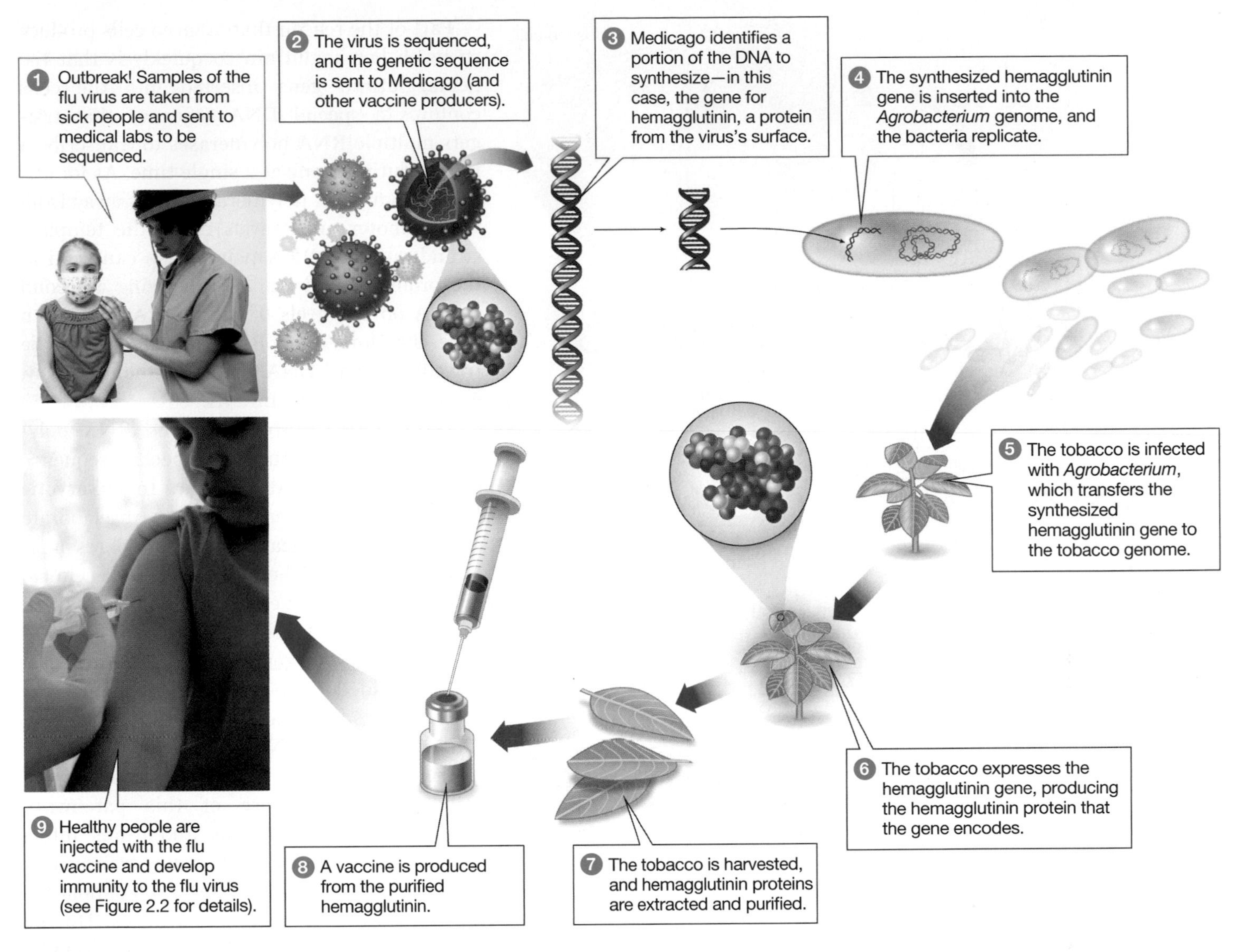

Figure 10.5

From outbreak to vaccine

During flu season or any other viral outbreak, a large network of medical professionals, scientists, and workers in the pharmaceutical industry quickly begin creating and distributing a vaccine.

Q1: In which of the step(s) illustrated here does DNA replication occur?

Q2: In which step(s) does gene expression occur?

Q3: Why do you think vaccine manufacturers produce a vaccine after the start of flu season rather than before it has begun?

See Appendix A for answers to the figure questions.

polymerase begins to move down the DNA template strand, constructing an RNA molecule. The RNA molecule is a strand of nucleotides complementary to the DNA, and the RNA polymerase constructs it from free nucleotides floating around in the nucleus. RNA does not have all the same bases as DNA: its four bases are adenine (A), cytosine (C), guanine (G), and uracil (U). Those bases pair with the four DNA bases according to the following rules: RNA's A with DNA's T, C with G, G with C, and U with A.

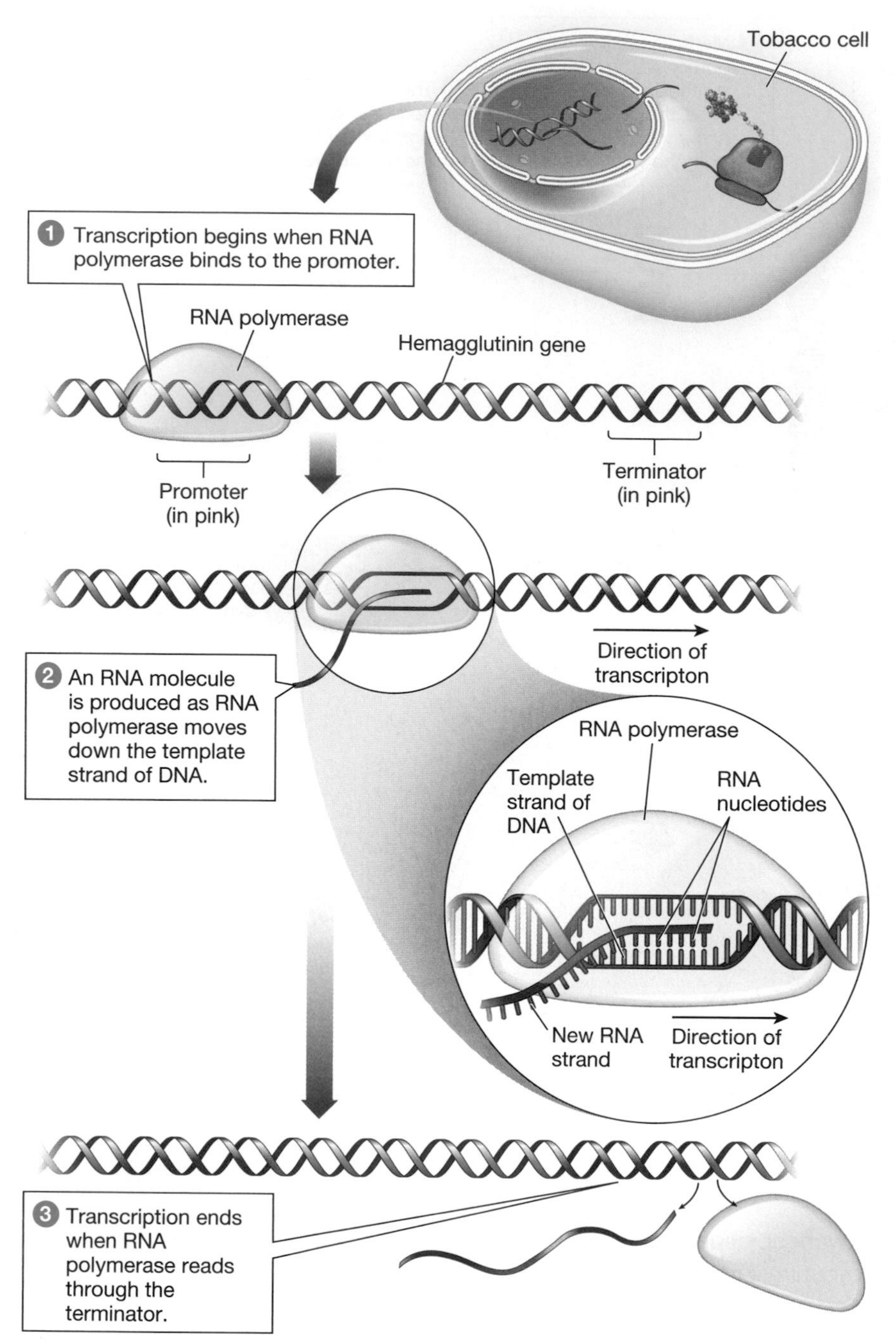

Part of the reason that tobacco cells produce so much hemagglutinin so quickly is that the hemagglutinin gene inserted into the cells contains a special DNA sequence that triggers multiple RNA polymerases to transcribe a hemagglutinin gene at a single time. As for any gene, as an RNA polymerase moves away from the promoter and travels down the template strand, another RNA polymerase can bind at the promoter and start synthesizing a second mRNA on the heels of the first. At any given time, therefore, many RNA polymerases can be traveling down a DNA template simultaneously, each synthesizing an mRNA.

Transcription stops when the RNA polymerase reads through a special sequence of bases called a **terminator**. In eukaryotic cells, the mRNA then undergoes an elaborate sequence of modifications that prepares it to leave the nucleus. These steps include chemical modification of both ends of the mRNA, as well as a process called RNA splicing.

Most eukaryotic genes (and many viral genes) are embedded with stretches of sequences that don't code for anything, called **introns**. The stretches of DNA in a gene that carry instructions for building the protein are called **exons**. Because of this patchwork construction, with introns and exons interspersed throughout a gene, newly transcribed RNA (pre-RNA) is also a patchwork of noncoding sequences and coding sequences. During **RNA splicing**, the introns are snipped out of a pre-RNA, and the remaining pieces of RNA—the exons—are joined to generate the mature RNA (**Figure 10.7**). This RNA is then ready to leave the nucleus.

Figure 10.6

Plants making proteins, I: Transcription

In the first step of protein creation, template DNA is transcribed. Here, RNA polymerase transcribes the hemagglutinin gene into a molecule of RNA.

Q1: Why does RNA polymerase use only one strand of DNA as a template for making RNA?

Q2: The template strand of part of a gene has the base sequence TGAGAAGACCAGGGTTGT. What is the sequence of RNA transcribed from this DNA, assuming that RNA polymerase travels from left to right on this strand?

Q3: If a mutation occurred within the promoter or terminator region, do you think it would affect the RNA transcribed? Why or why not?

See Appendix A for answers to the figure questions.

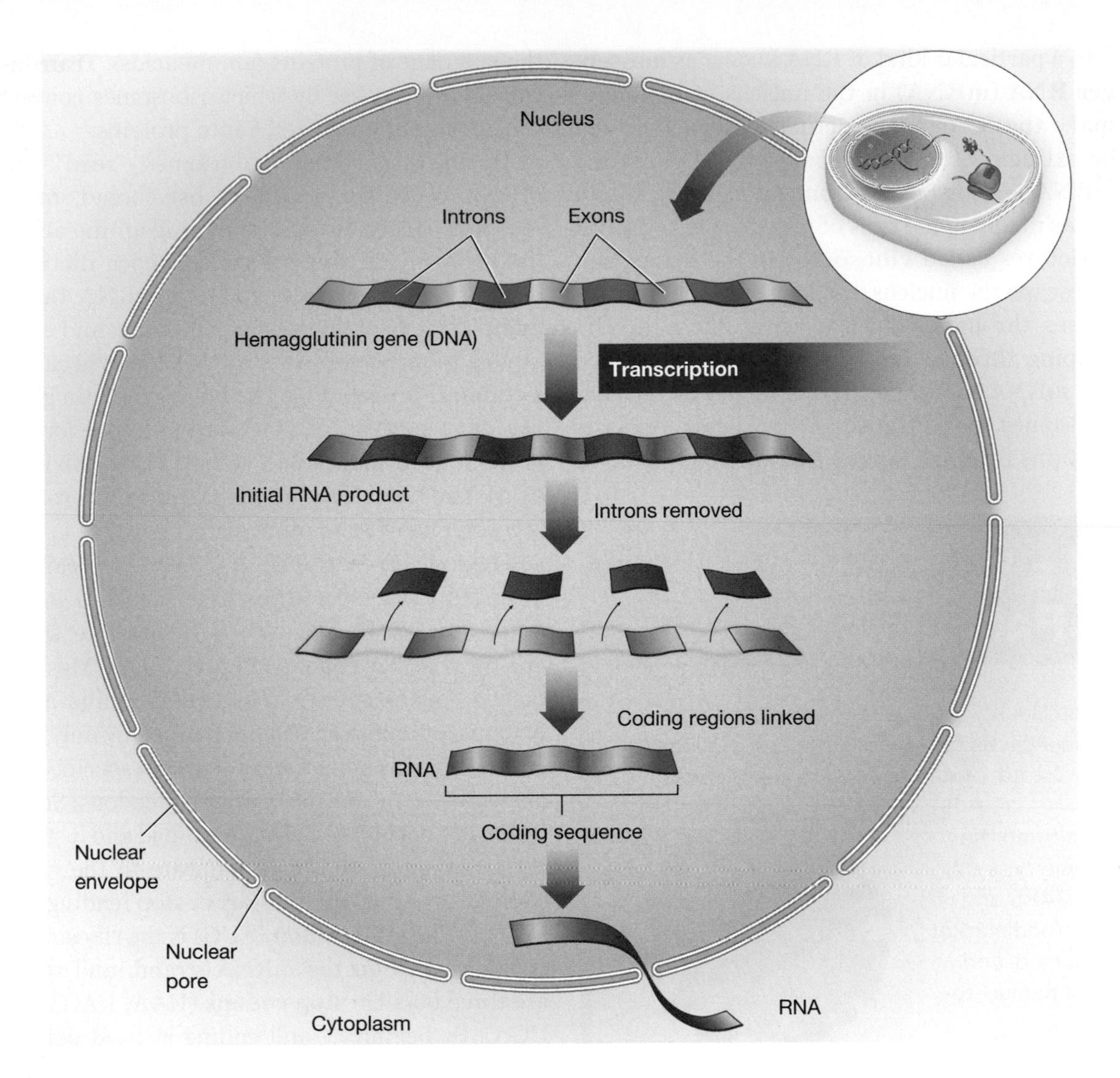

Figure 10.7

Processing RNA

In eukaryotes, introns must be removed before an mRNA leaves the nucleus.

Q1: In your own words, define RNA splicing. When during gene expression does it occur?

Q2: From where to where does the RNA transcript travel?

Q3: What do you predict would happen if the introns were not removed from RNA before translation? Why would it be a problem if the introns were not removed?

See Appendix A for answers to the figure questions.

To review, transcription occurs when RNA polymerase binds to a promoter, unzips the DNA helix, and constructs a strand of RNA based on the DNA template strand. Transcription ends at the terminator sequence, and the RNA is then processed, at which time noncoding introns are spliced out of the sequence. Voilà! RNA is created from a gene. The DNA itself remains unchanged.

Two-Step Dance, Translation: RNA to Protein

The microscopic molecular dance inside the tobacco cells continues with translation. Once the hemagglutinin gene has been transcribed into a particular kind of RNA known as **messenger RNA (mRNA)** in the nucleus, it is time to make the protein, the actual product that will be extracted from the tobacco leaves. First the mRNA is transported from the nucleus, where it was made, to the sites of protein synthesis: cell structures called **ribosomes** in the cytoplasm. To escape the nucleus, the long strand of mRNA passes through a nuclear pore, like a noodle slipping through the hole of a colander. Once the mRNA molecule arrives in the cytoplasm, ribosomes help translate the information it contains from the language of mRNA (bases) to the language of proteins (amino acids). **Translation** is the process by which ribosomes convert the information in mRNA into proteins.

During translation, ribosomes "read" the mRNA code like a grocery list (bread, milk, etc.) and collect the corresponding amino acids, linking them in the precise sequence dictated by mRNA. Ribosomes read the mRNA information in sets of three bases at a time, and each unique sequence of three mRNA bases is called a **codon** (**Figure 10.8**). The hemagglutinin gene has about 1,770 bases, of which 1,695 code for the protein. That makes 565 codons (1,695 divided by 3), and therefore the hemagglutinin protein is composed of 565 amino acids.

The four bases of mRNA (A, C, G, U) can be arranged to create a three-base sequence in 64 different ways (because $4^3 = 64$). Therefore, there are 64 possible codons (**Figure 10.9**). Most of the 64 codons specify a particular amino acid. A couple of amino acids are specified by only one codon, whereas other amino acids are specified by anywhere from two to six different codons. Some codons do not code for any amino acid and instead act as signposts that communicate to the ribosomes where they should start or stop reading the mRNA. The **start codon** (AUG) is the ribosome's starting point on the mRNA strand, and there are three possible **stop codons** (UAA, UAG, and UGA). By beginning and ending at fixed points, the cell ensures that the mRNA message is read in precisely the same way every time.

The information specified by all 64 possible codons is the **genetic code**, which the cell uses throughout translation. The genetic code has several significant characteristics. First, it is *unambiguous*: each codon specifies only one amino acid. It is also *redundant*: because there are 64 codons but only 20 amino acids, several different codons call for the same amino acid, as already mentioned. Finally, the genetic code is virtually *universal*: nearly every organism on Earth uses the same code, from agrobacteria to tobacco cells to human cells—a feature that illustrates the common descent of all organisms.

Making a protein from an mRNA strand requires two additional types of RNA. The first is **ribosomal RNA (rRNA)**, an important component of ribosomes. The second is **transfer RNA (tRNA)**, which is the caddy for the process, delivering specific amino acids to the ribosomes as the codons are read off the mRNA "list."

Racing toward a Vaccine during the COVID-19 Pandemic

On January 5, 2020, the World Health Organization (WHO) tweeted about a cluster of pneumonia cases detected in the city of Wuhan in the Hubei province of China. A total of 44 cases were linked to a seafood market in Wuhan.

On January 12, the WHO identified the cause as a coronavirus, the same type of virus responsible for the 2003 SARS and 2012 MERS outbreaks. The seafood market had been closed, the WHO reported, and there was "no clear evidence of human-to-human transmission."

Clearly, that was not the case. By the spring of 2020, countries around the world were on lockdown: schools, parks, and businesses closed; hospitals were running out of ventilators to care for patients; millions of people were out of work and scared. By the fall, more than 30 million cases were reported around the world, with over 960,000 deaths. By October 1, over 200,000 people in the U.S. alone had died from COVID-19, the disease caused by the virus SARS-CoV-2.

In response to the first reported cases in early January, scientists began working toward a vaccine to protect against the virus. At the National Institutes of Health, a team produced a vaccine candidate—a potential vaccine that still needed to be tested for safety and efficacy—in just 65 days. On March 16, that candidate became the first COVID-19 vaccine tested in humans. Many other candidates followed on its heels. Using their plant-based technology, Medicago produced a COVID-19 vaccine candidate quickly and saw an immune response in mice after a single dose. The company planned to begin human studies that summer.

By June 1, 2020, just 5 months after the outbreak began, ten candidate vaccines had begun human clinical trials—a shockingly fast beginning considering vaccine discovery usually takes 2 to 5 *years*. Meanwhile, another 115 candidates were being tested in the lab.

The race to a COVID-19 vaccine was on, with the world waiting and watching.

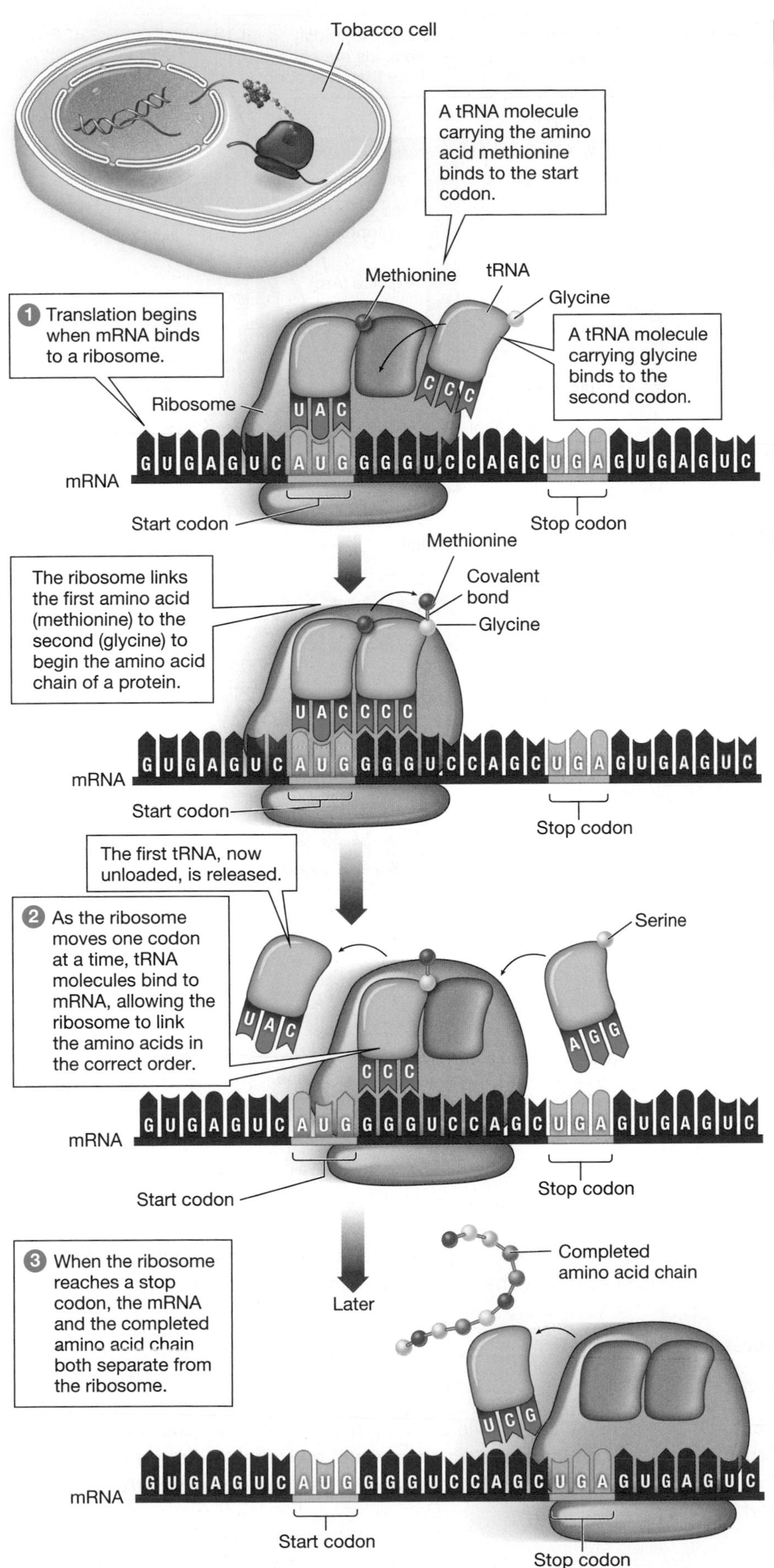

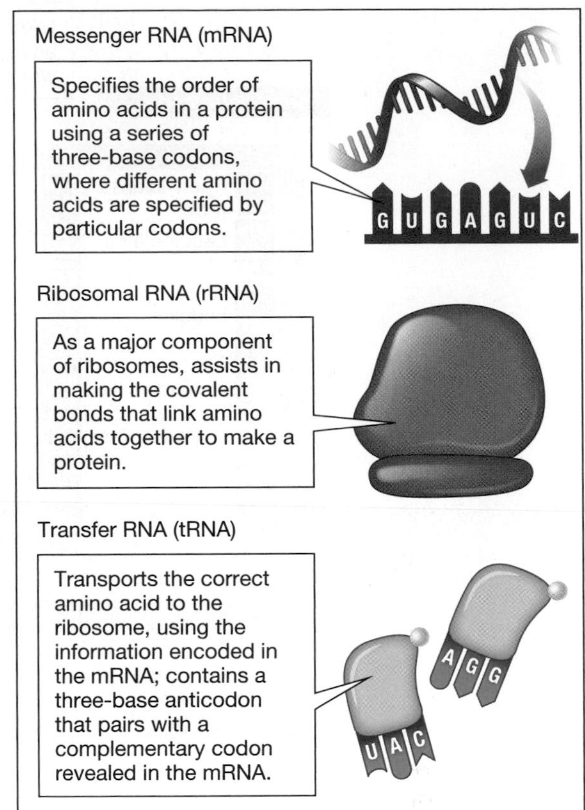

Messenger RNA (mRNA)
Specifies the order of amino acids in a protein using a series of three-base codons, where different amino acids are specified by particular codons.
GUGAGUC
Ribosomal RNA (rRNA)
As a major component of ribosomes, assists in making the covalent bonds that link amino acids together to make a protein.
Transfer RNA (tRNA)
Transports the correct amino acid to the ribosome, using the information encoded in the mRNA; contains a three-base anticodon that pairs with a complementary codon revealed in the mRNA.
AGG
UAC

Figure 10.8

Plants making proteins, II: Translation

This figure summarizes the process of translation. In the example used here, the hemagglutinin mRNA directs the synthesis of the hemagglutinin protein.

(You will need to finish reading the section on translation to answer the following three questions.)

Q1: Which amino acid always begins an amino acid chain? Which codon and anticodon are associated with that amino acid?

Q2: In the second question of Figure 10.6, you specified a partial mRNA sequence as being transcribed from the DNA template strand. What amino acid sequence would be translated from that mRNA sequence?

Q3: Each of the codons for stopping translation binds to a tRNA molecule that does not carry an amino acid. How would the binding of a stop codon cause the completed amino acid chain to be released?

See Appendix A for answers to the figure questions.

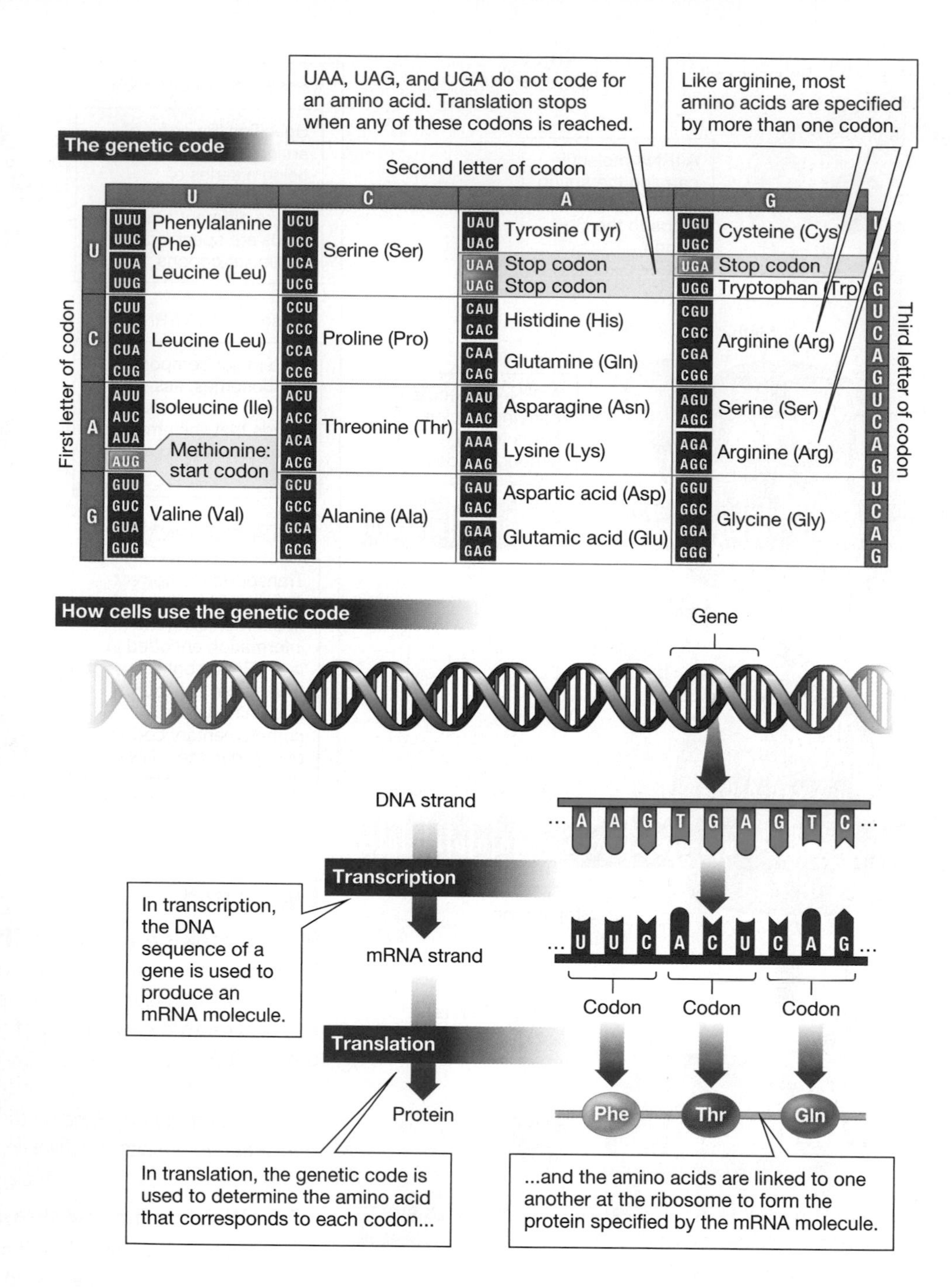

Figure 10.9

The genetic code

(Top) The genetic code is composed of the 64 possible codons found in the mRNA. Each codon specifies an amino acid or is a signal that starts or stops translation. (Bottom) The genetic code is used during the translation of mRNA to protein.

Q1: How many codons code for isoleucine? For tryptophan? For leucine?

Q2: What codons are associated with asparagine? With serine?

Q3: In the second question of Figure 10.6, you specified a partial mRNA sequence as being transcribed from the DNA template strand. From that sequence, remove only the first A. What amino acid sequence would be translated as a result of this change? How does that sequence compare to the amino acid sequence you translated from the original mRNA sequence? *Bonus:* What kind of mutation is this? (*Hint:* See Chapter 9.)

See Appendix A for answers to the figure questions.

Each tRNA specializes in binding to a specific amino acid and recognizes and pairs with a specific codon in the mRNA, like a puzzle piece that fits one amino acid on one end and one codon on the other. At one end of a tRNA molecule, a special sequence of three bases, called an **anticodon**, binds the correct codon on the mRNA. At the other end, the specific amino acid attaches.

Let's recap. For translation to occur, an mRNA molecule must first bind to a ribosome. The ribosomal machinery then "scans" through the mRNA until it finds a start codon (AUG). Next the ribosome recruits the appropriate tRNAs one by one, as determined by the codons read in the mRNA sequence. A special site on the ribosome facilitates the linking of one amino acid to another, like pearls on a necklace. Finally, the ribosome reaches a stop codon. The amino acid chain cannot be extended further, because none of the tRNAs will recognize and pair with any of the three stop codons. At this point the mRNA molecule and the completed amino acid chain separate from the ribosome. The new protein then folds into its compact, specific three-dimensional shape and is ready to go to work in the cell.

However, this process does not always go as planned. As we learned in Chapter 9, a mutation is a change in the sequence of DNA bases. Such a change affects an organism by disrupting or preventing the healthy formation of a protein. For example, a mutation can cause a DNA sequence not to be transcribed or translated, prompt the amino acid chain to end prematurely, or make the final protein fold incorrectly, among other possibilities.

Although single-base substitutions are not always a problem, single-base insertions and deletions cause a genetic "frameshift," shifting all subsequent codons "downstream" by one base (**Figure 10.10**). This shift scrambles the entire downstream DNA message, in turn scrambling the entire RNA message and causing the ribosomes to assemble a very different sequence of amino acids from the mutation point onward. For example, imagine if every letter in the sentence you're now reading was shifted to the left one space while the word length and spaces between words were retained: fo rexample , imagin ei fever ylette ri nth esentenc eyou'r eno wreadin gwa sshifte dt oth erigh ton espac ewhil eth ewor dlengt han dspace sbetwee nword swer eretaine d.

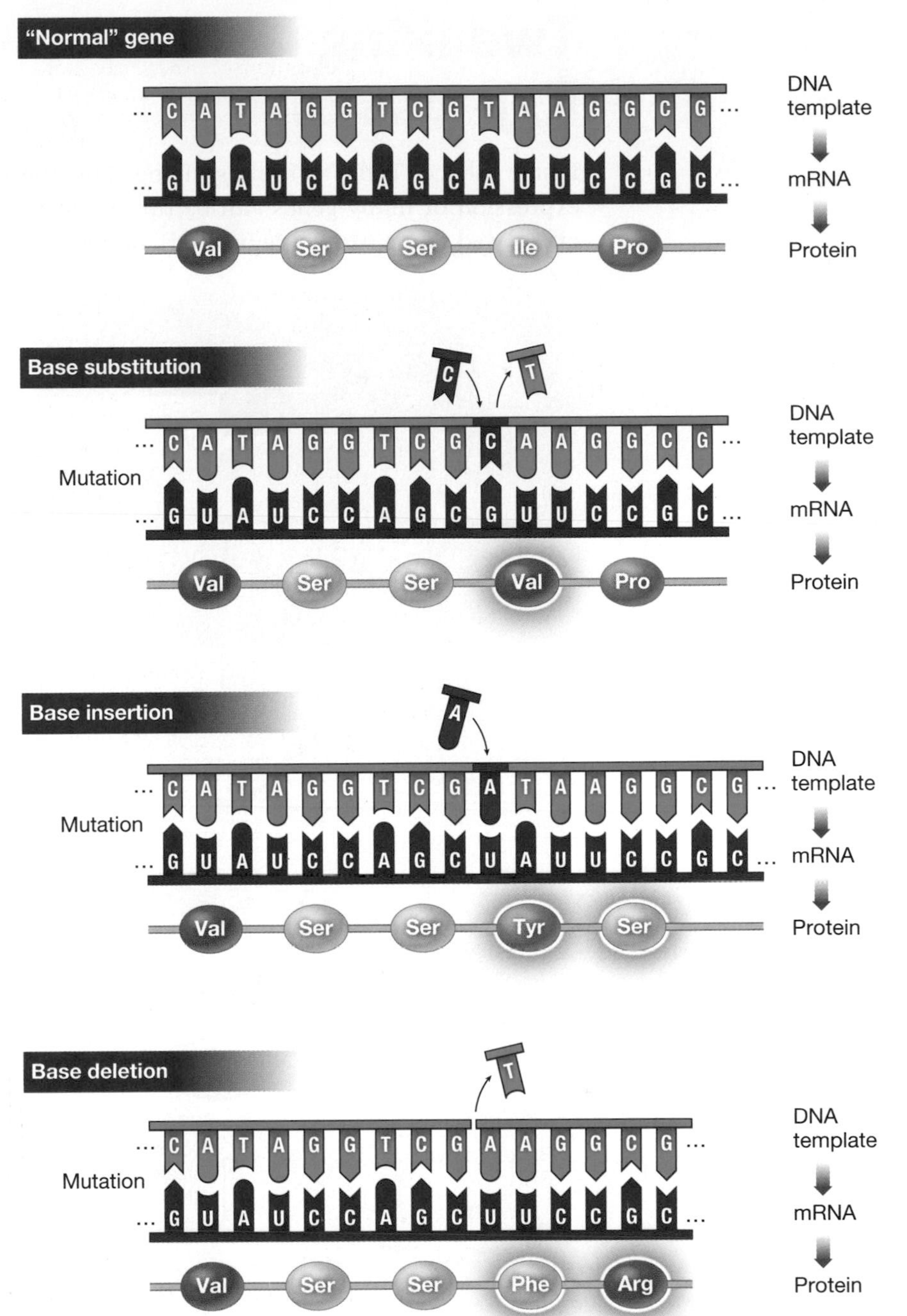

Figure 10.10

DNA mutations affect proteins

The different types of point mutations result in altered genetic expression.

Q1: Why is an insertion or a deletion in a gene more likely to alter the protein product than a substitution, such as A for C, would?

Q2: Which would you expect to have more impact on an organism: a point mutation as shown here or the insertion or deletion of a whole chromosome (discussed in Chapter 8)?

Q3: Which mechanisms in a cell prevent mutations? (*Hint:* Refer back to Chapter 6 if needed.)

See Appendix A for answers to the figure questions.

Tweaking Gene Expression

Inside tobacco cells, as in most living cells, the expression of many genes can be turned on or off, slowed down (**down-regulated**), or sped up (**up-regulated**). This **gene regulation** enables organisms to change which genes they express in response to internal signals (from inside the body) or to external cues in the environment. In this way, as different proteins are produced depending on circumstances, organisms can adjust to their surroundings.

All cells in a multicellular individual have essentially the same DNA, yet different cells express different sets of genes, and within a given cell the pattern of gene expression can change over time. Single-celled organisms, such as bacteria, face a more difficult challenge: they are directly exposed to their environment, and they have no specialized cells to help them deal with changes in that environment. One way they meet this challenge is to express different genes at different times.

The expression of most genes in prokaryotes and eukaryotes is regulated by both internal and external signals. Many genes are also developmentally regulated, meaning that their expression can change, sometimes dramatically, as an organism grows and develops. Gene expression is regulated at many different points in the cell, including DNA packing (the way DNA is compressed or unwound in the genome), transcription, mRNA processing, and several points during translation (**Figure 10.11**).

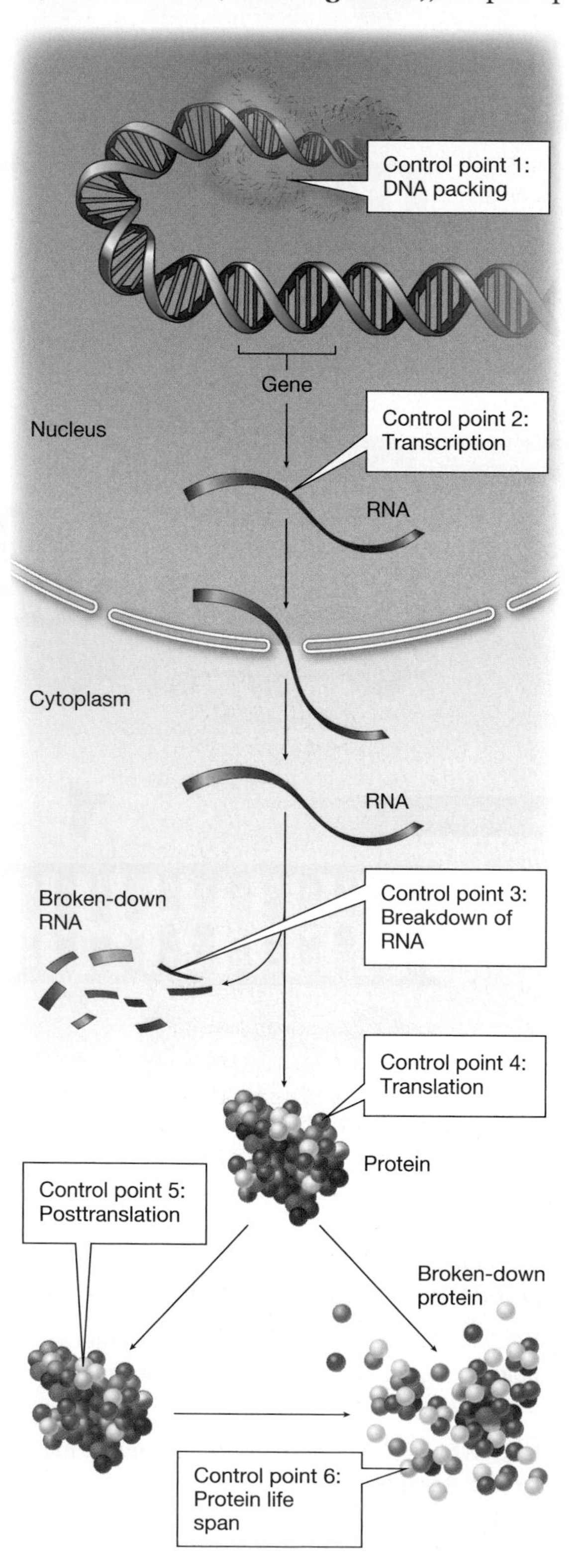

Figure 10.11

How gene expression is regulated

In eukaryotes, gene expression can be controlled at several points along the pathway from gene to protein to phenotype: before transcription, at transcription, during RNA processing, or at translation. Gene expression can also be regulated after translation by control of the activity or life span of the protein.

Q1: As illustrated here, at what control point is transcription regulated?

Q2: What is a possible advantage of regulating gene expression before transcription versus after transcription?

Q3: At which control point(s) is it possible to up-regulate production of the hemagglutinin protein in a tobacco plant carrying the hemagglutinin gene? Justify your reasoning.

See Appendix A for answers to the figure questions.

But for all living cells, a few genes are always expressed at a low level; their transcription is not regulated, because these genes are needed at all times.

At Medicago, the company takes advantage of gene regulation in tobacco cells to produce as much hemagglutinin protein as possible. After the agrobacteria are vacuum-sucked into the leaves and transcription and translation begin to actively occur in the tobacco cells, the plants are sent to grow in an incubation room, where technicians can alter the humidity, temperature, and level of light to maximize the amount of protein expressed by the plants.

"We've been able to tweak environmental conditions of the plants to boost gene expression," says Vézina. "You name it, we've tried it." It's an important step in the process, adds Wanner. "We determined what the best conditions for protein expression are."

To the Market

By the end of April, the final steps are taken to isolate and purify the hemagglutinin protein from the tobacco plants to see how much vaccine can be made in a month. After the plants have incubated for several days, the leaves are stripped and diced into green confetti, then digested with enzymes to break up the leaf material so that the desired proteins are released into solution. The resulting solution, which resembles green-pea soup, is filtered several times to isolate clusters of hemagglutinin, which will then be processed into a vaccine product that is safe to inject into people.

Medicago is not the first company to produce a human drug using a plant. The first humanlike enzyme was produced from tobacco in 1992 at Virginia Polytechnic Institute. Numerous other plant biopharming companies have sprung up since then, experimenting with various plant species, including corn, soybean, duckweed, and more. But the field is not without risks and controversy. "The main concern and risk is spread—that the gene will get out into nature and spread," says Roth. "But there are techniques to make sure it doesn't happen, and it is closely regulated."

One of those techniques is to grow the plants in contained environments where there is no risk of contaminating food crops. Israel-based biotech company Protalix Biotherapeutics, for example, grows carrot cells inside in large hanging bags of fluid and cells. In May 2012, the FDA approved the first biopharmed drug for humans, produced by Protalix. This therapy for Gaucher disease, a rare genetic disorder, was grown in the company's carrot cells. "This approval demonstrates a proof of concept for the power of this technology," the CEO of Protalix told the journal *Nature*.

When the April manufacturing test was completed, Medicago had produced an astounding 10 million doses of flu vaccine in a single month. It would have taken 5–6 months to produce the same amount using the traditional method of growing vaccines in chicken eggs. With that success under its belt, the company soon built a second production complex in Quebec City capable of delivering up to 50 million doses of seasonal flu vaccine.

Today, Medicago has completed safety trials for its pandemic flu vaccine and has had positive results in the first of two large clinical trials in humans that tested its seasonal flu vaccine's efficacy. In September 2018, the company began a phase 3 clinical trial, which is typically the final large human study before a drug or vaccine is approved. In October 2019, Medicago applied to the Canadian government for approval of the vaccine.

While also developing novel vaccines against rotavirus and rabies virus, Medicago immediately began research into a coronavirus vaccine in February 2020 during the worldwide outbreak of SARS-CoV-2, the virus that causes the COVID-19 disease. In just 20 days, the company was able to produce a COVID-19 vaccine candidate and planned to begin human clinical trials in the summer of 2020.

"It might have taken a bit longer than we thought for biopharming to be accepted, but it deserves the visibility and attraction it has now," says Vézina. "Biopharming is here to stay. I'm convinced of this."

DEBUNKED!

MYTH: You can catch the flu from the vaccine.

FACT: The flu shot is made from an inactivated virus that cannot cause infection. Because the flu shot takes a week or two to become effective, anyone who gets sick right after receiving a flu vaccination was going to get sick anyway.

Deadly Pandemics

The Spanish flu of 1918 devastated populations across Europe, killing not only the young and old, but also healthy adults. Since then, new strains of flu and other viruses causing respiratory infections—including the coronaviruses that caused COVID-19 and SARS—have emerged. Unlike SARS, which was quickly contained, the virus causing COVID-19 spread rapidly around the globe. Doctors and vaccine manufacturers struggle to anticipate the next outbreak.

Assessment available in smartwork

1918
H1N1
Spanish Flu

50M dead worldwide*	675K dead in U.S.

2019–present
COVID-19

1.0M dead worldwide**	208K dead in U.S.

1957
H2N2

1.1M dead worldwide	116K dead in U.S.

1968
H3N2

1.0M dead worldwide	100K dead in U.S.

2003
SARS

774 dead worldwide	0 dead in U.S.

2009
H1N1
Swine Flu

285K dead worldwide	12K dead in U.S.

*Some estimates are as high as 100M dead.

**Data taken from https://covid19.who.int/ on October 5, 2020

NEED-TO-KNOW SCIENCE

- Most genes contain instructions for building proteins. **Gene expression** (page 188) is the process by which genes are transcribed into RNA and then translated into a protein.
- During **transcription** (page 190), which occurs in the nucleus, **RNA polymerase** (page 190) binds the **promoter** (page 190) of a gene and produces an RNA version of the gene sequence from free nucleotides.
- Next, during **RNA splicing** (page 192), **introns** (page 192) are snipped out of the pre-mRNA sequence, and the remaining **exons** (page 192) are joined. The mRNA is transported out of the nucleus.
- During **translation** (page 194), which occurs in the cytoplasm, ribosomes convert the sequences of bases in a **messenger RNA (mRNA**, page 194) molecule to the sequence of amino acids in a protein, with the help of **ribosomes** (page 194) composed of **ribosomal RNA (rRNA**, page 194) and proteins, and associated with **transfer RNA (tRNA,** page 194).
- Ribosomes read the mRNA information in sets of three bases at a time called **codons** (page 194). There are 64 possible codons, including a **start codon** (AUG, page 194) and three possible **stop codons** (UAA, UAG, and UGA, page 194).
- The **genetic code** (page 194) is redundant because there are 64 codons and only 20 amino acids and universal because nearly all organisms on Earth use the same code.
- Gene expression is regulated at many different points in the pathway from gene to protein. Organisms rely on **gene regulation** (page 198) to respond to signals inside the body and to external cues in the environment.

THE QUESTIONS

See Appendix B for answers.

The Basics

1 Link each term with the correct definition.

gene expression — 1. The production of RNA, using the information in the DNA sequence of a gene.

gene regulation — 2. The flow of information from gene to protein.

transcription — 3. The control of gene expression in response to internal or environmental cues.

translation — 4. The production of a protein, by linking amino acids in the precise sequence dictated by an mRNA base sequence.

2 For each of the following items, identify the type of RNA involved (mRNA, rRNA, or tRNA).

_____ a. Transports the correct amino acid to the ribosome, using the information encoded in the mRNA.

_____ b. Is a major component of ribosomes.

_____ c. Specifies the order of amino acids in a protein, using a series of three-base codons, where different amino acids are specified by particular codons.

_____ d. Contains a three-base anticodon that pairs with a complementary codon revealed in the mRNA.

_____ e. Assists in making the bonds that link amino acids together to make a protein.

3 Select the correct terms: The genetic code is (**unambiguous** / **redundant**) because some amino acids are coded by more than one codon. At the same time, the genetic code is (**unambiguous** / **redundant**) because each codon codes for only one amino acid.

4 In the accompanying diagram of transcription, fill in the blanks with the appropriate terms: gene, promoter, terminator, RNA polymerase, RNA.

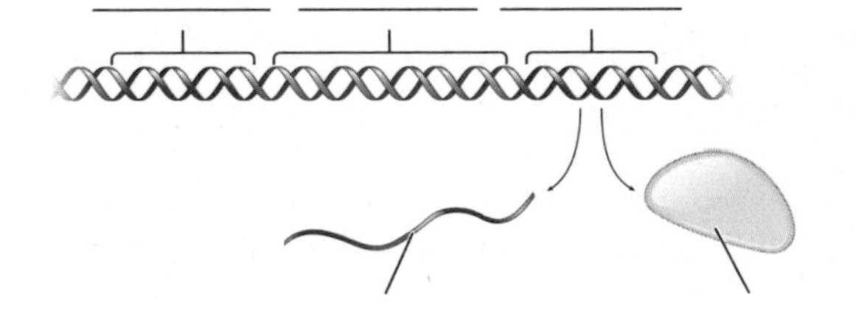

5 Place the following steps of translation in the correct order by numbering them from 1 to 9.

_____ a. A tRNA molecule carrying the amino acid methionine binds at its anticodon site to the mRNA start codon.

_____ b. The ribosome links the first amino acid to the second amino acid to begin the amino acid chain.

_____ c. The ribosome continues to link each amino acid to the growing amino acid chain.

_____ d. The ribosome reaches a stop codon.

_____ e. An mRNA binds to a ribosome.

_____ f. The mRNA as well as the completed amino acid chain separates from the ribosome.

_____ g. The first tRNA, separated from its amino acid, releases from the mRNA.

_____ h. A tRNA molecule carrying the second amino acid binds to the second mRNA codon.

_____ i. Each tRNA releases from the mRNA after it is separated from its amino acid.

6 Which of the following are possible reasons that a cell would regulate its expression of a gene? (Select all that apply.)

(a) an increased need for a particular enzyme
(b) a decreased need for a particular enzyme
(c) increasing temperature in the external environment
(d) changing needs as an organism ages
(e) death

Challenge Yourself

7 Using the genetic code shown in Figure 10.9, find the amino acid coded by each of the following codons.

(a) AAU
(b) UAA
(c) AUA
(d) GGG
(e) CCC

8 Using the genetic code shown in Figure 10.9, find a codon that codes for each of the following amino acids.

(a) arginine
(b) alanine
(c) methionine
(d) glycine

9 During transcription, what RNA molecule will be made from the DNA template CGTTACG?

(a) CGTTAGC
(b) GCAAUGC
(c) GCATTGC
(d) CGUUAGC

10 Which amino acid sequence will be generated during translation from the following small mRNA: ...CCC-AUG-UCU-UCG-UUA-UGA-UUG...? (*Hint:* Remember where translation starts and stops.)

(a) Met-Glu-Arg-Arg-Glu-Leu
(b) Met-Ser-Ser-Leu-Leu
(c) Pro-Met-Ser-Ser-Leu-Leu
(d) Pro-Met-Ser-Ser-Leu
(e) Met-Ser-Ser-Leu

11 The following nucleic acid is an entire primary transcript (pre-mRNA not yet processed): ACGCAUGCGaugaugccccucag GUCUguuuccgugaUGCCGUUGACCUGA. The nucleotides in capitals are exons; the nucleotides in lowercase type are introns. Appropriately splice this primary transcript to produce a mature mRNA.

(a) ACGCAUGCGGUCUUGCCGUUGACCUGA
(b) augccuuucagguuuccguga
(c) augccuuucagGUCUguuuccguga
(d) ACGCAUGCGaugGUCUUGCCGUUGACCUGA
(e) ACGCAUGCGaugccuagGUCUguuuccgugaUGCCGUUGACC UGA

Try Something New

12 How is gene expression similar to DNA replication, and how is it different? Give at least one similarity and one difference.

13 Some diseases, such as Huntington's and Parkinson's, appear to be related to increasing protein levels in brain cells, which lead eventually to cell death. At which of the control points in Figure 10.11 might a gene regulation error be occurring with these diseases? Identify one control point at which the error would result in up-regulation of gene expression and one control point at which the error would result in down-regulation of gene expression.

14 Malaria is a terrible mosquito-borne disease that kills more than 400,000 people every year. One of the most effective treatments is artemisinin, found in low quantities in the sweet wormwood shrub used in Chinese traditional medicine. The accompanying graph shows artemisinin levels in unaltered shrubs as well as in genetically engineered shrubs and tobacco plants (which are faster to grow to harvest than the shrubs). What is the percentage of artemisinin by dry weight in unaltered shrubs? Is genetic expression of artemisinin higher or lower in genetically engineered shrubs? In tobacco? By how much? *Bonus:* Why might a drug company choose to harvest artemisinin from tobacco rather than wormwood?

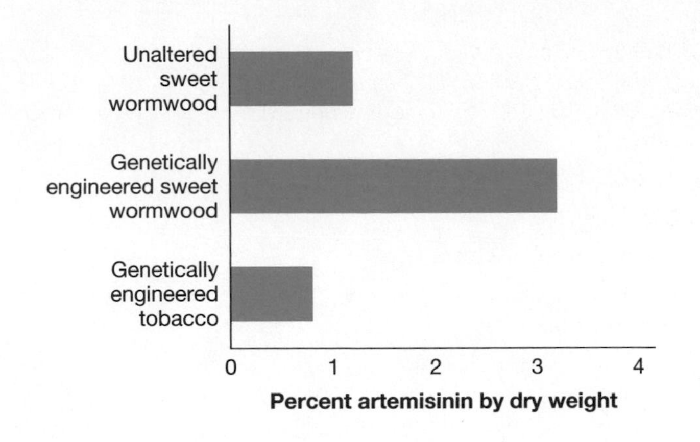

Leveling Up

15 ***Write Now*** **biology: genotype to phenotype** Your roommate, who is also taking a biology class, has become a little confused. He informs you that the genetic code is known to be ambiguous because a given genotype may give rise to a variety of phenotypes during gene expression (for example, his twin brother is an inch taller and more tan than he is). You like your roommate and would like him to pass his next biology exam, so you decide to help him. Write him a brief note explaining (a) why the genetic code is not, in fact, ambiguous and (b) how gene expression derives a phenotype from a genotype.

16 **What do *you* think?** Most people carry two copies of a normal gene that codes for an enzyme, glucosylceramidase, involved in breaking down lipids no longer needed in cells. (Enzymes are proteins that cause specific chemical changes; they are biological catalysts.) One in 100 people in the United States carries a recessive mutation that codes for a defective glucosylceramidase enzyme. And about one in 40,000 people carries two copies of the mutation and displays the symptoms of Gaucher disease. These symptoms, caused by the accumulation of lipids in cells, include anemia, enlarged organs, swollen glands and joints, and, in severe cases, neurological problems and early death.

Enzyme replacement therapy is effective but very expensive—about $200,000 annually—and must be continued, every 2 weeks, for life. The Israel-based biotech company Protalix Biotherapeutics, working with the U.S.-based Pfizer Pharmaceuticals, has developed a process to genetically modify carrots to produce a replacement enzyme. The biopharmed enzyme will cost about 25 percent less than the standard enzyme therapies, which are grown in mammalian cell lines. Protalix is now working on treatments for other enzyme deficiency diseases.

The FDA's May 2012 approval of the drug developed by Protalix alarmed some environmental activists and health advocates, who fear that the company's genetically modified carrot is just the beginning of a wedge that will lead to an underregulated and potentially dangerous industry. There is some legitimacy to their concerns: The U.S. Department of Agriculture (USDA) does not require an environmental impact assessment for biopharmed crops; nor does it require biotech companies to share the location of their test fields or the identity of the biopharmed molecules being produced. Furthermore, the USDA is not sufficiently staffed to effectively monitor companies involved in biopharming.

What do you think? Should biopharming be allowed in the United States? If so, under what conditions should it be allowed and with what limits? For example, should it be allowed to produce drugs for only life-threatening illnesses or only under highly controlled conditions? Be prepared to discuss your observations and reflections in class.

17 **Life choices** Go to the Centers for Disease Control and Prevention (CDC) influenza website (http://www.cdc.gov/flu) and read the pages "Key Facts About Influenza (Flu)" (under "About Flu") and "Prevent Seasonal Flu" (under "Prevent Flu"). You can also go to the Mayo Clinic's influenza website (http://www.mayoclinic.org/diseases-conditions/flu/home/ovc-20248057). Then answer the following questions.

(a) What is the flu? How is it passed on?

(b) What are the possible symptoms and complications of the flu?

(c) How can you decrease your chance of getting the flu, and what treatments are available if you become infected?

(d) What are the benefits and risks of the flu vaccine?

(e) Why is there a new flu vaccine every year?

(f) Why is the flu vaccine more effective in some years than in others?

(g) Who would you recommend should get the flu vaccine? Explain your reasoning.

(h) Do you get a flu vaccine every year? Why or why not?

Digital resources for your book are available online.

Whale Hunting

Fossil hunters discover Moby-Dick's earliest ancestor—a furry, four-legged land lover.

After reading this chapter you should be able to:

- Define evolution and list six types of evidence for evolution.
- Compare and contrast artificial selection and natural selection.
- Summarize the argument that the fossil record provides evidence in support of evolution.
- Give one example of a homologous trait and one example of a vestigial trait, and explain how such traits support the theory of common descent.
- Explain why even distantly related species have similar DNA.
- Use your knowledge of evolution and continental drift to make a prediction about the geographic location of a given set of fossils.
- Relate similarities in embryonic development among species to their shared evolutionary past.

EVOLUTION

CHAPTER 11

EVIDENCE FOR EVOLUTION

Fossils break all the time. This time, the 50-million-year-old ear bone of a small, deerlike mammal called *Indohyus* snapped clean off the skull. Sheepishly, the young laboratory technician cleaning the fossil handed the broken piece to his boss, paleontologist and embryologist J. G. M. "Hans" Thewissen at Northeast Ohio Medical University. Thewissen tenderly turned the preserved animal remains over in his hand. Then, as the tech reached for the fossil to glue it back onto the animal's skull, Thewissen went rigid.

"Wow, that is weird," said Thewissen. The *Indohyus* ear bone, which should have looked like the ear bone of every other land-living mammal—like half a hollow walnut shell, but smaller—was instead razor thin on one side and very thick on the other. "Wow," repeated Thewissen. This wasn't the ear bone of a deer or of any other land mammal. Thewissen squinted closer. "It looks just like a whale," he said (**Figure 11.1**).

Although they live in the ocean like fish, whales are mammals like us. So are dolphins and porpoises. All mammals share certain characteristics. For example, they maintain a stable body temperature, have backbones, breathe air, and nurse their young from mammary glands (**Figure 11.2**). Numerous fossils have been found documenting whales' unique transition from land-living mammals to the mammoths of the sea. During this very long process, populations of whale predecessors developed longer tails and shorter and shorter legs. But one crucial link in the fossil record was missing: the closest land-living relatives of whales. What did the ancestors of whales look like before they entered the water?

Staring at the strange fossil in his hand, Thewissen realized he could be holding the ear bone of that missing link.

Whales are only one of the many organisms that share our planet. Every species has evolved for life in its particular environment: whales in the open ocean, hawks streaking through the sky, tree frogs camouflaged in the green leaves of the rainforest. There is great diversity of life on Earth—animals, plants, fungi, and more—with each species well matched to its surroundings. This diversity of life is due to evolution.

Evolution, in everyday language, means change over time. In science, biological **evolution** is a change in the inherited characteristics of a group of organisms over generations. Whales, for example, evolved from four-legged land-living animals into sleek emperors of the ocean, slowly becoming suited for the water over tens of millions of years. Whales changed

Indohyus
When the *Indohyus* skull broke, Thewissen saw its very thick medial tympanic wall, like those found in all whales but no other living mammals.
Tympanic wall
Sediment filling middle ear
Whale

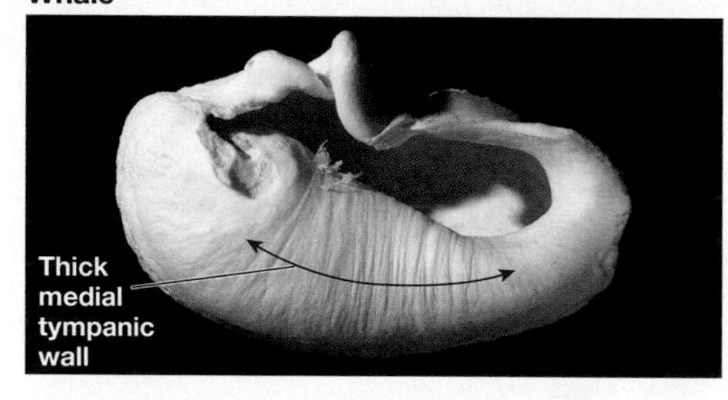
Thick medial tympanic wall
White-tailed deer

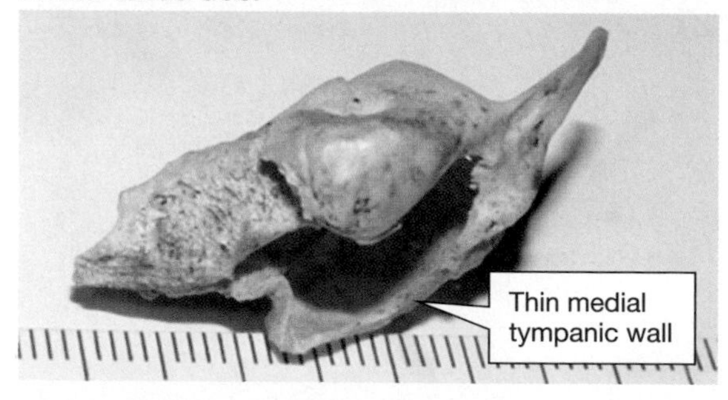
Thin medial tympanic wall

Figure 11.1

The mysterious ear bone

The *Indohyus* fossil ear bone (top) looks more like the ear bone of a whale (middle) than that of any modern land mammal (bottom). (Source: *Indohyus* and whale photos courtesy of Thewissen Lab, NEOMED.)

Figure 11.2

A montage of mammals

These photos represent only a fraction of the almost 5,500 species of mammals currently living on Earth. Within this diversity there is unity, however. Shared features mean these various species are all mammals.

Q1: What is a shared feature of mammals that you can see in these photos?

Q2: If dolphins and whales are mammals, can they breathe underwater like fish?

Q3: Name a mammal other than whales and dolphins that spends much of its life in the ocean.

See Appendix A for answers to the figure questions.

as a population. Populations evolve; individuals do not.

You may wonder how we can be sure of evolutionary change, especially for a transformation as extreme as that of a furry, four-legged beast into a whale. There is strong evidence for evolution in fossils, such as those that Thewissen studies. And there is equally strong evidence in features of existing organisms, common patterns of how embryos develop, DNA, geography, and even direct observation of organisms evolving today—including man's best friend.

Artificial to Natural

As we saw in Chapter 7, all dogs are a single species. That species is *Canis lupus familiaris*, a subspecies of the gray wolf. Scientists estimate that domestication of gray wolves began about 16,000 years ago, as humans introduced the animals to civilization. From there, people bred the wolves for desired qualities, such as decreased aggression and the ability to follow commands. Thousands of years later, the wolves

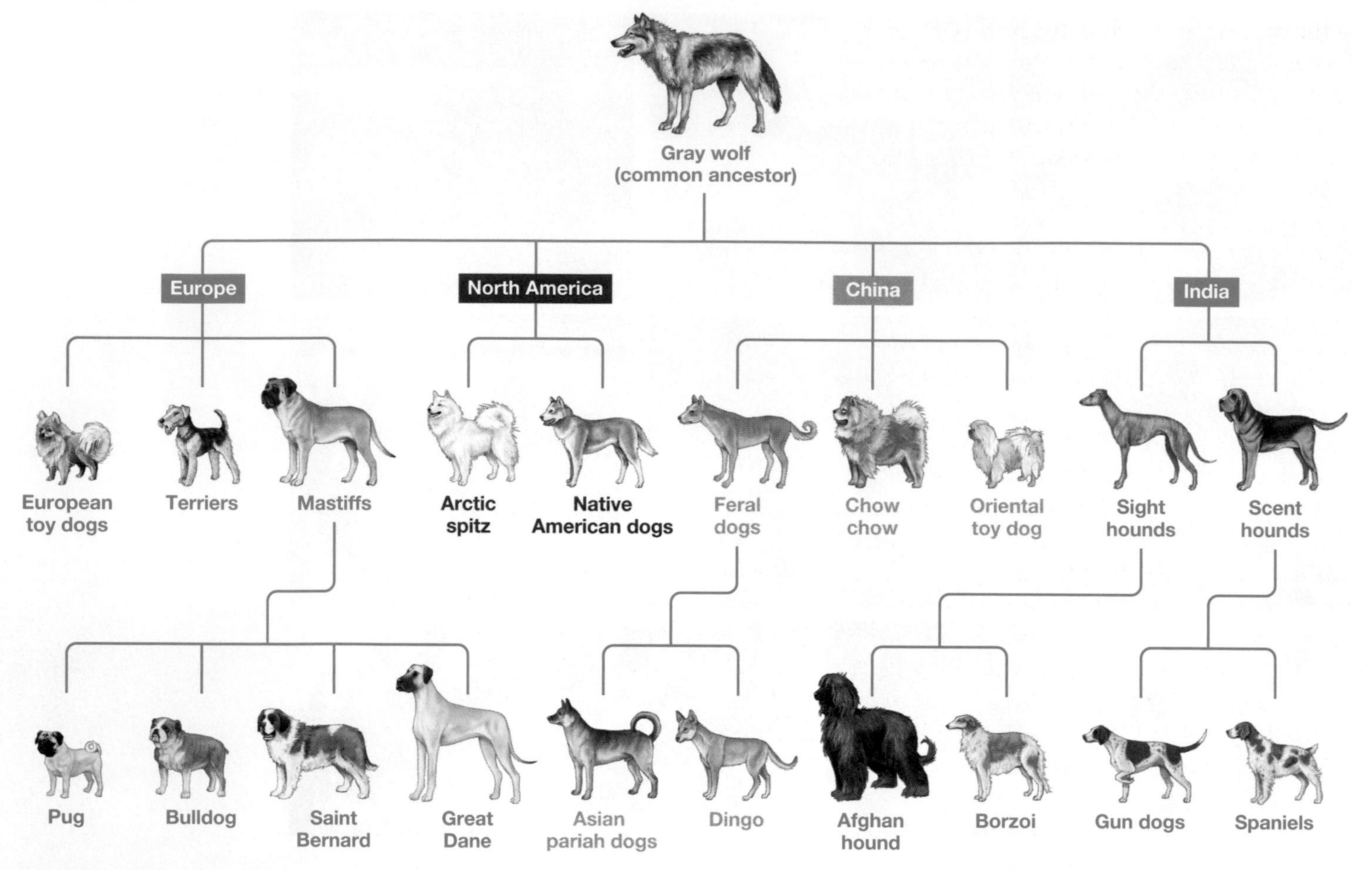

Figure 11.3

Selective breeding of dogs produces myriad traits

Dogs were domesticated only a few times and always from gray wolves. Thus, the remarkable diversity of dogs represents the effects of selective breeding on a small number of lineages of domesticated wolves.

Q1: What is selective breeding, and how does it work?

Q2: Explain how selective breeding leads to artificial selection.

Q3: Name as many organisms as you can that have characteristics due to artificial selection.

See Appendix A for answers to the figure questions.

had become recognizable dogs, which people then bred for specific traits, such as the short snout of a toy dog or the long legs of a hound. This selective process has led to incredible variation in the size and shape of dog breeds, from a 14-pound pug to a 200-pound Great Dane (**Figure 11.3**).

An example of evolution that we can directly observe, dogs evolved through artificial selection. **Artificial selection** is brought about by **selective breeding**, in which humans allow only individuals with certain inherited characteristics to mate. Through selective breeding, humans have crafted enormous evolutionary changes in not only dogs but also in many other domesticated organisms, including ornamental flowers, pet birds, and food crops.

Artificial selection happens when humans choose which individuals of a particular species are allowed to breed. But without humans, can the environment "choose" who survives and breeds? It does. In nature, evolution occurs mainly via natural selection as well as through other mechanisms we discuss later. **Natural selection** is the process by which individuals with advantageous inherited characteristics for a particular environment survive and reproduce at a higher rate than do individuals with other, less useful characteristics. In other words, whoever has the most kids wins!

After the environment "chooses" the winners—that is, those that successfully breed the most—the characteristics of those individuals become more common in successive generations because those individuals have produced more offspring. As with artificial selection, the overall changes are seen over time and across an entire population, not in individuals. For example, in 1977 a terrible drought struck the Galápagos Islands, off the coast of Ecuador. One species of small ground finches—petite birds with sharp, pointy beaks—starved as the small, tender seeds they ate became scarce. But some heat-loving, drought-resistant plants still produced large, hard seeds. The finches with larger beaks could eat those seeds. They survived and reproduced, and by 1978, in just one generation, the average beak size in the population had increased (**Figure 11.4**).

This is one of many examples of how a population can evolve via natural selection so that more and more individuals have beneficial traits, and fewer and fewer have disadvantageous traits. This change in traits is called **adaptation**—an evolutionary process by which a population becomes better matched to its environment over time. The finch population quickly adapted to its drier environment. Over time, the small-beaked finches died off, and the large-beaked birds survived and reproduced; as a result, the finch population adapted to its new environment in just a few years. Other adaptations take millions of years, such as whale ancestors adapting to aquatic life.

Six lines of evidence provide compelling support that species have evolved over time through adaptation and natural selection:

1. Direct observation of evolution through artificial selection
2. Fossil evidence
3. Shared characteristics among living organisms
4. Similarities and differences in DNA
5. Biogeographic evidence
6. Common patterns of embryo development

Figure 11.4

Natural selection results in larger beak size in finches

After a drought, only birds with larger beaks were able to eat the available food: large, hard seeds. In the span of just one generation, the average size of the species' beak was visibly larger. The bird in this photograph has the resulting large beak.

Q1: What is natural selection?

Q2: If heavy rains caused an abundance of small, tender seeds and fewer large seeds, what do you predict would happen to the average beak size of the finches?

Q3: Compare and contrast artificial selection and natural selection. Name two ways in which they are similar. How are they different?

See Appendix A for answers to the figure questions.

Nowhere is all this evidence more present and intriguing than in one of the most dramatic transitions to occur on Earth: the evolution of small, land-living mammals into dolphins, porpoises, and mighty whales.

Fossil Secrets

Fossils are often thought of as old, buried bones, but they are much more than that. **Fossils** are the mineralized remains—or even just the impressions, like footprints—of formerly living organisms (**Figure 11.5**). Scientists work long and hard to find and preserve fossils. For Thewissen, getting his hands on the *Indohyus* fossils was no easy task.

Soft-bodied animals such as this one dominated life on Earth 600 million years ago (mya).
This fossil is of a trilobite that lived between 410 and 355 mya.
This fossilized leaf is from a 300-million-year-old seed fern.
This 20-million-year-old termite is preserved in amber, the fossilized resin of a tree.
Two dinosaurs were fossilized together: a *Velociraptor*, on the left, is entangled with a *Protoceratops*, which bit down on the predator's claw, locking both in a death grip.
Once solid wood has fossilized into solid rock, it is known as petrified wood.

Figure 11.5

Fossils through the ages

Fossils range from imprints of organisms to preserved organisms to mineralized remains (such as bone or wood). Each fossil can be dated, and the combined results can tell the history of life on Earth.

Beginning in 2003, Thewissen made an annual pilgrimage to Dehradun, India, a city nestled in the foothills of the Himalayas. There he visited the widow of A. Ranga Rao, an Indian geologist who had hoarded piles of fossils excavated from Kashmir, a disputed border area between India and Pakistan. Most early whale fossils have come from the India-Pakistan region, where whales first evolved. But because of political tensions, in recent decades it has been too dangerous to travel to Kashmir, much less dig for fossils. Thewissen was frustrated by his inability to collect fossils in Kashmir, and he hoped Ranga Rao's widow would allow him access to her stockpile.

The fossil record enables biologists to reconstruct the history of life on Earth, and it provides some of the strongest evidence that species have evolved over time. The relative depth or distance from the surface of Earth at which a fossil is found indicates its *order* in the fossil record. The ages of fossils correspond to their order. Older fossils are found in deeper, older rock layers.

The fossil record includes numerous excellent examples of how major new groups of organisms arose from previously existing organisms. These **transitional fossils** are evidence of species with similarities to the ancestral group (in this case land-living mammals) and similarities to the descendant species (whales).

Thewissen spent decades studying these intermediates—from the oldest whale ancestor, the wolflike *Pakicetus*, which waded in shallow freshwater; to the larger crocodile-like *Ambulocetus*, which stalked its prey underwater; to the fully aquatic *Dorudon*, with its blowhole, flippers, and tail (**Figure 11.6**). Yet Thewissen and others had long been searching for the animal that preceded them all—the ancestor that lived on land. If Thewissen had to guess where those fossils might be, his guess would be Kashmir and potentially in Ranga Rao's basement.

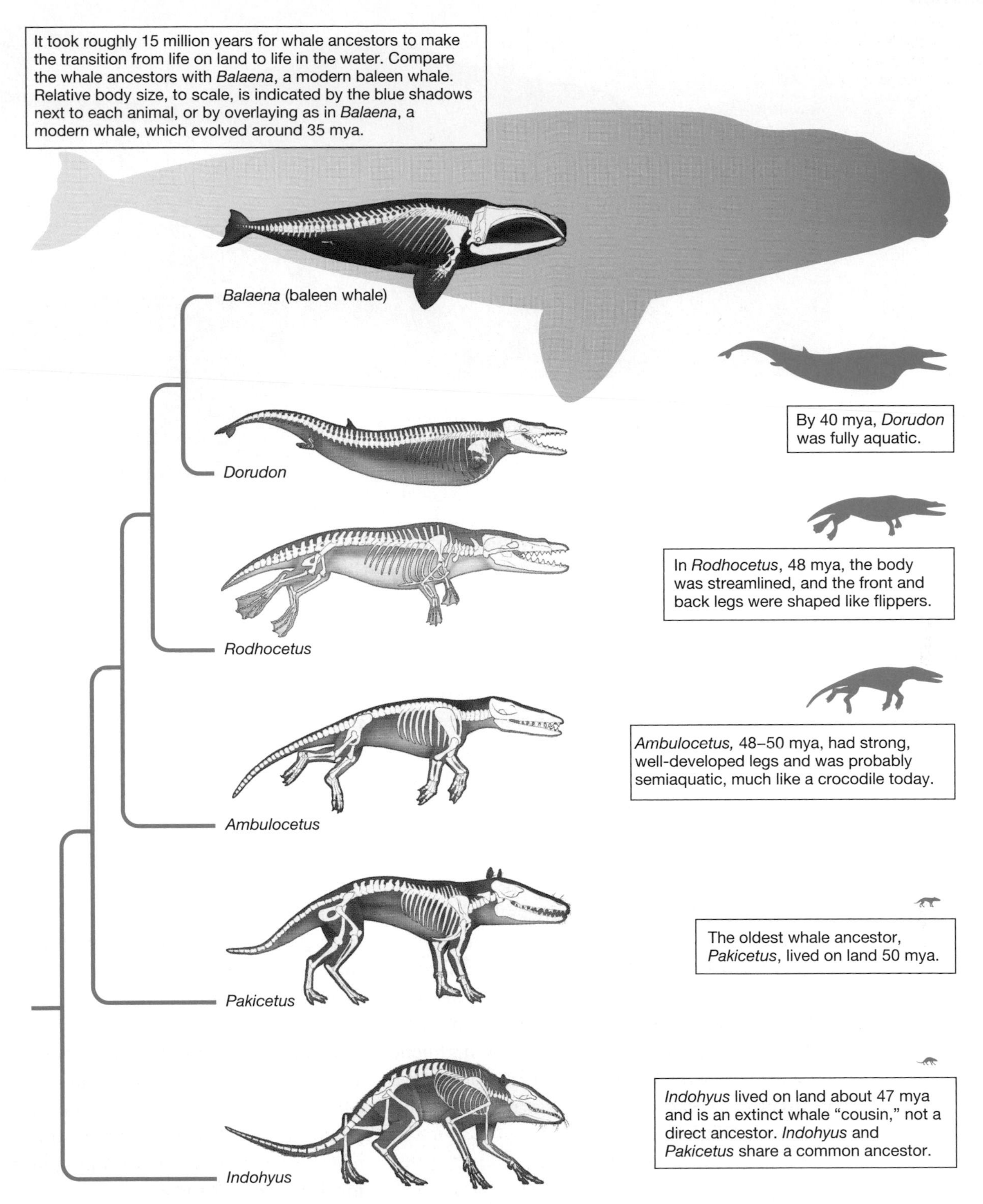

It took roughly 15 million years for whale ancestors to make the transition from life on land to life in the water. Compare the whale ancestors with *Balaena*, a modern baleen whale. Relative body size, to scale, is indicated by the blue shadows next to each animal, or by overlaying as in *Balaena*, a modern whale, which evolved around 35 mya.
Balaena (baleen whale)
By 40 mya, *Dorudon* was fully aquatic.
Dorudon
In *Rodhocetus*, 48 mya, the body was streamlined, and the front and back legs were shaped like flippers.
Rodhocetus
Ambulocetus, 48–50 mya, had strong, well-developed legs and was probably semiaquatic, much like a crocodile today.
Ambulocetus
The oldest whale ancestor, *Pakicetus*, lived on land 50 mya.
Pakicetus
Indohyus lived on land about 47 mya and is an extinct whale "cousin," not a direct ancestor. *Indohyus* and *Pakicetus* share a common ancestor.
Indohyus

Figure 11.6

Skeletons and body sizes of modern whales and fossil ancestors

Over roughly 15 million years, the ancestors of whales made the transition from life on land to life in the water. Here, from top to bottom, are reconstructed skeletons from a modern whale and various ancestors. Relative body size, to scale, is indicated by the blue shadow to the right of each animal or beneath it (*Balaena*).

Q1: What is the general definition of a fossil?

Q2: How do modern whales differ from their ancestors?

Q3: What is a transitional fossil?

See Appendix A for answers to the figure questions.

Figure 11.7

***Indohyus*, oldest cousin of whales**

The inset photo shows a reconstructed fossilized skeleton of *Indohyus*. The illustration is an artist's depiction of the living animal. (Source: Photo courtesy of Thewissen Lab, NEOMED.)

Unfortunately, Ranga Rao's widow was protective of the fossils, worried that someone might steal her husband's property and legacy. But each year, Thewissen visited her, chatted with her, and gained her trust. When she passed away in 2007, she made Thewissen cotrustee of her estate, and suddenly the fossils, which had sat in dusty piles for 30 years, were available for study.

"I focused on taking the rocks back to the U.S. and having my fossil preparers remove the fossils from the rocks, which is very difficult," says Thewissen. From Ranga Rao's collection, Thewissen identified more than 400 bones that belonged to *Indohyus*. By collecting a thighbone here and a jawbone there, his team compiled a Frankenstein-like skeleton of a representative *Indohyus* (**Figure 11.7**).

DEBUNKED!

MYTH: Evolution can't explain complex organs such as eyes.

FACT: Scientists have identified primitive, light-sensitive organs in animals such as snails and worms. Even without providing full vision, these simple structures would have been favored by natural selection. Eyes and other complex organs such as the placenta evolved from incremental changes over time.

After the discovery of the whalelike ear, the researchers looked even more carefully at the fossils and found additional evidence that *Indohyus* was a relative of whales. This unassuming little animal, with a pointy snout and slender legs tipped with hooves, lived near water and potentially spent a significant amount of time in the water.

The first clues about *Indohyus*'s lifestyle came from its teeth. Oxygen stored in the molecules that make up teeth comes from the water and food that an animal ingests, and the levels of oxygen isotopes in *Indohyus*'s teeth matched those of water-going mammals today. *Indohyus* also had large, crushing molars with levels of carbon isotopes that suggested it grazed on plants, as do some modern mammals that graze near and in water, such as hippopotamuses and muskrats (**Figure 11.8**).

Lisa Cooper, a graduate student in Thewissen's lab at the time, identified another adaptation to the water: *Indohyus*'s leg bones. From the outside, the limbs of *Indohyus* look like those of any other mammal walking around on land. But on the inside, it's another story. Cooper cut out a section of bone from a limb, ground it down until she could see light through it, and then looked at the bone under a microscope.

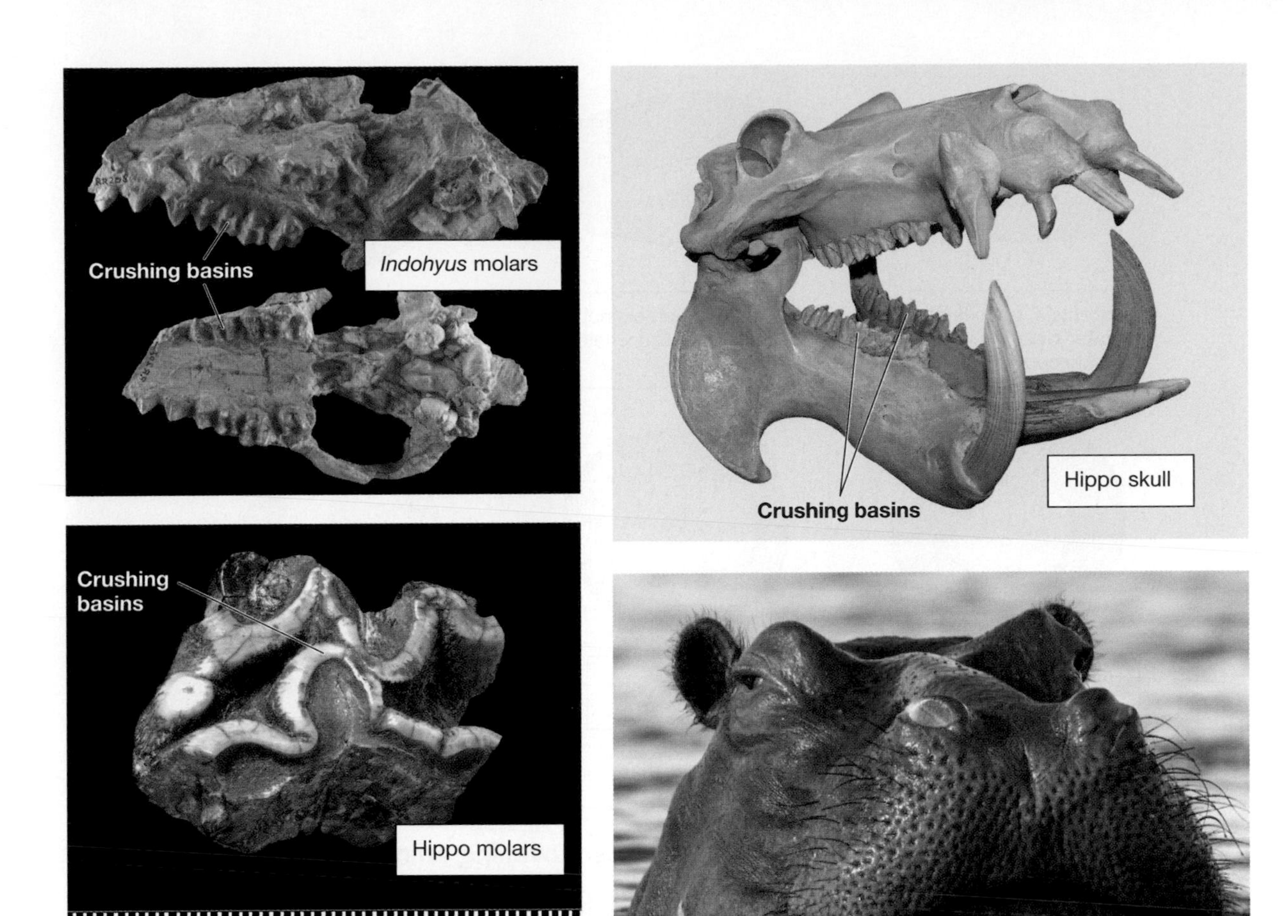

Figure 11.8

Comparing fossilized *Indohyus* and a modern hippopotamus

The molars of *Indohyus* are similar to those of contemporary aquatic plant-eating animals, such as the hippo. Crushing basins on teeth make it possible to grind up tough plant fibers for eating.

She saw a thick layer of bone wrapped around the bone marrow.

"Hans already had lots of bones of the earliest whales, and they all had extraordinarily thick bones," says Cooper, now an assistant professor at Northeast Ohio Medical University. Modern animals that live in shallow water, such as hippos and manatees, also have thick bones, which help prevent them from floating and enable them to dive quickly (**Figure 11.9**). "It isn't just isolated to whales," says Cooper. "Bones have thickened again and again as different groups of vertebrates entered the water. When you trace back through the fossil record, there is a pretty good correlation between thickness of bone and whether something was living in the water."

Indohyus's thick bones are an example of an **adaptive trait**, a feature that gives an individual better function than competitors. By being able to easily wade and dive in water, *Indohyus* had an advantage over other organisms in escaping predators and accessing edible plants underwater. Adaptive traits take many forms, from an anatomical feature such as *Indohyus*'s bones to behaviors to the functions of individual proteins. In bats, for example, echolocation—the physical ability to detect objects through sound waves—is an adaptation for catching insects in the dark. In stick insects, the ability to physically and behaviorally mimic the plants they live on is an adaptation for avoiding detection by predators (**Figure 11.10**).

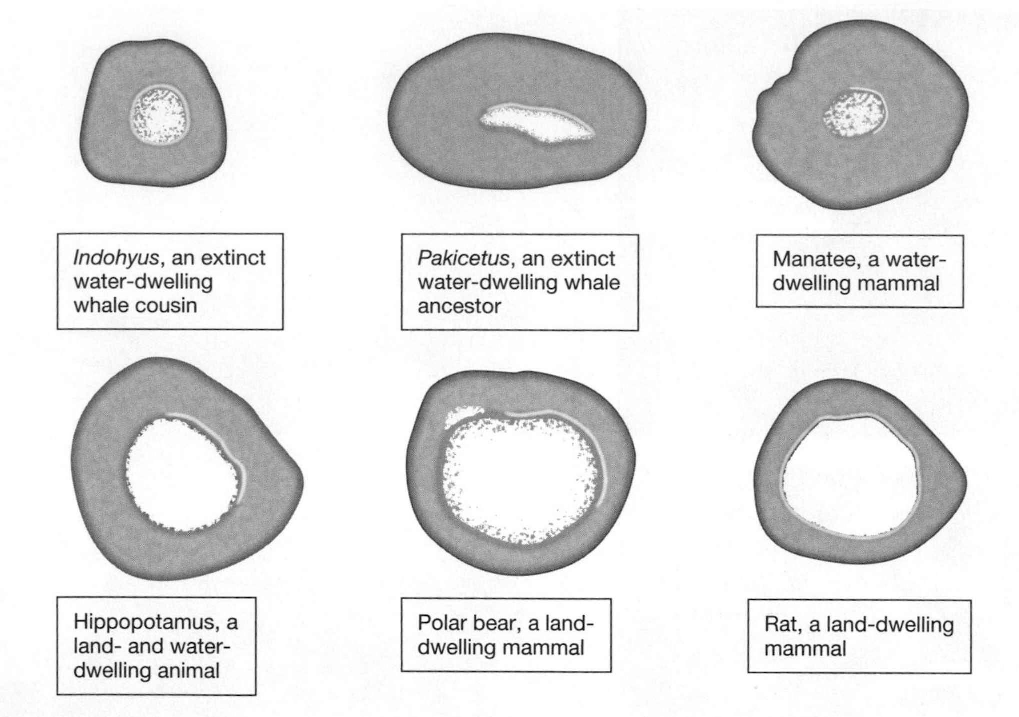

Figure 11.9

Cross sections of the femurs of *Indohyus*, water-dwelling animals, and land-dwelling animals

Aquatic organisms have thick, dense bone around a narrow marrow space, whereas terrestrial organisms have thin bone and a large marrow space. *Indohyus*'s bone structure is a trait that is shared with other water-dwelling animals.

Q1: **How would thick bones help water-dwelling animals?**

Q2: **Why does the adaptation of thick bones suggest a water-dwelling lifestyle?**

Q3: **How did this adaptation likely increase survival and reproduction in *Indohyus*?**

See Appendix A for answers to the figure questions.

Figure 11.10

Stick insects avoid detection by predators

Stick insects are well adapted to their environment. By looking like the branches they live on and moving slowly, they avoid detection by predators.

The Ultimate Family Tree

Thick bones are not just restricted to the ancestors of whales. They can also be seen in many other water-loving animals, such as hippos. This similarity across organisms is another type of evidence for evolution—namely, shared characteristics among species. Many shared characteristics—such as thick bones for animals that take to water, sexual reproduction via egg and sperm, or eukaryotic cells—result from organisms' having a **common ancestor**, a single organism from which many species have evolved. A group of organisms has **common descent** if the organisms share a common ancestor.

When one species splits into two or more species, the resulting species share similar features, or **homologous traits**, because they have common descent even though these features may begin to look different from one another over time. For example, whales and bats are so different from humans that it can be difficult to identify similarities, but we evolved from a common mammalian ancestor and do share homologous traits (**Figure 11.11**). In addition, humans, whales, and bats nurse their young and have a single lower jawbone because our common ancestor had those traits.

Vestigial traits are another type of trait that organisms have because of a common ancestor. These features are a piece of the evolutionary past, inherited from a common ancestor but no longer used. Vestigial traits may appear as reduced or degenerated parts whose function is hard to discern (**Figure 11.12**). For example, many modern whales have vestiges of thighbones, also called femurs, embedded in the skin next to the pelvis. In land mammals, birds, and other vertebrates with two pairs of limbs, these bones are critical for walking, running, and jumping. Aquatic whales have no need of this bone, yet its traces remain (**Figure 11.13**).

Whales also have small muscles devoted to nonexistent external ears, likely from a time when whales were able to move their ears as some land animals, such as dogs, do for directional hearing. Vestigial traits are not adaptations. In fact, they can be detrimental. Most humans no longer need wisdom teeth to replace

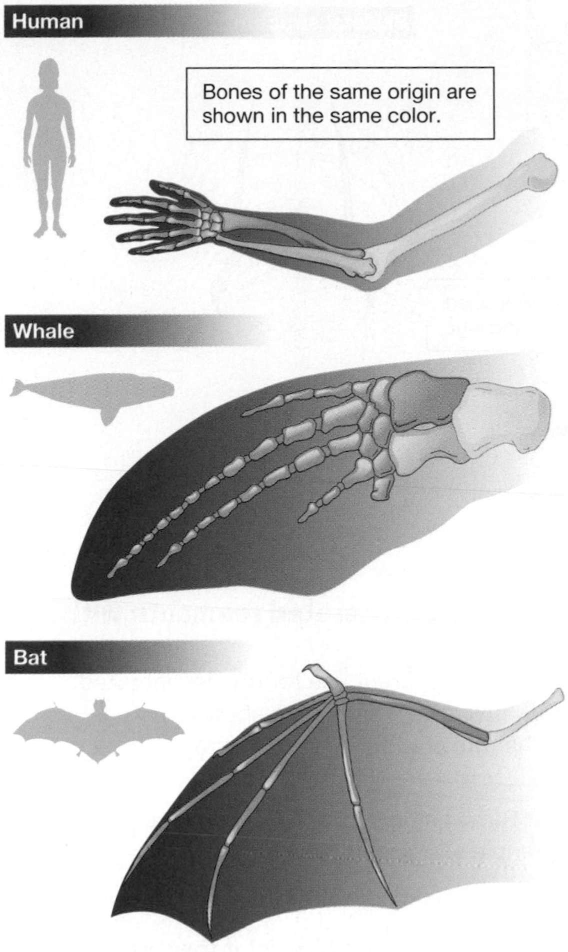

Human

Bones of the same origin are shown in the same color.

Whale

Bat

Figure 11.11

Homologous traits are shared characteristics inherited from a common ancestor

The human arm, the whale flipper, and the bat wing are homologous structures due to common descent. All three structures have a matching set of five digits and a matching set of arm bones, but these sets have been altered for different functions through evolution.

Q1: What is meant by the term "common ancestor"? Give an example.

Q2: Why are homologous structures among organisms evidence for evolution?

Q3: Aside from skeletal structural similarities, what other commonalities among organisms might be considered homologous?

See Appendix A for answers to the figure questions.

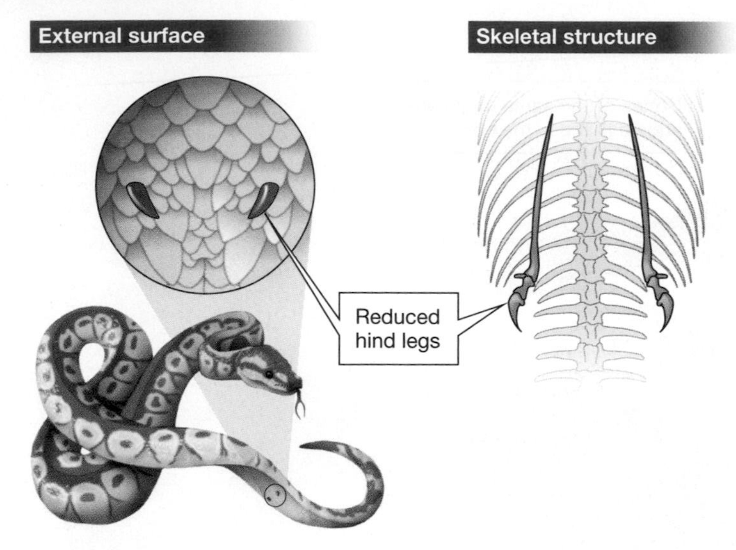

Figure 11.12

Vestigial traits are reduced or degenerated remnants with no apparent function

Snakes are limbless reptiles with no apparent use for the degenerated remnants of hind legs that they still have. The python shown here has extremely reduced hind legs that are only barely visible externally.

Q1: Why are vestigial structures among organisms evidence for evolution? Give another example of a vestigial structure.

Q2: Are vestigial structures also homologous structures? Explain.

Q3: Why do you think vestigial structures still exist if they are no longer useful?

See Appendix A for answers to the figure questions.

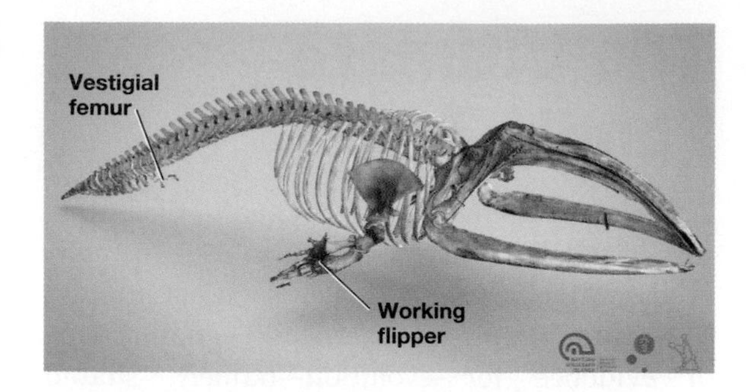

Figure 11.13

Whales still carry vestigial hind limbs

The front limbs, or flippers, of this whale skeleton have evolved to help it move through water, whereas its femurs have become much smaller and are nonfunctional.

lost teeth during adolescence, yet most people still have them. Wisdom teeth tend to erupt around the twentieth year of life, often causing severe pain and displacing other teeth, and they usually require removal.

Clues in the Code

Within the cells of every organism is one of the strongest pieces of evidence for evolution: DNA. Living things universally use DNA as hereditary, or genetic, material (see Chapter 8 for review). The fact that all organisms on Earth use the same genetic code—even organisms as different as bacteria, redwood trees, and humans—is further evidence that the great diversity of living things evolved from a common ancestor.

Researchers have analyzed the DNA sequences of whales and other animals and shown that, of all animals, whales are most closely related to even-toed ungulates—a group of hoofed mammals that includes modern deer, giraffes, camels, pigs, and hippos. *Indohyus* is an example of an extinct even-toed ungulate. Molecular studies confirm the prediction, backed by observations of fossils, that whales and *Indohyus* share a common ancestry.

According to DNA sequence similarity, hippos are whales' closest living relatives. Whale DNA is more similar to hippo DNA than to the DNA of other marine mammals, such as seals and sea lions. **DNA sequence similarity** is a measure of how closely related two DNA molecules are to each other. For example, in the DNA sequences of the same insulin gene in humans and mice, 83 percent of the nucleotides are identical at corresponding positions (**Figure 11.14**). We share a more recent common ancestor with mice than we do with chickens, for which the DNA sequence of the same gene is only 72 percent identical to ours. In fact, the insulin gene of our closest living relative on Earth, the chimpanzee, has 98 percent similarity to the human insulin gene. That comparability implies that humans and chimpanzees share a very recent common ancestor.

The fact that these separate lines of evidence—namely, anatomical features and DNA—yield the same result over and over again for diverse groups of organisms is strong evidence for evolution.

Whale evolution is "one of the best case studies documenting how a vertebrate can go from a terrestrial to an aquatic environment," says Cooper. Meanwhile, more evidence supporting whale evolution comes from the locations where whale fossils have been found.

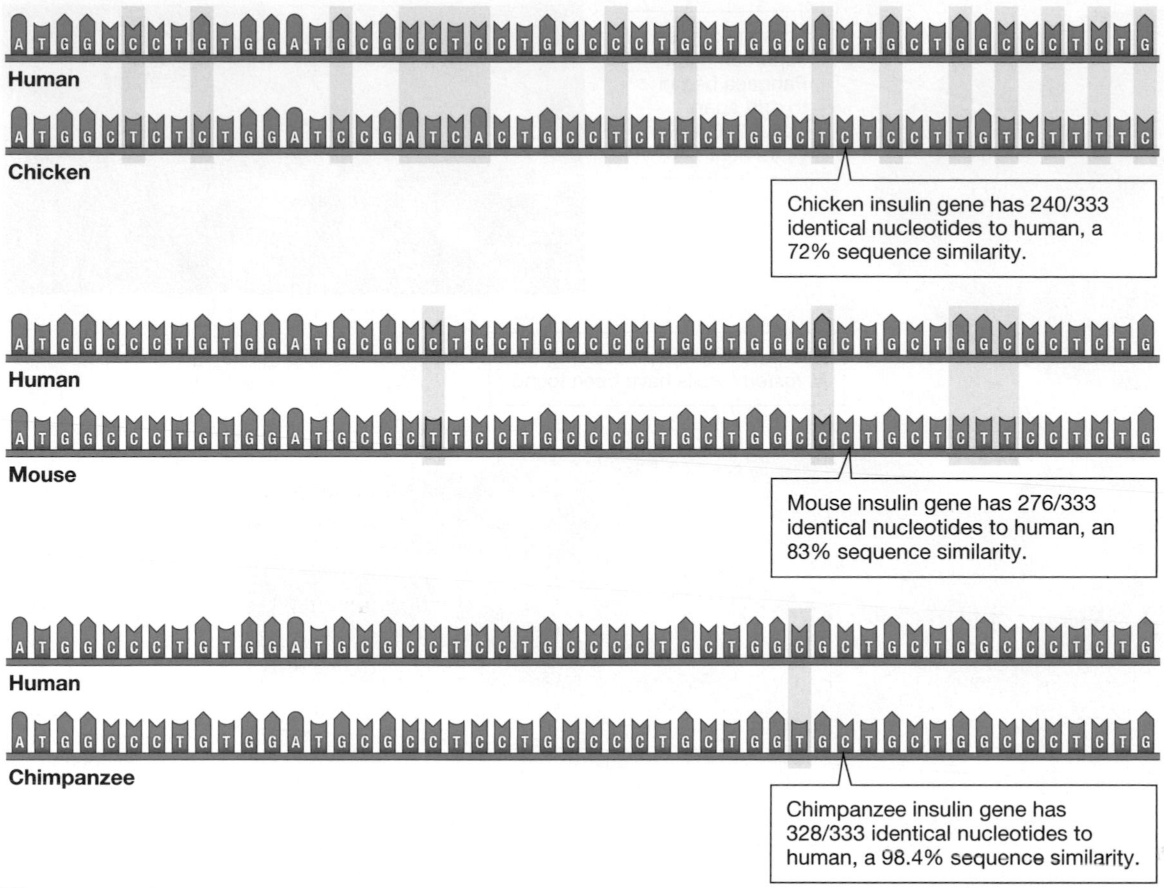

Figure 11.14

DNA sequence similarities of the insulin gene

The complete coding sequence of the human insulin gene contains 333 nucleotides. Only the first 50 nucleotides are shown here. Unshaded paired sequences are identical nucleotides at that position; those shaded in yellow are different.

Q1: **If a sequence from another species showed a 95 percent sequence similarity to humans, would that species be more or less closely related to humans than chimpanzees are?**

Q2: **Should similarities in the DNA sequences of genes be considered homologous traits? Explain your answer in terms of evolution.**

Q3: **How is the increased similarity in the DNA sequences of genes between more closely related organisms—and the decreased similarity between less closely related organisms—evidence for evolution? Use the examples in this figure to support your answer.**

See Appendix A for answers to the figure questions.

Birthplace of Whales

Earth's continents are on massive tectonic plates, which slowly move over time in a process called continental drift or plate tectonics. About 250 million years ago, South America, Africa, and all the other landmasses of Earth had drifted together to form one giant continent called Pangaea. About 200 million years ago, Pangaea slowly began to split up, separating to ultimately form the continents as we know them. That separation continues today; for example, each year South America and Africa drift farther apart by about an inch.

We can use knowledge of evolution and continental drift to make predictions about the **biogeography** of a species—that is, the geographic locations where its fossils will be

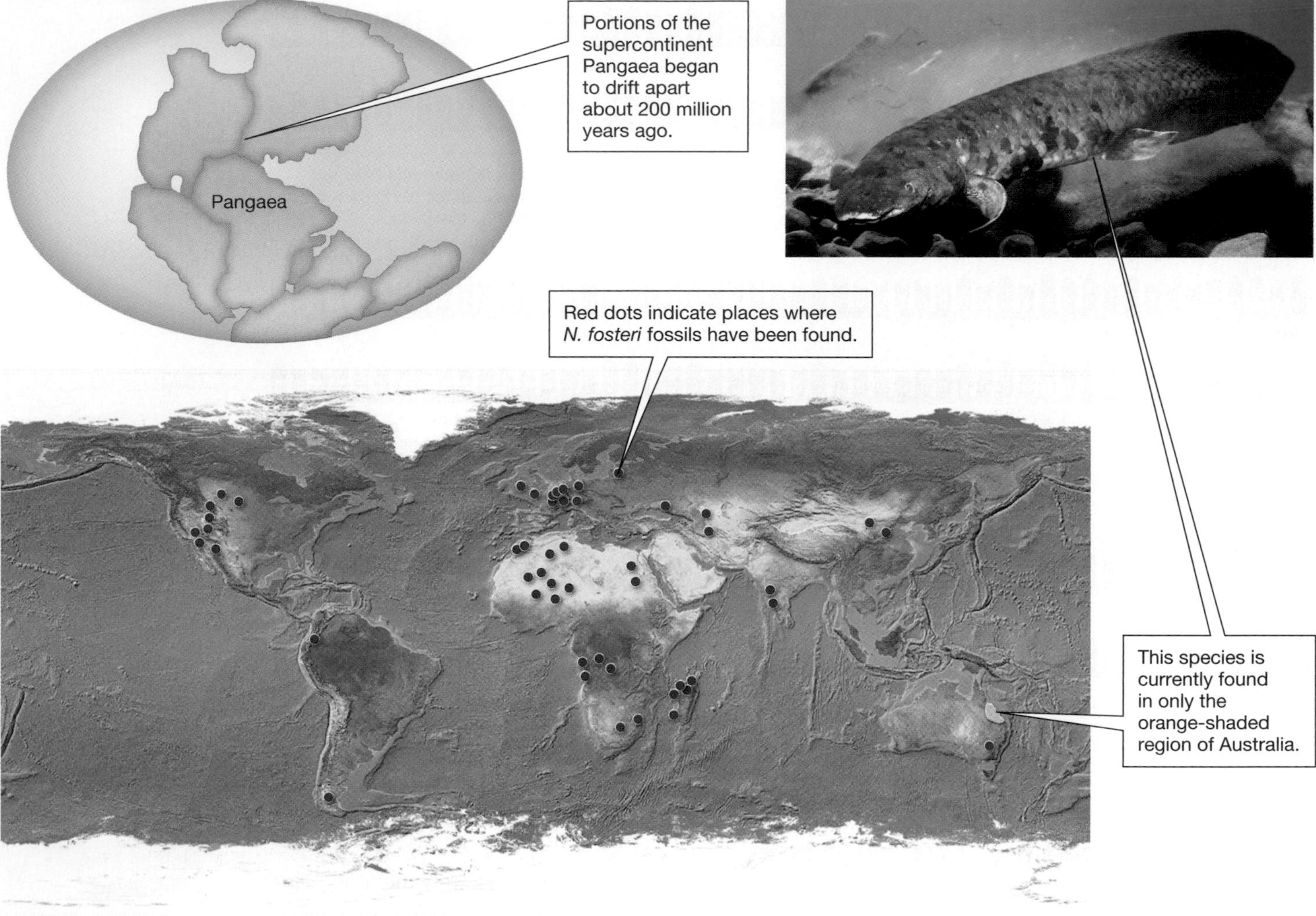

Portions of the supercontinent Pangaea began to drift apart about 200 million years ago.
Pangaea
Red dots indicate places where N. fosteri fossils have been found.
This species is currently found in only the orange-shaded region of Australia.

Figure 11.15

Biogeography can reveal a species' evolutionary past

Ancestors of the lungfish *Neoceratodus fosteri* lived during the time of Pangaea. *N. fosteri* fossils have been found on all continents except Antarctica.

Q1: Why should we expect to find *N. fosteri* fossils all over the world, given that it first evolved in Pangaea?

Q2: Can we use biogeographic evidence to support evolution without fossil evidence?

Q3: How might biogeographic evidence and DNA sequence similarities together support evolution?

See Appendix A for answers to the figure questions.

found. For example, today the freshwater lungfish *Neoceratodus fosteri* is found only in northeastern Australia, but its ancestors lived during the time of Pangaea. As predicted, therefore, fossils of those ancestors are found around the globe on all continents except Antarctica (**Figure 11.15**).

The biogeography of whale fossils matches the pattern predicted by evolution: all early species of whales, which lived in rivers and lakes but did not swim in the ocean, are found only near India and Pakistan. They were not capable of crossing the Atlantic. But fossils of fully aquatic mammals called protocetids, which emerged about 40 million years ago, are geographically much more widespread; they have been found as far away from Pakistan as Canada. "Protocetids are good swimmers, so we find their fossils all around the world," says Thewissen.

Growing Together

Though Thewissen has built a career on finding and describing whale fossils, he has recently become enamored with another vein of evolutionary evidence: embryology. According to the theory of evolution, organisms carry within themselves evidence of their evolutionary past. And, indeed, evidence of evolution can be observed in shared patterns of **embryonic development**.

As with homologous traits and DNA similarities, these common patterns are caused by descent from a common ancestor. Rather than evolving new organs "from scratch," new species inherit structures that may have been modified in form and sometimes even in function.

Upon fusion of sperm and egg, an animal embryo begins to grow and develop. The manner in which an embryo develops, especially at the early stages, may mirror early developmental stages of ancestral forms. For example, anteaters and some baleen whales do not have teeth as adults, but as fetuses they do. And the embryos of fishes, amphibians, reptiles, birds, and mammals (including humans) all develop pharyngeal pouches or gill slits (**Figure 11.16**). In fish, the pouches develop into gills that adult fish use to absorb oxygen underwater. In human embryos, these same features become parts of the ear and throat.

"I was interested to get embryos to look at some of these processes that we see happen in evolution, to see if they happen in development," says Thewissen. Upon obtaining whale and dolphin embryos (recall that dolphins are also mammals and are closely related to whales), the first trait he examined was hind limbs. "We know, from fossil evidence, that early whales lose their hind limbs," says Thewissen. So he wondered if hind limbs exist in whale or dolphin embryos. And if so, what makes them subsequently disappear before the animal is born.

Examining spotted dolphin embryos, Thewissen saw that when the embryos are the size of a pea, they do develop hind limb buds, but by the time they grow into the size of a bean, the limb buds are gone. In 2006, he and researchers at several other universities studied the genes that are active in whale and dolphin embryos and concluded that whales' hind limbs regressed over millions of years through small changes in a number of genes relatively late in embryonic development (**Figure 11.17**).

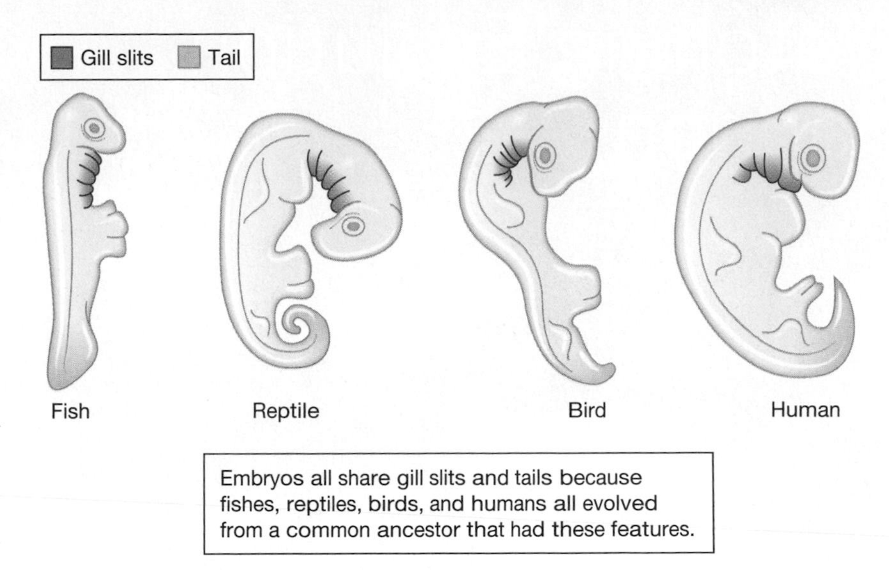

Figure 11.16

Evolutionary history can be extrapolated from similarities in embryo development

Complex structures in descendant species are generally further developments of structures that existed in their common ancestor.

Q1: Why are the similarities among organisms during early development evidence for evolution?

Q2: Are similar structures among species during early development homologous structures? Explain.

Q3: Why do you think embryonic structures still exist during early development if they are not used after birth?

See Appendix A for answers to the figure questions.

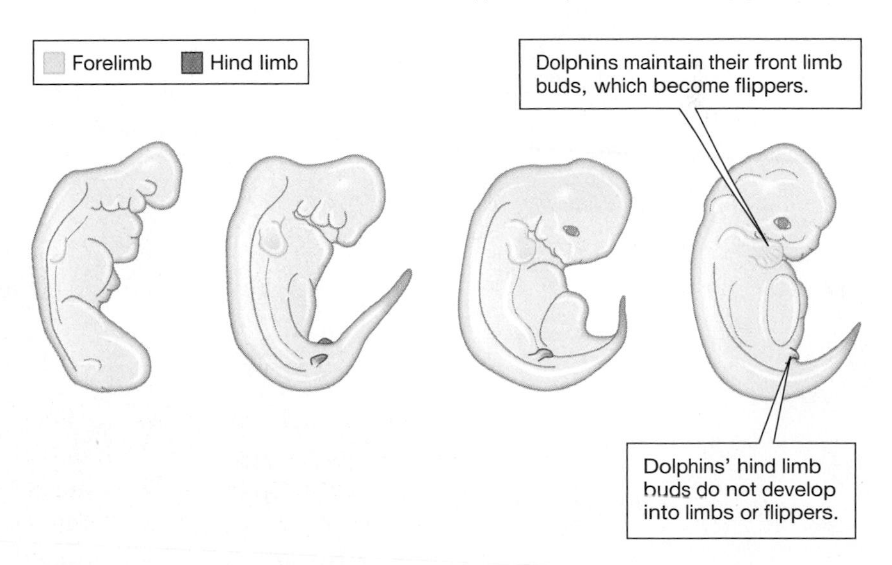

Figure 11.17

Dolphin embryonic development

The embryos of dolphins from weeks 4–9 of development show the formation and then subsequent loss of the hind limb buds. (Source: Based on a photo by Thewissen Lab, NEOMED, permission granted.)

Watching Evolution Happen

In 2012, researchers at UC Irvine watched evolution happen in real time via artificial selection. The scientists changed the growing environment of *Escherichia coli* bacteria, exposing them to far hotter temperatures than normal, to see if the microbes would adapt. Indeed, some of the bacteria evolved to be better able to survive in the hotter environment.

Assessment available in **smart**work

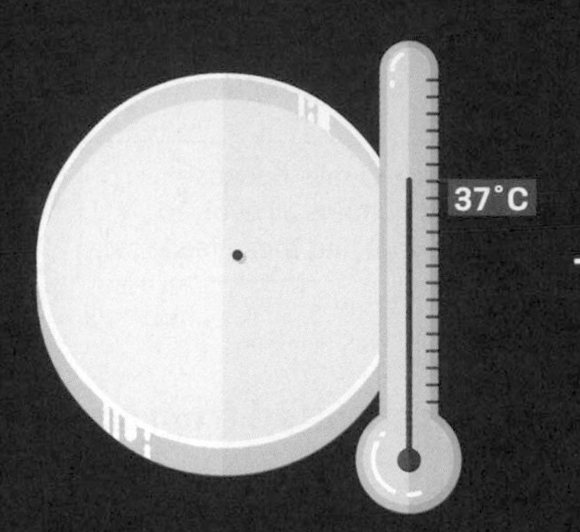

42.2°C

1 | Scientists grow **115** separate, genetically identical populations of *E. coli* at a comfortable temperature of **37**°C, or **98.6**°F.

2 | The populations reproduce through **2,000** generations while being subjected to an increased heat of **42.2**°C, or **107.96**°F.

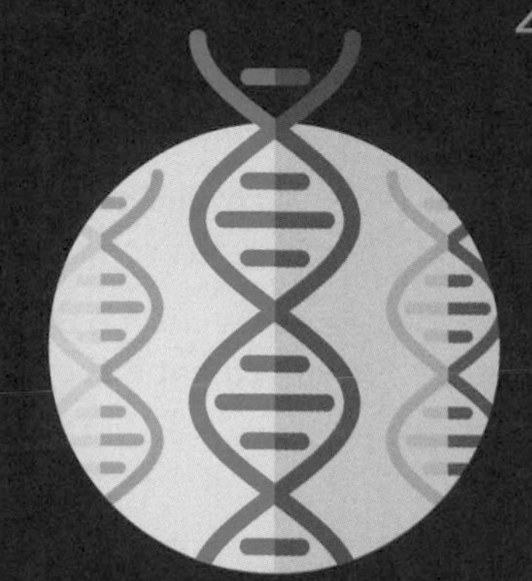

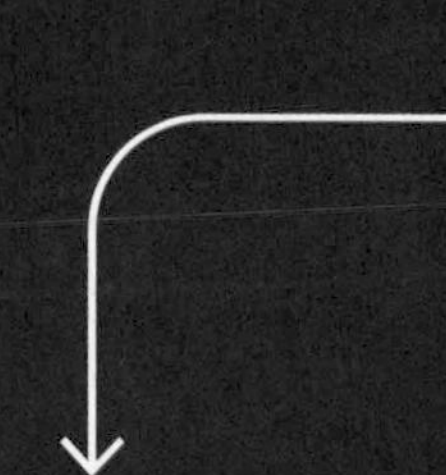

3 | One sample is taken from each of the **115** populations, and the genome is sequenced.

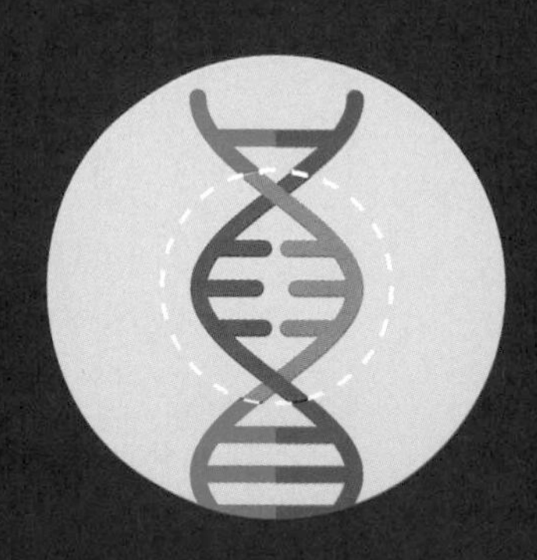

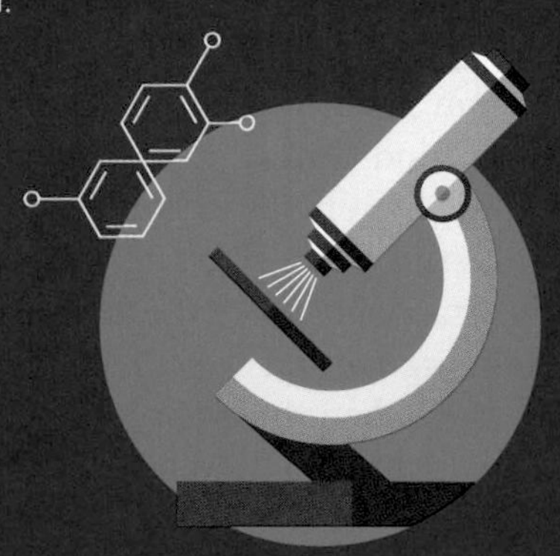

4 | **1,258** molecular changes, averaging **11** genetic mutations per clone, are detected. Two mutations in the *E. coli* enable it to survive the heat:

- Mutations in the RNA polymerase complex, an enzyme that transcribes RNA.
- Mutations in the *rho* gene, which encodes a protein that stops RNA transcription.

5 | The research continues: the next step is to figure out how the mutations in the RNA polymerase complex and in *rho* helped the *E. coli* survive the heat.

The loss of hind limbs in the embryo corresponds to the disappearance of hind limbs in the fossil record. "It was an awesome way to combine embryology with fossils," says Thewissen. In 2015, researchers at the Smithsonian Institution likewise found that the development of fetal ear bones in the womb paralleled changes observed in the whale fossil record.

With their discovery of *Indohyus*, Thewissen, Cooper, and their team bridged a 10-million-year gap in the fossil record, identifying an important transition species to whales. It is another rock in the mountain of evidence surrounding whale evolution, which was further fortified in April 2015 with the discovery in Panama of fossils from a previously unknown species of extinct pygmy sperm whale, and in April 2019, when scientists in Peru unearthed a 43-million-year-old fossil of a four-legged whale with a powerful tail and big feet for both walking and swimming. They named it *Peregocetus pacificus*, "the traveling whale that reached the Pacific Ocean."

It is clear that evolution is supported by mutually reinforcing, independent lines of evidence: direct observation, fossils, shared characteristics among living organisms, similarities and divergences in DNA, biogeographic evidence, and common patterns of embryo development. Just as the theory of gravity forms the foundation of physics, so evolution is the central tenet of biology.

Through scientific research, we know that whales descended from a series of land-living mammals, including *Indohyus*. But how did a population of small, furry animals become the mammoths of the sea? Next we investigate the strange and wonderful mechanisms of evolution—how and why it works.

NEED-TO-KNOW SCIENCE

- Biological **evolution** (page 206) is a change in the inherited characteristics of a population of organisms over generations.
- **Artificial selection** (page 208) results in biological evolution. Humans choose which organisms survive and reproduce—a process known as **selective breeding** (page 208).
- **Natural selection** (page 208) is the process by which individuals with advantageous genetic characteristics for a particular environment survive and reproduce at a higher rate than competing individuals with other, less useful characteristics.
- **Adaptation** (page 209) is an evolutionary process by which a population becomes better matched to its environment over time. An adaptation is also a trait—an **adaptive trait** (page 213)—that has evolved as a result of this process.
- **Fossils** (page 209) are the preserved remains (or just the impressions) of formerly living organisms. The fossil record enables biologists to reconstruct the history of life on Earth, and it provides some of the strongest evidence that species have evolved over time.

  Transitional fossils (page 210) are fossils of species with similarities to the ancestral group (for example, land-living mammals) and similarities to the descendant species (for example, whales).
- Many similarities among organisms are due to the fact that the organisms evolved via **common descent** (page 215) from a **common ancestor** (page 215). When one species splits into two, the resulting species share similar features, called **homologous traits** (page 215). If a homologous trait is no longer useful, it is called a **vestigial trait** (page 215).
- The fact that organisms as different as bacteria, redwood trees, and humans show **DNA sequence similarity** (page 216) is evidence that the great diversity of living things evolved from a common ancestor.
- **Biogeography** (page 217) uses knowledge about species and continental drift to make predictions about the geographic locations where fossils will be found.
- Similarities in **embryonic development** (page 219) of different organisms show that the characteristics of modern organisms arose through evolutionary modifications of traits inherited from common ancestors.

THE QUESTIONS

See Appendix B for answers.

The Basics

1 Transitional fossils

(a) share some similarities with their ancestral group.

(b) share some similarities with their descendant groups.

(c) both of the above

(d) none of the above

2 If two different organisms are closely related evolutionarily, then they will

(a) share a recent common ancestor.

(b) be similar in size.

(c) have very different DNA sequences in their genes.

(d) be randomly located throughout the world.

3 If two different organisms are very distantly related evolutionarily, then

(a) they should have extremely similar embryonic development.

(b) they must share a very recent common ancestor.

(c) the sequences of DNA in their genes should be less similar than those of two more closely related organisms.

(d) they should share more homologous traits than two more closely related organisms share.

4 Link each term with the correct definition.

biogeography	1. The preserved remains (or just the impressions) of once-living organisms.
fossil	2. The similarities in the nucleotide sequences among related organisms.
DNA sequence similarity	3. The similarities among organisms that are due to the fact that the organisms evolved from a common ancestor.
embryonic development	4. The geographic locations where related organisms and fossils are found.
homologous traits	5. The period of an organism's early growth in which structural developments can be observed that reveal links to evolutionarily related organisms.

Challenge Yourself

5 All mammals have tailbones and muscles for moving a tail. Even humans have a reduced tailbone and remnant tail-twitching muscles, though these features have no apparent usefulness. These traits in humans would best be described as

(a) convergent structures.

(b) fossil evidence.

(c) evidence from biogeography.

(d) vestigial traits.

6 Reduced tailbones and the associated remnant muscles in humans are an example of what type of evidence for common descent?

(a) artificial selection

(b) homologous traits

(c) biogeography

(d) fossil evidence

7 When a population evolves to be better fitted to its environment, this is an example of ________, which is brought about by the process of ________.

8 In one or two sentences, identify the similarities and the differences between artificial selection and natural selection.

Try Something New

9 Cat DNA is much more similar to dog DNA than to tortoise DNA. Why?

(a) Cats and dogs are both carnivores and take in similar nutrients.

(b) Cats and dogs have lived together with humans for a long period of time, so they have grown to be more similar to each other than to tortoises.

(c) Cats and dogs have more offspring during their lifetime than tortoises have, so their DNA changes less rapidly.

(d) Cats and dogs have a common ancestor that is more recent than the common ancestor of cats and tortoises.

10 DNA sequences were analyzed from humans and three other mammals: species X, Y, and Z. Of these mammals, which is most closely related to humans? (*Note:* Regions identical to human DNA are shown in bold type.)

Human: AATGCTTTGGGGATCGCGAGCGCAGCGC

Species X: GGGTT**TTTATCGCT**ATATATATATA

Species Y: AATGCTTTGGGGATCGCGAGCGCATATA

Species Z: AATGCGGGTTTTT**ATC**TATATATATATATA

11 Which of the following is *not* an example of artificial selection?

(a) Female fish in a natural pond have variable skill at depositing their eggs near the murky shore, where the eggs are better hidden from predators. Eggs that are not deposited near the murky shore are quickly devoured by fish. Female hatchlings from the eggs deposited near the murky shore grow up to be good at depositing eggs near the shore and therefore have a survival advantage.

(b) Your mother has been saving the best seeds from her lima bean plants every summer and replanting them the next year. She likes seeds that are plump and bright green and saves only these seeds each year. Within 10 years, almost all of her lima beans have become plump and bright green.

(c) The hamster you received as a birthday present turns out to be pregnant. Several of the offspring have long, silky hair, and you put them together in an enclosure to try to produce more baby hamsters with long, silky hair. You continue to breed long- and silky-haired hamsters to each other and now, several years later, your bedroom is full of cages containing long- and silky-haired hamsters.

(d) Farmer Brown has a duck that can type. He sets up an online dating profile for his duck to find a female duck that can also type. Seven years after the arranged wedding of the two typing ducks, Farmer Brown has an entire flock of ducks, most of which can type.

12 The fossil record shows that the first mammals evolved 220 million years ago. The supercontinent Pangaea began to break apart 200 million years ago. On which continents would you predict that fossils of the first mammals will be found?

13 The accompanying graph shows the number of whales hunted for food or other products both before and after an international ban on whaling. The ban excluded some scientific collecting and hunting by aboriginal groups.

(a) In what 5-year period were the most whales killed? How many whales would you estimate were killed in that time?

(b) Approximately how many whales have been killed since 2001?

(c) When would you estimate that the whale hunting was banned? Explain your reasoning.

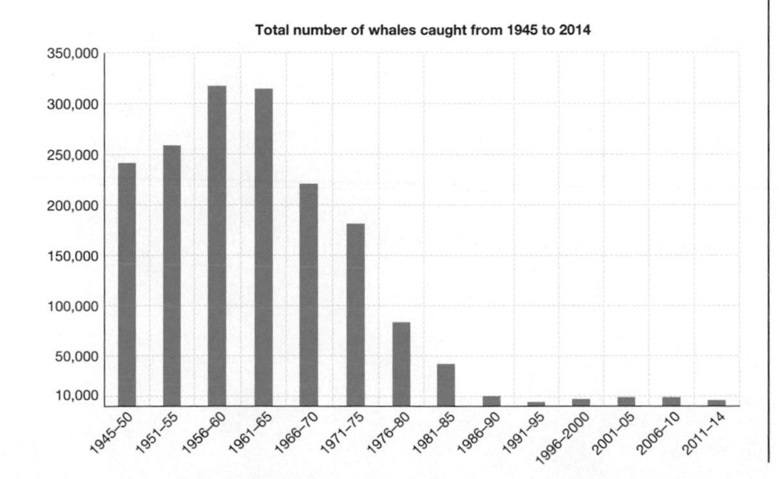

Leveling Up

14 **What do *you* think?** The prerequisites to apply for medical school always include courses in cell biology, genetics, and biochemistry but rarely include a formal course in evolution. Do you think medical schools should require a formal course in evolution as a prerequisite for admission? Why or why not? Research the issue and support your case using information you find from credible sources.

15 ***Write Now* biology: evidence for evolution** This assignment is designed to expand your knowledge of the evidence for evolution. View the following short videos, which can be found by going to pbslearningmedia.org and searching for the title of the video. Answer the questions below that accompany each video.

Video 1: "Isn't Evolution Just a Theory?"
Why is evolution not just a theory? Use specifics from the video to defend your answer.

Video 2: "Who Was Charles Darwin?"
Why do you think Charles Darwin's ideas and book *On the Origin of Species* were so groundbreaking and "revolutionary"? Use specifics from the video to defend your answer.

Video 3: "How Do We Know Evolution Happens?"
Describe how the video portrays whale evolution. Include specific examples of transitional fossils described in the video, and explain why the scientists at the time the video was made considered the fossils to be whale ancestors.

Video 5: "Did Humans Evolve?"
DNA sequences of different species can be used to provide evidence of common descent. Using examples from the video, explain why DNA sequence similarity is the best evidence of evolution on Earth.

Video 6: "Why Does Evolution Matter Now?"
Describe why the theory of evolution matters to the field of medicine and to individual doctors. Use the example of tuberculosis from the video to support your answer. How does this video affect your answer to question 14?

Digital resources for your book are available online.

Battling Resistance

Antibiotic-resistant bacteria are evolving to overcome our drugs of last defense. How do we beat superbugs?

After reading this chapter you should be able to:

- Clarify why evolution occurs only in populations, not in individuals.
- Distinguish among directional, stabilizing, and disruptive selection.
- Explain how natural selection brings about increased reproductive success of a population in its environment.
- Compare and contrast convergent evolution and evolution by common descent, and give an example of each.
- Describe how DNA mutations can create new alleles at random.
- Illustrate how gene flow works and how it may inhibit evolution.
- Relate the process of genetic drift to genetic bottlenecks and the founder effect, using examples.

EVOLUTION

CHAPTER 12

MECHANISMS OF EVOLUTION

Dawn Sievert vividly remembers June 14, 2002—the day disaster struck. Sievert is an epidemiologist, or "disease detective," someone who studies the patterns and causes of human disease. At the time, she was working at the Michigan Department of Community Health, monitoring reported cases of antibiotic-resistant bacteria, a major health care concern.

Sievert spent a significant amount of time investigating outbreaks of an increasingly common and worrisome microbe called MRSA, short for "methicillin-resistant *Staphylococcus aureus*." *S. aureus*, commonly known as "staph," is a small, round bacterium that usually lives benignly in our nostrils and on our skin (**Figure 12.1**). But on rare occasions, staph slips beneath the surface of a burn or cut and causes an infection, which can be especially dangerous, even deadly, for individuals with suppressed immune systems, such as the elderly and patients on chemotherapy.

Staph is one of the most common causes of hospital infections today, and it is treated with antibiotics, drugs that kill bacteria but not human cells. Penicillin (**Figure 12.2**) was the first antibiotic used against staph, but resistance to penicillin evolved in this wily microbe even before the drug became commercially available to the public in the 1940s.

When penicillin stopped working against staph, doctors switched to an antibiotic called methicillin. Methicillin worked for about 20 years, until populations of staph evolved widespread resistance to that antibiotic as well. To the chagrin of doctors and patients everywhere, bacteria are able to adapt rapidly to new antibiotics, thanks to their short generation time and ability to share resistance genes among themselves. These tiny microbes are the Navy SEALs of evolution—the best of the best at adapting. (Bacteria are discussed extensively in Chapter 15.)

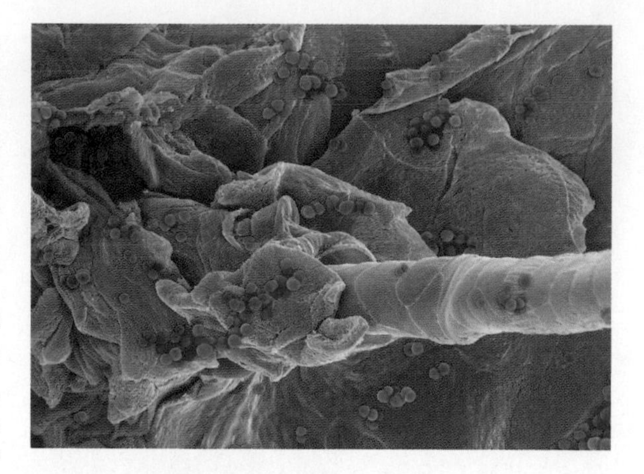

Figure 12.1

***Staphylococcus aureus*: friend or foe?**
These normally harmless bacteria, seen here on human skin at the base of a hair follicle, can become a source of deadly infection.

As we saw in Chapter 11, biological evolution is a change in the frequencies of inherited traits in a population over generations. Staph adapted to the presence of methicillin: only those microbes with genetic traits protecting them from the antibiotic survived. These traits were passed among populations and down from one generation to the next until methicillin resistance was frequent across staph populations.

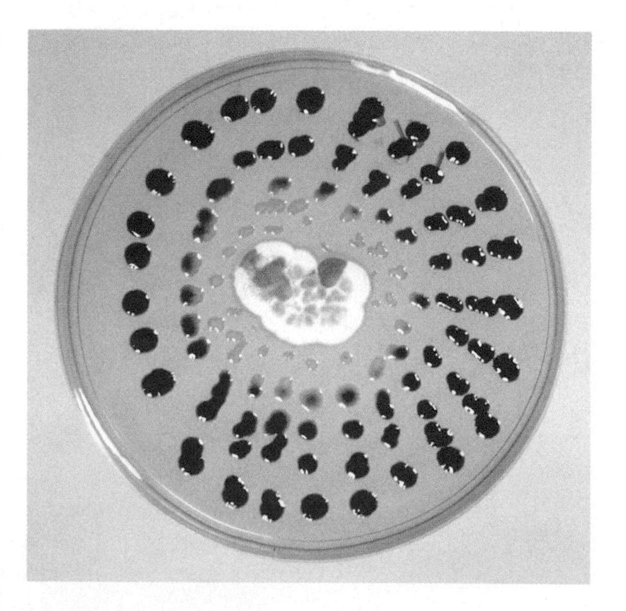

Figure 12.2

Penicillin, produced by mold, kills bacteria
The fuzzy white growth with blue spores in the center of this petri dish is *Penicillium*. This mold is secreting the antibiotic penicillin into the surrounding drops of liquid, killing the red bacteria in them.

DEBUNKED!

MYTH: Antibiotics can be used to treat colds and the flu.

FACT: Antibiotics kill only bacteria, so they're no help against the viruses that cause colds and the flu. Taking them won't help you get well any faster and could contribute to the growing problem of antibiotic-resistant bacteria.

Today, MRSA is rampant in hospitals, so doctors have been forced to turn to one of medicine's last lines of defense against the superbug. Vancomycin, a strong, blunt antibiotic that was first isolated from the mud of the Borneo jungle, is considered one of the drugs of last resort for fighting these serious infections. If the last-resort antibiotics don't kill these bacteria, no current antibiotic will, which brings us back to June 14, 2002, and Dawn Sievert.

On that day, a lab technician at a dialysis center in Detroit, Michigan, took two swabs of a foot ulcer infected with MRSA belonging to a 40-year-old diabetic woman (**Figure 12.3**). The patient had previously suffered from numerous foot infections, including MRSA, which had been treated with vancomycin for 6 weeks. Now she was in danger of losing her foot. The swabs of her latest infection were sent to a local laboratory, where technicians grew the bacteria in a petri dish to test its susceptibility to various antibiotics.

When the results of the first test came in, the laboratory staff immediately called Sievert's office. The bacteria, they told Sievert and the health department, appeared to be resistant to vancomycin.

"First, we needed laboratory confirmation and had to control any potential for panic," says Sievert. Her team asked the local lab to run its test again and send the health department a sample to test independently. Both teams waited.

The tests came back from each lab. Both were positive; the woman's foot was infected with the first reported case of vancomycin-resistant *S. aureus*. VRSA had arrived.

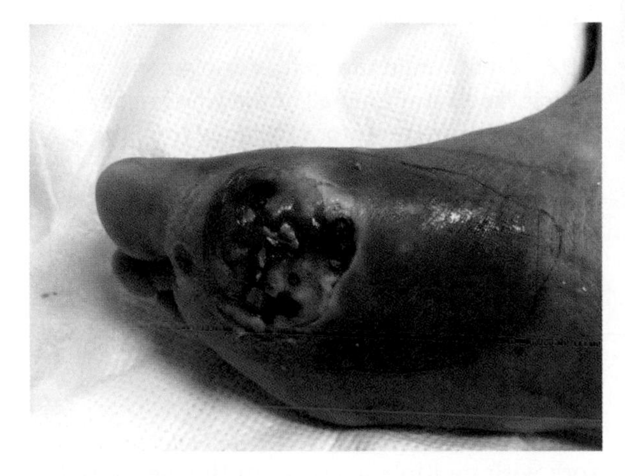

Figure 12.3

MRSA infection

This infected ulcer is on the foot of a patient with diabetes.

Birth of a Superbug

"At that point, it was the first-ever VRSA in the world," says Sievert. Unfortunately, it was not the last.

Through evolution, staph first survived penicillin, then methicillin, then vancomycin (**Figure 12.4**). The evolution of staph is a profound example of a species changing over time.

In the mid-1800s, two English biologists, Charles Darwin and Alfred Russel Wallace, separately studied the diversity of life. Each man concluded that existing species were not, as was generally thought at that time, the unchanging result of separate acts of creation. Instead, Darwin and Wallace came to the bold new conclusion that species "descend with modification"

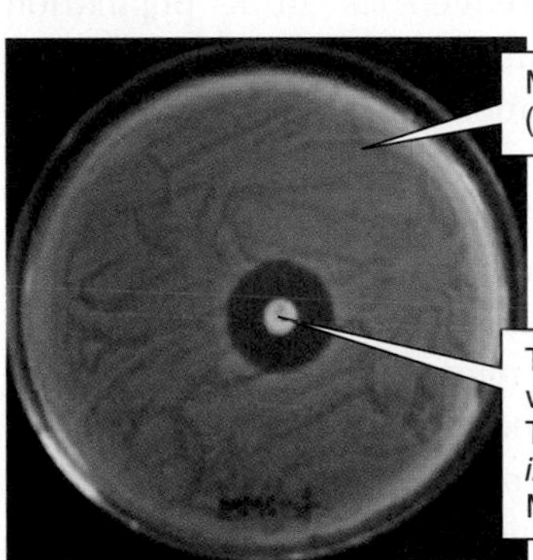

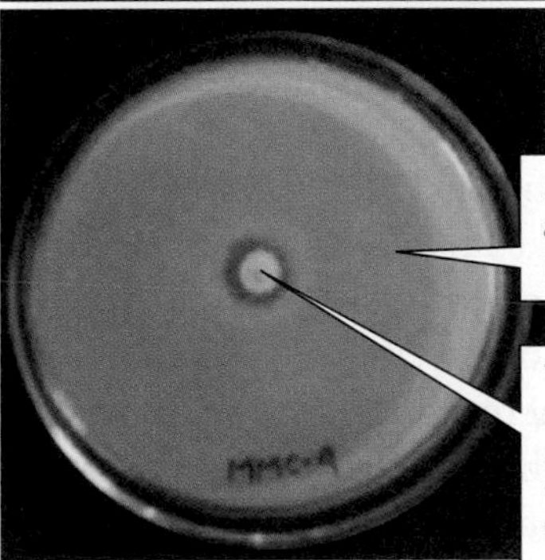

Now vancomycin-resistant *Staphylococcus aureus* (VRSA) bacteria are growing in a petri dish.
The paper disc is again infused with vancomycin. The zone of inhibition is much smaller than in the image above, indicating that the bacteria can now live and grow in the presence of the antibiotic.

Figure 12.4

MRSA versus VRSA in a vancomycin-resistance test

In the presence of vancomycin, MRSA (top) does not grow, whereas VRSA (bottom) grows relatively well.

Q1: What is the difference between MRSA and VRSA?

Q2: Explain in your own words the difference between the zone of inhibition in the top dish and the one in the bottom dish.

Q3: Why is the reduced zone of inhibition in the bottom dish so alarming?

See Appendix A for answers to the figure questions.

from ancestor species. In other words, new species arise from previous species.

Descent with modification, we know today, occurs not only in populations of large organisms such as whales but also in the tiniest single-celled bacteria and viruses. We generally think of evolution happening over millions of years, even hundreds of millions of years. But some evolutionary changes, such as adaptations for antibiotic resistance, take place over very short time spans as particular alleles spread rapidly through a population. Recall from Chapter 7 that *alleles* are different versions of the same gene (sequences of DNA) produced by random mutation, and therefore *allele frequencies* are the proportions of specific alleles in a population (**Figure 12.5**). Evolution corresponds to changes in a population's allele frequencies over time.

Because there are 15 mice, the gene pool has 30 allele possibilities (15 mice × 2 alleles per mouse).
Of the 30 total alleles in this population, 13 are white-fur-pigment alleles, so the white-allele frequency is 13/30 = 0.43, or 43%.
●●●●●●●●●●●●●●●●● 17/30 = 57%
○○○○○○○○○○○○○ 13/30 = 43%

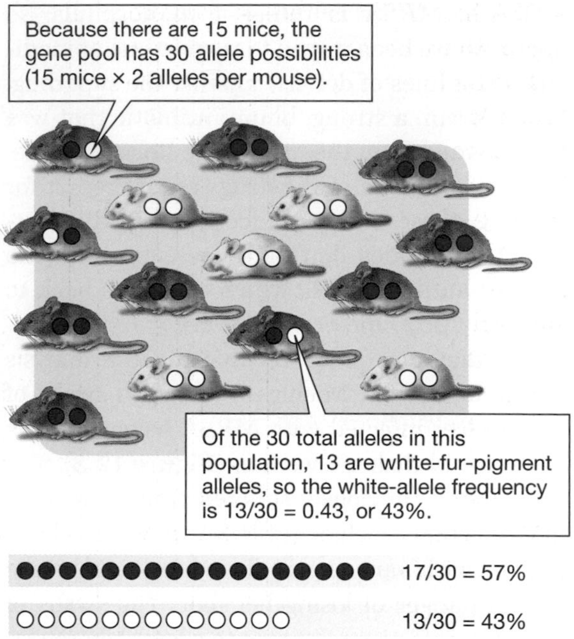

Figure 12.5

Allele frequencies are calculated as percentages in a population

These mice have two white-fur-pigment alleles and appear white, have two black-fur-pigment alleles and appear gray, or have one black and one white allele and appear gray. To calculate the white-fur-pigment allele frequency in the population, the number of white alleles is counted and divided by the total number of alleles.

Q1: What would the white-fur-pigment allele frequency be if three of the homozygous black allele mice (having two black alleles) were heterozygous (having one white and one black allele) instead?

Q2: What would the white-fur-pigment allele frequency be if all of the white mice died and were therefore removed from the population? Would the black-fur-pigment allele frequency be affected? If so, how would it be affected?

Q3: What would the white-fur-pigment allele frequency be if all of the gray mice died and were therefore removed from the population?

See Appendix A for answers to the figure questions.

When allele frequencies in a population change—that is, as particular alleles become more or less common—the phenotypes of the population change. In this way, the population evolves. As more and more staph containing the allele for methicillin resistance survived and reproduced, the whole population of staph evolved, becoming new, more powerful bacteria.

Sievert and other researchers and doctors experienced the emergence of a new allele and phenotype firsthand. When the results of the foot ulcer tests came in positive for VRSA, Sievert's team immediately called the Centers for Disease Control and Prevention (CDC), a government agency that investigates disease outbreaks and makes public health recommendations. Members of both the Michigan Health Department and the CDC converged on the dialysis center where VRSA had been found. They pored over the patient's medical history, examined her wound, took swabs from the nostrils and wounds of anyone who had come in contact with her, and then waited anxiously to see whether the new microbe had spread.

Thankfully, they found that the vancomycin-resistant bug had not infected anyone else. "We're lucky," says Sievert. "If that highly resistant organism had the ability to spread rapidly, we'd have some very sick people at risk in hospitals and other health care settings."

Yet it was not the end of VRSA. Between 2002 and 2020, there have been 13 additional cases of vancomycin-resistant staph infections in the United States: in the urine of a woman in New York with multiple sclerosis, in the toe wound of a man in Michigan with diabetes, in the triceps wound of a woman in Michigan, and more. So far, each infection has been isolated; the microbe has never been transmitted from person to person, as

MRSA has. VRSA, although dangerous, doesn't appear to spread through the human population.

But that's not to say it won't evolve to do so. When MRSA first emerged, it seemed restricted to medical facilities. Today, however, clinicians have been horrified to see cases of MRSA pop up from simple scrapes on a playground, suggesting that the microbe is out in communities where it can do widespread damage. In theory, the same is possible for VRSA, raising the bone-chilling specter of the superbug evolving into an "apocalyptic bug," as one reporter called it.

How did staph acquire vancomycin resistance 14 separate times? How likely is vancomycin resistance to become more widespread? We can find answers to these questions by understanding four mechanisms of evolution:

1. Natural selection
2. Mutation
3. Gene flow
4. Genetic drift

Bacteria are ideal organisms for examining these evolutionary mechanisms for the same reason that they are so dangerous: because they evolve incredibly fast.

Rising Resistance

Harvard Medical School microbiologist Michael Gilmore has long tracked the ways that bacteria evolve antibiotic resistance. After staph evolved widespread resistance to methicillin in the 1980s and vancomycin began to be used in hospitals, "we waited and waited and waited" for vancomycin resistance to emerge, says Gilmore.

He knew that once vancomycin was widely used by doctors to kill staph, it was extremely likely that an allele for vancomycin resistance would appear and spread through populations of staph. The process by which vancomycin-resistant alleles persist and spread in a population is called *natural selection*, as we saw in Chapter 11. Darwin and Wallace were the first to propose natural selection, and today we know that it is the central driver of evolution.

During natural selection, individuals with particular inherited characteristics survive and reproduce at a higher rate than other individuals in a population. Natural selection acts by favoring some phenotypes over others (**Figure 12.6**).

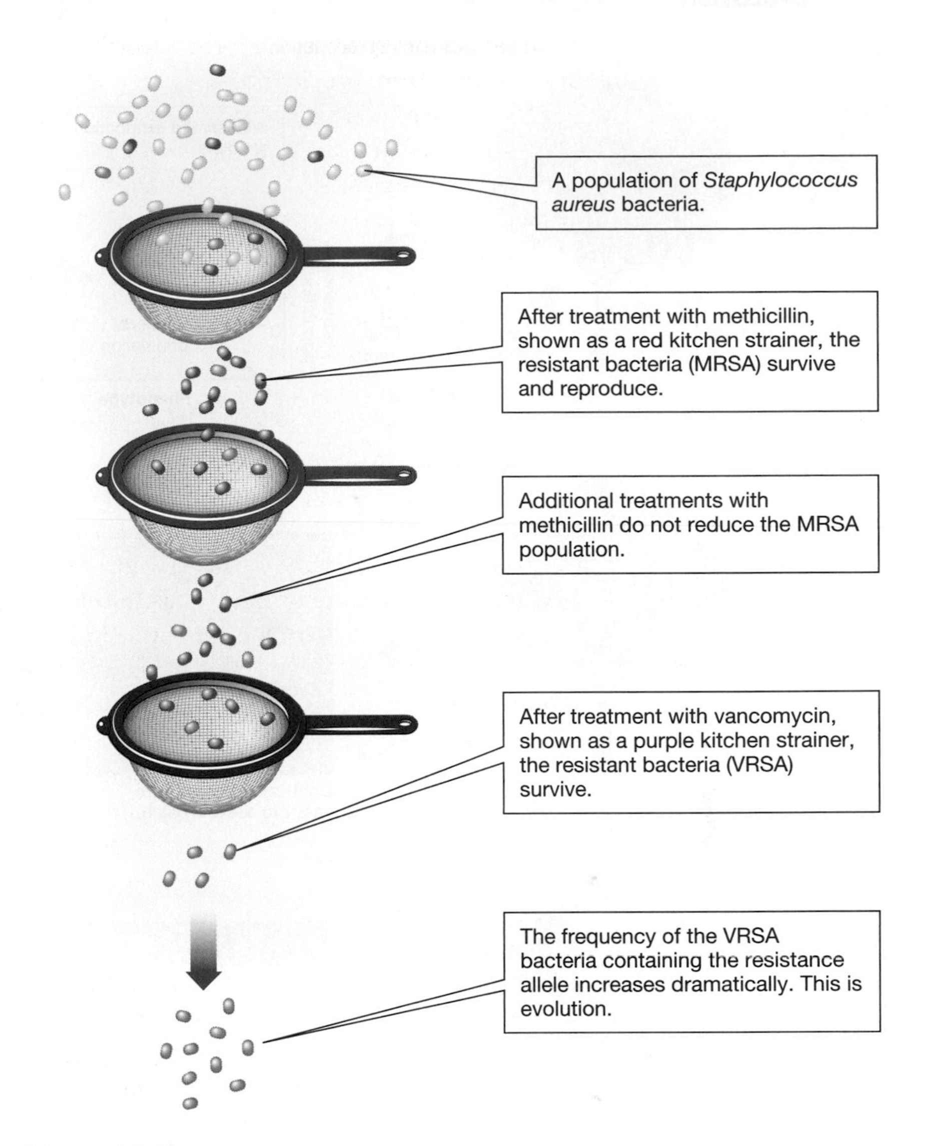

A population of *Staphylococcus aureus* bacteria.
After treatment with methicillin, shown as a red kitchen strainer, the resistant bacteria (MRSA) survive and reproduce.
Additional treatments with methicillin do not reduce the MRSA population.
After treatment with vancomycin, shown as a purple kitchen strainer, the resistant bacteria (VRSA) survive.
The frequency of the VRSA bacteria containing the resistance allele increases dramatically. This is evolution.

Figure 12.6

Evolution via natural selection in bacteria

Natural selection acts directly on the phenotype, not on the genotype, of a population. However, the alleles that code for a phenotype favored by natural selection tend to become increasingly common in future generations. Bacteria that survive an antibiotic attack, for example, pass on alleles that confer that resistance to their offspring. Here, *S. aureus* microbes survive multiple attacks from methicillin and vancomycin.

Q1: What is natural selection selecting for in this figure?

Q2: Why are the antibiotics represented by kitchen strainers in this figure?

Q3: Why do bacteria that are not resistant to antibiotics die when exposed to antibiotics?

See Appendix A for answers to the figure questions.

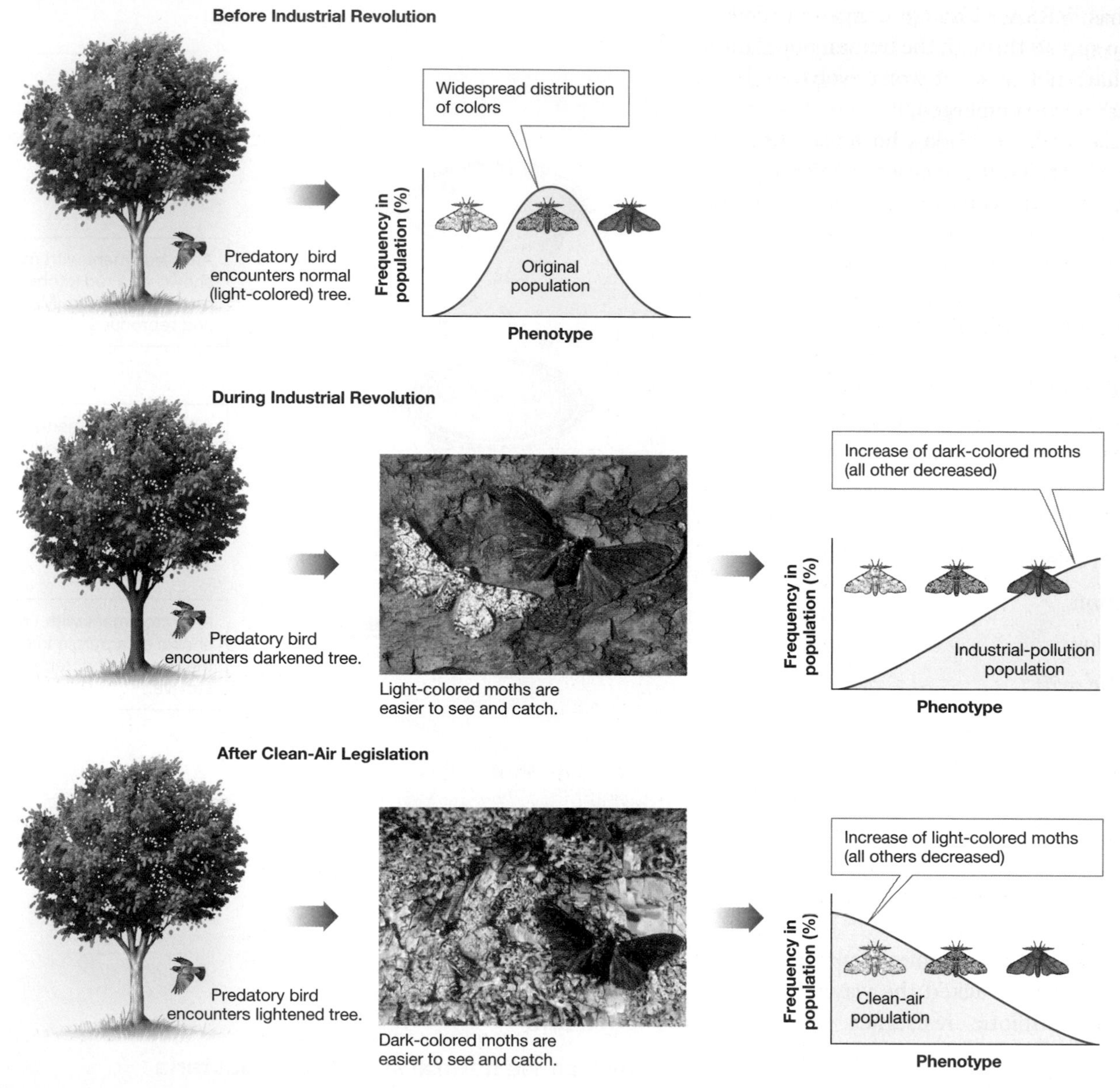

Figure 12.7

The peppered moth has undergone directional selection twice

During the industrial revolution of the nineteenth century, England and the United States experienced extreme air pollution, due mostly to soot from the mass burning of coal. Pollution blackened the bark of trees, causing dark-colored moths to be harder for predatory birds to find than light-colored moths. As a result, the proportion of dark-colored moths increased and the proportion of light-colored moths decreased. After clean-air legislation was enacted in 1956 in England and in 1963 in the United States, air pollution decreased, the bark of trees became lighter, light-colored moths became harder for predators to find than dark-colored moths, and the proportions reversed.

Q1: If one extreme phenotype makes up most of a population after directional selection, what happened to the individuals with the other phenotypes?

Q2: What do you predict would happen to the phenotypes of the peppered moth if the tree bark was significantly darkened again by disease or pollution?

Q3: What do you predict would happen to the phenotypes of the peppered moth if the tree bark became a medium color, that is, neither light nor dark?

See Appendix A for answers to the figure questions.

For example, when bacteria are exposed to vancomycin, the bacteria that can resist the antibiotic are likely to live and multiply, whereas those that cannot will perish.

Because staph's vancomycin-resistance adaptation is very recent, scientists have the opportunity to study how natural selection gives rise to VRSA. In nature, the three common patterns of natural selection are *directional selection, stabilizing selection,* and *disruptive selection.* All types of natural selection operate by the same principle: individuals with certain forms of an inherited trait have better survival rates and produce more offspring than do individuals with other forms of that trait.

Directional selection is the most common pattern of natural selection, in which individuals at *one extreme* of an inherited phenotypic trait have an advantage over other individuals in the population. The peppered moth provides a vivid example. Over the last two centuries, directional selection has caused the moth to darken in response to industrial pollution and then to lighten when that pollution decreased (**Figure 12.7**). Similarly, when methicillin became widely used to fight staph, MRSA evolved via directional selection: the bacteria that were resistant to the antibiotic survived, whereas those that were not resistant died.

In cases of **stabilizing selection**, individuals with *intermediate values* of an inherited phenotypic trait have an advantage over other individuals in the population. Birth weight in humans provides a classic example of this pattern of natural selection (**Figure 12.8**). Historically, light or heavy babies did not survive as well as babies of average weight, and as a result there was stabilizing selection for intermediate birth weights. Today, this stabilizing trend is not as strong, because advances in the care of low-birth-weight babies and an increase in the use of cesarean deliveries for large babies have allowed babies of all weights to thrive. However, even with these medical advances, babies at the extremes of newborn weight still survive at a lower rate than those closer to the median weight.

Finally, **disruptive selection** occurs when individuals with *either extreme of an inherited trait* have an advantage over individuals with an intermediate phenotype. This type of selection is the least commonly observed in nature, but one example of a trait affected by disruptive selection is the beak size within a population of birds called African seed crackers (**Figure 12.9**). During one dry season, almost all the birds starved. However, birds with small beaks were able to eat soft seeds, and birds with large beaks were able to crack and eat hard seeds, whereas birds with intermediate beaks fed inefficiently on both types of seeds. Through natural selection, some of the birds with large and small beaks lived, but all the birds with intermediate beaks died. As a result, beak sizes within the population tended to be of the extreme sizes.

Any of these patterns of natural selection can cause distantly related organisms to evolve

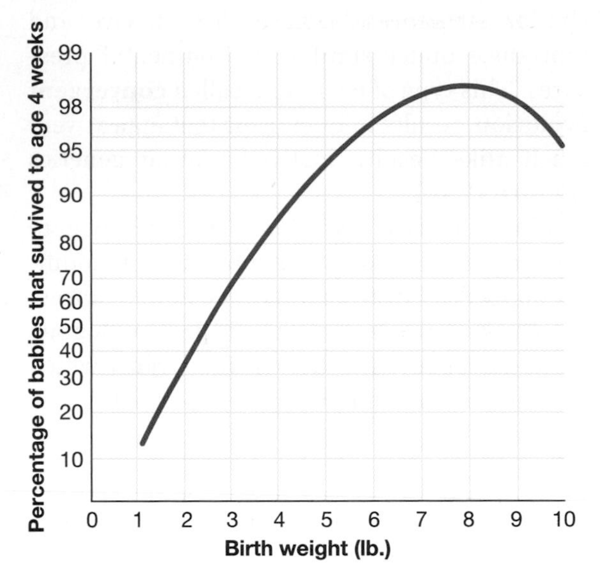

Figure 12.8

Stabilizing selection and human birth weight

This graph is based on data for 13,700 babies born between 1935 and 1946 in a hospital in London.

Q1: What does the curve in the graph indicate about the relationship between birth weight and survival?

Q2: Think of another example of stabilizing selection in human biology. Has modern technology or medicine changed its impact on the resulting phenotypes?

Q3: How do you think a graph of birth weight versus survival for a developing country with little health care would compare to the graph shown here? What about such a graph for the city of London today?

See Appendix A for answers to the figure questions.

similar structures because they survive and reproduce under similar environmental pressures. This type of evolution, called **convergent evolution**, results in organisms that appear very much alike despite vastly dissimilar genetics (**Figure 12.10**).

When species share characteristics due to convergent evolution and not because of modification by descent from a recent common ancestor, those characteristics are called **analogous traits** (as opposed to homologous traits, which are due to shared ancestry; see Chapter 11).

Enter *Enterococcus*

Intent on determining how VRSA was evolving after the first Michigan infection, Gilmore began to track the appearance of the bug. He and other researchers wanted to know where the allele for vancomycin resistance came from and how staph had acquired it 14 different times.

As Gilmore searched for patterns among the cases of VRSA infection, he noticed some peculiar things. First, most of the infections were turning up in people with diabetes, typically in bad foot wounds. Second, in most cases, when scientists looked closely at the samples, they saw not only staph but also a small spherical bacterium called *Enterococcus*, which had evolved resistance to vancomycin years earlier (**Figure 12.11**). Upon close observation, Gilmore and others noted that the enterococci, growing cozily side by side with staph, contained a vancomycin-resistance gene identical to the one in the staph. The presence of this gene suggested that staph had acquired vancomycin resistance from *Enterococcus*. So where did *Enterococcus* get it?

New alleles emerge in a species via mutation. A **mutation** is a change in the sequence of any segment of DNA in an organism, and it is the only way that new alleles are generated. DNA mutations create new alleles at random, thereby providing the raw material for evolution. In this sense, all evolutionary change depends on mutation. Mutation creates new genetic variations—differences in genotypes between individuals within a population. Then natural selection and other mechanisms of evolution act on the resulting phenotypes.

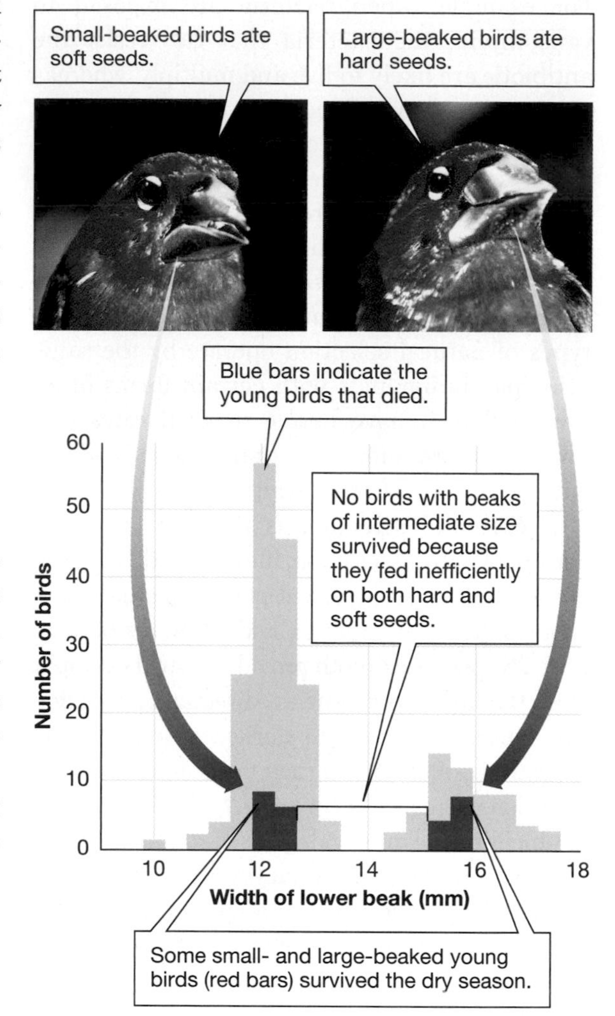

Figure 12.9

Disruptive selection for beak size

This graph shows survival rates among a group of young African seed crackers hatched in one year when seeds became scarce during the dry season.

Q1: What type of selection would have been present if only the intermediate-beaked birds had survived (instead of the small- and large-beaked birds)?

Q2: Why does disruptive selection, rather than directional selection or stabilizing selection, always result in two different phenotypes in the following generations?

Q3: Describe a scenario in which African seed crackers would experience directional selection for either smaller- or larger-beaked birds.

See Appendix A for answers to the figure questions.

Figure 12.10

Natural selection can result in convergent evolution

Plants and animals can undergo convergent evolution. (*Top*) Cacti are found in North American deserts, and euphorbias are found in African and Asian deserts. They are distantly related and have very different genetics, but they look alike and have evolved to function similarly in the desert. (*Bottom*) Sharks and dolphins are only distantly related—sharks are fish, and dolphins are mammals. Yet both evolved for success as predators in the ocean, and they share common characteristics, such as a streamlined body.

Q1: How is convergent evolution different from evolution by common descent?

Q2: What is the main difference between homologous traits and analogous traits?

Q3: Why are convergent traits considered evidence of evolution?

See Appendix A for answers to the figure questions.

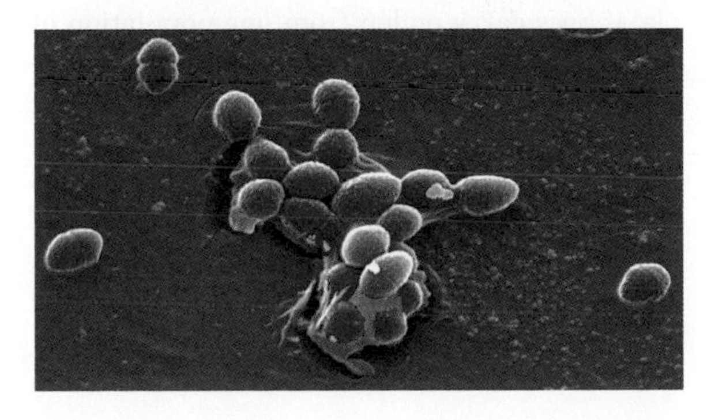

Figure 12.11

Enterococcus

Like *S. aureus*, these bacteria evolved resistance to the antibiotic vancomycin.

In species that reproduce sexually, genetic mutations in an individual's germ line cells can contribute to evolution because this cell lineage produces gametes (eggs and sperm). Mutations in other cells of the body, such as skin or blood cells, can affect the individual by causing cancer or other problems, but those mutations are not passed to the individual's offspring. If mutations are not passed to offspring, they cannot contribute to evolution.

The same does not hold true for bacteria, which are single cells and reproduce asexually. All genetic mutations in a bacterial cell are passed on to the offspring. (See Chapter 6 for a review of cellular replication.) If a mutation passed to an offspring increases the offspring's ability to reproduce, that mutation is favored by natural selection. These favored mutated genes are passed from parent to offspring, spreading through future generations in a way that alters the population as a whole.

But with bacteria, useful mutations don't always pop up randomly in the genome. Sometimes bacteria receive alleles from other bacteria. That appeared to be the case with VRSA, researchers found. "In most cases, VRSA has developed in a perfect storm," says Sievert. It emerges within "a very bad wound that's not healing and is a soup mix of organisms coming together and sharing genes."

Horizontal gene transfer is the process by which bacteria pass genes to one another (**Figure 12.12**). The genes are stored on small, circular pieces of DNA called *plasmids*. In a process called **conjugation**, the bacteria send plasmids to each other through small tubes.

Horizontal gene transfer is one example of another mechanism by which evolution occurs: gene flow. **Gene flow** is the exchange of alleles between populations. Gene flow can occur between two different species—in this case, between a population of staph and a population of *Enterococcus*—or between two populations of the same species, such as if a herd of deer carrying a mutation mixes and reproduces with another herd.

An individual may facilitate gene flow as well, such as when a fish or bird migrates between two otherwise isolated populations of a species (**Figure 12.13**). Gene flow can also occur when only gametes move from one population to another, as happens when wind or pollinators transport pollen from one population of plants to another. And although the examples discussed so far involve the one-way transfer of genes, gene flow can also occur in both directions.

Such two-way gene flow can have dramatic effects. It tends to make the genetic composition of different populations more similar. For example, if bees carry pollen from one population of flowers to another, then from the second population back to the first, the two populations may begin to look alike, even if they were initially very different. A mutual exchange of alleles through gene flow can counteract the effects of the other mechanisms, such as mutation, that tend to make populations more different from one another.

Donor bacterium
Recipient bacterium
Cell wall
Plasma membrane
Cytoplasm
Chromosomal DNA
Plasmid DNA
Conjugation tube
The donor bacterium attaches to a recipient.
The membranes of the two cells fuse to form a conjugation tube.
DNA is transferred to the recipient through the tube.

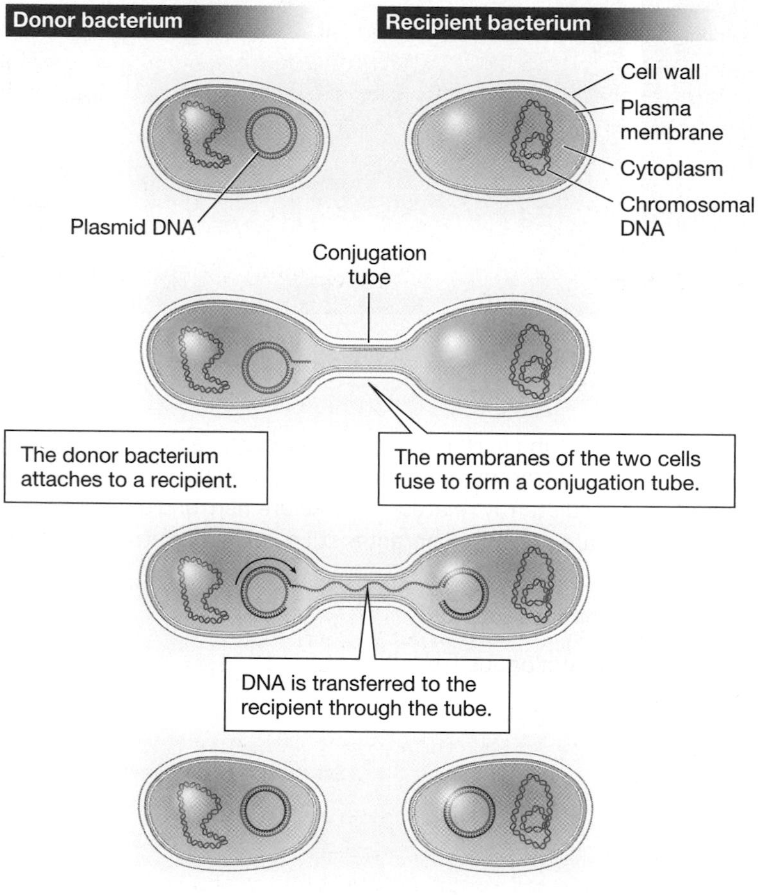

Figure 12.12

Horizontal gene transfer

Some bacteria share plasmid DNA through the physical process of conjugation. *S. aureus* acquired the vancomycin-resistance gene from *Enterococcus* through horizontal gene transfer.

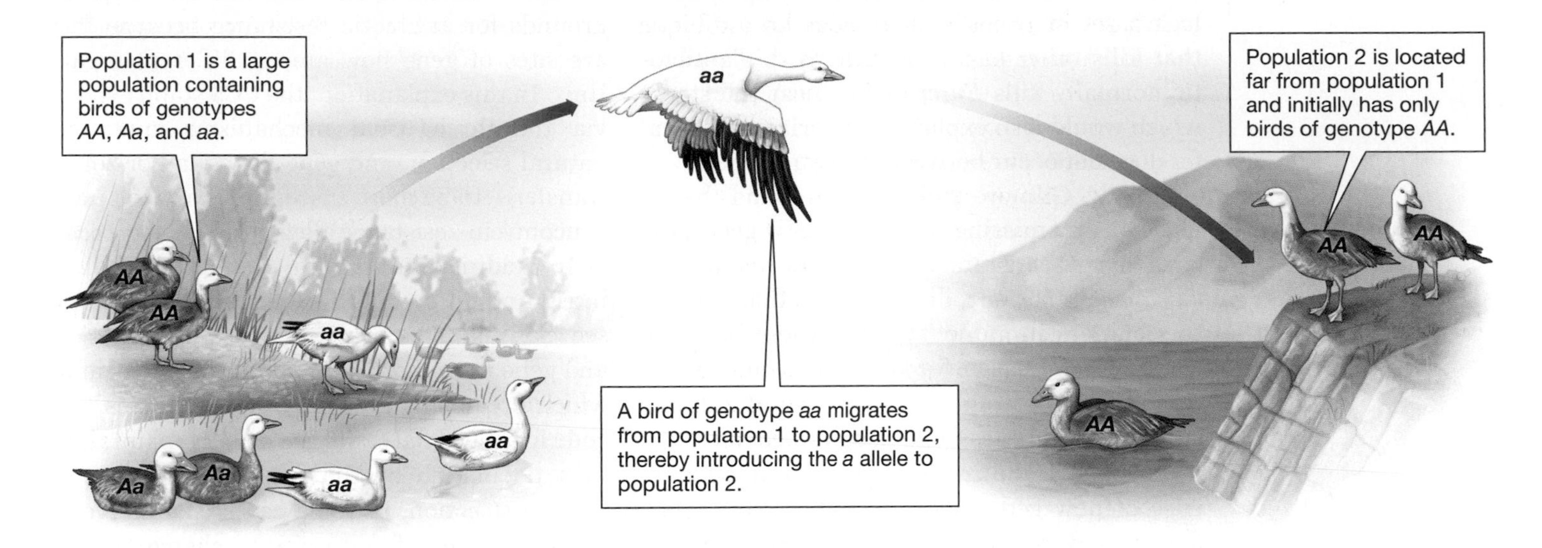

Figure 12.13

Migrants can move alleles from one population to another

As a result of migration, gene flow will occur between these two genetically disparate populations of geese. Mating between a migrant goose with genotype *aa* and a population 2 goose with genotype *AA* will result in offspring with genotype *Aa*. Continued mating between geese with genotype *Aa* will result in all three possible genotypes: *AA*, *Aa*, and *aa*. Eventually, the population 2 gene pool will look much more like the population 1 gene pool.

Q1: If a goose with genotype *AA* had migrated instead of the goose with genotype *aa*, would the scenario described here still be considered gene flow? Why or why not?

Q2: If a goose with genotype *Aa* had migrated instead of the goose with genotype *aa*, would the scenario still be considered gene flow? Why or why not?

Q3: If the goose with genotype *aa* had migrated to population 2 as shown but had failed to mate with any of the *AA* individuals, would the scenario still be considered gene flow? Why or why not?

See Appendix A for answers to the figure questions.

Gene flow appeared responsible for the emergence of VRSA; staph picked up an allele for vancomycin resistance from *Enterococcus*. But Gilmore made another observation that suggested why this horizontal gene transfer was happening now and not in the 1980s when vancomycin had first begun to be widely used.

Gilmore noted that all the VRSA samples taken from patients were clonal cluster 5 (CC5) strains of *S. aureus*. "There were implications that there was something special about clonal cluster 5 strains leading to this vancomycin resistance bubbling up," says Gilmore. Compared to other strains of staph CC5 strains appeared to have evolved the ability to easily take up and use vancomycin-resistance alleles from *Enterococcus*.

Primed for Pickup

In 2012, Gilmore and his team at Harvard analyzed the DNA of 11 of the 12 known cases of VRSA (samples from one of the 12 were not available, and the thirteenth and fourteenth cases had yet to occur). In the genome of every sample of VRSA, they found three traits that demonstrate how the CC5 strains of staph evolved to effectively pick up the allele for vancomycin resistance.

First, all the vancomycin-resistant CC5 staph bacteria have the same mutation in a gene called *DprA*. *DprA* appears to be involved in preventing horizontal gene transfer. A mutation in this gene, by enabling horizontal gene transfer, might make it easier for the staph to take up DNA from other bacteria, such as *Enterococcus*.

Second, the vancomycin-resistant CC5 strains lack a set of genes that encodes an antibiotic that kills other bacteria. Perhaps this antibiotic normally kills *Enterococcus* near the staph, which would also explain why horizontal transfer does not occur between the two species.

Finally, Gilmore and his team found that in place of that missing set of antibiotic genes, the vancomycin-resistant CC5 staph have a unique cluster of genes encoding proteins that confuse the human immune system. These proteins could make it easier for staph and other bacteria to grow in a wound because the host immune system would be less able to fight them off.

The lack of the antibiotic genes and the presence of new genes create the ideal conditions for a mixed infection, where different species of infectious agents mingle in a festering soup of contamination. Mixed infections are breeding grounds for antibiotic resistance because they are sites of gene flow among different organisms. In this explanation, the CC5 staph evolved via the three usual mechanisms—mutation, natural selection, and gene flow (via horizontal transfer)—to be more susceptible to take up the vancomycin-resistance allele from *Enterococcus*.

In addition to natural selection (and its intriguing permutation called *sexual selection*; see "Sex and Selection" on page 236), mutation, and gene flow, there is one other mechanism by which organisms evolve. Although alleles that code for beneficial traits are usually maintained in a population by the nonrandom action of natural selection, in some cases chance events can cause allele frequencies to change from a parent generation to the next generation.

Sex and Selection

Staphylococcus and other bacteria replicate asexually by copying their DNA and dividing in two. Other organisms reproduce sexually, and sex complicates evolution.

Another mechanism by which species evolve is called sexual selection. In **sexual selection**, nature selects a trait that increases an individual's chance of mating—even if that trait actually decreases the individual's chance of survival.

Sexual selection favors individuals that are good at getting mates, and it often helps explain differences between males and females in size, courtship behavior, and other traits. Species whose males and females are distinctly different in appearance—such as peacocks, lions, and ducks—are said to exhibit **sexual dimorphism**. In many species, the members of one sex, often females, are choosy about whether to mate. In birds, for example, brightly colored males may perform elaborate displays in their attempts to woo a mate. In other species, males may attract attention by calling vigorously. Females then select as their mates the males with the most impressive displays or the loudest calls.

Yet some characteristics that increase an individual's chance of mating can *decrease* its chance of survival. For example, male túngara frogs perform a complex mating call that may or may not end in one or more "chuck" sounds. Females prefer to mate with males that emit chucks. However, frog-eating bats hear that same sound and use it to help them locate their prey. As a result, in an evolutionary trade-off, a male frog's attempt to court a mate can end in his death.

A male peacock performs an elaborate mating display to a potential mate.

A male túngara frog has emitted the call to attract a female and has instead attracted a predatory bat.

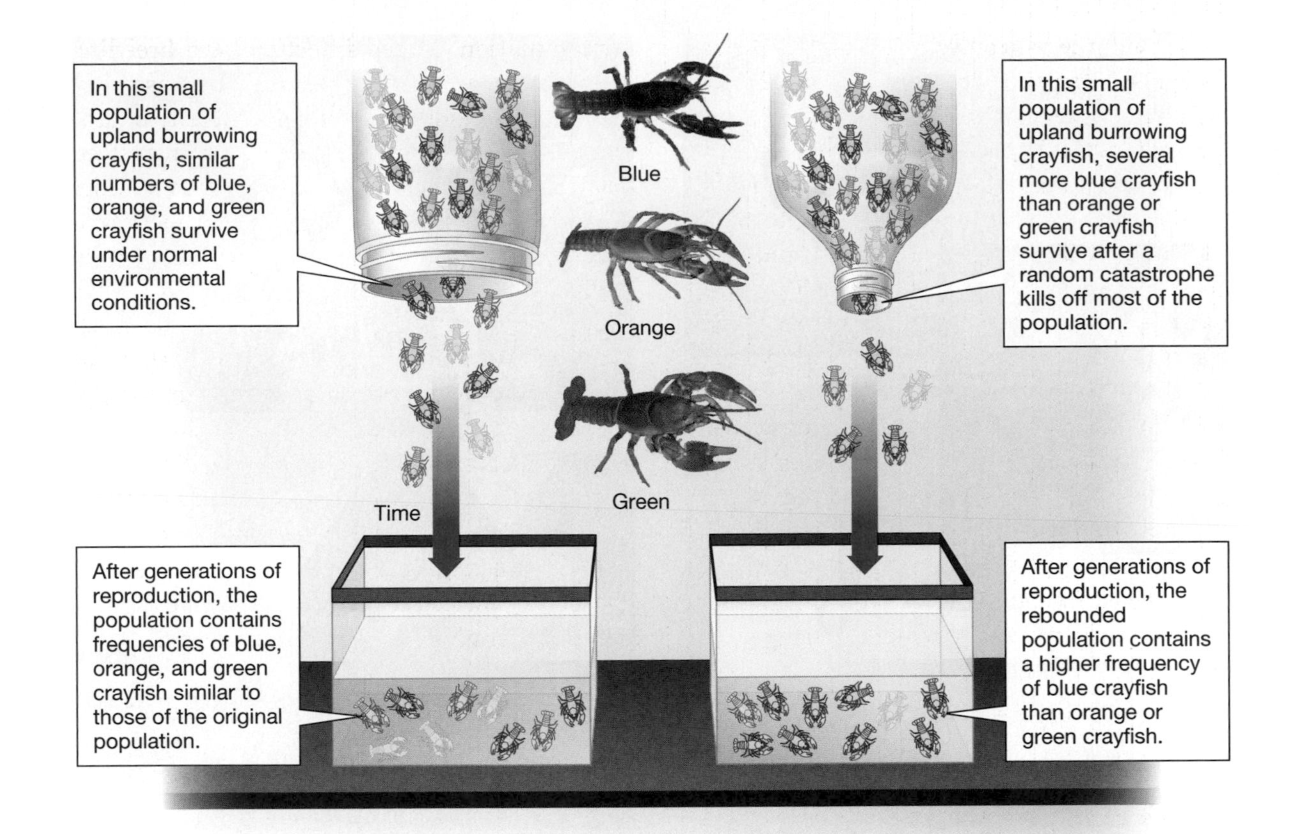

In this small population of upland burrowing crayfish, similar numbers of blue, orange, and green crayfish survive under normal environmental conditions.
Blue
Orange
Green
Time
In this small population of upland burrowing crayfish, several more blue crayfish than orange or green crayfish survive after a random catastrophe kills off most of the population.
After generations of reproduction, the population contains frequencies of blue, orange, and green crayfish similar to those of the original population.
After generations of reproduction, the rebounded population contains a higher frequency of blue crayfish than orange or green crayfish.

Figure 12.14

A genetic bottleneck is a type of genetic drift

On the left and right are two equivalent populations. The one on the right experiences a genetic bottleneck—a change in the frequency of a trait that is not associated with natural selection. In fact, although the blue crayfish on the right benefit from this type of genetic drift, they could be less well adapted to the environment than the other crayfish are.

Q1: Why do you think a genetic bottleneck is more likely to occur in a small population than in a large population?

Q2: List two chance events that could cause a genetic bottleneck.

Q3: Which resulting population has more genetic diversity? Why?

See Appendix A for answers to the figure questions.

Genetic drift is a change in allele frequencies caused by random differences in survival and reproduction among the individuals in a population. Just by chance, some individuals in a generation may leave behind more descendants than other individuals do. In this case, the genes of the next generation will be the genes of the "lucky" individuals, not necessarily the "better" individuals. Genetic drift occurs in all populations, including mammals and bacteria, but it is more likely to cause evolution in a small population than in a large one.

Strong genetic drift may occur through either a genetic bottleneck or the founder effect. A **genetic bottleneck** is a drop in the size of a population, for at least one generation, that causes a loss of genetic variation (**Figure 12.14**). A genetic bottleneck can threaten the survival of a population.

In the 1970s, for example, the population of the endangered Florida panther plummeted

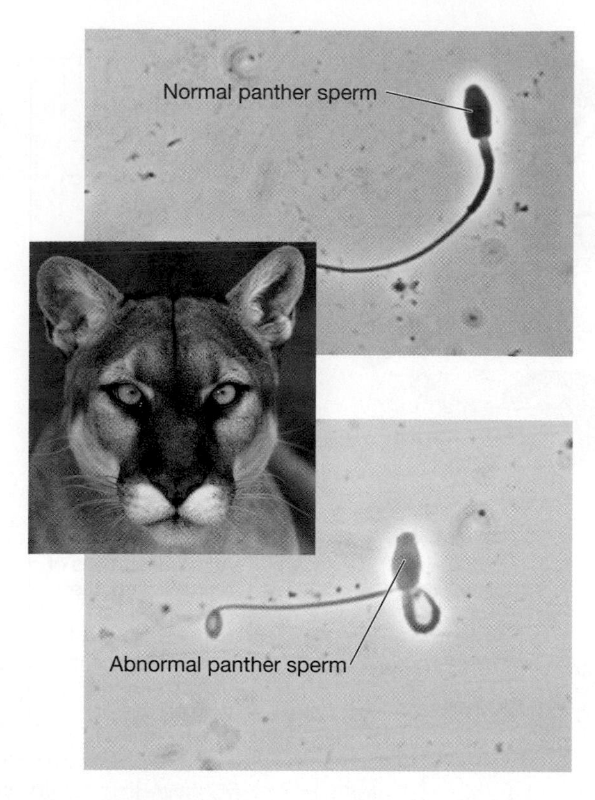

Figure 12.15

Probable effect of a genetic bottleneck
Florida panthers have more abnormally shaped sperm than do panthers from other populations—most likely because a genetic bottleneck promoted the increase of harmful alleles in the Florida panther population.

Figure 12.16

The founder effect
Early Dutch colonists in South Africa (top) had a high incidence of Huntington disease, which they passed on to their Afrikaner descendants (bottom).

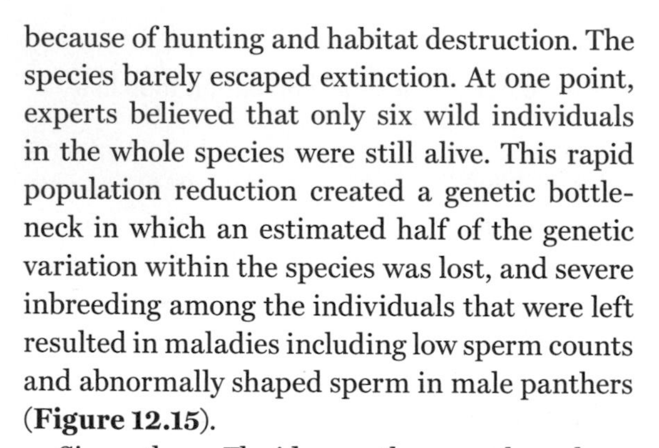

because of hunting and habitat destruction. The species barely escaped extinction. At one point, experts believed that only six wild individuals in the whole species were still alive. This rapid population reduction created a genetic bottleneck in which an estimated half of the genetic variation within the species was lost, and severe inbreeding among the individuals that were left resulted in maladies including low sperm counts and abnormally shaped sperm in male panthers (**Figure 12.15**).

Since then, Florida panther numbers have increased to an estimated number of 120–230 individuals, in part because of captive breeding programs. In an effort to restore genetic diversity to the population, in 1995 researchers released eight female Texas cougars, close genetic cousins to the Florida panthers, into the panthers' habitat. Five of those females produced eight litters of kittens—more genetically diverse, stronger, and longer-lived than their fathers.

The **founder effect** occurs when a small group of individuals establishes a new population isolated from its original, larger population. For example, in the 1600s a few Dutch colonists settled in South Africa (**Figure 12.16**). Today, their descendants are known as Afrikaners. The Afrikaner population has an unusually high frequency of the allele that causes Huntington disease because the original colonists by chance carried that allele.

After Vancomycin

When VRSA first appeared in 2002, the patient's foot was treated with an old topical drug that is rarely used today. The only other options were one or two new drugs that hadn't yet been released on the market. Today, "there are still some antibiotics we can use for treatment, but we recommend those be used very responsibly in order to maintain their effectiveness," says Sievert. "You don't want to give them out when not necessary because you don't want to see resistance to them. You save them as the end of the line, so something is available if the commonly used drugs end up failing because of resistance."

After 9 months of treatment, the patient's wound healed. Other cases of VRSA have been cured similarly but with the newer antibiotics.

But what happens when even these drugs no longer work against staph? Pharmaceutical companies are developing few new antibiotics, says Frank DeLeo, chief of the Laboratory of Bacteriology at the National Institute of Allergy and Infectious Diseases, and bacteria evolve more quickly than we can develop new antibiotics. It's a vicious cycle, he says. "As we continue to use antibiotics, populations of microbes will develop resistance to antibiotics."

Recently, microbes have appeared to be ahead in this race. Since 2015, researchers have been tracking the spread of a superbug resistant to colistin, an antibiotic of last resort, around the globe. They've identified nine alleles that confer colistin resistance popping up in a variety of microbes, including *E. coli* and *Salmonella*. Several of these alleles have appeared in the United States, including the rare *mcr-3*, announced in June 2019.

Thankfully, VRSA has not yet become as widespread as MRSA, though it has also appeared in isolated incidents in other countries, including Japan, Bangladesh, and Iran. Scientists believe its spread may be limited because staph does not handle the resistance allele for vancomycin very well; the gene from *Enterococcus* is bulky and difficult to manage, so the staph individuals that receive the gene are often outcompeted by other staph individuals in the environment. "Though it allows them to survive, they don't do well with it. There's a [survival] cost to having this resistance," says Gilmore. VRSA therefore remains "a rare, unstable organism in the environment that isn't very hardy, so it doesn't last and spread," adds Sievert. "That's the good news."

But there's still trepidation that staph may adapt and get comfortable with this resistance gene, just as staph adapted to methicillin resistance, says Gilmore. If staph survives and reproduces more easily with the resistance gene, the gene will likely spread to staph throughout the host/human population. "We don't know if that's a possibility," says Gilmore. "My suspicion is that it is." (See "How Can You Make a Difference? Help Prevent Antibiotic Resistance!" on page 239.)

In fact, the thirteenth VRSA case, in Delaware in July 2012, was not a CC5 strain. In other words, vancomycin resistance may be spreading to other strains of staph. "It's disconcerting," says Gilmore. "The main worry is that this would move into a strain that was highly transmissible in the community," adds DeLeo. "The potential is there."

How Can You Make a Difference? Help Prevent Antibiotic Resistance!

The U.S. Centers for Disease Control and Prevention (CDC) recommends the following guidelines:

- Take antibiotics exactly as the doctor prescribes. Do not skip doses. Complete the prescribed course of treatment, even when you start feeling better.
- Only take antibiotics prescribed for you; do not share or use leftover antibiotics. Antibiotics treat specific types of infections. Taking the wrong medicine may delay treatment and allow bacteria to multiply.
- Do not save antibiotics for the next illness. Discard any leftover medication once the prescribed course of treatment is completed.
- Do not ask for antibiotics when your doctor thinks you do not need them. Remember that antibiotics have side effects. When your doctor says you don't need an antibiotic, taking one might do more harm than good.
- Prevent infections by practicing good hand hygiene, covering your mouth and nose when you sneeze (preferably by sneezing into your elbow, not on your hands), and getting recommended vaccines.

SOURCE: https://www.cdc.gov/antibiotic-use/community/about/can-do.html.

Race Against Resistance

Methicillin-resistant *Staphylococcus aureus* (MRSA) is one of many microbial threats. Each year in the United States, more than 2 million people become infected with various types of antibiotic- and antifungal-resistant bacteria, resulting in at least 23,000 deaths. To combat antibiotic and antifungal resistance, doctors need novel antibiotics and antifungals, yet fewer and fewer new antibiotics and antifungals were coming to pharmacies each year, and in some cases, such as fluconazole, resistance was documented even before a drug was approved for widespread use. In 2011, Congress passed the Generating Antibiotic Incentives Now (GAIN) Act to stimulate the development of new antibiotics and antifungals, and it's working. So far, 12 new antibiotic and antifungal drugs have been approved through that program.

Assessment available in smartwork

Antibacterial drugs approved by the FDA

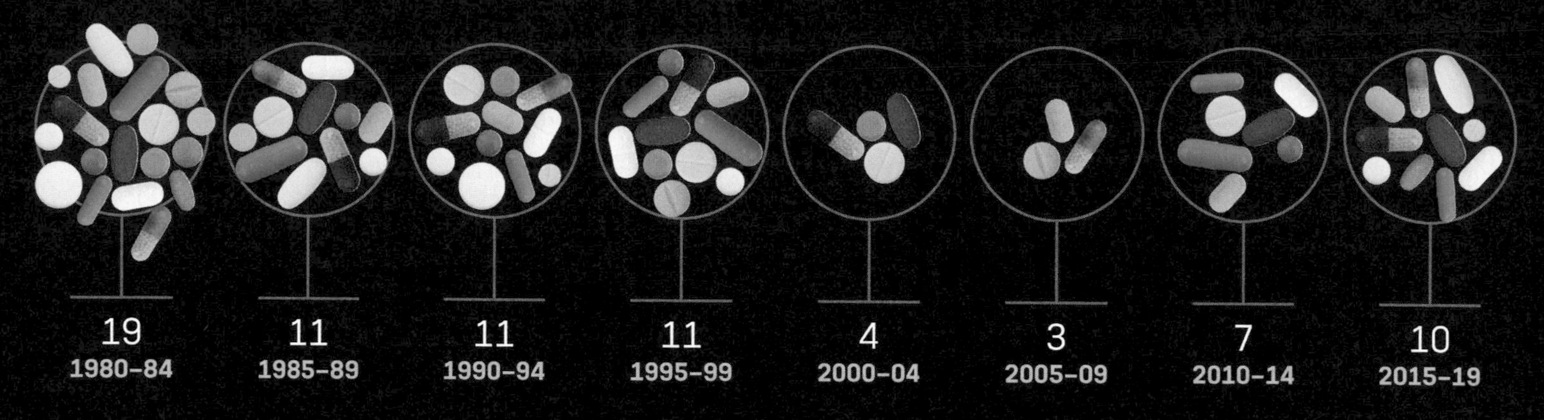

Timeline of microbe resistance

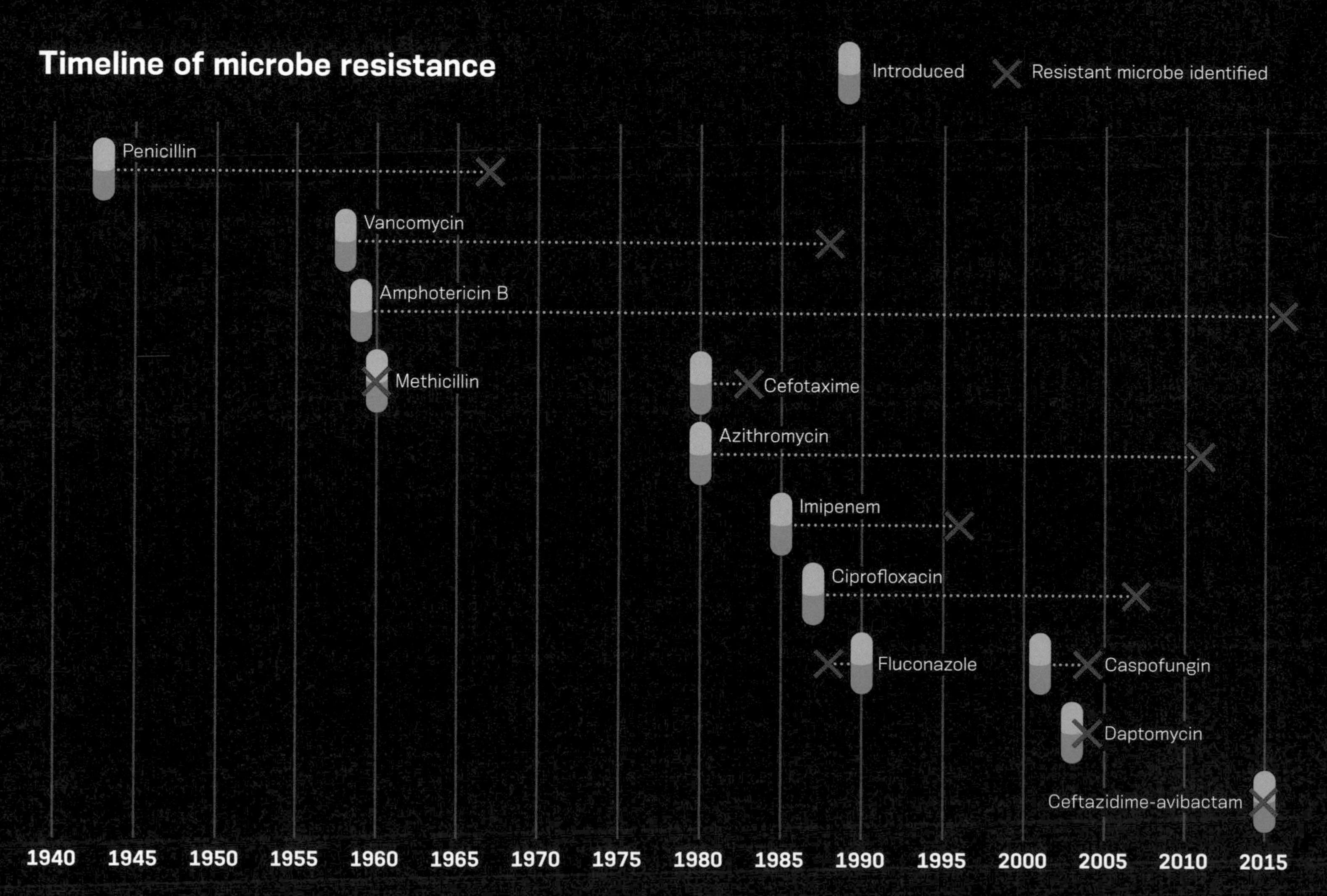

NEED-TO-KNOW SCIENCE

- Natural selection for inherited traits occurs in three common patterns:
    - In **directional selection** (page 231), individuals at one phenotypic extreme of a given genetic trait have an advantage over all others in the population.
    - In **stabilizing selection** (page 231), individuals with intermediate phenotypes have an advantage over all others in the population.
    - During **disruptive selection** (page 231), individuals with either extreme phenotype have an advantage over those with an intermediate phenotype.
- In **convergent evolution** (page 232), distantly related organisms (those without a recent common ancestor) evolve similar structures in response to similar environmental challenges.
- When species share characteristics due to convergent evolution and not because of modification by descent from a recent common ancestor, those shared characteristics are called **analogous traits** (as opposed to homologous traits, which are due to shared ancestry; page 232).
- All mechanisms of evolution depend on the genetic variation provided by **mutation** (page 232), a change in the sequence of any segment of DNA in an organism.
- **Horizontal gene transfer** (page 234) is the process by which some bacteria pass genes to one another.
- **Sexual selection** (page 236) occurs when a trait increases an individual's chance of mating even if it *decreases* that individual's chance of survival.
- **Gene flow** (page 234) is the exchange of alleles between populations.
- **Genetic drift** (page 237) is a change in allele frequencies produced by random differences in survival and reproduction in a small population. It occurs most dramatically in one of two ways:
    - A **genetic bottleneck** (page 237) occurs when a drop in the size of a population leads to a loss of genetic variation in the new, rebounded population.
    - The **founder effect** (page 238) occurs when a small group of individuals establishes a population isolated from its original, larger population, leading to a loss of genetic variation in the new population.

THE QUESTIONS

See Appendix B for answers.

The Basics

1 The founder effect is a type of (**genetic drift / gene flow**) in which a small group of individuals from a larger population (**establish a new, distant population / are the only survivors**) and then reproduce.

2 Unlike natural selection, _________ is not related to an individual's ability to survive and may result in offspring that are less well adapted to survive in a particular environment.

(a) sexual selection

(b) convergent evolution

(c) directional selection

(d) genetic drift

3 Which of the following statements about convergent evolution is true?

(a) It demonstrates how similar environments can lead to different physical structures.

(b) It demonstrates how similar environments can lead to the same physical structures.

(c) It demonstrates that similarity of structures is due to descent from a common ancestor.

(d) It demonstrates that similarity of structures is due to random chance.

4 Biological evolution is a change in allele frequencies in _________ over time.

(a) an individual

(b) a species

(c) a population

(d) a community

Challenge Yourself

5 In a population, which individuals are most likely to survive and reproduce?

(a) The individuals that are the most different from the others in the population.

(b) The individuals that are best adapted to the environment.

(c) The largest individuals in the population.

(d) The individuals that can catch the most prey.

6 Link each term with the correct definition.

genetic drift	1. Individuals with intermediate phenotypes survive and reproduce at higher rates than those with extreme phenotypes.
gene flow	2. Individuals with an extreme phenotype survive and reproduce at higher rates than those with intermediate phenotypes.
disruptive selection	3. Exchange of alleles between populations.
directional selection	4. Random changes in allele frequencies in a population.
stabilizing selection	5. Individuals at one extreme of a phenotype survive and reproduce at higher rates than other individuals in the population.

7 Explain how, because of sexual selection, an individual might be very successful at surviving but not pass on genes to the next generation.

Try Something New

8 Two large populations of the same species found in neighboring locations that have very different environments are observed to become genetically more similar over time. Which of the four main evolutionary mechanisms is the most likely cause of this trend? Justify your answer.

9 The Tasmanian devil, a marsupial indigenous to the island of Tasmania (and formerly mainland Australia as well), experienced a population bottleneck in the late 1800s, when farmers did their best to eradicate it. After it became a protected species, the population rebounded, but it is now experiencing a health crisis putting it at risk for disappearing again. Many current Tasmanian devil populations are plagued by a type of cancer called devil facial tumor disease, which occurs inside individual animals' mouths. Afflicted Tasmanian devils can actually pass their cancer cells from one animal to another during mating rituals that include vicious biting around the mouth.

Unlike the immune systems of other species, including humans, the Tasmanian devil's immune system does not reject the cells passed as foreign or nonself (as we reject a liver transplant from an unmatched donor) but accepts them as if they were their own cells. Why would a population bottleneck result in the inability of one devil's immune system to recognize another devil's cells as foreign?

10 A population of house mice, *Mus musculus*, lives in a garden shed. In a 10-year period, this population has experienced stabilizing selection. Which of the following scenarios might have occurred?

(a) Small mice could not reach the seed shelf, and large mice were easily seen by hawks circling above. Medium-sized mice therefore survived and reproduced better than both small and large mice.

(b) Small and medium-sized mice could not reach the seed shelf in the shed and therefore were at a disadvantage for finding food, so they did not survive and reproduce as well as large mice did.

(c) Small mice could easily cross the yard to the vegetable garden, and large mice could easily reach the seed shelf. Medium-sized mice had trouble with the seed shelf and were seen by hawks in the yard. Small and large mice survived and reproduced much better than medium-sized mice did.

(d) All of these are examples of stabilizing selection.

(e) None of these are examples of stabilizing selection.

11 The accompanying graph shows the five most common antibiotic-resistant strains of bacteria in Switzerland from 2004 to 2018.

(a) What was the most common antibiotic-resistant strain of bacteria in 2018? Was it always the most common?

(b) What percentage is MRSA, the strain we discussed in this chapter? Has it become more or less common over time?

(c) Why do you suppose VRSA isn't on this graph?

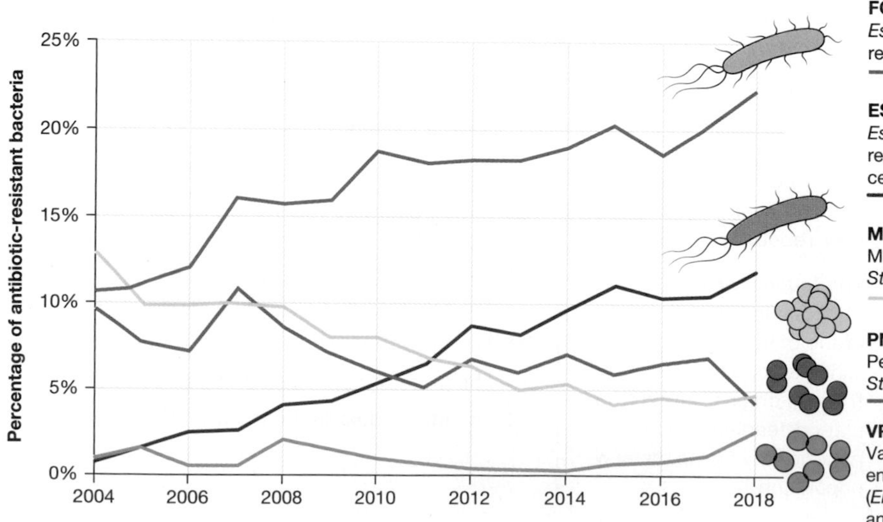

FQR-E. coli
Escherichia coli resistant to fluoroquinolones

ESCR-E. coli
Escherichia coli resistant to broad spectrum cephalosporins

MRSA
Methicillin-resistant *Staphylococcus aureus*

PNSP
Penicillin-resistant *Streptococcus pneumoniae*

VRE
Vancomycin-resistant enterococci (*Enterococcus faecalis* and *Enterococcus faecium*)

Leveling Up

12 **What do *you* think?** One way to prevent a small population of a plant or animal species from going extinct is to deliberately introduce some individuals from a large population of the same species into the smaller population. In terms of the evolutionary mechanisms discussed in this chapter, what are the potential benefits and drawbacks of transferring individuals from one population to another? Do you think biologists and concerned citizens should take such actions?

13 ***Write Now* biology: viral evolution** During the early months of 2020, COVID-19, the novel coronavirus, spread rapidly around the world. Read the review article at https://www.genengnews.com/news/coronavirus-evolved-naturally-and-is-not-a-laboratory-construct-genetic-study-shows/ about a study that convinced scientists that this coronavirus evolved naturally rather than being artificially created as a biological weapon, or bioweapon. Answer the following questions in about a paragraph for each question, using the article as well as other sources of information.

(a) What is a bioweapon? Give examples of bioweapons that have been used in the past. Do the United States and China (both accused of creating COVID-19 as a bioweapon) currently have bioweapons?

(b) COVID-19 is not the only coronavirus. How many others have been identified? Do they all cause severe illness?

(c) The authors of the study discuss two features of the virus that they feel demonstrate conclusively that it was not created in a laboratory. In your own words, what are those two features, and what is the argument for them evolving naturally rather than being artificially created?

(d) What is your informed opinion about the source of the coronavirus? Did it evolve naturally, or was it artificially created? Did it come directly from bats or through an intermediary host, such as a pangolin? Explain your reasoning.

Digital resources for your book are available online.

Penguins on Thin Ice

With bits of bone and bytes of data, scientists track the birth and loss of penguin species.

After reading this chapter you should be able to:

- Describe the relationships among adaptive traits, biological fitness, and evolution by natural selection.
- Explain the usefulness of the biological species concept and discuss its shortcomings.
- Define speciation and the role of genetic divergence among populations in creating new species.
- Differentiate between allopatric and sympatric speciation.
- Identify two kinds of reproductive isolating mechanisms and explain their role in speciation.
- Define and give an example of coevolution.

EVOLUTION

CHAPTER 13

ADAPTATION AND SPECIATION

On cool autumn days, tall Emperor penguins in Antarctica make their way across the frozen sea to the foot of the Brunt Ice Shelf, a floating sheet of ice that can grow as thick as a skyscraper is tall. After months of feasting at sea, these tens of thousands of penguins trek toward their frozen breeding ground to mate (**Figure 13.1**).

Two months after arriving, a female penguin lays an egg and carefully transfers it to her male partner, who will incubate it in a small pouch above his feet for three bitterly cold winter months while the mother returns to the sea to hunt. When she rejoins her family, both parents will take equal care of their fuzzy new chick as it grows into an adult (**Figure 13.2**).

At least, that's what supposed to happen. In 2016, as spring began and little gray chicks energetically emerged, Peter Fretwell and Philip Trathan watched with horror as disaster struck. As scientists with the British Antarctic Survey (BAS), a United Kingdom research organization that conducts scientific research in the Antarctic region, Fretwell and Trathan were using high-resolution satellite images to follow the movements of the Halley Bay colony, the second largest Emperor penguin colony in the world.

While reviewing the images from October 2016 (springtime in the southern hemisphere), the scientists expected to see thousands of breeding pairs and their chicks dotting the ice. Instead, all the penguins were gone.

Figure 13.2

A successful pair (so far) and their offspring of the year

The female rejoins her mate and young chick after she has restored her fat reserves that had been depleted from laying a very large egg. She is now ready to begin coparenting. *Photo by Christopher Michel.*

Figure 13.1

Emperor penguins advancing on the Brunt Ice Shelf

Well fed from months of feasting on ocean fish and invertebrates, Emperor penguins head to their breeding grounds to begin the challenging work of hatching and raising just one chick a year per couple.

"There were no penguins. Just open water," recalls Fretwell. Trathan—head of conservation biology at the BAS, who has studied penguins for decades and participated in 20 research trips to the Antarctic—couldn't believe that such a large, established colony could simply disappear. "We were pretty shocked," he recalls.

A storm had hit the Brunt Ice Shelf, breaking up the sea ice months early. Although adult penguins were protected from the cold water by their feathers and were strong enough to swim to land, the chicks were not. The ice broke up months before chicks would grow a full set of adult feathers to protect them. "We assume, then, that all the chicks in the Halley Bay colony that year died," says Fretwell.

The early sea ice breakup seemed like a crazy, chance catastrophe.

Then, it happened again.

And again.

In 2017 and 2018, the Halley Bay colony again experienced what Trathan and Fretwell would later call "catastrophic breeding failure"—a total loss of chicks each year. Although it is difficult to directly attribute the early sea ice breakup to global warming, says Fretwell, the Intergovernmental Panel on Climate Change, a United Nations organization, has predicted significant loss of Antarctic sea ice in the coming decades. In July 2019, NASA scientists announced that sea ice around Antarctica was at its lowest levels on record in 40 years of satellite monitoring.

Climate change is altering the quality and availability of food and nesting habitats for penguins around the world. On an island in the southern Indian Ocean, for example, a colony of 500,000 breeding pairs of King penguins, second largest in size after Emperor penguins, has lost 90 percent of its population since the 1980s (**Figure 13.3**). In parts of Antarctica, colonies of sleek, small Adélie penguins are declining in size where the sea surface is warming (**Figure 13.4**).

Research studies project that penguin numbers will fall dramatically this century as a result of climate change, despite penguins' unique adaptations to their environments. There are different species of penguin that live in habitats as varied as the tropics of the Galápagos Islands and the frigid ice shelves of Antarctica. In a warming world, Trathan says, understanding how penguins react and adapt to their environments will be critical to predicting the fate of these species.

Figure 13.4

Adélie penguins are small cousins of Emperor penguins

Emperor penguins (left) and Adélie penguins (right) in nature share the same habitat and breeding grounds.

Wonderfully Weird

While other birds soar through the sky and build snug nests in treetops, Emperor penguins dive deep into the ocean and breed on blocks of ice. How did these birds become so different from other feathered fowl around the world?

Figure 13.3

A King penguin and chick

King penguin chicks look larger than their parents before they molt their baby feathers and take on the sleek plumage of an adult.

Like all birds, penguins lay eggs, have wings, and are covered in feathers. Yet Emperor penguins are uniquely different from pigeons, robins, and other birds because Emperor penguins evolved in an extreme environment, and the *adaptive traits* that enabled them to survive and reproduce are unique to their environment.

Survival and successful reproduction are together referred to as an individual's **biological fitness**. If a trait helps an individual survive but interferes with its ability to reproduce, the individual would have low biological fitness. Similarly, if a trait makes an individual more successful at reproducing but that animal does not survive to reproductive age, the individual would have low biological fitness. Adaptive traits lead to higher biological fitness.

Adaptive traits can be structural features, biochemical traits, or behaviors. Penguins' feathers, for example, are a structural adaptive trait. Emperor penguins' quill-like contour feathers cover two types of insulating feathers in a complex network of protection against the harsh Antarctic climate. In addition, penguins' skin is thicker than that of other birds, and, like whales, penguins have an insulating layer of blubber as an added barrier against the cold (**Figure 13.5**).

Figure 13.5

Penguins have many adaptations to life as aquatic birds

The ancestors of modern penguins looked like albatrosses, which are adapted to life on and near the ocean. Penguins took things a step further by adapting to life *in* the ocean.

Q1: How might a salt gland serve as an adaptation for an aquatic bird such as a penguin?

Q2: Explain how the adaptive trait of muscles that store large amounts of oxygen increases the biological fitness of penguins.

Q3: Why would heavy bones be an adaptation for penguins but *not* an adaptation for most bird species?

See Appendix A for answers to the figure questions.

Biochemically, male Emperor penguins make "milk" in their esophagus that can be used to feed chicks until the mother penguins return from fishing. Only pigeons and flamingoes share this ability.

Penguins have useful adaptive behaviors too. Most famous, perhaps, is how Emperor penguins huddle into groups to protect themselves from cold, intense winds with speeds exceeding 100 miles per hour. The penguins swap positions inside and outside the huddle constantly, preventing heat loss of the overall community by an estimated 50 percent.

The term *adaptation* is commonly applied to adaptive traits or the process of evolution through natural selection that brings about adaptive traits, as discussed in Chapter 12. Therefore, an adaptation can be a trait that is advantageous to an individual or a population or, in broader terms, it can be the evolutionary process of natural selection that enables a good match between a population of organisms and its environment.

Yet natural selection does not always result in a perfect match between an organism and its

DEBUNKED!

MYTH: Penguins mate for life.

FACT: Contrary to popular belief, penguins are not monogamous. An estimated 85 percent of Emperor penguins, for example, find a new breeding partner each year.

environment. In many cases, organisms fail to adapt successfully. In fact, scientists estimate that 99 percent of all species that have ever lived are now extinct. Every extinct species is a silent testament to a failure to adapt in the face of adversity.

When evolution by natural selection does successfully enhance the adaptive fit of organisms to their environment and, thus, their biological fitness, it can occur over long periods of time, as with the evolution of whales from land-living mammals over 50 million years as described in Chapter 11. Evolution by natural selection can also occur over surprisingly short periods of time, such as the rise of antibiotic-resistant bacteria discussed in Chapter 12.

The speed at which a species adapts to an environment depends, among other things, on the species' generation time—that is, the average difference in age between a parent and offspring or the amount of time it takes to produce a new generation. Human generation time is roughly considered to be about 30 years. Emperor penguins have a generation time of 4–6 years, which is faster than humans but still slow compared to something like a fruit fly, with a generation time of 10–12 days, or the bacterium *E. coli*, with a generation time of just 20 minutes.

A slow generation time means that any population-wide adaptive change through evolution by natural selection will also occur slowly, which is not great news for penguins that are facing quickly warming temperatures. "If climate change is happening more rapidly than they've experienced throughout their history, it might be difficult for them to keep pace," says Trathan.

Figure 13.6

Penguins are most closely related to large seabirds like the albatross

Albatross share many traits with penguins, such as salt glands and water-resistant feathers. However, they do not have the adaptations for diving that penguins have.

What Makes a Species?

At the University of Otago in New Zealand, graduate student Theresa Cole began studying penguins because "they're just so weird," she recalls. "They've developed all these amazing adaptations to be able to exploit the marine environment as well as being tied to land."

The amazing diversity of penguins—from the regal Emperor penguin to the forest-dwelling, bushy-browed Fiordland penguin to the squat, tropical Galápagos penguin—got Cole wondering: Where did new penguin species come from, and why were they such an eclectic bunch?

Fossil records trace penguin history through more than 50 species across 60 million years, and the closest known living relative to a penguin is the albatross (**Figure 13.6**). Many people think of penguins in the wild only in Antarctica, says Cole, but most of the estimated 18 living species of penguins live elsewhere.

The term **species** is most commonly used to refer to members of a group that can mate with one another to produce fertile offspring. One way to define a species is using the **biological species concept**, which says a species is a group of natural populations that can interbreed to produce fertile offspring and cannot breed with other such groups; that is, they are **reproductively isolated** from other populations (**Figure 13.7**).

According to the biological species concept, if individuals from the two populations of penguins can still breed and produce fertile offspring, they are not different species. This same idea holds true for all sexually reproducing organisms. (Incidentally, it's not obvious why sexual reproduction is so common; see "Why Sex?" on page 250.)

Yet the definition of "species" is not a black-and-white issue; it is a frequently discussed, multifaceted topic in evolution, and the biological species concept doesn't always apply. For example, not all species can be defined by their ability to interbreed, including prokaryotes such as bacteria, which reproduce asexually. To handle these cases and others, scientists may use biogeographic information, DNA sequence similarity, and **morphology**—the organisms' physical characteristics—to identify and distinguish species (**Figure 13.8**). With that goal in mind, Cole and colleagues decided to use multiple types of data and information to build an evolutionary tree of both living (extant) and dead (extinct) penguin species.

First, they collected samples of skin and bone from penguins preserved in natural history museums and samples of blood from live penguins at zoos and in the wild (**Figure 13.9**). "They have pretty sharp beaks," admits Cole, who was pecked while taking samples.

Figure 13.7

A male and a female South Pacific rattlesnake confirm that they are the same species

After performing this mating ritual, these snakes successfully mated, producing viable offspring.

Q1: **How do we know that these rattlesnakes are members of the same species?**

Q2: **How would you design an experiment to determine whether two populations of snakes are distinct species according to the biological species concept?**

Q3: **For which types of populations does the biological species concept *not* work as a way of determining how they are related? (*Hint*: Read the next two paragraphs.)**

See Appendix A for answers to the figure questions.

Why Sex?

Sex is ubiquitous in the animal and plant kingdoms. An estimated 99 percent of multicellular eukaryotes are capable of sexual reproduction, which involves the joining of two haploid gametes produced through meiosis. Yet sex is very costly for individuals, so scientists have struggled to explain why it is so prevalent compared to asexual reproduction. And no, it's not because sex feels good; the first eukaryotes to engage in sex were single-celled protists some 2 billion years ago, long before animals developed neurons capable of giving an individual a sense of pleasure.

Costs of Sex

1. Time and energy must be invested to find or attract a mate.
2. Parents pass on only 50 percent of their genetic material to offspring, as opposed to the 100 percent that is passed on through asexual reproduction.
3. Gene combinations that have benefited the parents may be shuffled and broken apart during meiosis and recombination.

Possible Benefits of Sex

1. The genetic diversity created by sexual reproduction may be valuable for adaptation to new environments.
2. Sexual reproduction can help a population eliminate detrimental alleles and generate new beneficial alleles.
3. Rapid genetic change that occurs through sexual recombination may allow a population evolve resistance to infections.

From these samples, the scientists extracted mitochondrial DNA—that is, the genetic code inside the mitochondria of penguin cells. In many cases, it wasn't easy, says Cole. Extracting ancient DNA from the museum samples required a special sterile laboratory to prevent contamination, and the process was expensive and laborious, as one has to carefully extract the delicate DNA without damaging it.

Soon the team had gathered nearly complete mitochondrial genomes for all species. Next, Cole asked penguin morphology experts to add information to the database about what extant and extinct penguins look like. With all that information in one place, Cole and the team built an evolutionary tree of living and recently extinct penguins, which they published in a peer-reviewed scientific journal. Their evolutionary

Figure 13.8

One species or two?

Though their phenotypes differ, these tree frogs are genetically similar enough to be considered distinctly colored variations of the same species. Breeding these frogs with each other would determine their classification under the biological species concept.

Q1: In addition to reproductive isolation, what three kinds of information do scientists use to identify and distinguish between species?

Q2: What differences can you observe between the individuals in the photos? Why are these differences not enough to confirm that they are from two different species?

Q3: How is genetic divergence between two populations determined? (*Hint*: Read the next four paragraphs.)

See Appendix A for answers to the figure questions.

Figure 13.9

Research in the wild

Theresa Cole and her research team collect DNA samples from a penguin colony.

tree used the ages of penguin fossils to determine when species had diverged from one another.

Speciation is the process by which one species splits to form two species or more. Fundamentally, speciation occurs because of **genetic divergence**, the accumulation of differences in the DNA sequences of genes in two or more populations of organisms over time; as a result, the populations become more and more genetically dissimilar.

But what caused one species of penguin to split into others? Cole and her team compared their penguin family tree with a geological record of island formation. They found that penguin species known to live on a single island often form a new branch in the penguin family tree shortly after the formation of that species' native island: the Galápagos Islands and the Galápagos penguin, the Snares Islands and the Snares crested penguin, and so on. "In pretty much every case, that new penguin species diverged just after the island emerged," says Cole. In other words, the appearance of a new island spurred the birth of a new species.

"You get a whole pulse of evolution," says Cole. "An island forms, and then a [group of penguins] rapidly colonizes it. But then they're separated from the population they originally came from, so they'll develop new adaptations to the other plants and animals that live on that island."

That physical separation between populations, called **geographic isolation**, is one of the

The Kaibab squirrel is confined to the North Rim of the Grand Canyon.
Abert's squirrel lives on the South Rim and other southern locations, all the way into Mexico.

Figure 13.10

The Grand Canyon is a geographic barrier for squirrels

The Kaibab squirrel population became isolated from the Abert's squirrel population when the Colorado River cut the Grand Canyon, which is as deep as 6,000 feet in some places. With gene flow between them blocked, probably beginning about 5 million years ago, the two populations eventually accumulated enough genetic differences to become two distinct species.

Q1: What is the definition of gene flow? How was gene flow blocked between these species?

Q2: Name as many types of geographic barriers as you can. Which do you think would be the best at blocking gene flow?

Q3: Are geographic barriers universal for all species? If not, name a geographic barrier that might block gene flow for one species but not another.

See Appendix A for answers to the figure questions.

most common ways that new species form: individuals of a single population become geographically separated from one another. Geographic isolation can occur when a few members of a species colonize a region that is difficult to reach, such as an island located far outside the usual geographic range of the species, as in the case of the penguins. Alternatively, geographic isolation can begin when a newly formed geographic barrier, such as a river, canyon, or mountain chain, isolates two populations of a single species (**Figure 13.10**).

Geographically isolated populations are disconnected genetically; there is little or no gene flow between them. Without gene flow, the other mechanisms of evolution we saw in Chapter 12—namely, mutation, genetic drift, and natural selection—can more easily cause populations to diverge from one another. If populations remain isolated long enough, they can evolve into new species. The formation of new species from geographically isolated populations is called **allopatric speciation** (*allo*, "other"; *patric*, "country"), as shown in **Figure 13.11**.

Brand New Species

In a surprise finding of the study, Cole discovered that the allopatric speciation of penguins had occurred fairly recently on New Zealand's Chatham Islands. Her team found that two previously unknown penguin species—a unique crested penguin species and a subspecies of a yellow-eyed penguin—had lived on these islands within the last two centuries. But don't expect to find a photograph of these unique penguins—they went extinct shortly after humans arrived on the islands. It's likely that the humans ate the penguins to extinction, says Cole, as penguin remains were found in ancient trash heaps on the islands.

In that case, one species (humans) caused the extinction of another (penguins). But in other cases, quite the opposite is true: **coevolution** occurs when two species so strongly rely on each other for survival that they evolve in tandem. The term "coevolution" encompasses a wide variety of ways in which an adaptation in one species evolves alongside a complementary adaptation in another species. One example of coevolution is the relationship between hummingbirds and certain species of flowers (**Figure 13.12**).

Like adaptation, speciation can occur slowly or rapidly. One of the most famous cases of rapid speciation comes from the Galápagos Islands, a string of wildlife-rich islands in the Pacific Ocean made famous by one of the first scientists to propose evolution by natural selection, British naturalist Charles Darwin.

Darwin visited the Galápagos Islands in 1835 on a voyage on the HMS *Beagle*. He observed and collected small, colorful birds called finches from various islands in the archipelago. When he later examined those birds, Darwin realized they were all related but had observable differences. In his first book, *The Voyage of the Beagle* (1839), Darwin wrote, "Seeing this gradation and diversity of structure in one small, intimately related group of birds, one might really fancy that from an original paucity of birds in this archipelago, one species had been taken and modified for different ends."

Since Darwin, other scientists have traveled to the Galápagos to study his famed finches. In the 1980s, Rosemary and Peter Grant from

A single plant species is distributed over a broad geographic range.
Time
The sea level rises and isolates plant populations from one another. The populations may adapt to different environments on opposite sides of the barrier, indirectly causing genetic changes that reduce their ability to interbreed.
Time
When the barrier is removed, the plants recolonize the intervening area and mingle, but do not interbreed.
Range of overlap

Figure 13.11

Physical barriers can produce allopatric speciation by blocking gene flow

Allopatric speciation can occur when populations are separated by a geographic barrier, such as a rising sea.

Q1: What factor must be present for allopatric speciation to occur?

Q2: If a geographic barrier is removed and the two reunited populations intermingle and breed, what attributes must the offspring have for the two populations, according to the biological species concept, to be considered still the same species?

Q3: If the two populations in question 2 are determined to still be the same species, did allopatric speciation occur?

See Appendix A for answers to the figure questions.

Flowers that coevolved with hummingbirds produce nectar, are colored to attract the birds, and are shaped specifically to fit the hummingbirds' bills.

Figure 13.12

Coevolution in living color

This hummingbird's bill fits perfectly into this flower for easy access to the hummingbird's favorite food, nectar. The hummingbird feasts and then distributes the flower's pollen by carrying the pollen along to its next meal. It's a win-win situation!

Q1: Describe how coevolution, as seen with the hummingbird bill and hummingbird-pollinated flowers, is different from the kind of evolution described in Chapters 11 and 12.

Q2: Is coevolution the same thing as convergent evolution, described in Chapter 12? Why or why not?

Q3: Do you think one species' adapting over time to feed specifically and extremely successfully on another species is an example of coevolution? Why or why not?

See Appendix A for answers to the figure questions.

Princeton University, who had been carefully tracking finch populations, observed a new male bird appear on an island where three species of finches already lived. The intruder, which turned out to be a large cactus finch from an island more than 60 miles away, was much larger and sang a different song than the native finches. The newcomer mated with a resident medium ground finch and began a new "Big Bird" lineage, as the Grants called it.

The offspring of that breeding were bigger and sang an unusual song, which isolated them from other birds on the island. As a result, the offspring mated with members of their own lineage, leading to the development of a new species in just two generations (**Figure 13.13**).

Old Theory Meets New Theory

Cole's study was the first compelling evidence that penguin speciation is driven by island formation, a form of allopatric speciation. Although allopatric speciation is common with other types of animals, scientists did not think it occurs with penguins because the animals spend so much of their lives at sea where they are not geographically isolated. Instead, researchers proposed that ocean currents separate penguin populations with different groups of penguins preferring water and food sources at different temperatures. We call this ecological isolation.

Ecological isolation occurs when two closely related species in the same territory are reproductively isolated by slight differences in habitat, such as one penguin population preferring warmer ocean currents and the food available there, whereas a separate population migrates toward colder currents and food sources.

This formation of new species in the *absence* of geographic isolation is called **sympatric speciation** (from *sym*, "together"; **Figure 13.14**). Sympatric speciation is a particularly important process in plants (see "So Many Chromosomes" on page 255) and in the ocean, where plants (and other organisms) can drift around or swim extremely long distances. There are few physical barriers yet many separate, unique species.

It is likely that both geographic and ecological isolation contributed to penguin speciation, says Cole. For example, the three species of rockhopper penguins, known for their red eyes and the yellow and black spiky feathers on their heads, occupy three different ecosystems: warm, subtropical, and cold sub-Antarctic Ocean areas. In the case of rockhopper

Figure 13.13

New "Big Bird" Galápagos finch species

When a male large cactus finch arrived on Daphne Major Island in the Galápagos, he paired with a female medium ground finch. Their male offspring sang a different song from that of either parental species. As a result, females of the parental species were not attracted to these new males as potential mates, so the males instead paired with females of their own lineage. This immediate reproductive isolation caused speciation of the "Big Bird" large cactus finch/medium ground finch hybrids.

Note: Before answering these questions, read the next section.

Q1: Is this an example of allopatric or sympatric speciation?

Q2: Both large cactus finches and medium ground finches originally evolved from a shared ancestor that came from Central or South America and spread through the Galápagos Islands, speciating into different species on different islands. Is this an example of allopatric or sympatric speciation?

Q3: Finch species in the Galápagos differ in their beak size, among other things. How might the differences in beak size be an adaptation to the differing environments on the islands?

See Appendix A for answers to the figure questions.

So Many Chromosomes

New plant species can form in a single generation as a result of **polyploidy**, a condition in which an individual gains an extra full set or two (or three) of chromosomes. Humans and most other eukaryotes are *diploid* (having two sets of chromosomes), but some organisms are *triploid* (three sets) or *tetraploid* (four sets), or have an even higher number of chromosomes. Polyploidy is invariably fatal in people, but in many plant species and some other animals it is not lethal.

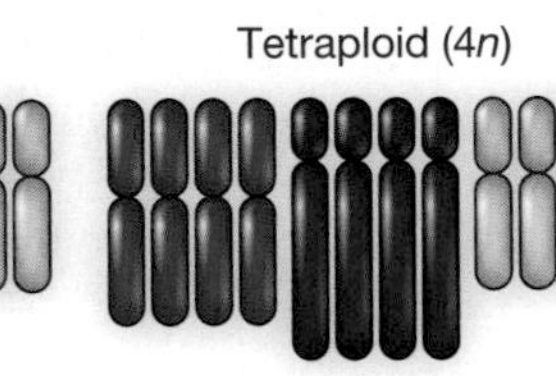

Polyploidy can occur when two species hybridize and produce an offspring with an odd number of chromosomes because of improper alignment of the chromosomes during mitosis (a form of reproductive cell division discussed in Chapter 6). This increase in chromosome sets can lead to reproductive isolation because the chromosome number in the gametes of the new polyploid individual no longer matches the number in the gametes of either of its parent species.

Polyploidy has had a large effect on life on Earth: more than half of all plant species alive today are descended from species that originated by polyploidy. A few animal species also appear to have originated by polyploidy, including several species of lizards, salamanders, and fish.

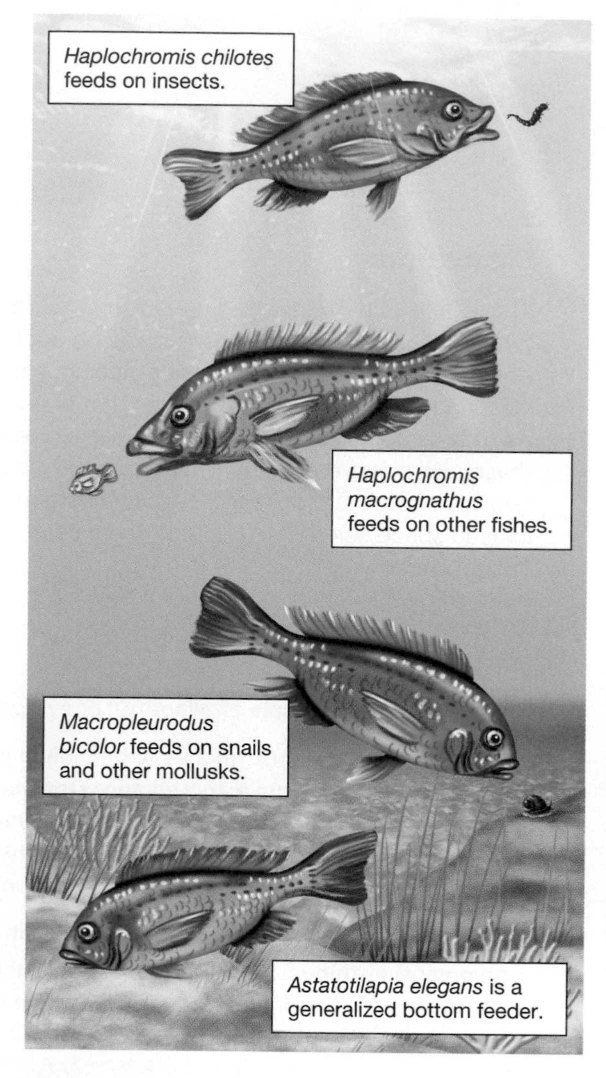
Haplochromis chilotes feeds on insects.
Haplochromis macrognathus feeds on other fishes.
Macropleurodus bicolor feeds on snails and other mollusks.
Astatotilapia elegans is a generalized bottom feeder.

Figure 13.14

Sympatric speciation drives diversity among Lake Victoria cichlid species

Scientists have described some 500 species of cichlid fishes in Lake Victoria. Genetic analyses indicate that they all descended from just two ancestor species over the past 100,000 years. These four species show some of the differences in feeding behavior and morphology.

Q1: What is the main difference between allopatric and sympatric speciation?

Q2: Name two events that must happen for both allopatric speciation and sympatric speciation to occur.

Q3: Do you think all of the 500 species in Lake Victoria arose through sympatric speciation? Why or why not?

See Appendix A for answers to the figure questions.

penguins, therefore, ecological isolation led to speciation.

When two species are prevented from reproducing with each other, we say that **reproductive barriers** exist between those species. Reproductive barriers are often divided into two categories: *prezygotic* and *postzygotic*.

Barriers that prevent a male gamete (such as a human sperm) and a female gamete (such as a human egg) from fusing to form a zygote are **prezygotic barriers**. Prezygotic barriers act *before* the zygote exists (**Figure 13.15**). Barriers that prevent zygotes from developing into fertile offspring are called **postzygotic barriers**, which

Blue-footed boobies point their beaks, wings, and tails upward in a mating dance called "sky pointing."

Figure 13.15

The blue-footed booby courtship dance is a prezygotic, behavioral reproductive barrier

This species of booby has a unique ritual dance that must be accurately completed before mating. Other booby species do not perform exactly the same dance and therefore do not mate with the blue-footed booby.

Q1: What does "prezygotic" mean?

Q2: How is the booby's ritual dance a prezygotic reproductive barrier?

Q3: What are some other prezygotic reproductive barriers besides a mating dance?

See Appendix A for answers to the figure questions.

Figure 13.16

A mule is the sterile offspring of a female horse and a male donkey

People breed horses and donkeys because the resulting mules are stronger and live longer than a horse and are smarter and more patient than a donkey. A mule inherits half of its horse mother's 64 chromosomes and half of its donkey father's 62 chromosomes, for a total of 63 chromosomes. Creating eggs or sperm requires an equal number of chromosomes, with rare exceptions, so both male and female mules are sterile.

Q1: Explain in your own words why this is an example of a postzygotic barrier that maintains species.

Q2: Occasionally a female mule will give birth, but her offspring are invariably infertile. Is this an example of a prezygotic or postzygotic barrier?

Q3: If a male mule was born fertile, but female horses and donkeys refused to mate with him, would this be an example of a prezygotic or postzygotic barrier?

See Appendix A for answers to the figure questions.

act *after* the zygote is formed. One of the best-known examples of a postzygotic barrier maintaining separate species is the mule, which is the sterile offspring of a female horse and a male donkey (**Figure 13.16**).

A wide variety of cellular, anatomical, physiological, and behavioral mechanisms generate prezygotic and postzygotic reproductive barriers, but they all have the same overall effect: little or no successful mating occurs and, therefore, few or no alleles are exchanged between species (**Table 13.1**). For example, in 2018 scientists discovered that two populations of the Northern cardinal, one of the most common backyard birds in the United States, in two adjacent deserts appear to be undergoing speciation in part due to differences in birdsong.

Song plays a critical role in a bird's ability to attract a mate, so if two birds sing different types of songs, it's like trying to communicate in different languages, and they are less likely to breed, as we saw with Darwin's finches and the interloping "Big Bird." Birdsong, therefore, is a prezygotic barrier. Although the Northern cardinal populations lived near each other and could potentially breed, each set of birds was more receptive to its own local song and did not respond to the song of the other population. Thus, we can directly observe speciation between two bird populations occurring due to behavioral isolation.

Table 13.1

Reproductive Barriers That Isolate Two Species in the Same Geographic Region

Type of Barrier	Description	Effect
PREZYGOTIC		
Ecological isolation	The two species breed in different portions of their habitat, in different seasons, or at different times of day.	Mating is prevented.
Behavioral isolation	The two species respond poorly to each other's courtship displays or other mating behaviors.	Mating is prevented.
Mechanical isolation	The two species are physically unable to mate.	Mating is prevented.
Gametic isolation	The gametes of the two species cannot fuse, or they survive poorly in the reproductive tract of the other species.	Fertilization is prevented.
POSTZYGOTIC		
Zygote death	Zygotes fail to develop properly, and they die before birth.	No offspring are produced.
Hybrid sterility	Hybrids survive but are unable to produce viable offspring.	No offspring are produced.
Hybrid performance	Hybrids survive poorly or reproduce poorly.	Hybrids are not successful.

Facing a Warmer Future

Although the catastrophic loss of chicks at Halley Bay for 3 years in a row does not bode well for Emperor penguins, there is a silver lining to this story: A smaller penguin colony 34 miles to the south, called the Dawson-Lambton colony, has grown more than tenfold over the past 3 years. Trathan and Fretwell speculate that many of the Halley Bay penguins relocated there (**Figure 13.17**).

"It seems most of the penguins breeding at Halley have moved," says Fretwell. "It shows that the Emperor penguins have got some resilience." That move suggests one way the Emperors are reacting to climate change and an increasing number of strong storms—by moving to safer breeding grounds.

But will that be enough to save the penguins? "Future climate change is predicted to cause species-wide extinction of some penguins," says Cole. "It's a major concern." The IUCN Red List has classified the species of penguins according to their vulnerability to extinction. Among all living penguin species, five are in danger of extinction, five are vulnerable, three are threatened, and only five are species of least concern. If we don't protect them now, there won't be any species left to evolve into new species, says Cole, who is now doing postdoctoral research on the genetics of penguin adaptive traits.

New species form and adapt to their environments in many different ways, and from those varied beginnings comes the vast diversity of life that exists on our planet, the result of billions of years of evolution. In the next section of this book, we explore the dramatic transitions and diversity of life-forms that have walked, crawled, and swum around our planet. And we'll start by investigating where penguins came from in the first place by diving deep into the history of life with a bird as our guide.

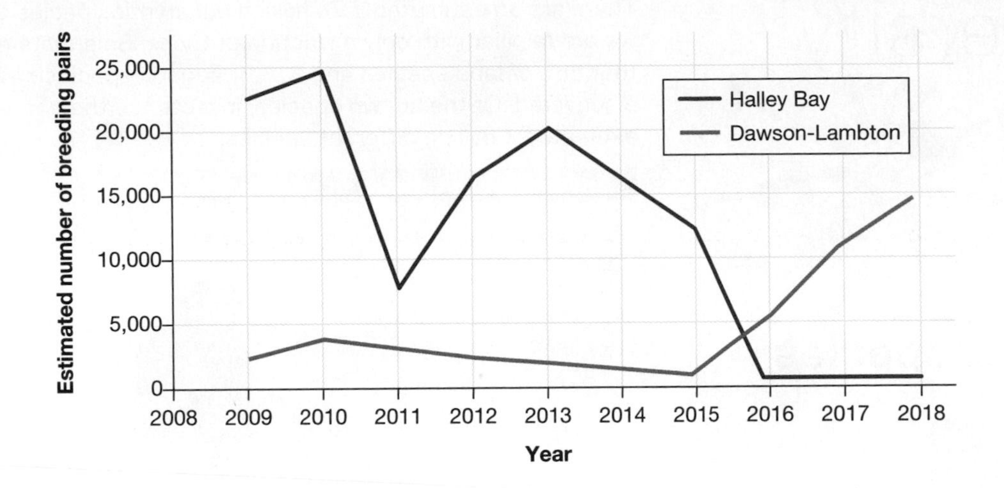

Figure 13.17

Emperor penguin breeding population changes over time

The number of penguin breeding pairs at the Halley Bay colony and the nearby Dawson-Lambton colony has varied enormously over 10 years.

Q1: According to this graph, in what year did the Halley Bay colony have its largest number of breeding pairs? In what year did the Dawson-Lambton colony have its largest number of breeding pairs?

Q2: How many total breeding pairs were there in Halley Bay and Dawson-Lambton colonies in 2018? How does this compare with 2010?

Q3: Given the numbers of breeding pairs in Halley Bay, what would you hypothesize about the change in biological fitness of this population of Emperor penguins over the 10-year period shown here?

See Appendix A for answers to the figure questions.

On the Diversity of Species

There are an estimated 8.74 million eukaryotic species on Earth, yet we are familiar with only a fraction of those. Scientists estimate that 86% of land species and 91% of aquatic species have not been discovered. Of the known species, insects top the chart with an estimated 1 million different species.

Assessment available in smartwork

Numbers of known species

Insects	1,000,000
Fish	31,200
Crustaceans	47,000
Mollusks	85,000
Fungi	99,000
Arachnids	102,200
Plants	310,100
Mammals	5,500
Sponges	6,000
Amphibians	6,500
Reptiles	8,700
Jellyfish and polyps	9,800
Birds	10,000
Millipedes and centipedes	16,100
Segmented worms	16,800
Flatworms	20,000
Roundworms	25,000

NEED-TO-KNOW SCIENCE

- An individual's survival and successful reproduction is termed its **biological fitness** (page 247). Natural selection leads to adaptation, genetic changes that produce adaptive traits, improving the biological fitness of organisms in their environment over time.
- According to the **biological species concept** (page 249), a **species** (page 249) is a group of populations that interbreed and can produce live and fertile offspring. Because this group cannot breed with other such groups, it is **reproductively isolated** (page 249) from other populations.
- To identify and distinguish species not addressed by the biological species concept, scientists may use biogeographic information, DNA sequence similarity, and **morphology** (page 250)—the organisms' physical characteristics.
- The process of **speciation** (page 251), one species splitting into two or more species, is the by-product of reproductive isolation and **genetic divergence** (page 251), the accumulation of differences in the DNA sequences of genes between populations of organisms over time.
- **Allopatric speciation** (page 252) occurs when populations of a species become **geographically isolated** (page 251), limiting gene flow and making genetic divergence more likely.
- Speciation that occurs between populations that are not geographically isolated is called **sympatric speciation** (page 254).
- In **ecological isolation** (page 254), two closely related species in the same area are reproductively isolated by slight differences in habitat.
- New plant species can form in a single generation as a result of **polyploidy** (page 255), a condition in which an individual gains an extra full set or two (or three) of chromosomes.
- **Reproductive barriers** (page 256) can occur either before individuals meet and produce a zygote, due to a **prezygotic barrier** (page 256), or after they meet and produce a zygote through a **postzygotic barrier** (page 256).
- When adaptive traits of one species evolve in concert with adaptive traits in another species, **coevolution** (page 252) has occurred.

THE QUESTIONS

See Appendix B for answers.

The Basics

1 When two populations are reproductively isolated, what else must occur for speciation to happen?

(a) gene flow
(b) genetic divergence
(c) coevolution
(d) convergent evolution
(e) none of the above

2 Traits that are inherited and improve an individual's biological fitness are called _________ traits.

(a) adaptive
(b) polymorphic
(c) biological
(d) sympatric
(e) allopatric

3 Prezygotic isolating mechanisms prevent hybrid offspring from occurring between species because

(a) the resulting offspring are not fertile and cannot reproduce.
(b) the egg and sperm fuse and form a zygote, but it does not survive.
(c) the egg and sperm do not ever meet or, if they do, cannot fuse to form a zygote.
(d) all of the above

4 Which of the following reproductive isolating mechanisms prevents mating because the two species are physically unable to mate?

(a) ecological isolation
(b) behavioral isolation
(c) mechanical isolation
(d) gametic isolation
(e) all of the above

5 Select the correct terms:
Speciation due to geographic separation and (**genetic divergence** / **gene flow**) of two populations is considered (**allopatric** / **sympatric**) speciation.

6 Select the correct phrases:
(**Prezygotic barriers** / **Postzygotic barriers**) are reproductive isolating mechanisms that occur after a zygote has been formed. An example would be (**an infertile hybrid** / **gametes that cannot fuse**).

Challenge Yourself

7 Natural selection

(a) leads to species being better adapted to their environment.
(b) may lead to speciation if there is no gene flow between populations.
(c) may lead to genetic divergence among populations in differing environments.
(d) all of the above
(e) none of the above

8 Which of the following examples refer to an adaptive trait? (Select all that apply.)

(a) A rainforest tree is vulnerable to storm damage.
(b) A male bird is more successful than others at attracting a female mate.
(c) A rabbit is better camouflaged in its environment.
(d) A desert plant is able to survive drought.
(e) A frog is more noticeable to predators.

9 Place the following events in the order in which they are most likely to occur by numbering them from 1 to 5.

_____ a. Speciation occurs.
_____ b. Genetic divergence takes place.
_____ c. Geographic barrier arises.
_____ d. Gene flow is discontinued.
_____ e. Species adapts to a new environment.

Try Something New

10 Distinct species that are able to interbreed in nature are said to "hybridize," and their offspring are called "hybrids." The gray oak and the Gambel oak can mate to produce fertile hybrids in regions where they co-occur. However, the gene flow in nature is sufficiently limited that, overall, the two species remain phenotypically distinct. If the hybrid offspring survive well and reproduce to the extent that there is a large population of hybrid individuals that breed between themselves but do not interbreed with either of the two original parent species (the gray and the Gambel), which of the following would you say most likely led to the new hybrid species?

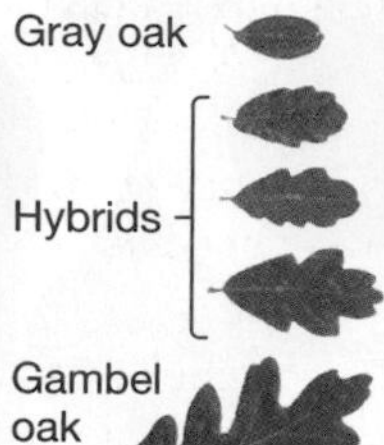

(a) sympatric speciation
(b) prezygotic reproductive barriers
(c) allopatric speciation
(d) postzygotic reproductive barriers
(e) none of the above

11 The four-eyed fish, *Anableps anableps*, really has only two eyes, which function as four, enabling the fish to see clearly through both air and water. *A. anableps* is a surface feeder, so the ability to see above water helps it locate prey such as insects. Its unique eyes also enable it to scan simultaneously for predators attacking from above (such as birds) or below (such as other fish). This species of fish __________ brought about through natural selection.

Anableps anableps

(a) is an example of coevolution
(b) has an evolutionary adaptation
(c) is an example of sympatric speciation
(d) is an example of allopatric speciation

12 Researchers from the Smithsonian Institution were startled to discover that the 3 species of *Starksia* blennies they had been studying in the Caribbean islands were really 10 different species. How could these researchers have thought that these 10 different reef fish species were only 3 species?

13 Adélie penguins, tiny cousins of Emperor penguins, are the subject of a long-term ecological study on Anvers Island, off the continent of Antarctica. This graph shows a steep decline over 20 years in the number of breeding pairs of Adélie penguins on the island.

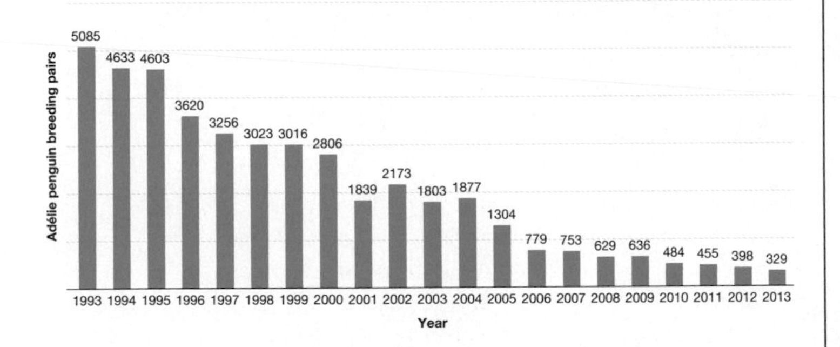

(a) What is the highest number of breeding pairs in the colony? In what year did this occur?

(b) Did the colony size decrease every year? In what years was the size greater than in the previous year?

(c) In 2016, a different Adélie colony of 18,000 breeding pairs suffered catastrophic breeding failure, with only two chicks fledged that year. Given that number of breeding pairs does not necessarily tell the whole picture of colony breeding success, draw a graph of your prediction of the number of Adélie chicks fledged over the same period of time as the graph that showed the number of mated pairs.

Leveling Up

14 ***Write Now* biology: adaptations** Select an organism (other than humans) that you find interesting. Research two adaptations of your organism and describe them in detail. Explain carefully why each of these features is considered an important adaptation for its species. Discuss how a change in the organism's environment might change the usefulness of these adaptations.

15 **Doing science** Eastern and western meadowlarks are very similar in morphology. Both are grassland species with some overlap in their range in the upper Midwest, and each sings a distinctly different song. Create a hypothesis as to whether these are two different species or two populations of the same species, and describe an experiment that would enable you to test your hypothesis. What result would support your hypothesis, and what result would cause you to reject your hypothesis?

16 **What do *you* think?** According to the Defenders of Wildlife organization, there were 20–30 million American range bison in the Old West. During the late 1800s, the American range bison were hunted to the brink of extinction, leaving behind a bottleneck population of only 1,091 individuals. The population has since rebounded to about 500,000 bison. Unfortunately, almost all of these bison are the descendants of these few individuals crossbred with domestic cattle by ranchers. Scientists and conservationists want to genetically test bison to find those of pure bison origin to preserve the species. Only these, they argue, should be called American range bison and be allowed to roam free in the national parks as bison. They think hybrids should be confined to farms and ranches, should be called "beefalo" rather than bison, and should not be afforded the protection that pure bison currently have. What do you think? Should bison carrying cattle genes be removed from free-range parks? Should the government spend scarce conservation monies on genetic testing and breeding efforts to preserve the pure bison population? Investigate conservation efforts and the costs of genetically testing and relocating bison to help you with your decision. Is speciation at the hands of human beings now part of evolution as we know it?

Digital resources for your book are available online.

Dinosaur Hunters

A fresh wave of fossils challenges the identity of the first dino-bird.

After reading this chapter you should be able to:

- Identify critical events on a timeline of life on Earth and explain their evolutionary importance.
- Interpret an evolutionary tree of a group of organisms.
- Explain how new scientific data may change an evolutionary tree without challenging our understanding of evolution.
- Classify a given species at each level of the Linnaean system.
- Give an example of an adaptation that enabled a group of organisms to transition from an aquatic existence to life on land.
- Describe the impact of prehistoric mass extinctions on Earth's biodiversity, and infer the probable impact of the current extinction rate on biodiversity.

BIODIVERSITY

CHAPTER 14

THE HISTORY OF LIFE

In 1861, quarry workers in Germany unearthed the fossil of a crow-sized bird. The 150-million-year-old preserved skeleton looked bizarre. It had feathers and clawed hands like a bird but teeth and a bony tail like a reptile.

After close examination by paleontologists, the fossil was hailed as a transitional form between birds and reptiles and labeled the oldest known bird. Named *Archaeopteryx*, the feathered dinosaur became famous around the world as the first solid evidence that birds descended from dinosaurs (**Figure 14.1**). Charles Darwin, who proposed the theory of evolution by natural selection, called the discovery "a grand case for me."

Over 100 years later, a young fossil hunter named Xu Xing bent over the dirt at a dig site in Liaoning province in northern China. It was Xu's first time in the field since finishing college. An excellent student, Xu had wanted to study economics at Peking University in Beijing, but at the time, students in China did not pick their majors, and Xu was assigned paleontology. Luckily, farmers in Liaoning had recently begun to uncover huge numbers of fossils, and research was booming.

So, as a grudging paleontologist, Xu began searching for dinosaurs in Liaoning. He had a particular interest in feathered dinosaurs. By the late twentieth century, the idea that birds descended from dinosaurs was no longer a hypothesis but a well-established scientific theory backed by a mountain of data. And atop that mountain of data sat *Archaeopteryx*, the earliest known bird.

Little did Xu know, as he dug through the dirt that day, that he would soon be the one to knock *Archaeopteryx* from its perch.

Dinosaurs and Domains

Although the *Archaeopteryx* fossil is old—150 million years old (mya), in fact—*much* older fossils have been found. Some of the oldest known rocks on Earth are 3.8 billion years old (bya) and contain carbon deposits that hint at life. Cell-like structures have been found in layered mounds of sedimentary rock called stromatolites that formed 3.7 bya, and projections based on DNA analysis also support the idea that life appeared on Earth at that time.

The question of how life arose from nonlife is one of the greatest riddles in biology, but scientists have little doubt that all life on Earth is related. As noted in Chapter 1, all living organisms are united by a basic set of characteristics. Living things share a set of common properties

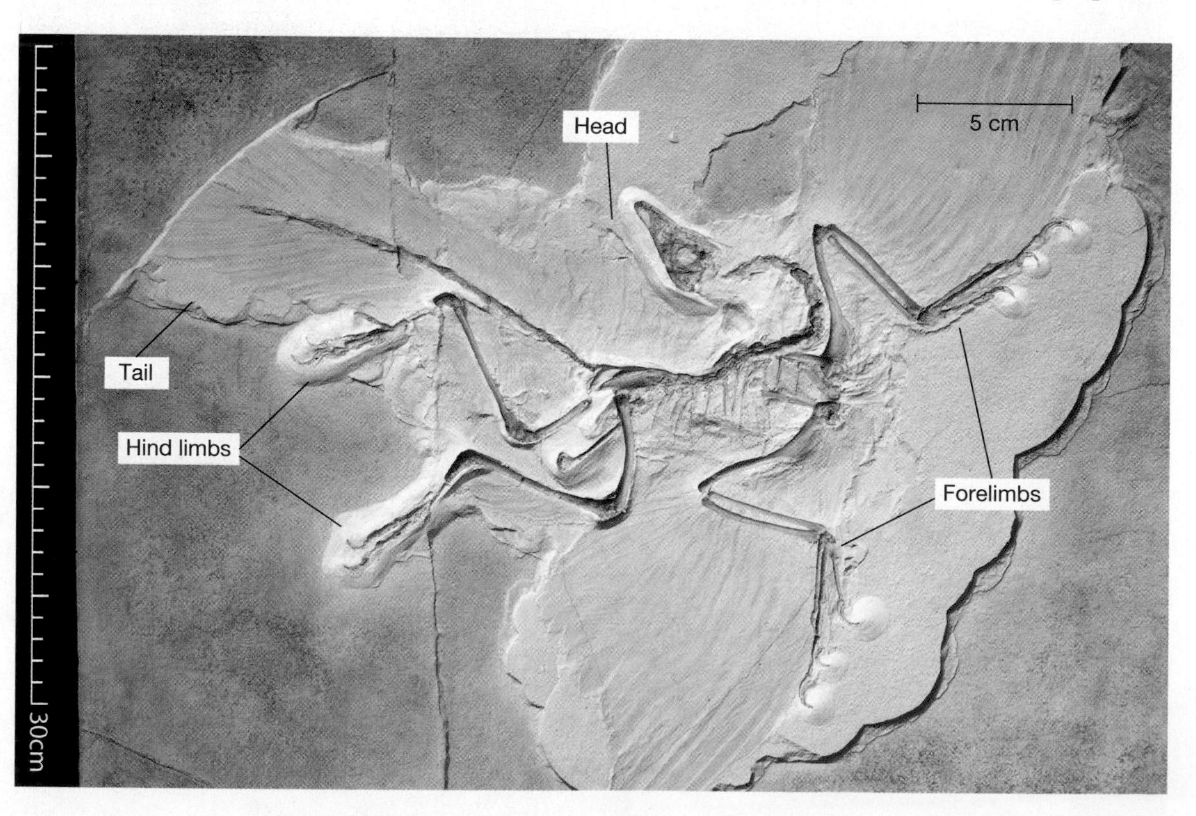

Figure 14.1

The original *Archaeopteryx* fossil

There are 11 complete *Archaeopteryx* fossils, but this one from Germany was the first to be found. Note the faint impressions of feathers on the forelimbs and tail.

because all life descended from a common ancestor, known as the *universal ancestor*.

This hypothetical ancestral cell is placed at the base of the tree of life. From that cell, all life emerged. Biological diversity, or **biodiversity**, embraces the variety of all the world's living things, as well as their interactions with each other and the ecosystems they inhabit. Biodiversity can be described at the level of genes, species, or entire ecosystems.

In spite of intense worldwide interest, scientists do not know the exact number of species alive today. Most traditional estimates have fallen in the range of 3–30 million species, but in 2016, scientists combined microbial, plant, and animal data sets from around the world, scaled up, and estimated that Earth could contain nearly a trillion species. So far, about 1.5 million species have been collected, named, and placed into an evolutionary tree, meaning it is possible that only one ten-thousandth of 1 percent of species are known to us. In other words, 99.9999 percent of species are yet undiscovered.

Life throughout history and today is so diverse that biologists created a classification system to organize it into categories. The **domains** form the highest, most inclusive hierarchical level in the organization of life, describing the most basic and ancient divisions among living organisms. There are three domains of life (**Figure 14.2**):

- **Bacteria**, which includes harmless inhabitants of our intestines and environment as well as disease-causing bacteria such as *E. coli*

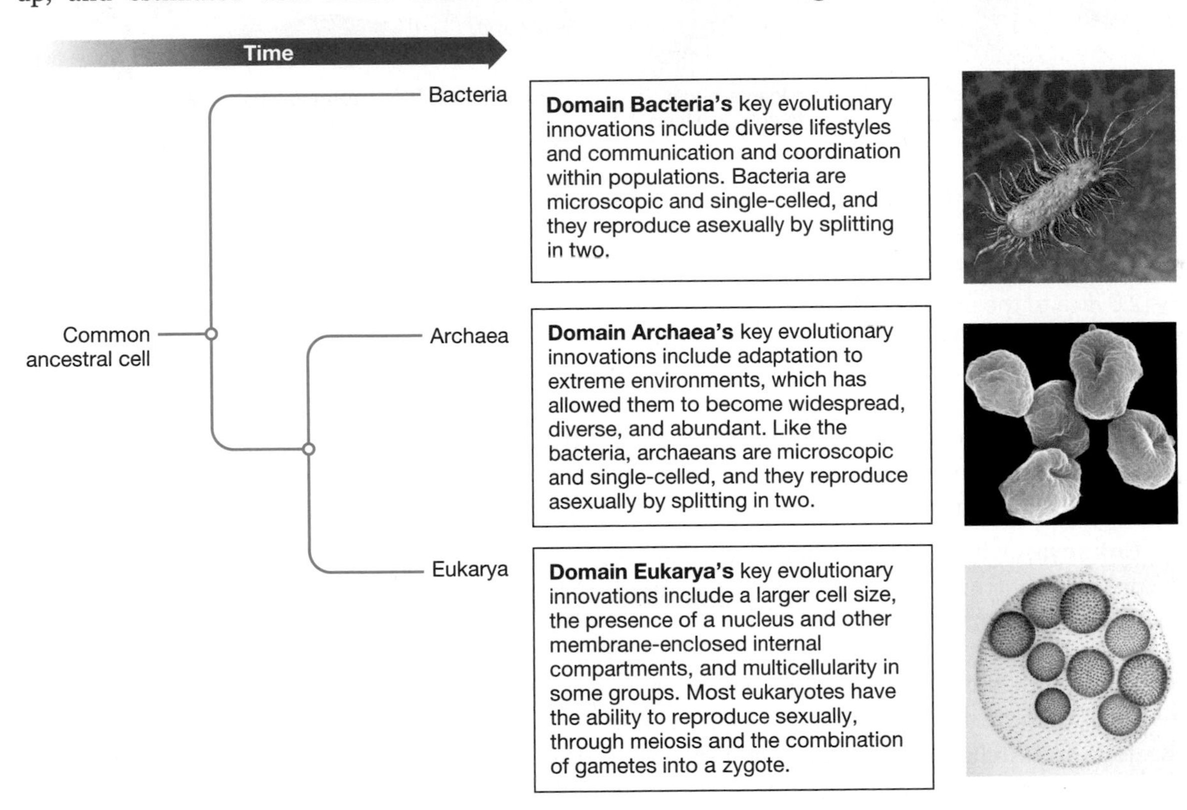

Figure 14.2

Three domains of life

This "tree of life" shows the relationships of the three domains of life. Each circle indicates a shared ancestor.

Q1: Why are Archaea and Eukarya connected more closely with each other than they are with Bacteria?

Q2: Where would multicellular organisms be found within this figure? What about birds and humans?

Q3: Which domains include disease-causing organisms? (*Hint*: Read the bulleted list after the figure callout.)

See Appendix A for answers to the figure questions.

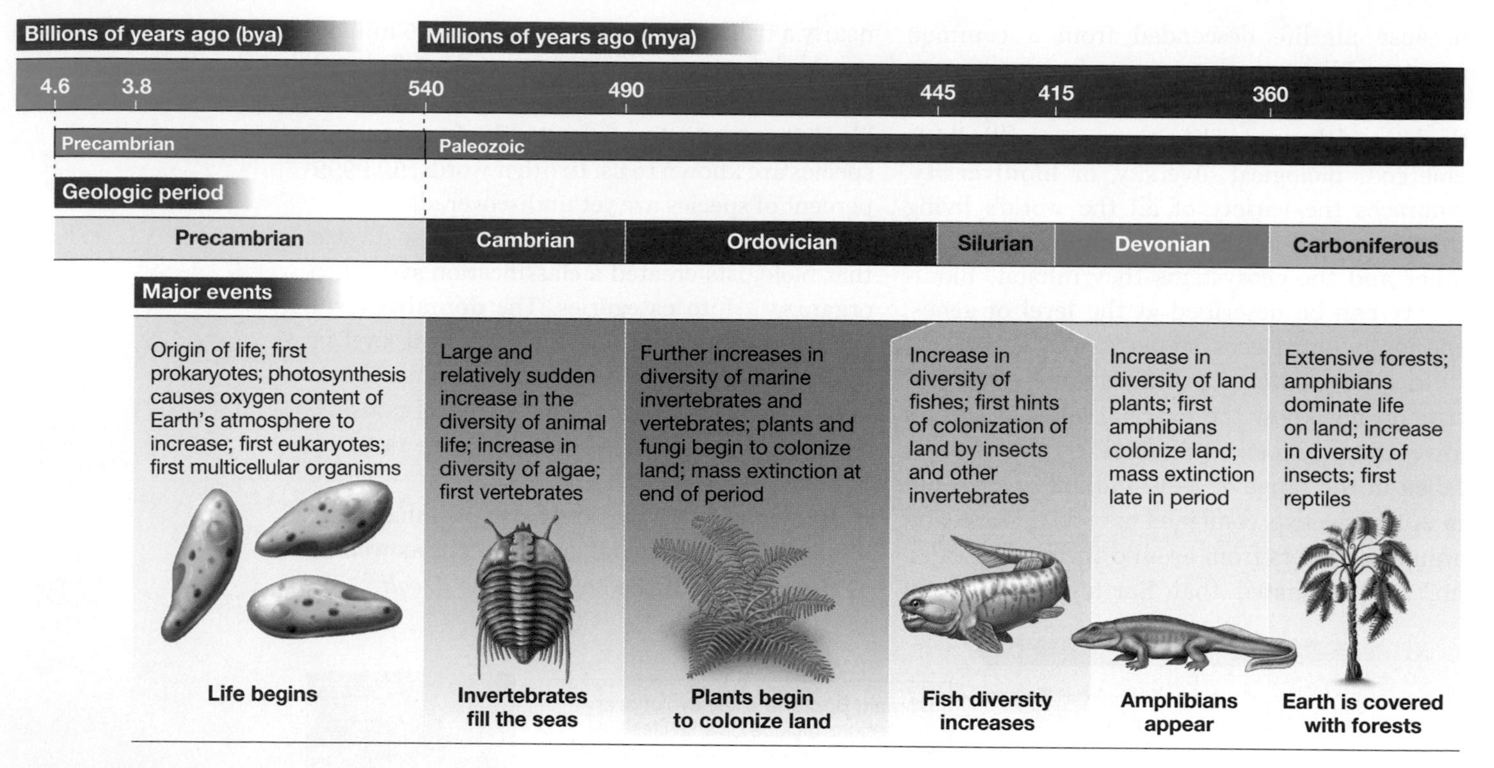

Billions of years ago (bya)
Millions of years ago (mya)
4.6
3.8
540
490
445
415
360
Precambrian
Paleozoic
Geologic period
Precambrian
Cambrian
Ordovician
Silurian
Devonian
Carboniferous
Major events
Origin of life; first prokaryotes; photosynthesis causes oxygen content of Earth's atmosphere to increase; first eukaryotes; first multicellular organisms
Large and relatively sudden increase in the diversity of animal life; increase in diversity of algae; first vertebrates
Further increases in diversity of marine invertebrates and vertebrates; plants and fungi begin to colonize land; mass extinction at end of period
Increase in diversity of fishes; first hints of colonization of land by insects and other invertebrates
Increase in diversity of land plants; first amphibians colonize land; mass extinction late in period
Extensive forests; amphibians dominate life on land; increase in diversity of insects; first reptiles
Life begins
Invertebrates fill the seas
Plants begin to colonize land
Fish diversity increases
Amphibians appear
Earth is covered with forests

Figure 14.3

The geologic timescale and major events in the history of life

The history of life can be divided into 12 major geologic time periods, beginning with the Precambrian (4.6 bya to 540 mya) and extending to the Quaternary (2.6 mya to the present). This time line is not drawn to scale; to do so would require extending the diagram off the book page to the left by more than 5 feet (1.5 meters).

- **Archaea**, which consists of single-celled organisms, some of which live in extremely harsh environments
- **Eukarya**, which includes all other living organisms, from amoebas to plants to fungi to animals

Dinosaurs, birds, and humans are all part of the Eukarya domain. They are *eukaryotes*. Bacteria and Archaea are two different domains—Archaea are more closely related and in some ways more similar to Eukarya than to Bacteria—yet because neither Bacteria nor Archaea are eukaryotes, the two have traditionally been lumped under a common label: *prokaryotes*.

Prokaryotes first appear in the fossil record at about 3.7 bya (**Figure 14.3**), but the first eukaryotes did not evolve until over a billion years later. Luckily for us, and all other eukaryotes, roughly 2.8 bya a group of bacteria evolved a type of photosynthesis that releases oxygen as a by-product. As a result of these prokaryotes releasing oxygen, the oxygen concentration in the atmosphere increased over time, and about 2.1 bya the first single-celled eukaryotes evolved. When the oxygen concentration reached its current level by about 650 mya, the evolution of larger, more complex multicellular organisms became possible, including fish, then land plants, then insects, amphibians, and reptiles.

One group of reptiles, which would eventually dominate most other species, was the dinosaurs. Dinosaurs first appeared about 230 mya, during the Triassic period, and they took over the planet.

Feathered Friends

Xu Xing may not have wanted to be a paleontologist when he went to college, but by the time he graduated, he was hooked on dinosaurs. Over the next 20 years, Xu became one of the most productive researchers in his field. Today, he is deputy director of the Institute of Vertebrate Paleontology and Paleoanthropology

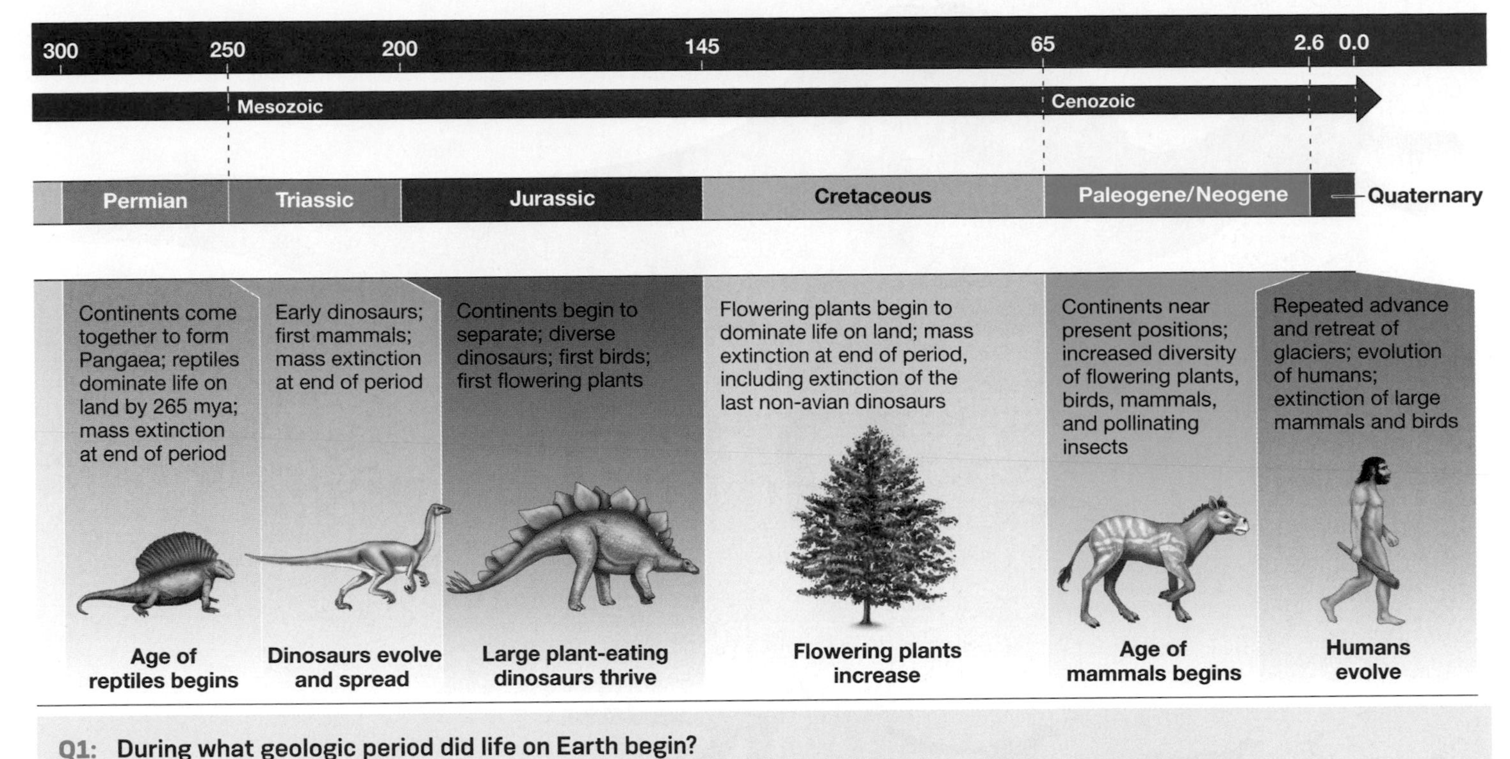
300
250
200
145
65
2.6
0.0
Mesozoic
Cenozoic
Permian
Triassic
Jurassic
Cretaceous
Paleogene/Neogene
Quaternary
Continents come together to form Pangaea; reptiles dominate life on land by 265 mya; mass extinction at end of period
Early dinosaurs; first mammals; mass extinction at end of period
Continents begin to separate; diverse dinosaurs; first birds; first flowering plants
Flowering plants begin to dominate life on land; mass extinction at end of period, including extinction of the last non-avian dinosaurs
Continents near present positions; increased diversity of flowering plants, birds, mammals, and pollinating insects
Repeated advance and retreat of glaciers; evolution of humans; extinction of large mammals and birds
Age of reptiles begins
Dinosaurs evolve and spread
Large plant-eating dinosaurs thrive
Flowering plants increase
Age of mammals begins
Humans evolve

Q1: **During what geologic period did life on Earth begin?**

Q2: **How long ago did species begin to move from water to land? What period was this?**

Q3: **In what period would *Archaeopteryx* have been alive?**

See Appendix A for answers to the figure questions.

at the Chinese Academy of Sciences and has discovered and named more than 70 extinct species—mostly dinosaurs but also a reptile and a salamander. And the majority of those dinosaur fossils had feathers (**Figure 14.4**).

As scientists traced back the **lineage**, or line of descent, from birds to dinosaurs, it became clear that birds are most closely related to theropods, fast-moving dinosaurs that ran on two legs and had hollow, thin-walled bones (as birds do). Theropods were a diverse group of dinosaurs (**Figure 14.5**). Most were carnivores, including insectivores, but a few were herbivores or omnivores. There were theropods that could swim and eat fish. Some theropod species boasted enlarged scales, and of course, many theropods had feathers.

Xu and other scientists map out how species are related using a diagram called an **evolutionary tree**, a model of evolutionary relationships among groups of organisms based on similarities and differences in their DNA, physical features, biochemical characteristics, or some

Figure 14.4

A feathered dinosaur tail, trapped in amber

In 2016, scientists discovered a 99-million-year-old dinosaur tail, complete with its feathers, in a piece of amber. Amber is fossilized tree resin, often used in jewelry; here it is the yellowish-white material surrounding the dinosaur parts.

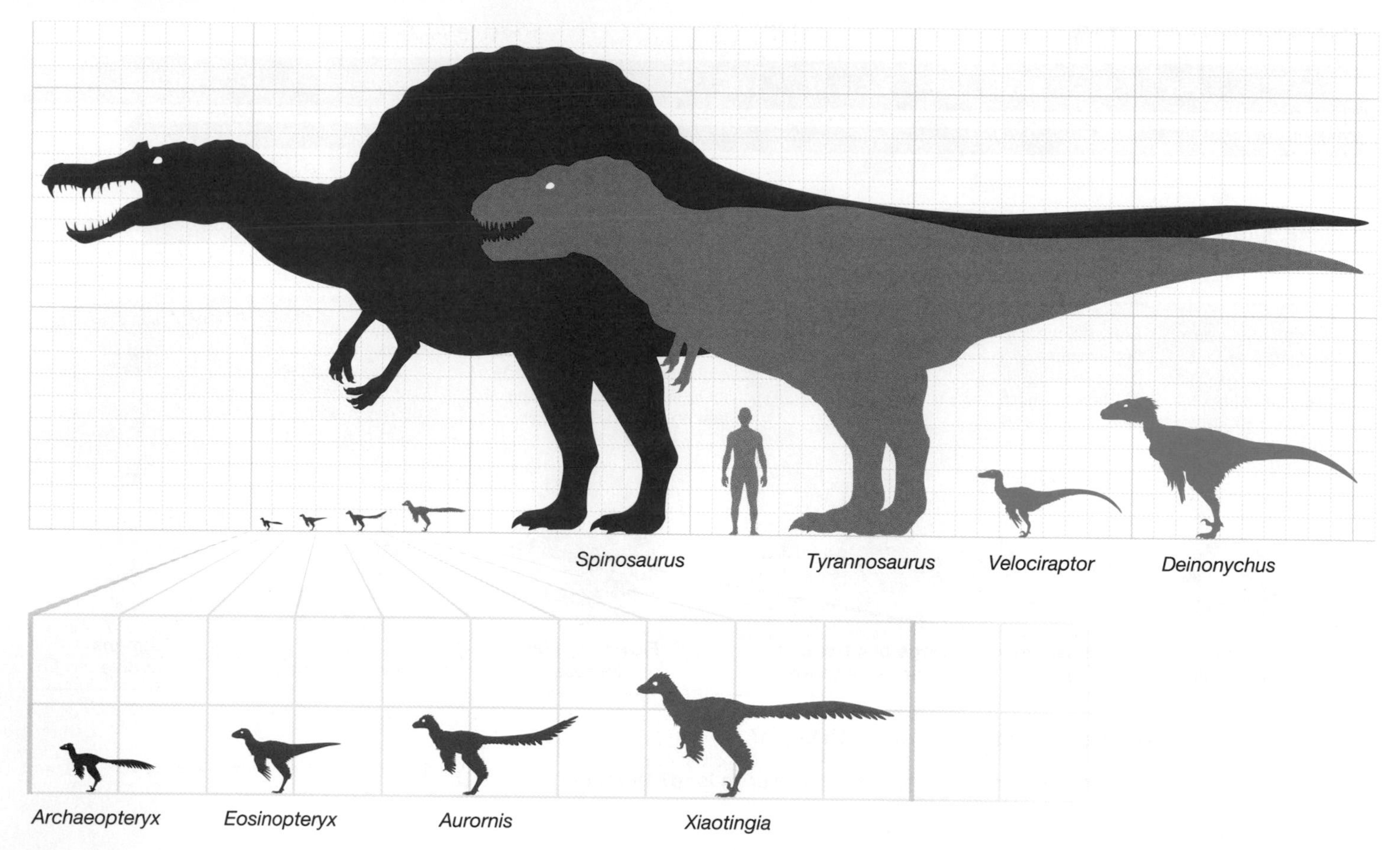

Figure 14.5

Dinosaurs large and small

Theropods ranged in size from tiny, like a chicken, to huge, like the group's most famous member, *Tyrannosaurus rex*.

Q1: **In what ways were theropods similar to modern birds? Give at least two similarities.**

Q2: **In what ways did theropods differ from modern birds? Give at least two differences. (*Hint*: Read the rest of the paragraph.)**

Q3: **Birds are often referred to as "living dinosaurs." Is this accurate? Why or why not?**

See Appendix A for answers to the figure questions.

combination of these. An evolutionary tree maps the relationships between ancestral groups and their descendants, and it clusters the most closely related groups on neighboring branches.

In an evolutionary tree, organisms are depicted as if they were leaves at the tips of the tree branches. A given ancestor and all its descendants make up a **clade**, or branch, on the evolutionary tree. *Archaeopteryx* and all subsequent animals that evolved from it are considered a clade.

A **node** marks the moment in time when an ancestral group split, or diverged, into two separate lineages. A node represents the **most recent common ancestor** of two lineages in question—that is, the most *immediate* ancestor that *both* lineages share. For over 100 years, researchers considered *Archaeopteryx* the most recent common ancestor of both birds and dinosaurs and thus placed it at the root of the avian clade—the first bird (**Figure 14.6**, top).

The first bird, that is, until Xu stumbled across a fossil that threw the field into controversy. In 2008, Xu visited the Shandong Tianyu Museum of Nature, a dinosaur museum

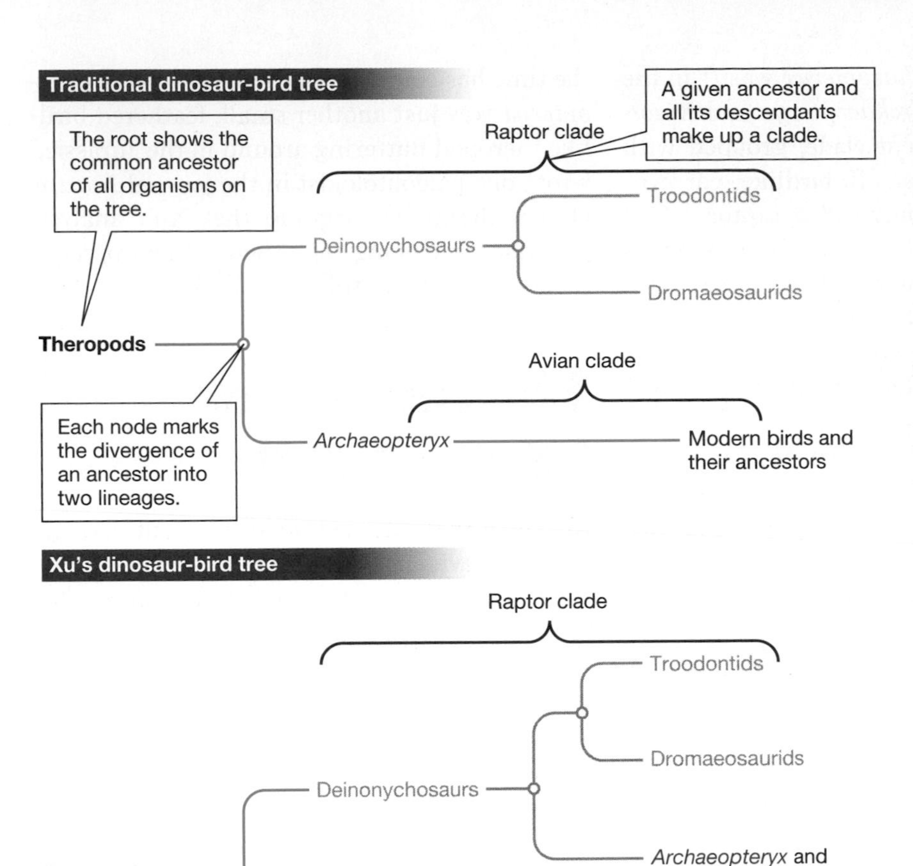

Figure 14.6

The evolutionary origins of birds

(*Top*) The traditional evolutionary tree, showing *Archaeopteryx* as an early bird that split off from the deinonychosaurs (birdlike, carnivorous dinosaurs). (*Bottom*) The evolutionary tree proposed by Xu after the discovery of a new fossil, *Xiaotingia*. This discovery, discussed in the following paragraphs, led to the classification of *Archaeopteryx* as a deinonychosaur rather than an early bird.

Q1: In the traditional tree, identify the node showing the common ancestor for early birds and dinosaurs.

Q2: What do both the traditional tree and Xu's tree suggest about troodontids and dromaeosaurids?

Q3: In both trees, identify the node for the common ancestor of *Archaeopteryx* and other birds. In what way are the nodes different in the two trees?

See Appendix A for answers to the figure questions.

in eastern China. There, he happened upon a unique fossil that had been collected by a farmer in Liaoning and sold through a dealer to the museum. The fossil, entombed in yellowish rock, shows a small, birdlike dinosaur seemingly craning its neck forward and spreading its short wings (**Figure 14.7**). "I saw it and said, 'Oh, this is an important species,'" recalls Xu. He asked the museum to let him study it.

As Xu examined the fossil, which he later named *Xiaotingia zhengi*, he wondered where it belonged on the dinosaur-bird evolutionary tree. To find out, he analyzed *shared derived traits* of the fossil and similar early birds. **Shared derived traits** are unique features common to all members of a group that originated in the group's most recent common ancestor and then were passed down in the group (but not in groups that are not direct descendants of that ancestor). In this case, the original ancestor in question was assumed to be *Archaeopteryx*, and the shared traits included feathers, clawed hands, and a long, bony tail.

By comparing *Xiaotingia* to *Archaeopteryx* and other related species, Xu created a new evolutionary tree of early birds (**Figure 14.6**,

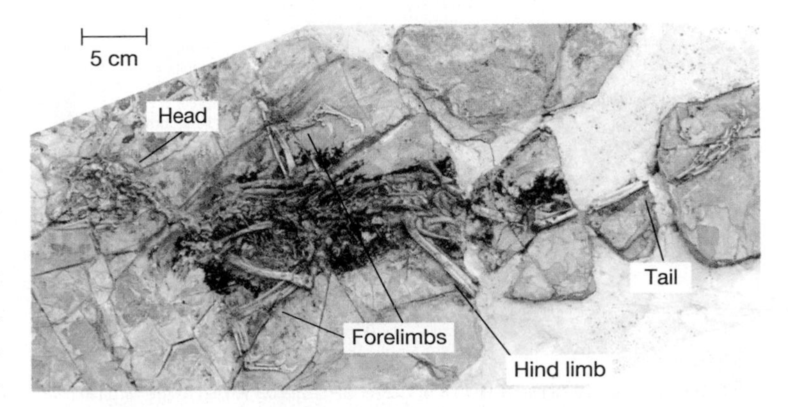

Figure 14.7

Xiaotingia zhengi

After its discovery, this rock-encased fossil added a controversial new leaf to the dinosaur-bird evolutionary tree.

bottom). Suddenly, *Archaeopteryx* wasn't in the avian clade. Instead, *Archaeopteryx* and *Xiaotingia* were in a different clade, grouped with deinonychosaurs—the small, birdlike, carnivorous dinosaurs commonly called raptors. (The term "raptor" refers to carnivorous modern birds such as hawks and owls, but it is also used informally to describe this group of dinosaurs.) *Archaeopteryx*, Xu believed, was not the first bird but a raptor. It had feathers, but its descendants did not evolve into birds.

"It was a big change," says Xu. Yet, he adds, it wasn't entirely unexpected. In the last 20 years, more and more fossils of early avian species have been discovered—and the more Xu compared *Archaeopteryx*, discovered in Europe, to other early birds discovered in China, the less it looked like a bird and the more it looked like a raptor. Early birds have small, thick skulls and two toes on each foot; *Archaeopteryx*, on the other hand, has a long, almost pointy skull and three toes on its feet. "*Archaeopteryx* is just so different from other early birds," says Xu.

Xu published his revised evolutionary tree in 2011. It took the scientific world by storm. Some researchers embraced the idea: "Perhaps the time has come to finally accept that *Archaeopteryx* was just another small, feathered birdlike theropod fluttering around in the Jurassic," wrote one paleontologist in the journal *Nature*. Others disagreed, arguing that Xu's analysis was not convincing. For them, *Archaeopteryx* remained the first bird.

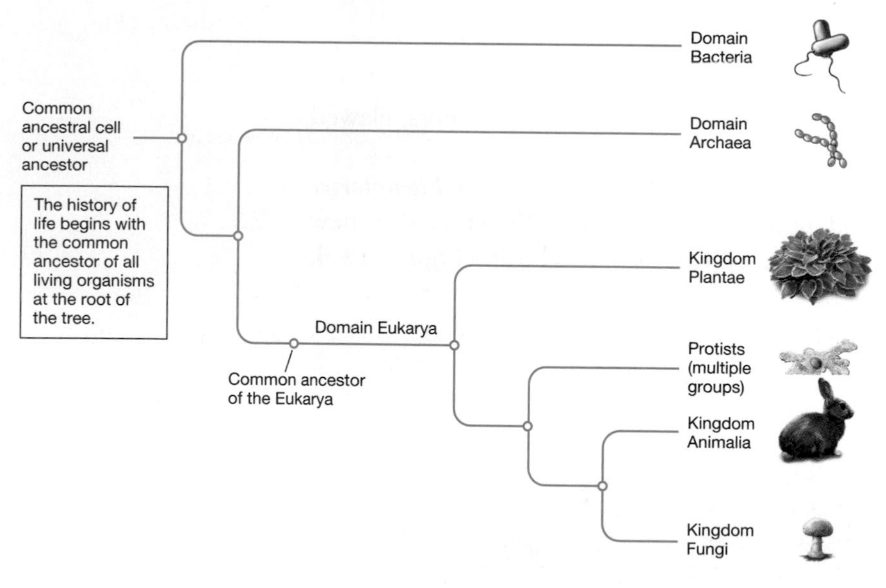

Figure 14.8

The tree of life

The domains Bacteria and Archaea do not have separate kingdoms within them. The domain Eukarya encompasses three kingdoms: Plantae consists of plants. Fungi includes yeasts and mushroom-producing species. Animalia consists of animals. Eukarya also includes protists, an artificial grouping defined by what its members are not—protists are eukaryotes but they are not plants, animals, fungi.

The History of Life on Earth

Xu knew that one new fossil was but a single shred of evidence and therefore unlikely, by itself, to convince the scientific community. "The evidence is not very strong," he admits, "but it was a question we wanted to discuss and add more analyses to, including old and new fossils." The identity of the first bird is important because our understanding of how flight evolved is based on the classification of *Archaeopteryx*. If *Archaeopteryx* is not the species from which flight evolved, then our understanding of flight is wrong.

Xu's redesign of the early bird evolutionary tree raised another important question: Did flight evolve more than once? If *Archaeopteryx*, which looked capable of flight, was a bird, then flight probably evolved just once among dinosaurs, in the avian lineage. If *Archaeopteryx* was not a bird and was instead simply a raptor that could fly, then flight evolved at least twice among dinosaurs, once in raptors and once in avians—but only avians went on to evolve into modern birds.

Biological classification helps us answer important evolutionary questions such as these. In addition to recognizing three broad domains of life, biologists group the Eukarya into three distinct **kingdoms**, the second-highest level in the hierarchical classification of life, and **protists** (meaning "very first" or "primal"), an artificial group that includes amoebas and algae (**Figure 14.8**). The kingdoms are **Plantae**, which encompasses all plants; **Fungi**, which includes mushrooms, molds, and yeasts; and **Animalia**, which encompasses all animals, including dinosaurs, birds, and humans. As an evolutionary tree shows, these kingdoms share common ancestors but diverge at various points (**Figure 14.9**).

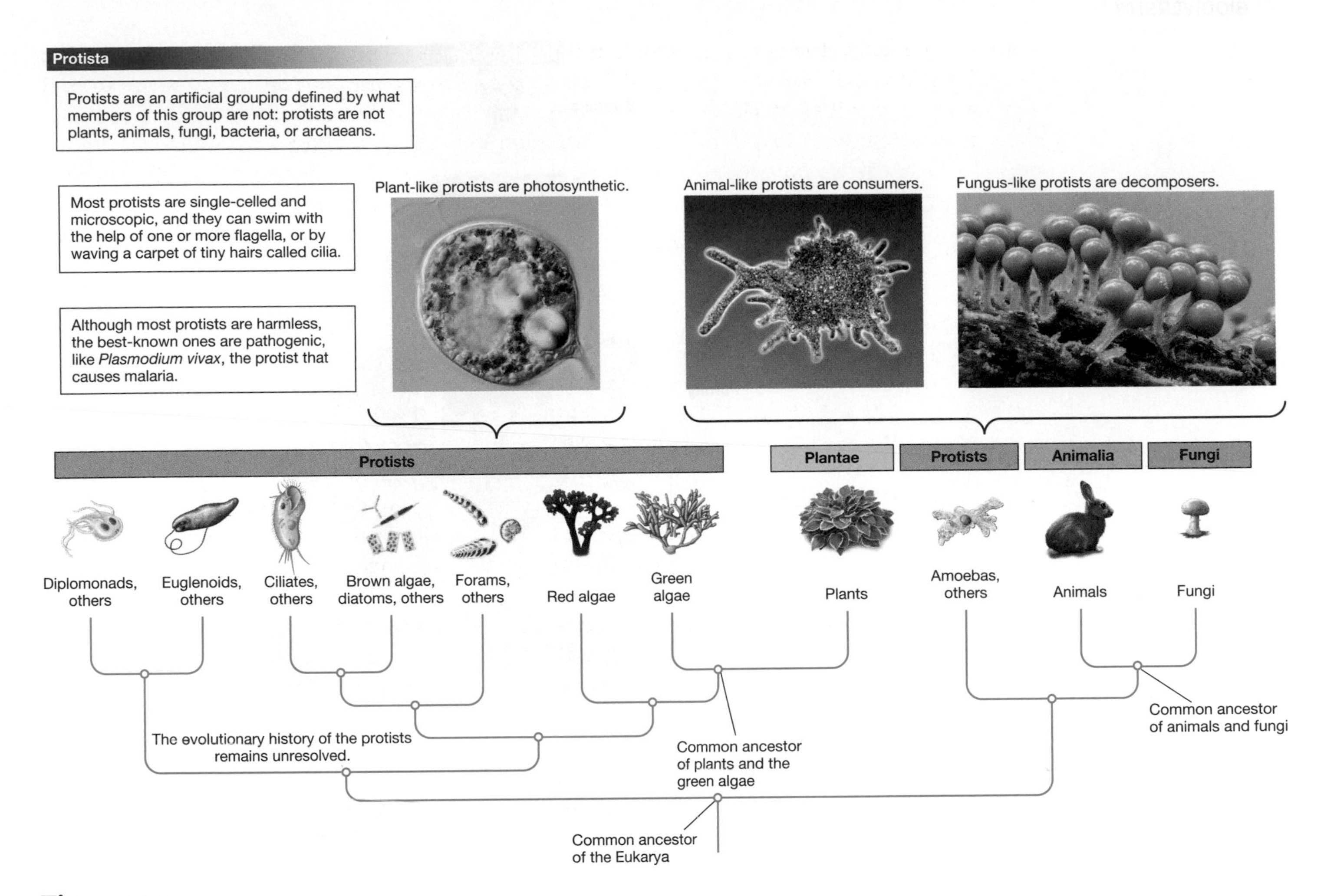
Protista
Protists are an artificial grouping defined by what members of this group are not: protists are not plants, animals, fungi, bacteria, or archaeans.
Most protists are single-celled and microscopic, and they can swim with the help of one or more flagella, or by waving a carpet of tiny hairs called cilia.
Although most protists are harmless, the best-known ones are pathogenic, like Plasmodium vivax, the protist that causes malaria.
Plant-like protists are photosynthetic.
Animal-like protists are consumers.
Fungus-like protists are decomposers.
Protists
Plantae
Protists
Animalia
Fungi
Diplomonads, others
Euglenoids, others
Ciliates, others
Brown algae, diatoms, others
Forams, others
Red algae
Green algae
Plants
Amoebas, others
Animals
Fungi
The evolutionary history of the protists remains unresolved.
Common ancestor of plants and the green algae
Common ancestor of animals and fungi
Common ancestor of the Eukarya

Figure 14.9

The domain Eukarya

This tree traces evolutionary relationships among the organisms within Eukarya. The protists are a kind of biological catchall. (The three kingdoms—Plantae, Fungi, and Animalia—are explored in Figures 14.12–14.14.)

Q1: What group of organisms shares the most recent common ancestor with plants?

Q2: Are fungi more closely related to plants or to animals? Does the answer surprise you? Why or why not?

Q3: If you were to create an evolutionary tree in which amoebas were included within the kingdom of organisms to which they were the most closely related (rather than with protists, where they are currently placed), where would you put them?

See Appendix A for answers to the figure questions.

Below the level of kingdom, biological classification gets even more specific using the **Linnaean hierarchy**, a system of biological classification devised in the eighteenth century by the Swedish naturalist Carolus Linnaeus and revised by later scientists (**Figure 14.10**). The smallest unit of classification in the Linnaean hierarchy—the **species**—reflects individuals that are the most related to each other. The most closely related species are grouped together to form a **genus** (plural "genera"). Using these two categories in the hierarchy, every species is given a unique, two-word Latin name, called its **scientific name**. The first word of the name identifies the genus to which the organism belongs; the second word defines the species. Comparing our own species name, *Homo sapiens*, to that of the Neanderthals, *Homo neanderthalensis*, shows that the two are classified as separate species but belong to the same genus, *Homo*.

In the Linnaean hierarchy, each species is placed in successively larger and more inclusive

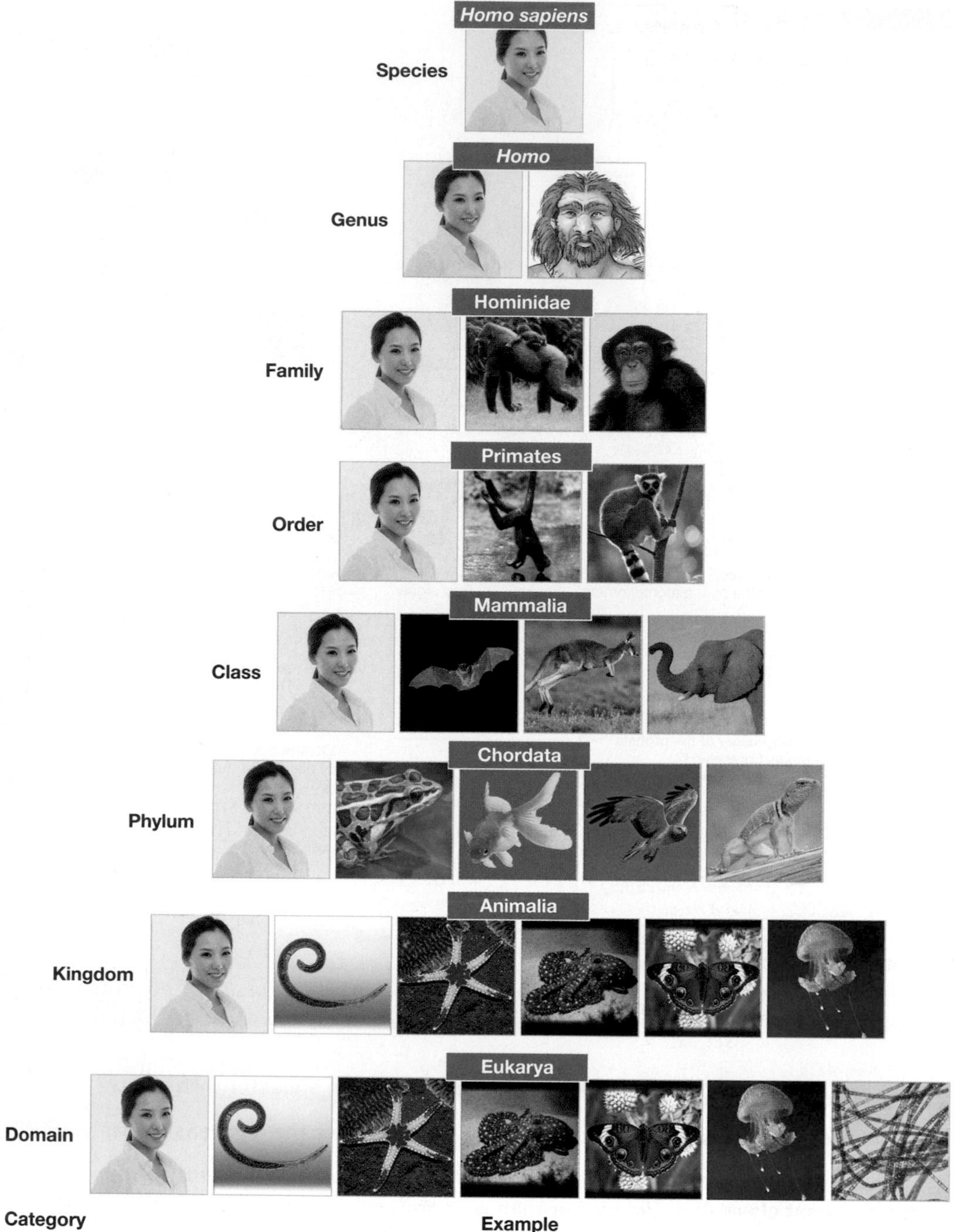

Figure 14.10

The Linnaean hierarchy of classification

According to the system created by Linnaeus, each species is part of a genus, each genus is part of a family, each family is part of an order, and so on, building up to a kingdom.

Q1: Within which category are the organisms shown most closely related to one another?

Q2: Within which category are the organisms shown most distantly related?

Q3: Are individual species within the same order or within the same family more closely related to each other?

See Appendix A for answers to the figure questions.

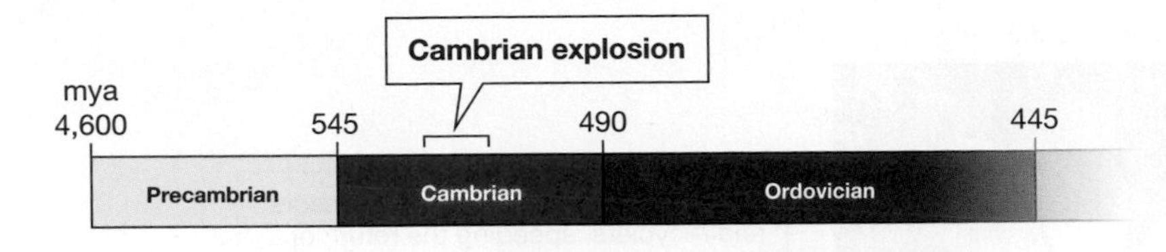

Figure 14.11

Cambrian biodiversity

During the Cambrian period, animal diversity increased dramatically. Shown here are representations of some of the life-forms, all aquatic, that lived during this time. The remains of many of these species have been found in Canada (the Burgess Shale), China (the Maotianshan Shale), and fossil beds in Greenland and Sweden. Some of the fossils look familiar, resembling sponges and clams, but do not appear to be related to any living groups of animals.

categories beyond genus. Closely related genera are grouped into a **family**. Closely related families are grouped into an **order**. Closely related orders are grouped into a **class**. Closely related classes are grouped into a **phylum** (plural "phyla"). And, you guessed it, closely related phyla are grouped together into a kingdom.

So where did all these carefully labeled organisms come from? The earliest forms of life evolved in water. About 650 mya (during the Precambrian period; see Figure 4.3), the number of organisms appearing in the fossil record increased. At that time, much of Earth was covered by shallow seas filled with small, mostly single-celled organisms floating freely in the water.

Then, about 540 mya, the world experienced an astonishing burst of evolutionary activity with a dramatic increase in biodiversity. Most of the major living animal groups first appear in the fossil record during this time, popularly known as the **Cambrian explosion**. The Cambrian explosion changed the face of life on Earth from a world of relatively simple, slow-moving, soft-bodied scavengers and herbivores to a world dominated by large, fast-moving predators. The presence of predators sped up the evolution of Cambrian herbivores, judging by the variety of scales and shells and other protective body coverings typical of many Cambrian, but not Precambrian, fossils. The Cambrian explosion, however, occurred primarily in the oceans (**Figure 14.11**).

Because life first evolved in water, the colonization of land by living organisms posed enormous challenges. Indeed, many of the functions basic to life, including physical support, movement, reproduction, and the regulation of heat, had to be handled very differently on land than in water. About 480 mya, near the beginning of the Ordovician period, algae-like plants were the first organisms to meet these challenges. These early terrestrial colonists were single-celled or had just a few cells.

Fungi are thought to have made their way onto land next, according to fossil evidence. For example, scientists have found fossils of terrestrial fungi that are 455–460 million years old. Reconstructing the evolutionary history of eukaryotes from DNA data, scientists estimate

Zygomycetes (molds)

Many fungi act as garbage processors and recyclers, speeding the return of the nutrients in dead and dying organisms to the ecosystem.

Ascomycetes (sac fungi)

Fungi digest organic material outside their bodies and absorb the molecules that are released as breakdown products.

Basidiomycetes (club fungi)

Fungi are similar to animals in that they don't photosynthesize and they store surplus food energy in the form of glycogen. Like some animals, such as insects and lobsters, fungi produce a tough material, chitin, that strengthens and protects the body. Unlike animals, fungal cells have a protective cell wall that wraps around the plasma membrane and encases the cells.

Figure 14.12

Fungi up close

Fungi, such as the three types shown here, play several roles in terrestrial ecosystems. Many are decomposers (extractors of nutrients from the remains of dead or dying organisms and from waste products such as urine and feces).

that the common ancestor of fungi and animals diverged from all other eukaryotes about 1.5 bya, and fungi diverged from their closest cousins, the animals, about 10 million years after that (**Figure 14.12**).

Next, land plants evolved and diversified from the original green algae that made it to land. By 360 mya, at the end of the Devonian period, Earth was covered with plants (**Figure 14.13**). The first land animals likely emerged from an aquatic environment 400 mya. Many of the early animal colonists on land were carnivores; others fed on living plants or decaying plant material. Every one of the main animal groups except sponges—namely, cnidarians, protostomes, and deuterostomes—has some terrestrial representatives (**Figure 14.14**).

The first vertebrates to colonize land were amphibians, the earliest fossils of which date to about 365 mya (see Chapter 17 for more on vertebrates). Early amphibians descended from lobe-finned fishes. Amphibians were the most abundant large organisms on land for about 100 million years. Then, in the late Permian period, reptiles took over as the most common vertebrate group on land. Reptiles were the first group of vertebrates that could reproduce without returning to open water, because they lay amniotic eggs that have a built-in food

Bryophytes (mosses and liverworts)
Plants have a waxy covering, the *cuticle*, that covers their above-ground parts. A waxy cuticle holds in moisture—an important adaptation to life on land.

Ferns

Gymnosperms
Gymnosperms were the first plants to evolve pollen, a microscopic structure that contains sperm cells, which freed them from a dependence on water for fertilization. Gymnosperms were also the first to evolve seeds, which can be disseminated so they will not compete with the mother plant for sunlight, or for water and nutrients in the soil.

Angiosperms
Flowering plants, or angiosperms, are the most diverse group of plants on our planet. The keys to the success of anglosperms are the flower, a structure that evolved through modification of early plant reproductive organs, and the fruit, a fleshy ovary wall that protects and helps disperse the seeds inside it.

Figure 14.13

Plants up close

Plants, such as the four types shown here, are multicellular autotrophs (primary producers of organic compounds, through photosynthesis) and are mostly terrestrial. Because plants are producers, they form the basis of essentially all food webs on land.

source and are protected from drying out by a hardened shell as compared to the jellylike sac that encloses the eggs of other vertebrates. The evolution of the amniotic egg was a major event in the history of life because it established a new evolutionary branch, the amniotes, which later included all reptiles, birds, and mammals.

Sponges

Cnidarians

The sponges are the most ancient animal lineage. Cnidarians, a group that includes jellyfish and corals, evolved next. The remaining animal phyla fall into two groups: the *protostomes* and *deuterostomes*, distinguished by different patterns of embryonic development. Protostomes comprise more than 20 separate subgroups, including mollusks (such as snails), annelids (segmented worms), and arthropods (including spiders and insects). Deuterostomes include echinoderms (sea stars and their relatives) and the chordates. The chordates are a large group that includes all animals with backbones, such as fish, birds, and humans.

Mollusks

Arthropods

Echinoderms

Chordates

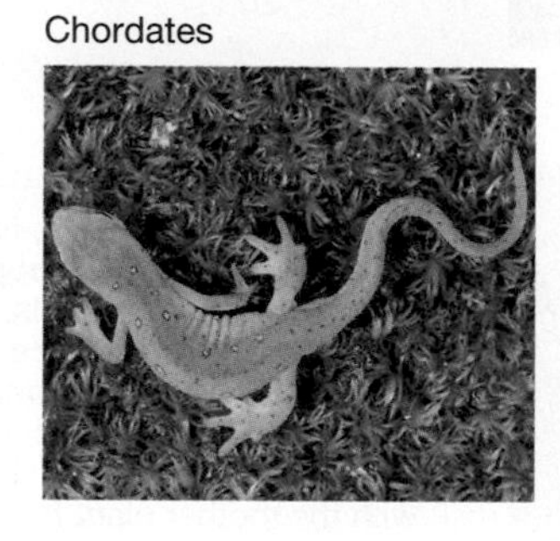

Animal cells differ from those of plants and fungi in that they lack cell walls. Instead, many cells in the animal body are enveloped in, or attached to, a felt-like layer known as the *extracellular matrix*. An important evolutionary innovation of animals is the development of true tissues. Most animals have two or three main tissue layers that give rise to a structurally complex body.

Figure 14.14

Animals up close

Animals are multicellular ingestive heterotrophs (obtainers of energy and carbon by ingesting food). All animals are consumers, and some are important decomposers.

And so, with the rise of reptiles 230 mya, the age of dinosaurs began.

Repeated Revisions

In January 2013, a year and a half after Xu published his controversial paper about *Archaeopteryx* and *Xiaotingia*, his work received unexpected support. Pascal Godefroit, a paleontologist at the Royal Belgian Institute of Natural Sciences, reported the discovery of another feathered dinosaur, *Eosinopteryx brevipenna*. A commercial collector in northeastern China had dug up *Eosinopteryx* in the same area where *Xiaotingia* was discovered. The tiny 161-million-year-old dinosaur was preserved as a virtually complete skeleton, with its legs bent and arms out, as if about to jump (**Figure 14.15**).

When Godefroit and his team added *Eosinopteryx* to the evolutionary tree of feathered dinosaurs, they came to the same surprising conclusion that Xu had: *Archaeopteryx* was not a bird. Instead, they hypothesized that *Archaeopteryx* was a raptor along with *Eosinopteryx* and *Xiaotingia*. All three species share traits that early birds did not have, such as arms longer than their legs, reduced tail plumage, and primitive feather development.

But science is a process of continual revision, and only 4 months later, Godefroit published a description of another birdlike fossil that caused him to revise his hypothesis. This time it was a feathered dinosaur that Godefroit found collecting dust in the archives of a Chinese museum. The 18-inch-long fossil had small, sharp teeth and long forelimbs. Godefroit believed the dinosaur, which his team named *Aurornis xui*

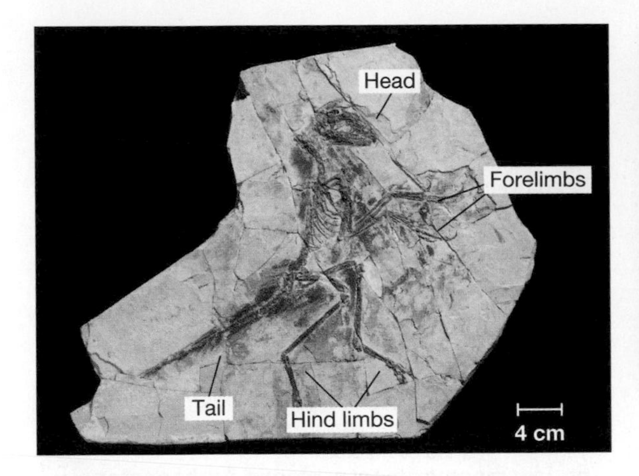

Figure 14.15

Eosinopteryx brevipenna
The discovery of this feathered-dinosaur fossil supported Xu's new tree.

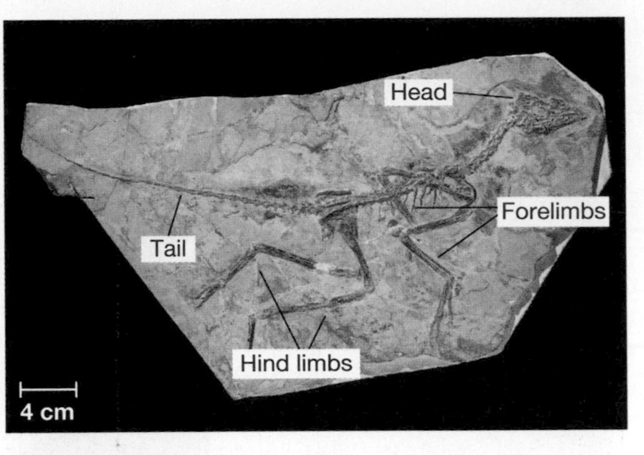

Figure 14.16

Aurornis xui
This fossil illuminates the transition from birdlike dinosaurs into birds.

to honor Xu's work, probably couldn't fly but instead used its wings to glide from tree to tree. But the fossil's other features, including the hip bones, are clearly shared by modern birds (**Figure 14.16**).

Using *Aurornis*, Godefroit constructed another evolutionary tree. This time he started from scratch, compiling data on almost a thousand characteristics of skeletons of 101 species of dinosaurs and birds. "It's very impressive," paleontologist Mike Lee at the South Australian Museum told *National Geographic*. "They considered more than twice the anatomical information as even the best previous analyses."

Contrary to the tree that arose from his first study, Godefroit's second tree places *Archaeopteryx* back on its roost in the bird family, although no longer as the oldest bird. That place belongs to *Aurornis*, says Godefroit (**Figure 14.17**).

Yet the debate is not over. "Of course, we need more evidence and more work," says Xu. "Many of these new species are a possible candidate for the earliest bird." There are likely to be more fossils that will shake up the bird tree, says Xu, but that's a good problem to have in science. "There are so many new species it just makes it difficult for us all to agree."

Xu, Godefroit, and others are sure to continue to dig up dinosaur fossils for some time yet. Dinosaurs arose about 230 mya and dominated the planet from about 200 mya to about 65 mya. Then the majority of them went extinct, except for those that evolved into birds. As the fossil record shows, species have regularly gone extinct throughout the history of life. The rate at which this has happened—that is, the number of species that have gone extinct during a given period—has varied over time, from low to very high. At the upper end of this scale, the fossil record shows that there have been five **mass extinctions**, periods of time during which great numbers of species go extinct (**Figure 14.18**).

Although difficult to determine, the causes of the five mass extinctions are thought to include such factors as climate change, massive volcanic eruptions, changes in the composition of marine and atmospheric gases, and sea level changes. The Cretaceous extinction event occurring about 65 mya wiped out three-fourths of plant and animal species on Earth, including non-avian dinosaurs. Researchers suspect that a massive comet or asteroid slammed into the Gulf of Mexico, choking the skies around the planet with debris and

DEBUNKED!

MYTH: Dinosaurs are extinct.

FACT: Although there was a mass extinction of dinosaurs 65 mya, not *all* dinosaur species went extinct. A handful of small feathered dinosaurs survived, eventually evolving into roughly 10,000 species of birds.

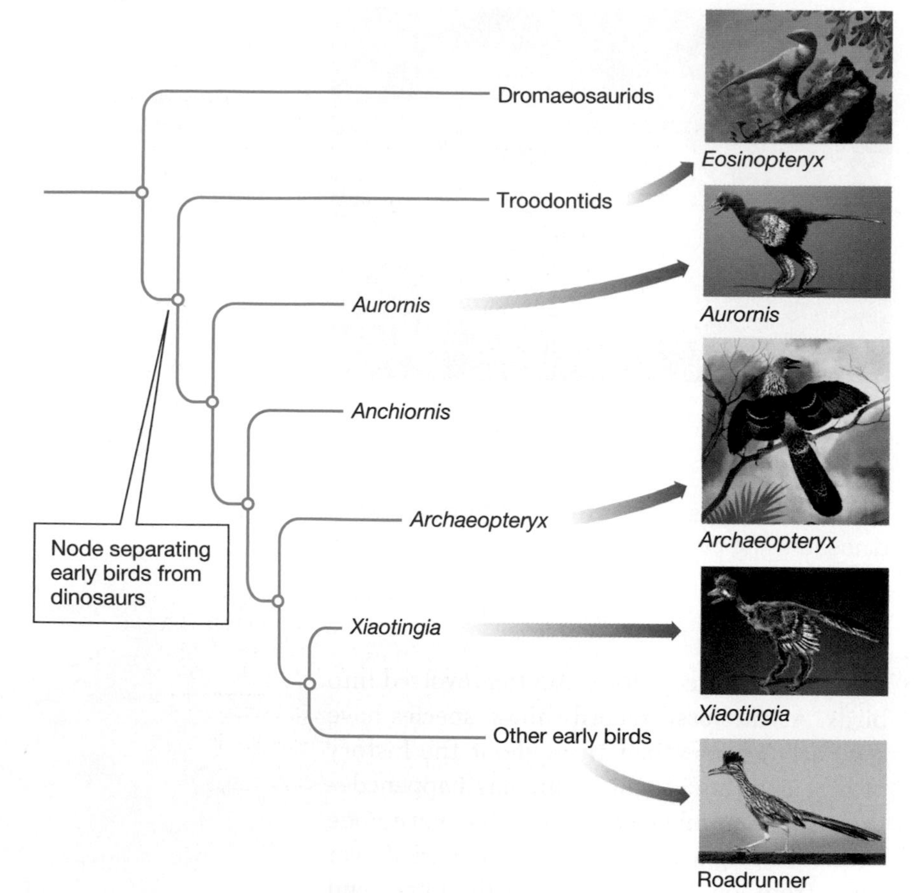

Figure 14.17

The early birds

Godefroit's second 2013 study places *Archaeopteryx* and *Xiaotingia* with birds rather than dinosaurs, as in the traditional dinosaur-bird evolutionary tree (see Figure 14.6, top). But it places *Aurornis* as the earliest known bird on the evolutionary tree.

Q1: Is *Xiaotingia* an earlier or later bird than *Archaeopteryx* in this tree?

Q2: If a future study, based on more fossils or new measurements, placed *Archaeopteryx* back with dinosaurs, would this suggest that birds are not related to dinosaurs? Why or why not?

Q3: If you were to create an evolutionary tree of modern birds, where would you expect to place the roadrunner (judging by its appearance in this figure) as compared to a house sparrow or pigeon?

See Appendix A for answers to the figure questions.

decreasing the ability of plants to photosynthesize. As the plants died, so did animals further up the food chain.

Mass extinctions affect the diversity of life in two main ways: First, entire groups of organisms perish, changing the history of life forever. Second, the extinction of one or more dominant groups of organisms can provide new opportunities for groups previously of minor importance, thereby dramatically altering the course of evolution. When a group of organisms expands to take on new ecological roles and to form new species and higher taxonomic groups, that group is said to have undergone an **adaptive radiation**. Some of the great adaptive radiations in the history of life occurred after mass extinctions, such as when the mammals diversified after the extinction of the dinosaurs.

Today, one species of mammal, *Homo sapiens*, dominates life on land, and our impact on biodiversity is unprecedented. Because of human activities, the world is losing species at an alarming rate. In July 2016, researchers at the United Nations Environment Programme World Conservation Monitoring Centre in the United Kingdom announced that worldwide biodiversity had fallen below predetermined "safe" levels, thresholds below which ecological function is likely to be negatively affected. An estimated 58 percent of the world's land has lost more than 10 percent of its biodiversity, the researchers found, with grasslands and biodiversity hot spots such as the Amazon rainforest hit the hardest.

That is alarming news because nonhuman organisms provide even the most basic requirements for human life. Plants and plantlike prokaryotes and protists such as algae produce the oxygen we breathe and provide us with food. Whole ecosystems provide so-called "ecosystem services," environmental benefits that humans rely on. For example, coastal redwood trees in northern California intercept fog, mist, and rain, channeling the water onto and into the ground.

Biologists today assert that we are on our way toward a new mass extinction and the cause is clear: the activities of the ever-increasing number of humans living on, and exploiting, Earth.

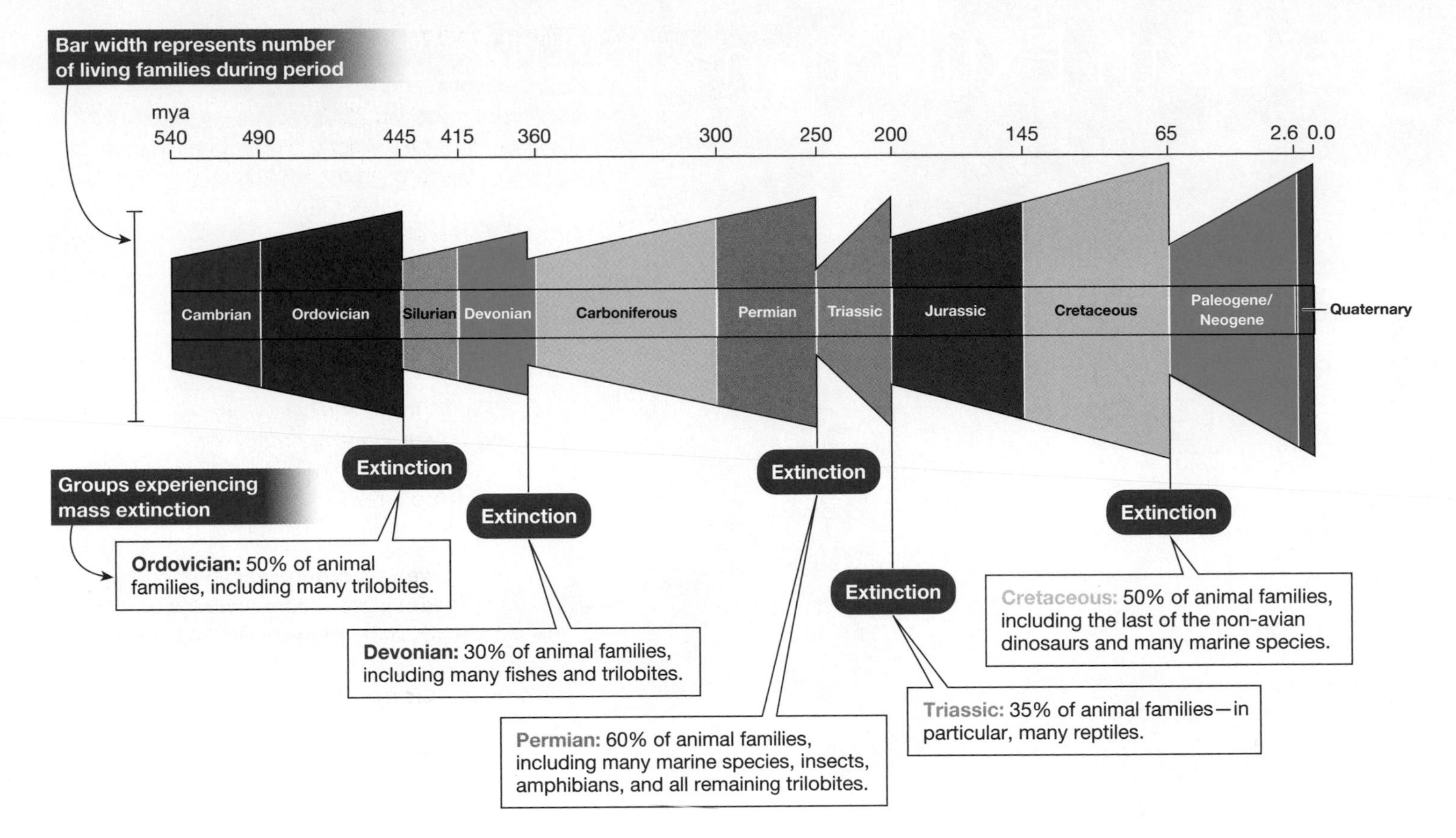

Figure 14.18

Mass extinctions and biodiversity in animals

In addition to the marine and terrestrial animal groups shown here, other groups of living organisms were severely affected by the five mass extinctions that have occurred in Earth's history. After each of the extinctions, life again diversified.

Q1: What extinction event occurred about 200 mya? What animal groups were most affected by this event?

Q2: Which of the mass extinctions appears to have removed the most animal groups? How long ago did this extinction occur?

Q3: The best studied of the mass extinctions is the Cretaceous extinction. Why do you think it has been better studied than the other extinctions?

See Appendix A for answers to the figure questions.

The Sixth Extinction

On at least five occasions, mass extinctions have occurred across the globe, caused at different times by climate change, volcanic eruptions, and possible asteroids. Today, scientists agree we are in the midst of a sixth extinction, and this time, we, the human race, are the cause.

Assessment available in smartwork

Ghosts of species past

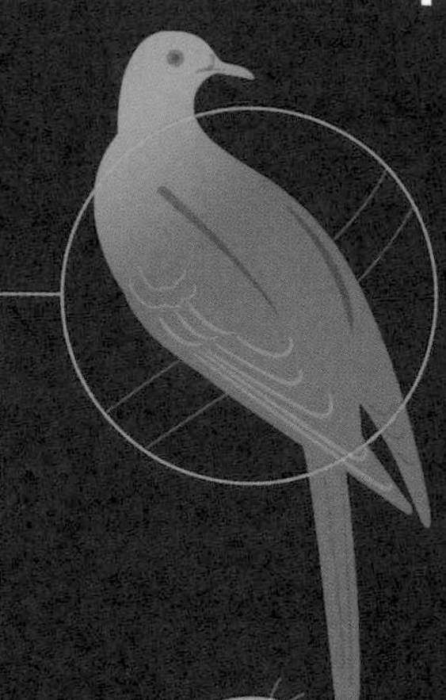

Passenger Pigeon

The passenger pigeon, the most common bird in North America 200 years ago, was hunted to extinction in the nineteenth century. The last wild bird was shot in 1900, and the last captive bird died in 1914.

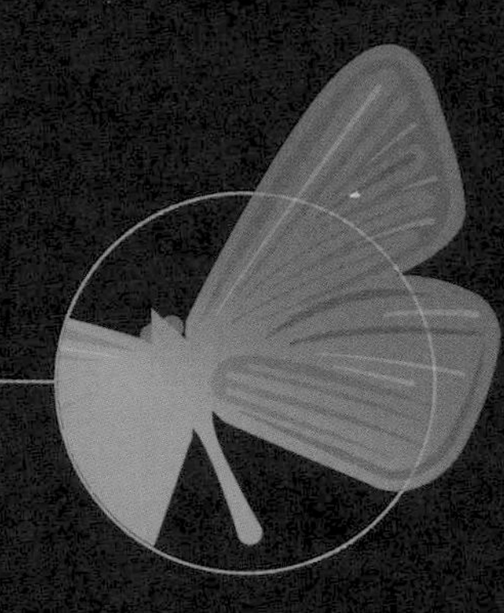

Xerces Blue Butterfly

The Xerces blue butterfly once lived on sand dunes around San Francisco—until its habitat was destroyed by urban development. It was last seen in 1943.

Caribbean Monk Seal

The Caribbean monk seal was the only seal native to the Caribbean Sea and Gulf of Mexico. It was overhunted for oil, and its food sources were overfished. It was last sighted in 1952.

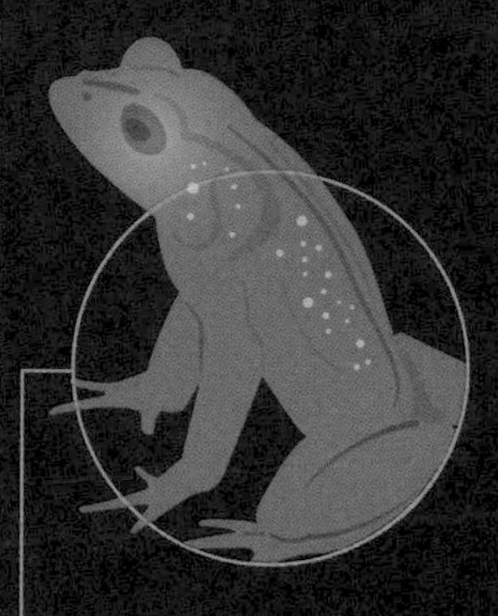

Golden Toad

Once common in the cloud forests of Monteverde, Costa Rica, the golden toad has not been seen since 1989. Pollution and global warming likely contributed to its extinction.

Number of species known to be extinct or extinct in the wild since 1500

Arachnids	9
Crustaceans	12
Reptiles	22
Amphibians	36
Insects	58
Fishes	71
Mammals	79
Plants	134
Birds	145
Mollusks	324

NEED-TO-KNOW SCIENCE

- The first single-celled organisms resembled modern prokaryotes and probably evolved by 3.7 bya.
- The most basic and ancient branches of the tree of life define three **domains** (page 267): **Bacteria** (page 267), **Archaea** (page 268), and **Eukarya** (page 268). All life-forms fall into one of these three domains. The Eukarya are further divided into three kingdoms: **Fungi** (page 272), **Plantae** (page 272), and **Animalia** (page 272). The **protists** (page 272) are an artificial grouping of eukaryotes, defined by their not belonging to any of the three kingdoms. The variety of these life-forms, as well as their interactions with each other and the ecosystems they inhabit, is **biodiversity** (page 267).
- Scientists use an **evolutionary tree** (page 269) to model ancestor-descendant relationships among different organisms. The tips of branches represent existing groups of organisms, and each **node** (page 270) represents the moment when an ancestor split into two descendant groups. A **clade** (page 270) is an ancestral species and all its descendants.
- Closely related groups of organisms share distinctive features that originated in their **most recent common ancestor** (page 270). These **shared derived traits** (page 271) are used to identify each **lineage** (page 269), line of descent, of a species.
- The **Linnaean hierarchy** (page 273) places each **species** (page 273) in successively larger and more inclusive categories. Closely related species are grouped together into a **genus** (page 273), related genera are grouped into a **family** (page 275), related families into an **order** (page 275), related orders into a **class** (page 275), related classes into a **phylum** (page 275), and finally, related phyla into a **kingdom** (page 272).
- Organisms are identified by their genus and species names, together referred to as their Latin **scientific name** (page 273).
- The release of oxygen by photosynthetic prokaryotes caused oxygen concentrations in the atmosphere to increase. Rising oxygen concentrations made possible the evolution of single-celled eukaryotes about 2.1 bya. Multicellular eukaryotes followed about 650 mya.
- Life in the oceans changed dramatically during the **Cambrian explosion** (page 275), when large predators and well-defended herbivores suddenly appear in the fossil record. The Cambrian explosion is an example of **adaptive radiation** (page 280), in which a group of organisms takes on new ecological roles and forms new species and higher taxonomic groups.
- The land was first colonized by plants (about 480 mya), fungi (455–460 mya), and then animals (e.g., insects about 400 mya), with vertebrates the last to come ashore (about 365 mya). Each group evolved unique adaptations that enabled life to flourish on land.
- There have been five **mass extinctions** (page 279) during the history of life on Earth. The extinction of a dominant group of organisms may provide new opportunities for other groups.

THE QUESTIONS

See Appendix B for answers.

The Basics

1 The first single-celled organisms on Earth were likely prokaryotes and probably evolved

(a) 3,700 years ago
(b) 3.7 mya
(c) 3.7 bya
(d) We have no idea when the first life on Earth appeared.

2 The production of _________ by prokaryotes increased its atmospheric concentration and enabled more complex forms of life to evolve.

(a) carbon dioxide
(b) oxygen
(c) nitrogen
(d) all of the above

3 The term "Cambrian explosion" refers to

(a) an increase in biodiversity.
(b) a mass extinction.
(c) an atmospheric detonation due to increasing gas concentrations.
(d) an oceanic eruption caused by underwater volcanic activity.

4 Which of the following terms most specifically describes what occurs when a group of organisms expands to take on new ecological roles, forming new species and higher taxonomic groups in the process?

(a) speciation

(b) mass extinction

(c) evolution

(d) adaptive radiation

5 Match each term with its definition.

clade	1. A distinctive feature that originated in two groups' most recent common ancestor.
node	2. An ancestor and all its descendants.
lineage	3. The point at which an ancestral group splits into two separate lineages.
evolutionary tree	4. A diagram showing the evolutionary relationships among a related group of organisms.
shared derived trait	5. The line of descent of a group of organisms.

6 Select the correct terms:

The domains Archaea and Bacteria are referred to as (**prokaryotes** / **eukaryotes**). The domain (**Eukarya** / **Prokarya**) includes three kingdoms. The kingdom (**Plantae** / **Fungi**) was first to make the transition to land. The kingdom (**Animalia** / **Plantae**) is most closely related to the kingdom Fungi.

7 All of the following are considered possible causes of the five mass extinctions *except*

(a) climate change.

(b) change in the composition of atmospheric or marine gases.

(c) comet or asteroid strike.

(d) worldwide thunderstorms.

(e) volcanic eruptions.

Challenge Yourself

8 *Archaeopteryx*

(a) belongs to the domain Prokarya.

(b) belongs to the protist group.

(c) lived during the Precambrian period.

(d) was much larger than modern birds.

(e) is an early example of the evolution of birds.

9 Identify the domain and kingdom for each of the following organisms.

(a) *Archaeopteryx*

(b) chanterelle mushrooms

(c) palm tree

(d) green algae

(e) you

10 "King Philip came over for good soup" is a mnemonic to remember the levels of the Linnaean hierarchy.

(a) List each level of the Linnaean hierarchy in the order indicated by the mnemonic.

(b) What level of the Linnaean hierarchy is missing from the mnemonic?

(c) Create your own mnemonic to remember the Linnaean hierarchy.

11 Place the following evolutionary events in the correct order from earliest to most recent by numbering them from 1 to 5.

____ a. Cambrian explosion

____ b. origin of life on Earth

____ c. plants' transition to land

____ d. oxygen-rich environment created by plantlike prokaryotes

____ e. evolution of birds from dinosaurs

12 Fill in the tree below to show the evolutionary relationships among the domains and kingdoms of life (and protists!).

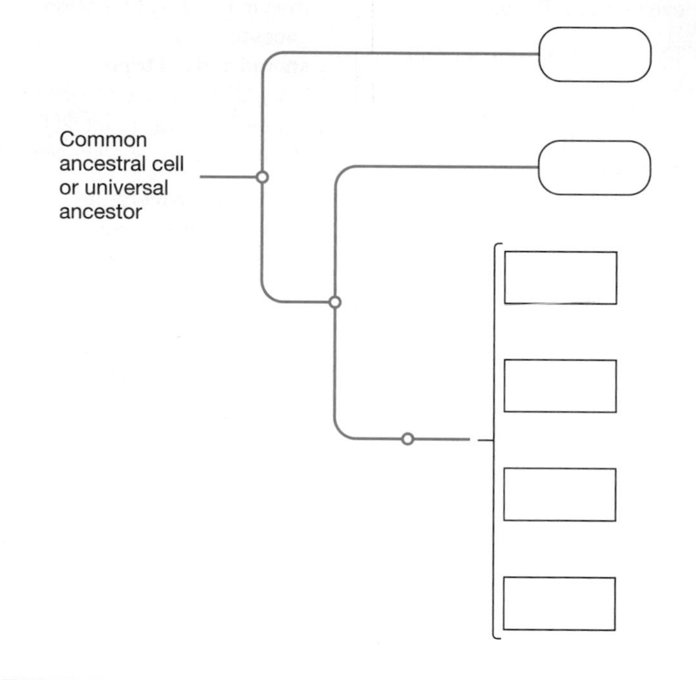

Try Something New

13 In each of the following cases, which domain(s), kingdom(s), or both might contain organisms with the traits described? (You may need to review **Figures 14.2, 14.9**, and **14.12–14.14** to answer.)

(a) movement

(b) single-celled organisms only within the domain or kingdom

(c) found in extreme environments (for example, high temperature, low oxygen, high salt)

(d) multicellular with organ systems

(e) photosynthetic

14 The traditional dinosaur-bird tree (Figure 14.6, top) can be restated as a hypothesis: "We hypothesize that *Archaeopteryx* is an early bird and that birds split off from the closely related dinosaur group deinonychosaurs, which then split into troodontids and dromaeosaurids." Restate Xu's tree (Figure 14.6, bottom) and Godefroit's tree (Figure 14.17) as hypotheses involving *Archaeopteryx*, *Xiaotingia*, and *Aurornis*.

15 The accompanying figure shows the taxonomic diversity of terrestrial (land-living) vertebrates over time.

(a) What is the unit of taxonomic diversity in the graph?

(b) Why were there no terrestrial vertebrates in the Silurian?

(c) What was the first class of vertebrates to make the water-to-land transition?

(d) What class of terrestrial vertebrates is the most diverse, i.e., has the most orders in the present day?

(e) Describe how amphibian diversity has changed over time. How is that different from the status of other terrestrial vertebrates?

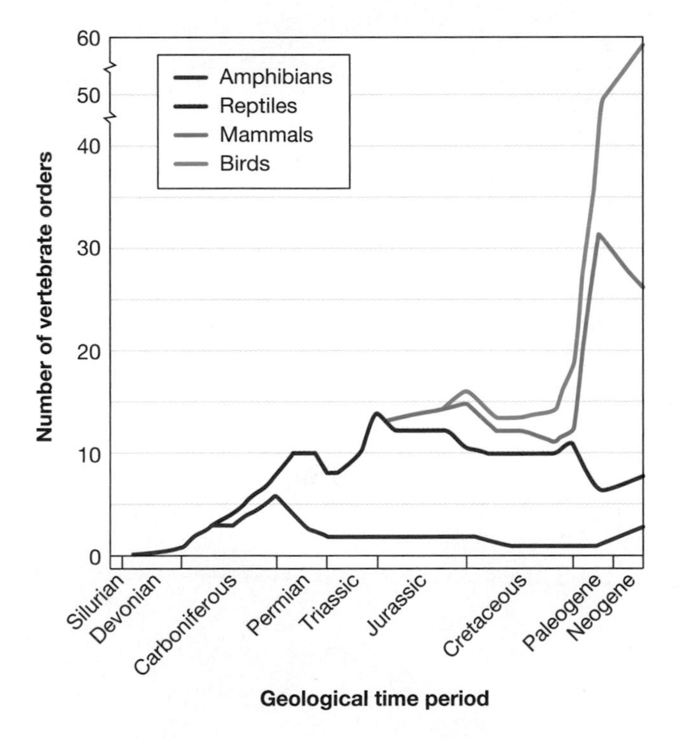

Leveling Up

16 ***Write Now*** **biology: early life on Earth** Read the *New York Times* article "World's Oldest Fossils Found in Greenland" (http://www.nytimes.com/2016/09/01/science/oldest-fossils-on-earth.html). Answer the following questions, writing one paragraph for each.

(a) Summarize the findings of the study reported in the *New York Times* in your own words.

(b) Some scientists have argued these objects are not fossils but were instead caused by normal tectonic activity (https://www.nationalgeographic.com/science/2018/10/news-oldest-stromatolite-fossilized-life-rocks-greenland/). If the scientific community reached a consensus that these are *not* fossils, how would this conclusion affect our understanding of early life on Earth? Explain why such a conclusion would be not a failure of the scientific method but rather a natural result of it.

(c) Define the Late Heavy Bombardment (LHB). Describe how our understanding of the LHB has changed, and how this change has influenced our hypotheses about early life on Earth.

17 **Is it science?** Some people became quite upset when scientists agreed that the dinosaur *Brontosaurus* was in fact *Apatosaurus*, *Triceratops* was only a juvenile *Torosaurus*, and Pluto was not a planet. They clung to their *Brontosaurus*, or *Triceratops*, or planet Pluto, dismissing the scientists' arguments by saying, "Scientists are always changing their minds." Explain how scientists "changing their minds" is not a problem but instead a necessary and useful feature of science, using the science described in this chapter as an example.

Digital resources for your book are available online.

Navel Gazing

Scientists discover hundreds of new species of microbes—burrowed deep in our belly buttons.

After reading this chapter you should be able to:

- Differentiate between bacteria and archaea, and explain why they are together referred to as prokaryotes.
- Explain why prokaryotes can reproduce more rapidly than eukaryotes and why rapid reproduction can be advantageous.
- Give examples of the environments in which archaeans are found.
- Illustrate through example the breadth of metabolic diversity found in bacteria.
- Explain the function of at least one cellular structure that is unique to prokaryotes.
- Explain how citizen science contributes to scientific research.

BIODIVERSITY

CHAPTER 15

BACTERIA AND ARCHAEA

The gathering crowd jostled for position around the table. The scientists behind the table smiled as they handed out long, thin cotton swabs. Holly Menninger took a swab and then stared down at her shirt. Like other volunteers at the conference, she'd been told what to do: lift your shirt, insert the swab into your belly button, swirl it around, place it into a test tube, seal the top. "Oh, the things we do for science," Holly thought. Then she swirled.

After returning her tube, Menninger—an entomologist by training and, at the time, coordinator of the New York Invasive Species Research Institute at Cornell University—struck up a conversation with the researchers manning the table, led by Rob Dunn of North Carolina State University. Dunn, an applied ecologist and writer with an infectious passion for science, initiated the swabbing project to answer a simple question: What lives on our skin? The belly button, as small and strange as it is, was an excellent place to look.

If you pick away the lint, the leftover schmutz in your belly button isn't dirt; it's alive. Our bodies are ecosystems, home to an estimated 39 trillion resident **microbes** (microscopic organisms), primarily bacteria. Considering that the average human body is made up of 30 trillion human cells, that's about a 1:1 ratio of microbial cells to human cells. (The old 10:1 ratio is a miscalculation that has unfortunately become enshrined in popular culture.) The **human microbiome**—the complete collection of microbes that live in and on our cells and bodies—affects human gut health, brains, and even body odor.

Microbes live on most surfaces of our bodies, inside and out: in our guts, on our eyelashes, and yes, in our belly "holes," as Dunn calls them. In fact, the belly "hole" is an ideal place for a scientific study of resident microbes because it is a protected, moist patch of skin and one of the few areas that individuals don't regularly wash. Dunn knew that if his students could determine which microbes were swimming around in individual volunteers' navels, then they could dig into the more burning question: *Why* do we each have the microbes we do?

"We know that which microbes you have on your skin influences your risk of infection, how attractive you are to other people, and how attractive you are to mosquitoes," says Dunn. "So this question of 'What determines which microbes are on your skin?' is super intriguing."

Of the three domains of life on Earth, Bacteria was the first to split off from the shared ancestor of the Archaea and Eukarya (**Figure 15.1**). Some fossil evidence places that split at about 3.48 billion years ago (bya), yet a 2016 discovery in Greenland suggests that bacteria existed as early as 3.7 bya, close to a period of time when Earth was being bombarded by asteroids (**Figure 15.2**).

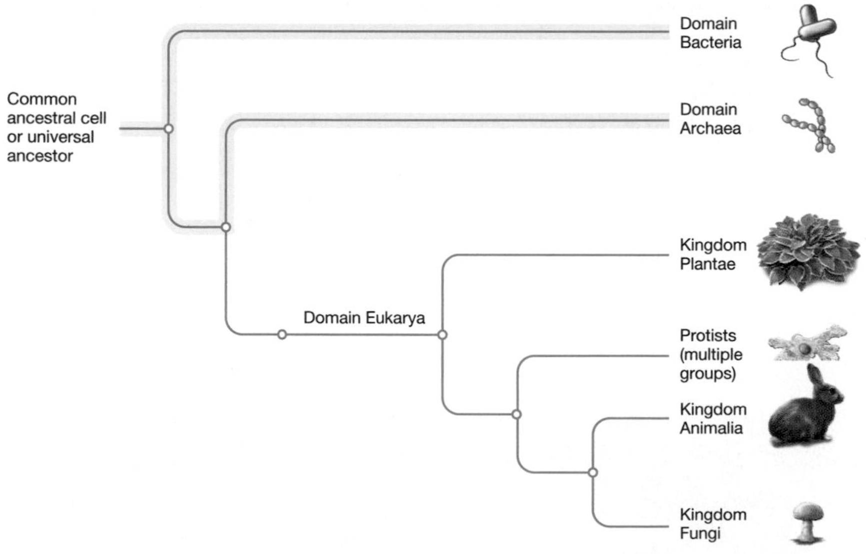

Figure 15.1

Two of the three domains of life

Bacteria and Archaea are two distinct domains. They share many characteristics that are not found in the Eukarya.

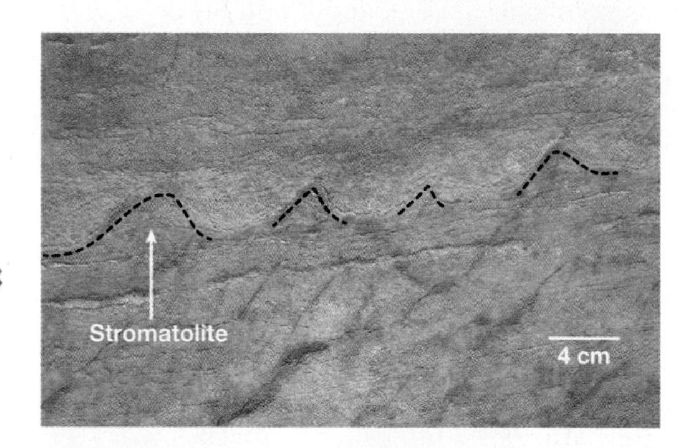

Figure 15.2

Did earth's first life leave its mark in Greenland rocks?

In 2016, Australian scientists found stromatolites, cell-like structures that may be the remains of early life, in rocks formed 3.7 billion years ago in Greenland. This finding is controversial, but similar sites in Australia document life on earth at least 3.45 billion years ago.

Archaea most likely split from Eukarya much later, around 2.7 bya. The microbes that make up Bacteria and Archaea display many small but significant differences in their DNA, plasma membrane structure, and metabolism. They also share several important characteristics and so they are traditionally grouped together as prokaryotes.

Prokaryotes are single-celled organisms, with a single loop of DNA floating free in the cytoplasm of the cell (check out Figure 4.8 for a comparison of prokaryotic and eukaryotic cells). Unlike cells found in the Eukarya, prokaryotes do not have membrane-enclosed organelles. Prokaryotic cells are not only simpler than eukaryotic cells but also they are smaller—almost exclusively microscopic and invisible to the naked eye because of their diminutive size. The simple structure and single loop of DNA enable prokaryotes to reproduce at a much more rapid rate than eukaryotes, doubling in number in as few as 10 to 30 minutes (**Figure 15.3**).

Prokaryotes in both domains are widespread, extremely abundant, and display an astonishing diversity in metabolism. The vast majority of life on Earth is single-celled and prokaryotic. Scientists estimate that the number of prokaryotes on Earth is about 5,000,000,000,000,000,000,000,000,000,000 (5 nonillion, or 5×10^{30}). For example, prokaryotes—not fishes or algae—are the most abundant organisms in the open ocean where they play a crucial role in the ecology of our biosphere.

The success of prokaryotes is due in part to how quickly a prokaryotic population reproduces. It is also due to the fact that prokaryotes can live practically anywhere, including places where few other forms of life can survive. For example, prokaryotes exist in deep, hot thermal vents as well as in the acidic environment of the human intestine. Humans are saturated in microbes. Each of us is a microbial zoo.

Menninger was so intrigued by Dunn's project that when a job became available in his lab a few months after her swab test, she immediately applied for and got the position. By then, the team had gathered about 60 swabs at local events, including the conference where Menninger swabbed, and now they had bigger ambitions: Dunn hired Menninger to spearhead

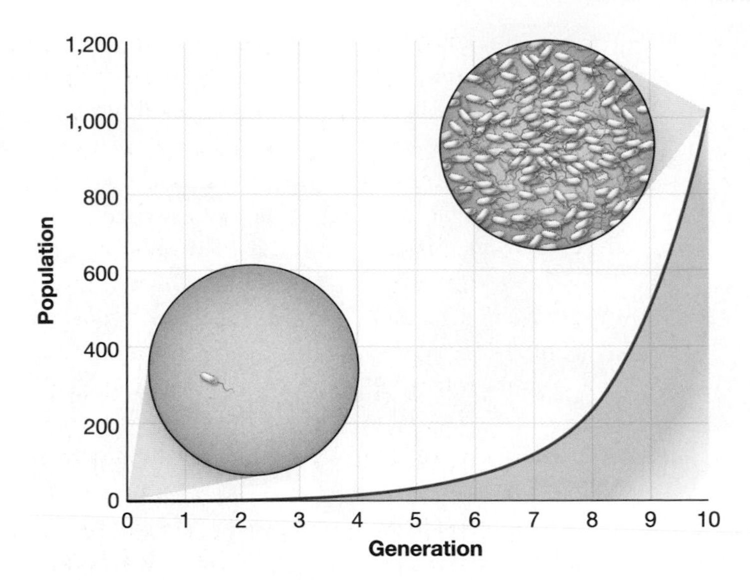

Figure 15.3

Prokaryotes are capable of extremely rapid population growth

An individual prokaryote is able to divide in two within 10–30 minutes. Those two can each divide in two in the same amount of time. This means that a single bacterium or archaean can become a large population in a very short period of time.

Q1: If an individual prokaryote divides every 20 minutes, how many individuals will there be after an hour?

Q2: If the generation time is 20 minutes, how much time will have gone by when the final generation shown has doubled?

Q3: Many bacteria are able to reproduce more quickly in warmer conditions. What does this suggest to you about the importance of refrigerating foods?

See Appendix A for answers to the figure questions.

DEBUNKED!

MYTH: Five-second rule! Food dropped on the floor and left for less than 5 seconds is safe to eat because microbes need time to transfer.

FACT: Some bacteria transfer onto food instantaneously, in less than 1 second, so think twice about eating that cookie off the floor.

Figure 15.4

Professional scientists and scientists-in-training

Rob Dunn and Holly Menninger are pictured here with students in the laboratory.

a massive public outreach effort to acquire belly button swabs from all around the country (**Figure 15.4**). In **citizen science** projects such as this, the public plays a part in research by contributing, collecting, and sometimes even analyzing data in cooperation with professional scientists (**Figure 15.5**).

"At the time, we were right at the forefront of an explosion in research, learning about how important the microorganisms that live on and in our bodies are to our health and well-being," says Menninger. "And this was a really great project to get people talking about the skin microbiome."

Dutch tradesman Antonie van Leeuwenhoek first opened our eyes to the microbial world back in 1668, when he used simple, handcrafted microscopes to discover bacteria swimming around in pond water. In the intervening 350 years, our knowledge of the microbial world has come a long way.

Today, we know that bacteria and archaeans make up more than two-thirds of the species found on Earth. In 2016, microbiologists at UC Berkeley published a revised evolutionary tree of life that, for the first time, included the vast diversity of bacteria and archaeans lurking in Earth's nooks and crannies. With the data used to generate that tree, they showed that life is clearly dominated by prokaryotes (**Figure 15.6**).

Data collection

Citizen scientists from the Republic of the Congo are learning to geo-tag key resources in their forest as part of the Extreme Citizen Science Intelligent Maps project.

Data analysis

A citizen scientist in North America uses an app to upload and compare observations of when local plants begin to bud to a large database of plant budding dates.

Sebastian Seung, a professor at Princeton University, works with citizen scientists around the world to map the brain through a video game.

Figure 15.5

Citizen scientists work alongside professional scientists to increase our knowledge about the natural world

You can contribute to science by collecting data, analyzing data, or even being an experimental participant like the video game players who work with Sebastian Seung. Citizen scientists learn new things, often have fun, and report feeling a sense of purpose.

Q1: In which of the three ways listed above did the navel microbiome participants contribute?

Q2: Which of the advantages listed above do you think the navel microbiome citizen scientists received?

Q3: Would you be willing to contribute to the navel microbiome project? Why or why not?

See Appendix A for answers to the figure questions.

Merry Microbes

In 2010, as a gag, an undergraduate student in Dunn's lab decided to make a Christmas card with streaks of microbes taken from people in the lab. She asked for belly button swabs from her coworkers, spread each sample on a petri dish, and grew the bacteria into colonies. As she grew bacteria from multiple individuals, it quickly became clear that there was more variety among the microbes growing on people in the lab than had been expected.

"It went from being a fun lab project to a serious question," says Dunn. The Christmas gag had raised an important question: Why do the microbes of one person differ from those of someone else? From his experience studying ecology, Dunn knew that microbes are critical in our lives. But neither Dunn nor the others knew what factors determine which particular skin microbes a person has or doesn't have.

Dunn and his team began to study the locally collected swabs. An early part of the experiment involved visually depicting each person's microbial menagerie. The team plated each sample onto a petri dish and then took a picture of the dish for the microbes' owner. Although prokaryotes are single-celled organisms, some prokaryotes form colonies or long chains of identical cells, produced by repeated splitting begun by one original cell. This bacterial pattern of reproduction resulted in bright, dramatic patterns on the plates (**Figure 15.7**).

Under the microscope, the individual microbes appeared even more diverse and elaborate. Bacteria and Archaea cells are quite variable in shape, ranging from rods to spheres to spirals (**Figure 15.8**). Still, they all have a basic structural plan. Most bacteria and many archaeans have a protective cell wall that surrounds the plasma membrane. Some have an additional wrapping around that cell wall called a **capsule**. The capsule, made of slippery biomolecules, works like an invisibility cloak to help disease-causing bacteria evade the immune system that protects organisms like us from foreign invaders.

Dunn's team observed many bacteria whose surface was covered in short, hairlike projections called **pili** (singular "pilus"), a common bacterial feature. Bacteria use pili to link together to form

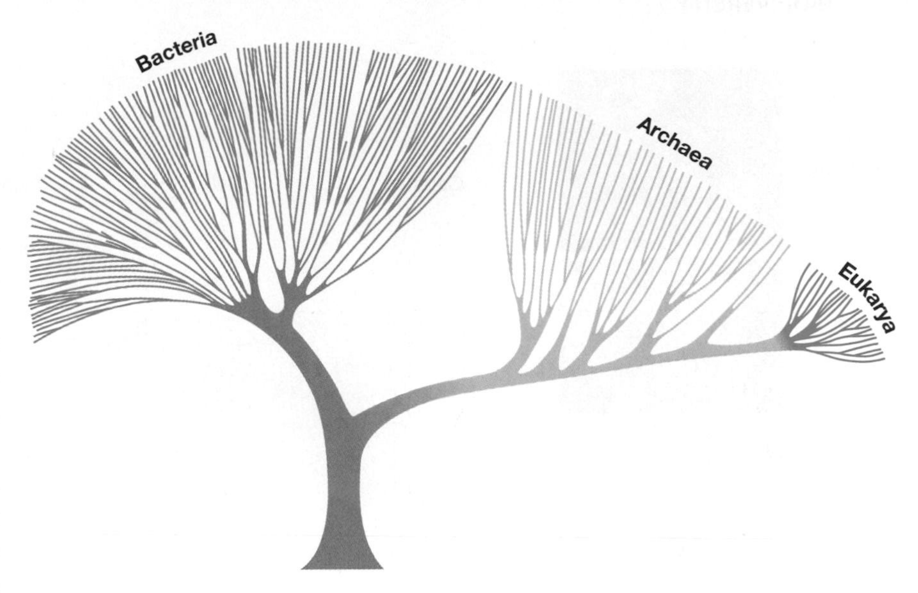

Figure 15.6

The most recent tree of life

To create this graphic, researchers sequenced DNA from individuals in 3,083 genera across all three domains. This information was combined by phylum, and the placement of each phylum in the tree was based on its relatedness to other phyla, as measured by DNA similarity.

Q1: Where in the figure would you place the first life on Earth?

Q2: Where in the figure did Bacteria split off from the ancestor of Archaea and Eukarya?

Q3: The figure (and thus the study) demonstrates that Archaea and Eukarya are more closely related to each other than to Bacteria. How is that illustrated?

See Appendix A for answers to the figure questions.

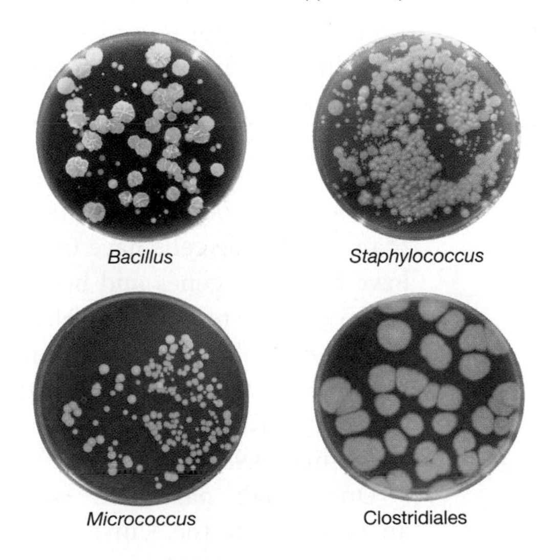

Figure 15.7

Belly button bacterial biodiversity

Each of these petri dishes contains colonies of the bacteria collected from one person's belly button sample.

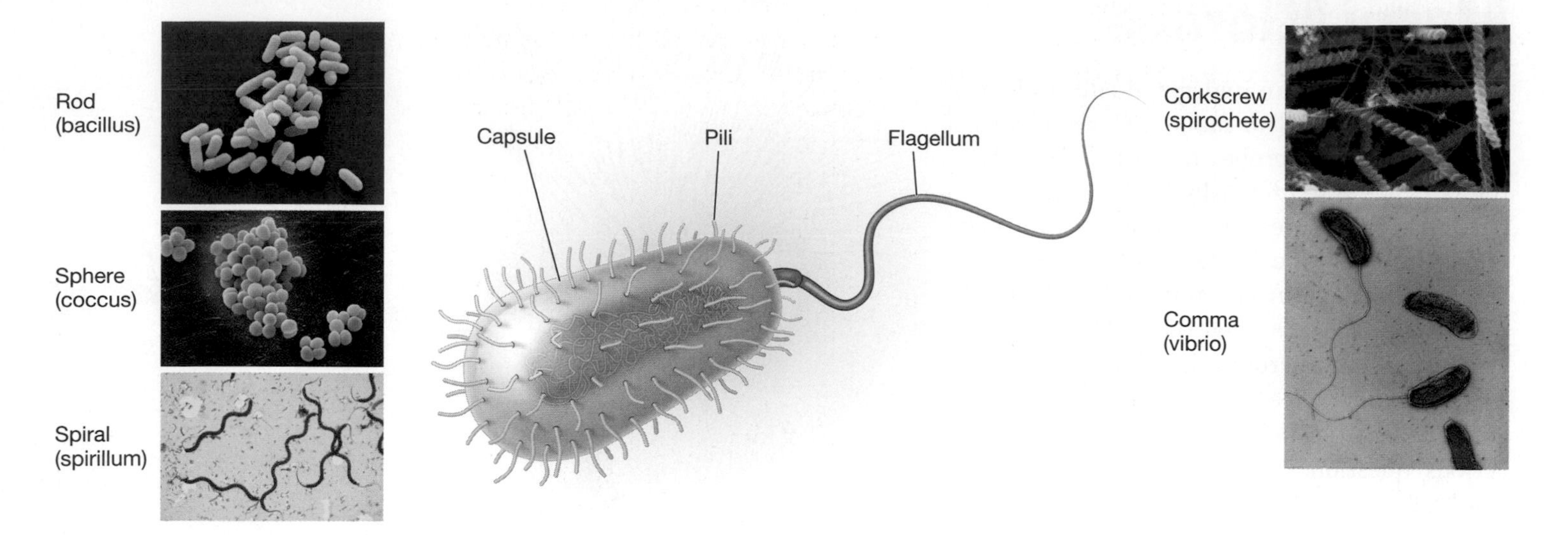

Figure 15.8

A simple structure, a diversity of forms

Bacteria and Archaea share a simple cell structure compared to eukaryotes, but they display a striking diversity of shapes, and some have additional structures that perform special functions.

Q1: Which of these shapes do you think *Streptococcus* would take?

Q2: From the micrographs here, does it appear that all prokaryotes have a flagellum?

Q3: Which one of these shapes is most clearly capable of moving by its own volition? Why?

See Appendix A for answers to the figure questions.

bacterial mats or to attach to surfaces in their environment, such as the cells of the human intestine. Some bacteria have one or more long, whiplike structures called **flagella** (singular "flagellum"), which spin like a propeller to push the bacterium through liquid.

Under the microscope, one obvious cellular feature is missing: prokaryotes do not have a true nucleus. They typically have much less DNA than eukaryotic cells have because prokaryotes have far fewer genes and because they contain relatively little noncoding DNA (DNA that is not used to construct proteins but may have regulatory or other functions). In contrast, eukaryotes generally have many more genes and far more noncoding DNA.

Dunn's lab partnered with Noah Fierer, an ecologist at the University of Colorado, to sequence the DNA from each belly button sample to determine the identity of the resident microbes, especially those that don't grow in a laboratory culture. To do so, lab members first accessed that DNA by mechanically crushing the microbes and chemically dissolving their cell membranes with detergent, similar to how dish soap dissolves grease clinging to a dirty dish.

Next, they ran each sample through a silica strainer, called a column, to capture the DNA while allowing other debris to pass through. Once the DNA was isolated, the team amplified it using the polymerase chain reaction technique (see Figure 9.9) and then sequenced one short section of it, about 250 nucleotides of a well-studied gene called 16S ribosomal RNA (16S rRNA). Although 16S rRNA is found in all bacteria and archaeans, it varies from one species to another, providing an identifying tag, like a fingerprint, to distinguish one species of microbe from another.

In 2012, the team published results from the initial set of swabs. Among 92 belly buttons, the team identified 1,400 species of bacteria. "About 600 or so don't match up in obvious ways with known species, which is to say either they are new to science or we don't know them well enough," Dunn said in a media interview after the paper was published.

They had discovered new species of bacteria living in our navels. Still, Dunn was most interested in finding patterns in the data. The first pattern the scientists identified was that the microbes in belly buttons are somewhat predictable. Six species showed up in about 80 percent of people, and when present, those species tended to be most widespread. "If I go to a cocktail party, I can tell you which species will be most abundant in the room," says Menninger with a laugh. That distribution pattern reminded Dunn of rainforest ecology: in any given forest, the types of plants might vary, but an ecologist can depend on a certain few tree types dominating the landscape.

Likewise, infrequent species tended to be present only in small quantities in belly buttons. Some were familiar: *Bacillus subtilis*, known to cause foot odor, also lives in some belly buttons (where it likely causes the same smell). Others were quite rare: one participant hosted a species previously found only in the soil of Japan, yet the man had never been to Japan. Dunn found a species on himself known solely for eating pesticides. Why it was on his skin remains a mystery.

Amazingly, the team found more than just bacteria. On one particularly fragrant individual, who said he had not washed in years, they detected two species of archaeans. Although some bacteria thrive in unusual environments, the domain Archaea is well known for the extreme lifestyles of its members (**Figure 15.9**). Some are extreme **thermophiles** ("heat lovers") that live in geysers, hot springs, and hydrothermal vents—cracks in the seafloor that spew boiling water. The cells of most organisms cannot function at such high temperatures, but evolutionary innovations enable thermophiles to succeed where others cannot. Other archaeans, classified as **halophiles** ("salt lovers"), thrive in very salty environments where nothing else can live.

Archaeans have also been found in less exotic locations. In another citizen science project from Dunn and Fierer, students analyzed the microbial DNA swabs from home surfaces such as kitchen counters, door frames, television screens, and more. Once again, they found several species of archaeans. "They are actually a lot more common than people tend to realize," says Menninger. "We're starting to have the molecular tools that allow us to find them." Dunn thinks the discovery is a bit ironic: "We spent so much time finding archaeans in the first place, in hard-to-reach, faraway places, and somebody could have just had some introspection with their own belly hole."

Yet back in the belly hole, Dunn was unable to answer his main question about the bacteria: Why do individuals have the bacteria they do? The team looked at gender, ethnicity, how often participants washed their belly buttons, age, and more, yet they were unable to attribute any of the variation to biological or lifestyle factors.

Archaea found in mineral hot springs are able to withstand extreme acidity *and* high temperatures.

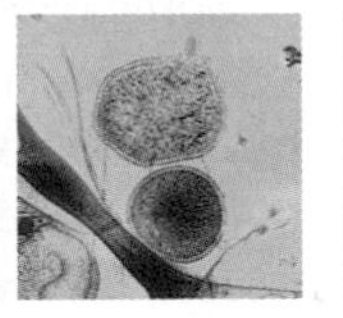

A new archaean species was recently found in an abandoned copper mine. Others have been found living in acidic drainage from mines.

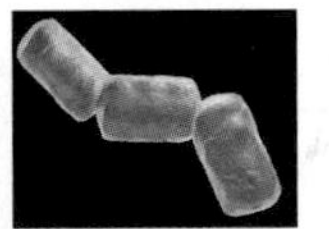

Hydrogen-eating, methane-producing archaeans have been found in deep-sea thermal vents.

Figure 15.9

Archaeans are everywhere

Archaeans, and bacteria to a lesser extent, thrive in many environments, including places where humans are incapable of surviving. They are found at extremely high or low temperatures or in extremely salty or acid environments.

Q1: Which shape in the previous Figure 15.8 corresponds to the archaeans from deep-sea thermal vents in Figure 15.9?

Q2: Why are many archaeans referred to as "extremophiles"?

Q3: Is there anywhere you think archaeans could *not* survive? Justify your answer.

See Appendix A for answers to the figure questions.

So, as scientists tend to do, Dunn went looking for more samples. When Menninger joined the team, she initiated a program to have students and volunteers from around the country send in belly button swabs. Thanks to Menninger's outreach efforts, volunteers nationwide swabbed, swirled, and sealed. Soon the lab had over 600 samples to work with. That's a lot of microbes.

Talk but No Sex

Like the plants and animals in any environment, such as a rainforest, microbes in an ecosystem communicate with each other, reproduce, and more. Researchers used to think prokaryotes were the ultimate loners—self-sufficient and maintaining strict single-celled lifestyles. Then they discovered that bacteria physically interact with each other in many ways.

Some bacteria (and archaeans) have a unique system of cell-to-cell communication called **quorum sensing**, which enables them to sense and respond to other bacteria in the area in accordance with the density of the population. Disease-causing bacteria, for instance, may begin to multiply rapidly after sensing that their numbers are high enough to overwhelm the host organism's immune system. Other bacteria coordinate their behavior by forming tough aggregates called *biofilms*, which are made up of the same or different species (**Figure 15.10**). One well-known example of biofilms is dental plaque.

It's strange to think about, but all these activities are happening on our skin: "Everything you can imagine life doing happens in you," says Dunn, including reproduction. Prokaryotes typically reproduce by splitting in two in a process called **binary fission**, a form of asexual reproduction (see Figure 6.3). The DNA in the parent cell is copied before fission, and one copy is transferred to each of the resulting daughter cells. The genetic information in the daughter cells is virtually identical to that of the parent cell, as in asexual reproduction.

Although sexual reproduction has not been seen in prokaryotes, they are adept DNA pickpockets. Microbes can capture bits of DNA from their environment or other bacteria and

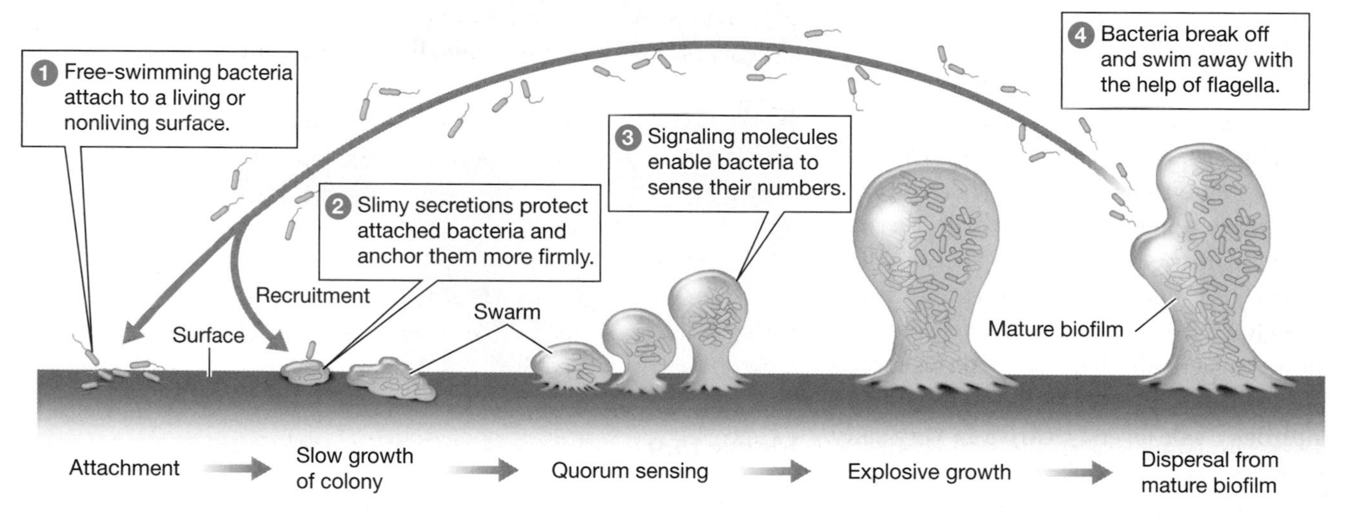

1 Free-swimming bacteria attach to a living or nonliving surface.
2 Slimy secretions protect attached bacteria and anchor them more firmly.
3 Signaling molecules enable bacteria to sense their numbers.
4 Bacteria break off and swim away with the help of flagella.
Recruitment
Surface
Swarm
Mature biofilm
Attachment
Slow growth of colony
Quorum sensing
Explosive growth
Dispersal from mature biofilm

Figure 15.10

Quorum sensing enables coordinated behavior in bacteria

With quorum sensing, a bacterial population can increase its virulence (harmfulness), reproductive rate, or antibiotic resistance under the appropriate conditions. Biofilms are produced via quorum sensing to protect bacteria from environmental hazards.

Q1: How do individual bacteria know that they have a "quorum"?

Q2: There is a well-known biofilm found in your mouth. What is it?

Q3: Under what conditions might bacteria want to coordinate (via quorum sensing) to increase their reproductive rate?

See Appendix A for answers to the figure questions.

incorporate them into their own genetic material. The transfer of genetic material between microbes is known as *horizontal gene transfer*, and it involves plasmids, loops of non-chromosomal DNA in the cytoplasm of prokaryotes (see Figure 12.12). Horizontal gene transfer has been seen in some eukaryotes as well. Bdelloid rotifers, for example, a type of microscopic freshwater animal, lifted about 8 percent of the DNA in their genome from bacteria.

A bacterium can directly trade DNA with another bacterium through a process known as *bacterial conjugation*. Alternatively, when a bacterium dies, the cell may burst open and another bacterium may simply take up the released DNA.

Some types of bacteria—but no archaeans—can undergo **sporulation**, the formation of thick-walled dormant structures called spores. Spores are the bomb shelters of microbes: they can survive boiling and freezing, thereby allowing the microbes to hang out for a long period of time until the conditions are again favorable to reproduce. In many cases, even antibiotics can't kill bacterial spores.

All Hands on Deck

Dunn, Menninger, and the team analyzed samples flooding in from citizen scientists around the country. Still, they were unable to explain the differences among belly button microbes. They could not identify any factors affecting the number and types of species a person might host.

They decided to take a step that more and more scientists are taking—that is, they made all their data freely available online for other researchers to use (with all identifying information about the participants removed). "We asked other researchers to let us know what their insights are, because we're all in this together," says Menninger. "We need all hands on deck."

The gambit worked. Mathematician Sharon Berwick at the University of Maryland took a new approach to the data. Instead of focusing on characteristics of the participants, such as gender or location, she looked at the characteristics of the microbes and found that people tend to be dominated by one of two types of belly button bacteria: *aerobic* or *anaerobic*.

Aerobes are prokaryotes that need oxygen to survive. *Micrococcus* species, which Dunn found in the navel, are aerobes; they need oxygen. Therefore, species of *Micrococcus* are unlikely to do well deep inside a belly button, says Dunn, but they appear to thrive on the surface.

Another type of navel resident, species of *Clostridia*, do not use oxygen. They are **anaerobes**, prokaryotes that survive without oxygen. In fact, some anaerobes may actually be poisoned by oxygen. Among the anaerobic archaeans are several species of **methanogens**, which feed on hydrogen and produce methane gas as a by-product of their metabolism. The ability to exist in both oxygen-rich and oxygen-free environments is another reason why prokaryotes can be found in most habitats.

Some prokaryotes can even switch between the two. One of the most common bacteria Dunn's team found in belly buttons was *Staphylococcus*. Although it is best known as a pathogen, staph on your skin is typically a good guy. On the skin, it is beneficial for your health, fighting off other pathogens that want to crowd in. Some species of staph, such as *Staphylococcus epidermidis*, are typically aerobes, but when oxygen is in short supply, they switch over to a special type of anaerobic metabolism known as **fermentation**, involving the breakdown of sugars (yes, the same process that is used to ferment beer and wine; to review fermentation, see Figure 5.12). In other words, as Dunn wrote in one description of *S. epidermidis*, "Right now you might be making a teeny tiny bit of navel wine."

Like staph, *Bacillus subtilis*, a rod-shaped, spore-forming bacterium, can grow in both aerobic and anaerobic conditions. *B. subtilis* is one of the warrior clans in our belly buttons and other areas of the skin, producing antibiotics that kill off other bacteria and even foot fungi, says Dunn. Prokaryotes are either consumers, called **heterotrophs**, or producers, called **autotrophs** (**Figure 15.11**). Autotrophs make food on their own, but heterotrophs, like *B. subtilis*, obtain energy by taking it from other sources. Specifically, *B. subtilis* is a **chemoheterotroph**, an organism that consumes organic molecules to get energy (in the form of chemical bonds) and carbon (in the form of

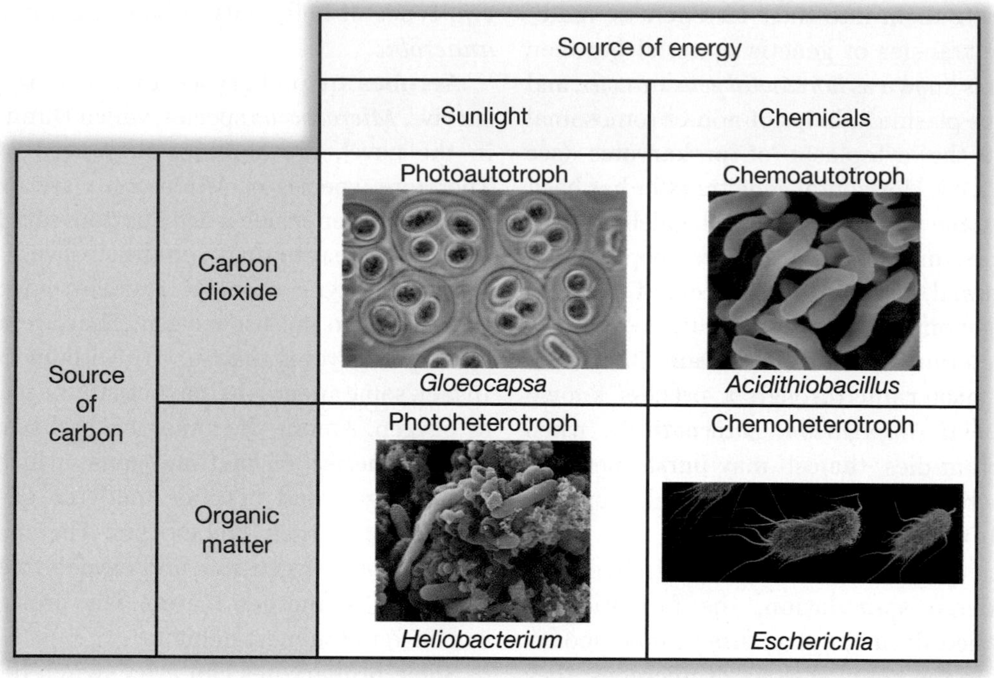

		Source of energy	
		Sunlight	Chemicals
Source of carbon	Carbon dioxide	Photoautotroph *Gloeocapsa*	Chemoautotroph *Acidithiobacillus*
	Organic matter	Photoheterotroph *Heliobacterium*	Chemoheterotroph *Escherichia*

Figure 15.11

How prokaryotes feed themselves

Prokaryotes have many more ways to feed themselves than do eukaryotes. They are categorized by their energy source and by the source of carbon they use. To answer the following questions, read the rest of this section.

Q1: What source of energy would you expect a cave-dwelling prokaryote to use?

Q2: In which of these categories would you place the bacteria responsible for nitrogen fixation? Why?

Q3: In which of these categories do decomposers belong? Explain your reasoning.

See Appendix A for answers to the figure questions.

carbon-containing molecules). Some species of bacteria, like the heliobacteria abundant in rice paddies, are **photoheterotrophs**, which acquire carbon from organic sources but their energy from sunlight.

Some autotrophs, called **photoautotrophs**, absorb the energy of sunlight and take in carbon dioxide to conduct photosynthesis. Others, called **chemoautotrophs**, get their energy from inorganic chemicals in their environment, including iron ore, hydrogen sulfide, and ammonia, instead of from sunlight.

In fact, the very first photoautotrophs on the planet were prokaryotes. One type of aquatic bacteria, the cyanobacteria, is believed to be responsible for changing Earth's chemistry by evolving photosynthesis and producing oxygen gas as a by-product. Oxygen gas accumulated in the air and water, and the levels rose from next to nothing to almost 10 percent about 2.1 bya. At that time in the fossil record, eukaryotes appear, suggesting that the oxygen generated by cyanobacteria may have facilitated the evolution of eukaryotes, especially multicellular forms.

Today, prokaryotes continue to aid eukaryotes. Bacteria directly help plants through a process known as **nitrogen fixation**. Plants need nitrogen in the form of ammonia or nitrate, which they cannot make themselves. Bacteria, however, can take nitrogen gas from the air—our atmosphere is 78 percent nitrogen—and convert it to ammonia, making it available for plants.

Autotrophs aren't the only microbes that enable life on Earth. Many heterotrophic bacteria and archaeans are decomposers, consumers that extract nutrients from the remains of dead or dying organisms and from waste products such as urine and feces. Decomposers play a crucial role in **nutrient cycling**. By breaking down dead or dying organisms or waste products, decomposers release the chemical elements locked in the biological material and return them to the environment. Those released elements, such as potassium or nitrogen or phosphorus, are used then by autotrophs and eventually by heterotrophs as well.

Healthy Balance

Ever since van Leeuwenhoek first looked through his microscope, people have often regarded microbes with suspicion and fear (**Figure 15.12**). We've vilified microbes as the bad guys that our bodies were constantly at war with. But now scientists recognize that our ecosystems wouldn't function without microbes. Humans would have no oxygen to breathe, no plants to eat. And microbes in our gut and on our skin, when in healthy balance, promote human health.

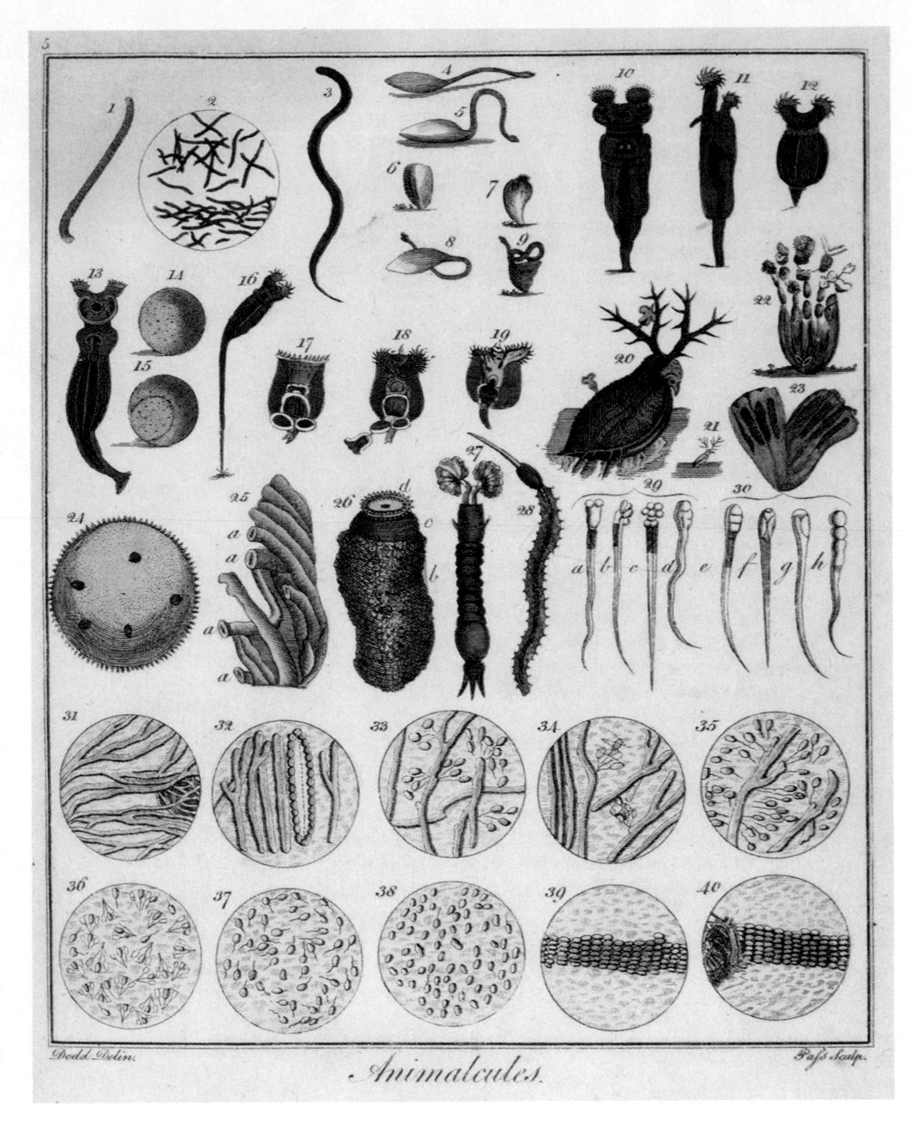

Figure 15.12

The first illustrations of microbes

Antonie van Leeuwenhoek's descriptions and illustrations of microscopic life, such as the array above, led to his being considered the "father of microbiology."

Q1: **From the prokaryotic structures shown in Figure 15.8, what shape would you assign to drawing number 8 in van Leeuwenhoek's illustration in Figure 15.12?**

Q2: **Which of the large prokaryote drawings (numbers 24–30) has the coccus shape?**

Q3: **Do you think all of these "animalcules" drawn by van Leeuwenhoek are prokaryotes? Why or why not?**

See Appendix A for answers to the figure questions.

We emphasize a healthy balance because it is possible for communities in the human microbiome to shift out of balance—a phenomenon called *dysbiosis*, which may cause illness. Alternatively, pathogenic bacteria can make us sick as well. Although the great majority of bacteria are harmless and many are actually beneficial to humans, some cause mild to deadly disease. Interestingly, archaeans are not known to be pathogens of any organism.

The biodiversity of microbes on our skin actually helps keep pathogens away. When a dangerous microbe lands on the skin, even before it meets the immune system it meets other microbes. If the skin has a diversity of microbes, odds are "one of them has the ability to kick the butt" of the pathogen, says Dunn.

After the overwhelming response to the belly button project, the lab kicked off a new citizen science spectacular: "Armpit-pa-looza" (**Figure 15.13**). As with the belly button, the team began collecting swabs of microbes living in people's armpits. In a preliminary study, the team found that the use of antiperspirants or deodorants dramatically

The Bugs in Your Belly Button

There's no way around it: Your belly button is full of bugs. According to research from Rob Dunn's lab at North Carolina State University, the most abundant species in a belly button—those with the biggest populations, such as *Corynebacterium* and *Staphylococcus*—are found in most people. Yet no two people have exactly the same set of microbes. In other words, our belly button microbiomes look alike, but we each have a unique navel fingerprint.

Assessment available in smartwork

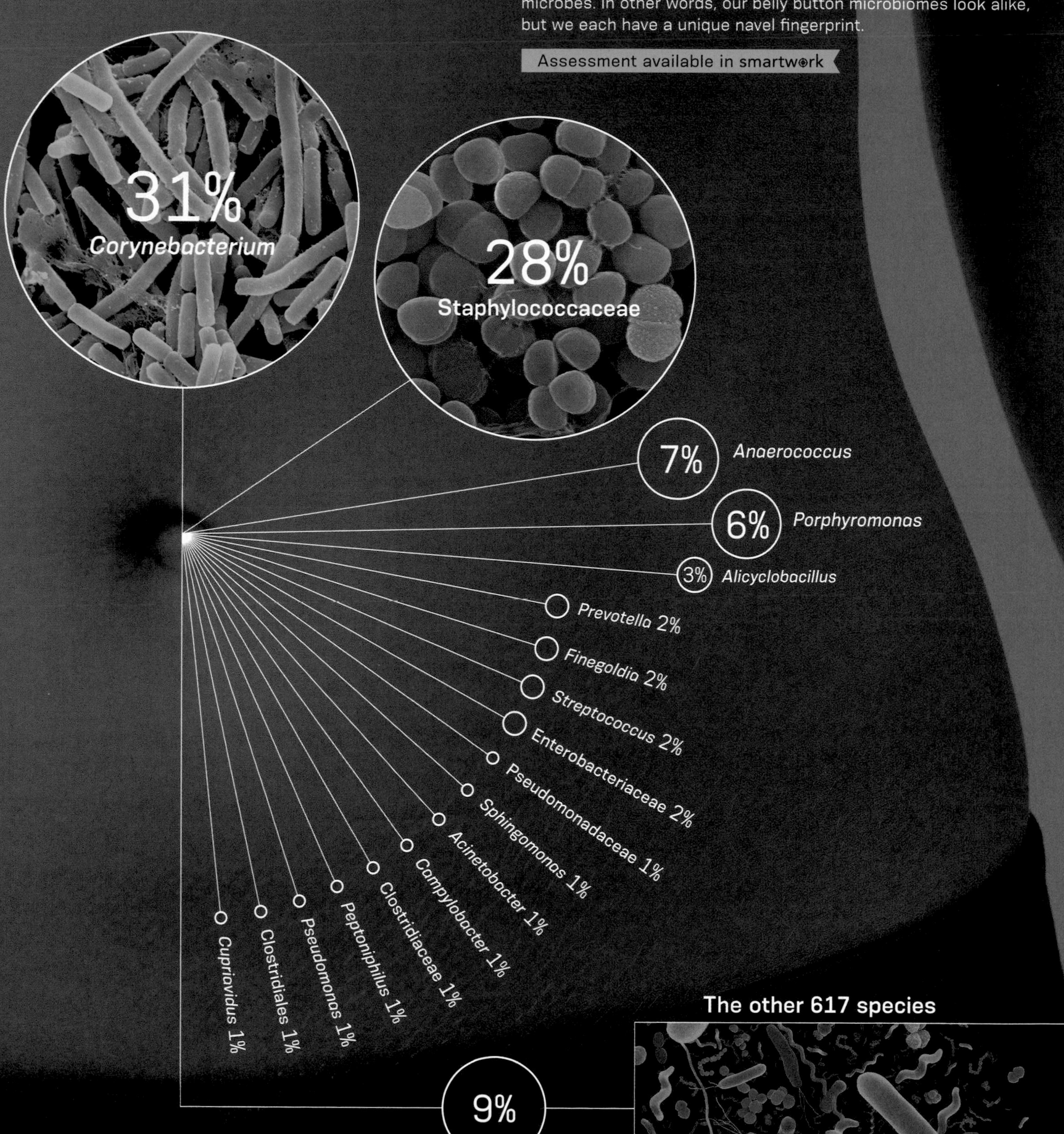

inhibits bacterial growth, affecting the composition of the microbial populations on the skin more strongly than any other factor. "It now looks like one of the biggest things affecting skin microbes in general is the use of antiperspirant or anti-odor products," says Dunn.

Now the team is taking a look at microbes on the skin of dogs. Skin wounds on pet pooches actually heal about four times faster than human skin wounds do, says Dunn, and he suspects that skin microbes are involved. "We're starting to figure out if we can predict wound healing rates as a function of which microbes are there to start with," says Dunn. "I'm pretty excited about that."

Figure 15.13

Sampling armpits for science
Citizen scientists volunteer to have samples taken of the prokaryotes living under their arms.

NEED-TO-KNOW SCIENCE

- The non-Eukarya, or prokaryotes, fall into two domains: Bacteria and Archaea.
- All prokaryotes are **microbes** (page 288)—microscopic, single-celled organisms—but the Bacteria and Archaea differ in significant ways, such as in their DNA, plasma membrane structure, and metabolism.
- Prokaryotes can reproduce extremely rapidly and are the most numerous life-forms on Earth. They also have the most widespread distribution.
- Many prokaryotes, particularly bacteria, have specialized structures. Bacteria with a **capsule** (page 291) surrounding the cell wall can avoid detection by organisms' immune systems. Short, hairlike **pili** (page 291) help bacteria attach to surfaces and to each other. Whiplike **flagella** (page 292) assist in locomotion.
- Some prokaryotes, particularly archaeans, thrive in extreme environments. For example, **thermophiles** (page 293) live in extremely hot places, and **halophiles** (page 293) live in very salty places.
- Prokaryotes typically reproduce by **binary fission** (page 294), and they may acquire DNA from each other or their environment via horizontal gene transfer.
- Some prokaryotes use **quorum sensing** (page 294) to communicate with each other, and some undergo **sporulation** (page 295) for protection.
- Prokaryotes exhibit unmatched diversity in methods of getting and using nutrients and energy. **Autotrophs** (page 295) make their own food, whereas **heterotrophs** (page 295) obtain food from other sources. **Chemoautotrophs** (page 296) and **chemoheterotrophs** (page 295) use inorganic chemicals as their energy source; **photoautotrophs** (page 296) and **photoheterotrophs** (page 296) use sunlight.
- Prokaryotes perform key tasks in ecosystems, including photosynthesis, **nitrogen fixation** (page 296) that provides nitrogen to plants, and **nutrient cycling** (page 297) that involves breaking down dead or dying organisms to release chemical elements.
- The **human microbiome** (page 288)—the complete collection of microbes that live in and on our cells and bodies—affects human gut health, brains, and even body odor. Although prokaryotes are useful to humanity in many ways, some cause deadly diseases.
- In **citizen science** (page 290) projects, the public plays a part in research by contributing, collecting, and sometimes even analyzing data in cooperation with professional scientists.

THE QUESTIONS

See Appendix B for answers.

The Basics

1 Prokaryotes include

(a) archaeans.

(b) bacteria.

(c) fungi.

(d) both a and b

(e) both b and c

2 Some prokaryotes can

(a) break down chemicals for energy.

(b) use sunlight for energy.

(c) create their own energy.

(d) harvest energy from other organisms.

(e) all of the above

3 Prokaryotes are extremely abundant because

(a) they can survive in a narrow range of environments.

(b) they have a single loop of DNA.

(c) they reproduce very rapidly.

(d) they are able to form biofilms.

(e) all of the above

4 Quorum sensing

(a) is the transfer of plasmid DNA from one bacterium to another.

(b) enables bacteria to form biofilms.

(c) is the formation of thick-walled dormant structures, called spores, under conditions unfavorable for growth.

(d) enables bacteria to switch from cellular respiration to fermentation when they sense that oxygen levels are low.

5 Select the correct terms:

Of the two types of prokaryotes, (**Archaea** / **Bacteria**) are more closely related to eukaryotes. The domain (**Archaea** / **Bacteria**) contains some species that are pathogens. (**Prokaryotes** / **Eukaryotes**) have smaller, less complex cells and are able to reproduce at a more rapid rate. Some (**prokaryotes** / **eukaryotes**) are multicellular.

Challenge Yourself

6 Which of the following can prokaryotes *not* do?

(a) communicate with each other about environmental conditions

(b) reproduce sexually

(c) live in extreme environments, including high salt, high temperature, and high atmospheric pressure

(d) share DNA with other prokaryotes

(e) double in population size two or more times an hour

7 Which of the following descriptions is true *only* of prokaryotes?

(a) They reproduce through binary fission.

(b) Their genetic material is DNA.

(c) They are unicellular.

(d) They contain organelles within the cell.

(e) none of the above

8 Place the following steps of the belly button research described in this chapter in the correct order by numbering them from 1 to 5.

_____ a. An observation was made that belly button bacteria seemed to differ across individuals.

_____ b. The DNA was sequenced.

_____ c. DNA was isolated from the samples.

_____ d. Volunteers had their navels swabbed to collect bacterial samples.

_____ e. The bacterial samples were grown in petri dishes.

Try Something New

9 Which of the following is an example of citizen science?

(a) reporting when the first hummingbirds arrive at your feeders in the spring

(b) completing a Facebook quiz on your horoscope

(c) filling out your medical history at the doctor's office

(d) all of the above

(e) none of the above

10 There are more genera of bacteria and archaeans in the world than of eukaryotes and more individuals. However, the total mass of prokaryotes is thought to be approximately equal to that of eukaryotes. How can this be?

11 Why are prokaryotes able to replicate so much more quickly than eukaryotes? Why is this difference in replication rate an important part of our vulnerability to bacterial pathogens?

Leveling Up

12 **Doing science** The research discussed in this chapter would not have been possible without the many citizen scientists who shared samples of their belly button microbes with the research team. Choose a citizen science project. Perform an internet search on the term "citizen science project," or find regularly updated lists of projects at the U.S. government site Citizen Science (http://www.citizenscience.gov) or Your Wild Life (http://www.yourwildlife.org). Write a one-paragraph summary of the goals of the project you have chosen. Would you participate in this project? Why or why not? If not, describe a citizen science project that you *would* want to contribute to.

13 **Looking at data** The table here is based on the published results of the study discussed in the chapter. It shows the number of bacterial phylotypes (kinds of bacteria) that were found on a particular percentage of people.

(a) How many bacterial phylotypes were found on only 10 percent or fewer of the people sampled? How many were found on more than 90 percent of people?

(b) How many phylotypes were found on more than half of the people sampled?

(c) Describe in one or two sentences the frequency of phylotypes found in the study. Were most phylotypes rare or common?

Phylotypes	Percentage of Human Samples Where Found
2,188	1–10
97	11–20
31	21–30
12	31–40
17	41–50
11	51–60
4	61–70
2	71–80
2	81–90
4	91–100

Digital resources for your book are available online.

The Dirt on Black-Market Plants

Nighttime poaching and illegal trade threaten extinction.

After reading this chapter you should be able to:

- Compare and contrast prokaryotes and eukaryotes.
- Explain what is distinctive about protists.
- Outline the key evolutionary innovations of plants.
- Connect structure and function in how fungi obtain energy from the environment.
- Identify key characteristics and give an example organism for protists, plants, and fungi.

BIODIVERSITY

CHAPTER 16

PLANTS, FUNGI, AND PROTISTS

Jacob Phelps spent a lot of his time in Thailand trying not to arouse suspicion. Walking among vendor stalls at wildlife markets in Bangkok, the tall, thin graduate student stopped occasionally to help a trader trim dead leaves off his plants or to chat about the weather. Wherever Phelps walked, plants surrounded him—hanging from the ceiling, stuffed into boxes, piled into mounds.

Phelps was careful about the questions he asked, because he was at the wildlife market to document illegal activity. The market traders had tens of thousands of plants for sale, most of which weren't supposed to be there. But when Phelps cautiously asked whether the plants were wild or rare, the traders were surprisingly relaxed. Concern for plant conservation and enforcement of trading laws is so limited that illegal sales occur openly at public markets across Southeast Asia.

Even Thailand's wildlife trade management authority at the time claimed that illegal trade in ornamental flowers was limited, "found in small case[s] in some parties." But Phelps was familiar enough with plants to know that most of the flowers he saw in the Thailand markets were wild, protected plants that were illegal to sell: the charismatic pink blossoms of *Dendrobium* orchids; the dense, fragrant plumes of *Rhynchostylis* flowers; the delicate, wavy petals of the coveted lady slipper orchid, *Paphiopedilum*. Plant enthusiasts prize such species for their beauty, fragrance, and rarity. Coveted wild orchids make up more than 80 percent of the plants traded at these unregulated markets (**Figure 16.1**).

Figure 16.1

A plant market in Thailand

Jacob Phelps visited plant markets like this one to conduct research on illegal trade in endangered orchid species.

As a graduate student at the National University of Singapore, Phelps received the blessing of his PhD supervisor, plant ecologist Ted Webb, to conduct an extensive survey of the illegal plant trade at the four largest plant markets in Thailand. It was the first time such a survey had been conducted of plant markets anywhere in Southeast Asia. In fact, the illegal trade of plants is often called the "invisible wildlife trade" because it is so rarely discussed or documented, in contrast to the widely publicized illegal trade of animals and animal products such as elephant ivory and rhinoceros horns.

Plants often take second-tier status behind animals—it's easier to get the public to care about a fuzzy baby tiger than a sprouting redwood tree—but their biology is no less wondrous. In Chapter 15 we explored the microscopic worlds of the Bacteria and Archaea domains of life. Here, we begin to meet the Eukarya, consisting of the artificial grouping of protists and three kingdoms: Plantae, Fungi, and Animalia. Animalia will be covered in detail in Chapter 17.

The defining feature of the Eukarya is a true nucleus. Instead of floating free in the cytoplasm, eukaryotic DNA is enclosed in two concentric layers of membranes that form a nuclear envelope. In addition to the nucleus, eukaryotes have a great variety of membrane-enclosed subcellular compartments, many of which are specialized for various tasks—such as sending messages, converting energy to useable form, or cleaning up—so that the cell can function efficiently (review Figure 4.8 for a comparison of prokaryotic and eukaryotic cells, and Figures 4.9–4.14 for an overview of eukaryotic organelles). All of these organelles take up space, so the diameter of a eukaryotic cell is, on average, 10 times larger than that of a prokaryotic cell, and the eukaryotic cell volume is a thousand times greater.

This compartmentalization of the cell interior enables eukaryotic cells to do things that most prokaryotic cells cannot do. For example, some eukaryotes engulf their prey and digest them internally. That's the way many single-celled eukaryotes, such as the blob-like amoebas, eat: They extend their gooey cytoplasmic arms to engulf other cells whole and then digest their prey with an elaborate system of internal compartments, ridding the cell of waste and storing surplus food (**Figure 16.2**).

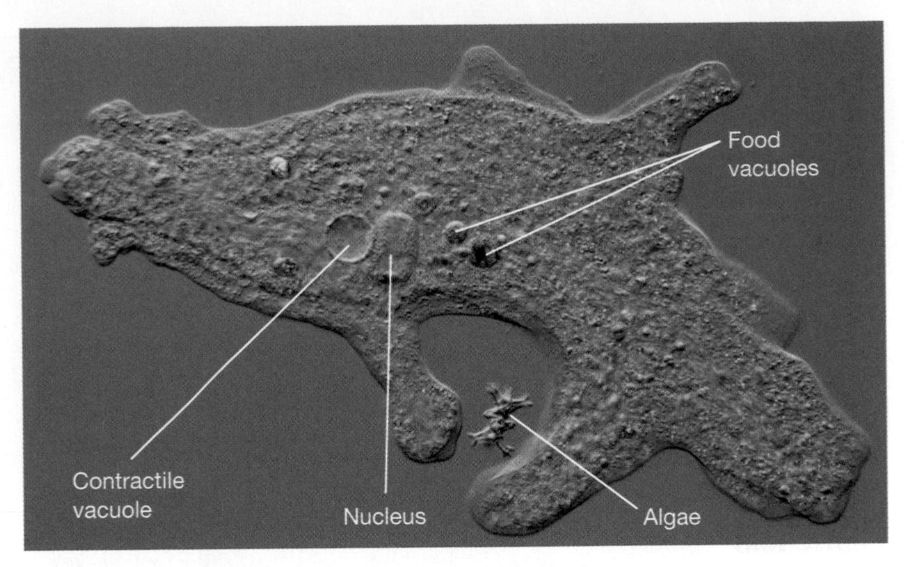

Figure 16.2

Eukaryotes have a true nucleus and compartmentalized cells
The more complex cell structure of eukaryotes enables them to perform functions that are impossible for prokaryotes. For example, intracellular compartments such as food vacuoles enable this amoeba to digest its food, single-celled algae. The amoeba expels excess water with the help of its contractile vacuole, another type of intracellular organelle.

Peculiar Protists

Amoebas are single-celled eukaryotes. Most single-celled eukaryotes are **protists**, a strange group that is defined by exclusion instead of inclusion. Although all protists are eukaryotes, they are grouped together simply because they are *not* plants, animals, or fungi (**Figure 16.3**).

Researchers have proposed several classification schemes to split the protists into multiple kingdoms; however, no consensus exists on the best way to do so. For now, protists remain divided into two traditional, broad categories: the **protozoans**, which are nonphotosynthetic and motile (capable of moving); and the **algae**, which are photosynthetic and may or may not be motile.

Protists are diverse in size, shape, cellular organization, and mode of nutrition. Most protists are single-celled and microscopic, but some protists have evolved from free-living single cells into multicellular associations, such as slime molds and kelp. Certain single-celled protists are bound by nothing more than a flexible plasma membrane, whereas others are covered in protective sheets, heavy coats, or other types of armor made of protein or silica.

Most protists are motile and can swim with the help of one or more flagella or by waving a carpet of tiny hairlike extensions called cilia. Others can crawl on a solid surface with the help of cellular projections called *pseudopodia* (false feet).

Many protists are heterotrophs and eat other organisms. Some of these heterotrophs function as decomposers, breaking down waste material and releasing nutrients into the environment to be taken up by producers and cycled back into the food chain. Other protists are nutritional opportunists, or *mixotrophs*, organisms that use energy and carbon from a variety of sources to fuel their growth and reproduction.

Although most protists are harmless, many of the best-known protists are pathogens (disease-causing agents), such as the genus *Plasmodium*, which is transmitted by mosquitoes and causes malaria, and *Toxoplasma gondii*, passed to humans through poorly cooked food or cat feces. *T. gondii* causes toxoplasmosis, which typically causes mild flulike symptoms in adults but more severe symptoms for a fetus.

Two Cells Are Better Than One

A multicellular organism is a well-integrated assemblage of genetically identical cells, in which different groups of cells perform distinctly specialized functions. This functional compartmentalization makes it easier for multicellular eukaryotes to sense and respond to the external environment through the use of complex structures such as leaves, fruit, eyes, and wings.

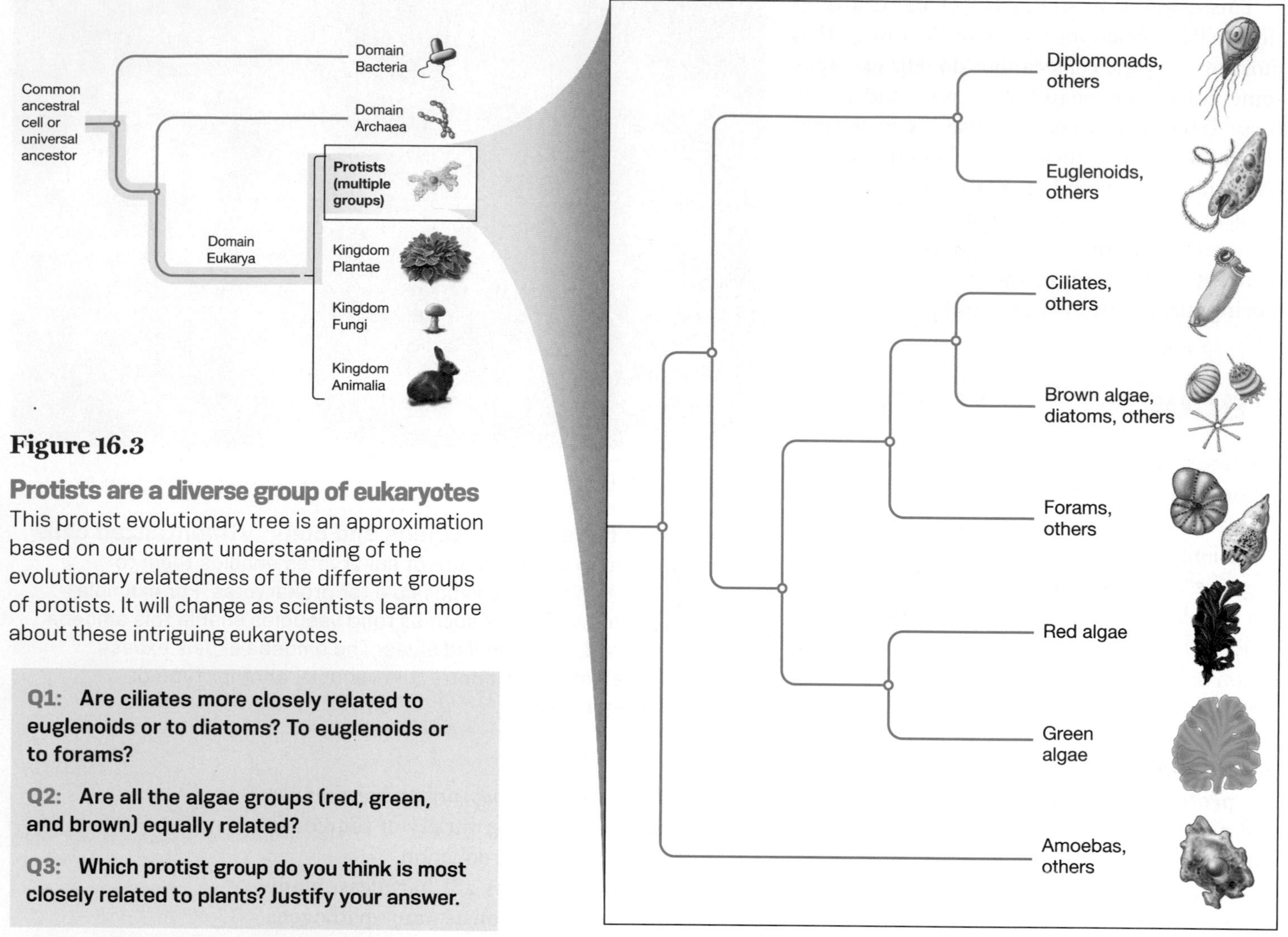

Figure 16.3

Protists are a diverse group of eukaryotes

This protist evolutionary tree is an approximation based on our current understanding of the evolutionary relatedness of the different groups of protists. It will change as scientists learn more about these intriguing eukaryotes.

Q1: Are ciliates more closely related to euglenoids or to diatoms? To euglenoids or to forams?

Q2: Are all the algae groups (red, green, and brown) equally related?

Q3: Which protist group do you think is most closely related to plants? Justify your answer.

See Appendix A for answers to the figure questions.

Multicellular forms evolved several times among different lineages of the eukaryotes. Multicellularity enables an individual organism to grow large, which can be advantageous for evading potential predators. A bigger individual can also gather resources from its environment more effectively than can a smaller individual. Having more resources, such as light or food, may translate into producing more surviving offspring, the ultimate measure of biological success.

To make those offspring, eukaryotes reproduce via either asexual or sexual reproduction. Asexual reproduction is common among species in this domain and generates genetically identical offspring. Protists, for example, can split into two in a process similar to binary fission in prokaryotes. Many plants reproduce asexually by fragmenting into pieces, each piece developing into a new individual.

But it is through the second type of reproduction that eukaryotes have made an indelible mark on our planet. By combining genetic information from two parents, sexual reproduction produces offspring that are genetically different from each other and from both parents. Although their life cycle is distinctly different from that of animals, plants similarly produce embryos. The fusion of egg and sperm produces a single cell, called a zygote, which then divides to produce a multicellular structure called an embryo. Sexual reproduction is one means by which natural populations become genetically diverse; it is why there are an estimated 400,000 species of plants on Earth.

Fungi Play Well with Others

Plants probably would not have been as successful on land if they had not entered into a mutualistic relationship with fungi almost immediately on their arrival. Today, the vast majority of plants in the wild have mutualistic fungi, known as **mycorrhizal fungi**, associated with their root systems. (The mutualistic relationships between plants and mycorrhizal fungi are called "mycorrhizae.") Truffles, morels, and chanterelles, all beloved by gourmets, are the reproductive structures of mycorrhizal fungi.

Mycorrhizal fungi form thick, spongy mats of mycelium on and in the roots of their plant hosts and also extend into the surroundings, sometimes permeating several acres of the soil around the root. Mycelia are thinner, more extensively branched, and in closer contact with the soil than even the thinnest branches on a plant root. As a result, a mycelial mat plumbs far more water and mineral nutrients, such as phosphorus and nitrogen, than the plant's root system could absorb on its own. In return for sharing absorbed water and mineral nutrients, the fungus obtains sugars that the plant manufactures through photosynthesis.

Mycorrhizal fungi assist in providing nutrients to orchid seeds, which are tiny and lack stored food. The embryo within a newly sprouted orchid seed could not survive without the mycorrhizal network linking the seedling to mature photosynthesizing plants, from which the seedling draws nourishment until it can photosynthesize on its own.

Fungal mutualisms are not found only with other eukaryotes. A **lichen** is a mutualistic association between a photosynthetic microbe and a fungus. The fungus receives sugars and other carbon compounds from its photosynthetic partner, usually a green alga or a cyanobacterium. In return, the fungus produces lichen acids, a mixture of chemicals that scientists believe may function to protect both the fungus and its partner from being eaten by predators.

Lichens grow very slowly and are often pioneers in barren environments. Lichen acids wear down a rocky surface, facilitating soil formation. Soil particles build up from the slow weathering of rock and over time other life-forms, including plants, gain a toehold in the newly made soil.

Plants that have no mycorrhizal fungi associated with their roots do not grow as well...
...as those that host the beneficial fungi.

Through these unique traits—namely, a true nucleus, cellular compartmentalization, multicellularity, and sexual reproduction—eukaryotes have evolved into amazingly diverse and dynamic species. Of these species, Phelps was particularly concerned about orchids. In addition to spending time in Bangkok, the budding ecologist spent months traveling along Thailand's borders with Laos and Myanmar, wandering hours upon hours through wildlife markets. Over time, he built local networks that allowed him to interview more than 150 plant harvesters and intermediaries, including market traders, online traders, nursery owners, and more.

Orchids are, on paper, one of the most heavily protected families, representing almost 75 percent of all species, plant or animal, for which international trade is regulated. Artificially propagated orchids can be traded legally, but wild orchid species in Southeast Asia are

protected, says Phelps, who now works at Lancaster University in the United Kingdom: "None should be collected from the wild, unless you have specialized permissions."

Yet, as Phelps came to discover, there is no enforcement of those protections in Thailand or across the region. Orchids and other ornamental and medicinal plants were not only being traded but also poached from nearby countries, threatening the very existence of certain species and the overall biodiversity of the region.

Though it may sound exotic, the plant black market is not exclusive to far-flung countries. It is alive and well in the United States, from the wild forests of North Carolina to cultivated botanical gardens in California.

Green-Fingered Thieves

Under the dark of night, the thieves worked swiftly. Armed with shovels, they snuck past the closed gift shop and made a beeline for their quarry. They knew exactly what to take.

Figure 16.4

Cycads: ancient, rare, and hard to move

Several workers struggle to move a cycad. In spite of their size, many cycads have been stolen from public and private gardens and then sold on the black market.

The next morning, the staff at the Quail Botanical Gardens in Encinitas, California, discovered the theft. In a frantic effort, they spread details of the heist to local newspapers, television stations, and online community message boards. Their quick action paid off; the loot became too hot. Within days, an anonymous tip led authorities to a rural road where the thieves had dumped their haul in a sorry-looking pile: 21 stocky plants, specimens with thick trunks covered in woody scales and a crown of long, green palm-like leaves, like a pineapple.

The thieves had attempted to steal a group of rare African cycads, an ancient type of plant that once lived alongside dinosaurs in the Jurassic period (**Figure 16.4**). As with the Thai orchids, horticulturalists around the world highly prize cycads as ornamental plants. Today, a rare, mature cycad can fetch $20,000 or more on the international black market. That's right—enough to fund a year of college. There are about 300 species of cycads; most are threatened with extinction.

Ironically, 4 of the 21 cycads stolen from the Quail Botanical Gardens (now the San Diego Botanic Garden) were part of a rescue program; the plants had been illegally brought into the United States, where they were seized upon import by authorities, and the botanical garden had taken over their care. Some of the plants survived the ordeal and were replanted. But they weren't the only cycads stolen that year; a nursery in San Diego lost almost 40 large cycads to thieves, and many homes in the Long Beach area had plants swiped right out of their front yards.

Plants are multicellular autotrophs that are mostly terrestrial (**Figure 16.5**). Like the green algae from which they evolved, plants use chloroplasts in order to photosynthesize. Most photosynthesis in plants takes place in their leaves, which typically have a broad, flat surface—a design that maximizes light interception. Because plants are producers, they form the basis of essentially all food webs on land. They may not be as cute, fuzzy, or exciting as animals, but animals wouldn't exist without them. Nearly all organisms on land ultimately depend on plants for food, either directly by eating plants or indirectly by eating other organisms that eat plants.

In addition to being food, plants are valuable for many other reasons. Many organisms live on or in plants or in soils largely made up

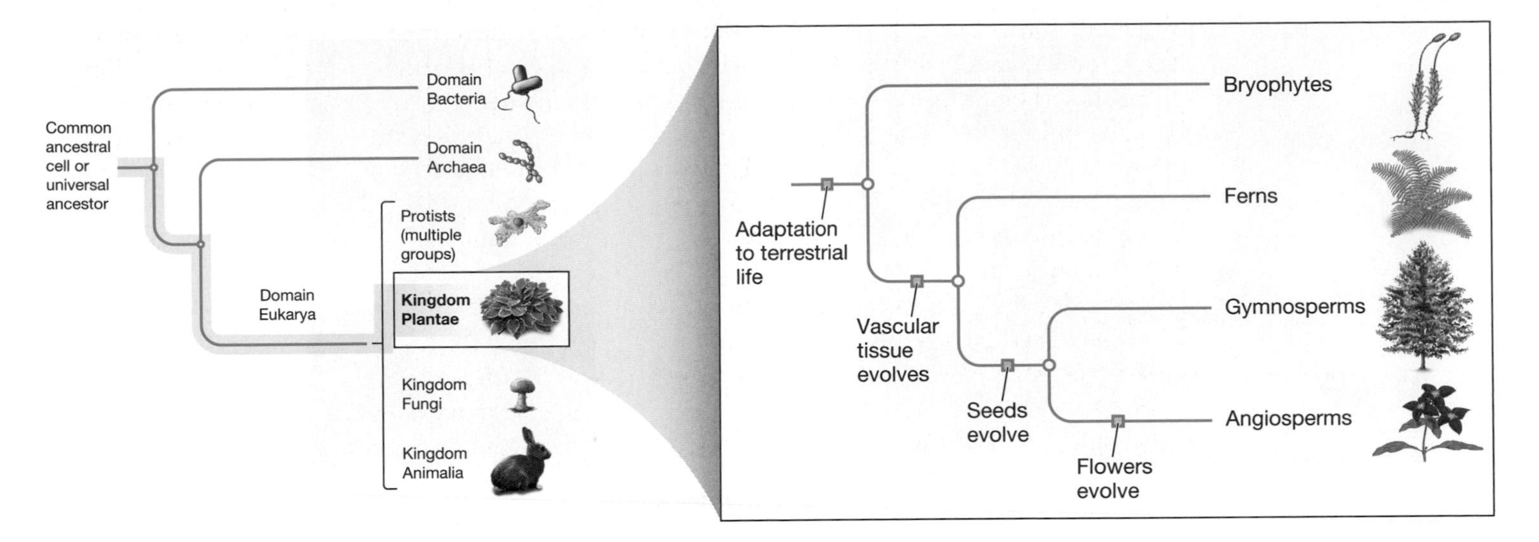

Figure 16.5

Plants are eukaryotic autotrophs

Plants are photosynthetic, making their own food from sunlight, and are almost exclusively found on land. They are at the base of almost all terrestrial food webs.

Q1: What evolutionary innovation separates all land plants from their aquatic ancestors?

Q2: How do ferns differ from bryophytes? Do they share this difference with other plant groups? (You will need to read ahead to answer this question.)

Q3: What group(s) might a plant with seeds belong to? What about a plant with flowers?

See Appendix A for answers to the figure questions.

of decomposed plants. By soaking up rainwater in their roots and other tissues, plants prevent runoff and erosion that can contaminate streams. Plants (and algae) also recycle carbon dioxide and produce the oxygen we breathe.

Many of these traits are adaptations to the challenges of living on land (**Figure 16.6**). The greatest of those challenges is how to obtain and conserve water. The first plants, the ancestors of present-day **bryophytes** (liverworts and mosses), grew as ground-hugging carpets of greenery. These simple plants had relatively thin bodies, often just a few cells thick, and absorbed water through a wicking action. However, absorption by direct contact cannot transport fluids effectively in a plant that rises a foot or more aboveground, so more complex plants, including cycads, have a network of tissues, called the **vascular system**, that is made up of tubelike structures specialized for transporting fluids. Roots, found in most plants, also have an extensive vascular system. The first land plant with a vascular system was the ancestor of present-day **ferns**.

Like green algae, plant cells have strong but flexible cell walls composed of **cellulose**, which gives structural strength to the cells, including those of the low-growing bryophytes. Beginning with the ferns, plants evolved yet another type of strengthening material called **lignin**, which allowed them to grow even taller. As one of the strongest materials in nature, lignin links together cellulose fibers in the cell wall to create a rigid network. Lignin is the reason that some cycads in Japan can grow over 20 feet (7 meters) tall, though it takes them about 50–100 years to achieve such heights.

One reason rare cycads are so prized is that it can take a long time to raise and grow them. Cycads are **gymnosperms**, the first plants to evolve **pollen**, a microscopic structure containing sperm cells that can be released into the air in massive quantities. The evolution of pollen freed gymnosperms from a dependence on water for fertilization. Instead of having to transport sperm cells through water, they could do it through air.

Algae (aquatic)

- Photosynthesis occurs, and CO_2 is absorbed throughout the organism.
- Water and minerals are absorbed by the whole organism.
- Water supports the whole organism.
- Dehydration is not an issue in an aquatic environment.

Plants (terrestrial)

- Photosynthesis occurs, and CO_2 is absorbed primarily in the leaves. Air enters leaf cells through minute openings in the leaf's surface. Many plants have more elaborate pores (*stomata*) that open and close to regulate the flow of gases into and out of the leaf.
- Roots absorb water and minerals from the soil.
- Roots anchor and support the plant within the ground. Lignin and vascular tissues help support the plant aboveground.
- The waxy cuticle holds in moisture to keep plant tissues from drying out, even when exposed to sunlight and air throughout the day.

Figure 16.6

Moving to land brought unique challenges to plants

Terrestrial plants evolved in response to challenges that their aquatic ancestors had not faced, which resulted in adaptations designed to slow dehydration, provide support and anchoring, and enable photosynthesis and nutrient uptake.

Q1: In what ways are terrestrial plants and their aquatic ancestors similar? Give at least two similarities.

Q2: In what ways do terrestrial plants and their aquatic ancestors differ? Give at least two differences.

Q3: Would you predict that aquatic plants, which have secondarily evolved to live in water (in other words, their ancestors were terrestrial plants), would be more like plants in a rainforest or more like desert plants? Explain your reasoning.

See Appendix A for answers to the figure questions.

Gymnosperms were also the first plants to evolve **seeds**, each of which consists of the plant embryo and a supply of stored food, all encased in a protective covering. The embryo uses this stored food to grow until it is able to make its own food via photosynthesis. Seeds also provide embryos with protection from drying and from attack by predators. Unfortunately, they don't provide protection from poachers. Today, the San Diego Botanic Garden keeps not only its prized cycads but also their valuable seeds in a greenhouse under lock and key.

Searching for Flowers

Botanical gardens also go to extreme measures to keep flowering plants, or **angiosperms**, safe for long-term conservation. Orchids, as we have seen, are the most common illegally traded angiosperms. Others include *Galanthus*, or "snowdrop" plants, with delicate white flowers. In 2012, a single bulb of a rare variety of snowdrop fetched $945 at auction.

Compared to gymnosperms, angiosperms are a relatively recent development in the history of life. With about 250,000 species today, angiosperms are the most dominant and diverse group of plants on our planet. Nearly all agricultural crops are flowering plants. These plants also provide humans with materials such as cotton and pharmaceuticals.

Angiosperms' key evolutionary innovation is the flower, a structure that evolved through modification of the conelike reproductive organs of gymnosperms. **Flowers** are structures that enhance sexual reproduction in angiosperms by bringing male gametes (sperm cells) to the female gametes (egg cells) in highly efficient ways by attracting animal pollinators through scent, shape, and color (**Figure 16.7**).

Figure 16.7

Bumblebees visit a purple coneflower

Pollination in the angiosperms is more efficient than in the wind-pollinated gymnosperms. Flowers attract animal pollinators that carry pollen to and from individual plants of the same species.

Gymnosperms produce "naked" seeds that sit bare, unwrapped in any additional layers, on the modified leaf (the scale in a pinecone). In angiosperms, the modified leaf evolved into the ovary wall, which consists of tissue layers that enclose and protect the egg-bearing structures, or **ovules**. After fertilization, the ovules develop into seeds, and the ovary wall that enclosed them becomes the fruit wall.

At the Thailand plant markets, Phelps had trouble identifying orchid species until they flowered, so he went back again and again—four times per year to four different markets—looking for newly opened flowers, listing species he recognized, taking photos of others, and even occasionally asking for a flower off a plant he did not recognize, which he quickly stored in a vial of alcohol to take back for identification. In the end, Phelps gathered evidence of 348 orchid species in 93 genera, representing 13–22 percent of the area's known orchid flora, and tens of thousands of individual plants, including several new species.

Phelps's results were shockingly different from those published in preexisting government reports on plant trades among Thailand, Laos, and Myanmar. The Convention on International Trade in Endangered Species of Wild Fauna and Flora (CITES) is an international treaty with more than 175 countries to monitor and regulate the international trade of plants and animals. Member countries are required to produce permits for wildlife trade of species protected under the agreement, including all wild orchids. These permits are used to guarantee plants are legally harvested in a sustainable way that does not endanger either the species or the environment.

"CITES is about conservation and sustainable use," says Anne St. John, a biologist with the U.S. Fish and Wildlife Service's Division of Management Authority, which implements CITES in the United States. "The goal is to ensure that these species are around for our grandchildren. That includes not only tigers and elephants, but also bigleaf mahogany and Brazilian rosewood."

And the orchids of Southeast Asia are protected as well. Over 9 years, Laos reported permits for the export of just 20 wild-collected orchids into Thailand; Myanmar reported none. Yet during just one day with a single market trader at the border between Laos and Thailand, Phelps documented that the woman sold at least 168 plants of eight different genera. In one day, she sold eight times more plants than the government reported as sold over 9 years. "It's totally anecdotal, but incredibly illustrative of the problem," says Phelps. "This trade is completely unacknowledged. It's an open secret."

Other countries are working hard to crack down on the plant black market. American ginseng, a short leafy plant with a tan, gnarled root commonly used in Chinese herbal medicine, is the largest CITES-regulated plant export

Figure 16.8

Ginseng plants carpet a forest floor

After pollination, ginseng flowers develop a bright-red seed head, which helps ginseng hunters find the plants to collect their roots in the filtered light of the forest.

Q1: What feature(s) of the ginseng plant tell you it is not a bryophyte?

Q2: What feature(s) of the ginseng plant tell you it is not a fern or gymnosperm?

Q3: Because of the CITES classification of ginseng, you are not allowed to sell plants younger than 5 years even if they grew on your own land. Do you agree with that law? Why or why not?

See Appendix A for answers to the figure questions.

of the United States (**Figure 16.8**). A pound of quality, dried ginseng can sell for up to $900, so some people try to bypass CITES permits, harvesting plants that are too young (legally traded roots must be 5 years or older), out of season, or poached from federal lands.

"As you would expect, species that are high value and in high demand are the subject of illegal trade," says St. John. The illegal trade of ginseng even spawned a reality show on the History television channel: *Appalachian Outlaws*. To combat the criminal activity, some states have taken to spraying ginseng plants on state and federal lands with paint, so if a marked plant shows up in a batch of ginseng to be exported to China, authorities will know where it came from. Those caught selling protected plants face fines and jail time, and violating CITES is a federal crime.

Truffle Trouble

The illegal wildlife trade is a problem not just for animals and plants but also for **fungi**, none of which have been listed as CITES-protected species so far. Fungi are absorptive heterotrophs. They secrete digestive enzymes that break down organic material from dead or dying organisms and then absorb this material for food. The majority of fungal species fall into three main groups: **zygomycetes**, which contains many species of molds; **ascomycetes**, a diverse group informally known as sac fungi; and the more familiar **basidiomycetes**, or club fungi (**Figure 16.9**). Each of these groups differs in—and is named for—its unique reproductive structures.

Fungi have properties in common with both plants and animals. Like plant cells, all fungal cells have a protective cell wall that wraps around the plasma membrane and encases the cell. However, fungi are similar to animals in that they store surplus food energy in the form of glycogen.

In 2012, two bandits broke through security gates and stole an estimated $60,000 of fungi from a locked warehouse. The loot was truffles, the fruiting body of a particular group of ascomycetes. Certain white European truffles can sell for as much as $3,600 per pound, making them the most expensive food in the world. That high price also makes them a target for thieves,

DEBUNKED!

MYTH: Animals won't eat poisonous mushrooms.

FACT: Just because a snail or other wild animal has taken a bite out of a mushroom does not mean it's safe to eat. Toxins in a mushroom may be harmless to other organisms and still dangerous for humans.

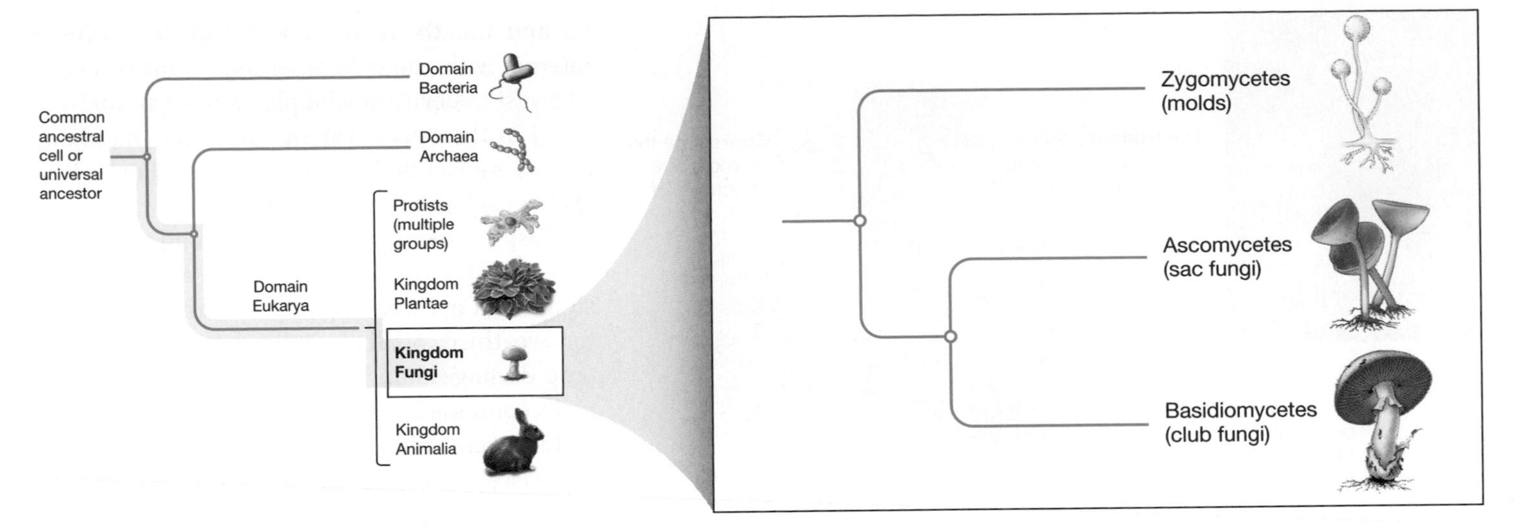

Figure 16.9

Fungi are eukaryotic absorptive heterotrophs

Fungi must take their food from other organisms, and they do this in a unique way. Instead of ingesting their food, as most heterotrophs do, fungi use chemicals to break down their food outside of their bodies and then absorb the nutrients.

Q1: What group of fungi most resembles the mushrooms you buy in a grocery store?

Q2: Are sac fungi (ascomycetes) more closely related to molds (zygomycetes) or to club fungi (basidiomycetes)?

Q3: How do we know that fungi are eukaryotes rather than prokaryotes?

See Appendix A for answers to the figure questions.

Figure 16.10

A truffle hunter and his dog search for a fungus beloved by gourmets

Although farmers continue attempts to grow truffles commercially, their success has been limited. For now, we must rely on truffle hunters and their dogs (or pigs, which can also be trained to sniff out truffles) for this fungal delicacy.

especially because white truffles cannot be cultivated in greenhouses as other fungi can.

There's danger in the forests too. Truffle hunters use dogs and pigs to sniff out wild truffles (**Figure 16.10**), and some have reported competitors planting spiked traps or poisoned meatballs in forests to eliminate trained dogs. Other hounds have been stolen from their owners, never to be seen again.

That's a lot of drama for fungi. Most fungi are multicellular, but there are some single-celled species collectively known as **yeasts.** Many of us are familiar with yeasts thanks to two important products they produce: alcohol and carbon

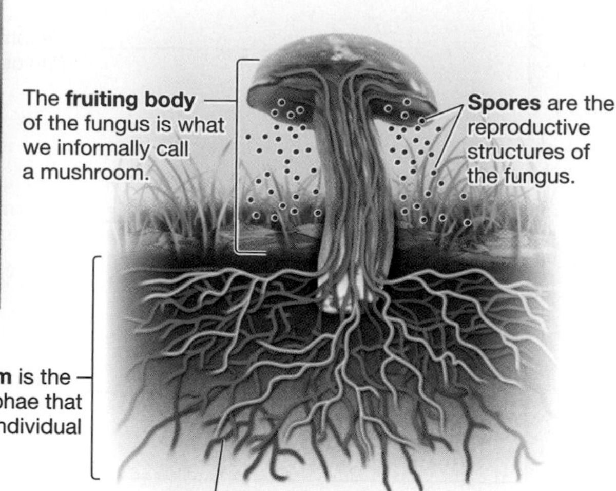

The **fruiting body** of the fungus is what we informally call a mushroom.
Spores are the reproductive structures of the fungus.
The **mycelium** is the bundle of hyphae that make up an individual fungus.
A **hypha** is a single fungal thread.

Figure 16.11

A fungus lives both belowground and aboveground
The main body of a fungus is unseen belowground. To reproduce, the fungus generates a fruiting body that is usually aboveground and releases spores. Spores travel by wind and develop into new fungi.

Q1: Why is it important that the fruiting body of a fungus is aboveground?

Q2: What part of a fungus is the mushroom that you can buy in the grocery store?

Q3: Write a sentence in your own words that uses the terms "mycelium," "fruiting body," and "spore" correctly.

See Appendix A for answers to the figure questions.

dioxide, crucial to the rising of bread, the brewing of beer, and the fermenting of wine.

Fungi's key evolutionary innovation is their body form, which is well suited for absorptive heterotrophy (**Figure 16.11**). They are made up of a network of fine, colorless, branching, hair-like threads called **hyphae** (singular "hypha"), which absorb nutrients from the environment. The entire bundle of hyphae, composing the main body of the fungus, is called the **mycelium** (plural "mycelia").

Because of this unique body form, many fungi are decomposers and consume nonliving organic material. As they eat, these fungi release back into the environment inorganic chemicals that had previously been trapped in the bodies of dead organisms. Once these chemicals are back in the environment, plants and algae scoop them up and use them to manufacture food. Fungi interact with plants in other important ways too. The vast majority of wild plants have mutualistic fungi in their root systems that help the plants absorb nutrients from the soil (see "Fungi Play Well with Others" on page 307).

Despite that assistance, fungi and plants are not always close friends. Fungi are the most significant parasite of plants and are responsible for two-thirds of all plant diseases, causing more crop damage than bacteria, viruses, and insect pests combined.

Like many plants (and animals and protists), fungi can reproduce both asexually and sexually. Some species appear to multiply only asexually, and most multicellular fungi can reproduce asexually through fragmentation—that is, by simply breaking off from the mother colony. When fungi do reproduce sexually, they do not have distinct male and female individuals. Instead, a sexually reproducing mycelium belongs to one of two (or more!) mating types. Each mating type can mate successfully with only one of the other types. After mating, a **fruiting body** is formed that may be large enough to be readily observed.

Fungal fruiting bodies release offspring as sexual spores. A **spore** is a reproductive structure that can survive for long periods of time in a dormant state and will sprout under favorable conditions to produce the body of the organism. Spores released from a fruiting body that is raised up in the air are better able to catch a ride on wind currents or to attract animals that can carry them far and wide. Many fungi also produce asexual spores.

Fighting for the Future

The illegal sale of plants and fungi could drive species to extinction. In 2015, for example, the International Union for Conservation of Nature (IUCN), a global environmental authority, announced that 31 percent of cactus species—renowned for their unique forms and beautiful flowers—are in danger of extinction and that the greatest threat to these plants is illegal trade (**Figure 16.12**).

During his studies, Phelps observed traders selling plants that are highly endangered in the wild. He published results from his thesis in

Food Banks

Over millennia, farmers have saved seeds from crops that were easiest to grow, process, or store, and have used those seeds and their offspring year after year. The results of this domestication process are staple crops we rely on today: corn, soybeans, wheat. Yet to feed a growing human population, we're going to have to look beyond what we're familiar with, to the smorgasbord of edible plants that pepper the planet.

Assessment available in smartwork

30,000

terrestrial plants are known to be edible.

7,000

are cultivated or collected by humans for food.

30

crops provide 95% of the food energy taken in by the world's human population.

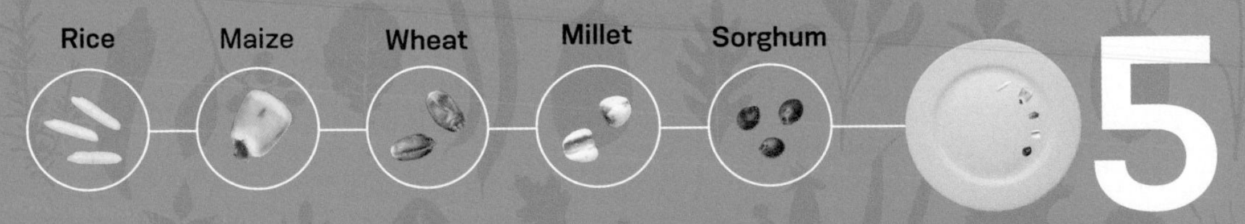

5

cereal crops provide 60% of the food energy taken in by the world's human population.

Figure 16.12

The greatest threat to endangered cacti is illegal trade

As with orchids and cycads, the main threat to most species of cacti is illegal trade. Collectors pay high prices to smugglers who are willing to risk fines and prison time to meet the demand.

2015 in the journal *Biological Conservation*, pleading with scientific and policy communities for a greater focus on these plants and calling the illegal plant trade "a major conservation challenge that has been almost completely overlooked."

"It's hard to get people to care about plants, and nobody can care if we don't have data to show it's a problem," says Phelps, who is now trying to engage governments and experts to raise awareness and initiate policies that will better regulate plant trade. At the CITES conference in August 2019, country delegates increased protections for several plant species, including Malawi's national tree, the rare Mulanje cedar, which is being cut down at an unsustainable rate for its highly sought-after wood.

"All these plants are part of a larger ecosystem," says St. John. "They serve important purposes in temperature regulation, in creating oxygen, and more. We need to recognize that plants are an integral part of a healthy ecosystem and key to the diversity of life."

NEED-TO-KNOW SCIENCE

- The domain Eukarya is divided into the artificial group called protists and three major kingdoms: Plantae, Fungi, and Animalia. Eukaryotes possess a true nucleus; they have complex subcellular compartments, which enable larger cell size. Sexual reproduction and multicellularity are key evolutionary innovations of the Eukarya.
- The **protists** (page 305) lump together many evolutionarily distinct lineages under one highly diverse grouping. Most protists are single-celled and microscopic.
- Protists are traditionally divided into two categories: **protozoans** (page 305), which are nonphotosynthetic and motile; and **algae** (page 305), which are photosynthetic and may or may not be motile.
- **Plants** (page 308) are descended from green algae and have evolved numerous evolutionary innovations to adapt to life on land. All plants photosynthesize and use **cellulose** (page 309) to strengthen their cell walls.
- The first plants, the ancestors of present-day **bryophytes** (liverworts and mosses, page 309), had relatively thin bodies and absorbed water through a wicking action.
- **Ferns** (page 309) were the earliest plant group to grow larger and taller than the bryophytes. This extended growth required both cellulose and **lignin** (page 309) to provide structure to plant bodies, as well as a network of fluid-transporting tissues called the **vascular system** (page 309).
- **Pollen** (page 309) and **seeds** (page 310) first evolved among the **gymnosperms** (page 309), the plant group that includes cycads.
- **Angiosperms** (page 310) evolved **flowers** (page 311), and they enclose their seeds in the fruit. Many angiosperms attract animals, which then deliver the plants' pollen and disperse their seeds.
- The **fungi** (page 312) are distinguished by their mode of nutrition. They acquire their nutrients by absorption, digesting their food outside of their

bodies. To accomplish this, their bodies consist of a **mycelium** (page 314) composed of hairlike threads called **hyphae** (page 314).

- Most fungi are multicellular, but there are some single-celled species collectively known as **yeasts** (page 313).
- Fungal reproduction may be sexual or asexual. During reproduction, a **fruiting body** (page 314) is formed that releases **spores** (page 314) into the environment. Much as seeds do, these spores then develop into new individuals.
- There are at least three main groups of fungi—**zygomycetes** (page 312), **ascomycetes** (page 312), and **basidiomycetes** (page 312)—and each group is characterized by distinctive reproductive structures (fruiting bodies).
- Most plant roots in natural habitats form close associations with beneficial fungi, called **mycorrhizal fungi** (page 307). A **lichen** (page 307) is a mutually beneficial association between a fungus and a photosynthetic microbe, usually a green algae or a cyanobacterium.

THE QUESTIONS

See Appendix B for answers.

The Basics

1 Which of the following descriptions is *not* true of all eukaryotes?

(a) They are multicellular.
(b) They have cellular organelles.
(c) They have a larger cell size than prokaryotes have.
(d) They have a true nucleus.
(e) All of the above are characteristics of all eukaryotes.

2 Protists

(a) are the largest prokaryotes.
(b) are all single-celled.
(c) include plants and fungi.
(d) are an artificial grouping, placed together for convenience.
(e) are all photosynthetic.

3 Fungi

(a) reproduce only sexually.
(b) reproduce only asexually.
(c) may have multiple mating types within a single species.
(d) are more closely related to plants than to animals.
(e) are more closely related to protists than to animals.

4 Which of these evolutionary innovations enabled larger cell size?

(a) autotrophic mode of nutrition
(b) multicellularity
(c) sexual reproduction
(d) subcellular compartmentalization
(e) all of the above

5 Select the correct terms:

Green algae are (**plants** / **protists**). They are (**aquatic** / **terrestrial**) and are (**autotrophs** / **heterotrophs**).

Challenge Yourself

6 Place the following adaptations in the correct order from earliest to most recent by numbering them from 1 to 5.

_____ a. vascular tissue
_____ b. flowers
_____ c. seeds
_____ d. multicellularity
_____ e. movement to land

7 Which of the following groups contains only multicellular species?

(a) algae
(b) protists
(c) eukaryotes
(d) fungi
(e) angiosperms

8 Of the kingdoms covered in this chapter, which, if any, are composed of only autotrophic species? Which, if any, have only heterotrophs?

Try Something New

9 Symbiosis is a long-term and intimate association between two different types of organisms. A symbiotic organism may live on or inside another species. For each of the following symbiotic relationships, (1) define the relationship in one to three sentences, (2) identify the domain—and for Eukarya, the group or kingdom—of each partner in the relationship, and (3) discuss whether the relationship evolved as a mutualism (both benefit), commensalism (one benefits while the other is not affected), or parasitism (one benefits to the detriment of the other). You may need to read more than your textbook to answer this question.

(a) mycorrhizae

(b) lichens

(c) hermit crabs/shells

(d) malaria

10 Which plant groups produce pollen, and what is the adaptive value of pollen?

11 Plants are not the only organisms susceptible to fungal infections. Give an example of a fungal infection to which humans are susceptible.

12 Mycoremediation js the use of fungi to inactivate pollutants in the soil. The accompanying graph shows the results of a mycoremediation study. Scientists applied local fungi to soil from a closed sawmill polluted with chlorophenols, which are used as disinfectants, pesticides, and herbicides.

(a) What is the highest percentage of inactivated chlorophenol in this graph? On which day of the experiment did this occur?

(b) Which condition had the lowest percentage inactivated?

(c) Which fungus do you predict would have the most effect after a year of treatment?

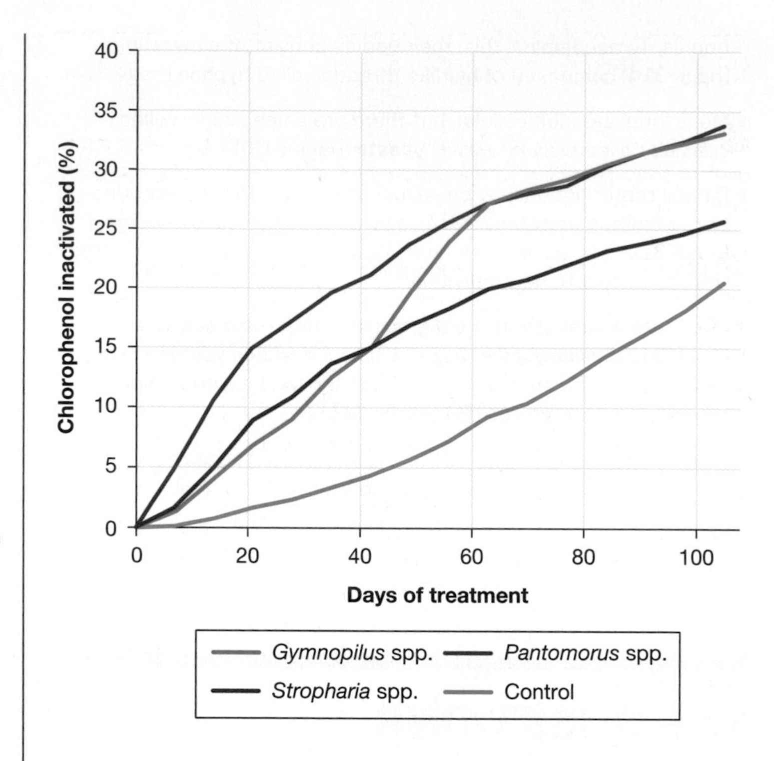

Leveling Up

13 **Life choices** If fungi are more closely related to animals than to plants, should vegetarians and vegans refrain from eating fungi? Why or why not?

14 ***Write Now* biology: Is a mass extinction under way?** The International Union for Conservation of Nature (IUCN) maintains its Red List, which identifies the world's threatened species. To be defined as such, a species must face a high to extremely high risk of extinction in the wild. The 2019 Red List contains more than 28,000 species threatened with extinction out of the approximately 105,500 species assessed. Because this assessment accounts for only a small percentage of the world's 1.7 million described species, the total number of species threatened with extinction worldwide may actually be much larger.

The Red List is based on an easy-to-understand system for categorizing extinction risk. It is also objective, yielding consistent results when used by different people. These two attributes have earned the Red List international recognition as an effective tool for assessing extinction risk.

(a) If the threatened species listed by the IUCN do become extinct and the percentages of species under threat in other taxonomic groups turn out to be similar to those listed, then the percentages of species that will go extinct will approach the proportions lost in some of the previous mass extinctions. Does this mean that a mass extinction is under way? Why or why not? Explain your reasoning, using information about mass extinctions from Chapter 14 (in particular, Figure 14.18) to support your argument.

(b) What are the causes of the current high extinction rates? Compare the causes of the current situation with those of previous mass extinction events. In what ways are they similar, and in what ways do they differ?

(c) Why are some groups in more danger of extinction than others? Choose one group that has a large proportion of species threatened with extinction, and find out more about it (you can begin at http://www.iucnredlist.org):

- What kinds of habitat are they found in?
- Does a particular aspect of their ecology—for example, feeding or reproduction—make them more vulnerable to extinction?
- Is anything being done to protect them and/or their habitat?
- Has there been an increase or decrease in the number of species within the group that have been identified as vulnerable to extinction?

Digital resources for your book are available online.

Neanderthal Sex

The relationship between modern humans and Neanderthals just got a whole lot spicier.

After reading this chapter you should be able to:

- Identify the key characteristics of animals, and list the major groups within the kingdom Animalia.
- Explain the significance of symmetry and segmentation for some animal groups.
- Compare and contrast the three types of mammals.
- Interpret an evolutionary tree of the primates.
- Illustrate the inheritance of mitochondrial DNA as compared to nuclear DNA.
- Describe the evidence suggesting that prehistoric hominin species reproduced with our direct ancestors.

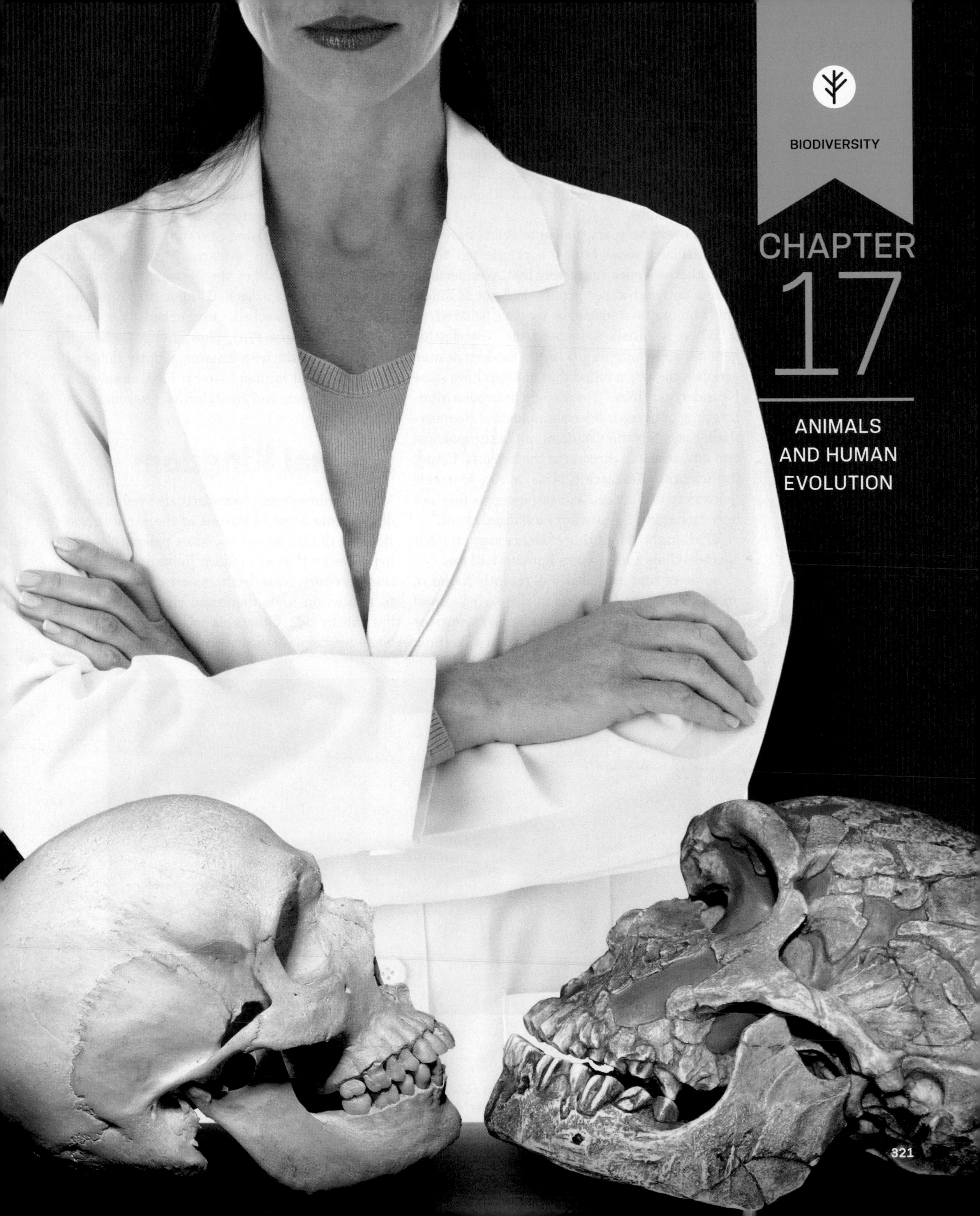

BIODIVERSITY

CHAPTER 17

ANIMALS AND HUMAN EVOLUTION

In popular culture, Neanderthals are ugly, hairy cavemen with big brains but no wits. Since the first Neanderthal bones were discovered in Germany in 1856, we've cultivated an image of our closest extinct human relatives as hulking brutes who communicated by grunting, walked like chimps, and hit each other over the head with clubs.

Yet, over the years, paleontologists have dug up fossilized vocal bones, sophisticated tools, and other evidence suggesting that Neanderthals were a fairly advanced group—and not as different from our own species as we once believed.

When scientists sequenced the Neanderthal genome and compared it to the modern human genome, they even found that humans have some Neanderthal DNA. "There could have been interbreeding between modern humans and Neanderthals," says Silvana Condemi, an anthropologist and now research director at the National Center for Scientific Research (CNRS) at Aix-Marseille University in France. "We can imagine they not only exchanged culture but exchanged genes."

That's right: Mounting evidence suggests that Neanderthals and modern humans had sex.

Some of that evidence was recently found in an old pile of dusty bones. In the early years of the twenty-first century, Laura Longo, a curator at the Civic Natural History Museum of Verona, Italy, decided to take a second look at a group of fossils excavated from a rock shelter called Riparo Mezzena. Riparo Mezzena is nestled in the Lessini Mountains in northern Italy, a wide-open landscape speckled with large rocks and evergreen trees. The region is snowy and silent in the winter but green and thriving in the summer when paleontologists come to work.

Neanderthals existed 300,000–28,000 years ago and lived in Riparo Mezzena about 35,000 years ago. Yet the Neanderthal fossils collected in that area had sat at the museum where Longo worked, untouched, for over 50 years (**Figure 17.1**). Now, Longo believed, the fossils could help answer a hotly debated question about human history: How closely did modern humans and Neanderthals interact?

Animal Kingdom

Modern humans and Neanderthals lived in some of the same areas of Europe at the same times. Because of this proximity, some paleontologists hypothesize that as modern humans expanded their territory, Neanderthals were quickly driven to extinction and, therefore, the two did not live side by side. Others claim the opposite: that Neanderthals, *Homo neanderthalensis*, were

Figure 17.1

Fossil remains

A jawbone like this one was found at the Riparo Mezzena rock shelter in the Lessini Mountains in Italy. The individual it belonged to lived between 40,000 and 30,000 years ago.

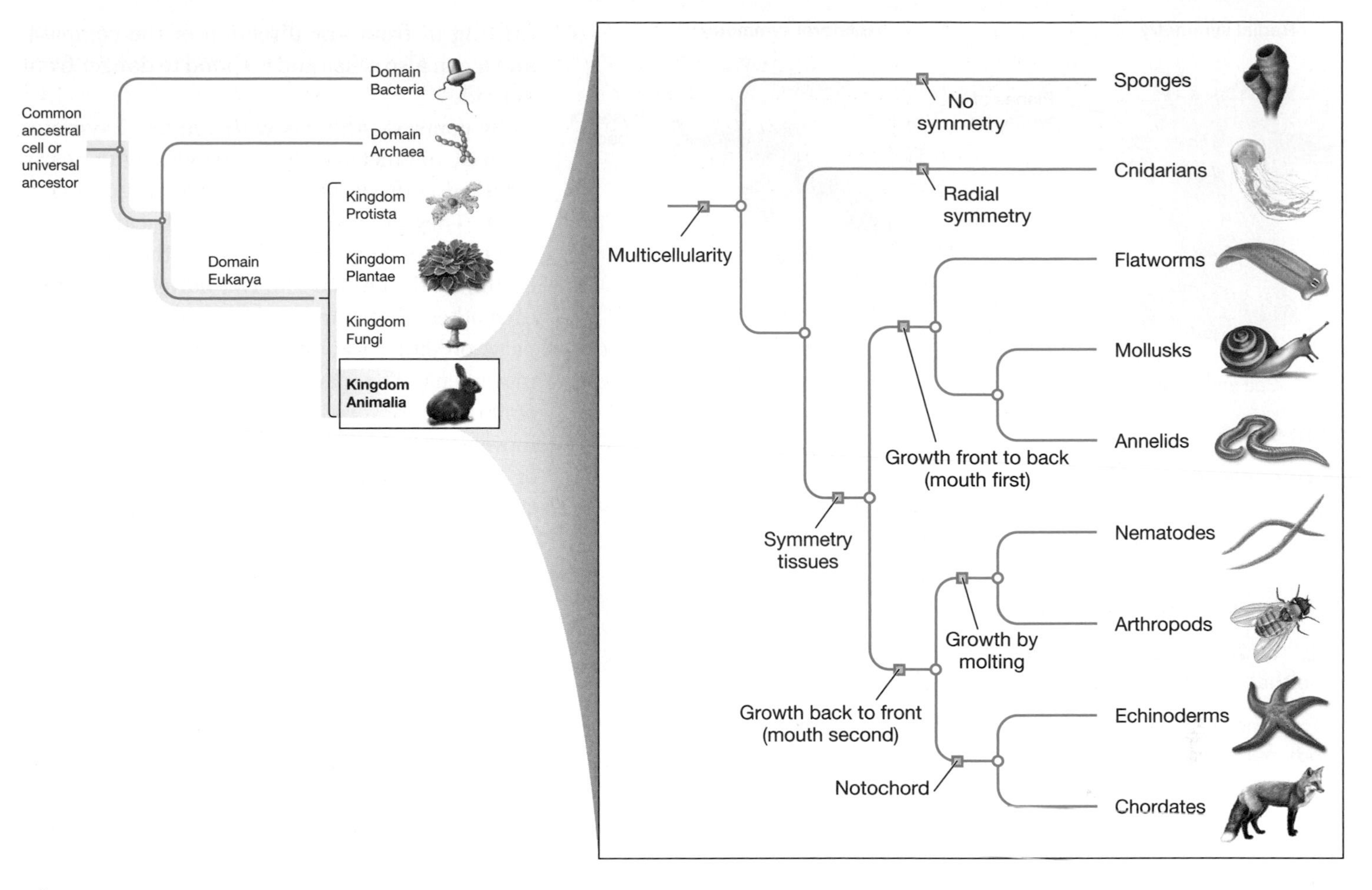

Figure 17.2

Animals are heterotrophic eukaryotes

On these evolutionary trees, circles indicate ancestors, and squares indicate developments. From a common ancestor, three organism domains emerged: Eukarya, Bacteria, and Archaea. Eukaryotes are further divided into four kingdoms: Protista, Plantae, Fungi, and Animalia. Animals ingest their food, which can be organisms from any domain or kingdom, including their own, and then digest it internally. The multicellularity of these organisms led to the development of tissues and organs and all the subsequent diversity that emerged.

Q1: Are mollusks more closely related to flatworms or to annelids? Explain your reasoning.

Q2: If you found an animal with no symmetry, to which group do you think it would belong? What about an animal with radial symmetry? One with bilateral symmetry?

Q3: If an animal shows growth back to front (or mouth second), what kind of symmetry does it have?

See Appendix A for answers to the figure questions.

slowly incorporated into the population of newly incoming humans, *Homo sapiens*. But to know how humans evolved, it is important first to realize where we came from.

All species in the *Homo* genus, of which *H. sapiens* is the only one alive today, are in the kingdom Animalia. **Animals** are multicellular ingestive heterotrophs; that is, we are complex organisms, eukaryotes, and obtain energy and carbon by ingesting food into our bodies and digesting it internally. Animals first evolved some 700 million years ago, descended from flagellated protists, single-celled organisms with whiplike tails that thrived in wet habitats and began to use oxygen to break down food into energy. From those humble origins, countless animal species evolved, beginning with sponges and including mollusks such as snails and clams, the annelids (segmented worms), and arthropods such as crustaceans, spiders, and insects (**Figure 17.2**).

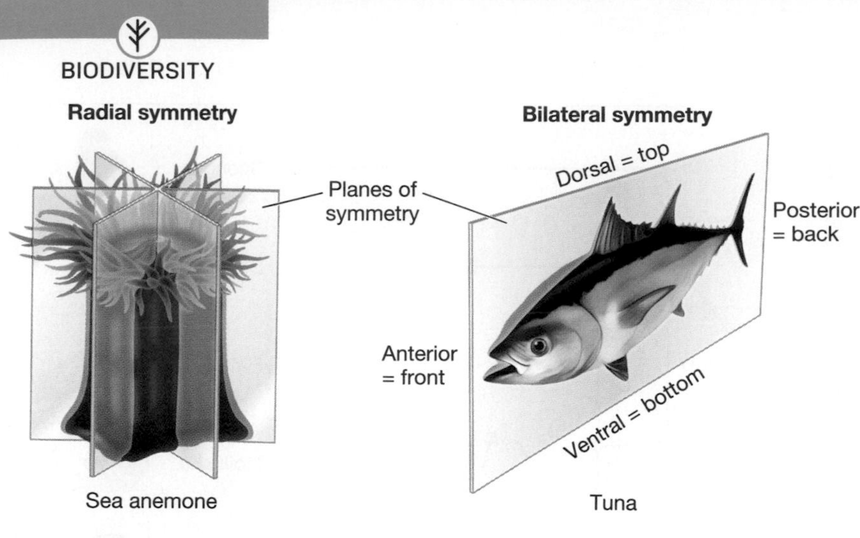

Figure 17.3

Body symmetry in animals

All animals other than sponges have symmetrical bodies. This arrangement helps them sense the world around them and respond to it.

Q1: Is a sea star (starfish) radially or bilaterally symmetrical?

Q2: What advantage might a bilaterally symmetrical animal have over one that is radially symmetrical, and vice versa?

Q3: What kind of symmetry do you (a human) have? What external body parts indicate this arrangement?

See Appendix A for answers to the figure questions.

A key factor that distinguishes one group of animals from another is body plan. All animals except sponges, the most ancient of the animal lineages, have a distinct body symmetry and can be divided into two main groups: those with radial symmetry and those with bilateral symmetry.

Animals with **radial symmetry** (**Figure 17.3**, left)—including cnidarians such as jellyfish, sea anemones, and corals— have bodies that can be divided symmetrically along any number of vertical planes that pass through the center of the animal, like cutting a pie. Radial symmetry gives an animal sweeping, 360-degree access to its environment. The animal can snare food drifting in from any direction of the compass, and it can also sense and respond to danger from any side.

By contrast, animals with **bilateral symmetry** have distinct right and left sides, with nearly identical body parts on each side (**Figure 17.3**, right). They can be divided symmetrically by just one plane passing vertically from the top to the bottom of the animal into two halves that mirror each other. The symmetrical arrangement of body parts on either side of a central body facilitates movement. The paired arrangement of limbs or fins, for example, enables quick and efficient movement on land or in water. Whether in radial or bilateral animals, locomotion is a key evolutionary innovation and has sparked a wide range of behaviors, including ways of capturing prey, eating prey, avoiding capture, attracting mates, caring for young, and migrating to new habitats.

DEBUNKED!

MYTH: Neanderthals lived alongside dinosaurs.

FACT: The last dinosaurs died out 65 million years ago (mya), and the earliest human ancestors emerged just 6 mya. Dinosaurs and Neanderthals—and dinosaurs and humans, for that matter—coexisted only in fiction.

Get a Backbone!

All of the chordates, including us, are bilateral animals. **Chordates** are a large phylum of bilateral animals that includes all animals with a dorsal **notochord**, a flexible rod along the center of the body that is critical for development, and a dorsal **nerve cord**, a solid strand of nervous tissue that we call the spinal cord in humans (**Figure 17.4**).

Humans belong to the phylum of chordates and the subphylum of vertebrates, meaning chordates with backbones. In vertebrates, the notochord has evolved to become the cushioning discs between **vertebrae** (singular "vertebra"), which are strong, hollow sections of the backbone (see "Repeat after Me" on page 326). Vertebrates include fish, amphibians (frogs and salamanders), reptiles (snakes, lizards, turtles, and crocodiles), birds, and mammals.

The chordate phylum also includes several subgroups of less familiar animals, such as sea squirts and lancelets (**Figure 17.4**), that also have a nerve cord along the back of the body but have no backbone. These chordates and all other phyla of animals are informally lumped together as invertebrates. But keep in mind that "invertebrates" is not an evolutionarily meaningful label; it is an artificial grouping (like that of protists) of animals with varied evolutionary histories and different sets of evolutionary adaptations.

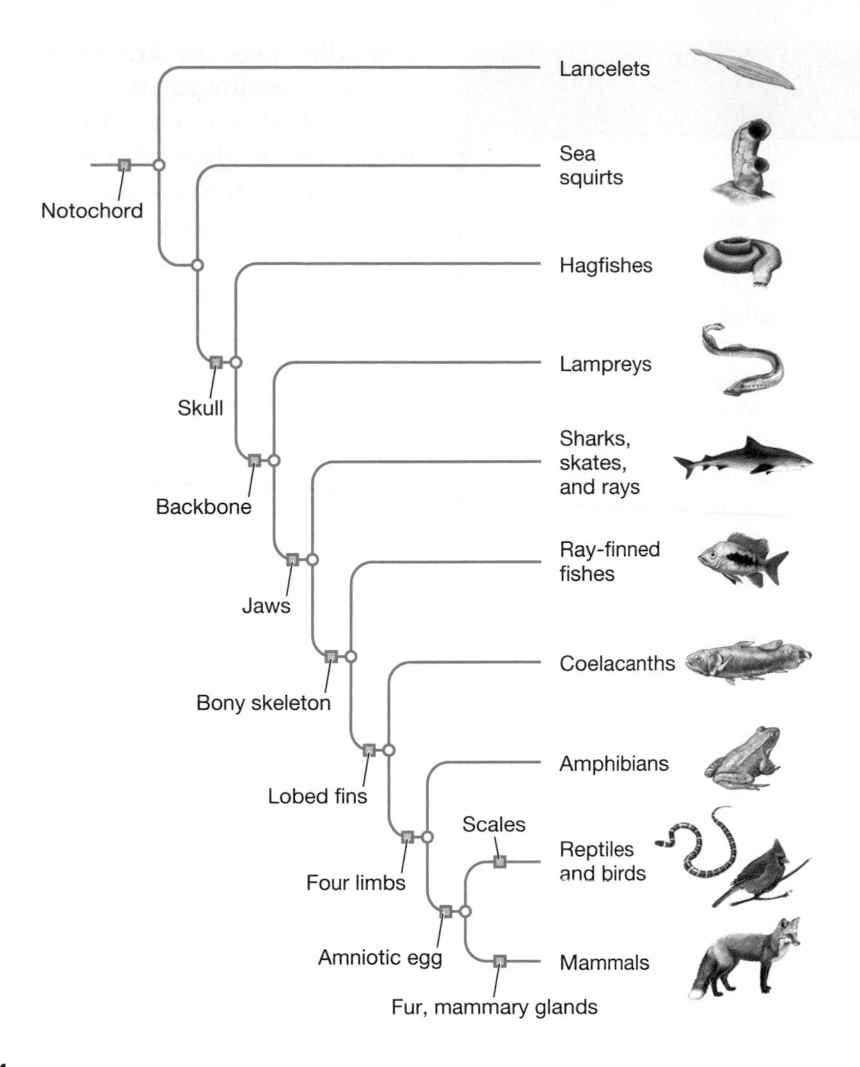

Figure 17.4

Chordate features

As with all eukaryotes except the protists (see Chapter 14), the evolutionary relatedness of the chordates is very well known. Not all chordates have a skull or backbone, but all have a notochord, which is the precursor to part of the vertebrate backbone.

Q1: Do amphibians have amniotic eggs?

Q2: What group of animals has jaws but not a bony skeleton?

Q3: When people talk about animals, they are sometimes referring only to mammals. How would you explain to them their error?

See Appendix A for answers to the figure questions.

We can trace the evolution of our species—and figure out when and where humans intersected with Neanderthals—by walking through the evolution of vertebrates.

It starts with fish. Jawless fish were the first vertebrates to evolve. Their skeletons, including the backbone, were made from a strong but flexible tissue called **cartilage**. Only a few groups of jawless fish have survived to the present day, most notably the lampreys (**Figure 17.4**). The next great leap in vertebrate evolution was hinged jaws, which enabled predators to grab and swallow prey efficiently. The evolution of teeth made jaws even more effective because teeth enabled animals to seize and tear food.

Repeat after Me

Three large groups of animals—arthropods (such as insects), annelids (such as earthworms), and vertebrates—have segmented bodies; that is, their body plan consists of repeated identical units, known as **segments**, running from the front to the rear. The three groups are not closely related to one another, but segmentation seems to offer significant advantages in terms of diversity, longevity, and overall evolutionary success. Arthropods, for example, make up nearly 40 percent of animal biomass on Earth, and vertebrates are highly diverse and widespread.

Nearly the entire body of an annelid consists of identical segments. In a vertebrate, segmentation is harder to see. It is hidden inside the body, in the vertebrae of the backbone and in the muscles and nerves that spread from the spinal cord. At some point in their lives, all arthropods have bodies that are internally and externally segmented. Each body segment tends to repeat the same type of structures: a pair of legs, a set of breathing organs, and a set of nerves. The head, thorax, and abdomen of arthropods are sets of segments grouped into sections.

The evolution of just the posterior segments of arthropods illustrates how evolution can take a basic body plan and modify it to produce many variations over time. Consider that the last segment in arthropods has evolved into the delicate abdomen of the butterfly, the piercing abdomen of the wasp, and the delicious tail of the lobster. **Appendages**, body parts with specialized functions, develop in pairs from particular segments of the body. The front appendage of vertebrates has evolved as an arm in humans, a wing in birds, a flipper in whales, and a front leg in salamanders and lizards, whereas the hind legs in snakes have been reduced to almost nonexistent nubs (see Figure 11.12). Over evolutionary time, the segments and the appendages that spring from them have evolved diverse form and function, enabling the animal body to adapt to new habitats or acquire new modes of life.

Another major step in the evolution of vertebrates was the replacement of the cartilage-based skeleton with a denser tissue strengthened by calcium salts: bone. Although the descendants of cartilaginous fish—sharks, skates, and rays (**Figure 17.4**)—are still with us today, bony fish are far more diversified and widespread in both saltwater and freshwater environments. With more than 30,000 species, bony fish are the most diverse vertebrates today.

The advent of lungs was a crucial milestone in the transition of vertebrates onto land. Amphibians made this transition only partially; they can live on land but must return to the water to lay eggs and breed. There are several thousand species living the amphibious lifestyle.

Reptiles were the first vertebrates to head into drier environments, and they evolved a number of adaptive traits to deal with the risk of dehydration. These adaptations included skin covered in waterproof scales, a water-conserving system for excretion, and the amniotic egg with its calcium-rich protective shell, which slows the loss of moisture while allowing the entry of life-giving oxygen and the release of waste carbon dioxide for the developing embryo. Reptiles dominated Earth during the age of the dinosaurs, and the dinosaurs' descendants (as discussed in Chapter 14) remain with us today in the form of birds. Like mammals, birds are warm-blooded, but they have feathers instead of fur for insulation. At least 10,000 species of birds are living today.

Mammals R Us

To get back to our egocentric focus—that is, humans—we are part of the kingdom Animalia and the class Mammalia. All **mammals** share specific features, including body hair, sweat glands, and milk produced by mammary glands. And at the risk of tooting our own horn, mammals have been a highly successful class of animals, with over 5,000 species living in a variety of habitats. This success is largely thanks to the extinction of dinosaurs; if dinosaurs still roamed Earth, it is likely that mammals would be nothing but dinner. Lucky for us, mammals have replaced dinosaurs as the top predators in most terrestrial habitats, and they thrive in both saltwater and freshwater environments. Only one type of mammal can fly—bats—although a few others, such as some species of squirrels, can glide through the air.

Mammals can be divided into three broad categories: eutherians, marsupials, and monotremes, all of which feed their offspring with milk (**Figure 17.5**). More than 95 percent of mammals alive today are **eutherians**, including humans. A unifying characteristic of eutherians is that the offspring are nourished inside the mother's body through a special organ called the *placenta* and are therefore born in a relatively well-developed state.

Marsupials have a simple placenta, resulting in offspring born early that then complete development in an external pocket or pouch.

Figure 17.5

Three kinds of mammals

Eutherians, such as polar bears, give birth to well-developed young. Marsupials, such as kangaroos, give birth to immature young that finish developing in a pouch. Monotremes, such as platypuses, lay eggs from which well-developed young hatch.

Q1: The Virginia opossum, or possum, is the only North American marsupial. How would the birth and development of its young compare with the birth and development of the young of eutherians and monotremes?

Q2: Do monotremes produce milk and nurse their young?

Q3: What kind of mammal is a cow? How about a human? How do you know?

See Appendix A for answers to the figure questions.

Marsupials are found mainly in Australia and New Zealand, with a few species in the Americas.

Monotremes are egg-laying mammals that have no placenta. The only living species of monotremes are just one platypus species and several echidna species, all in Australia and New Guinea.

In addition to being in the class Mammalia, humans are in the order **primates**. Like all other primates, we have flexible shoulder and elbow joints, five functional fingers and toes, thumbs that are **opposable** (that is, they can be placed opposite each of the other four fingers), flat nails (instead of claws), and brains that are large in relation to our body size (**Figure 17.6**).

Within the order primates we are members of the ape family, the **hominids**. We are not just closely related to apes; we *are* apes. As such, we share many characteristics with other apes, especially chimpanzees, including the use of tools, a capacity for symbolic language, and the performance of deliberate acts of deception. But we are part of a distinct branch of apes called **hominins**. This is the "human" branch of the ape family, consisting of *H. sapiens* and our extinct relatives. The members of the hominin lineage have one or more humanlike features—for example, thick tooth enamel or upright posture—that set them apart from the other apes, such as gorillas and chimpanzees. One member of this lineage is the Neanderthals.

Rise of the Apes

It is important to realize that biological evolution includes human evolution. Surveys taken since 2013 reveal that about 40 percent of adults in the United States do not believe that humans evolved from earlier species of animals. In scientific circles, evolution has been a settled issue for nearly 150 years. Scientists like Condemi go to work every day and see evolution in action. In fact, the vast majority of scientists of all nations agree that the evidence for evolution is overwhelming.

A major step in human evolution and the main feature that distinguishes hominins from

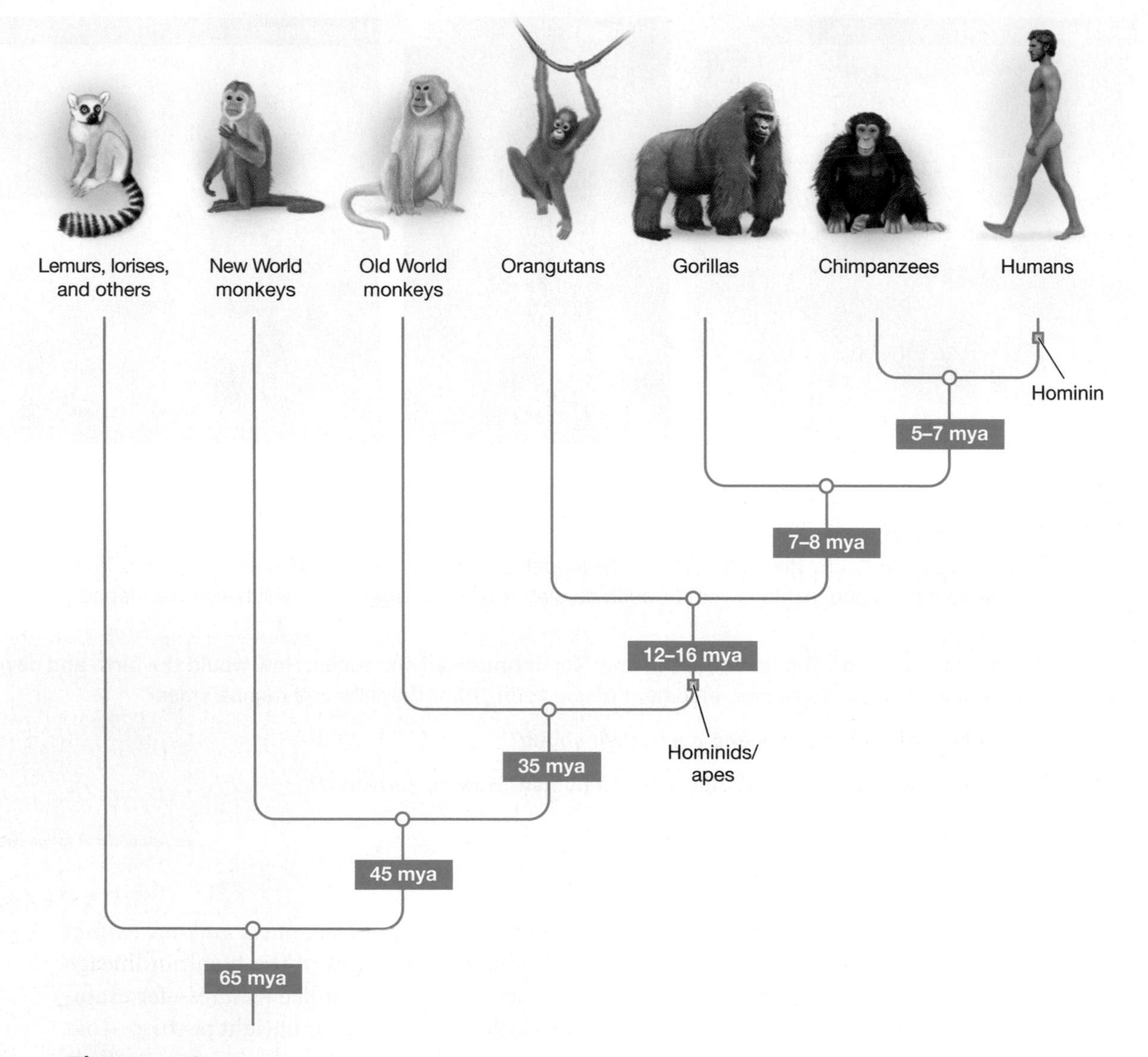

Figure 17.6

Primate groups

This tree illustrates our current understanding of primates' evolutionary relationships, based on genetic analyses and a series of spectacular fossil discoveries. Each circle indicates where scientists propose that the lineage leading to humans diverged from other primate groups.

Q1: According to this evolutionary tree, which primate group is most closely related to humans?

Q2: According to this evolutionary tree, which primate group is most distantly related to humans?

Q3: What characteristics are common to all primates, including humans?

See Appendix A for answers to the figure questions.

Because of a shared common ancestor 1.6 billion years ago, humans share DNA with all animals, plants, and fungi. But how much? Take a look to see how much genetic material you have in common with other organisms.

Assessment available in **smartwork**

Percentage of genes shared with humans

Humans
100%

Neanderthals
99%

Chimpanzees
96%

Mice
88%

Dogs
84%

Zebrafish
73%

Platypuses
69%

Chickens
65%

Honeybees
44%

Roundworms
38%

Grapes
24%

Baker's yeast
18%

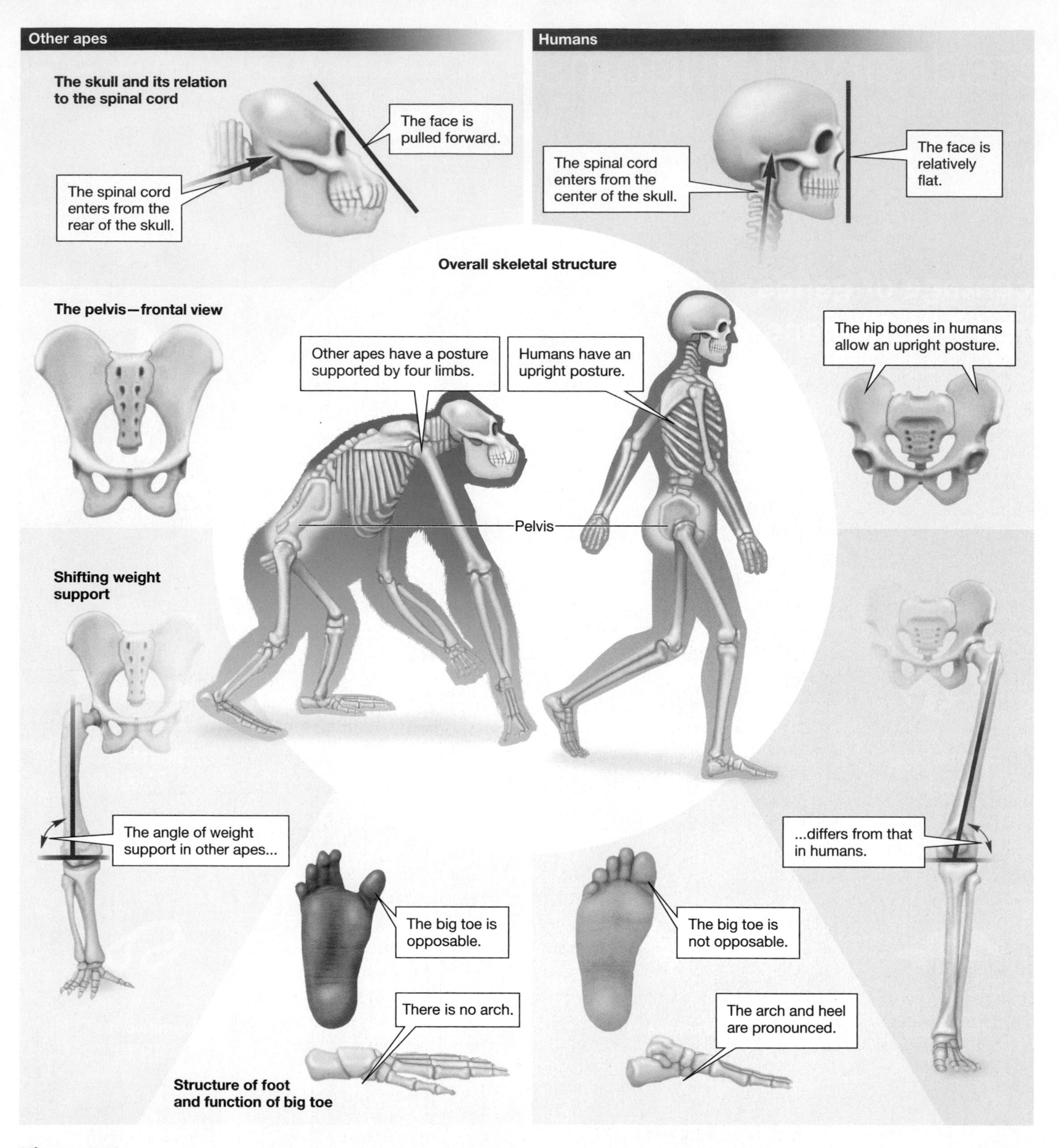

Figure 17.7

Evolutionary differences between humans and other apes

The switch to walking upright required a drastic reorganization of primate anatomy, especially of the hip bones. See figure questions on page 331.

other hominids was the shift from moving on four legs to being **bipedal**, walking upright on two legs (**Figure 17.7**). Many skeletal changes accompanied the switch to walking upright, including the loss of opposable toes, as you will notice if you try touching your little toe with the big toe on the same foot.

The loss of opposable toes that accompanied walking upright would have been a handicap in trees, as opposable toes help grasp branches during climbing. It is therefore likely that bipedalism was an adaptation for living on the ground. Walking on two feet freed the hands to carry food, tools, and weapons, and it also elevated the head, enabling the walker to see farther and over more things.

The shift to life on the ground was probably not sudden or complete. The skeletal structure of some of the oldest fossil hominins (3–3.5 million years old) indicates that they walked upright. However, foot bones and fossilized footprints show that the hominins living at that time still had partially opposable big toes (**Figure 17.8**). Perhaps they still occasionally climbed trees.

The earliest known hominin is *Sahelanthropus tchadensis*, identified from a 6- to 7-million-year-old skull discovered in 2002. Other early hominins include *Ardipithecus ramidus*, who lived 4.4 million years ago, and several *Australopithecus* species that are 3–4.2 million years old, including the first full-time walker with the first modern foot: *Australopithecus afarensis*. All of these hominins are thought to have walked upright. Their brains were still relatively small (less than 400 cubic centimeters in volume), and their skulls and teeth were more similar to those of other apes than to those of humans (**Figure 17.9**). On the other hand, the other *Homo* species were more similar to modern humans than to their ape relatives. A typical modern human has a brain volume of about 1,400 cubic centimeters, about the same volume as a 1.5-liter soda bottle.

Within the hominin branch is the *Homo* genus. The fossils identified at Riparo Mezzena were believed to be *Homo neanderthalensis* bones, but they had never been closely studied. So, Longo, the curator from Verona, assembled a team of researchers to analyze them, including Condemi, the anthropologist.

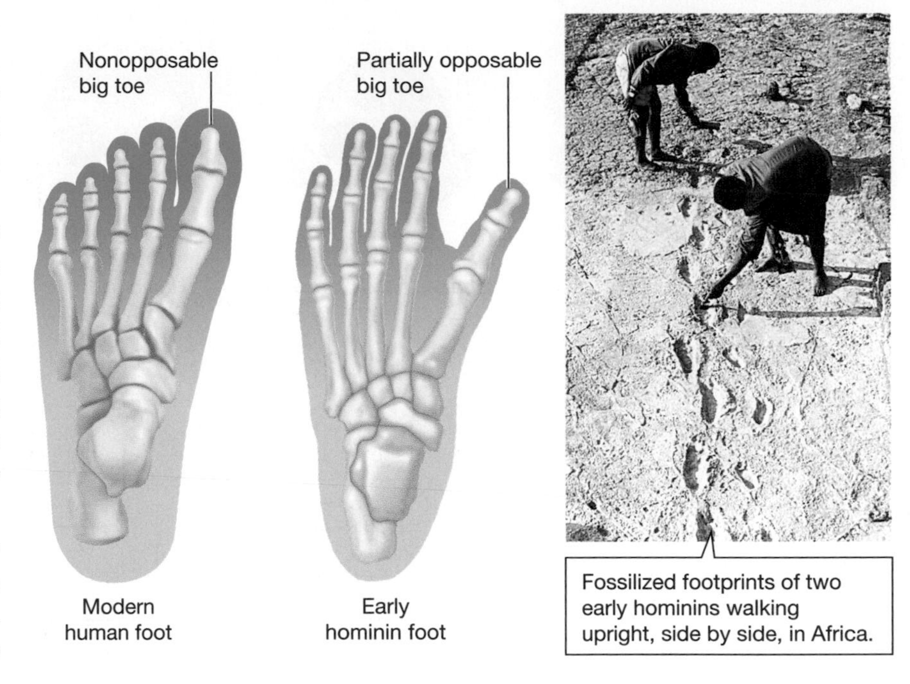

Figure 17.8

Early hominin locomotion

Fossilized foot bones show that some hominins living between 3.5 and 3 mya walked upright but had partially opposable big toes.

Q1: What other reason besides continuing to spend time in trees might explain why early hominins had partially opposable big toes?

Q2: In what way does the pattern of footprints in this figure suggest that the print makers were walking upright?

Q3: Why do you think we no longer have partially opposable big toes?

See Appendix A for answers to the figure questions.

Condemi had long been interested in the movement of Neanderthals across Europe and how their populations overlapped with modern *H. sapiens* populations, and she wanted to compare the Riparo Mezzena fossils to fossils from Neanderthal groups that had been dug up elsewhere around Europe. She hoped the fossils might illuminate interactions between Neanderthals and modern humans.

Q1: Through natural selection, deleterious traits will tend to disappear from a population over time. Which traits might have been deleterious for ground-dwelling early hominins that shifted from living in trees to living on the ground?

Q2: Through natural selection, advantageous traits will tend to persist in a population over time. Which traits might have been advantageous for ground-dwelling early hominins?

Q3: The adaptation to upright walking means that human females have more difficulty giving birth than do females of other species. What adaptation would you predict has had the greatest impact on this difficulty?

See Appendix A for answers to the figure questions.

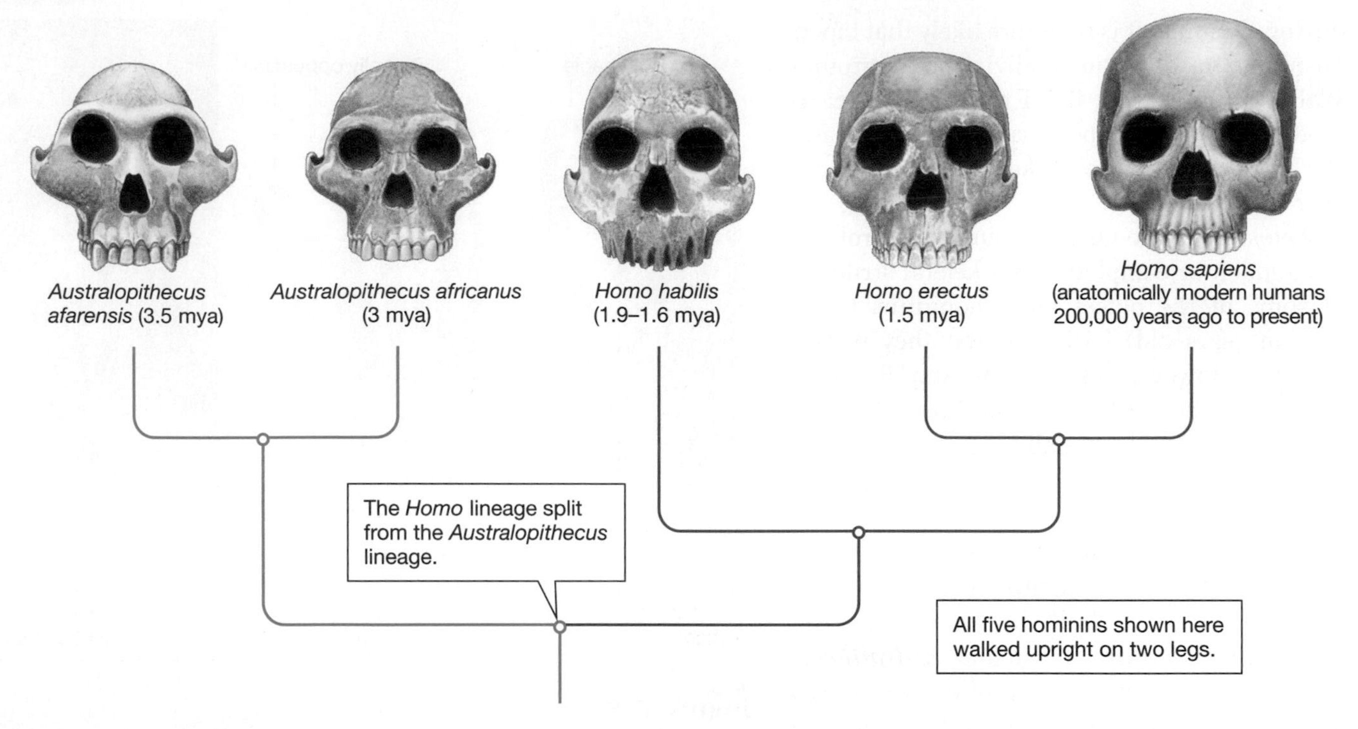

Figure 17.9

A gallery of hominin skulls

This tree shows the evolutionary relationships and the skulls of five hominin species. A complete evolutionary tree of hominins would be "bushier," with multiple side branches emerging at different times.

Q1: Would the Neanderthal species branch be on the *Homo* lineage side or the *Australopithicus* side of this tree?

Q2: Do you think *Homo habilis* and Neanderthal teeth would be more similar to apes or modern humans?

Q3: Which of these species has the smallest braincase?

See Appendix A for answers to the figure questions.

Hominins United

"For years, it was a very simple story: Neanderthals either disappeared quickly when modern humans came or they integrated with humans," says Condemi. But she suspected that the dynamic was more complex. "In some regions, Neanderthals disappeared very quickly. In other regions, we have evidence the two [species] could have overlapped."

Researchers, including Condemi, had often suggested that during the overlap, Neanderthals and humans interbred. But there was little physical evidence—until the Riparo Mezzena bones.

Even DNA evidence initially suggested that there was no interbreeding. DNA from the mitochondria of Neanderthals (isolated first from a single Neanderthal fossil in 1997 and then from four Neanderthal fossils in 2004) was compared to modern human mitochondrial DNA, and the tests showed that there was no genetic overlap between the species. **Mitochondrial-DNA inheritance** occurs only from the egg, not from the sperm (**Figure 17.10**). In fact, mitochondrial DNA (mtDNA) is unique because it is passed down virtually unchanged from mother to child, so it can be tracked from one generation, or one species, to another. But modern *H. sapiens* did not have Neanderthal mitochondrial DNA, so it appeared there had been no interbreeding, at least not involving female Neanderthals.

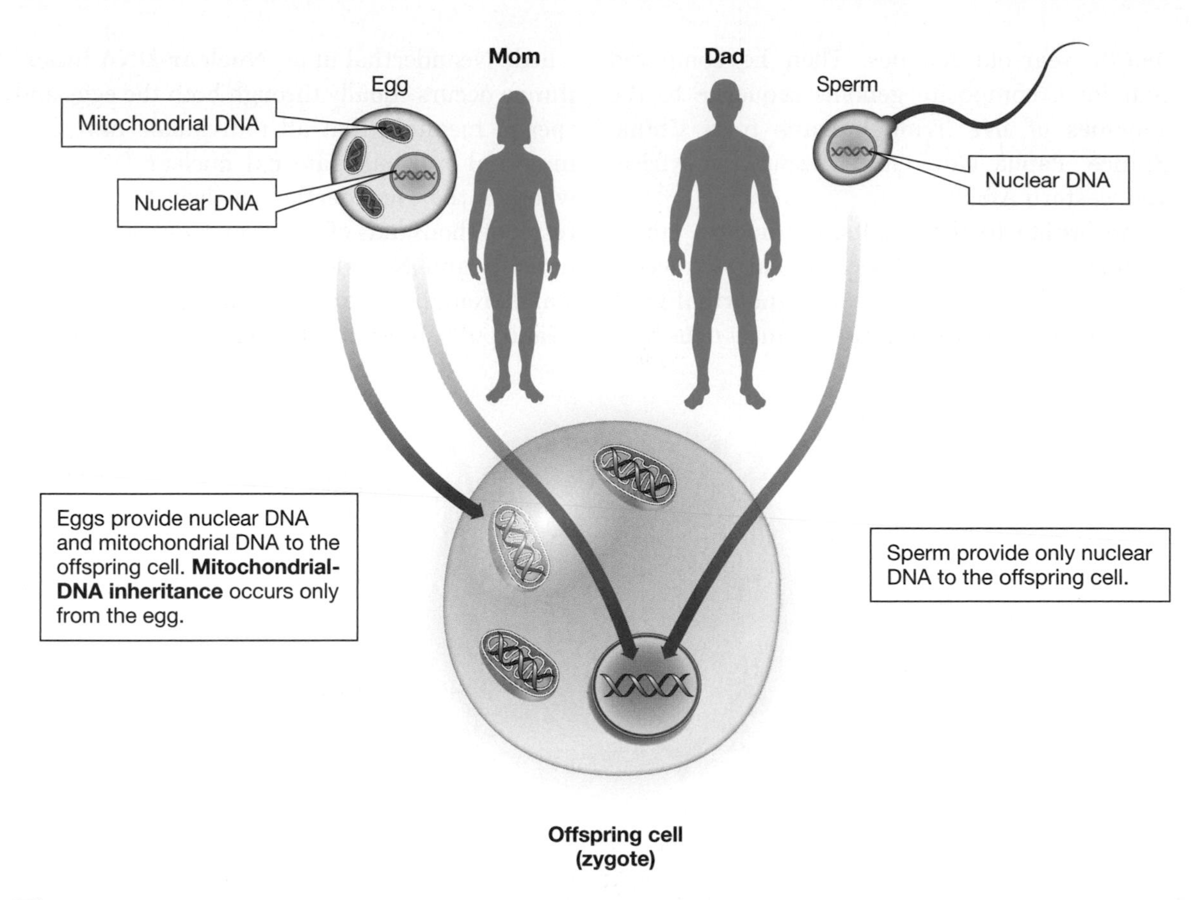
Mom
Dad
Egg
Sperm
Mitochondrial DNA
Nuclear DNA
Nuclear DNA
Eggs provide nuclear DNA and mitochondrial DNA to the offspring cell. Mitochondrial-DNA inheritance occurs only from the egg.
Sperm provide only nuclear DNA to the offspring cell.
Offspring cell (zygote)

Figure 17.10

Mitochondrial DNA comes only from your mom (not from your dad)
In addition to the chromosomes in the nucleus (nuclear DNA), cells also have DNA in their mitochondria. An offspring's mitochondria and their resident mtDNA all come from the mother.

Q1: Why does mitochondrial DNA come only from the mother?

Q2: If a child had a modern human mother and a Neanderthal father, could you tell that hybrid parentage by mitochondrial-DNA sequencing? Why or why not?

Q3: If a child had a modern human father and a Neanderthal mother, could you tell that hybrid parentage by mitochondrial-DNA sequencing? Why or why not?

See Appendix A for answers to the figure questions.

That mitochondrial work was performed by Svante Pääbo and researchers at the Max Planck Institute for Evolutionary Anthropology in Leipzig, Germany. Pääbo is one of the founders of the effort to use genetics to study early humans and other ancient populations, and he has pioneered numerous techniques to extract delicate DNA from even the tiniest slivers of fossilized bones. At the time that he began his work, it was easier to find and extract mitochondrial DNA because cells contain hundreds of copies of this type of DNA and only one copy of nuclear DNA. In addition, mitochondrial DNA can be isolated from cells and tissues that aren't so well preserved and from damaged DNA. Whole-genome sequencing, by contrast, requires well-preserved cells or tissues with fully intact nuclear DNA.

But despite his initial findings that modern humans and Neanderthals did not share mtDNA, Pääbo still thought there might be room for some small contribution. Being a diligent scientist, he decided to look even deeper—this time at the whole Neanderthal genome.

Pääbo spent 4 years sequencing the 1.5 billion base pairs in the Neanderthal genome, using DNA extracted from the femur bones of three

38,000-year-old females. Then he compared that long, composite genome sequence to the genomes of five living humans from China, France, Papua New Guinea, southern Africa, and western Africa.

According to the results, all modern ethnic groups, other than Africans, carry traces—between 1 and 4 percent—of Neanderthal DNA in their genomes. In other words, most of us have a little Neanderthal in us. **Nuclear-DNA inheritance** occurs equally through both the eggs and sperm, meaning that all individuals have half maternal and half paternal nuclear DNA. But whether the Neanderthal nuclear DNA is the result of thousands of sexual encounters between humans and Neanderthals or a few one-night stands remains unknown (**Figure 17.11**). When Pääbo published his work in 2010, he and others

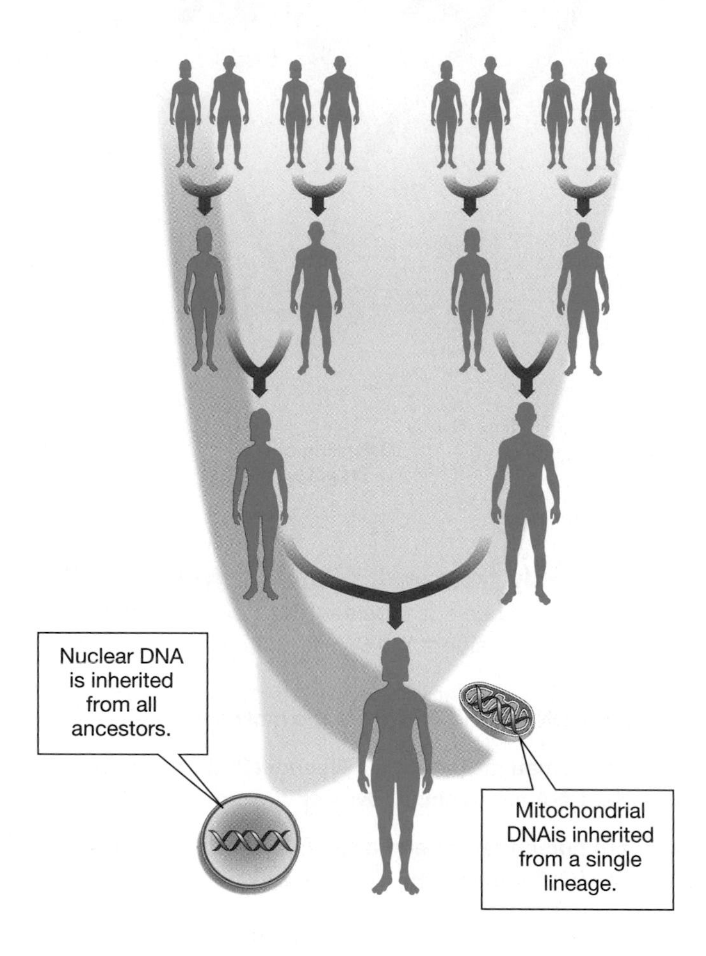

Figure 17.11

Two types of DNA inheritance

Mitochondrial-DNA sequencing can determine how related an individual is to only the female ancestors on its mother's side. By contrast, nuclear-DNA sequencing can determine how related an individual is to both its male and its female ancestors.

Q1: If a child had a modern human mother and a Neanderthal father, could you tell that hybrid parentage by whole-genome DNA sequencing? Why or why not?

Q2: If a child had a Neanderthal mother and a modern human father, could you tell that hybrid parentage by whole-genome DNA sequencing? Why or why not?

Q3: Under what circumstances are scientists able to do whole-genome sequencing, and when are they restricted to mitochondrial-DNA sequencing?

See Appendix A for answers to the figure questions.

admitted it was possible that the shared DNA wasn't necessarily a product of interbreeding. It could have been a remnant of DNA from a common shared ancestor.

Who would that ancestor have been? Anthropological research indicates that there were numerous species of *Homo* and that several of these species existed in the same places and times. More research and evidence will be necessary before general agreement is reached regarding the exact number of early *Homo* species and their evolutionary relationships.

The oldest *Homo* fossil fragments were found in Ethiopia in 2015 and date between 2.80 and 2.75 mya, suggesting that the earliest members of the genus *Homo* originated in Africa 2–3 mya. More complete early *Homo* fossils exist from 1.9–1.6 mya; these fossils have been given the species name *Homo habilis*. The oldest *H. habilis* fossils resemble those of *Australopithecus africanus*, the oldest known early human from southern Africa. More recent *H. habilis* fossils show a more rounded skull and a face that isn't pulled as far forward (see **Figure 17.9**). *H. habilis* fossils therefore provide an excellent record of the evolutionary shift from ancestral hominins (in *Australopithecus*) to more recent species, such as *Homo erectus*, the most likely candidate for a shared common ancestor of Neanderthals and modern humans (**Figure 17.12**).

Taller and more robust than *H. habilis*, *H. erectus* also had a larger brain and a skull more like that of modern humans (see **Figure 17.9**). It is likely that by 500,000 years ago, *H. erectus* could use, but not necessarily make, fire. In addition, *H. erectus* probably hunted large animals, as suggested by a remarkable 2010 discovery in Germany of three 400,000-year-old spears, each about 2 meters long and designed for throwing with a forward center of gravity (like a modern javelin).

Uniquely Human?

The frontal lobe of the brain is unique to human beings, and it enables us to reason like no other animal. But what about other attributes that we so commonly consider unique to us? As humans we pride ourselves on our intelligence and deep emotional connections to others, but are these really only human traits?

- *Language.* Researchers once believed language was an exclusively human trait. And it is, if we're referring to the use of words to represent things. But what about other forms of language? Chimpanzees in the wild use sign language, with approximately 70 different signs for distinct words. Other primates, birds, whales, and bats have distinct, learned vocalizations that they use to communicate.
- *Memory.* Some have suggested humans alone possess the ability to store memories. But dogs easily learn and remember many commands, whereas crows can learn and remember shapes better than human adults can, and they can use causal reasoning, not trial and error, to unlock doors and find hidden objects.
- *Social culture.* Once thought to be strictly human, social culture is a learned trait that chimpanzees, Japanese macaques, and killer whales pass on throughout their populations. Tool use by dolphins, elephants, and octopi varies in its specifics from population to population—a sure sign of learned behavior.
- *Emotions.* Our emotions make us human, right? Others in the animal kingdom have been documented expressing empathy (elephants), grief (elephants, dolphins), jealousy (apes), curiosity (cats, lizards), altruism (apes), and gratitude (whales). Apes have been seen laughing at clumsy fellow apes and using deception to outwit family members.
- *Self-awareness.* The ability to recognize oneself in the mirror was once thought to be ours alone. It was also thought to show self-awareness. As it turns out, all the apes and some gibbons, elephants, magpies, some dolphins and whales, and many fish pass the mirror test. However, researchers debate whether this ability truly reflects self-awareness. In this regard, we simply don't know what's going on in nonhuman animals' brains.
- *Altruism.* Finally, what about behavior that might not benefit oneself but might benefit others? Altruism has been well documented in nonhuman animals. For example, monkeys and rats will not accept offered food if, in doing so, a fellow member of their species receives an electric shock.

To be sure, there *is* something unique about humans that lies at the intersection of all these abilities. However, our expanding knowledge of animal behavior can't help but make us feel more closely connected to the other species with which we share this planet.

Figure 17.12

Meet the folks

These cartoons depict the presumed features of seven hominin species, giving their average height in centimeters. *H. sapiens* is represented as a 6-foot-tall male for reference.

Q1: Are you surprised by the interpretations of the hominins in this picture? Why or why not?

Q2: Describe the main differences that distinguish the hominin species.

Q3: From what you've learned about these species, do you think these representations are accurate? How can you find more information about each species to help you answer this question?

See Appendix A for answers to the figure questions.

H. erectus, or one of the other *Homo* ancestors, migrated from Africa about 2 mya. From there, this ancestor species spread around the Middle East and into Asia. *Homo* fossils dating from 1.9–1.7 mya have been found in the central Asian republic of Georgia, in China, and in Indonesia.

So, was the Neanderthal DNA found in modern human genomes simply a remnant of a common ancestor and not indicative of interbreeding? In 2012, Pääbo's team and others were able to determine the age of the pieces of Neanderthal DNA in the human genome. They

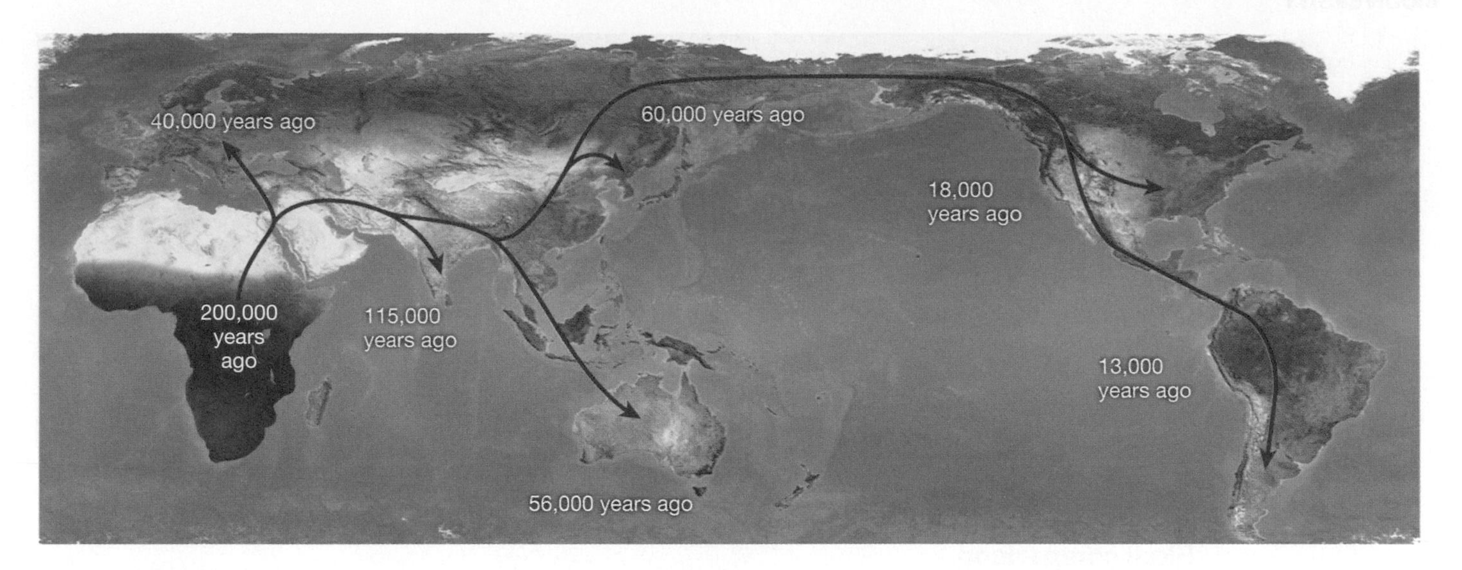

Figure 17.13

Anatomically modern humans evolved in Africa

This map traces the age of the earliest archaeological evidence that anatomically modern humans (*H. sapiens*) lived in different regions of the world. According to the earliest known archaeological specimens of modern humans, *H. sapiens* originated in Africa and traveled these routes.

Q1: When was Neanderthal DNA introduced into modern human DNA?

Q2: What conclusion can you make from the fact that all modern human populations besides Africans have a small amount of Neanderthal DNA?

Q3: What species of hominins other than the Neanderthals may have commingled with modern humans (see Figure 17.12)?

See Appendix A for answers to the figure questions.

found that the DNA had been introduced into our genome between 90,000 and 40,000 years ago, around the same time that modern humans spread out of Africa and met the Neanderthals. A remnant of DNA from a common ancestor would have been 10 times older.

All in the Family

The fossil record indicates that the first *Homo sapiens*, called archaic *H. sapiens*, originated 400,000–300,000 years ago. Archaic *H. sapiens* bore features intermediate between those of *H. erectus* and those of "anatomically modern" *H. sapiens*—our species. According to the out-of-Africa hypothesis, anatomically modern humans first evolved in Africa about 195,000–200,000 years ago from a unique population of archaic *H. sapiens* and then spread into other continents to live alongside other hominins (**Figure 17.13**). These ancestors of anatomically modern humans developed new tools and new ways of making tools, ate new foods, and built complex shelters. (But humans are not the only organisms to do many of these things; see "Uniquely Human?" on page 335.)

Early populations of archaic *H. sapiens* eventually gave rise to both the Neanderthals (who lived from 300,000 to 28,000 years ago) and us—that is, anatomically modern humans. In fact, there is some debate as to whether Neanderthals were simply an odd form of archaic *H. sapiens* or their own distinct species.

Evidence from the fossil record indicates that anatomically modern humans overlapped in time with *H. erectus* and Neanderthal populations yet remained distinct from them. Neanderthals and modern humans coexisted in western Asia for about 80,000 years and in Europe for some 10,000 years, until modern humans completely replaced all other *Homo* populations.

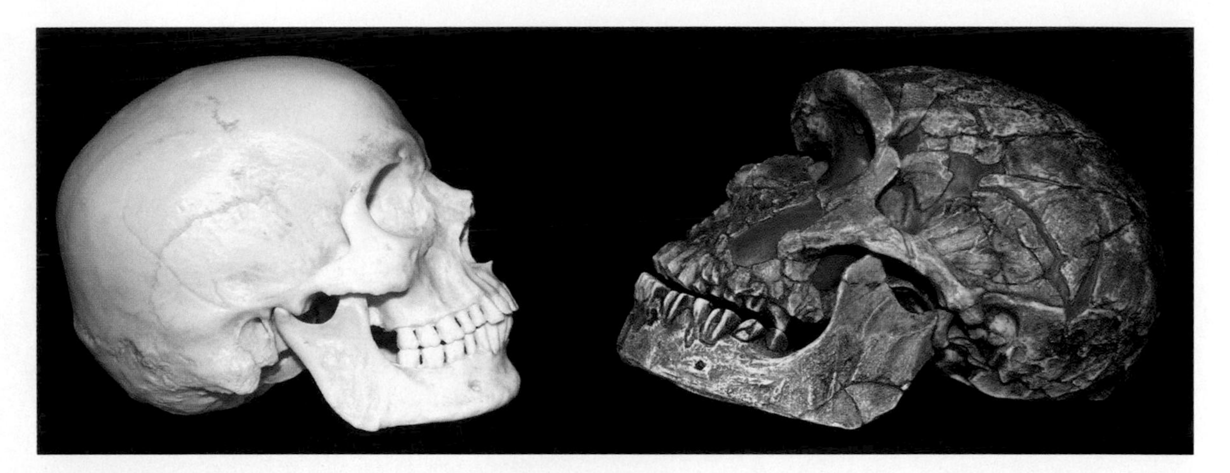

Figure 17.14

Skull comparison

The skull on the left is from a modern human; the one on the right is from a Neanderthal.

Q1: How do the lower jaws differ between the modern human skull and the Neanderthal skull?

Q2: How do the eyebrow ridges and foreheads differ between the two skulls?

Q3: What would you expect a hybrid of Neanderthals and modern humans to look like?

See Appendix A for answers to the figure questions.

But what happened in that intervening time? Were modern humans and Neanderthals friendly neighbors, or were the latter quickly wiped out by the former?

With a previous team, Condemi used fossil evidence from southern Italy to determine that modern humans arrived on the Italian peninsula 45,000–43,000 years ago, *before* the disappearance of Neanderthals. So the two populations likely made contact in Italy.

During this period, it's possible that "there was a kind of interbreeding," says Condemi. But if the two species interbred and had children, what did those children look like? And why hadn't Condemi's team found their remains?

Then Condemi examined the Riparo Mezzena bones. One caught her attention. It was a jawbone, a mandible, from a late Neanderthal living in Italy at the same time that modern humans had already made their way into Europe (see **Figure 17.1**). But the jawbone didn't look like a Neanderthal's, which has no chin (**Figure 17.14**). Instead, the face of the Riparo Mezzena individual, when reconstructed with three-dimensional imaging, had an intermediate jaw, something between no chin and a strongly projected chin. Because chins are a feature unique to modern humans, the jaw appeared to be something of a hybrid between a Neanderthal and a modern human. "This, in my view, could only be a sign of interbreeding," says Condemi.

To test her hypothesis, Condemi's team analyzed the fossil's DNA. The fossil had Neanderthal mtDNA, confirming that at least the individual's mother was a Neanderthal. From the DNA and imaging evidence, Condemi and her team concluded that it was the child of a "female Neanderthal who mated with a male *Homo sapiens*" (**Figure 17.15**). What's more, she added, this evidence supports the idea of a slow transition from Neanderthals to anatomically modern humans, in which the two species intermingled in both culture and sex, rather than the abrupt extinction of Neanderthals when modern humans arrived.

Since Condemi published her results, additional studies have led to a growing belief that there was an interbreeding bonanza among ancient human species, including between early modern humans and Neanderthals, between humans and Denisovans (a subspecies of archaic humans), between Denisovans and Neanderthals, and others. Each of these species likely interbred with the others on multiple occasions over the past 100,000 years, according to the scientific journal *Nature*.

These findings leave an open question for future scientists: While modern humans continued to develop their culture and populate the planet, Neanderthals became extinct. Why? Perhaps another pile of bones will someday reveal the answer.

Figure 17.15

Neanderthal-human hybrid
This is a paleontological reconstruction of what a Neanderthal-human hybrid might have looked like.

NEED-TO-KNOW SCIENCE

- **Animals** (page 323) are multicellular ingestive heterotrophs. The **chordates** (page 324) include all animals with a dorsal nerve cord. Chordates with a backbone are **vertebrates** (page 324), and all other phyla of animals are informally designated as invertebrates.
- All animals except sponges, the most ancient of the animal lineages, have a distinct body symmetry. Some invertebrates, such as the cnidarians, have **radial symmetry** (page 324); all other animals, including all chordates, have **bilateral symmetry** (page 324).
- Chordates have a dorsal **notochord** (page 324) and a dorsal **nerve cord** (page 324).
- The first vertebrates had skeletons made from tissue called **cartilage** (page 325).
- Many animal groups have a body plan consisting of repeated units known as **segments** (page 326). **Appendages** (page 326), body parts with specialized functions, develop in pairs from particular segments of the body. The evolution of segmentation and appendages has allowed animals to exploit new habitats and food sources in novel ways.
- **Mammals** (page 326) share specific features, including body hair, sweat glands, and milk-producing mammary glands. They can be grouped into three categories: **Eutherians** (page 326) have a placenta and deliver well-developed offspring. **Marsupials** (page 326) have a simple placenta and deliver partly developed offspring that finish developing in a pouch. **Monotremes** (page 327) do not have a placenta and instead lay eggs, from which their offspring hatch.
- Humans belong to the class of mammals. More specifically, humans belong to the order **primates** (page 327) and have flexible shoulder and elbow joints, five functional fingers and toes, **opposable** (page 327) thumbs, flat nails (instead of claws), and large brains.
- **Hominins** (page 327) are characterized by being **bipedal** (page 330), having the ability to walk upright on two legs. They are a branch of the apes, the family of **hominids** (page 327).
- **Mitochondrial-DNA inheritance** (page 332) occurs exclusively through the maternal egg; the sperm contributes essentially no mitochondria and none of the mitochondrial DNA. **Nuclear-DNA inheritance** (page 334) occurs equally through both the eggs and sperm; all individuals have half maternal and half paternal nuclear DNA.

THE QUESTIONS

See Appendix B for answers.

The Basics

1 Which of the following statements is *not* true?

(a) A single evolutionary line led from *Ardipithecus ramidus* to modern humans.

(b) Some hominid traits evolved more rapidly than others.

(c) Brain size increased greatly from early hominids to *Homo sapiens*.

(d) Toolmaking technology has improved greatly over the past 300,000 years.

2 The out-of-Africa hypothesis states that

(a) all new species of hominins arose in Africa and then migrated to the rest of the world.

(b) many new species of hominins arose outside of Africa and then migrated back to Africa later.

(c) all hominin speciation events occurred outside of Africa.

(d) all speciation events occurred in Africa, but only *Homo sapiens* distributed itself across the globe.

3 Which of the following features do sponges lack that some other animals have?

(a) segmented body plan

(b) symmetrical body plan

(c) backbone

(d) all of the above

4 Select the correct terms:

(**Mitochondrial DNA** / **Nuclear DNA**) is passed on only from the maternal line. (**Mitochondrial DNA** / **Nuclear DNA**) is inherited from both the mother and the father.

5 For each of the following cases, identify whether the animal group has bilateral symmetry (B), radial symmetry (R), or no distinct symmetry (N).

_____ a. sponges

_____ b. cnidarians

_____ c. arthropods

_____ d. chordates

_____ e. primates

Challenge Yourself

6 Examine the accompanying feline phylogenetic tree. Of the saber-toothed cats, which are the most closely related? The least related? Which of the modern cats is most closely related to the saber-toothed cats? Which of the modern cats is most closely related to the domestic cat? Explain your answers.

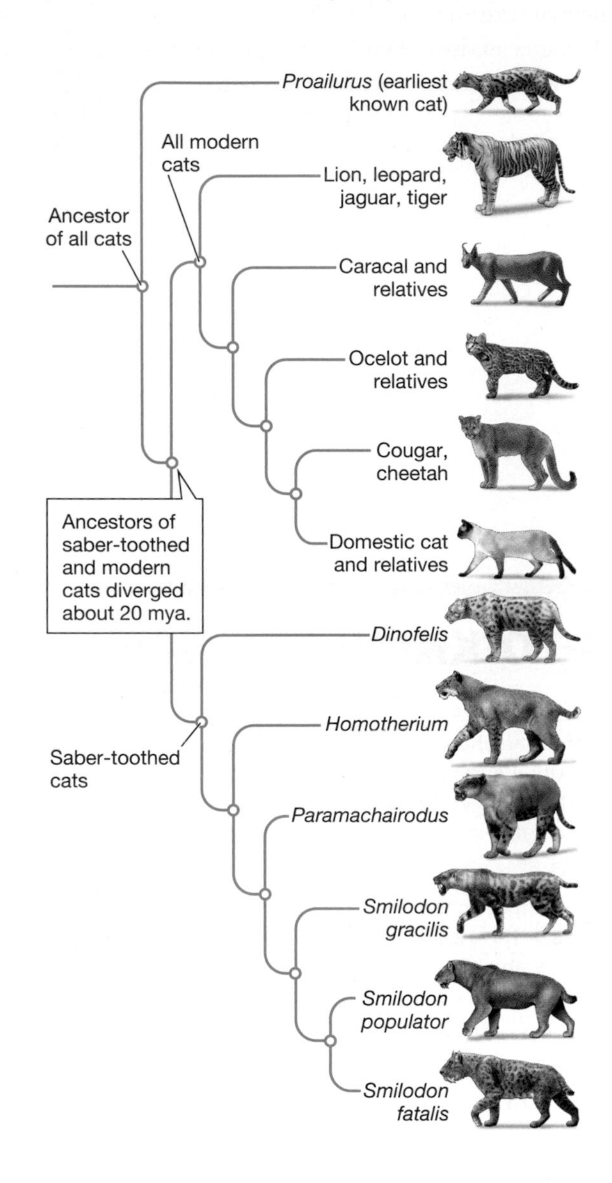

7 ________ specimens have features that are intermediate between those of *Australopithecus africanus* and *Homo erectus* and provide an amazing record of the evolutionary shift from ancestral hominin characteristics seen in *Australopithecus* fossils to more recent ones seen in *H. erectus* fossils.

(a) *Homo sapiens*

(b) *Homo neanderthalensis*

(c) *Homo habilis*

(d) *Ardipithecus ramidus*

8 Briefly describe the key differences that distinguish monotreme, marsupial, and eutherian mammals.

9 Place the following hominids in the correct order from earliest to most recent by numbering them from 1 to 5.

_____ a. archaic *Homo sapiens*

_____ b. *Australopithecus afarensis*

_____ c. modern *Homo sapiens*

_____ d. *Homo habilis*

_____ e. *Homo erectus*

Try Something New

10 You visit your local museum of natural history and come upon an exhibit showcasing hominin fossils dating back 300,000–400,000 years. You notice that these fossils have features intermediate between those of *Homo erectus* and those of modern *Homo sapiens*. To which species did these fossils belong?

(a) archaic *Homo erectus*

(b) archaic *Homo sapiens*

(c) *Homo neanderthalensis*

(d) *Homo habilis*

11 In 2004, scientists discovered the fossilized remains of an extinct species they named *Tiktaalik roseae*. The fossil appears to be a transitional species between (aquatic) fish and (terrestrial) four-legged amphibians. In what type of environment do you predict this animal lived?

12 Do some research to find an animal that has at least one of the following key adaptations of chordates: (a) backbone, (b) skull, (c) amniotic egg, (d) bony skeleton. Choose a different animal for each adaptation.

13 How does the fact that all ethnic groups *except* Africans contain some Neanderthal DNA (1–4 percent of their DNA) support the out-of-Africa hypothesis for the origin of modern humans (*Homo sapiens*)?

Leveling Up

14 ***Write Now biology:* if we were not alone** Fossil evidence indicates that in the relatively recent past (about 30,000 years ago), anatomically modern humans, or *Homo sapiens*, may have shared the planet with at least three other distinct hominins: *H. erectus*, *H. neanderthalensis*, and *H. floresiensis*. If one or more of these species were alive today, how would their existence affect the world as we know it?

15 **Doing science** Many free apps available for Android devices and iPhones enable you to do "citizen science." Find a biodiversity-project app and add to the collective knowledge base about aspects of the natural world. Your phone and the world are all you need to become a citizen scientist in your own backyard, on a hike, on vacation, on a city block, and beyond. Blog about your experiences and spread the word about your biodiversity app on social media. Go out and make a difference in the world.

16 **Is it science?** Watch the original *Planet of the Apes* movie from 1968. As it plays, list the scientific problems you spot. For example, based on your understanding of apes on Earth today, which of the apes' adaptations would be biologically possible, and which ones would be impossible? If these species did evolve to have the adaptive traits of *Homo*, would their members still be called apes?

Digital resources for your book are available online.

Climate Meltdown

NASA's "OMG" scientists track Greenland's rapidly melting ice sheet, which could single-handedly raise sea levels by 23 feet.

After reading this chapter you should be able to:

- Define the biosphere and the role that humans play in it.
- Differentiate between a biotic factor and an abiotic factor in the study of ecology.
- Distinguish between climate and weather.
- Describe the greenhouse effect, and identify the gas most responsible for it.
- Compare and contrast the hydrologic cycle and the carbon cycle.
- Explain how global warming contributes to climate change.
- List at least five consequences of climate change, and describe one consequence in detail.
- Articulate the need for a measure such as an ecological footprint, and describe your own ecological footprint.

ECOLOGY

CHAPTER 18

INTRODUCTION TO ECOLOGY

"Three, two, one . . . drop!" With a shove, Josh Willis launches a 3-foot-long canister from a hole in the bottom of the small propeller plane (**Figure 18.1**). The canister rockets 500 feet downward and splashes into Greenland's icy coastal waters. Willis and the rest of the scientific team aboard the plane speed ahead to the next drop zone.

Where the canister hit, a buoy floats to the surface and a specialized probe, which measures the depth, temperature, and saltiness of the water, sinks to the ocean floor. A thin metal wire connects the probe to the buoy on the surface where a radio transmitter sends data to the plane in real time.

Ten minutes later, the data transmission stops. The whole instrument sinks to the ocean floor where it biodegrades over time. The plane continues on its expedition along the shore of Greenland, an island northeast of Canada in the Arctic. Willis, a NASA oceanographer, prepares to drop the next canister.

It is August 2019, and Willis has launched nearly a thousand probes into Greenland's coastal waters, wrapping up the fourth year of NASA's Oceans Melting Greenland (OMG) project. Willis, head of the project, intentionally named it OMG to catch the public's attention (and because he's got a good sense of humor). OMG's goal is to better understand the role of the ocean in Greenland's massive and rapid ice loss. Three-quarters of Greenland is covered by a permanent ice sheet—the largest in the Northern Hemisphere and over a mile thick in some places. This ice sheet contains enough ice to raise the global sea level by 7 meters (or about 23 feet). And because of climate change, that ice is melting faster than ever before.

The OMG project is one of many efforts being undertaken around the world to quantify where and how fast ice sheets are melting and what effects that melting will have on the biosphere. The **biosphere** consists of Earth's living organisms plus the physical spaces we all inhabit. It includes inorganic chemicals such as water, our nitrogen-rich atmosphere, and more (**Figure 18.2**). Simply put, the biosphere is the integration of all the environments on Earth. The biosphere is crucial to our survival and well-being because humans depend on the biosphere for food and raw materials.

Our biosphere is rapidly being altered by climate change. Sea level rise, for instance, is occurring at a rate of about 3.5 millimeters

Figure 18.1

Probing for data

To determine the impact of rising ocean temperatures on how quickly Greenland glaciers are melting, scientists fly along the coastline and drop biodegradable probes to measure the depth, temperature, and salinity of coastal waters. Pictured here is NASA oceanographer Josh Willis.

Figure 18.2

The biosphere

This is a view of Earth from space. The biosphere consists of all the environments on the planet and their elements—the surface, the atmosphere, and all the living organisms.

per year. By 2100, when today's toddlers are grandparents, rising seas could displace entire populations in cities such as New Orleans, contaminate freshwater drinking supplies, and lead to chronic flooding in coastal areas. Eventually, if all the ice on the planet melts, raising sea levels by a whopping 216 feet, cities such as San Diego, New York, Buenos Aires, and London will simply be gone.

Greenland sits at the core of concerns about sea level rise. The island's ice sheet is now the single largest source of new water added to the ocean every year. "Greenland is melting," says Willis, who has studied sea level rise at NASA for more than 15 years. "The only question is how fast."

Scientists once predicted that due to global warming, the ice sheet would be fully melted in 1,000 years. But as our planet continues to warm, Willis and others say, the melting could take less time than that—far less.

Canary in a Coal Mine

Changes to Earth's physical environment have a major global effect on organisms living in various habitats. **Ecology** is the scientific study of interactions between organisms and their environment, where the environment of an organism includes **abiotic** (nonliving) factors and **biotic** factors (other living organisms). In Greenland, for example, abiotic factors include glaciers, bedrock beneath the ice, sunlight, and air temperature. Biotic factors include polar bears, narwhals, willow trees, ferns, and seagrasses (**Figure 18.3**). Warming in the Arctic has already begun to affect many of these species. Polar bears, for example, are losing the sea ice on which they hunt and give birth. The system of interacting abiotic factors and biotic factors is an **ecosystem** (we cover ecosystems in more depth in Chapter 21).

In August 2019, as Willis drops his probes, Greenland is in the middle of a heat wave. Temperatures rise 40 degrees Fahrenheit above normal on some days, and the surface of more than half the ice sheet is melting, including some areas typically frozen year-round. Massive icebergs break off the ice sheet, including one the size of lower Manhattan measuring 4 miles wide and half a mile thick.

Figure 18.3

Ecology and ecosystem: physical environment plus living inhabitants

Ecology is the study of how living organisms (biotic factors) interact with other organisms and with their nonliving environment (abiotic factors). The interaction of biotic factors and abiotic factors is an ecosystem.

Q1: List as many biotic and abiotic factors in this photograph as you can.

Q2: Is the tundra ecosystem part of the biotic or abiotic environment? Explain.

Q3: Name a part of the tundra ecosystem that is not also part of the surrounding ocean ecosystem. Is it biotic or abiotic?

See Appendix A for answers to the figure questions.

Yet Willis and his team are not here to observe Greenland's weather but to track the effects of climate. "Weather" and "climate" do not mean the same thing. **Weather** refers to short-term atmospheric conditions in a limited geographic area, such as today's temperature, precipitation, wind, humidity, and cloud cover. **Climate** describes the prevailing weather of a region over relatively long periods of time (30 years or more) and is determined by incoming solar radiation, global movements of air and water, and major features of Earth's surface. Organisms are more strongly influenced by climate than by any other feature of their environment. On land, for example, features of climate such as temperature and precipitation determine whether a particular region is desert, grassland, or tropical forest.

Figure 18.4

Consequences of climate change

Increasing global temperatures have led to melting of glaciers (top) and polar ice, causing sea level rise, increase in ocean temperatures, and coral reef bleaching and death. Extreme weather and new rainfall patterns caused by climate change have led to increased flooding in some areas (middle left) and drought in others (lower left), bringing about ecosystem and habitat destruction and species extinction. Effects on our food sources include increased crop failure (middle right) and depletion of important fisheries (lower right).

Q1: Name two ways in which climate change affects the frequency and severity of floods.

Q2: How has climate change caused a rise in sea level?

Q3: Give an example of an environmental effect of climate change in your state or region.

See Appendix A for answers to the figure questions.

Climate change, then, is a large-scale and long-term *alteration* in Earth's climate, and it includes such phenomena as global warming, change in rainfall patterns, and increased frequency of violent storms. Although Earth has gone through many changes in its climate over its 4.6-billion-year history, the speed of the change that has taken place in the past 100 years is without precedent in the climate record.

Climate change in recent history has been caused to a large extent by human actions, and its consequences are likely to be negative for people and ecosystems around the world (**Figure 18.4**). In addition to sea level rise affecting coastal regions around the globe, climate change threatens the world's food supplies, as extreme weather such as floods, droughts, and erratic rainfall disrupt food production and the availability of farmland.

With so many effects from climate change, one might wonder why scientists focus on Greenland, an autonomous territory of Denmark with a population of only about 56,000 citizens, roughly the same size as Idaho Falls, Idaho, or Terre Haute, Indiana. It's because changes in Greenland are an advance warning of the dangers of climate change. Temperatures in the extreme north and south of the planet—that is, the Arctic and Antarctica—are rising twice as fast as the global average, so the effects of climate change on the biosphere are particularly evident in these places.

Where Water Meets Ice

Willis and his crew of three—two additional scientists and a pilot—are in Greenland to measure how ocean water contributes to the melting of Greenland's glaciers. Traditionally, people think of ice sheets melting like an ice cube under a hair dryer—warm air causes ice to melt.

But an ice sheet touches more than just air, says Willis. The Greenland ice sheet is surrounded by ocean water roughly 1,000 meters deep, supported underneath by bedrock. In addition, ice along the edge of its glaciers is in constant contact with the ocean, breaking off and dumping chunks of ice into the water at the edges. Suspecting the ocean has just as much, if not more, effect on melting as air temperature does, Willis and other scientists at NASA set out to investigate.

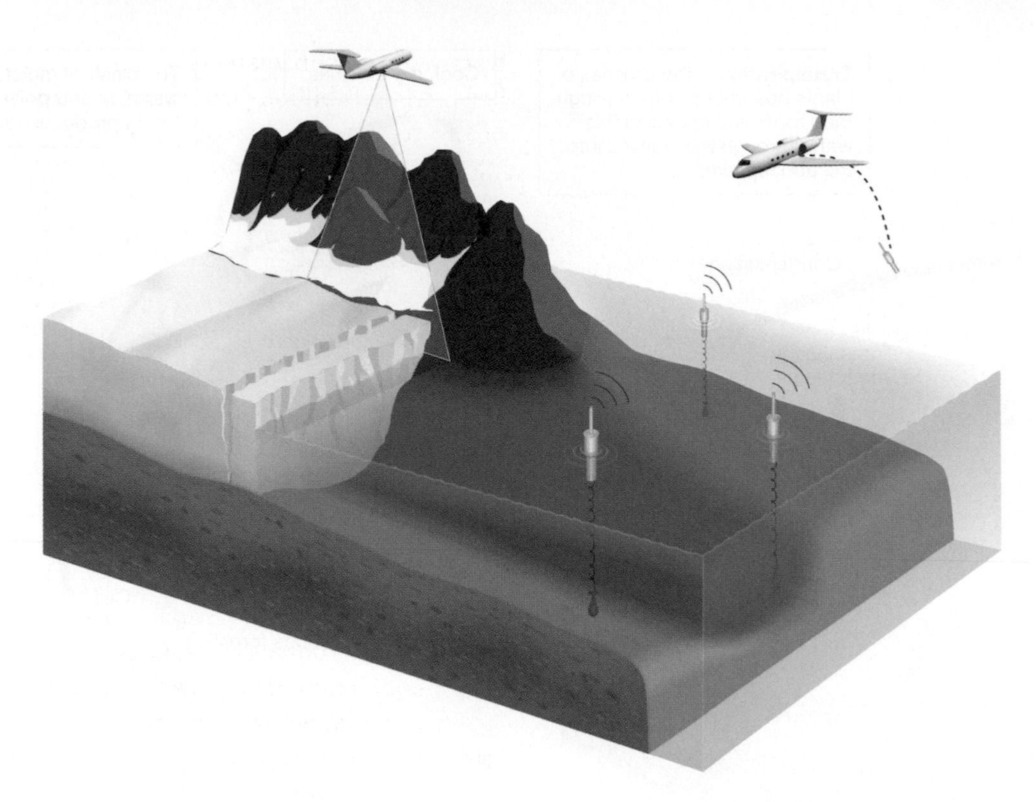

Figure 18.5

Collecting data on the extent and height of glacial ice in Greenland
To determine how much the glaciers of Greenland are melting each year, scientists drop temperature probes in summer coastal waters (right) and use radar to measure the height and extent of spring glacial ice (left).

To do so, the OMG team decided to measure ocean temperatures and ice thickness to see how the two affect each other. The team visits Greenland twice each year. In the summer, when there is minimal sea ice, the team drops probes into water along the coast to measure the temperature of the water touching the glaciers. In the spring, the team flies at a height of 40,000 feet and, sweeping radar along the surface, maps the height and extent of the ice (**Figure 18.5**). "We can fly over the edges of all these glaciers and map in great detail how they are advancing, retreating, thinning or thickening," says Willis.

In March 2016, Willis and the team flew over the Jakobshavn glacier for the first time and took radar measurements of the ice thickness. In cold areas of the world, such as Greenland and Antarctica, fallen snow gradually becomes compressed by additional layers of snow and turns into ice. That's how a glacier grows. If there is more ice formation than melting, the glacier thickens. If there is more melting than ice formation, the glacier thins.

The amount of snow that falls is determined by the **hydrologic cycle**, the circulation of water from the land to the sky and back again (**Figure 18.6**). Snow falling in Greenland and rain falling in Ohio are each part of the hydrologic cycle.

The hydrologic cycle is a never-ending global process of water circulation. This process happens most substantially near the equator where the greatest amount of direct sunlight causes water to evaporate from the planet's surface. The warm, moist air rises because heat causes it to expand, making it less dense and lighter than air that has not been heated. But as it rises, the warm, moist air begins to cool. Because cool air cannot hold as much water as warm air can, much of the moisture from a cooling air mass is "wrung out" and falls as rain.

For this reason, most tropical regions receive ample rainfall. Earth has four giant **convection cells**, in which warm, moist air rises and cools, releasing moisture as rain or snow depending on temperature and then sinks back to the ground as

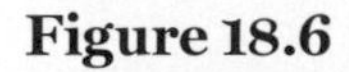
Arrows pointing toward the atmosphere show where evaporation occurs from water sources or is released from plants through transpiration to form humidity or clouds.
Transpiration is the process of plants absorbing water through their roots and releasing this water through their leaves into the atmosphere.
Precipitation
Condensation
Transpiration from plants
Evaporation
Uptake by plants
Runoff
Water percolation into soil
Groundwater flow
Ocean
Arrows pointing toward Earth show where water is returned as precipitation.

Figure 18.6

The hydrologic cycle

Arrows indicate the direction of water flow.

Q1: What is transpiration?

Q2: Why is transpiration important to the hydrologic cycle?

Q3: If plants and therefore transpiration decrease in a given area, what will happen to the humidity or cloud cover in this area?

See Appendix A for answers to the figure questions.

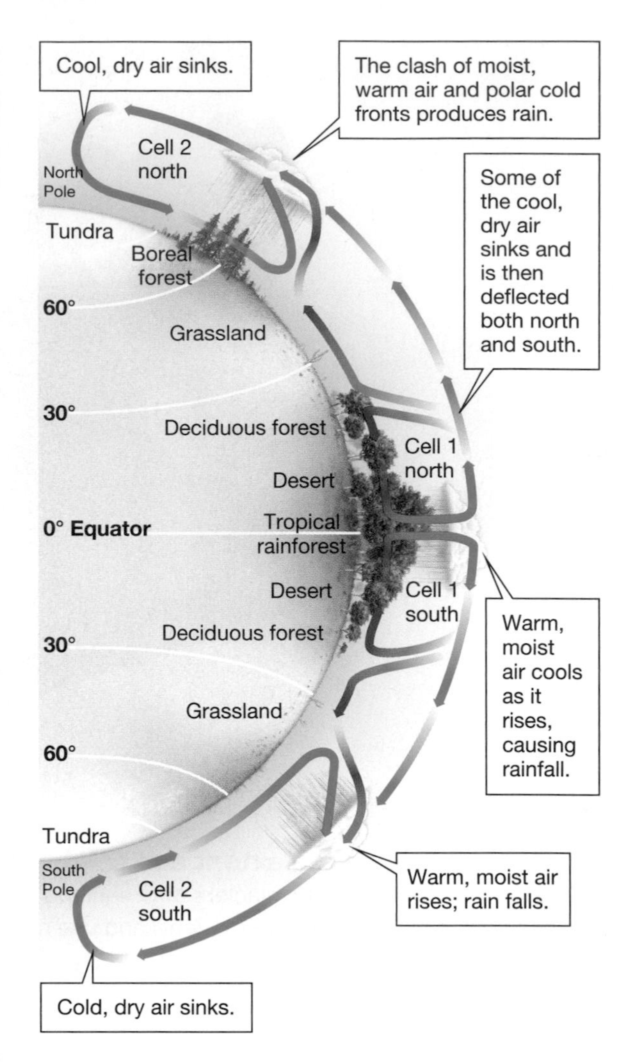
Cool, dry air sinks.
The clash of moist, warm air and polar cold fronts produces rain.
Cell 2 north
North Pole
Tundra
Boreal forest
60°
Grassland
Some of the cool, dry air sinks and is then deflected both north and south.
30°
Deciduous forest
Cell 1 north
Desert
0° Equator
Tropical rainforest
Desert
Cell 1 south
30°
Deciduous forest
Warm, moist air cools as it rises, causing rainfall.
Grassland
60°
Tundra
South Pole
Cell 2 south
Warm, moist air rises; rain falls.
Cold, dry air sinks.

Figure 18.7

Earth's convection cells

Two giant convection cells are in the Northern Hemisphere, and two are in the Southern Hemisphere.

Q1: How do the patterns of rainfall in the Northern and Southern Hemispheres compare?

Q2: How do the patterns in the kinds of environments shown in the Northern and Southern Hemispheres compare?

Q3: What happens at the equator to make this region so wet?

See Appendix A for answers to the figure questions.

dry air (**Figure 18.7**). These convection cells, in combination with the different angles of sunlight striking Earth, play a large role in the creation of regional environments on Earth, such as rainforests and deserts. They even affect regions far from the equator, including Greenland.

Since 1997, the Jakobshavn glacier had been doing one thing—thinning. The glacier was flowing fast, retreating away from the coastline, and thinning dramatically. From 2000 to 2010, Jakobshavn alone contributed about 1 millimeter of sea level rise. Although 1 millimeter may sound small, we're talking about raising the *entire* sea level around the globe.

When the OMG scientists took measurements with their new radar instruments in March 2017, the readings said the surface elevation was higher, not lower, than the previous year—that is, that Jakobshavn was thickening.

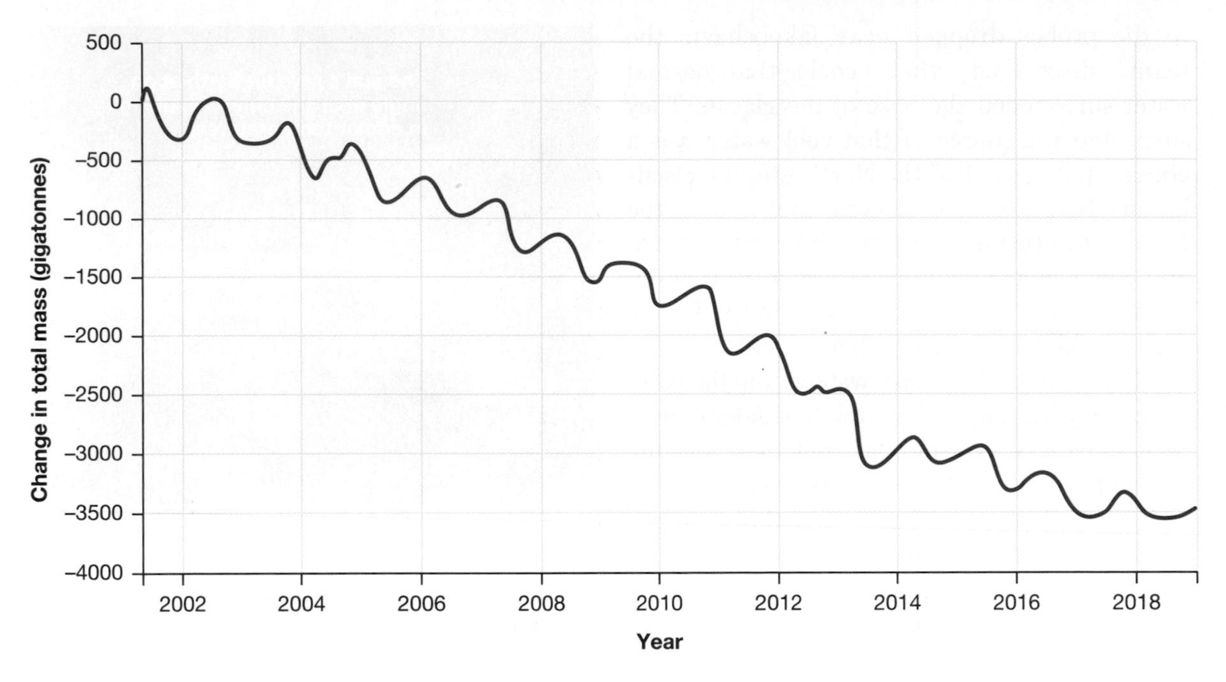

Figure 18.8

Greenland is melting

The Greenland ice sheet has been losing mass steadily at least since 2002, when these measurements began. A gigatonne is equal to 1 trillion (1,000,000,000) kilograms, or 2.2 trillion pounds.

Q1: How many gigatonnes of ice have been lost since 2002?

Q2: Why does the amount of ice loss change so dramatically over the course of each year?

Q3: About how much ice was lost between summer 2014 and summer 2016?

See Appendix A for answers to the figure questions.

"Our first reaction was that there must be something wrong," recalls Ala Khazendar, a scientist with NASA for 11 years and part of the OMG project. Khazendar had long studied what happens where ice meets ocean—two abiotic factors in the biosphere.

Where a glacier meets the sea, it loses ice in one of two ways, says Khazendar: either an iceberg breaks off and melts in the ocean or water melts the ice sheet from underneath. "These processes in general are part of the natural cycle of an ice sheet," says Khazendar. But over the past few decades, the processes have been out of balance in both the Arctic and Antarctic, with far more melting than ice formation, especially in Greenland, he adds (**Figure 18.8**).

Except, apparently, on Jakobshavn. Initially, Khazendar suspected there was something wrong with the instruments or the computer algorithms used to crunch the data. But after checking every possible part of the experiment, the team couldn't find any errors. Then, in 2017 and 2018, the team recorded additional increases in the height of the ice—Jakobshavn was definitely getting thicker, not thinner.

"It turns out, it was the first time in 20 years that the glacier was slowing down and thickening," says Khazendar. Although the Greenland ice sheet overall continued to lose mass, the Jakobshavn glacier was building up. But why?

Changing Climate

Willis, Khazendar, and the team began to investigate why a glacier that had been thinning for 20 years would suddenly start to thicken, especially as global temperatures were rising.

It turned out that ocean water—not the air—was causing the glacier to grow. Thanks to the probes dropped near Jakobshavn, the team discovered that cooler-than-normal water surrounded the base of the glacier. They suspected the source of that cool water was a climate pattern called the North Atlantic Oscillation (NAO), a natural cycle that causes the North Atlantic Ocean to switch between warm and cold waters about once every 5–20 years in a climate event similar to El Niño in the Pacific Ocean, which impacts weather patterns.

They concluded that cool waters from the NAO were likely leading to the growth of Jakobshavn. So could the scientists relax, put their feet up, and stop worrying about sea level rise?

Unfortunately, no. First, Greenland overall continues to lose a tremendous amount of ice mass each year, contributing to sea level rise. And, second, "We fully expect that this is not going to last," says Khazendar. "Jakobshavn is getting a temporary break" from warm waters, adds Willis. "But in the long run, the oceans are warming. And seeing the oceans have such a huge impact on the glaciers is bad news for Greenland's ice sheet." In a few years, the NAO will switch back, and all that cool North Atlantic Ocean water will be replaced with a current even warmer than normal due to the long-term trend of global warming.

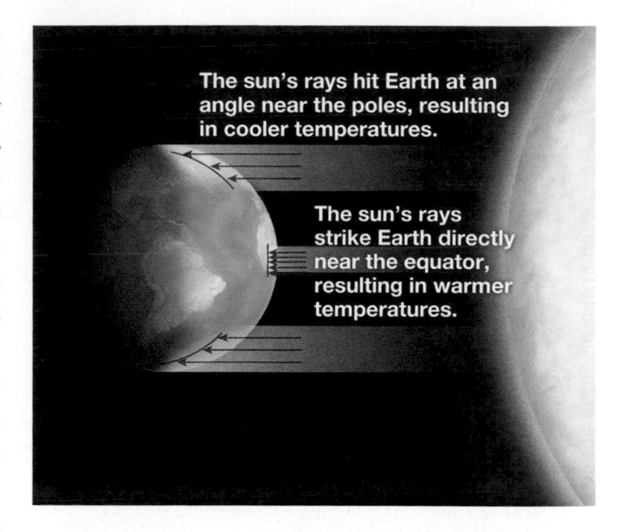

The sun's rays hit Earth at an angle near the poles, resulting in cooler temperatures.
The sun's rays strike Earth directly near the equator, resulting in warmer temperatures.

Figure 18.9

Sunlight strikes Earth most directly at the equator

The angle at which the rays of the sun strike Earth determines how much energy or heat reaches Earth's surface. The more direct the rays are when they strike Earth, the more heat they deliver.

Q1: Why is it colder at the poles than at the equator?

Q2: Why is it warmer at the equator than at the poles?

Q3: Why is it cooler at night than during the daytime?

See Appendix A for answers to the figure questions.

Stuck in a Greenhouse

The terms "global warming" and "climate change" are related but not synonymous. Global warming is one type of climate change among many. **Global warming** is a significant increase in the average surface temperature of Earth over decades or more.

Temperature on Earth is generally determined by the angle at which sunlight strikes the planet. Sunlight strikes Earth most directly at the equator but at a more slanted angle near the North and South Poles (**Figure 18.9**). Where sunlight hits the atmosphere more directly, its solar energy is more concentrated. For this reason, more solar energy reaches the equator than the poles, making the equator and neighboring tropical regions much warmer.

In addition, because Earth is tilted on its axis as it revolves around the sun in its annual orbit, sunlight hits different areas of the planet at different angles over the course of the year. This is the cause of our seasons: the Northern Hemisphere experiences summer when Earth is tilted toward the sun, while the Southern Hemisphere simultaneously experiences winter because it is tilted away from the sun (**Figure 18.10**).

Global warming, however, is not dependent on how sunlight strikes Earth. Instead, it is caused by an increase in **greenhouse gases**. Some gases in Earth's atmosphere, such as carbon dioxide (CO_2), water vapor (H_2O), methane (CH_4), and nitrous oxide (N_2O), absorb heat that radiates away from Earth's surface, preventing that heat from being released into space. These gases are called greenhouse gases because they function much as the walls of a greenhouse or the windows of a car. They let in sunlight and

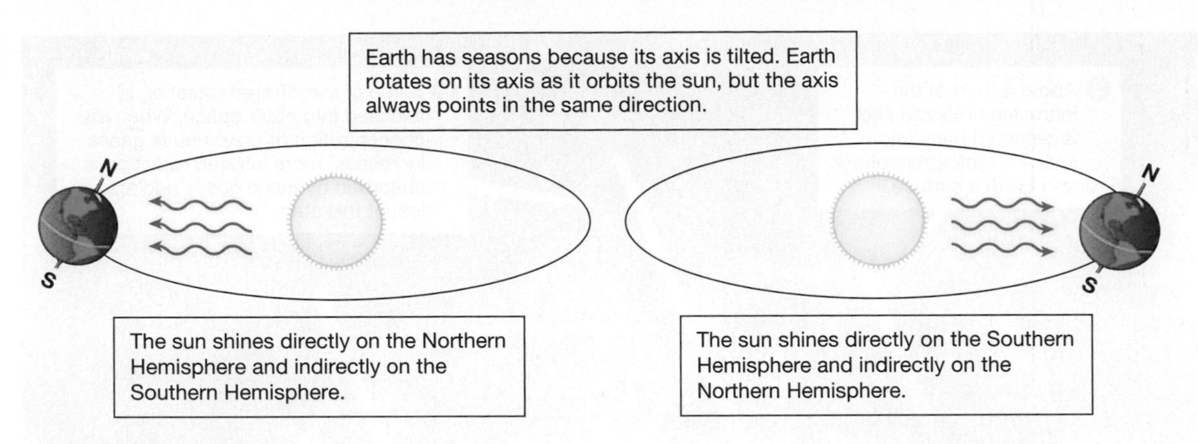

Earth has seasons because its axis is tilted. Earth rotates on its axis as it orbits the sun, but the axis always points in the same direction.
N
S
N
S
The sun shines directly on the Northern Hemisphere and indirectly on the Southern Hemisphere.
The sun shines directly on the Southern Hemisphere and indirectly on the Northern Hemisphere.

Figure 18.10

Seasons are caused by the tilt of Earth and its annual journey around the sun

As Earth moves around the sun, its tilt means that the Northern and Southern Hemispheres trade off receiving more direct sunlight.

Q1: When the Northern Hemisphere is tilted most directly toward the sun, what season is experienced there?

Q2: When the Northern Hemisphere is tilted the furthest away from the sun, what season is experienced in the Southern Hemisphere?

Q3: Why are the seasons less distinct at the equator than at the North and South Poles?

See Appendix A for answers to the figure questions.

trap heat in a process known as the **greenhouse effect** (**Figure 18.11**).

Greenhouse gases are not inherently bad; in fact, they have existed in Earth's atmosphere for more than 4 billion years, and they play an important role in maintaining temperatures that are warm enough for life to thrive on Earth. Yet human activities, primarily the burning of fossil fuels, have released an excess of greenhouse gases into the atmosphere, especially the infamous king of greenhouse gases: **carbon dioxide (CO_2)**.

The Great Carbon Rise

Carbon, in the form of CO_2 gas, makes up only a small part of Earth's atmosphere, about 0.04 percent. Carbon is also found in Earth's crust where carbon-rich sediments and rocks formed from the remains of ancient marine and terrestrial organisms. Carbon is present in every living thing.

Living cells are built mostly from organic molecules that contain carbon atoms bonded to hydrogen atoms. After oxygen, carbon is the most abundant element in cells by weight; every one of the large biomolecules in an organism has a backbone of carbon atoms. Living organisms, in both aquatic and terrestrial ecosystems, acquire carbon mostly through photosynthesis. Aquatic producers, such as photosynthetic bacteria and algae, absorb dissolved CO_2 and convert it into organic molecules using sunlight as a source of energy. Plants, the most important producers in terrestrial ecosystems, absorb CO_2 from the atmosphere and transform it into food with the help of sunlight and water.

The transfer of carbon within biotic communities and their physical surroundings, the abiotic world, is known as the global **carbon cycle** (**Figure 18.12**). (For more on nutrient cycling, see Chapter 21.) The biotic elements act as carbon reservoirs until they decompose and release carbon into the surrounding environment. Another way that carbon is transferred between the biotic and abiotic worlds is through combustion—that is, the burning of carbon-rich materials, living or not.

Some of the organic matter from ancient organisms has been transformed by geologic

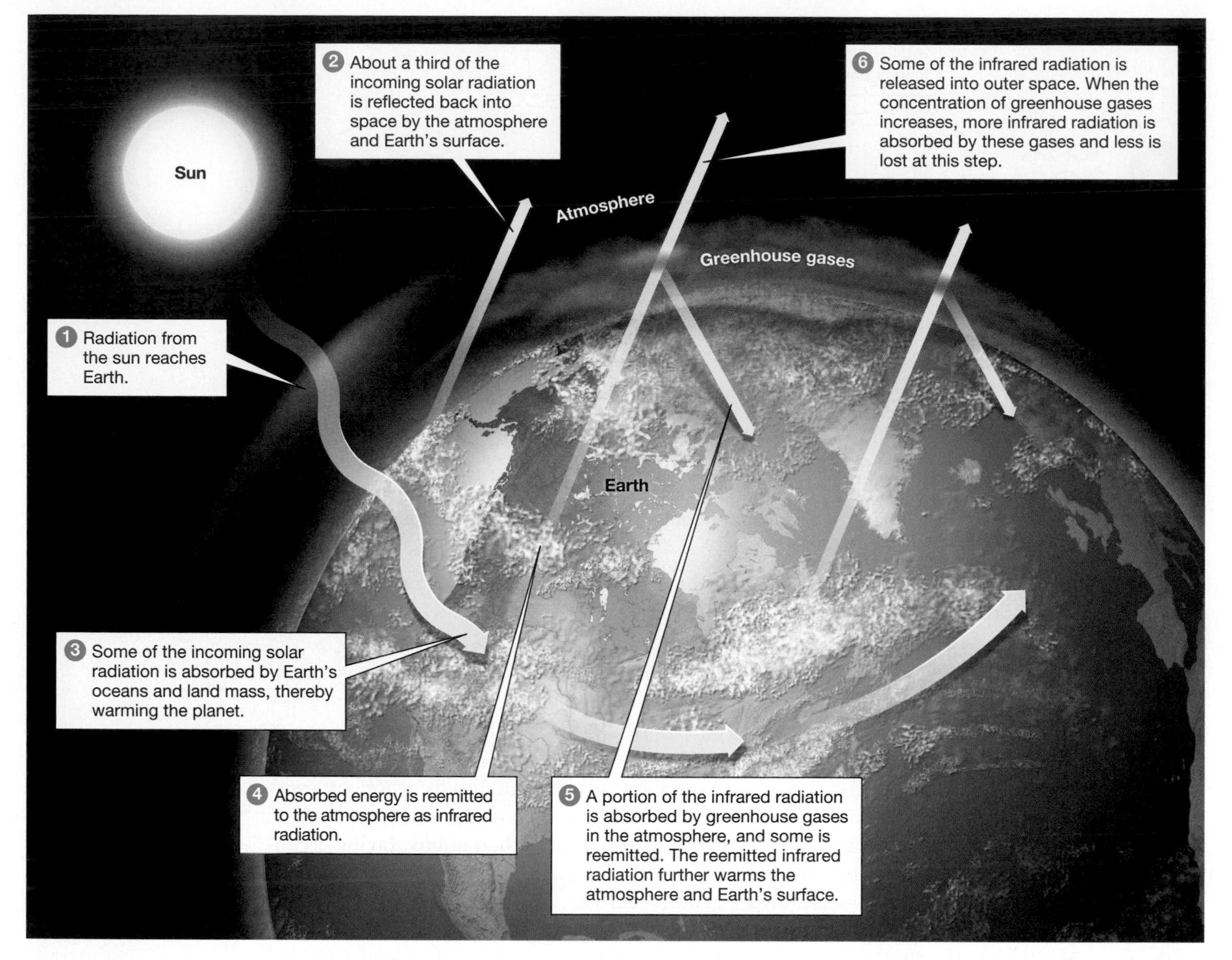

Sun
Atmosphere
Greenhouse gases
Earth
1 Radiation from the sun reaches Earth.
2 About a third of the incoming solar radiation is reflected back into space by the atmosphere and Earth's surface.
3 Some of the incoming solar radiation is absorbed by Earth's oceans and land mass, thereby warming the planet.
4 Absorbed energy is reemitted to the atmosphere as infrared radiation.
5 A portion of the infrared radiation is absorbed by greenhouse gases in the atmosphere, and some is reemitted. The reemitted infrared radiation further warms the atmosphere and Earth's surface.
6 Some of the infrared radiation is released into outer space. When the concentration of greenhouse gases increases, more infrared radiation is absorbed by these gases and less is lost at this step.

Figure 18.11

How greenhouse gases warm the surface of Earth

Carbon dioxide (CO_2), water vapor (H_2O), methane (CH_4), and nitrous oxide (N_2O) are known as greenhouse gases because they absorb and trap heat that would otherwise radiate away from Earth.

Q1: How much of the incoming solar energy is reflected back to outer space?

Q2: What kind of energy is reemitted to the atmosphere after being absorbed by Earth's surface?

Q3: How are greenhouse gases like a blanket on your bed at night?

See Appendix A for answers to the figure questions.

processes into deposits of fossil fuels such as petroleum, coal, and natural gas. When we extract these fossil fuels and burn them to meet our energy needs, the carbon that was locked in these deposits for hundreds of millions of years is released into the atmosphere as carbon dioxide. Plants also release carbon back into the atmosphere when they are burned.

Scientists estimate atmospheric CO_2 levels for both the recent and the relatively distant past, up to hundreds of thousands of years ago, by measuring CO_2 concentrations in air bubbles trapped

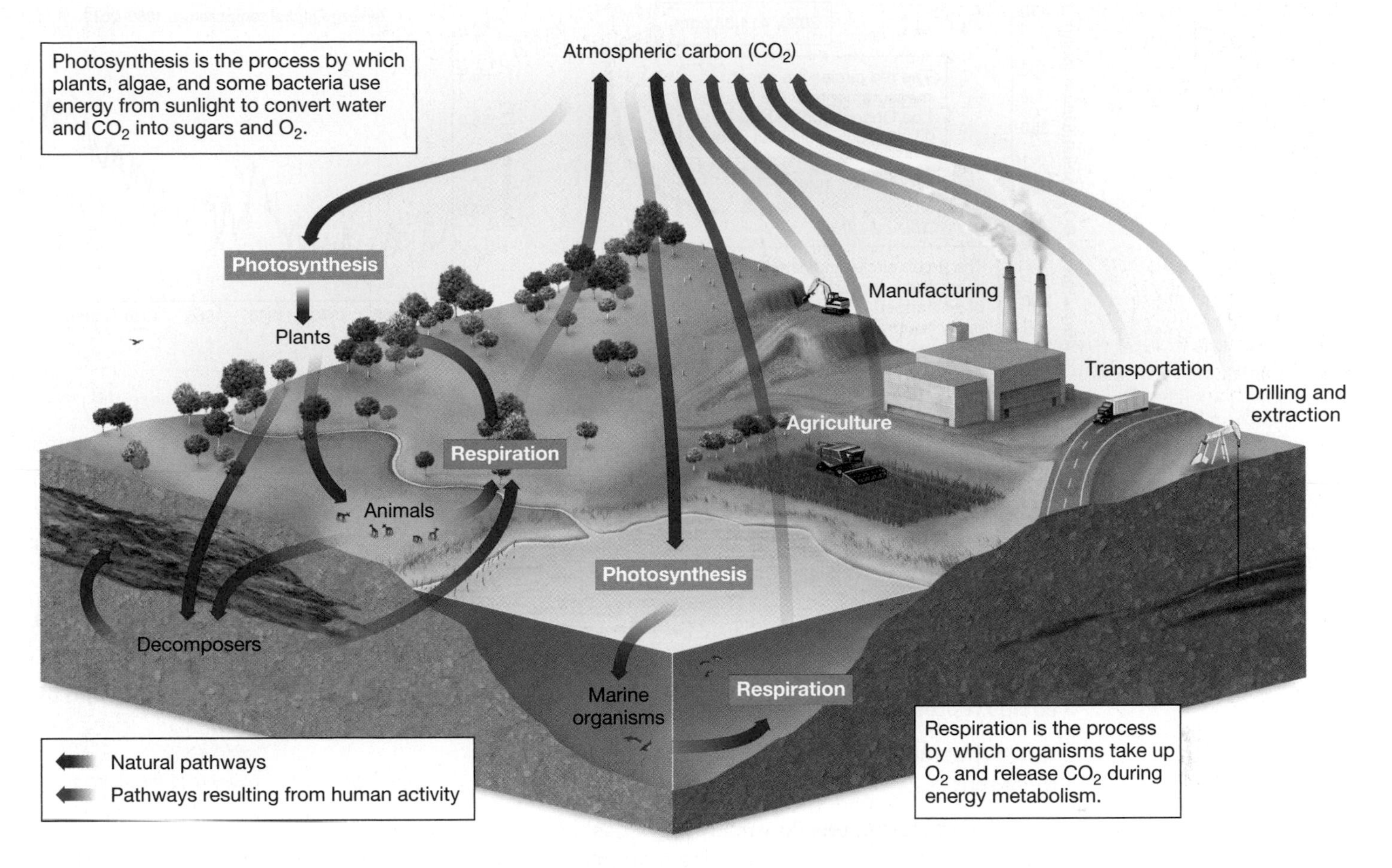

Figure 18.12

The carbon cycle
Arrows indicate the direction of carbon flow.

Q1: What are three ways that carbon is released into the atmosphere?

Q2: Are all the pathways you listed for question 1 affected by human activity?

Q3: What are two biotic reservoirs of carbon?

See Appendix A for answers to the figure questions.

in ice. Using these data, they can observe that CO_2 levels in the atmosphere did not rise above a certain concentration, 310 parts per million (ppm), for millennia, all the way back to 800,000 years ago. But starting in 1950, the atmosphere surpassed that threshold.

During the last 200 years, levels of atmospheric CO_2 have risen from roughly 280 to over 400 parts per million (**Figure 18.13**). Measurements from ice bubbles show that this rate of increase is greater than even the most sudden increase that occurred naturally during the past 420,000 years. Carbon dioxide levels are now higher than those estimated for any time during that period.

About 75 percent of the current yearly increase in atmospheric CO_2 is due to the burning of fossil fuels. Logging and burning of forests are responsible for most of the remaining 25 percent of the increase, but industrial processes also make a significant contribution.

Methane, another potent greenhouse gas, is also increasing in the atmosphere due to human activity. Methane is released into the air as a by-product of the extraction and processing of natural gas and, in smaller volumes, from

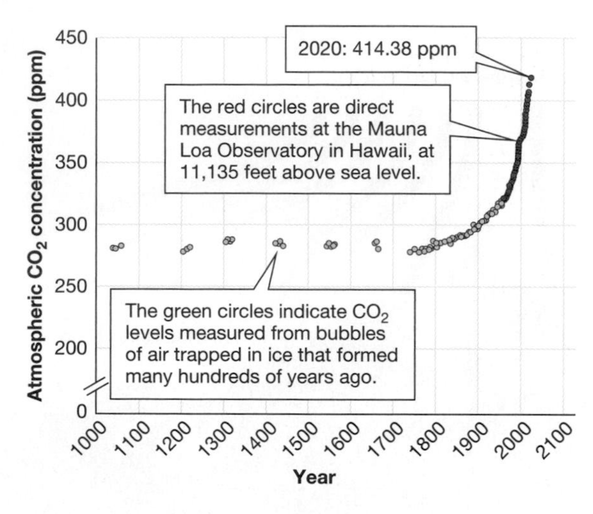

Figure 18.13

CO_2 in the atmosphere

Atmospheric CO_2 levels (measured in parts per million, or ppm) have increased greatly in the past 200 years.

Q1: What measurements do the green circles represent?

Q2: What measurements do the red circles represent?

Q3: For approximately how many years has the Mauna Loa Observatory been recording CO_2 levels?

See Appendix A for answers to the figure questions.

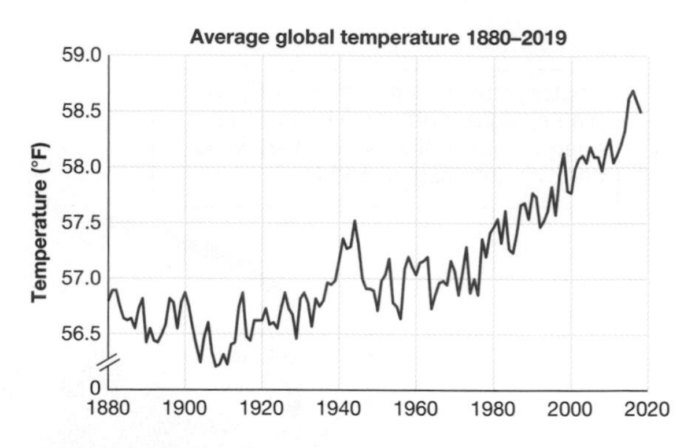

Figure 18.14

Global temperatures

Average global temperature has increased greatly over the 140 years since it has been recorded.

Q1: In what years were global temperatures the lowest?

Q2: In what years were global temperatures the highest?

Q3: What trend is apparent in this graph of actual global temperatures?

See Appendix A for answers to the figure questions.

livestock manure and waste decaying in landfills. (For more on the environmental impact of personal choices, see "How Big Is Your Ecological Footprint?" on page 356.)

All this is to say that human burning of fossil fuels has led to an increase in greenhouse gases in our atmosphere. These gases trap additional heat as it leaves Earth's surface, causing atmospheric temperatures to rise. Since the 1880s, when people started recording surface temperatures, scientists have documented a near-perfect historical correlation between CO_2 levels and the surface temperature on Earth. Since the early twentieth century, Earth's mean surface temperature has increased by about 1.62°F (0.9°C), and it is estimated to rise another 2°F–11.5°F (1.1°C–6.4°C) in the future (**Figure 18.14**). As of 2020, the hottest year ever recorded was 2016, and 2010–19 was the hottest decade ever recorded.

What happens to all that carbon once it's in the atmosphere? This is a critical point for Willis and other oceanographers: most of it goes into the oceans. Oceans are a **carbon sink**, a natural or artificial reservoir that absorbs more carbon than it releases. The opposite of a carbon sink is a **carbon source**, a reservoir that releases more carbon than it absorbs (**Figure 18.15**).

Oceans absorb carbon dioxide wherever air meets water, and since the 1950s, the oceans have absorbed an estimated 20–35 percent of fossil fuel emissions. This intake is increasing the acidity of the water, which impacts the species living there. In addition, the oceans are directly absorbing the heat. According to a comprehensive study of ocean warming

DEBUNKED!

MYTH: Scientists can't agree on Earth's temperature changes.

FACT: They can and they do. Records from four international science institutions show similar temperature peaks and valleys since 1880, and all show rapid warming since 1980.

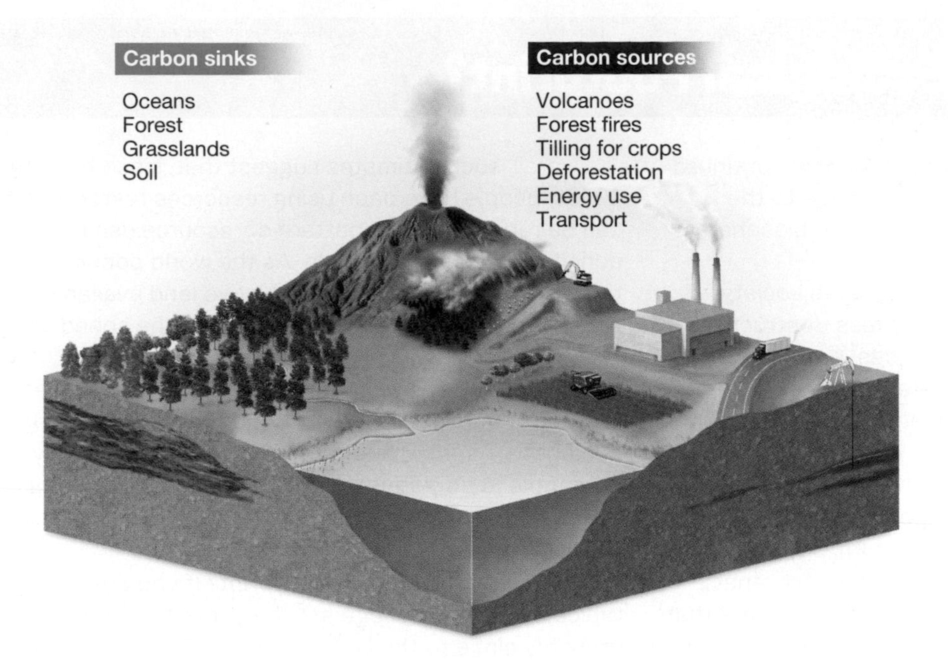

Figure 18.15

Carbon sinks and sources

There are both natural and human-created sources of atmospheric carbon, but to date only natural pathways exist to remove carbon from the atmosphere.

Q1: How does a carbon source contribute to global warming?

Q2: How does a carbon sink protect against global warming?

Q3: How can trees act as both a source and a sink?

See Appendix A for answers to the figure questions.

published in 2016, the world's oceans absorb an estimated 93 percent of the heat trapped from greenhouse gases. The rising temperatures also affect marine species. For example, researchers have been tracking widespread and severe episodes of coral bleaching, a deadly stress response to high water temperatures, in reefs around the world.

Oceans are warming and, according to the OMG results, glaciers are more sensitive to the ocean than we previously thought. This means as the oceans continue to warm, they will dramatically affect the rate of melting. "If the oceans are important, then our old idea of the ice cube under the hair dryer is wrong," says Willis. "And our estimates for how fast the whole ice sheet is going to melt have to be revised upward." In other words, Greenland's ice sheet is likely to melt faster than we previously thought.

To the Core

A second key piece of evidence for the accelerated melting of Greenland comes from a team at Woods Hole Oceanographic Institute in Massachusetts. In 2015, glaciologist Sarah Das and postdoctoral researcher Luke Trusel traveled to three remote field sites in Greenland, accessible only by plane or helicopter.

The sites were chosen to be high enough on the glacier that when the snow melts in the summer, it does not run off into the ocean but simply refreezes, creating a dense layer of ice on the ground. Over years, layers of ice and snow pile up, creating a literal record of melt, the way layers of sedimentary rock create a record of Earth's sediments.

The team wanted to harvest and examine those records. At each of the three sites, they set up a large orange and white tent to house and

How Big Is Your Ecological Footprint?

An action or process is **sustainable** if it can be continued indefinitely without causing serious damage to the environment. The current human impact on the biosphere is *not* sustainable.

Each of us can help build a more sustainable society. We can advocate legislation that fosters less destructive and more efficient use of natural resources, patronize businesses that take measures to lessen their negative impact on the planet, support sustainable agriculture, and modify our own lifestyles. For example, we can increase our use of renewable energy and energy-efficient appliances, reduce all unnecessary use of fossil fuels (for instance, by biking to work or using public transportation), buy seafood from sustainable fisheries, use "green" building materials, and reduce, reuse, and recycle waste. Experts estimate that more than 200 million women around the world wish to limit their family size but have no access to family planning. Those of us who live in developed countries can support aid efforts that provide education, health care, and family-planning services in developing countries.

One measure of sustainability is an **ecological footprint**, which is the area of biologically productive land and water that an individual or a population requires to produce the resources it consumes and to absorb the waste it produces. Scientists compute an ecological footprint using standardized mathematical procedures and express it in *global hectares* (gha): 1 gha is equivalent to 1 hectare (2.47 acres) of *biologically productive* space. Approximately one-fourth of Earth's surface is considered biologically productive; this definition excludes areas such as glaciers, deserts, and the open ocean.

According to recent estimates, the ecological footprint of the average person in the world is 2.7 gha, which is about 60 percent higher than the 1.7 gha that would be needed to support each of the world's 7.4 billion people in a sustainable manner. An ecological footprint can also be expressed in **Earth equivalents**, the number of planet Earths needed to provide the resources we use and absorb the wastes we produce. Currently, the global population uses 1.6 Earth equivalents each year, as shown in the bottom row of the accompanying figure.

Overall, such estimates suggest that, since the late 1970s, people have been using resources faster than they can be replenished—a pattern of resource use that, by definition, is not sustainable. As the world population grows, the amount of biologically productive land available per person continues to decline, increasing the speed at which Earth's resources are consumed.

The per capita consumption of Earth's resources by different countries is most directly related to energy demand, affluence, and a technology-driven lifestyle. As people in populous countries such as China and India become wealthier, their ecological footprints are growing rapidly.

What is *your* ecological footprint? If you are a typical American college student, your footprint is probably close to the U.S. average of 8.2 gha. It would take nearly five planet Earths to support the human population if everyone on Earth enjoyed the same lifestyle that you do (see the top row of the accompanying figure). Your ecological footprint depends on four main types of resource use:

1. *Carbon footprint*, or energy use
2. *Food footprint*, or the land, energy, and water needed to grow what you eat and drink
3. *Built-up land footprint*, which includes the building infrastructure (from schools to malls) that supports your lifestyle
4. *Goods-and-services footprint*, which includes your use of everything from home appliances to paper products

If you drive a gas guzzler, live in a large suburban house, routinely eat higher up on the food chain (more beef than grains or veggies and fruits), and do not recycle much, your footprint is likely to be higher than that of a person who uses public transportation, shares an apartment, eats mostly plant-based foods, and sends relatively little to the local landfill. Most of us can significantly reduce our ecological footprint with little reduction in our quality of life while bestowing an outsized benefit on our planet.

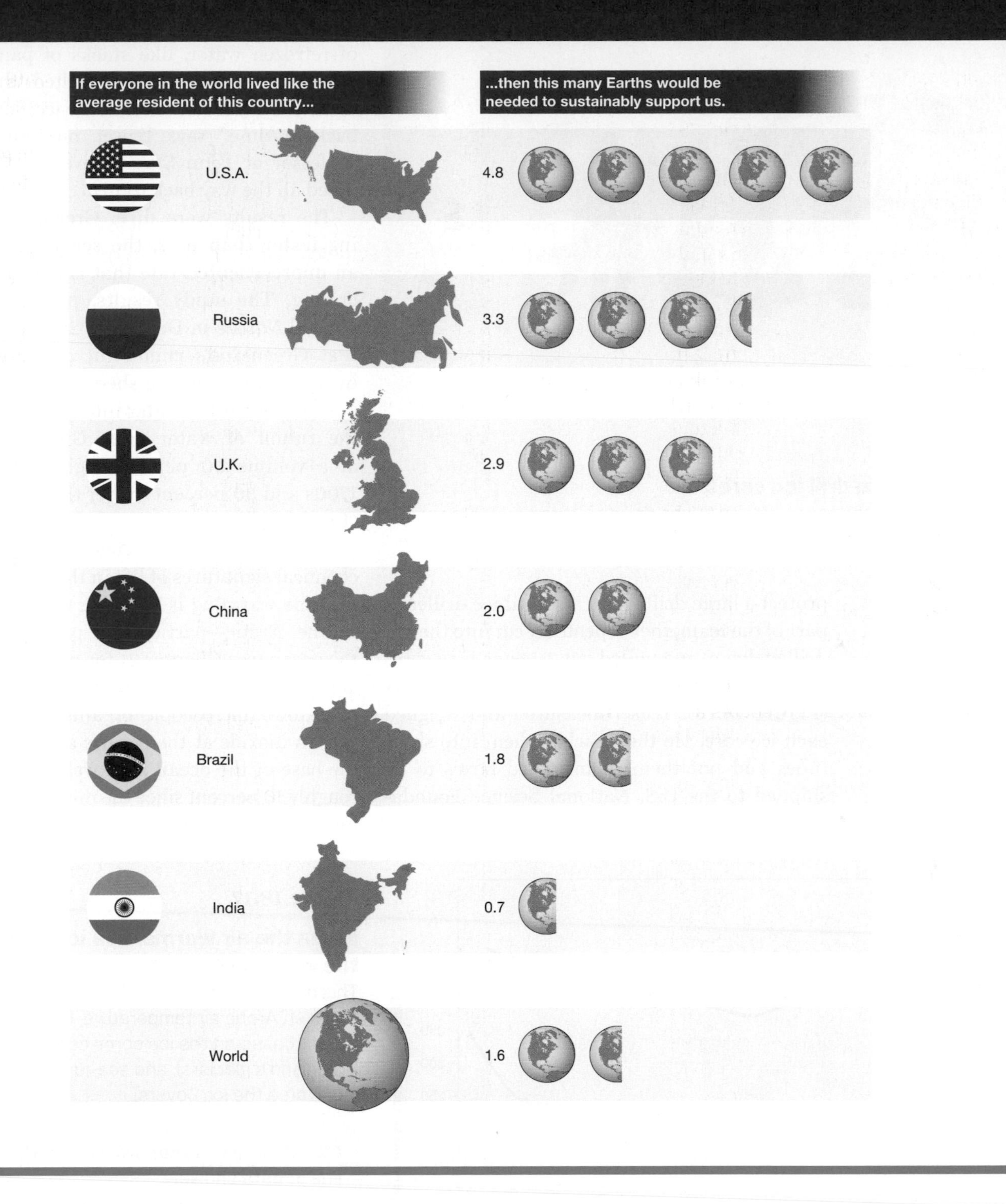
If everyone in the world lived like the average resident of this country...
...then this many Earths would be needed to sustainably support us.
U.S.A.
4.8
Russia
3.3
U.K.
2.9
China
2.0
Brazil
1.8
India
0.7
World
1.6

Figure 18.16

Setting up camp to drill ice cores

Scientists at a Greenland research site will be drilling a three-meter ice core that will document hundreds of years of melting and refreezing of glacial ice.

protect a large drill. With a skilled ice driller as part of the team, they repeatedly cut into the ice, drilled down, and pulled out 3-meter-long cores of ice, until they reached depths of 140 meters (**Figure 18.16**). Trusel measured and weighed each ice core. He then packed them into shiny tubes and put them in insulated boxes to be shipped to the U.S. National Science Foundation's Ice Core Facility in Colorado.

There technicians scanned the cores with high-resolution imaging systems. From those images, Das and Trusel visually identified bands of refrozen water, like stacks of pancakes, to determine when the ice melted throughout history. "As you go down in the ice, you're going back in time," says Trusel, now an assistant professor at Penn State University. The cores dated all the way back to 1675.

The results were dire: Greenland is melting faster than ever, the scientists found, at an unprecedented rate that is currently accelerating. The study results, published in the journal *Nature* in December 2018, determined that Greenland's runoff hit a 350-year high in 2012, when the ice sheet dumped a record 600 gigatonnes of water into the ocean. Today, the runoff of water from Greenland occurs at a volume 50 percent higher than in the 1700s and 30 percent higher than in the 1900s (**Figure 18.17**).

In follow-up work, scientists examined the chemical signatures of life in the ice cores to see how the warming is affecting the biotic factors in the Arctic—particularly phytoplankton in the ocean (see Chapter 21 for more on plankton and ecosystems). They found that marine phytoplankton—microscopic organisms that absorb carbon dioxide at the ocean's surface and form the base of the ocean food web—have declined roughly 10 percent since the mid-1800s, coinciding with the warming temperatures.

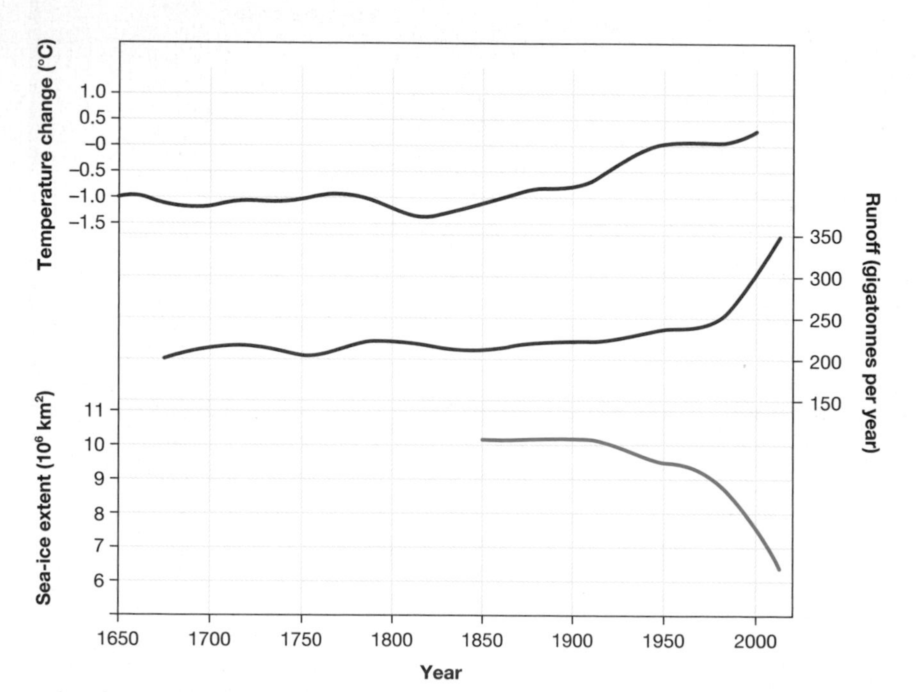

Figure 18.17

When the air warms, sea ice melts and the ice sheet shrinks

This graph shows the relationship over time between Arctic air temperature, ice-melt runoff (based on the ice cores collected from Greenland's glaciers), and sea-ice extent (how much area the ice covers).

Q1: **Describe in your own words what each line of data shows.**

Q2: **When did runoff seem to increase significantly?**

Q3: **How would you explain the clear relationship among temperature, runoff, and sea-ice extent?**

See Appendix A for answers to the figure questions.

As World Warms, Extinction Risks Soar

As the planet warms, animals will continue to be impacted by shifting seasons, fluctuating rainfall patterns, changing access to food, and new exposures to predators and diseases. To gain a broad understanding of how climate change might affect species around the globe, a biologist at the University of Connecticut analyzed 131 published predictions about extinction risks. Shockingly, he found that if climate change proceeds as expected with 4 degrees Celsius of warming, one in six species faces extinction.

Assessment available in smartwork

Global extinction risk from climate change

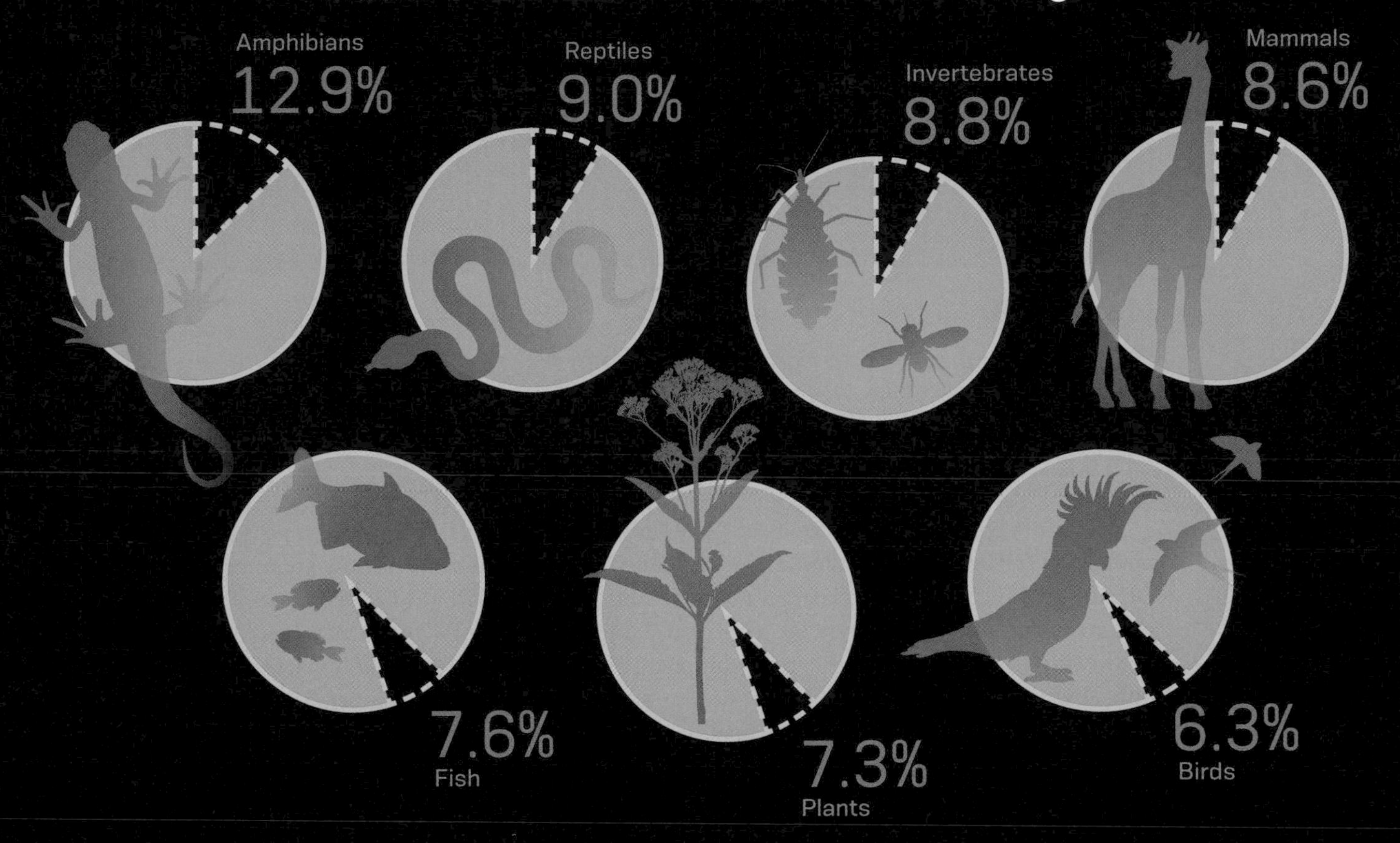

Percent of total species in danger of extinction by region

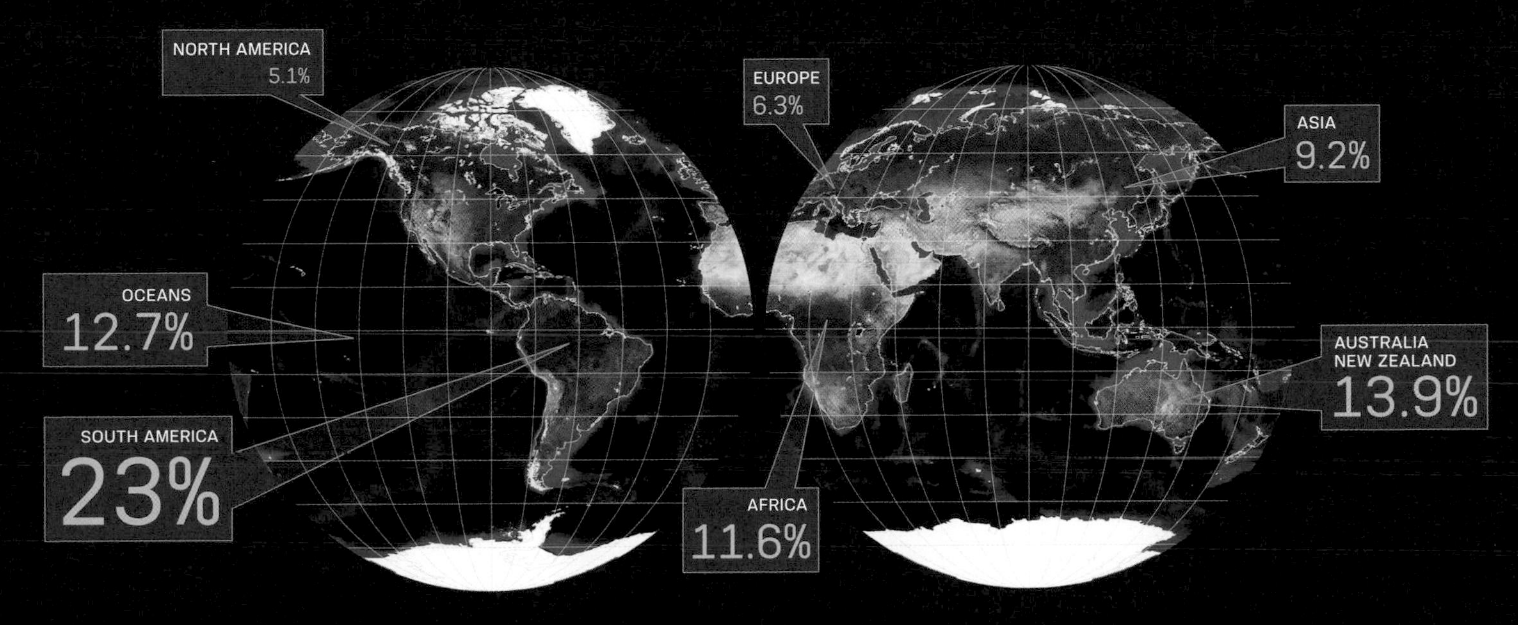

Figure 18.18

Sled dogs haul scientific equipment through frigid meltwater
Sea-ice melt came early to Greenland in 2019, making the annual trip to collect over-wintering equipment more challenging than usual.

The authors hypothesize that as Greenland melts, cold freshwater flows into the North Atlantic Ocean, pushing down the warmer, nutrient-rich water that phytoplankton rely on, leading to loss of the organisms. "There are really important links between the physical climate system and the biological side of things," says Trusel.

Predicting the Future

The news continues to be dismal: Greenland's melt, says Trusel, has become a feedback loop in which melting begets more melting. When the ice sheet's top layer melts, bright snow is replaced by darker patches of ice. Those darker patches absorb more heat from the sun, warming Greenland even further. As temperatures continue to rise, scientists predict that feedback loop will become even stronger. Greenland's melt will have a huge impact on the biosphere, and it's just a question of how soon and how large that impact will be (**Figure 18.18**).

The OMG team was headed back to Greenland in 2020 to see if warm water has returned to Jakobshavn glacier. Before the project ends, Willis hopes to come up with a formula to better predict how fast Greenland is melting by taking oceans into account. "It will help us predict future sea level rise better," says Willis.

As we think about that future, the best way to rescue Greenland's ice sheet, and protect cities—including New Orleans, Miami, and Houston—from rising sea levels is to reduce greenhouse gas emissions, says Trusel. By taking steps to reduce one's carbon footprint, such as by recycling, driving less, and eating less meat, as well as supporting government initiatives to reduce carbon emissions, we can still affect the future of climate change.

"It's not set in stone yet," says Trusel. "Ultimately, what happens to Greenland in the future is up to us."

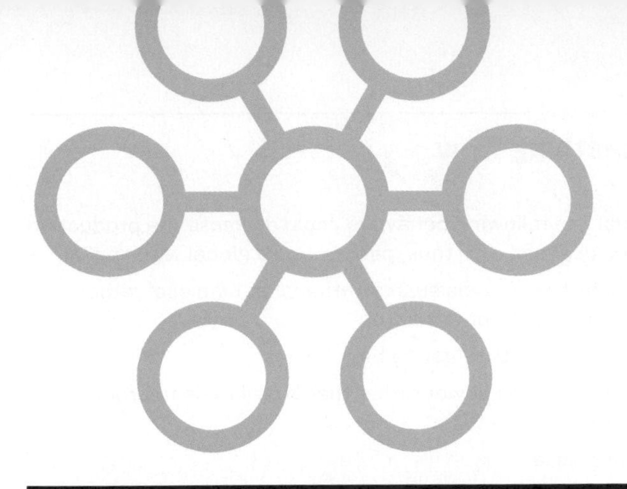

NEED-TO-KNOW SCIENCE

- **Ecology** (page 345) is the scientific study of interactions between organisms and their environment. All ecological interactions occur in the **biosphere** (page 344), which consists of **biotic** factors (all living organisms on Earth, page 345) together with **abiotic** factors (the physical environments they inhabit, page 345). An interacting system of living organisms and their physical environments is an **ecosystem** (page 345).
- **Weather** (page 345) consists of the short-term atmospheric conditions in a given area, such as temperature, precipitation, wind, humidity, and cloud cover. **Climate** (page 345), the prevailing weather of a specific region over relatively long periods of time, has a major effect on the biosphere. Climate is determined by incoming solar radiation, global movements of air and water, and major features of Earth's surface. Earth's regional environments are greatly affected by its four giant **convection cells** (page 347).
- **Climate change** (page 346) is a large-scale and long-term alteration in Earth's climate, and it includes such phenomena as global warming, change in rainfall patterns, and increased frequency of violent storms.
- **Greenhouse gases** (page 350) contribute to the **greenhouse effect** (page 351) by absorbing heat released at Earth's surface, rather than allowing it to leave the atmosphere. The greenhouse effect has led to a rapidly rising average global temperature, a phenomenon known as **global warming** (page 350).
- The atmospheric concentration of **carbon dioxide** (**CO_2**, page 351), the most significant greenhouse gas, is increasing at a dramatic rate because of an increase in **carbon sources** (page 354), such as the release of CO_2 through the burning of fossil fuels, and a loss of **carbon sinks** (page 354), such as the absorption of CO_2 through photosynthesis in large forests.
- Both the **hydrologic cycle** (page 347) and the **carbon cycle** (page 351) are affected by global warming, which disrupts the natural cycling of water and carbon molecules in the biosphere.
- The area of the biosphere required to produce the resources and to absorb the waste produced by an individual or population is known as the **ecological footprint** (page 356). The ecological footprint is often expressed in **Earth equivalents** (page 356), the number of planet Earths needed to provide the resources and absorb the waste of an individual or a population.

THE QUESTIONS

See Appendix B for answers.

The Basics

1 The biosphere is

(a) all organisms on Earth, together with their physical environments.
(b) crucial to human survival and well-being.
(c) a source of food and raw materials for human society.
(d) a web of interconnected ecosystems.
(e) all of the above

2 Greenhouse gases function by

(a) blocking sunlight but letting out heat from Earth to outer space.
(b) absorbing heat that radiates from Earth that would otherwise escape to outer space.
(c) absorbing heat that radiates from the sun toward Earth.
(d) releasing heat that radiates from Earth into outer space.

3 Which of the following is an abiotic component of the biosphere?

(a) algae
(b) insects
(c) lichen
(d) water
(e) none of the above

4 Place the following components of the hydrologic cycle in the correct order by numbering them from 1 to 5, beginning with precipitation.

_____ a. precipitation
_____ b. uptake by plants
_____ c. transpiration
_____ d. water percolation into soil
_____ e. condensation

5 Select the correct terms:

(Climate / Weather) is the short-term atmospheric conditions in an area, whereas (climate / weather) is the average atmospheric conditions in a given area over a long period of time. (Climate change / Global warming) consists of the long-term and large-scale changes in atmospheric conditions, which have been brought about by (climate change / global warming), an increase in the average global temperature.

Challenge Yourself

6 The carbon cycle and the hydrologic cycle are similar in that

(a) both cycle molecules between abiotic and biotic components of the environment.

(b) photosynthesis is a critical process in both.

(c) both are involved in global warming and climate change.

(d) all of the above

(e) none of the above

7 An ecologist is studying an area of forested land that experiences a forest fire every 5 years. In the years between forest fires, the area of study absorbs 300 million tons of CO_2. If a forest fire releases 200 million tons of CO_2 in this area, does this forested area act as a carbon sink or a carbon source? Why?

8 Place the following events in the correct order of their onset by numbering them from 1 to 5, beginning with the earliest.

____ a. Climate change, including more extreme weather

____ b. Increase in human population and activities

____ c. Increase in release of CO_2 and other greenhouse gases into the atmosphere

____ d. Increase in calls to control the production and release of greenhouse gases

____ e. Global warming

Try Something New

9 Which of the following behaviors could *decrease* the production of greenhouse gases and, thus, perhaps curb global warming?

(a) using plant-based fuels such as ethanol or biodiesel rather than fossil fuels to power autos

(b) using coal rather than gas to heat homes

(c) using solar or wind power rather than fossil fuels to produce electricity

(d) all of the above

(e) none of the above

10 Reviewing Figure 18.7, identify the type of environment at your latitude, and explain how it is influenced by Earth's convection cells.

11 Discuss in one or two sentences how the more extreme weather caused by global warming may affect where you live.

Leveling Up

12 **Looking at data** Review Figures 18.13 and 18.14. Describe the change in atmospheric CO_2 levels from 1880 to the present and the change in average global temperature for the same time period. What were the CO_2 level and the average temperature in 1880? In 1960? In 2020? Is there a relationship between the two variables? How would you explain the pattern you see here?

13 **Life choices** You can estimate your impact on the planet by using one of the many "footprint calculators" on the internet. Take three online quizzes available from organizations such as the Global Footprint Network, the Nature Conservancy, and the U.S. Environmental Protection Agency. Each site calculates your ecological footprint a little differently. Compare and contrast the different sites you used. Which one do you think is the most accurate? Why? Which one do you think is the most superficial? Why? Write a list of ways you can decrease your ecological footprint, and try to adhere to your new lifestyle!

14 ***Write Now* biology: climate change** Watch one of the following documentaries: *Ice on Fire* (2019), *True North* (2018), *Before the Flood* (2016), *Chasing Ice* (2012), or Frontline's *Climate of Doubt* (2012). Write an essay reflecting on what you learned from the film and from this chapter about global warming. Use knowledge of your ecological footprint to reflect on how important it is for all citizens to be aware of this global crisis.

Digital resources for your book are available online.

Zika-Busting Mosquitoes

Genetically engineered insects help stop the spread of dengue, malaria, and Zika.

After reading this chapter you should be able to:

- Articulate the difference between population size and population density.
- Interpret graphical population data to determine whether the population exhibits logistic or exponential growth.
- Explain the concept of carrying capacity, and give examples of factors that can increase or decrease carrying capacity for a population.
- Differentiate between density-dependent and density-independent changes in population size.
- Illustrate population cycles of predators and prey.

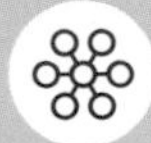

ECOLOGY

CHAPTER 19

GROWTH OF POPULATIONS

It is impossible to know when an infected mosquito bit the first person in Miami-Dade County, Florida. But on Friday, July 29, 2016, health officials confirmed that a woman and three men in a small area north of downtown Miami had been infected with the Zika virus through the bite of a local mosquito. It was the moment everyone had been dreading: the first known cases of the virus being transmitted by mosquitoes in the United States. "As we have anticipated, Zika is now here," Tom Frieden, the director of the U.S. Centers for Disease Control and Prevention (CDC), told reporters that afternoon (**Figure 19.1**).

By July 2016, Zika had become a household name. Zika fever, the illness that results from infection with Zika virus, is a mild sickness that typically lasts less than a week and involves fever, red eyes, headache, and a possible rash. But Zika has another, much more dire effect on health. A year before the Miami outbreak, doctors in Brazil had begun noticing an increasing number of infants born with a serious birth defect known as microcephaly, in which the head and brain are smaller than expected; they have not developed properly. Microcephaly can cause many problems in growing infants, including seizures, speech and intellectual delays, feeding and movement problems, and hearing and vision loss. Brazil's new cases of microcephaly were soon linked to a rapidly spreading virus passed through the bite of a mosquito: Zika.

The scientific evidence linking Zika virus and microcephaly is strong: infected mothers pass the virus to their fetuses, where it may stunt

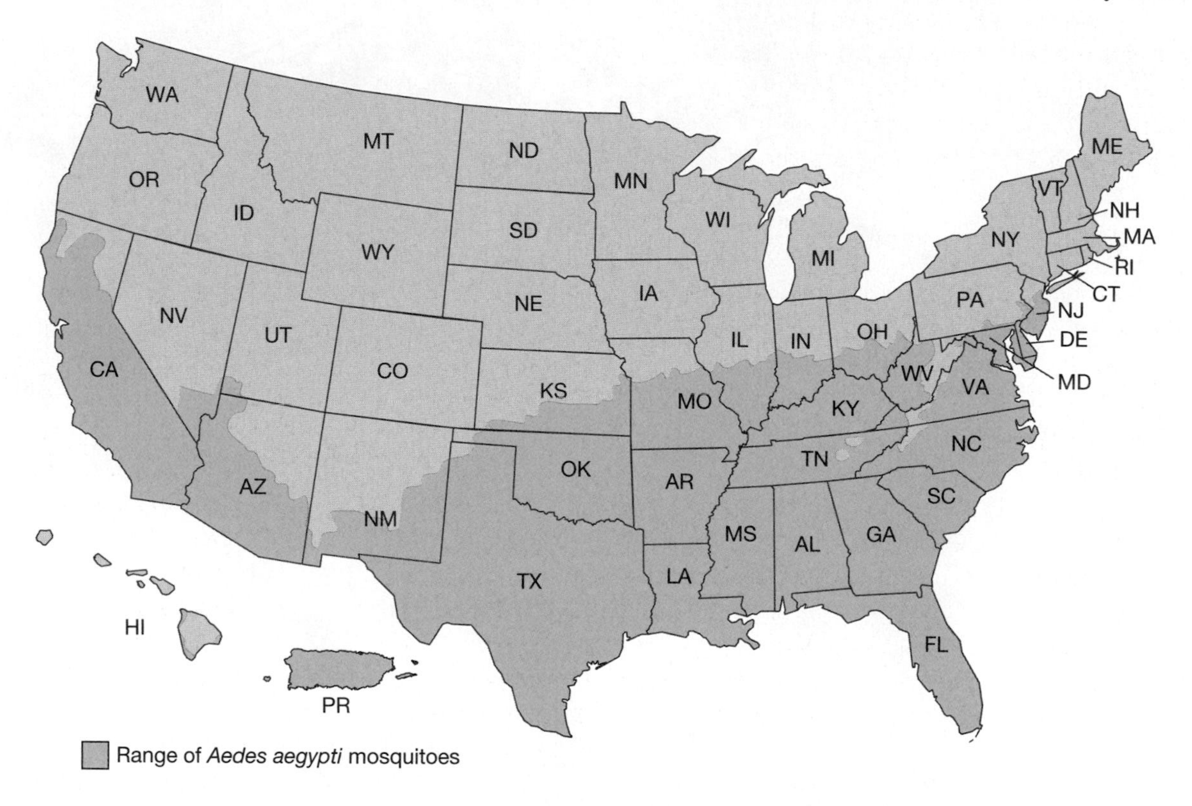

Figure 19.1

U.S. range of the mosquito species most likely to carry the Zika virus

Aedes aegypti mosquitoes spread viruses such as Zika, dengue, and chikungunya more often than do other mosquito species.

Q1: What parts of the United States are within the range of the mosquito that carries Zika?

Q2: What areas are *not* within the mosquito's range? Why do you think that is?

Q3: Find your own state and your location in the state. Are you at risk of contracting Zika?

See Appendix A for answers to the figure questions.

brain growth (**Figure 19.2**). Studies suggest that infants exposed to the Zika virus in utero are more than 50 times more likely to be born with microcephaly than those who are not exposed. In February 2016, the World Health Organization declared the spread of Zika virus to be a "public health emergency of international concern."

Since the outbreak in Brazil, governments around the world have begun looking for ways to stop Zika, and the local transmission in Florida galvanized U.S. officials and scientists. "It's scary," says Matthew DeGennaro, a biologist who studies mosquito genetics and behavior at Florida International University in Miami. "But people in other countries face these things all the time, and not much is being done about it. It is possible this [outbreak] will focus Americans' attention on mosquito-borne illness."

The human species has a long history of battling mosquito-borne viruses, including dengue virus and West Nile virus. Along the way, we've learned that when a vaccine is lacking, the best way to stop the virus is to stop the insect. Yet despite efforts to eradicate mosquito populations around the world, the pests continue to roar back no matter what we throw at them. Mosquitoes quickly develop resistance to common insecticides, for example, and in at least one documented case in Senegal, the pests showed an adapted behavior, attacking farmers outdoors in the early morning rather than indoors during the night when the farmers were protected by mosquito nets.

But now there's a new weapon in the arsenal against mosquito-borne diseases. Taking a technological approach to the problem, scientists have begun to target mosquito populations by using the insects themselves as weapons. It's a new kind of war—mosquito versus mosquito.

Figure 19.2

A Brazilian mother holds her microcephalic baby

The incidence of microcephaly in newborns has skyrocketed, and the increase has been linked to Zika infection of the mother during pregnancy.

Population Control

One month after the first four reported cases of Zika transmission in Florida, officials captured Zika-infected mosquitoes at seven different locations on Miami Beach—the smoking gun to show that local disease transmission was indeed occurring. By October 2016, mosquitoes bearing Zika virus had spread another 3 miles north.

Zika is spread by females of both the *Aedes aegypti* species and the less common *Aedes albopictus* species, although the latter is less likely to bite humans because it breeds in rural areas and feeds on other animals in addition to humans. These two types of mosquitoes also serve as carriers, or *vectors*, for dengue, chikungunya, and other viruses. (Malaria, which is caused by parasitic protists, is transmitted to humans by female mosquitoes of the genus *Anopheles*.) Active primarily during the day, a female mosquito feeds on human blood via a tubelike mouthpart (which male mosquitoes lack) that pierces the skin of a host. During that moment, virus particles in the mosquito's saliva can be transferred to the person's skin. Once on the skin, the virus is able to replicate in skin cells and then can spread to the lymph nodes and bloodstream.

In addition to being transmitted by mosquitoes, Zika can spread from person to person via body fluids: blood, tears, semen, and vaginal fluids. In some cases, the virus can remain in those

DEBUNKED!

MYTH: A mosquito dies after it bites you.

FACT: Mosquitoes are capable of biting more than once. After a blood meal, a female mosquito may lay up to 200 eggs and then seek another meal. Males do not feed on blood; they drink nectar.

fluids for months after infection (**Figure 19.3**). Slowing the spread of Zika, therefore, involves the screening of donated blood and the proper use of condoms during sex. But true prevention ultimately requires decreasing the incidence of mosquito bites, which can be accomplished by getting rid of the mosquitoes.

But what effects might that have on the planet? Mosquitoes live on every continent and have been on Earth for more than 100 million years. Birds, fish, and other animals eat mosquitoes, and mosquitoes pollinate some plants. Yet ecosystems could adapt to the loss of mosquitoes, according to scientists who study mosquito

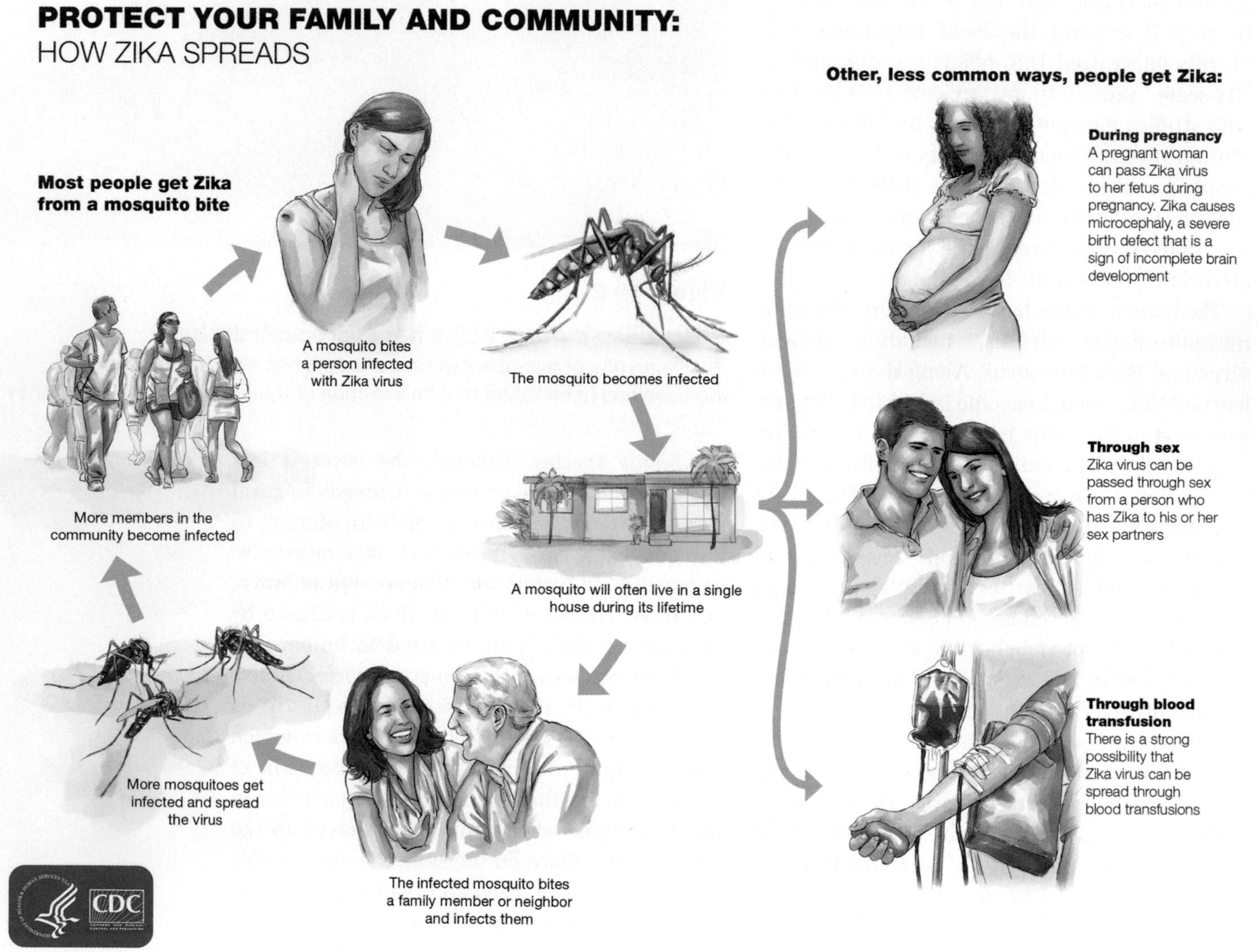

PROTECT YOUR FAMILY AND COMMUNITY:
HOW ZIKA SPREADS

Most people get Zika from a mosquito bite

A mosquito bites a person infected with Zika virus

The mosquito becomes infected

A mosquito will often live in a single house during its lifetime

The infected mosquito bites a family member or neighbor and infects them

More mosquitoes get infected and spread the virus

More members in the community become infected

Other, less common ways, people get Zika:

During pregnancy
A pregnant woman can pass Zika virus to her fetus during pregnancy. Zika causes microcephaly, a severe birth defect that is a sign of incomplete brain development

Through sex
Zika virus can be passed through sex from a person who has Zika to his or her sex partners

Through blood transfusion
There is a strong possibility that Zika virus can be spread through blood transfusions

CDC

Figure 19.3

How Zika is transmitted

The CDC produces posters such as this for public health issues including Zika, sexually transmitted diseases, and the flu.

Q1: What is the main way that someone is infected with the Zika virus?

Q2: Judging by the poster and your knowledge of mosquito behavior, what can you do to decrease your risk of being infected with the Zika virus?

Q3: Besides the transmission methods shown on the poster, what are some other ways you could become infected with the Zika virus?

See Appendix A for answers to the figure questions.

biology and ecology. If mosquitoes were gone, it is likely that mosquito predators would adapt and eat different insects as prey, according to some experts. In addition, mosquito pollination is not known to be crucial for any crops on which humans depend. Although one cannot predict all possible outcomes of eradicating an organism, the health benefits of eradicating mosquito populations near human settlements are clear. "The elimination of *Anopheles* would be very significant for mankind," entomologist Carlos Brisola Marcondes from the Federal University of Santa Catarina in Brazil told *Nature* magazine.

Because of that potential, scientists have tried many approaches to reducing mosquitoes' **population size**, the total number of individuals in the population. A **population** is a group of organisms of the same species in a defined area. Population size tends to change over time—sometimes increasing, sometimes decreasing. Whether the size of a population increases or decreases depends on the number of births and deaths in the population, as well as on the number of individuals that enter (immigrate) or leave (emigrate) the population. Birth and immigration increase population size; death and emigration reduce it. Environmental factors also have a strong impact on population size. Mosquito populations, for example, increase in warmer, wetter weather, when the conditions are ripe for reproduction.

To target mosquito populations, scientists rely on **population ecology**, the study of the size and structure of populations and how they change over time and space. Population ecology can help us determine where mosquitoes live, eat, and breed. Mosquitoes congregate and lay their eggs at water sources, and many prefer urban environments with trash receptacles and concentrated groups of people. At these locations mosquitoes have their highest **population density**, or number of individuals per unit of area.

To calculate population density, total population size is divided by the corresponding area of interest. For example, in an urban neighborhood of Rio de Janeiro, Brazil, where people were being regularly infected with dengue virus, scientists captured mosquitoes in mosquito traps and calculated the neighborhood's population of female mosquitoes to be 3,505 pregnant females. The neighborhood has an area of 911 acres, so the female mosquito population density was 3.85 mosquitoes per acre (3,505 mosquitoes divided by 911 acres), a relatively high density, according to the scientists who conducted the study, which may help explain the high number of dengue cases in the area. Keep in mind that population density is often difficult to measure because it depends on an accurate count of the population size. Individuals may be hard to detect, may move between populations, or may inhabit a complex, hard-to-define area.

Scientists can track population density to see how well eradication efforts work against mosquitoes. As mentioned earlier, one of the most popular ways of reducing mosquito populations is poisoning them with insecticides. But scientists agree that poisoning is not ideal. "Insecticides affect the entire insect population usually, which leads to overall imbalances in the environment," says DeGennaro. "So putting a chemical into the environment is something you need to do very carefully."

An alternative approach, thanks to advanced genetic technologies, is to tweak the genes of the mosquitoes to sabotage their reproduction. If the mosquitoes can't reproduce, their population size will plummet, which means there will be fewer mosquitoes to bite and infect people. This gene tweaking called **genetic modification (GM)** alters the genes of an organism for a specific purpose, creating a **genetically modified organism (GMO)**. It is also referred to as genetic engineering (GE). In contrast to insecticides, GM mosquitoes can directly target just one insect species, says DeGennaro. "You're only affecting one species of mosquito, the species that is the primary vector for Zika, dengue, and more."

Early field tests of GM mosquitoes in Brazil appeared to stop the spread of dengue. But could they do the same for Zika?

Rapid Spread

Before it hit Florida, Zika virus struck Brazil like a tsunami. In a span of 3 months in early 2015, Brazil documented nearly 7,000 cases of mild illness similar to dengue fever, though no one yet suspected Zika. That's the problem with a newly emerging infectious disease: no one is looking for it.

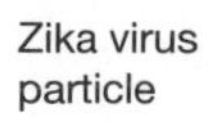

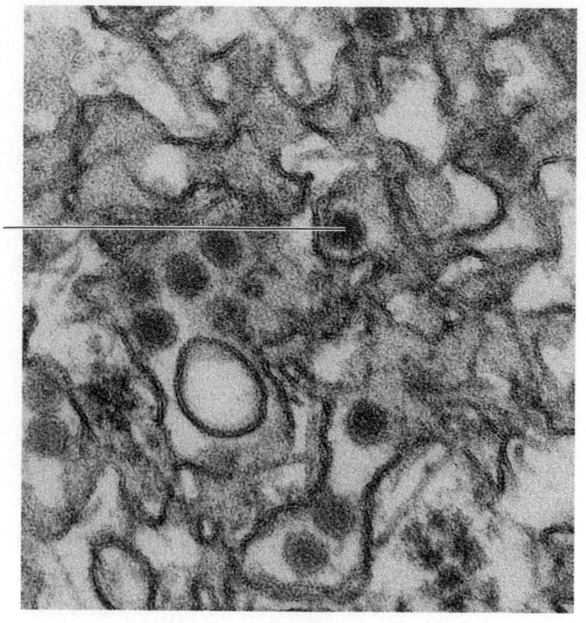

Figure 19.4

Virus under the microscope

Zika is a spherical RNA virus.

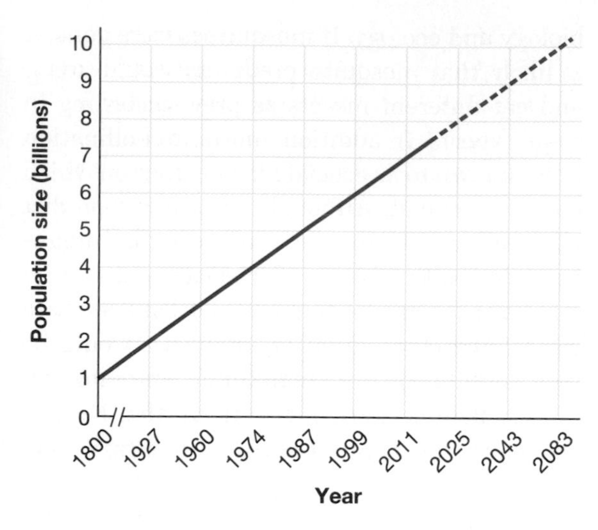

Figure 19.5

Doubling time for the global human population

Although human populations grow much more slowly than those of mosquitoes (whose populations double in months rather than decades), our growth has an unusually large impact on the global environment.

> **Q1: Over what years did the human population double from 1 billion to 2 billion people? How many years did that take?**
>
> **Q2: How many years did the second population doubling take (from 2 billion to 4 billion)? When did that occur?**
>
> **Q3: When is it predicted that the *next* population doubling will occur? What will be the size of the human population then?**

See Appendix A for answers.

Then, in May, a national laboratory reported that Zika virus was circulating around the country. Two months later, Brazilian health workers began reporting neurological disorders associated with Zika infection, including brain inflammation. In October came the worst news of all: reports of an unusual spike in newborns with microcephaly.

That dramatic spread and link to microcephaly put Zika on the map, but it isn't actually a new virus. Zika virus, a round particle with a dense core packed with RNA (**Figure 19.4**), was first isolated in 1947 in the Zika Forest of Uganda, for which it is named. The first human cases of Zika were detected in Uganda in 1952, and the little studied virus subsequently spread through central Asia, across the South Pacific, and into South and Central America and the Caribbean.

Humans are no strangers to mosquito-borne diseases (see "Mosquito-Borne Diseases" on page 371). Although Zika and other diseases were once isolated to specific areas of the globe, global travel and human population growth, along with climate change, have increased mosquitoes' ability to proliferate and spread disease on a grand scale. In the second half of the twentieth century, our planet saw the fastest rate of human population growth in its history. **Population doubling time**—that is, the time it takes a population to double in size—is a good measure of how fast a population is growing. For example, the doubling time for the U.S. human population at our current reproductive rate is about 100 years, and it is even faster worldwide (**Figure 19.5**). In stark contrast, the population doubling time for mosquitoes can be as little as 30 days in an optimal environment.

Reaching Capacity

Now back to mosquitoes and how to get rid of them. Scientists argue strongly that understanding mosquito population ecology is a prerequisite for eliminating mosquito-borne diseases. For example, many mosquitoes lay their eggs directly on the surface of water. That knowledge led to one of the most important public health efforts to reduce mosquito-borne diseases: the removal of pools of standing water. Eliminating these breeding sites makes it possible to reduce the **carrying capacity** of the environment, the maximum population size that can be sustained in a given area.

In most species, the growth rate of a population decreases as the population size nears the carrying capacity, because resources such as food and water begin to run out. If there are only a few locations of standing water in a given area, for example, a limited number of mosquito eggs can be laid in that area, thereby limiting the size of the population. Any limiting resource needed for survival, such as habitat, food, or water, will determine an environment's carrying capacity for a specific population. Because different species have unique needs, the same environment can have different carrying capacities for different resident species. For the same reason, two different environments in which the same species lives will have different carrying capacities. At the carrying capacity, the population growth rate is zero.

If a population has no constraints on its resources, it will experience **exponential growth**, which occurs when a population increases by a constant proportion over a constant time interval, such as 1 year. Exponential growth is represented by a **J-shaped growth curve** (**Figure 19.6**). A population that approaches its carrying capacity because of constrained resources will experience **logistic growth**, in

Mosquito-Borne Diseases

Mosquitoes cause more human suffering than any other organism (with the exception of our own species). Here are just a few of the deadly or debilitating diseases they carry:

- *Dengue virus* is a huge problem for the population of Brazil. Spread by the same mosquitoes as Zika, dengue also has no vaccine or cure, and it causes a disease chillingly known as "breakbone fever." Symptoms include continuous joint pain, vomiting, high fever, and severe rashes. In 2015, Brazil chalked up a record 1.6 million cases and 863 deaths to dengue—an 82.5 percent increase over 2014.

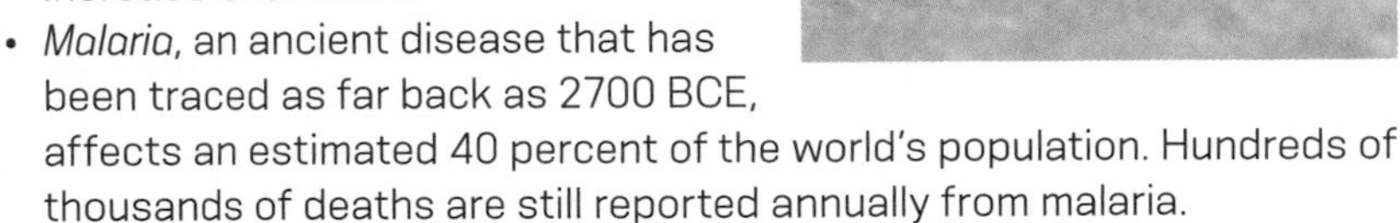

- *Malaria*, an ancient disease that has been traced as far back as 2700 BCE, affects an estimated 40 percent of the world's population. Hundreds of thousands of deaths are still reported annually from malaria.
- *Yellow fever*, against which there is a vaccine, still leads to an estimated 30,000 deaths per year.
- *West Nile virus*, which emerged in the 1930s, can invade the nervous system and cause death.
- *Chikungunya* is found mainly in Africa and Asia, but cases have been reported in Florida as well. The main symptoms are fever, rash, and joint pain that can last for years. Only one in 1,000 cases results in death.

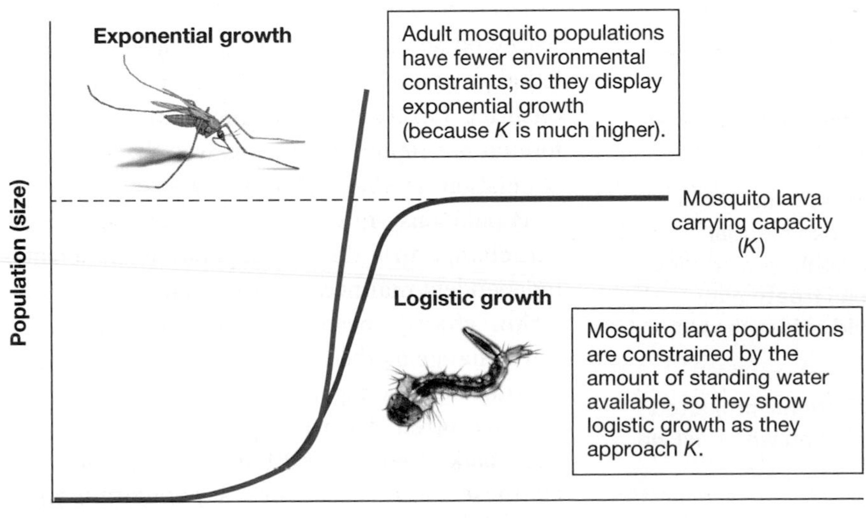

Exponential growth
Adult mosquito populations have fewer environmental constraints, so they display exponential growth (because K is much higher).
Mosquito larva carrying capacity (K)
Logistic growth
Mosquito larva populations are constrained by the amount of standing water available, so they show logistic growth as they approach K.
Population (size)
Time

Figure 19.6

Populations can experience exponential growth or logistic growth

A population that is not constrained by resources or by the environment can grow exponentially, whereas one that is constrained by a set carrying capacity shows logistic growth.

Q1: Which form of population growth displays a J-shaped curve?

Q2: Which form of population growth displays an S-shaped curve?

Q3: Describe a situation in which a population initially shows exponential growth and later shows logistic growth.

See Appendix A for answers to the figure questions.

which the population grows nearly exponentially at first but stabilizes at the maximum population size that can be supported indefinitely by the environment. Logistic growth is represented by an **S-shaped growth curve**.

Little research has been done on growth curves for mosquito populations, primarily because the insects are small and airborne, making them hard to track. But we do know how the human population has changed. Over the last 500 years, Earth's human population exhibited both logistic and exponential growth. At the end of the last ice age, in approximately 10,000 BCE, there were only 5 million people on Earth. With the advent of agriculture in about 8000 BCE, the world population began to rise logistically, until about 200 years ago. Then, alongside the use of fossil fuels and during the industrial revolution, human population growth exploded exponentially. Modern populations have continued to show exponential growth to the detriment of the environment. Current estimates of the carrying capacity of Earth range from 2 billion to over 1,000 billion people, with the majority of studies insisting that the maximum number that Earth can support is 9–10 billion people. At current population growth rates, we will reach this number by the year 2050 (**Figure 19.7**).

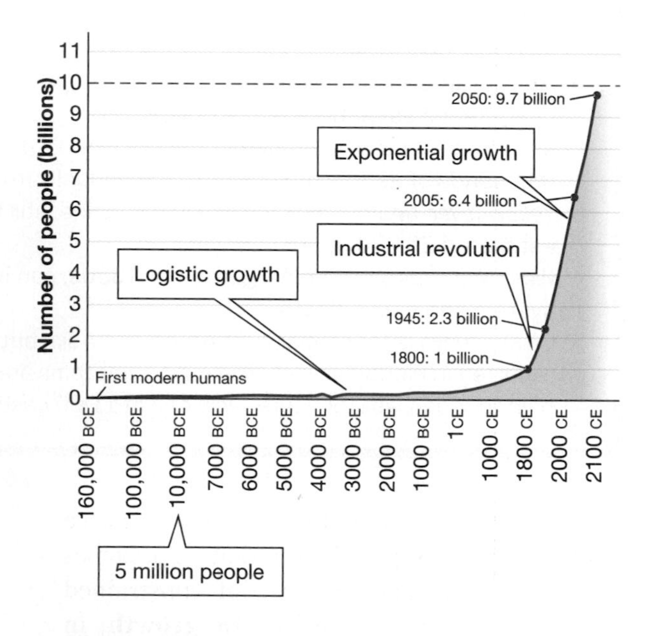

Figure 19.7

Curves of logistic and exponential growth in the world human population

The dashed line indicates the United Nations' estimated carrying capacity of Earth.

Q1: According to this graph, approximately when did exponential growth begin?

Q2: What milestone corresponds to the transition from logistic to exponential population growth? How would you predict that modern medicine (e.g., treatment, hygiene, antibiotics) would affect the slope of the line?

Q3: What is the United Nations' projected carrying capacity of Earth, and when will we reach it?

See Appendix A for answers to the figure questions.

Seeking Change

With human populations increasing in number and concentrating in urban areas, we can expect the continued spread of mosquito-borne illnesses. Thankfully, recent results from field tests of GM mosquitoes in Brazil suggest that we may have a new way to stop that spread.

Starting in 2002, British company Oxitec began producing a GM line of the *Aedes aegypti* mosquito. The company inserts a single gene into a line of lab-bred mosquitoes. This gene, designed to work only in insect cells, produces a protein that prevents the mosquito from transitioning between two of its life stages: larva and adult. In the lab, the mosquitoes are exposed to an antibiotic (tetracycline) that neutralizes the protein so that the mosquitoes can grow to be adults. Then only males are released into the wild because the males don't bite and so will not contribute to the spread of disease. Once free, the male GM mosquitoes mate with wild females and produce offspring with the lethal gene—and no access to antibiotics—so the offspring die off before becoming adults and therefore do not reproduce. In this way the size of the population is reduced (**Figure 19.8**).

Populations—mosquito, human, or otherwise—can change in a density-dependent or density-independent manner. **Density-dependent population change** occurs when birth and death rates change as the population density changes. The number of offspring produced and the death rate are often density dependent. Food shortages, lack of space, and habitat deterioration—all these factors influence a population more strongly as it increases in density (**Figure 19.9**).

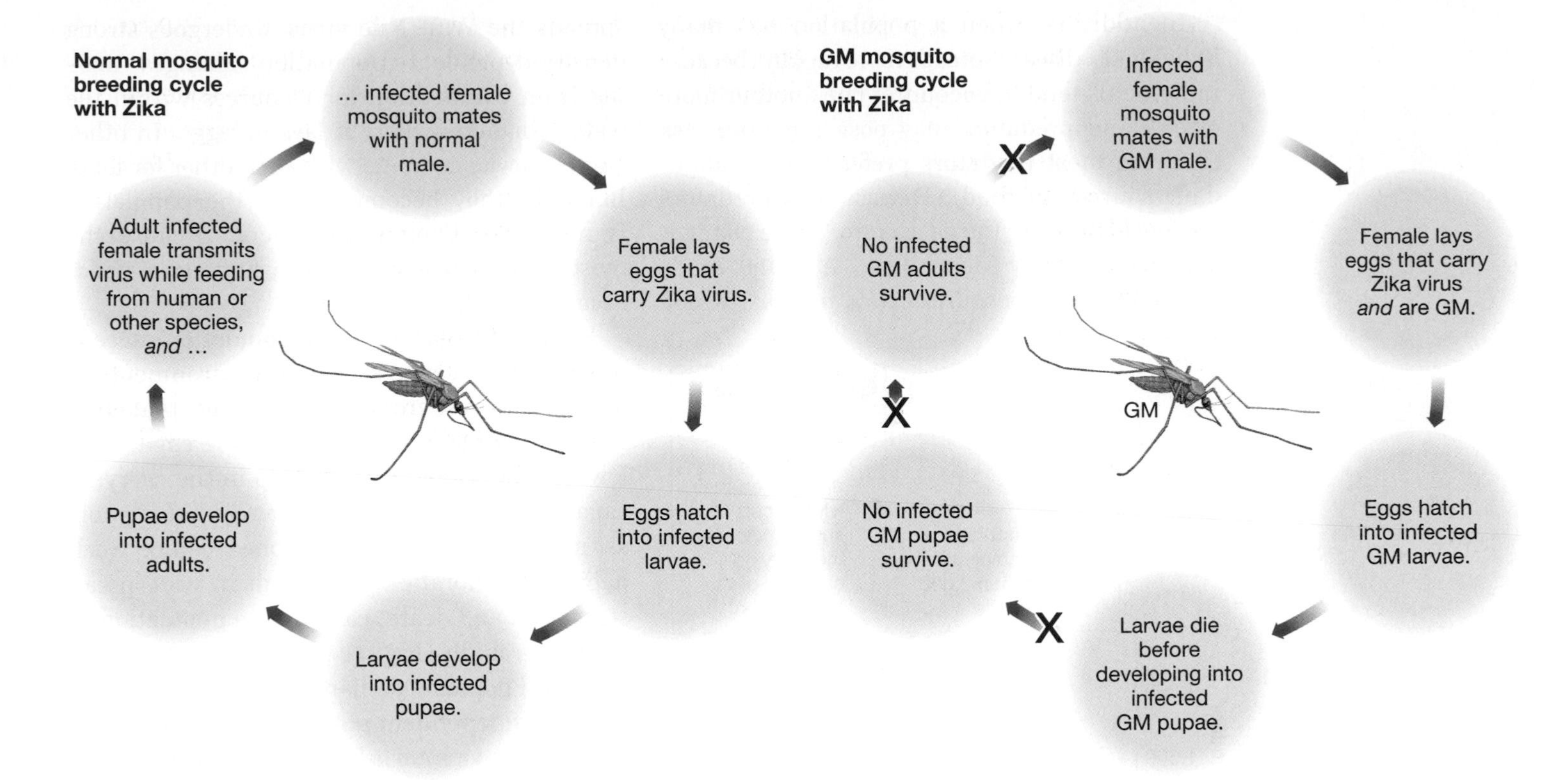
Normal mosquito breeding cycle with Zika
... infected female mosquito mates with normal male.
Female lays eggs that carry Zika virus.
Eggs hatch into infected larvae.
Larvae develop into infected pupae.
Pupae develop into infected adults.
Adult infected female transmits virus while feeding from human or other species, *and* ...
GM mosquito breeding cycle with Zika
Infected female mosquito mates with GM male.
Female lays eggs that carry Zika virus *and* are GM.
Eggs hatch into infected GM larvae.
Larvae die before developing into infected GM pupae.
No infected GM pupae survive.
No infected GM adults survive.
GM
X

Figure 19.8

How mosquitoes are genetically modified to stop the Zika virus

Genetically modified (GM) male mosquitoes have been created and released in Brazil, where they mate with wild females and produce offspring that cannot grow to adulthood.

Q1: List all the stages of the mosquito life cycle.

Q2: Which life cycle stage is vulnerable to the GM treatment?

Q3: Why are only male GM mosquitoes released into the wild rather than both males and females?

See Appendix A for answers to the figure questions.

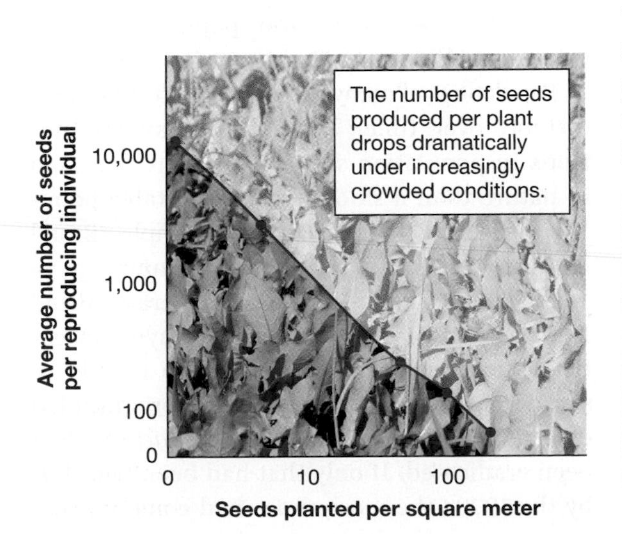
The number of seeds produced per plant drops dramatically under increasingly crowded conditions.
Average number of seeds per reproducing individual
10,000
1,000
100
0
0
10
100
Seeds planted per square meter

Figure 19.9

Overcrowded conditions result in density-dependent population change

Overcrowding affects many species. The plantain, shown here, is a small, herbaceous plant that has decreased reproduction in crowded conditions.

Q1: What factors may be limiting growth and reproduction in the plantain's crowded conditions?

Q2: Why are overcrowded conditions considered density-dependent population changes?

Q3: Relate this example of overcrowded conditions to the human population growth shown in Figure 19.7. How do you think the situations are similar? How are they different?

See Appendix A for answers to the figure questions.

In addition, when a population has many individuals, disease spreads more rapidly (because individuals tend to encounter one another more often), and predators may pose a greater risk (because most predators prefer to hunt abundant sources of food). Disease and predators obviously increase the death rate. These changes are also density dependent. A 2014 study found that the mosquito *Culex pipiens*, which spreads the West Nile virus, undergoes strong density-dependent population changes arising from competition for resources within the watery habitats where it lays its eggs. In other words, larvae compete with each other for food. But once they become adults, the population is no longer density dependent, because the mosquitoes become airborne and have wider access to food.

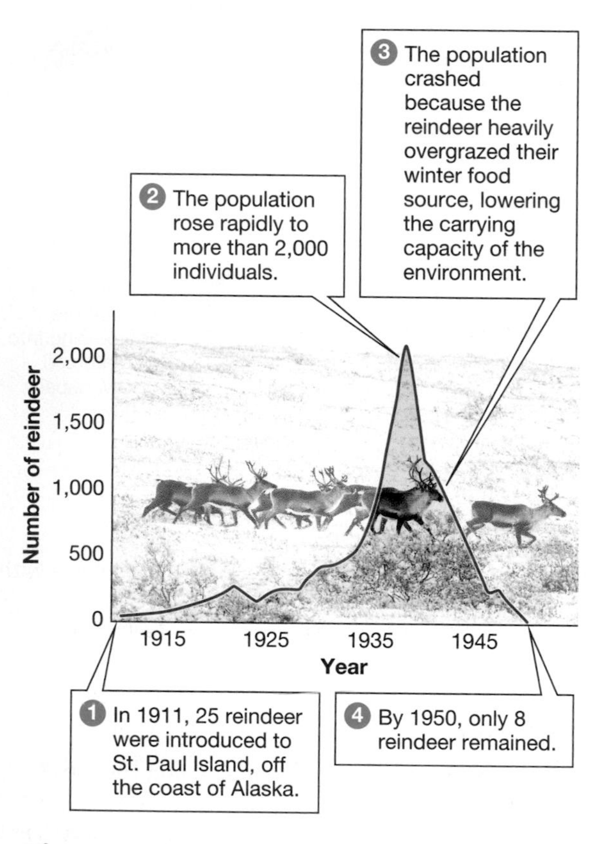
3 The population crashed because the reindeer heavily overgrazed their winter food source, lowering the carrying capacity of the environment.
2 The population rose rapidly to more than 2,000 individuals.
Number of reindeer
2,000
1,500
1,000
500
0
1915
1925
1935
1945
Year
1 In 1911, 25 reindeer were introduced to St. Paul Island, off the coast of Alaska.
4 By 1950, only 8 reindeer remained.

Figure 19.10

Habitat destruction results in density-dependent population change

Like overcrowding, habitat destruction affects many species. The reindeer on St. Paul Island are one example.

Q1: In what year did the reindeer's numbers begin to rise exponentially?

Q2: In what years was the reindeer's population growth logistic?

Q3: How do you predict the graph (population size) would change if someone had begun bringing in supplemental food for the reindeer in 1940? Draw a sketch of your prediction.

See Appendix A for answers to the figure questions.

For some organisms, if a population exceeds the carrying capacity of its environment by depleting its resources, it may damage that environment so badly that the carrying capacity is lowered for a long time. A drop in the carrying capacity means that the habitat cannot support as many individuals as it once could. Such habitat deterioration may result in widespread starvation and death, causing the population to decrease rapidly (**Figure 19.10**).

Not all population changes are due to density. **Density-independent population change** occurs when populations are held in check by factors unrelated to population density. Density-independent factors can prevent populations from reaching high densities in the first place. Year-to-year variation in weather, for example, may cause conditions to be suitable for population growth in an especially warm year. Or poor weather conditions may reduce the growth of a population directly (by freezing the eggs of a mosquito, for example) or indirectly (such as by decreasing the number of food plants available to an animal).

Other natural disturbances, such as fires and floods, can also limit the growth of populations in a density-independent way. Finally, the effects of environmental pollutants, such as the insecticide DDT, are density independent; such pollutants can threaten natural populations with extinction (**Figure 19.11**).

Populations of many species rise and fall unpredictably over time. These **irregular fluctuations** in population size are far more common in nature than a smooth rise to a stable population size. In the 1950s, for example, Brazil mounted a massive anti-mosquito campaign to combat yellow fever—a disease also transmitted by *Aedes aegypti*—that included spraying insecticides and encouraging citizens to get rid of standing water. The success of the program led officials to declare in 1958 that *Aedes aegypti* had been eradicated. If only that had been true, but by the 1970s the mosquitoes had come roaring

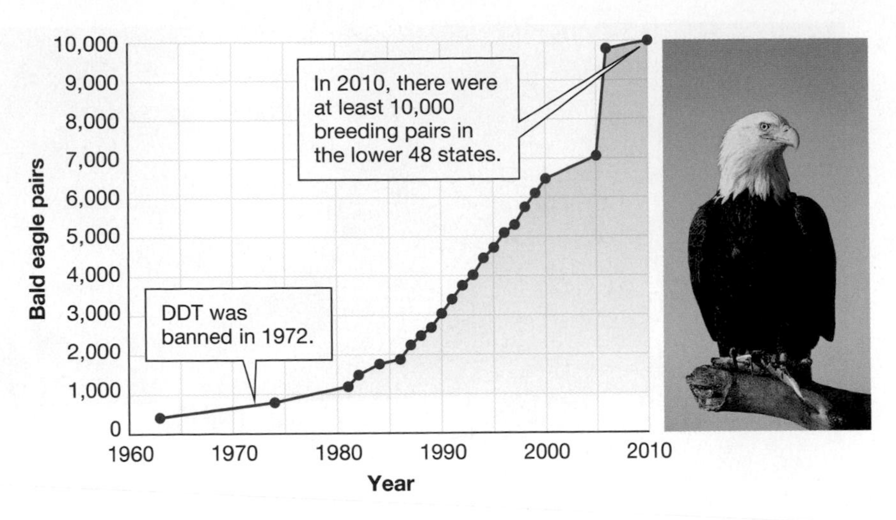

Figure 19.11

Banning the use of the pesticide DDT removed a density-independent population limit

DDT poisoning was directly responsible for declining eagle populations by the middle of the twentieth century. By the early 1960s, population counts indicated that only 417 breeding pairs of bald eagles remained in the lower 48 states—a huge drop from the estimated 100,000 breeding pairs present in 1800. Bald eagle populations increased dramatically after DDT was banned.

Q1: In what year did the bald eagle population rise to more than 2,000 breeding pairs?

Q2: Give some examples of possible density-dependent limits on bald eagle populations.

Q3: Is the population growth of bald eagles more like logistic or exponential growth? Explain why you think so.

See Appendix A for answers to the figure questions.

back, breeding and spreading like crazy—an unwelcome irregular fluctuation. And now the pests are resistant to many chemical attacks.

Populations can also exhibit **cyclical fluctuations**, predictable patterns that occur with seasonal changes in temperature and precipitation or when at least one of two species is strongly influenced by the other. The Canadian lynx, for example, depends on the snowshoe hare for food, so lynx populations increase when hare populations rise, and they decrease when hare populations drop. In this example, the population cycles are also density-dependent population changes because each population is affected by the other's numbers. They cycle together in response to each other's density (**Figure 19.12**).

Friendly Fight

The Oxitec GM mosquitoes—trademarked "Friendly" *Aedes aegypti*—were first released in Brazil in February 2011, under the direction of biochemist Margareth Capurro of the University of São Paulo. In a densely populated suburb in northeastern Brazil, the released mosquitoes reduced the wild population of *Aedes aegypti* by 85 percent, says Capurro. In 2013, a similar test in the village of Mandacaru resulted in a 96 percent reduction of the wild mosquito population in the area after only 6 months.

"For the first time, we demonstrated that transgenic mosquitoes can work," says Capurro, who is now developing alternate GM mosquito lines in her own laboratory, attempting to make the GM insects even more potent against their wild counterparts. For example, Capurro's lab is trying out different genetic mutations to create a line of mosquitoes that does not require the use of an antibiotic to be kept alive in the laboratory, which would be less expensive. "We are finishing the lines now. They are very promising," says Capurro. Other genetic modification techniques are being attempted in labs elsewhere to make *Anopheles* incapable of carrying the malaria parasite, for example.

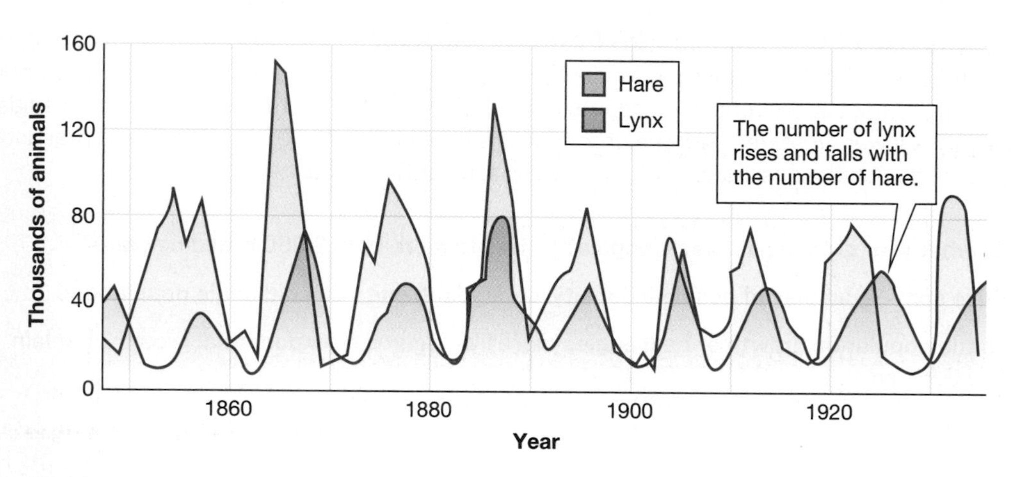

Figure 19.12

Populations of two species may increase and decrease together

The Canadian lynx depends on the snowshoe hare for food, so the number of lynx is strongly influenced by the number of hare.

Q1: During which years did the hare likely have the greatest food supply?

Q2: Besides the number of hare, what other factors might contribute to the number of lynx?

Q3: Can you draw an average carrying-capacity line on the graph? Why or why not?

See Appendix A for answers to the figure questions.

In 2016, Oxitec released Friendly *Aedes aegypti* in a neighborhood of 5,000 residents in the city of Piracicaba, Brazil, to combat the spread of dengue virus. The company released 3–4 million mosquitoes per month over a period of 10 months. The experiment reduced the number of wild *A. aegypti* larvae in the area by 82 percent, and the neighborhood experienced a 91 percent reduction in cases of dengue, according to local officials.

In August 2016, shortly after the first cases of Zika infection from local mosquitoes were reported, the U.S. Food and Drug Administration (FDA) approved a field test of the Oxitec mosquitoes in Key Haven, Florida. Critics argued that the approach is too expensive, because mosquitoes had to be bred and released multiple times. There were also concerns that the engineered gene wouldn't correctly insert in the genome or would become somehow deactivated in the wild. If either of those situations occurred, it would exacerbate the problem by adding reproducing mosquitoes into the environment, critics argued.

After examining such claims, the FDA concluded that the proposed field trial in Florida "would not result in significant impacts on the environment." In a 3–2 vote in November 2016, the Florida Keys Mosquito Control District approved a trial of genetically modified mosquitoes, but in the 2016 general election, 65 percent of Key Haven residents voted to reject the proposal and the trial was not carried out.

In September 2019, Oxitec applied to the government yet again to release mosquitoes into the Florida Keys. This time, Oxitec proposed releasing a second generation of GM *Aedes aegypti* mosquitoes. This generation would not be exposed to antibiotics and can be distributed as eggs rather than as adults, which makes the distribution process easier and cheaper, according to Oxitec. In May 2020, the federal government approved a 2-year field test of Oxitec's mosquitoes in Florida subject to approval by state and local authorities.

Some reject the plan solely on the grounds that it involves genetic modification (**Figure 19.13**). "The public fears genetic engineering. Nearly all politicians don't understand it," said bioethicist Arthur Caplan at the New York University School of Medicine, as quoted in a 2016 article in the *Atlantic*. "I don't think the issue is economic. It is ignorance, distrust, fear of the unknown." Recent polls suggest that the tide may be changing. Three independent polls found that 53–78 percent of Americans support the release of GM mosquitoes in the United States. "I think it will work well," says DeGennaro. "It's time we use modern methods here in Florida like they are using in Brazil to reduce mosquito populations."

Figure 19.13

Some Florida residents oppose the release of GM mosquitoes in their community

Misinformation about genetically modified organisms, including mosquitoes, has caused concerns about their use.

Just the Beginning?

In 2016 and 2017, the Florida Department of Health made a concerted effort to control mosquito populations using insecticides and eliminating standing water. By the end of 2017, the virus was eradicated in local mosquito populations. By 2019, there was no longer any local transmission of Zika virus in the United States, according to the CDC. Around the world, the incidence of Zika followed a similar pattern: Zika infection peaked in 2016 and then declined "substantially" throughout 2017 and 2018, according to the World Health Organization.

Still, the risk of another Zika epidemic remains. There is evidence of continuing Zika virus transmission across 87 countries and territories around the globe. The CDC continues to warn pregnant women away from traveling to any area with Zika. As of 2020, there were no approved vaccines or therapies for Zika virus infection, but it is not for lack of trying. At least six different vaccines are being tested in clinical trials to see if they will be safe for people and effective at protecting against Zika virus.

It remains too soon to tell whether GM mosquitoes can stop the spread of Zika or other mosquito-borne diseases. In December 2019, for example, Miami-Dade County, finally free of local Zika transmission, reported 14 confirmed cases of locally transmitted dengue fever, spread by the same mosquitoes that spread the Zika virus. Eliminating these viruses will depend on population ecology: on mosquito populations, the human population, and how much virus is circulating among both. But history suggests that in the fight against mosquitoes, it's time to try something new.

World's Deadliest Animals

You may have heard that humans are the deadliest animals on the planet. It's true that we, as a species, kill hundreds of thousands of humans. But there's one family of animals that has us beat: mosquitoes. Many species of these small, pesky insects transmit harmful infections, including Zika fever, malaria, West Nile disease, dengue fever, and many more.

Assessment available in smartwork

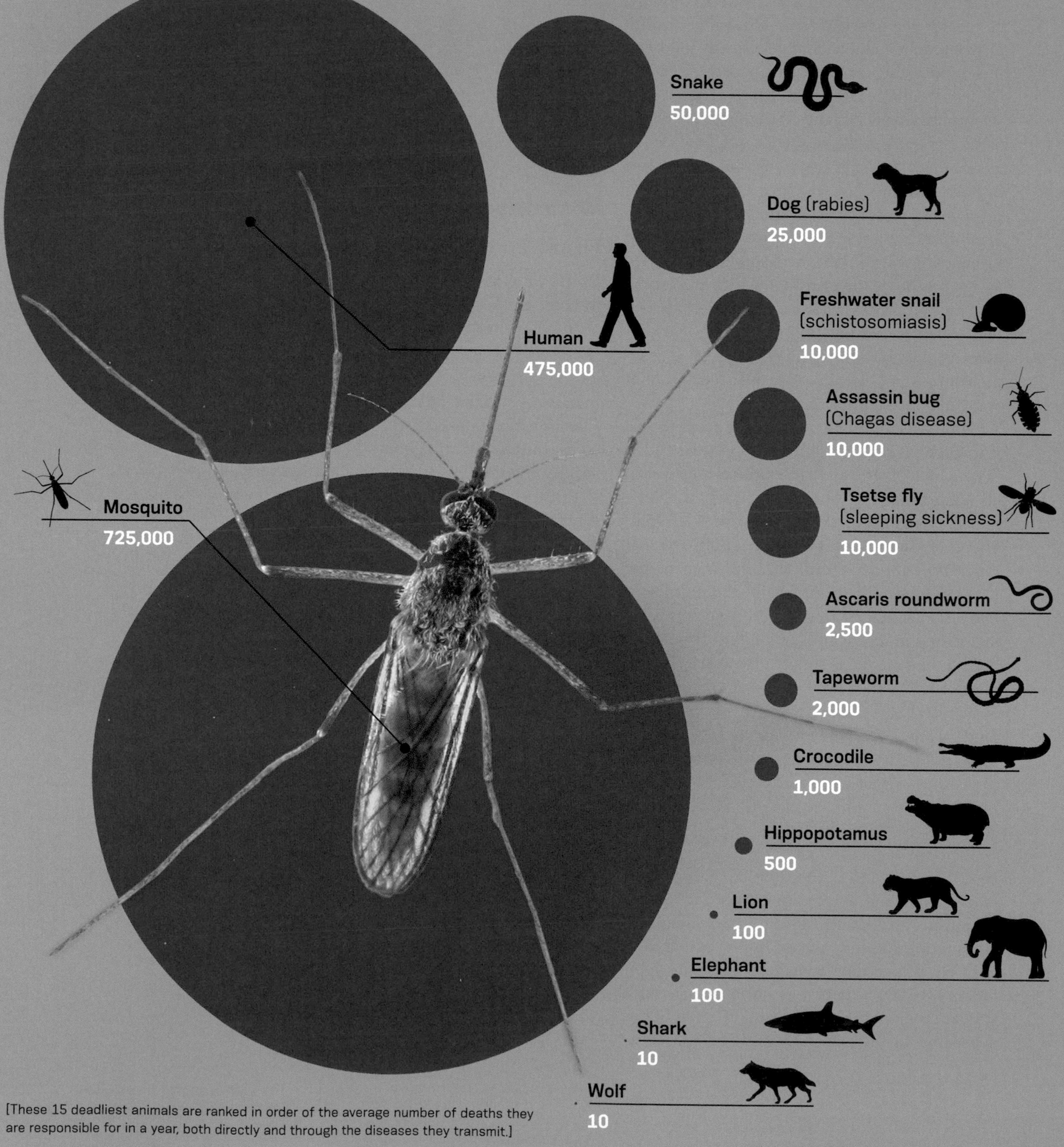

Mosquito
725,000

Human
475,000

Snake
50,000

Dog (rabies)
25,000

Freshwater snail (schistosomiasis)
10,000

Assassin bug (Chagas disease)
10,000

Tsetse fly (sleeping sickness)
10,000

Ascaris roundworm
2,500

Tapeworm
2,000

Crocodile
1,000

Hippopotamus
500

Lion
100

Elephant
100

Shark
10

Wolf
10

[These 15 deadliest animals are ranked in order of the average number of deaths they are responsible for in a year, both directly and through the diseases they transmit.]

NEED-TO-KNOW SCIENCE

- A **population** (page 369) is a group of organisms of a single species located within a particular area.
- **Population size** (page 369) is the total number of individuals in a population. **Population density** (page 369) is the population size divided by the area covered by that population.
- **Population ecology** (page 369) is the study of the size and structure of populations and how they change over time and space.
- **Genetic modification** (**GM**, page 369) is altering the genes of an organism for a specific purpose resulting in a **genetically modified organism** (**GMO**, page 369). It is also referred to as genetic engineering (GE).
- **Population doubling time** (page 370), or the time it takes a population to double in size, is a good measure of how fast a population is growing.
- Environmental factors such as lack of space, food shortages, predators, disease, and habitat deterioration limit populations. These factors affect the **carrying capacity** (page 371), the number of individuals that can live in an environment indefinitely.
- In **exponential growth** (page 371), a population increases by a constant proportion from one generation to the next generation. It is associated with a **J-shaped growth curve** (page 371) when graphed.
- In **logistic growth** (page 371), a population grows exponentially at first but then stabilizes after reaching the carrying capacity. It is associated with an **S-shaped growth curve** (page 372) when graphed.
- **Density-dependent population change** (page 372) occurs when birth and death rates are affected by population density, which is the case when many individuals occupy the same space and therefore compete for resources.
- **Density-independent population change** (page 374) occurs when population size is affected by factors that have nothing to do with the density of the population.
- Populations may rise and fall over time, exhibiting **irregular fluctuations** (page 374) or **cyclical fluctuations** (page 375).

THE QUESTIONS

See Appendix B for answers.

The Basics

1 A group of individuals of a single species located within a particular area is called

(a) a biosphere.
(b) an ecosystem.
(c) a community.
(d) a population.

2 A population that is growing exponentially increases

(a) by the same number of individuals in each generation.
(b) by a constant proportion in each generation.
(c) in some years and decreases in other years.
(d) none of the above

3 In a population with an S-shaped (logistic) growth curve, after an initial period of rapid increase, the number of individuals

(a) continues to increase exponentially.
(b) drops rapidly.
(c) remains near the carrying capacity.
(d) cycles regularly.

4 The growth of populations can be limited by

(a) natural disturbances.
(b) weather.
(c) food shortages.
(d) all of the above

5 The maximum number of individuals in a population that can be supported indefinitely by the population's environment is called the

(a) carrying capacity.
(b) J-shaped curve.
(c) sustainable size.
(d) exponential size.

6 Select the correct terms:
(Density-independent / Density-dependent) factors limit the growth of populations more strongly at high densities than at low densities. One example of a (density-independent / density-dependent) factor is a natural disaster.

Challenge Yourself

7 A population of plants has a density of 12 plants per square meter and covers an area of 100 square meters. What is the population size?

(a) 120
(b) 1,200
(c) 12
(d) 0.12
(e) none of the above

8 Which of the following would *not* cause the carrying capacity of a mosquito population to change?

(a) lower availability of standing water in which to lay eggs
(b) fewer animals for female mosquitoes to feed on
(c) warmer weather that increases overwintering survivorship
(d) All of the above would change carrying capacity.
(e) None of the above would change carrying capacity.

9 Place the following elements in the strategy to control mosquito populations in the correct order from earliest to latest by numbering them from 1 to 5.

_____ a. Scientists genetically modify male mosquitoes in the lab.
_____ b. Mosquito offspring produced by the matings do not survive to adulthood.
_____ c. GM male mosquitoes mate with normal wild females.
_____ d. The mosquito population decreases dramatically.
_____ e. Scientists release GM male mosquitoes into the wild.

Try Something New

10 Draw a diagram that illustrates the possible fluctuating population cycles of a predator, such as the mountain lion, and its prey species, such as deer. The prey regularly experiences population crashes when it overgrazes its food source. Label your axes.

11 An artificial pond at a college campus was populated with several mating pairs of pond slider turtles. After many years of slow population growth, campus administrators installed a coin-operated food dispenser at the pond's edge. Students and community members bought and fed the turtles an enormous amount of food each day, causing the population of turtles in the pond to increase exponentially. How did the installation of the food dispenser affect the carrying capacity of the turtle population at this pond?

12 Suppose that population ecologists at the college in question 11 determined that the number of turtles in the pond had increased from 6 individuals to 24 individuals in just the first year after installation of the coin-operated food dispenser. What is the population doubling time for the turtles in this population?

(a) 6 years
(b) 6 months
(c) 1 year
(d) 1 month

Leveling Up

13 **What do *you* think?** Some Florida residents do not want to allow genetically modified (GM) mosquitoes to be released, even though they have been demonstrated to significantly reduce mosquito-borne illnesses in Brazil.

(a) Should Floridians be allowed to decide not to have GM mosquitoes released in their neighborhoods? Should this question be decided by vote? Explain your reasoning.
(b) Currently, the Florida Department of Health may require people to remove standing water from their yards in order to decrease breeding habitat for mosquitoes. Should people be allowed not to comply with this order, given that the yard is their private property? Explain your reasoning.

14 **Life choices** Do you feel that people should choose to limit the number of children they have, given the increasing human population? Alternatively, should people focus on decreasing their own ecological footprint—given that, for example, one American uses the same amount of resources as do four Chinese citizens? Explain your reasoning.

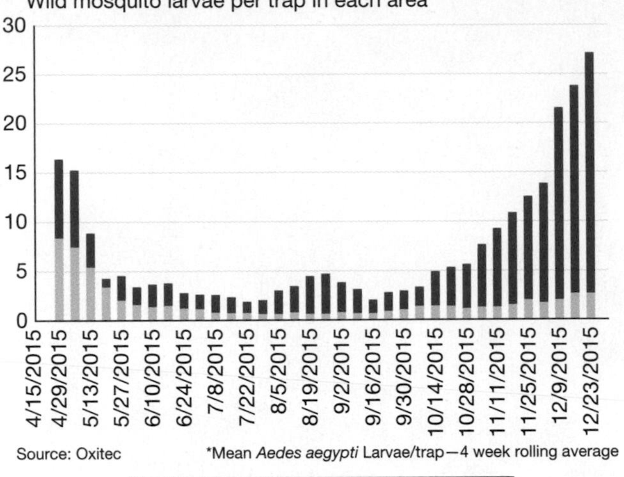

15 **Is it science?** Your best friend found a Reddit thread stating that GM mosquitoes do not suppress the population size of Zika-bearing mosquitoes and instead claiming that Zika was *caused* by the release of GM mosquitoes. Putting aside the idea that Zika was caused by GM mosquitoes, which is refuted by multiple studies as discussed on many legitimate websites, you decide to simply show your friend some data comparing mosquito population sizes in areas with and without GM mosquitoes. In your own words, write a paragraph that describes the results shown in the accompanying figure and explains to your friend how they relate to the Reddit claims.

Digital resources for your book are available online.

Of Wolves and Trees

The extermination of wolves in Yellowstone National Park had unforeseen effects on the park's ecosystem. Can the return of this top predator restore order?

After reading this chapter you should be able to:

- Explain the concept of an ecological community and how it relates to ecological diversity.
- Articulate the role of relative species abundance and species richness in defining ecological diversity.
- Distinguish between a food chain and a food web.
- Illustrate how the removal of a keystone species disrupts an ecological community.
- Describe the four major kinds of species interactions, and give an example of each.
- Compare and contrast primary and secondary succession.

ECOLOGY

CHAPTER 20

COMMUNITIES OF ORGANISMS

Robert Beschta will always remember the day he visited Yellowstone National Park's Lamar Valley in 1996. Beschta, a hydrologist, was there to observe the Lamar River, which winds through the valley's lush lowland of grass and sage (**Figure 20.1**). But on that day, as he walked into the valley, he noticed something odd. All around the area, there were not many trees. The few tall, white aspens he saw looked haggard, their bark eaten away. And there were no young saplings.

Beschta knew what a healthy valley was supposed to look like. As a professor of forestry, he had spent decades studying forests, rivers, and wildlife. This valley, Beschta observed, was not healthy.

He approached the river and saw that its banks were also devoid of trees. The leafy green cottonwoods and wide willows that had once arched gracefully over the riverbank were absent. And with no tree roots to hold the soil in place, the banks themselves were jagged and eroding. "I was dumbstruck," recalls Beschta. Something unprecedented was happening in Yellowstone.

Beschta returned to Oregon State University (OSU), where he worked, and described his observations in a seminar. He showed pictures of aspens with their bark stripped away and empty riverbanks where saplings should have been growing. In the audience, William Ripple sat up a little straighter. Also a scientist at OSU, Ripple studied forest ecology. He was particularly interested in aspen trees, which grow as tall as 70 feet and live up to 150 years. From what Beschta was saying, aspens were no longer growing in Yellowstone.

"It was a scientific mystery as to what was the cause of the decline," he recalls. In that moment, listening to Beschta, Ripple knew exactly what his next research project would be: to document the extent of the aspen decline and determine why it was occurring.

Within a year, Ripple and one of his graduate students, Eric Larsen, traveled to the Lamar Valley. There, they drilled small holes into the centers of aspens and removed from each a plug of wood about the diameter of a pencil. Then they counted the growth rings in each plug—one for every year the tree has been in existence—to determine the age of each tree. They found that most of the aspens had begun to grow *prior* to 1920. After 1920, almost no new trees had begun growing.

"We started scratching our heads at that point," says Ripple. He, Beschta, and Larsen began to brainstorm reasons why the trees might have stopped regenerating. They looked for environmental changes in the 1920s that could have done it: a fire that killed off saplings or a change in climate that reduced the trees' ability to reproduce. But nothing lined up—until an ordinary moment gave Ripple an extraordinary idea.

Standing in a gift shop in Grand Teton National Park, just south of Yellowstone, Ripple looked up at a poster on the wall. It featured a grove of tall, white aspen trees in the winter. In the middle of the trees, its paws covered in snow, stood a large gray wolf. "That was an 'aha' moment," says Ripple. "I thought, 'Maybe the wolf protects the aspen.'"

The idea of wolves protecting aspens initially seems nonsensical. How would meat-eating predators protect trees? They wouldn't—at least not directly. Ripple surmised that wolves in Yellowstone might have had an indirect effect on the community. As an ecologist, Ripple had long studied **ecological communities**, associations of species that live in the same area. Communities vary in size and complexity, from a small group of microorganisms in a puddle of water to the whole of Yellowstone Park, home

Figure 20.1

River valley

The Lamar River flows through Yellowstone National Park.

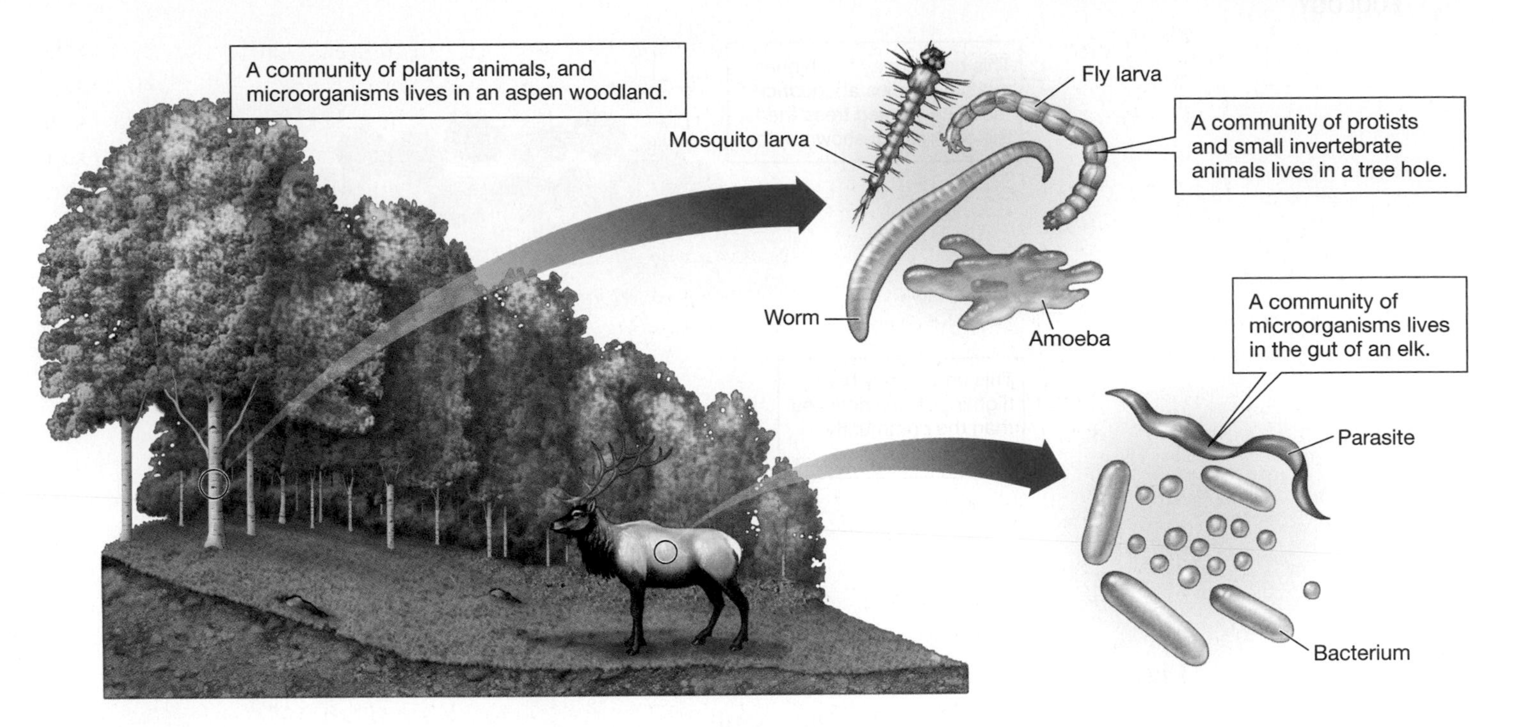

Figure 20.2

Ecological communities come in all sizes

Smaller communities can be nested within a larger community. This aspen woodland community contains the smaller communities of a temporary pool of water in a tree hole and an elk's gut, among others.

Q1: List another species that is part of this community.

Q2: Of which community could this aspen woodland be a smaller part?

Q3: Which other small communities could be found within this larger community?

See Appendix A for answers to the figure questions.

to an estimated 322 species of birds, 67 species of mammals, 1,386 species of plants, and an uncounted number of insects (**Figure 20.2**).

Ripple knew that an ecological community is characterized by the diversity of species that live there, and that diversity is governed by two things: the **relative species abundance** (how common one species is when compared to another) and the **species richness** (the total number of different species that live in the community; **Figure 20.3**). Ripple also knew that communities are subject to constant change and suspected that something important changed in Yellowstone starting in the 1920s.

Ecological communities change naturally as a result of interactions between and among species and as a result of interactions between species and their physical environment. Ripple knew that both the relative abundance and the richness of species had changed in Yellowstone during the previous century. The relative species abundance had changed because aspen trees and willows were in decline, elk and coyote numbers had increased, and bison and beaver populations had decreased. Many of the common inhabitants of Yellowstone had changed in abundance; therefore, the relative species abundance of this community had changed.

The species richness had also changed because early in the twentieth century, a keystone species, the gray wolf, had gone missing. The total number of species in the community had gone down by only one, but that seemingly small change had a big impact. A **keystone species** is a species that has a disproportionately large effect on a community, relative to the species' abundance. There are few wolves compared to, say, rabbits, yet the wolves have a stronger effect on the community. Keystone species in other communities include prairie dogs in the American western plains,

This community has higher relative species abundance for white-barked trees than the community shown below.

This community has higher species richness than the community shown above.

Figure 20.3

Two measures of species diversity

The diversity of an ecological community is determined by relative species abundance and also species richness. High relative species abundance means that one species dominates the community (so the community is less diverse), and high species richness means that many species are present in the community (so it is more diverse).

Q1: If relative species abundance increased in the first community, how would the figure look different than it does now?

Q2: If species richness decreased in the second community, how would the figure look different than it does now?

Q3: How do relative species abundance and species richness define the species diversity of a forest community?

See Appendix A for answers to the figure questions.

hummingbird pollinators in the Sonoran Desert, and sea stars (also called starfish) in intertidal waters (**Figure 20.4**).

Keystone species are often recognized only when they go missing and their disappearance results in dramatic changes to the rest of the community. And that is exactly what happened in Yellowstone in the 1920s. The mighty gray wolf disappeared, or more accurately the mighty gray wolf was exterminated.

A Key Loss

In the early 1900s, ranchers and homesteaders killed wolves all across the United States and eliminated them in many eastern states. Then, in 1915, the U.S. government began subsidizing wolf extermination programs all over the country, and a systematic slaughter began. Under the national program, states paid bounties of up to $150 for individual wolf pelts. Wolves, feared and hated by private landowners for killing livestock, were trapped, shot, and skinned.

The extermination happened quickly. The last known wolf den in Yellowstone was destroyed in 1926. At least 136 wolves, maybe more, were killed during the eradication campaign in Yellowstone. Once it was done, the park was wolf free.

Seven decades later, staring at the poster of a wolf standing among the aspens, Ripple wondered whether the loss of that keystone species had contributed to the decline of aspen

How a predator maintains diversity
A *Pisaster* sea star feeding on a mussel. Without this keystone species, the mussels crowd out other species in their community.
Loss of a keystone species reduces diversity
Number of species
25
20
15
10
5
0
1962
'64
'66
'68
'70
'72
'74
Year
Sea star not removed
Sea star removed
Sea stars completely eliminated mussels in submerged areas of this marine community, enabling other intertidal species to thrive there.
When the sea star *Pisaster* was removed from a community experimentally, the number of species dropped from 18 to 1, a mussel.

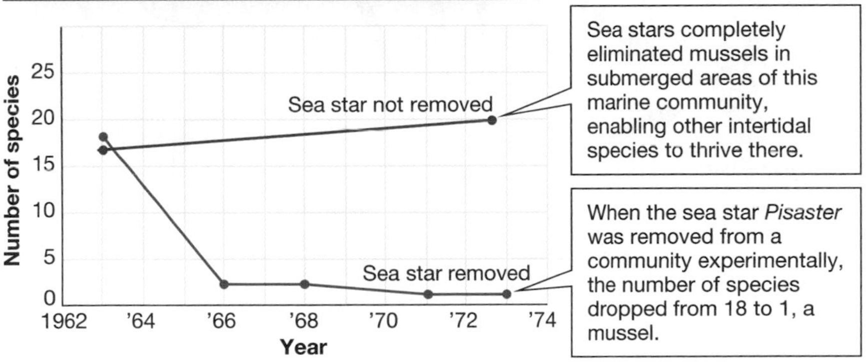

Figure 20.4

The star of the community

The sea star *Pisaster ochraceus* is an example of a keystone species. In a classic experiment conducted along the Pacific coast of Washington State in 1963, sea stars were removed from one site while an adjacent site was left undisturbed.

Q1: How many species were left in 1966 in the community where sea stars were *not* removed?

Q2: How many species were left in 1966 in the community where sea stars *were* removed?

Q3: How do your answers to questions 1 and 2 demonstrate the importance of a keystone species for the maintenance of diversity in a community?

See Appendix A for answers to the figure questions.

trees. He looked up the historical records to see whether the timing matched. Lo and behold, the last wolf had been killed at about the same time the aspens stopped regenerating in the mid-1920s. Suddenly, it seemed obvious why the aspens had declined: wolves kill elk, and elk eat aspens—three species in the same food chain.

A **food chain** is a simple list of who eats whom. In scientific terms, it is the direct path by which nutrients are transferred through the community. A **food web**, by contrast, is a

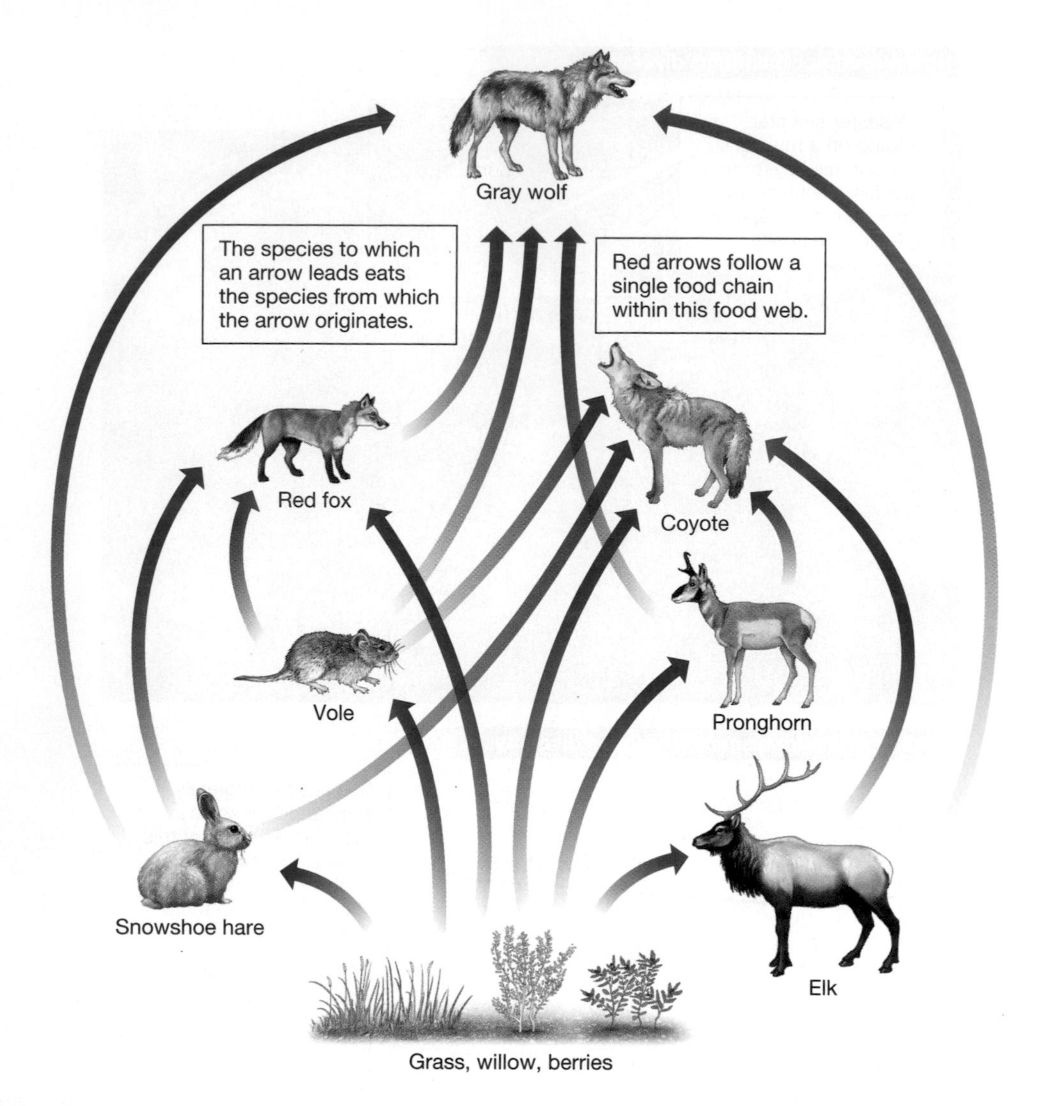
Gray wolf
The species to which an arrow leads eats the species from which the arrow originates.
Red arrows follow a single food chain within this food web.
Red fox
Coyote
Vole
Pronghorn
Snowshoe hare
Elk
Grass, willow, berries

Figure 20.5

Food webs show how energy moves through a community

Food webs are composed of many food chains that show one species eating another.

Q1: What species eats the coyote?

Q2: What species does the coyote eat?

Q3: What do you think would happen to a community that lost its coyotes?

See Appendix A for answers to the Figure Questions.

more complex diagram of all the food chains in a single ecosystem and how they interact and overlap (**Figure 20.5**).

In the wolf-elk-aspen food chain, aspen are the **producers**, the organisms at the bottom of the food chain that use energy from the sun to produce their own food through photosynthesis. In Yellowstone and on land all over Earth, photosynthetic plants such as trees, grasses, and shrubs are the major producers (**Figure 20.6**). In aquatic biomes, the major producers are photosynthetic plankton, as we see in Chapter 21.

Further up the food chain are **consumers**, organisms that obtain energy by eating all or parts of other organisms or their remains. Elk and wolves are both consumers: elk eat aspens, and wolves eat elk. In the Yellowstone food chain, elk are **primary consumers**: they eat producers.

In tropical rainforests, the abundant producers (plants) store chemical energy harvested from sunlight, which in turn is readily available to consumers.

In deserts, the sparseness of plant life means that relatively little chemical energy is available for consumers.

Figure 20.6

Producers are the energy base of a food chain

All communities and ecosystems have producers, but they vary in abundance, depending on the circumstances.

Q1: Where do producers acquire the energy they need to perform their function in the food chain?

Q2: Why are producers necessary for life on Earth?

Q3: Given the lower abundance of producers in desert regions compared to tropical rainforests, what would you predict about the abundance of consumers in the two environments?

See Appendix A for answers to the figure questions.

Wolves are **secondary consumers** because they eat primary consumers. This sequence of organisms eating organisms can continue: a bird that eats a spider that ate a beetle that ate a plant is a **tertiary consumer**; a killer whale that eats a leopard seal that ate a sea bass that ate a krill that ate a phytoplankton is a **quaternary consumer**. We explore the flow of energy up the food chain in Chapter 21.

The more they discussed it, the more Ripple, Larsen, and Beschta believed that the loss of wolves in Yellowstone had allowed the elk population to flourish and eat so many young trees that the aspen population could not regenerate. "We developed a hypothesis that maybe the killing of wolves actually affected the reproduction of aspen trees," says Ripple.

Any change in species diversity will have a ripple effect (no pun intended) throughout the community, and the wolves of Yellowstone were no exception. Ripple and Larsen published their hypothesis in 2000, suggesting that the loss of wolves had led to increased elk populations and altered elk movements and feeding patterns. In other words, with wolves gone, elk were free to find and eat young aspen shoots whenever they wanted, with no fear of wolves.

In 2001, Beschta returned to the Lamar Valley and collected data on another species of tree, the cottonwood, which can live more than 200 years. He documented the same trend as the aspens showed. In the 1920s, cottonwoods suddenly stopped generating new, young trees. In fact, since the 1970s, not a single new cottonwood had been established. "This was dramatic," recalls Beschta. "It's a big deal when you can't have a single cottonwood in this large valley make it to a mature tree."

It is possible for a consumer to eat a species to extinction, so if elk populations had continued to grow unchecked, they could have reduced the aspen and cottonwood populations to zero, disrupting the ecological community permanently. But before that could happen, something dramatic occurred in Yellowstone. Humans brought the wolves back.

A Second Ripple Effect

In 1973, wolves became the first animals to be protected under the Endangered Species Act. It was the dawn of the modern conservation movement, and the idea of returning wolves to Yellowstone grew in popularity. It took some time, but eventually, lawmakers agreed to the plan, and between 1995 and 1997, 41 wild wolves were captured in Canada and released in Yellowstone. It didn't take long for the wolf populations to recover. By 2007, an estimated 170 wolves lived in and around Yellowstone. As of January 2020, the wolf population in the park hovered around 80 wolves in nine packs (**Figure 20.7**). The lower recent number is likely due to outbreaks of disease and packs moving out of the park, said Doug Smith, project leader for the Wolf Restoration Project in Yellowstone.

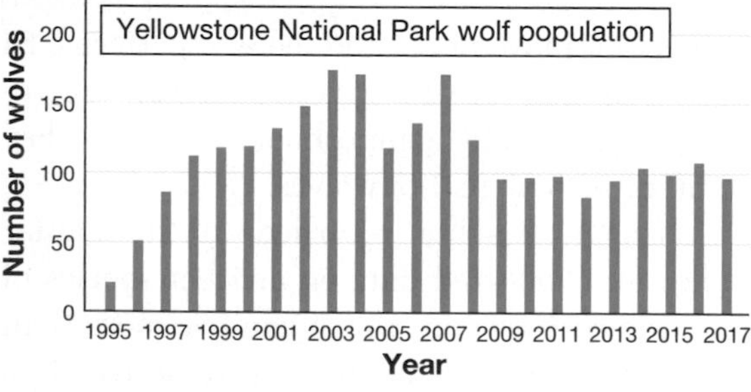

Figure 20.7

Wolves reintroduced

Today wolves run freely in Yellowstone.

Q1: When were wolves first reintroduced to Yellowstone?

Q2: What was the highest number of wolves ever observed in Yellowstone? In what year did that occur?

Q3: Why did it take a few years after the reintroduction of wolves for the aspen population to increase as well as the populations of beavers and bears?

See Appendix A for answers to the figure questions.

The loss, and then return, of the wolves had significant impacts on **species interactions** in the park. There are four central ways in which species in a community interact: mutualism, commensalism, predation, and competition. The classification is based on whether the interaction is beneficial, harmful, or neutral to each species involved. These interactions affect where organisms live and how large their populations grow. Species interactions also drive natural selection and evolution, thereby changing the composition of communities over short and long periods of time.

With wolves back in the park, Ripple suddenly had a way to test his hypothesis about aspen decline. If the loss of wolves was responsible for aspen loss, then the return of the wolves should incite a revival of aspens (and other woody plants, including cottonwoods). But he needed a way to quantify that change. "It's a scientifically difficult task to connect a wolf to a plant. Obviously wolves don't consume plants, so instead we had to connect the dots. There are data that wolves affect elk, and other data that elk affect plants," says Ripple.

Ripple and Beschta went into the field in 2006 and again in 2010 to take measurements. In addition to recording the ages of the trees, they looked for and documented signs of elk feeding on aspens—such as scars where branches or buds had been bitten off—and measured the heights of young aspens. Beschta did the same with the cottonwoods. They were eager to see which species interactions would occur now that wolves were back.

The first type of species interaction, **mutualism**, occurs when two species interact and both benefit. For example, Yellowstone is home to 4,600 bison, the largest land mammals in North America. Bison have a mutualistic relationship with the black-billed magpie. Pests such as ticks burrow into a bison's short, dense hair to suck the beast's blood, but hungry

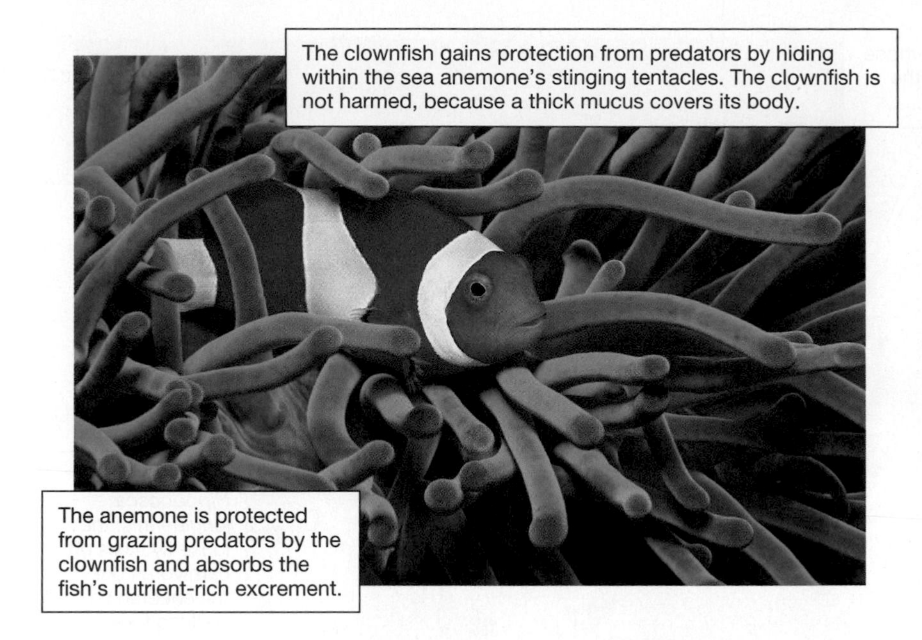

The clownfish gains protection from predators by hiding within the sea anemone's stinging tentacles. The clownfish is not harmed, because a thick mucus covers its body.
The anemone is protected from grazing predators by the clownfish and absorbs the fish's nutrient-rich excrement.

Figure 20.8

Mutualism: friends in need

The clownfish and the sea anemone both benefit from their relationship.

Q1: **What might happen to an anemone without a resident clownfish?**

Q2: **What would happen to a clownfish that did not produce a mucous coating?**

Q3: **In the example of mutualism given in the text, what would happen if ticks were no longer able to feed on bison?**

See Appendix A for answers to the figure questions.

little magpies perch on top of the bison and eat those ticks. Thus, both the bison and the magpie benefit from close interaction with one another. Mutualism is common and important in ecosystems all over Earth. Many species receive benefits from, and provide benefits to, other species, as the clownfish-anemone partnership in **Figure 20.8** illustrates. These benefits increase the survival and reproduction of both interacting species.

When they aren't perched atop bison, black-billed magpies can be found in large nests atop trees, where they reproduce once a year. These trees, another member of the community, share a commensal relationship with the magpies. **Commensalism** happens when one partner benefits while the other is neither helped nor harmed—in this case, the magpie benefits from having a safe place to lay eggs, and the interaction has no effect on the tree. Another example of a commensal relationship is barnacles living on a whale (**Figure 20.9**).

As you might have guessed, not all species interactions are as pleasant as those among bison, birds, and trees. In two other types of interactions, at least one of the two species is harmed: *competition* (which we return to in a moment) and *predation*.

In **predation**, one species benefits and the other is harmed, and **predators** are defined as consumers that eat part or all of other organisms. A **parasite** is a kind of predator that

The barnacles on a whale's snout enjoy a continuous stream of nutrient-rich water flowing across them from which they can feed. The whale is neither harmed nor helped by these stowaways.

Figure 20.9

Commensalism: a whale of a ride!

This gray whale's snout is covered in barnacles.

Q1: **How do barnacles benefit from living on a whale?**

Q2: **Do you think a whale could avoid being colonized by barnacles? Why or why not?**

Q3: **Explain why detritivores are considered commensal to the organisms they consume. (You will need to read ahead to answer this question.)**

See Appendix A for answers to the figure questions.

This cheetah is a predator whose many prey species include the African hare.
The parasite *Hippobosca longipennis*, the louse fly, lives off carnivores like the cheetah by sucking their blood, but does not kill its host.

Figure 20.10

Predators can also be prey

We traditionally think of the cheetah as a predator, but it can also be prey to parasites and eventually to decomposers and detritivores.

Q1: What kind of predator is the cheetah in the figure?

Q2: What kind of predator is the louse fly in the figure?

Q3: What kind of predator are the elk that graze on aspen tree saplings?

See Appendix A for answers to the figure questions.

lives in or on the organism it harms, its **host**. An important group of parasites is pathogens, which cause disease in their hosts. The bacteria that cause strep throat, tuberculosis, and pneumonia are pathogens, for example. Many organisms have evolved mechanisms to avoid being hosts, such as immune systems to help fight off parasitic diseases and infections.

Other types of predators are distinguished by what they eat. Elk are **herbivores**, animals that eat plants. Yellowstone elk feed on the shoots, saplings, and new branches of woody plants such as aspen, cottonwood, and willow, especially in winter, when other plants are scarce. Wolves are **carnivores**, animals (and, in rare cases, plants) that kill other animals for food. Yellowstone wolves predominantly eat elk, especially in winter, but they also eat deer and any small mammals they can catch, notably beaver. Other animals, such as raccoons and coyotes, eat both animals and plants, so we call them **omnivores**. Both raccoons and coyotes are also **scavengers**, animals that eat dead or dying plants and animals. Of the species that are scavengers, **decomposers**, such as fungi, dissolve their food, and **detritivores**, including worms and millipedes, mechanically break apart and consume their food.

The animals eaten by predators are called **prey**. All of the animal residents of Yellowstone (and most plants) are eaten by other species, except for grizzly bears, mountain lions, eagles, and gray wolves. These four animals are all at the top of the food chain. Although they are top predators, they are themselves prey to parasites (**Figure 20.10**).

Cause and Effect

Wolves were reintroduced to Yellowstone National Park in 1995, and their return had a major impact on other species in the park, especially elk and aspens. Elk began avoiding areas where wolves could easily prey upon them, such as near riverbanks, especially riverbanks with downed logs, where escape would be difficult. With decreasing elk presence, aspens flourished in these areas. The effects depicted here are a powerful example of the influence of a keystone species on an ecological community.

Assessment available in smartwork

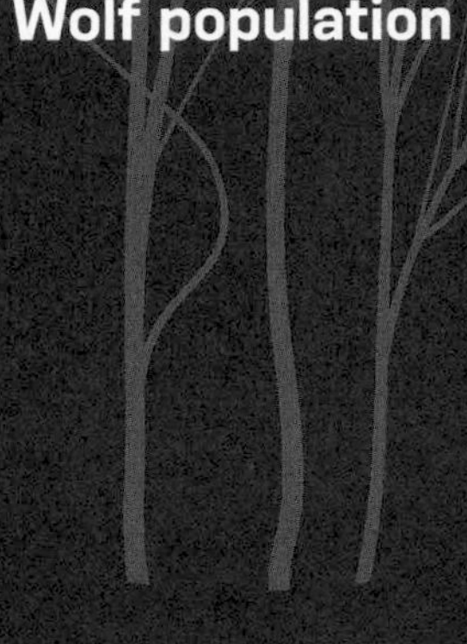

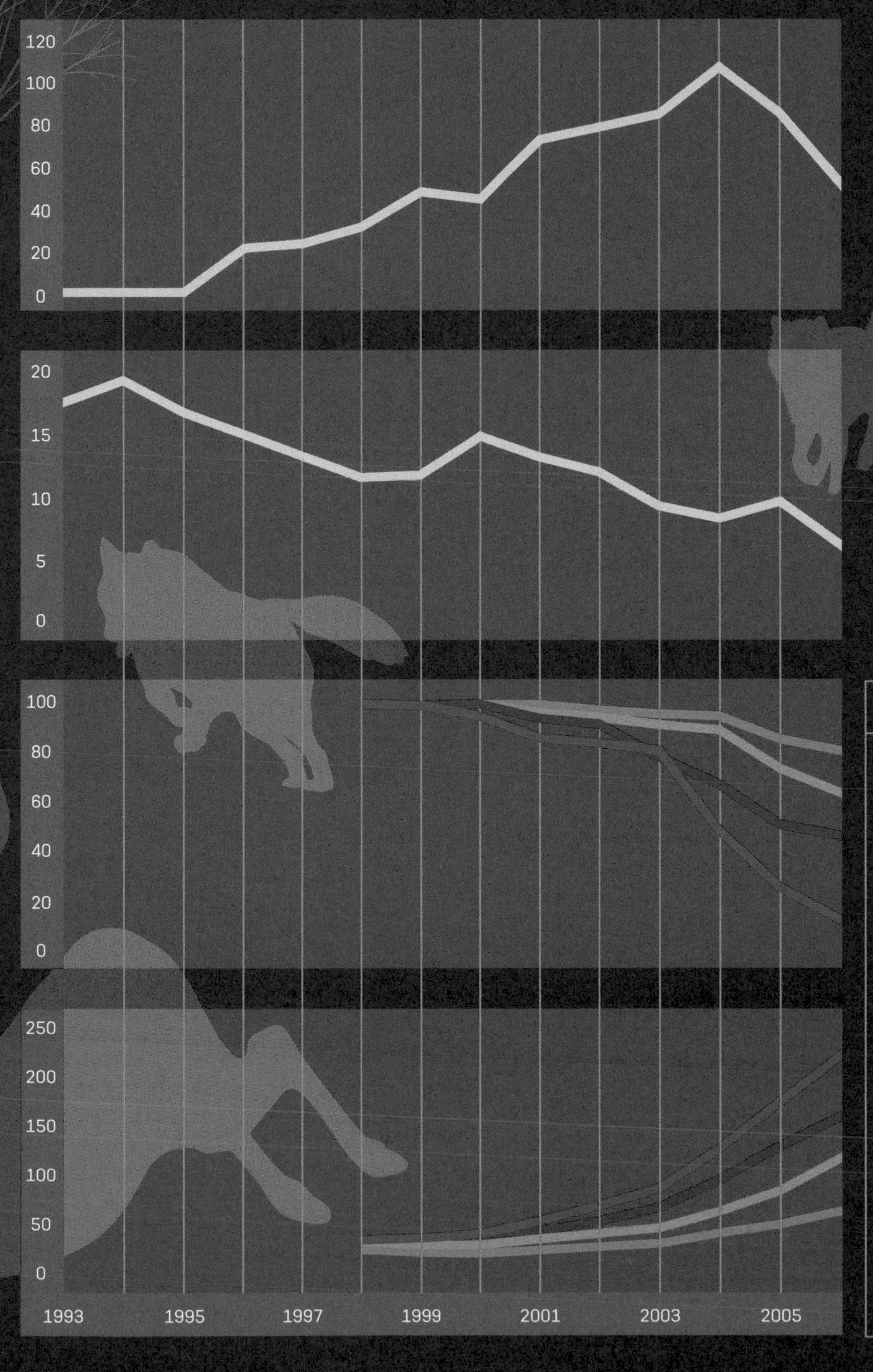

Back in the Park

Now that wolves, a top predator in the community, were back in Yellowstone, how would their prey, the elk, react? And how would that reaction affect the elk's food, the trees?

Using the plant measurements taken in the park and comparing those measurements to historical data, Beschta and Ripple found that between 1998 and 2010, as the wolf population in the park grew, the elk population decreased and therefore fewer aspens were eaten. In 1998, essentially 100 percent of the young aspen plants were being preyed upon, but by 2010, only 18–24 percent of young aspens were being eaten. In addition, average aspen heights increased for all areas that the scientists observed. Cottonwoods experienced the same revival. In the 1970s, cottonwoods had entirely stopped adding new young saplings, but by 2012, some 4,660 young cottonwoods had grown to over 2 meters high.

Together, the aspen and cottonwood data sets convinced Ripple and Beschta that the reintroduction of wolves was responsible for a cascade of species interactions leading to the restoration of aspen and cottonwood populations. "With wolves back, young, woody plants are doing better and growing taller," says Ripple. "Plant communities are beginning to recover." Different species are growing and spreading at different rates of recovery, he adds, "but there's enough new growth that we suggest it is in support of our basic hypothesis, that the presence or absence of the top predator—the wolf in this case—makes a difference in these plant populations."

In a follow-up paper in 2018, the two scientists, along with Larsen and OSU ecologist Luke Painter, compared aspen data from inside the park to data collected in three areas outside the park. The team showed that aspens are also recovering in areas around the park and that wolf predation on elk is the major reason for the new tree growth. "It's a restoration success story," Painter said in a press release about the study.

Since Ripple and Beschta's initial discovery, the scientific community has been debating two potential reasons why the plants are flourishing with the return of the wolves. The most straightforward possibility is that the elk population has decreased. Wolves kill elk, so there are fewer elk to consume the plants. Yet some of the tree populations seemed to recover faster than the drop in the elk populations would suggest. So a second possibility is that the presence of wolves led to a change in elk behavior called a *fear effect*. Often, the presence of a predator in an area can affect the behavior of its prey. In this case, it is possible that elk stopped grazing in areas where they could easily be seen by wolves, such as along the banks of the Lamar River.

"These two mechanisms, the population density and fear behavior, are difficult to tease apart, and we're working on that," says Ripple. "Many today believe it [the change in plant populations in Yellowstone] is due to a combination of the two." Beschta agrees: "In my opinion, they've both been going on."

Safety in Numbers and Colors

Elk may avoid lingering at streams as a way to evade their predators, but other types of prey have far more elaborate strategies to avoid being consumed. The poison dart frog, for example, is among the most toxic animals on Earth, and it evolved bright colors as **warning coloration** to alert potential predators to the dangerous chemicals in its tissues (**Figure 20.11**, top). Such warning coloration can be highly effective. Young blue jays, for example, quickly learn not to eat brightly colored monarch butterflies, which contain chemicals that cause nausea and, at high doses, death.

Then there are species that, though not poisonous, evolved coloration to make them look as if they were. Through **mimicry**, the viceroy butterfly, which is not poisonous, imitates the color and pattern of the monarch butterfly (**Figure 20.11**, middle). That "borrowed" coloration scares away blue jays and other birds that may have felt sick the last time they ate a monarch. Another mechanism to avoid being eaten is **camouflage**, any type of coloration or appearance that makes an organism hard to find or hard to catch (**Figure 20.11**, bottom). Finally, many prey, from musk oxen to wood pigeons, have evolved a different strategy to avoid becoming dinner: living together. By group living, these animals are able to act together to warn each other when a predator is

Warning coloration

The bright colors of the poison dart frog warn potential predators of the deadly chemicals contained in its tissues.

Mimicry

The viceroy butterfly (left) mimics the color and pattern of the monarch butterfly (right), which contains toxic compounds.

Camouflage

With its long legs outstretched, this lichen-mimic katydid, a relative of the cricket, sits motionless on lichen-covered branches to escape detection by predators.

Figure 20.11

Adaptive coloration responses to predation

Prey species have adapted many elaborate strategies to avoid being eaten by predators, including warning coloration, mimicry, and camouflage.

Q1: How do predators know that brightly colored prey are usually toxic?

Q2: Do you think mimicry works if the toxic species is in low abundance? Why or why not?

Q3: Why is camouflage considered an adaptive response to predation?

See Appendix A for answers to the figure questions.

Although a single musk ox may be vulnerable to predators such as wolves, a group forming a circle makes a difficult target.

Figure 20.12

Safety in numbers

Animals that live in groups are better able to warn each other and sometimes fend off attacking predators.

Q1: What percentage of pigeons are caught when they are alone and not in a flock?

Q2: For wood pigeons, what is the minimum number of individuals that provides protection from goshawks?

Q3: Why do you think a group of musk oxen versus a lone musk ox would be safer from a pack of wolves?

See Appendix A for answers to the figure questions.

about to attack and even to repel attacks as a united front (**Figure 20.12**).

In predation, it is the predator that benefits. And each of the other two types of species interactions we've discussed—mutualism and commensalism—benefits at least one of the species involved. But in one final type of species interaction, **competition**, no one benefits. Instead, both interacting species are negatively affected.

Competition most often occurs when two species share an important but limited resource, such as food or space. In Yellowstone, both beavers and elk eat woody plants. For both species, woody plants are part of the **ecological niche**, the set of conditions and resources that a population needs in order to survive and reproduce in its habitat. Because the niches of beavers and elk overlap, these species compete. When two (or three or more) species compete, each has a negative effect on the other because one is using resources that the other then cannot access. (If resources are abundant, however, there may be no competition between species, even if their niches overlap.)

There are two main categories of competition: exploitative and interference. In **exploitative competition**, species compete indirectly for shared resources, such as food. In this case, each species reduces the amount of the resource that is available for the other species, but the species do not directly interact or come in contact with each other (**Figure 20.13**). When wolves returned to Yellowstone and elk populations declined as a result, beavers had less competition for food, especially willow trees, and the number of beaver colonies in the park rose from 1 in 1996 to 12 in 2009. By 2015, Yellowstone had an estimated 100 beaver colonies.

Elk and bears also interact through exploitative competition. Elk eat the leaves and branches on shrubs, resulting in a decrease in the number of berries the shrubs produce. That's not good for grizzly bears, which love to eat berries. Knowing

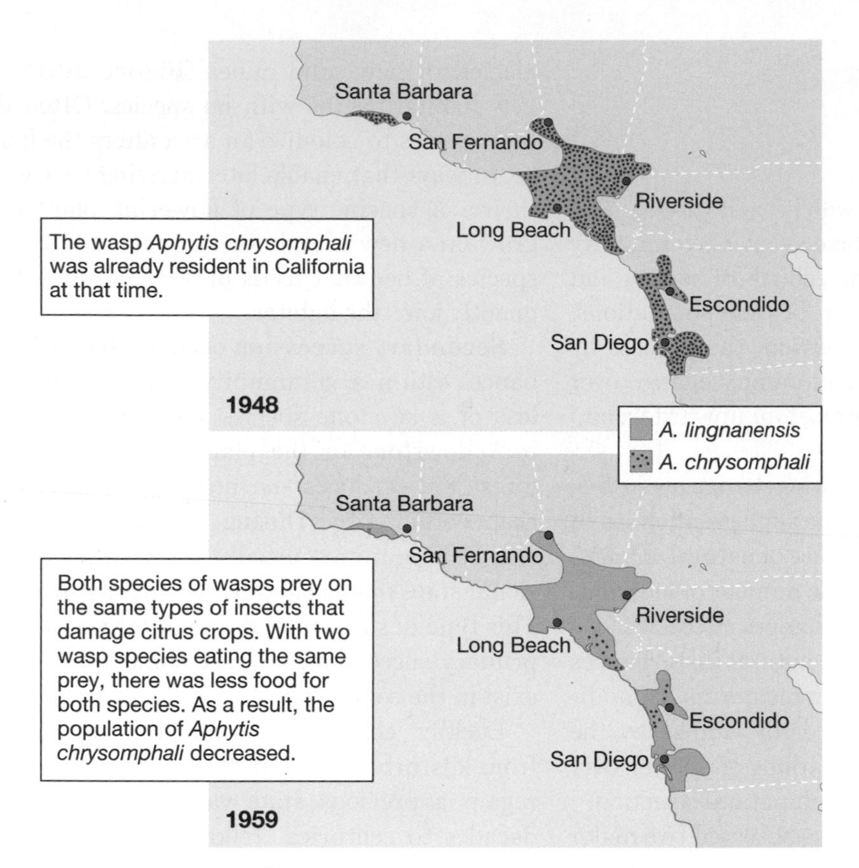
Santa Barbara
San Fernando
Riverside
Long Beach
Escondido
San Diego
The wasp Aphytis chrysomphali was already resident in California at that time.
1948
A. lingnanensis
A. chrysomphali
Santa Barbara
San Fernando
Riverside
Long Beach
Escondido
San Diego
Both species of wasps prey on the same types of insects that damage citrus crops. With two wasp species eating the same prey, there was less food for both species. As a result, the population of Aphytis chrysomphali decreased.
1959

Figure 20.13

Exploitative competition: a new species moves in

Two different species of wasps—*Aphytis lingnanensis* and *Aphytis chrysomphali*—feed on the same resource but do not directly compete for access to it.

Q1: Which species appears to be the superior competitor?

Q2: Why is this example considered exploitative competition?

Q3: What would you predict if these species had undergone competitive exclusion? (You will need to read the next paragraph to answer this question.)

See Appendix A for answers to the figure questions.

of this relationship, Ripple hypothesized that a decrease in elk would result in an increase in berry-producing shrubs and that bears would be eating more berries.

To test this hypothesis, Ripple, Beschta, and others spent 2 years analyzing grizzly scat that had been collected in the park. They compared the percentage of fruit in current scat to that of scat data that had been collected and saved before 1995, prior to the wolf return. Over a 19-year period, they found that the percentage of berries in the grizzly diet went up as elk populations went down.

This is an example of the **competitive exclusion principle**, which predicts that different species that use the same resource can coexist only if one of the species adapts to using other resources. In this case, grizzlies adapted to the presence of elk by finding food sources other than berries, and in the absence of elk they increased their consumption of berries. Organisms also compete through **interference competition**, in which one organism physically and directly excludes another from the use of a resource (**Figure 20.14**).

Figure 20.14

Interference competition can be dramatic

In Yellowstone bears and wolves often fight over the carcass of an animal.

A Community Restored

Today, the return of the wolf is having a clear and significant impact on the ecological community of Yellowstone, from the rebirth of aspens and cottonwoods to growing beaver populations. These are signs of **succession**, the process by which the species in a community change over time. "We're on a very important upward trend," says Beschta.

All ecological communities change over time, sometimes because of human intervention, as in Yellowstone, but also because of natural changes in species composition—the number of individuals in a population often changes as the seasons change, for example—and natural disturbances such as fires, floods, and windstorms. In addition, communities can broadly change by the slow loss or gain of populations of species over long periods of time through natural selection.

In nature, ecologists have observed two major types of succession: primary and secondary. **Primary succession** occurs in newly created habitats, such as an island that just emerged from the sea, the soil left behind a retreating glacier, or new sand dunes (**Figure 20.15**). A new habitat begins with no species. Often the first species to colonize an area alters the habitat in ways that enable later-arriving species to thrive. A specific type of flowering plant may grow on a new island, for example, and then a species of bee that feeds on that flower subsequently joins the habitat.

Secondary succession occurs after a disturbance within a community—for example, the loss of a keystone species such as the wolves in Yellowstone or the plant loss caused by a forest fire—reduces the number of species in that community. During secondary succession, communities usually regain the successional state that existed before the disturbance. This type of succession does not take as long as primary succession, because some species still exist in the community.

Luckily, communities can and do recover from disturbances, but the time required to regain a previous state varies from years to decades to centuries. Yellowstone is currently experiencing a secondary succession as wolf, aspen, and beaver populations slowly return. It will likely still be a while before the park returns to its previous state, says Ripple. "We were

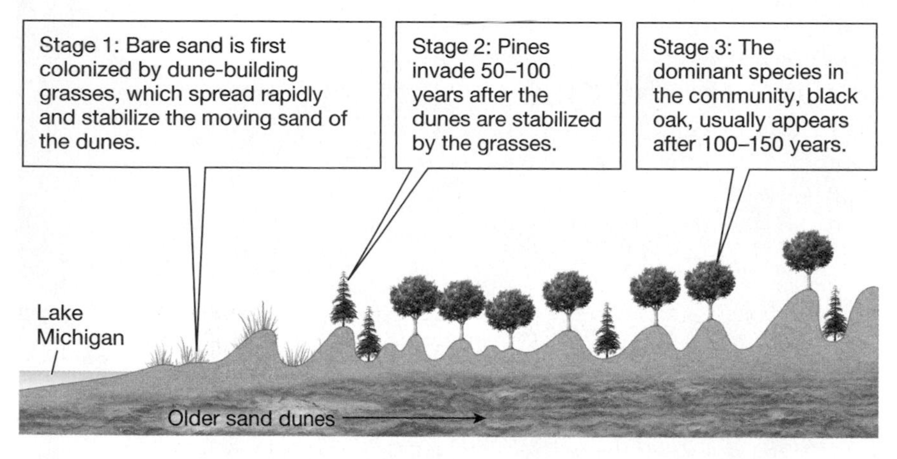

Stage 1: Bare sand is first colonized by dune-building grasses, which spread rapidly and stabilize the moving sand of the dunes.
Stage 2: Pines invade 50–100 years after the dunes are stabilized by the grasses.
Stage 3: The dominant species in the community, black oak, usually appears after 100–150 years.
Lake Michigan
Older sand dunes

Stage 4: Climax communities of black oak have lasted up to 12,000 years.

Figure 20.15

Primary succession: from nothing to climax community

Over thousands of years, sand becomes woodland near Lake Michigan.

Q1: What species represents the first colonizers of the sand dunes?

Q2: What species is the intermediate species, and how does it become the dominant species?

Q3: What species is the mature, climax community species, and how does it become the dominant species?

See Appendix A for answers to the figure questions.

In 1988, a large fire destroyed a portion of the mature lodgepole-pine forest in Yellowstone National Park.

By 1992, the lodgepole-pine forest regrowth was gaining momentum.

An example of a mature, climax community lodgepole-pine forest.

Figure 20.16

Secondary succession: from disturbance to climax community

Lodgepole-pine forest has been slowly but steadily growing again in Yellowstone National Park.

Q1: What other types of disturbances could you imagine destroying a forest?

Q2: How is secondary succession different from primary succession?

Q3: What is a climax community?

See Appendix A for answers to the figure questions.

70 years without wolves, and now we're less than 20 years since wolves have returned, so this is going to take time." But he and others are hopeful that Yellowstone will return to its status as a **climax community**, a mature community whose species composition remains stable over long periods of time (**Figure 20.16**).

Yet even as Yellowstone's recovery is under way, ecological communities around the planet are being threatened by the loss of other keystone species, especially large carnivores. In 2013, Ripple and colleagues analyzed 31 carnivore species around the globe, including leopards, lions, cougars, and sea otters. They found that more than 75 percent of those 31 species are declining and that 17 of them now occupy less than half of their former ranges. In 2016, they found that bushmeat hunting, in particular, is driving many mammal species to extinction, including 113 species in Southeast Asia.

Because of all the species interactions discussed here, it is clear that those changes will have major effects on ecological communities. "Humans are affecting predators around the globe in a major way," says Ripple. "It's a worldwide issue."

DEBUNKED!

MYTH: Wolves kill for sport.

FACT: Wolves kill for survival. If wolves kill more prey than they can eat at one time, they will come back and finish off that food source later.

NEED-TO-KNOW SCIENCE

- An **ecological community** (page 384) can be characterized by its species composition or diversity. This diversity has two components: **relative species abundance** (how common each species is when compared to others, page 385) and **species richness** (the total number of different species that live in the community, page 385).
- **Keystone species** (page 385) have a disproportionately large effect, relative to their own abundance, on the richness and abundances of the other species in a community. The removal or disappearance of these keystone species results in dramatic changes to the rest of the community.
- A **food chain** (page 387) is a single direct line of who eats whom among species in a community. A **food web** (page 387) depicts how overlapping food chains of a community are connected.
- **Producers** (page 388), organisms found at the bottom of a food chain that use light energy to produce their own food, are eaten by **consumers** (page 388). After **primary consumers** (page 388) eat producers, they are eaten by **secondary consumers** (page 389), which are eaten by **tertiary consumers** (page 389), which are eaten by **quaternary consumers** (page 389).
- Consumers are classified as **carnivores** (page 392), **herbivores** (page 392), or **omnivores** (page 392), depending on whether they eat animals or plants or both, respectively.
- Dead and dying organisms are **prey** (page 392) for many kinds of consumers. **Scavengers** (page 392) are omnivores or carnivores that hunt for dead and dying prey. **Decomposers** (page 392) dissolve their food, and **detritivores** (page 392) mechanically break apart and consume their food.
- **Species interactions** (page 390) in a community can be beneficial, harmful, or without benefit or harm to each of the interacting species.
- In **mutualism** (page 390), both species benefit.
- In **commensalism** (page 391), one species benefits at no cost to the other.
- In **predation** (page 391), one species benefits and the other is harmed. **Parasites** (page 391) are predators that live in or on their **hosts** (page 392).
- Strategies to avoid being consumed include **warning coloration** (page 394), **mimicry** (page 394), and **camouflage** (page 394).
- In **competition** (page 396), both species are harmed. Competition occurs when **ecological niches** (page 396) overlap and includes **exploitative competition** (page 396) and **interference competition** (page 397).

  The **competitive exclusion principle** (page 397) predicts that different species that use the same resource can coexist only if one of the species adapts to using other resources.
- **Succession** (page 398) establishes new communities (**primary succession**, page 398) and replaces disturbed communities (**secondary succession**, page 398). In stable environments without disturbances, called mature communities or **climax communities** (page 399), species composition remains stable over long periods of time.

THE QUESTIONS

See Appendix B for answers.

The Basics

1 A single sequence of feeding relationships describing who eats whom in a community is a

(a) life history.

(b) keystone relationship.

(c) food web.

(d) food chain.

2 The process of species replacement over time in a community is called

(a) global climate change.

(b) succession.

(c) competition.

(d) community change.

3 Organisms that can produce their own food from an external source of energy without having to eat other organisms are called

(a) suppliers.

(b) consumers.

(c) producers.

(d) keystone species.

4 A low-abundance species that has a large effect on the composition of an ecological community, especially when removed from that community, is called a

(a) predator.

(b) herbivore.

(c) keystone species.

(d) dominant species.

5 Select the correct terms:

A cheetah eats an antelope that ate some grass. The cheetah is a (**secondary consumer** / **primary consumer**), whereas the antelope is a (**secondary consumer** / **primary consumer**). The grass is a (**tertiary consumer** / **producer**).

Challenge Yourself

6 Wolves are considered a keystone species in Yellowstone because

(a) their removal in the early twentieth century caused many changes to the Yellowstone ecological community.

(b) their reintroduction in the late twentieth century caused many changes to the Yellowstone ecological community.

(c) when they were removed, elk populations increased, leading to increased competition with beavers and bears, which then declined.

(d) when they were reintroduced, aspen populations began to increase because of decreased predation from elk.

(e) all of the above

7 Link each species interaction with an example of the interaction.

mutualism	1. Elk graze on aspen.
commensalism	2. Bison allow magpies to perch on them; the birds eat ticks they find on the bison's bodies.
predation	3. Birds nest in aspen trees.
competition	4. Beavers and elk eat the same trees.

8 Place the following elements of the scientists' study of the relationship between aspen populations and wolf populations in the correct order from earliest to latest by numbering them from 1 to 5.

_____ a. Beschta hypothesized that the decimation of wolf populations in the early twentieth century allowed elk populations to grow, thus increasing grazing of the aspen by elk.

_____ b. Beschta observed there were few trees in the river valley.

_____ c. Ripple heard Beschta describe his observations in a talk.

_____ d. Ripple and his graduate student collected data showing that no new trees had grown in the valley since the 1920s.

_____ e. With the reintroduction of wolves into the park, elk populations declined and aspen populations rebounded.

9 Which ecological community would be more diverse: one with high relative species abundance or one with low relative species abundance? Which community would be more diverse: one with high species diversity or one with low species diversity?

Try Something New

10 When a female cat comes into heat and is ready to mate, she urinates more frequently and in a large number of places. Male cats from the neighborhood congregate near urine deposits and fight with each other for the female's attention and breeding rights. In what type of interaction are the male cats engaging?

(a) commensalism

(b) predation

(c) interference competition

(d) exploitative competition

(e) mutualism

11 Rabbits can eat many plants, but they prefer some plants over others. Assume that the rabbits in a grassland community containing many plant species prefer to eat a species of grass that happens to be a superior competitor. If the rabbits are removed from the region, predict whether relative species abundance will increase or decrease and whether species richness will increase or decrease.

12 Identify whether each of the following is an example of primary succession (P) or secondary succession (S) in Rocky Mountain National Park.

_____ a. A mountain slope was cleared of evergreen trees and is now sprouting aspen trees.

_____ b. A mountainside with mature evergreen trees has dominated the landscape for generations.

_____ c. An area of soil, sand, and rocks remains after a dam burst and flooded the area.

_____ d. Lichens and mosses grow on bare rock at upper elevations because of an increase in average yearly temperatures.

13 Analyze the food web shown in the accompanying figure and answer the following questions:

(a) Which species do *not* have a predator shown on the food web?

(b) Which species shown on the food web performs photosynthesis (captures light energy to make its own food)?

(c) Which species have only one predator and only one prey shown on the web?

(d) Which species has the most predators shown on the web?

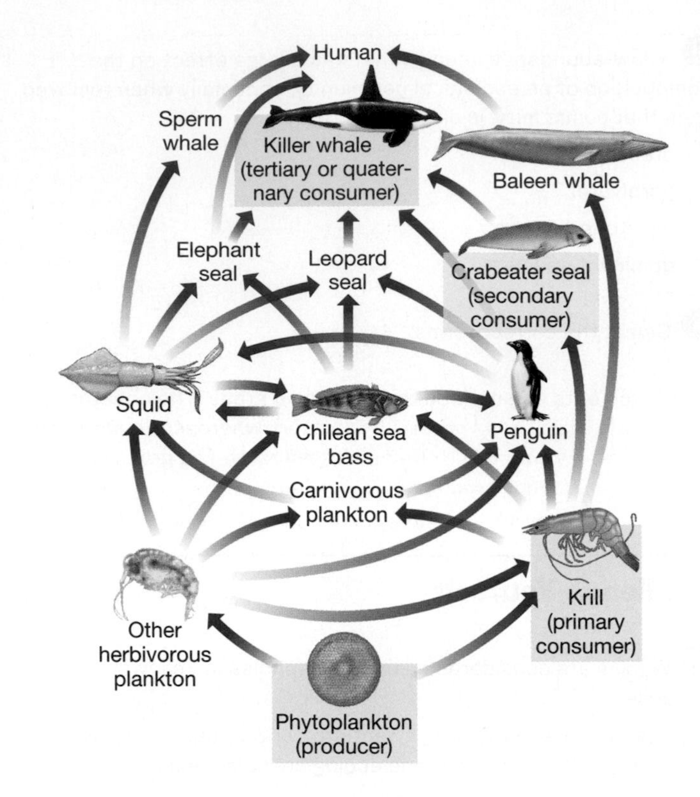

Leveling Up

14 **Looking at data** The accompanying graphs are from a study of the impact of insect predators on a food web involving an herbivore (willow beetles) and its food source (willow trees, just like in the chapter story!). The diversity (number of different kinds) of predators—wasps, predatory flies, and "crawlers"—was varied across experimental conditions. Please answer the following questions based on these data.

(a) What happened to beetle survival as the diversity of predators increased?

(b) What happened to the amount of willow ("plant biomass") eaten as the diversity of predators increased?

(c) How would you explain this result? Please compare it to the chapter story about the relationship between wolves and willows.

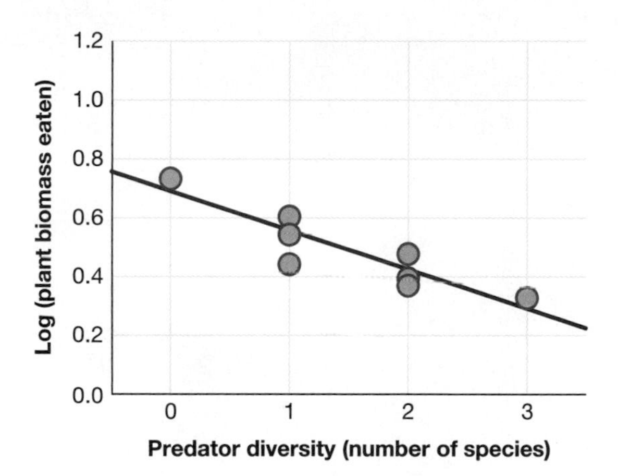

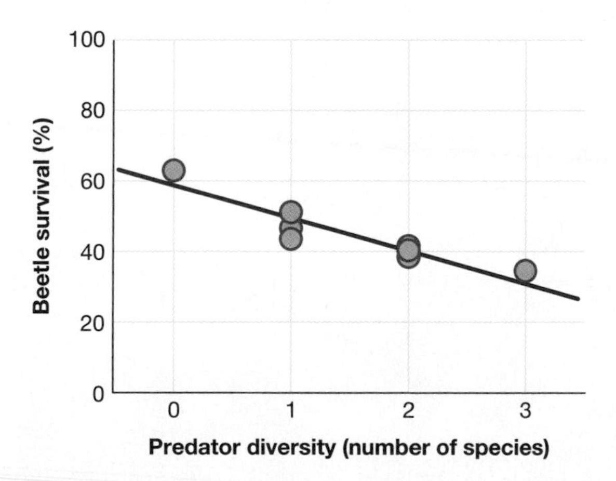

15 **Doing science** Citizen science is an amazing way for anyone to get involved in scientific research. Search the internet for citizen science projects relevant to this chapter's topics, using keywords such as "parasite," "predator," "prey," and "group living." Examples include Project Monarch Health (http://www.monarchparasites.org), in which volunteers sample wild monarch butterflies for a protozoan parasite to track its prevalence across North America. Participate in a project and, in writing, reflect on what you learned.

16 ***Write Now* biology: saving a species** Write a letter to your congressperson in which you use the concepts learned in this chapter to explain how wolves (or a different species near you) are beneficial to our wilderness ecosystems. Explain what role the species plays in the environment, what has caused it to be endangered or at risk now, what the likely effect will be if it goes extinct (locally or globally), and what could be done to ensure its survival.

Digital resources for your book are available online.

Here and Gone

Researchers discover an alarming decline of plankton in the ocean.

After reading this chapter you should be able to:

- Define an ecosystem and explain how it differs from an ecological community.
- Compare and contrast how nutrients and energy move through the environment.
- Place the members of a specified ecosystem into an energy pyramid at the appropriate trophic level.
- Identify where on Earth a specified biome is found, and compare its relative productivity to other biomes.
- Define primary productivity and explain its significance to studies of ecosystems.

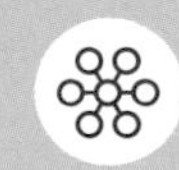

ECOLOGY

CHAPTER 21

ECOSYSTEMS

As a teenager, Daniel Boyce made extra cash working as a deckhand on fishing boats, so he grew comfortable on the ocean. Eventually, Boyce turned his interest in marine environments into a career, and in 2007 he joined the lab of marine biologist Boris Worm at Dalhousie University in Nova Scotia, Canada.

Worm had been investigating the decline of some fish communities. Several years earlier, the scientist had discovered that an industrialized fishing boom that began in the 1950s had decimated predatory fish communities. Boyce decided to follow up on his mentor's work by studying how that loss of predatory fish reverberated down the food chain. Had the disappearance of ocean predators affected plankton? Plankton, a diverse group of free-floating organisms, are the base of the ocean food web, supporting virtually all marine animals. Although they are tiny, they play a mighty role in marine environments.

When Boyce proposed the study idea, little did he know that it would uncover a profound shift in oceans around the world—a finding so shocking that just suggesting it would plunge him and his collaborators into a heated public controversy.

Going Green

Boyce's initial goal, similar to that of William Ripple's work studying wolves in Yellowstone National Park (see Chapter 20), was to show how the loss of a top predator affects the environment in which the predator lives. In Ripple's story, a variety of species living in Yellowstone interacted to form an ecological community. A large community of organisms interacting with one another and with the physical environment they share is an **ecosystem** (**Figure 21.1**). To say it another way, an ecosystem is characterized by interactions of organisms in the **biotic** (living) world with each other as well as with the **abiotic** (nonliving) world.

An ecosystem may be small or large; a puddle teeming with protists is an ecosystem, as is the

Figure 21.1

Ecosystem in Nova Scotia, Canada

Overfishing has decimated fish populations in Nova Scotia, affecting the entire ecosystem in which these fish reside. Both biotic and abiotic elements of the ecosystem have been affected.

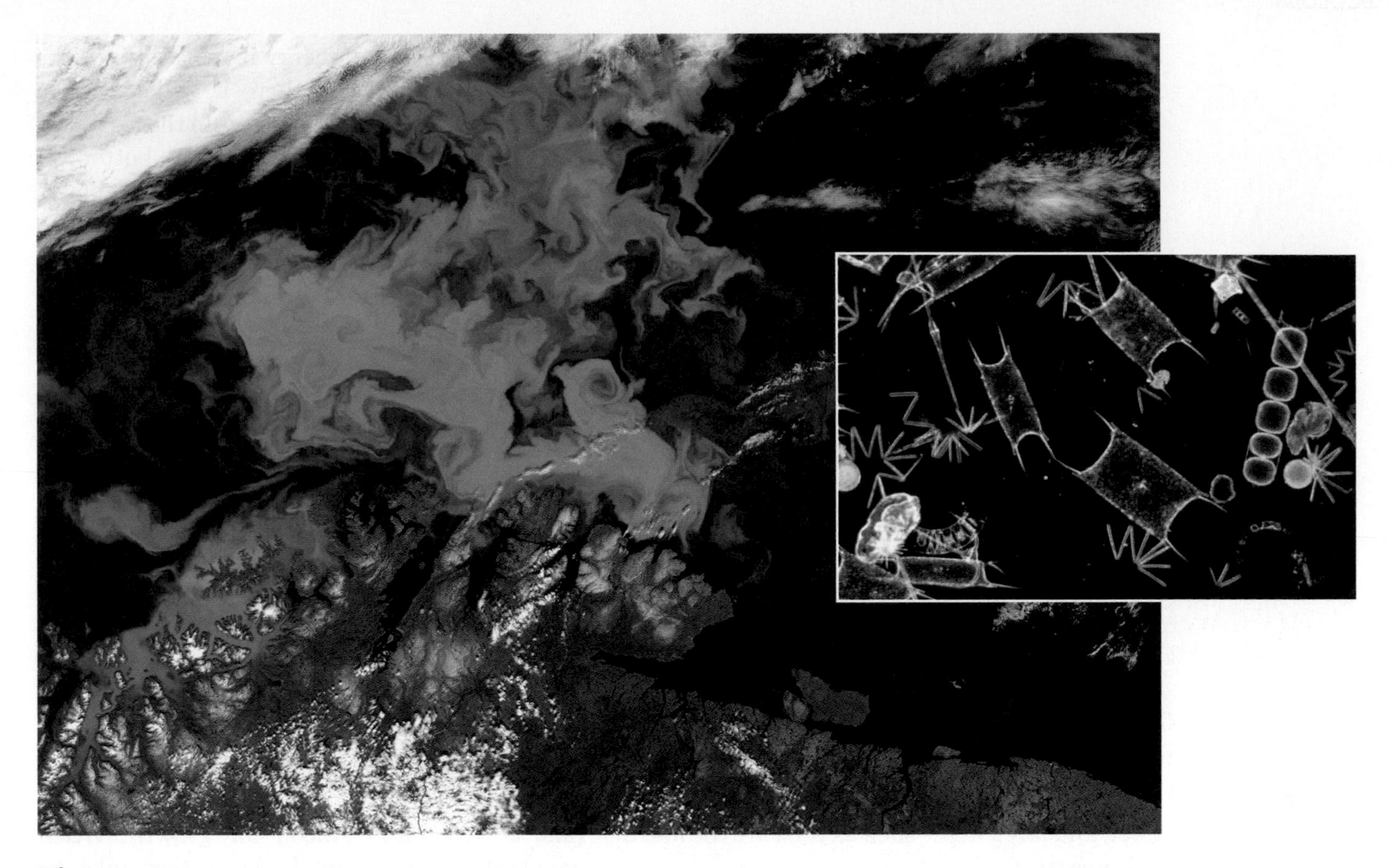

Figure 21.2

Phytoplankton bloom

The turquoise area in this aerial photo is a phytoplankton bloom occurring off the coast of Norway. When a population of phytoplankton (inset) increases rapidly, it discolors the water in which it resides.

Atlantic Ocean. And smaller ecosystems can be nested inside larger, more complex ecosystems. This variety means that ecosystems do not always have sharply defined physical boundaries. Instead, ecologists often define an ecosystem by the distinctive ways in which it functions, especially the means by which energy and nutrients are acquired and distributed by the biotic community.

The activity of primary producers, in particular, profoundly influences the characteristics of an ecosystem. Ecologists often describe an ecosystem according to the types of producers it contains and the consumers that the producers support. A duckweed-covered pond, a tallgrass prairie, and a beech-maple woodland are all examples of ecosystems that can be defined by the specific types of producers that capture and supply energy to consumers.

To see how overfishing was affecting food chains in the ocean, Boyce first looked to the primary producers of the ocean ecosystem: phytoplankton. These small, floating microalgae come in a fantastic array of shapes and sizes, from smooth orbs to segmented spirals to pointy crescents. Phytoplankton are primarily microscopic, but in large groups they form the green color often seen in water. The more phytoplankton in the water, the greener the water is; the less phytoplankton, the clearer it is (**Figure 21.2**).

Phytoplankton are green because they are photosynthetic. They convert light energy from the sun into chemical energy using chlorophyll, the green pigment critical to the process of photosynthesis. Because they are photosynthetic, these water-living organisms inhabit the top layer of water in the ocean (and almost every body of freshwater as well), a location that gives them easy access to sunlight. Thanks to their ability to photosynthesize using sunlight, phytoplankton are primary producers and the central means through which energy enters the ocean ecosystem. If overfishing was affecting phytoplankton levels, Boyce worried, the whole ocean ecosystem could be in danger.

Bottom of the Pyramid

Energy and nutrients flow through ecosystems in distinctive patterns. First, let's consider the path of energy. Primary producers such as phytoplankton capture energy from the sun and transform it into fuel energy. (There are primary producers in ecosystems centered on deep sea volcanic vents, and they convert chemicals released from the vents into energy.)

The fuel produced by primary producers is passed up the food chain as one organism eats another. An **energy pyramid** represents the amount of energy available to organisms in an ecosystem. Each level of the pyramid corresponds to a step in the food chain and is called a **trophic level**. In the ocean, for instance, phytoplankton are the first trophic level. Larger, multicellular plankton that feed on phytoplankton, known as zooplankton, are the second trophic level. Small fishes such as herring are the third level, and large fishes such as tuna are the fourth (**Figure 21.3**).

At each trophic level, a portion of the energy captured by producers is lost as **metabolic heat**, the heat released as a by-product of chemical reactions within a cell, especially during cellular respiration. Organisms lose a lot of energy as metabolic heat, as revealed by the fact that a small room crowded with people rapidly becomes hot; that warmth is the result of metabolic heat leaving our bodies. On average, roughly 10 percent of the energy at one trophic level is transferred to the next trophic level. The remaining 90 percent of the energy that is not transferred is either not consumed (for example, when we eat an apple, we eat only a small part of the apple tree), is not taken up by the consumer's body (for example, we cannot digest the cellulose contained in the apple), or is lost as metabolic heat.

Because of this steady loss of heat, energy flows in *only one direction* through ecosystems. It enters Earth's ecosystems from the sun (in most cases) and leaves them as heat. Therefore, energy cannot be recycled within an ecosystem. It travels up an energy pyramid, never down.

But we know that energy cannot be created or destroyed, so what happens to energy as it leaves an ecosystem? Heat energy is released into the atmosphere, and from there it will leave the planet. Other energy is held within the fossilized remains of organisms buried at the bottom of oceans or swamps and converted over geologic time into oil, gas, or coal (fossil fuels).

In contrast, **nutrients**—chemical elements required by living organisms—are recycled and reused within and across ecosystems. Although Earth receives a constant stream of light energy from the sun, our planet does not acquire more nutrients on a daily basis; rather, a constant and finite pool of nutrients cycles through the land, water, and air. If nutrients were not cycled between organisms and the physical environment, life on Earth would not exist.

Nutrients pass through the abiotic world, from rocks and mineral deposits into soil, water, and air, and then on to the biotic world via absorption by producers. Once in the biotic world, they are cycled among consumers for varying lengths of time. Phytoplankton, for example, require the nutrients nitrogen, phosphorus, iron, and silicon for growth. When zooplankton eat phytoplankton, they take up those nutrients, and so on up the food chain (**Figure 21.4**).

Nutrients are eventually returned to the abiotic world when **decomposers** break down the dead bodies of other organisms. In some ecosystems, decomposers break down 80 percent of the biomass, or biological material, made by producers. Without decomposers, nutrients could not be repeatedly reused, and life would cease because all essential nutrients would remain locked up in the bodies of dead organisms. In this way, decomposers are the "cleaners" of an ecosystem. Bacteria and fungi are important decomposers in the ocean, as are hagfishes, worms, and others.

Ecologists and earth scientists use the term "nutrient cycle" to describe the passage of a chemical element through an ecosystem (**Figure 21.5**). The nutrient cycle and the flow of energy are two of four processes that link the biotic and abiotic worlds in an ecosystem. These **ecosystem processes** also include the water cycle (see Figure 18.6) and succession, the process by which species in a community change over time (as discussed in Chapter 20).

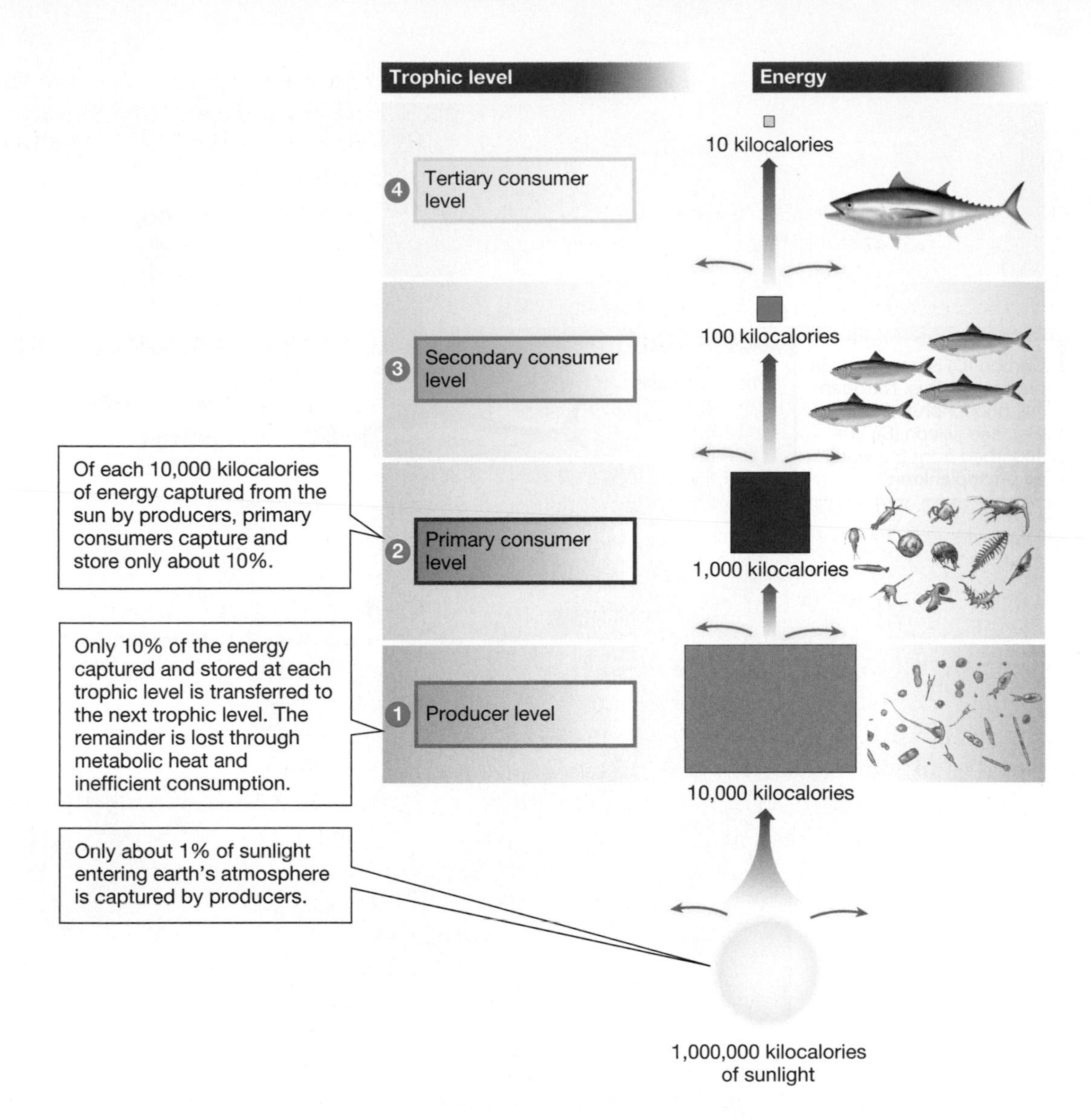
Trophic level
Energy
4 Tertiary consumer level
3 Secondary consumer level
2 Primary consumer level
1 Producer level
10 kilocalories
100 kilocalories
1,000 kilocalories
10,000 kilocalories
1,000,000 kilocalories of sunlight
Of each 10,000 kilocalories of energy captured from the sun by producers, primary consumers capture and store only about 10%.
Only 10% of the energy captured and stored at each trophic level is transferred to the next trophic level. The remainder is lost through metabolic heat and inefficient consumption.
Only about 1% of sunlight entering earth's atmosphere is captured by producers.

Figure 21.3

Energy pyramid

The levels of the energy pyramid correspond to steps in a food chain.

Q1: What percentage of the 10,000 kilocalories in the first trophic level will be available to a shark that might eat the tuna in the fourth trophic level?

Q2: What trophic level and term would describe a predator of tuna?

Q3: Give an example of a primary consumer in a terrestrial environment.

See Appendix A for answers to the figure questions.

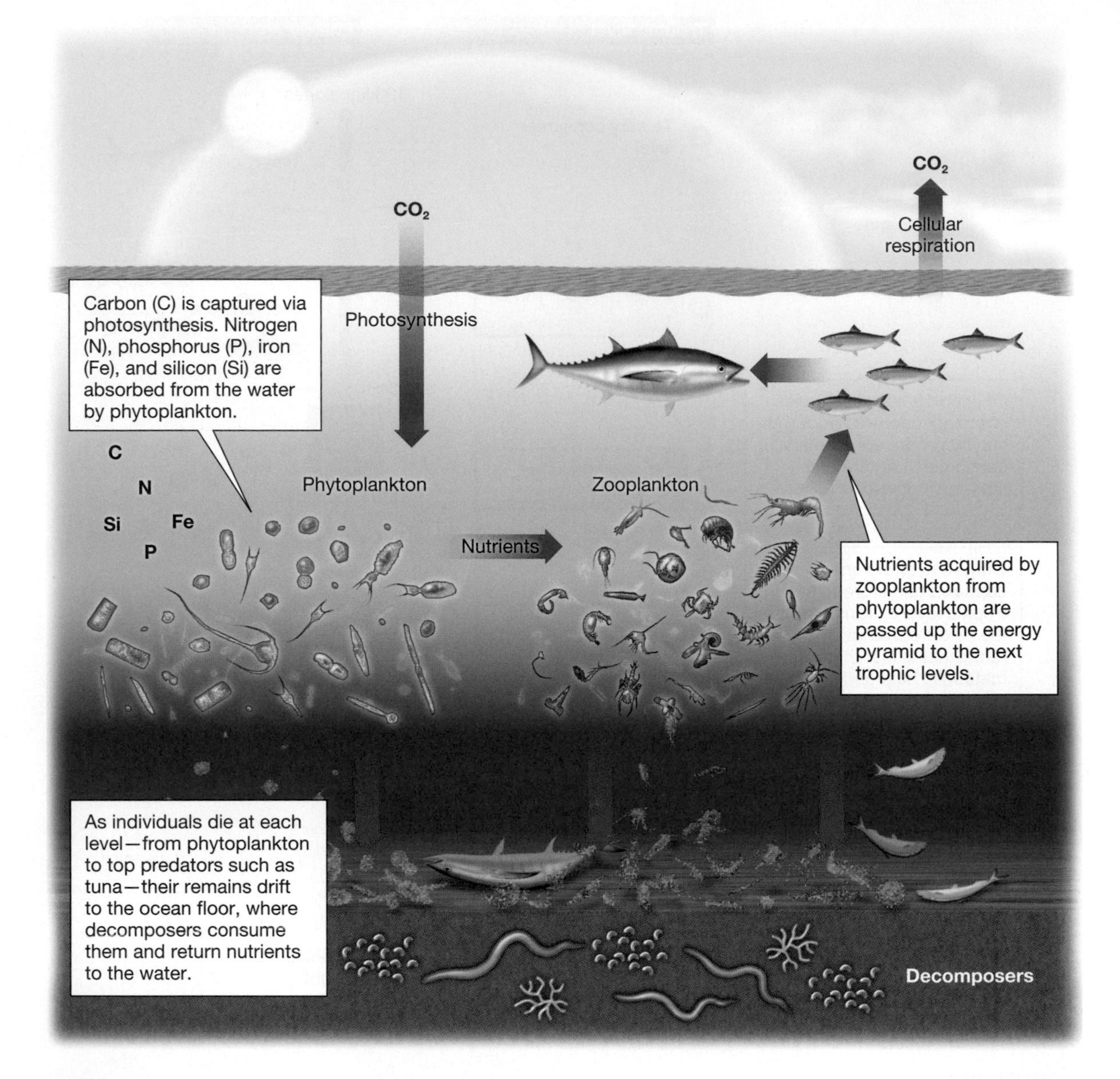

Figure 21.4

Nutrients cycle within and beyond the ocean ecosystem

In the ocean, phytoplankton absorb nutrients from the abiotic world. Within the biotic world, these nutrients are cycled among consumers, beginning with zooplankton and moving up through trophic levels. The nutrients then return to the abiotic world when dead or dying organisms are broken down by decomposers into their constituent elements.

Q1: Which organisms are the producers in this ecosystem?

Q2: Which organisms are responsible for nutrient flow from the biotic world to the abiotic world? (*Hint*: Read the next paragraph.)

Q3: How do nitrogen and phosphorus move from abiotic to biotic elements of this ecosystem? How would that differ in a terrestrial system?

See Appendix A for answers to the figure questions.

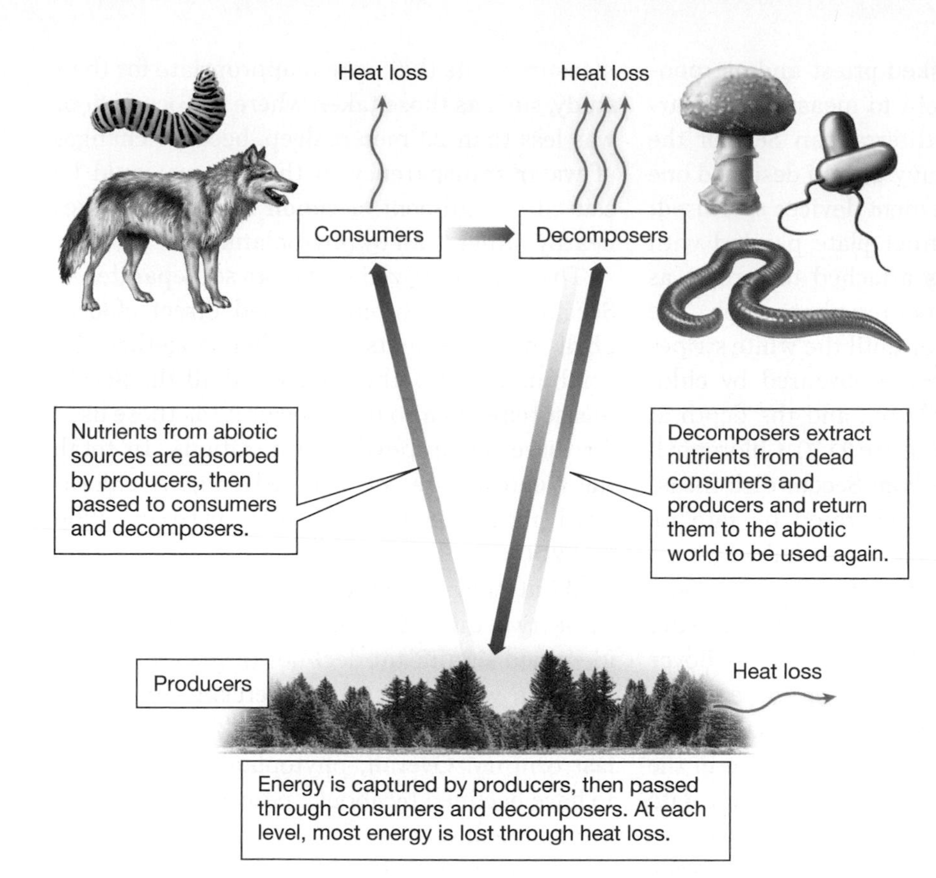

Heat loss
Heat loss
Consumers
Decomposers
Nutrients from abiotic sources are absorbed by producers, then passed to consumers and decomposers.
Decomposers extract nutrients from dead consumers and producers and return them to the abiotic world to be used again.
Producers
Heat loss
Energy is captured by producers, then passed through consumers and decomposers. At each level, most energy is lost through heat loss.

Figure 21.5

Energy flow and nutrient cycling

Unlike energy, which moves up and out of ecosystems, nutrients are constantly cycled between the abiotic and biotic worlds. Important nutrients for the biotic world include carbon (C), potassium (K), phosphorus (P), and nitrogen (N).

Q1: How is a decomposer different from a more typical consumer?

Q2: What would happen to nutrient cycles if decomposers did not exist?

Q3: Describe all the points at which heat is lost from the ecosystem in this figure.

See Appendix A for answers to the figure questions.

A Multitude of Measurements

For over 100 years, researchers around the globe have studied ecosystems containing phytoplankton. Boyce tapped into that wealth of research to document past and present levels of phytoplankton in the oceans.

The amount of phytoplankton biomass in a given area can be estimated by the concentration of chlorophyll found there (**Figure 21.6**). For decades, nearly all ocean studies have used chlorophyll concentration as a reliable metric of phytoplankton biomass. Chlorophyll concentration is measured by detecting the color of water. Water takes on deeper shades of green as the amount of chlorophyll increases. When there is no chlorophyll, water appears clear.

Ideally, Boyce would have used satellite data to detect chlorophyll and thus phytoplankton concentrations, as satellites today take high-resolution color measurements of the ocean surface. Yet Boyce planned to review phytoplankton levels over the past 100 years, and high-quality satellite data have been available for only the last decade. He needed another source of data.

In a first-of-its-kind analysis, Boyce, together with his adviser Boris Worm and the oceanographer Marlon Lewis, combined two types of chlorophyll measurements. The first type, dating all the way back to 1899, was recorded with nothing more than a rope and a disk.

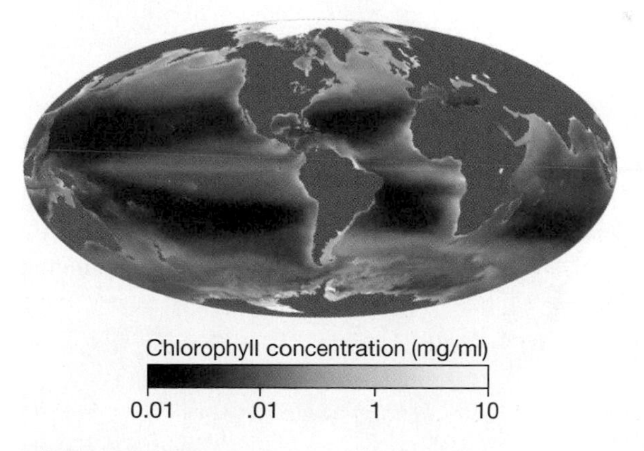

Figure 21.6

Average chlorophyll concentration in the oceans

As indicated by the yellow areas on this map, phytoplankton are most abundant in high latitudes, along coastlines and continental shelves, and along the equator in the Pacific and Atlantic Oceans. As indicated by the dark blue patches, these organisms are scarce in remote oceans.

In 1865, the pope asked priest and astronomer Pietro Angelo Secchi to measure the clarity of water in the Mediterranean Sea for the purposes of the papal navy. Secchi designed one of the simplest measurement devices ever used: a disk the size of a dinner plate painted with black and white stripes attached to a rope, as seen in this chapter's opening photograph. The disk is lowered into water until the white stripes disappear (as they become obscured by chlorophyll from phytoplankton), and the depth at that point is recorded (**Figure 21.7**). Chlorophyll concentrations derived from Secchi disk measurements have been corroborated by satellite data, so scientists know they are reliable.

In addition to gathering Secchi disk data, scientists at sea use lab tools to directly measure the quantity of chlorophyll in the water. Boyce found hundreds of thousands of these direct chlorophyll measurements online in open-source databases. "There's been a huge increase in the amount of publicly available oceanographic data out there," says Boyce.

But to use the data, Boyce first had to separate the wheat from the chaff. "With any big database, there are bound to be measurements that are entered incorrectly, for whatever reason," says Boyce. He, Worm, and Lewis ruled out measurements that were inappropriate for their study, such as those taken where the ocean floor was less than 25 meters deep, because changes of water transparency in those cases could be caused by sediment or runoff from landmasses nearby rather than by phytoplankton.

The team analyzed each data set separately—Secchi disk measurements and direct chlorophyll measurements—and then together. To combine the two, they converted all the Secchi measurements into the same units as those used for direct chlorophyll concentrations. In total, the blended data set included 445,237 chlorophyll measurements collected between 1899 and 2008.

What they found made them pause. With those two different methods of analysis, they identified significant declines in phytoplankton levels—a whopping 60–80 percent—in Earth's oceans where data were available during the last century. Overall, phytoplankton appeared to have declined by about 1 percent of the global average each year.

One percent sounds like a small number, but that amount every year just since 1950 translates into a staggering total phytoplankton decrease of 50 percent in the world's oceans.

Figure 21.7

Indirectly measuring chlorophyll concentration

The Secchi disk is lowered into water until its white stripes become obscured by the chlorophyll in phytoplankton. The length of line underwater is used as an estimate of chlorophyl concentration.

Net Loss

A 50 percent loss in the main producer in any ecosystem is a worrisome number but especially with respect to phytoplankton. Phytoplankton support fisheries, produce half the oxygen we breathe, and take in carbon dioxide from the atmosphere, which helps offset the greenhouse effect and global warming.

An ocean with less phytoplankton will function differently because ecosystems depend on **energy capture**, the trapping and storing of solar energy by the producers at the base of the ecosystem's energy pyramid. Herbivores, carnivores, and detritivores all depend indirectly on energy capture. If an ecosystem has an abundance of producers, it can often support more consumers at higher trophic levels. In a tropical forest, for example, an abundance of plants captures energy from the sun, and the forest teems with life.

On the flip side, relatively little energy is captured in an environment with few producers.

In tundra or desert regions, for example, less food is available, and fewer animals can live there. These significant differences have prompted ecologists to categorize large areas of Earth's surface into distinct regions, called **biomes**, which are defined by their unique climatic and ecological features. Biomes do not always begin and end abruptly but rather transition into one another. Terrestrial biomes are categorized by temperature, precipitation, and altitude; aquatic biomes are determined by proximity to shorelines (**Table 21.1**).

Assessing the overall amount of energy captured by producers is important in determining how an ecosystem works because energy capture influences the amount of food available to other organisms. The **primary productivity** of an ecosystem is the energy, acquired through photosynthesis over a particular time period, available for the growth and reproduction of producers. Primary productivity is typically determined by estimating the amount of carbon captured during photosynthesis. This can be done by measuring the amount of new biomass produced by the photosynthetic organisms in a given area during a specified period of time.

According to scientists' estimates, the productivity of all producers on Earth exceeds 100 billion tons of carbon biomass per year. Roughly half of this productivity comes from

Table 21.1

Amazing Biomes

Biome	Type	Location	Description	Typical Vegetation	Typical Animals
Tundra	Terrestrial	Polar, mountaintops	Very cold and dry year-round	Low-growing flowering plants, mosses, lichens	Herbivores: rodents Carnivores: foxes, wolves Large mammals: bears, musk oxen
Boreal forest	Terrestrial	Northern Hemisphere, 50–60° N	Cold, dry winters and mild summers	Coniferous trees, low plant diversity	Large herbivores: elk, moose Small carnivores: weasels, wolverines, martens Larger carnivores: lynx, wolves
Temperate forest	Terrestrial	25–50° latitude in both hemispheres, rich soils	Snowy winters and humid, warm summers	Greater diversity than tundra and boreal forest: oak, maple, hickory, beech, elm	Herbivores: squirrels, rabbits, deer, raccoons, beavers Carnivores: bobcats, mountain lions, bears Amphibians and reptiles
Grassland	Terrestrial	Across latitudes in the middle of continents	Arid but less so than desert; many areas converted to agriculture	Grasses and herbaceous plants with scattered trees	Burrowing rodents: voles, prairie dogs
Chaparral	Terrestrial	Western coasts of continents; Mediterranean	Cool, rainy winters and hot, dry summers	Evergreen shrubs and small trees	Small mammals: jackrabbits, gophers Lizards and snakes

Continued

Table 21.1

Amazing Biomes—Continued

Biome	Type	Location	Description	Typical Vegetation	Typical Animals
Desert	Terrestrial	Extremely arid regions	High daytime and low nighttime temperatues	Plants with small leaves that minimize heat loss and store water in fleshy stems or leaves	Nocturnal animals such as tarantulas, kangaroo rats, owls, and coyotes—hiding in burrows during day and emerging to feed at night
Tropical forest	Terrestrial	Near equator	Seasonally heavy or year-round rain	Rich variety that locks up nutrients (so soil is nutrient poor)	Rich diversity including poison dart frogs, parrots, boa constrictors, and jaguars; tropical rainforests are among Earth's most productive ecosystems
Freshwater	Aquatic	Nonsaline water bordering or flowing through terrestrial biomes	Lakes: landlocked bodies of standing water Rivers: continuously moving Wetlands: shallow standing water Bogs: wetlands with stagnant, oxygen-poor water	Diverse, with low productivity in bogs but high productivity in wetlands such as grassy marshes and tree-filled swamps	Range from low species diversity in bogs to high species diversity in wetlands such as marshes and swamps; besides fish, there are snails, turtles, beavers, and alligators
Estuary	Aquatic	Where rivers flow into the ocean	Tidal ecosystem, with constant ebb and flow of fresh and salt water	Rich and diverse photosynthesizers due to plentiful light, abundant nutrients from rivers, and regular stirring of nutrient-rich sediments; primarily grasses and sedges	Organisms able to tolerate daily changes in saltwater concentrations, such as salmon, crabs and oysters, and otters
Marine	Aquatic	Largest biome, consisting of saltwater areas across the planet	Coastal: from shoreline to edge of continental shelf Intertidal: closest to shore Open ocean: begins about 40 miles (64 km) offshore	Coastal: highly productive because rich in nutrients and oxygen Intertidal: plants that endure being submerged and exposed to dry air twice daily Open ocean: less productive because nutrient poor	Coastal: most marine species live here Intertidal: seaweeds, worms, crabs, sea stars (starfish), sea anemones, mussels, etc. Open ocean: whales, dolphins, pelagic fishes such as tuna

Data on net primary production (NPP)—the total energy available in an ecosystem for the growth and reproduction of primary producers—is sparse for whole biomes, but researchers have been able to estimate the NPP of those listed here using information about different vegetation types and carbon sources in each area.

Assessment available in smartwork

Total energy available from primary producers

(in grams per square meter per year)

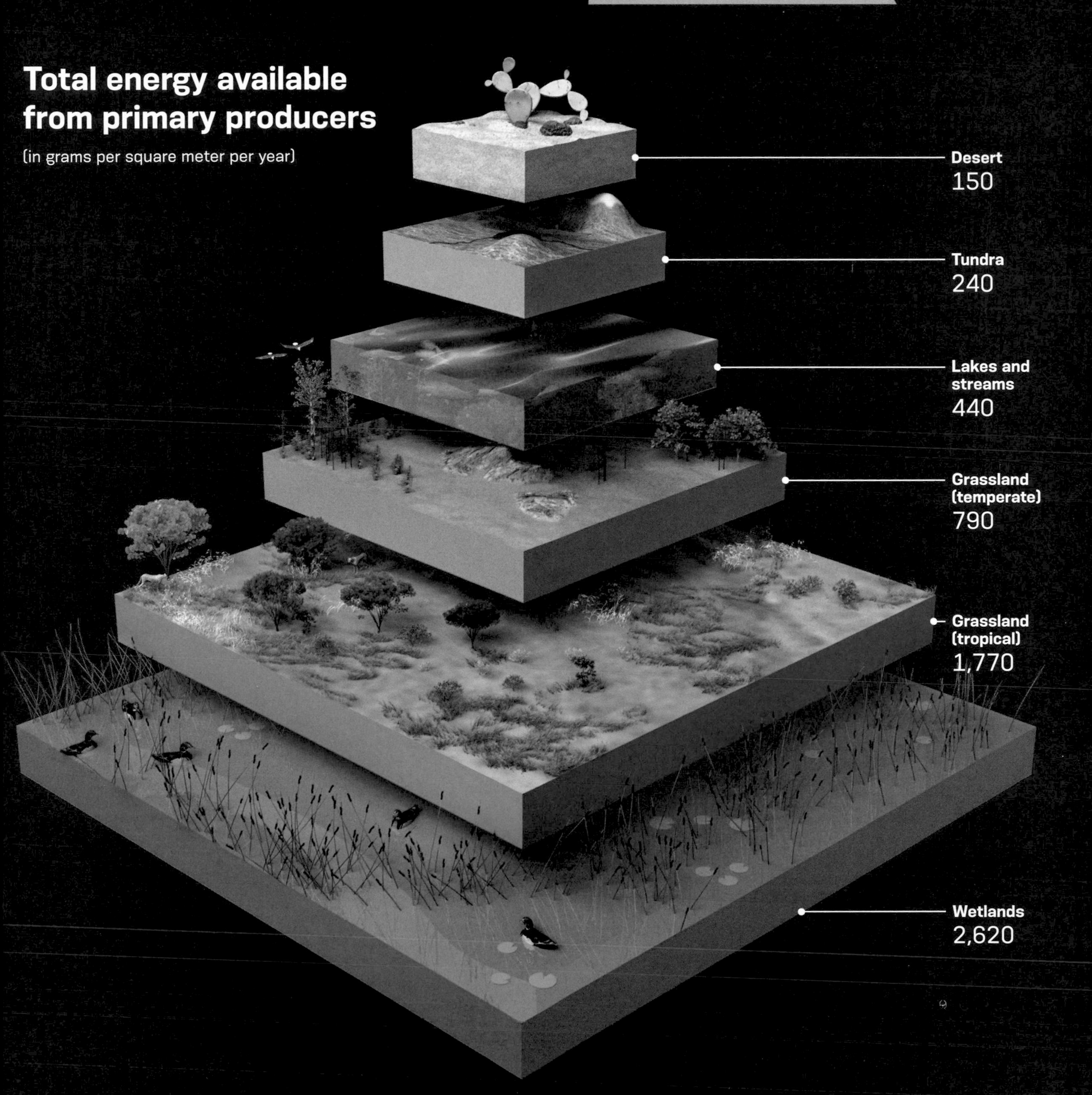

phytoplankton in the ocean. Therefore, phytoplankton capture approximately 50 billion tons of carbon per year. So if Boyce's calculations are right, a loss of 1 percent of that biomass is 500 million tons of organic matter lost from the oceans *each year*. That's a lot of biomass to lose.

Primary productivity relies on four things: sunlight, water, temperature, and the availability of nutrients. The most productive ecosystems on land are tropical forests; the least productive are deserts and tundra (including some mountaintop communities). The most productive ecosystems in water are estuaries—regions where rivers empty into the sea—because nutrients drained off the land stimulate the growth and reproduction of phytoplankton and other producers, which in turn nourish large populations of consumers. The least productive aquatic biome is the deep ocean, where sunlight does not penetrate.

Despite similarities between the productivity requirements on land and in the ocean, the global pattern of productivity differs between the two. On land, productivity is highest at the equator and decreases toward the poles. But in the ocean, the general pattern relates not to latitude but to distance from shore. The productivity of marine ecosystems is typically high in ocean regions close to land and relatively low in the open ocean (**Figure 21.8**). This is because nutrients needed by aquatic photosynthetic organisms are in better supply near land, thanks to delivery from streams and rivers. Wetlands such as swamps and marshes, which trap soil sediments rich in nutrients and organic matter, can be so productive that they match the productivity levels of tropical forests.

A loss of 500 million tons of phytoplankton each year could potentially affect the ocean's productivity and ocean life. "Almost all biological life in the ocean depends on phytoplankton. A reduction in the biomass of phytoplankton will result in less secondary production in the oceans," says Boyce. That means fewer sharks, whales, fishes, eels—you name it.

The team's discovery was shocking, to say the least. No one else had documented a global decline in phytoplankton before. So, to be confident in their results, Boyce and Worm did several more rounds of data analysis, checking again and again to make sure they were using the right numbers in ways that correctly represented what was happening in the natural world. And over and over, they came back with the same results: global phytoplankton declined over the last century. In 2010, they published that finding in the peer-reviewed scientific journal *Nature*.

Phyto-Fight

The scientific community reacted immediately. Some researchers doubted Boyce's conclusions; others were outright incredulous. Paul Falkowski at Rutgers University told a *New York Times* reporter that he had not found the same trend in a long-term analysis of the North Pacific (though Boyce contends that the teams' trends were very similar), and another team had actually seen an increase in phytoplankton starting around 1978 in the central North Pacific. Falkowski called Boyce's paper "provocative" but said he would "wait another several years" to see whether satellite data would back up the finding.

Then, in 2011, three separate research teams went so far as to publish formal critiques of the work. One team suggested that the declining trend was an error resulting from the use of two different types of measurements: Secchi disk readings and direct chlorophyll measurements. A second team echoed that idea, reanalyzed the data in a way that showed an increase in phytoplankton, and then bluntly concluded, "Our results indicate that much, if not all, of the century-long decline reported [by Boyce] is attributable to this [sampling bias] and not to a global decrease in phytoplankton biomass."

The third team noted that Boyce's finding conflicted with eight decades of data on phytoplankton biomass collected by a large project called the Continuous Plankton Recorder (CPR) survey, which monitors the Northeast Atlantic Ocean. The CPR survey, started in 1931, employs a unique instrument pulled through the ocean by commercial fishing vessels to collect millions

DEBUNKED!

MYTH: The deep sea is totally dark.

FACT: Although sunlight penetrates only about 3,200 feet into the ocean depths, there is another type of light further down—glimmering blue and green bioluminescence produced by a special enzyme in the bodies of bacteria and some deep-sea fish.

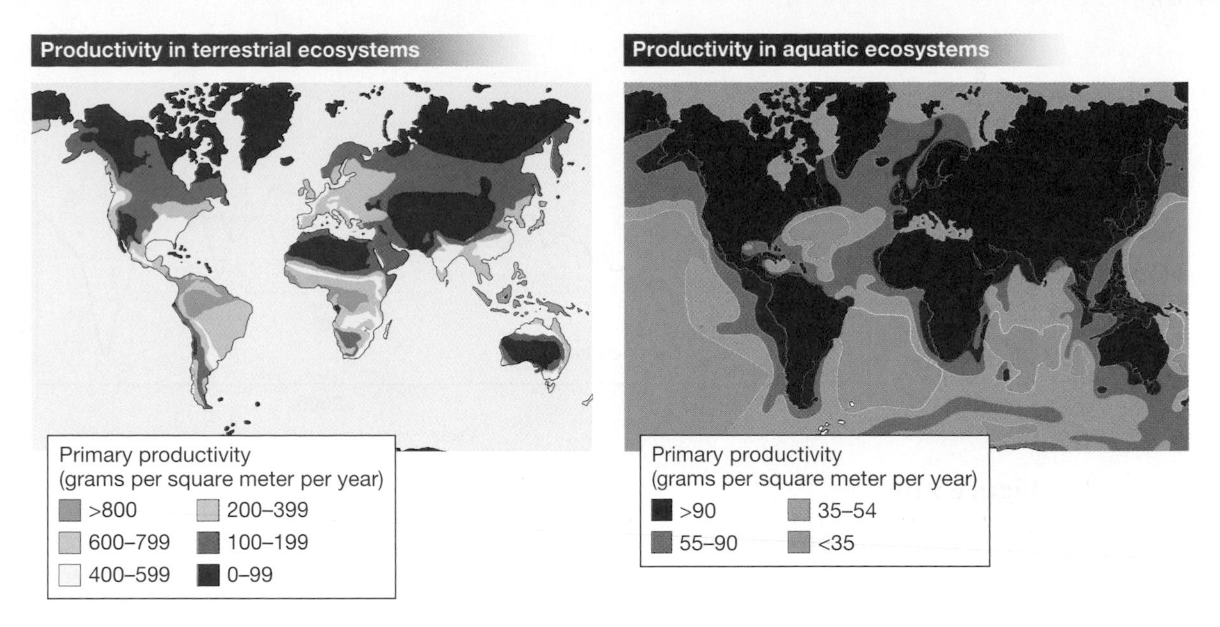

Figure 21.8

Global variation in primary productivity

Primary productivity can be measured as the number of grams of new biomass made by producers each year in a square meter of each biome's area. Productivity varies greatly across both terrestrial and aquatic ecosystems.

Q1: Which terrestrial biome has the lowest productivity? Which aquatic biome has the lowest productivity?

Q2: Where are the most productive terrestrial biomes located?

Q3: Give a possible reason for your answer to question 2.

See Appendix A for answers to the figure questions.

of samples of plankton. The CPR survey found that over the last 20–50 years, phytoplankton biomass increased in the Northeast Atlantic, says Abigail McQuatters-Gollop, a former researcher at the Sir Alister Hardy Foundation for Ocean Science, which operates the survey.

In response, Boyce, Worm, and Lewis didn't get angry; they got focused. After reading the critiques, the three researchers went back to the data. First they applied a correction factor suggested by the critics, in the hopes of removing any bias between the two types of data. "We did that, and the trends remained similar," says Boyce. Next, they again estimated changes over time individually for the two data sources. "That didn't change the trends either," says Boyce. After that, they incorporated additional suggestions by their peers and created a new, expanded database of chlorophyll measurements. Finally, they reestimated changes in chlorophyll using this new database and their revised analysis methods, but the phytoplankton still seemed to be declining, independently of the type of data or how the data were analyzed. The three researchers published their reanalysis in a series of three papers in 2011, 2012, and 2014, demonstrating the same decline again and again.

But the additional work has not silenced the critics. "It's still pretty hotly debated," admits Boyce. "The story is not over by any means." In 2011, in fact, Worm traveled to an international plankton conference where he, McQuatters-Gollop, and others debated the topic in front of a live audience. "It was an amicable meeting that generated loads of discussion," says McQuatters-Gollop. Still, they could not agree on whether phytoplankton populations have increased or decreased in the ocean. At the meeting, the researchers agreed that the best thing to do was to combine as many data sets as possible.

Heating Up

In 2015, the long-awaited satellite data came in—and they weren't good. Ocean color measurements from two NASA satellites led

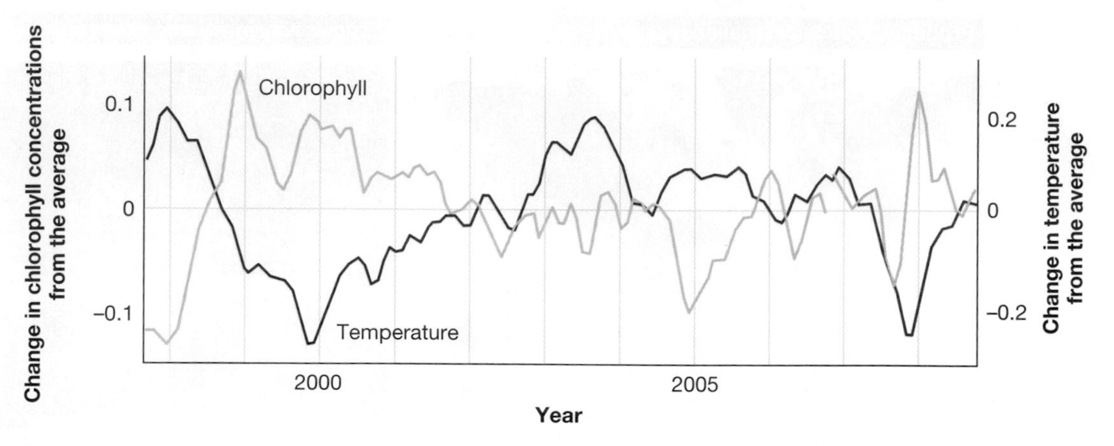

Figure 21.9

As ocean temperature increases, chlorophyll decreases

Between late 1997 and mid-2008, satellites observed that warmer-than-average temperatures (red line) were correlated with below-average chlorophyll concentrations (green line).

Q1: In what years were chlorophyll levels the highest?

Q2: In what years were the temperature changes from the average the greatest?

Q3: How do you predict this graph will look 10 years from now? Explain your reasoning.

See Appendix A for answers to the figure questions.

scientists to conclude that diatoms, the largest type of phytoplankton, had declined more than 1 percent per year from 1998 to 2012 globally, with major losses in the North Pacific, North Indian, and Equatorial Indian Oceans. And in 2019, a team at Massachusetts Institute of Technology and Woods Hole Oceanographic Institution documented a steady decline of phytoplankton's productivity in the North Atlantic, one of the world's most productive marine basins. The decline in productivity, the authors note, coincides with rising sea surface temperatures.

If the documented decline in phytoplankton trend is correct, a key question remains: Why is the decline occurring? As part of his research, Boyce investigated possible causes of the downward trend. In one study, he compared changes in sea surface temperature to changes in chlorophyll levels and noted a strong correlation. Over the last 100 years, chlorophyll concentrations declined *and* ocean temperatures increased, in line with increases in global warning. This correlation has been closely followed in recent years (**Figure 21.9**).

But correlation does not prove causation, and much more work needs to be done to test the idea that global warming caused the decline. It is a logical hypothesis, however, because as the planet warms, water in the oceans mixes less, limiting the nutrients delivered to the surface from decomposers in the deep sea. As a result, phytoplankton will not receive the nutrients they need for growth and reproduction. In 2014, Boyce published experimental data supporting the hypothesis. Working with marine scientists in Germany, he found that warming the water in a controlled, experimental ocean water enclosure led to reduced phytoplankton biomass.

A 2014 study by a large international group of researchers predicted that an increase in ocean temperature due to global warming would cause phytoplankton biomass to decrease by 6 percent by the end of this century. Consequences of continued phytoplankton decline could include altering the carbon cycle between the ocean and the atmosphere, changing heat distribution in the ocean, and causing a decrease in the supply of food in the ocean. Whatever the case, if phytoplankton populations are decreasing, says Boyce, our planet will experience profound effects, which have only just begun.

NEED-TO-KNOW SCIENCE

- An **ecosystem** (page 406) consists of a large community of organisms and the physical environment in which those organisms live. It is the sum of all **biotic** (page 406) elements interacting with each other and with all **abiotic** (page 406) elements. Energy, materials, and organisms can move from one ecosystem to another.
- An **energy pyramid** (page 408) represents the amount of energy available to each **trophic level** (page 408) of a food chain in an ecosystem.
- Energy enters an ecosystem when producers capture it from an external source, primarily the sun. A portion of the energy captured by producers is lost as **metabolic heat** (page 408) at each trophic level and is eventually released into the atmosphere and then the planet. As a result, energy flows in only one direction through ecosystems.
- **Nutrients** (page 408) are the chemical elements required by living organisms. Unlike energy, nutrients are recycled and reused within and across ecosystems. Earth has a fixed amount of nutrients.
- **Decomposers** (page 408) break down the dead bodies of other organisms, both consumers and producers, releasing the nutrients in the bodies of dead organisms back to the physical environment.
- Four **ecosystem processes** (page 408) link the biotic and abiotic worlds in an ecosystem: nutrient cycling, energy flow, water cycling, and succession.
- Ecosystems depend on **energy capture** (page 412), the trapping of solar energy by producers via photosynthesis, and the storage of that energy as chemical compounds in their bodies.
- Earth is categorized into 10 major **biomes** (page 413), regions defined by their climatic and ecological features.
- **Primary productivity** (page 413) is the energy acquired through photosynthesis that is available for growth and reproduction to producers in an ecosystem. It is estimated by the amount of biomass produced in a given area during a specified period of time.

THE QUESTIONS

See Appendix B for answers.

The Basics

1 The movement of nutrients between organisms and the physical environment is called

(a) nutrient cycling.
(b) ecosystem services.
(c) primary productivity.
(d) decomposition.

2 How much energy is transferred up the energy pyramid from one trophic level to the next?

(a) 90 percent
(b) 50 percent
(c) 10 percent
(d) 10–50 percent

3 Which organisms are considered the "recyclers" of our planet?

(a) consumers
(b) producers
(c) phytoplankton
(d) decomposers

4 The terrestrial biome that receives the most consistent year-round rainfall is

(a) wetland.
(b) boreal forest.
(c) tropical forest.
(d) chaparral.

5 Link each term with the correct definition.

biome	1. The energy acquired through photosynthesis by producers of an ecosystem.
ecosystem	2. A large community of organisms interacting with one another and with the physical environment they share.
primary productivity	3. A large, distinct region defined by its unique climatic and ecological features.
ecological community	4. The populations of different species that live and interact with one another in a particular place.

Challenge Yourself

6 Select the correct terms:

The biome characterized by shrubs and small trees that grow in regions with cool, rainy winters and hot, dry summers is (**tundra** / **chaparral**). Another biome with few trees, but in this case dominated by grasses, is (**grassland** / **tundra**). The largest biome is (**deserts** / **marine**).

7 Which of the following is a component of an ecosystem but *not* of an ecological community?

(a) a producer

(b) water

(c) a secondary consumer

(d) a primary consumer

8 Place the following elements of the scientists' study of oceanic phytoplankton levels in the correct order from earliest to latest by numbering them from 1 to 5.

____ a. Boyce analyzed data from multiple sources showing that ocean phytoplankton levels have declined dramatically over the last 100 years.

____ b. Worm demonstrated that industrialized fishing had decimated predatory fish communities.

____ c. Secchi designed a device to measure phytoplankton levels in water.

____ d. Boyce's finding of declining phytoplankton levels was challenged by fellow scientists because of his methodology.

____ e. Boyce, a graduate student in Worm's lab, decided to study how the loss of predatory fish reverberated through the oceanic ecosystem—in particular, its phytoplankton populations.

9 How is an ecosystem different from an ecological community?

Try Something New

10 In the energy pyramid shown here, an owl, a cardinal, and a grasshopper are the fourth, third, and second trophic levels, respectively.

(a) If each grasshopper passes 100 kilocalories to the cardinal when eaten, how many grasshoppers would the cardinal have to eat to obtain 10,000 kilocalories?

(b) How many kilocalories of grass are required to produce 10,000 calories' worth of grasshopper?

(c) Where does the lost energy go?

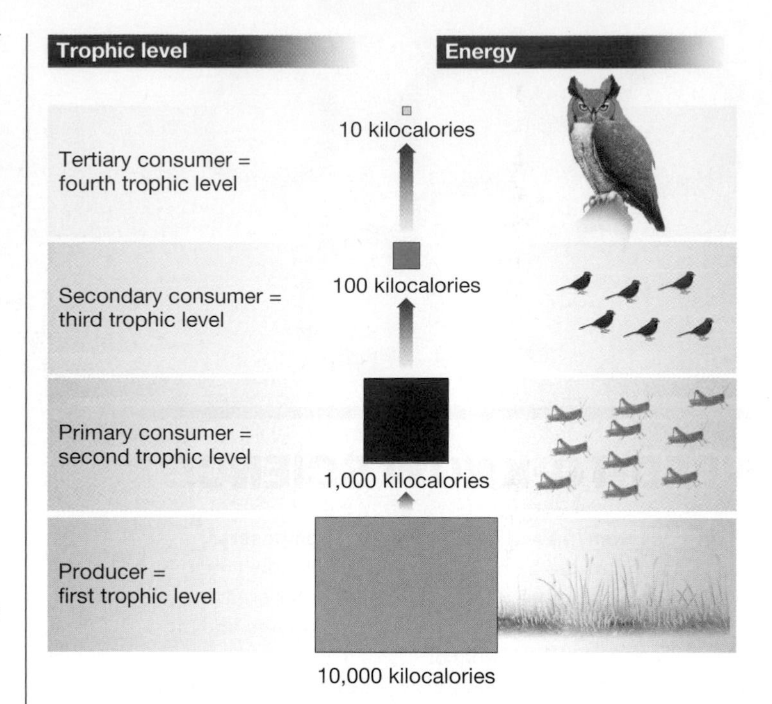

11 Is the water cycle (see Figure 18.6) more similar to the movement of energy or to the movement of nutrients through an ecosystem? Justify your answer.

12 Introduced species can disrupt energy flow through ecosystems. The mud crab *Rhithropanopeus harrissii* is a native of the Atlantic coast of North America, but it has been introduced all over the world. When introduced, it is able to rapidly decimate native invertebrates such as mussels and clams. These invertebrates—primary consumers of chlorophyll-producing, photosynthesizing phytoplankton—are critical to maintaining energy balance in coastal ecosystems; without them, phytoplankton blooms can block light to organisms lower in the water and clog fish gills, among other things. Answer the following questions based on the data in the accompanying graphs from the Baltic Sea, where *R. harrissii* has been introduced.

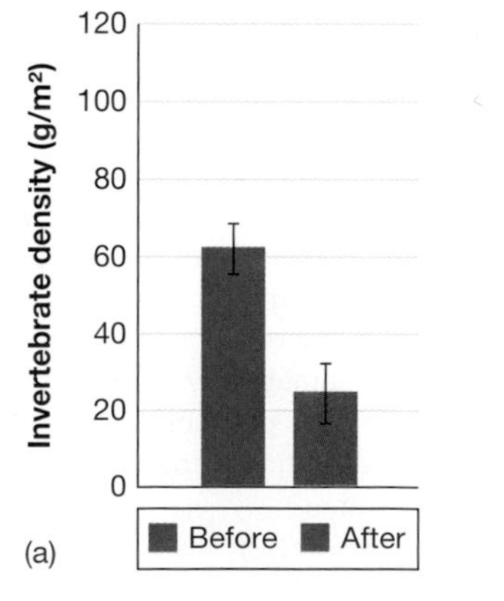

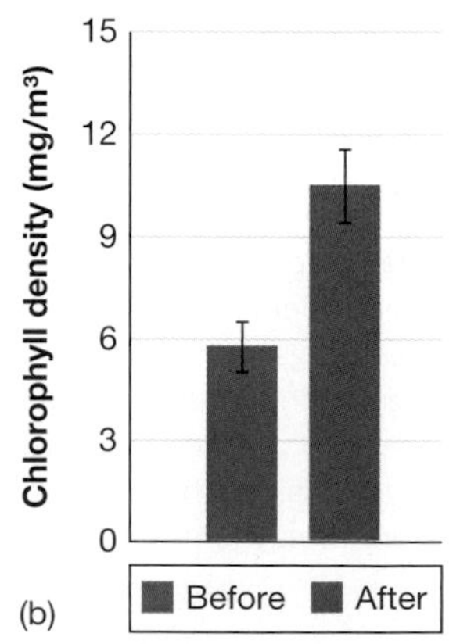

(a) What was the average invertebrate density before and after the mud crabs were introduced? What does this tell you about the effect of mud crabs on the amount of native invertebrates in this ecosystem?

(b) What was the average chlorophyll density before and after the mud crabs were introduced? What does this tell you about the effect of mud crabs on the amount of phytoplankton in this ecosystem?

(c) How has energy flow changed in the Baltic Sea ecosystem since mud crabs were introduced?

Leveling Up

13 ***Write Now* biology: human-caused biome shifts** The locations of Earth's different biomes depend on climate and altitude, for the most part. However, human activities play a role in the conversion of one biome to another, as has been seen many times in history. Research one major change in a biome category based on human activity and describe how and why this change happened. Speculate on how this change could have been specifically avoided. (*Note*: Do not analyze a change via deforestation to agricultural land, as agricultural land is not a natural biome. *Hint*: Take a look at Easter Island as one example.)

14 **Doing science** Join forces with millions of others by classifying phytoplankton on your computer. Do an internet search for "citizen science phytoplankton" and sign in as a citizen scientist. Complete the tutorial and start helping researchers quantify the phytoplankton in our oceans.

15 **What do *you* think?** Some people think the current U.S. Endangered Species Act should be replaced with a law designed to protect ecosystems, not species. The intent of such a law would be to focus conservation efforts on what its advocates think really matters in nature: whole ecosystems. Given how ecosystems are defined, do you think it would be easy or hard to determine the boundaries of what should and should not be protected if such a law were enacted? Give reasons for your answer.

Digital resources for your book are available online.

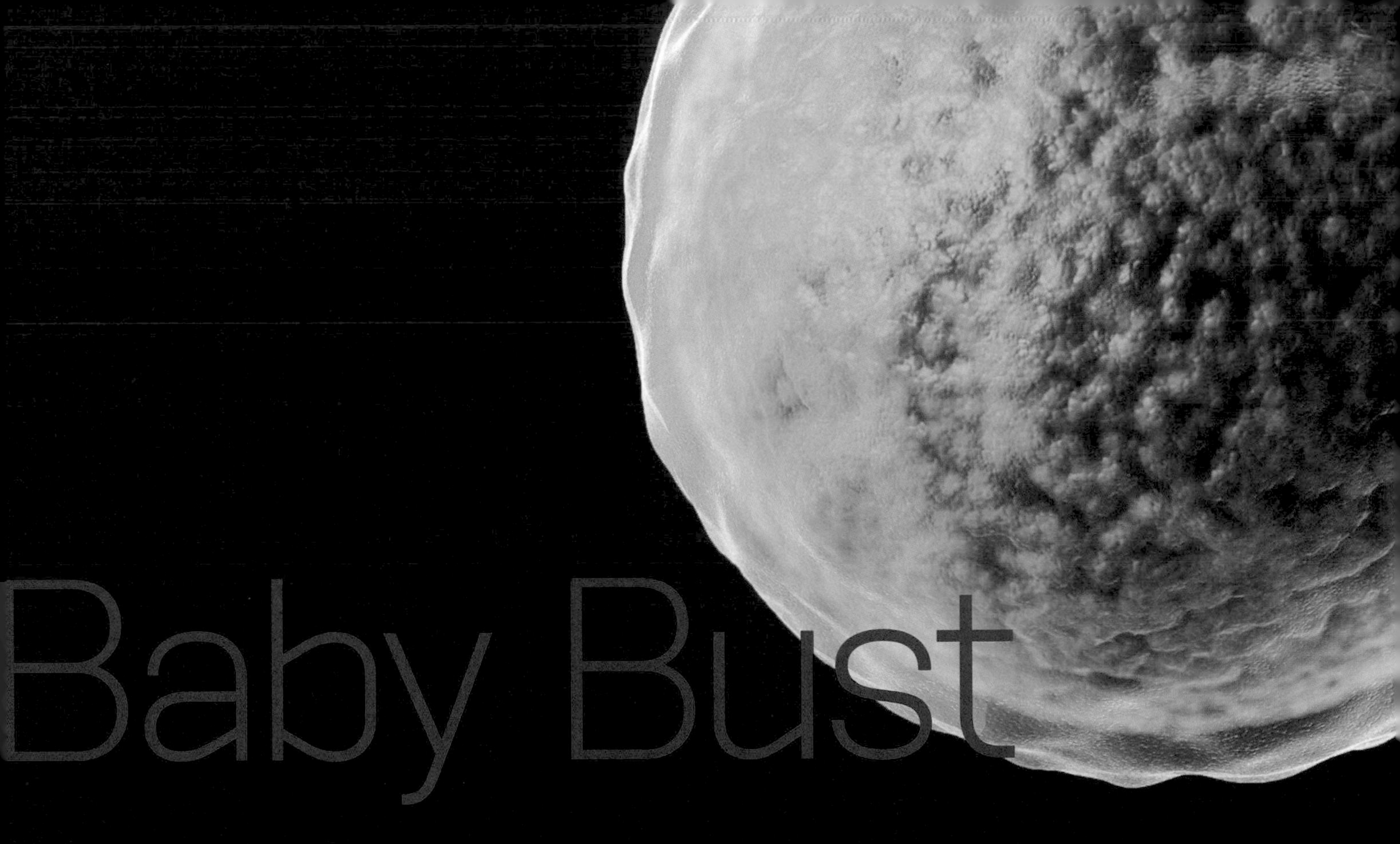

Baby Bust

Facing dwindling births, Denmark searches to resolve problems of infertility.

After reading this chapter you should be able to:

- Distinguish among tissues, organs, and organ systems.
- Explain the importance of homeostasis for life.
- Outline the steps of gamete formation in males and females.
- Create a flowchart of the steps of fertilization in humans.
- Describe the stages of prenatal development in humans.
- Explain the role of a given hormone in the reproductive system.

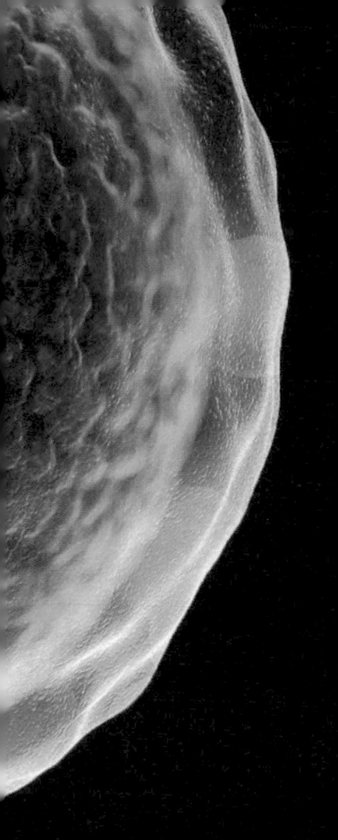

PHYSIOLOGY

CHAPTER 22

HOMEOSTASIS, REPRODUCTION, AND DEVELOPMENT

A woman slides a bra strap off her naked shoulder. In Danish, a seductive man's voice asks, "Can sex save Denmark's future?"

The commercial, aired on Danish television, then switches to a view of an empty playground. Produced by a travel agency, the ad encourages responsible Danes to book a romantic holiday with the company's "ovulation discount" and try to conceive while on vacation. It concludes with a large banner proclaiming, "Do It for Denmark" (**Figure 22.1**).

Birth rates reached a 23-year low in the year 2013 in Denmark, a Nordic country in northern Europe. When asked, most Danish couples said they would like to have two or three children, yet the present fertility rate is only 1.67 children per family—not high enough to maintain Denmark's current population. Infertility is now considered an epidemic in the country. In fact, one in 10 children in Denmark is conceived using reproductive technologies.

"We see more and more couples needing to get assisted fertility treatment," Bjarne Christensen, secretary general of Sex and Society, Denmark's leading family-planning association, told Bloomberg News. "We see a lot of people who don't succeed in having children."

And it's not just Denmark. A survey of fertility trends across 195 countries from 1950 to 2017 found that nearly half of countries are facing a "baby bust," in which not enough children are being born to maintain the country's population size. Countries where women are having the fewest children include Taiwan, Puerto Rico, and South Korea. Since 1950, the global fertility rate has essentially been cut in half. And that includes the United States. In 2018, the United States hit an all-time low in fertility rates in a decline that has affected women of almost all ages and races (with the exception of Native Hawaiians and Pacific Islanders). Then, in 2020, the coronavirus pandemic led to an even further drop in fertility rates.

Commercials like the one that ran in Denmark make a patriotic, if whimsical, appeal to citizens to have more children, but experts debate whether falling fertility rates are a cause for concern or for celebration. Some applaud the decrease as a way to slow global population growth and reduce human consumption of limited natural resources. But others in countries facing declining fertility rates worry about the economic burdens that will be placed on a generation smaller than the one before it (**Figure 22.2**). Government officials fear that this trend, along with the coronavirus pandemic resulting in an estimated 500,000 fewer babies being born in the U.S. alone, will produce a smaller workforce and fewer young people to care for retirees.

Figure 22.1

Making Denmark mate again

This print advertisement was part of the "Do It for Denmark" campaign.

To counteract falling birth rates before the coronavirus pandemic, some governments had begun taking profertility stances, including providing free postnatal care and subsidized day care. In Denmark, Sex and Society, a nongovernmental organization, provides sex education materials for most of Denmark's schools, and it recently unveiled a new series of lesson plans entitled "This Is How You Have Children!" Instead of focusing on contraception and how to avoid becoming pregnant, these new classes educate students about what fertility is, how aging affects fertility, and when may be the best times to have children.

"We have for many years addressed the very important issues of how to avoid becoming pregnant, how to avoid sexual diseases, how they have a right to their own bodies, but we totally forgot to tell the kids that we cannot have children forever," Søren Ziebe, head of Copenhagen University Hospital's fertility clinic, told reporters. "There is a biological limit."

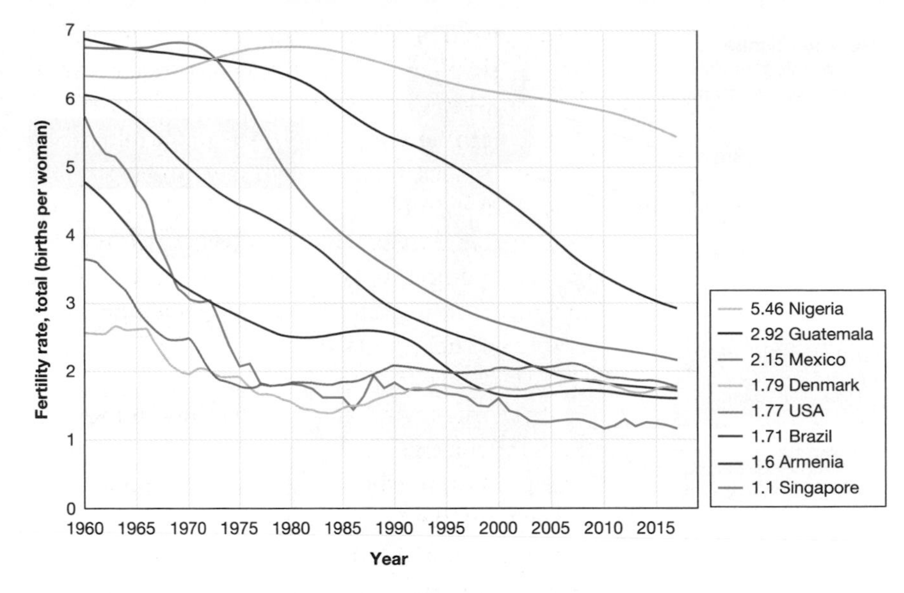

Figure 22.2

Declining fertility rates in countries all over the world

The number of births per woman has been declining in most countries. Fertility data from 1960 to 2017 from several countries are shown here.

Q1: List the fertility rates in 2017 of the countries shown from highest to lowest. China (not shown) had a fertility rate of 1.63. Where on your list would China's data fit?

Q2: List the approximate fertility rates in 1960 of the countries shown from highest to lowest. China (not shown) had a fertility rate of 5.75. Where on your list would China's data fit?

Q3: Which country showed the greatest decrease in fertility over the 57 years depicted in the graph? How many births decreased per woman over this time? Which country showed the least fertility decrease, and how many births decreased per woman over these years?

See Appendix A for answers to the figure questions.

Seeking Stability

That biological limit may be one key reason the global birth rate was falling. Pre-pandemic population studies showed that, on average, couples were waiting longer to have children, and conception can be more difficult after age 30.

But other causes were also possible. Fewer childhood deaths due to improved health care means women may choose to have fewer children. In addition, greater access to contraception means women have more choice in preventing pregnancy. The coronavirus pandemic has disrupted access to contraception in lower-income countries and could lead to millions of unplanned pregnancies, increasing fertility rates in these countries. On the other hand, assisted reproductive technologies have also been disrupted in higher-income countries, further lowering the fertility rate in these countries. For those trying but struggling to have children, it is also possible that chemicals or other environmental factors might be affecting the human **reproductive system**, the parts of the body responsible for producing offspring.

Our bodies are highly efficient and well-coordinated communities of over 200 specialized cell types. **Anatomy** is the study of the structures that make up a complex multicellular body, and **physiology** is the science of the functions of anatomical structures. Through physiological research, scientists hope to determine why fertility rates are falling in Denmark and whether anything can be done about it.

Their first step has been to examine the reproductive system, which consists of cells, tissues, and organs. **Tissues** are made up of cells that, in an integrated manner, perform a shared set of functions. The different types of tissues found in the vertebrate animal body can be placed into four broad categories: **epithelial**, **connective**, **muscle**, and **nervous tissue** (**Figure 22.3**). A tissue may be composed of just one cell type, or it may contain multiple cell types; in either case, the cells that compose

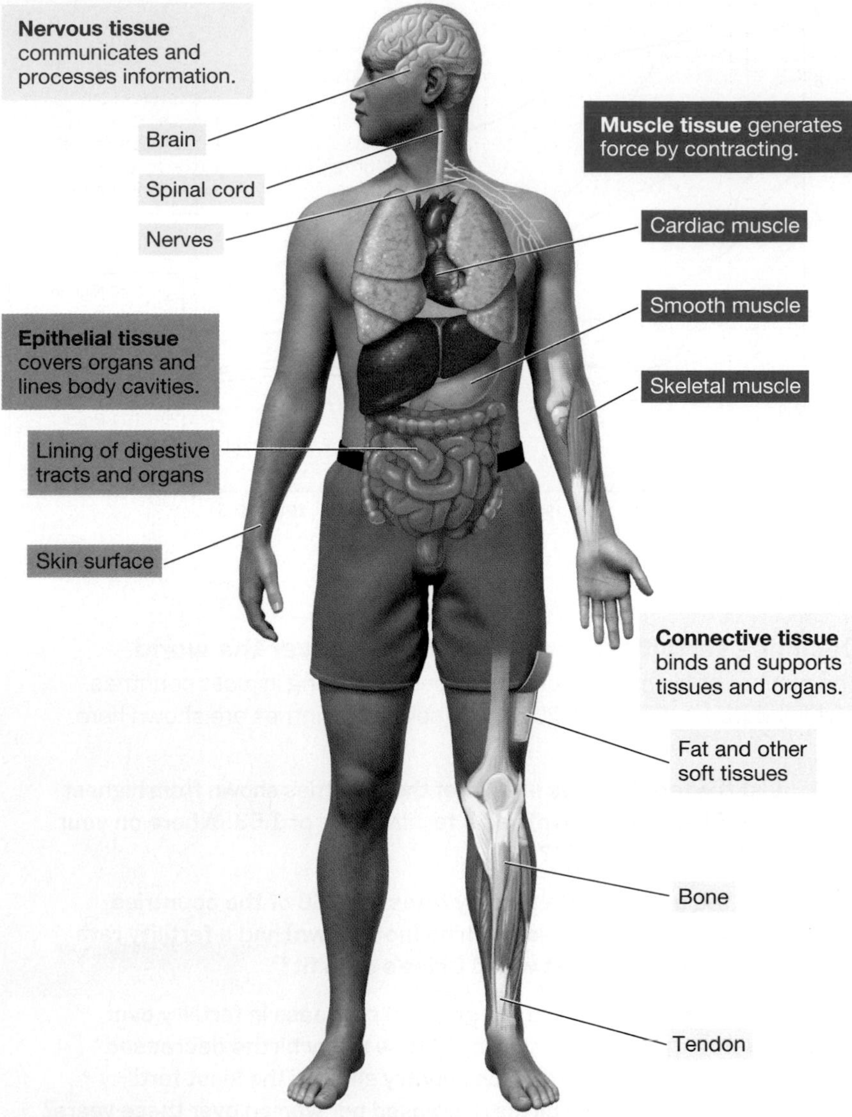

Figure 22.3

Animal tissue

The four types of animal tissue have different functions.

Q1: Which tissue type is the primary component of skin?

Q2: Which tissue type is the primary component of bone?

Q3: Which tissue types do you think the hand contains?

See Appendix A for answers to the figure questions.

a tissue cooperate to perform the distinctive functions of that tissue. An **organ** has more than one tissue type and forms a functional unit with a distinctive function, shape, and location in the body. An **organ system** is composed of two or more organs that, in close coordination, perform a set of functions.

The human body has 11 major organ systems, including the reproductive system (**Figure 22.4**). Each system is covered in more detail in the following chapters, and plant organ systems are discussed in Chapter 26. Each organ system and its organs are unique in function and form—from the beating, blood-filled heart of the circulatory system to the electrical, threadlike nerves of the nervous system—but they all share one important commonality: for proper function, an organ system requires proper structure and a stable internal environment.

Most biological processes take place within only a certain temperature range, with the right amount of water, at the appropriate pH, and at a particular concentration of chemicals. In the male reproductive system, for example, the testes need to maintain a temperature around 95°F (35°C), about 3–4 degrees cooler than the body temperature of 98.6°F (37°C), to produce and store functional sperm. In the female reproductive system, the vagina requires a pH of 3.8–4.5 to keep out bacteria while maintaining a healthy environment for fertility.

These environments are regulated through **homeostasis**, the body's way of maintaining a relatively constant internal state despite changes in the external environment. Homeostasis enables the body to continually sense its internal state and rapidly adjust. In this way, despite large fluctuations in the outside world, homeostasis maintains the internal conditions best suited for life processes.

Homeostasis occurs via **homeostatic pathways**, sequences of steps that reestablish homeostasis if there is any departure from the normal state (also called the **set point**). Homeostatic pathways continually monitor the physical and chemical characteristics of the internal environment and trigger regulatory processes within the body if this monitoring system detects any deviation from the set point.

Homeostatic pathways depend on **feedback loops**. In a **negative feedback** loop, the results of a process cause that process to slow down or stop. For example, if a person drinks a large milk shake, the level of glucose in the blood rises. In response, cells in the pancreas produce insulin, which allows glucose to enter cells, and blood glucose level declines. The regulation of body temperature—called *thermoregulation*, or the control of heat gain and loss—also relies on negative feedback loops (**Figure 22.5**), as does *osmoregulation*, or the control of internal water content and solute concentration.

The *urinary system* removes excess fluid from the body, along with waste products and toxins.

In the *digestive system*, large molecules of food are broken down in the mouth, stomach, and small intestine, and nutrients are absorbed in the small and large intestines.

The *circulatory system* diffuses oxygen from the lungs to the heart, which then pumps oxygen-rich blood to the rest of the body through a closed network of vessels.

The *respiratory system* brings in oxygen and expels carbon dioxide through the lungs.

The *endocrine system* works closely with the nervous system to regulate all other organ systems. It consists of a number of glands and secretory tissue.

The *integumentary system* is the largest organ system in the human body, covering and protecting the surface of the body.

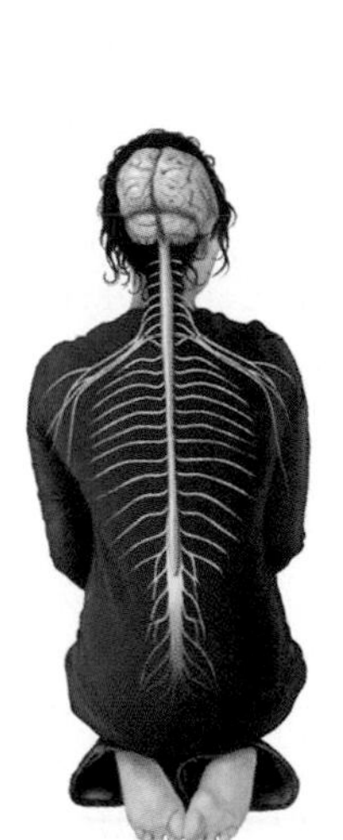

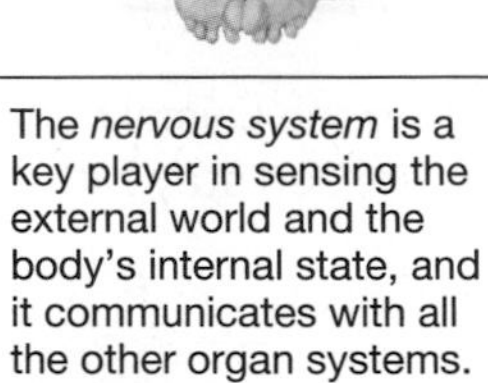

The *nervous system* is a key player in sensing the external world and the body's internal state, and it communicates with all the other organ systems.

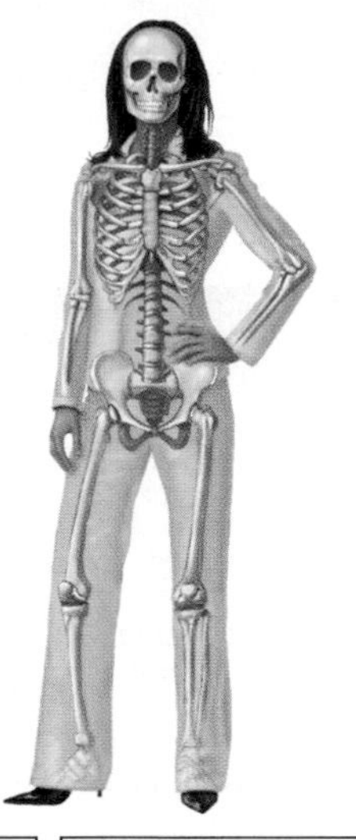

The *skeletal system* provides an internal framework to support the body in vertebrates. It consists of bone, cartilage, and ligaments.

The *muscular system* produces the force that moves structures within the body. It works closely with the skeletal system.

The *immune system* defends the body from invaders such as viruses, bacteria, and fungi.

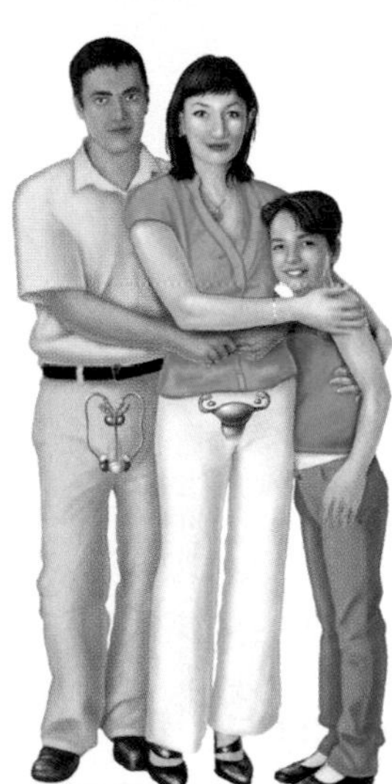

The *reproductive system* generates gametes and, depending on the animal group, may also support fertilization and prenatal development.

Figure 22.4

Organ systems

The 11 major organ systems of the human body work in an integrated manner to maintain the body's overall health.

Q1: Which organ system defends the body from infectious diseases such as the common cold or flu?

Q2: Which two organ systems together transport oxygen to cells?

Q3: Which two organ systems together regulate the activities of the other organ systems?

See Appendix A for answers to the figure questions.

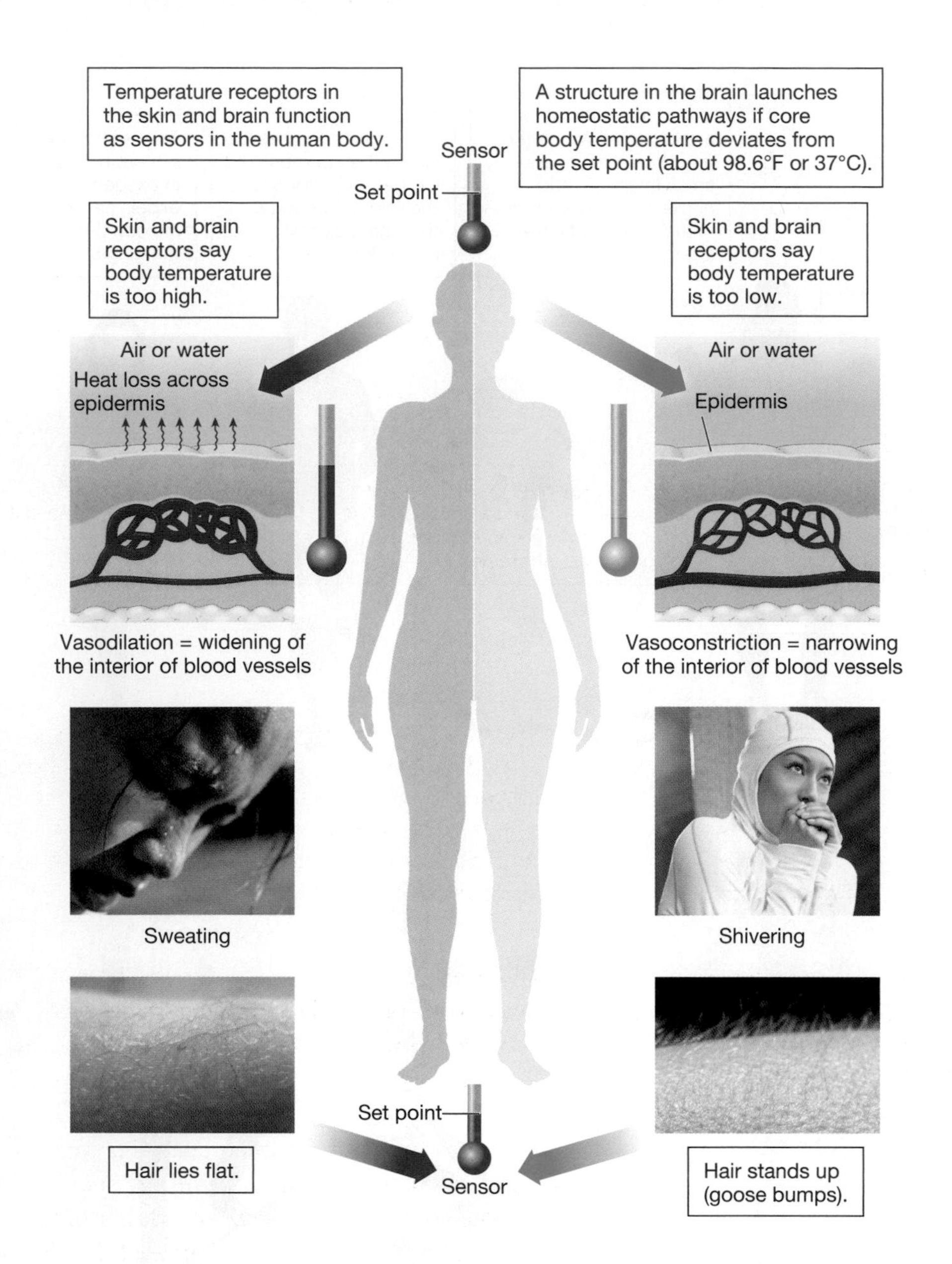

Figure 22.5

Through homeostasis, the body maintains stable internal conditions

Because of homeostatic pathways, fluctuations in external conditions may produce little or no overall change in the animal body. For example, even when the outside temperature is very hot or very cold, the human body's internal temperature can stay within the narrow range required for survival.

Q1: Why is it important to maintain a stable body temperature?

Q2: Which organ system is mainly involved in sensing the body's external and internal states? (See Figure 22.4 for an overview of organ systems.)

Q3: Give another example of a homeostatic pathway in humans that is discussed in this chapter.

See Appendix A for answers to the figure questions.

In a **positive feedback** loop, by contrast, the results of a process cause it to speed up until an endpoint is reached. Blood clotting triggered by a broken blood vessel, for example, is a positive feedback loop: the process of clotting releases chemicals that accelerate clotting, increasing the amount of clotting until the clot plugs the break in the blood vessel.

Sea stars can reproduce asexually by breaking off an arm, which then regenerates into a new individual. Alternatively, they can reproduce sexually, with females releasing eggs and males releasing sperm and fertilization occurring in the water.

As with sea stars, frog fertilization occurs in the water. The male takes position on the female so their gametes can join, but he doesn't penetrate her.

Clownfish begin life as males, but the largest fish in a group will change to female.

Figure 22.6

Animals display a rich variety of reproductive systems

Some animals reproduce by cloning themselves exclusively through asexual reproduction. Many asexually reproducing species can switch to sexual reproduction, depending on environmental conditions (left). Sexually reproducing species fertilize gametes externally or internally. External fertilization is common among aquatic animals (middle), whereas internal fertilization is more common among land animals. Still other species are both male and female, either simultaneously or sequentially (right). Individuals with functional reproductive organs of both sexes are called *hermaphrodites*.

In homeostatic pathways, negative feedback loops are more common than positive feedback loops. However, both types are critical to the maintenance of our organ systems, including the reproductive system—which brings us back to doing it in Denmark.

All in the Timing

Danish researchers have long suspected that the primary reason for declining fertility rates is that women are waiting longer to have children.

In the 1970s, on average, a Danish woman gave birth to her first child at 24 years old. But in 2014, that average age was 29, with more and more women waiting until they were over 35 to have a first child. And the older a woman gets, the more difficult it is for her to conceive.

Humans reproduce through *sexual reproduction*, in which male (**sperm**) and female (**egg** or **ovum**) haploid gametes produced through *meiosis* combine to form a diploid *zygote*. Through the process of mitosis, the zygote develops into a multicellular individual that is genetically unique and different from either parent. As the processes of meiosis and mitosis were detailed in Chapter 6, we concentrate primarily on human sexual reproduction processes here. Keep in mind that sex in other animals is more variable than our human perspective might lead us to expect. In particular, some animals can also reproduce via *asexual reproduction*, in which cells from only one individual produce the offspring, so all of the offspring's genes come from that parent (**Figure 22.6**). (For more on the benefits and pitfalls of sexual and asexual reproduction, see "Why Sex?" on page 250 of Chapter 13.)

Human eggs develop through **oogenesis**, cell divisions that produce mature eggs capable of being fertilized (**Figure 22.7**, left). Oogenesis begins before birth, when the germ line cells in

DEBUNKED!

MYTH: Women can't get pregnant if they have sex while on their period.

FACT: It's unlikely, but if the timing is right, a woman can become pregnant while menstruating (see Figure 22.8). Because sperm can live inside the human body for up to 5 days, if a woman ovulates early, or more than once per cycle as happens in approximately 10 percent of women, sperm could find and fertilize an egg.

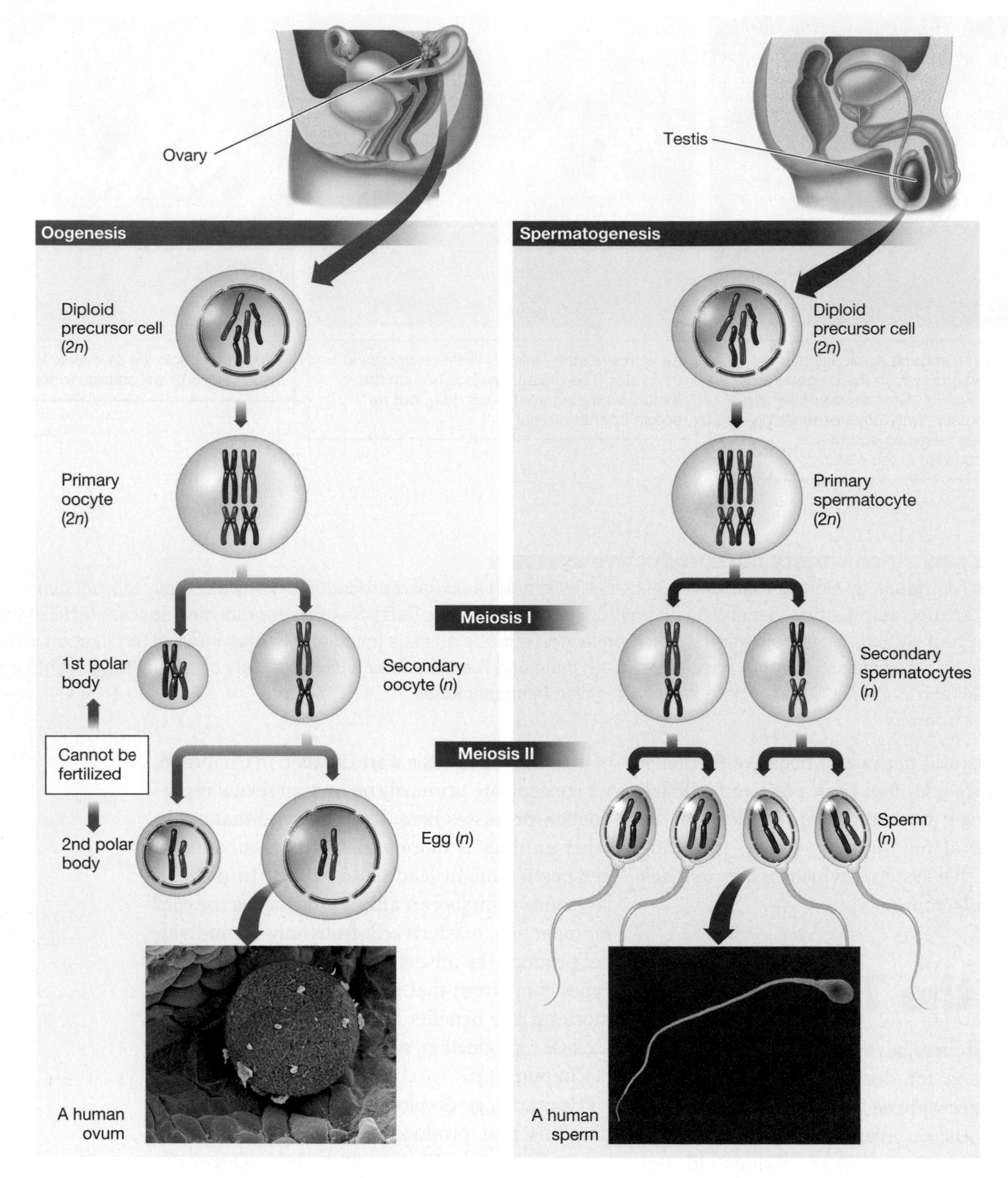

Figure 22.7

Sexual reproduction requires the production of haploid gametes

Oogenesis produces haploid eggs (ova), and spermatogenesis produces haploid sperm.

Q1: Identify one way in which oogenesis and spermatogenesis are similar.

Q2: How many eggs are produced from each precursor cell? How many sperm?

Q3: How much time elapses between the appearance of a precursor cell and the formation of an egg? How does this process differ for a sperm?

See Appendix A for answers to the figure questions.

the ovaries multiply and develop into immature, diploid egg cells called **primary oocytes**. At birth, the female's **ovaries** already contain her entire lifetime supply of primary oocytes, about 1–2 million cells. These cells remain in a state of suspended development until the production of *hormones* at puberty stimulates one, or occasionally two, of them to mature each month in preparation for ovulation. By the time a female reaches puberty, at about 10–12 years of age, approximately 400,000 viable primary oocytes remain—still more than she will use in her lifetime.

Human females do not produce mature eggs continuously. Instead, individual eggs mature and are released in a hormone-driven sequence of events known as the **menstrual cycle** (**Figure 22.8**). The menstrual cycle averages about 28 days, but cycle lengths from 21 days to 35 days are considered normal.

A woman has more primary oocytes than she will use in her lifetime, but evidence suggests that those eggs decline in quality as a woman ages—a conclusion supported by the increased risk of birth defects in children born to older mothers. In addition, if eggs from younger women are implanted into women over 40, the pregnancy rate equals the rate associated with the younger women who donated the eggs. In other words, young eggs result in young pregnancy rates.

When a woman passes 40 years of age, she shows a clear drop in her ability to produce normal eggs and bear children. Human females reach menopause—the end of their reproductive phase—around age 50, although this is highly variable among women.

A decline in egg quality is one suggested cause of the global decline in fertility. Because women are waiting longer to have children, their eggs are older and their fertility has decreased. Age also affects the fertility of men, although males do not undergo the clearly identifiable menopause that is characteristic of females. Instead, their sex drive and their ability to produce sperm slowly decrease as they age. In addition, men produce fewer and lower-quality sperm as they age, decreasing their chances of fertilizing an egg.

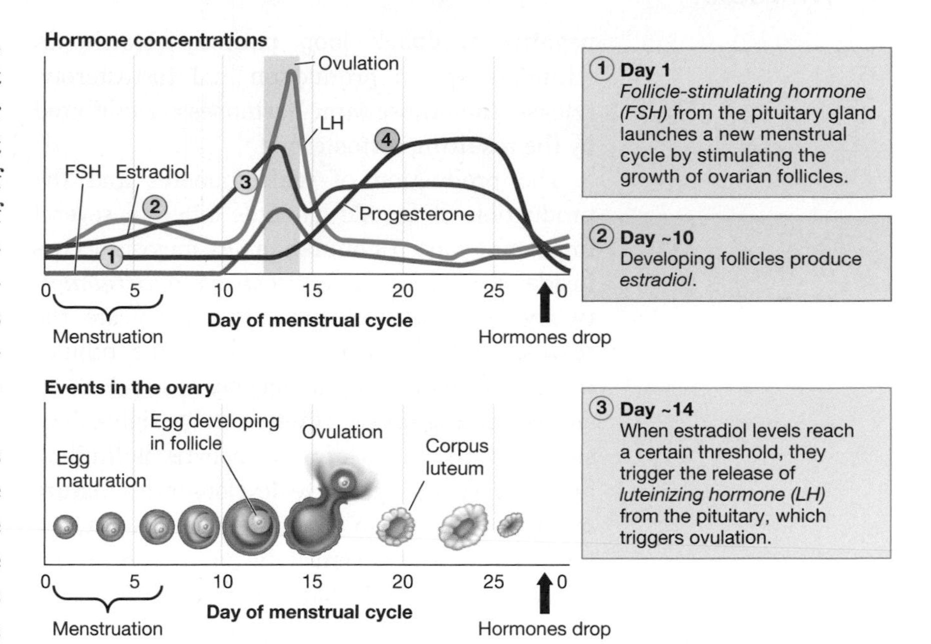

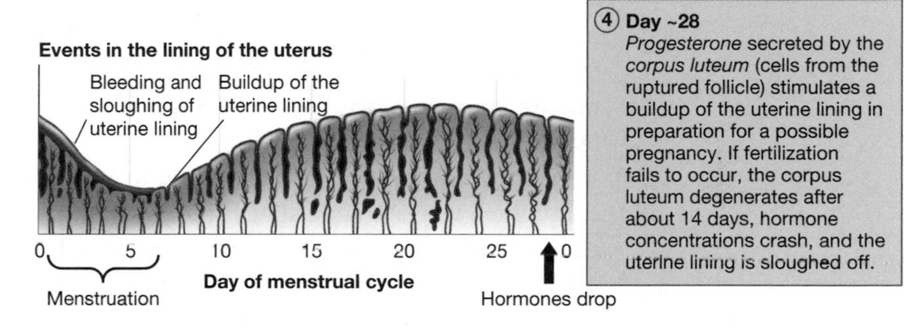

Events in the lining of the uterus

Bleeding and sloughing of uterine lining

Buildup of the uterine lining

0 5 10 15 20 25 0

Menstruation

Day of menstrual cycle

Hormones drop

④ **Day ~28**
Progesterone secreted by the *corpus luteum* (cells from the ruptured follicle) stimulates a buildup of the uterine lining in preparation for a possible pregnancy. If fertilization fails to occur, the corpus luteum degenerates after about 14 days, hormone concentrations crash, and the uterine lining is sloughed off.

Figure 22.8

The human menstrual cycle

A menstrual cycle begins with the first day of bleeding, which marks the end of the previous cycle. Over the next few weeks, a succession of hormones stimulates the release of an egg and signals the uterine lining to grow and thicken in preparation for a potential pregnancy. If pregnancy does not occur, hormone levels plummet and the lining is sloughed off as menstrual flow.

Q1: Which hormones important for the menstrual cycle are produced in the pituitary gland?

Q2: Which hormone is involved in producing the uterine lining?

Q3: How is the egg follicle involved in producing hormones?

See Appendix A for answers to the figure questions.

Just Keep Swimming

Although parents' increasing age at conception likely contributes to fertility problems, researchers suspect that other factors are also at play—particularly, the quality of male sperm.

At the onset of puberty, a surge of FSH and LH brings on the production of sperm and the release of testosterone (discussed in the next section). Unlike in females, hormones in males are relatively constant due to a homeostatic negative feedback loop pathway. Hormones stimulate sperm production and testosterone release; then those same hormones are inhibited by the resulting testosterone.

The production of male gametes and the production of female gametes differ in several important ways. In human males, meiosis occurs inside structures called *seminiferous tubules*: twisty, spaghetti-like tubes that permeate the **testes**. Diploid germ line cells in the tubules divide to form sperm in a sequence of steps known as **spermatogenesis** (**Figure 22.7**, right). The supply of a female's primary oocytes is limited, and once a primary oocyte develops into a mature ovum, it is lost from the supply. In contrast, the testes constantly replenish the pool of sperm. It is noteworthy to recall that in a normal menstrual cycle, a human female begins maturing only one egg cell—in other words, only one egg per month—while an average male produces millions of sperm cells every day. Surplus sperm that accumulate over time are degraded and reabsorbed by the cells that line the tubules.

Another difference is that an ovum is typically much larger than a sperm cell. The human egg is visible (just barely) to the naked eye, but an individual sperm can be seen only under a microscope. A sperm contains little substance beyond chromosomes and the cellular machinery needed to move up the female reproductive tract, attach to an egg, and propel the sperm's chromosomes into the egg's cytoplasm. A sperm is a simple package containing valuable genetic information. An ovum, on the other hand, is a plump, complex cell full of genetic material and organelles.

The first reports of falling sperm counts came in 1992 from Niels Skakkebæk at the University of Copenhagen. Looking at 61 different studies on the semen quality of almost 15,000 men, Skakkebæk and his team found that sperm counts around the world had dropped by 50 percent from about 1940 to 1990. In a peer-reviewed paper they concluded, "As male fertility is to some extent correlated with sperm count, the results may reflect an overall reduction in male fertility." In other words, a decrease in the quality of semen could be lowering male fertility.

Skakkebæk's 1992 study of declining semen quality sparked a lot of interest. Hundreds of studies followed, as researchers sought the cause for the sperm decline. Many suspected an environmental cause, such as a toxin.

"We all wanted to study it in more detail," says Tina Kold Jensen, a researcher at the University of Southern Denmark. But instead of looking at past studies, Jensen and her collaborators determined that it was necessary to track these trends in real time. "We decided if we wanted to get closer to the truth, we had to collect our own data."

Jensen partnered with Skakkebæk and Niels Jørgensen at Copenhagen University Hospital. Starting around 1995, the scientists began recruiting young male volunteers at government-required physical exams, asking each to provide a sperm sample. By 2010, the team had amassed sperm from more than 5,000 volunteers. Surprisingly, when analyzed, the data seemed to contradict Skakkebæk's initial conclusions from the earlier studies. In fact, over the 15 years of this new study, semen quality did not decline; in fact, it improved slightly. Sperm concentrations rose from 43 million per milliliter in 1996–2000 to 48 million per milliliter in 2006–10. How could this be? Looking retrospectively at studies from 1940 to 1990 that had all different sample sizes, control populations, and experimental parameters could have skewed the interpretations of their results.

Still, the slight improvement in sperm quality over the retrospective study was not enough to suggest that fertility improved during that period. "Although we see a slight rise, only 23 percent of the young men had optimal semen quality," said Jørgensen in a statement when the study was released. "In fact, the semen quality of 27 percent of the men was so poor, it will probably take these men longer to make their partner pregnant," he added. "Furthermore, for 15 percent of the men, semen quality was so poor, they are likely to need fertility treatment in order to conceive."

Driven by Hormones

Jørgensen and Jensen have gone on to study the factors that might have caused such low levels of semen quality. In May 2014, their team found that 98 percent of 308 young men had detectable urinary levels of bisphenol A (BPA), a chemical found in plastics that disrupts the body's endocrine system (see Chapter 6 for more on BPA). Men with higher BPA levels, they discovered,

Preventing Pregnancy

The Food and Drug Administration approved the first birth control pill for sale in the United States in 1960. It became immediately popular as a way for women to control their fertility, and studies found access to contraception contributed to a sharp increase in college attendance and graduation rates for women. Here is an overview of some of today's most popular forms of birth control, with data on the effectiveness of each type. It's still too early to say for sure, but initial trends suggest spikes in some parts of the world and declines in others. Broadly speaking, birthrates should continue to drop in many higher-income countries and climb in many poor and middle-income nations, where the U.N. Population Fund (UNFPA) projects that pandemic-driven disruptions in access to contraception could lead to millions of unplanned pregnancies.

Assessment available in smartwork

Pregnancies per 100 women in a year

Intrauterine Device
T-shaped plastic device must be positioned inside the uterus by a health care professional.

Hormonal

Nonhormonal

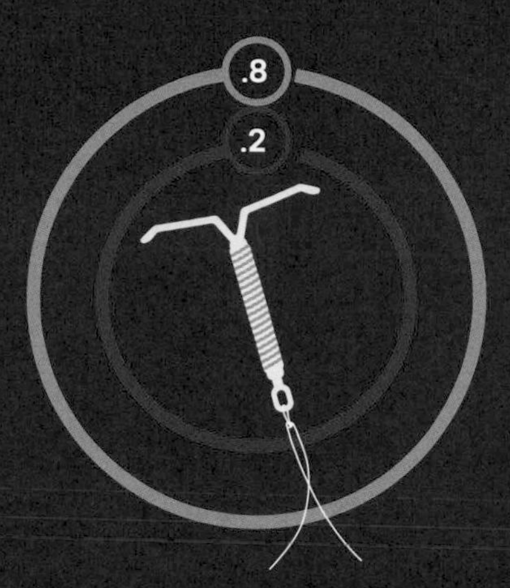

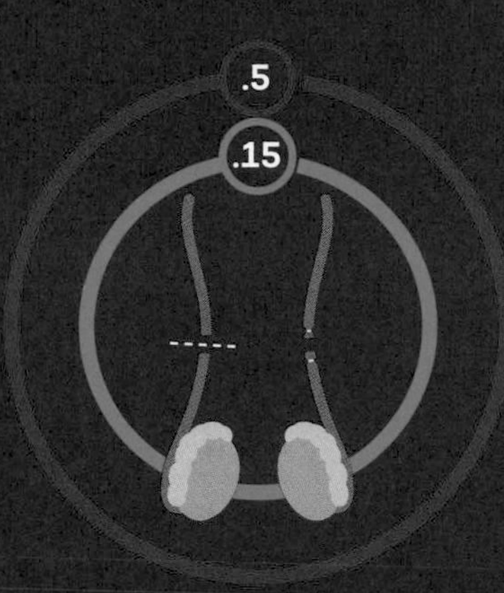

Female Sterilization
The oviducts are sealed with clamps or by other surgical means.

Male Sterilization
The tubes that carry sperm are sealed surgically by a health care professional.

The Pill
Hormones suppress ovulation, preventing the release of an egg, by mimicking the hormone levels of pregnancy.

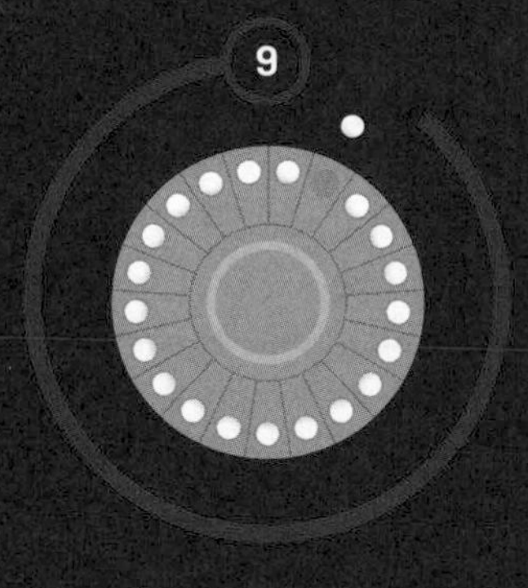

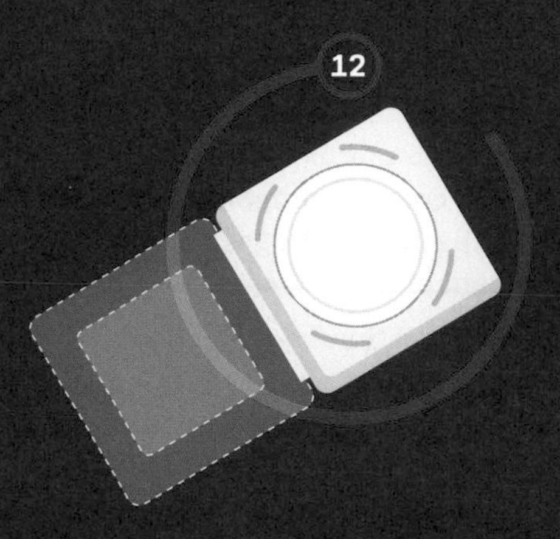

Diaphragm
Dome-shaped latex cup filled with spermicide, inserted before intercourse, covers the cervix and keeps sperm out of the uterus.

Female Condom
Plastic pouch, inserted before intercourse, lines the vagina to prevent sperm from entering.

Male Condom
Plastic or latex pouch covers the penis to keep sperm from entering the vagina.

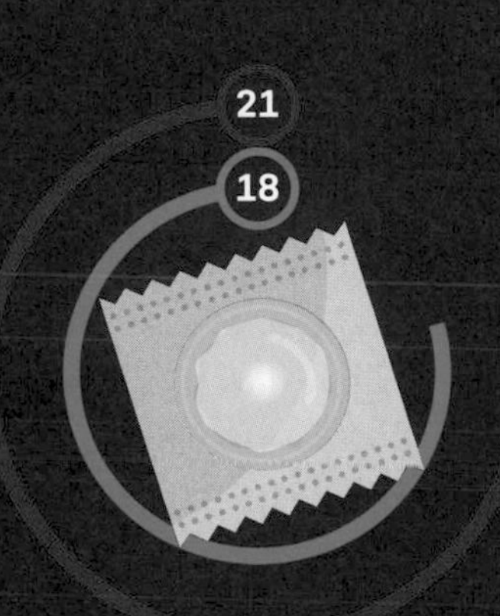

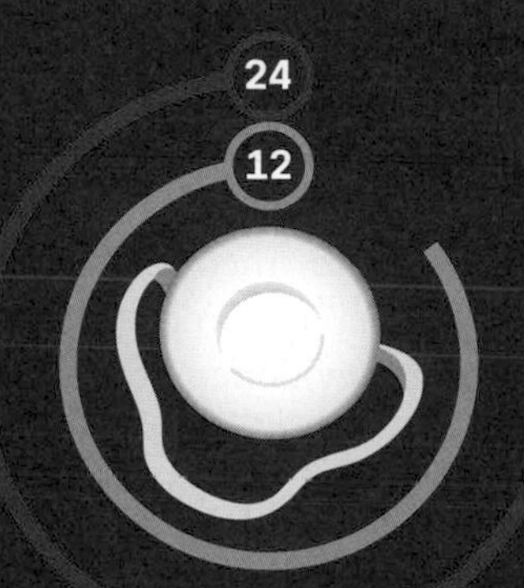

Sponge
A sponge containing spermicide is inserted deep in the vagina prior to intercourse.

Women who have given birth before

Women who have never given birth

also had higher levels of testosterone and other hormones. That correlation suggests that BPA could be affecting hormone feedback loops, but additional research is needed to identify possible mechanisms for such an effect.

Hormones regulate nearly all aspects of reproduction in animals, from mating behaviors to the development and birth of offspring. The emergence of sex-specific characteristics in the fetus and the maturation of reproductive organs during puberty are examples of long-term aspects of reproduction controlled by hormones. As mentioned previously, hormones also regulate the monthly cycle of menstruation in females and the continuous stimulation of sperm production in males (see **Figure 22.8** for the role of hormones in the menstrual cycle and Chapter 25 for more on hormones in general).

Testes and ovaries produce three major types of hormones: estrogens, progestogens, and androgens. Both males and females produce all three but in different ratios; for example, males have more androgens than estrogens, and females have more estrogens than androgens.

- **Estrogens** play a role in determining female characteristics such as wide hips, a voice that is pitched higher than that of males, and the development of breast tissues. The primary estrogen is **estradiol**.
- **Progestogens** have a number of functions in the female body, including thickening the lining of the uterus and increasing its blood supply to create a suitable environment for a developing fetus. **Progesterone** is the most important of the progestogens.
- **Androgens** stimulate the body to develop male characteristics, such as beard growth and sperm production. The primary androgen is **testosterone**. Together with another closely related androgen, testosterone directs the development of internal reproductive structures, such as the sperm ducts and prostate gland. A third androgen directs the development of external structures, such as the penis.

In addition to uncovering the link between BPA and hormone levels, the Danish research team found evidence that regular alcohol consumption may affect semen quality, possibly by changing testosterone levels. In particular, large amounts of alcohol significantly lowered semen quality: men who consumed 40 or more drinks in a week had a 33 percent reduction in sperm as compared with men who drank just 1–5 drinks per week. In their conclusion, the authors went so far as to warn young men to "avoid habitual alcohol intake." And a 2019 study in Denmark found that smoking is also associated with lower sperm count and reduced sperm concentration.

"Do It for Denmark"

Jørgensen, Jensen, and others continue to seek the causes for declining semen quality. At the same time, demographers—that is, researchers who study the statistics of human populations—track social reasons for decreasing fertility rates, such as the availability of contraceptives, the increasing age of mothers, and a global pandemic. "Of course, there are a lot of factors involved, including social factors," says Jensen. "But it's not all due to social factors. When we talk about fertility rates, we need to think about biology as well."

Details, details. A woman releases an egg from her ovaries, or *ovulates*, about once every 28 days (see **Figure 22.8**, middle panel). The egg moves down the **oviduct**, or *fallopian tube*, a 4-inch-long tube that connects the ovary to the **uterus** (**Figure 22.9**). If the woman has unprotected sexual intercourse, the man's **penis** may ejaculate almost 300 million sperm into her **vagina**. The sperm swim from the vagina into the uterus through an opening called the **cervix**, and then up the oviducts in response to a chemical signal released by the ovary.

Only a few hundred sperm manage to reach the egg in the oviduct, and only one of those lucky sperm fertilizes the egg to create a zygote. Although both parents contribute equally to the genetic material of the zygote, its organelles and other cellular machinery come almost entirely from the female.

Human development in the uterus averages about 38 weeks and is divided into three stages known as **trimesters**, each about 3 months long (**Figure 22.10**). During the first trimester, the zygote develops from a single cell into an **embryo** possessing all the main tissue types. All organ systems are established by the third month,

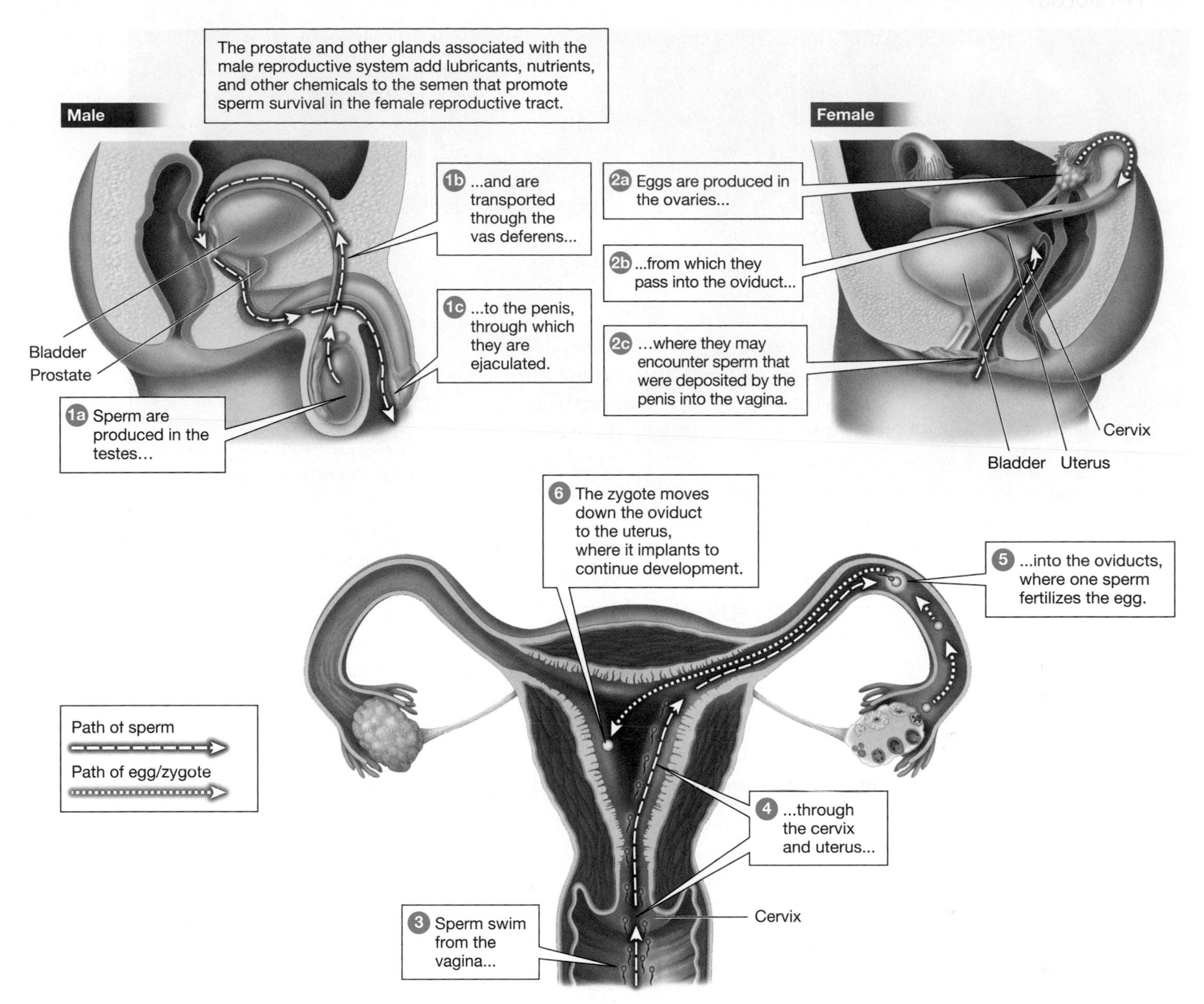

Male
The prostate and other glands associated with the male reproductive system add lubricants, nutrients, and other chemicals to the semen that promote sperm survival in the female reproductive tract.
Female
1b ...and are transported through the vas deferens...
2a Eggs are produced in the ovaries...
2b ...from which they pass into the oviduct...
1c ...to the penis, through which they are ejaculated.
2c ...where they may encounter sperm that were deposited by the penis into the vagina.
Bladder
Prostate
1a Sperm are produced in the testes...
Cervix
Bladder
Uterus
6 The zygote moves down the oviduct to the uterus, where it implants to continue development.
5 ...into the oviducts, where one sperm fertilizes the egg.
Path of sperm
Path of egg/zygote
4 ...through the cervix and uterus...
Cervix
3 Sperm swim from the vagina...

Figure 22.9

Fertilization takes place in the oviduct

Fertilization results in a zygote that can develop in the sheltered environment of the uterus.

Q1: If an egg is released but no sperm enter the oviduct, what is likely to occur? (*Hint*: Refer to Figure 22.8.)

Q2: If sperm enter the oviduct but no egg is present, what is likely to occur?

Q3: Given the timing of egg release and the life span of sperm, when is it safe to have unprotected intercourse without risking pregnancy?

See Appendix A for answers to the figure questions.

and the developing individual is now known as a **fetus**. It is during these first 3 months, when the fetus's organs are initially formed, that most birth defects occur, many of them severe enough to cause miscarriage.

Birth defects, the leading cause of infant mortality in the United States, are structural changes that affect how the body forms or works. Physicians know the cause of some birth defects, such as fetal alcohol syndrome resulting

PHYSIOLOGY

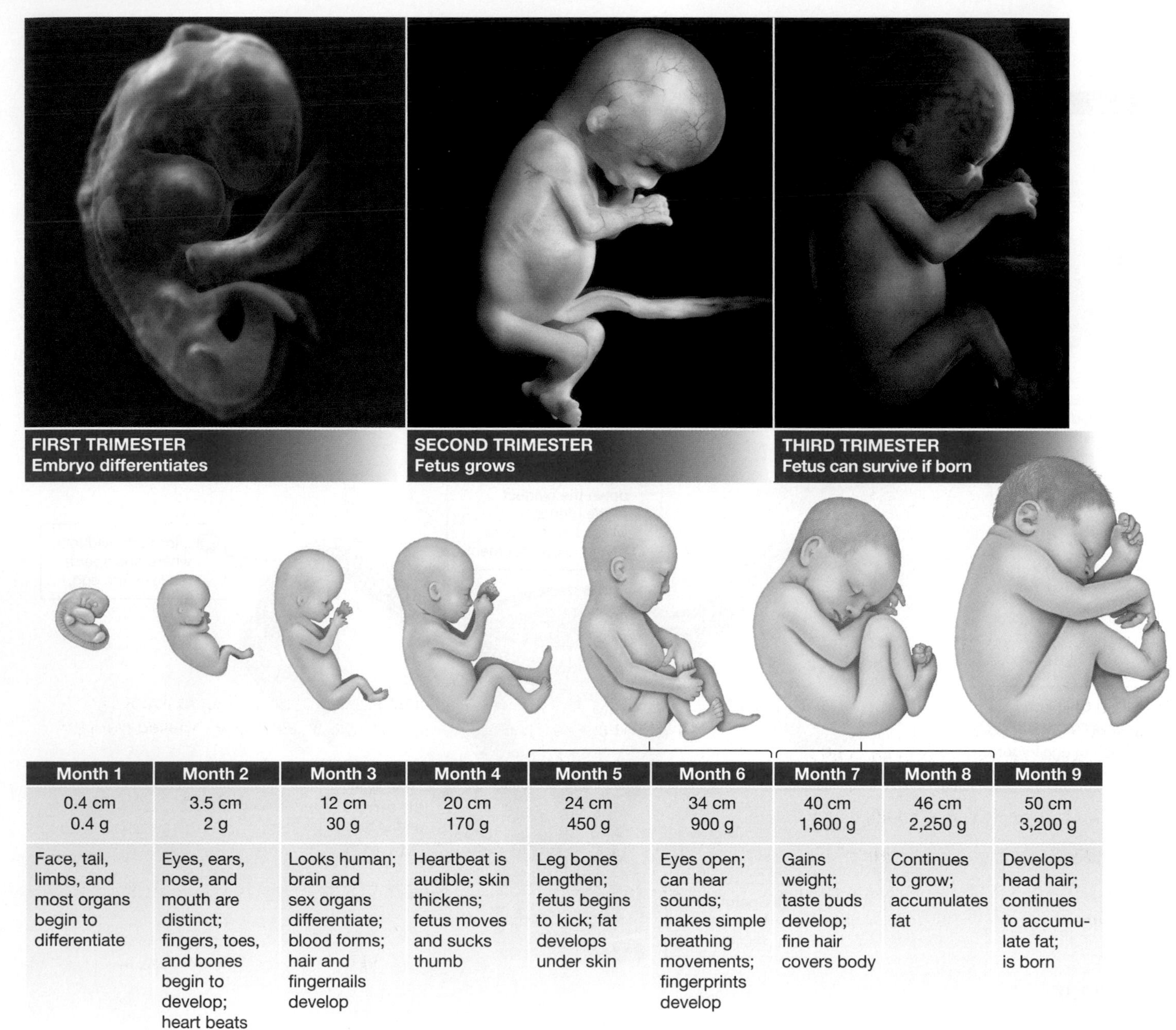

Month 1	Month 2	Month 3	Month 4	Month 5	Month 6	Month 7	Month 8	Month 9
0.4 cm 0.4 g	3.5 cm 2 g	12 cm 30 g	20 cm 170 g	24 cm 450 g	34 cm 900 g	40 cm 1,600 g	46 cm 2,250 g	50 cm 3,200 g
Face, tail, limbs, and most organs begin to differentiate	Eyes, ears, nose, and mouth are distinct; fingers, toes, and bones begin to develop; heart beats	Looks human; brain and sex organs differentiate; blood forms; hair and fingernails develop	Heartbeat is audible; skin thickens; fetus moves and sucks thumb	Leg bones lengthen; fetus begins to kick; fat develops under skin	Eyes open; can hear sounds; makes simple breathing movements; fingerprints develop	Gains weight; taste buds develop; fine hair covers body	Continues to grow; accumulates fat	Develops head hair; continues to accumulate fat; is born

Figure 22.10

Nine months in the womb

Fertilization of an ovum leads to a 9-month-long period of development within the mother's uterus.

Q1: Place these terms in the correct order of development: embryo, fetus, infant, zygote.

Q2: In what trimester is the fetus most likely to survive outside its mother's body?

Q3: The first trimester is the most sensitive time for exposure to mutagens (factors, such as chemicals, that may produce or increase mutations). Why might that be?

See Appendix A for answers to the figure questions.

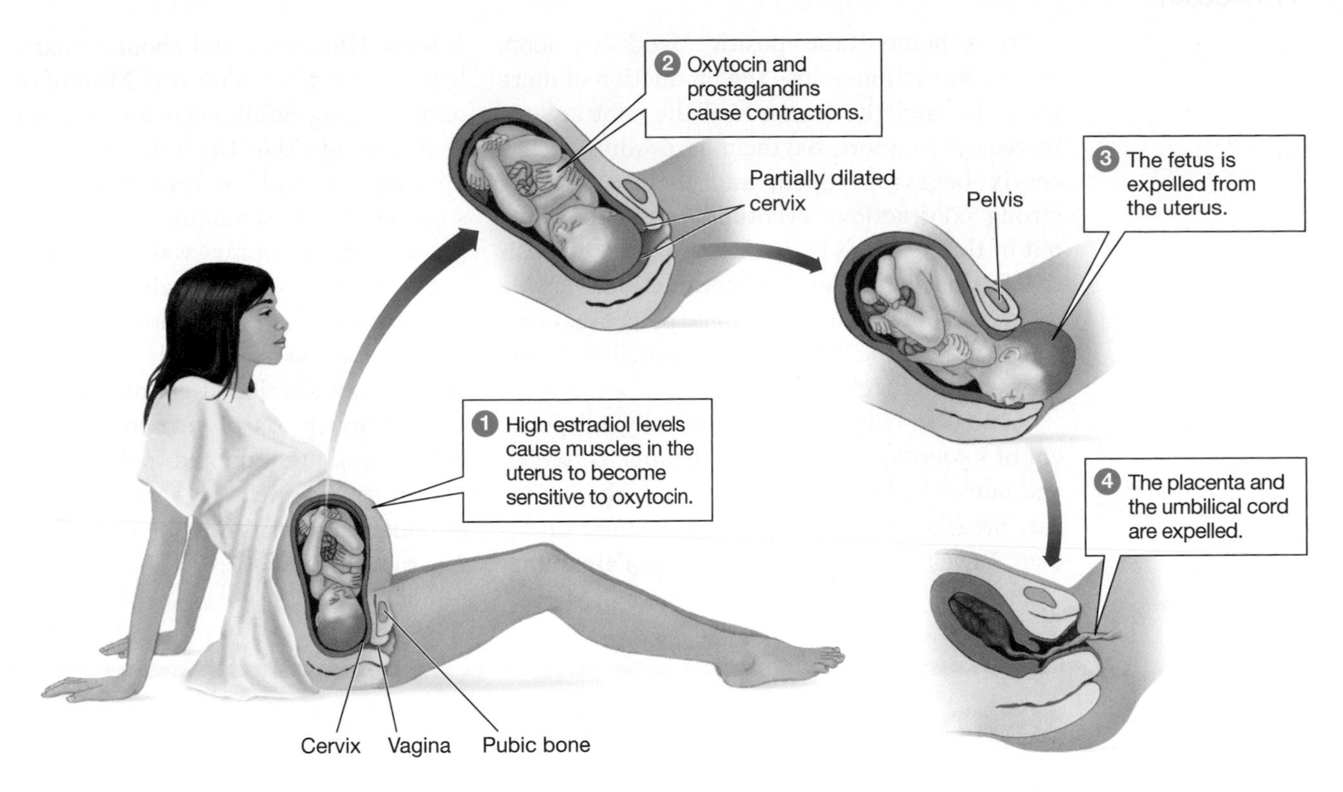

Figure 22.11

Childbirth is orchestrated by hormones

Childbirth occurs in stages, driven by the hormone oxytocin. Oxytocin signals uterine muscles to contract. The contractions become stronger as the amount of oxytocin increases. The mother's cervix opens, and the fetus is eventually expelled from the uterus, followed by the placenta soon after.

Q1: What is the role of estradiol in childbirth?

Q2: Explain how the involvement of hormones in childbirth is an example of a positive feedback loop.

Q3: If a woman has been pregnant for more than 40 weeks, her doctor might give her an injection of oxytocin to induce labor. How would that bring about labor?

See Appendix A for answers to the figure questions.

from heavy drinking by the mother during pregnancy. However, the causes of most birth defects remain unknown, though they are likely due to a mix of genetic and environmental factors.

During the next 3 months, the second trimester, the organs develop and the fetus increases in size. By the start of the third and final trimester, fetal development has progressed to the point that, with the help of modern technology, the fetus has reasonably good odds of surviving outside the mother's body. It gains a good deal of weight during the third trimester, and its circulatory and respiratory systems prepare for living in a gaseous atmosphere rather than the watery world of the amniotic fluid.

By the end of the third trimester, the fetus is ready for its sudden transition from the uterus to the outside world: childbirth (**Figure 22.11**). The last few weeks of pregnancy are marked by hormonal changes. Specifically, higher estrogen levels in the mother's blood make the muscles of her uterus more sensitive to **oxytocin**, a hormone secreted by the fetus and, later in the birth process, by the mother's pituitary gland. Oxytocin stimulates the uterine muscles and causes the placenta to secrete prostaglandins, which reinforce the contractions. Labor begins when the muscles of the uterus begin to contract in response to these hormones.

In a homeostatic positive feedback loop, more contractions cause the production of more oxytocin, and the strength of the contractions increases as more oxytocin is produced. The cervix begins to open, and the increasingly strong contractions eventually push the fetus out of the mother's body. At this point the positive feedback ends, and the contractions subside as oxytocin levels decrease. The placenta, often referred to as the "afterbirth," is expelled during the last stage of childbirth.

At birth, a baby becomes physically independent of its mother. It no longer obtains its oxygen and nutrients directly from her blood, and it must eat, breathe, and maintain homeostasis on its own. Development does not end when an animal is born. Humans spend about a quarter of their lives reaching full adult size. Most of this growth occurs during childhood before sexual maturity.

Amazingly, the "Do It for Denmark" ad blitz appeared to work, at least temporarily. In the summer of 2016, 9 months after the commercials premiered, the country was preparing for a baby boom, with 1,200 more babies due to be born in that summer than in the previous summer.

All signs point to birth rates continuing to fall in most developed countries, including the United States. As of yet, however, the United States has not adopted policies designed to encourage women to have more children, although the idea is not far-fetched. Is "Do It for America" in our future?

NEED-TO-KNOW SCIENCE

- **Anatomy** (page 425) is the study of the structures that make up a complex multicellular body, and **physiology** (page 425) is the science of the functions of anatomical structures.
- **Tissue** (page 425) consists of cells that, in an integrated manner, perform a shared set of functions. Four main types of tissues are found in vertebrates: **epithelial** (page 425), **connective** (page 425), **muscle** (page 425), and **nervous** (page 425).
- In an **organ** (page 426), more than one tissue type forms a unit with a function, shape, and location in the body. In an **organ system** (page 426), two or more organs, in close coordination, perform a set of functions.
- **Homeostasis** (page 426) is the body's way of maintaining a relatively constant internal state despite changes in the external environment. **Homeostatic pathways** (page 426) have two basic features: sensors that monitor the internal environment, and regulatory processes that attempt to restore the normal internal state (the **set point**, page 426) when deviations from optimal conditions are detected.
- Many homeostatic pathways are controlled by **feedback loops** (page 426). In **negative feedback** (page 426) loops, the results of a process cause that process to slow down or stop. In **positive feedback** (page 428) loops, the results of a process cause it to speed up until an endpoint is reached.
- Sexual reproduction involves the organs of the **reproductive system** (page 425) and the union of male and female gametes (**sperm**, page 429, and **eggs**, page 429). Most animals produce offspring in this way, some can individually produce genetically identical offspring via asexual reproduction, and some can reproduce sexually and asexually.
- **Oogenesis** (page 429) and **spermatogenesis** (page 432) are the production of (haploid) eggs and (haploid) sperm, respectively. At birth, the human female's **ovaries** (page 431) already contain her entire lifetime supply of **primary oocytes** (page 431). Males produce sperm in **testes** (page 432).
- Three major types of hormones are **estrogens** (page 434), such as **estradiol** (page 434), which play a role in determining female characteristics; **progestogens** (page 434), such as **progesterone** (page 434), which have a number of functions in the female body; and **androgens** (page 434), such as **testosterone** (page 434), which stimulate cells to develop male characteristics.
- Approximately monthly, one egg released from a woman's ovary moves into the **oviduct** (page 434), where it can be fertilized. During sexual intercourse, the man's **penis** (page 434) releases into the woman's **vagina** (page 434) up to nearly 300 million sperm, which travel through the **cervix** (page 434) and **uterus** (page 434) to the oviduct, where only one of them can fertilize the egg to produce a diploid zygote.
- During the first **trimester** (page 434) of human development, cells of the **embryo** (page 434) rapidly differentiate into the various organs and structures present at birth. From the ninth week of development on, the developing human is called a **fetus** (page 435). During the second and third trimesters, the fetus grows rapidly.
- Childbirth occurs in stages. The hormone **oxytocin** (page 437) signals uterine muscles to contract. The contractions become stronger as positive feedback increases the amount of oxytocin produced.

THE QUESTIONS

See Appendix B for answers.

The Basics

1 Denmark's birth rate is dropping because

(a) women are now older when they begin a family.

(b) men are now older when they begin a family.

(c) exposure to chemicals is decreasing the sperm count in men.

(d) all of the above

2 Tissues

(a) are composed of cells that work in an integrated manner.

(b) have a distinctive shape and location in the body.

(c) are composed of multiple organs.

(d) are composed of only one cell type.

3 Homeostasis does *not* maintain

(a) cellular pH.

(b) body temperature.

(c) environmental temperature.

(d) blood oxygen levels.

4 Which of the following stimulate the body to develop the characteristics of maleness?

(a) estrogens

(b) spermatogens

(c) progestogens

(d) androgens

5 What is the typical order of events that produce an embryo?

(a) gamete development, ovum release, intercourse, fertilization, implantation

(b) gamete development, ovum release, fertilization, implantation, intercourse

(c) gamete development, ovum release, intercourse, implantation, fertilization

(d) ovum release, gamete development, intercourse, fertilization, implantation

6 Link each term with the correct definition.

spermatogenesis	1. Cells from only one parent produce offspring.
oogenesis	2. The process of producing sperm.
sexual reproduction	3. The process of producing eggs (ova).
asexual reproduction	4. Gametes from two parents combine to produce offspring.

7 Select the correct terms:

The body maintains a constant internal state through (**homeopathic** / **homeostatic**) pathways that trigger regulatory processes when there is movement away from the body's (**set point** / **initiator**). For example, body temperature is maintained through a (**negative** / **positive**) feedback loop—a process known as (**thermoregulation** / **osmoregulation**).

Challenge Yourself

8 Which of the following is an example of a homeostatic positive feedback loop?

(a) the interaction of oxytocin and prostaglandins during labor

(b) the interaction of estradiol and progesterone during the menstrual cycle

(c) the interaction of glucose and insulin during eating

(d) both a and b

(e) both a and c

9 Is this an example of asexual or sexual reproduction? California blackworms are hermaphroditic, meaning that they have both male and female reproductive parts. These worms can break apart and each worm fragment can become a new worm.

10 American and European women aged 40–54 were asked, "What do you think is the ideal number of children for a family to have?" Their answers were consistently higher than actual birth rates. In which two countries shown in the accompanying graph did women want the most children, on average? The least? Which country has the largest gap between the number of children that women would like to have and the number they actually have?

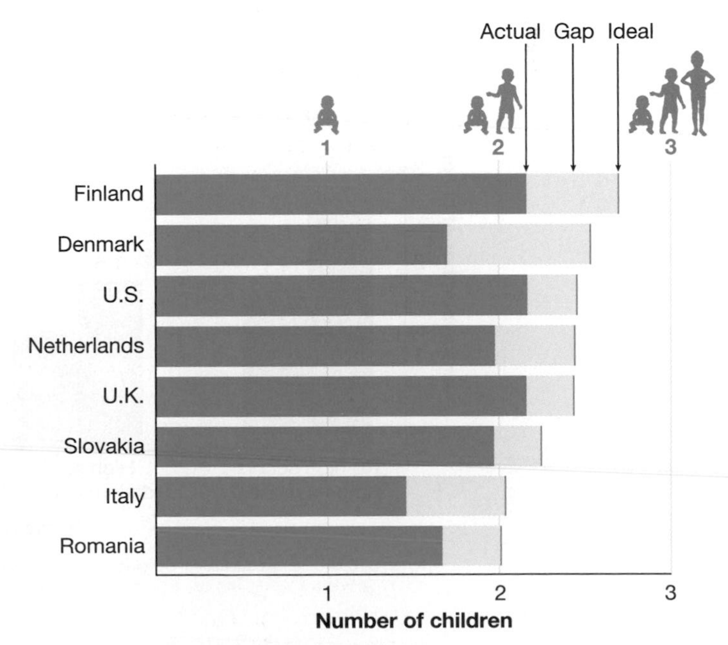

Sources: European data from 2011 Eurobarometer, U.S. data from 2006 and 2008 General Social Survey

11 Place the following events of the human menstrual cycle in the correct order by numbering them from 1 to 5.

_____ a. The corpus luteum secretes progesterone.

_____ b. Follicle-stimulating hormone (FSH) stimulates the growth of ovarian follicles.

_____ c. The egg is released from the follicle.

_____ d. With decreased progesterone, the uterine lining sloughs off (menstruation occurs).

_____ e. Estrogen levels increase and trigger luteinizing hormone (LH).

Try Something New

12 By producing the hormone leptin, fat cells signal to the body that it has sufficient energy stores. Higher leptin levels then decrease the feeling of being hungry. With fewer fat cells, less leptin is produced, and hunger levels are higher. This is an example of

(a) oogenesis.

(b) an organ system.

(c) spermatogenesis.

(d) homeostasis.

(e) contraception.

13 The actual effectiveness of birth control is often significantly lower than the theoretical effectiveness, or "effectiveness if used as directed." For example, there is a 9% annual pregnancy rate of women on the Pill, although its effectiveness is stated as over 99%. Similarly, although the condom is considered 98% effective, the associated annual pregnancy rate is between 10% and 18%. Would you expect a difference between theoretical and actual effectiveness for sterilization (female tubal ligation and male vasectomy; see "Preventing Pregnancy" on page 433)? If not, why not? If so, what would cause the difference?

14 In maintaining homeostasis, small animals face more challenges than do large animals. Small animals tend to exchange water, solutes, and heat with their environments more quickly than large animals do, because a small animal has a larger surface area relative to its volume than a larger animal has. The ratio between these two quantities—surface area and volume—determines how quickly or slowly an animal can gain or lose water, solutes, or heat. When the ratio of surface area to volume is relatively high (small animals), gains and losses are rapid; when the ratio is relatively low (large animals), gains and losses are slower. How does this ratio relate to newborn humans and temperature extremes? What else does the ratio of surface area to volume suggest that we need to consider when caring for a newborn?

Leveling Up

15 **Looking at data** To answer the following questions, refer to the accompanying graph, which shows data collected for the National Survey of Family Growth by the U.S. Centers for Disease Control and Prevention (CDC).

(a) What does the *y*-axis show?

(b) How many children, on average, does a woman with a bachelor's degree (or higher) have?

(c) How many children, on average, does a man with a high school diploma or GED have?

(d) Describe in your own words what the graph shows about the relationship between education and reproduction in the United States.

(e) State a scientific hypothesis that might explain the relationship you described in part (d).

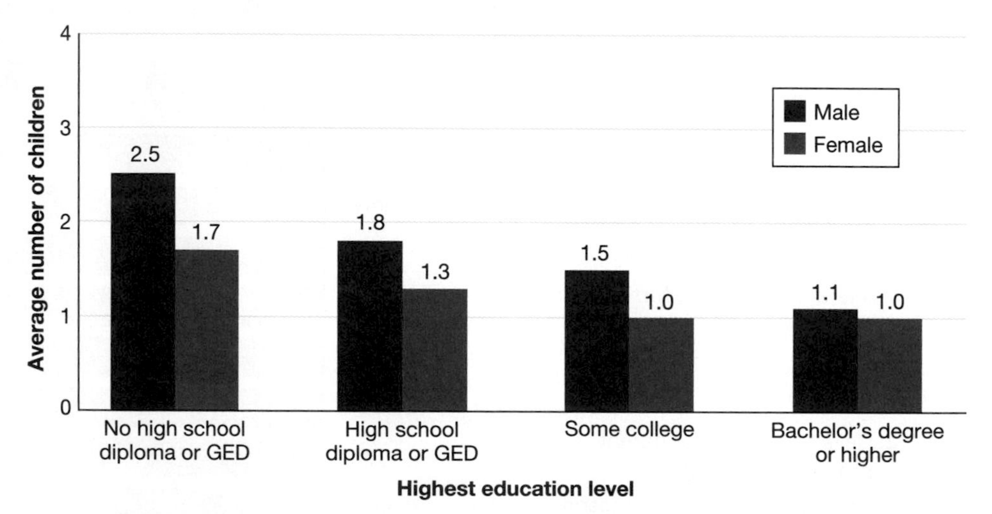

NOTE: GED is General Educational Development high school equivalency diploma.

16 ***Write Now*** **biology: birth control and fertility rates** You are an aide to an extremely busy U.S. senator, who neglected to take a biology class in college. The senator has received an email from a constituent, which included a link to an article about fertility rates: https://www.bloomberg.com/graphics/2019-global-fertility-crash/. The constituent is concerned that people are having fewer children than in the past and thinks that the Senate should pass a law making it illegal to sell birth control to married couples. The senator asks you to write a two-page "white paper" (500 words) that she will use to respond to the email and possibly also to propose legislation. She is relying on you to clearly, concisely, and accurately summarize the issue raised by the email, as well as of possible actions to take. After reading the article, write a position paper addressing the following points (using about half a page, or 125 words, on each point). You may use other resources, including this textbook.

(a) Summarize the main points of the article, defining terms (for example, "fertility rate," "replacement rate") as needed.

(b) Find the fertility rate for your state, and compare it to the U.S. fertility rate. How do the two rates differ, and what is your best hypothesis for why they differ? Do the points made in the article hold for your state?

(c) What are some potential challenges to the legislation proposed in the email? Is there alternative legislation—or another action—that might have the same effect with fewer challenges?

(d) In your final paragraph, advise the senator on how she should respond to the letter and whether she should propose the legislation recommended in the email. Also provide an alternative recommendation, which does not have to involve increasing reproduction. Justify your opinion with data and logic.

Digital resources for your book are available online.

Made to Move

Lack of exercise could be more deadly than smoking, diabetes, even heart disease.

After reading this chapter you should be able to:

- Understand the different roles of involuntary and voluntary muscle contractions.
- Compare and contrast the functions of cardiac, smooth, and skeletal muscle.
- Describe how the components of skeletal muscle work together to contract.
- Separate the skeletal system into axial and appendicular categories by function.
- Articulate the features and functions of cartilage and bone.
- List the elements of a functional joint.
- List the parts of the integumentary system, including the layers of skin.
- Relate the structures to the functions of the individual digestive system components.
- Name and give examples of the two key groups of nutrients essential for maintaining a healthy body.

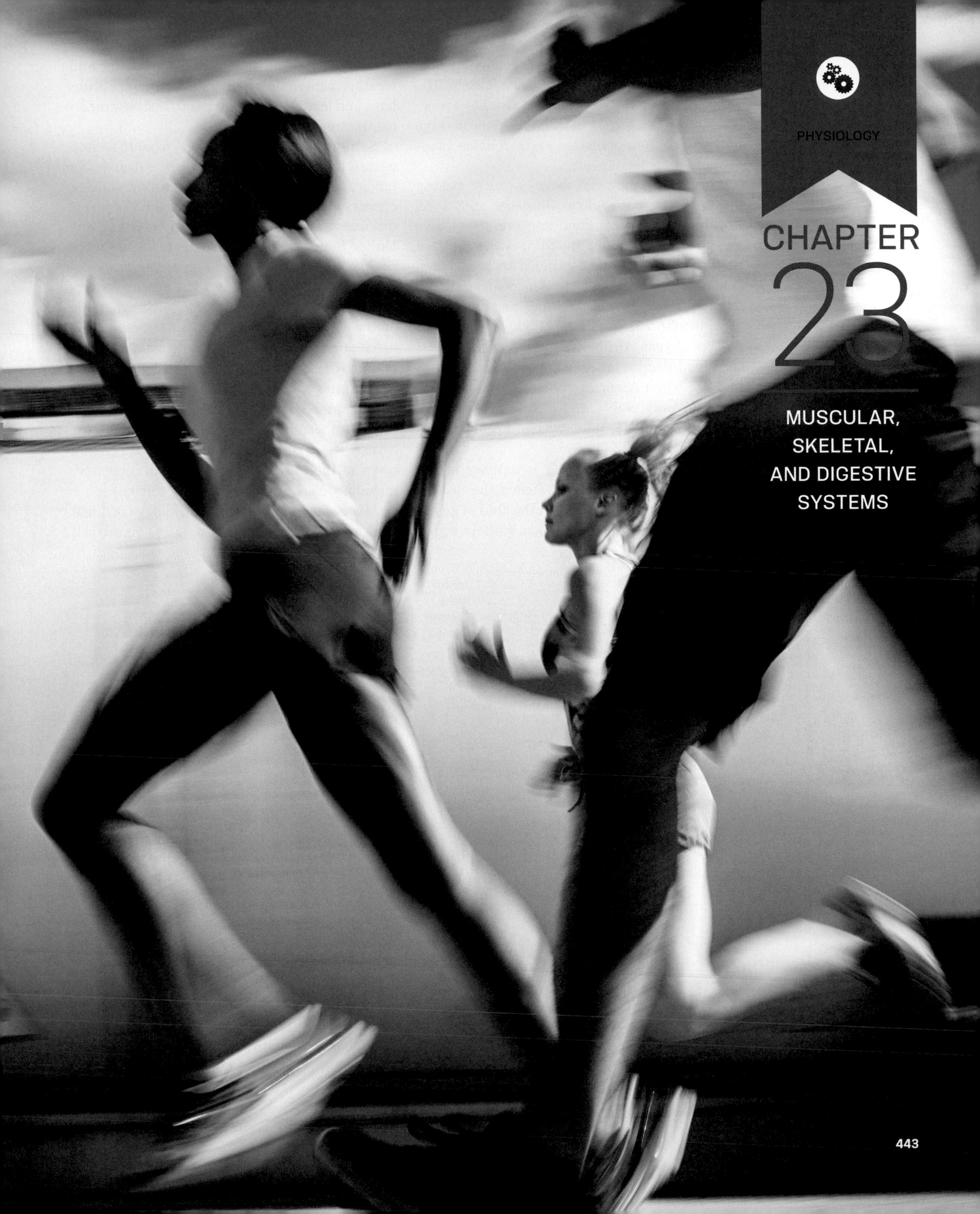

PHYSIOLOGY

CHAPTER 23

MUSCULAR, SKELETAL, AND DIGESTIVE SYSTEMS

Wael Jaber was worried about a line sloping downward on a graph. The slim line represented the life expectancy of Americans—a calculation of the average life span of the population and a reliable snapshot of the society's health.

Life expectancy has been steadily ticking upward across the globe since the 1960s. As health care and sanitation improve, people live longer. Yet beginning in 2015, while most of the world population continued to live longer, Americans did the opposite. Americans shaved a month off their average lifetimes, and the decline continued in 2016 and 2017 (**Figure 23.1**).

The 3-year drop was America's longest sustained decline in life expectancy since 1915 to 1918, when Americans died by the millions during World War I and a flu pandemic. Yet from 2015 to 2018, there was no world war, no deadly pandemic.

Jaber, a cardiologist and researcher at the Cleveland Clinic in Ohio, watched the decline with great concern. Jaber isn't the kind of guy to sit around and observe a problem without trying to fix it. His typical day involves diagnosing and treating patients with complicated heart problems and leading clinical trials to test new medical treatments.

Seeing the decline, Jaber couldn't stop wondering why it was happening. Advances in medical technology were flourishing, curing cancers and genetic diseases (see the Chapter 8 discussion on gene therapy). In addition, the United States spends thousands of dollars more per person on health care than any other country in the world. So why was the country's average life expectancy dropping? Why was that line on the graph going *down*?

The Centers for Disease Control and Prevention (CDC) attributed part of the decline to rising suicide rates and the country's ongoing drug crisis (see Chapter 3 on the opioid epidemic). But other factors appeared to be at play. During the same 3-year period as the decline in life expectancy, seven of the top ten causes of death increased, including heart disease—Jaber's specialty.

Jaber sat down with cardiology trainees working in his lab and said, "Let's figure out

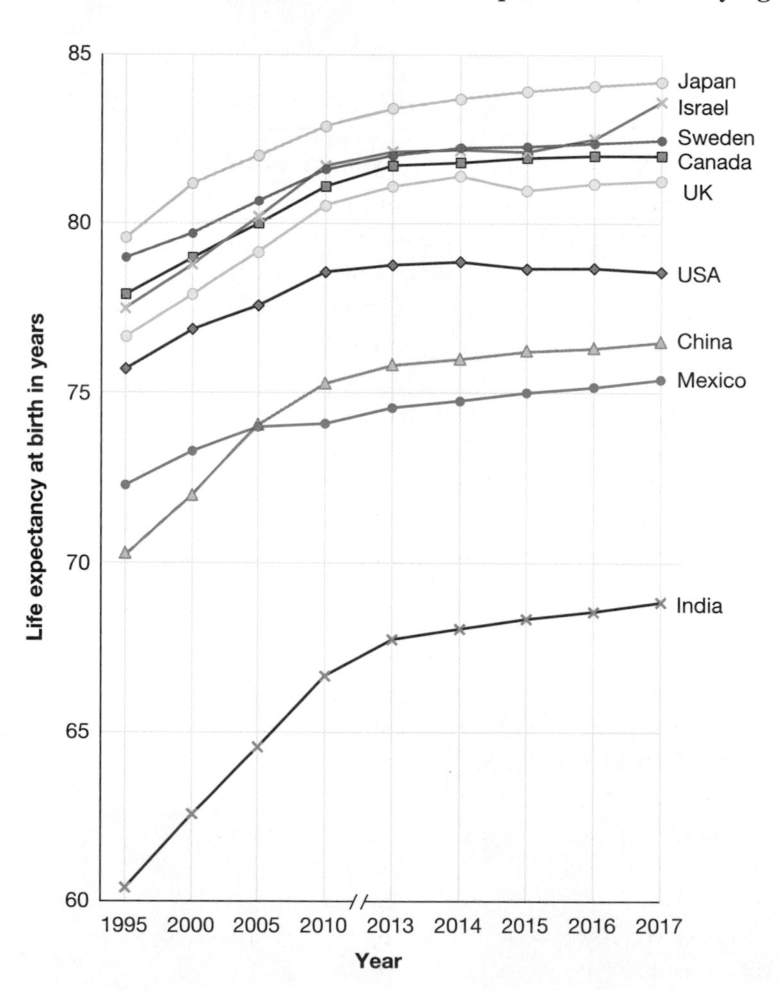

Figure 23.1

United States life expectancy lags behind that of other developed countries

This graph shows life expectancy by birth year in 5-year increments from 1995 to 2010, then by year from 2013 to 2017. Life expectancy increased for all the countries shown except for the United States, whose life expectancy by birth year began declining in 2015.

Q1: **In which two countries did people born in 2017 have the longest life expectancy? In which two countries did people born in 2017 have the shortest life expectancy?**

Q2: **In which two countries did people born in 1995 have the longest life expectancy? In which two countries did people born in 1995 have the shortest life expectancy?**

Q3: **In what year did China's life expectancy exceed that of Mexico? In what year did Israel's life expectancy first exceed that of Canada? In what year did Israel's life expectancy first exceed that of Sweden?**

See Appendix A for answers to the figure questions.

if there's an easy way to look at this." At the Cleveland Clinic, he directs the Stress Lab, a facility that conducts tests to measure heart health. And since 1990, the facility has stored data, with permission, from tens of thousands of patients who underwent stress testing. During that testing, the lab technicians gathered information on age, gender, blood pressure, cholesterol, heart health, smoking habits, and other conditions.

In other words, Jaber had a trove of human health data at his fingertips. And he was going to use it to investigate the question of why Americans are dying at younger ages than they used to.

Little did he know, he would be very surprised by the answer.

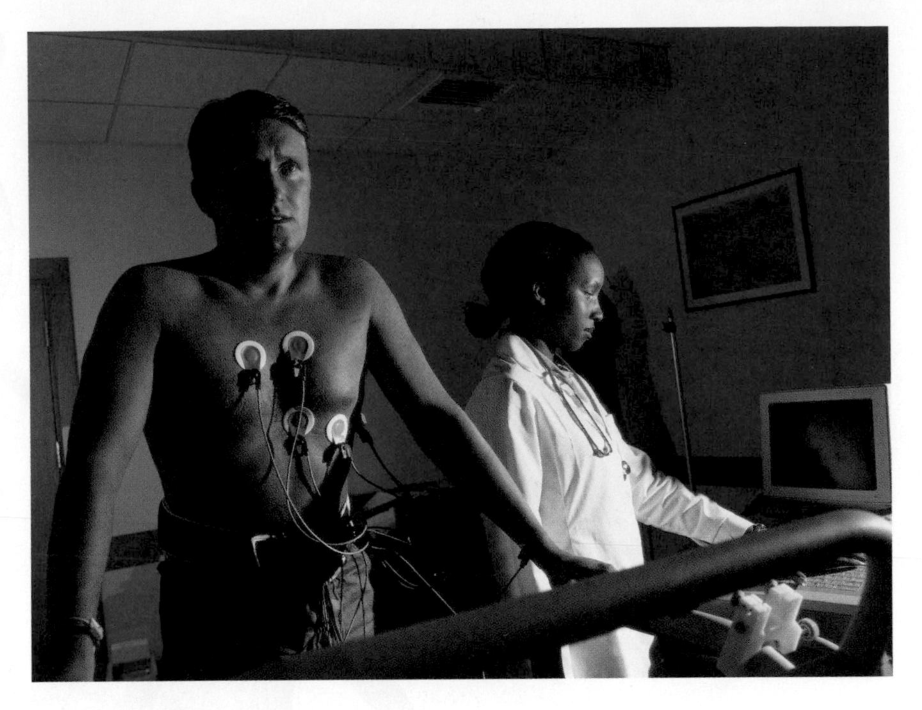

Figure 23.2

Stress test

A stress test involves aerobic exercise on a treadmill as a way to objectively measure cardiorespiratory fitness.

Show of Strength

Typically performed on a treadmill, a stress test evaluates how the heart and rest of the body respond to a challenging workload (**Figure 23.2**). Jaber became director of the Cleveland Clinic Stress Lab in 2000, and he has since examined the hearts and cardiorespiratory fitness levels of thousands of people, from professional basketball players to senior citizens with heart disease, to determine how well their heart, lungs, and muscles perform during exercise.

When an individual enters the Cleveland Clinic for a stress test—say, a pilot who must pass the test for certification, a person with suspected heart disease, or a teenager with an irregular heart valve—the person is asked to walk or run on a treadmill while hooked up to various monitors. During the test, the treadmill progressively increases in speed and ramp angle until the patient says stop. In some cases, physicians inject the patient with a medical dye that enables them to take pictures of blood flow in and around the heart before and after the treadmill test.

The heart is the hardest-working muscle in the human body, pumping out 2 ounces of blood with every heartbeat at an average of 80 times per minute. With the ability to beat over 3 billion times in a lifetime, the heart is a central part of the body's **muscular system**. Muscle tissue is unique to animals and provides the power necessary for movement. Muscle tissue possesses a crucial property: it can contract and relax.

Even when we sit still, the muscular system is hard at work. *Involuntary* muscles operate constantly: the heart pumps blood, the lungs move air, and muscular contractions move food along the digestive tract. Involuntary muscles do their work without our having to think about them.

Two specialized types of muscle engage in involuntary contraction: cardiac muscle and smooth muscle (**Figure 23.3**). The vertebrate heart is the only organ that contains **cardiac muscle**. During a stress test, Jaber and his team can watch the contractions of that cardiac muscle as it powers the movement of blood through the lungs and into the body.

Smooth muscle, by contrast, can be found all around the body. It forms the digestive tract, the walls of blood vessels, the respiratory tract, the uterus, and the urinary bladder. Unlike other types of muscle, smooth muscle cells appear spindle-shaped: wide in the middle and tapered at both ends, like the pupil of a cat's eye. Some smooth muscles contract and release rapidly, whereas others can sustain a contraction for hours.

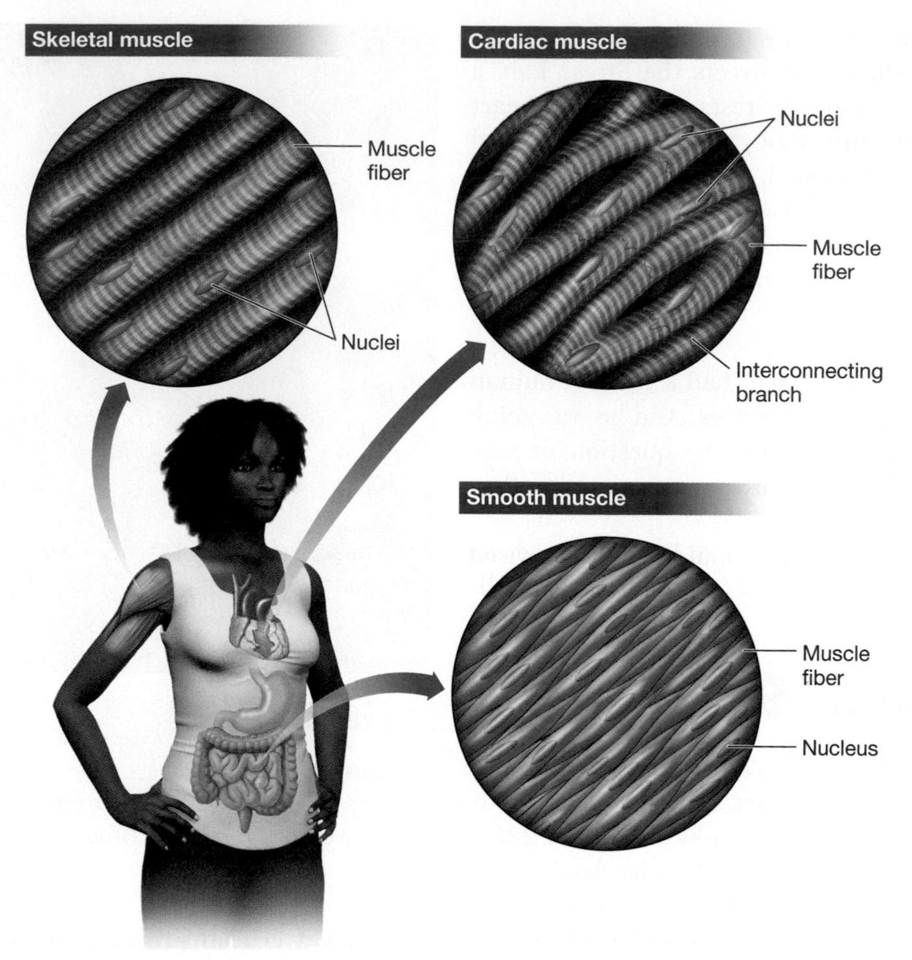

Figure 23.3

Specialized types of muscles for different types of movement

Skeletal muscle has a distinctive banded appearance, brought about by the presence of the sarcomeres (see Figure 23.5). Cardiac muscle is also banded, and its muscle fibers are branched, which helps produce the coordinated contractions known as heartbeats. Because it lacks sarcomeres, smooth muscle has no visible bands.

Q1: Which type(s) of muscles can you voluntarily contract?

Q2: Do the muscles in the heart contract voluntarily or involuntarily?

Q3: You do not have to think about breathing (otherwise, sleeping would be dangerous!), but you *can* increase or decrease your rate of breathing. Are the muscles involved in breathing, then, under voluntary or involuntary control?

See Appendix A for answers to the figure questions.

The muscular system also works at our command. When we walk, run, or jump, we are using *voluntary* contractions. These types of contractions are controlled by **skeletal muscle**. Skeletal muscle makes up your biceps, your calves, and even the facial muscles enabling you to smile or grimace right now (depending on how excited you are to be reading a textbook…).

Skeletal muscle consists of bundles of muscle fibers. A **muscle fiber** is a long, narrow cell that can span the length of an entire muscle because it is made up of many muscle cells that fused together during development (**Figure 23.4**). Each muscle fiber is packed with cylindrical structures, like straws in a tube. These cylinders known as **myofibrils** contain proteins that contract by bracing against each other.

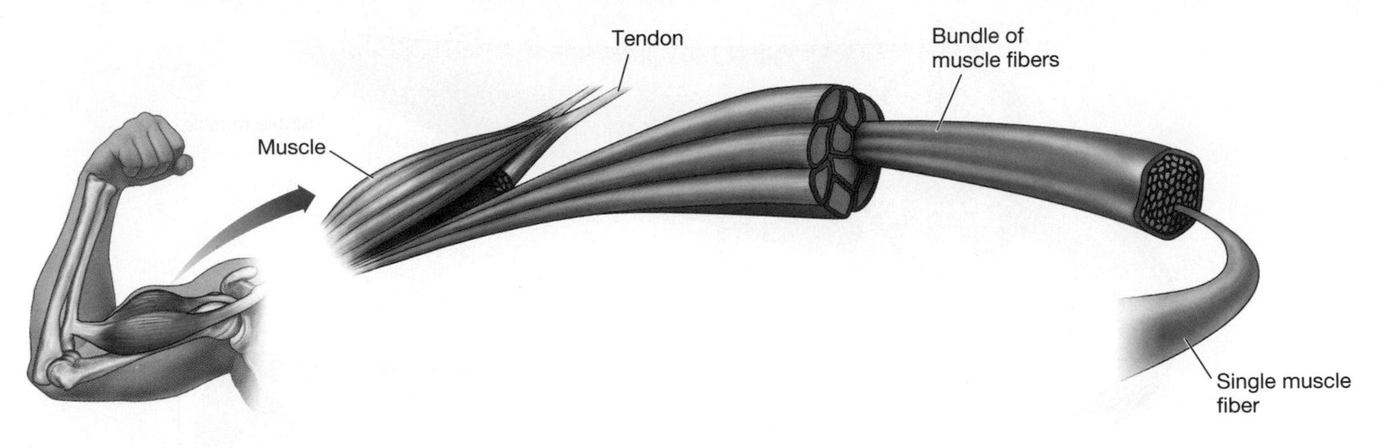

Figure 23.4

The muscles that move you

Both ends of a skeletal muscle are anchored by tendons to nearby support structures, such as bones. A muscle contains bundles of muscle fibers, each one running the length of the entire muscle.

Q1: How is skeletal muscle attached to bones in the skeleton?

Q2: Are skeletal muscles under voluntary or involuntary control?

Q3: Is a muscle fiber one cell or multiple cells?

See Appendix A for answers to the figure questions.

Myofibrils are organized into series of contractile units called **sarcomeres** (**Figure 23.5**).

The simultaneous contraction of sarcomeres, usually taking no more than a tenth of a second, produces the contraction of a whole muscle. Sarcomeres are visible as bands when seen through a microscope. At extremely high magnification, it is possible to see two kinds of protein filaments arranged in a specific manner inside sarcomeres: **actin filaments**, made up of molecules of the protein actin, and **myosin filaments**, consisting of the protein myosin.

Under a microscope, each end of a sarcomere appears as a dark line called the **Z disc**. The two Z discs that mark the ends of the sarcomere contain a large protein that provides anchor points for actin filaments. Between the free ends of the actin filaments, thick myosin filaments attach to the middle of the sarcomere. The sliding of the myosin filaments along the actin filaments enables the sarcomere to contract, like interlocking your fingers together.

Because we've convinced you that skeletal muscle is fascinating, go ahead and smile. See? You just made a slew of sarcomeres contract. Now imagine running on a treadmill for a stress test. Millions of sarcomeres contract and release as myosin filaments slide along actin filaments.

That's what happened in the bodies of all the patients who underwent Jaber's stress tests while machines recorded cardiac muscle contractions, blood pressure, and breathing. The more fit a patient was, the longer amount of time that person was able to spend on a treadmill during the stress test.

Starting in 2017, Jaber and his team began comparing the overall cardiorespiratory fitness level of individuals in the Stress Lab database to their mortality—that is, how long each person lived—based on medical charts and death records. The team analyzed nearly 25 years of data from 122,007 patients and categorized each patient into one of five levels of fitness performance from lowest to highest.

Right away, Jaber saw the correlation. Across all ages and genders, one health factor stood out: time on the treadmill. The longer individuals could last on a treadmill—an indicator of their level of physical fitness—the longer they were likely to live. Fitness level was clearly

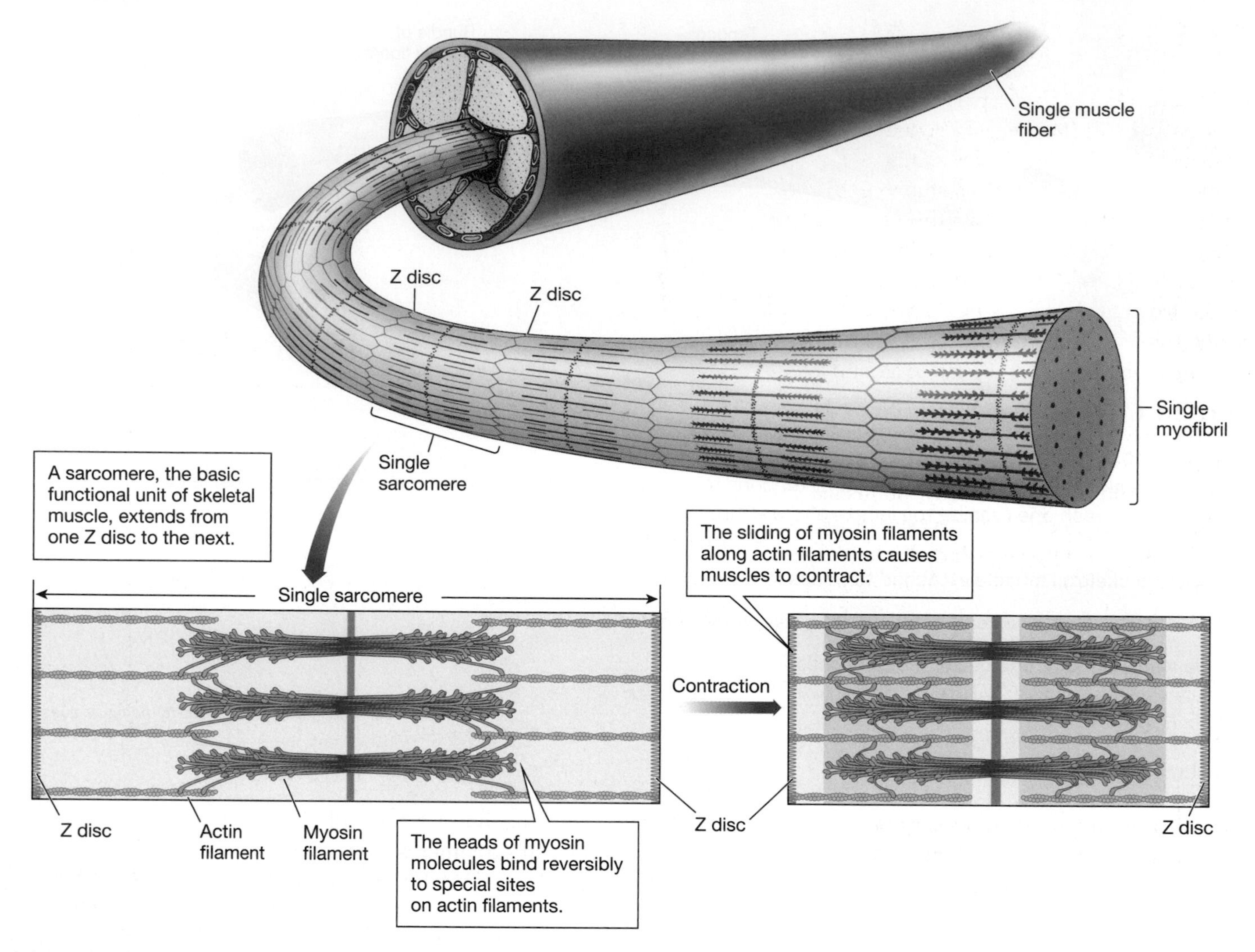
Single muscle fiber
Z disc
Z disc
Single myofibril
Single sarcomere
A sarcomere, the basic functional unit of skeletal muscle, extends from one Z disc to the next.
The sliding of myosin filaments along actin filaments causes muscles to contract.
Single sarcomere
Contraction
Z disc
Actin filament
Myosin filament
The heads of myosin molecules bind reversibly to special sites on actin filaments.
Z disc
Z disc

Figure 23.5

The microscopic structure of muscle

Each muscle fiber contains myofibrils made up of sarcomeres. Muscle contraction depends on the movement of actin filaments and myosin filaments within each sarcomere. At each end of a sarcomere is a Z disc; each of the two Z discs contains a large protein that provides anchor points for the actin filaments.

Q1: List muscle structures from smallest to largest, beginning with sarcomeres.

Q2: What are the components of the sarcomere?

Q3: Across animal species, the microscopic structure of muscles is the same. Why, do you think, are there differences in strength among animals?

See Appendix A for answers to the figure questions.

associated with living longer (**Figure 23.6**). "Even in advanced age, if you take patients in their 70s who can do really well on a treadmill, they are almost invincible. They live and live," says Jaber.

Jaber hypothesized that lack of exercise, therefore, was contributing to early death in Americans. But just how important a risk factor was it? Was sitting around as bad as eating too much pizza? Or having high cholesterol? Or smoking?

Sit Less, Move More

Jaber's team set to work comparing the increased risk of early death from being sedentary—that is, not moving much—to other traditional risk factors for early death, such as smoking, diabetes, high cholesterol, even heart disease.

They listed all the risk factors side by side and included the impact of each factor on survival. "It was extremely surprising," says Jaber. "I thought smoking would be the worst, probably, and being diabetic maybe second to smoking."

Poor fitness level blew the other health factors out of the water. It raised one's risk of an early death to a whopping 3.5 times the increased risk from smoking. Jaber's study, published in 2018 in *JAMA Network Open*, a peer-reviewed medical journal, also concluded that there was no ceiling to how fit a person could be: there is no such thing as too much fitness. Ultrafit athletes in the study lived the longest. Exercise, it turns out, is exceptionally good at keeping us alive.

Fitness can be thought of like a financial retirement plan, says Jaber. One does not enter their 60s, 70s, or 80s in good physical shape by just sitting around. Instead, they need to "deposit" in a fitness "bank" over the years by being physically active and exercising. The more one deposits over time, the higher the return later in life, he says.

Jaber's conclusion—that poor fitness level is more deadly than known risk factors such as smoking, diabetes, and heart disease—has been supported by numerous other studies. A 2015 report from the University of Cambridge in the United Kingdom, which analyzed data from 334,000 men and women over 12 years, found that a moderate amount of physical activity was key to lowering the chances of early death. Even just a 20-minute brisk walk per day, the authors concluded, was enough to reduce the risk of premature death by as much as 30 percent.

In addition, not exercising was twice as bad for a person as being obese. Obesity, a complex health issue resulting from both genetic and environmental factors, increases a person's risk for other potentially serious health conditions, including type 2 diabetes, certain cancers, sleep apnea, and heart disease.

In 2019, the same Cambridge team published a paper combining data from eight studies that

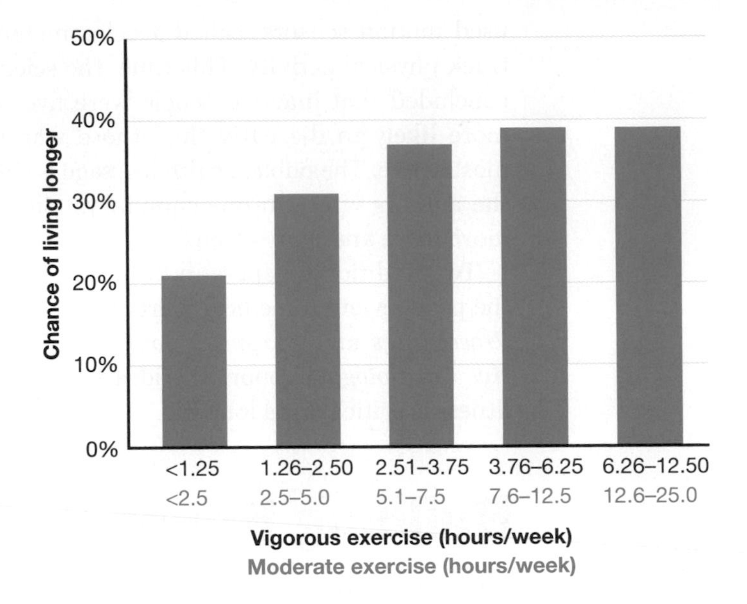

Figure 23.6

Even minimal exercise can substantially increase longevity
In 2008, the U.S. Department of Health and Human Services included this graph with guidelines for the types and amounts of exercise that offer health benefits. The guidelines recommended at least 1.25 hours of vigorous exercise or at least 2.50 hours of moderate exercise per week. Vigorous activities included running, swimming, playing basketball, and any activity that included sprinting, such as indoor cycling. Moderate activities included walking, hiking, bicycle riding, and other activities that do not include sprinting or extremely heavy breathing.

Q1: Participants who vigorously exercised 1.25–2.5 hours per week or moderately exercised 2.5–5.0 hours per week, as recommended by the guidelines, increased their chance of living longer by what percentage?

Q2: How much moderate exercise increased participants' chance of living longer by 20 percent? How much moderate exercise maxed out participants' increased chance of living longer? What was the maximum percentage chance?

Q3: Previous studies suggested that 12.5 hours of vigorous exercise per week, such as elite athletes commonly perform, can be harmful. Does this study support that conclusion? Why?

See Appendix A for answers to the figure questions.

DEBUNKED!

MYTH: Crunches trim belly fat.

FACT: Eating less and exercising more burns fat stores evenly around the body. One cannot reduce fat at a single spot by targeting it with a specific exercise.

used motion sensors, called accelerometers, to track physical activity. This time, the scientists concluded that inactive people were five times more likely to die early than those who were most active. The public health message is simple, the authors wrote in one summary: "Sit less—move more and more often."

Two additional papers in 2019, published in the peer-reviewed medical journals *Mayo Clinic Proceedings* and *European Journal of Preventive Cardiology*, support the idea that physical fitness is critical for a long life.

Building Bones

Although there is no known, clear-cut reason why exercise helps us live longer, for now, there are hypotheses.

For example, people who are fit spend less time being chronically ill with conditions such as diabetes, high cholesterol, or high blood pressure. Chronic conditions cause stress on the body and increase the likelihood of other illnesses. Exercise also improves the strength and function of our muscles and bones. "Exercise affects the muscular and skeletal systems," says Jaber. People who are fit typically have fewer falls, fewer broken hips, and fewer knee problems, he explains.

The **skeletal system** supports the body, gives it shape, and protects soft tissues and organs (**Figure 23.7**). Like most other vertebrates, humans have an internal skeleton made of bones. This skeleton is often divided into two categories.

First, the **axial skeleton**, made up of 80 bones, supports the long axis of the body. It includes the skull, the ribs, and the long, bony spinal column. Although the axial skeleton plays a role in movement, its primary purpose is to protect the body's vital organs.

Second, the **appendicular skeleton** consists of the 126 bones of the arms, shoulders, legs, and pelvis ("appendicular" means "relating to an appendage or limb"). The appendicular skeleton is key for motion.

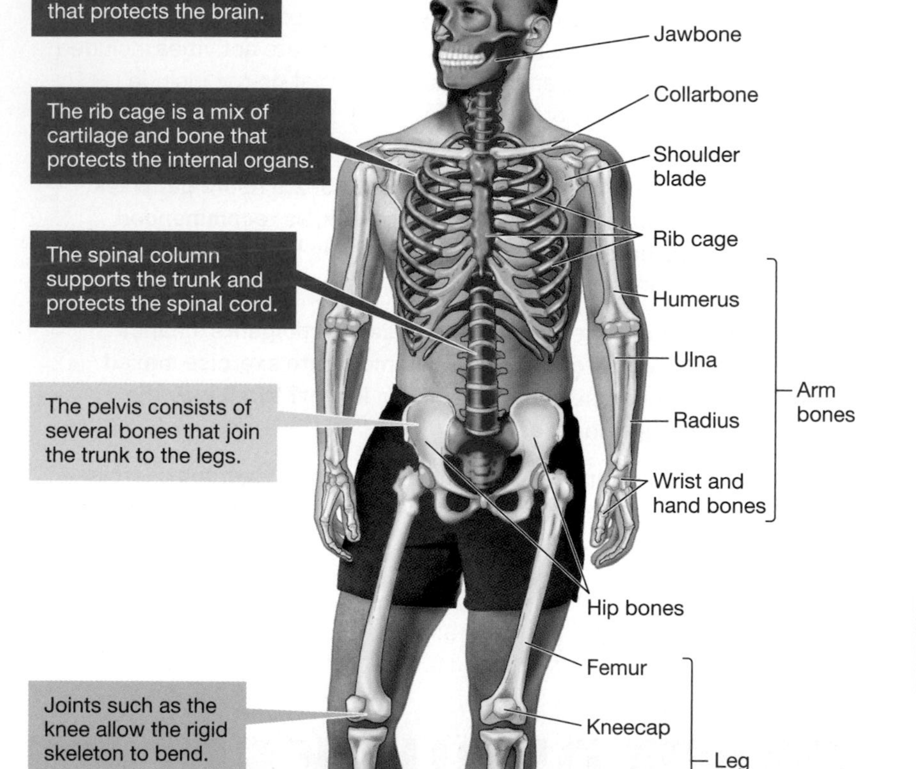

The skull forms a part of the axial skeleton that protects the brain.
The rib cage is a mix of cartilage and bone that protects the internal organs.
The spinal column supports the trunk and protects the spinal cord.
The pelvis consists of several bones that join the trunk to the legs.
Joints such as the knee allow the rigid skeleton to bend.
Jawbone
Collarbone
Shoulder blade
Rib cage
Humerus
Ulna
Radius
Wrist and hand bones
Arm bones
Hip bones
Femur
Kneecap
Tibia
Fibula
Leg bones
Foot bones
Axial skeleton
Appendicular skeleton
Cartilage

Figure 23.7

The human skeleton

In humans, the axial skeleton protects vital organs; the appendicular skeleton facilitates movement.

Q1: **The collarbone is part of which skeleton: axial or appendicular?**

Q2: **Which parts of the skeleton are made of cartilage?**

Q3: **Which skeleton—axial or appendicular—protects the central nervous system (the brain and spinal cord)?**

See Appendix A for answers to the figure questions.

Another important part of our skeleton is **cartilage**, dense tissue that combines strength with flexibility. In the human skeleton, cartilage gives form to the nose, the ears, and part of the rib cage. In addition, cartilage is found at nearly every point in the body where two bones meet (**Figure 23.7**; see also **Figure 23.10**, right side); it creates a smooth surface that prevents the two bony surfaces from grinding against one another.

Cartilage contains cells, but it consists primarily of nonliving, extracellular material—bundles of **collagen**, a tough but pliable protein found in a great variety of tissues, including skin, blood vessels, bones, teeth, and the lens of the eye.

Remember when we noted that most vertebrates have a bony skeleton? Sharks belong to a small class of vertebrates that do not have a skeleton made of bones; their skeleton consists entirely of cartilage and connective tissue. Keep in mind, too, that not all animals have their skeleton on the inside. Although humans and other vertebrates have an internal skeleton, or **endoskeleton**, many other animals, such as lobsters and insects, have an **exoskeleton**, an external skeleton that surrounds and encloses the soft tissues it supports (**Figure 23.8**).

Like cartilage, much of bone is made up of nonliving material. Yet bone is not solid, inactive tissue; it is living tissue with blood vessels and nerve cells. Specialized bone cells called **osteocytes** surround themselves with a hard, mineral matrix of calcium and phosphate. Although they are just single cells, osteocytes can live as long as the organism whose skeleton they belong to. Studies have shown that exercise, especially in growing children, stimulates osteocytes and strengthens bones. Activities that put force on bones, such as running, jumping, and gymnastics, are good at increasing bone strength and density. Even in adults, exercise can have moderate bone-building effects.

Bones are made of two major types of bone tissue. **Compact bone** forms the hard, white outer region. **Spongy bone**, honeycombed with numerous tiny cavities, lies inside the compact bone and is most abundant at the knobby ends of our long bones. Long bones and some others, such as the ribs and breastbone, have a hollow interior, which makes them light but strong. Cavities inside hollow bones contain **marrow**, a living tissue that, depending on the type of bone, stores fat or produces blood cells (**Figure 23.9**).

Figure 23.8

A newly molted cicada emerges from its exoskeleton

Exoskeletons provide protective armor for many animals and also protect terrestrial invertebrates from excessive moisture loss. The rigidity of an exoskeleton means that animals that outgrow their exoskeletons must shed them periodically—a process known as molting.

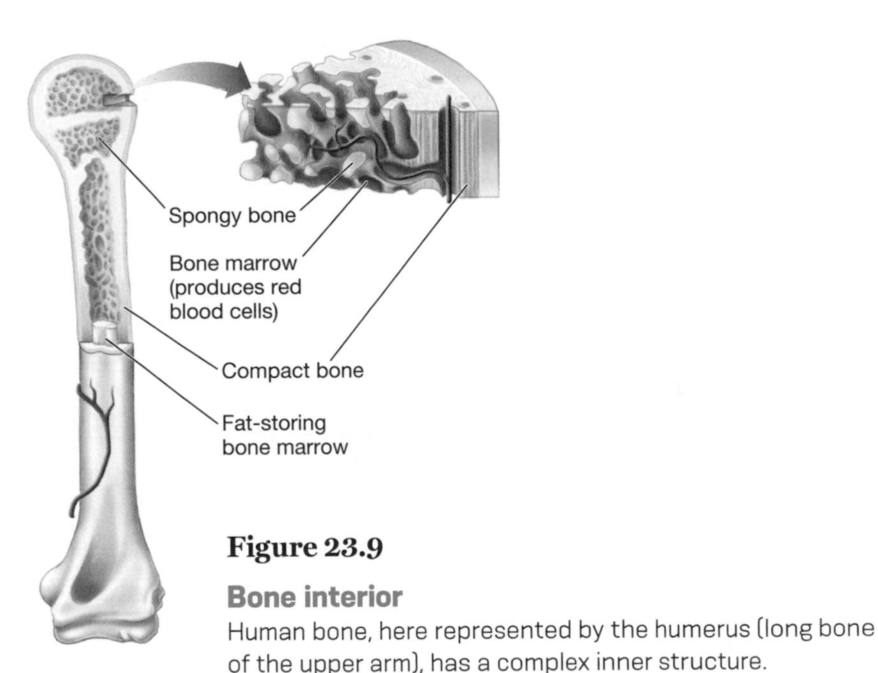

Figure 23.9

Bone interior

Human bone, here represented by the humerus (long bone of the upper arm), has a complex inner structure.

Q1: How does spongy bone differ from compact bone?

Q2: What parts of a bone are made of living cells or living tissue?

Q3: What characteristics of hollow bones make them important for organisms' survival?

See Appendix A for answers to the figure questions.

Bones meet at joints, the junctions in the skeletal system that let the skeleton move in specific ways. Walking, for example, requires movement at the hips and knees, as well as at other joints. The lower jaw connects to the skull at a joint so that it can move relative to the rest of the skull, enabling us to chew and talk.

Joints are held together by collagen-rich ligaments and tendons. **Ligaments** are flexible bands of connective tissue that join bone to bone, whereas **tendons** connect muscle to bone. Wherever two moving parts rub against each other, as in most joints, wear can erode bone, and friction can waste energy. So, moveable joints are lined with sheets of tissue (synovial membranes) that form cavities called **synovial sacs**. The space inside each synovial sac is filled with a lubricating fluid that reduces friction between the two bony surfaces (**Figure 23.10**).

Altogether, these five components—bone, cartilage, ligaments, tendons, and synovial sacs—work to move a joint safely and with precise control. Being fit involves regular, active movement of these components and protects against damage to the skeletal system, leading into a positive feedback loop, says Jaber. "Being fit makes you steadier on your feet for a longer time, so you don't fall. And, if you're steady on your feet for a long time, you keep walking or running," he says.

Exercise also has positive effects on many other body systems, including the brain and nervous system, as discussed in the next chapter. In 2019, scientists at the University of Iowa found that exercise can lead to improvements in brain function and connectivity after just a single workout.

Exercise even affects the skin—the main part of a critical organ system (see "The Skin We're

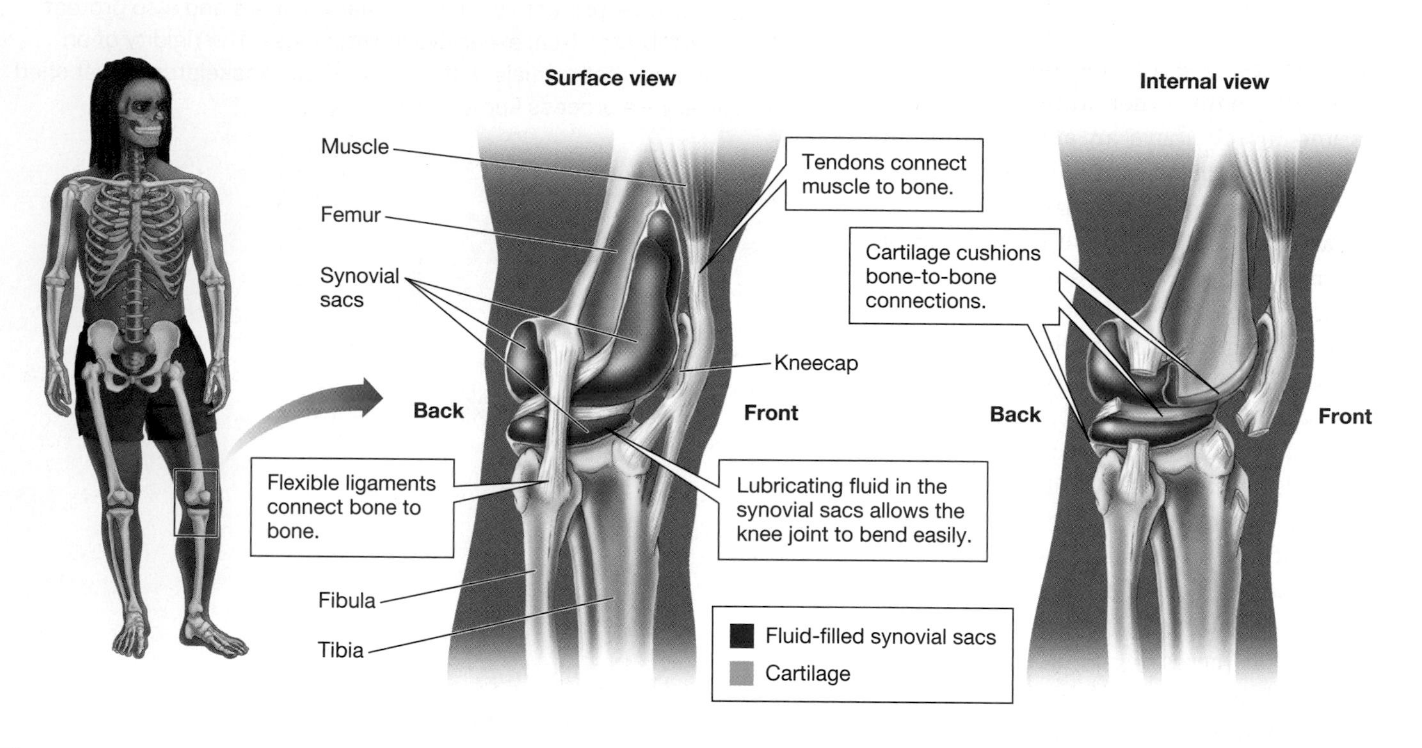

Figure 23.10

Rigidity and flexibility in joints

Although the human knee differs in detail from other joints in the body, it illustrates how rigid and flexible materials are combined in a joint to allow controlled motion.

Q1: What is the function of the synovial sacs?

Q2: Compare and contrast ligaments and tendons.

Q3: Knee injuries are some of the most common sports injuries. Why do you think that is?

See Appendix A for answers to the figure questions.

In" on page 453). Scientists in Canada gathered a group of volunteers, aged 65 or older who did not exercise much, and made an odd request. They asked each volunteer to reveal a buttock, so the scientists could examine skin that hadn't been frequently exposed to the sun. Then, under a microscope, they examined a small skin biopsy taken from each volunteer buttock. Next, the volunteers were put on an endurance training program, jogging or cycling twice per week.

After 3 months, another buttock skin biopsy was taken and examined. Under a microscope, the volunteers' skin "looked like that of a much younger person and all that they had done differently was exercise," the authors told the *New York Times*.

As for Jaber, "I practice what I preach," he says. Every day of the year, Jaber bikes or runs to work, and he goes for long runs on weekends.

The Skin We're In

The **integumentary system** is the largest organ system in the human body, accounting for almost 15 percent of our weight. It covers the body and protects it from environmental hazards such as extreme temperatures and dangerous pathogens. It also prevents water loss and protects the body from physical damage.

The integumentary system consists of the skin and structures embedded in the skin, such as hair and nails in humans, or feathers, hooves, and scales in other vertebrates. As shown in the accompanying figure, the skin is made up of three layers. Moving inward from the outermost layer, the layers are the epidermis, dermis, and hypodermis.

In addition, the skin contains multiple tissue types. The epidermis is an example of epithelial tissue. The nerve endings are examples of nervous tissue. The arrector pili muscle is composed of smooth muscle tissue and involuntarily causes hair to stand on end when stimulated by cold or fear. Much of the dermis is made up of connective tissue. And adipose tissue dominates in the hypodermis, a thick insulating sheet under the other layers of the skin.

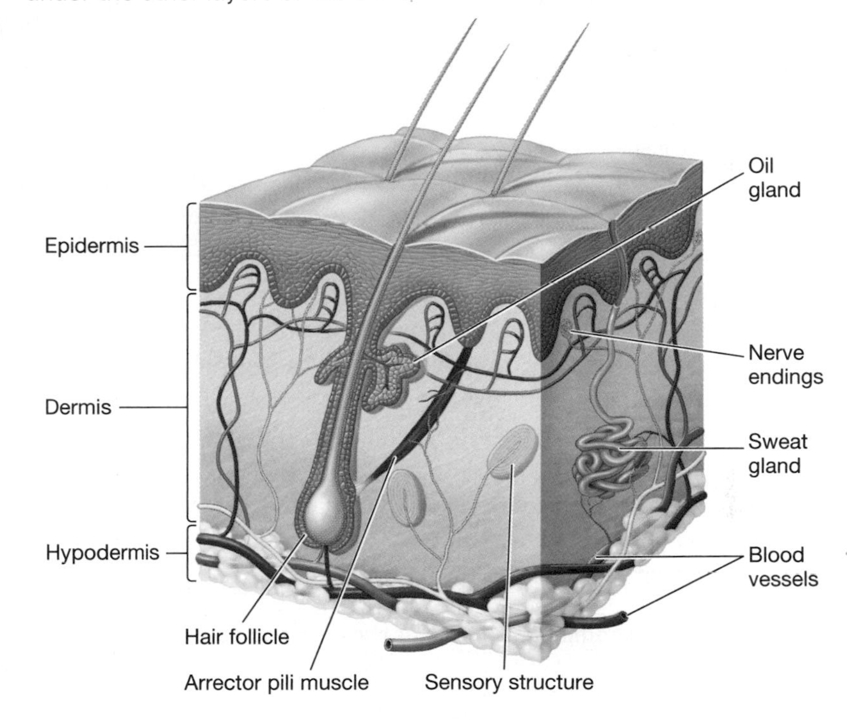

You Are What You Eat

Overwhelming scientific evidence supports the idea that exercise is good for the human body—for muscles, bones, skin, and even the gut.

The **digestive system**, also known as the gut or the gastrointestinal (GI) tract, processes food, absorbs vitamins and minerals, and eliminates unusable waste. The digestive system for most mammals consists of a long, hollow passageway known as the digestive tract and a number of accessory organs, such as the pancreas and liver (**Figure 23.11**).

In 2018, Stephen Carter, an exercise physiologist at the University of Alabama at Birmingham, was studying the effects of exercise training on breast cancer survivors. He became intrigued with two new studies describing a link between exercise and gut microbiota—that is, the population of microbes, including fungi and bacteria—living in our gut.

"There were two papers reporting similar findings. They agreed that a relationship existed between the gut microbiota and aerobic fitness level," says Carter. "Basically, if a person had a higher fitness level, this corresponded with greater gut microbiota diversity."

Growing evidence suggested that having a wide variety of different gut microbes was associated with better overall health. Consumption of a healthy diet with regular exercise is associated with a greater variety of gut microbe species. However, the digestive system tends to lack microbe species diversity in obese persons and those with chronic inflammatory diseases. Low microbe species diversity is also correlated with many other diseases and cancers.

Carter and colleagues enrolled a group of 37 breast cancer survivors who had recently

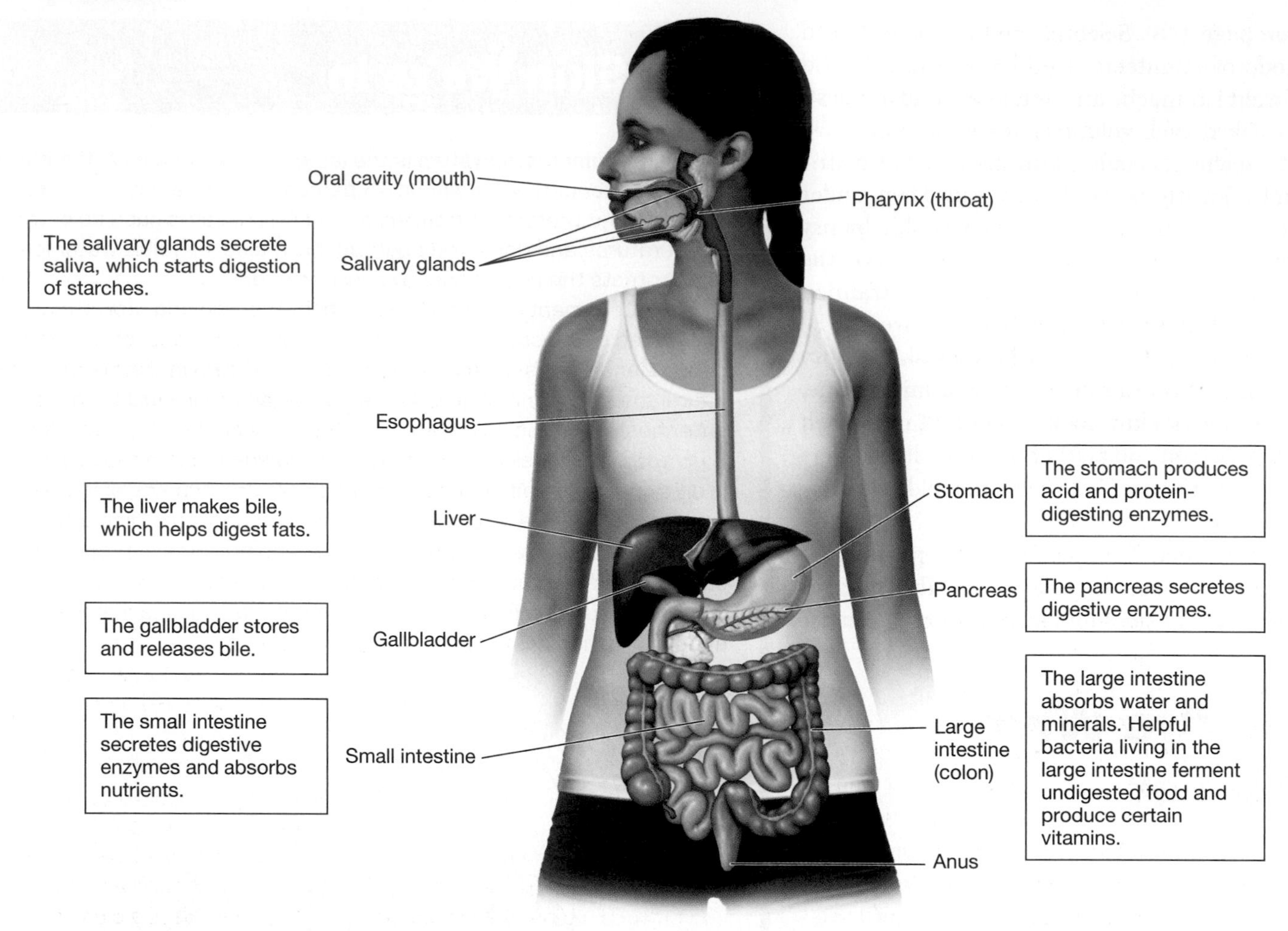

Oral cavity (mouth)
Pharynx (throat)
The salivary glands secrete saliva, which starts digestion of starches.
Salivary glands
Esophagus
The stomach produces acid and protein-digesting enzymes.
Stomach
The liver makes bile, which helps digest fats.
Liver
Pancreas
The pancreas secretes digestive enzymes.
The gallbladder stores and releases bile.
Gallbladder
The large intestine absorbs water and minerals. Helpful bacteria living in the large intestine ferment undigested food and produce certain vitamins.
The small intestine secretes digestive enzymes and absorbs nutrients.
Small intestine
Large intestine (colon)
Anus

Figure 23.11

The digestive system converts food into absorbable nutrients

As food moves through the digestive system, it is broken down into small molecules that can be absorbed by the lining of the intestines. Unabsorbed food passes through the intestines and is excreted.

Q1: List, in order, the structures of the digestive system that a piece of swallowed food would pass through, beginning with the oral cavity.

Q2: What part of the digestive system hosts bacteria that produce vitamins?

Q3: What is the common function of the liver and gallbladder?

See Appendix A for answers to the figure questions.

completed treatment for their illness. Breast cancer survivors often become less fit over time because the disease and treatment can leave them feeling persistently fatigued.

Carter compared the aerobic fitness level of each participant—again based on treadmill tests—to the person's gut microbiota, analyzed using fecal samples. He and his colleagues found that individuals with higher aerobic fitness levels had significantly greater gut microbiota diversity than participants with lower fitness levels.

"Fitness is correlated with the gut microbiota diversity," says Carter, who now works at Indiana University Bloomington. Carter believes the results, combined with the previous papers, extend beyond cancer survivors to all people. "It's becoming an accepted idea that fitness and the gut microbiome are related."

An individual's digestive system begins to be colonized by microbes during birth and soon includes microbes the baby ingests from breast milk or formula. Eating, or **ingestion**, is the first stage in the processing of food by the digestive system. **Digestion**, the mechanical and chemical breakdown of food, begins almost immediately after ingestion in many species.

During ingestion, a bite of food—say, a spoonful of cornflakes—is deposited in the **oral cavity**, the mouth. There, an array of different types of teeth, which are shaped to cut, crush, or grind food into smaller pieces, begin to break apart the cornflakes. Many small pieces of food provide a greater surface area for digestive enzymes to work on than do fewer large pieces.

The muscular tongue mixes the crushed cereal particles with saliva. **Saliva** contains enzymes that start to break down starches—or any carbohydrate—into sugars. If you chew a piece of bread long enough, for example, its starches are digested to sugar and it will begin to taste sweet. Saliva is also important for turning the crunchy cereal into a moist mass that can slip easily down the throat.

The tongue assists in pushing the now moist cereal into the throat, or **pharynx**, where the back of the mouth and the nasal cavity come together. The pharynx is the common entryway for both the air tube (the trachea) and the food tube (the **esophagus**). That is why, on occasion, a person might cough up food or liquid that "went down the wrong tube"; it accidentally went down the trachea instead of the esophagus.

Normally, the pharynx is very good at separating air and food. When the mushy bite of cereal makes contact with the wall of the pharynx, it stimulates nerves that launch the **swallowing reflex**, in which a flap of tissue, called the epiglottis, seals off the entry into the trachea. The cereal is then pushed into the esophagus.

Waves of muscular contractions carry the cereal down the esophagus and into the **stomach**. The digestion of nutrients begins in the stomach. **Nutrients** are components of foods that an organism needs to survive and grow. These include micronutrients (vitamins and minerals) and macronutrients (see "Necessary Nutrients" on page 455).

The stomach secretes acid and enzymes to break down macronutrients. Muscles in the wall of the stomach alternately contract and relax to mix the food particles with the acid and enzymes. The resulting watery mixture is stored in the stomach until it can move into the **small intestine**, a highly coiled tube about 3–4 centimeters in diameter. If straightened, the small intestine would extend about 20 feet (6 meters).

The upper region of the small intestine, which lies nearer to the stomach, uses enzymes secreted by the **pancreas** and the intestine to break down large molecules into simpler forms that the body can absorb. Here, the digestion of proteins, carbohydrates, and lipids, including fats, is essentially completed. The resulting small molecules include amino acids, simple sugars, and fatty acids and are absorbed here. The upper region of the small intestine also absorbs vitamins, electrolytes, and copious amounts of water.

The digestion of fats poses a particular problem because fats are not soluble in water, yet they need to be broken down and made to mix with the watery contents of the digestive tract. **Bile** is a fluid that helps digest fats by creating a coating that enables the fat globules to interact

Necessary Nutrients

Two key types of nutrients are essential for the normal functioning of our bodies.

Macronutrients are large organic molecules classified into three main categories: carbohydrates, lipids, and proteins. These biomolecules serve as sources of energy and furnish the body with chemical building blocks such as sugars, fatty acids, and amino acids. Although an adult human can synthesize some of the 20 amino acids needed to make proteins, we must get nine of them, called **essential amino acids**, from food. Another type of macronutrient is dietary fiber, which does not contribute amino acids or energy to the body but is critical for survival because it affects how other nutrients are absorbed in the gut.

Micronutrients include vitamins and minerals. **Vitamins** are small, organic nutrients needed by our bodies but only in tiny amounts (**Figure 23.12**). They participate in a great variety of essential metabolic processes, such as helping blood cells form and maintaining brain function. Some vitamins bind to enzymes to speed up chemical reactions within a cell. Some act as a delivery service, supplying chemical groups needed in important metabolic reactions. Others act as signaling molecules. And some are even believed to work as antioxidants, substances that protect body tissues from destructive chemicals known as free radicals. **Minerals** are inorganic chemicals that have critical biological functions. Carbon, hydrogen, oxygen, and nitrogen make up about 93 percent of the animal body so, by convention, these four elements are excluded from the category of dietary minerals. But more than 20 other elements, such as fluoride, sodium, and iodine, are classified as dietary minerals. Calcium is the most abundant dietary mineral in the body; it makes up a large proportion of your bones.

Vitamin C is abundant in fruits and vegetables, and it assists in the maintenance of teeth, bones, and other tissues. Deficiency of this vitamin leads to scurvy, in which teeth and bones degenerate.

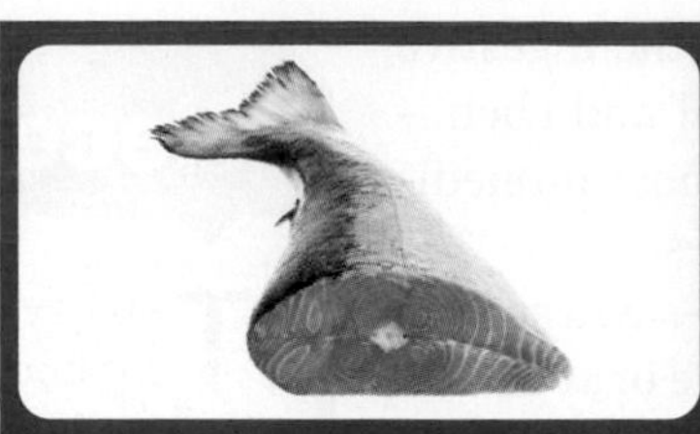

Fish is the richest source of *vitamin D*; fortified foods (such as milk, soy milk, and breakfast cereals) are important sources for most people. A deficiency in vitamin D leads to poor formation of bones and teeth.

Folic acid, a *B vitamin*, is abundant in green vegetables, legumes, and whole grains. B_{12}, another B vitamin, is scarce in plant foods but abundant in milk, meat, fish, and poultry. It is important for the maintenance of teeth, bones, and other tissues.

Vitamin E is abundant in nuts, vegetable oils, whole grains, and egg yolk. It protects lipids in cell membranes and other cell components.

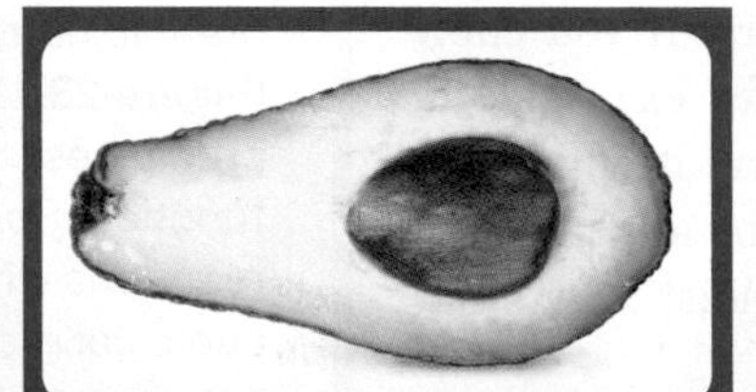

Leafy green vegetables and some fruits (e.g., avocado and kiwi) are rich in *vitamin K*, which is also manufactured by intestinal bacteria. A deficiency can cause prolonged bleeding and slow wound healing.

Carotene is responsible for the color of yellow and orange fruits and vegetables. It is converted into *vitamin A* within our bodies. Vitamin A aids in production of the visual pigment needed for good eyesight and is used in making bone.

Figure 23.12

Vitamins needed in the human diet

Humans need nine **water-soluble vitamins** (C and eight different B vitamins) and four **fat-soluble vitamins** (A, D, E, and K) from their diet. Because water-soluble vitamins are easily excreted in urine, they tend not to accumulate in body tissues, meaning we must obtain these vitamins from food on a regular basis. Fat-soluble vitamins are not excreted as readily and tend to accumulate in body fat, so excessive consumption can lead to overdosing.

Q1: Which vitamins described in the figure are important for healthy bones?

Q2: Which vitamins are you more likely to store and accumulate more than you need?

Q3: In your own diet, are there any vitamins that you may not be eating enough of?

See Appendix A for answers to the figure questions.

with water molecules to partially dissolve them. Bile is produced by the **liver**, an organ that serves a multitude of functions. Some of the bile made by the liver is stored in the **gallbladder**, which dispenses the bile into the small intestine as needed.

The main function of the lower region of the small intestine is absorption of vitamin B_{12}, bile salts, and any nutrients not previously absorbed. **Absorption** occurs by cells lining the cavity of the digestive tract. The innermost lining of the small intestine presents a large surface area for

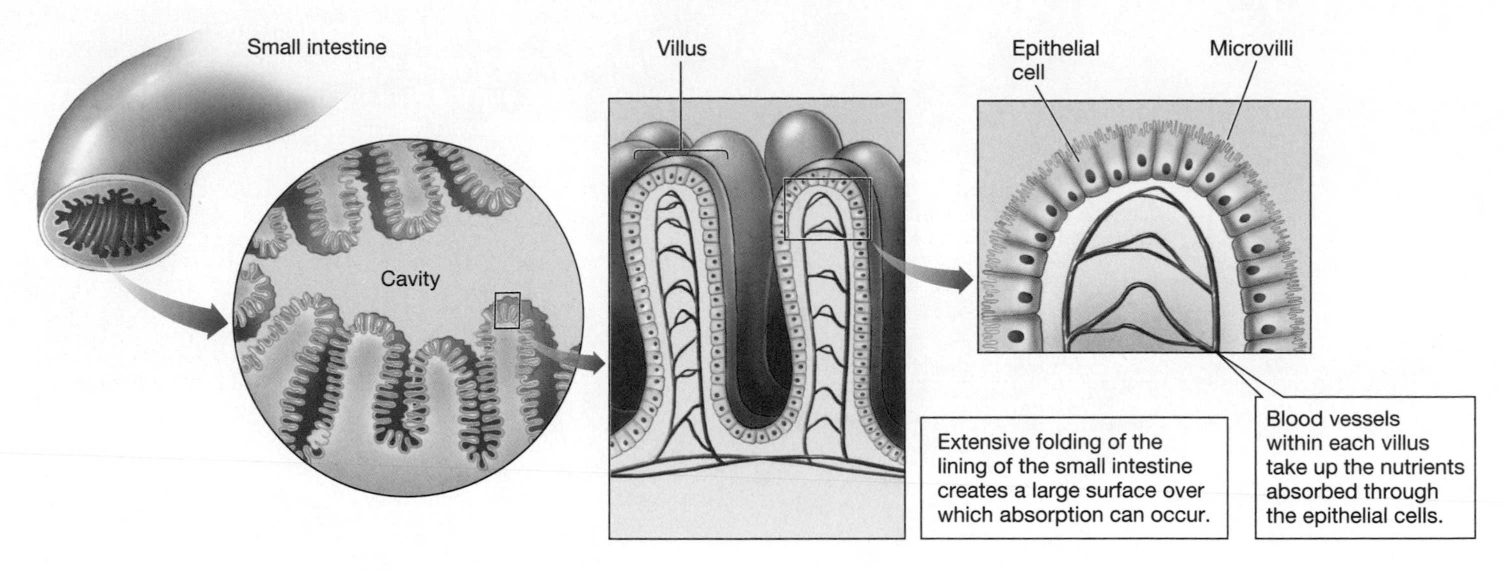
Small intestine
Cavity
Villus
Epithelial cell
Microvilli
Extensive folding of the lining of the small intestine creates a large surface over which absorption can occur.
Blood vessels within each villus take up the nutrients absorbed through the epithelial cells.

Figure 23.13

The small intestine is specialized for absorption

Nutrients are absorbed in the small intestine by large numbers of fingerlike projections called villi. Each villus is about 1 millimeter long, with a surface consisting of epithelial cells specialized for nutrient absorption. The plasma membrane of each of these cells also has many tiny projections called microvilli. This complex folding of the intestinal lining produces almost 300 square meters of surface area for absorption.

Q1: Why is a larger surface area important for absorption?

Q2: What feature of the epithelial cells lining the villi increases absorption?

Q3: In your own words, explain the role of the blood vessels within each villus.

See Appendix A for answers to the figure questions.

that process (**Figure 23.13**). Most of the nutrients absorbed by the digestive tract are sent to the bloodstream, which eventually delivers them to cells in the body.

Our original bite of cereal contains very few nutrients by the time it arrives in the final segment of the digestive tract, the **large intestine**, or **colon**. Large numbers of bacteria living in the colon break down a specific carbohydrate called fiber that our bodies cannot digest without help. These bacteria also produce certain vitamins that are absorbed with all remaining minerals, electrolytes, and water from the undigested residual matter to squeeze out the very last nutrients. This waste is prepared for **elimination**—the removal from the body of solid waste, consisting mostly of indigestible material and microbes that inhabit the digestive tract—during passage through the colon. The waste, or **feces**, leaves the body through the **anus**, a muscle-lined opening.

Keep Moving

Even in the colon, one can observe the benefits of exercise. There is strong scientific evidence that physical activity reduces the risk of colon cancer by up to 50 percent.

Carter is now investigating exactly how exercise improves the diversity of microbes in the gut, and he will soon begin a study in which participants eat a standardized diet for 10 weeks while performing varying amounts of exercise.

Jaber continues his daily work as a cardiologist and advises everyone he meets, patient or not, to "Keep moving!" It could, quite literally, save your life.

Nutritional Needs

Vitamins and minerals are essential nutrients that our bodies require in small amounts. Supplement manufacturers sell a range of different multivitamins, but the broad consensus from nutrition experts is that nothing substitutes for a healthy diet. Check out the data below to see how much of each nutrient you should be ingesting in a day, and some of the best foods in which to find them.

Assessment available in smartwork

- Men's multivitamin
- Women's multivitamin
- RDA male
- RDA female
- Example food source

The recommended daily allowance (RDA) is the average daily dietary nutrient intake level that is sufficient to meet the nutrient requirements of nearly all healthy individuals (97%–98%) in a particular life stage and gender group.

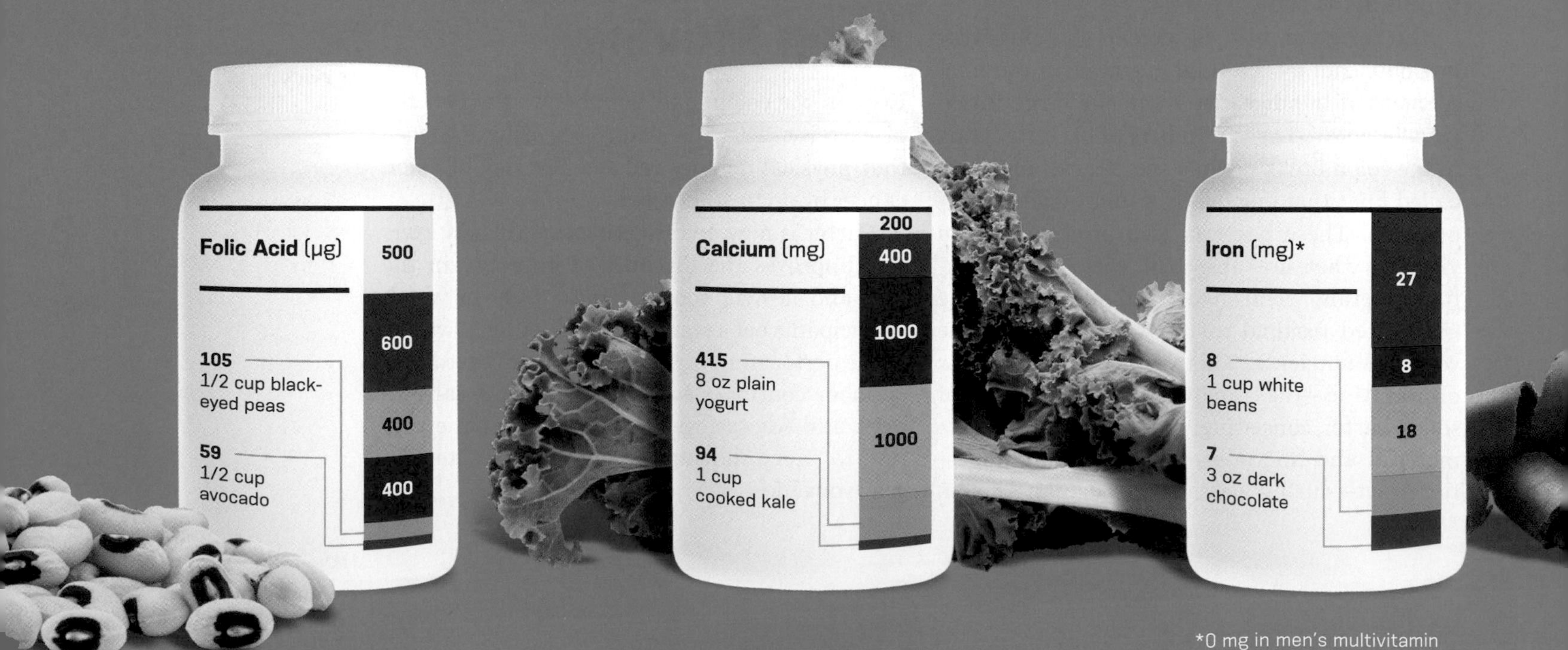

*0 mg in men's multivitamin

NEED-TO-KNOW SCIENCE

- The power necessary for movements is created by specialized muscle types. **Cardiac muscle** (page 445) is found in the heart where its contractions pump blood. **Smooth muscle** (page 445), which is found in the digestive tract and blood vessels, contracts in waves. **Skeletal muscle** (page 446) makes up the rest of our **muscular system** (page 445), has a banded appearance, and is required to move all our body parts.
- Involuntary contractions of cardiac muscle and smooth muscle occur outside of our control, whereas voluntary contractions of skeletal muscle occur at our command.
- Muscles consist of **muscle fibers** (page 446), each of which is packed with **myofibrils** (page 446). Myofibrils contain repeating units called **sarcomeres** (page 447) that contract when **myosin filaments** (page 447) slide along **actin filaments** (page 447). Each end of a sarcomere has a **Z disc** (page 447) where a large protein anchors actin to the sarcomere.
- The **skeletal system** (page 450) supports the body, gives it shape, and protects soft tissues and organs. One part of that system, the **axial skeleton** (page 450), supports and protects vital organs along the long axis of the body. The second part of the system, the **appendicular skeleton** (page 450), is composed of the bones in the arms, legs, and pelvis and is primarily involved with movement.
- The skeleton of an organism may be inside the body (**endoskeleton**, page 451) or outside the body (**exoskeleton**, page 451).
- Bones are living tissue containing specialized cells called **osteocytes** (page 451) that excrete and surround themselves with a hard, mineral matrix of calcium and phosphate. Most bones are made of **spongy bone** (page 451) surrounded by harder **compact bone** (page 451). Hollow bones contain **marrow** (page 451).
- **Ligaments** (page 452) connect bone to bone. **Tendons** (page 452) connect muscle to bone. Tendons and ligaments are made of **collagen** (page 451). **Cartilage** (page 451) is also made of collagen but is found inside the joints where bones meet and in other areas of the skeletal system where strength and flexibility are needed. **Synovial sacs** (page 452) are also found inside moveable joints and provide lubricating fluid for smooth motion.
- The **integumentary system** (page 453) is the largest organ system in the human body, covering and protecting the surface of the body. It consists of the skin layers found from the surface inward—namely, epidermis, dermis, hypodermis—and the structures embedded in the skin, such as hair and nails.
- The **digestive system** (page 453) is a tubular passageway that, in conjunction with accessory organs, processes ingested food.
- **Ingestion** (page 454) occurs through the **oral cavity** (page 455), where the grinding action of teeth breaks food down into smaller pieces and **saliva** (page 455) moistens the food and begins the chemical breakdown of carbohydrates.
- Food passes from the oral cavity to the **pharynx** (page 455) and then down the **esophagus** (page 455). The **swallowing reflex** (page 455) prevents food from entering the airway by sealing off the entry into the trachea with a flap of skin.
- In the acidic environment of the **stomach** (page 455), **digestion** (page 455) of **nutrients** (page 455) begins. Animals rely on nutrients for chemical building blocks and energy.
- Partially digested food moves from the stomach into the upper region of the **small intestine** (page 455) where enzymes secreted by the **pancreas** (page 455), the intestine, and the **liver** (page 456) complete digestion of the food. The liver produces **bile** (page 455), which is stored and delivered by the **gallbladder** (page 456) and helps digest fats. The lining of the small intestine is highly folded and bears villi, finger-like projections that present a large surface area for absorbing nutrients. The digestion and **absorption** (page 456) of proteins, carbohydrates, and lipids, including fats, is essentially completed here.
- The lower region of the small intestine is responsible for absorption of vitamin B_{12}, bile salts, and any nutrients not previously absorbed.
- From the small intestine, the unabsorbed material moves into the **large intestine**, or **colon** (page 457), where remaining water, electrolytes, and minerals are absorbed. Here intestinal bacteria help break down indigestible carbohydrates (fiber) to release extra nutrients and produce vitamins that the colon absorbs.
- The remaining waste, or **feces** (page 457), leaves the body through a muscle-lined opening called the **anus** (page 457). This process is called **elimination** (page 457).
- Macronutrients are carbohydrates, proteins, and lipids. These biomolecules serve as sources of energy and furnish the body with chemical building blocks such as sugars, fatty acids, and amino acids, including the nine **essential amino acids** (which come from food, page 455).
- Micronutrients include **vitamins** (organic compounds obtained from food that regulate metabolic processes in the animal body, page 455) and **minerals** (inorganic molecules needed by the body in small amounts, page 455).

THE QUESTIONS

See Appendix B for answers.

The Basics

1 Which nutrient does saliva in the oral cavity begin to break down?

(a) proteins

(b) lipids

(c) carbohydrates

(d) vitamins

2 Which of the following statements is *not* true of bones?

(a) They are a mineral-rich, nonliving tissue.

(b) They can contain marrow.

(c) They can be spongy, compact, or hollow.

(d) They make up our endoskeleton.

3 Cardiac muscle is

(a) part of the voluntary muscle system.

(b) found only in heart tissue.

(c) important for contractile waves in the digestive system.

(d) made up of osteocytes.

4 Which of the following statements is *not* true of the small intestine?

(a) It is specialized for absorption.

(b) It absorbs nutrients through villi and microvilli.

(c) It secretes digestive enzymes.

(d) It is specialized for elimination of waste.

5 Use these terms correctly in the following sentences: adipose, dermis, epithelial. _______ tissue is found in the hypodermis and insulates us from temperature extremes. One level up, the _______ is composed mainly of connective tissue. And protecting us from the external environment is the epidermis, made up of _______ tissue.

6 Link each term with the best definition.

skeletal muscle	1. Muscle that is consciously controlled.
smooth muscle	2. Type of muscle found in the heart.
cardiac muscle	3. Type of muscle used for walking and running.
voluntary muscle	4. Type of muscle found in the digestive system.
involuntary muscle	5. Muscle that works without conscious control.

Challenge Yourself

7 Which of the following are important for controlled movement of the knee?

(a) the axial and appendicular skeletons

(b) the ulna and the radius

(c) rigidity and flexibility

(d) the femur and the tibia

8 Select the correct terms:

Within a (**muscle** / **tendon**), sarcomeres shorten when their (**actin** / **myosin**) filaments slide along (**actin** / **myosin**) filaments joined to (**Z** / **X**) discs on either end of the (**muscle fiber** / **sarcomere**).

9 According to unpublished data from Purdue University, working out at college gyms results in better grades. Data published for grade school kids support the hypothesis that exercise increases grades. The accompanying figure shows the results of the FitnessGram testing of 3 million grade school kids from California and Texas split into the number of standards (six total) that were passed. The Standard Achievement Test scores were averaged for all the students in each FitnessGram category. For example, the average Standard Achievement Test score for all the students who passed three of the six FitnessGram standards was 300. From this graph, you can conclude that

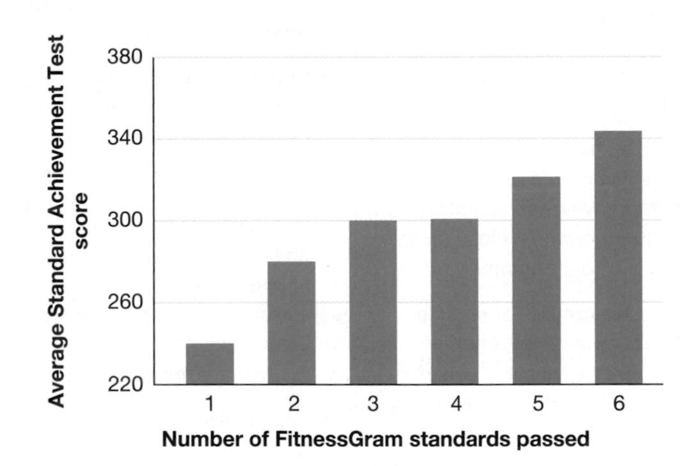

(a) fewer FitnessGram standards passed correlates with higher Standard Achievement Test scores.

(b) fewer FitnessGram standards passed correlates with lower Standard Achievement Test scores.

(c) more FitnessGram standards passed correlates with lower Standard Achievement Test scores.

(d) there is no correlation between FitnessGram standards passed and Standard Achievement Test scores.

10 Place the following events of food movement through the digestive system in the correct order by numbering them from 1 to 5.

_____ a. Nutrients are absorbed through cells of the villi.

_____ b. Food is broken into smaller pieces through chewing.

_____ c. Digestive enzymes are released by accessory organs.

_____ d. Bacteria help digest food and produce some vitamins before sending waste out of the body.

_____ e. Acids break down proteins for further digestion.

Try Something New

11 Which of the following is *not* involved in supporting movement of a joint?

(a) bone

(b) smooth muscle

(c) ligaments

(d) synovial sacs

(e) All of these are involved in joint movement.

12 Our bones are constantly changing in response to how we live. Physical activity builds stronger bones. The bones in a pitcher's throwing arm, for example, are stronger and have larger ridges to which muscles can attach than do the same bones in the other arm. Inactivity leads to weaker bones because special osteocytes step up the rate at which they remove tissue from the bone when the skeleton is not under physical stress; such weakening is seen, for example, in a person confined to bed by an injury or illness. Which of the following activities would be the least effective in increasing bone strength? Explain why.

(a) walking

(b) dancing

(c) weight lifting

(d) swimming

(e) running/jogging

13 A January 2013 Tumblr post that went viral on the internet read as follows:

> eating is so bad*** i mean you put something in a cavity where you smash and destroy it with 32 protruding bones and then a meat tentacle pushes it into a vat of acid and after a few hours later you absorb its essence and transform it in[to] energy just wow.

Identify each element of the digestive system discussed in the post. Do you think this is an accurate description of digestion?

Leveling Up

14 **Looking at data** The recommended daily allowance (RDA) for nutrients is the average daily amount necessary to meet from 97 percent to 98 percent of the nutrient requirements of most healthy adults. The two kinds of necessary mineral nutrients are macrominerals, which your body needs in larger amounts, and trace minerals, which your body needs in smaller amounts. Macrominerals include calcium and phosphorus. Trace minerals include iron, copper, iodine, zinc, and fluoride.

(a) According to the accompanying graph, what is the dividing line between macrominerals and trace minerals?

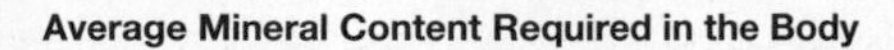

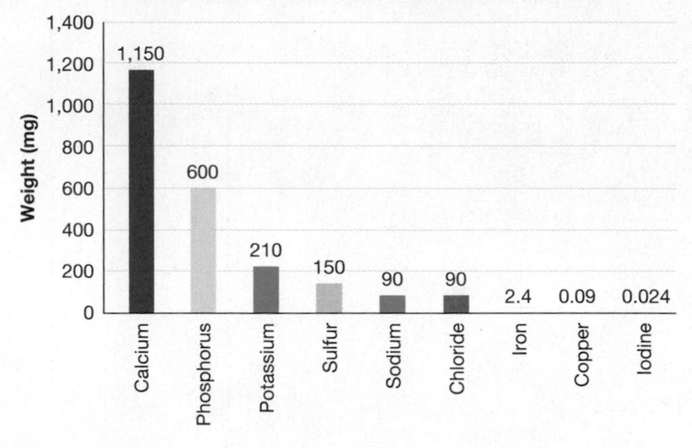

(b) What does the *y*-axis show?

(c) Describe in your own words what the graph shows.

(d) How much potassium is needed by an average person?

(e) Is the amount of potassium you stated in part (d) the same amount that we need to eat daily?

15 **Life choices** Nutritionists agree that we can obtain all our necessary nutrients in optimal amounts from natural sources provided we eat a well-balanced diet that includes a variety of foods. Recognizing that relatively few people meet this ideal in reality, some physicians advise their patients to take a multivitamin and mineral supplement as "added insurance." The use of this kind of dietary supplement raises the specter of toxicity from excessive intake, particularly of the fat-soluble vitamins. Most manufacturers, but by no means all, attempt to avoid high levels of the fat-soluble vitamins in their vitamin preparations, but the burden is largely on us to use supplements wisely, because this industry is not highly regulated by the government. Review the information provided on dietary supplements by the National Institutes of Health Office of Dietary Supplements (http://ods.od.nih.gov/HealthInformation/DS_WhatYouNeedToKnow.aspx) and the U.S. Food and Drug Administration, or FDA (http://www.fda.gov/Food/DietarySupplements/UsingDietarySupplements/ucm110567.htm). Take notes so that you can answer the following questions.

(a) What aspects of the sale of dietary supplements does the government regulate?

(b) Are you confident that this regulation is sufficient to make supplements safe to use as labeled? Why or why not?

(c) List three recommendations that you found informative or helpful. Will they change your use of dietary supplements? Explain your reasoning.

(d) Do you think a mobile app for dietary supplements would be helpful to have? Why or why not?

(e) Identify one dietary supplement that you feel would be helpful for you to take. Will you take it? Why or why not?

Digital resources for your book are available online.

Body (Re)Building

Could engineered human tissues, brought to life in the lab, replace failing organs in people?

After reading this chapter you should be able to:

- Create a flowchart depicting the movement of blood through the cardiovascular system.
- Describe the functions of the different components of blood, and compare the three types of blood vessels.
- Diagram the elements of the respiratory system, and show how air moves through it.
- Explain how gases are exchanged in the lungs, and why.
- Describe how the structure of a nephron relates to its function in the urinary system.
- Distinguish between the central and peripheral nervous systems.
- Identify the sensory receptors in humans.

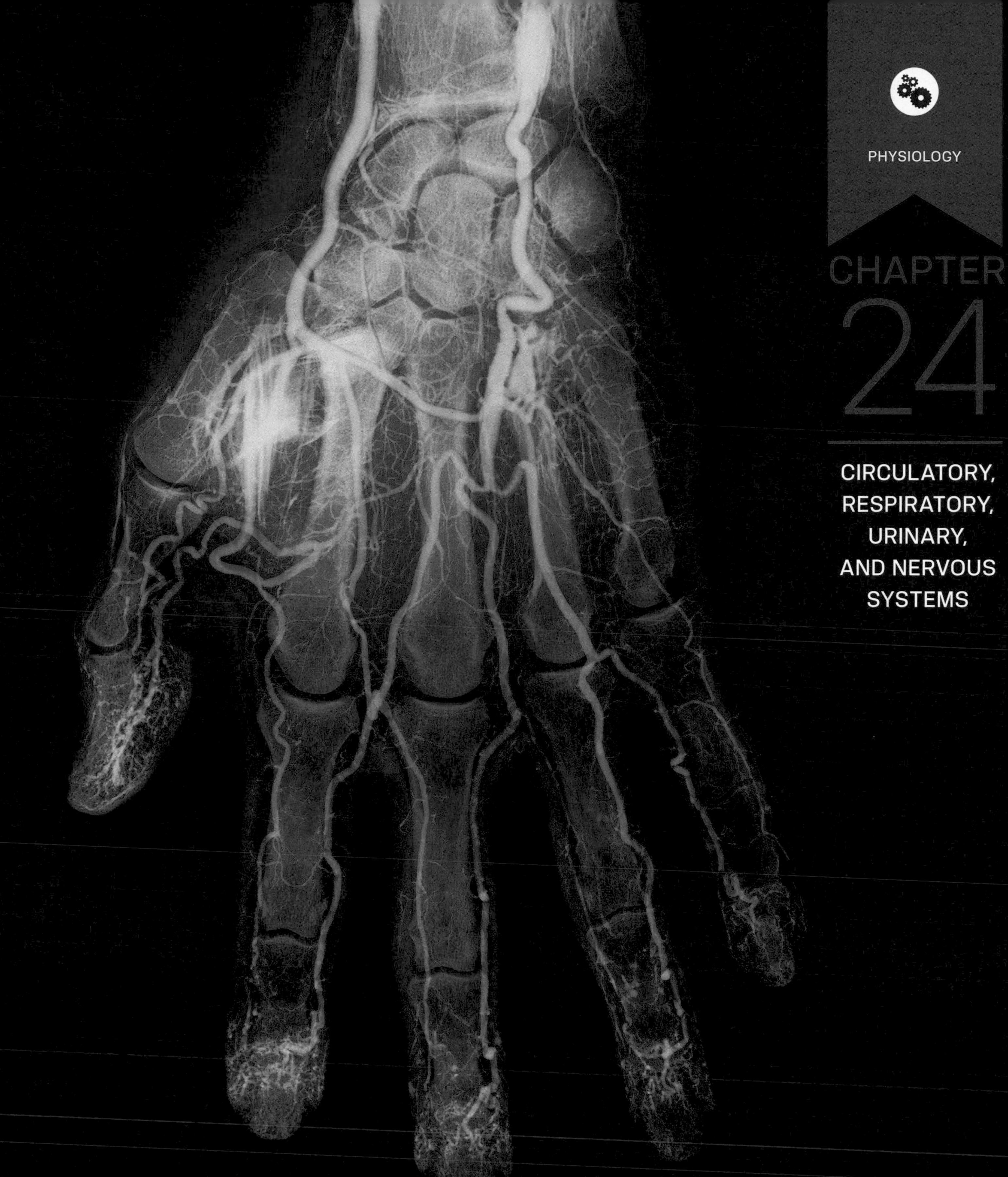

PHYSIOLOGY

CHAPTER 24

CIRCULATORY, RESPIRATORY, URINARY, AND NERVOUS SYSTEMS

Standing over the operating table, covered head to toe in teal scrubs, physician Jeffrey Lawson lifts a long, white tube out of a bath of clear liquid. Carefully, slowly, he threads the tube through the unconscious patient's upper arm and stitches each side of the tube to an exposed blood vessel.

Bioengineer Laura Niklason stands to the side of the operating table, watching closely. She and Lawson created the tube being implanted in the patient. Once the operation is complete, Lawson steps back, and Niklason gives him a hug. "Congratulations to you," says Niklason happily. "We're saving the world!"

On June 5, 2013, Lawson, Niklason, and their team at Duke University in Durham, North Carolina, transplanted a laboratory-grown blood vessel into a human (**Figure 24.1**). It was the first such procedure in the United States and a major feat for the field of *tissue engineering*, the effort to grow or regenerate tissues or organs using engineering materials and principles. Other engineered tissues implanted into humans have included nerves, bladders, and windpipes.

Lawson and Niklason imagine a future in which any type of organ can be constructed in the lab. In an ideal world, "we'll be able to grow all sorts of tissues for patients, so a surgeon can literally reach up on a shelf, pull down a tissue graft, and implant it in a patient," Niklason said in 2013, shortly after the first blood vessel transplant. "That's really going to be a revolution—being able to grow replacement parts for patients that they don't have to wait for."

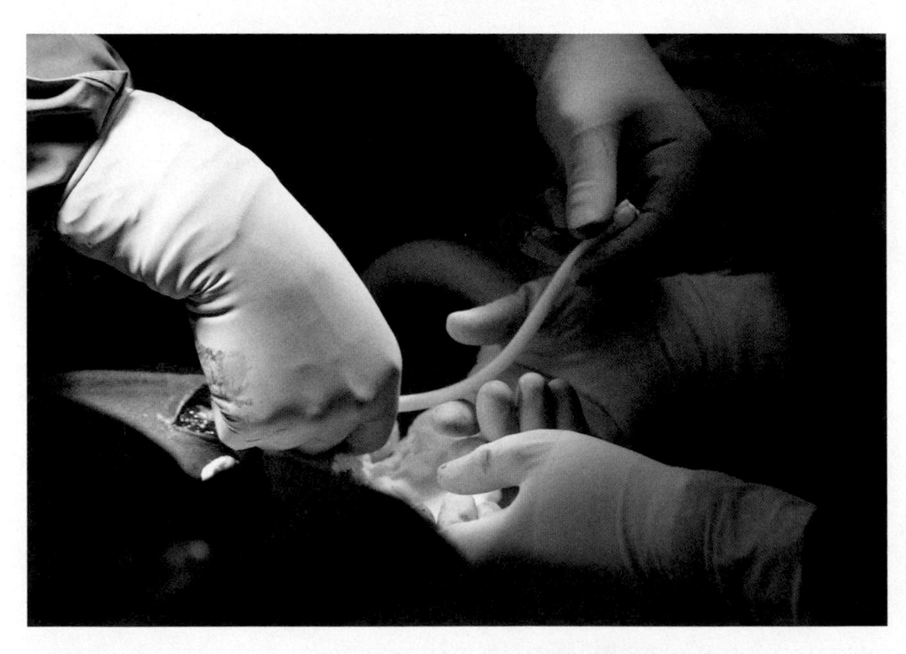

Figure 24.1

A bioengineered blood vessel

In 2013, a team of doctors at Duke University was the first in the United States to implant a bioengineered blood vessel into a patient. The white vessel in this image is composed entirely of proteins formed by living cells in the lab.

Organ shortage is a major concern in health care (**Figure 24.2**). According to the U.S. government, the number of people waiting for an organ would fill a football stadium—twice. Every day, an average of 20 people die waiting for transplants of kidneys, hearts, livers, lungs, and more.

As discussed in Chapter 9, one possible solution to this organ crisis is to harvest compatible organs from animals such as pigs. But a more versatile, potentially faster option is the Frankenstein-like approach of tissue engineering: building organs in the lab.

It sounds far-fetched, but research teams have already begun to engineer complicated organs such as kidneys, lungs, and even whole hearts. These are just a few of the organs that make up the major organ systems in our bodies, including the circulatory, respiratory, urinary, and nervous systems—each one critical to the healthy functioning of our bodies (see Figure 22.4 for an overview of these systems).

That healthy functioning, including growth and homeostasis (see Figure 22.5), depends on the internal transfer of essential proteins, waste products, and other substances in the body. Distances inside the bodies of most multicellular organisms are far too great for diffusion to be an effective means of distributing these materials, so elaborate organ systems exist to transport them around the body. The **circulatory system** moves oxygen (O_2) from the lungs to the heart, which then pumps oxygen-rich blood to the rest of the body. The **respiratory system** brings in oxygen and expels carbon dioxide (CO_2) to support cellular respiration. The circulatory system also moves CO_2 to the lungs and substances such as nutrients, hormones, and blood cells throughout the body. The **urinary system** removes excess fluid from the circulatory system, along with waste products, toxins, and other water-soluble substances that are harmful or not needed by the body.

These three systems are interconnected in the body: Carbon dioxide collected by the circulatory system is delivered to the respiratory system for exchange with the outside environment. The circulatory system brings substances dissolved in blood to the urinary system for discharge into the environment. And a high-speed communication system called the **nervous system** coordinates the many muscles involved in the functioning of organs, including the heart, lungs, and bladder. The nervous system directs the rapid contractions of muscles and processes information received by the senses, such as touch, sound, and sight. In this way, the nervous system enables an animal to detect food, find a mate, avoid predators, and respond to extremes of heat and cold.

These organ systems, along with all the other organ systems introduced in Chapter 22, govern how your body operates. If one of your organs fails, a whole system can shut down. That's why researchers are exploring how to build new organs, writing a recipe for flesh and blood.

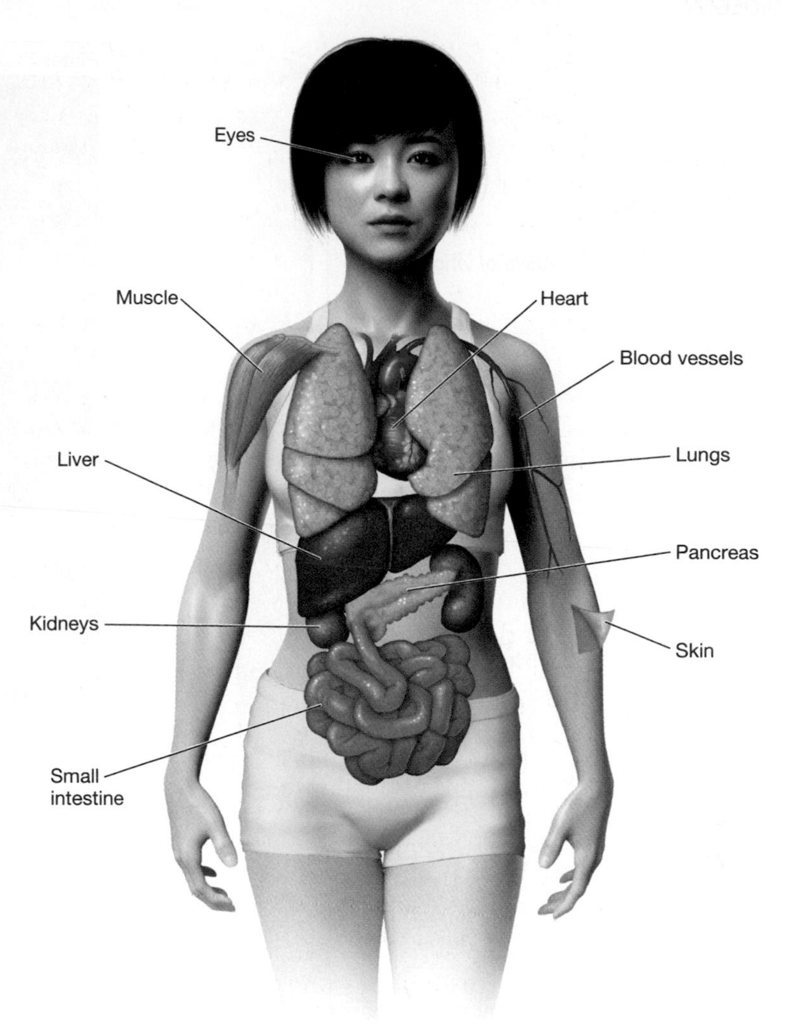

Figure 24.2

Organs needed for donation

The organs highlighted here can be transplanted from one human body to another.

Totally Tubular

Lawson and Niklason met at Duke University Medical Center in 1999 while performing surgery together. Waiting in the operating room to move their patient, the two struck up a conversation. Within moments, they discovered they shared strikingly similar interests: Lawson was a surgeon who spent his days suturing blood vessels in the hospital and growing different types of blood cells in his lab, and Niklason was an anesthesiologist and bioengineer who had recently opened a lab to build blood vessel tubes. Both dreamed of the same goal—to create a blood vessel from scratch—and they had complementary skills. "It made us a perfect partnership," says Lawson.

That first conversation bloomed into a friendship and working partnership. Together, Lawson and Niklason started building prototypes of blood vessels, using his knowledge of blood cells and her knowledge of the engineering forces at work in blood vessels. **Blood vessels** are a critical part of the **cardiovascular system**, a closed circulatory system consisting of a muscular heart, a complex network of blood vessels that collectively form a closed loop, and blood that circulates through the heart and blood vessels. Almost all of our cells lie within 0.03 millimeter of a blood vessel with which they exchange materials by diffusion. Carrying blood so close to all the trillions of cells in our bodies requires an extensive network of vessels.

Lawson and Niklason built blood vessels out of various materials and cell types. They implanted these vessels in rats to see whether blood would successfully flow through the tubes. **Blood** is composed of cells and cell fragments that float in a fluid known as **plasma** (**Figure 24.3**). Plasma has a low capacity for transporting dissolved oxygen, but **red blood cells** in the plasma carry significant amounts of oxygen, greatly increasing the oxygen-carrying capacity of blood.

In 2005, the duo hit upon a technique that seemed to work. They collected smooth muscle cells from the blood vessels of organ donors

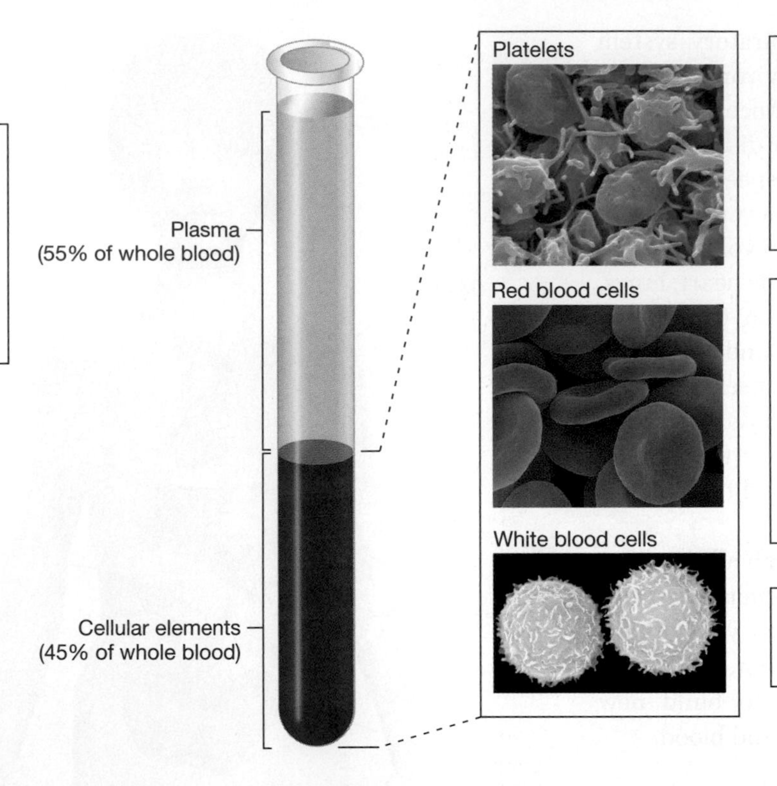

Blood plasma is 92% water and contains dissolved gases, ions, and molecules that are critical for homeostasis, as nutrients, or as signaling molecules. Most of the carbon dioxide carried in the blood is dissolved in plasma.
Plasma (55% of whole blood)
Cellular elements (45% of whole blood)
Platelets
Red blood cells
White blood cells
Platelets are small cell fragments. They can clump together to help stop the loss of blood if a blood vessel is damaged. Platelets release substances that stimulate plasma proteins to create a meshwork of protein strands, platelets, and blood cells to collectively form a blood clot.
A mature red blood cell has no nucleus, and its cytoplasm is packed with oxygen-binding proteins called hemoglobin. Each hemoglobin molecule can carry up to four oxygen molecules. Because each human blood cell contains about 280 million hemoglobin molecules, a single one of these cells can bind over a billion molecules of oxygen.
Several different kinds of white blood cells help defend the body from invading organisms.

Figure 24.3

The ingredients of human blood

Whole blood consists of plasma and different kinds of cells and cell fragments, three of which are shown here. Red blood cells account for about 95 percent of the cells in blood.

Q1: Where in the blood is the majority of carbon dioxide carried?

Q2: Where in the blood is the majority of oxygen carried?

Q3: What would happen if your red blood cells carried a mutation that made the hemoglobin less effective at binding to oxygen (as in sickle cell disease)?

See Appendix A for answers to the figure questions.

and grew those muscle cells on a biodegradable frame shaped like a blood vessel. The cells worked like little machines, churning out proteins that formed a three-dimensional scaffold of connective tissue called the *extracellular matrix*. Then the original, biodegradable frame dissolved, leaving behind a sturdy tube of cells and extracellular matrix.

There was still a catch. The immune system of a human body rejects cells from another person (see Chapter 25 for more on the immune system), so Lawson and Niklason had to wash away the original donor muscle cells to leave behind just the tubular extracellular matrix. "It's like we kept the mortar but the bricks all washed away," says Lawson. They tested their creation in rats. It worked. They had made a functioning blood vessel.

The vessels were first tested in humans by researchers in Poland in December 2012. The researchers found that once an engineered tube is implanted into a patient, the patient's own blood and muscle cells take up residence in the tube, filling in the cracks. "What started off as our structure now becomes your tissue," says Lawson. "And it all seals together so it doesn't leak." In 2019, the team got the results of a long-term study tracking the progress of their vessels. Many of the first bioengineered vessels implanted 6 years ago remained healthy and robust in patients receiving dialysis three times a week. The vessels took on the recipient's cells, matured, and essentially became the patient's own living tissue.

A critical feature of the engineered vessel is that it has the structural strength of a natural

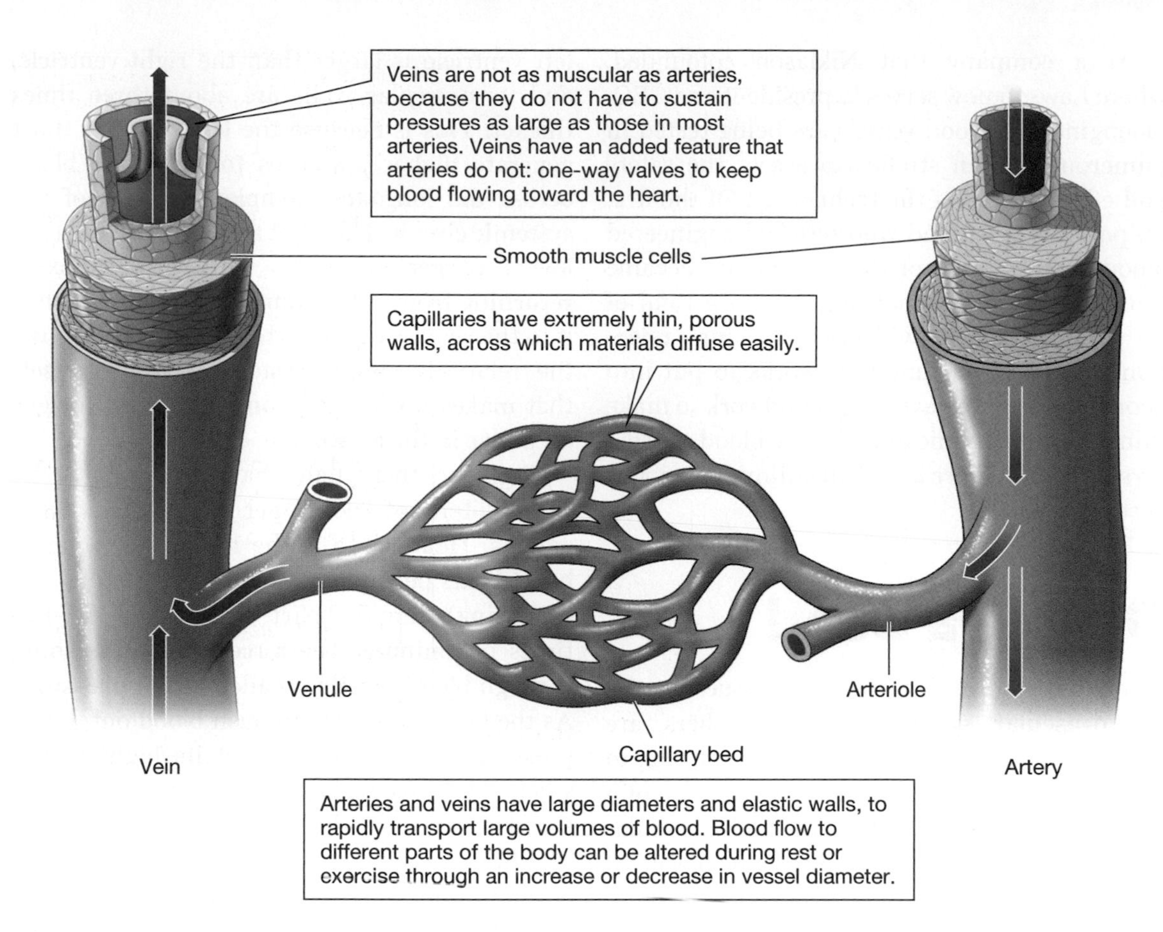

Veins are not as muscular as arteries, because they do not have to sustain pressures as large as those in most arteries. Veins have an added feature that arteries do not: one-way valves to keep blood flowing toward the heart.
Smooth muscle cells
Capillaries have extremely thin, porous walls, across which materials diffuse easily.
Venule
Arteriole
Vein
Capillary bed
Artery
Arteries and veins have large diameters and elastic walls, to rapidly transport large volumes of blood. Blood flow to different parts of the body can be altered during rest or exercise through an increase or decrease in vessel diameter.

Figure 24.4

Arteries, veins, and capillaries

Arteries carry blood away from the heart. Veins carry blood toward the heart. Arterioles and venules are smaller arteries and veins, respectively. The narrowest arteries and veins connect with each other in a fine network known as a capillary bed.

Q1: Why are arteries more muscular than veins?

Q2: What two structural features of capillaries enable easier diffusion into and out of surrounding tissues?

Q3: Why do you think capillaries are not typically transplanted?

See Appendix A for answers to the figure questions.

blood vessel, so it can withstand the force of the blood pulsing through it. The human body has three major kinds of blood vessels: arteries, veins, and capillaries (**Figure 24.4**). **Arteries** are large vessels (0.1–10 mm in diameter) that transport blood away from the heart. **Veins** are large vessels (0.1–2 mm in diameter) that carry blood back to the heart. **Capillaries**, the smallest vessels (0.005–0.01 mm in diameter), exchange materials by diffusion with nearby cells.

The large vessels—arteries and veins—are built for mass transport of blood. Niklason, now a professor of anesthesiology and biomedical engineering at Yale University, and Lawson have built blood vessels with diameters of 6 millimeters (about the width of a pencil) and 3 millimeters. But they also expect to be able to make and transplant larger vessels, such as the aorta, the main artery of the body. Capillaries, however, are very small and are not typically transplanted. They are built for slower movement of blood. Because there are many of them, they have a large surface area, which facilitates the exchange of materials with surrounding cells.

At a company that Niklason cofounded, where Lawson now serves as president and CEO, bioengineered blood vessels are being tested in numerous human studies to assess the safety and effectiveness of the technology. Of the first 20 patients in Poland who received engineered blood vessels, none of the transplants became infected, and additional implants in a total of 60 patients saw no evidence of immune rejection. "We've got a tube that works to put into people, so now we have the groundwork to make things more complicated than a blood vessel," says Lawson. "There is still an unlimited amount of science to do."

Gimme a Beat!

Blood vessels are not the only tissues in the cardiovascular system that researchers are trying to engineer. Scientists are attempting to grow the **heart**, a muscular organ the size of a fist in humans that works as the body's circulatory pump (**Figure 24.5**). Like the hearts of all other mammals, the human heart is divided into four chambers that form two distinct pumping units, which are independent but coordinated. The right and left sides of the heart function as two separate pumps. However, the two upper chambers, or **atria** (singular "atrium"), contract in unison, as do the two lower chambers, the **ventricles**. This unified contraction begins with a signal from the *pacemaker*, or *sinoatrial (SA) node*. The signal causes both atria to contract; it also causes the *atrioventricular (AV) node* to pass the signal on to the ventricles about a tenth of a second later. The short delay allows the atria to empty completely.

The left atrium receives oxygenated blood from the lungs and pumps it to the left ventricle, which pumps it through the **systemic circuit** to body cells performing cellular respiration. The left ventricle is larger than the right ventricle, and its muscular walls are about three times thicker. This is because the left ventricle must generate higher pressures to distribute blood across the long and complex network of the systemic circuit. The right atrium receives blood low in oxygen and laden with carbon dioxide returning from the systemic circuit and pumps it to the right ventricle, which pumps it through the relatively simple system of blood vessels that makes up the **pulmonary circuit**, for gas exchange in the lungs.

Together, these four chambers pump some 7,000 liters of blood per day—about 1,850 gallons. **Heart rate** is the number of times a heart beats per minute; an average resting (not exercising) human heart beats about 60–100 times per minute. The force of blood pushing through blood vessels is called **blood pressure**. As the heart contracts to push blood out, blood pressure in the arteries is at its highest, and when the heart relaxes after each contraction, blood pressure in the arteries is at its lowest. The first pressure is referred to as **systolic** and is the top number in a blood pressure reading. The second, **diastolic**, is the bottom number in a blood pressure reading. The human circulatory system adjusts the heart rate and patterns of blood distribution according to the body's needs.

The heart might seem too complex an organ to bioengineer, but researchers have had some success with a technique similar to the one Niklason and Lawson used with blood vessels. The researchers started with an entire organ as a scaffold, such as a heart from a rat donor. After using detergents to strip away all the original cells that would cause an immune reaction, they repopulated the heart with cells that were a better match for the recipient. Although the resulting hearts pumped in a dish, they were too primitive to work when transplanted into an animal.

DEBUNKED!

MYTH: Blood in veins is blue.

FACT: Human blood is always red. It can look blue through our skin due to an optical illusion resulting from how tissues absorb light and how our eyes see color. Some animals do have blue blood, however, such as lobsters and squid.

Breathe In, Breathe Out

In 2010, after she had moved from Duke to Yale University, Niklason expanded her research program from creating simple blood vessels to attempting to build a lung. **Lungs** are the main organs of the respiratory system, which carries

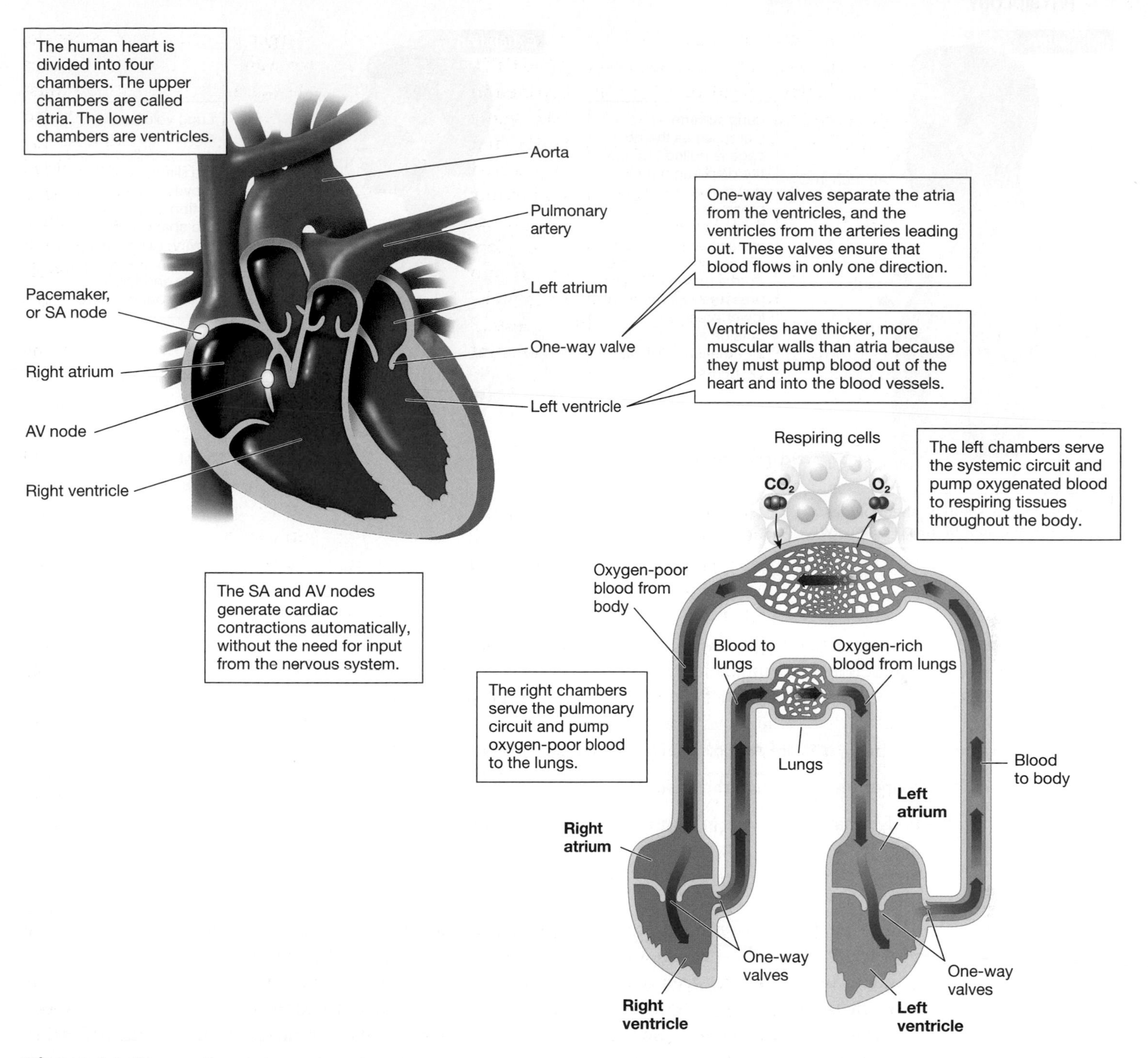

Figure 24.5

The human heart

The heart is shown from the front of the body, so the left atrium is on the right side of the diagram, and so forth.

Q1: Beginning with the left atrium, list the locations of blood as it moves through the circulatory system.

Q2: Why is the left ventricle larger than the right ventricle, and why are its walls thicker and more muscular?

Q3: What do you think is the function of an artificial pacemaker implanted in the heart?

See Appendix A for answers to the figure questions.

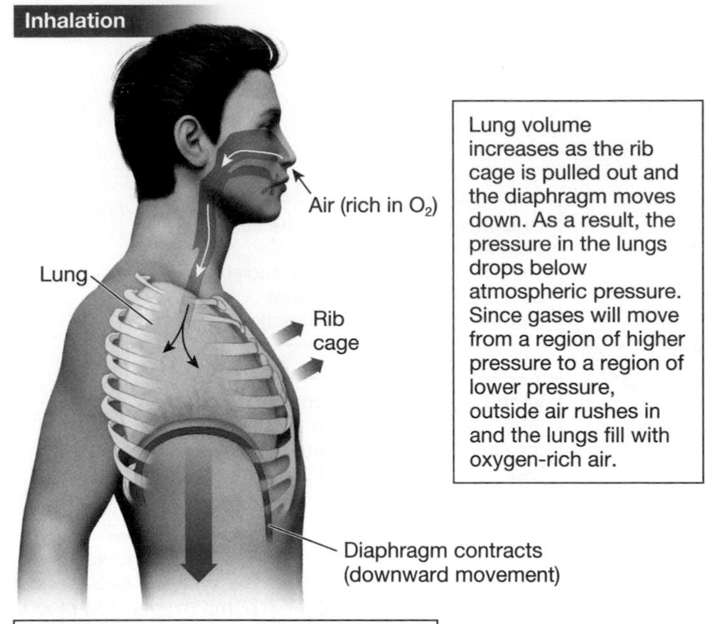
Inhalation
Air (rich in O_2)
Lung
Rib cage
Lung volume increases as the rib cage is pulled out and the diaphragm moves down. As a result, the pressure in the lungs drops below atmospheric pressure. Since gases will move from a region of higher pressure to a region of lower pressure, outside air rushes in and the lungs fill with oxygen-rich air.
Diaphragm contracts (downward movement)
The diaphragm is a thick sheet of muscle that forms the floor of the chest cavity.

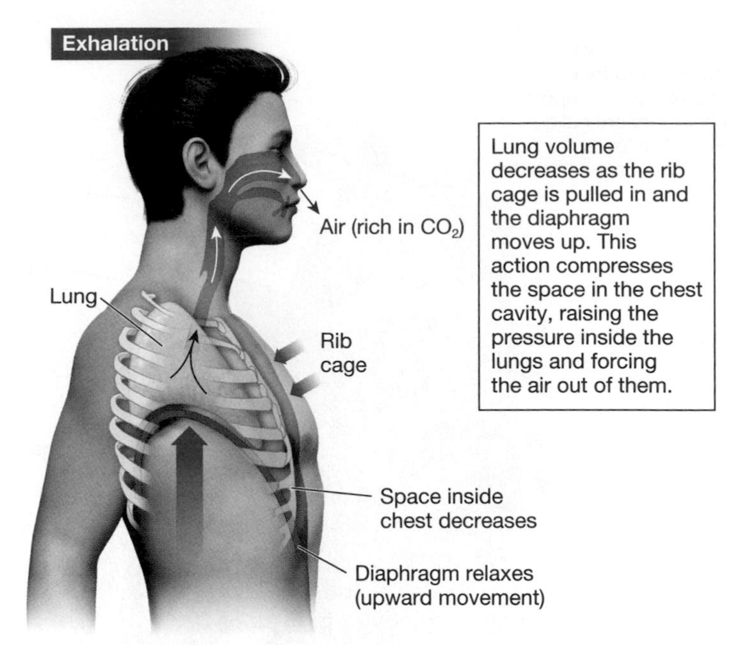
Exhalation
Air (rich in CO_2)
Lung
Rib cage
Lung volume decreases as the rib cage is pulled in and the diaphragm moves up. This action compresses the space in the chest cavity, raising the pressure inside the lungs and forcing the air out of them.
Space inside chest decreases
Diaphragm relaxes (upward movement)

Figure 24.6

Breathing

Breathing involves two main steps: inhalation, when air is pulled into the lungs, and exhalation, when air is pushed out of the lungs. Most of the time, breathing is controlled automatically by sensory systems located in the heart and brain. If we choose, we can also control our breathing with a system of muscles, the most important of which are the rib muscles and the diaphragm.

Q1: When air enters the body, is it richer in oxygen or carbon dioxide? What about when it exits the body?

Q2: The figure shows air entering through the nose. Where else can air enter the respiratory system?

Q3: Explain how the body creates a change in air pressure during breathing.

See Appendix A for answers to the figure questions.

air from the nose (or mouth) to the lungs through a series of tubular passageways. These airways allow air to move between the external environment and the inside of the body—specifically, the gas exchange surfaces in the lungs.

The process of taking air into the lungs (inhaling) and expelling air from them (exhaling) is called **breathing** (**Figure 24.6**). The air we inhale is about 21 percent oxygen and contains little carbon dioxide and water vapor. The air we exhale has less oxygen (15 percent) and contains about 4 percent each of carbon dioxide and water vapor. That's because these gases are exchanged at the surface of the cells that line our lungs: oxygen moves from the inhaled air and enters the bloodstream, while carbon dioxide and water vapor move from the bloodstream and enter the air that is exhaled. As discussed in Chapter 5, our bodies need oxygen to obtain energy via cellular respiration (see Figure 5.11).

At Yale, Niklason's team built a rat lung and transplanted it into a rat. There, it briefly supported gas exchange for the animal before filling up with fluid, suggesting that the technology needs further development. A lung needs to be highly reliable, as lungs are vital to transporting oxygen throughout the body. Although humans can live for more than a week without food and for a few days without water, a mere 4 minutes without oxygen results in irreversible brain damage. And lung tissue is notoriously bad at repairing itself, so lung transplants are often the only option in the case of lung damage from disease or trauma.

The respiratory system can be divided into two parts (**Figure 24.7**). The **upper respiratory system** includes airways in the nose, mouth, and

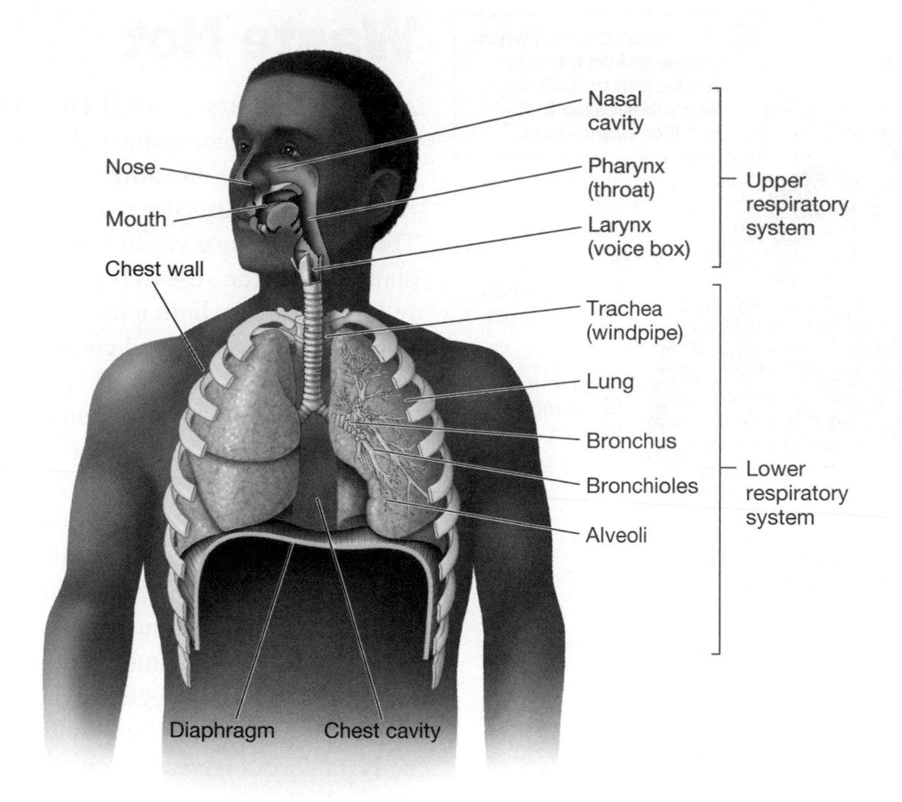

Figure 24.7

The human respiratory system

The respiratory system consists of two parts, upper and lower, each of which has specific components.

Q1: Beginning with the nose, list the locations of oxygen as it moves through the upper respiratory system. Then beginning with the mouth, list the locations of oxygen as it moves through the upper respiratory system.

Q2: List the locations of oxygen as it moves through the lower respiratory system.

Q3: From the bronchus, where does the oxygen move?

See Appendix A for answers to the figure questions.

throat. When we inhale, air enters through the nostrils and moves into each nasal cavity. It can also enter through the mouth and bypass the nasal cavities on its way to the throat, or **pharynx**, an area where the back of the mouth and the two nasal cavities join into a single passageway. From the pharynx, air moves into the **larynx**, or voice box, which forms the entryway to the **trachea**, or windpipe.

The trachea is the start of the **lower respiratory system**. Within the chest, the trachea branches into two smaller tubes called **bronchi** (singular "bronchus"). Each bronchus leads to one of the paired lungs, the organs where gases are exchanged. Inside the lungs is where the respiratory system gets intricate: The bronchi divide into smaller bronchi and then **bronchioles**, a series of branching, ever-smaller tubes. The tiniest bronchioles open into the **alveoli** (singular "alveolus"), small clusters of sacs that resemble a bunch of grapes. Gases are exchanged across the moist surface of the thin layer of cells that forms the surface of each alveolar sac and the surrounding capillaries. Together, the trachea, bronchi, and lungs make up the lower respiratory system (**Figure 24.8**). Niklason continues to try to achieve this gas exchange in an engineered lung. So far, her rat lung transplants have managed this exchange for only about 2 hours before failing.

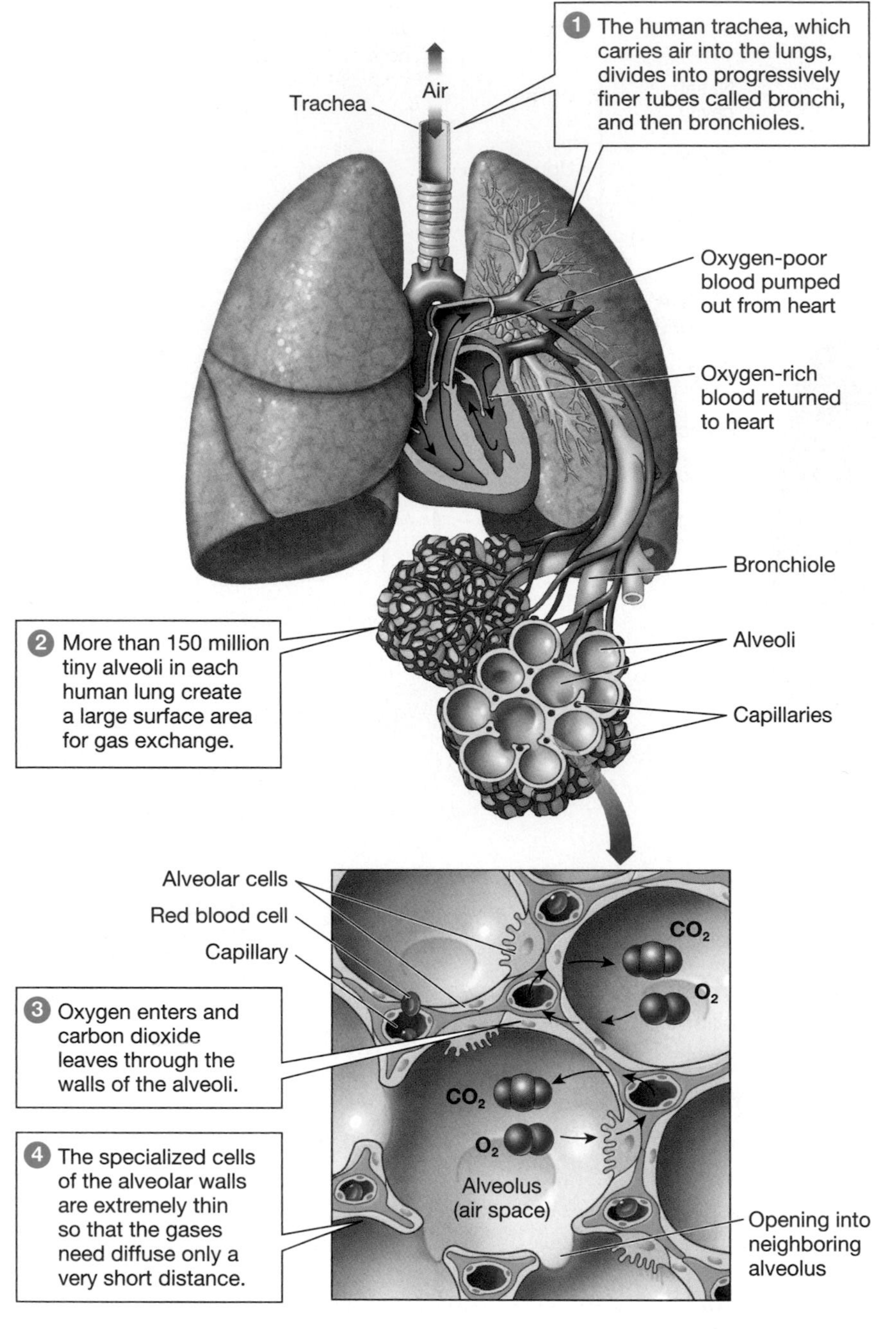

Figure 24.8

Gas exchange in the alveoli

The structure of the lungs speeds the diffusion of oxygen and carbon dioxide into and out of the body.

Q1: Why is the large surface area created by alveoli important for gas exchange? How do the thin walls of alveoli help with gas exchange?

Q2: Does carbon dioxide move into or out of alveoli? Does it move into or out of capillaries at the surface of the alveoli?

Q3: When a person has pneumonia, the alveoli may fill with fluids. Why would this be a problem?

See Appendix A for answers to the figure questions.

Waste Not

At Massachusetts General Hospital in Boston, Harald Ott has bioengineered whole rat, pig, and human lungs by cleaning donated organs and repopulating them with new cells (**Figure 24.9**). These organs have yet to be successfully transplanted, but the research is advancing rapidly. In 2019, his team built a fully automated system to help them quickly and consistently build and monitor bioengineered lung tissues.

Ott, a thoracic surgeon and bioengineer, also has an extensive tissue engineering program beyond lungs. In 2013, Ott's lab used the scaffolding technique to engineer rat kidneys, which were then transplanted into rats. In vertebrates, **kidneys** are a set of paired organs that maintain water and solute homeostasis, serving as key components of the urinary system.

All animals must regulate the concentrations of solutes, such as sodium and calcium, in their body fluids, but terrestrial animals face the additional challenges of conserving water and retaining vital solutes. The solute composition of an animal is affected by metabolic activities within the body: as cells metabolize biomolecules, they use up chemicals dissolved in body fluids and produce new ones. This process results in waste products that must be removed from the body.

Kidneys filter and regulate the composition of the blood as it moves through them. As blood passes through the kidneys, it is cleansed of metabolic wastes. The volume of blood leaving

Figure 24.9

A bioengineered rat lung

Researchers at Massachusetts General Hospital have bioengineered the lungs of rats, pigs, and humans. None have been capable of functioning within a living individual.

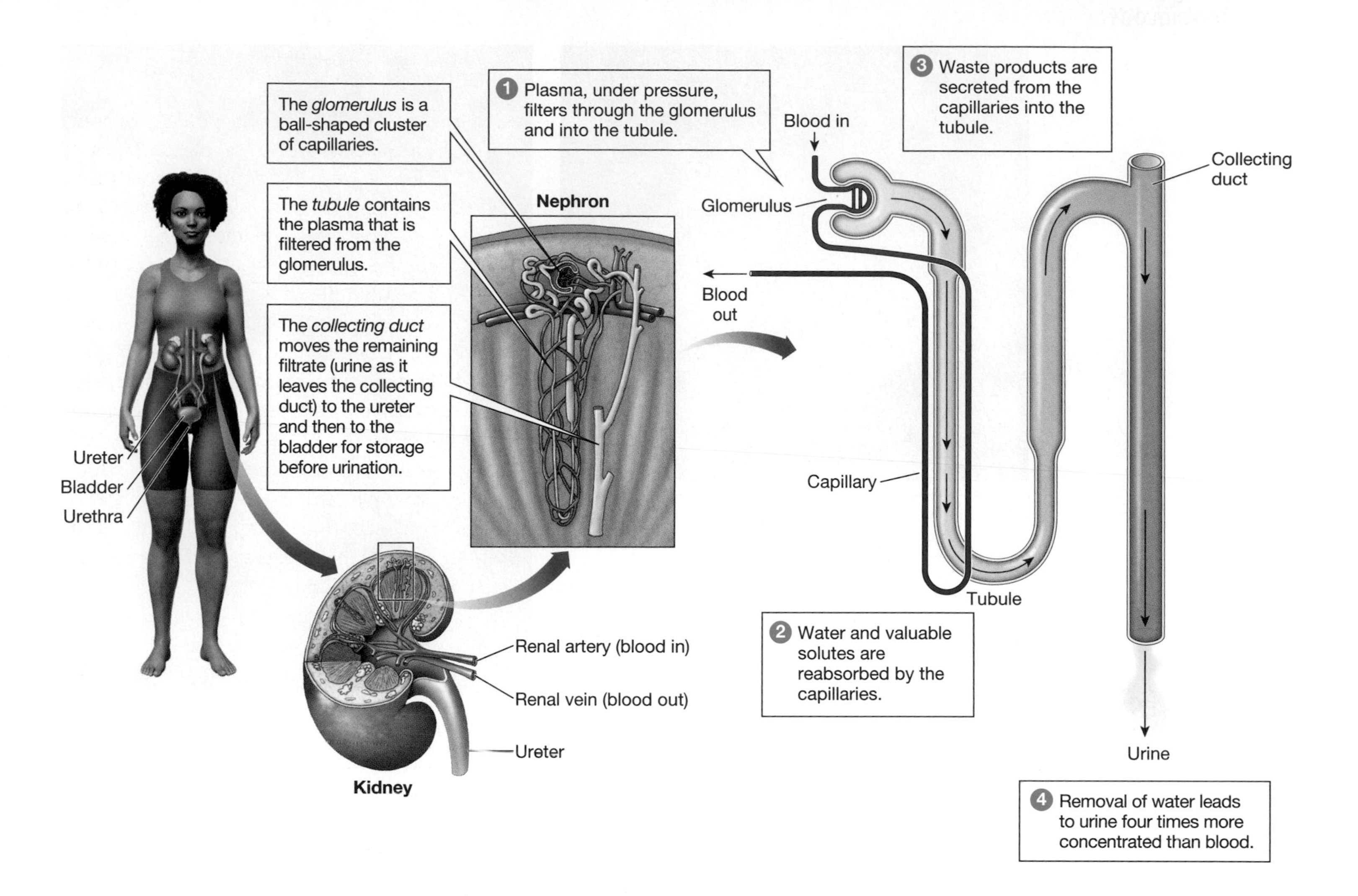

Figure 24.10

The kidney, a critical component of the human urinary system

The kidney regulates internal water content, balances solute concentrations, and removes toxic wastes. All of this work occurs within the nephron.

Q1: What is drained from the many collecting ducts of the kidney, what does it drain into, and then where does it go?

Q2: What is the difference between reabsorption and secretion?

Q3: Alcohol suppresses the kidney's ability to reabsorb water through the capillaries. What common consequence of drinking alcohol is related to this fact, and how might you alleviate this problem?

See Appendix A for answers to the figure questions.

the kidneys is slightly smaller than the entering volume because some water is lost to make **urine**, the waste-carrying solution that is expelled from the body.

The blood-cleansing work of the kidneys—**filtration**—is performed by the kidney's basic functional unit, the **nephron** (**Figure 24.10**). Each human kidney has about a million of these tiny filtration units, tightly integrated with surrounding capillaries. When the kidneys fail, an individual can no longer filter waste. Approximately 100,000 individuals in the United States currently await kidney transplantation. As they wait, these patients generally receive dialysis three times per week, during which a machine filters their blood for them. This lifesaving technology requires a significant time commitment, and dialysis patients have a high risk of infection.

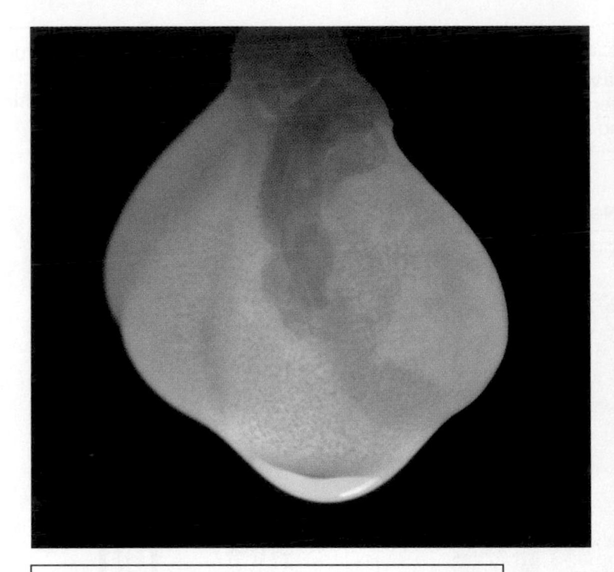

A rat kidney with all its cells removed, leaving only the extracellular matrix made of collagen.

The collagen "scaffold" has now been reseeded with cells.

Figure 24.11

A bioengineered kidney: before and after
The Ott lab at Massachusetts General Hospital bioengineered rat kidneys that function when transplanted. Before implanting, the kidney is flushed of its donor cells and repopulated (seeded) with recipient cells so that it will not be rejected by the recipient's immune system when transplanted.

An engineered kidney would provide a permanent solution for these patients. But an engineered kidney would need to do more than just remove waste from the blood, as filtration is only one of three parts of the kidney's job. A second important function is the **reabsorption** of water and valuable solutes (such as sodium, sugars, and small proteins) before they leave the kidneys. A third function is **secretion**, which involves the kidney in actively transporting excess quantities of substances (such as potassium and hydrogen ions, and some medications and toxins) from the blood into the liquid passing through the kidney.

The concentrated fluid that results from the combination of filtration, reabsorption, and secretion is urine. Urine from the many collecting ducts in a kidney drains into a long tube, the *ureter*, which delivers the fluid to the urinary bladder for storage. In urination, the bladder empties through a tube called the *urethra*. In Ott's lab in Boston, the bioengineered rat kidneys successfully produced urine when transplanted into rats (**Figure 24.11**).

Coming to Your Senses

To date, the kidney is the most complex organ recreated in the lab and successfully transplanted into an animal. Meanwhile, bioengineering research in a different organ system—the nervous system—has yielded astonishing improvements in patients' lives, including the life of U.S. Navy corpsman Edward Bonfiglio Jr.

At 5:00 a.m. on August 27, 2009, Edward Bonfiglio awoke to the sound of the phone ringing. It was his son, Edward Jr., a member of the U.S. Navy serving on active duty in Afghanistan. "Dad?" Edward said, "I got shot. Don't tell Mom."

During a routine foot patrol, Edward's unit had been ambushed, and he'd been shot in his left leg. The bullet hit his sciatic nerve, a long, thick bundle of nerve fibers that runs from the lower back down the leg. Edward immediately lost all feeling and function below his left knee. Doctors found that the bullet had sliced a 5-centimeter gap in the sciatic nerve. Without that nerve intact, the leg would not function.

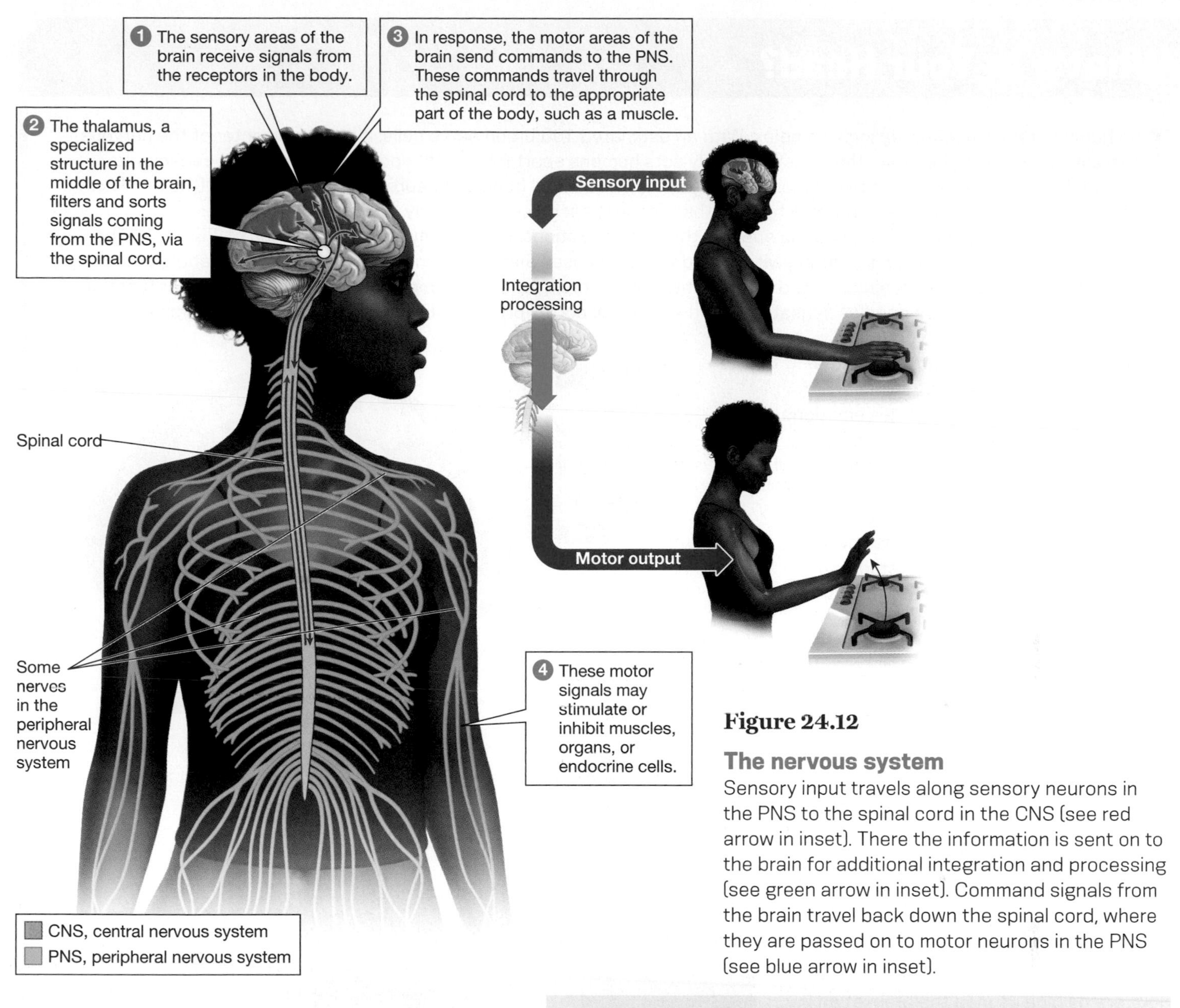

1 The sensory areas of the brain receive signals from the receptors in the body.
2 The thalamus, a specialized structure in the middle of the brain, filters and sorts signals coming from the PNS, via the spinal cord.
3 In response, the motor areas of the brain send commands to the PNS. These commands travel through the spinal cord to the appropriate part of the body, such as a muscle.
Sensory input
Integration processing
Motor output
Spinal cord
Some nerves in the peripheral nervous system
4 These motor signals may stimulate or inhibit muscles, organs, or endocrine cells.
CNS, central nervous system
PNS, peripheral nervous system

Figure 24.12

The nervous system

Sensory input travels along sensory neurons in the PNS to the spinal cord in the CNS (see red arrow in inset). There the information is sent on to the brain for additional integration and processing (see green arrow in inset). Command signals from the brain travel back down the spinal cord, where they are passed on to motor neurons in the PNS (see blue arrow in inset).

Q1: To which part of the nervous system—CNS or PNS—does the retina in your eye belong?

Q2: To which part of the nervous system—CNS or PNS—does the spinal cord belong?

Q3: In a *reflex arc*, sensory input is processed in the spinal cord and an immediate signal is sent to activate motor output without first routing the input signal to the brain. The "knee jerk" is one example of a reflex arc. Give another example of this kind of reflexive reaction to sensory input.

See Appendix A for answers to the figure questions.

After Edward arrived home in the United States, doctors gave him two options. The nonfunctioning leg could be amputated, or the damaged nerve might be repaired with a new kind of nerve graft, a bioengineered tube that could potentially reconnect the two ends of his severed nerve and bring feeling back to his leg. Edward chose the latter option to try to repair the nerve in his leg, part of his peripheral nervous system.

The nervous system of vertebrates is a communication system that transmits signals among various parts of the body. It can be divided into two main units: the **central nervous system**, or **CNS**, and the **peripheral nervous system**, or **PNS** (**Figure 24.12**).

PHYSIOLOGY

What's in Your Head?

The human brain is mind-bogglingly complex. With an estimated 100 billion nerve cells, it is the epicenter of the nervous system. The brain is the organ that most distinctly sets humans apart from other species, giving us our capacity to reason, feel, and remember. The brain is made up of *gray matter* (the cell bodies of neurons) and *white matter* (the *dendrites* and *axons*, a branching network of winding tendrils that connect each neuron to many, many other neurons).

The human brain weighs about 3 pounds and has three main sections: *forebrain*, *midbrain*, and *hindbrain* (Figure 1). The forebrain contains the *thalamus*, a central switchboard that processes and directs incoming sensory information, and the *cerebrum*, whose outer layer (*cerebral cortex*) handles most of the brain's information processing, as well as higher functions. In other words, the thalamus handles sensation, but the cerebral cortex produces perception, thought, and complex behavior. The forebrain also contains the *hypothalamus*, the *amygdala*, and the *hippocampus*. These structures make up the *limbic system*, which regulates emotions and motivations and participates in memory formation.

The midbrain and hindbrain make up the *brain stem*. The midbrain coordinates sensory information from the thalamus with the peripheral nervous system to bring about simple physical movement. The main structure in the midbrain is the *reticular formation*, which regulates consciousness. Finally, the hindbrain, evolutionarily the oldest part of the brain, consists of the *pons*, the *medulla oblongata*, and the *cerebellum*. The hindbrain controls the body's most basic functions, including heart rate, breathing, and balance.

Neuroscience is an extremely active area of research. Scanning technologies such as diffusion MRI, which tracks the diffusion of water through white matter, and functional MRI, which detects brain activity by measuring changes in blood flow, are helping us map the brain at an unprecedented level of detail. Figure 2 is a scan from a diffusion MRI (artificially colored) depicting the elegant structure of white-matter fibers twisting through the brain.

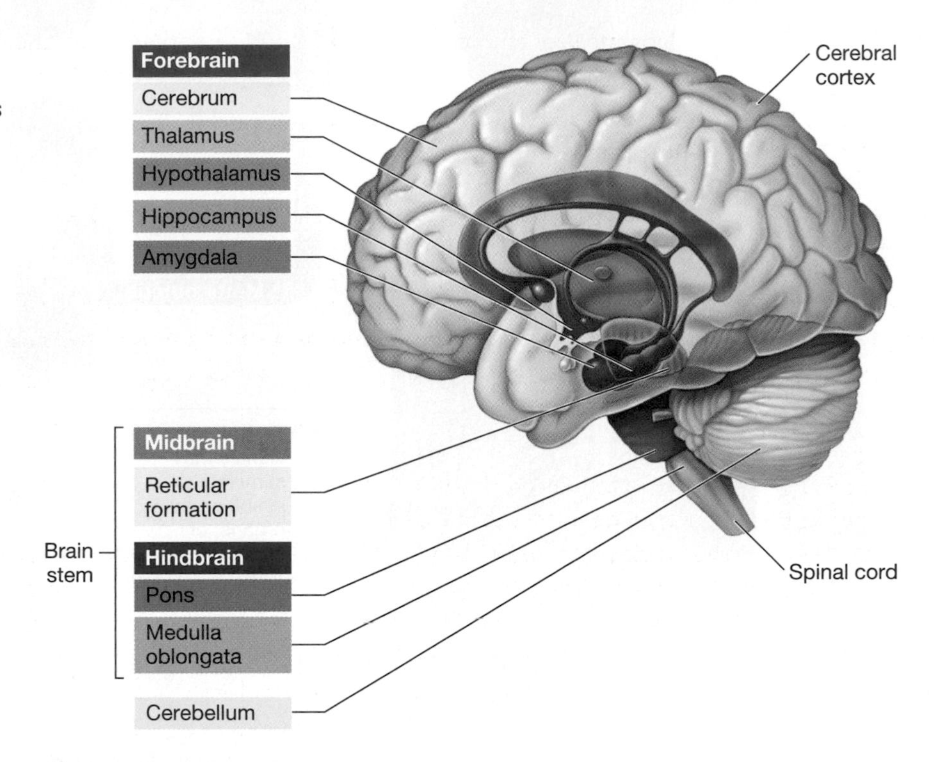

Figure 1

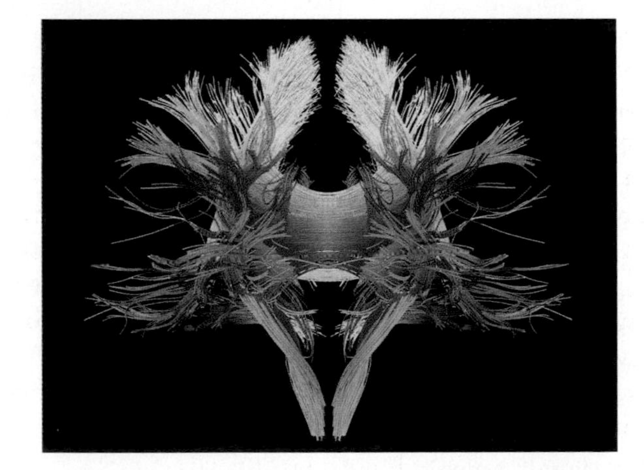

Figure 2

Table 24.1

Sensory Receptors

Receptor Type	Stimulus	Sense(s)
Chemoreceptors	Chemicals	Taste, smell, concentration of chemicals in the body
Photoreceptors	Light	Vision
Mechanoreceptors	Physical changes	Touch, hearing, proprioception (body position), balance
Thermoreceptors	Moderate heat and cold	Thermoreception (gradations of heat and cold)
Pain receptors	Injury, noxious chemicals, chemical and physical irritants	Pain, itch
Electroreceptors*	Electrical fields (especially those generated by muscle contractions of other animals)	Electrical sense
Magnetoreceptors*	Magnetic fields	Magnetic sense

*Electroreceptors and magnetoreceptors are found in many animals, including most vertebrates. They are not known to be active in humans.

The CNS consists of the brain and the spinal cord. The **brain** has a large capacity for processing diverse types of sensory information, and it controls and coordinates nerve signals throughout the body (see "What's in Your Head?" on page 476). The brain is also responsible for higher functions such as learning, memory, and conscience. The **spinal cord** is a thick central nerve cord that is continuous with the brain, acting as a relay between the brain and the rest of the body, but it can also process spinal cord reflexes—simple movements directed entirely within the spinal cord. The retina—a sheath of nerves connected directly to the optic nerve—and the optic and olfactory nerves are also considered part of the CNS, because they lead directly to the brain without connective nerve tissue.

The PNS consists of **sensory organs**—such as the eyes and ears—plus the nerves (except for the retinal, optic, and olfactory nerves). For example, the nonfunctioning nerve in Edward Bonfiglio's leg was part of his PNS. The PNS converts stimuli from the sensory organs into **sensory input**—signals that are received, transmitted, and processed by the CNS. Animals are constantly bombarded with sensory input from their external and internal environments. Five main classes of sensory receptors in humans receive this input; some additional categories, not known to be active in humans, are found in other animals (**Table 24.1**).

Through sensory input, the PNS gathers information from the external and internal environments and sends it to the CNS. For example, imagine placing your hand near a hot stove; the heat input to your skin is transmitted as a signal from the PNS to the CNS. The CNS integrates and processes the information and generates a signal in response, dictating a particular action, or **motor output**, such as moving your hand away from the stove. The PNS then relays that signal to the body part that will complete the action: the hand moves away from the stove.

All this action in the nervous system is conducted by **nerves**. Nerves—made up of one or more specialized cells called **neurons**—transmit signals from one part of the body to another in a fraction of a second. The structure of a neuron

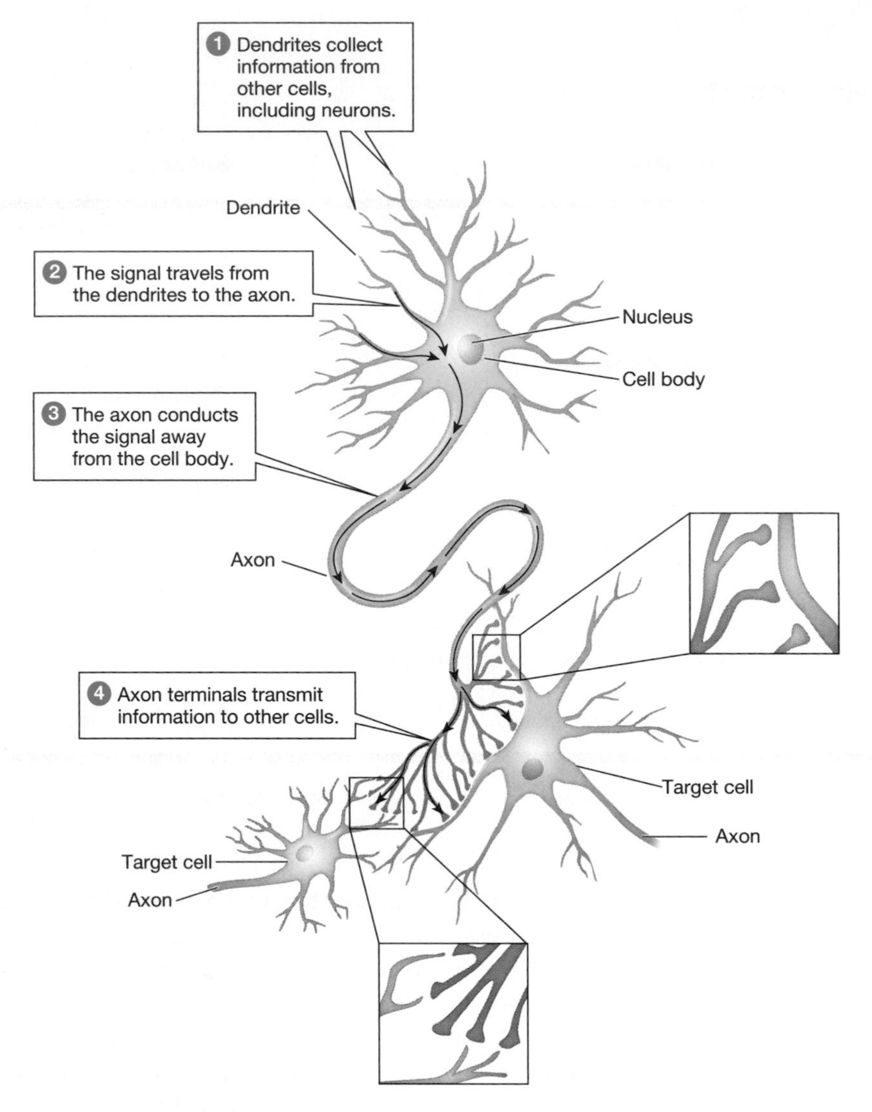

1 Dendrites collect information from other cells, including neurons.
Dendrite
2 The signal travels from the dendrites to the axon.
Nucleus
Cell body
3 The axon conducts the signal away from the cell body.
Axon
4 Axon terminals transmit information to other cells.
Target cell
Axon
Target cell
Axon

Figure 24.13

Neurons transmit signals from one cell to another

A neuron receives information from other cells, including other neurons, through one or more dendrites. The neuron pictured here has a single long axon with branched endings. Axons carry signals away from the cell body and to another cell, such as the two target cells shown here.

Q1: The cell body of a neuron contains all the structures common to animal cells. How do neurons look different from other cells you've learned about in this book?

Q2: What is the function of these differences?

Q3: Some people are born without the capacity to feel pain. Although it might initially sound nice not to feel pain, it is actually quite dangerous. Describe a situation in which it would be dangerous not to feel pain.

See Appendix A for answers to the figure questions.

reflects this unique function (**Figure 24.13**). Throughout the body, different types of neurons respond to different types of stimuli.

Nerves in the PNS are fragile and can be damaged easily. For instance, if you've ever hit your "funny bone," you've actually caused trauma to your ulnar nerve at the elbow, resulting in numbness and tingling. A more severe blow—such as trauma from a broken bone, or a bullet wound as in Edward's

Heart to Heart

Over 100,000 people in the United States are right now waiting for an organ transplant. Each year, the number of people on the waiting list is larger than both the number of donors and transplants, so many of those awaiting a transplant never receive one. On average, 20 people die each day in the U.S. while awaiting a transplant. Consider registering to be an organ donor: a single organ, eye, and tissue donor can save or improve as many as 75 lives.

Assessment available in **smart**work

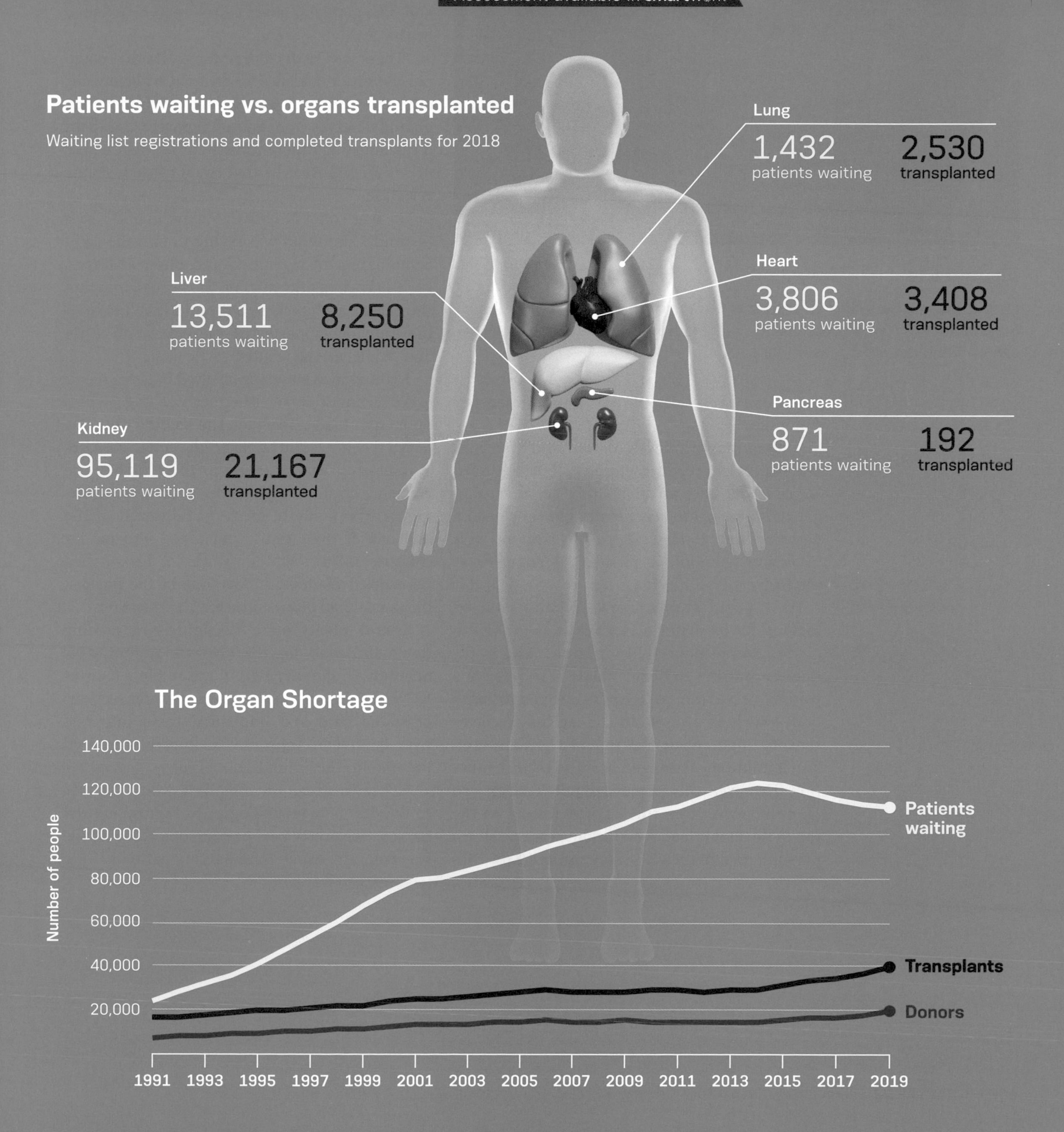

Figure 24.14

Using bioengineering to repair a damaged nerve
Edward Bonfiglio chose experimental surgery over amputation, and today he can walk and run without assistance.

case—can cause loss of motor and/or sensory function.

Unlike lungs, kidneys, or the heart, peripheral nerves can repair themselves, but they need help. "If you have an injury to a nerve, we can get those nerves to regenerate, but they need a pathway, like a sidewalk, to migrate on," says Christine Schmidt, a biomedical engineer at the University of Florida.

As a postdoctoral fellow, Schmidt started looking for biomaterials to act as sidewalks to guide severed neurons to migrate. The traditional approach to nerve repair is to remove a piece of nerve from somewhere else in a patient's body—such as a nerve in the leg that receives sensory input from the top of the foot—and stitch it into the damaged nerve to restore function. However, this process requires surgery to take the nerve from the leg, and feeling on the top of the foot is lost. And in some cases, using a person's own nerves is not an option. Edward's gap was 5 centimeters wide—far too big for a nerve from another part of his body to work.

Another option is a synthetic graft, a hollow plastic tube inserted to provide a way for nerve cells to grow from one location to another. Yet these tubes can be used only in very small places and, like artificial blood vessels, can be rejected by the immune system.

"There was really a need for grafts that could provide better guidance for neurons, without using one's own nerves," says Schmidt. Schmidt's lab obtained nerves from animal cadavers and spent 3 years testing different methods of washing the nerves to remove cells that could cause immune rejection.

"It was like a cooking experiment," says Schmidt. Her team was trying to find the right combination of chemicals and physical forces, such as rinsing and swirling the nerves, to retain the nerve's architecture but remove all the original cells. They perfected the process in 2004 and published the results.

Right away, a company involved in nerve repair asked Schmidt about using her washing process on human tissue. Together, she and the company tested the process on human cadaver tissues. After a successful demonstration in those tissues, the company began selling off-the-shelf nerve grafts for use in hospitals. In late 2009, Edward Bonfiglio had one of these grafts implanted into his leg (**Figure 24.14**). Within months of his surgery, he wiggled his toes. It was "one of the greatest moments I had in my entire life," he later said.

There's plenty more bioengineering coming down the road for the nervous system, says Schmidt. In January 2019, for example, a team at UC San Diego 3D-printed 2-millimeter sections of spinal cord and implanted them into rats with spinal cord injuries. The implants restored function to the animals' hind limbs.

"It's rewarding," says Schmidt, who continues to develop new technologies to regrow injured nerves. "It's pretty neat to take something all the way from the bench to impacting patients."

NEED-TO-KNOW SCIENCE

- In humans and other vertebrates, the **circulatory system** (page 464) moves oxygen from the lungs to the heart, which then pumps oxygen-rich **blood** (page 465) to the rest of the body.
- Blood is composed of cells and cell fragments that float in a fluid known as **plasma** (page 465). **Red blood cells** (page 465) are the main oxygen carriers in the blood.
- The **cardiovascular system** (page 465) is a closed circulatory system with a chambered heart that pumps blood through a complex network of **blood vessels** (page 465). In the **pulmonary circuit** (page 468), oxygen-deficient blood is pumped to the lungs. In the **systemic circuit** (page 468), oxygenated blood returning from the lungs is pumped out to body tissues.
- **Arteries** (page 467) are large vessels that transport blood away from the heart. **Veins** (page 467) are large vessels that carry blood back to the heart. **Capillaries** (page 467), the smallest vessels, facilitate the exchange of materials with surrounding cells.
- The mammalian **heart** (page 468) is composed of four chambers that make up two separate muscular pumps, each composed of an **atrium** (page 468) and a **ventricle** (page 468). The left atrium and ventricle pump blood to the body; the right atrium and ventricle pump blood to the lungs.
- **Heart rate** (page 468) is the number of times a heart beats per minute. The force of blood pushing through blood vessels is called **blood pressure** (page 468). The pressure of the heart contracting to push blood out is **systole** (page 468). The pressure of the heart relaxing after contracting is **diastole** (page 468).
- The human **respiratory system** (page 464) consists of two parts. The **upper respiratory system** (page 470)—including airways such as the **pharynx** (page 471) and the **larynx** (page 471)—carries air from the nose (or mouth) to the **lower respiratory system** (page 471), which consists of the **trachea** (page 471), **bronchi** (page 471), **bronchioles** (page 471), and **lungs** (page 468). The air eventually reaches **alveoli** (page 471), where gas exchange occurs (oxygen diffuses into the blood and carbon dioxide diffuses out of it).
- **Breathing** (page 470) consists of inhalation and exhalation, which are controlled by the contraction of muscles, especially those of the diaphragm and the rib cage.
- In the **urinary system** (page 464) of many animals, including humans, **kidneys** (page 472) regulate concentrations of both body water and solutes. The kidney's basic unit, the **nephron** (page 473), performs three functions: **filtration** (page 473), **reabsorption** (page 474), and **secretion** (page 474). The resulting concentrated solution, **urine** (page 473), is carried by ducts to the bladder and excreted from the body.
- The vertebrate **nervous system** (page 465) is divided into the **central nervous system** (**CNS**, page 475), consisting of the **brain** (page 477) and the **spinal cord** (page 477), and the **peripheral nervous system** (**PNS**, page 475), consisting of the **sensory organs** (page 477) and all the remaining nerves.
- Sensory organs convert environmental stimuli into **nerve** (page 477) impulses that are carried by sensory **neurons** (page 477) to the CNS. All human senses rely on **sensory input** (page 477) from just five types of sensory receptors.
- The CNS integrates sensory information and sends an output signal, often **motor output** (page 477), through the PNS to the appropriate body part.

THE QUESTIONS

See Appendix B for answers.

The Basics

1 Blood plasma transports
(a) waste products.
(b) water.
(c) solutes.
(d) all of the above

2 In humans, where is the gas exchange surface located?
(a) pharynx
(b) bronchus
(c) alveolus
(d) bronchiole

3 Which blood vessels carry blood back toward the heart?
(a) veins
(b) arteries
(c) ventricles
(d) capillaries

4 In a neuron, the ___________ conducts signals to other cells.
(a) dendrite
(b) axon
(c) nucleus
(d) cell body

5 Which of the following functions is *not* performed by the nephron?
(a) filtration
(b) reabsorption
(c) secretion
(d) deletion

6 Identify each of the following items as belonging to either the central nervous system (CNS) or the peripheral nervous system (PNS).
_____ a. brain
_____ b. sensory organs
_____ c. spinal column
_____ d. temperature-sensing nerves in the skin

7 Link each term with the correct definition.

SA node	1. Sends blood to the right ventricle.
left ventricle	2. Pumps blood to the systemic circuit.
right atrium	3. Serves as the heart's pacemaker.
pulmonary circuit	4. Pumps oxygen-poor blood to the lungs.

Challenge Yourself

8 Which of the following sensory receptors is *not* active in humans?
(a) pain receptor
(b) electroreceptor
(c) mechanoreceptor
(d) chemoreceptor

9 Select the correct terms:
Kidneys conduct their primary work of filtering the blood through their basic functional unit, the (**neuron** / **nephron**). Besides filtration, the kidney must (**reabsorb** / **secrete**) water and important solutes and (**reabsorb** / **secrete**) toxins and overabundant substances. Urine drains from the kidneys into the bladder via the (**urethra** / **ureter**) and empties from the bladder via the (**urethra** / **ureter**).

10 According to the U.S. Centers for Disease Control and Prevention (CDC), there are seven criteria for staying heart healthy to avoid heart attack and stroke: be active, keep a healthy weight, learn about cholesterol, don't smoke or use smokeless tobacco, eat a heart-healthy diet, keep blood pressure healthy, and learn about blood sugar and diabetes mellitus. According to the accompanying graph, about what total percentage of kids aged

12–19 years old met all seven criteria for staying heart healthy? About what total percentage met six of the seven criteria? Was the percentage higher for males or females of the kids who met three of the seven criteria?

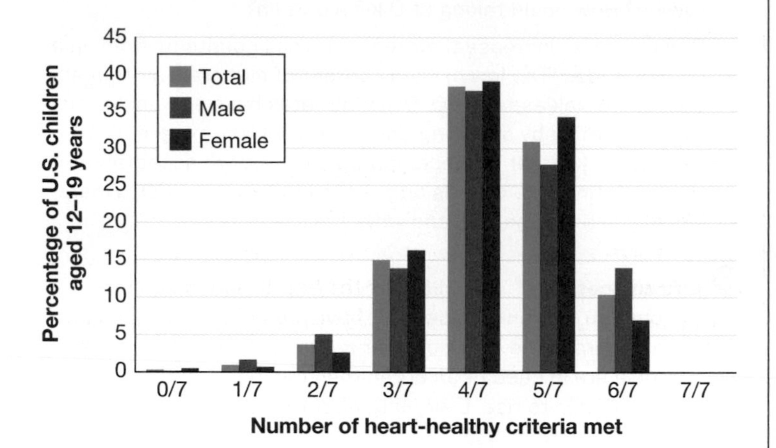

11 Beginning with sensory input from a mechanoreceptor in your toe, place the response events of your nervous system in the correct order by numbering them from 1 to 5.

_____ a. Command signals travel from the brain to the spinal cord.

_____ b. Input is sent to the brain.

_____ c. A sensory signal is sent to the spinal cord.

_____ d. The muscle responds to the signal.

_____ e. Processing occurs in the brain.

12 How do veins and capillaries differ in both structure and function?

Try Something New

13 Which of the following consequences would most likely result from kidney failure?

(a) Heart rate would be uncontrolled.

(b) Toxin levels in the bloodstream would increase.

(c) Incorrect sensory signals would be sent to the brain.

(d) Oxygen levels would fluctuate dramatically.

(e) None of these would occur because of kidney failure.

14 If an organism has a greater concentration of carbon dioxide (CO_2) in its lungs than in its blood, will there be (a) net transport of CO_2 from the lung air space to the alveolar capillaries, or (b) net transport of CO_2 from the alveolar capillaries to the lung air space? Explain the reasoning for your answer.

15 The number of times our hearts beat per minute is referred to as heart rate. Each heartbeat lasts a little less than 1 second and consists of a series of events called the *cardiac cycle*. The blood pressure measured in a doctor's office reflects the pressure in the arteries leading to the body from the left ventricle. A blood pressure reading of 120/80 (systole/diastole), for example, means that contraction of the left ventricle generates 120 millimeters of mercury (mm Hg) of pressure in the arteries, followed by a drop to 80 mm Hg when the ventricles relax and refill.

(a) Blood vessels become less flexible with age. How might this change affect the body's ability to respond to environmental changes with changes in blood pressure or heart rate?

(b) How would you predict that heart rate changes with exercise in the short term and in the long term?

(c) "White coat hypertension" refers to the higher blood pressure displayed at the doctor's office than in normal daily life. How would you explain this phenomenon?

Leveling Up

16 **What do *you* think?** In the highly competitive world of endurance sports such as marathon running, cross-country skiing, and bicycle racing, any means of improving performance offers a significant advantage. In recent years, many cyclists have admitted to injecting the hormone erythropoietin (EPO), also known as blood doping. Made naturally in the kidneys, EPO increases the oxygen-carrying capacity of blood by stimulating red blood cell production. In addition to increasing the production of red blood cells, EPO stimulates the growth of capillaries that carry oxygen to tissues.

A synthetic form of EPO developed by drug companies is used to treat patients with anemia, kidney damage, and malaria. Athletes engaging in "blood doping" subject themselves to many health risks. An excess of red blood cells can make the blood so thick that it clots or fails to flow easily through the heart. Nearly two dozen endurance athletes are thought to have died of heart attacks caused by doping with EPO. Because EPO is a naturally occurring hormone, identifying synthetic EPO in the blood or urine samples of athletes has been difficult.

(a) What is EPO, and why does it offer a performance advantage in sports, especially endurance events such as cycling and rowing? How could taking EPO kill a person?

(b) Some cyclists increase their red blood cell counts by training at high altitudes. The low oxygen content of mountain air triggers the natural release of EPO. Other athletes have accomplished the same thing by spending time in special low-oxygen tents. Do you think either of these approaches is more acceptable than injecting EPO or cells engineered to express EPO? Where would you draw the line, and why?

17 **Life choices** As of 2019, 95 percent of U.S. adults support organ donation, but only 58 percent have identified themselves as organ donors in the event of their death. Unfortunately, many more people are in need of an organ than can be helped, and this number continues to rise. Every day, about 80 people receive

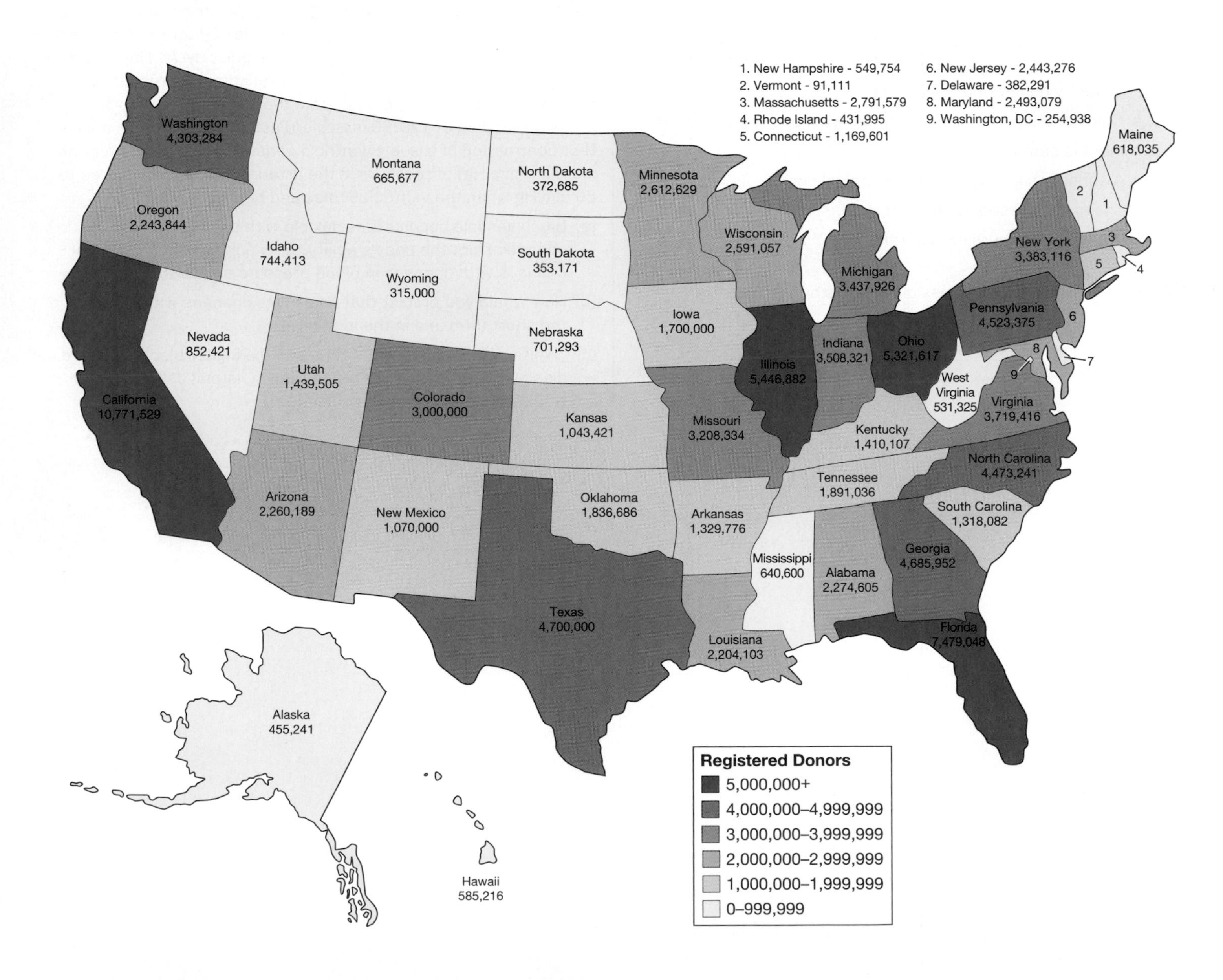

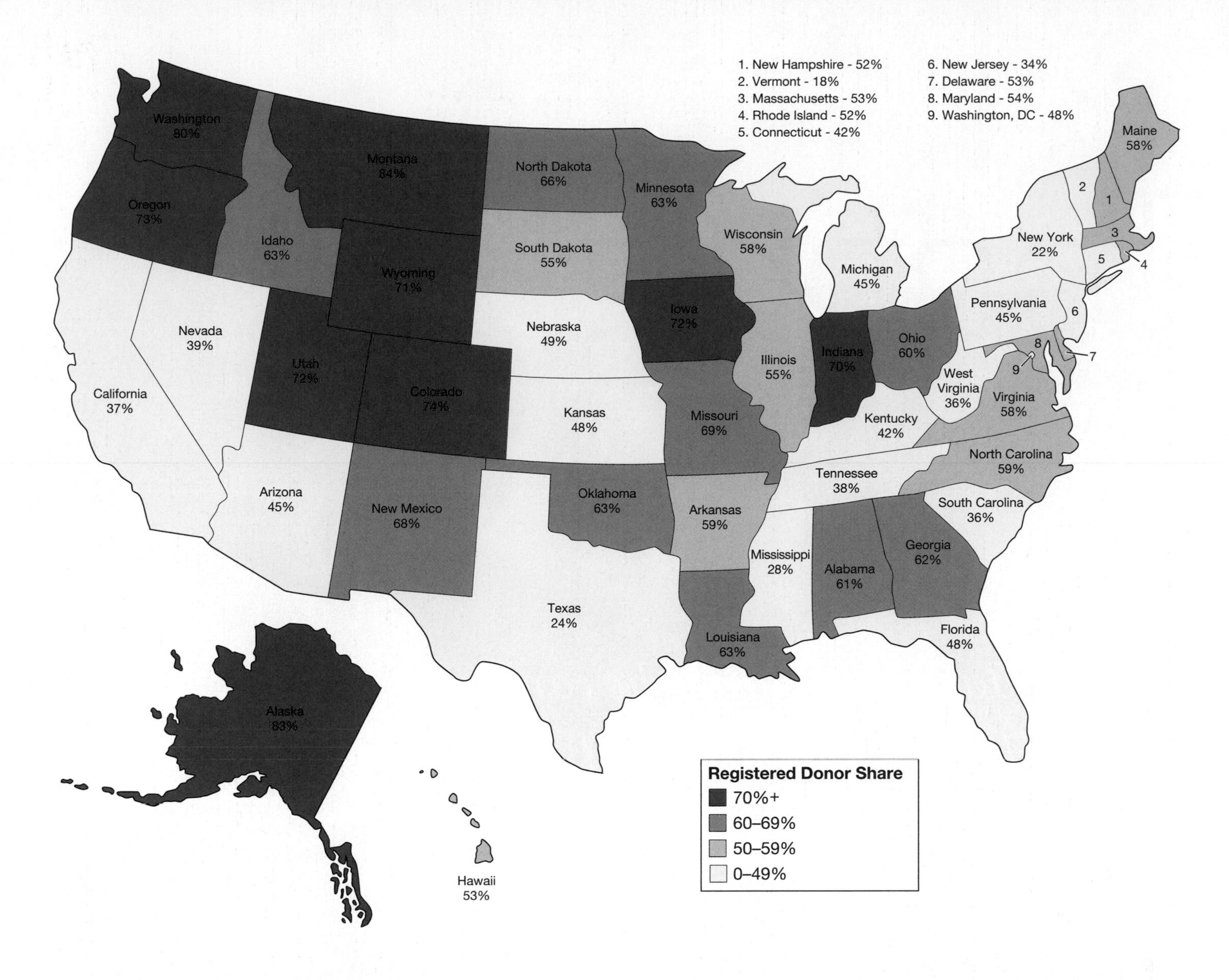

organ transplants but another 20 people die while waiting for a transplant that never came. Refer to the following two maps to answer the first three questions here.

(a) Which state has the highest number of organ donors? Which state has the lowest number?

(b) Which state has the highest percentage of organ donors? Which state has the lowest percentage?

(c) Where does your state fall in terms of the percentage of organ donors? Does this surprise you? Why or why not?

(d) Are you an organ donor? Why or why not?

(e) Suggest one way that a state (yours or another) could increase the number of its residents who identify themselves as organ donors.

Digital resources for your book are available online.

Testing the Iceman

A Dutch daredevil claims he can fend off disease with his mind. Two skeptical scientists take the case.

- Explain how cells communicate with each other via the endocrine system.
- Describe how a hormone can act on a target cell.
- Identify the immune system's first, second, and third lines of defense.
- Compare and contrast the role of white blood cells in the innate and adaptive immune systems.
- Relate the processes of inflammation and blood clotting.
- Distinguish between a primary and a secondary adaptive immune response.
- Create a flowchart depicting the sequence of events as a vertebrate immune system responds to a pathogen.

PHYSIOLOGY

CHAPTER 25

ENDOCRINE AND IMMUNE SYSTEMS

The scantily clad young men lie on the ground, their hands behind their heads and their faces pointed toward the sky. Wearing only swim trunks and sunglasses, they look as if they're tanning at the beach—but there are no piña coladas or warm sand here. Instead, these 18 volunteers are lying on cold, white snow in the mountains of Poland. And lying with them is the Iceman.

Wim Hof, a Dutch daredevil known as the "Iceman," who holds numerous world records for cold exposure, breathes deeply and leads the youths in an exercise. Over 4 days, he will train them to tolerate extreme cold. During his rigorous program, they will swim in near-freezing water every day and climb a snow-covered mountain wearing just shorts (**Figure 25.1**). Hof claims that exposure to the cold combined with meditation and breathing exercises will enable the young men to fend off illness and disease.

Figure 25.1

Iceman in action
Wim Hof (third from left) trains volunteers under extreme conditions.

Matthijs Kox stands to the side of the Iceman's trainees, taking notes. Kox, a researcher in intensive-care medicine at Radboud University Medical Center in the Netherlands, first met Hof in 2010, when the Iceman was visiting another laboratory at the university. A team in the physiology department was measuring Hof's ability to regulate his core temperature while standing in an ice bath (**Figure 25.2**). The scientists were surprised to find that rather than decreasing as expected, Hof's core temperature actually increased, and his metabolism climbed. While standing in the ice bath talking to his examiners, Hof mentioned that he could also consciously modulate his autonomic nervous system and immune system.

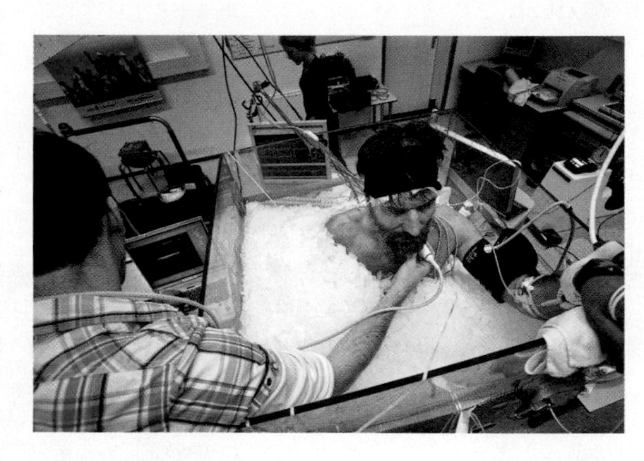

Figure 25.2

Iceman at rest
Hof's vital signs, including his core temperature, are monitored while he is immersed in ice.

It was an unbelievable claim. The autonomic nervous system is a major component of the peripheral nervous system (PNS, discussed in Chapter 24). It operates body functions that humans cannot voluntarily control, such as heartbeat and blood pressure. The **immune system**, a remarkable defense system that protects us against most infectious agents, has also long been known to be involuntary. But Hof had a history of doing the unbelievable. He claimed the Guinness World Record for longest ice bath by staying immersed in ice for 1 hour, 52 minutes, and 42 seconds. He climbed part of Mount Everest wearing nothing but shorts. He ran a half marathon through the snow at –4°F (–20°C), again wearing only shorts.

Hof's testers in the physiology unit told him that a Radboud University professor named Peter Pickkers had a way to measure a person's immune response. So Hof hoofed it to Pickkers's office, shook his hand, and said, "I can modulate my immune system. I heard you can measure it. Will you measure mine?"

Hormonal Changes

Pickkers was skeptical of Hof's claim, which had a whiff of pseudoscience. But Hof was an interesting character, so Pickkers went online and watched videos of his feats. "There were remarkable things I did not know of—things that, if you had asked me beforehand, I would have said, 'That's not possible. It's not possible to run half a marathon barefoot in the snow,'" says Pickkers. "But he did that."

Pickkers raised the idea of testing Hof to Kox, who was one of Pickkers's PhD students at the time, studying how the brain and immune system interact. Pickkers and Kox discussed the possibility at length, and they decided to give Hof a chance to document his claim. But they would do it while adhering strictly to the principles of the scientific process. "You can imagine some people wondered what we were doing with this guy," says Kox. "So we really focused on doing this in a very sound, precise manner, with no doubt about the scientific integrity of the project."

Hof claimed that the regimen for consciously controlling his immune system required three components: cold exposure, meditation, and breathing exercises. So the team tested Hof's blood before and after an 80-minute full-body ice bath while Hof performed breathing and meditation exercises.

Each time the scientists took blood, they went back to the lab and exposed the blood cells to molecules of endotoxin, a substance found in the cell walls of bacteria that activates an immune response in the human body. They wanted to see how Hof's immune cells in the blood would react to the endotoxin. After the regimen of ice, breathing, and meditation, Hof's cells had a far more subdued immune system response, showing very low levels of proteins associated with activation of the immune system, compared to similar cells before the regimen.

The cause of that subdued immune response was unclear, but the researchers suspected that stress hormones played a role. As you learned in Chapter 22, **hormones** are signaling molecules produced by certain cells that tell other cells what to do under specific situations or at certain times in the life cycle of the individual. Hormones are produced by specialized secretory cells of the **endocrine system** (**Figure 25.3**).

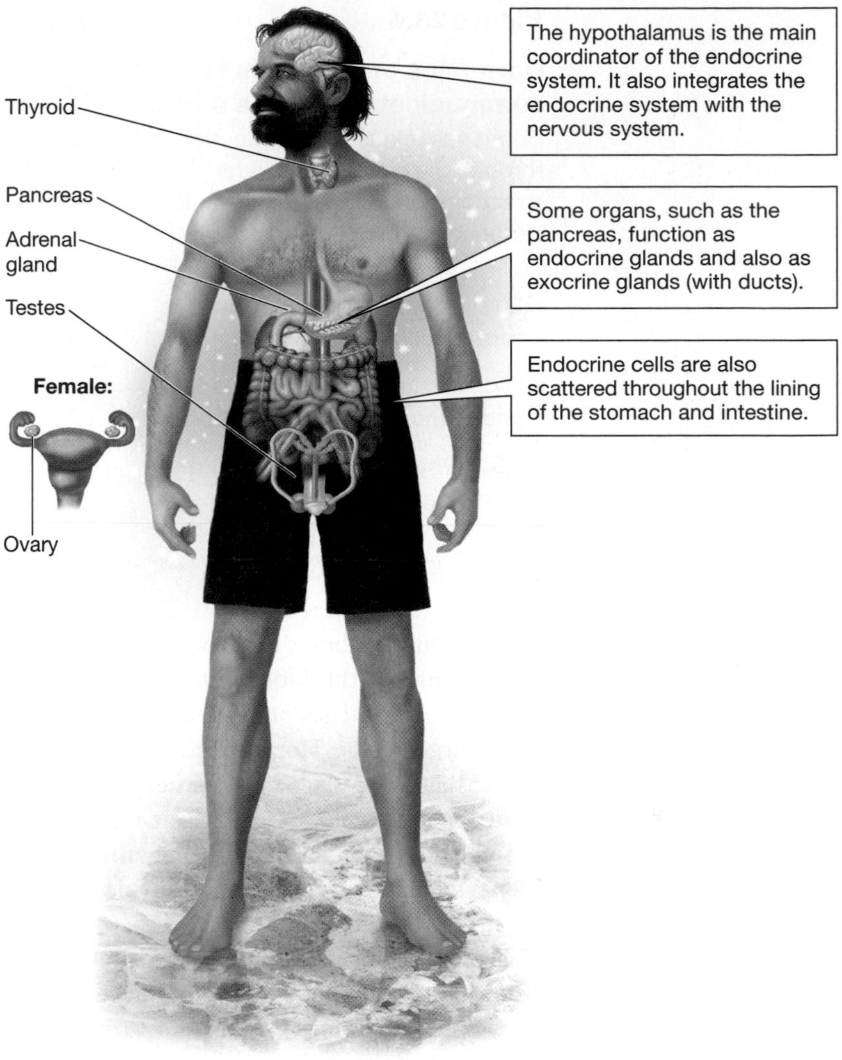

Secretory cells are often organized into discrete organs called **endocrine glands**. Major endocrine glands are located throughout the human body. Unlike *exocrine glands*, such as tear ducts, endocrine glands do not have ducts or tubes that deliver secretions from the gland to the site of action. Instead, endocrine glands release hormones into body fluids such as blood, which carries these chemical messengers throughout the body (**Figure 25.4**).

In a subsequent experiment, Kox measured the levels of a stress hormone called cortisol in Hof's blood. After the ice, breathing, and meditation regimen, Hof's blood contained far higher levels of cortisol than before.

Figure 25.3

The endocrine system is composed of hormone-secreting cells

The endocrine system consists of cells organized into ductless glands, plus scattered endocrine cells embedded in other tissues or organs. These cells all release hormones directly into the circulatory system.

Q1: What brain region coordinates the endocrine system?

Q2: What is the relationship between the endocrine system, endocrine glands, and endocrine cells?

Q3: What are the main male and female endocrine organs?

See Appendix A for answers to the figure questions.

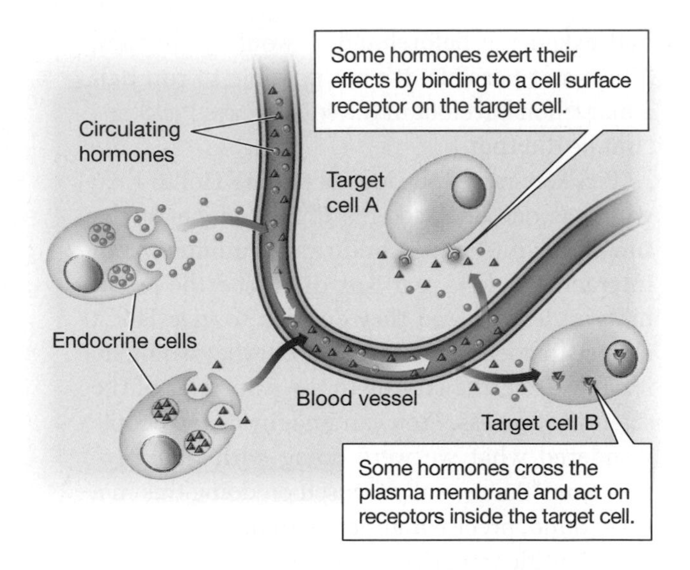

Figure 25.4

Hormones enable cells to communicate with one another

Hormones released by endocrine cells travel through the circulatory system to produce a response in target cells often located at a distance in the body.

Q1: **How do hormones travel to target cells?**

Q2: **Distinguish between an endocrine cell and a target cell.**

Q3: **Why is a hormone called a signaling molecule?**

See Appendix A for answers to the figure questions.

In the human body, most hormones can travel only as fast as the blood moves, which means they take several seconds or more to arrive at their target cells. Hormones coordinate functions that take place over timescales ranging from seconds (such as quickly increasing one's heartbeat in reaction to fear) to months (such as preparing a uterus to contract during the birthing process; see Figure 22.11).

Typically, hormones become greatly diluted after they are released into the circulatory system. They must therefore be able to exercise their effects at very low concentrations. Hormones are effective in small amounts because they bind to their targets with great specificity. Cortisol, for example, binds to a very specific receptor present on the surface of almost every cell in the body.

Cortisol, adrenaline (also called epinephrine), and noradrenaline (norepinephrine) are three hormones produced by the **adrenal glands**, a pair of endocrine glands sitting atop the kidneys. The release of these hormones launches a number of rapid physiological responses, including boosting blood glucose levels to increase energy levels in times of stress.

When a single hormone molecule binds to its receptor, it sets in motion a chain of events that may ultimately activate thousands of protein molecules in the target cell (**Figure 25.5**). Consider that when cortisol binds to its receptor, it initiates a pathway that results in the regulation of genes involved in development, metabolism, and immune response. This signal amplification—a single hormone molecule binding to a receptor triggers the activation of many proteins and genes inside the cell—means that just a few hormone molecules can have a substantial impact on target cells. Through its effects on many cells, a hormone can exert a profound influence on the body as a whole. In some cases, those influences can be negative. Sustained elevated levels of cortisol in the bloodstream, for example, have been linked to anxiety, depression, digestive problems, and more.

Some hormone-secreting cells are not organized into distinct glands like the adrenal glands but are instead embedded as single cells or clusters of cells within other specialized tissues and organs. For instance, the main role of the kidneys is to filter blood, yet some cells in the kidneys produce hormones that stimulate red blood cell production.

Altogether, the endocrine glands and the endocrine cells embedded in other organs, such as the kidneys, make up the endocrine system.

Brain-Body Connection

Simply testing Hof's cells alone was not enough for Kox and Pickkers. They not only wanted to measure his entire body's immune response but also wanted to know if the breathing and meditation techniques made any difference.

After the ice bath, the scientists asked Hof to perform his meditation and breathing techniques while they injected him directly with endotoxin.

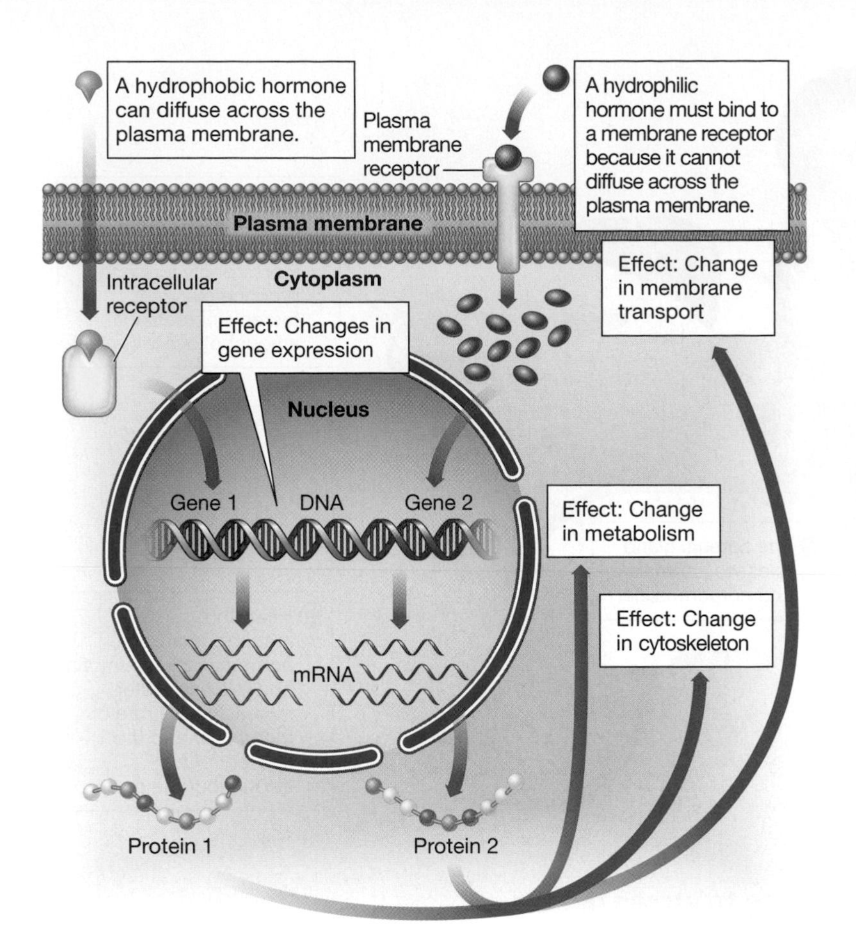

Figure 25.5

Hormonal signals are amplified within the cell

Hormones are effective at low concentrations because of their specificity and because tiny amounts of a hormone can generate a large internal signal within a target cell.

Q1: Describe the two ways that a hormone outside a cell can exert its effect on a cell.

Q2: Within the cell, how does a hormone bring about a change in cell activity?

Q3: It takes very little of the hormone cortisol to have large effects throughout the body. Explain why.

See Appendix A for answers to the figure questions.

In previous experiments, healthy volunteers injected with endotoxin had experienced fever, headaches, and shivering, accompanied by high levels of signaling proteins, called **cytokines**, that immune system cells secrete to coordinate an attack when an invader is present. Like hormones, they rely on the receptors on target cells.

At various times before and after the injection, Kox measured Hof's blood levels for hormones and cytokines. Kox then compared Hof's results to those of a control group of 112 healthy volunteers who had previously taken the same test.

To the scientists' surprise, as soon as Hof began practicing his breathing techniques, his cortisol and adrenaline levels skyrocketed. And unlike the other volunteers, Hof reported almost no flu-like symptoms. Topping it off, the concentration of cytokines in his blood—indicative of an immune response—was less than half that of the control group. Hof appeared to have suppressed his immune system—voluntarily.

How was that possible? Kox considered the possibilities. First, a tiny region at the base of the vertebrate brain, the **hypothalamus**, coordinates the endocrine system and integrates it with the nervous system (refer back to **Figure 25.3**). The hypothalamus contains neurons, which interact with other neurons in the brain, and endocrine cells, which produce hormones. It is a literal brain-body connection.

One well-known part of that connection involves the adrenal glands. In response to stress messages from the brain, the adrenal glands release adrenaline into the blood. If you see a rattlesnake in front of you, for example, you are likely to jump back or at least freeze in place, your heart racing. This quick response is due to the connection between the nervous system and the adrenal glands (**Figure 25.6**): the nervous system processes visual information (*snake!*) and transmits an alarm signal to the adrenal glands within a fraction of a second. The adrenal glands kick in right away, pouring adrenaline and noradrenaline into the blood. Adrenaline stimulates glycogen breakdown in liver and skeletal muscle cells, causing glucose to be released into the bloodstream. It also speeds up the heartbeat and increases the force with which the heart contracts, so that glucose is delivered throughout the body more rapidly.

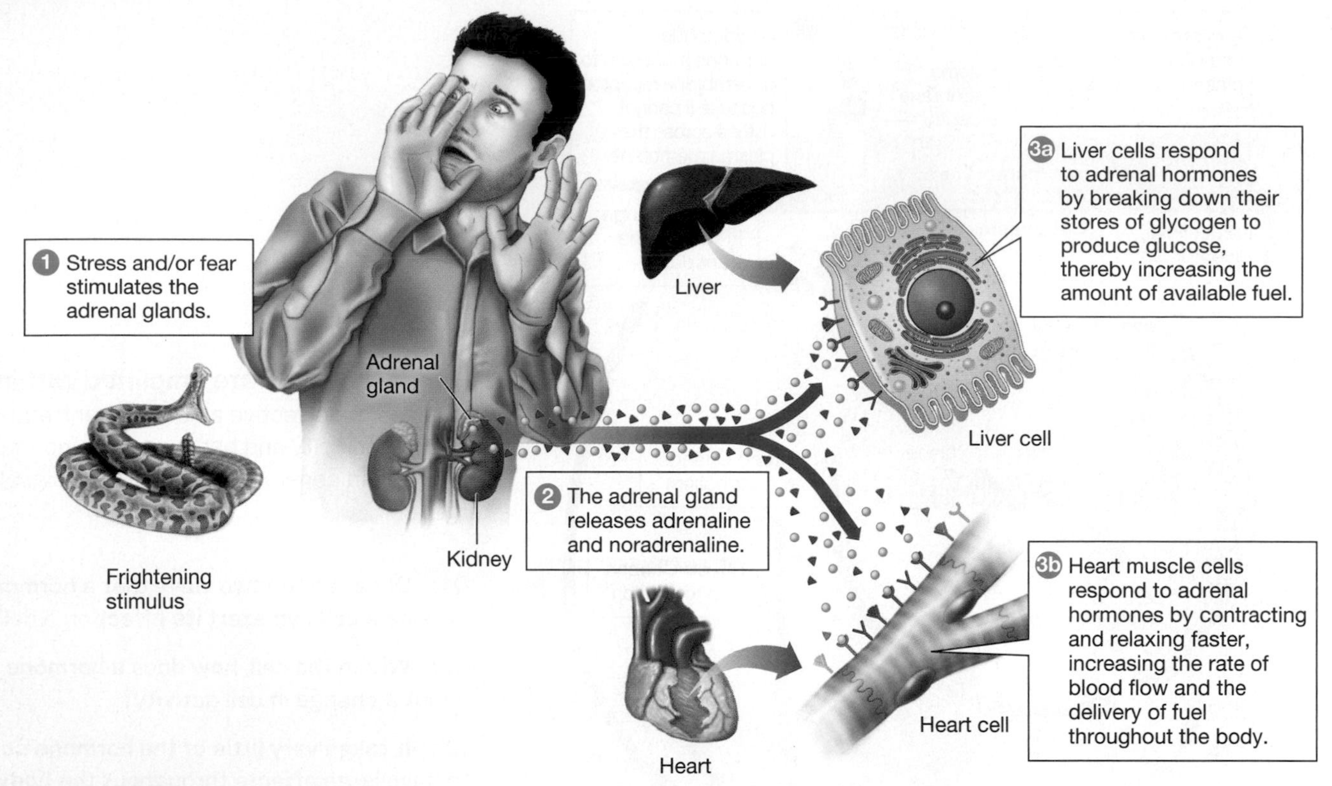

1 Stress and/or fear stimulates the adrenal glands.
Frightening stimulus
Adrenal gland
Kidney
Liver
2 The adrenal gland releases adrenaline and noradrenaline.
3a Liver cells respond to adrenal hormones by breaking down their stores of glycogen to produce glucose, thereby increasing the amount of available fuel.
Liver cell
Heart
Heart cell
3b Heart muscle cells respond to adrenal hormones by contracting and relaxing faster, increasing the rate of blood flow and the delivery of fuel throughout the body.

Figure 25.6

Adrenal hormones produce a rapid response to stress or fear

The adrenal glands produce adrenaline (epinephrine) and noradrenaline (norepinephrine), which trigger the rapid release and delivery of stored energy.

Q1: Describe an event (other than the one illustrated in the figure) that might cause the release of adrenaline.

Q2: What organs does adrenaline affect?

Q3: What do you think would happen if your adrenal glands were constantly releasing adrenaline?

See Appendix A for answers to the figure questions.

In this way, glucose becomes available to fuel a rapid response to a stressful situation.

Within just a few seconds, then, these hormones increase the pumping of blood and trigger the release of glucose, all of which support the next move: fight or flight. In the case of an encounter with a snake, that may mean either arming yourself with a hefty stick or running away. As it turns out, however, adrenaline and cortisol play another role aside from triggering glucose delivery. Research has shown that the hormones also subdue the activity of immune system cells.

Hof appeared to be able to consciously activate his nervous system (in the absence of real stress) to prompt the release of cortisol and adrenaline, thereby suppressing his immune response to the endotoxin voluntarily.

It was a "remarkable" finding, says Pickkers, but he wasn't ready to jump to conclusions. The case study of a single individual is weak scientific evidence for any phenomenon. Perhaps Hof was simply an outlier: "Everyone can play a little baseball, but there is only one Derek Jeter," says Pickkers. Perhaps Hof had a unique genetic mutation or another factor that enabled him to control his autonomic nervous and immune systems.

But Hof insisted that he was not an outlier and that he could teach his technique to anyone. "I'm sure everybody is able to do this," Hof told Pickkers. Pickkers challenged him to prove it. For scientific validation, Hof would need to teach his method to a group of healthy volunteers so that Pickkers could then compare that group's

Driven by Hormones

Hormones regulate not only the fight-or-flight response but also the sleep and wake cycles of all people. These signaling molecules in the blood also become elevated or depressed during periods of stress (cramming for a test, anyone?) and exercise. What are your levels of melatonin, adrenaline, and cortisol right now?

Assessment available in smartwork

Hormones in the human body

Melatonin

Regulates sleep timing, blood pressure, and more. Levels fluctuate over a 24-hour cycle, peaking at night.

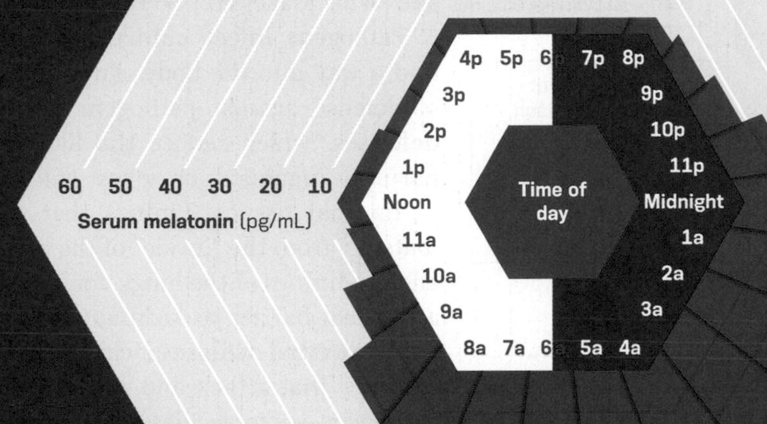

Stress hormone that quickens the heartbeat, among other effects. Levels increase as exercise intensity goes up.

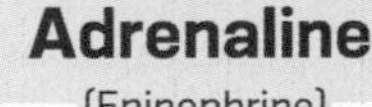

Adrenaline

(Epinephrine)

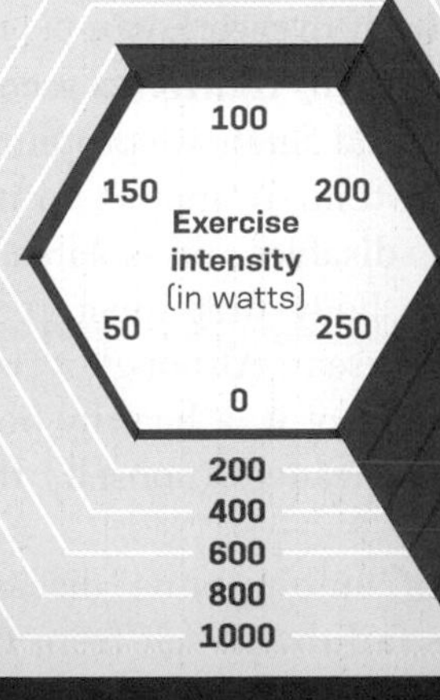

Plasma epinephrine (pg/mL)

Stress hormone that regulates homeostasis in the body. Levels drop during prolonged high-intensity exercise.

Cortisol

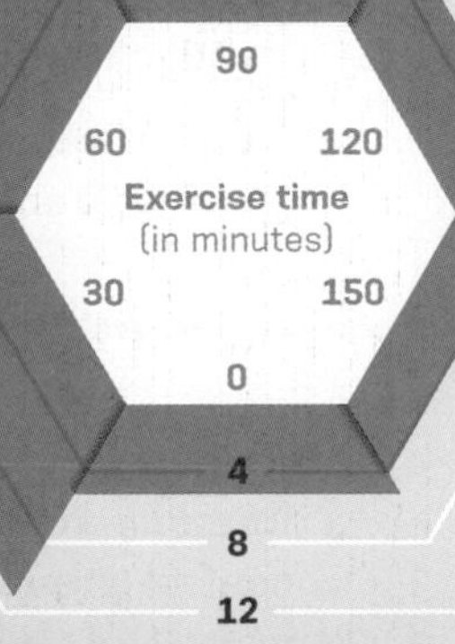

Cortisol (μg/dL)

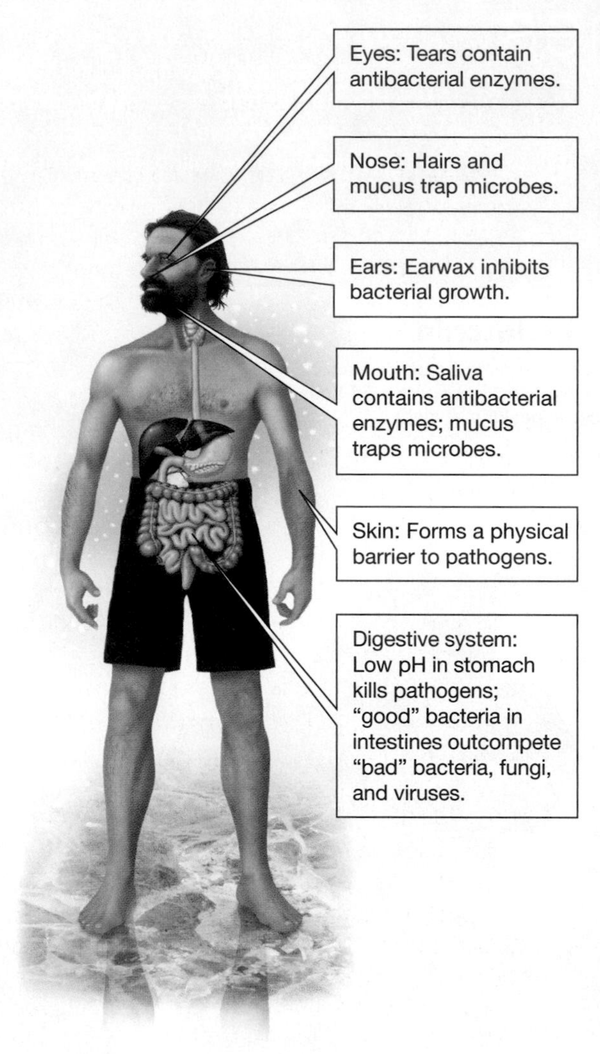
Eyes: Tears contain antibacterial enzymes.
Nose: Hairs and mucus trap microbes.
Ears: Earwax inhibits bacterial growth.
Mouth: Saliva contains antibacterial enzymes; mucus traps microbes.
Skin: Forms a physical barrier to pathogens.
Digestive system: Low pH in stomach kills pathogens; "good" bacteria in intestines outcompete "bad" bacteria, fungi, and viruses.

Figure 25.7

The immune system's first line of defense is to prevent the entry of pathogens

Our external defenses, the skin and the linings of our respiratory and digestive systems, form physical and chemical barriers against pathogens.

Q1: In animals, what is the main physical barrier that keeps out pathogens?

Q2: Give an example of a chemical defense within the digestive system.

Q3: Explain why touching your eyes and nose during flu and cold season is not recommended.

See Appendix A for answers to the figure questions.

immune responses to those of an untrained control group of volunteers. In this controlled way, Hof might produce stronger scientific evidence for his claim.

Innate Defenders

If Hof is right—if it really is possible to voluntarily control the immune system—the discovery would do more than change our understanding of the immune system; it would offer hope to people with autoimmune diseases, individuals in whom the immune system is out of control.

Normally, it is a good thing for the immune system to be active and on high alert, ruthlessly fighting off infectious agents called **pathogens**. Human pathogens include viruses, bacteria, and protists, as well as some fungi and multicellular animals such as parasitic worms. A well-known example of a pathogen is HIV, the virus responsible for AIDS (see "What Makes HIV So Deadly?" on page 495).

Pathogens infect animals only if they can find a way into the body. An animal's first line of defense against pathogens is its **external defenses**, which reduce the likelihood that a harmful organism or virus will gain access to internal tissues. Linings that separate the "outside" from the "inside" of the body—the skin and the linings of the lungs, for example—act as a physical barrier to keep out most pathogens. Other external defenses include proteins such as enzymes that attack and destroy molecules on pathogens—or molecules like mucus that prevent pathogens from binding to a surface—and chemical environments like acidic conditions that prevent pathogens from growing (**Figure 25.7**).

Although external defenses do a good job of keeping out most pathogens, the body is still vulnerable. Mucous membranes, found in all parts of the body that open to the outside world, such as the eyes, nasal cavity, and mouth, are a common point of entry for pathogens. Wounds, in the form of cuts, abrasions, and punctures, are common, and many pathogens will take advantage of breaks in the skin to gain entry to their hosts.

Once inside, pathogens confront a second line of defense: the cells and defensive proteins of the **innate immune system**. To mount an internal defense that kills, disables, or isolates invading pathogens, the body first must recognize that an invader is present. Although a person is not consciously aware of it, a healthy body can distinguish foreign invaders (nonself) from its own cells (self).

But in cases of autoimmune disease, the internal defenders fail to tell self from nonself. They mistakenly attack the body's own cells, leading to conditions such as rheumatoid arthritis (in which immune cells attack the lining of the membranes that surround joints) or type 1 diabetes (in which immune cells attack the pancreas, which makes insulin). In these cases, doctors prescribe medications to dampen the overactive immune system.

If there were a way, as Hof claims, to subdue the immune system at will, people with autoimmune diseases might have another avenue, aside from expensive medications, by which to control their rebellious immune systems.

Team Effort

Despite Hof's personal achievement, Pickkers and Kox didn't really think he would be able to teach others to voluntarily control their innate immune systems. "We thought it would be a negative result," says Pickkers. If nothing else, Hof had been performing his technique for 30 years, so even if he could teach it to others, Pickkers doubted he could teach novices enough to influence their own immune system in just a few days. Hof disagreed, arguing that a short training regimen would be sufficient to impart the ability. Kox didn't think most of the volunteers would even make it through the training.

It was no easy study for the participants. Over 4 days, 18 healthy, young, male volunteers were taken into the mountains of Poland and exposed to the cold in various ways: standing in the snow barefoot for 30 minutes, lying in the snow bare-chested for 20 minutes, swimming in ice-cold water each day for several minutes, hiking up a snowy mountain in nothing but shorts and shoes. Hof also taught them his meditation and breathing techniques, including deep inhalations and exhalations.

Contrary to Kox's expectation, all 18 participants completed the training, and 12 of them were then randomly selected to come back to the lab for the final part of the experiment: exposure to endotoxin to test whether they could consciously regulate their immune response. Their results would then be compared to those of 12 healthy controls who had also been exposed to endotoxin but had not received Hof's training. Now the scientists would finally find out whether Hof could teach others to control the activity of their innate immune system.

What Makes HIV So Deadly?

In the early 1980s, doctors in the United States began to notice that gay men were dying of a variety of rare diseases, including a skin cancer called Kaposi's sarcoma, an unusual kind of pneumonia, and other infections that most people ordinarily shake off. By the mid-1980s it was clear that these patients had broken immune systems. Their condition was named acquired immunodeficiency syndrome (AIDS), and it resulted from infection by human immunodeficiency virus (HIV).

In North America and Europe, the number of new cases rapidly increased, claiming the lives of tens of thousands of people each year. Initially, most new cases were limited to gay men, intravenous drug users, and people who had received blood transfusions. The common denominator was contact with the blood or body fluids of others: couples during sex, drug users when sharing used needles, and surgical and hemophilia patients who received blood transfusions contaminated with HIV.

In time, safe-sex education, clean-needle programs, and screening blood for HIV contamination reduced the rates of infections among gay men and blood transfusion patients, but the virus spread to other populations. Current at-risk populations include sex workers, people in prison, and transgender people. Globally, an estimated 32 million people have died of AIDS-related illnesses since the start of the epidemic, and 74.9 million people have become infected with HIV. In 2018 alone, 770,000 people died of AIDS-related illnesses, and 37.9 million were living with HIV.

In the bloodstream, HIV enters immune system cells and reproduces inside them. In the short term, remaining immune system cells track down HIV-infected cells and destroy them. Because the immune cells do such a good job of killing HIV-infected cells in the blood, most people with HIV have about a decade of normal health before they become ill, even without any treatment.

Over time, however, the HIV particles in the body evolve, and the immune system cells no longer recognize and kill the virus. The population of HIV particles increases and begins destroying immune system cells faster than they can multiply. The body no longer has the immune system cells it needs to fight off infections by bacteria, yeasts, and other viruses. Once the immune system collapses, a person is vulnerable to almost any infection.

So far, there is no effective vaccine or cure for HIV. But a variety of drugs enable people with HIV to live years longer with fewer symptoms than they used to. Antiretroviral therapy (ART) is a combination of HIV medicines that are taken together and called an HIV treatment regimen. This standard mixture of therapeutic drugs prevents the viral genetic material from replicating and prevents the virus from merging with plasma membranes and entering cells. Globally, only 62 percent of HIV patients are on ART. Much of the problem is that many patients do not know they are infected with HIV, and for those who do, the drugs can cost thousands of dollars a month. Because of this, only about half the HIV patients in much of Africa, Asia, and the United States receive effective treatment. The best ways to slow the spread of the disease is advocating for more resources for HIV testing and treatment and to promote safe-sex education, the free availability of condoms, and clean-needle programs.

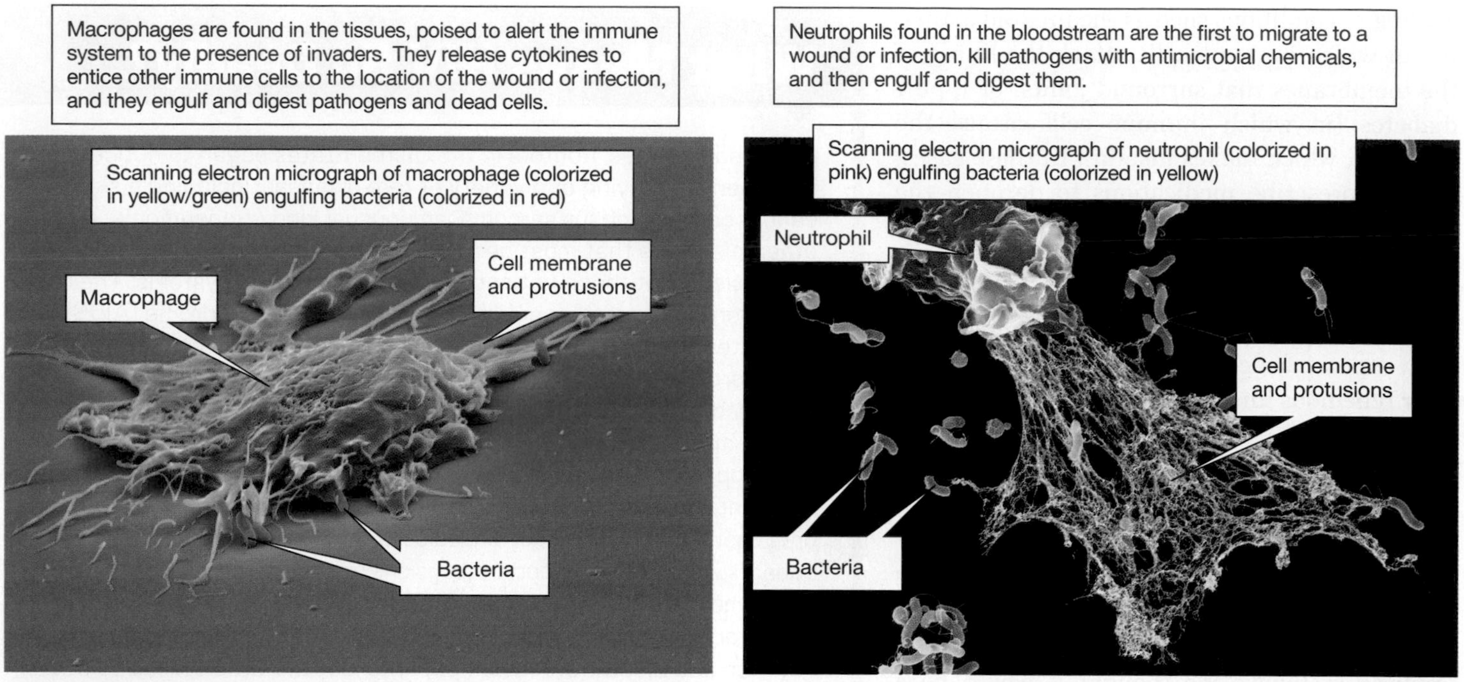

Macrophages are found in the tissues, poised to alert the immune system to the presence of invaders. They release cytokines to entice other immune cells to the location of the wound or infection, and they engulf and digest pathogens and dead cells.

Scanning electron micrograph of macrophage (colorized in yellow/green) engulfing bacteria (colorized in red)

Macrophage

Cell membrane and protrusions

Bacteria

Neutrophils found in the bloodstream are the first to migrate to a wound or infection, kill pathogens with antimicrobial chemicals, and then engulf and digest them.

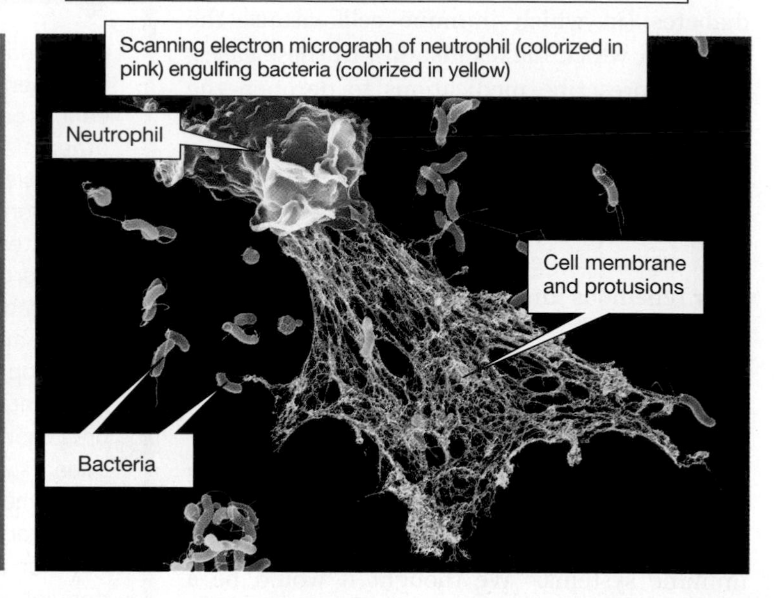

Figure 25.8

Phagocytes destroy pathogens by engulfing them

Phagocytes are a kind of white blood cell, a family of defense cells found in body fluids including blood, where they intercept invading pathogens. Two different kinds of phagocytes—a macrophage and a neutrophil—are seen in these colorized transmission electron micrographs (TEMs).

Q1: Place these terms in order from most to least inclusive: neutrophil, white blood cell, phagocyte, innate immune system.

Q2: Name a function that is similar between macrophages and neutrophils. Name a difference between these two cells.

Q3: Why would it be a problem if your innate immune system identified the insulin-producing cells in your pancreas as "nonself"?

See Appendix A for answers to the figure questions.

The innate immune system reacts to cells or molecules that do not belong in the body by activating defense cells and proteins to eliminate the unwelcome guests. A suite of pathogen-recognizing cells called **phagocytes**, a type of white blood cell, destroy foreign invaders by engulfing and digesting them (**Figure 25.8**).

DEBUNKED!

MYTH: Getting a flu shot can give you the flu.

FACT: The influenza vaccine is made up of killed or weakened viral particles that do not cause illness. Some people mistake the short-lived side effects of the vaccine, including a slight fever and aches, for the actual flu.

This immune response is said to be innate (inherent) because the necessary components are constantly at the ready for deployment against an invading pathogen. The innate response can be local, occurring at the point of entry, or global, involving the whole body. Like the external defense system, innate immunity is indiscriminate as to which foreign invaders it repels, so it is considered a **nonspecific response**. The innate immune system is an ancient defense mechanism found in both invertebrates and vertebrates.

In addition to defending against invaders, the innate immune system plays two other critical roles. First, it responds to tissue damage from a pathogen invasion or wound by mounting an immediate and coordinated sequence of

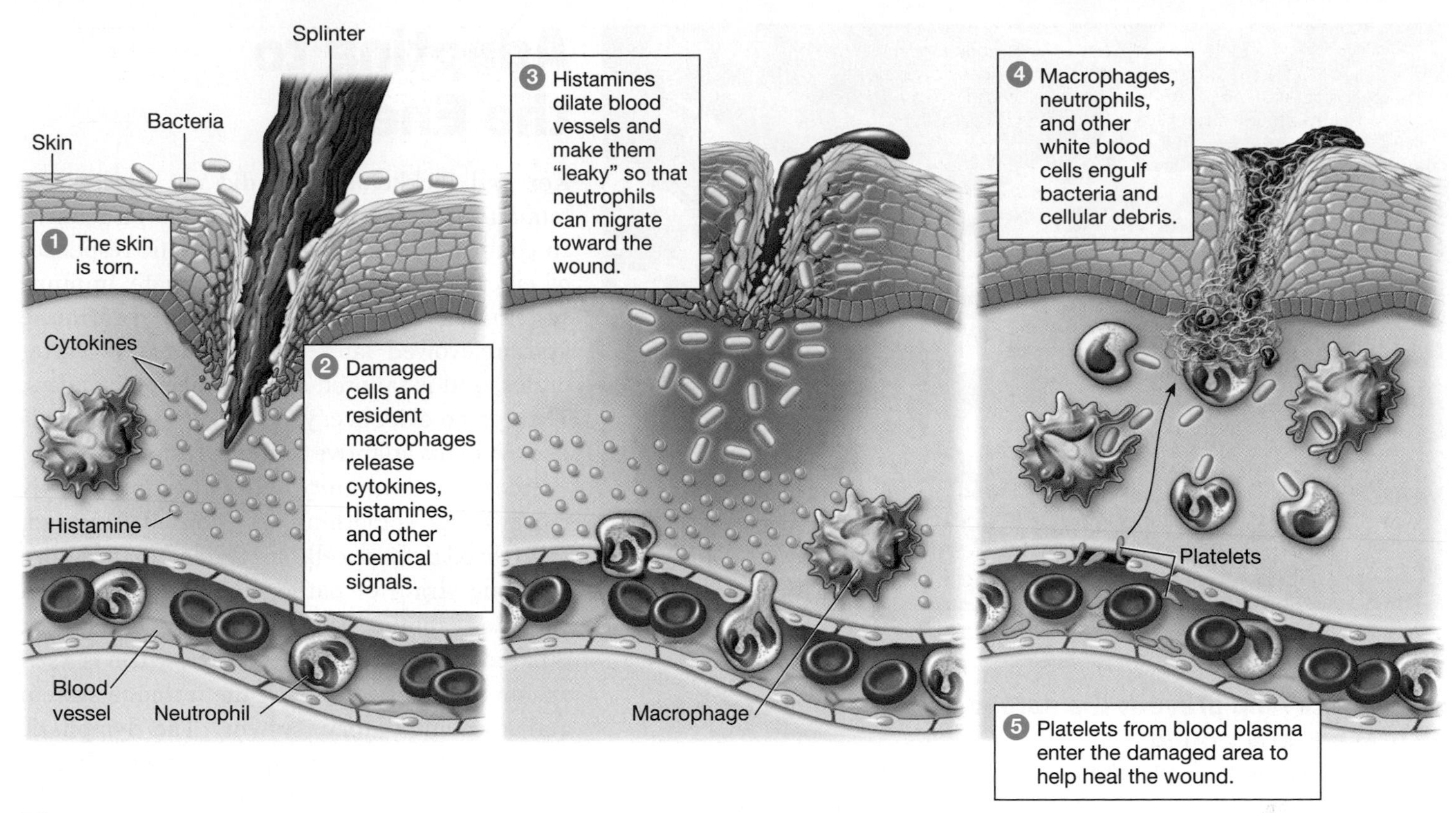

Figure 25.9

The inflammatory response acts against invading pathogens

Inflammation occurs when the innate immune system swings into action after cellular damage is detected, cleaning up damaged tissues and preventing the spread of pathogens. Inflammation can occur anywhere inside the body. Here an inflammatory response follows a puncture wound to the skin.

Q1: What is the role of white blood cells in inflammation?

Q2: What would happen if histamines were not produced during inflammation?

Q3: Why is inflammation called a "nonspecific" immune response?

See Appendix A for answers to the figure questions.

events known as **inflammation** (**Figure 25.9**). Cytokines are a clear marker of inflammation and, thus, a good way to measure the action of the immune system. In rare cases, like in some COVID-19 patients, cytokines are over-induced and create a "cytokine storm," causing widespread tissue damage and, consequently, multi-organ failure and death. A second role of the innate immune system is **blood clotting** to close a wound. Sealing an open wound reduces blood loss and restores the integrity of external defense barriers (**Figure 25.10**).

To test the innate immune response of the 24 study participants (12 trained volunteers and 12 untrained controls), the scientists injected the participants with endotoxin and monitored them for 6 hours. Hof visited his trainees during the experiment, coaching them through his breathing techniques.

The results were clear: after being injected, while performing the breathing techniques, the trainees showed higher adrenaline levels than the controls—higher even than the adrenaline produced by a person's first bungee jump. "They produced more adrenaline just lying in bed than somebody standing in front of an abyss going to jump in fear for the first time," says Hof. "That means direct control of your hormone system, and your hormones have a direct relationship to the immune system."

In addition, the trainees had fewer flu-like symptoms and lower fevers, and their cytokines—

Figure 25.10

Blood clots help prevent the spread of pathogens that may be present in a wound

Sticky cell fragments, or **platelets** (shown here in light blue), and clotting proteins (yellow) form a gel-like mesh that traps blood cells, creating a blood clot that seals broken skin. Clotting can begin as quickly as 15 seconds after tissue damage occurs. Growth of new tissue eventually repairs the wound more permanently.

Q1: **Why is blood clotting an important immune response?**

Q2: **How are the inflammatory response and blood clotting similar?**

Q3: **Some people have a genetic disorder in which their blood cannot clot. Why would this be a problem?**

See Appendix A for answers to the figure questions.

the signaling proteins of the immune system and markers of inflammation—were at less than half the level of the control group. "We were very surprised," says Kox. "It was impressive that these guys could do all this cold exposure training, but I still thought the chances were slim they'd be able to modulate their immune systems. But the results were so convincing."

In the future, the immune-suppressing ability of Hof and his trainees may help researchers identify new, less expensive treatments for autoimmune diseases. In the meantime, it's important to not yet try Hof's method at home. Humans have an innate immune system for a reason—to protect against pathogens—so voluntarily subduing it without the guidance of a doctor is not a good idea.

Adapting to the Enemy

Kox and Pickkers's study did not address the human immune system's third line of defense. In contrast to the broad, nonspecific responses of external defenses and the innate immune system, the more complex **adaptive immune system** evolved later, is found only in vertebrates, and is tailored against specific invaders. There is no evidence, yet, that Wim Hof is able to control his adaptive immune response.

Adaptive immunity goes beyond simply recognizing something as nonself. Instead, specialized defense cells are trained to recognize only one strain of pathogen and to activate a **specific response**. The adaptive immune system is based in the **lymphatic system**, which is itself an important part of both the immune system and the circulatory system. The lymphatic system consists of lymphatic ducts, lymph nodes, and associated organs, and it operates by moving a type of white blood cell called **lymphocytes** (**Figure 25.11**). Lymphocytes operate in two main weapon systems: antibody-mediated immunity and cell-mediated immunity.

In **antibody-mediated immunity**, powerful Y-shaped membrane proteins called **antibodies** recognize and attack invaders. Antibodies recognize nonself markers, or **antigens**, on the pathogen and mark the pathogen for destruction by phagocytes. **B cells**, which are specialized lymphocytes that originate and mature in the bone marrow, produce thousands of antibodies per second. The antibodies are aimed specifically at the pathogen that has been recognized.

In **cell-mediated immunity**, by contrast, the body's own cells are targeted because they have been infected by a pathogen, such as a virus. Cancer cells may also be targeted (see "Cancer: Uncontrolled Cell Division" on page 113 of Chapter 6). **T cells**, which are lymphocytes that originate in the bone marrow and mature in the thymus, recognize markers on the surface of infected cells. The T cells then kill the infected cells so that they cannot spread the disease to other cells.

Compared to innate immunity, adaptive immunity is slow to mobilize. However, it is the most sophisticated and effective animal defense

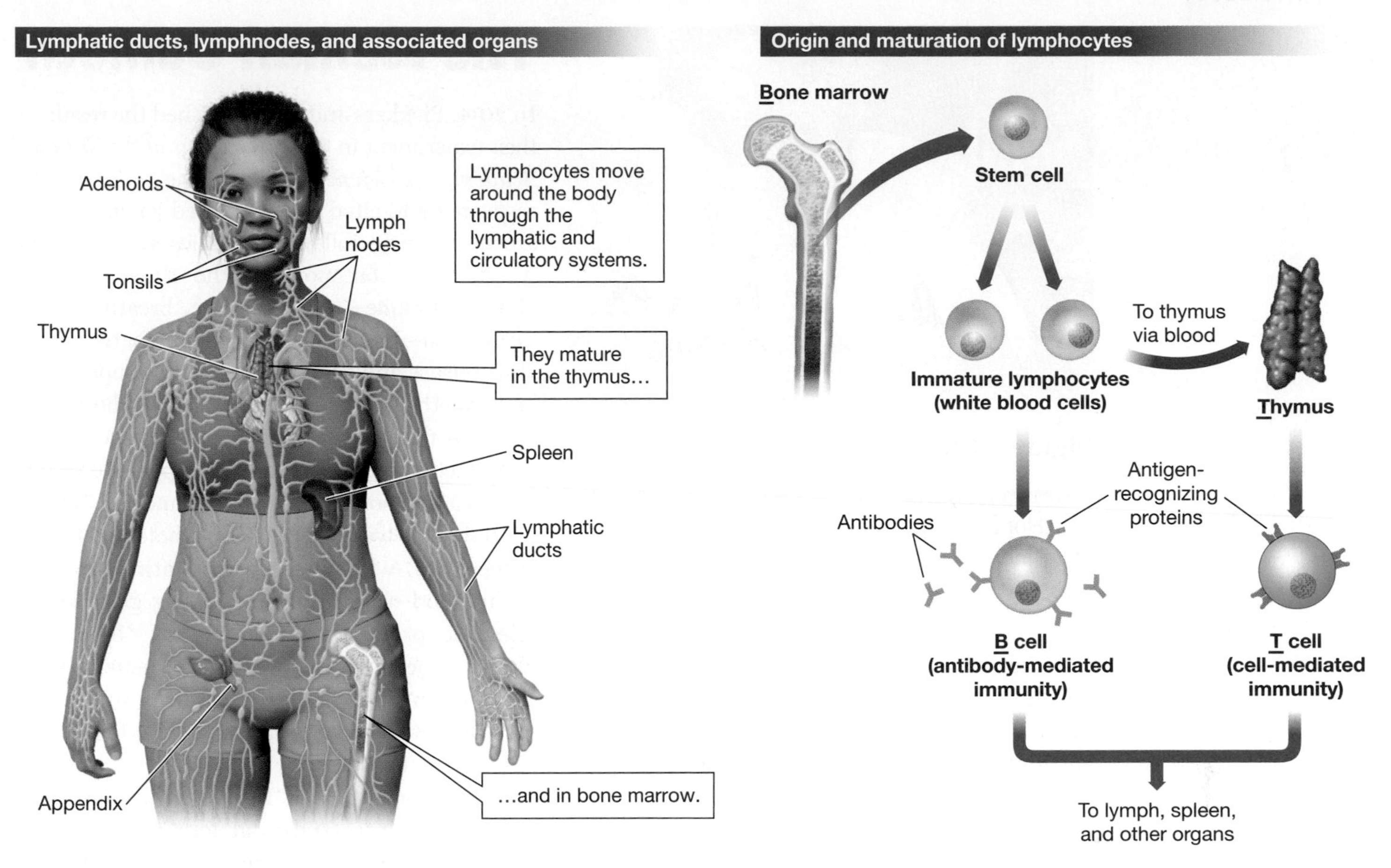

Figure 25.11

Adaptive immunity resides in the lymphatic system

(Left) Lymphocytes circulate in the lymphatic and circulatory systems and accumulate in lymphatic ducts, lymph nodes, and associated organs, such as the tonsils, spleen, and appendix. (Right) Lymphocytes originate from stem cells in bone marrow. B cells mature in the bone marrow; T cells mature in the thymus.

Q1: Why are B and T cells so named?

Q2: In what way is this immune system "adaptive"?

Q3: Why is the adaptive immune response considered the third layer of the immune system?

See Appendix A for answers to the figure questions.

system because of its amazing selectivity in attacking a particular invader.

Adaptive immunity occurs in two stages. The first time a person is exposed to a particular pathogen, the **primary immune response** is activated. This response takes time—more than 2 weeks sometimes—to reach full steam. Because of that slow start and because pathogens multiply so rapidly, people infected with an aggressive pathogen for the first time can sometimes lose the race, becoming ill and dying. Therefore, any pathogen that is new to humans is particularly dangerous, and the nonspecific response of innate immunity may be more effective in combating it.

A distinctive feature of the adaptive immune system is **immune memory**, the capacity to remember the system's first encounter with a specific pathogen and to mobilize a speedy and targeted response to future infection by the same strain. This "memory" enables the body to become immune to attacks by a disease after it first experiences that strain. Once you've had measles, for example, you never get sick from

Figure 25.12

Iceman in sand

Wim Hof continues to train volunteers to use his novel method. Here, women and men of various ages exercise on a beach.

the measles virus again, because the adaptive immune system recognizes the virus and quickly eradicates it the next time. The second encounter, when the adaptive immune system is poised and ready to respond, is the **secondary immune response**. Each individual must build up immune memory over time by being exposed and responding to various pathogens.

Immunity may be acquired actively or passively. **Active immunity** to a particular pathogen develops when the body produces antibodies against that pathogen; they are not received from an outside source. This happens naturally when you're exposed to certain pathogens, such as the measles virus. You can also acquire active immunity to certain diseases through vaccination, which introduces antigens to the body in a harmless form (vaccination is discussed extensively in Chapter 2).

Passive immunity develops when the body receives antibodies that it did not make. A human fetus acquires antibodies from exchanges between its blood and its mother's blood. This antibody sharing continues after birth. Mother's milk is rich in antibodies because the mother's immune system has encountered many antigens and made many antibodies in her lifetime. Thanks to that antibody-rich milk, a nursing baby receives passive immunity to a broad range of pathogens. Passive immunity produces no memory cells, so it wears off as the received antibodies degrade, usually within a few weeks or months.

The Iceman Cometh

In 2014, Pickkers and Kox published the results of their experiment in the *Proceedings of the National Academy of Sciences*, one of the world's most respected and cited peer-reviewed journals. They are conducting follow-up studies to determine whether one or more of the three parts of Hof's technique—cold exposure, breathing, and meditation—is primarily responsible for the adrenaline release and subsequent immune suppression and exactly how these effects come about. Kox suspects that the breathing techniques are the main factor, as Hof's breathing appears to trigger the release of hormones, but he cannot yet be sure.

"It needs to be studied a whole lot more," agrees Hof, who is eager to continue putting his method under the magnifying glass of the scientific process (**Figure 25.12**). "By meticulous experiments and measurements—not speculation—we want to show this works. I'm very thankful to the professors at Radboud who dared to go into this."

Since working with Kox and Pickkers, Hof has entered into studies at four other institutions, investigating the impact of his technique on areas such as inflammation and pain and further exploring how it works. A team at Wayne State University School of Medicine in Michigan even scanned Hof's brain while subjecting him to bouts of mild hypothermia and found certain brain regions to be highly activated during the breathing exercises. In May 2018, they concluded that Hof's brain, rather than his body, is key to mediating his responses to cold exposure.

But whether Hof's technique can help individuals with autoimmune disorders is still up for debate. Although the training worked for young, physically fit men, scientists do not know whether it will work for older people with autoimmune diseases who already have compromised organ systems. Until further research indicates that the method is safe, "We would not advise people to do this," says Pickkers. "We have to be careful there are no unwanted side effects or risks."

But he is optimistic about the future and still sounds surprised about how well the training worked. "We confirmed that, indeed, using the techniques of Wim Hof, humans are able to modulate their autonomic nervous system and influence their immune response," says Pickkers. "It is remarkable."

NEED-TO-KNOW SCIENCE

- A **hormone** (page 489) is a signaling molecule distributed throughout the body by the circulatory system. Because hormones move only as quickly as the blood moves, they tend to coordinate functions that are slower and longer-lasting than those under the influence of the nervous system.
- A single hormone may affect many different kinds of target cells, potentially triggering a different response in each. Hormones act on target cells either by moving through the plasma membrane to the cell's interior or by acting on receptors embedded in the plasma membrane.
- The **endocrine system** (page 489) is made up of the glands and specialized cells that produce hormones. The **hypothalamus** (page 491), a tiny region at the base of the vertebrate brain, coordinates the endocrine system and integrates it with the nervous system. The **adrenal glands** (page 490), located atop the kidneys, produce hormones responsible for rapid physiological responses, such as the fight-or-flight response.
- The vertebrate **immune system** (page 488) uses signaling proteins called **cytokines** (page 491) to communicate when an invader is present. The system possesses three layers of defenses against **pathogens** (page 494):
  - The first layer consists of **external defenses** (page 494): physical and chemical barriers, including the skin and the linings of the respiratory and digestive systems.
  - The second line of defense is the **innate immune system** (page 494). Several types of blood cells and molecules produce the **nonspecific responses** (page 496) of the innate immune system, including **phagocytes** (page 496) such as macrophages and neutrophils, which engulf and destroy pathogens. Tissue damage stimulates **inflammation** (page 497) and **blood clotting** (page 497).
  - The third line of defense is the **adaptive immune system** (page 498), providing long-term protection in the form of **specific responses** (page 498) to pathogens and parasites. These responses are mediated either by powerful proteins called **antibodies** (page 498) or by cells.
- The **lymphatic system** (page 498) provides the primary sites for adaptive immunity. White blood cells called **lymphocytes** (page 498) confer specific immunity. Immature lymphocytes differentiate into **B cells** (which confer **antibody-mediated immunity**, page 498) in the bone marrow and **T cells** (which confer **cell-mediated immunity**, page 498) in the thymus. Each lymphocyte has special membrane proteins that bind to only a specific **antigen** (page 498) of a specific pathogen.
- The **primary immune response** (page 499) from the adaptive immune system is relatively slow and mild. The **secondary immune response** (page 500) is a faster, stronger response to a pathogen that has been encountered one or more times, creating **immune memory** (page 499).
- **Active immunity** (page 500) can be acquired through natural exposure to a pathogen or through a vaccine. **Passive immunity** (page 500) comes from receiving antibodies that were not made by our own bodies, such as when a fetus acquires antibodies from its mother.

THE QUESTIONS

See Appendix B for answers.

The Basics

1 Hormones are

(a) secretory cells.

(b) endocrine glands.

(c) signaling molecules.

(d) target cells.

2 Which of the following is *not* true of hormones?

(a) They are distributed through body fluids.

(b) They must be present in large amounts to be effective.

(c) They are produced by specialized cells.

(d) They act on target cells.

3 Adrenaline

(a) is produced in the adrenal glands.

(b) increases the amount of glucose in the bloodstream.

(c) suppresses immune system activity.

(d) all of the above

4 The ________ is the immune system's second line of defense against pathogens.

(a) innate immune system

(b) adaptive immune system

(c) combination of physical and chemical barriers to pathogen entry

(d) all of the above

5 Which of the following is/are *not* a part of the innate immune system?

(a) phagocytes

(b) antibodies

(c) inflammation

(d) clotting

6 Link each term with the correct definition.

adaptive immune response	1. The glands and specialized cells that produce hormones.
endocrine system	2. The blood cells and molecules that provide a nonspecific response to pathogens.
innate immune response	3. The region of the brain that coordinates the endocrine system and integrates it with the nervous system.
hypothalamus	4. Long-term defense against pathogens, centered in the lymphatic system.

Challenge Yourself

7 Which of the following is *not* true of a B cell?

(a) It is a kind of lymphocyte.

(b) It is produced in the bone marrow.

(c) It matures in the thymus.

(d) It is part of the adaptive immune response.

(e) All of the above are true of B cells.

8 Identify whether each of the following is characteristic of either antibody-mediated (A) or cell-mediated (C) immunity.

_____ a. The immune response relies on Y-shaped membrane proteins to identify pathogens.

_____ b. B cells produce proteins specific to a pathogen.

_____ c. Lymphocytes that matured in the thymus identify infected cells.

_____ d. Antigens on the pathogen allow it to be identified as nonself.

_____ e. Infected cells are destroyed so that an infection cannot spread to other cells.

9 Select the correct terms:

The first time you are exposed to a pathogen, the (**primary** / **secondary**) immune response is activated. The (**primary** / **secondary**) immune response to a pathogen is stronger and more rapid. You acquire (**active** / **passive**) immunity to a pathogen when your own body creates the antibodies against that pathogen. (**Active** / **Passive**) immunity comes from the antibodies produced by another person, such as your mother when you were in utero or nursing. The immunity conferred by vaccines is an example of (**active** / **passive**) immunity.

10 Scenario (albeit unlikely): A person is exposed to SARS-CoV-2 (the coronavirus that causes COVID-19) after touching a surface covered in viral particles and then touching their face. They do not get sick and test negative for SARS-CoV-2 in their nasal cavity. Further, a negative blood test indicates that there is no coronavirus in this person's bloodstream. A serology test shows no antibodies against SARS-CoV-2, either. How did this person remain uninfected when they were clearly in contact with viral particles? Of the following, choose all possible reasons they were not infected after exposure:

(a) Mucus in the nasal cavity successfully blocked viral infection.

(b) B cells were activated, and antibodies marked the virus for destruction by phagocytes.

(c) A "cytokine storm" effectively blocked viral infection.

(d) Antimicrobial chemicals in saliva effectively blocked viral infection.

11 Beginning with a perceived threat (for example, a spider), identify the correct order of events in the stress response by numbering the following events from 1 to 5.

_____ a. Target cells amplify the hormonal signal to produce a response.

_____ b. The liver breaks down glycogen to glucose and the heart increases its rate and the force of its contractions.

_____ c. Adrenaline reaches target cells in the liver and heart.

_____ d. The hypothalamus signals the adrenal glands that a threat is present.

_____ e. The adrenal glands release adrenaline into the bloodstream.

Try Something New

12 Wim Hof and his trainees had increased levels of the stress hormone adrenaline and decreased immune function during the experiments described in this chapter. How might these changes negatively affect the endocrine and immune systems over the long term? How do you think these changes might affect their response(s) to exposure and infection by SARS-CoV-2, the virus that causes COVID-19?

13 Allergies, in extremely simplified terms, are an overreaction of the immune system to harmless foreign materials, including pollen, molds, and food items such as peanuts. According to the Children's Hospital of Chicago, one in 13 kids has at least one food allergy. That's 7.6 percent of U.S. children. The accompanying graph shows the top nine allergy-inducing foods and the percentage of kids allergic to these foods. Why do these percentages *not* add up to 100 percent? What are the three most common child food allergies in the United States? What are the three least common child food allergies shown in the graph? To which food are about 0.9 percent of kids allergic?

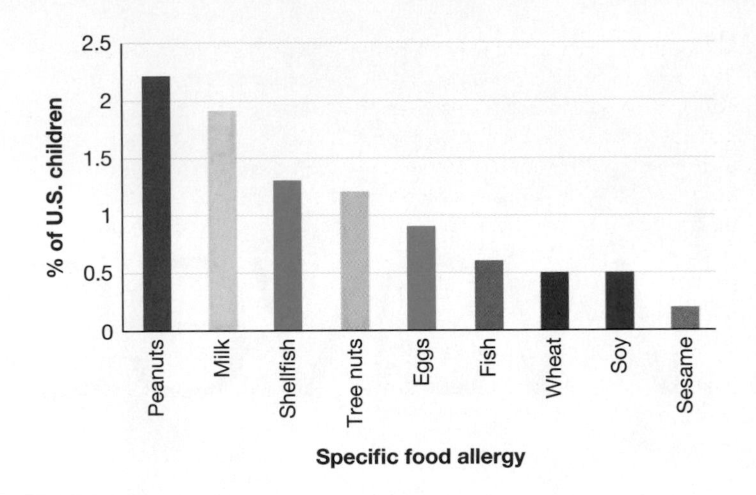

14 Increased body temperature (fever) is part of the body's innate immune response. Fever is uncomfortable and can be dangerous if very high. It is often treated with over-the-counter medicines such as acetaminophen, ibuprofen, naproxen, or aspirin. What are possible negative effects of this treatment?

Leveling Up

15 **Life choices** Although clotting is an important component of the innate immune response, it can also be dangerous. For example, a blood clot could block an artery to the heart or brain, leading to a heart attack or stroke. Aspirin reduces blood clotting by interfering with the body's production of a lipid called thromboxane A2. This lipid normally helps platelets clump together (see Figure 25.10), so aspirin, by inhibiting its production, reduces clotting and "thins the blood." Some doctors may prescribe a daily dose of aspirin for patients who are at risk of heart attack or stroke. Review the U.S. Preventive Services Task Force 2016 recommendations on daily aspirin therapy (https://www.uspreventiveservicestaskforce.org/Page/Document/RecommendationStatementFinal/aspirin-to-prevent-cardiovascular-disease-and-cancer), and then answer the following questions.

(a) Do you fall into one of the categories for which daily aspirin therapy is recommended? If yes, which one? If no, is there an aspirin therapy category that you think you'll be in eventually?

(b) How strong is the evidence supporting aspirin therapy in the category you identified in the previous question, if any? (See the "Grade" column in the task force recommendations.)

(c) With this information in hand, do you plan to take aspirin daily at some point in your life?

(d) Will you speak with your doctor before taking aspirin daily? Why or why not?

(e) Do you know anyone who has had a heart attack or stroke? Does that person take aspirin daily?

16 **Doing science** Pickkers and Kox say that their next experiment will attempt to determine whether one or more of Hof's techniques—cold exposure, meditation, and breathing exercises—are primarily responsible for the increased adrenaline release and decreased immune response. Kox predicts that the breathing exercises will prove to be most important. (You can view a video of cold-exposure training at https://www.youtube.com/watch?v=ziXm9oWJm6A.) Imagine that it is your responsibility to design the next experiment for Pickkers and Kox. Include answers to the following questions in the description of your experimental design.

(a) What are your experimental hypotheses?

(b) Give at least one prediction for each hypothesis.

(c) Identify your control group and treatment group(s). How many participants will be in each group? Justify your sample size.

(d) Give a detailed description of the treatment for each group.

Digital resources for your book are available online.

Amber Waves of Grain

We've been growing the same domesticated crops for thousands of years. To survive the future, we're going to need new ones.

After reading this chapter you should be able to:

- Identify the structures and explain the functions of plant tissues, organs, and organ systems.
- Give an example of how plants use chemicals to survive, grow, and reproduce.
- Explain how a plant obtains a given nutrient.
- Diagram the alternation of generations that is a plant life cycle.
- Explain the ways in which plants are pollinated, and how this variety relates to the structures of flowers.
- Describe how plants disperse their seeds and how fruits have evolved to assist this effort.

PHYSIOLOGY

CHAPTER 26

PLANT PHYSIOLOGY

Lee DeHaan wanders through a field of grain, golden knee-high stalks brushing against his faded jeans. Tall, with an angular face and goatee, DeHaan looks the part of a farmer, with a floppy hat and tanned arms to boot. But though he grew up on a corn farm in Minnesota, DeHaan is now an *agronomist*, a researcher studying the science of producing and using plants for food, energy, land preservation, and more. In these fields around his office at the Land Institute in Salina, Kansas, DeHaan is domesticating a new crop with the potential to transform agriculture—a grain that is bred from a humble prairie grass (**Figure 26.1**).

The way we plant and harvest crops is not sustainable. Nearly 70 percent of the freshwater used by humans is put toward irrigation. Farmland is losing productivity because of agricultural practices that lead to erosion, pesticide and herbicide pollution, and the accumulation of too much salt in the soil from poor irrigation. Staple crop production worldwide has leveled off. To make matters worse, we are losing 1.5 million acres of U.S. farmland a year to development, according to American Farmland Trust, a national agricultural organization. All this does not bode well when the world population is projected to grow by another 3 billion people over the next half century.

We need new crops. The common crops grown in the United States today—corn, wheat, and soybeans—were domesticated by our ancestors thousands of years ago to produce high yields and be easily harvested and replanted. Yet these crops are failing to meet our sustainability needs, due to a lack of genetic diversity. They require large amounts of water, fertilizer, and pesticides, and they are vulnerable to weather changes, pests, and diseases. They are inefficient and delicate; we need food crops that are hardy and resilient.

Unfortunately, humans stopped domesticating new crop plants long ago. So, despite Earth's rich diversity of plants—over 300,000 known species, including more than 50,000 species of edible plants—we rely on just 30 of them to provide 95 percent of our food (see "Food Banks" on page 315 in Chapter 16).

More than 250,000 of Earth's plant species are *flowering plants* (also called *angiosperms*). Worldwide, people get more than 80 percent of their calories from flowering plants such as grasses (wheat, rice, and corn), legumes (peas, beans, and peanuts), potatoes, and sweet potatoes. But as climate change affects agriculture—increasing temperatures and severe weather and causing more disease—and as the global population grows, we're going to need tough, plentiful crops. That's no easy requirement to meet.

Figure 26.1

Harvest at the Land Institute
Researchers gather ripe stalks of a prairie grass that holds hope for feeding our growing population.

Perfecting Plants

In 2001, as a young and ambitious plant breeder, DeHaan joined a research team at the Land Institute, a 600-acre research center devoted to developing sustainable alternatives in agriculture. At the time, he was the young scientist on the team, so his bosses handed him an ambitious long-term project: to create a new type of grain by domesticating a wild grass that grows year after year.

Plants can be grouped into three categories on the basis of their life cycle: annuals, biennials, and perennials.

- **Annuals** complete their entire life cycle in one year. In flowering plants, an annual has 1 year to grow from a seed into a mature plant, produce flowers, and make the seeds that will start the next generation. Annual crops must be replanted every year.

- **Biennials** grow and mature for a year but do not initially reproduce. Reproduction takes place in the second year of growth. After the second year, biennial crops must be replanted.
- **Perennials** live 3 years or more and sometimes for hundreds or even thousands of years. Perennials, which once made up much of the natural grasslands that dominated Earth, are alive year-round and are efficient at nutrient cycling and water management.

Wheat, corn, rice, soybeans—these are all annuals. Our ancestors saw the advantages of breeding annuals. Compared with perennials, annuals produce more seed, and replanting them every season speeds the process of domestication. But perennials have advantages too, and it was those advantages that DeHaan wanted to tap into by breeding a perennial as a food crop.

First, perennials do not waste energy by growing all new **roots** each year, as annuals do. Instead, they grow long, deep roots that anchor in the soil and remain there year after year. These roots enable perennials to absorb water and nutrients more efficiently than annuals do. The roots also outcompete weeds, so less weed killer is needed to grow perennials. Deep roots also hold carbon in the soil, acting as a *carbon sink*.

Flowering plants can be classified as either dicots or monocots (**Figure 26.2**). The **dicots** are the larger of these two informal categories and encompass about 175,000 species, including beans, squash, oak trees, and roses. The name "dicot" comes from the presence of two cotyledons in dicot seeds. **Cotyledons** are food-storing organs that are part of the embryonic seedling inside a plant seed. **Monocots** include all the grasses, members of the lily family, palm trees, banana plants, and grain-producing plants (such as wheat, rice, and corn).

Both types of plants have relatively simple bodies compared to the bodies of vertebrate animals (**Figure 26.3**). Plant bodies are made of three basic tissue types:

- **Dermal tissues**, which form the outermost layer of the plant, protect the plant from the outside environment and control the flow of materials into and out of the plant.
- **Ground tissues**, which form the intermediate layer, make up the bulk of the plant body and perform a wide range of functions, including physical support, wound repair, and photosynthesis.
- **Vascular tissues** contain stacks of long cells forming continuous tubes that run throughout the plant body, linking all organs of the root and shoot systems. The vascular tissue known as **phloem** transports sugars from

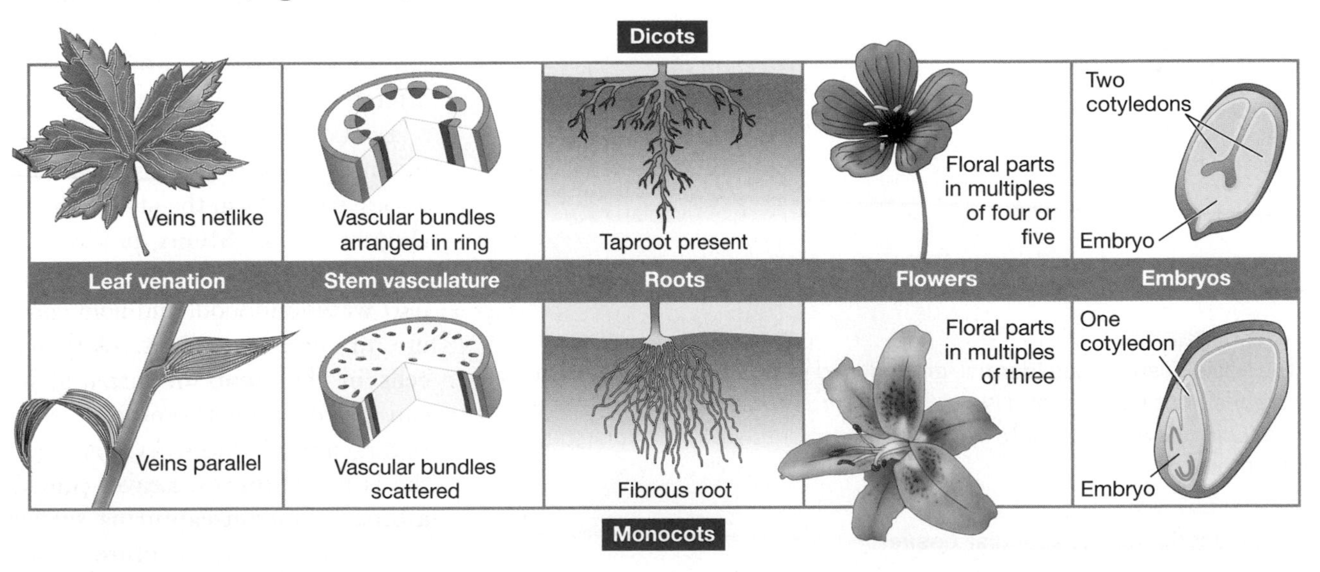

Figure 26.2

Dicots and monocots

Flowering plants have traditionally been classified into two main groups, dicots and monocots, on the basis of their external form and internal structure. For example, dicots grow straight, thick taproots. Monocots grow fibrous, branching roots.

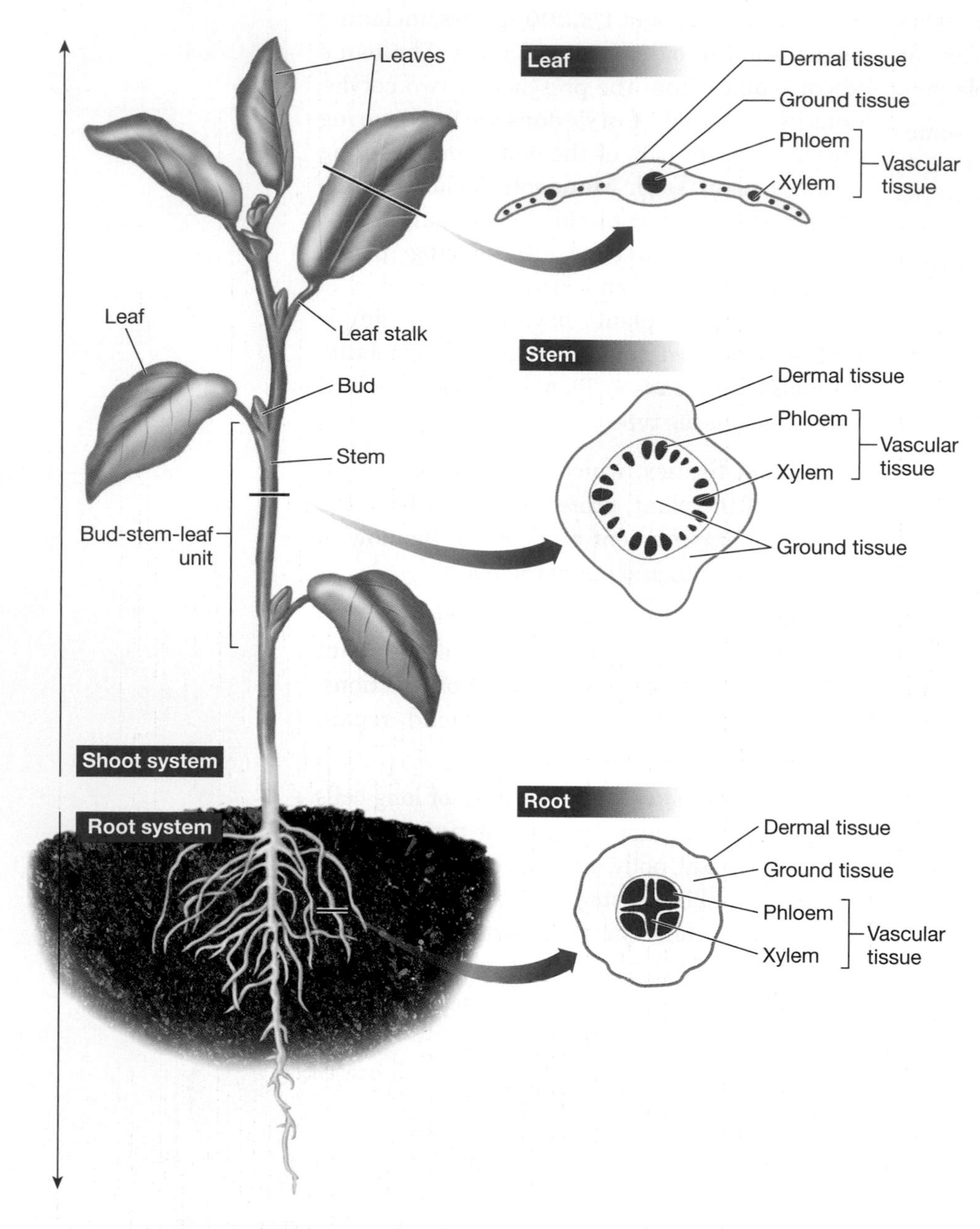

Figure 26.3

How plants are organized

Plants have three basic tissue types (dermal, ground, and vascular), which make up three types of organs (roots, stems, and leaves). Belowground, plants grow by extending old roots and producing new lateral roots. Aboveground, plants grow by adding new bud-stem-leaf units.

Q1: What is the function of the vascular tissue?

Q2: A plant organ is green if the cells within it contain chloroplasts, which carry out photosynthesis. In the figure, which of the plant organs do not contain chloroplasts, and why do you think that is?

Q3: Which tissue type has chloroplast-containing cells? Why?

See Appendix A for answers to the figure questions.

the leaves, where they are produced, to living cells in every part of the plant. **Xylem**, in contrast, transports water and minerals, absorbed from the soil, upward from the roots and outward to the leaves.

DeHaan aimed to domesticate a flowering perennial called intermediate wheatgrass (*Thinopyrum intermedium*). Widely used for hay and pasture, intermediate wheatgrass grows wild across the western United States and Canada. It has tall, thin shoots that grow green in the fall and turn golden brown in the spring and summer, and long, deep roots that stretch belowground.

As shown in **Figure 26.3**, the body of a flowering plant can be divided into those two basic organ systems: roots in soil, shoots in air. The **root system** anchors the plant, absorbs water and nutrients from the soil, transports water and food, and may store food (**Figure 26.4**). *Root hairs* greatly increase the surface area through which plants can absorb water and mineral nutrients. Annuals typically grow short roots. Perennials grow longer roots, which last year after year, depending on the species.

Roots are one of plants' three basic organs. The other two are stems and leaves, which form the **shoot system** (**Figure 26.5**). **Stems** provide the plant with structural support, transport water and food, and hold leaves up to intercept light. Although cells in the stems of many plants can perform photosynthesis, most sugars are produced by photosynthesis in the **leaves**. Leaves provide a broad, sunlight-capturing surface to maximize energy capture. Wheat and intermediate wheatgrass have long, pointy leaves that grow from the stems. At the tip of each plant shoot and at the base of many leaves is a bud. Under the right conditions, buds produce new shoots or flowers.

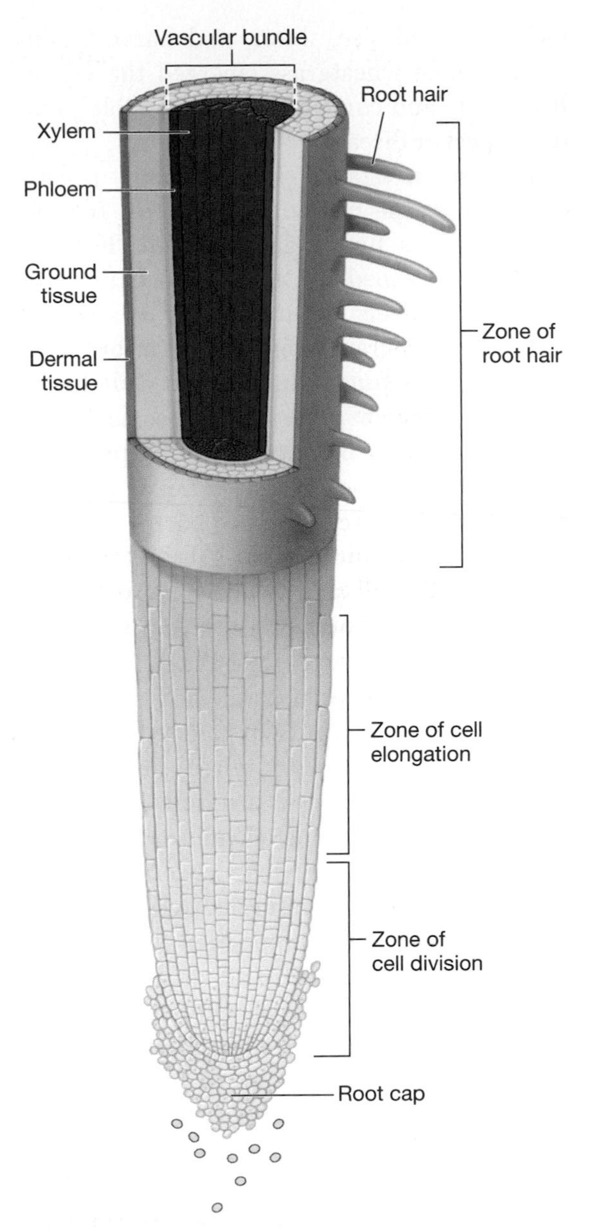

Figure 26.4

The root system

The dermal tissues of roots produce numerous outgrowths called *root hairs*, which aid in the absorption of water and nutrients. Plant roots have a region of active cell division, protected by the *root cap*, a region of cell elongation, in which cells increase in size, and a root hair zone where they complete their development.

Q1: How do root hairs increase a plant's ability to absorb water and nutrients?

Q2: How do roots make a plant more stable?

Q3: Given question 2, which do you predict would be more stable in harsh weather: annuals or perennials?

See Appendix A for answers to the figure questions.

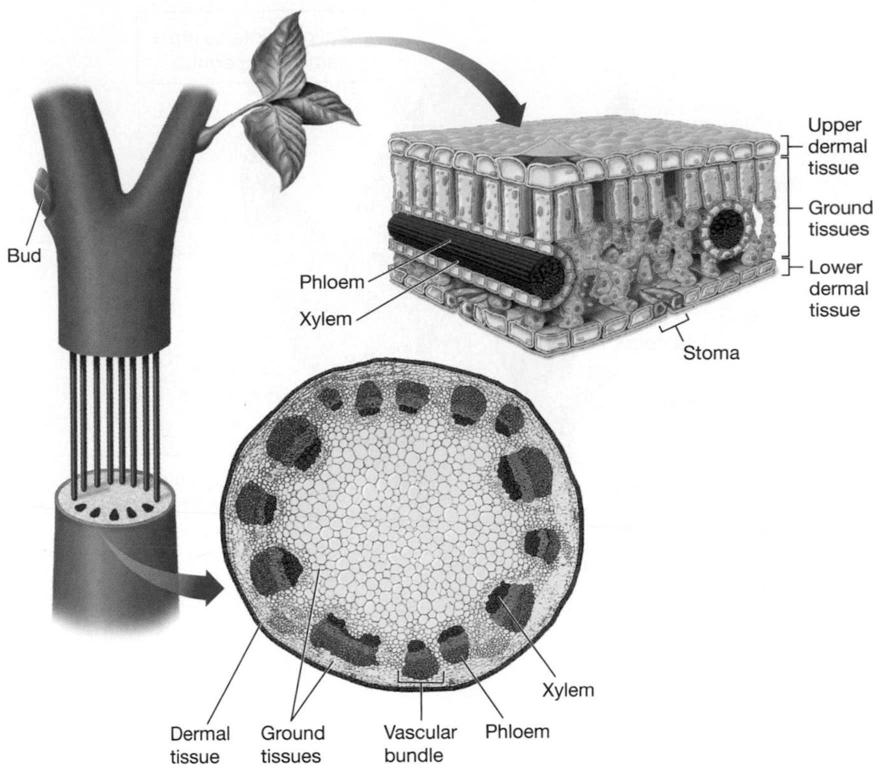

Figure 26.5

The shoot system

Stems provide structural support and may perform a limited amount of photosynthesis. Leaves produce the majority of the plant's food through photosynthesis. Buds are the focal points of plants' growth.

Q1: What nutrients does phloem move from the leaves to other parts of the plant?

Q2: Which part of the shoot system—the bud, stem, or leaf—produces flowers?

Q3: How is the location of stomata important for their function?

See Appendix A for answers to the figure questions.

Flowers house the structures that produce male and female gametes and, in many species, also facilitate the delivery of the sperm-bearing pollen to the female reproductive organs (**Figure 26.6**). After fertilization, the ovary of the flower develops into fruit, which helps disperse seeds in highly effective ways.

The outer layer of leaves is made up of dermal tissues, which include pores known as **stomata** (singular "stoma," see **Figure 26.5**). Plants open stomata to let in carbon dioxide (CO_2) needed for photosynthesis. When open, the stomata release

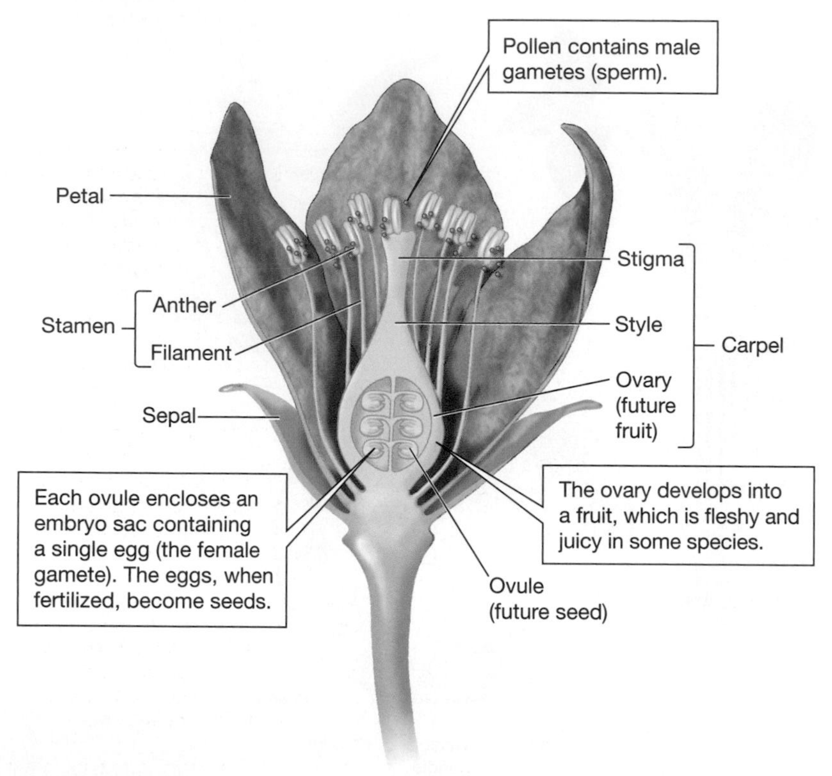
Pollen contains male gametes (sperm).
Petal
Stigma
Stamen
Anther
Filament
Style
Carpel
Ovary (future fruit)
Sepal
Each ovule encloses an embryo sac containing a single egg (the female gamete). The eggs, when fertilized, become seeds.
The ovary develops into a fruit, which is fleshy and juicy in some species.
Ovule (future seed)

Figure 26.6

Four whorls make a flower

The various parts of a flower are arranged in concentric rings, or *whorls*. From the outermost whorl inward, a typical flower consists of four whorls: *sepals, petals, stamens*, and *carpels*. (All the petals in a flower are collectively known as the *corolla*. The collective term for all of the sepals is the *calyx*.)

oxygen and lose water through evaporation. Most plants open their stomata in the daytime and close them at night, conserving water when photosynthesis is not an option. A plant experiencing water stress because of an inadequate water supply will close its stomata to conserve water, no matter the time of day.

Intermediate wheatgrass and other deep-rooted perennials are less likely to experience water stress, because of their extensive underground network of roots to extract water from the ground. Yet they do often become dormant and close their stomata during the hot summer.

Breeding Begins

Intermediate wheatgrass first came to the attention of agronomists in 1983, when the Rodale Institute, a Pennsylvania research facility that studies organic agriculture, evaluated close to 100 species of perennial grasses for traits including seed size, flavor, and harvestability. Intermediate wheatgrass emerged the winner. It is hardy and currently not susceptible to any major pest or disease.

Disease resistance is important because crops are at the mercy of pathogens. In 1999, for example, a fungal disease called "Fusarium head blight" wiped out Minnesota's barley crop. *Fusarium* infects wheat, barley, oats, corn—essentially any grain crop. Every year for the last 20 years, this fungus has caused a damaging epidemic somewhere in the US. Because plants are under attack by pathogens and predators, they have developed a rich variety of mechanisms to deter attacks (**Figure 26.7**, left).

Plants have physical defenses such as thorns, as well as chemical substances that are toxic to herbivores. Nicotine, the addictive chemical in cigarette smoke, protects tobacco plant leaves from insect predators. Caffeine protects the leaves and seeds of coffee and tea plants from potential predators. Plants also circulate and secrete antimicrobial chemicals to kill pathogens that could infect them.

In addition to using chemicals to protect themselves, plants produce chemicals called *hormones* to coordinate internal activities necessary for growth and reproduction (**Figure 26.7**, right). Hormones are active at very low concentrations (see Chapter 25 for more on hormones) and include **auxins**, **cytokinins**, **gibberellins**, **abscisic acid (ABA)**, and **ethylene**.

After intermediate wheatgrass proved to be resistant to pests and produced better seeds than other perennial grasses do, the Rodale Institute began collecting strains of the plant from around the world. Over 12 years, researchers bred the strains, selected the best 20, and then identified the top 14 individual plants.

At the Land Institute, DeHaan received seed from the 14 Rodale plants and began breeding them, tinkering with the possibilities of intermediate wheatgrass. In his first year, DeHaan planted some 3,000 plants in an effort to create a large group of diverse individuals. Then he watched the grass grow.

Plants can increase in length by **primary growth**, in which the plant's shoots and roots lengthen. They can also grow in thickness, called **secondary growth**. Secondary growth includes the thickening of both stems and roots.

Unlike most animals, which grow early in their lives and then level off in a pattern

Plants have three lines of defense against the external environment:

1. A tough outer surface protects the plant from water loss and stress due to hot and cold temperatures and serves to block pathogens from entry. Thorns and other defenses keep predators away.

2. Plants deploy nonspecific chemical defenses, which involve a range of broadly targeted antipathogen and antiherbivore chemicals, including hormones that signal other parts of the plant to manufacture chemical weapons and have them "battle ready."

3. Plants also have specific chemical defenses. For example, they have a number of genes that respond to complementary genes in a pathogen. If a specific pathogen is detected, protective defenses including toxic chemicals are unleashed.

Five main hormones control the internal environment of a plant:

1. **Auxins** are necessary for cell division and the formation of organs such as roots. They are involved in *phototropism*, the growth of shoots toward the light, and in *gravitropism*, the growth of roots toward the ground and the growth of shoots away from the ground.

2. **Cytokinins** are necessary for cell division, and they promote shoot formation. The levels of both cytokinins and auxins decline sharply just before plants drop their leaves in the fall season.

3. **Gibberellins** bring about stem growth through both cell elongation and cell division. They also stimulate seeds to germinate.

4. **Abscisic acid (ABA)** mediates adaptive responses to drought, cold, heat, and other stresses.

5. **Ethylene** stimulates the ripening of some fruits, including apples, bananas, avocadoes, and tomatoes. Ethylene activates enzymes that convert starches into sugars, resulting in a sweeter fruit.

Figure 26.7

Plants use chemicals to survive, grow, and reproduce

(Left) Plants use chemicals to protect themselves from both living and nonliving threats in their environment. (Right) Like animals, plants use hormones to coordinate internal activities necessary for growth and reproduction.

Q1: **Give an example of how plants use chemicals to defend themselves.**

Q2: **Of the plant hormones listed in this figure, which three promote growth?**

Q3: **What is a plant's first line of defense? Is this a chemical defense?**

See Appendix A for answers to the Figure Questions.

termed **determinate growth**, most plants grow throughout their lives—a pattern called **indeterminate growth**. Also unlike most animals, plants are able to grow through **modular growth**, the repeated addition of "modules" of bud-stem-leaf units aboveground or new lateral roots belowground.

Indeterminate and modular growth habits give plants great flexibility to respond to changing environmental conditions, such as high and low levels of sunlight, water, or nutrients. Plants tend to add many new parts when conditions are favorable and few new parts when conditions are not. This flexibility also enables plants to replace damaged tissues and organs. In fact, plants are so flexible in their development that most living cells in the adult plant body have the potential to generate whole new plants.

DeHaan and his team frequented their field and recorded traits of the plants, such as their heights and how early they flowered. When it was time to harvest the intermediate wheatgrass, the team plucked the seed heads off the plants and placed them in bar-coded bags.

Back in the lab, technicians analyzed the seeds from each plant. They recorded size and weight, because the larger the seed, the more grain can be produced for food. They also noted how easily the seeds shed their outer husks, because the less sticky a grain's husk is, the easier it is to process that grain.

DeHaan's first experimental crop grew plants of many sizes and traits. "It's like scratching off lottery tickets," DeHaan told National Public Radio in 2009. "Maybe there's something amazing in there. We'll see." As he recorded traits, DeHaan found that most of his best performers—those with larger seeds and deeper roots—came from a few families, so DeHaan went back into the field and uprooted those specific plants. These plants were brought into the warm greenhouse for the winter and bred with each other. Their offspring were then planted the following fall, and the process was repeated.

But the more DeHaan bred these particular plants, the more he restricted their gene pool. In fact, by excluding certain alleles of genes that had been in the original population of 3,000 plants, he created a genetic bottleneck (see Chapter 12). He began to worry that he might have lost genes needed to improve certain traits. One particular trait he struggled to identify and retain in his crop was the ability to mature early in the season.

Many plants, especially species native to temperate regions, perceive the seasons by sensing the length of the day. This is possible because day length varies with the season: days are shorter in winter and longer in summer.

This sensing of the duration of light and dark in a 24-hour cycle is known as **photoperiodism**. Plants use night length to sense when conditions are favorable for flowering and seed germination. The dormancy of buds through fall and winter, as well as their regrowth in spring, is also influenced by photoperiodism in addition to changes in temperature and precipitation. DeHaan wanted to identify wheatgrass that flowered early in the spring, so from the U.S. Department of Agriculture (USDA) he obtained hundreds of wild collections of wheatgrass that mature early, and those that flowered early he crossed with his own plants.

By 2010, DeHaan had something good. He had doubled both the seed yield and the weight of seeds of his intermediate wheatgrass. "It was succeeding a lot faster than we thought," says DeHaan. "It was really easy to grow." Unfortunately, "intermediate wheatgrass" doesn't sound like a tasty grain one might like in cereal or bread. So DeHaan and his team renamed the newly domesticated grain "Kernza," a combination of the word "kernel" and the name of a native tribe of the region, Kanza (which also inspired the state name "Kansas").

Wheat seeds are larger and heavier.
Kernza seeds are smaller and lighter.

Figure 26.8

Wheat and Kernza seeds, side by side

Kernza growers hope to eventually breed plants with seeds as large as those of wheat. They still have a long way to go.

Kernza® perennial grain is different from wild intermediate wheatgrass in several ways. First, the seeds are larger. "When I started working with it, the typical seed weighed 3.5 milligrams," DeHaan said in 2010. "Now, our best seeds are 10 milligrams." That's progress, but there's more to be done: an average wheat seed weighs 35 milligrams (**Figure 26.8**).

Then there are the roots. Kernza's roots grow down to 10 feet (**Figure 26.9**). The deep roots make the plant very efficient at sucking water out of the soil. That makes it more resistant to climate change than annual crops are, and the plant has already been shown to fare better in drought conditions than traditional wheat does. Furthermore, the long roots hold soil together, preventing erosion that carries away fertile soil, fertilizers, and pesticides with it. Major crop-producing countries around the world lose tens of billions of tons of topsoil every year to erosion.

Wheat roots are shorter.
Kernza roots grow deeper.

Figure 26.9

Wheat and Kernza roots, side by side

Kernza's longer roots make the plant better able to collect water from surrounding soil and store carbon and nitrogen that would otherwise be lost to the air or waterways.

Kernza's long roots have additional benefits, such as outdoing wheat at accessing groundwater, accumulating carbon in the soil, and absorbing nitrogen fertilizer. To grow, plants need CO_2, water, and mineral nutrients, especially nitrogen, phosphorus, and potassium. Most of the dry weight of a plant comes from CO_2 absorbed from the air and converted into carbohydrates by photosynthesis.

Plants require **macronutrients** (nitrogen, phosphorus, potassium, calcium, sulfur, and magnesium) in relatively large amounts and **micronutrients** (including iron, zinc, and copper) in small amounts. Carbon, oxygen, and hydrogen are obtained from air or water; the rest of the nutrients that plants need must be obtained from soil. In agriculture, farmers add most of the nutrients to the soil as fertilizer, but fertilizer can run off into water, polluting it. Perennial plants make better use of fertilizer than do annuals by efficiently retaining and recycling it. In the same study mentioned earlier, Kernza reduced the amount of nitrogen leached into the nearby ecosystem by up to 99 percent compared with wheat.

Crop Collaboration

Large seeds and deep roots are good traits, but they aren't enough to make Kernza a successful crop. Other specialty grain crops yield about 1,000 pounds per acre. In Kansas, Kernza was yielding less than 300 pounds per acre. So DeHaan sought help. Around 2007, DeHaan took his Kernza plants north to the University of Minnesota, where he had once been a student. He began working with agronomist Donald Wyse and other wheat breeders at the university. Planting in Minnesota turned out to be a boon. Kernza had larger yields there because intermediate wheatgrass is a cool-season grass and performs better in a cooler place.

Wyse was an ideal collaborator. He had spent years breeding perennial and winter annual crops for agriculture in Minnesota, including perennial flaxseed and perennial sunflower. One of the great values of a perennial, in addition to having deep roots that prevent erosion and requiring less fertilizer and pesticide, says Wyse, is that the roots and ground tissues are active during a longer period of the year, absorbing more solar energy and thus increasing an area's net primary productivity (NPP). This, he says, is "high-efficiency agriculture" (see Chapter 21 for a discussion of NPP).

In Minnesota, for example, the roots of corn and soybeans are active for just 2½ months a year, usually after the heaviest rains. That means all the sunlight from the other 9½ months of the year is wasted. But perennials can perform photosynthesis from the moment the snow melts in the spring until the first snowfall of the winter, says Wyse. With perennials, he says, "we're harvesting more energy."

In 2011, DeHaan, Wyse, and the Minnesota team planted a field of more than 2,000 intermediate-wheatgrass plants from 69 families and then measured traits such as biomass, grain yield, and seed shape. They've continued this process every year, sometimes pairing complementary strengths and weaknesses. In addition, they've begun recruiting farmers to grow and harvest Kernza in small fields of 40 acres or so in different states.

Yet the overall domestication process continues, and it is not easy. DeHaan estimates it will be another 10 years before Kernza is ready to compete economically with wheat. One of the challenges of breeding intermediate wheatgrass has to do with how it reproduces. Many plants, such as dandelions and poplar trees, can reproduce asexually, when a parent plant forms a genetically identical clone. Some crops, including potatoes, apples, and grapes, are also propagated as clones. But staple crops, including grains, reproduce sexually.

The overall principle of sexual reproduction in plants is similar to that in animals. A haploid male gamete (sperm) fuses with a haploid female gamete (egg) to give rise to a diploid cell, the zygote, which undergoes mitosis to create a multicellular diploid embryo. In time, the embryo develops into an individual offspring, which represents the next generation (**Figure 26.10**).

Plant and animal life cycles differ in one key respect. In animals, meiosis creates gametes and nothing but gametes. In plants, meiosis generates haploid cells called **spores**. Each spore undergoes mitotic divisions to create a haploid, multicellular structure called a **gametophyte**. Specialized cells in the gametophyte differentiate to produce sperm or egg cells.

After fertilization in flowering plants, the zygote develops into an embryo inside the ovule, which is contained within the ovary, at the base of the flower (see **Figure 26.10**, top right). The outer layers of the ovule harden into a protective seed coat. Each seed contains the ingredients for growing a young plant of the next generation: a mature embryo, a food source, and the seed coat.

The ovary surrounding the seed forms a *fruit*, yet another plant organ. Once a fruit is formed, the embryo enters dormancy, and the seed is dispersed from its parent. Seeds are often dispersed by wind or water, by attaching to the fur of animals, or by being eaten (and excreted) by animals (**Figure 26.11**). Dormancy ends when the embryo is stimulated by favorable conditions to start growing again. Then the seed germinates, and a seedling grows into a plant that will mature and produce flowers and fruit, the beginning of another generation. This plant is called a **sporophyte**, analogous to an individual animal in that both are diploid, multicellular organisms.

Annual wheat plants *self-pollinate*: that is, they reproduce by fertilizing themselves with their own pollen to create a zygote before their flowers even open. But intermediate wheatgrass and many other plants do not self-pollinate. They require a mate. Given that plants cannot travel to find mates, how does sperm-containing pollen from one plant reach the eggs of another plant?

In some species, such as grasses and pine trees, pollen is transported by wind. Wind can carry pollen long distances, but most pollen blown by wind lands in inhospitable places (such as parking lots or lakes), not on the flower of another plant of the same species. For many flowering plants, the solution to this problem is having their pollen transported by animals such as insects, birds, and even mammals

DEBUNKED!

MYTH: Tomatoes are vegetables.

FACT: Tomatoes are considered both a fruit *and* a vegetable. They contain seeds, as a fruit does, but have low sugar content, as a vegetable does. The USDA food guidelines list tomatoes as vegetables. (Other "vegetables" that count as fruit include cucumbers, squash, and peppers. And did you know a banana is a berry?)

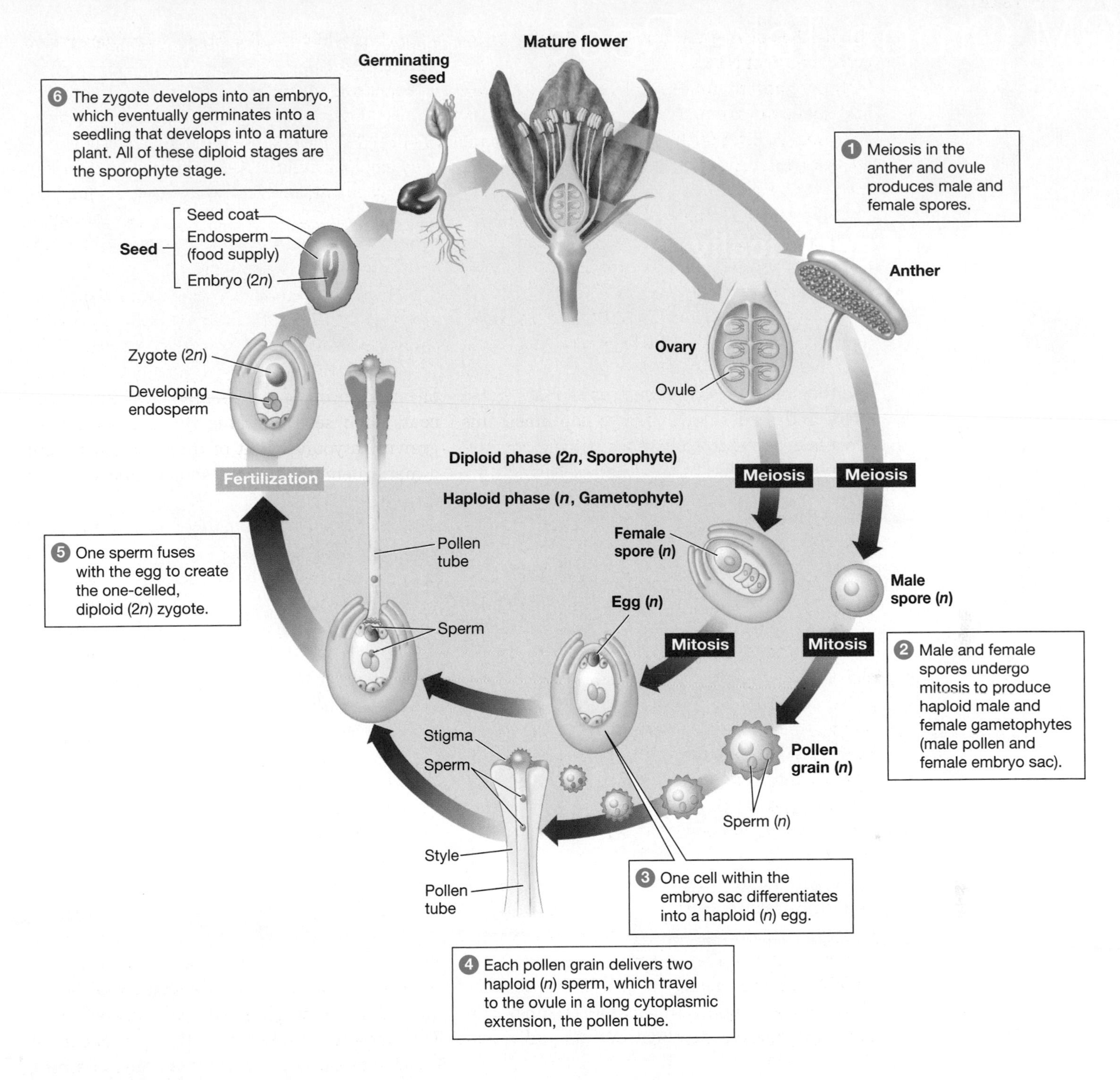
Mature flower
Germinating seed
6 The zygote develops into an embryo, which eventually germinates into a seedling that develops into a mature plant. All of these diploid stages are the sporophyte stage.
1 Meiosis in the anther and ovule produces male and female spores.
Seed
Seed coat
Endosperm (food supply)
Embryo (2n)
Anther
Ovary
Ovule
Zygote (2n)
Developing endosperm
Diploid phase (2n, Sporophyte)
Fertilization
Meiosis
Meiosis
Haploid phase (n, Gametophyte)
Female spore (n)
Male spore (n)
5 One sperm fuses with the egg to create the one-celled, diploid (2n) zygote.
Pollen tube
Egg (n)
Sperm
Mitosis
Mitosis
2 Male and female spores undergo mitosis to produce haploid male and female gametophytes (male pollen and female embryo sac).
Stigma
Sperm
Pollen grain (n)
Sperm (n)
Style
Pollen tube
3 One cell within the embryo sac differentiates into a haploid (n) egg.
4 Each pollen grain delivers two haploid (n) sperm, which travel to the ovule in a long cytoplasmic extension, the pollen tube.

Figure 26.10

From generation to generation

The life cycle of flowering plants consists of the **alternation of generations**. This means that haploid stages of the life cycle (purple) alternate with diploid stages (orange).

Q1: Are eggs and sperm haploid or diploid? Are they part of the sporophyte or gametophyte generation?

Q2: Why is the plant life cycle, but not the animal life cycle, called an "alternation of generations"?

Q3: How does asexual reproduction differ from the plant sexual reproduction life cycle diagrammed here?

See Appendix A for answers to the figure questions.

GM Crops Take Root

Since 1996, the total land area used to grow genetically modified (GM) crops has expanded dramatically. Most of these crops, including the varieties depicted here, are enhanced with the insertion of a gene that imparts insect resistance or herbicide tolerance. Broad scientific consensus has concluded that food derived from GM crops poses no greater risk to health than food from nonmodified crops.

Assessment available in smartwork

Global area of genetically modified crops

Area shown in million hectares

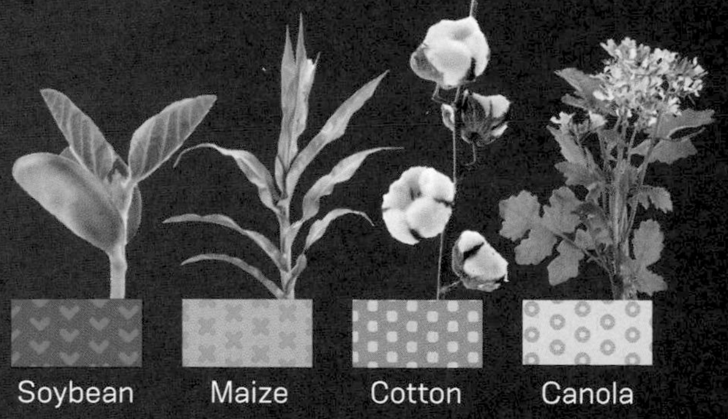

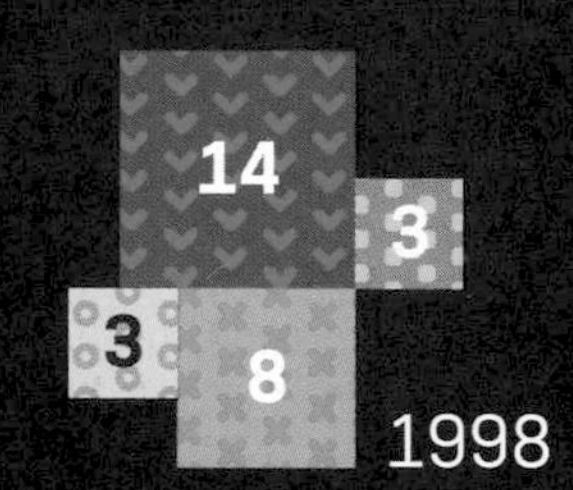

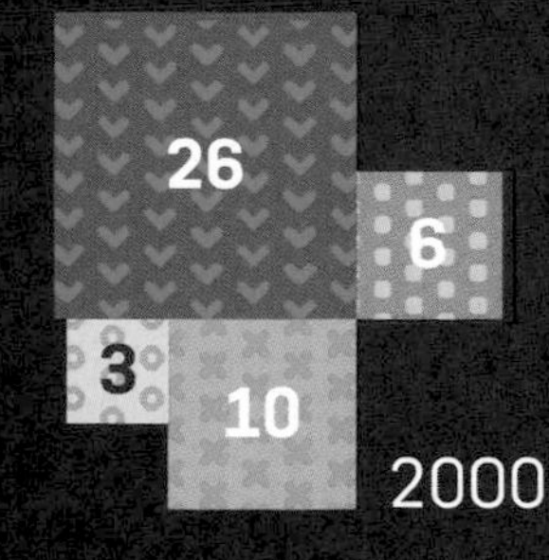

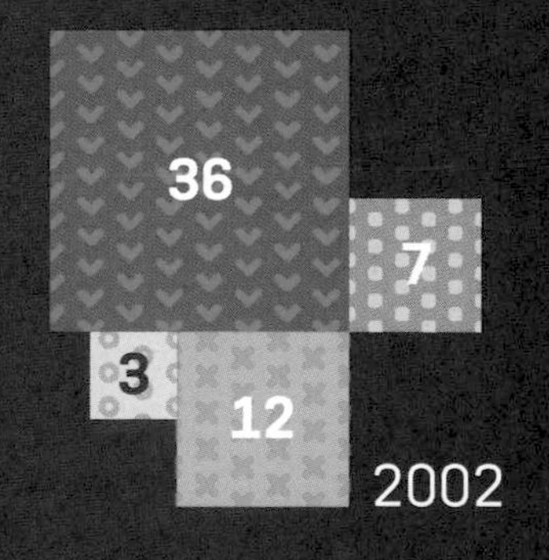

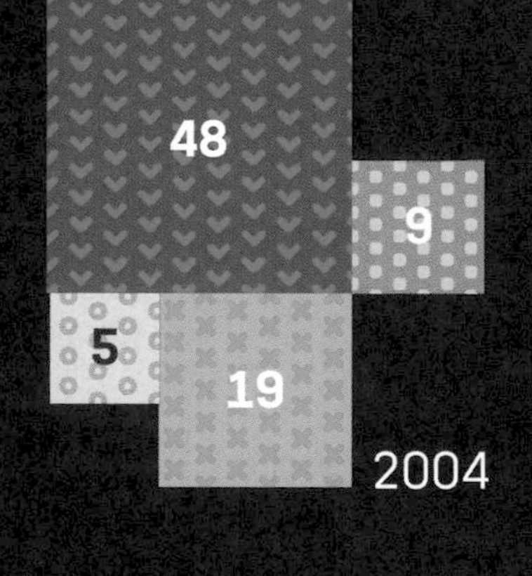

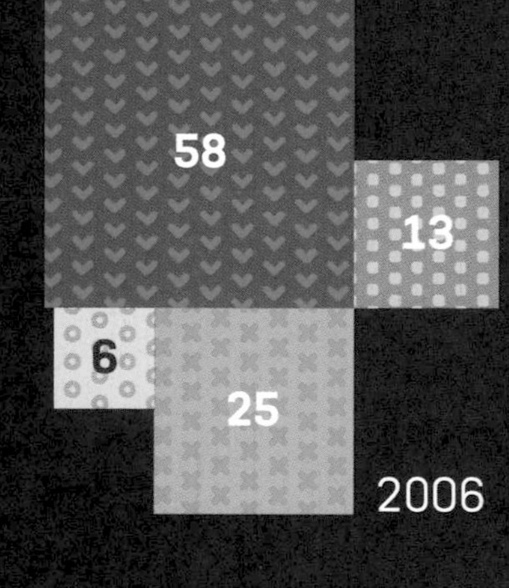

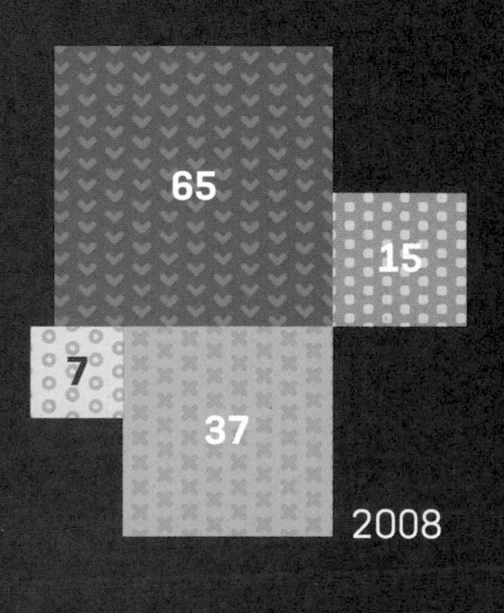

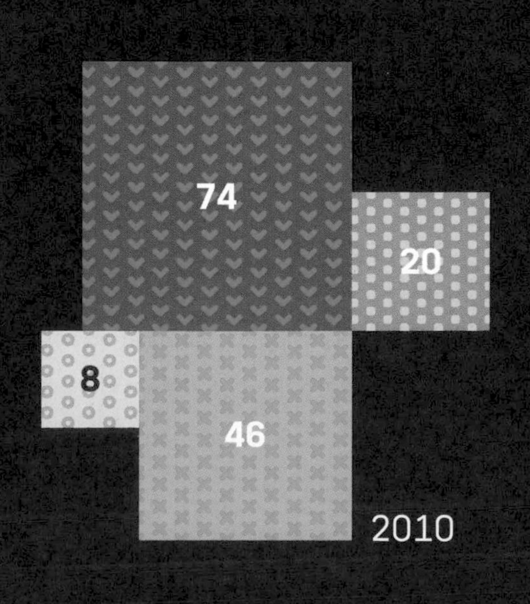

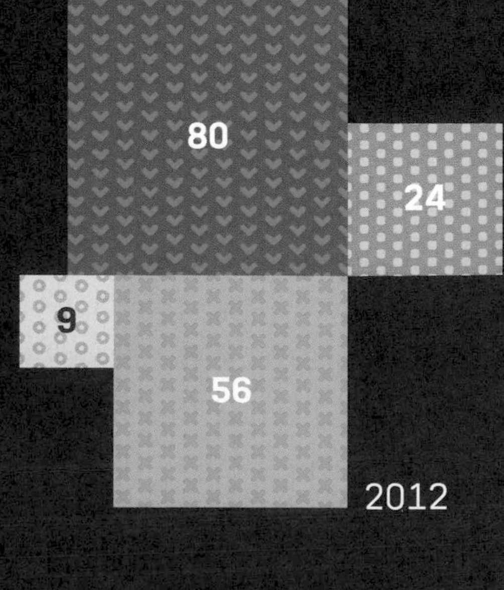

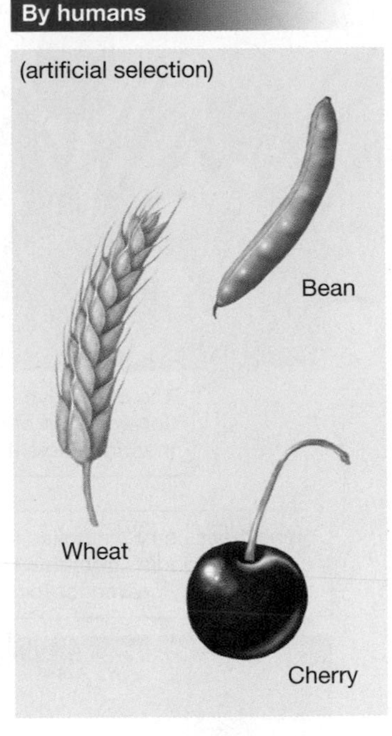

Figure 26.11

Plant seeds spread in various ways

The seeds of many plants spread via wind or water. Others attract animals to eat (and excrete) them or attach to their fur. Humans have artificially selected many plant species to make the fruit, seeds, or both more edible.

Q1: What kinds of seeds would you expect to be contained within a sugary fruit: seeds spread by water or by animals? Why?

Q2: What kinds of seeds would you expect to weigh less: those spread by air or those artificially selected by humans? Why?

Q3: How does the height of a tree affect the spread of its seeds?

See Appendix A for answers to the figure questions.

(**Figure 26.12**). Plants attract these **pollinators** with brightly colored, sweet-smelling flowers filled with sugary nectar. It is a mutualistic relationship, in which both species benefit.

DeHaan and his team need to mate specific plants to breed for specific traits. So, instead of relying on wind or pollinators, they put a bag over the flowering stalks of two plants they want to breed so that pollen from one plant passes directly to the second plant in the bag and no others (**Figure 26.13**). "We call them 'plant condoms,'" says DeHaan with a laugh. "The bags keep crosses we don't want to happen from happening."

And because an intermediate-wheatgrass plant can breed with only another plant—rather than with itself, as annual wheat can do—the offspring are genetically diverse; no offspring is *exactly* like its parent. In fact, an offspring can be very different from its parents because there is a great deal of variation in the polyploid genome of intermediate wheatgrass. Each gene may have six or more versions, or alleles, in a single plant, says DeHaan. So, when breeding new plants, the team can never match the exact genotype and phenotype of a parent but rather must constantly juggle traits to try to capture the right set of traits in a single line of plants. For example, DeHaan is currently working to grow shorter plants because taller plants tend to tip over, making them difficult to harvest.

Perennial Pancakes

Early on, DeHaan estimated it would take 50 years for a new plant to be domesticated to match the yield of wheat, but now USDA and University of Minnesota researchers are supplementing his hands-on breeding work with gene

The distinctive colors, shapes, and smells of flowers often attract highly specific pollinators that are efficient pollen dispersal agents.
Honeybees carry the pollen that dusts their bodies from flower to flower as they search for food.

Birds have good color vision and favor red flowers with long floral tubes.

Figure 26.12

Bribing animals to do the work

Pollinators provide stationary plants with a way of transporting sperm to eggs. The spectacular colors, shapes, and odors of flowers, in combination with food rewards such as nectar, lure pollinating animals into visiting several flowers of the same species, incidentally transferring pollen in the process.

Q1: Many crops, including apples, peppers, and tomatoes, depend on insect pollinators. What would happen if no pollinator visited such a plant?

Q2: Flower petals are actually modified leaves, although they look very different from leaves. Do flowers still perform the main function of leaves? Explain your reasoning.

Q3: Some flowers look like an insect. How might this resemblance attract pollinators?

See Appendix A for answers to the figure questions.

Figure 26.13

Breeding the next generation of Kernza plants

The flowering stalks of Kernza plants are enclosed in bags to prevent the pollen from spreading to any other plants. Scientists identify plants with the traits they want to pass on to the next generation and selectively breed them in this way.

Q1: What traits are scientists selecting for in each new generation of Kernza?

Q2: What would happen if the Kernza plants were allowed to freely pollinate rather than being selectively pollinated by hand?

Q3: After choosing two plants to breed, researchers put the plant heads together under a bag as shown in the photo. Why do they do this before the plants begin to pollinate?

See Appendix A for answers to the figure questions.

sequencing efforts. These scientists have been associating genetic markers with particular traits, such as seed size. Once they have enough traits mapped on the genome, they will no longer have to wait for a plant to mature to observe its phenotype; instead they will simply sample the DNA of young seedlings and predict the plant's traits on the basis of genetic markers.

"Within a very short period of time, we'll be able to identify key genes in the domestication process at a very low cost," says Wyse. With these tools, Wyse thinks the breeders won't take too long to match the yield of wheat. "It may only take half a decade to see dramatic improvements in these new crops," says Wyse. "We think crops of this type are going to have great value."

But there's more to a food crop than how well it grows and can be harvested. It also has to taste good. DeHaan has used Kernza flour to make cookies, cakes, scones, bread, and more (**Figure 26.14**). He and other Land Institute staff have enjoyed them, though at least one food scientist says the flour will need some flavor improvement to go mainstream. In 2019, food manufacturer General Mills produced 6,000 small boxes of cereal made from Kernza and handed them out as samples to food writers and environmentalists. The cereal tasted like Wheaties, according to one reporter. The company says it hopes to scale up production to eventually sell Kernza-based cereal at grocery stores.

Other food companies have expressed an interest in using the new grain, as have beer brewers. DeHaan and Wyse also imagine that the crop by-product could be used as hay or to produce biofuel. "We'd love this to be a dual-use crop," says DeHaan. Farmers, for example, could harvest the grain for a food company and then use the "residue" plant material in the field, typically the leftover bottom of the stalk, to make biofuel. "Those two things together could result in an economically viable crop," he says.

Figure 26.14

Kernza flour

Scientists are finding uses for Kernza in traditional baking and also using it to make alcohol, animal feedstock, and hopefully even biofuel in the future.

NEED-TO-KNOW SCIENCE

- Plants can be grouped into three categories. **Annuals** (page 506) complete their entire life cycle in 1 year. **Biennials** (page 507) grow and mature for a year and then reproduce in the second year. **Perennials** (page 507) live 3 years or more.
- Plant bodies are made of three basic tissue types. **Dermal tissues** (page 507) protect the plant from attack and control the flow of materials into and out of the plant. **Ground tissues** (page 507) make up the bulk of the plant body and participate in support, wound repair, and photosynthesis. The two types of **vascular tissue** (page 507) are **phloem** (page 507), which transports sugars throughout the plant, and **xylem** (page 508), which transports water and minerals from soil.
- The plant body contains two basic organ systems: the belowground **root system** (page 508) and the aboveground **shoot system** (page 508). Plant organs include roots, stems, leaves, flowers, and fruits. **Roots** (page 507) absorb water and mineral nutrients, anchor the plant, and may store food. **Stems** (page 508) support the plant, transport water and food, hold leaves up to intercept light, and may perform some photosynthesis. **Leaves** (page 508) produce the majority of the plant's food through photosynthesis and include pores known as **stomata** (page 509), which are central to the plant's gas exchange with the atmosphere. **Flowers** (page 509) house the structures that produce male and female gametes and, in many species, also facilitate the delivery of the sperm-bearing pollen to the female reproductive organs.
- Plants defend themselves against the external environment with a tough outer covering, physical weapons such as thorns, and an arsenal of toxic chemicals.
- Like animals, plants use hormones, chemicals produced at low concentrations, to coordinate growth and reproduction.
- Whereas animals are limited to **determinate growth** (page 511) that levels off, plants have **indeterminate growth** (page 511). That is, they can grow throughout their lives, adding repeating bud-stem-leaf units aboveground and root units below in a pattern called **modular growth** (page 511). Aboveground, plants increase in length by **primary growth** (page 510), the lengthening of shoots and roots. Plants increase in thickness by **secondary growth** (page 510), the thickening of stems and roots.

- Plants experience **photoperiodism** (page 512), the sensing of the duration of light and dark in a 24-hour cycle.
- To grow, plants need CO_2, water, and mineral nutrients. Plants require **macronutrients** (page 513), such as nitrogen, phosphorus, potassium, calcium, sulfur, and magnesium, in relatively large amounts and **micronutrients** (page 513), including iron, zinc, and copper, in small amounts.
- Many plants can reproduce asexually, but most can also reproduce sexually. Meiosis in plants produces single-celled **spores** (page 514), which divides through mitosis to create a haploid, multicellular **gametophyte** (page 514). Cells in the gametophyte differentiate into egg or sperm. The fertilization of the egg by sperm generates a diploid zygote, which gives rise to the diploid, multicellular plant. All diploid stages are a **sporophyte** (page 514). This **alternation of generations** (page 515), that is, of gametophytes and sporophytes, is the plant life cycle.
- Fertilization produces a zygote, which divides by mitosis and develops into an embryo. The embryo is located within the ovule, which hardens into a protective seed coat. Each seed contains the ingredients for growing a young plant of the next generation: a mature embryo, a food source, and the seed coat.
- In flowering plants, male and female reproductive parts are contained in flowers. Flowers attract animal **pollinators** (page 517), which provide immobile plants with a way of transporting sperm-containing pollen to eggs.

THE QUESTIONS

See Appendix B for answers.

The Basics

1 Water is transported throughout the plant body by

(a) xylem.

(b) phloem.

(c) stomata.

(d) root hairs.

(e) flowers.

2 Which of the following is a chemical defense strategy used by plants?

(a) maintaining a tough dermal layer

(b) storing toxic chemicals in leaves

(c) signaling plant shoots to grow toward the light

(d) stimulating the ripening of fruit

(e) none of the above

3 Which plant tissue controls the flow of materials into and out of the plant and protects the plant from the outside environment?

(a) vascular tissue

(b) ground tissue

(c) dermal tissue

(d) phloem

(e) xylem

4 What kind of seed will be dispersed most widely by wind?

(a) dandelion

(b) apple

(c) coconut

(d) Kernza

(e) a seed held within a sticky bur

5 Alternation of generations

(a) involves a haploid gametophyte.

(b) involves a diploid sporophyte.

(c) is the plant life cycle.

(d) all of the above

(e) none of the above

6 Link each term with the correct definition.

modular growth	1. Increase in length by the division of cells located at the tip of each stem and each root
indeterminate growth	2. Growth throughout life
primary growth	3. Growth in size by repeatedly adding the same basic bud-stem-leaf unit
secondary growth	4. Increase in thickness

Challenge Yourself

7 Which of the following is *not* a plant organ?

(a) stem

(b) fruit

(c) vascular bundles

(d) root

(e) flower

8 Which *micro*nutrient do plants need to absorb from the environment to survive and grow?

(a) carbon

(b) sugar

(c) potassium

(d) nitrogen

(e) zinc

9 Select the correct terms:

(**Annual** / **Biennial**) plants complete their life cycle in 1 year, whereas (**biennial** / **perennial**) plants live for 3 or more years. Wheat is a(n) (**annual** / **perennial**), whereas Kernza is a(n) (**annual** / **perennial**). Kernza is also a (**monocot** / **dicot**), which has a (**fibrous root** / **taproot**).

10 Beginning with the stage following spores, place the following steps of the plant life cycle in the correct order by numbering them from 1 to 4.

_____ a. Zygote develops into an embryo.

_____ b. Gametophyte develops.

_____ c. Seed germinates.

_____ d. Egg is fertilized.

Try Something New

11 Birds pollinate flowering plants during the day. Some flowering plants are pollinated by bats, which are active at night. What would you predict about how bat-pollinated flowers differ from bird-pollinated flowers?

12 Several plants domesticated by humans have highly modified organs. For example, potatoes are modified stems, and carrots and sweet potatoes are modified roots. From which plant organ do you think onions and tomatoes are modified? Why?

13 Ethylene is a plant hormone that causes fruit to ripen by converting starches into sugars. It also signals flowers to open, seeds to germinate, and leaves to shed. Consider that placing a ripe piece of fruit next to but not touching an unripe piece will cause the second piece to ripen more quickly. How would you explain that by placing fruit in a paper bag (rather than exposed on the counter), it ripens more quickly?

14 Although bees are well-known pollinators, butterflies are also important pollinators because they can travel longer distances, pollinating flowering plants over larger areas. Unfortunately, one important species native to North America, the monarch butterfly, has been in a population decline over the past 20 years. Examine the accompanying graph. What years had the highest and lowest monarch butterfly counts? What was the average monarch count in 2008–12? In 2013–17? From 2008–12 to 2013–17, what was the average decline in monarch butterflies?

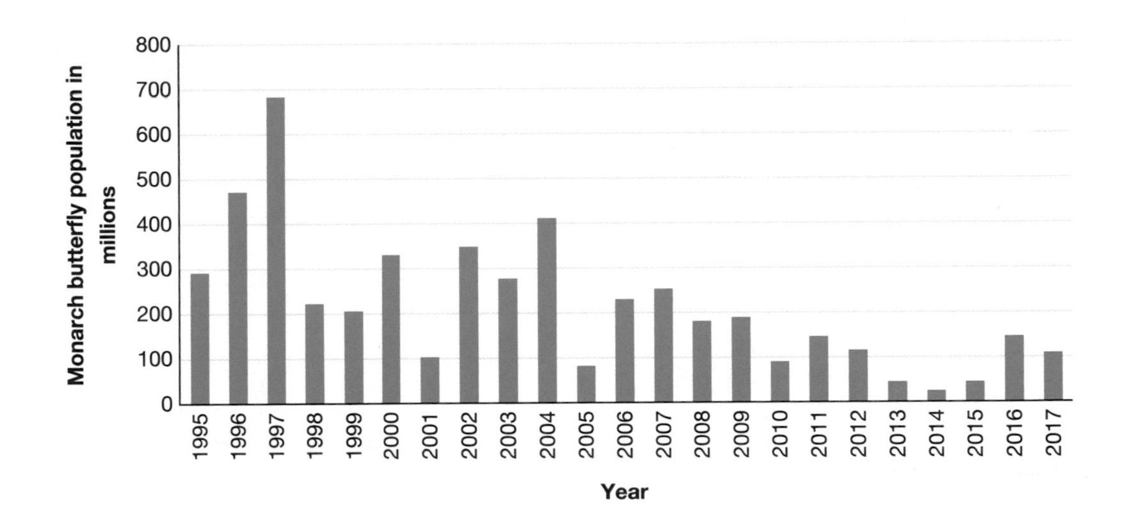

Leveling Up

15 **Looking at data** The debate over genetically modified (GM) crops and genetically modified organisms (GMOs) has been impassioned but not always informed. Genetically modified crops are those with DNA altered through biotechnology to increase productivity or resistance to disease.

Approximately 93 percent of the soybeans harvested in the United States, fed almost exclusively to domesticated animals, are genetically modified for pesticide and herbicide resistance. Examine the accompanying graph, which shows the yield in bushels per acre of soybeans from 1980 to 2019. The red circle is the yield for organic (i.e., unmodified) soybeans in 2007, which is based on information from 1,331 organic farms.

(a) What does the *x*-axis represent? What does the *y*-axis show?

(b) Describe the trend in soybean productivity from 1980 to 2019.

(c) The yield for organic soybeans in 2007 is only 66 percent of the average yield for that year. Organic soybean yield in 2011 is 80 percent of the average yield for that year. What years of conventional soybean production do these productions match?

(d) Critics say that the widespread growth and consumption of GM crops may have unexpected health and environmental costs. Proponents argue that genetic modification is the only way to feed a rapidly growing population. What do you think?

16 **Doing science** Although you are an expert on Kernza® perennial grain, you have been asked to contribute your scientific expertise to developing another perennial crop, *Silphium integrifolium* (sometimes referred to as a "perennial sunflower"), which could replace annual crops grown as sources of cooking oil.

(a) Read the Land Institute's overview of its research on *Silphium*: https://landinstitute.org/our-work/perennial-crops/perennial-oilseeds/.

(b) The *Silphium* researchers say that they aim to identify critical ways in which *Silphium* responds to changes in the environment each year but that they will need to follow up such observations with experiments. Are the scientists conducting observational or experimental research on *Silphium*? Explain your answer.

(c) Propose a hypothesis about an environmental variable that you believe might increase oil production in *Silphium*, and then identify a testable prediction from that hypothesis. Design an experiment to test the prediction, identifying your independent and dependent variables, the control group, and the treatment conditions. Create a graph to show the results you expect to find (1) if your hypothesis is supported and (2) if your hypothesis is not supported.

Digital resources for your book are available online.

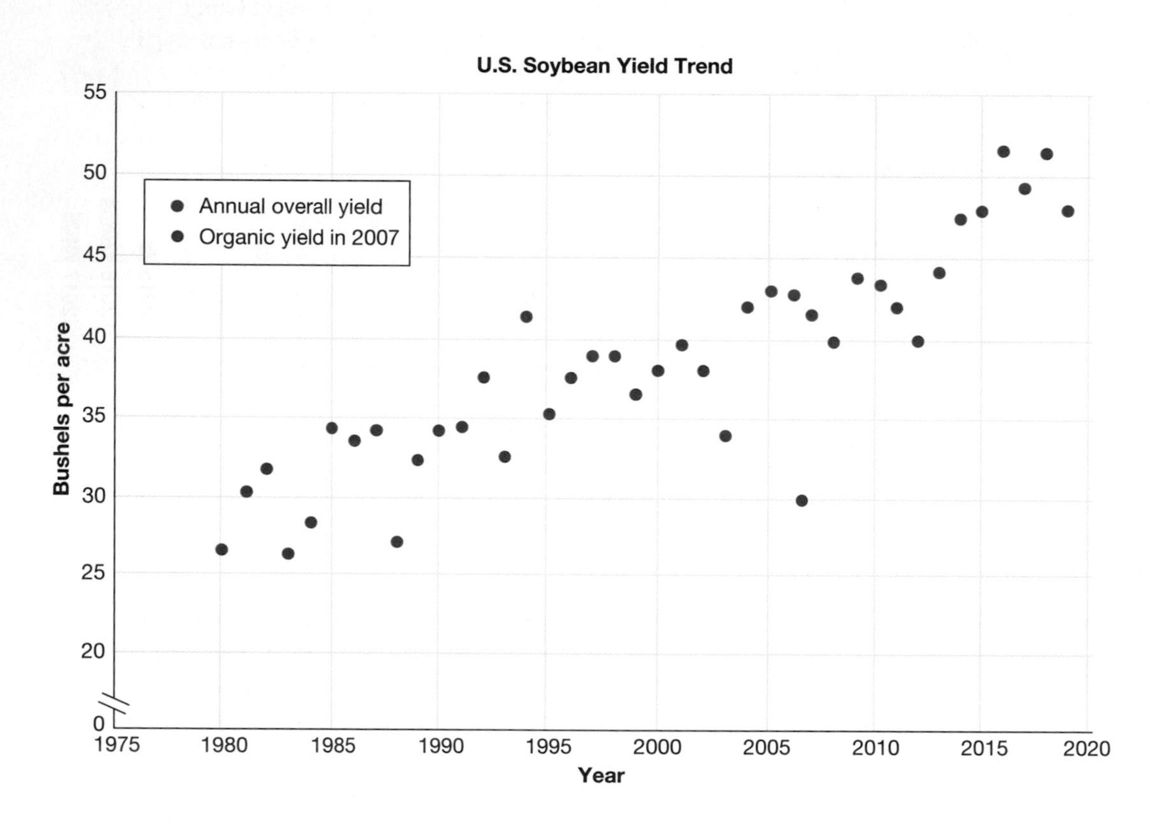

Appendix A: Answers to Figure Questions

CHAPTER 1

Figure 1.1

Q1: What were the original observation and question of the scientists studying the sick bats?
A1: They observed many dead bats and many bats with white noses. They questioned whether the high death rates were in some way related to the white noses.

Q2: At what point in the scientific method would a scientist decide on the methods to use to test a hypothesis?
A2: A scientist would design tests after predictions are generated from a hypothesis and before those predictions are tested (by observing or measuring or by designing and running experiments).

Q3: How might you explain the process of science to someone who complains that "scientists are always changing their minds; how can we trust what they say?"
A3: Science is a process and, because nothing is ever "proved" in science, we have to expect that our best understanding of the natural world will change as science proceeds.

Figure 1.2

Q1: Which step(s) in the scientific method does this photograph illustrate?
A1: The photograph illustrates making observations and testing the predictions of the hypothesis through observational study (either descriptive or analytical).

Q2: What types of environmental data might the researchers have collected in the cave?
A2: Environmental data might have included temperature and humidity readings within the cave, as well as soil and air samples to see whether fungal spores were present.

Q3: Why would the researchers in this photograph have worn protective gear?
A3: The researchers would have worn protective gear to avoid coming into contact with and perhaps being infected by whatever caused WNS.

Figure 1.4

Q1: According to this advertisement, what hypothesis was scientifically tested?
A1: Lucky Strike cigarettes taste milder than other leading brands of cigarettes.

Q2: What prediction comes from this hypothesis? Is this prediction testable? Why or why not?
A2: Prediction: People who smoke a Lucky Strike cigarette will say it is milder than cigarettes from other brands they are asked to smoke. Yes, this prediction is testable, because it can be measured and repeated.

Q3: Why can't the hypothesis in this ad be "proved"?
A3: As noted in the text, science is open to challenge by anyone at any time on the basis of evidence. Therefore, nothing in science can be proved. In this case, the evidence would be based on only the research participants who are tested. There might always be people (who weren't tested) who think another brand is milder.

Figure 1.5

Q1: State a possible hypothesis that could be tested with the analytical approach shown here.
A1: Hypothesis: Bats with WNS weigh less than uninfected bats.

Q2: State the hypothesis being tested with the experimental approach.
A2: Hypothesis: Healthy bats injected with a fungicide will have lower rates of infection by WNS than will bats that are not injected with a fungicide.

Q3: Why can't an observational study show a cause-and-effect relationship? Why is an experimental study the only research method that can show a cause-and-effect relationship?
A3: One possible explanation: Even if you find a relationship between two variables in an observational study, you can't know whether one variable causes the other (or whether another variable causes both of them). In an experimental study, you manipulate one variable to see whether that manipulation causes the second variable to change.

Figure 1.6

Q1: Explain how each treatment group differs from the control group. What is the independent variable for each group? What is the dependent variable for all groups?

A1: The independent variable for group 2 is airborne exposure to WNS. For group 3, it is physical contact with bats with WNS. For group 4, it is the fungus applied directly to their wings. For all groups, the dependent variable is the number of bats with WNS.

Q2: What hypothesis was being tested in this experiment?

A2: Hypothesis: WNS is caused by contact with the fungus *Geomyces destructans*.

Q3: In one or two sentences, state the conclusions you can draw from the experiment. Did the results support the hypothesis? Why or why not?

A3: Bats that come into physical contact with *Geomyces destructans* are highly likely to develop WNS, whereas those exposed only through the air do not develop WNS. The hypothesis was supported in part (physical versus air contact).

Figure 1.7

Q1: What is the control group in this experiment, and what are the two treatment groups?

A1: Control group: received sham injection. Treatment groups: injected with the white-nose fungus (*Geomyces destructans*) from North America; injected with a fungus related to *G. destructans* from Europe.

Q2: At day 40, approximately how many individuals were alive in each treatment group? At day 80? At day 100?

A2: Day 40: 100 percent of all groups survived; 18 in each. Day 80: 100 percent of control and North American *Geomyces destructans* (*Gd*) groups survived; 14 of 18 in the European *Gd* -related group survived. Day 100: 100 percent of control group survived; 13 of 18 in the North American *Gd* group survived; only 2 of 18 in the European *Gd* -related group survived (although the European *Gd*-related experiment ended at about day 90).

Q3: In one or two sentences, state the conclusions you can draw from the experiment. Did the results support the hypothesis? Why or why not?

A3: Conclusion: *Geomyces destructans* causes WNS and leads to higher mortality. The hypothesis was supported because the study found that WNS was caused by the fungus.

Figure 1.8

Q1: Give one *fact* about bats that you learned from this chapter.

A1: One possible fact: Bats can develop fungal infections.

Q2: What is another example of evidence for the *germ theory of disease*? (*Hint*: Think about human diseases.)

A2: One example would be strep throat, caused by *Streptococcus* bacteria (and cured by antibiotics).

Q3: Explain in your own words the difference between a fact and a hypothesis and between a hypothesis and a theory.

A3: Facts are objective observations of things in the physical world. They can be used to form hypotheses. Hypotheses are not as certain and are more complex than facts; they are simpler and less well documented than theories.

Figure 1.9

Q1: Give examples of other kinds of organs that mammals such as bats have. (*Hint*: Think of the organs in your own body.)

A1: Examples include kidney, liver, heart, and lungs.

Q2: Are bats in California part of the community of bats in upstate New York if they are of the same species? Why or why not?

A2: No, they are not, because they do not interact with each other.

Q3: Is the soil in a cave where bats live a part of the bats' population, community, or ecosystem? Explain your reasoning.

A3: Soil is part of the ecosystem; populations and communities are composed only of living things, and soil is part of the physical environment.

CHAPTER 2

Figure 2.2

Q1: How does a vaccine create immunity to a virus?

A1: A vaccine causes the immune system to produce antibodies in response to an inactive or harmless amount of a virus. If a live virus of that type enters the body, those antibodies then immediately identify and attack the virus.

Q2: Why is it important that the virus in the vaccine is inactive or harmless?

A2: If a person is exposed to the live virus without first being vaccinated with the inactive or harmless version, the person may become infected and have no antibodies to fight the infection.

Q3: Natural immunity occurs without a vaccine, just by exposure to a particular illness. For example, why will someone who has had chicken pox not get it a second time?

A3: Your body creates antibodies when you are exposed to, for example, chicken pox virus. If you are exposed a second time, those antibodies immediately identify and attack the chicken pox virus (just as in question Q1). Natural immunity doesn't always last for a lifetime. For example, antibodies for COVID-19 may only be produced for a few months. Note, however, that the chicken pox virus can remain in the body and emerge in later adulthood as shingles, a viral inflammation. (See end-of-chapter Try Something New question 12.)

Figure 2.3

Q1: Before vaccinations, which diseases had the highest and lowest mortality rates?

A1: Before vaccinations, tetanus had the highest mortality rate (90 percent), and smallpox had the lowest (3.17 percent).

Q2: After vaccinations, which diseases had the highest and lowest mortality rates?

A2: After vaccinations, tetanus still had the highest mortality rate (14 percent). Polio, diphtheria, and smallpox all had the lowest (0 percent).

Q3: When you look at the all numbers and percentages in this figure, what is your takeaway?

A3: In every case, postvaccination numbers and percentages are lower than prevaccination numbers and percentages. Therefore, vaccines are helping eradicate these diseases, in some cases completely.

Figure 2.4

Q1: Which vaccine has the highest uptake rate (percentage of population vaccinated), based on the most recent year? Which has the lowest? What are the approximate uptake rates in each case?

A1: The highest uptake rate is for Tdap at 90 percent. The lowest is for HPV at 65 percent.

Q2: Why are there no data for HPV vaccine uptake before 2011?

A2: The ACIP did not recommend the vaccine for teens before 2011.

Q3: What do you predict the uptake rate will be for the HPV vaccine in the United States 10 years from now? Explain your reasoning.

A3: Possible answers include the prediction that the rate won't rise any further because parents see cancers related to HPV as a distant possibility in their children's future (compared with, for example, meningitis or pertussis). Or students may predict that the rate will rise as more parents and teenagers become aware of HPV, associated cancers, and the vaccine. In addition, the rate might rise as schools require students to be vaccinated.

Figure 2.6

Q1: Why do the manufacturers of vaccines begin with tests on animals or cell lines before moving on to adult human subjects?

A1: They need to be sure of the safety of the vaccine before exposing people to it.

Q2: What types of ongoing testing and reporting are vaccines subjected to?

A2: Manufacturers test all vaccine lots, the FDA regularly inspects manufacturing facilities, the ACIP and the CDC review all test results before approving vaccines, and vaccine safety is continually monitored through the VSD and the VAERS.

Q3: What do FDA, ACIP, and CDC stand for, and what is the role of each in evaluating vaccines?

A3: The FDA, or Food and Drug Administration, is involved in the initial approval of vaccines. The ACIP, or Advisory Committee on Immunization Practices, and the CDC, or Centers for Disease Control and Prevention, are involved in the ongoing monitoring of vaccines.

Figure 2.7

Q1: Why are we less confident of scientific claims made over social media?

A1: Claims made over social media are not subject to peer review before being "published." (See Chapter 1 for a refresher on peer-reviewed publications.) Therefore, people can and do make ridiculous scientific claims—and claims purporting to be scientific—in social media.

Q2: Where would you place a blog in this figure? Would it matter whether the blog was written by a practicing scientist? Explain your reasoning.

A2: You could list a blog under social media or secondary literature, depending on the quality of the content. In the case of a science blog written by a practicing scientist, the choice would depend on the subject matter, but it would likely be listed under secondary literature. (However, a physicist who thinks global climate change is a scientific conspiracy should make you suspicious of that scientist's writing.)

Q3: Give an example of when you would rely on secondary literature to evaluate a scientific claim and an example of when you would go to the primary literature. What is the basis of that decision?

A3: You might go to the secondary literature only for something that is not life-threatening—for example, what kind of exercise to do or whether to turn down the thermostat in your bedroom. Primary literature is challenging reading, so it is usually left for life-critical choices—for example, whether to vaccinate your children or possibly whether to eat a vegetarian or vegan diet.

Figure 2.9

Q1: Why do all the green arrows continue down to "real science," whereas each red arrow points directly to "pseudoscience"?

A1: All of the criteria listed in boxes 1–6 must be met for a study to meet the expectations of the scientific method and therefore be real science. By contrast, any single item in a red box indicates pseudoscience.

Q2: Return to the main text and read the rest of this section and the following section, "Real or Pseudo?" Then come back and answer this question: What part(s) of the figure show where Wakefield's study failed to meet the standards of the scientific method?

A2: Wakefield's study really falls apart in steps 4–6. It was not carefully designed and definitely not reproducible. The sample studied was not of sufficient size and did not include a random control group. The conclusions did not follow from the analysis of the experimental results, and the study was published without adequate review from scientists in the field.

Q3: What is the scientific claim behind the vaccine-autism controversy? What is an alternative scientific claim?

A3: The original scientific claim was that the MMR vaccine leads to the development of autism (specifically, some said, by harming a child's immune system). One alternative claim is that autism is caused by multiple factors that include a genetic predisposition but do *not* include vaccination.

Figure 2.10

Q1: How much did organic food sales grow during the period covered in the graph? How much did autism diagnoses grow?

A1: Organic food sales grew from $5 billion to $25 billion (a fivefold increase). Autism diagnoses grew from about 50,000 individuals to over 250,000 (also a fivefold increase).

Q2: Why might both organic food sales and autism diagnoses have increased during this time period? A Reddit user in the original discussion thread suggested that both might be affected by increasing wealth in the United States. How might an increase in wealth affect these variables?

A2: People with more disposable income are able to spend more on food (hence, the rise in organic food sales), and they are also better able to take their children in for advanced medical care (possibly the reason for increased identification of autism spectrum disorder).

Q3: How has the vaccine-autism debate been confused by people misinterpreting correlation as causation?

A3: Because vaccination was increasing at the same time that autism rates were rising (correlation), people suggested that the former caused the latter (causation). In addition, the time at which children are typically vaccinated is about the same age at which autism symptoms often appear.

Figure 2.11

Q1: Why is it important to know the education and expertise of a person making a scientific claim?

A1: The appropriate education and expertise are likely to give a person the knowledge necessary to make a valid claim. The opinion of a person who doesn't understand the science behind a claim is not a valid opinion.

Q2: List at least five biases that people making scientific claims might have.

A2: Possibilities include the biases that people could make money if you buy a product related to the claim, people might win a lawsuit if the judge believes the claim, people could become famous if others accept the claim, people's religious beliefs might be supported if the claim is true, and people's political beliefs might be supported if the claim is true.

Q3: Describe a situation in which you might *not* dismiss the scientific claim of a person who did not have appropriate credentials or who had a bias toward the claim.

A3: One possible scenario is that an independent researcher, someone without an advanced degree, creates a new vaccine. That vaccine is appropriately tested and adopted for use by the public, and its creator makes a lot of money by selling the product.

Figure 2.13

Q1: What happens to an immunized person when a disease spreads through a population? (*Hint*: In the graphic, follow an immunized individual before and after a disease spreads.)

A1: The immunized person does not, or is much less likely to, contract the disease.

Q2: How does vaccination of an individual help that person's community?

A2: Because an individual who is vaccinated is much less likely to become ill, that person is less likely to pass on a disease to others in the community.

Q3: Explain why a disease is less likely to spread to vulnerable members of a population if most people are immunized.

A3: "Herd immunity" means that fewer people contract the disease and, therefore, vulnerable people are less likely to be exposed to a contagious person.

CHAPTER 3

Figure 3.2

Q1: How many protons, neutrons, and electrons does the hydrogen atom shown here have? What are the atomic number and the atomic mass number of the hydrogen atom?

A1: As shown at the bottom of the figure, the hydrogen atom has 1 proton, 1 electron, and 0 neutrons. The atomic number (number of protons) and the atomic mass number (number of protons plus neutrons) are both 1.

Q2: What are the atomic number and the atomic mass number of the carbon isotope shown?

A2: The atomic number is 6 (6 protons). The atomic mass number is 12 (6 protons plus 6 neutrons).

Q3: Nitrogen-11 is an isotope of nitrogen that has 7 protons and 4 neutrons. What are the atomic number and atomic mass number of nitrogen-11?

A3: The atomic number is 7 (7 protons). The atomic mass number is 11 (7 protons plus 4 neutrons).

Figure 3.5

Q1: Which type of bond results in one positive ion and one negative ion?

A1: In ionic bonding, one atom becomes a positive ion and one atom becomes a negative ion.

Q2: Which type of bond is not a covalent bond but requires the atoms to be covalently bonded to another atom?
A2: In hydrogen bonding molecules must have a covalent bond between a partially positive hydrogen atom and a partially negative atom. Two such molecules are electrically attracted to each other: the partial positive charge of one molecule to the partial negative charge of the other.

Q3: What is a common feature of the valence shells of both atoms in covalent and ionic bonds?
A3: The common factor is electrons, which are shared by the valence shells in covalent bonds and exchanged by the valence shells in ionic bonds.

Figure 3.6

Q1: In methane gas, how many electrons is each hydrogen atom sharing? How many is the carbon atom sharing?
A1: Hydrogen is sharing one electron with carbon. Carbon is sharing four electrons, one with each of the four hydrogen atoms.

Q2: Carbon dioxide is not technically considered an organic compound, even though it contains a carbon atom. What essential atom is found in the organic compounds shown here that is not included in carbon dioxide?
A2: Hydrogen is not found in CO_2.

Q3: Draw a molecule of formaldehyde (CH_2O). How many electrons is the oxygen atom sharing with the carbon atom? How many is the carbon atom sharing with the oxygen atom and with each hydrogen atom?
A3: Oxygen is sharing two electrons with carbon. Carbon is sharing two electrons with oxygen and one electron with each of the hydrogens.

Figure 3.7

Q1: Which element has the most electrons in its valence shell? The least?
A1: Sulfur has the most electrons (6) in its valence shell. Hydrogen has the least (1).

Q2: Which elements have three electron shells? Which elements have two? Which element has only one?
A2: Phosphorus and sulfur have three electron shells. Carbon, nitrogen, and oxygen have two. Hydrogen has only one.

Q3: To fill the valence shell of each element, how many electrons are needed?
A3: Hydrogen needs two electrons to fill its valence shell. Carbon needs four. Nitrogen needs three. Oxygen needs two. Phosphorous needs three. Sulfur needs two.

Figure 3.8

Q1: Which monomers make up proteins, and how many are there?
A1: There are 20 amino acid monomers that make up proteins.

Q2: Describe the four levels of protein structure.
A2: Level 1: the sequence of amino acids; level 2: the chain of amino acids forms coils or sheets; level 3: the chain in coils and sheets folds further onto itself to form a functional 3D form; level 4: multiple 3D proteins group together for additional functionality.

Q3: Which of the function(s) listed in the caption do you think opioid receptors in brain cell membranes perform?
A3: Because opiates and opioids must bind to opioid receptors lodged in brain cell membranes to cause their responses of pain relief, suppressed breathing, and euphoria, the listed function of "transmitting signals through cell membranes" is their main function.

Figure 3.9

Q1: Which of the big six elements on Earth are found in carbohydrates?
A1: Carbons, hydrogens, and oxygens make up carbohydrates.

Q2: Which carbohydrate is used for energy storage in animal muscle tissue? Plant tissue?
A2: Animal muscle tissue uses glycogen for energy storage. Plant tissue uses starch.

Q3: Name two carbohydrates used as structural support.
A3: Possibilities include cellulose, chitin, and peptidoglycan.

Figure 3.10

Q1: Which nucleic acid is built of two strands of joined nucleotides?
A1: DNA consists of two strands of joined nucleotides.

Q2: How many different types of ribonucleotide monomers make up RNA molecules?
A2: RNA molecules consist of four types of ribonucleotide monomers.

Q3: Chromosomes are made up of which type of nucleotide monomers?
A3: Nucleotides are the monomers that make up DNA, which makes up chromosomes.

Figure 3.11

Q1: How many fatty acid chains attach to glycerol to form a triglyceride? How many are needed to form a phospholipid?
A1: There are three fatty acid chains in a triglyceride. There are two in a phospholipid.

Q2: What type of lipid is formed from hydrocarbon rings?
A2: Steroids are formed from hydrocarbon rings.

Q3: What type of molecules do chains of hydrocarbons form?
A3: They form fatty acids.

Figure 3.12

Q1: Where are the covalent bonds in this figure?

A1: There are two covalent bonds located at the electrons shared between the oxygen atom and each hydrogen atom.

Q2: This figure shows a water molecule. A hydrogen molecule (H_2) consists of two hydrogen nuclei that share two electrons. Draw a simple diagram of a hydrogen molecule indicating the positions of the two electrons.

A2: The diagram looks like this:

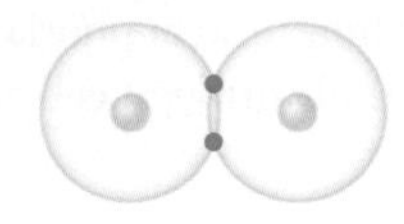

The electrons are equidistant from the two hydrogen nuclei because there is no difference in the strength of their attraction to the electrons.

Q3: When table salt (sodium chloride, NaCl) dissolves in water, it separates into a sodium ion (Na^+) and a chloride ion (Cl^-). Which portion of a water molecule would attract the Na^+ ion, and which portion would attract the Cl^- ion?

A3: The partial negative charge on the oxygen atom (O) would attract the Na^+ ion, and the partial positive charge on the hydrogen atoms (H) would attract the Cl^- ion.

Figure 3.13

Q1: Describe what will happen to the molecules of olive oil if you shake the bottle and then let it sit for an hour. What about the molecules of vinegar?

A1: When you shake the bottle, the olive oil molecules will not combine with the vinegar and water molecules. They will be broken up from their current form (into small beads of oil) and then will rejoin into a layer of oil at the top of the bottle; that is, they will return to their starting condition. Shaking the bottle will distribute the vinegar molecules through the water, and they will remain in solution.

Q2: What will happen if you add another fat to the bottle, such as warm bacon grease, and shake it?

A2: Like the olive oil molecules, the newly added fat molecules will break up (again into small beads) and then rejoin into a layer at the top of the bottle.

Q3: Given how sugar behaves when it is mixed into coffee or tea, would you predict that it is hydrophobic or hydrophilic?

A3: Sugar dissolves in coffee or tea; therefore, it is hydrophilic.

Figure 3.14

Q1: Identify where in the picture water can be seen in its liquid, solid, and gas states.

A1: The water in the hot springs is in a liquid state. The snow and ice are solid water. Water vapor (steam) in the air is a gas.

Q2: In the gas state, water molecules move too rapidly and are too far apart to form hydrogen bonds. Compare the volumes occupied by an equal number of water molecules in the liquid, solid, and gas states.

A2: The same number of molecules will occupy more volume in both ice (solid) and water vapor (gas) than in liquid water. The volume of ice is defined (9 percent greater than the volume in water) because the water molecules form an ordered array. The volume of water vapor depends on its temperature.

Q3: Explain in your own words how ice floats on water.

A3: Water molecules are farther apart in ice than they are in liquid water. Because the same number of water molecules occupies a greater volume in ice than in water, the density of ice is lower than that of water, and ice floats.

Figure 3.16

Q1: Medically administered opioids—namely, prescription drugs—are meant to target which pathway(s)?

A1: Medically administered opioids are meant to target pain relief but also cause suppressed breathing, euphoria, and addiction.

Q2: Most if not all opioid abusers were originally prescribed opioids for medical pain relief. These people were "hooked" into illicit drug abuse by which pathway(s)?

A2: Addictive euphoria hooked many opioid users, who subsequently became opioid abusers.

Q3: Opioid receptor proteins are lodged in cell membranes. Their centers interact with hydrophobic fatty acids of phospholipids, and their ends interact with the hydrophilic phosphate groups of phospholipids. Are opioid receptors hydrophilic or hydrophobic? Explain.

A3: Opioid receptors must be both hydrophilic to interact with the phosphate groups of phospholipids and hydrophobic to interact with fatty acids. When proteins have regions that are hydrophilic and other regions that are hydrophobic, like opioid receptors, they are called amphipathic proteins.

Figure 3.17

Q1: List one reason why carbon ring structures might be diagrammed without the hydrogens and carbons in the structures of complex molecules.

A1: Answers may vary, but all should mention simplifying the complex molecule. By assuming all corners are carbon atoms without labeling with a "C" and knowing that four covalent bonds must form to fill its valence shell (4), we know the number of covalently bonded H^+ are associated with each carbon atom without drawing them on the structure. This keeps the structure from becoming overly cluttered with H^+. In this way, only noncarbon atoms are written out as letters, keeping the structures as simple and uncluttered as possible.

Q2: Single covalent bonds are represented by one line between atoms, whereas double covalent bonds between atoms are represented by two lines. How many double covalent bonds are in the chemical structure of fentanyl? How many are in the chemical structure of NFEPP?

A2: Seven double covalent bonds are in the chemical structure of fentanyl. Seven double covalent bonds are in the chemical structure of NFEPP.

Q3: If NFEPP is approved and marketed as a safe and potent pain reliever, do you think fentanyl drug addictions and overdoses will decrease, increase, or stay the same? Why do you think so? What about oxycodone (OxyContin) addictions and overdoses?

A3: Answers will vary, and you could logically support any of the three possibilities. Because fentanyl is rarely prescribed as a pain medication now but is a popular street drug, you could argue that NFEPP approval will not change the use of fentanyl. You could then argue that oxycodone (OxyContin) use will decrease because NFEPP will be safer to use medically. Patients will not become addicted to NFEPP from a medical issue, and they will not fall into the cycle of abuse. You could also argue that NFEPP medical use will decrease fentanyl use for a similar reason. Patients given medically administered oxycodone (OxyContin) become addicted and then purchase fentanyl on the street to fulfill their addiction after their medically approved prescriptions expire. Breaking the cycle of addiction to medically administered opioids of any kind would then break the cycle of using street drugs.

Figure 3.18

Q1: Which has a higher concentration of free hydrogen (H^+) ions: vinegar with a pH of 2.8 or milk with a pH of 6.5?

A1: A *high* concentration of free H^+ ions corresponds to a *low* pH. Thus, vinegar has a higher concentration of free H^+ ions than milk has.

Q2: What happens to the concentration of free H^+ ions in your stomach when you drink milk?

A2: The concentration of free H^+ ions decreases (that is, the pH increases, as milk is more alkaline than stomach acid).

Q3: Black coffee has a pH of 5. Does adding coffee to water (pH of 7) increase or decrease the concentration of free H^+ ions in the liquid?

A3: The concentration of free H^+ ions increases.

CHAPTER 4

Figure 4.2

Q1: Why did the researchers insert the gene that codes for blue pigment into the synthetic DNA?

A1: The blue color identified the *M. capricolum* cells that contained the synthetic DNA from *M. mycoides*.

Q2: What part of the transformed bacterium was synthetic?

A2: Only the DNA was synthetic; all the structural components of the transformed cells are from the *M. capricolum* cells into which the synthetic DNA was inserted.

Q3: Do you think this experiment created life?

A3: This is an opinion question. Answers could support either side. Against: Although some articles in the popular press referred to the synthetic bacterium as a new life-form, it is better described as "repackaged life." The DNA was synthetic, but all the intracellular components that enabled the DNA to function were already present in the cell. For: With all newly synthesized DNA, this is a new life-form. All genetic information stored in the DNA will drive the existence and survival of this new life-form.

Figure 4.3

Q1: Why is it important that the phosphate head of a phospholipid is hydrophilic?

A1: The fact that the phosphate head is attracted to water (hydrophilic) and also to other phosphate heads means that a bilayer will form.

Q2: What genetic component of most organic cells do liposomes lack? What does that omission mean for liposomes?

A2: Liposomes lack genetic material (DNA). As a result, the characteristics of a liposome are not transmitted to its descendants.

Q3: Could the tendency of phospholipid bilayers to spontaneously form spheres have played a role in the origin of life? (*Hint*: Refer to "The Characteristics of Living Organisms" on page 67.)

A3: Yes. Once phospholipids formed (how that happened is still an open question), they would have formed liposomes spontaneously, trapping substances in their interiors.

Figure 4.4

Q1: How is the plasma membrane a barrier? How is it a gatekeeper?

A1: It is a barrier in that it keeps out many molecules. It is a gatekeeper in that it selectively allows in other molecules.

Q2: Why can't ions (such as Na^+) cross the plasma membrane without the help of a transport protein?

A2: Ions are hydrophilic (they are electrically charged [+] or [–]), so they cannot cross the lipid portion of the bilayer, which is hydrophobic, without the help of a transport protein.

Q3: If no energy were available to the cell, what forms of transport would not be able to occur? What forms of transport could occur?

A3: No form of active transport is possible without the input of energy. Any form of passive transport, including diffusion and osmosis, could occur even in the absence of an energy source.

Figure 4.5

Q1: Is the dye at equilibrium in any of these glasses? Describe how the first glass will look when the dye is at equilibrium with the water.

A1: Yes, the dye is at equilibrium in the third glass. If the first glass reaches equilibrium, it will look like the third glass.

Q2: Will diffusion mix the molecules of dye evenly through the water, or is it necessary to shake the container to get a uniform mixture?

A2: Diffusion is sufficient to mix the dye thoroughly, but it is a slow process and you would probably get tired of waiting.

Q3: Will diffusion mix the dye faster in hot water than in cold water? Why or why not? (*Hint*: Review the discussion of the behavior of water molecules at different temperatures in Chapter 3.)

A3: Diffusion is faster at higher temperatures because the water molecules have more energy, so they form and break hydrogen bonds at a higher rate and, hence, move around more rapidly.

Figure 4.6

Q1: What would the second diagram look like if the pores in the semipermeable membrane were too small to allow water molecules to pass through?

A1: The second diagram would look the same as the first diagram, as neither the sugar molecules nor the water molecules would be able to pass through the membrane.

Q2: What would the second diagram look like if the pores were large enough to let both water molecules and sugar molecules pass through?

A2: If the pores were large enough for sugar molecules to pass through, diffusion would cause the concentration of sugar molecules to decrease on the left side and increase on the right. At the same time, it would cause the concentration of water molecules to decrease on the right and increase on the left, moving down the concentration gradient of water. At equilibrium, the concentration of sugar and water would be the same on both sides of the membrane, and the depth of the solution would also be the same on both sides.

Q3: The fluid in an IV bag is isotonic to blood. What change would you see in the red blood cells of a patient if a bag of hypertonic solution was used in error?

A3: A hypertonic solution is less dilute than blood, so an IV of hypertonic solution would decrease the concentration of water in the blood. Now the red blood cells would have a higher concentration of water than the blood surrounding them, so water would move by osmosis out of the red blood cells, causing them to shrivel.

Figure 4.7

Q1: If endocytosis itself is nonspecific, how does receptor-mediated endocytosis bring only certain molecules into a cell?

A1: The receptor protein embedded in the plasma membrane attracts and holds only specific molecules. When they attach, the plasma membrane bulges inward to engulf the molecules.

Q2: What sorts of molecules could be moved by endocytosis or exocytosis but not by diffusion?

A2: By endocytosis or exocytosis, cells move molecules that are too large to pass through the plasma membrane by diffusion.

Q3: How does the fluid that enters a cell via pinocytosis differ from the fluid that enters by osmosis?

A3: The fluid that fills the inward bulge of a cell during pinocytosis may contain solutes, molecules that are in solution, whereas osmosis allows only water molecules to enter a cell.

Figure 4.8

Q1: What structures do prokaryotic and eukaryotic cells have in common?

A1: Both prokaryotic and eukaryotic cells contain genetic material (DNA), a plasma membrane, ribosomes, and cytoplasm.

Q2: What structures are unique to prokaryotic cells? To eukaryotic cells?

A2: Prokaryotic cells have cell walls, and some have flagella as shown here. Eukaryotic cells have all the organelles including a nucleus.

Q3: Both plants and animals are eukaryotes, but there are differences in their cellular structure. What are those differences?

A3: A plant cell has a cell wall, chloroplasts, and a central vacuole, but an animal cell does not.

Figure 4.9

Q1: What is the main function of the nucleus?

A1: It compartmentalizes the genetic material, the DNA, of the cell.

Q2: What are nuclear pores, and what do they do?

A2: Nuclear pores are openings in the nuclear envelope that control passage of molecules into and out of the nucleus.

Q3: If prokaryotic cells do not have a nucleus, where do they keep their genetic material?

A3: Prokaryotic genetic material is found in the cytoplasm.

Figure 4.10

Q1: What are the functions of the smooth ER and the rough ER?

A1: The smooth ER manufactures lipids and hormones. The rough ER manufactures proteins destined for membranes of other organelles or for the plasma membrane.

Q2: Where is the ER located in the cell?

A2: The ER is contiguous with the nuclear membrane.

Q3: If prokaryotic cells do not have an ER, where do you think they perform the functions mentioned in Q1?

A3: All of these functions occur in the cytoplasm of prokaryotic cells.

Figure 4.11

Q1: What is the function of the Golgi apparatus?

A1: The Golgi apparatus packages and directs proteins and lipids produced by the ER to other membranes of cellular organelles or to the plasma membrane of the cell.

Q2: Where is the Golgi apparatus located in the cell? What is the significance of that location?
A2: The Golgi apparatus is found adjacent to the rough and smooth ER. It must be next to the ER because vesicles budding from the ER fuse with its membranes.

Q3: Why do you think prokaryotic cells can survive without a Golgi apparatus? (*Hint*: Eukaryotic cells are very large, and prokaryotic cells are very small.)
A3: All processes occur in the plasma membrane and the cytoplasm, and because prokaryotic cells are very small, they can rely on diffusion to move molecules around, into, and out of the cell.

Figure 4.12

Q1: What are the functions of lysosomes? What is the main function of vacuoles?
A1: Lysosomes break down biomolecules and organelles for recycling or disposal. Among the many functions vacuoles perform, the main function is water storage and water balance in plant cells.

Q2: Which types of cells have lysosomes? Which types have vacuoles?
A2: Both plant cells and animal cells have lysosomes. Only plant cells, fungi, and protists have some lysosomes that we call vacuoles.

Q3: Lysosomes and vacuoles were once thought to have analogous functions in animal and plant cells, respectively. Now that you have learned about both of these types of organelles, explain why they are not analogous structures.
A3: Analogous structures perform the same functions in different cells. Lysosomes and vacuoles do not perform the same functions.

Figure 4.13

Q1: What is the main function of mitochondria? Of chloroplasts?
A1: Mitochondria generate cellular energy, ATP, from the chemical energy of biomolecules. Chloroplasts generate chemical energy, sugar, from the solar energy of the sun.

Q2: Which types of cells have mitochondria? Which types have chloroplasts?
A2: Animal cells and plant cells have mitochondria. Only plant cells and some protists have chloroplasts.

Q3: According to the endosymbiotic theory, mitochondria and chloroplasts were once free-living single-celled organisms that were absorbed by another single-celled organism. Over the millennia, they evolved into the organelles of today. Before they were absorbed, what processes must these two single-celled organisms have been able to perform on their own?
A3: The premitochondria must have been able to convert sugar into cellular energy, ATP. The prechloroplast must have been able to convert solar energy into chemical energy, sugar.

Figure 4.14

Q1: What is the main function of the cytoskeleton?
A1: The main function of the cytoskeleton, a network of protein cylinders and filaments, is to structurally support the cell.

Q2: Which types of cells have a cytoskeleton?
A2: Both plant cells and animal cells have a cytoskeleton, as do all other eukaryotes.

Q3: Postulate: The cytoskeleton of a cell is similar in function to the skeleton of a human. Give an example in support and one in rebuttal of this statement.
A3: Support: Both the cytoskeleton and the skeleton provide structural support. Rebuttal: Only the plant cell wall gives a similar amount of support to the plant cell as the skeleton does to the human body.

CHAPTER 5

Figure 5.2

Q1: Why is the pink line showing cellular respiration originating at both the giraffe and the tree? (*Hint*: Before answering this question and the following ones for this figure, read the next paragraph in the text.)
A1: The pink line indicates that both animals and plants perform cellular respiration to generate ATP for cellular work.

Q2: How does animal life depend on photosynthesis?
A2: Only photosynthetic organisms convert sunlight into chemical energy through photosynthesis. Animals cannot carry out photosynthesis. Instead, they eat plants or other animals and get their energy from that food. (There are a very few exceptions to these generalizations.)

Q3: Explain how photosynthesis and cellular respiration are "complementary" processes.
A3: They use each other's products and reactants.

Figure 5.3

Q1: What source of energy would plants use for anabolic reactions? Would an animal use the same kind of energy?
A1: Both plants and animals use chemical energy for anabolic reactions. Plants generate chemical energy through photosynthesis, while animals must consume chemical energy from food.

Q2: What source of energy would plants release in catabolic reactions? Would an animal release the same kind of energy? (*Hint*: To answer this question, read the rest of this paragraph and the next one in the text.)
A2: All living things release energy carriers, usually ATP, during catabolic reactions.

Q3: What would happen to a plant if catabolism stopped in all its cells? If catabolism stopped in all of an animal's cells, would the same result occur? (*Hint*: To answer this question, read the rest of this paragraph and the next one in the text.)

A3: Because catabolism provides energy to all living things, the lack of catabolism would cause the plant cells and the plant to die. Yes, the same thing would happen to an animal.

Figure 5.4

Q1: Define ATP in your own words.

A1: You might say something like this: "ATP is a molecule that cells use as their energy to do work. It is the universal energy carrier for cells."

Q2: When ATP is transformed to ADP, what two things are released in the process?

A2: Energy and a free phosphate group are released.

Q3: Arsenic disrupts the production of ATP from ADP. Why would this characteristic cause arsenic to be a potent poison?

A3: If ATP stops being produced, then cells lack the energy to function, and the organism quickly dies.

Figure 5.6

Q1: How does embedding chlorophyll in the membranes of thylakoids versus the membrane of the chloroplast increase the amount of chlorophyll each chloroplast can accommodate?

A1: The thylakoids provide increased membrane surface area for embedding chlorophyll.

Q2: Do all photosynthetic organisms have their chlorophyll embedded in the membranes of thylakoids?

A2: No, bacteria have chlorophyll embedded in their plasma membranes.

Q3: What would cells with higher numbers of chloroplasts be better at absorbing?

A3: They would be better at absorbing light energy through photosynthesis.

Figure 5.7

Q1: What are the three reactants (types of input) needed for photosynthesis to occur?

A1: The three types of input needed are sunlight, water (H_2O), and carbon dioxide (CO_2) from the air.

Q2: Which products of the light-dependent reactions of photosynthesis do the light-independent reactions use?

A2: They use ATP and NADPH.

Q3: What are the two major overall products (types of output) of photosynthesis?

A3: The light-dependent reactions produce oxygen (O_2), and the light-independent reactions produce glucose.

Figure 5.8

Q1: Why is it important that enzymes are not permanently altered when they bind with substrate molecules?

A1: The fact that they are not permanently altered means that they can continue to perform their function. This recycling of enzymes means they do not have to exist in great numbers or be continually produced.

Q2: If the active site of an enzyme were blocked, how would the production of a product be affected?

A2: If the active site were blocked, no induced fit would occur, substrate molecules would not bind, and no product would be produced.

Q3: If a cell could not produce more of a particular enzyme and a poison blocked this enzyme from resuming its original shape after an induced fit, what would happen to the production of that enzyme's product?

A3: These problems would bring the metabolic pathway to a complete halt, so the cell would not make that product.

Figure 5.11

Q1: What are the end products of cellular respiration after all three steps are complete?

A1: The end products are ATP, carbon dioxide, and water.

Q2: Considering the inputs and products of each process, why is cellular respiration considered the reciprocal process to photosynthesis?

A2: Photosynthesis uses carbon dioxide and produces oxygen, whereas cellular respiration does the opposite, using oxygen and producing carbon dioxide.

Q3: Which of the three stages of cellular respiration—namely, glycolysis, the Krebs cycle, or oxidative phosphorylation—could organisms have used 4 billion years ago before photosynthesis by cyanobacteria released oxygen into the atmosphere?

A3: Organisms could have used glycolysis because it does not depend on oxygen. As the text says, "Glycolysis was probably the earliest means of producing ATP from food molecules. It is still the primary means of energy production in many prokaryotes."

Figure 5.12

Q1: Which product released by fermentation accounts for the bubbles in beer?

A1: Carbon dioxide (CO_2), released as a gas, produces the bubbles. The other product, ethanol, is a liquid at room temperature.

Q2: Bread makers rely on yeast to cause their dough to rise before baking. What must be included in the dough for fermentation to occur?

A2: The yeast needs glucose (or another carbohydrate) to perform glycolysis and subsequently fermentation.

Q3: Explain in your own words why lactic acid builds up in your muscles during strenuous physical activity.
A3: You might say something like this: "During strenuous exercise, my muscles can't get enough oxygen to produce all the needed ATP. Glycolysis produces ATP anaerobically, which generates lactic acid as a by-product."

CHAPTER 6

Figure 6.2

Q1: When is DNA replicated during the cell cycle?
A1: DNA is replicated during interphase.

Q2: When in the cell cycle does the cell separate into two genetically identical daughter cells?
A2: The cell separates into daughter cells during cell division.

Q3: If a cell does not complete the cell cycle, what phase does it enter? Is this part of the cell cycle?
A3: It enters the G_0 phase, which is not part of the cell cycle.

Figure 6.3

Q1: When single-celled organisms go through cell division, why is it called asexual reproduction?
A1: Both single-celled eukaryotes and prokaryotes go through cell division that results in genetically identical offspring. By definition, asexual reproduction is the process of creating offspring that are genetically identical to the parent.

Q2: What is asexual reproduction referred to specifically in prokaryotes?
A2: Prokaryotic asexual reproduction is called binary fission.

Q3: Although prokaryotes do not have an "interphase" during cell division like eukaryotes do, they must perform one of the main processes of eukaryotic interphase. What is this process in prokaryotes that must occur before the cell physically divides in two?
A3: Prokaryotes must replicate their one loop of DNA before physically dividing into two cells.

Figure 6.4

Q1: The image at far left in the figure is a cross section of an onion root tip stained to display the chromosomes. From the arrangement of the chromosomes, it is clear that all of these cells are not in the same stage of the cell cycle. What stages do you see, and how do you know?
A1: You see two cells in metaphase, with chromosomes aligned in the middle of the cell; a cell in anaphase, with chromosomes pulling apart toward opposite sides of the cell; several cells in prophase, with condensed chromosomes not yet aligned; and several cells in interphase, with chromsomes not yet condensed.

Q2: What happens between the end of interphase and early prophase that changes the appearance of the chromosomes?
A2: The DNA within the chromosomes is condensed to prepare for division.

Q3: Explain in your own words the role of the mitotic spindle in mitosis.
A3: The mitotic spindle aligns the replicated, condensed chromosomes and then separates these chromatids into daughter cells.

Figure 6.5

Q1: Why is it important for a chromosome to be copied before mitosis?
A1: There need to be two copies of each chromosome, so that each daughter cell has identical genetic material.

Q2: Are sister chromatids attached at the centromere considered to be one or two chromosomes?
A2: Sister chromatids are, when attached, one chromosome. When split apart, they are two separate chromosomes.

Q3: Why is the chromosome's DNA tightly packed and condensed with proteins for mitosis and cytokinesis instead of remaining exceptionally long and loose?
A3: Tightly packed chromosomes are more easily aligned in metaphase and separated during anaphase into the two sides of the cell. Without being tightly packed—if they were unpacked as they are during interphase—the chromatin fibers would become entangled and be impossible to align and separate without breakage.

Figure 6.6

Q1: What could happen if the cell's checkpoints are disabled?
A1: The cell cycle could occur more quickly than normal. That is, the cycle might not pause to let the regulatory proteins ensure that the checkpoint conditions are met. If checkpoint conditions are not met and the cell cycle continues, DNA damage leading to cancer could occur.

Q2: What is the advantage of stopping the cell cycle if the cell's DNA is damaged?
A2: If the cell's DNA is damaged, the daughter cells will inherit damaged DNA. By stopping the cell cycle, the parent cell can prevent parts of the organism from developing abnormally and from developing cancer.

Q3: Which cell cycle checkpoint may have been influenced in Soto and Sonnenschein's breast cancer cell experiments?
A3: If a checkpoint had been influenced, it must have been G_1, as the cell cycle was abnormally committed to begin.

Figure 6.7

Q1: Is a zygote haploid or diploid?
A1: The zygote is diploid.

Q2: What cellular process creates a baby from a zygote?
A2: A zygote develops into a baby through mitosis.

Q3: How might long-term, significant exposure to BPA experienced by a mother or father prior to conceiving a child explain potential birth defects in the fetus?

A3: If BPA exposure damaged the parent's sex cells or the genetic material within the cells (eggs or sperm), it could lead to a birth defect in the fetus.

Figure 6.8

Q1: Is a daughter cell haploid or diploid after the first meiotic division? How about after the second meiotic division?

A1: After meiosis I, a daughter cell is haploid (has one of each homologous chromosome). After meiosis II, each daughter cell is still haploid (has one of each homologous chromosome), and sister chromatids have split into separate daughter cells.

Q2: What is the difference between homologous chromosomes and sister chromatids?

A2: A homologous chromosome pair consists of the two copies—namely, one maternally derived from the egg and the other paternally derived from the sperm—of the same type of chromosome. The pair is present at all times in diploid cells, but a haploid cell has just one or the other copy. Each homologous chromosome contains two sister chromatids. Sister chromatids are identical DNA molecules, replicated from the single DNA molecule, that remain bound to each other. They exist in a cell only from the S phase until anaphase of mitosis or anaphase II of meiosis.

Q3: If the skin cells of house cats contain 19 homologous pairs of chromosomes, how many chromosomes are present in the egg cells they produce?

A3: The egg cells contain 19 chromosomes.

Figure 6.9

Q1: Why is the term "crossing-over" appropriate for the exchange of DNA segments between homologous chromosomes?

A1: Segments of DNA physically "cross over" between homologous chromosomes.

Q2: At what stage of meiosis (I or II) does crossing-over occur?

A2: Crossing-over occurs during meiosis I.

Q3: What would be the effect of crossing-over between two sister chromatids?

A3: There would be no effect because sister chromatids are genetically identical.

Figure 6.10

Q1: During meiosis, does independent assortment occur before or after crossing-over?

A1: Independent assortment occurs after crossing-over.

Q2: What would be the effect on genetic diversity if homologous chromosomes did not randomly separate into the daughter cells during meiosis?

A2: Genetic diversity would decrease.

Q3: With two pairs of homologous chromosomes, four kinds of gametes can be produced (2^2). In the same way, a mosquito (*Aedes aegypti*) with three pairs of homologous chromosomes can produce eight kinds of gametes (2^3). How many kinds of gametes can a spinach plant produce with its six pairs of homologous chromosomes? How many kinds of gametes can be produced with the 23 homologous pairs of chromosomes in human cells?

A3: A spinach plant can produce 64 kinds of gametes (2^6). Therefore, 23 homologous pairs could produce huge variation in gametes. To be more precise, they could produce 2^{23} (8,388,608) different combinations of chromosomes in gametes.

CHAPTER 7

Figure 7.2

Q1: How are chromosomes related to DNA and genes?

A1: Chromosomes are made of DNA and proteins, and a gene is a segment of DNA that contributes to an organism's phenotype or function.

Q2: How many copies of each gene are found in the diploid cells in a woman's body? (*Hint*: See Chapter 6 for a refresher on diploid versus haploid cells.)

A2: Each diploid cell has two copies of each gene.

Q3: There are 46 chromosomes in a human diploid cell; how many chromosomes come from the person's mother and how many from the father?

A3: Each parent contributes half of the chromosomes, so 23 chromosomes come from the mother and 23 come from the father.

Figure 7.3

Q1: Which might you observe directly: the genotype or the phenotype?

A1: You might observe the phenotype directly because it is an observable physical characteristic of an individual. To observe the genotype, you would need to be able to see the DNA.

Q2: Which poodle could be heterozygous: the one with the black coat or the one with the brown coat?

A2: Only the black-coated poodle may be heterozygous.

Q3: Can you identify with certainty the genotype of a black poodle? A brown poodle? In each case, why?

A3: No, you can't identify the genotype of a black poodle, because the *B* allele is dominant. Yes, you can identify the genotype of a brown poodle, because the *b* allele is recessive.

Figure 7.4

Q1: What would you predict about the color of the F_1 plants' flowers?

A1: The flowers would be purple rather than white because the darker color would be dominant over the lighter one.

Q2: Why was it important for Mendel to begin with pea plants that he knew bred true for flower color? Why couldn't he simply cross a purple-flowered plant and a white-flowered plant?

A2: The purple-flowered plant might be heterozygous *or* homozygous, so the results of the cross-breeding would not be predictable.

Q3: Over the years, Mendel experimented with more than 30,000 pea plants. Why did he collect data on so many plants? Why didn't he study just one cross? (*Hint*: Read "What Are the Odds?" on page 136 before answering.)

A3: With data from more plants, there was a better chance that Mendel's results would accurately reflect reality.

Figure 7.5

Q1: Why did Mendel's entire F_1 generation look the same?

A1: All of the F_1 plants were heterozygous (*Pp*), so the purple allele was dominant and they all had a purple phenotype.

Q2: The phenotypic ratio in the F_2 generation is 3:1 purple-to-white flowers. What is the genotypic ratio?

A2: The genotypic ratio is 1:2:1 (for *PP* to *Pp* to *pp*).

Q3: Draw a Punnett square for a genetic cross of two heterozygous, black-coated dogs. What are the phenotypic and genotypic ratios of their offspring?

A3: Phenotypic ratio is 3:1 black to brown. Genotypic ratio is 1:2:1 *BB* to *Bb* to *bb*.

Figure 7.6

Q1: List all the possible offspring phenotypes and genotypes.

A1: The phenotypes and genotypes are round, yellow (dominant dominant): *RRYY, RrYY, RRYy, RrYy* ; round, green (dominant recessive): *RRyy, Rryy* ; wrinkled, yellow (recessive dominant): *rrYY, rrYy* ; and wrinkled, green (recessive recessive): *rryy*.

Q2: What is the offspring phenotypic ratio?

A2: The phenotypic ratio is 9:3:3:1.

Q3: Complete a Punnett square for a genetic cross of two true-breeding Portuguese water dogs—one with a black, wavy coat (homozygous dominant, *BBWW*) and one with a brown, curly coat (homozygous recessive, *bbww*). What is the phenotypic ratio of their offspring (F_1)? Now fill out another Punnett square, crossing two of the offspring. What is the phenotypic ratio of the F_2 generation?

A3: F_1 generation phenotypic ratio is 3:1 black wavy to brown curly. F_2 generation phenotypic ratio is 9:3:3:1 black wavy to black curly to brown wavy to brown curly.

Figure 7.7

Q1: Boxers are far more inbred than poodles. Why does that inbreeding make boxers a better target for genetic studies of disease than poodles are?

A1: Inbreeding leads to less genetic variation. Because they are more inbred, boxers are more likely than poodles to be homozygous for traits of interest.

Q2: Explain why a geneticist interested in finding a gene linked to cancer would want to look at the DNA of senior golden retrievers with *and* without cancer.

A2: To identify the genetic difference between goldens with and without cancer, the geneticist would need to compare the DNA of dogs that are cancer-free with that of dogs with cancer.

Q3: Obsessive-compulsive disorder (OCD) in humans is characterized by obsessive thoughts and compulsive behavior, such as pacing. Canine compulsive disorder (CCD) is characterized by compulsive behavior such as "flank sucking," sometimes seen in Doberman pinschers. Would you predict that the medications given to humans with OCD would decrease compulsive behaviors in CCD dogs? Why or why not?

A3: You might predict that the medications would have that effect, as dogs and humans share many genes and, therefore, OCD and CCD may share a common genetic basis and be treatable by the same means.

Figure 7.8

Q1: What are the genotypes of a large dog and a small dog?

A1: The genotypes are *II* for large and *BB* for small.

Q2: Is it possible to have a heterozygous large dog? Explain why or why not.

A2: No, a heterozygous large dog is not possible, because *IB* results in a medium-sized dog.

Q3: Crossing a Great Dane and a Chihuahua is likely to be unsuccessful, even though they are members of the same species (and thus have compatible sperm and egg). Why is that? What are some potential risks of such a cross?

A3: Although conception is possible, it would be difficult for a Great Dane and a Chihuahua to mate because of the size difference. One risk would be that if a female Chihuahua were to become pregnant this way, the pups would be too large for her to carry safely to term.

Figure 7.9

Q1: What are the possible genotypes of the black dog? The brown dog? The yellow dog?

A1: For the black dog, the possible genotypes are *BBEE, BbEE, BBEe, BbEe*. For the brown dog: *bbEe, bbEE*. For the yellow dog: *BBee, Bbee, bbee*.

Q2: Draw a Punnett square showing possible matings between the black dog (assuming it is heterozygous at both genes) and the

yellow dog (assuming it is heterozygous at the *B* gene). List all the possible phenotypes of their offspring. (See Figure 7.6 for an example of a Punnett square made with two traits.)
A2: The three possible offspring phenotypes are black, yellow, and brown.

Q3: If you wanted the most variable litter possible, what colors of Labrador retrievers would you cross? Assume that your Labradors are true-breeding for color.
A3: If they were true-breeding (homozygous) for color, you would cross yellow and brown. The resulting offspring could be black, brown, or yellow. Because the *B* allele is dominant, any dog bred from a black dog could be black or yellow but not brown.

Figure 7.10

Q1: The gene that brings about the pale Siamese body fur is also partly responsible for the typical blue eyes of the species. What is the term for this type of inheritance?
A1: This type of inheritance is called pleiotropy.

Q2: Siamese kittens that weigh more tend to have darker fur on their bodies. Why might this be?
A2: Larger kittens may have lower core temperatures, which allow more melanin to be produced in their bodies.

Q3: The Siamese cat pictured is called a "seal point" because it has seal-colored (dark brown) extremities. Some Siamese cats show the same color pattern, but the dark areas are of a lighter color or even a different shade—for example, lilac point, red point, blue point. What results would you predict if the tests described in the text (shaving the cat and then increasing or decreasing temperature) were performed on cats with these color patterns? Why?
A3: The results should be the same. Even though the fur color has changed, the mechanism for amount of melanin produced (based on temperature) would not change.

CHAPTER 8

Figure 8.4

Q1: Is this illustration the karyotype of a male or a female?
A1: It is the karyotype of a male, because its sex chromosomes are XY, not XX.

Q2: How would the karyotype of a person with Down syndrome differ from this karyotype?
A2: The karyotype of a person with Down syndrome would have three copies of chromosome 21.

Q3: The size of a chromosome correlates roughly with the number of genes residing on it. Why are an extra copy of chromosome 21 and a missing Y chromosome two of the least damaging chromosomal abnormalities?
A3: Because chromosome 21 and the Y chromosome are two of the smallest chromosomes, fewer genes would be affected by there being too many or too few of them (compared to a larger chromosome).

Figure 8.5

Q1: Why are changes in chromosome *number* almost always more severe than changes in chromosome structure?
A1: There are so many genes on an individual chromosome that deleting or adding an entire chromosome has massive effects on an individual.

Q2: In which part of meiosis would you predict that chromosomal abnormalities are produced? (Refer back to Chapter 6 if necessary.)
A2: Chromosomal abnormalities are produced during metaphase.

Q3: Create a mnemonic (a code) to help remember the four kinds of structural changes to chromosomes.
A3: Answers will vary. One possibility for recalling deletion, inversion, translocation, and duplication is "doctors improve treatment daily."

Figure 8.6

Q1: How do we know whether two chromosomes are homologous?
A1: If the chromosomes carry alleles of the same genes and align during cell division, they are homologous.

Q2: In one sentence, explain how the terms "gene," "locus," and "chromosome" are related.
A2: A gene is found at a particular location—a locus—on a chromosome.

Q3: If hair color were determined by a single gene for brown or blonde hair, what would be an example of the gene's alleles?
A3: An example is *B* for brown hair, *b* for blonde hair.

Figure 8.7

Q1: Which chromosome contains the gene for cystic fibrosis? For Tay-Sachs disease? For sickle cell disease?
A1: The gene for cystic fibrosis is on chromosome 7; for Tay-Sachs, on chromosome 15; for sickle cell, on chromosome 11.

Q2: No known genetic disorders are encoded on the Y chromosome. Why do you think this is?
A2: The Y chromosome has very few genes; further, any disorder would always be expressed and therefore selected against.

Q3: In your own words, explain why most single-gene disorders are recessive rather than dominant.
A3: Dominant, single-gene disorders experience heavier selection than recessive disorders because they are always expressed (that is, there are no carriers).

Figure 8.8

Q1: Which of the children in this Punnett square represents Arabella? What is her genotype?

A1: Arabella is the "affected child" with genotype *aa*.

Q2: If Arabella's parents had another child, what is the probability that the child would have SMA1? What is the probability that the child would be a carrier of SMA1?

A2: The child would have a 25 percent (1/4) probability of having SMA1 and a 50 percent (1/2) probability of being a carrier.

Q3: If Arabella is able to have a child of her own someday, and the other parent is not a carrier of SMA1 (he would likely be tested before they chose to have children; see "Prenatal Genetic Screening" on page 157), what is the probability that the child would have SMA1? What is the probability that the child would be a carrier?

A3: The child would have a 0 percent probability of having SMA1 and a 100 percent probability of being a carrier.

Figure 8.9

Q1: What is the probability that a child with one parent who has an autosomal dominant disorder will inherit the disease?

A1: The probability is 50 percent.

Q2: Why are there no carriers with a dominant genetic disorder?

A2: Anyone with the gene would express the disorder.

Q3: Because dominant genetic disorders are rare, it is extremely rare for both parents to have the condition (genotype *Aa*). Draw a Punnett square with two *Aa* parents. What proportion of the offspring would have the disorder? What proportion would not have it?

A3: The proportion of the offspring that would have the disorder is 75 percent, whereas 25 percent would not have the disorder.

Figure 8.11

Q1: What are the odds that a given egg cell will contain an X chromosome? A Y chromosome? What are those odds for a sperm cell?

A1: An egg cell has a 100 percent chance of containing an X chromosome and a 0 percent chance of containing a Y chromosome. For the sperm cell, the odds are 50/50 for both cases.

Q2: If a couple has two daughters, does that mean their next two children are more likely to be sons? Explain your reasoning. (*Hint:* Refer back to "What Are the Odds?" on page 136.)

A2: No, the probability for each event is independent of prior and future events.

Q3: Sisters share the same X chromosome inherited from their father, but they may inherit different X chromosomes from their mother. What is the probability that brothers share the same Y chromosome? What is the probability that brothers share the same X chromosome?

A3: Brothers have a 100 percent probability of sharing the same Y chromosome. They have a 50 percent probability of sharing the same X chromosome.

Figure 8.12

Q1: Which of the children specified in this Punnett square represents Samuel? What is his genotype?

A1: Samuel is the "affected son," with genotype X^aY.

Q2: Explain why Samuel is neither homozygous nor heterozygous for the *XSCID* gene.

A2: The *XSCID* gene is found on the X chromosome, and Samuel carries only one X chromosome because he is a male. Homozygous and heterozygous genotypes are possible only when two chromosomes are present.

Q3: Create a Punnett square to illustrate the offspring that could result if Samuel had children with a noncarrier woman. What is the probability that a son would have XSCID? What is the probability that a daughter would be a carrier of XSCID?

A3: A son, with the genotype $X^A Y$, would have a 0 percent probability of having XSCID. A daughter, with the genotype $X^A X^a$, would have a 100 percent probability of being a carrier.

Figure 8.13

Q1: Which two children in this pedigree have sickle cell disease? How do you know?

A1: Individuals III-2 and III-3 have sickle cell disease. They are depicted by filled (here, red) symbols.

Q2: Does either parent have sickle cell disease? If so, which one(s)? How do you know?

A2: Neither parent, II-1 nor II-2, has sickle cell disease. They are both depicted by open (here, blue) symbols.

Q3: Do any of the grandparents have sickle cell disease? If so, which one(s)? How do you know?

A3: Yes. I-2, the paternal grandmother, has sickle cell disease. She is depicted by a filled (red) symbol.

Figure 8.14

Q1: How many male and how many female descendants (individuals who did not join the family by marriage) does generation IV of the pedigree contain?

A1: Generation IV includes 8 male descendants and 4 female descendants.

Q2: What proportions of the male and female descendants in generation IV were affected by the disorder?

A2: In generation IV, 50 percent (4/8, or 0.50) of males and 0% (0/4, or 0.00) of females were affected.

Q3: Why would a geneticist hypothesize that the disease was X-linked?

A3: The disease was observed only in males, but their mothers' male relatives were also affected.

CHAPTER 9

Figure 9.3

Q1: Name two base pairs.

A1: Two base pairs are adenine-thymine (A-T) and cytosine-guanine (C-G).

Q2: Why is the DNA structure referred to as a "ladder"? What part of the DNA represents the rungs of the ladder? What part represents the sides?

A2: The structure looks like a ladder. The rungs are base pairs, and the sides are alternating sugar and phosphate molecules.

Q3: Is the hydrogen bond that holds the base pairs together a strong or weak chemical bond? Why is that important?

A3: It is a weak chemical bond, which allows the two strands of DNA to separate for transcription and replication.

Figure 9.5

Q1: If all genes are composed of just four nucleotides, how can different genes carry different types of information?

A1: Each gene is composed of different numbers of nucleotides in different arrangements, thus allowing for many different types of information to be conveyed.

Q2: Would you expect to see more variation in the sequence of DNA bases between two members of the same species (such as humans) or between two individuals of different species (for example, humans and pigs)? Explain your reasoning.

A2: More variation would be expected between two individuals of different species. Within a species, individuals have the same genes but different alleles; across species, there are different genes *and* different alleles.

Q3: Do different alleles of a gene have the same DNA sequence or different DNA sequences?

A3: They have different DNA sequences, causing the alleles to encode different versions of a protein.

Figure 9.6

Q1: What common mechanism is employed by the guide RNA to find its target DNA sequence?

A1: The guide RNA uses complementary base-pairing between itself and one strand of the target DNA.

Q2: How many strands of DNA must Cas9 cut to be effective?

A2: Both strands must be cut.

Q3: Does Cas9 also cause the deletion of DNA from the genome?

A3: No. Normal DNA repair proteins cause the deletion of DNA from the genome.

Figure 9.7

Q1: What structures result from the first level of DNA coiling around proteins?

A1: Nucleosomes result from that initial coiling.

Q2: What makes up a "bead" and what makes up a "string" in the beads-on-a-string structure of DNA?

A2: The beads are the nucleosomes; the string is the double-stranded DNA linking them together.

Q3: What structure is more compact than the beads-on-a-string structure but less compact than an actual chromosome?

A3: Chromatin fiber fits that description.

Figure 9.8

Q1: Where do the DNA template strands come from? Why are they called "template" strands?

A1: The template strands come from the original double helix. They are so called because new DNA is built using the information from them as a template.

Q2: What must be broken before replication can begin?

A2: Hydrogen bonds between base pairs must first be broken.

Q3: In your own words, explain why replication is described as "semiconservative."

A3: It is semiconservative because each new double helix of DNA is composed of one "conserved" strand from the original DNA molecule and one new strand.

Figure 9.9

Q1: PCR replicates DNA many times to increase the amount available for analysis. Why is this process called "amplification"?

A1: This replication is called amplification because PCR substantially increases, or "amplifies," the quantity of DNA.

Q2: During the PCR cycle, what causes the DNA strands to separate?

A2: Heat causes them to separate, by breaking the hydrogen bonds between the base pairs.

Q3: What is one major difference between PCR and DNA replication? (*Hint*: Think about what is being copied.)

A3: PCR amplifies a specific targeted region of DNA, whereas DNA replication duplicates an entire chromosome.

Figure 9.10

Q1: Summarize how DNA repair works and why the repair mechanisms are essential for the normal functioning of cells and of whole organisms.

A1: Once an error is detected and tagged, the damaged section of DNA is removed and replaced. Without DNA repair, mutations would persist and potentially result in the death of cells and of organisms.

Q2: Is DNA repair 100 percent effective?

A2: No, some mismatch errors can remain.

Q3: What would happen to an organism if its DNA repair became less effective?

A3: Its cells would have more trouble operating properly as the DNA's genetic instructions became less accurate.

Figure 9.11

Q1: What are the three types of point mutations? Which one of these causes sickle cell disease?

A1: The three types of point mutations are substitution, insertion, and deletion. In sickle cell disease, one nucleotide in hemoglobin DNA is substituted for another.

Q2: Sickle cell disease is an autosomal recessive genetic disorder. How many mutated hemoglobin alleles do people with sickle cell disease have?

A2: People with sickle cell disease have two mutated hemoglobin alleles.

Q3: In previous decades, sickle cell disease was often fatal in childhood. Because of improved treatments, individuals with sickle cell disease are now living into their forties, fifties, or longer. How might this extension of life span affect the prevalence of sickle cell disease in the population?

A3: Sickle cell disease will increase because people will be able to survive long enough to reproduce and pass on the sickle cell allele.

Figure 9.13

Q1: What step in this process is similar to one in the original CRISPR method that removed the PERVs from the pig genome? (*Hint:* Review Figure 9.6.)

A1: Here, a kidney development gene is removed from the genome. In the original CRISPR method, a *pol* gene was removed from the genome.

Q2: What step in this process was not included in the original CRISPR method that removed the PERVs from the pig genome?

A2: Here, human stem cells are added to the pig embryo.

Q3: What parts of this process would scientists need to change in order to develop several different human organs in a single pig?

A3: Scientists would need to use CRISPR to target only the genes that are required for the development of each organ in the pig. Those organs would then be replaced by human organs via use of the same human stem cells added to create a single human kidney in the pig.

CHAPTER 10

Figure 10.3

Q1: Why is it faster to produce vaccines through biopharming with plants than by injecting viruses in eggs? Why is speed important?

A1: Biopharming with plants is faster because plants can be grown in vast quantities more quickly than viruses can multiply in eggs. This speed makes it possible to bring greater quantities of vaccine to the public before an outbreak becomes an epidemic or a pandemic.

Q2: How much cheaper is biopharming with plants than creating vaccines in eggs? Why is cost important?

A2: Biopharming with plants is $364 million cheaper than creating vaccines in eggs. The lower cost means that the vaccines will be marketed to the public at a much lower price. In addition, more insurance companies will cover the lower costs, and more individuals will receive the vaccine.

Q3: The FDA is responsible for ensuring the safety of drugs and food. Why must tobacco-derived vaccines, like any new medications, be approved by the FDA?

A3: FDA approval is necessary because vaccines are drugs, so they must be safe for use in people.

Figure 10.5

Q1: In which of the step(s) illustrated here does DNA replication occur?

A1: DNA replication occurs in steps 4–5.

Q2: In which step(s) does gene expression occur?

A2: Gene expression occurs in step 6.

Q3: Why do you think vaccine manufacturers produce a vaccine after the start of flu season rather than before it has begun?

A3: Because the flu virus changes every year, vaccine manufacturers have to wait for the season's flu virus to appear to know what proteins to replicate to create the vaccine.

Figure 10.6

Q1: Why does RNA polymerase use only one strand of DNA as a template for making mRNA?

A1: RNA polymerase uses only one strand of DNA because the other strand would code for the *opposite* mRNA.

Q2: The template strand of part of a gene has the base sequence TGAGAAGACCAGGGTTGT. What is the sequence of RNA transcribed from this DNA, assuming that RNA polymerase travels from left to right on this strand?

A2: The sequence is ACUCUUCUGGUCCCAACA.

Q3: If a mutation occurred within the promoter or terminator region, do you think it would affect the mRNA transcribed? Why or why not?

A3: Yes, a mutation would affect transcription, because it would make unclear where the gene started or stopped, so the mRNA might not be transcribed at all, or it would grow too long.

Figure 10.7

Q1: In your own words, define RNA splicing. When during gene expression does it occur?

A1: RNA splicing removes the introns from the mRNA transcript. It occurs after transcription but before the mRNA transcript is translated into a protein.

Q2: From where to where does the mRNA transcript travel?

A2: It travels from the nucleus to the cytoplasm.

Q3: What do you predict would happen if the introns were not removed from mRNA before translation? Why would it be a problem if the introns were not removed?

A3: If the introns were not removed, they would be translated, and the protein would be much larger and also nonfunctional. Alternatively, there could be stop codons in the introns, causing translation to stop prematurely.

Figure 10.8

Q1: Which amino acid always begins an amino acid chain? Which codon and anticodon are associated with that amino acid?

A1: Methionine always begins the chain. The associated codon and anticodon are AUG and UAC, respectively.

Q2: In the second question of Figure 10.6, you specified a partial mRNA sequence as being transcribed from the DNA template strand. What amino acid sequence would be translated from that mRNA sequence?

A2: The partial mRNA sequence ACUCUUCUGGUCCCAACA yields the amino acid sequence threonine-leucine-leucine-valine-proline-threonine.

Q3: Each of the codons for stopping translation binds to a tRNA molecule that does not carry an amino acid. How would the binding of a stop codon cause the completed amino acid chain to be released?

A3: Because there is no amino acid, the growing protein becomes detached from the mRNA.

Figure 10.9

Q1: How many codons code for isoleucine? For tryptophan? For leucine?

A1: Three codons code for isoleucine. One codes for tryptophan. Six code for leucine.

Q2: What codons are associated with asparagine? With serine?

A2: AAU and AAC are associated with asparagine. UCU, UCC, UCA, UCG, AGU, and AGC are associated with serine.

Q3: In the second question of Figure 10.6, you specified a partial mRNA sequence as being transcribed from the DNA template strand. From that sequence, remove only the first A. What amino acid sequence would be translated as a result of this change? How does that sequence compare to the amino acid sequence you translated from the original mRNA sequence? *Bonus:* What kind of mutation is this? (*Hint:* See Chapter 9.)

A3: The partial mRNA sequence CUCUUCUGGUCCCAACA yields the amino acid sequence leucine-phenylalanine-tryptophan-serine-glutamine. There is one fewer amino acid, and all the amino acids are different. *Bonus:* This mutation is a deletion.

Figure 10.10

Q1: Why is an insertion or a deletion in a gene more likely to alter the protein product than a substitution, such as A for C, would?

A1: An insertion or deletion causes a "frameshift," so every single amino acid from that point on is likely to be different, resulting in an entirely different protein. By contrast, substitution typically changes at most one amino acid, often resulting in the same protein.

Q2: Which would you expect to have more impact on an organism: a point mutation as shown here or the insertion or deletion of a whole chromosome (discussed in Chapter 8)?

A2: The loss or addition of an entire chromosome, with all the genes on it, is likely to have far more impact on an organism than a mutation within a single gene.

Q3: Which mechanisms in a cell prevent mutations? (*Hint:* Refer back to Chapter 6 if needed.)

A3: Checkpoints in the cell cycle prevent (or signal for repair of) mutations.

Figure 10.11

Q1: As illustrated here, at what control point is transcription regulated?

A1: Transcription is regulated at control point 2.

Q2: What is a possible advantage of regulating gene expression before transcription versus after transcription?

A2: The cell does not waste time and energy producing mRNA transcripts that it will not use.

Q3: At which control point(s) is it possible to up-regulate production of the hemagglutinin protein in a tobacco plant carrying the hemagglutinin gene? Justify your reasoning.

A3: Control points 1–4 all could have an impact on the amount of hemagglutinin being produced by a cell. The levels of DNA compaction, transcription, mRNA degradation, and translation all work together to increase or decrease the production of hemagglutinin. Modifying any one of these or any in combination can have a large impact on production.

CHAPTER 11

Figure 11.2

Q1: What is a shared feature of mammals that you can see in these photos?

A1: All the mammals in the photos clearly have a backbone and are breathing in the air.

Q2: If dolphins and whales are mammals, can they breathe underwater like fish?

A2: No, they breathe air like all mammals.

Q3: Name a mammal other than whales and dolphins that spends much of its life in the ocean.

A3: Other mammals that spend much of their lives in the ocean are seals and sea lions. Also manatees and dugongs!

Figure 11.3

Q1: What is selective breeding, and how does it work?

A1: Selective breeding is the process by which humans allow only individuals with certain inherited characteristics to breed. Generation after generation, the selective breeder chooses which individuals mate and pass their traits to offspring.

Q2: Explain how selective breeding leads to artificial selection.

A2: Over time, the population resulting from generations of selective breeding can change significantly, and we can then say that artificial selection has occurred.

Q3: Name as many organisms as you can that have characteristics due to artificial selection.

A3: Examples include any domesticated animal, including common pets and agriculturally important animals. All agricultural plants are also products of artificial selection.

Figure 11.4

Q1: What is natural selection?

A1: Natural selection is the process by which individuals with genetic characteristics that are advantageous for a particular environment survive and reproduce at a higher rate than do individuals that have other, less useful characteristics.

Q2: If heavy rains caused an abundance of small, tender seeds and fewer large seeds, what do you predict would happen to the average beak size of the finches?

A2: Average beak size would become smaller.

Q3: Compare and contrast artificial selection and natural selection. Name two ways in which they are similar. How are they different?

A3: They are similar in that both occur in populations, not individuals; that there must be a change or changes in the population; and that they occur over time—usually many, many generations. They are different in that artificial selection results from selective breeding performed by human beings, whereas natural selection occurs by the breeding of individuals in a population that survives in a particular environment.

Figure 11.6

Q1: What is the general definition of a fossil?

A1: Fossils are the mineralized remains of formerly living organisms or the impressions of formerly living organisms.

Q2: How do modern whales differ from their ancestors?

A2: They are larger, they don't have back limbs, and their front limbs are proportionally smaller.

Q3: What is a transitional fossil?

A3: A transitional fossil is evidence of a species with similarities to the ancestral group and similarities to a descendant species. It is an intermediate form and can be thought of as a (non-missing) "missing link" in evolution.

Figure 11.9

Q1: How would thick bones help water-dwelling animals?

A1: The heaviness of thick bones would help water-dwelling animals control their ability to submerge and move in the water.

Q2: Why does the adaptation of thick bones suggest a water-dwelling lifestyle?

A2: Fossil mammals with thick bones are presumed to have been water dwellers because almost all currently living water-dwelling mammals have thick bones.

Q3: How did this adaptation likely increase survival and reproduction in *Indohyus* ?

A3: Thick bones probably enabled *Indohyus* to escape predators and to forage on the bottoms of lakes or ponds more efficiently than could lighter-boned species, which had to work harder to stay submerged. Hiding and feeding advantages could have enabled *Indohyus* to eat more, live longer, and have a higher reproductive rate (producing more babies that survived) than those with lighter bones.

Figure 11.11

Q1: What is meant by the term "common ancestor"? Give an example.

A1: A common ancestor is the species from which at least two species descended. The new species arose from the ancestral species through changes in their populations' traits over time. For example, the gray wolf is the common ancestor of modern wolves and modern dogs.

Q2: Why are homologous structures among organisms evidence for evolution?

A2: Homologous structures are similar parts of different organisms. These structures have changed in size or form over time, but they are still easily determined to be the same structure in the ancestral species from which the organisms evolved. So species with homologous traits are all related, having evolved from an ancestor with those structures.

Q3: Aside from skeletal structural similarities, what other commonalities among organisms might be considered homologous?

A3: Any traits that are shared by related organisms and also shared in an ancestor could be homologous traits. Examples include mammary glands, egg laying, structures to extract oxygen from air or water, and the use of DNA as the genetic material.

Figure 11.12

Q1: Why are vestigial structures among organisms evidence for evolution? Give another example of a vestigial structure.

A1: Vestigial structures are evidence for evolution because they are shared among species that have a common ancestor. One example is goose bumps in humans. In our furry ancestors, goose bumps fluffed the fur, thereby increasing its insulating effects and helping keep the animals warm.

Q2: Are vestigial structures also homologous structures? Explain.

A2: Yes. Vestigial traits are homologous structures because they are shared in organisms that have a common ancestor. For example, all organisms that descended from a furry ancestor have goose bumps when they are cold.

Q3: Why do you think vestigial structures still exist if they are no longer useful?

A3: The full (nonvestigial) structure was selected against because it was harmful to the individuals carrying it. Individuals carrying reduced versions of the structure had a survival and/or reproductive advantage over those with the full structure. Once the structure was so reduced as to be termed vestigial, it would no longer be subject to selection.

Figure 11.14

Q1: If a sequence from another species showed a 95 percent sequence similarity to humans, would that species be more or less closely related to humans than chimpanzees are?

A1: The hypothetical species would be less similar to humans than chimpanzees are (95 percent similarity versus 98.4 percent in chimps).

Q2: Should similarities in the DNA sequences of genes be considered homologous traits? Explain your answer in terms of evolution.

A2: All living organisms use DNA as their genetic material, suggesting that the first true ancestral cell used DNA as its genetic material. As species evolved from that common ancestor, their DNA changed in ways that separated them, but sequence similarities remained. Therefore, sequence similarities are homologous traits.

Q3: How is the increased similarity in the DNA sequences of genes between more closely related organisms—and the decreased similarity between less closely related organisms—evidence for evolution? Use the examples in this figure to support your answer.

A3: DNA sequence similarity is the gold standard for determining species relatedness—it is the best homologous trait that exists. Because DNA is the genetic material in all cells on Earth, the changes in DNA sequences in populations over time create the changes in traits that drive evolution. We can map the changes that occur in populations or species by looking at sequence similarity and recreating a family tree. The more related a species is to another species, the more similar the DNA sequences are. Humans and chimps are primates, mammals, and vertebrates (these classifications are discussed in Chapter 14), and they are more similar in DNA sequence than are humans and mice, which are both vertebrates and mammals, but are not both primates. Of the examples in this figure, chickens are the least similar to humans because although they are vertebrates, they are not mammals or primates. A nonvertebrate animal, such as a jellyfish or a worm, would be even less similar to humans than the species named here, but it would likely still show some similarity.

Figure 11.15

Q1: Why should we expect to find *N. fosteri* fossils all over the world, given that it first evolved in Pangaea?

A1: Ancestors of *N. fosteri* existed on the supercontinent. When that landmass broke up, these organisms traveled on the resulting continental portions, and their fossils can be found on all the continents.

Q2: Can we use biogeographic evidence to support evolution without fossil evidence?

A2: Yes, we can. Biogeographic evidence without fossil evidence would consist of the current locations of living organisms that are related. Consider that members of the primate family are found on almost all the continents. Their similarity indicates that they had a common ancestor, and their widespread presence suggests that their common ancestor lived at the time of Pangaea. In short, they evolved into their current forms.

Q3: How might biogeographic evidence and DNA sequence similarities together support evolution?

A3: Biogeographic evidence would consist of the locations of fossils or living organisms. DNA sequence analysis would reveal the genetic similarities and differences among the living organisms and, where possible, the fossils. Together, the locations and genetic information would make it possible to interpret species' evolutionary histories. For example, we can work out the evolutionary histories of every primate on Earth.

Figure 11.16

Q1: Why are the similarities among organisms during early development evidence for evolution?

A1: Similarities between organisms during early development suggest that they have a common ancestor whose early development occurred in the same or similar manner. For example, all vertebrates go through similar stages of development in the early embryo. Many invertebrate organisms also share the same steps in embryonic development.

Q2: Are similar structures among species during early development homologous structures? Explain.

A2: Yes. The similar structures are homologous traits because they are shared with a common ancestor.

Q3: Why do you think embryonic structures still exist during early development if they are not used after birth?

A3: These structures can be considered vestigial traits, as they are now useless to the organism in which they still exist embryonically.

Remember, vestigial traits still exist because they do not harm the organism's ability to survive and reproduce.

CHAPTER 12

Figure 12.4

Q1: What is the difference between MRSA and VRSA?

A1: The difference between these two populations of *S. aureus* is that MRSA survives in the presence of the antibiotic methicillin but can be killed by vancomycin, whereas VRSA survives in the presence of both methicillin and vancomycin.

Q2: Explain in your own words the difference between the zone of inhibition in the top dish and the one in the bottom dish.

A2: The zone of inhibition in each dish represents the area where the antibiotic has killed off the bacteria. The zone in the top dish is larger because more bacteria have been killed there.

Q3: Why is the reduced zone of inhibition in the bottom dish so alarming?

A3: The reduced zone of inhibition in the bottom dish indicates that the antibiotic of last resort, vancomycin, cannot kill the bacteria, and they grow in the antibiotic area as in the areas away from the vancomycin-soaked paper tab. VRSA is a deadly bacteria against which we have no good defense.

Figure 12.5

Q1: What would the white-fur-pigment allele frequency be if three of the homozygous black allele mice (having two black alleles) were heterozygous (having one white and one black allele) instead?

A1: The white-fur-pigment allele frequency would be 16/30 = 53%.

Q2: What would the white-fur-pigment allele frequency be if all of the white mice died and were therefore removed from the population? Would the black-fur-pigment allele frequency be affected? If so, how would it be affected?

A2: The white-fur-pigment allele frequency would be 3/20 = 15%. Yes, the black-fur-pigment allele frequency would be affected; there would then be 17 black alleles out of a total of only 20 alleles: 17/20 = 85%.

Q3: What would the white-fur-pigment allele frequency be if all of the gray mice died and were therefore removed from the population?

A3: The white-fur-pigment allele frequency would then be 10/10 = 100%.

Figure 12.6

Q1: What is natural selection selecting for in this figure?

A1: Natural selection means that those organisms most fit to survive in an environment are most likely to survive. In this case, during the initial treatment (red strainers), methicillin-resistant *S. aureus* (MRSA) bacteria are most fit and, therefore, are being selected for naturally. In the final treatment (purple strainer), vancomycin-resistant *S. aureus* (VRSA) are most fit and, therefore, are being selected for naturally.

Q2: Why are the antibiotics represented by kitchen strainers in this figure?

A2: The antibiotics are depicted as kitchen strainers because antibiotics can "catch," or kill, bacteria in a population. However, at least one bacterium will always survive the antibiotic assault and slip through the strainer.

Q3: Why do bacteria that are not genetically resistant to antibiotics die out when exposed to antibiotics?

A3: After entering a bacterium, an antibiotic generally blocks or poisons one or more processes of the bacterium's life cycle so that it cannot survive or reproduce. (Bacteria that have a mechanism to pump out the poison generally survive the poison and live to reproduce; they are termed "resistant.")

Figure 12.7

Q1: If one extreme phenotype makes up most of a population after directional selection, what happened to the individuals with the other phenotypes?

A1: They declined or disappeared. In this case, individuals with other phenotypes were killed by predators.

Q2: What do you predict would happen to the phenotypes of the peppered moth if the tree bark was significantly darkened again by disease or pollution?

A2: Darker moths would once again be more similar to the color of the bark than would lighter moths, so the darker moths would be protected from predators. As a result, darker moth phenotypes would increase and other phenotypes would decrease.

Q3: What do you predict would happen to the phenotypes of the peppered moth if the tree bark became a medium color, that is, neither light nor dark?

A3: Medium-colored moths would be protected from predators. As a result, the medium-color phenotype would increase and other phenotypes would decrease. (This pattern, stabilizing selection, is discussed in the next paragraph.)

Figure 12.8

Q1: What does the curve in the graph indicate about the relationship between birth weight and survival?

A1: Babies that weighed 8 pounds at birth survived at a higher rate than did lighter or heavier babies.

Q2: Think of another example of stabilizing selection in human biology. Has modern technology or medicine changed its impact on the resulting phenotypes?

A2: Before modern technology and medicine played a major role in survival and quality of life, stabilizing selection probably affected many human traits. Examples include adult height and weight, which would be affected by many hormone levels and overall metabolism.

Q3: How do you think a graph of birth weight versus survival for a developing country with little health care would compare to the graph shown here? What about such a graph for the city of London today?

A3: Compared to the graph shown here, the curve in a graph for a developing country with little health care would be even sharper showing less survival at either end, whereas the curve in a graph for London today would be much wider showing more survival at both ends.

Figure 12.9

Q1: What type of selection would have been present if only the intermediate-beaked birds had survived (instead of the small- and large-beaked birds)?

A1: Survival of only the intermediate-beaked birds would have represented stabilizing selection.

Q2: Why does disruptive selection, rather than directional selection or stabilizing selection, always result in two different phenotypes in the following generations?

A2: Disruptive selection favors both extreme ends of a particular trait, whereas directional selection favors one extreme end and stabilizing selection favors intermediate versions.

Q3: Describe a scenario in which African seed crackers would experience directional selection for either smaller- or larger-beaked birds.

A3: If the only seeds produced in a particular year were so small they could be picked up only by small beaks, then only small-beaked birds would survive, and the population would evolve toward having smaller beaks. Similarly, if only large seeds were produced, then only large-beaked birds would survive, and the population would evolve toward having larger beaks.

Figure 12.10

Q1: How is convergent evolution different from evolution by common descent?

A1: Convergent evolution is essentially the opposite of evolution by common descent. Convergent evolution begins with two distantly related organisms that, over many generations, end up with similar phenotypes because they have adapted to similar environments. Evolution by common descent begins with an original common ancestor and, over many generations, may split into many different populations that are phenotypically different.

Q2: What is the main difference between homologous traits and analogous traits?

A2: Homologous traits are similarities between organisms that exist because a common ancestor had a trait that served as a basis for evolution. By contrast, analogous traits perform similar functions in different organisms, but they formed through convergent evolution rather than evolving from a trait of a common ancestor.

Q3: Why are convergent traits considered evidence of evolution?

A3: As discussed in Chapter 11, evolution reflects changes in allele frequencies over time as populations become better adapted to their environments. Convergent evolution shows phenotypic results of these processes in separate organisms.

Figure 12.13

Q1: If a goose with genotype *AA* had migrated instead of the goose with genotype *aa*, would the scenario described here still be considered gene flow? Why or why not?

A1: No, this is technically not gene flow. Although alleles are being exchanged, they are the same as the existing alleles in the population and will not change allele frequencies over time and many generations.

Q2: If a goose with genotype *Aa* had migrated instead of the goose with genotype *aa*, would the scenario still be considered gene flow? Why or why not?

A2: Yes, this is gene flow. Although the effect is not as extreme as with the *aa* genotype, the *Aa* genotype introduces a new allele into an existing population, creating offspring that can be *Aa*, and thereby changing allele frequencies over time and many generations.

Q3: If the goose with genotype *aa* had migrated to population 2 as shown but had failed to mate with any of the *AA* individuals, would the scenario still be considered gene flow? Why or why not?

A3: No, this is not gene flow. Just adding an individual with different alleles to a population does not count as gene flow. There must be an exchange of alleles between the newcomer and an existing individual.

Figure 12.14

Q1: Why do you think a genetic bottleneck is more likely to occur in a small population than in a large population?

A1: In a large population, it is less likely that a chance event can kill off almost all of the individuals, leaving only a few behind that randomly represent only one of multiple phenotypes. In a small population, a chance event could easily kill off all but a few individuals. All subsequent offspring would arise from these few individuals, whatever phenotype they might have, regardless of which phenotypes in the original population were best adapted.

Q2: List two chance events that could cause a genetic bottleneck.

A2: Chance events that could cause a genetic bottleneck include disease, famine, drought, arrival of more predators, habitat loss, tsunami, hurricane, volcanic eruption, or other natural disaster.

Q3: Which resulting population has more genetic diversity? Why?

A3: The population on the left has greater diversity, because its gene pool was not decreased by a genetic bottleneck.

CHAPTER 13

Figure 13.5

Q1: How might a salt gland serve as an adaptation for an aquatic bird such as a penguin?

A1: Because a penguin's diet includes eating oceanic fish, a salt gland would help the penguin pump any excess salt from its body.

Q2: Explain how the adaptive trait of muscles that store large amounts of oxygen increases the biological fitness of penguins.

A2: Muscles that store a large amount of oxygen would allow the penguins to make deep dives for food, which would help them better survive and then reproduce (= biological fitness).

Q3: Why would heavy bones be an adaptation for penguins but *not* an adaptation for most bird species?

A3: Penguins need heavy bones to dive and swim underwater. Most birds fly, so they need to have light bones.

Figure 13.7

Q1: How do we know that these rattlesnakes are members of the same species?

A1: The caption specifies that they successfully mated and the resulting offspring survived and reproduced.

Q2: How would you design an experiment to determine whether two populations of snakes are distinct species according to the biological species concept?

A2: Your experiment would require mixing individuals from the two populations under conditions conducive to sexual reproduction. If the individuals did mate, the offspring would then need to be raised to maturity and, to test their fertility, be set up to also mate and produce live offspring.

Q3: For which types of populations does the biological species concept *not* work as a way of determining how they are related? (*Hint*: Read the next two paragraphs.)

A3: Because it requires sexual reproduction, the biological species concept cannot be applied to populations that reproduce asexually (such as bacteria).

Figure 13.8

Q1: In addition to reproductive isolation, what three kinds of information do scientists use to identify and distinguish between species?

A1: Scientists use biogeographic information, DNA sequence similarity, and morphology (physical characteristics).

Q2: What differences can you observe between the individuals in the photos? Why are these differences not enough to confirm that they are from two different species?

A2: These frogs display differences in coloration, most strikingly along the side of the body. These differences don't mean the frogs belong to different species because the differences could be—and in fact are—within-species variation, much like hair-color and eye-color differences between humans (who, of course, all belong to the same species).

Q3: How is genetic divergence between two populations determined? (*Hint*: Read the next four paragraphs.)

A3: Genetic divergence between two populations is determined by examination of the DNA sequences of many individuals in the two populations. A lot of similarity among DNA sequences suggests little genetic divergence, whereas many differences among DNA sequences suggest much genetic divergence.

Figure 13.10

Q1: What is the definition of gene flow? How was gene flow blocked between these species?

A1: As discussed in Chapter 12, gene flow is the passing of alleles between populations of different species or of the same species. After the river created a geographic barrier between them, squirrels from the population on one side of the canyon could not mate with squirrels from the population on the other side; thus, gene flow was blocked.

Q2: Name as many types of geographic barriers as you can. Which do you think would be the best at blocking gene flow?

A2: Examples of geographic barriers include but are not limited to rivers, lakes, oceans, glaciers, mountains, canyons, brick walls, freeways, and fences. The larger the barrier, the better it blocks two individuals from finding each other and mating.

Q3: Are geographic barriers universal for all species? If not, name a geographic barrier that might block gene flow for one species but not another.

A3: No, geographic barriers are not universal. A river, for example, might block gene flow between two lizard populations but not two bird populations.

Figure 13.11

Q1: What factor must be present for allopatric speciation to occur?

A1: A geographic barrier must be present.

Q2: If a geographic barrier is removed and the two reunited populations intermingle and breed, what attributes must the offspring have for the two populations, according to the biological species concept, to be considered still the same species?

A2: The offspring must be viable (alive) and fertile (be able to reproduce).

Q3: If the two populations in question 2 are determined to still be the same species, did allopatric speciation occur?

A3: If they are still the same species, no speciation occurred.

Figure 13.12

Q1: Describe how coevolution, as with the hummingbird bill and hummingbird-pollinated flowers, is different from the kind of evolution described in Chapters 11 and 12.

A1: In coevolution, a species evolves directly to interact better with another species. Coevolution can be both species evolving to interact better with each other, or it can be just one of the species adapting to the other species.

Q2: Is coevolution the same thing as convergent evolution, described in Chapter 12? Why or why not?

A2: No. In convergent evolution, two genetically different species look more alike over time because they are adapting to similar environments. In coevolution, two different species adapt to each other's adaptations over time.

Q3: Do you think one species' adapting over time to feed specifically and extremely successfully on another species is an example of coevolution? Why or why not?

A3: Yes. Coevolution does not have to be reciprocal (shared).

Figure 13.13

Q1: Is this an example of allopatric or sympatric speciation?

A1: This is an example of sympatric speciation, because speciation occurred in the same location as the founding species.

Q2: Both large cactus finches and medium ground finches originally evolved from a shared ancestor that came from Central or South America and spread through the Galápagos Islands, speciating into different species on different islands. Is this an example of allopatric or sympatric speciation?

A2: This is an example of allopatric speciation, because the finches evolved on different islands.

Q3: Finch species in the Galápagos differ in their beak size, among other things. How might the differences in beak size be an adaptation to the differing environments on the islands?

A3: There could be different food sources on each island that made a particular beak size more successful there.

Figure 13.14

Q1: What is the main difference between allopatric and sympatric speciation?

A1: Allopatric speciation requires a geographic barrier, and sympatric speciation cannot include a geographic barrier.

Q2: Name two events that must happen for both allopatric speciation and sympatric speciation to occur.

A2: The two populations must be reproductively isolated, and genetic change must occur.

Q3: Do you think all of the 500 species in Lake Victoria arose through sympatric speciation? Why or why not?

A3: Yes, all of the 500 species in Lake Victoria arose through sympatric speciation because there are no geographic barriers to separate the populations in the lake. There would have to be a human-made wall or fence for allopatric speciation to occur.

Figure 13.15

Q1: What does "prezygotic" mean?

A1: "Prezygotic" means "before a zygote" or "before fertilization of an egg by a sperm"—in other words, no fusion of egg and sperm.

Q2: How is the booby's ritual dance a prezygotic reproductive barrier?

A2: This dance happens before mating. If the dance is not performed correctly, no mating happens. No mating means no fusion of egg and sperm.

Q3: What are some other prezygotic reproductive barriers besides a mating dance?

A3: Examples include all geographic barriers, inability of egg and sperm to fuse for genetic reasons (gamete incompatibility and isolation), inability to mate because the genitalia are physically incompatible (mechanical incompatibility and isolation), and ecological isolation in which two species breed in different portions of their habitat, at different seasons, or at different times of day.

Figure 13.16

Q1: Explain in your own words why this is an example of a postzygotic barrier that maintains species.

A1: It is postzygotic, because the mating of the horse and donkey created a zygote. The zygote survived and grew to adulthood but is unable to reproduce.

Q2: Occasionally a female mule will give birth, but her offspring are invariably infertile. Is this an example of a prezygotic or postzygotic barrier?

A2: This is an example of a postzygotic barrier. A horse and donkey created a zygote, which produced an (extremely rare!) fertile mule, which then created zygotes that became infertile mules.

Q3: If a male mule was born fertile, but female horses and donkeys refused to mate with him, would this be an example of a prezygotic or postzygotic barrier?

A3: This would be an example of a prezygotic barrier, as he would be unable to fertilize an egg and, thus, no zygote would be produced.

Figure 13.17

Q1: According to this graph, in what year did the Halley Bay colony have its largest number of breeding pairs? In what year did the Dawson-Lambton colony have its largest number of breeding pairs?

A1: Between 2008 and 2018, Halley Bay had its largest number of breeding pairs in 2010. Dawson-Lambton had its largest number in 2018.

Q2: How many total breeding pairs were there in Halley Bay and Dawson-Lambton in 2018? How does this compare with 2010?

A2: There were approximately 15,000 breeding pairs in 2018 compared with about 29,000 in 2010.

Q3: Given the numbers of breeding pairs in Halley Bay, what would you hypothesize about the change in biological fitness of this population of Emperor penguins over the 10-year period shown here?

A3: This population of Emperor penguins has suffered a significant decrease in biological fitness over the last 10 years.

CHAPTER 14

Figure 14.2

Q1: Why are Archaea and Eukarya connected more closely with each other than they are with Bacteria?

A1: Archaea and Eukarya share a common ancestor with each other more recently than either one does with Bacteria.

Q2: Where would multicellular organisms be found within this figure? What about birds and humans?

A2: Multicellular organisms, birds, and humans would be found as branches of Eukarya.

Q3: Which domains include disease-causing organisms? (*Hint*: Read the bulleted list after the figure callout.)

A3: Disease-causing organisms exist within Bacteria and within Eukarya (protists plus the kingdoms Fungi and Animalia).

Figure 14.3

Q1: During what geologic period did life on Earth begin?

A1: Life on Earth began during the Precambrian.

Q2: How long ago did species begin to move from water to land? What period was this?

A2: Species began to move from water to land between 490 and 445 mya in the Ordovician.

Q3: In what period would *Archaeopteryx* have been alive?

A3: The first birds, including *Archaeopteryx*, existed during the Jurassic.

Figure 14.5

Q1: In what ways were theropods similar to modern birds? Give at least two similarities.

A1: They ran on two legs and had hollow, thin-walled bones.

Q2: In what ways did theropods differ from modern birds? Give at least two differences. (*Hint*: Read the rest of the paragraph.)

A2: Theropods were more variable in size and in skin covering than modern birds.

Q3: Birds are often referred to as "living dinosaurs." Is this accurate? Why or why not?

A3: Birds are direct descendants of dinosaurs, so it could be argued they *are* dinosaurs.

Figure 14.6

Q1: In the traditional tree, identify the node showing the common ancestor for early birds and dinosaurs.

A1: The common ancestor came after the split from the theropods.

Q2: What do both the traditional tree and Xu's tree suggest about troodontids and dromaeosaurids?

A2: Both trees suggest that these two groups were closely related.

Q3: In both trees, identify the node for the common ancestor of *Archaeopteryx* and other birds. In what way are the nodes different in the two trees?

A3: The traditional tree shows *Archaeopteryx* on the bird side of the split between birds and dinosaurs; the common-ancestor node is the point where the birds split into two groups. Xu's tree shows *Archaeopteryx* on the dinosaur side of the split; the common-ancestor node is the theropods.

Figure 14.9

Q1: What group of organisms shares the most recent common ancestor with plants?

A1: As noted in the figure, green algae and plants share a recent common ancestor.

Q2: Are fungi more closely related to plants or to animals? Does the answer surprise you? Why or why not?

A2: Fungi are more closely related to animals.

Q3: If you were to create an evolutionary tree in which amoebas were included within the kingdom of organisms to which they were the most closely related (rather than with protists, where they are currently placed), where would you put them?

A3: You could put them with either Animalia or Fungi.

Figure 14.10

Q1: Within which category are the organisms shown most closely related to one another?

A1: Organisms are most closely related within a species.

Q2: Within which category are the organisms shown most distantly related?

A2: Organisms are most distantly related within a kingdom.

Q3: Are individual species within the same order or within the same family more closely related to each other?

A3: Species are more closely related within a family than within an order.

Figure 14.17

Q1: Is *Xiaotingia* an earlier or later bird than *Archaeopteryx* in this tree?

A1: *Xiaotingia* is later than *Archaeopteryx*.

Q2: If a future study, based on more fossils or new measurements, placed *Archaeopteryx* back with dinosaurs, would this suggest that birds are not related to dinosaurs? Why or why not?

A2: No. What is in question is not *whether* birds are related to dinosaurs but when exactly they split off from related dinosaur species and which species is the first example of that split.

Q3: If you were to create an evolutionary tree of modern birds, where would you expect to place the roadrunner (judging by its appearance in this figure) as compared to a house sparrow or pigeon?

A3: The roadrunner looks more like a dinosaur, so you might argue that it belongs closer to the base of the tree.

Figure 14.18

Q1: What extinction event occurred about 200 mya? What animal groups were most affected by this event?

A1: The Triassic extinction occurred about 200 mya; reptiles were most affected by it.

Q2: Which of the mass extinctions appears to have removed the most animal groups? How long ago did this extinction occur?

A2: The Permian extinction appears to have removed the most animal groups; it occurred about 250 mya.

Q3: The best studied of the mass extinctions is the Cretaceous extinction. Why do you think it has been better studied than the other extinctions?

A3: Because it is more recent, more fossils may be available from this extinction than from earlier ones.

CHAPTER 15

Figure 15.3

Q1: If an individual prokaryote divides every 20 minutes, how many individuals will there be after an hour?

A1: There will be eight prokaryotes.

Q2: If the generation time is 20 minutes, how much time will have gone by when the final generation shown has doubled?

A2: Another 20 minutes will have passed.

Q3: Many bacteria are able to reproduce more quickly in warmer conditions. What does this suggest to you about the importance of refrigerating foods?

A3: Refrigerating food slows generation time, so dangerous bacteria such as some strains of *E. coli* have less of a chance to increase to threatening levels.

Figure 15.5

Q1: In which of the three ways listed above did the navel microbiome participants contribute?

A1: They were experimental participants.

Q2: Which of the advantages listed above do you think the navel microbiome citizen scientists received?

A2: They may have felt a sense of contribution and purpose.

Q3: Would you be willing to contribute to the navel microbiome project? Why or why not?

A3: Answers will vary; some might express privacy concerns.

Figure 15.6

Q1: Where in the figure would you place the first life on Earth?

A1: It belongs at the base of the tree.

Q2: Where in the figure did Bacteria split off from the ancestor of Archaea and Eukarya?

A2: This occurred at the first split from the base of the tree.

Q3: The figure (and thus the study) demonstrates that Archaea and Eukarya are more closely related to each other than to Bacteria. How is that illustrated?

A3: The first split seen is Bacteria, pictured apart from the other groups.

Figure 15.8

Q1: Which of these shapes do you think *Streptococcus* would take?

A1: It would be a sphere, or coccus.

Q2: From the micrographs here, does it appear that all prokaryotes have a flagellum?

A2: No, of the bacteria shown here, only the comma, or vibrio, has a flagellum.

Q3: Which one of these shapes is most clearly capable of moving by its own volition? Why? (*Hint*: Read the next paragraph.)

A3: The comma, or vibrio, can move itself because of the flagellum.

Figure 15.9

Q1: Which shape in Figure 15.8 corresponds to the archaeans from deep-sea thermal vents?

A1: The rod, or bacillus, is the shape of these archaeans.

Q2: Why are many archaeans referred to as "extremophiles"?

A2: They are found in extreme environments (at extremes of heat, acidity, and salt level). "Philes" means "lovers of."

Q3: Is there anywhere you think archaeans could *not* survive? Justify your answer.

A3: Various answers are possible.

Figure 15.10

Q1: How do individual bacteria know that they have a "quorum"?

A1: They detect increasing concentrations of signaling molecules.

Q2: There is a well-known biofilm found in your mouth. What is it?

A2: It is dental plaque, which causes cavities and gum disease.

Q3: Under what conditions might bacteria want to coordinate (via quorum sensing) to increase their reproductive rate?

A3: If environmental conditions are favorable for the bacteria, it makes sense for the population to grow as quickly as possible to take advantage of those conditions.

Figure 15.11

Q1: What source of energy would you expect a cave-dwelling prokaryote to use?

A1: The prokaryote could not rely on light, so it would most likely use chemical energy.

Q2: In which of these categories would you place the bacteria responsible for nitrogen fixation? Why?

A2: The bacteria would be chemoautotrophs, using nitrogen from the atmosphere and carbon from carbon dioxide.

Q3: In which of these categories do decomposers belong? Explain your reasoning.

A3: Decomposers are chemoheterotrophs, because they receive both carbon and energy from dead or dying organisms.

Figure 15.12

Q1: From the prokaryotic structures shown in Figure 15.8, what shape would you assign to drawing number 8 in van Leeuwenhoek's illustration?

A1: Its shape is the comma, or vibrio, because of the flagellum.

Q2: Which of the large prokaryote drawings (numbers 24–30) has the coccus shape?

A2: Number 24 has that circular shape.

Q3: Do you think all of these "animalcules" drawn by van Leeuwenhoek are prokaryotes? Why or why not?

A3: No. Some have such an elaborate structure that they are most likely eukaryotes.

CHAPTER 16

Figure 16.3

Q1: Are ciliates more closely related to euglenoids or to diatoms? To euglenoids or to forams?

A1: Ciliates are more closely related to both diatoms and forams than they are to euglenoids.

Q2: Are all the algae groups (red, green, and brown) equally related?

A2: No. Red and green algae are more closely related to each other than either is to brown algae.

Q3: Which protist group do you think is most closely related to plants? Justify your answer.

A3: Green algae are most closely related to plants because they share a common ancestor with plants.

Figure 16.5

Q1: What evolutionary innovation separates all land plants from their aquatic ancestors?

A1: They are terrestrial.

Q2: How do ferns differ from bryophytes? Do they share this difference with other plant groups? (You will need to read ahead to answer this question.)

A2: Ferns have lignin and vascular systems, whereas bryophytes do not. Gymnosperms and angiosperms also have these adaptations.

Q3: What group(s) might a plant with seeds belong to? What about a plant with flowers?

A3: Seeds are found in gymnosperms and angiosperms; flowers are found only in angiosperms.

Figure 16.6

Q1: In what ways are terrestrial plants and their aquatic ancestors similar? Give at least two similarities.

A1: Both photosynthesize and must absorb nutrients.

Q2: In what ways do terrestrial plants and their aquatic ancestors differ? Give at least two differences.

A2: Unlike their aquatic ancestors, terrestrial plants must protect themselves from dehydration and absorb nutrients from the soil.

Q3: Would you predict that aquatic plants (which have secondarily evolved to live in water; in other words, their ancestors were terrestrial plants) would be more like plants in a rainforest or more like desert plants? Explain your reasoning.

A3: They should be more like rainforest plants because water is not as limiting a resource (limiting the plants' distribution) in the rainforest as in the desert.

Figure 16.8

Q1: What feature(s) of the ginseng plant tell you it is not a bryophyte?

A1: It has roots, and also it is able to grow taller than a bryophyte.

Q2: What feature(s) of the ginseng plant tell you it is not a fern or gymnosperm?

A2: It has flowers, which only angiosperms have.

Q3: Because of the CITES classification of ginseng, you are not allowed to sell plants younger than 5 years even if they grew on your own land. Do you agree with that law? Why or why not?

A3: Answers will vary. You may feel you should be able to do what you want on your own land, or you may agree it is best to maintain the species.

Figure 16.9

Q1: What group of fungi most resembles the mushrooms you buy in a grocery store?

A1: Basidiomycetes (club fungi) most resemble the mushrooms sold in grocery stores.

Q2: Are sac fungi (ascomycetes) more closely related to molds (zygomycetes) or to club fungi (basidiomycetes)?

A2: Sac fungi are more closely related to club fungi than to molds.

Q3: How do we know that fungi are eukaryotes rather than prokaryotes?

A3: The simplest reason is that fungi are multicellular, but they also are composed of eukaryotic cells (with nuclei and other organelles, for example, and larger size).

Figure 16.11

Q1: Why is it important that the fruiting body of a fungus is aboveground?

A1: It is important that the fruiting body of a fungus is aboveground because, if belowground, spores would not be able to travel by wind.

Q2: What part of a fungus is the mushroom that you can buy in the grocery store?

A2: You can buy the fruiting body.

Q3: Write a sentence in your own words that uses the terms "mycelium," "fruiting body," and "spore" correctly.

A3: One possible answer: The mycelium is the main body of a mushroom, whereas the fruiting body develops to produce and release spores into the environment for reproduction.

CHAPTER 17

Figure 17.2

Q1: Are mollusks more closely related to flatworms or to annelids? Explain your reasoning.

A1: Mollusks share a more recent common ancestor with annelids than with flatworms, so they are more closely related to annelids.

Q2: If you found an animal with no symmetry, to which group do you think it would belong? What about an animal with radial symmetry? One with bilateral symmetry?

A2: Animals with no symmetry are sponges. Animals with radial symmetry include cnidarians and echinoderms (secondarily). Animals with bilateral symmetry include flatworms, mollusks, annelids, nematodes, arthropods, and chordates.

Q3: If an animal shows growth back to front (or mouth second), what kind of symmetry does it have?

A3: The animal has bilateral symmetry.

Figure 17.3

Q1: Is a sea star (starfish) radially or bilaterally symmetrical?

A1: It is radially symmetrical.

Q2: What advantage might a bilaterally symmetrical animal have over one that is radially symmetrical, and vice versa?

A2: Examples of advantages include (1) animals with bilateral symmetry move more efficiently and (2) those with radial symmetry have 360-degree access to their environment.

Q3: What kind of symmetry do you (a human) have? What external body parts indicate this arrangement?

A3: The human body has bilateral symmetry. Ideally, the body's left and right sides divide paired features related to sensing and moving: eyes, ears, arms, hands, legs, and feet.

Figure 17.4

Q1: Do amphibians have amniotic eggs?

A1: No, only reptiles, birds, and mammals have amniotic eggs.

Q2: What group of animals has jaws but not a bony skeleton?

A2: Sharks and rays meet this description.

Q3: When people talk about animals, they are sometimes referring only to mammals. How would you explain to them their error?

A3: Animals are a very large group that can be broken up into smaller groups that include chordates and nonchordates. Chordates include mammals, birds, and lizards, among others. But nonchordates, such as invertebrates and sponges, are also animals.

Figure 17.5

Q1: The Virginia opossum, or possum, is the only North American marsupial. How would the birth and development of its young compare with the birth and development of the young of eutherians and monotremes?

A1: Because possums are marsupials, their young are born relatively undeveloped and crawl into the mother's pouch, where they nurse and grow. Eutherians' young develop fully inside the womb and are born well developed. Monotremes' young hatch from eggs well developed.

Q2: Do monotremes produce milk and nurse their young?

A2: Yes; all mammals produce milk and nurse their young, even egg-laying monotremes.

Q3: What kind of mammal is a cow? How about a human? How do you know?

A3: Cows and humans are eutherians. Unlike marsupials, they don't have pouches. Unlike monotremes, they don't lay eggs.

Figure 17.6

Q1: According to this evolutionary tree, which primate group is most closely related to humans?
A1: Chimpanzees are most closely related to humans. The lines of chimpanzees and humans diverged 5–7 mya.

Q2: According to this evolutionary tree, which primate group is most distantly related to humans?
A2: Lemurs, lorises, and others in the same branch are most distantly related to humans. Their line and the line of humans diverged 65 mya.

Q3: What characteristics are common to all primates, including humans? (*Hint*: You will need to read the rest of the paragraph in the text to answer this question.)
A3: All primates have flexible shoulder and elbow joints, five functional fingers and toes, opposable thumbs, flat nails, and large brains.

Figure 17.7

Q1: Through natural selection, deleterious traits will tend to disappear from a population over time. Which traits might have been deleterious for ground-dwelling early hominins that shifted from living in trees to living on the ground?
A1: Natural selection favors traits that provide an advantage for survival and reproduction. Walking on four limbs would have been an obstacle for ground-dwelling hominins, and opposable toes would have further hindered walking on land.

Q2: Through natural selection, advantageous traits will tend to persist in a population over time. Which traits might have been advantageous for ground-dwelling early hominins?
A2: An upright posture would have freed their hands for other activities, including the use of tools. Nonopposable toes and arched feet would have provided walking and running advantages.

Q3: The adaptation to upright walking means that human females have more difficulty giving birth than do females of other species. What adaptation would you predict has had the greatest impact on this difficulty?
A3: The reorganization of the pelvic structure, brought about by the need to balance on two feet, brought with it a narrowness that makes it difficult for a baby to emerge from the mother's body.

Figure 17.8

Q1: What other reason besides continuing to spend time in trees might explain why early hominins had partially opposable big toes?
A1: This trait may have taken a long, long time to lose. A major change in structure cannot occur within a few generations. It was likely a gradual change over thousands of generations.

Q2: In what way does the pattern of footprints in this figure suggest that the print makers were walking upright?
A2: No hand or knuckle prints accompany the footprints, so these hominins were not moving on four limbs.

Q3: Why do you think we no longer have partially opposable big toes?
A3: Even partially opposable big toes would have made walking upright or running more difficult. If we no longer returned to the trees, there was no selective advantage to having them. In fact, individuals with fully opposable toes were at a disadvantage.

Figure 17.9

Q1: Would the Neanderthal species branch be on the *Homo* lineage side or the *Australopithecus* side of this tree?
A1: The *Homo* lineage side. The *Homo neanderthalensis* lineage would branch either from the common ancestor of *H. erectus* and *H. sapiens* or from *H. erectus* along with *H. sapiens*. The exact hereditary line for these two closely related species is still unclear.

Q2: Do you think *Homo habilis* and Neanderthal teeth would be more similar to apes or modern humans?
A2: *Homo habilis* and Neanderthal teeth would be more similar to modern humans because they are all *Homo* species.

Q3: Which of these species has the smallest braincase?
A3: *Australopithecus afarensis* has the smallest braincase.

Figure 17.10

Q1: Why does mitochondrial DNA come only from the mother?
A1: Mitochondria are found in the egg (female gamete) but not in the sperm (male gamete).

Q2: If a child had a modern human mother and a Neanderthal father, could you tell that hybrid parentage by mitochondrial-DNA sequencing? Why or why not?
A2: You could not, because all of the child's mtDNA would be from *H. sapiens*.

Q3: If a child had a modern human father and a Neanderthal mother, could you tell that hybrid parentage by mitochondrial-DNA sequencing? Why or why not?
A3: You could, because all of the child's mtDNA would be from *H. neanderthalensis*.

Figure 17.11

Q1: If a child had a modern human mother and a Neanderthal father, could you tell that hybrid parentage by whole-genome DNA sequencing? Why or why not?
A1: You could, because the child would have nuclear DNA from both the egg and the sperm.

Q2: If a child had a Neanderthal mother and a modern human father, could you tell that hybrid parentage by whole-genome DNA sequencing? Why or why not?

A2: You could, because the child would have nuclear DNA from both the egg and the sperm.

Q3: Under what circumstances are scientists able to do whole-genome sequencing, and when are they restricted to mitochondrial-DNA sequencing?

A3: Whole-genome sequencing requires well-preserved cells or tissues with fully intact DNA. Mitochondrial DNA can be isolated from cells and tissues that aren't so well preserved and from damaged DNA.

Figure 17.12

Q1: Are you surprised by the interpretations of the hominins in this picture?' Why or why not?

A1: This question asks for an opinion. One possible answer: *H. sapiens* looks more primitive than expected, and the other species looks strikingly like us.

Q2: Describe the main differences that distinguish the hominin species.

A2: The species are differentiated by mainly by height, musculature, and size of skull.

Q3: From what you've learned about these species, do you think these representations are accurate? How can you find more information about each species to help you answer this question?

A3: The first part of the question asks for an opinion. One possible answer for the second part: I can do much more extensive research about all of our family members online and at museums of natural history.

Figure 17.13

Q1: When was Neanderthal DNA introduced into modern human DNA?

A1: Neanderthal DNA was introduced into modern human DNA between 90,000 and 40,000 years ago.

Q2: What conclusion can you make from the fact that all modern human populations besides Africans have a small amount of Neanderthal DNA?

A2: Neanderthals did not evolve in Africa. Modern humans evolved from archaic humans in Africa and spread to the rest of the world. Neanderthals evolved from archaic humans in the Middle East and spread to Europe.

Q3: What species of hominins other than the Neanderthals may have commingled with modern humans?

A3: *H. erectus* overlapped in time with modern humans and so may have interacted with them.

Figure 17.14

Q1: How do the lower jaws differ between the modern human skull and the Neanderthal skull?

A1: The modern human skull has a chin, a ridge that is higher and juts out further than the bottom of the Neanderthal jaw.

Q2: How do the eyebrow ridges and foreheads differ between the two skulls?

A2: The Neanderthal skull has a more pronounced eyebrow ridge and a more sloping forehead.

Q3: What would you expect a hybrid of Neanderthals and modern humans to look like?

A3: In a hybrid, half the DNA would be from *H. sapiens* and half from *H. neanderthalensis*. A hybrid has features of both parents. It might have intermediate features, or it might have mixed features (such as a sloping forehead but a prominent chin).

CHAPTER 18

Figure 18.3

Q1: List as many biotic and abiotic factors in this photograph as you can.

A1: Biotic: all living things, such as plants, as well as all animals, including microscopic bacteria and algae. Abiotic: all nonliving things, such as rocks and snow (frozen water).

Q2: Is the tundra ecosystem part of the biotic or abiotic environment? Explain.

A2: It is both. The plants and animals and microorganisms are biotic, whereas the rocks and snow are abiotic.

Q3: Name a part of the tundra ecosystem that is not also part of the surrounding ocean ecosystem. Is it biotic or abiotic?

A3: A terrestrial plant is part of the tundra ecosystem that isn't part of the ocean ecosystem. It is biotic. (Polar bears are arguably both, as they hunt on land and sea.)

Figure 18.4

Q1: Name two ways in which climate change affects the frequency and severity of floods.

A1: Changes in rainfall patterns and melting ice can cause rivers and lakes to overflow.

Q2: How has climate change caused a rise in sea level?

A2: Melting of glaciers and polar ice has caused sea levels to rise.

Q3: Give an example of an environmental effect of climate change in your state or region.

A3: Answers may include drought or flooding, crop failure, and depletion of fisheries.

Figure 18.6

Q1: What is transpiration?

A1: Transpiration is the process of plants absorbing water through their roots and releasing this water through their leaves to the atmosphere.

Q2: Why is transpiration important to the hydrologic cycle?

A2: Transpiration returns water from the soil to the atmosphere to form clouds and eventually precipitation.

Q3: If plants and therefore transpiration decrease in a given area, what will happen to the humidity or cloud cover in this area?

A3: The humidity and cloud cover will decrease where there is a substantial decrease in transpiration.

Figure 18.7

Q1: How do the patterns of rainfall in the Northern and Southern Hemispheres compare?

A1: The same patterns emerge as you move away from the equator either northward or southward.

Q2: How do the patterns in the kinds of environments shown in the Northern and Southern Hemispheres compare?

A2: As you move away from the equator, the major biomes are equivalent distances from the equator either north or south.

Q3: What happens at the equator to make this region so wet?

A3: The density of plants is very high in the equatorial tropical rainforest and, therefore, the amount of transpiration is very high, resulting in cloud cover and high precipitation. High precipitation results in high plant growth and high transpiration rates.

Figure 18.8

Q1: How many gigatonnes of ice have been lost since 2002?

A1: 3,250 gigatonnes have been lost.

Q2: Why does the amount of ice loss change so dramatically over the course of each year?

A2: There is less ice loss in the winter than in the summer.

Q3: About how much ice was lost between summer 2014 and summer 2016?

A3: About 300 gigatonnes of ice were lost between summer 2014 and summer 2016.

Figure 18.9

Q1: Why is it colder at the poles than at the equator?

A1: The sun's rays are spread more widely at the poles and strike Earth less directly than at the equator. Less direct sunlight results in cooler temperatures.

Q2: Why is it warmer at the equator than at the poles?

A2: The sun's rays strike Earth directly at the equator. The result is more intense heat at the equator.

Q3: Why is it cooler at night than during the daytime?

A3: During the planet's 24-hour rotation, it is nighttime for the part of Earth that it is rotated away from the sun, so it is not receiving direct rays from the sun and becomes cooler than the daytime.

Figure 18.10

Q1: When the Northern Hemisphere is tilted most directly toward the sun, what season is experienced there?

A1: When the Northern Hemisphere is tilted most directly toward the sun, it is summer in the Northern Hemisphere.

Q2: When the Northern Hemisphere is tilted the furthest away from the sun, what season is experienced in the Southern Hemisphere?

A2: When the Northern Hemisphere is tilted the furthest away from the sun, it is summer in the Southern Hemisphere.

Q3: Why are the seasons less distinct at the equator than at the North and South Poles?

A3: The equator remains relatively equidistant to the sun throughout the year, so the temperature does not change much, whereas the poles experience the most significant changes in tilt toward the sun.

Figure 18.11

Q1: How much of the incoming solar energy is reflected back to outer space?

A1: About a third is reflected back.

Q2: What kind of energy is reemitted to the atmosphere after being absorbed by Earth's surface?

A2: Infrared radiation is reemitted.

Q3: How are greenhouse gases like a blanket on your bed at night?

A3: Greenhouse gases absorb the heat around Earth and hold it near the surface just as a blanket absorbs body heat, prevents it from escaping, and holds it near your body.

Figure 18.12

Q1: What are three ways that carbon is released into the atmosphere?

A1: Carbon is released through respiration from organisms, the burning of organic matter including fossil fuels and wood, and the decomposition of dead organic material.

Q2: Are all the pathways you listed for question 1 affected by human activity?

A2: Almost everything on our planet is affected by human activity in some way. Of the three pathways of carbon release, the one most affected by humans is the burning of organic materials for energy.

Q3: What are two biotic reservoirs of carbon?

A3: Plants and animals are biotic reservoirs of carbon.

Figure 18.13

Q1: What measurements do the green circles represent?

A1: The green circles represent CO_2 levels measured from air bubbles in ice that formed hundreds of years ago.

Q2: What measurements do the red circles represent?

A2: The red circles represent direct measurements of CO_2 levels at the Mauna Loa Observatory in Hawaii.

Q3: For approximately how many years has the Mauna Loa Observatory been recording CO_2 levels?

A3: It has been recording them directly for about 60 years.

Figure 18.14

Q1: In what years were global temperatures the lowest?

A1: They were lowest in the years around 1910.

Q2: In what years were global temperatures the highest?

A2: They were highest in the years around the late 1990s through the present.

Q3: What trend is apparent in this graph of actual global temperatures?

A3: Average global temperatures are rising.

Figure 18.15

Q1: How does a carbon source contribute to global warming?

A1: A carbon source is a reservoir of carbon that releases more than it absorbs, thereby dumping CO_2 into the environment, increasing greenhouse gases, and causing global warming.

Q2: How does a carbon sink protect against global warming?

A2: A carbon sink absorbs more carbon than it releases, thereby removing CO_2 from the environment, decreasing greenhouse gases, and blocking global warming.

Q3: How can trees act as both a source and a sink?

A3: Trees act as a carbon sink when photosynthesizing and absorbing CO_2 and as a carbon source when they are burned for fuel or in a wildfire.

Figure 18.17

Q1: Describe in your own words what each line of data shows.

A1: The air temperature is increasing over time, as is sea-ice runoff. The sea-ice extent is decreasing over time.

Q2: When did runoff seem to increase significantly?

A2: Runoff began to increase substantially in the late 1990s or early 2000s.

Q3: How would you explain the clear relationship among temperature, runoff, and sea-ice extent?

A3: As the air temperature increases, ice melt and runoff increase as well, leading to a decreased area of Arctic Ocean covered by sea ice.

CHAPTER 19

Figure 19.1

Q1: What parts of the United States are within the range of the mosquito that carries Zika?

A1: Roughly, the southern United States and the eastern coastline are within the range of the mosquito.

Q2: What areas are *not* within the mosquito's range? Why do you think that is?

A2: The northern states are not within the mosquito's range, probably because it is too cold for mosquitoes to overwinter there.

Q3: Find your own state and your location in the state. Are you at risk of contracting Zika?

A3: Answers depend on student location. If you travel to areas with Zika-infected mosquitoes, your risk increases. If local Zika infections have not yet been reported in your state, you are at less risk than, for example, citizens of Florida.

Figure 19.3

Q1: What is the main way that someone is infected with the Zika virus?

A1: The main way is being bitten by an infected mosquito.

Q2: Judging by the poster and your knowledge of mosquito behavior, what can you do to decrease your risk of being infected with the Zika virus?

A2: Cover up when you're outside, don't go out at dusk, wear insect repellent, and use protection when having sex with someone who has been exposed.

Q3: Besides the transmission methods shown on the poster, what are some other ways you could become infected with the Zika virus?

A3: Possibilities include kissing an infected individual or possibly sharing a razor or toothbrush with someone.

Figure 19.5

Q1: Over what years did the human population double from 1 billion to 2 billion people? How many years did that growth take?

A1: That doubling occurred from 1800 to 1927; the span was 127 years.

Q2: How many years did the second population doubling take (from 2 billion)? When did that occur?
A2: That doubling took 47 years; it happened in 1974.

Q3: When is it predicted that the *next* population doubling will occur? What will be the human population then?
A3: The next one will happen in 2025; the population is expected to be 8 billion.

Figure 19.6

Q1: Which form of population growth displays a J-shaped curve?
A1: Exponential growth displays a J-shaped curve.

Q2: Which form of population growth displays an S-shaped curve?
A2: Logistic growth displays an S-shaped curve.

Q3: Describe a situation in which a population initially shows exponential growth and later shows logistic growth.
A3: If a population enters a new area—an island, for example—it will initially grow exponentially as it expands into the environment. Later, as it reaches the carrying capacity of the new area, it will begin to display logistic growth.

Figure 19.7

Q1: According to this graph, approximately when did exponential growth begin?
A1: It began around the year 1800 CE.

Q2: What milestone corresponds to the transition from logistic to exponential population growth? How would you predict that modern medicine (e.g., treatment, hygiene, antibiotics) would affect the slope of the line?
A2: The onset of the industrial revolution corresponds to that transition. Modern medicine would increase the slope of the line because there would be lower mortality.

Q3: What is the United Nations' projected carrying capacity of Earth, and when will we reach it?
A3: The UN projects Earth's carrying capacity to be 9–10 billion people, which we will reach in about 2050 CE.

Figure 19.8

Q1: List all the stages of the mosquito life cycle.
A1: The stages are egg, larva, pupa, and adult.

Q2: Which life cycle stage is vulnerable to the GM treatment?
A2: The larvae are vulnerable in that they won't survive to become pupae.

Q3: Why are only male GM mosquitoes released into the wild rather than both males and females?
A3: Males do not feed on blood, so they are not able to transmit Zika virus, whereas females do feed on blood and thus can transmit Zika. It's better not to release more mosquitoes that feed on blood into the population.

Figure 19.9

Q1: What factors may be limiting growth and reproduction in the plantain's crowded conditions?
A1: Possible factors include availability of nutrients, water, sunlight, and room for root and shoot growth.

Q2: Why are overcrowded conditions considered density-dependent population changes?
A2: The more organisms there are in the environment or the more densely packed they are, the more competition there is for resources and the less each individual will likely receive.

Q3: Relate this example of overcrowded conditions to the human population growth shown in Figure 19.7. How do you think the situations are similar? How are they different?
A3: Increasing human population may lead to an increase in competition for jobs, housing, and food. So far, humans have been able to increase their carrying capacity through technology, but that may no longer be possible.

Figure 19.10

Q1: In what year did the reindeer's numbers begin to rise exponentially?
A1: That happened in about 1933.

Q2: In what years was the reindeer's population growth logistic?
A2: That occurred between about 1911 and 1932.

Q3: How do you predict the graph (population size) would change if someone had begun bringing in supplemental food for the reindeer in 1940? Draw a sketch of your prediction.
A3: Supplemental food would have caused the population decline to end, and the population would have either remained constant or again increased, depending on the amount of supplemental food provided.

Figure 19.11

Q1: In what year did the bald eagle population rise to more than 2,000 breeding pairs?
A1: That happened in about 1986 or 1987.

Q2: Give some examples of possible density-dependent limits on bald eagle populations.
A2: Examples include numbers of prey available, adequate habitat for nesting and hunting, and availability of water.

Q3: Is the population growth of bald eagles more like logistic or exponential growth? Explain why you think so.
A3: The population growth of bald eagles is more like exponential growth, since it has continued to climb for several decades.

Figure 19.12

Q1: During which years did the hare likely have the greatest food supply?
A1: That happened around 1865 and 1888.

Q2: Besides the number of hare, what other factors might contribute to the number of lynx?

A2: Other possible factors include the quality of habitat for building dens and raising young, the availability of freshwater, and the competition with other lynx and other predators of the hare.

Q3: Can you draw an average carrying-capacity line on the graph? Why or why not?

A3: It would be difficult to draw a line representing a carrying capacity for both of these animals because of the way they go through cycles of "boom and bust" in population size. They do not show logistic growth, where the leveling off of the S-shaped curve provides an obvious carrying capacity.

CHAPTER 20

Figure 20.2

Q1: List another species that is part of this community.

A1: One possible answer is the wolf.

Q2: Of which community could this aspen woodland be a smaller part?

A2: It could be part of a larger deciduous forest community.

Q3: Which other small communities could be found within this larger community?

A3: Many other communities could exist, including soil communities (soil-dwelling insects, other invertebrates, and microbes), communities of plants and animals residing in the forest undergrowth, and canopy communities of animals that live in the treetops.

Figure 20.3

Q1: If relative species abundance increased in the first community, how would the figure look different than it does now?

A1: There would be even more trees of the white-trunked species and fewer of the other species.

Q2: If species richness decreased in the second community, how would the figure look different than it does now?

A2: There would be fewer different species of trees.

Q3: How do relative species abundance and species richness define the species diversity of a forest community?

A3: High species diversity relies on having each species in reasonable abundance and not on one species taking up the majority of space in an area (relative abundance) and on there being many different species in a given area (richness).

Figure 20.4

Q1: How many species were left in 1966 in the community where sea stars were *not* removed?

A1: About 17 or 18 species were left.

Q2: How many species were left in 1966 in the community where sea stars *were* removed?

A2: Only two or three species were left.

Q3: How do your answers to questions 1 and 2 demonstrate the importance of a keystone species for the maintenance of diversity in a community?

A3: The species diversity plummeted as a result of the loss of the sea stars. Without this species, the entire intertidal community changed from having many species to having only a few.

Figure 20.5

Q1: What species eats the coyote?

A1: The gray wolf sometimes eats coyotes (though generally they are competitors for food).

Q2: What species does the coyote eat?

A2: The coyote eats the vole, the snowshoe hare, berries, grass, willows, the pronghorn, and the elk.

Q3: What do you think would happen to a community that lost its coyotes?

A3: The snowshoe hare and the vole would overeat the grasses and other plants; once those food sources were destroyed, their unavailability would cause the snowshoe hare and vole populations to crash.

Figure 20.6

Q1: Where do producers acquire the energy they need to perform their function in the food chain?

A1: They acquire it from the sun. In combination with CO_2 and water, they produce glucose through photosynthesis.

Q2: Why are producers necessary for life on Earth?

A2: Without producers, there would be no influx of energy into Earth's biosphere—no energy source for consumers to acquire.

Q3: Given the lower abundance of producers in desert regions compared to tropical rainforests, what would you predict about the abundance of consumers in the two environments?

A3: Consumer species would be less abundant in deserts than in rainforests.

Figure 20.7

Q1: When were wolves first reintroduced to Yellowstone?

A1: They were reintroduced in 1995.

Q2: What was the highest number of wolves ever observed in Yellowstone? In what year did that occur?

A2: The highest number was about 175 individuals in 2003.

Q3: Why did it take a few years after the reintroduction of wolves for the aspen population to increase as well as the populations of beavers and bears?

A3: Each individual parent plant can produce only a certain number of offspring at a time, and then those offspring have to reach maturity before they themselves can reproduce.

Figure 20.8

Q1: What might happen to an anemone without a resident clownfish?

A1: An anemone unaccompanied by clownfish and, therefore, unprotected from anemone-eating fish could be grazed extensively and be unable to support its growth needs without the nutrients in the clownfish excrement.

Q2: What would happen to a clownfish that did not produce a mucous coating?

A2: It would be stung by the anemone's tentacles, so the mutualism would fall apart.

Q3: In the example of mutualism given in the text, what would happen if ticks were no longer able to feed on bison?

A3: There would be nothing for the birds to eat off the bison and no reason for the bison to allow the birds on them. The mutualism would end.

Figure 20.9

Q1: How do barnacles benefit from living on a whale?

A1: The whale provides a home and a constant stream of water passing over the barnacles to bring them the tiny particles of food that they filter from the sea.

Q2: Do you think a whale could avoid being colonized by barnacles? Why or why not?

A2: It is probably unlikely that a whale could avoid being covered by barnacles, although because the barnacles do not help or hurt the whale, there would be no reason for a whale to try to avoid them.

Q3: Explain why detritivores are considered commensal to the organisms they consume. (You will need to read ahead to answer this question.)

A3: Because the organism is already dead, the detritivore is not harming it by eating it. The detritivore, though, benefits from the association.

Figure 20.10

Q1: What kind of predator is the cheetah in the figure?

A1: It is a carnivore.

Q2: What kind of predator is the louse fly in the figure?

A2: It is a parasite.

Q3: What kind of predator are the elk that graze on aspen tree saplings?

A3: They are herbivores.

Figure 20.11

Q1: How do predators know that brightly colored prey are usually toxic?

A1: Most predators have experienced through the trial and error of tasting prey and becoming sick that brightly colored prey are toxic. Predators must learn by trying one to avoid others later. If the organism were so toxic that the predator died from eating one, then bright displays would not help prey avoid predation.

Q2: Do you think mimicry works if the toxic species is in low abundance? Why or why not?

A2: Mimicry is similar to warning coloration in that its benefit is usually accomplished through trial and error. If there are many more nontoxic mimics than real toxic individuals, predators that successfully eat the mimics will not learn to avoid them. Only if it is more likely that a predator will encounter an actual toxic and bad-tasting prey will it learn to avoid anything that looks similar to the prey it encountered.

Q3: Why is camouflage considered an adaptive response to predation?

A3: Camouflage enables prey species to blend in with their surroundings so that they're difficult for predators to detect. Random variation in coloration of a population enables those that are better hidden from predators to survive and reproduce, whereas those that do not match the surroundings well are easily seen and eaten. Over time, only well-camouflaged prey will survive to reproduce, passing the camouflage coloration on to their offspring.

Figure 20.12

Q1: What percentage of pigeons are caught when they are alone and not in a flock?

A1: Very close to 80 percent are caught.

Q2: For wood pigeons, what is the minimum number of individuals that provides protection from goshawks?

A2: As few as 2 to 10 individuals provide protection, but 11 or more have a strong effect on predation.

Q3: Why do you think a group of musk oxen versus a lone musk ox would be safer from a pack of wolves?

A3: The group of musk oxen could form a circle in which they face out and use their large horns to impale the wolves, keeping their sides and rear ends protected from attack. A lone ox would be completely unprotected from a pack of wolves working together.

Figure 20.13

Q1: Which species appears to be the superior competitor?

A1: *Aphytis lingnanensis* appears superior, because it occupied the same territory from 1948 to 1959, whereas the extent of *Aphytis chrysomphali* decreased.

Q2: Why is this example considered exploitative competition?

A2: These wasp species feed on the same foods in the same place—they directly compete for the resource—but they never physically interact or come in contact with each other.

Q3: What would you predict if these species had undergone competitive exclusion? (You will need to read the next paragraph to answer this question.)

A3: The first species would now be exploiting another resource—for example, apples rather than citrus crops.

Figure 20.15

Q1: What species represents the first colonizers of the sand dunes?

A1: The first colonizers were dune-building grasses.

Q2: What species is the intermediate species, and how does it become the dominant species?

A2: The pine is the intermediate species. It likely becomes dominant by outcompeting the grasses for water, nutrients, and sunlight. Once established, the pines' shade further inhibits the growth of grasses.

Q3: What species is the mature, climax community species, and how does it become the dominant species?

A3: The black oak is the mature, climax community species. It likely becomes the dominant species because it outcompetes the pines for water, nutrients, and sunlight.

Figure 20.16

Q1: What other types of disturbances could you imagine destroying a forest?

A1: Other types of disturbances could include clear-cutting or heavy logging, burning to create agricultural fields, flooding, clearing by tornado or hurricanes, and the creation of empty lots in a newly built subdivision.

Q2: How is secondary succession different from primary succession?

A2: Secondary succession starts with some existing producers that survived a disturbance, whereas primary succession starts from nothing—pure rock or sand. A disturbance that would result in primary succession would be a lava flow that covered an entire area; succession would have to start from pure lava rock and no existing producers.

Q3: What is a climax community?

A3: A climax community is the final step in succession, in which the species in the community remain and are not replaced by any other species as time goes by.

CHAPTER 21

Figure 21.3

Q1: What percentage of the 10,000 kilocalories in the first trophic level will be available to a shark that might eat the tuna in the fourth trophic level?

A1: At that point, 0.1 percent of the energy will be available.

Q2: What trophic level and term would describe a predator of tuna?

A2: A predator of tuna would be at the fifth trophic level and therefore a quaternary consumer.

Q3: Give an example of a primary consumer in a terrestrial environment.

A3: Possible answers include a deer, a grasshopper, a seed-eating bird, and a mouse.

Figure 21.4

Q1: Which organisms are the producers in this ecosystem?

A1: Phytoplankton are the producers.

Q2: Which organisms are responsible for nutrient flow from the biotic world to the abiotic world? (*Hint*: Read the next paragraph.)

A2: Decomposers are responsible for this flow of nutrients.

Q3: How do nitrogen and phosphorus move from abiotic to biotic elements of this ecosystem? How would that differ in a terrestrial system?

A3: These nutrients are taken up from the water by phytoplankton. On land, they would be taken up from the air or soil by plants.

Figure 21.5

Q1: How is a decomposer different from a more typical consumer?

A1: Decomposers feed off only dead organic matter (dead organisms). All other types of consumers feed off live plants and animals.

Q2: What would happen to nutrient cycles if decomposers did not exist?

A2: The nutrient cycles would stop, because the nutrients would not move back into the abiotic world and be available for other organisms.

Q3: Describe all the points at which heat is lost from the ecosystem in this figure.

A3: Heat is lost—to the atmosphere and eventually escaping it—as producers, consumers, and decomposers grow and reproduce.

Figure 21.8

Q1: Which terrestrial biome has the lowest productivity? Which aquatic biome has the lowest productivity?

A1: The lowest terrestrial productivity appears to occur in the desert and tundra biomes in the terrestrial map and in the marine biome of the open ocean in the aquatic map.

Q2: Where are the most productive terrestrial biomes located?

A2: They appear to be mainly near the equator.

Q3: Give a possible reason for your answer to question 2.

A3: More sunlight reaches Earth at the equator, so larger populations of producers can be supported.

Figure 21.9

Q1: In what years were chlorophyll levels the highest?

A1: They peaked in 1999 and 2008.

Q2: In what years were the temperature changes from the average the greatest?

A2: This happened in 2000 (low), 2003–4 (high), and 2008 (low).

Q3: How do you predict this graph will look 10 years from now? Explain your reasoning.

A3: If the world's oceans are warming, then the chlorophyll levels may continue to decrease.

CHAPTER 22

Figure 22.2

Q1: List the fertility rates in 2017 of the countries shown from highest to lowest. China (not shown) had a fertility rate of 1.63. Where on your list would China's data fit?

A1: Your list should read: Nigeria, 5.46; Guatemala, 2.92; Mexico, 2.15; Denmark, 1.79; USA, 1.77; Brazil, 1.71; China, 1.63; Armenia, 1.6; Singapore, 1.1.

Q2: List the approximate fertility rates in 1960 of the countries shown from highest to lowest. China (not shown) had a fertility rate of 5.75. Where on your list would China's data fit?

A2: Your list should read: Guatemala, 6.9; Mexico, 6.8; Nigeria, 6.4; Brazil, 6.1; Singapore, 5.8; China, 5.75; Armenia, 4.8; USA, 3.7; Denmark, 2.5.

Q3: Which country showed the greatest decrease in fertility over the 57 years depicted in the graph? How many births decreased per woman over this time? Which showed the least fertility decrease, and how many births decreased per woman over these years?

A3: Singapore experienced the greatest decrease, by 4.7 births per woman. Denmark experienced the least decrease, by 0.71 births per woman.

Figure 22.3

Q1: Which tissue type is the primary component of skin?

A1: Skin is primarily composed of epithelial tissue.

Q2: Which tissue type is the primary component of bone?

A2: Bone is primarily composed of connective tissue.

Q3: Which tissue types do you think the hand contains?

A3: Together, all four tissue types give the hand structure and enable it to function.

Figure 22.4

Q1: Which organ system defends the body from infectious diseases, such as the common cold or flu?

A1: The immune system performs that function.

Q2: Which two organ systems together transport oxygen to cells?

A2: The respiratory and circulatory systems perform that function.

Q3: Which two organ systems together regulate the activities of the other organ systems?

A3: The nervous and endocrine systems perform that function.

Figure 22.5

Q1: Why is it important to maintain a stable body temperature?

A1: The processes that occur in living things have an optimal temperature range and will be less efficient (or unable to occur) if body temperature deviates too much from that range.

Q2: Which organ system is mainly involved in sensing the body's external and internal states? (See Figure 22.4 for an overview of organ systems.)

A2: The nervous system performs that function.

Q3: Give another example of a homeostatic pathway in humans that is discussed in this chapter.

A3: Possible answers include the system for control of internal water content and solute concentration and the system for control of internal pH levels.

Figure 22.7

Q1: Identify one way in which oogenesis and spermatogenesis are similar.

A1: One possible answer is they each produce haploid gametes through meiosis.

Q2: How many eggs are produced from each precursor cell? How many sperm?

A2: One egg is produced. Four sperm are produced.

Q3: How much time elapses between the appearance of a precursor cell and the formation of an egg? How does this process differ for a sperm?

A3: Whereas only one egg completes development per month, new sperm are produced daily.

Figure 22.8

Q1: Which hormones important for the menstrual cycle are produced in the pituitary gland?

A1: Both follicle-stimulating hormone (FSH) and luteinizing hormone (LH) are produced in the pituitary gland.

Q2: Which hormone is involved in producing the uterine lining?

A2: Progesterone, secreted by the corpus luteum, stimulates a buildup of the uterine lining.

Q3: How is the egg follicle involved in producing hormones?

A3: The egg is carried within a follicle, which produces estradiol. After the egg is released from the follicle, which ruptures, the remaining corpus luteum produces progesterone.

Figure 22.9

Q1: If an egg is released but no sperm enter the oviduct, what is likely to occur? (*Hint*: Refer to Figure 22.8.)

A1: The unfertilized egg will travel into the uterus and then out of the body, and the menstrual cycle will continue.

Q2: If sperm enter the oviduct but no egg is present, what is likely to occur?

A2: The sperm will eventually die, and no pregnancy will occur.

Q3: Given the timing of egg release and the life span of sperm, when is it safe to have unprotected intercourse without risking pregnancy?

A3: It is essentially never safe to have unprotected intercourse without risking pregnancy, as sperm can live for 5 days inside the human body and eggs may be released at any point. In fact, the act of intercourse can cause the release of an egg(s).

Figure 22.10

Q1: Place these terms in the correct order of development: embryo, fetus, infant, zygote.

A1: The order of development is zygote, embryo, fetus, infant.

Q2: In what trimester is the fetus most likely to survive outside its mother's body? (*Hint*: Read the next two paragraphs of the text.)

A2: The fetus is most likely to survive independently during the third trimester.

Q3: The first trimester is the most sensitive time for exposure to mutagens (factors, such as chemicals, that may produce or increase mutations). Why might that be?

A3: This is when most organ systems develop, so any mutations could have profound effects on the viability of the developing fetus.

Figure 22.11

Q1: What is the role of estradiol in childbirth?

A1: It makes uterine muscles more sensitive to oxytocin.

Q2: Explain how the involvement of hormones in childbirth is an example of a positive feedback loop.

A2: Contractions of the uterus are caused by oxytocin and increasing contractions, in turn, cause more oxytocin to be produced.

Q3: If a woman has been pregnant for more than 40 weeks, her doctor might give her an injection of oxytocin to precipitate labor. How would that bring about labor?

A3: Because of the positive feedback loop with oxytocin and uterine contractions, the initial injection of oxytocin should cause contractions—which would then kick off the feedback loop, with the woman's body now releasing more oxytocin.

CHAPTER 23

Figure 23.1

Q1: In which two countries did people born in 2017 have the longest life expectancy? In which two countries did people born in 2017 have the shortest life expectancy?

A1: Japan and Israel had the longest life expectancy in 2017. Mexico and India had the shortest.

Q2: In which two countries did people born in 1995 have the longest life expectancy? In which two countries did people born in 1995 have the shortest life expectancy?

A2: Japan and Sweden had the longest life expectancy in 1995. China and India had the shortest.

Q3: In what year did China's life expectancy exceed that of Mexico? In what year did Israel's life expectancy first exceed that of Canada? In what year did Israel's life expectancy first exceed that of Sweden?

A3: China's life expectancy exceeded Mexico's life expectancy in 2005. Israel's life expectancy first exceeded that of Canada in 2005 and that of Sweden in 2010.

Figure 23.3

Q1: Which type(s) of muscles can you voluntarily contract?

A1: You can voluntarily contract skeletal muscles.

Q2: Do the muscles in the heart contract voluntarily or involuntarily?

A2: Heart muscle contraction is under involuntary control.

Q3: You do not have to think about breathing (otherwise, sleeping would be dangerous!), but you *can* increase or decrease your rate of breathing. Are the muscles involved in breathing, then, under voluntary or involuntary control?

A3: The muscles involved in breathing are under involuntary control, but we can override them to consciously control breathing.

Figure 23.4

Q1: How is skeletal muscle attached to bones in the skeleton?

A1: It is attached via tendons.

Q2: Are skeletal muscles under voluntary or involuntary control?

A2: For the most part, skeletal muscles are involved in voluntary contractions. However, they can sometimes contract involuntarily (for example, when your leg twitches during sleep or any reflex action).

Q3: Is a muscle fiber one cell or multiple cells?

A3: A muscle fiber is considered one cell, but it is created by the fusion of several cells.

Figure 23.5

Q1: List muscle structures from smallest to largest, beginning with sarcomeres.
A1: From smallest to largest, the muscle structures are sarcomeres, myofibrils, muscle fibers, muscle fiber bundles, and muscles.

Q2: What are the components of the sarcomere?
A2: The sarcomere consists of myosin filaments, actin filaments, and Z discs.

Q3: Across animal species, the microscopic structure of muscles is the same. Why, do you think, are there differences in strength among animals?
A3: Animals are of different sizes and shapes, so their muscles are too. If a mouse were as large as an elephant, it would likely be as strong.

Figure 23.6

Q1: Participants who vigorously exercised 1.25–2.5 hours per week or moderately exercised 2.5–5.0 hours per week, as recommended by the guidelines, increased their chance of living longer by what percentage?
A1: Participants increased their chance of living longer by 31 percent.

Q2: How much moderate exercise increased participants' chance of living longer by 20 percent? How much moderate exercise maxed out participants' increased chance of living longer? What was the maximum percentage chance?
A2: Even minimal moderate exercise, anything over zero minutes, increased participants' chance of living longer by 20 percent. Participants maxed out their increased chance of living longer at 7.6 hours of moderate exercise. The maximum chance of living longer was 39 percent.

Q3: Previous studies suggested that 12.5 hours of vigorous exercise per week, such as elite athletes commonly perform, can be harmful. Does this study support that conclusion? Why?
A3: This study doesn't appear to support the conclusion that 12.5 hours of vigorous exercise per week can be harmful. The graph shows no decrease in longevity at the highest number of hours of vigorous or moderate activity. However, we need to be careful about extending the conclusions of these data. The graph does not show percentages beyond 12.5 hours of vigorous exercise or 25 hours of moderate exercise per week.

Figure 23.7

Q1: The collarbone is part of which skeleton: axial or appendicular?
A1: It is part of the appendicular skeleton.

Q2: Which parts of the skeleton are made of cartilage?
A2: Cartilage is a component of joints, such as elbows, knees, and shoulders, as well as parts of most ribs, the nose, and the ears.

Q3: Which skeleton—axial or appendicular—protects the central nervous system (the brain and spinal cord)?
A3: The axial skeleton protects the central nervous system.

Figure 23.9

Q1: How does spongy bone differ from compact bone?
A1: Compact bone forms the hard, white outer region without any cavities. Spongy bone is honeycombed with numerous tiny cavities and lies inside the compact bone.

Q2: What parts of a bone are made of living cells or living tissue?
A2: Osteocytes are living cells that surround themselves with a hard, mineral matrix of calcium and phosphate. Bone marrow is a living tissue that, depending on the type of bone, stores fat or produces blood cells. Bone also contains living blood vessels and nerve cells.

Q3: What characteristics of hollow bones make them important for organisms' survival?
A3: Hollow bones are the site of bone marrow, which is essential for forming new blood cells. In addition, the combined lightness and strength of hollow bones makes them especially important for flying animals.

Figure 23.10

Q1: What is the function of the synovial sacs?
A1: They produce lubricating fluid that allows joints to move more easily.

Q2: Compare and contrast ligaments and tendons.
A2: Ligaments connect bone to bone; tendons connect muscle to bone.

Q3: Knee injuries are some of the most common sports injuries. Why do you think that is?
A3: Knees are critical for running and jumping, and many sports involve both. In addition, so many components of the knee must function together that damage to one affects functioning of the entire knee.

Figure 23.11

Q1: List, in order, the structures of the digestive system that a piece of swallowed food would pass through, beginning with the oral cavity.
A1: Food would pass through the oral cavity, pharynx, esophagus, stomach, small intestine, large intestine, and anus.

Q2: What part of the digestive system hosts bacteria that produce vitamins?
A2: The large intestine performs this function.

Q3: What is the common function of the liver and gallbladder?
A3: They produce, store, and release bile to digest fats.

Figure 23.12

Q1: Which vitamins described in the figure are important for healthy bones?

A1: Vitamins A, B_{12}, C, and D contribute to bone formation and maintenance.

Q2: Which vitamins are you more likely to store and accumulate more than you need?

A2: You are more likely to store the fat-soluble vitamins: A, D, E, and K.

Q3: In your own diet, are there any vitamins that you may not be eating enough of?

A3: Possible answers include vitamin A if no yellow vegetables are eaten and B_{12} if no meat is consumed.

Figure 23.13

Q1: Why is a larger surface area important for absorption?

A1: With a larger surface area, a higher proportion of nutrients traveling through the small intestine will be absorbed into the body, as they are more likely to come into contact with the cells lining the small intestine.

Q2: What feature of the epithelial cells lining the villi increases absorption?

A2: The microvilli of the cells increase surface area even further and therefore increase absorption.

Q3: In your own words, explain the role of the blood vessels within each villus.

A3: The blood vessels are important for moving nutrients from the villi to the rest of the body.

CHAPTER 24

Figure 24.3

Q1: Where in the blood is the majority of carbon dioxide carried?

A1: Most of the carbon dioxide is carried in the plasma.

Q2: Where in the blood is the majority of oxygen carried?

A2: Most of the oxygen is carried in the red blood cells, attached to hemoglobin molecules.

Q3: What would happen if your red blood cells carried a mutation that made the hemoglobin less effective at binding to oxygen (as in sickle cell disease)?

A3: Less oxygen would be available to the body, which could have many negative effects; for example, one complication far less serious than sickle cell disease is that with less oxygen available you would have trouble exercising.

Figure 24.4

Q1: Why are arteries more muscular than veins?

A1: Arteries need to sustain higher pressures to move blood throughout the body.

Q2: What two structural features of capillaries enable easier diffusion into and out of surrounding tissues?

A2: Capillaries are very narrow, and they have thin, porous walls.

Q3: Why do you think capillaries are not typically transplanted?

A3: The small size of capillaries makes them harder to transplant than veins and arteries.

Figure 24.5

Q1: Beginning with the left atrium, list the locations of blood as it moves through the circulatory system.

A1: Blood moves from the left atrium to the left ventricle, arteries, capillaries, veins, right atrium, right ventricle, lungs, and then back to the left atrium.

Q2: Why is the left ventricle larger than the right ventricle, and why are its walls thicker and more muscular?

A2: The left ventricle must generate higher pressures to drive blood through the long systemic circuit that runs the length of the body. The right ventricle pushes blood through the much shorter pulmonary circuit.

Q3: What do you think is the function of an artificial pacemaker implanted in the heart?

A3: If someone's natural pacemaker (SA node) is not working, then the heart will not receive the necessary signal for it to contract in unison. An artificial pacemaker ensures that the heart receives the signal it needs.

Figure 24.6

Q1: When air enters the body, is it richer in oxygen or carbon dioxide? What about when it exits the body?

A1: When air enters the body, it is rich in oxygen. When it exits the body, it is rich in carbon dioxide.

Q2: The figure shows air entering through the nose. Where else can air enter the respiratory system?

A2: It can enter through the mouth.

Q3: Explain how the body creates a change in air pressure during breathing.

A3: During inhalation, the diaphragm pulls down and rib muscles push out to create a larger volume and thus decrease air pressure in the lungs, bringing air into the lungs from outside the body. During exhalation, the diaphragm moves up and rib muscles pull in, decreasing volume and increasing pressure so that air moves out.

Figure 24.7

Q1: Beginning with the nose, list the locations of oxygen as it moves through the upper respiratory system. Then beginning with the mouth, list the locations of oxygen as it moves through the upper respiratory system.

A1: From the nose, oxygen moves through the nasal cavity, pharynx, and larynx. From the mouth, it bypasses the nasal cavity, moving through the pharynx and larynx.

Q2: List the locations of oxygen as it moves through the lower respiratory system.

A2: It moves through the trachea into the bronchus of the lung.

Q3: From the bronchus, where does the oxygen move?

A3: From the bronchus, oxygen moves through bronchioles into alveoli, where it undergoes gas exchange.

Figure 24.8

Q1: Why is the large surface area created by alveoli important for gas exchange? How do the thin walls of alveoli help with gas exchange?

A1: The more surface area there is, the greater is the rate at which gases are exchanged by diffusion through the walls of the alveoli. The thinness of those walls makes it easier for gases to diffuse.

Q2: Does carbon dioxide move into or out of alveoli? Does it move into or out of capillaries at the surface of the alveoli?

A2: Carbon dioxide moves into alveoli. It moves out of capillaries.

Q3: When a person has pneumonia, the alveoli may fill with fluids. Why would this be a problem?

A3: Fluid inside the alveoli would make gas exchange of oxygen and carbon dioxide more difficult.

Figure 24.10

Q1: What is drained from the many collecting ducts of the kidney, what does it drain into, and then where does it go?

A1: Urine is drained into the ureter, enters the bladder, and is then excreted from the body.

Q2: What is the difference between reabsorption and secretion?

A2: Reabsorption is bringing substances (such as water and valuable solutes) back into the body; secretion is removing substances from the body (in urine).

Q3: Alcohol suppresses the kidney's ability to reabsorb water through the capillaries. What common consequence of drinking alcohol is related to this fact, and how might you alleviate this problem?

A3: Decreased reabsorption can lead to dehydration, a major cause of hangovers. You can alleviate this problem by increasing your intake of nonalcoholic fluids, such as water, before, during, and after you drink alcohol.

Figure 24.12

Q1: To which part of the nervous system—CNS or PNS—does the retina in your eye belong?

A1: Like all sensory neurons, the retina is part of the PNS.

Q2: To which part of the nervous system—CNS or PNS—does the spinal cord belong?

A2: It is part of the CNS.

Q3: In a *reflex arc*, sensory input is processed in the spinal cord and an immediate signal is sent to activate motor output without first routing the input signal to the brain. The "knee jerk" is one example of a reflex arc. Give another example of this kind of reflexive reaction to sensory input.

A3: Examples include squinting in bright light and/or flinching at a loud sound or rapidly moving object.

Figure 24.13

Q1: The cell body of a neuron contains all the structures common to animal cells. How do neurons look different from other cells you've learned about in this book?

A1: Neurons have long extensions called dendrites and axons.

Q2: What is the function of these differences?

A2: Dendrites and axons receive and send information rapidly from cell to cell.

Q3: Some people are born without the capacity to feel pain. Although it might initially sound nice not to feel pain, it is actually quite dangerous. Describe a situation in which it would be dangerous not to feel pain.

A3: One possible answer is that if you couldn't tell that the water was too hot in a shower or bath you could be badly burned.

CHAPTER 25

Figure 25.3

Q1: What brain region coordinates the endocrine system?

A1: The hypothalamus, a region on the underside of the brain, is the main coordinator of the endocrine system.

Q2: What is the relationship between the endocrine system, endocrine glands, and endocrine cells?

A2: The endocrine system consists of endocrine glands, which are made of endocrine cells (plus other types of cells), and endocrine cells embedded in other tissues or organs. Both the glands and the scattered cells can release hormones directly into the circulatory system.

Q3: What are the main male and female endocrine organs?

A3: Different sex organs contribute to the endocrine system: in males the testes, in females the ovaries.

Figure 25.4

Q1: How do hormones travel to target cells?

A1: Hormones travel through body fluids, especially the blood.

Q2: Distinguish between an endocrine cell and a target cell.

A2: Endocrine cells produce and secrete hormones, which then travel to target cells; target cells change in response to the presence of hormones.

Q3: Why is a hormone called a signaling molecule?

A3: A hormone signals that something has changed in the organism or its environment, triggering a cellular response.

Figure 25.5

Q1: Describe the two ways that a hormone outside a cell can exert its effect on a cell.

A1: The hormone may cross the plasma membrane, or it may bind to a receptor embedded in the membrane.

Q2: Within the cell, how does a hormone bring about a change in cell activity?

A2: The hormone may influence gene expression, metabolism, cytoskeletal organization, or membrane transport.

Q3: It takes very little of the hormone cortisol to have large effects throughout the body. Explain why.

A3: The binding of cortisol to specific cells (with the appropriate receptors) activates many proteins within those cells.

Figure 25.6

Q1: Describe an event (other than the one illustrated in the figure) that might cause the release of adrenaline.

A1: One possible answer: taking an exam.

Q2: What organs does adrenaline affect?

A2: It affects the liver and heart (also the immune system).

Q3: What do you think would happen if your adrenal glands were constantly releasing adrenaline?

A3: The resulting breakdown of glycogen to glucose, increase in heart rate, and suppression of the immune system could be damaging over a long time (bringing on the health problems associated with constant stress).

Figure 25.7

Q1: In animals, what is the main physical barrier that keeps out pathogens?

A1: Skin is animals' main physical barrier against pathogens.

Q2: Give an example of a chemical defense within the digestive system.

A2: One example is low pH in the stomach.

Q3: Explain why touching your eyes and nose during flu and cold season is not recommended.

A3: Viruses from your hands can move into your eyes and nose, where it is easier for them to gain access and infect you.

Figure 25.8

Q1: Place these terms in order from most to least inclusive: neutrophil, white blood cell, phagocyte, innate immune system.

A1: The correct order is innate immune system, white blood cell, phagocyte, and neutrophil.

Q2: Name a function that is similar between macrophages and neutrophils. Name a difference between these two cells.

A2: Both macrophages and neutrophils engulf pathogens (they are both phagocytes, a type of white blood cell). However, macrophages are found in the tissues, whereas neutrophils are found in the bloodstream until enticed by cytokines to enter the tissues. Neutrophils also differ from macrophages by using antimicrobial chemicals to attack. Macrophages release cytokines at sites of wounds and infections in tissues, but neutrophils do not.

Q3: Why would it be a problem if your innate immune system identified the insulin-producing cells in your pancreas as "nonself"?

A3: If your immune system thought that your insulin-producing cells were pathogens, it would attack and destroy them. (This is what occurs in type 1 diabetes.)

Figure 25.9

Q1: What is the role of white blood cells in inflammation?

A1: They destroy pathogens and engulf cellular debris.

Q2: What would happen if histamines were not produced during inflammation?

A2: Neutrophils could not leave the bloodstream to attack invading pathogens at the site of cellular damage.

Q3: Why is inflammation called a "nonspecific" immune response?

A3: It responds in the same way to any invading pathogen (and to any cellular damage).

Figure 25.10

Q1: Why is blood clotting an important immune response?

A1: It seals a potential point of entry for pathogens and also reduces blood loss from the wound.

Q2: How are the inflammatory response and blood clotting similar?

A2: Both are components of the innate immune system, and both are rapid responses to a wound.

Q3: Some people have a genetic disorder in which their blood cannot clot. Why would this be a problem?

A3: If your blood can't clot, even the smallest wound can cause enormous loss of blood.

Figure 25.11

Q1: Why are B and T cells so named?

A1: B cells mature in the bone marrow; T cells mature in the thymus.

Q2: In what way is this immune system "adaptive"?

A2: It adapts to specific invading pathogens, rather than having a generalized response to all pathogens as in the innate immune system.

Q3: Why is the adaptive immune response considered the third layer of the immune system?

A3: It responds more slowly to pathogens than do either the external defenses or the innate immune system. It also evolved later and is found only in vertebrates.

CHAPTER 26

Figure 26.3

Q1: What is the function of the vascular tissue?

A1: It moves water, sugars, and minerals throughout the plant.

Q2: A plant organ is green if the cells within it contain chloroplasts, which carry out photosynthesis. In the figure, which of the plant organs do not contain chloroplasts, and why do you think that is?

A2: The roots do not contain chloroplasts because the roots are not exposed to sunlight and so cannot photosynthesize.

Q3: Which tissue type has chloroplast-containing cells? Why?

A3: Ground tissue has chloroplast-containing cells because this is where photosynthesis occurs.

Figure 26.4

Q1: How do root hairs increase a plant's ability to absorb water and nutrients?

A1: Root hairs increase the plant surface area exposed to the soil, enabling the plant to absorb more water and nutrients.

Q2: How do roots make a plant more stable?

A2: Plants without broad and deep roots are more easily knocked over by wind, rain, animals, and so on.

Q3: Given question 2, which do you predict would be more stable in harsh weather: annuals or perennials?

A3: Perennials should be more stable than annuals as their roots are longer (because they continue to grow over multiple years).

Figure 26.5

Q1: What nutrients does phloem move from the leaves to other parts of the plant?

A1: Phloem moves sugars created through photosynthesis.

Q2: Which part of the shoot system—the bud, stem, or leaf—produces flowers?

A2: Buds produce flowers.

Q3: How is the location of stomata important for their function?

A3: Stomata take in carbon dioxide and release water and oxygen, so they need to be on the surface of the leaf.

Figure 26.7

Q1: Give an example of how plants use chemicals to defend themselves.

A1: One possible answer: Plants store toxic chemicals in their leaves so that herbivores won't eat them.

Q2: Of the plant hormones listed in this figure, which three promote growth?

A2: Auxins, cytokinins, and gibberellins promote growth.

Q3: What is a plant's first line of defense? Is this a chemical defense?

A3: A plant's first line of defense is its tough outer surface and defensive organs (for example, thorns); these are mainly physical rather than chemical defenses.

Figure 26.10

Q1: Are eggs and sperm haploid or diploid? Are they part of the sporophyte or gametophyte generation?

A1: Eggs and sperm are haploid. They are part of the gametophyte generation.

Q2: Why is the plant life cycle, but not the animal life cycle, called an "alternation of generations"?

A2: The plant life cycle, but not the animal life cycle, is called an "alternation of generations" because plants have a complete haploid structure, the gametophyte, whereas in animals, only the gametes are haploid.

Q3: How does asexual reproduction differ from the plant sexual reproductive life cycle diagrammed here?

A3: In asexual reproduction, no haploid gamete is produced, so the entire gametophyte portion of the life cycle (bottom half of the figure) does not occur.

Figure 26.11

Q1: What kinds of seeds would you expect to be contained within a sugary fruit: seeds spread by water or by animals? Why?

A1: The sugary fruit most likely contains seeds spread by animals. We can guess that natural selection resulted in sweet fruit that attracts animals to eat it (and excrete the seeds).

Q2: What kinds of seeds would you expect to weigh less: those spread by air or those artificially selected by humans? Why?

A2: Seeds spread by air are likely to be lighter than seeds artificially selected by humans. Natural selection will have produced lighter seeds that can be carried farther and dispersed more widely by wind. Humans tend to artificially select for larger fruit to create more food, and larger fruit tends to have larger and therefore heavier seeds.

Q3: How does the height of a tree affect the spread of its seeds?

A3: Seeds from taller trees should be dispersed more widely because they are higher up and better able to be lifted and carried long distances by wind.

Figure 26.12

Q1: Many crops, including apples, peppers, and tomatoes, depend on insect pollinators. What would happen if no pollinator visited such a plant?

A1: If a plant was not pollinated, the fruit wouldn't develop.

Q2: Flower petals are actually modified leaves, although they look very different from leaves. Do flowers still perform the main function of leaves? Explain your reasoning.

A2: Most flowers aren't green, suggesting that they don't have chloroplasts and therefore cannot produce food via photosynthesis, which is the main function of leaves.

Q3: Some flowers look like an insect. How might this resemblance attract pollinators?

A3: Male insects in search of a mate might be attracted to the flower, thinking it is a female of their own species. Alternatively, insect predators might be attracted to the flower, thinking it is food. Both the insects and their predators could serve as pollinators.

Figure 26.13

Q1: What traits are scientists selecting for in each new generation of Kernza?

A1: They are selecting for larger seeds, better flavor, and deeper roots.

Q2: What would happen if the Kernza plants were allowed to freely pollinate rather than being selectively pollinated by hand?

A2: Pollination would be random, so the desirable traits could be lost in the next generation of plants.

Q3: After choosing two plants to breed, researchers put the plant heads together under a bag as shown in the photo. Why do they do this before the plants begin to pollinate?

A3: They need to ensure that no inadvertent pollination occurs by wind or by an insect pollinator.

Appendix B: Answers to End-of-Chapter Questions

CHAPTER 1

1. d
2. e
3. observation, hypothesis, predictions
4. (a) 1, (b) 4, (c) 5, (d) 3, (e) 6 or 7, (f) 2, (g) 6 or 7
5. (a) organ, (b) organism, (c) population, (d) ecosystem, (e) organ system, (f) community
6. Observation: Bats have been observed with white noses.
   Hypothesis: Bats with white noses are infected with a fungus.
   Experiment: Inject bats with a fungicide and observe whether they are less likely to develop white-nose syndrome than are bats that are given sham injections.
7. a
8. c
9. d; even if this observation were true, it would not necessarily relate to a prion being the cause of this disease.
10. (a) hypothesis, (b) result, (c) experiment, (d) observation, (e) result
11. Possible answers:
    Cellular level: The fungus is a single-celled organism.
    Organism level: The fungal infections were infecting individual bats.
    Population level: Entire populations of bats were infected and died.
12. (a) No; a question is not a hypothesis. (b) Yes; this is a plausible and falsifiable explanation. (c) Yes; this is a plausible and falsifiable explanation. (d) No; a question is not a hypothesis. (e) No; proposing the existence of a "mysterious cloud" does not generate testable predictions.

CHAPTER 2

1. b
2. c
3. d
4. scientific literacy: 2, basic research: 5, applied research: 1, secondary literature: 4, primary literature: 3
5. credentials, bias, secondary literature
6. (a) 1, (b) 6, (c) 3, (d) 2, (e) 4, (f) 5
7. a
8. (a) primary, (b) neither, (c) neither, (d) secondary
9. c, because the doctor's knowledge may influence her results
10. (a) Pseudoscience. Dr. Oz has relevant credentials, but he has earned a reputation for offering nonscientific advice unsupported by research. In addition, the idea of a "fat-burning" dietary aid does not align with current scientific understanding.

    (b) Pseudoscience. The idea of a scientific conspiracy is a red flag, as is the rejection of scientific consensus. And though the *Wall Street Journal* is a respected newspaper, it tends to have a conservative bias.

    (c) Real science. Practicing scientists have reported experimental findings to their peers, but they do not yet appear to have published in a peer-reviewed journal, which will need to be their next step.

    (d) Pseudoscience. Astrologers are not scientists. There is no scientific evidence that date of birth has any effect on personality.

    (e) Real science. The journal *Diabetes* is a peer-reviewed and well-established scientific journal.
11. (a) The first graph shows the incidence of pertussis (whooping cough) cases in the United States from 1922 to 2017, with the inset showing a zoomed-in view of 1990–2017. The x -axis shows the year from which the pertussis cases were counted; the y -axis shows the counts. Any point on the line shows the total number of cases reported in the corresponding year on the x -axis. The general trend of the graph is a decrease in cases of pertussis over time.

(b) The reason for the increase in cases in the first decades of the twenty-first century is probably that parents were choosing not to have their children vaccinated.

(c) The second graph shows the incidence of pertussis (whooping cough) cases in the United States by age group from 1990 to 2017. The x -axis shows again the year from which the pertussis cases were counted, and the y -axis shows the counts. This graph's y -axis shows not the total number of cases but rather the number of cases per 100,000 individuals in the population of that particular age group. One point on each line shows the number of cases per 100,000 individuals in the population of that particular age group reported in the corresponding year on the x -axis. The general trend is an increase in cases of pertussis by age group over time. The lines differ by age group. The younger the age group, the greater the increase in pertussis cases per year.

(d) The incidence of pertussis cases per 100,000 individuals in the population of children under 1 year old is very high, and in people over 20 years old it is very low. Increase over time is much greater in children under 1 year old and is very minimal in people over 20 years old.

(e) Answers may vary.

12. (a) Evaluate the credentials of the people recommending the vaccine (in this case, it would be the ACIP and the CDC).

    (b) Assess any possible biases of the people in step 1.

    (c) Read the secondary literature (for example, recent articles in *Popular Science* or the *New York Times*) for an overview of the issue.

    (d) Read recent primary literature that was cited in the secondary literature (for example, articles in *JAMA*, *New England Journal of Medicine*, or *Vaccine*).

    (e) Review papers from the primary literature for credentials and biases of the authors, good research design, reasonable sample size, and conclusions.

13. Possible answers include the following: Dunning-Kruger effect—although you have little knowledge about how vaccines work, you assume that you know more than highly trained medical professionals and scientists. Confirmation bias—seeing a Facebook post from an anti-vaccine friend or relative supports your own concerns, which strengthens those concerns. Neglect of probability—you overestimate the likelihood of a reaction to a vaccine (highly unlikely) versus the chance of being infected with the particular disease if unvaccinated (increasingly likely). Omission bias—you prefer not acting (e.g., not being vaccinated) to acting in uncertainty (e.g., being vaccinated despite your worries about effects). Illusory correlation—because autism shows up developmentally around the same time that children are being vaccinated, you assume that the vaccine causes autism (see Figure 2.10 for an example of illusory correlation).

CHAPTER 3

1. a
2. d
3. ion: 3, matter: 6, solution: 9, element: 5, chemical compound: 8, molecule: 4, isotope: 7, polymer: 2, atom: 1
4. a
5. polymers, sugar, nucleotides, are not, carbon
6. a, c, e
7. Carbon is the basis of life on Earth because, thanks to its ability to form four bonds, it can form large molecules that contain thousands of atoms. In addition, it can form more complex molecules than can any other element, including hydrogen or oxygen.
8. (a) 18%, 1%, (b) nucleic acids, (c) carbohydrate, 5%, (d) protein, 18%
9. The recommended food categories that include more than one biomolecule class as nutrients are the vegetables (carbohydrates and, for some types, proteins), grains (proteins, carbohydrates), dairy (protein, carbohydrates, lipids), and protein foods (protein, carbohydrates, lipids). The recommended food categories that include only one biomolecule class as nutrients are the fruits (carbohydrates) and oils (lipids).
10. b
11. b
12. (a) grains, (b) none, (c) vegetables, dairy, fruit, (d) meat, eggs, and nuts
13. (a) oven, (b) coffee maker, (c) neither, (d) oven, (e) coffee maker
14. The nonpolar end of the detergent will bond to the oil in the salad dressing while the polar end bonds to the water molecules, lifting the oil into the wash water. Vinegar is a polar molecule, so it will dissolve in the wash water; you don't need detergent to remove vinegar.

CHAPTER 4

1. a, c, e. Viruses meet no criteria except for evolving as groups. The diamond meets no criteria.
2. c
3. d
4. receptor-mediated endocytosis: 2, phagocytosis: 1, pinocytosis: 4, exocytosis: 3

5. chloroplast: 7, Golgi apparatus: 4, lysosome: 5, mitochondrion: 6, nucleus: 1, rough endoplasmic reticulum: 2, smooth endoplasmic reticulum: 3

6.

Component	Prokaryotes	Eukaryotes	
		Animals	**Plants**
PLASMA MEMBRANE	X	X	X
CELLULOSE CELL WALL			X
NUCLEUS		X	X
ENDOPLASMIC RETICULUM		X	X
GOLGI APPARATUS		X	X
RIBOSOMES		X	X
CYTOSKELETON		X	X
MITOCHONDRIA		X	X
CHLOROPLASTS			X

7. a

8. The purple curve represents simple diffusion. The key to figuring this out is the dashed line, which indicates when the transport proteins are saturated (fully in use). According to the purple curve, the concentration of the dissolved substance and the rate of transport increase proportionally. This occurs even past the point of transport protein saturation. Why does that continuance matter? It matters because, in fact, transport proteins don't matter for this process—they are not involved in simple diffusion. By contrast, the green curve represents facilitated diffusion. Here, the curve reaches the transport protein saturation line but cannot exceed it, because facilitated diffusion relies on transport proteins to send the dissolved substance across the membrane. If the transport proteins are saturated, adding a higher concentration of solute will not increase the rate of transport.

9. b

10. (a) reproducing autonomously, (b) responding to the environment, (c) obtaining energy from the environment, (d) evolving as a group, (e) consisting of one or more cells

11. right side, more, fewer

12. c, because ATP is cellular fuel and active transport is the one process listed here that requires energy

13. (a) isotonic, (b) neither gain nor lose, (c) equal to, (d) hypertonic, (e) lose, (f) higher than, (g) hypotonic, (h) gain, (i) lower than

CHAPTER 5

1. c
2. a
3. c
4. b
5. d
6. opposite, catabolism, produces
7. The correct labels are the following:

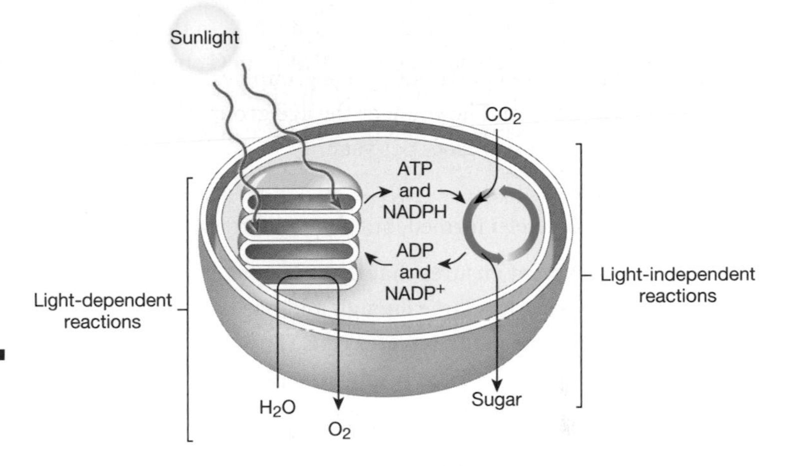

8. (a) 3, (b) 2, (c) 1, (d) 4
9. If all three enzymes are functional, the fur will be black. If only enzymes 1 and 2 are functional, it will be chocolate (brown). If only enzyme 1 is functional, it will be yellow (blonde).
10. e
11. (a) The reaction without an enzyme requires more energy to proceed. Its line on the graph representing how much energy is needed shows a higher peak on the y -axis.

    (b) This reaction is catabolic because there is less energy in the products than in the reactants, as shown by the line's lower position on the y -axis at the end of the reaction. Catabolic reactions release energy; therefore, their products have less energy than their reactants.
12. c
13. If metabolic processes occur at a higher rate than can be sustained by cellular respiration, then it is possible to die from lack of usable energy.
14. a

CHAPTER 6

1. b
2. a

3. cytokinesis, 4; S phase, 1; G_1 phase, 2; G_0 phase, 3
4. Meiosis, binary fission, homologous chromosomes, sister chromatids
5. (a) 1, (b) 2, (c) 5, (d) 3, (e) 4
6. b
7. b
8. d
9. During the G_1 phase, the cell has its original number of chromosomes. Chromosomes replicate during the S phase where there will be the original DNA strands and increasing amounts of DNA from newly synthesized strands of DNA. In the G_2 phase, the cell will contain twice as much DNA as in G_1, because S phase doubles the amount of DNA. In the M phase, the cell will contain that same amount of DNA until cytokinesis occurs.
10. (a) 54, (b) 27, (c) 27
11. c (so multiple nuclei formed, but the cell did not divide)
12. The G_1 checkpoint ensures that the cell is ready to divide—for example, that it is large enough and has enough energy to produce two normal daughter cells. The G_2 checkpoint ensures that the cell's DNA has been replicated and packed into pairs of sister chromatids. Bypassing the G_1 checkpoint could allow cells to divide before they're ready; bypassing the G_2 checkpoint could lead to the production of daughter cells with defective chromosomes.
13. Normal stomach tissue and ovary tissue have the highest number of dividing cells. Ovary tissue has the highest number of dividing cells for cancer. Ovary tissue shows the greatest increase in dividing cells between normal and cancerous cells.

CHAPTER 7

1. genotype: 4, phenotype: 5, heterozygote: 1, homozygote: 2, dominant: 6, recessive: 3
2. gene, alleles
3. meiosis, segregation, independent assortment
4. (a) M, (b) C, (c) C, (d) C, (e) M
5. a
6. e
7. incomplete dominance
8. d
9. pleiotropic (influencing multiple traits)
10. b, c, e
11. The first-generation result suggests that the round-shape allele is dominant to the oval-shape allele. The next cross should be to breed offspring with themselves to create an F_2 generation, and the proportion of ovals would be 1 in 4, or 25 percent (3:1 round to oval).
12. Orange color must be dominant to black color. You could breed their offspring to test the hypothesis.
13. (a) *llff* (homozygous recessive for both long hair and no furnishings)

    (b) *LlFf, LLFF, LlFF, LLFf*

    (c) 9:3:3:1 short furnished to short unfurnished to long furnished to long unfurnished.

CHAPTER 8

1. chromosomes, genes, loci, alleles
2. d
3. gene therapy: 3, in vitro fertilization: 5, preimplantation genetic diagnosis: 4, chorionic villus sampling: 1, amniocentesis: 2
4. The sex chromosomes are the two in the final pair. Because there is one large sex chromosome and one small one, the individual is a male.
5.

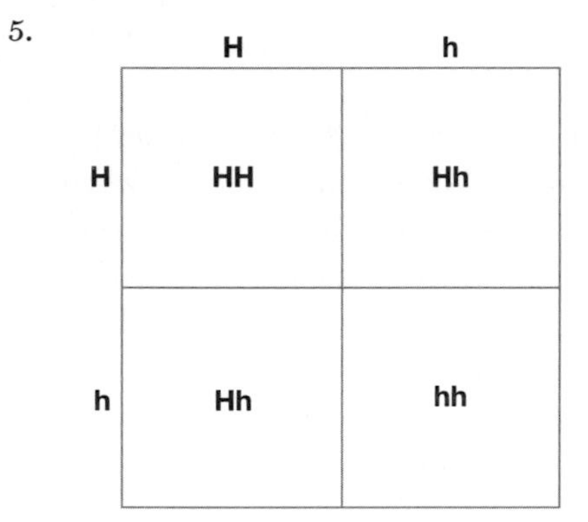

6. a
7. e
8. c
9. b, d, e
10. (a) pie graph; viral vectors; 68 percent of trials; naked DNA was the second most common vector (at nearly 18 percent of trials)

    (b) bar graph; x = year, y = number of approved gene therapy clinical trials; least: 1989, most: 2008; number of annual trials has increased since 1989
11. (a) from their mother, (b) no, (c) yes, (d) two types, (e) no, no, yes, female only
12. (a) $X^g X^g$, (b) $X^G X^g$, (c) $X^g Y$, (d) $X^g Y$, (e) $X^G X^g$
13. (a) *gg*, (b) *Gg*, (c) *Gg*, (d) *gg*

CHAPTER 9

1. c
2. d
3. a
4. d
5. nucleotide, 4; base pair, 1; DNA molecule, 3; base, 2
6. The correct labels are the following:

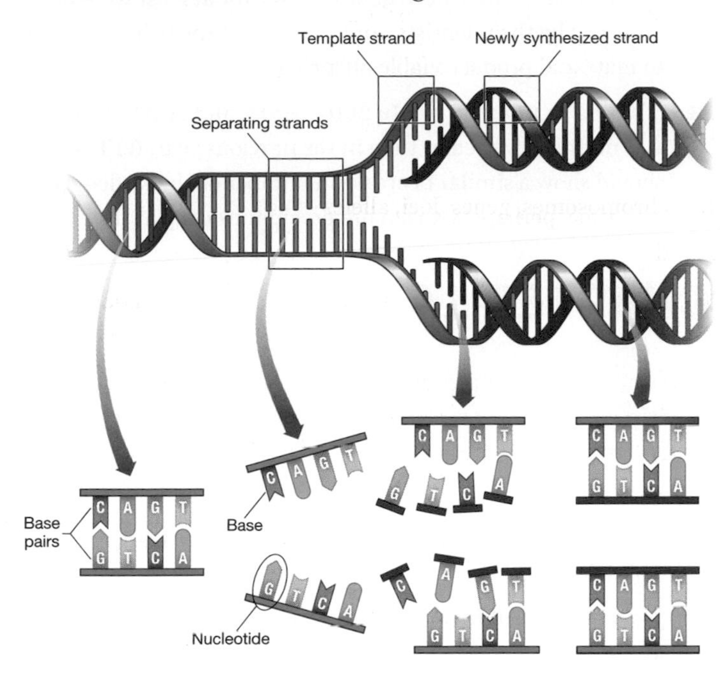

7. PCR, CRISPR
8. c
9. (a) 1, (b) 5, (c) 3, (d) 2, (e) 4
10. (a) deletion, (b) substitution, (c) insertion
11. ATGCAAATCCTGG and TACGTTTAGGACC
12. 20 percent T, 30 percent G, and 30 percent C (because T needs to be the same percentage as A, and the remaining 60 percent needs to be divided evenly between G and C)
13. (a) 72 percent, (b) 38 percent, (c) making a baby more intelligent was considered the least appropriate use of gene editing.
14. Yes, it matters. Noncoding DNA sequences still perform important functions and can easily be disrupted by mutations. Regulatory regions rely on specific nucleotide sequences, and they could become nonfunctional if those sequences are changed.

CHAPTER 10

1. gene expression: 2, gene regulation: 3, transcription: 1, translation: 4
2. (a) tRNA, (b) rRNA, (c) mRNA, (d) tRNA, (e) rRNA
3. redundant, unambiguous
4. The correct placements are:

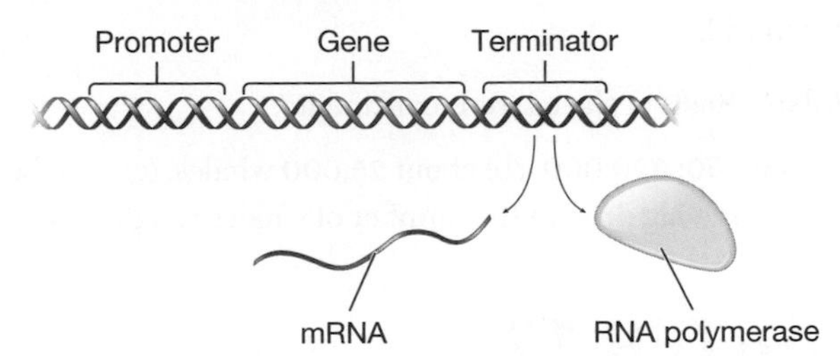

5. (a) 2, (b) 4, (c) 6, (d) 8, (e) 1, (f) 9, (g) 5, (h) 3, (i) 7
6. a, b, c, d
7. (a) asparagine, (b) stop codon, (c) isoleucine, (d) glycine, (e) proline
8. (a) CGU, CGC, CGA, CGG, AGA, or AGG; (b) GCU, GCC, GCA, or GCG; (c) AUG; (d) GGU, GGC, GGA, or GGG
9. b
10. e
11. a (removing the introns, as occurs during RNA splicing)
12. They are similar in that both involve building one molecule using another molecule as a template. They differ in that gene expression is DNA to RNA, whereas DNA replication is RNA to protein.
13. At control point 2 (transcription), there might be an error in upregulation (speeding up). At control point 3 (breakdown of mRNA), there might be an error in down-regulation (slowing down).
14. Unaltered sweet wormwood has 1.2 percent artemisinin. The genetic expression is higher in genetically engineered sweet wormwood by 2 percent and lower in genetically engineered tobacco by 0.4 percent. *Bonus*: Even though tobacco does not have as high a production of artemisinin by dry weight, it can be harvested more quickly and could be more profitable.

CHAPTER 11

1. c
2. a
3. c
4. biogeography: 4, fossil: 1, DNA sequence similarity: 2, embryonic development: 5, homologous traits: 3
5. d
6. b
7. adaptation, natural selection
8. In artificial selection, humans choose which organisms survive and reproduce. In natural selection, fitness for the particular environment determines which organisms survive and reproduce, as individuals with more beneficial inherited traits have a competitive advantage over individuals with less useful ones.

9. d
10. species Y
11. a
12. All (or most) of the current continents.
13. (a) 1956–60; 320,000, (b) about 25,000 whales, (c) 1986 because there was a big drop in the number of whales caught that year

CHAPTER 12

1. genetic drift; establish a new, distant population
2. a
3. b
4. c
5. b
6. genetic drift: 4, gene flow: 3, disruptive selection: 2, directional selection: 5, stabilizing selection: 1
7. An individual that survives well (natural selection) but is unattractive to potential mates or is unable to compete for access to mates will not reproduce and will not pass on genes (sexual selection).
8. Gene flow is the most likely mechanism because of these facts about the other three mechanisms: mutations are random (so they could have produced these results but aren't likely to have), natural selection would have caused populations in different environments to diverge, and genetic drift is most relevant for small populations.
9. Population bottlenecks cause individuals in the resulting population to be more genetically similar to each other. In this case, the two individuals are so similar that each individual does not distinguish another devil's cells as different from its own cells.
10. a
11. a. FQR-E.coli was the most common in 2018, as it has been since 2005. In 2004, it was the second most common.

    b. MRSA was at 5 percent in 2018; it has decreased in percentage over time.

    c. VRSA is likely not on the graph because it is not very common; these are the five most common antibiotic-resistant bacterial strains.

CHAPTER 13

1. b
2. a
3. c
4. c
5. genetic divergence, allopatric
6. Postzygotic barriers, an infertile hybrid
7. d
8. b, c, d
9. (a) 5, (b) 4, (c) 1, (d) 2, (e) 3
10. a
11. b
12. The 10 species must have differed significantly in a combination of traits that were unobservable (or at least were not observed by the scientists) and that meant the fish were unable to mate and produce viable offspring.
13. (a) 5,085 in 1993; (b) No. In 2002, 2004, and 2009, the size was marginally greater than in the previous year; (c) The graph should show a similar decrease in number of chicks fledged over time, perhaps with a more precipitous drop at the end.

CHAPTER 14

1. c
2. b
3. a
4. d
5. clade: 2, node: 3, lineage: 5, evolutionary tree: 4, shared derived trait: 1
6. prokaryotes, Eukarya, Plantae, Animalia
7. d
8. e
9. (a) Eukarya, Animalia; (b) Eukarya, Fungi; (c) Eukarya, Plantae; (d) Eukarya, Protists; (e) Eukarya, Animalia
10. (a) kingdom, phylum, class, order, family, genus, species; (b) domain; (c) answers will vary
11. (a) 3, (b) 1, (c) 4, (d) 2, (e) 5
12. (See figure below.)

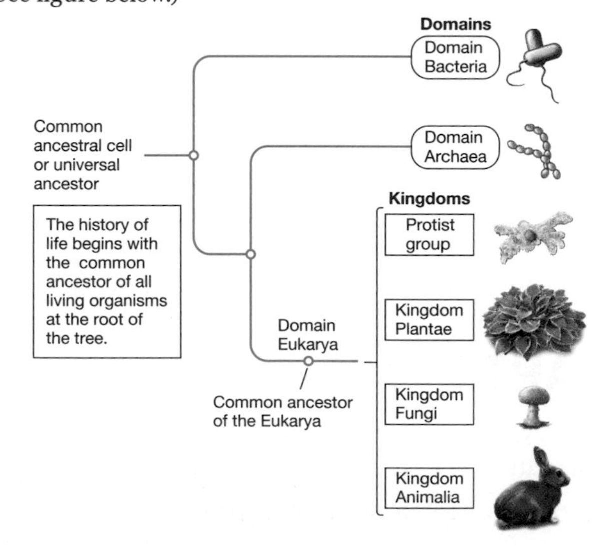

13. (a) domains: Bacteria, Archaea; kingdom: Animalia (also protists); (b) domains: Bacteria, Archaea; (c) domain: Archaea; (d) kingdom: Animalia; (e) domain: Bacteria, kingdom: Plantae (also protists)

14. Xu's tree: "We hypothesize that *Archaeopteryx* and *Xiaotingia* are dinosaurs, closely related to the deinonychosaurs." Godefroit's tree: "We hypothesize that *Archaeopteryx* and *Xiaotingia* are early birds, and *Aurornis* is an earlier bird."

15. (a) orders; (b) no vertebrates had yet made the water-to-land transition during the Silurian; (c) amphibians were the first; (d) birds are the most diverse; (e) amphibian diversity increased until the Permian and has since declined slowly; it has not increased nearly as much as other terrestrial vertebrates over time.

CHAPTER 15

1. d
2. e
3. c
4. b
5. Archaea, Bacteria, Prokaryotes, eukaryotes
6. b
7. a
8. (a) 1, (b) 5, (c) 4, (d) 2, (e) 3
9. a
10. Almost all individual eukaryotes weigh much more than almost any individual prokaryote.
11. Prokaryotes have a single circular chromosome and divide by binary fission, a simpler and quicker process than mitosis. Because they replicate so much more quickly, it is hard for our immune systems to respond to them in a timely way.

CHAPTER 16

1. a
2. d
3. c
4. d
5. Protists, aquatic, autotrophs
6. (a) 3, (b) 5, (c) 4, (d) 1, (e) 2
7. e
8. The kingdom Plantae contains only autotrophs. The kingdom Fungi contains only heterotrophs.
9. For each case, answers to the first part of the question will vary; here's an example for part (b): Lichens are a symbiotic relationship between a fungus and intracellular bacteria or alga. The photosynthetic bacteria produce food for the fungus, and the fungus provides protection and anchoring for the bacteria. Answers to the second and third parts of the question are as follows:

   (a) The partners of mycorrhizae are plants (domain Eukarya, kingdom Plantae) and fungi (domain Eukarya, kingdom Fungi). The relationship is a mutualism.

   (b) The partners of a lichen are fungi (domain Eukarya, kingdom Fungi) and either algae (domain Eukarya, Protists) or cyanobacteria (domain Bacteria). The relationship is a mutualism.

   (c) The partners in this relationship are hermit crabs and the (nonliving) shells of various species of snails (both domain Eukarya, kingdom Animalia). The relationship is a commensalism.

   (d) The partners in this relationship include the malaria protozoan, its insect host, and its vertebrate host (all domain Eukarya; protists for the malaria protozoan, kingdom Animalia for the two host types). It is a parasitic relationship.
10. Gymnosperms and angiosperms produce pollen, which is able to travel via wind or a pollinator to other individuals of the same species, thereby enabling sexual reproduction in a nonaquatic environment.
11. Possible answers include athlete's foot, fungal pneumonia, and yeast infections.
12. a. The highest percentage is about 34 percent, which occurred on day 100.

    b. The control condition had the lowest percentage inactivated.

    c. Either *Gymnopilus* spp. or *Pantomorus* spp. would have the most effect after a year of treatment.

CHAPTER 17

1. a
2. a
3. d
4. Mitochondrial DNA, Nuclear DNA
5. (a) N, (b) R, (c) B, (d) B, (e) B
6. Of the saber-toothed cats, *Smilodon fatalis* and *Smilodon populator* are the most closely related, because they are placed closest on the tree and equally branched off at the end. *Smilodon fatalis* or *Smilodon populator* and *Dinofelis* are the least related, because they are branched off the farthest apart among the saber-toothed cat branch. Of the modern cats, the

lion, leopard, jaguar, and tiger are most closely related to the saber-toothed cats, because their group branches off first from the ancestors of saber-toothed and modern cats branch. The cougar and cheetah are most closely related to the domestic cat, because their group is closest to it and equally branched off the end of the tree.

7. c
8. Monotremes lay eggs rather than developing young internally via a placenta. Marsupials have a simple placenta and so give birth to relatively undeveloped young, which then develop externally in a pouch. Eutherians have a well-developed placenta and so can support their young internally until they are more fully developed.
9. (a) 4, (b) 1, (c) 5, (d) 2, (e) 3
10. b
11. *Tiktaalik roseae* most likely lived in an environment with shallow water. (Specifically, it most likely lived in an area with oxygen-poor and nutrient-poor water but nutrient-rich earth, such as brought about by seasonal flooding.)
12. Possible answers include (a) shark, (b) hagfish, (c) bird, and (d) frog.
13. It suggests that Neanderthals evolved and left Africa before modern humans evolved in Africa and that modern humans intermingled with Neanderthal populations only after leaving Africa.

CHAPTER 18

1. e
2. b
3. d
4. (a) 1, (b) 3, (c) 4, (d) 2, (e) 5
5. Weather, climate, Climate change, global warming
6. d
7. Carbon sink because more carbon is absorbed in the years between the fires than is produced by the fires.
8. (a) 4, (b) 1, (c) 2, (d) 5, (e) 3
9. c
10. Answers will depend on student location.
11. Answers will depend on student location.

CHAPTER 19

1. d
2. b
3. c
4. d
5. a
6. Density-dependent, density-independent
7. b
8. c
9. (a) 1, (b) 4, (c) 3, (d) 5, (e) 2
10. One possibility is shown here:

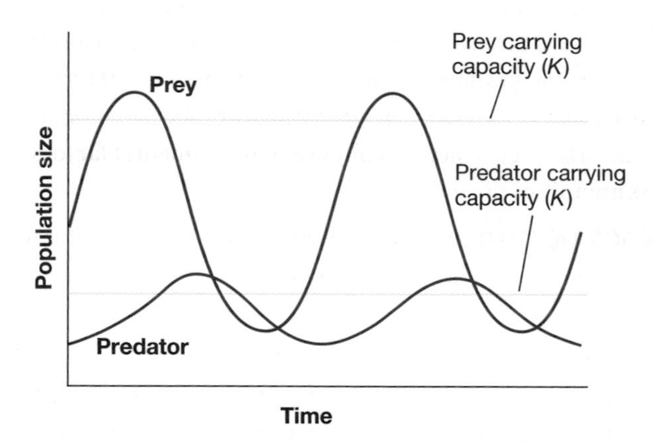

11. The food dispenser increased the carrying capacity dramatically.
12. b

CHAPTER 20

1. d
2. b
3. c
4. c
5. secondary consumer, primary consumer, producer
6. e
7. mutualism: 2, commensalism: 3, predation: 1, competition: 4
8. (a) 4, (b) 1, (c) 2, (d) 3, (e) 5
9. More diverse communities have low relative species abundance and high species diversity.
10. c
11. Relative species abundance will decrease, and species richness will increase.
12. (a) S, (b) S, (c) P, (d) P
13. (a) humans and killer whales, (b) phytoplankton, (c) baleen whale, sperm whale, and crabeater seal, (d) other herbivorous plankton and krill

CHAPTER 21

1. a
2. c
3. d
4. c
5. biome: 3, ecosystem: 2, primary productivity: 1, ecological community: 4
6. chaparral, grassland, marine
7. b
8. (a) 4, (b) 2, (c) 1, (d) 5, (e) 3
9. An ecosystem is composed of multiple ecological communities, and it also includes physical environmental factors (communities do not).
10. (a) 100. (b) 100,000. (c) The energy is expended in metabolism, and because predation is not 100 percent efficient (that is, predators don't eat every last bit of the calories in their prey, leaving, for example, the bones), some energy is lost there too.
11. The water cycle is more similar to the movement of nutrients, because water cycles through the biotic and abiotic components of an ecosystem, whereas energy flows in only one direction through ecosystems because of the steady loss of heat.
12. (a) Average invertebrate density decreased from about 60 g/m^2 to about 20 g/m^2. Mud crabs have caused the density of native invertebrates to drop to a third of its normal amount. (b) Average chlorophyll density increased from about 5 mg/m^3 to about 10 mg/m^3. Mud crabs are associated with a doubling of phytoplankton in this area. (c) Double the amount of energy is being brought into the ecosystem, based on the doubling of phytoplankton. However, that energy is not moving up through the ecosystem because the primary consumer populations have dropped so much.

CHAPTER 22

1. d
2. a
3. c
4. d
5. a
6. spermatogenesis: 2, oogenesis: 3, sexual reproduction: 4, asexual reproduction: 1
7. homeostatic, set point, negative, thermoregulation
8. a
9. asexual reproduction
10. Most: Finland and Denmark, least: Italy and Romania, largest gap: Denmark
11. (a) 4, (b) 1, (c) 3, (d) 5, (e) 2
12. d
13. Yes, there is a difference between theoretical and actual effectiveness, and there are two possible causes. First, intercourse could happen too soon after surgery, when an egg is past the point of ligation or (more commonly) sperm are still in the vas deferens (see Figure 22.9). Second, the tubes can grow back together, as happens in about one out of 200 surgeries for women (0.5% failure) and one out of 2,000 surgeries for men (0.05% failure).
14. It is vital to protect newborns from temperature extremes. Because they have a higher surface area–to–volume (SA-to-V) ratio than older and larger people do, they can chill or overheat more quickly. They also need to be fed more often, and they become dehydrated more quickly because of their high SA-to-V ratio.

CHAPTER 23

1. c
2. a
3. b
4. d
5. Adipose, dermis, epithelial
6. skeletal muscle: 3, smooth muscle: 4, cardiac muscle: 2, voluntary muscle: 1, involuntary muscle: 5
7. c
8. muscle, myosin, actin, Z, sarcomere
9. b
10. (a) 4, (b) 1, (c) 3, (d) 5, (e) 2
11. b
12. d; swimming is not weight-bearing and does not directly build muscle (as in weight lifting), so it is the least effective for increasing bone strength
13. "Cavity" = mouth; "32 protruding bones" = teeth; "meat tentacle" = tongue; "vat of acid" = stomach; "absorb its essence and transform it in[to] energy" = action of the small intestine (mainly)

CHAPTER 24

1. d
2. c
3. a
4. b
5. d
6. (a) CNS, (b) PNS, (c) CNS, (d) PNS

7. SA node: 3, left ventricle: 2, right atrium: 1, pulmonary circuit: 4
8. b
9. nephron, reabsorb, secrete, ureter, urethra
10. About 0 percent met all seven criteria. About 10–11 percent met six of the seven criteria. A higher percentage of females met three of the seven criteria.
11. (a) 4, (b) 2, (c) 1, (d) 5, (e) 3
12. Veins are larger to transport large quantities of blood. Capillaries are smaller to facilitate the transfer of oxygen and nutrients into surrounding cells.
13. b
14. The net transport of CO_2 will be from the lung air space to the alveolar capillaries. Gas exchange occurs through diffusion, which is the movement of a substance from an area of higher concentration to an area of lower concentration.
15. (a) The decrease in blood vessel flexibility that comes with age makes it harder for the body to respond to environmental changes by increasing blood pressure or heart rate. Blood vessels lose the ability to respond to the body's signals for such changes. In addition, such changes can damage an increasingly inflexible cardiovascular system. (b) In the short term, heart rate increases. In the long term, resting heart rate decreases (because the heart is stronger from exercise). (c) Some people become anxious in the doctor's office, and anxiety can increase blood pressure.

CHAPTER 25

1. c
2. b
3. d
4. a
5. b
6. adaptive immune response: 4, endocrine system: 1, innate immune response: 2, hypothalamus: 3
7. c
8. (a) A, (b) A, (c) C, (d) A, (e) C
9. primary, secondary, active, Passive, active
10. a, d
11. (a) 4, (b) 5, (c) 3, (d) 1, (e) 2
12. Sustained elevated levels of cortisol in the bloodstream, for example, have been linked to anxiety, depression, digestive problems, and more. A long-term suppressed immune system increases the risks of viral, bacterial, parasitic, and fungal infection, cancer, and heart disease, and causes a decrease in the bone marrow's ability to produce blood cells. Individuals with decreased immune function could be more susceptible to infection by SARS-CoV-2, the COVID-19 virus. Interestingly, it is possible that a cytokine storm, often associated with COVID-19, would be less likely to occur in these individuals with suppressed immune systems.
13. Percentages do not add up to 100 percent (they add up to more than 100 percent) because some children are allergic to more than one of these items. The top three child food allergies in the United States are peanuts, milk, and shellfish. The three least common child food allergies are wheat, soy, and sesame. Some 0.9 percent of kids are allergic to eggs.
14. If you "turn off" one of the body's immune response mechanisms, it may not be able to fight a pathogen/infection as well. In addition, some people are allergic to one or more of these medications.

CHAPTER 26

1. a
2. b
3. c
4. a
5. d
6. modular growth: 3, indeterminate growth: 2, primary growth: 1, secondary growth: 4
7. c
8. e
9. Annual, perennial, annual, perennial, monocot, fibrous root
10. (a) 3, (b) 1, (c) 4, (d) 2
11. Because bats are pollinating flowers at night, they will not be as attracted to colorful flowers as birds are, and they will be more drawn to strong-smelling flowers. In addition, because pollinating bats are generally larger than pollinating birds, the bats will tend to seek sturdier flowers and ones that have larger nectar rewards.
12. Onions are modified leaves, enveloping a bud at the tip of the stem (the green parts at the top are also leaves). Tomatoes (often considered vegetables) are a fruit, as suggested by their often plentiful seeds and their sweetness when ripe.
13. Ethylene is clearly released as a gas into the air because it doesn't require physical contact to act. This would explain why it works more effectively in an enclosed space, such as a bag, which would prevent it from dispersing through the air as quickly.
14. The highest monarch count was in 1997 at 682 million butterflies. The lowest was in 2014 at 25 million butterflies. The average monarch count in 2008–12 was 138.8 million. In 2013–17, the average was 74.2 million. Between these two periods, monarch butterflies decreased by 64.6 million, a loss of 47 percent.

Glossary

A

ABA See **abscisic acid**.

abiotic Nonliving. Compare *biotic*.

abscisic acid (ABA) A plant hormone that mediates adaptive responses to drought, cold, heat, and other stresses.

absorption The uptake of mineral ions and small molecules by cells lining the cavity of the digestive tract.

acid A chemical compound that loses hydrogen ions (H^+) in aqueous surroundings. Compare *base* (definition 1).

acidic Having the properties of an acid, or containing acid; having a pH below 7. Often contrasted with *basic*.

actin filament One of the two kinds of protein filaments, made of the protein actin, that is found in sarcomeres. The sliding of *myosin filaments* against actin filaments enables sarcomeres to contract.

active immunity Immunity to a particular pathogen that is conferred by antibodies made by the body itself. Compare *passive immunity*.

active site The location within an enzyme where substrates are bound.

active transport The movement of a substance in response to an input of energy. Compare *passive transport*.

adaptation 1. An evolutionary process by which a population becomes better matched to its environment over time. 2. See **adaptive trait**.

adaptive immune system The immune system's third line of defense against pathogens, which mounts responses against specific invaders via antibodies (antibody-mediated immunity) and phagocytes and other specialized cells (cell-mediated immunity). Compare *external defenses* and *innate immune system*.

adaptive radiation The expansion of a group of organisms to take on new ecological roles and to form new species and higher taxonomic groups.

adaptive trait Also called an *adaptation*. A feature that gives an individual improved function in a competitive environment.

adenine (A) One of the four nucleotides that make up DNA. The other three are thymine (T), guanine (G), and cytosine (C).

ADP Adenosine diphosphate.

adrenal glands The paired glands, located atop the kidneys, that release the hormones cortisol, adrenaline (epinephrine), and noradrenaline (norepinephrine), which launch a number of rapid physiological responses, including boosting blood glucose levels.

aerobes Prokaryotes that need oxygen to survive.

aerobic Requiring oxygen. Compare *anaerobic*.

algae One of two main groups of protists, whose members are photosynthetic and may or may not be motile. Compare *protozoans*.

alleles Different versions of a given gene.

allopatric speciation The formation of new species from geographically isolated populations. Compare *sympatric speciation*.

alternation of generations The life cycle of flowering plants, in which haploid stages (gametophytes) alternate with diploid stages (sporophytes).

alveoli (sing. alveolus) Small clusters of minute sacs resembling a bunch of grapes where gases are exchanged in the lungs.

amino acid Any of a class of small molecules that are the building blocks of proteins.

amniocentesis A prenatal genetic screening technique in which amniotic fluid is extracted from the pregnancy sac that surrounds a fetus by means of a needle that is inserted through the abdomen into the uterus. Compare *chorionic villus sampling*.

anabolism Metabolic pathways that create complex biomolecules from smaller organic compounds. Compare *catabolism*.

anaerobes Prokaryotes that survive without oxygen.

anaerobic Not requiring oxygen. Compare *aerobic*.

analogous trait A feature that is shared across species because of convergent evolution, not because of modification by descent from a recent common ancestor. Compare *homologous trait*.

analytical study An observational study that looks for patterns in the information collected and addresses how or why those patterns came to exist. Compare *descriptive study*.

anatomy The study of the structures that make up a complex multicellular body. Compare *physiology*.

anchorage dependence The phenomenon in which cells stop dividing when detached from their surroundings.

androgens Any of the hormones produced in the testes that stimulate cells to develop characteristics of maleness, such as beard growth and the production of sperm. Compare *estrogens* and *progestogens*.

angiogenesis The formation of new blood vessels.

angiosperms Flowering plants. One of two main groups of seed-bearing plants, characterized by seeds enclosed in an ovary and by flowers. Compare *gymnosperms*.

Animalia The animals. One of the five kingdoms of life, in the domain Eukarya, encompassing all animals, including humans, birds, and dinosaurs.

animals Multicellular ingestive heterotrophs; that is, complex organisms (eukaryotes) that obtain energy and carbon by ingesting food into their bodies and digesting it internally.

annual A plant that completes its life cycle in one year. Compare *biennial* and *perennial*.

antibody Any of various Y-shaped proteins that recognize invaders by their *antigens* and attack them in the antibody-mediated response of the adaptive immune system.

antibody-mediated immunity A response mounted by the adaptive immune system in which specific invaders are attacked by antibodies. Compare *cell-mediated immunity*.

anticodon A unique sequence of three nitrogenous bases at one end of a tRNA molecule that binds to the corresponding *codon* on an mRNA molecule.

antigen Any of various nonself markers found on pathogens that mark them for destruction by *antibodies*.

anus The muscle-lined opening through which solid waste is eliminated from the body.

appendages Body parts with specialized functions that develop in pairs from particular segments of an animal's body.

appendicular skeleton The part of the skeleton that has to do with motion. It is made up of the arms, legs, and pelvis. Compare *axial skeleton*.

applied research Research in which scientific knowledge is applied to human issues and often commercial applications. Compare *basic research*.

Archaea One of the three domains of life (compare *Bacteria* and *Eukarya*) and also one of the five kingdoms of life. The domain and kingdom Archaea consists of single-celled organisms best known for living in extremely harsh environments.

artery Any of the large vessels that transport blood away from the heart. Compare *capillary* and *vein*.

artificial selection The process by which individuals display specific traits through selective breeding. Compare *natural selection*.

ascomycetes Sac fungi, one of three main groups of fungi. Compare *basidiomycetes* and *zygomycetes*.

asexual reproduction The process by which clones, offspring that are genetically identical to the parent, are generated. Compare *sexual reproduction*.

atom The smallest unit of an element that retains the element's distinctive properties. Atoms are the building blocks of all matter.

atomic mass number The sum of the number of protons and the number of neutrons in an atom's nucleus. Compare *atomic number*.

atomic number The number of protons in an atom's nucleus. Compare *atomic mass number*.

ATP Adenosine triphosphate, a small, energy-rich organic molecule that is used to store energy and to move it from one part of a cell to another. Every living cell uses ATP.

atrium (pl. atria) Either of the two upper chambers of the heart. Compare *ventricle*.

autosome Any chromosome that is not one of the *sex chromosomes*.

autotroph A metabolic producer, an organism that makes food on its own. *Chemoautotrophs* acquire their energy from inorganic chemicals in their environment. *Photoautotrophs* absorb the energy of sunlight and take in carbon dioxide to conduct photosynthesis. Compare *heterotroph*.

auxin Any of a group of plant hormones that are necessary for cell division and promote root formation. Compare *cytokinin* and *gibberellin*.

axial skeleton The part of the skeleton that supports and protects the long axis of the body. It includes the skull, the ribs, and a long, bony spinal column. Compare *appendicular skeleton*.

axon The part of a neuron that sends information to other cells. Compare *dendrite*.

B

B cell A type of lymphocyte that matures in the bone marrow. B cells are involved in antibody-mediated immunity. Compare *T cell*.

Bacteria One of the three domains of life (compare *Archaea* and *Eukarya*) and also one of the five kingdoms of life. The domain and kingdom Bacteria includes familiar disease-causing bacteria.

base 1. A chemical compound that accepts hydrogen ions (H^+) in aqueous surroundings. Compare *acid*.[+] 2. The nitrogen-containing component of nucleotides. The four nitrogenous bases of DNA are adenine, cytosine, guanine, and thymine. In RNA, thymine is replaced by uracil.

base pair Also called *nucleotide pair*. Two nucleotides that form one rung of the DNA ladder.

base-pairing rules The rules that govern the pairing of nucleotides in DNA. Adenine (A) on one strand can pair only with thymine (T) on the other strand, and cytosine (C) can pair only with guanine (G).

basic Having the properties of a base, or containing a base; having a pH greater than 7. Often contrasted with *acidic*.

basic research Research that is intended to expand the fundamental knowledge base of science. Compare *applied research*.

basidiomycetes Club fungi, one of three main groups of fungi. Compare *ascomycetes* and *zygomycetes*.

behavioral trait A characteristic of an individual's behavior, such as shyness or extroversion. Compare *biochemical trait* and *physical trait*.

benign Referring to a tumor, confined to one site.

bias A prejudice or opinion for or against something.

biennial A plant that takes two years to complete its life cycle, growing and maturing in the first year and reproducing in the second year. Compare *annual* and *perennial*.

bilateral symmetry An animal body plan in which the body can be divided by just one plane passing vertically from the top to the bottom of the animal into two halves that mirror each other. Compare *radial symmetry*.

bile A fluid, produced by the liver, that helps digest fats by creating a coating that enables the fat globules to interact with water molecules to partially dissolve them.

binary fission A type of cell division in which a cell simply divides into two equal halves, resulting in daughter cells that are genetically identical to each other and to the parent cell.

biochemical trait A characteristic due to specific chemical processes of an individual, such as the level of a particular enzyme. Compare *behavioral trait* and *physical trait*.

biodiversity The variety of all the world's life-forms, as well as their interactions with each other and the ecosystems they inhabit.

biogeography The geographic locations where organisms or the fossils of a particular species are found.

biological fitness An individual's survival and successful reproduction.

biological hierarchy A way to visualize the breadth and scope of life, from the smallest structures to the broadest interactions between living and nonliving systems that we can comprehend.

biological species concept The idea that a species is defined as a group of natural populations that can interbreed to produce fertile offspring and cannot breed with other such groups.

biome A large region of the world defined by shared physical characteristics (especially climate) and a distinctive community of organisms.

biomolecule Also called *macromolecule*. A large organic molecule that is critical for living cells. Four major classes of biomolecules are proteins, carbohydrates, nucleic acids, and lipids.

biosphere All the world's living organisms and the physical spaces where they live.

biotic Living. Compare *abiotic*.

bipedal Walking upright on two legs.

blood The cells, cell fragments, and plasma that circulate through the heart and blood vessels.

blood clotting A role of the innate immune system to close a wound by clotting blood to reduce blood loss.

blood pressure The force of blood pushing from the heart through blood vessels.

blood vessel A vessel (for example, vein, capillary, artery) that transports blood throughout the body.

brain The part of the central nervous system that controls and coordinates nerve signals throughout the body and has a large capacity for processing diverse types of sensory information.

breathing The process of taking air into the lungs (inhaling) and expelling air from them (exhaling).

bronchioles A series of ever-smaller tubes that branch from the bronchi and open into the alveoli.

bronchus (pl. bronchi) Either of two small tubes that the trachea branches into in the chest and that lead to the lungs.

bryophytes A group of nonflowering plants that includes liverworts and mosses.

C

Calvin cycle See **light-independent reactions**.

Cambrian explosion The burst of evolutionary activity, occurring about 540 million years ago, that resulted in a dramatic increase in the diversity of life. Most of the major living animal groups first appear in the fossil record during this time.

camouflage Any type of coloration or appearance that makes an organism hard to find or hard to catch. Compare *mimicry* and *warning coloration.*

cancer cell Also called *malignant cell.* A tumor cell gains anchorage independence and starts invading other tissues.

capillary Any of the tiny blood vessels that exchange materials by diffusion with nearby cells. Compare *artery* and *vein.*

capsule An additional protective layer that surrounds the cell wall in prokaryotes.

carbohydrate Any of a major class of biomolecules, including sugars and starches, built of repeating units of carbon, hydrogen, and oxygen.

carbon cycle The movement of carbon within biotic communities, between living organisms and their physical surroundings, and within the abiotic world.

carbon dioxide (CO_2) The most abundant and consequential of the greenhouse gases.

carbon fixation See **light-independent reactions**.

carbon sink A natural or artificial reservoir that absorbs more carbon than it releases. Compare *carbon source.*

carbon source A natural or artificial reservoir that releases more carbon than it absorbs. Compare *carbon sink.*

cardiac muscle The specialized muscle that helps produce the coordinated contractions known as heartbeats. It has a banded appearance and branched muscle fibers, and its contractions are entirely involuntary. Compare *skeletal muscle* and *smooth muscle.*

cardiovascular system A closed circulatory system consisting of a muscular heart, a complex network of blood vessels that collectively form a closed loop, and blood that circulates through the heart and blood vessels.

carnivore An animal (or, rarely, plant) that eats other animals. Compare *herbivore* and *omnivore.*

carrying capacity The maximum population size that can be sustained in a given environment.

cartilage A dense tissue of the skeleton that combines strength with flexibility. It is found almost everywhere that two bones meet and prevents them from grinding together.

catabolism Metabolic pathways that release chemical energy in the process of breaking down complex biomolecules. Compare *anabolism.*

causation A statistical relation indicating that a change in one aspect of the natural world causes a change in another aspect. Compare *correlation.*

cell The smallest and most basic unit of life—a microscopic, self-contained unit enclosed by a water-repelling membrane.

cell cycle The sequence of events that make up the life of a typical eukaryotic cell, from the moment of its origin to the time it divides to produce two daughter cells.

cell division The final stage of the cell cycle. Cell division includes the transfer of DNA from the parent cell to the daughter cells.

cell-mediated immunity A response mounted by the adaptive immune system in which specific invaders are attacked by phagocytes and other specialized cells. Compare *antibody-mediated immunity.*

cell theory One of the unifying principles of biology; a theory stating that every living organism is composed of one or more cells, and that all cells living today came from a preexisting cell.

cellular respiration The reciprocal process to *photosynthesis,* in which sugars are broken down into energy usable by the cell.

cellulose A chemical substance that provides structural strength in plant cells. Compare *lignin.*

central nervous system (CNS) One of two main parts of the nervous system, consisting of the brain and spinal cord. Compare *peripheral nervous system.*

centromere The central region of a chromosome that attaches sister chromatids together.

cervix The lower portion of the uterus, which narrows and connects the uterus to the vagina.

chemical bond A force that holds two atoms together.

chemical compound A molecule that contains atoms from two or more different elements.

chemical reaction The process of breaking existing chemical bonds and creating new ones.

chemoautotroph See **autotroph**.

chemoheterotroph See **heterotroph**.

chlorophyll A green pigment that is specialized for absorbing light energy.

chloroplast An organelle of plant cells and some protist cells that captures energy from sunlight and uses it to manufacture food molecules via photosynthesis.

chordates A large phylum that encompasses all animals with backbones, such as fishes, birds, and mammals.

chorionic villus sampling (CVS) A prenatal genetic screening technique in which cells are extracted by gentle suction from the villi (a cluster of cells that attaches the pregnancy sac to the wall of the uterus). Ultrasound is used to guide the narrow, flexible suction tube through a woman's vagina and into her uterus. Compare *amniocentesis.*

chromatin fiber A chromosome in relaxed form, made up of DNA and nucleosomes.

chromosomal abnormality Any change in the chromosome number or structure, compared to what is typical for a species.

chromosome A DNA double helix wrapped around spools of proteins.

chromosome theory of inheritance The theory that genes are located on chromosomes, and that these chromosomes are the basis for all inheritance.

circulatory system The organ system that moves oxygen from the respiratory system to the heart, which then pumps oxygen-rich blood to the rest of the body.

citizen science Research that is assisted by members of the public, who participate by collecting and sometimes analyzing data in cooperation with professional scientists.

citric acid cycle See **Krebs cycle**.

clade A branch of an evolutionary tree, consisting of an ancestor and all its descendants.

class The unit of classification in the Linnaean hierarchy above order and below phylum.

climate The prevailing *weather* of a specific place over relatively long periods of time (30 years or more).

climate change A large-scale and long-term alteration in Earth's climate, including such phenomena as *global warming*, change in rainfall patterns, and increased frequency of violent storms.

climax community A mature community whose species composition remains stable over long periods of time.

CNS See **central nervous system**.

codominance An interaction between two alleles of a gene that causes a heterozygote to display a phenotype that clearly displays the effects of both alleles. Compare *incomplete dominance*.

codon A unique sequence of three mRNA bases that either specifies a particular amino acid during translation or signals the ribosomes where to start or stop translation. Compare *anticodon*.

coevolution The tandem evolution of two species that results because interaction between the two so strongly influences their survival.

collagen A tough but pliable protein that, in addition to being the main component of cartilage, is found in a great variety of tissues.

colon Also called *large intestine*. The final portion of the digestive system, where water and nutrients are absorbed before the remaining solid waste is expelled.

commensalism A species interaction in one species benefits at no cost to the other species. Compare *competition*, *mutualism*, and *predation*.

common ancestor An organism from which many species have evolved.

common descent The sharing of a common ancestor by two or more different species.

community The populations of different species that live and interact with one another in a particular place.

compact bone One of the two major types of bone tissue. It forms the hard, white outer region of bones. Compare *spongy bone*.

competition A species interaction in which both species may be harmed. Compare *commensalism*, *mutualism*, and *predation*.

competitive exclusion principle The idea that different species that use the same resource can coexist only if one of the species adapts to using other resources.

complementary base-pairing The relationship between two nucleic acid strands in which each purine on one strand hydrogen-bonds with a specific pyrimidine on the opposite strand. A pairs with T or U, and C pairs with G.

complex trait A genetic trait whose pattern of inheritance cannot be predicted by Mendel's laws of inheritance.

condensation The transition from gas to liquid. Compare *evaporation*.

conjugation The physical process of transferring genetic material through direct contact.

connective tissue A tissue that binds and supports tissues and organs.

consumer An organism that obtains energy by eating all or parts of other organisms or their remains. Compare *producer*.

contraceptive Any means of preventing pregnancy.

control group The group of subjects in an experiment that is maintained under a standard set of conditions with no change in the independent variable. Compare *treatment group*.

controlled experiment An experiment that measures the value of a dependent variable for two groups of subjects that are comparable in all respects except that one group (the treatment group) is exposed to a change in the independent variable and the other group (the control group) is not.

convection cell A large and consistent atmospheric circulation pattern in which warm, moist air rises and cool, dry air sinks. Earth has four giant convection cells.

convergent evolution Evolution that results in organisms that have different genetics but appear very much alike.

corpus luteum The cells of the ruptured follicle that remain behind in the ovary after ovulation to produce the hormone progesterone.

correlation A statistical relation indicating that two or more aspects of the natural world behave in an interrelated manner:

if one shows a particular value, we can predict a particular value for the other aspect. Compare *causation*.

cotyledon A food-storing organ that is part of the tiny, embryonic seedling that lies inside a plant seed.

covalent bond The sharing of electrons between two atoms. Compare *hydrogen bond* and *ionic bond*.

credentials Evidence of qualifications and competence to be recognized as an authority on a subject. Such evidence would include education and accomplishments.

CRISPR Clustered regularly interspaced short palindromic repeats. An RNA sequence that guides molecular machinery to a DNA sequence through complementary base-pairing.

cross See **genetic cross**.

crossing-over The physical exchange of chromosomal segments between homologous chromosomes. Compare *genetic recombination*.

CVS See **chorionic villus sampling**.

cyclical fluctuation A relatively predictable pattern of change in population size that occurs when at least one of two species is strongly influenced by the other. Compare *irregular fluctuation*.

cytokine Any of a group of signaling proteins that immune system cells use to communicate when an invader is present.

cytokinesis Division of the cytoplasm, the second step of mitotic division, resulting in two self-contained daughter cells. Compare *mitosis*.

cytokinin Any of a group of plant hormones that are necessary for cell division and promote shoot formation. Compare *auxin* and *gibberellin*.

cytosine (C) One of the four nucleotides that make up DNA. The other three are adenine (A), thymine (T), and guanine (G).

cytoskeleton The network of protein cylinders and filaments that forms the framework of a cell.

D

data (sing. datum) Information collected in a scientific study.

decomposer A scavenger that dissolves the dead bodies of other organisms to consume them. Compare *detritivore*.

deletion A mutation in which a base is deleted from the DNA sequence of a gene. Compare *insertion* and *substitution*.

dendrite The part of a neuron that receives information from other cells. Compare *axon*.

density-dependent population change A change in population size that occurs when birth and death rates change as the population density changes. Compare *density-independent population change*.

density-independent population change A change in population size that occurs when populations are held in check by factors that are not related to the density of the population. Compare *density-dependent population change*.

deoxyribonucleic acid See **DNA**.

dependent variable Any variable that responds, or could potentially respond, to changes in an *independent variable*.

dermal tissue One of three types of plant tissue. Forming the outermost layer of the plant, dermal tissue protects the plant from the outside environment and controls the flow of materials into and out of the plant. Compare *ground tissue* and *vascular tissue*.

descriptive study An observational study that reports information about what is found in nature. Compare *analytical study*.

determinate growth The general growth pattern of animals, in which they grow early in their lives and then level off. Compare *indeterminate growth*.

detritivore A scavenger that mechanically breaks apart the dead bodies of other organisms to consume them. Compare *decomposer*.

diastole The second pressure, or bottom number, in a blood pressure reading—the pressure at the point the heart relaxes after a contraction. Compare *systole*.

dicot Any plant that has two cotyledons in each of its seeds. Dicots are characterized by parallel veins in the leaves, scattered vascular bundles, fibrous roots, and floral parts in multiples of three. Examples include beans, squash, oak trees, and roses. Compare *monocot*.

diffusion The movement of a substance from a region of higher concentration to a region of lower concentration.

digestion The chemical breakdown of food.

digestive system The organ system that breaks down food in the mouth, stomach, and small intestine and absorbs nutrients in the small and large intestine.

dihybrid cross A controlled mating experiment involving organisms that are heterozygous for two traits.

diploid Possessing a double set of genetic information, represented by $2n$. Somatic cells are diploid. Compare *haploid*.

directional selection The most common pattern of natural selection, in which individuals at one extreme of an inherited phenotypic trait have an advantage over other individuals in the population. Compare *disruptive selection* and *stabilizing selection*.

disruptive selection The least common pattern of natural selection, in which individuals with either extreme of an inherited trait have an advantage over individuals with an intermediate phenotype. Compare *directional selection* and *stabilizing selection*.

DNA Deoxyribonucleic acid, the genetic code of life, consisting of two parallel strands of nucleotides twisted into a double helix. DNA is the genetic material that transfers information from parents to offspring.

DNA polymerase The enzyme that builds new strands of DNA in DNA replication.

DNA replication The duplication of a DNA molecule.

DNA sequence similarity The degree to which the sequences of two different DNA molecules are the same—a measure of how closely related two DNA molecules are to each other.

domain The highest hierarchical level in the organization of life, describing the most basic and ancient divisions among living organisms. The three domains of life are Bacteria, Archaea, and Eukarya.

dominant allele An allele that prevents a second allele from affecting the phenotype when the two alleles are paired together. Compare *recessive allele.*

dominant genetic disorder A genetic disorder that is inherited as a dominant trait on an autosome. Compare *recessive genetic disorder.*

double helix The spiral formed by two complementary strands of nucleotides that is the backbone of DNA.

down-regulation The slowing down of gene expression. Compare *up-regulation.*

E

Earth equivalent The number of planet Earths needed to provide the resources we use and absorb the wastes we produce.

ecological community An association of species that live in the same area.

ecological footprint The area of biologically productive land and water that an individual or a population requires to produce the resources it consumes and to absorb the waste it produces.

ecological isolation The condition in which closely related species in the same territory are reproductively isolated by slight differences in habitat. Compare *geographic isolation* and *reproductive isolation.*

ecological niche The set of conditions and resources that a population needs in order to survive and reproduce in its habitat.

ecology The scientific study of interactions between organisms and their environment, where the environment of an organism includes both biotic factors (other living organisms) and abiotic (nonliving) factors.

ecosystem A particular physical environment and all the communities in it.

ecosystem process Any of four processes—nutrient cycling, energy flow, water cycling, and succession—that link the biotic and abiotic worlds in an ecosystem.

egg Also called *ovum.* The female gamete. Compare *sperm.*

electron A negatively charged particle found outside the nucleus of an atom. Compare *neutron* and *proton.*

electron transport chain An elaborate chain of chemical events in which electrons and protons (H^+) are handed over to other molecules that ultimately generates ATP and NADPH.

element A pure substance that has distinctive physical and chemical properties, and that cannot be broken down into other substances by ordinary chemical methods.

elimination The removal from the body of solid waste, consisting mostly of indigestible material and bacteria that inhabit the digestive tract.

embryo The earliest stage of development of an individual after fertilization, up to 2 months of age in humans. Compare *fetus.*

embryonic development The process by which an embryo develops. Common patterns of embryonic development across species provide evidence of evolution.

endocrine gland A gland that releases hormones into body fluids, such as the bloodstream, for transport to target cells throughout the body.

endocrine system The organ system, consisting of a number of glands and secretory tissues that produce and secrete hormones, that works closely with the nervous system to regulate all other organ systems.

endocytosis The process by which materials are transported into a cell via vesicles. Compare *exocytosis.*

endoplasmic reticulum (ER) An extensive and interconnected network of sacs made of a single membrane that is continuous with the outer membrane of the nuclear envelope. See also **rough ER** and **smooth ER.**

endoskeleton An internal skeleton. Compare *exoskeleton.*

endosymbiotic theory An evolutionary theory stating that eukaryotic cells evolved through a process in which a larger prokaryotic cell engulfed smaller free-living prokaryotes that were not digested and eventually developed into mitochondria, chloroplasts, and possibly other organelles.

energy The capacity of any object to do work, which is the capacity to bring about a change in a defined system.

energy capture The trapping and storing of solar energy by the producers at the base of an ecosystem's energy pyramid.

energy carrier A molecule that can store and deliver usable energy.

energy pyramid A pyramid-shaped representation of the amount of energy available to organisms in a food chain.

enzyme Any of a class of small molecules that speed up chemical reactions.

epistasis A form of inheritance in which the phenotypic effect of the alleles of one gene depends on the presence of alleles for another, independently inherited gene.

epithelial tissue A tissue that covers organs and lines body cavities.

ER See **endoplasmic reticulum**.

esophagus The food tube that connects the pharynx to the stomach.

essential amino acid Any of the eight amino acids that can be obtained only from food.

estradiol The primary estrogen. Compare *progesterone* and *testosterone*.

estrogens Any of the hormones produced in the ovaries that play a role in determining female characteristics such as wide hips, a voice that is pitched higher than that of males, and the development of breast tissues. Compare *progestogens* and *androgens*.

ethylene A plant hormone that stimulates the ripening of some fruits, including apples, bananas, avocadoes, and tomatoes.

Eukarya One of the three domains of life, including all the living organisms that do not fit into the domains *Archaea* or *Bacteria*, from amoebas to plants to fungi to animals.

eukaryote An organism that belongs to the Eukarya. Animals, plants, fungi, and protists are all eukaryotes. Compare *prokaryote*.

eutherians One of three main groups of mammals, whose members have a placenta and produce offspring that are born in a well-developed state. Compare *marsupials* and *monotremes*.

evaporation The transition from liquid to gas. Compare *condensation*.

evolution A change in the overall inherited characteristics of a group of organisms over multiple generations.

evolutionary tree A model of evolutionary relationships among groups of organisms that is based on similarities and differences in their DNA, physical features, biochemical characteristics, or some combination of these. It maps the relationships between ancestral groups and their descendants, and it clusters the most closely related groups on neighboring branches.

exocytosis The process by which materials are exported out of a cell via vesicles. Compare *endocytosis*.

exon A stretch of DNA that carries instructions for building a protein. Compare *intron*.

exoskeleton An external skeleton. Compare *endoskeleton*.

experiment A repeatable manipulation of one or more aspects of the natural world.

experimental group See **treatment group**.

exploitative competition Competition between species in which the two species indirectly compete for shared resources, such as food. Compare *interference competition*.

exponential growth A pattern of population growth in which the population increases by a constant proportion over a constant time interval, such as one year. Exponential growth occurs when there are no constraints on resources and is represented by a J-shaped curve. Compare *logistic growth*.

external defenses The immune system's first line of defense against pathogens, consisting of physical and chemical barriers that reduce the likelihood of harmful organisms or viruses gaining access to internal tissues. Compare *adaptive immune system* and *innate immune system*.

F

F_1 generation The first generation of offspring in a series of genetic crosses. Compare *F2 generation* and *P generation*.

F_2 generation The second generation of offspring in a series of genetic crosses. Compare *F1 generation* and *P generation*.

facilitated diffusion Diffusion that requires transport proteins. Compare *simple diffusion*.

fact A direct and repeatable observation of any aspect of the natural world. Compare *theory*.

fallopian tube See **oviduct**.

falsifiable Able to be refuted.

family The unit of classification in the Linnaean hierarchy above genus and below order.

fat-soluble vitamin A vitamin that cannot dissolve in water and therefore tends to accumulate in body fat because it cannot be so easily excreted in urine. Compare *water-soluble vitamin*.

feces The solid waste produced by digestion.

feedback loop The steps of a process that either decrease (*negative feedback*) or increase (*positive feedback*) the output of that process.

fermentation A metabolic pathway by which most anaerobic organisms extract energy from organic molecules. It begins with glycolysis and is followed by a special set of reactions whose only role is to help perpetuate glycolysis. Fermentation enables organisms to generate ATP anaerobically.

ferns A group of vascular, nonflowering plants that reproduce via spores.

fertilization The fusion of two gametes, resulting in a zygote.

fetus The second stage of development of an individual, from 2 months to birth in humans. Compare *embryo*.

filtration The blood-cleansing work of the kidneys.

flagellum (pl. flagella) A long, whiplike structure that assists bacteria in locomotion.

flower A structure in angiosperms that enhances sexual reproduction by bringing male gametes (sperm cells) to the female gametes (egg cells) in highly efficient ways, by attracting animal pollinators through scent, shape, and color.

follicle-stimulating hormone (FSH) A hormone, produced by the pituitary gland, that launches each menstrual cycle by stimulating the growth of ovarian follicles.

food chain A single direct line of who eats whom among species in a community. Compare *food web*.

food web The way in which various and often overlapping *food chains* of a community are connected.

fossil The mineralized remains or impression of a formerly living organism.

founder effect A form of genetic drift that occurs when a small group of individuals establishes a new population isolated from its original, larger population.

fruiting body In fungi, the structure resulting from mating that releases offspring as sexual spores.

FSH See **follicle-stimulating hormone**.

Fungi; the fungi (sing. fungus) One of the five kingdoms of life, in the domain Eukarya, distinguished by their modes of reproduction. Fungi are absorptive heterotrophs.

G

G_0 phase A nondividing state of the cell.

G_1 phase "Gap 1," the first phase in the life of a newborn cell.

G_2 phase "Gap 2," the phase of the cell cycle between the S phase and cell division.

gallbladder An organ of the digestive system that stores bile made by the liver and dispenses the bile into the small intestine as needed.

gamete A sex cell. Male gametes are sperm; female gametes are eggs. Compare *somatic cell*.

gametophyte A haploid, multicellular structure produced by mitotic division of a plant's spore. Cells in the gametophyte differentiate into egg or sperm. Compare *Sporophyte*.

gene The basic unit of information, consisting of a stretch of DNA, that codes for a distinct genetic characteristic.

gene expression The process by which genes are transcribed into RNA and then translated to make proteins.

gene flow The exchange of alleles between populations.

gene regulation The changing of the genes that are expressed in response to internal signals or external cues that allows organisms to adapt to their surroundings by producing different proteins as needed.

gene therapy A genetic engineering technique for correcting defective genes responsible for disease development.

genetic bottleneck A form of genetic drift that occurs when a drop in the size of a population causes a loss of genetic variation.

genetic carrier An individual who has only one copy of a recessive allele for a particular disease and therefore can pass on the disorder allele but does not have the disease.

genetic code The information specified by each of the 64 possible codons.

genetic cross A controlled mating experiment performed to examine how a particular trait may be inherited.

genetic disorder A disease caused by an inherited mutation, passed down from a parent to a child.

genetic divergence The presence of differences in the DNA sequences of genes.

genetic drift A change in allele frequencies produced by random differences in survival and reproduction among the individuals in a population.

genetic engineering The permanent introduction of one or more genes into a cell, tissue, or organism.

genetic modification (GM) Altering the genes of an organism for a specific purpose.

genetic recombination The exchange of DNA between homologous chromosomes brought about by *crossing-over*, contributing to variation in gametes.

genetic trait Any inherited characteristic of an organism that can be observed or detected in some manner.

genetically modified organism (GMO) Any organism with genetic material that has been altered using genetic engineering techniques.

genome The complete set of genes of an organism.

genotype The allelic makeup of a specific individual with respect to a specific genetic trait. Compare *phenotype*.

genus (pl. genera) The unit of classification in the Linnaean hierarchy above species and below family.

geographic isolation The condition in which populations are separated by physical barriers. Compare *ecological isolation* and *reproductive isolation*.

gibberellin Any of a group of plant hormones that bring about stem growth through both cell elongation and cell division. Compare *auxin* and *cytokinin*.

global warming A significant increase in the average surface temperature of Earth over decades or more. Compare *climate change*.

glycolysis The first of three stages of cellular respiration. During glycolysis, sugars (mainly glucose) are split to make the three-carbon compound pyruvate. For each glucose molecule that is split, two molecules of ATP and two molecules of NADH are released. Compare *Krebs cycle* and *oxidative phosphorylation*.

GM See **genetic modification**.

GMO See **genetically modified organism**.

Golgi apparatus A collection of flattened membranes that packages and directs proteins and lipids produced by the ER to their final destinations either inside or outside the cell.

greenhouse effect The process by which greenhouse gases let in sunlight and trap heat.

greenhouse gas A gas in Earth's atmosphere that absorbs heat that radiates away from Earth's surface. Examples include carbon dioxide (CO_2), water vapor (H_2O), methane (CH_4), and nitrous oxide (N_2O).

ground tissue One of three types of plant tissue. Forming the intermediate layer of the plant, ground tissue makes up the bulk of the plant body and performs a wide range of functions, including support, wound repair, and photosynthesis. Compare *dermal tissue* and *vascular tissue*.

guanine (G) One of the four nucleotides that make up DNA. The other three are adenine (A), thymine (T), and cytosine (C).

gymnosperms One of two main groups of seed-bearing plants, characterized by naked seeds. Compare *angiosperms*.

H

halophile A prokaryote, usually archaean, that can live in extremely salty environments.

haploid Possessing a single set of genetic information, represented by $2n$. Gametes are haploid. Compare *diploid*.

heart A muscular organ the size of a fist that works as the body's circulatory pump.

heart rate The number of times a heart beats per minute.

herbivore An animal that eats plants. Compare *carnivore* and *omnivore*.

herd immunity Protection against disease that is brought about by vaccination of a critical portion of a population.

heterotroph A metabolic consumer, an organism that obtains energy by taking it from other sources. *Chemoheterotrophs* consume organic molecules as a source of energy and carbon. *Photoheterotrophs* absorb the energy of sunlight but require an organic source of carbon. Compare *autotroph*.

heterozygous Carrying two different alleles for a given phenotype (*Bb*). Compare *homozygous*.

histone protein One of a class of specific proteins that, together with DNA, form nucleosomes.

homeostasis The process of maintaining constant internal conditions.

homeostatic pathway The sequence of steps that reestablishes homeostasis if there is any departure from the genetically determined normal state of a particular internal characteristic.

hominids The ape family, which includes humans and chimpanzees. All hominids are capable of tool use, symbolic language, and deliberate acts of deception. Compare *hominins*.

hominins The "human" branch of the *hominids*, including modern humans and extinct relatives such as Neanderthals.

homologous pair A pair of chromosomes consisting of one chromosome received from the father and one from the mother.

homologous trait A feature that is similar across species because of common descent. Homologous traits may begin to look different from one another over time. Compare *analogous trait*.

homozygous Carrying two copies of the same allele (such as *BB* or *bb*) for a particular gene. Compare *heterozygous*.

horizontal gene transfer The transfer of genes on plasmids from one bacterium to another.

hormone A signaling molecule that coordinates internal activities necessary for growth and reproduction.

host An organism in which a *parasite* lives.

human microbiome The complete collection of microbes that live in and on our cells and bodies.

hydrogen bond The weak electrical attraction between a hydrogen atom with a partial positive charge and a neighboring atom with a partial negative charge. Compare *covalent bond* and *ionic bond*.

hydrologic cycle The movement of water as it circulates from the land to the sky and back again.

hydrophilic Literally, "water-loving." Soluble in water. Compare *hydrophobic*.

hydrophobic Literally, "water-fearing." Excluded from water. Compare *hydrophilic*.

hypertonic Describing a fluid that has a solute concentration higher than that of the cell it surrounds. Compare *hypotonic* and *isotonic*.

hyphae (sing. hypha) The fine, branching threads of fungi that absorb nutrients from the environment.

hypothalamus (pl. hypothalami) A small organ at the base of the vertebrate brain that coordinates the endocrine system and integrates it with the nervous system.

hypothesis (pl. hypotheses) An informed, logical, and plausible explanation for observations of the natural world.

hypotonic Describing a fluid that has a solute concentration lower than that of the cell it surrounds. Compare *hypertonic* and *isotonic*.

I

immune memory The capacity of the adaptive immune system to remember a first encounter with a specific pathogen and to mobilize a speedy and targeted response to future infection by the same strain.

immune system The organ system that defends the body from invaders such as viruses, bacteria, and fungi.

in vitro fertilization (IVF) Fertilization of an egg by a sperm in a petri dish, followed by implantation of one or more embryos into a woman's uterus.

incomplete dominance An interaction between two alleles of a gene in which neither one can exert its full effect, causing a heterozygote to display an intermediate phenotype. Compare *codominance*.

independent assortment The random distribution of the homologous chromosomes into daughter cells during meiosis I.

independent variable The variable that is manipulated by the researcher in a scientific experiment. Compare *dependent variable*.

indeterminate growth The general growth pattern of plants, in which they grow throughout their lives. Compare *determinate growth*.

induced fit The way an enzyme changes shape when molecules bind to its active site.

inflammation The immediate and coordinated sequence of events mounted by cytokines in the innate immune system in response to tissue damage from a pathogen invasion or wound.

ingestion The taking in of food, the first stage in the processing of food by the digestive system.

innate immune system The immune system's second line of defense against pathogens, consisting of cells and proteins that recognize the presence of an invader and mount an internal defense to kill, disable, or isolate it. Compare *adaptive immune system* and *external defenses*.

insertion A mutation in which a base is inserted into the DNA sequence of a gene. Compare *deletion* and *substitution*.

integumentary system The largest organ system in the human body, covering and protecting the surface of the body.

interference competition Competition between species in which one organism directly excludes another from the use of a resource. Compare *exploitative competition*.

interphase The longest stage of the cell cycle. Most cells spend 90 percent or more of their life span in interphase.

intron A stretch of DNA that does not code for anything. Compare *exon*.

invariant trait A trait that is the same in all individuals of a species. Compare *variable trait*.

ion An atom that has lost or gained electrons and therefore is either negatively or positively charged.

ionic bond The chemical attraction between a negatively charged ion and a positively charged ion. Compare *covalent bond* and *hydrogen bond*.

irregular fluctuation An unpredictable pattern of change in population size. Compare *cyclical fluctuation*.

isotonic Describing a fluid that has a solute concentration equal to that of the cell it surrounds. Compare *hypertonic* and *hypotonic*.

isotopes Two or more forms of an element that have the same number of protons but different numbers of neutrons.

IVF See **in vitro fertilization**.

J

J-shaped growth curve The type of graphical curve that represents exponential growth. Compare *S-shaped growth curve*.

joint A junction in the skeletal system that lets the skeleton move in specific ways.

K

karyotype A depiction showing all the chromosomes of a particular individual or species arranged in homologous pairs.

keystone species A species that has a disproportionately large effect on a community, relative to the species' abundance.

kidneys The paired organs that maintain water and solute homeostasis.

kingdom The second-highest hierarchical level in the organization of life; the unit of classification in the Linnaean hierarchy above phylum and below domain. The five kingdoms of life are Bacteria, Archaea, Protista, Fungi, and Animalia.

Krebs cycle Also called *citric acid cycle*. The second of three stages of cellular respiration. In this sequence of enzyme-driven reactions, the pyruvate made in glycolysis is broken down, releasing CO_2 and producing large amounts of energy carriers, including ATP, NADH, and $FADH_2$. Compare *glycolysis* and *oxidative phosphorylation*.

L

large intestine See **colon**.

larynx Also called *voice box*. The breathing and sound-producing structure that forms the entryway to the trachea.

law of independent assortment The law, proposed by Gregor Mendel, stating (in modern terms) that when gametes form, the two alleles of any given gene segregate during meiosis independently of any two alleles of other genes. Compare *law of segregation*.

law of segregation The law, proposed by Gregor Mendel, stating (in modern terms) that the two alleles of a gene are separated during meiosis and end up in different gametes. Compare *law of independent assortment*.

leaf A structure in plants that produces the majority of the plant's food through photosynthesis.

LH See **luteinizing hormone**.

lichen A mutualistic association between a photosynthetic microbe (usually a green alga or cyanobacterium) and a fungus.

ligament A specialized, flexible band of tissue that joins bone to bone. Compare *tendon*.

light-dependent reactions The first of two principal stages of photosynthesis, in which chlorophyll molecules absorb energy from sunlight and use that energy for the splitting of water, which in turn produces oxygen gas (O_2) as a by-product that is released into the atmosphere. Compare *light-independent reactions*.

light-independent reactions Also called the *Calvin cycle* or *carbon fixation*. The second of two principal stages of photosynthesis, in which a series of enzyme-catalyzed chemical reactions converts carbon dioxide (CO_2) into sugar, using energy delivered by ATP and electrons and hydrogen ions donated by NADPH. Compare *light-dependent reactions*.

lignin A strengthening substance that links together *cellulose* fibers in plant cells to create a rigid network.

lineage A single line of descent.

Linnaean hierarchy A system of biological classification devised by the Swedish naturalist Carolus Linnaeus in the eighteenth century.

lipid Any of a major class of biomolecules built of fatty acids and insoluble in water.

liposome A sphere formed by a phospholipid bilayer.

liver A large organ of the digestive system that produces bile, stores glycogen, and detoxifies dangerous chemicals in the body.

locus (pl. loci) The physical location of a gene on a chromosome.

logistic growth A pattern of population growth in which the population grows nearly exponentially at first but then stabilizes at the maximum population size that can be supported indefinitely by the environment. Logistic growth is represented by an S-shaped curve. Compare *exponential growth*.

lower respiratory system The part of the respiratory system made up of the trachea, bronchi, and lungs. Compare *upper respiratory system*.

lungs The paired organs in which gases (oxygen and carbon dioxide) are exchanged.

luteinizing hormone (LH) A hormone, produced by the pituitary gland, that triggers ovulation.

lymphatic system The network of ducts, lymph nodes, and associated organs that are the primary sites for action by the adaptive immune system.

lymphocyte A type of white blood cell that confers specific immunity as part of the adaptive immune system. Immature lymphocytes differentiate into B cells and T cells.

lysosome An organelle in animal cells that acts as a garbage or recycling center. Compare *vacuole*.

M

macromolecule See **biomolecule**.

macronutrient Any of nine nutrients—carbon, oxygen, hydrogen, nitrogen, phosphorus, potassium, calcium, sulfur, and magnesium—that plants need in relatively large amounts. Compare *micronutrient*.

malignant cell See **cancer cell**.

mammals A large class of animals that have body hair, sweat glands, and milk produced by mammary glands.

marrow A tissue found in the cavities of hollow bones that, depending on the type of bone, stores fat or produces blood cells.

marsupials One of three main groups of mammals, whose members have a simple placenta and produce offspring that complete development in their mother's pouch. Compare *eutherians* and *monotremes*.

mass extinction A period of time during which a great number of Earth's species goes extinct. The fossil record shows that there have been five mass extinctions in the history of Earth.

matter Anything that has mass and occupies a volume of space.

meiosis A specialized type of cell division that kicks off sexual reproduction. It occurs in two stages: meiosis I and meiosis II, each involving one round of nuclear division followed by cytokinesis. Compare *mitosis*.

meiosis I The first stage of meiosis, in which the chromosome set is reduced by the separation of each homologous pair into two different daughter cells. Each homologous chromosome lines up with its partner and then separates to the two ends of the cells. Compare *meiosis II*.

meiosis II The second stage of meiosis, in which sister chromatids are separated into two new daughter cells. Compare *meiosis I*.

Mendelian trait A trait that is controlled by a single gene and unaffected by environmental conditions.

menstrual cycle The process in human females by which individual eggs mature and are released in a hormone-driven sequence of events approximately monthly.

messenger RNA (mRNA) A type of RNA that is complementary to a DNA template strand. Compare *ribosomal RNA* and *transfer RNA*.

meta-analysis Work that combines results from different studies.

metabolic heat The heat released as a by-product of chemical reactions within a cell, typically during cellular respiration.

metabolic pathway Any of various chains of linked events that produce key biological molecules in a cell, including important chemical building blocks like amino acids and nucleotides.

metabolism All the chemical reactions that occur inside living cells, including those that release and store energy.

metastasis The spread of a disease from one organ to another.

methanogen An anaerobic archaean that feeds on hydrogen and produces methane gas as a by-product of its metabolism.

microbe A microscopic, single-celled organism.

micronutrient Any of a variety of nutrients (including iron, zinc, and copper) that plants need in relatively small amounts. Compare *macronutrient*.

mimicry Coloration of a nonpoisonous animal that resembles the coloration of a toxic species. Compare *camouflage* and *warning coloration*.

mineral Any of various small, inorganic nutrients needed by the human body for critical biological function, but only in small amounts. Compare *vitamin*.

mitochondrial-DNA inheritance The passing down of DNA from the mitochondria in an egg cell to a new generation. Mitochondrial DNA passes virtually unchanged from mother to child, so it can be tracked from one generation, or one species, to another. Sequencing of mitochondrial DNA can determine how related an individual is to its female ancestors on its mother's side. Compare *nuclear-DNA inheritance*.

mitochondrion (pl. mitochondria) An organelle that is a tiny power plant fueling cellular activities. Mitochondria are the main source of energy in eukaryotic cells.

mitosis Division of the nucleus, the first step of mitotic division. Mitosis is divided into four main phases: prophase, metaphase, anaphase, and telophase. Compare *cytokinesis* and *meiosis*.

mitotic division A type of cell division that generates two genetically identical daughter cells from a single parent cell in eukaryotes. It consists of two steps: mitosis and cytokinesis.

modular growth The way plants grow by repeatedly adding the same module of bud-stem-leaf unit aboveground or new lateral roots belowground.

molecule An association of atoms held together by chemical bonds.

monocot Any plant that has one cotyledon in each of its seeds. Monocots are characterized by netlike veins in the leaves, vascular bundles arranged in rings, taproots, and floral parts in multiples of four or five. Examples include all the grasses, members of the lily family, palm trees, and banana plants. Compare *dicot*.

monomer A small molecule that is the repeating unit of a *polymer*. For example, amino acids are the monomers that make up protein polymers.

monotremes One of three main groups of mammals, whose members lack a placenta and lay eggs. Compare *eutherians* and *marsupials*.

morphology An organism's physical characteristics.

most recent common ancestor The most immediate ancestor that two lineages share.

motor output A particular action of the body.

mRNA See **messenger RNA**.

muscle fiber A long, narrow cell that can span the length of an entire muscle because it is made up of several muscle cells that fused together during development.

muscle tissue A tissue that generates force by contracting.

muscular system The organ system that, working closely with the skeletal system, produces the force that moves structures within the body.

mutation A random change to the sequence of bases in an organism's DNA.

mutualism A species interaction in which both species benefit. Compare *commensalism*, *competition*, and *predation*.

mycelium (pl. mycelia) The entire bundle of hyphae that composes the main body of a fungus.

mycorrhizal fungi Fungi that form mutualistic associations with the root systems of plants, that help the plants absorb more water and nutrients from the soil.

myofibril Any of the cylindrical structures packed inside of muscle fibers containing proteins that contract by bracing against each other.

myosin filament One of the two kinds of protein filaments, made of the protein myosin, that is found in sarcomeres. The sliding of myosin filaments against *actin filaments* enables sarcomeres to contract.

N

natural selection The process by which individuals with advantageous genetic characteristics for a particular environment survive and reproduce at a higher rate than do individuals with other, less useful characteristics. Compare *artificial selection*.

negative feedback The steps of a process that decrease its output. Compare *positive feedback*.

nephron The basic functional unit of the kidney.

nerve One or more neurons that transmit signals to and from the central nervous system.

nerve cord A solid strand of nervous tissue that we call the spinal cord in humans.

nervous system The organ system that directs the rapid contractions of muscles and processes information received by the senses, such as touch, sound, and sight.

nervous tissue A tissue that communicates and processes information.

neuron A specialized cell of the nervous system that transmits signals from one part of the body to another in a fraction of a second.

neutron An electrically neutral particle found in the nucleus of an atom. Compare *electron* and *proton*.

nitrogen fixation The process, carried out by bacteria, of taking nitrogen gas from the air and converting it to ammonia, making it available for plants.

node The point on an evolutionary tree indicating the moment in time when an ancestral group split, or diverged, into two separate lineages. The node represents the most recent common ancestor of the two lineages in question.

nonspecific response A response mounted by the immune system against pathogens that is indiscriminate as to the invaders it repels. External defenses and the innate immune system are both nonspecific responses. Compare *specific response*.

notochord In chordates, a flexible yet rigid rod along the center of the body that is critical for development.

nuclear-DNA inheritance The passing down of DNA from the nucleus in an egg or sperm cell to a new generation. Sequencing of nuclear DNA can determine how related an individual is to all of its ancestors, both male and female. Compare *mitochondrial-DNA inheritance*.

nuclear envelope The boundary of a cell's nucleus, consisting of two concentric phospholipid bilayers.

nuclear pore Any of many small openings in the nuclear envelope that allow chemical messages to enter and exit the nucleus.

nucleic acid Any of a major class of biomolecules, including DNA and RNA, built of chains of nucleotides.

nucleosome DNA wrapped around histone proteins that, in multiples, form the beads-on-a-string complex in a chromatin fiber.

nucleotide The basic repeating subunit of DNA, composed of the sugar deoxyribose, a phosphate group, and one of four bases: adenine (A), cytosine (C), guanine (G), or thymine (T).

nucleotide pair See **base pair**.

nucleus (pl. nuclei) 1. The dense core of an atom, which contains protons and neutrons. 2. The control center of the eukaryotic cell, containing all of the cell's DNA and occupying up to 10 percent of the space inside the cell.

nutrient A chemical element that is required by a living organism.

nutrient cycling The process by which decomposers break down dead organisms or waste products, release the chemical elements locked in the biological material, and return them to the environment.

nutrition The scientific study of what we eat—that is, our diet.

O

observation A description, measurement, or record of any object or phenomenon.

omnivore An animal that eats both animals and plants. Compare *carnivore* and *herbivore*.

oogenesis The series of cell divisions in human females that results in an egg. Compare *spermatogenesis*.

opposable Able to be placed opposite other digits of the hand or foot. For example, opposable thumbs can be placed opposite each of the other four fingers.

oral cavity The mouth.

order The unit of classification in the Linnaean hierarchy above family and below class.

organ A collection of different types of tissues that form a functional unit with a distinctive shape and location in the body.

organ system A network of organs that work in a closely coordinated manner to perform a distinct set of functions in the body.

organelle Any of the membrane-enclosed subcellular compartments found in eukaryotic cells.

organic compound A compound that is made up of multiple organic molecules bound together.

organic molecule A molecule that includes at least one carbon-hydrogen bond.

organism An individual living thing composed of interdependent parts.

origin of replication A DNA sequence where DNA replication is initiated.

osmosis A form of simple diffusion in which water moves in and out of cells (and compartments inside cells).

osteocyte A specialized bone cell that surrounds itself with a hard, nonliving mineral matrix composed largely of calcium and phosphate.

ovary Either of a pair of female reproductive organs that produce eggs and estrogens in vertebrates. Compare *testis*.

oviduct Also called *fallopian tube*. The tube through which an egg travels from the ovary to the uterus.

ovule The egg-bearing structure in plants.

ovum (pl. ova) See **egg**.

oxidative phosphorylation The third of three stages of cellular respiration. During this process, the chemical energy of NADH and $FADH_2$ is converted into the chemical energy of ATP, while electrons and hydrogen atoms removed from NADH and $FADH_2$ are handed over to molecular O_2, creating water (H_2O). A large amount of ATP is generated. Compare *glycolysis* and *Krebs cycle*.

oxytocin A hormone—secreted by the fetus and, later in the birth process, by the mother's pituitary gland—that stimulates the uterine muscles and causes the placenta to secrete prostaglandins, which reinforce contractions.

P

P generation The first set of parents in a series of genetic crosses. Compare *F1 generation* and *F2 generation*.

pancreas A gland that produces insulin and secretes fluids that aid in the digestion of food.

parasite An organism that lives in or on another species and harms it by stealing resources. For example, some parasites suck blood or live off the food in our intestines. Compare *host*.

passive immunity Immunity to a particular pathogen that is conferred by antibodies not made by the body, but received from an outside source. Compare *active immunity*.

passive transport The movement of a substance without the addition of energy. Compare *active transport*.

pathogen An infectious agent.

PCR See **polymerase chain reaction**.

pedigree A chart similar to a family tree that shows genetic relationships among family members over two or more generations of a family's medical history.

peer-reviewed publication The publishing of original research only after it has passed the scrutiny of experts who have no direct involvement in the research under review, or a scientific journal that follows this standard.

penis The male reproductive organ that introduces sperm into a female or hermaphrodite sexual partner. The penis is also involved in urination in mammals.

perennial A plant that lives three or more years. Compare *annual* and *biennial*.

peripheral nervous system (PNS) One of two main parts of the nervous system, consisting of the sensory nerves plus all the nerves that are not part of the *central nervous system*.

PGD See **preimplantation genetic diagnosis**.

pH scale A logarithmic scale that indicates the concentration of hydrogen ions. The scale goes from 0 to 14, with 0 representing an extremely high concentration of free H^+ ions and 14 representing the lowest concentration.

phagocyte A type of white blood cell that functions as part of the innate immune system to mark and destroy foreign invaders by engulfing and digesting them.

phagocytosis Literally, "cellular eating." A large-scale version of endocytosis in which particles considerably larger than biomolecules are ingested. Compare *pinocytosis*.

pharynx Also called *throat*. An area where the back of the mouth and the two nasal cavities join together into a single passageway.

phenotype The physical, biochemical, or behavioral expression of a particular version of a trait. Compare *genotype*.

phloem One of two types of vascular tissue in plants. Phloem transports sugars from the leaves, where they are produced, to living cells in every part of the plant. Compare *xylem*.

phospholipid An organic molecule with a hydrophilic head and a hydrophobic tail.

phospholipid bilayer A double layer of phospholipids in which the heads face out and the tails face in. Plasma membranes are phospholipid bilayers.

photoautotroph See **autotroph**.

photoheterotroph See **heterotroph**.

photoperiodism The ability of plants to sense the duration of light and dark in a 24-hour cycle.

photosynthesis The process by which organisms capture energy from the sun and use it to create sugars from carbon dioxide and water, thereby transforming light energy into chemical energy stored in the covalent bonds of sugar molecules. Compare *cellular respiration*.

phylum (pl. phyla) The unit of classification in the Linnaean hierarchy above class and below kingdom.

physical trait An anatomical or physiological characteristic of an individual, such as the shape of an animal's head. Compare *behavioral trait* and *biochemical trait*.

physiology The science that focuses on the functions of anatomical structures. Compare *anatomy*.

pili (sing. pilus) Short, hairlike projections that cover the surface of many bacteria.

pinocytosis Literally, "cellular drinking." A large-scale version of endocytosis in which fluids are ingested. Compare *phagocytosis*.

Plantae; the plants One of the five kingdoms of life, in the domain Eukarya, encompassing all plants, which are multicellular photosynthetic autotrophs.

plasma The fluid portion of the blood.

plasma membrane A barrier consisting of a phospholipid bilayer that separates a cell from its external environment.

platelet A type of sticky cell fragment found in circulating blood that helps form blood clots.

pleiotropy The pattern of inheritance in which a single gene influences multiple different traits. Compare *polygenic trait*.

PNS See **peripheral nervous system.**

point mutation A mutation in which only a single base is altered.

polar molecule A molecule whose electrical charge is shared unevenly, with some regions being electrically negative and others electrically positive.

pollen In plants, a microscopic structure containing sperm cells that can be lofted into the air in massive quantities.

pollinator An animal that transports pollen from one plant to another.

polygenic trait A genetic trait that is governed by the action of more than one gene. Compare *pleiotropy*.

polymer A long strand of repeating units of small molecules called *monomers*. For example, proteins are polymers made up of amino acid monomers.

polymerase chain reaction (PCR) A technique for replicating DNA that can produce millions of copies of a DNA sequence in just a few hours from a small initial amount of DNA.

polyploidy The condition in which an individual's somatic cells have more than two sets of chromosomes.

population A group of individuals of the same species living and interacting in a shared environment.

population density The number of individuals per unit of area.

population doubling time The time it takes a population to double in size, as a measure of how fast a population is growing.

population ecology The study of the number of organisms in a particular place.

population size The total number of individuals in a population.

positive feedback The steps of a process that increase its output. Compare *negative feedback*.

postzygotic barrier A barrier that prevents a zygote from developing into a healthy and fertile individual—that is, a reproductive barrier that acts after a zygote exists. Compare *prezygotic barrier*.

predation A species interaction in which one species benefits and the other species is harmed. Compare *commensalism*, *competition*, and *mutualism*.

predator A consumer that eats either plants or animals. Compare *prey*.

preimplantation genetic diagnosis (PGD) The removal of one or two cells from an embryo developing in a petri dish, usually 3 days after fertilization, followed by testing for genetic disorders. One or more embryos that are free of disorders are then implanted into a woman's uterus.

prey An animal that is eaten by a *predator*.

prezygotic barrier A barrier that prevents a male gamete and a female gamete from fusing to form a zygote—that is, a reproductive barrier that acts before a zygote exists. Compare *postzygotic barrier*.

primary consumer An organism that eats producers. Compare *secondary consumer*, *tertiary consumer*, and *quaternary consumer*.

primary growth In plants, an increase in length by the division of cells at the tips of shoots, stems, and roots. Compare *secondary growth*.

primary immune response The slow response mounted by the adaptive immune system against an invading pathogen the very first time a person is exposed to that pathogen. Compare *secondary immune response*.

primary literature Scientific literature in which research is first published. Compare *secondary literature*.

primary oocyte An immature egg cell.

primary productivity The energy acquired through photosynthesis that is available for the growth and reproduction of producers within an ecosystem. It is the amount of energy captured by photosynthetic organisms, minus the amount they expend on cellular respiration and other maintenance processes.

primary succession Succession that occurs in a newly created habitat, usually from bare rock or sand. Compare *secondary succession.*

primates The order of mammals to which humans belong. All primates have flexible shoulder and elbow joints, five functional fingers and toes, opposable thumbs, flat nails, and brains that are large in relation to the body.

primer A short stretch of RNA or DNA that is complementary-base-paired to a DNA template to provide a 3'-hydroxyl group for polymerase to initiate DNA replication or the polymerase chain reaction (PCR), respectively.

process of science See **scientific method**.

producer An organism at the bottom of a food chain that uses energy from the sun to produce its own food. Compare *consumer.*

product A substance that results from a chemical reaction. Compare *reactant.*

progesterone The primary progestogen. Compare *estradiol* and *testosterone.*

progestogens Any of the hormones produced in the ovaries that have a number of functions in the female body, including thickening the lining of the uterus and increasing the blood supply to it to create a suitable environment for a developing fetus. Compare *estrogens* and *androgens.*

prokaryote An organism that belongs to either the Bacteria or the Archaea. Compare *eukaryote.*

promoter A segment of DNA near the beginning of a gene that RNA polymerase recognizes and binds to begin transcription. Compare *terminator.*

protein Any of a major class of biomolecules built of amino acids.

protists An artificial grouping of eukaryotes, defined by not belonging to any of the three kingdoms of Eukarya, composed of mainly single-celled, microscopic organisms grouped together simply because they are not plants, animals, or fungi and include amoebas and algae.

proton A positively charged particle found in the nucleus of an atom. Compare *electron* and *neutron.*

protozoans One of two main groups of protists, whose members are nonphotosynthetic and motile. Compare *algae.*

pseudoscience Scientific-sounding statements, beliefs, or practices that are not actually based on the scientific method.

pulmonary circuit The circuit in the heart, consisting of the two chambers on the right side, that receives blood low in oxygen and pumps it to the lungs for gas exchange. Compare *systemic circuit.*

Punnett square A grid-like diagram showing all possible ways that two alleles can be brought together through fertilization.

Q

quaternary consumer An organism that eats *tertiary consumers.* Compare *primary consumer* and *secondary consumer.*

quorum sensing A system of cell-to-cell communication used by prokaryotes that enables them to sense and respond to other bacteria in the area in accordance with the density of the population.

R

radial symmetry An animal body plan in which the body can be sliced symmetrically along any number of vertical planes that pass through the center of the animal. Compare *bilateral symmetry.*

reabsorption The removal of valuable solutes such as sodium, chloride, and sugars from the fluid filtered by the kidneys so that they don't exit the body in the urine.

reactant A substance that undergoes change in a chemical reaction. Compare *product.*

receptor-mediated endocytosis A form of specific endocytosis in which receptor proteins embedded in the membrane recognize specific surface characteristics of substances that will be incorporated into the cell.

receptor protein A site where a molecule from another cell can bind.

recessive allele An allele that has no effect on the phenotype when paired with a *dominant allele.*

recessive genetic disorder A genetic disorder that is inherited as a recessive trait on an autosome. Compare *dominant genetic disorder.*

red blood cell A cell in the blood that greatly increases the oxygen-carrying capacity of blood.

relative species abundance How common one species is when compared to another.

reproduction The making of a new individual like oneself.

reproductive barrier A barrier that prevents two species from interbreeding, making them reproductively isolated.

reproductive isolation The condition in which barriers prevent populations from interbreeding. Compare *ecological isolation* and *geographic isolation.*

reproductive system The organ system that generates gametes and may also support fertilization and prenatal development.

respiratory system The organ system that brings in oxygen and expels carbon dioxide to support cellular respiration.

ribonucleic acid See **RNA**.

ribosomal RNA (rRNA) A type of RNA that is an important component of ribosomes. Compare *messenger RNA* and *transfer RNA.*

ribosome The site of protein synthesis (translation) in the cytoplasm. Ribosomes are embedded in the rough endoplasmic reticulum.

RNA Ribonucleic acid, a single-stranded nucleic acid transcribed from DNA and consisting of the ribonucleotides adenine, guanine, cytosine, and uracil.

RNA polymerase An enzyme that recognizes and binds a gene's promoter sequence and then separates the two strands of DNA during transcription.

RNA splicing Processing of mRNA in which the introns are snipped out of a pre-mRNA and the remaining pieces of mRNA—the exons—are joined to generate the mature mRNA.

root A structure in plants that absorbs water and mineral nutrients, anchors the plant, and stores food.

root system One of two plant organ systems. It anchors the plant, absorbs water and nutrients from the soil, transports food and water, and may store food. Compare *shoot system.*

rough ER A part of the endoplasmic reticulum, having a knobby appearance because of embedded ribosomes, where proteins are assembled. Compare *smooth ER.*

rRNA See **ribosomal RNA.**

rubisco The enzyme that catalyzes the first step in the light-independent reactions of photosynthesis, fixing a carbon molecule from CO_2.

S

S phase The "synthesis" phase of the cell cycle, in which preparations for cell division begin. A critical event during this phase is the replication of all the cell's DNA molecules.

S-shaped growth curve The type of graphical curve that represents logistic growth. Compare *J-shaped growth curve.*

saliva A fluid secreted into the oral cavity to aid in the digestion of food.

sarcomere Any of the contractile units of the muscular system that make up each myofibril.

scavenger An animal that eats dead or dying animals and plants. Scavengers are categorized as either *decomposers* or *detritivores.*

science A body of knowledge about the natural world, and an evidence-based process for acquiring that knowledge.

scientific claim A statement about how the world works that can be tested using the scientific method.

scientific literacy An understanding of the basics of science and the scientific process.

scientific method Also called *process of science.* The practices that produce scientific knowledge.

scientific name The unique two-word Latin name, consisting of the genus and species names, that is assigned to a species in the Linnaean hierarchy.

secondary consumer An organism that eats *primary consumers.* Compare *tertiary consumer* and *quaternary consumer.*

secondary growth In plants, an increase in thickness in either stems or roots. Compare *primary growth.*

secondary immune response The rapid response mounted by the adaptive immune system against an invading pathogen the second and subsequent times a person is exposed to that pathogen. Compare *primary immune response.*

secondary literature Scientific literature that summarizes and synthesizes an area of research. Compare *primary literature.*

secondary succession Succession that occurs after a disturbance in a community. Compare *primary succession.*

secretion The active transport by the kidneys of excess quantities of substances such as potassium and hydrogen ions, and some toxins, from the blood into the liquid passing through the kidney and out of the body in the urine.

seed In plants, the embryo and a supply of stored food, all encased in a protective covering.

segments Repeated identical units that make up the body plan of arthropods, annelids, and vertebrates.

selective breeding The process by which humans allow only individuals with certain inherited characteristics to mate.

selective permeability The quality of plasma membranes by which some substances are allowed to cross the membrane at all times, others are excluded at all times, and still others can pass through the membrane when they are aided by transport proteins.

semiconservative replication The mode of replication by which DNA is duplicated, where one "old" strand (the template strand) is retained (conserved) in each new double helix.

sense To perceive the world through a sensory system such as sight, touch, or smell.

sensory input Signals that are received, transmitted, and processed by the central nervous system.

sensory organ An organ of the body, such as the eyes or ears, that receives sensory input.

serotype See **viral strain.**

set point The genetically determined normal state of any physical or chemical characteristic of the body's internal environment.

sex chromosome One of the two chromosomes (X and Y) that determine gender. Compare *autosome.*

sex-linked Found solely on the X or Y chromosome. See also **X-linked** and **Y-linked**.

sexual dimorphism A distinct difference in appearance between the males and females of a species.

sexual reproduction The process by which genetic information from two individuals is combined to produce offspring. It has two steps: cell division through meiosis, followed by fertilization. Compare *asexual reproduction*.

sexual selection Natural selection in which a trait increases an individual's chance of mating even if it decreases the individual's chance of survival.

shared derived trait A unique feature common to all members of a group that originated in the group's most recent common ancestor and then were passed down in the group.

shoot system One of two plant organ systems, consisting of stems and leaves. Stems provide the plant with structural support, transport food and water, and hold leaves up to intercept light so that they can perform photosynthesis. Compare *root system*.

simple diffusion Diffusion in which substances such as the small, uncharged molecules of water, oxygen, or carbon dioxide, slip between the large molecules in the phospholipid bilayer without much hindrance. Compare *facilitated diffusion*.

sister chromatids The two identical DNA molecules produced by the replication of a chromosome.

skeletal muscle The specialized muscle that is associated with the skeleton. It has a banded appearance, and its contractions are mainly voluntary. Compare *cardiac muscle* and *smooth muscle*.

skeletal system The organ system, consisting of bones, cartilage, and ligaments, that provides an internal framework to support the body of vertebrates.

small intestine The highly coiled tube, specialized for absorption, into which food moves from the stomach during digestion.

smooth ER A part of the endoplasmic reticulum, having a smooth appearance, where lipids and hormones are manufactured. Compare *rough ER*.

smooth muscle The specialized muscle found in the walls of the digestive system and blood vessels. It has no visible bands, and its contractions are entirely involuntary. Compare *cardiac muscle* and *skeletal muscle*.

soluble Able to mix completely with water.

solute A dissolved substance, such as sugar in water. Compare *solvent*.

solution Any combination of a solute and a solvent.

solvent The fluid, such as water, into which a substance has dissolved. Compare *solute*.

somatic cell A non-sex cell. Compare *gamete*.

speciation The process by which one species splits to form two species or more.

species 1. Members of a group that can mate with one another to produce fertile offspring. 2. The smallest unit of classification in the Linnaean hierarchy.

species interaction Any of four ecological interactions—mutualism, commensalism, predation, and competition—that occur between different species.

species richness The total number of different species that live in an ecological community.

specific response A response mounted by the adaptive immune system against a specific strain of pathogen. Compare *nonspecific response*.

sperm The male gamete. Compare *egg*.

spermatogenesis The series of cell divisions in human males that results in sperm. Compare *oogenesis*.

spinal cord A thick central nerve cord that is continuous with the brain, acting as a filter between the brain and the sensory neurons.

spongy bone One of the two major types of bone tissue. Honeycombed with numerous tiny cavities, it lies inside the *compact bone*. Spongy bone is most abundant at the knobby ends of our long bones.

spore 1. In fungi, a reproductive structure that can survive for long periods of time in a dormant state and will sprout under favorable conditions to produce the body of the organism. 2. In plants, a haploid cell produced by meiosis that divides mitotically to produce the gametophyte.

sporophyte A diploid, multicellular individual that arises from the zygote in plants. Compare *gametophyte*.

sporulation The formation of thick-walled dormant structures called spores.

stabilizing selection The pattern of natural selection in which individuals with intermediate values of an inherited phenotypic trait have an advantage over other individuals in the population. Compare *directional selection* and *disruptive selection*.

start codon The codon AUG; the point on an mRNA strand at which the ribosomes begin translation. Compare *stop codon*.

statistics A branch of mathematics that estimates the reliability of data.

stem A structure in plants that provides structural support, transports food and water, and holds leaves up to intercept light.

stem cells Unique, unspecialized cells that can make identical copies of themselves for long periods of time and can be used in gene therapy.

stoma (pl. stomata) An air pore in the leaf of a plant that controls gas exchange.

stomach The organ of the digestive system, located between the esophagus and intestines, in which most digestion occurs, through mechanical and chemical means.

stop codon The codon UAA, UAG, or UGA; the point on an mRNA strand at which the ribosomes end translation. Compare *start codon.*

substitution A mutation in which one base is substituted for another in the DNA sequence of a gene. Compare *deletion* and *insertion.*

substrate A molecule that will react to form a new product.

succession The process by which the species in a community change over time.

sustainable Able to be continued indefinitely without causing serious damage to the environment.

swallowing reflex The reaction of the digestive system when food comes into contact with the pharynx, in which the epiglottis seals off the entry into the trachea and food is pushed into the esophagus.

sympatric speciation The formation of new species in the absence of geographic isolation. Compare *allopatric speciation.*

synovial sac The cavity inside a joint that is filled with a lubricating fluid that reduces friction between two bony surfaces.

systemic circuit The circuit in the heart, consisting of the two chambers on the left side, that receives oxygenated blood from the lungs and pumps it to the body. Compare *pulmonary circuit.*

systole The first pressure, or top number, in a blood pressure reading—the pressure at the point the heart contracts to push blood out. Compare *diastole.*

T

T cell A type of lymphocyte that matures in the thymus. T cells are involved in cell-mediated immunity. Compare *B cell.*

template strand The strand of DNA that is used as a template to make a new strand of DNA.

tendon A specialized, flexible band of tissue, rich in collagen, that joins muscle to bone. Compare *ligament.*

terminator A segment of DNA that, when reached by RNA polymerase, stops transcription. Compare *promoter.*

tertiary consumer An organism that eats *secondary consumers.* Compare *primary consumer* and *quaternary consumer.*

testis (pl. testes) Either of a pair of male reproductive organs that produce sperm and androgens in vertebrates. Compare *ovary.*

testosterone The primary androgen. Compare *estradiol* and *progesterone.*

theory A hypothesis, or a group of related hypotheses, that has received substantial confirmation through diverse lines of investigation by independent researchers. Compare *fact.*

thermophile A prokaryote, usually archaean, that can live in extremely hot environments, such as geysers, hot springs, and hydrothermal vents.

throat See **pharynx.**

thymine (T) One of the four nucleotides that make up DNA. The other three are adenine (A), guanine (G), and cytosine (C).

tissue A group of cells that function in an integrated manner to perform a unique set of tasks in the body.

trachea Also called *windpipe.* The structure that connects the pharynx to the bronchi.

transcription The synthesis of RNA based on a DNA template. Compare *translation.*

transfer RNA (tRNA) A type of RNA that facilitates translation by delivering specific amino acids to the ribosomes as codons are read off of an mRNA. Compare *messenger RNA* and *ribosomal RNA.*

translation The process by which ribosomes convert the information in mRNA into proteins. Compare *transcription.*

transitional fossil A fossil that displays physical characteristics in between those of two known fossils in a family tree.

transpiration The process of plants absorbing water through their roots and releasing this water through their leaves into the atmosphere.

transport protein A protein that acts like a gate, channel, or pump that allows molecules to move into and out of a cell.

treatment group Also called *experimental group.* The group of subjects in an experiment that is maintained under the same standard set of conditions as the control group but is subjected to manipulation of the independent variable. Compare *control group.*

trimester Any of the three defined stages of human pregnancy. Each trimester is about 3 months long.

tRNA See **transfer RNA.**

trophic level Each level of the energy pyramid, corresponding to a step in a food chain.

tumor A cell mass that results from runaway cell division.

U

up-regulation The speeding up of gene expression. Compare *down-regulation.*

upper respiratory system The part of the respiratory system that includes airways in the nose, mouth, and throat. Compare *lower respiratory system.*

urinary system The organ system that removes excess fluid from the body, along with waste products, toxins, and other water-soluble substances that are not needed.

urine The waste-carrying watery solution that is expelled from our bodies.

uterus The female reproductive organ in which a fertilized egg implants and develops until birth.

V

vacuole An organelle in plant cells that acts as a garbage or recycling center and that stores water. Compare *lysosome.*

vagina The female reproductive organ that connects the uterus to the external genitalia.

variable A characteristic of any object or individual organism that can change.

variable trait A trait that is different in different individuals of a species. Compare *invariant trait.*

vascular system In plants, a network of tissues that is made up of tubelike structures specialized for transporting fluids.

vascular tissue One of three types of plant tissue. Forming the innermost layer of the plant, vascular tissue (consisting of phloem and xylem) contains stacks of long cells forming continuous tubes that run throughout the plant body and transport materials throughout the plant. Compare *dermal tissue* and *ground tissue.*

vein Any of the large vessels that carry blood back to the heart. Compare *artery* and *capillary.*

ventricle Either of the two lower chambers of the heart. Compare *atrium.*

vertebrae (sing. vertebra) The strong, hollow sections of the backbone, or vertebral column.

vertebrates Chordates that possess a backbone.

vesicle A sac, formed by the bulging inward or outward of a section of the plasma membrane, that moves molecules from place to place inside a cell but also may transport substances into and out of the cell.

vestigial trait A feature that is inherited from a common ancestor but no longer used. Vestigial traits may appear as reduced or degenerated parts whose function is hard to discern.

villus (pl. villi) Any of the large number of fingerlike projections in the small intestine that are specialized for nutrient absorption.

viral strain Also called *serotype.* Any of the variant forms of a particular type of virus.

virus A small, infectious agent that can replicate only inside a living cell.

vitamin Any of various small, organic nutrients needed by the human body, but only in tiny amounts. Compare *mineral.*

voice box See **larynx**.

W

warning coloration Bright coloring of an animal that alerts a potential predator to dangerous chemicals in the animal's tissues. Compare *camouflage* and *mimicry.*

water-soluble vitamin A vitamin that can dissolve in water and therefore tends not to accumulate in body tissues because it can be easily excreted in urine. Compare *fat-soluble vitamin.*

weather Short-term atmospheric conditions, such as today's temperature, precipitation, wind, humidity, and cloud cover. Compare *climate.*

windpipe See **trachea**.

X

X-linked Found solely on the X chromosome. Compare *Y-linked.*

xylem One of two types of vascular tissue in plants. Xylem transports water and minerals, absorbed from the soil, upward from the roots and outward from the central stem to the leaves. Compare *phloem.*

Y

Y-linked Found solely on the Y chromosome. Compare *X-linked.*

yeasts Single-celled fungi that belong in the group zygomycetes and are important in the rising of bread, the brewing of beer, and the fermenting of wine.

Z

Z disc A structure, found at each end of a sarcomere, that contains a large protein that provides an anchor point for actin filaments.

zygomycetes One of three main groups of fungi, containing many species of molds. Compare *ascomycetes* and *basidiomycetes.*

zygote The single cell that results from fertilization.

References for Debunked!

Cited papers provide a starting point into the evidence debunking each myth. Additional peer-reviewed studies support the claims.

Chapter 1, p. 4

Eklöf, Johan, Jurģis Šuba, Gunars Petersons, and Jens Rydell. 2014. "Visual acuity and eye size in five European bat species in relation to foraging and migration strategies." *Environmental and Experimental Biology* 12:1–6. http://eeb.lu.lv/EEB/201403/EEB_12_Eklof.pdf.

Chapter 2, p. 26

Vetter, Volker, Gülhan Denizer, Leonard R. Friedland, Jyothsna Krishnan, and Marla Shapiro. 2018. "Understanding modern-day vaccines: What you need to know." *Annals of Medicine* 50 (2): 110–20. https://doi.org/10.1080/07853890.2017.1407035.

Chapter 3, p. 49

Kolodny, Andrew, David T. Courtwright, Catherine S. Hwang, Peter Kreiner, John L. Eadie, Thomas W. Clark, and G. Caleb Alexander. 2015. "The prescription opioid and heroin crisis: A public health approach to an epidemic of addiction." *Annual Review of Public Health* 36:559–74. https://doi.org/10.1146/annurev-publhealth-031914-122957.

Chapter 4, p. 68

Sender, Ron, Shai Fuchs, and Ron Milo. 2016. "Revised estimates for the number of human and bacteria cells in the body." *PLoS Biology* 14 (8): e1002533. https://doi.org/10.1371/journal.pbio.1002533.

Chapter 5, p. 91

Rich, P. R. 2003. "The molecular machinery of Keilin's respiratory chain." *Biochemical Society Transactions* 6 (31): 1095–105. https://doi.org/10.1042/bst0311095.

Chapter 6, p. 115

Spalding, Kirsty L., Olaf Bergmann, Kanar Alkass, Samuel Bernard, Mehran Salehpour, Hagen B. Huttner, Emil Boström, et al. 2013. "Dynamics of hippocampal neurogenesis in adult humans." *Cell* 153 (6): 1219–27. https://doi.org/10.1016/j.cell.2013.05.002.

Chapter 7, p. 136

Sturm, Richard A., and Mats Larsson. 2009. "Genetics of human iris color and patterns." *Pigment Cell and Melanoma Research* 22 (5): 544–62. https://doi.org/10.1111/j.1755-148X.2009.00606.x.

Chapter 8, p. 149

Vilain, Eric, and Edward R. B. McCabe. 1998. "Mammalian sex determination: From gonads to brain." *Molecular Genetics and Metabolism* 65 (2): 74–84. https://doi.org/10.1006/mgme.1998.2749.

Chapter 9, p. 178

Takaoka, Miho, and Yoshio Miki. 2018. "BRCA1 gene: Function and deficiency." *International Journal of Clinical Oncology* 23:36–44. https://doi.org/10.1007/s10147-017-1182-2.

Chapter 10, p. 199

Centers for Disease Control and Prevention. 2019. "Influenza (flu) vaccine (inactivated or recombinant): What you need to know." https://www.cdc.gov/vaccines/hcp/vis/vis-statements/flu.pdf.

Chapter 11, p. 212

Lamb, Trevor D., Shaun P. Collin, and Edward N. Pugh, Jr. 2007. "Evolution of the vertebrate eye: Opsins, photoreceptors, retina and eye cup." *Nature Reviews Neuroscience* 8 (12): 960–76. https://doi.org/10.1038/nrn2283.

Chapter 12, p. 226

Mohr, Kathrin I. 2016. "History of antibiotics research." In *How to Overcome the Antibiotic Crisis*, edited by Marc Stadler and Petra Dersch. *Current Topics in Microbiology and Immunology* 398:237–72. https://doi.org/10.1007/82_2016_499.

Chapter 13, p. 248

Williams, Tony D. 1995. *The Penguins*. Oxford: Oxford University Press.

Chapter 14, p. 279

Brusatte, Stephen L., Jingmai K. O'Connor, and Erich D. Jarvis. 2015. "The origin and diversification of birds." *Current Biology* 25 (19): R888–98. https://doi.org/10.1016/j.cub.2015.08.003.

Chapter 15, p. 289

Miranda, Robyn C., and Donald W. Schaffner. 2016. "Longer contact times increase cross-contamination of *Enterobacter aerogenes* from surfaces to food." *Applied and Environmental Microbiology* 82 (21): 6490–96. https://doi.org/10.1128/AEM.01838-16.

Chapter 16, p. 312

Camazine, Scott. 1983. "Mushroom chemical defense: Food aversion learning induced by hallucinogenic toxin, muscimol." *Journal of Chemical Ecology* 9 (11): 1473–81. https://doi.org/10.1007/BF00988513.

Chapter 17, p. 324

Brusatte, Stephen L., Richard J. Butler, Paul M. Barrett, Matthew T. Carrano, David C. Evans, Graeme T. Lloyd, Philip D. Mannion, et al. 2015. "The extinction of the dinosaurs." *Biological Reviews Cambridge Philosophical Society* 90 (2): 628–42. https://doi.org/10.1111/brv.12128.

Chapter 18, p. 354

Lockwood, Mike. 2009. "Solar change and climate: An update in the light of the current exceptional solar minimum." *Proceedings of the Royal Society A* 466 (2114). https://doi.org/10.1098/rspa.2009.0519.

Chapter 19, p. 367

Klowden, Marc J., and Arden O. Lea. 1978. "Blood meal size as a factor affecting continued host-seeking by *Aedes Aegypti* (L.)." *American Journal of Tropical Medicine and Hygiene* 27 (4): 827–31. https://doi.org/10.4269/ajtmh.1978.27.827.

Chapter 20, p. 399

Kruuk, Hans. 1972. "Surplus killing by carnivores." *Journal of Zoology* 166 (2): 233–44. https://doi.org/10.1111/j.1469-7998.1972.tb04087.x.

Chapter 21, p. 416

Tamburini, C., Miquel Canals, Xavier Durrieu de Madron, Loïc Houpert, Dominique Lefèvre, Séverine Martini, Fabrizio D'Ortenzio, et al. 2013. "Deep-sea bioluminescence blooms after dense water formation at the ocean surface." *PLoS One* 8 (7): e67523. https://doi.org/10.1371/journal.pone.0067523.

Chapter 22, p. 429

Ferreira-Poblete, A. 1997. "The probability of conception on different days of the cycle with respect to ovulation: An overview." *Advances in Contraception* 13 (2-3): 83–95. https://doi.org/10.1023/A:1006527232605.

Chapter 23, p. 449

Gwinup, Grant, Reg Chelvam, and Terry Steinberg. 1971. "Thickness of subcutaneous fat and activity of underlying muscles." *Annals of Internal Medicine* 74 (3): 408–11. https://10.7326/0003-4819-74-3-408.

Chapter 24, p. 468

Kienle, Alwin, Lothar Lilge, I. Alex Vitkin, Michael S. Patterson, Brian C. Wilson, Raimund Hibst, and Rudolf Steiner. 1996. "Why do veins appear blue? A new look at an old question." *Applied Optics* 35 (7): 1151–60. https://doi.org/10.1364/AO.35.001151.

Chapter 25, p. 496

Trombetta, Claudia Maria, Elena Gianchecchi, and Emanuele Montomoli. 2018. "Influenza vaccines: Evaluation of the safety profile." *Human Vaccines and Immunotherapeutics* 14 (3): 657–70. https://doi.org/10.1080/21645515.2017.1423153.

Chapter 26, p. 514

Van Eck, Joyce, Dwayne D. Kirk, and Amanda M. Walmsley. 2006. "Tomato (*Lycopersicum esculentum*)." In *Agrobacterium Protocols*, edited by K. Wang. *Methods in Molecular Biology*, vol. 343 (pp. 459–74). Humana Press. https://doi.org/10.1385/1-59745-130-4:459.

Credits

Text

Ch.1 Review Question 13, p. 18: Figure 1 from Greenspan, S.E., Lambertini, C., Carvalho, T. et al. Hybrids of amphibian chytrid show high virulence in native hosts. Sci Rep 8, 9600 (2018). https://doi.org/10.1038/s41598-018-27828-w. Under CC by 4.0 License.

Fig. 2.10, p. 31: Graph: "The real cause of increasing autism prevalence?" by J. Emory Parker. Reprinted by permission of the author.

Ch. 8 Review Question 10, p. 164: Figure 3 republished with permission of the Royal Society of Chemistry, from "Biomaterials at the interface of nano- and micro-scale vector-cellular interactions in genetic vaccine design," by Charles H. Jones, Anders P. Hakansson, Blaine A. Pfeifer, *Journal of Materials Chemistry B*, Issue 46. © 2014; permission conveyed through Copyright Clearance Center, Inc.

Ch. 9 Review Question 13, p. 184: Figure from "Public Views of Gene Editing for Babies Depend on How It Would Be Used," Pew Research Center, Washington, D.C. (June 2018). https://www.pewresearch.org/science/2018/07/26/public-views-of-gene-editing-for-babies-depend-on-how-it-would-be-used/.

Ch. 11 Review question 13, p. 223: "Total number of whales killed from 1945 to 2018, i.e. before and after the moratorium." Graph reprinted courtesy of OceanCare. OceanCare.org.

Ch. 12 Review Question 11, p. 242: Graph reprinted courtesy of Swiss Federal Office of Public Health (design) and www.anresis.ch (data).

Fig. 13.5, p. 248: Text reprinted courtesy of the New England Aquarium.

Ch. 13 Review Question 13, p. 263: Figure reprinted from Gallacher C (2019) Possible causes for the decline in Adélie Penguin population numbers at anvers island, Western antarctic Peninsula. Ann Mar Sci 3(1): 006-010. DOI: 10.17352/ams.000013. Under CC by 4.0 License.

Fig. 18.5, p. 347: Figure adapted from https://omg.jpl.nasa.gov/portal/about-en. Reprinted courtesy of NASA/JPL-Caltech.

Ch. 19 Review Question 14, p. 381: Graph: "Oxitec Approach Suppresses Mosquitoes in Piracicaba, Brazil." Appeared in "No GM Mosquitoes Didn't Start the Zika Outbreak," Discovermagazine.com, January 31, 2016. Reprinted by permission of Oxitec Ltd.

Ch. 20 Review Question 14, p. 403: Figure 3 from Sonja B. Otto et. al, Predator Diversity and Identity Drive Interaction Strength and Trophic Cascades in a Food Web, *Ecological Society of America*, Vol. 89, Issue 1, Jan. 2008, pp. 134-144. © Ecological Society of America. Reprinted by permission of John Wiley & Sons, Inc.

Ch. 21 Review Question 12, p. 420: Figure 1 from Kotta, J., Wernberg, T., Jänes, H. et al. Novel crab predator causes marine ecosystem regime shift. Sci Rep 8, 4956 (2018). https://doi.org/10.1038/s41598-018-23282-w. Under CC by 4.0 License.

Fig. 22.2, p. 425: Graph adapted from the World Bank. Under CC by 4.0 License. https://data.worldbank.org/indicator/sp.dyn.tfrt.in?end=2017&start=1960.

Fig. 23.1, p. 444: Graph adapted from the World Bank. Under CC by 4.0 License. https://data.worldbank.org/indicator/SP.DYN.LE00.IN.

Ch. 26 Review Question 14, p. 521: "Monarch Population in Millions," graph by Tierra Curry, Center for Biological Diversity, February 9, 2017. Reprinted by permission.

Photo

Front Matter

Page vii top: Jonathan Mays Wildlife Biologist Maine Department of Inland Fisheries and Wildlife/USFWS, bottom: BSIP SA/Alamy Stock Photo; **p. viii** top: Towfiqu Photography, bottom: (puzzle) George Diebold/ Getty Images, (cell) Mopic/ Shutterstock; **p. ix** top to bottom: Michael Melford/National Geographic, Tsuneo Yamashita/Getty Images, (room) Iaroslav Neliubov/Shutterstock, (dogs) Eric Isselee/Shutterstock; **p. x** top: FS Productions/Getty images, bottom: Mike Kemp/Rubberball/Getty Images; **p. xi** top to bottom: (background) shironosov/iStock/Getty Images Plus, (tobacco leaf) joannawnuk/Shutterstock, Andrey Nekrasov/Alamy, (background) muss/Shutterstock, (IV bag) Thinkstock/Getty Images, (VRSA) Science Source; **p. xii** top: David Tipling Photo Library/Alamy Stock Photo, bottom: Lou Linwei/ Alamy; **p. xiii** top: Elke Van de Velde/Photodisc/Getty Images Plus, center: JimmyWrangles/iStock/Getty Images Plus, bottom: (TV screen): thomaslenne/iStock/Getty Images Plus, (karyotype): Leonard Lessin/Science Source, (steel lab table): Piotr Adamowicz/iStock/Getty Images Plus, (scientist): LattaPictures/iStock/Getty Images Plus, (skulls): Sabena Jane Blackbird/Alamy; **p. xiv** top: Paul Souders/Getty Images, Paul Starosta/Corbis Documentary/Getty Images, Holly Kuchera/ Shutterstock, bottom: SCIEPRO/Science Photo Library/ Getty Images; **p. xv** top: SECCHI Disk, bottom: TFoxFoto/ Shutterstock; **p. xvi** top: poba/Getty Images, bottom: Vince Michaels/Getty Images; **p. xvii** top: Henny Boogert, bottom: Adam Gault/OJO Images/Getty Images.

Chapter 1

Pages 2–3 Jonathan Mays, Wildlife Biologist Maine Department of Inland Fisheries and Wildlife/USFWS; **p. 6**: Photo by Dr. Kimberli Miller courtesy USGS National Wildlife Health Center; **p. 7**: clockwise from top left: Photo courtesy Ryan von Linden. Photo used with permission from New York State Department of Environmental Conservation. All rights reserved., Kevin Wenner/Pennsylvania Game Commission, David S. Blehert, U.S. Geological Survey - National Wildlife Health Center; **p. 8**: Image Courtesy of The Advertising Archives; **p. 9** (all): Amy Smotherman Burgess/Knoxville News Sentinel/ZUMAPRESS.com; **p. 12** clockwise from top left: Photo used with permission from New York State Department of Environmental Conservation, Deborah Springer, Design Pics Inc/Alamy, Dr. Mary Hausbeck, Photo courtesy Ryan von Linden. Photo used with permission from New York State Department of Environmental Conservation. All rights reserved., Greg Turner/Pennsylvania Game Commission; **p. 14** clockwise from top left: Vitalii Hulai/Shutterstock, Valeriy Vladimirovich Kirsanov/Shutterstock, Valeriy Vladimirovich Kirsanov/Shutterstock, Rosa Jay/Shutterstock, nico99/ Shutterstock, ASA - Carlos Asanuma/Getty Images; **p. 17**: Lindsey Heffernan PA Game Commission/Patty Stevens USGS/USFWS.

Chapter 2

Pages 20–21: BSIP SA/Alamy Stock Photo; **p. 22**: Photo courtesy of the Brennan family; **p. 24** clockwise from top left: BSIP SA/Alamy Stock Photo, South West News Service, CDC, AP Photo, Science Source/Colorization by: Mary Madsen; **p. 26**: Facebook; **p. 29**: https://twitter.com/laura_jbrennan/ status/1016699474838720512; **p. 33**: Republished with permission of Elsevier Science and Technology Journals from Ileal-lymphoid-nodular hyperplasia non-specific colitis and pervasive developmental disorder in children. Wakefield AJ et al. Lancet v351(9103) 1998.; **p. 35**: Mara008/Shutterstock; **p. 36**: Courtesy of the Brennan family and the HSE Immunisation Office, Dublin.

Chapter 3

Page 40–41: Towfiqu Photography; **p. 42**: Edwin Chindongo; **p. 44**: Lawrence K. Ho/Los Angeles Times via Getty Images; **p. 48** clockwise from top right: Deep OV/Shutterstock, designelements/Shutterstock, Africa Studio/Shutterstock, Andrey Starostin/Shutterstock, iofoto/Shutterstock, Photo Melon/Shutterstock, antonprado/Panther Media GmbH/ Alamy Stock Photo; **p. 52** left to right: tryton2011/Shutterstock, Jacqui Hurst/Corbis Documentary via Getty Images, Abramova Elena/Shutterstock; **p. 54**: Photolukacs/Shutterstock, **p. 55**: Scott Wiseman/Scripps Research; **p. 58** top and bottom: Helen Sessions/Alamy; **p. 59**: majcot/Shutterstock.

Chapter 4

Pages 64–65: (puzzle) George Diebold/ Getty Images, (cell) Mopic/Shutterstock; **p. 66** clockwise from top left: A. Barry Dowsett/Science Source, Biophoto Associates/ Science Source Colorization by: Mary Martin, CNRI/ Science Source, EMBL/Johanna Höög, **p. 67** left to right: Judith Collins/Alamy, Scott Camazine/Science Source, AS Food studio/Shutterstock, Lev Kropotov/Shutterstock, Rosa Jay/Shutterstock; **p. 68**: Photos courtesy J. Craig Venter Institute. from Gibson DG et al. Creation of a bacterial cell controlled by a chemically synthesized genome. Science. 2010 Jul 2;329(5987):52-6. doi: 10.1126/science.1190719; **p. 69**: Roger Harris/Science Source; **p. 72**: Photo Researchers Inc./Composition by: Eric Cohen/Science Source; **p. 74** top to bottom: Don W. Fawcett/Science Source, SPL/ Science Source, Dennis Kunkel/Science Photo Library/ Science Source, Dennis Kunkel Microscopy/Science Photo Library/Science Source; **p. 76** left: Hybrid Medical Animation/Science Source, right: Biophoto Associates/Science Source; **p. 77**: Dennis Kunkel Microscopy/Science Photo Library/Science Source; **p. 78** top and bottom: Russell Kightley; **p. 79**: Russell Kightley; **p. 80**: top: Biophoto Associates/Science Source, bottom: Biophoto Associates/Science Source Colorization by: Mary Martin; **p. 81** top: Photo Researchers/Science History Images/Alamy Stock Photo, bottom: Dr. David Furness Keele University/Science Source; **p. 82**: Dr. Torsten Wittmann/Science Source; **p. 83**: Larry Downing/Reuters/Newscom; **p. 86**: Susumu Nishinaga/

Science Source; **p. 87** top: Dennis Kunkel Microscopy/ Science Source, bottom: David M. Phillips/Science Source.

Chapter 5

Pages 88–89: Michael Melford/National Geographic; **pp. 90, 94, 98**: Mike T. Booth; **p. 101** left: Ruslan Semichev/ Shutterstock, right: ostill/Shutterstock; **p. 103**: Dr. Karl Stetter; **p. 105** left to right: Ernie Janes/Alamy Stock Photo, Life on white/Alamy Stock Photo, Claire Bryant/Alamy Stock Photo.

Chapter 6

Pages 108–109: Tsuneo Yamashita/Getty Images; **p. 110** top: Steve Gschmeissner/Science Source, bottom: Martin Oeggerli/Science Source; **p. 114** left: Manfred Kage/Science Source, **pp. 114–115** top: Thomas Deerinck, NCMIR/Science Source; **p. 123**: David McNew/Getty Images; **p. 124**: Graphic design/Shutterstock.

Chapter 7

Pages 128–129: (room) Iaroslav Neliubov/Shutterstock, (dogs) Eric Isselee/ Shutterstock; **p. 130**: Lynda McFaul/ Shutterstock; **p. 132**: Amir Paz/PhotoStock-Israel/Alamy; **p. 136**: Tereza Huclova/Shutterstock; **p. 137**: clockwise from top left: GK Hart/Vikki Hart/Getty Images, NHGRI/Broad Institute, HelenaQueen/istock/Getty Images Plus, ESB Professional/Shutterstock, GlobalP/istock/Getty Images Plus; **p. 138** left to right: MPG/Getty Images, jinga80/istockGetty Images Plus, standby/istockphoto/Getty Images Plus; **p. 140**: adogslifephoto/istock/Getty Images Plus; **p. 141** left: Vasiliy Koval/ Dreamstime.com, right: Vincent J. Musi/Nat Geo Creative; **p. 143**: Sean Graff Photography; **p. 144** left: cynoclub/Shutterstock, right: William Mullins/Alamy Stock Photo; **p. 145**: left to right: AleksVF/iStock/Getty Images Plus, Irochka_T/iStock/Getty Images Plus, GlobalP/iStock/ Getty Images Plus, bottom: Klein-Hubert/KimballStock.

Chapter 8

Pages 146–147: FS Productions/Getty Images; **p. 148**: NIH Clinical Center; 149: Mickie Gelsinger; **p. 150**: CNRI/Science Source; **p. 151**: clockwise from top left: recep-bg/Getty Images, Mainardi, P. Cri du Chat Syndrome. Orphanet Journal of Rare Diseases 20061:33 DOI: 10.1186/1750-1172-1-33; licensee BioMed Central Ltd. 2006/CC by 2.0, Mid Essex Hospital Services NHS Trust/Science Photo Library via Science Source; **p. 157**: Baylor College of Medicine Photo Archives; **p. 163**: National Human Genome Research Institute.

Chapter 9

Pages 166–167: Mike Kemp/Rubberball/Getty Images; **p. 168**: Ashley Rasys and Hannah Schriever, "CRISPR-Cas9 Gene Editing in Lizards through Microinjection of Unfertilized Oocytes" in Cell Reports; **p. 170**: Professor Enzo Di Fabrizio, IIT/Science Source; **p. 177** left: Dr. Tony Brain/SPL/ Science Source, right: Meckes Ottawa/Science Source; **p. 178**: daniel san martin/Alamy Stock Photo; **p. 181**: Science Source.

Chapter 10

Pages 186–187: (background) shironosov/iStock/Getty Images Plus, (tobacco leaf) joannawnuk/Shutterstock; **p. 188**: Mathieu Belanger/Reuters/Newscom; **p. 189** left: Zbigniew Guzowski/Shutterstock, right: dra_schwartz/iStock/Getty Images Plus; **p. 190**: Dong yanjun/Imaginechina/AP Photo; **p. 191** top: Image Source/Getty Image, bottom: Blend Images/ REB Images/Getty Images; **p. 194**: CDC/ Alissa Eckert, MS; Dan Higgins, MAMS; **p. 200**: CDC/ Alissa Eckert, MS; Dan Higgins, MAMS.

Chapter 11

Pages 204–205: Andrey Nekrasov/Alamy; **p. 206**: top to bottom: Thewissen Lab NEOMED, Robert Bowman for the Custom House Maritime Museum New London Connecticut, Alton Dooley; **p. 207** clockwise from top left: David Tipling/Getty Images, Steve Bloom Images/Alamy, Kerstin Meyer/Getty Images, Volodymyr Burdiak/Alamy Stock Photo, digitalskillet/Getty Images, Image Source/Alamy, Sergey Uryadnikov/Alamy Stock Photo, Momatiuk-Eastcott/ Corbis via Getty Images, Konrad Wothe/imageBROKER/ Alamy, Rudmer Zwerver/Alamy Stock Photo; **p. 209**: Wilfred Marissen/Shutterstock; **p. 210** clockwise from top left: O. Louis Mazzatenta/National Geographic Creative, James L. Amos/National Geographic Creative, Scott Camazine/ Science Source, WitGorski/Shutterstock, Francois Gohier/ Science Source, Francois Gohier/Science Source; **p. 212**: (inset) Thewissen Lab NEOMED; **p. 213** clockwise from top left: Reprinted by permission from Macmillan Publishers Ltd: Nature 450:1190-1194 Thewissen J.G.M. et. al. Whales originated from aquatic artiodactyls in the Eocene epoch of India. ©2007, Trinacria photo/Shutterstock, Shem Compion/ Getty Images, The Natural History Museum, London/ Science Source; **p. 214**: Brian Lasenby/Shutterstock; **p. 216**: The Icelandic Museum of Natural History and The Natural History Museum of Denmark; **p. 218**: Paulo Oliveira/Alamy Stock Photo.

Chapter 12

Pages 224–225: (background) muss/Shutterstock, (IV bag) Thinkstock/Getty Images, (VRSA) Science Source; **p. 226** top: Dennis Kunkel Microscopy/ Science Source, bottom: Biophoto Associates/Science Source; **p. 227** left: Dr. David Armstrong SALSA Technologies LLC, right: Reprinted from Chakraborty et. al. In vitro antimicrobial activity of nanoconjugated vancomycin against drug resistant Staphylococcus aureus. International Journal of Pharmaceuticals Volume 436 Issues 1-2 15 October 2012 Pages 659-676 with permission from Elsevier and Dr. Somenath Roy; **p. 231** top and bottom: Breck P. Kent/Animals Animals-Earth Scenes; **p. 232**: Courtesy of Tom Smith, UCLA; **p. 233** clockwise from top left: Richard Wong/Alamy Stock Photo, Christian Musat/Alamy Stock Photo, Flip Nicklin/Minden Pictures/National Geographic Images, George Grall/National Geographic Image Collection; **p. 234**: Science Source; **p. 236** left: F.J. Hiersche/Science Source, right: Merlin D. Tuttle/Bat Conservation Int'l/Science Source; **p. 237** top to bottom: Roger F. Thoma; **p. 238** top and bottom left: Courtesy of JoGayle Howard Department of Reproductive Sciences Smithsonian's National Zoological Park, center left: Thomas & Pat Leeson/Science Source, top right: Swaziland Digital Archives, bottom right: Staphane De Sakutin/AFP via Getty Images; **p. 239**: JGI/Jamie Grill/Blend Images/Corbis; **p. 240**: Sergiy Kuzmin/Shutterstock.

Chapter 13

Pages 224–225: David Tipling Photo Library/Alamy Stock Photo; **p. 246** left: Stuart Holroyd/Alamy Stock Photo, right: Photo by Christopher Michel; **p. 247** left: Rich Lindie/Shutterstock, right: Frans Lanting/MINT Images/Science Source; **p. 249**: Roberta Olenick/All Canada Photos/Alamy Stock Photo; **p. 250**: Photo by Marc C. Perkins; **p. 251** top: Juniors Bildarchiv GmbH/Alamy, bottom: Peter Clayton Photography/Getty Images, right: Chris Long; **p. 252** left to right: Thomas & Pat Leeson/Science Source, Larry Geddis/Alamy, Gerald C. Kelley/Science Source; **p. 254**: Anthony Mercieca/Science Source; **p. 255** top to bottom: ©B.R. Grant, ©K.T. Grant, ©P.R. Grant; **p. 256**: Sohns/imageBROKER/Alamy; **p. 257** top to bottom: Eric Isselee/Shutterstock, Yuri Kevhiev/Alamy Stock Photo, Coprid/Shutterstock; **p. 260** clockwise from top left: Martina_L/Shutterstock. Eric Isselee/Shutterstock, Blinka/Shutterstock, IrinaK/Shutterstock, Mike Truchon/Shutterstock, BOONROONG/Shutterstock, oksana2010/Shutterstock, Boule/Shutterstock, Sakdinon Kadchiangsaen/Shutterstock, StockPhotoAstur/Shutterstock; **p. 262** top left: Robert Sivinski, right: Michael Quinn/NPS, bottom: gallimaufry/Shutterstock.

Chapter 14

Pages 264–265: Lou Linwei/Alamy; **p. 266**: The Natural History Museum/Alamy Stock Photo; **p. 267** top to bottom: MichaelTaylor3d/Shutterstock, Eye of Science/Science Source, blickwinkel/Fox/Alamy Stock Photo; **p. 269**: Royal Saskatchewan Museum (RSM/R.C. McKellar); **p. 271**: Image courtesy of Dr. Xu Xing. Reprinted by permission from Macmillan Publishers Ltd: Nature 475:465-470 Xing X. et. al. An Archaeopteryx-like theropod from China and the origin of Avialae. ©2011; **p. 273** left to right: Lebendkulturen.de/Shutterstock, Roland Birke/Getty Images, Rainer Fuhrmann/Shutterstock; **p. 274**: (modern human) Hans Kim/Shutterstock, (gorilla) Image Source/Alamy Stock Photo, (chimpanzee) Eric Isselee/Shutterstock, (spider monkey) Wegner/Arco Images/Alamy, (lemur) Bernd Rohrschneider/FLPA/Science Source, (bat) Valeriy Vladimirovich Kirsanov/Shutterstock, (kangaroo) FLPA/Alamy, (elephant) AfriPics.com/Alamy Stock Photo, (frog) Gary Meszaros/Science Source, (goldfish) Irochka_T/iStock/Getty Images Plus, (bird) Anthony Mercieca/Science Source, (lizard) James DeBoer/Shutterstock, (nematode) Science Source (Colorization by: Eric Cohen), (starfish) BSIP/Science Source, (octopus) Georges Antoni/Hemis/Alamy, (butterfly) Gary Meszaros/Science Source, (jellyfish) Picture Partners/Science Source, (algae) Ray Simons/Science Source; **p. 276** top to bottom: Andrew Syred/Science Source, Fletcher & Baylis/Science Source, Ingeborg Knol/imageBROKER/Corbis; **p. 277** top to bottom: Chris Mattison/FLPA/Science Source; Robert and Jean Pollock/Science Source, Dennis Flaherty/Science Source, Schmitz Olaf/Getty Images; **p. 278** clockwise from top left: WaterFrame/Alamy Stock Photo, Stuart Wilson/Science Source, Andrew J. Martinez/Science Source, Martin Shields/Alamy, Marius Dobilas/Shutterstock, Andrew J. Martinez/Science Source; **p. 279** left and right: The Royal Belgian Institute of Natural Sciences Brussels; **p. 280** top to bottom: Emily Willoughby/Stocktrek Images/Alamy, Masato Hattori, Natural History Museum London, Masato Hattori, Laurie E Wilson/Shutterstock.

Chapter 15

Pages 286–287: Elke Van de Velde/Photodisc/Getty Images Plus; **p. 288**: image courtesy Allen Nutman. University of Wollongong. A.P. Nutman et al. Rapid emergence of life shown by discovery of 3 700-million-year-old microbial structures. Nature. Published online August 31 2016. doi: 10.1038/nature19355; **p. 290** clockwise from top: NC State University, Photograph courtesy of Gardenista, Jim Davis/The Boston Globe via Getty Images, Gillian Conquest, Extreme Citizen Science (ExCiteS) research group University College London; **p. 291**: Dunn Lab; **p. 292**: clockwise from top left: Eye of Science/Science Source, Scimat/Science Source, Michael Abbey/Science Source, Eye of Science/Science Source, David McCarthy/Science Source; **p. 293** clockwise from top left: David Henderson/caia image/Alamy Stock Photo, Eye of Science/Science Source, image courtesy Luis R. Comolli. Comolli L. R; Baker B; Downing K.H; Siegerist C. E; and Banfield J. (2008). Three-dimensional analysis of the structure and ecology of a novel ultra-small archaeon. ISME J 3(2) 159-167, MichaelTaylor3d/ Shutterstock, IFE URI-IAO UW

Lost City Science Party; NOAA/OAR/OER; The Lost City 2005 Expedition; **p. 293**: Jose Arcos Aguilar/Shutterstock; **p. 296** clockwise from top left: Dr. Ron Dengler, Dennis Kunkel Microscopy/Science Source, CDC/James Archer Illustrators: Alissa Eckert and Jennifer Oosthuizen, Eye of Science/Science Source; **p. 297**: Universal History Archive/REX/Shutterstock; **p. 298** clockwise from top left: Dennis Kunkel Microscopy/Science Source, Dennis Kunkel Microscopy/Science Source, Dennis Kunkel Microscopy/Science Source, (background) Michael Heim/EyeEm/Getty Images; **p. 299**: photo by Lauren Nichols Your Wild Life/Dunn Lab.

Chapter 16

Pages 302–303: JimmyWrangles/iStock/Getty Images Plus; **p. 304**: Jacob Phelps; **p. 305**: Thomas Deerinck, NCMIR/Science Source; **p. 307** left: Plant-Success, right: NPS/Jacob W. Frank; **p. 308**: Leon Neal/AFP/Getty Images; **p. 310** left: Michael Patrick O'Neill/Science Source, right: Casther/Shutterstock; **p. 311**: Lance Cheung/USDA; **p. 312** top: Carroll & Carroll/AgStock/Design Pics Inc/Alamy Stock Photo, bottom: Shawn Poynter/The New York Times/Redux; **p. 313** left and right: Stefano Rellandini/Reuters/Newscom; **p. 314**: Laszlo Podor/Moment Open/Getty Images; **p. 315**: clockwise from left: Le Do/Shutterstock, Suzifoo/iStock/Getty Images Plus, Egor Rodynchenko/Shutterstock, Alexander Ruiz/Alamy Stock Photo, Vitaly Korovin/Shutterstock, tarapong srichaiyos/Shutterstock, kzww/Shutterstock, xpixcl/Shutterstock, Amazing snapshot/Shutterstock; **p. 316**: Glass and Nature/Shutterstock.

Chapter 17

Pages 320–321 (TV screen): thomaslenne/iStock/Getty Images Plus, (karyotype): Leonard Lessin/Science Source, (steel lab table): Piotr Adamowicz/iStock/Getty Images Plus, (scientist): LattaPictures/iStock/Getty Images Plus, (skulls): Sabena Jane Blackbird/Alamy; **p. 322**: Mandible of Homo sp. from Mala Balanica Serbia. Photo by Dr. Mirjana Roksandic; **p. 327** left to right: Steve Bloom Images/Alamy, John W Banagan/Getty Images, Dave Watts/Nature Picture Library/Alamy Stock Photo; **p. 331**: John Reader/Science Source; **p. 338**: Sabena Jane Blackbird/Alamy, **p. 339**: Mark Thiessen/National Geographic Creative.

Chapter 18

Pages 342–343: Paul Souders/Getty Images; **p. 344**: left and right: NASA; **p. 345**: chbaum/Shutterstock; **p. 346** clockwise from top: Greenpeace/Ges.oek.Forschung, John Sommers II/Reuters/Newscom, AP Photo/Nick Ut, James Steinberg/Science Source, David Greedy/Getty Images; **p. 358**: photo courtesy of Dr. Sarah Das; **p. 359** top: nikiteev_konstantin/Shutterstock, bottom: NASA Goddard Institute for Space Studies; **p. 360**: Steffen M. Olsen and the Danish Meteorological Institute.

Chapter 19

Pages 364–365: Paul Starosta/Corbis Documentary/Getty Images; **p. 367**: Estefan Radovicz/Agencia o Dia/AGENCIA ESTADO/Xinhua/Alamy Live News; **p. 368**: CDC; **p. 370**: CDC/Cynthia Goldsmith; **p. 371**: 7th Son Studio/Shutterstock; **p. 373**: Dyakova Yulia/Shutterstock; **p. 374**: Mark Newman/Tom Stack & Assoc.; **p. 375**: Tom Brakefield/Getty Images; **p. 376**: Alan G. Nelson/Dembinsky Photo Associates; **p. 377**: Angel Valentin/The New York Times/Redux; **p. 378**: (all silhouettes) nikiteev_konstantin/Shutterstock, (mosquito) Antagain/E+/Getty Images Plus.

Chapter 20

Pages 382–383: Holly Kuchera/Shutterstock; **p. 384**: Pat & Chuck Blackley /Alamy; **p. 387**: Don Johnston_IH/Alamy Stock Photo, inset: Kevin Ebi/Alamy Stock Photo; **p. 389** left: Martin Harvey/Corbis via Getty Images, right: Bill Lawson/Shutterstock; **p. 390**: Daniel Stahler/NPS; **p. 391** left: Helmut Corneli/imageBROKER/Alamy Stock Photo, right: David Fleetham/Alamy Stock Photo; **p. 392**: PhotoStock-Israel/Alamy; **p. 393** clockwise from top left: Vertyr/Shutterstock, rachisan alexandra/Shutterstock, KatarinaF/Shutterstock, Ksanawo/Shutterstock; **p. 395**: clockwise from top: Daniel Borzynski/Alamy Stock Photo, CHAINFOTO24/Shutterstock, Premaphotos/Alamy Stock Photo, Nancy Nehring/Getty Images; **p. 396**: Norbert Rosing/National Geographic Creative; **p. 397**: Anne-Marie Kalus; **p. 398**: Adam Burton/Alamy; **p. 399** left to right: Stan Osolinski/Dembinsky Photo Associates, Howard Garrett/ Dembinsky Photo Associates, Michael P. Gadomski/Science Source.

Chapter 21

Pages 404–405: (disk) SECCHI Disk, (water) TFoxFoto/Shutterstock; **p. 406**: Gary Corbett/Alamy; **p. 407** left: Jacques Descloitres MODIS Rapid Response Team NASA/GSFC, right: D.P. Wilson/FLPA/Science Source; **p. 411**: NASA image by Jesse Allen & Robert Simmon; **p. 412**: Susan Biddle/for the Washington Post/Getty Images; **p. 413** top to bottom: Bernd Zoller/imagebroker/Alamy, John E Marriott/Alamy, Michael P. Gadomski/Earth Sciences/Animals Animals, Mark De Fraeye /SPL /Science Source, meganopierson/Shutterstock; **p. 414** top to bottom: Willard Clay/Dembinsky Photo Associates, Pixtal Images/Media Bakery, Design Pics Inc/Photolibrary/Getty Images, Mark Goodreau/Alamy, Jim Zipp/National Audubon Society Collection/Science Source.

Chapter 22

Pages 422–423: SCIEPRO/Science Photo Library/Getty Images; **p. 424:** Spies 2015; **p. 428** clockwise from top left: Werayuth Tes/Shutterstock, Motofish Images The Image Bank/Getty Images, Bele Olmez/imageBROKER/Alamy Stock Photo, Boligolov Andrew/Shutterstock; **p. 429** left to right: Reinhard Dirscherl/agefotostock, Matteo photos/Shutterstock, cbpix/Alamy; **p. 430** left: Clouds Hill Imaging Ltd/Getty Images, right: Dennis Kunkel Microscopy/Science Source; **p. 436:** left to right: MedicalRF.com/Corbis, Joo Lee/Corbis Documentary via Getty Images, Anatomical Travelogue/Science Source.

Chapter 23

Pages 422–423: poba/Getty Images; **p. 445:** Stockbyte/Getty Images; **p. 451:** James L. Amos/Science Source; **p. 456** clockwise from top left: Gtranquillity/Shutterstock, Juliya Shangarey/Shutterstock, Media Bakery, Shutterstock, D7INAMI7S/Shutterstock, Nik Merkulov/Shutterstock; **p. 458:** clockwise from top left: jeehyun/Shutterstock, Maks Narodenko/Shutterstock, Petr Malyshev/Shutterstock, Nattika/Shutterstock, Viktor1/Shutterstock, Binh Thanh Bui/Shutterstock, Levent Konuk/Shutterstock(bottle) HeinzTeh/Shutterstock.

Chapter 24

Pages 462–463: Vince Michaels/Getty Images; **p. 464:** Shawn Rocco/Duke Medicine; **p. 466:** top to bottom: Dennis Kunkel/Science Source, Dennis Kunkel/Science Source, Dr. Gopal Murti/Science Photo Library/Alamy Stock Photo; **p. 472:** The Ott Lab/Massachusetts General Hospital Boston MA; **p. 474** left and right: The Ott Lab/Massachusetts General Hospital Boston MA; **p. 476:** Alfred Pasieka/Science Source; **p. 480:** Scott Lewis Media.

Chapter 25

Pages 486–487: Henny Boogert; **p. 488** top: Matthijs Kox PhD, bottom: Henny Boogert; **p. 496** left: Eye of Science/Science Source, right: Science Photo Library/Science Source; **p. 498:** Volker Steger/Science Source; **p. 500:** Henny Boogert.

Chapter 26

Pages 504–505: Adam Gault/OJO Images/Getty Images; **p. 506:** courtesy of The Land Institute; **p. 512** left: Dragomir Radovanovic/Shutterstock, right: Photograph by Kathleen Bauer GoodStuffNW.com; **p. 513:** courtesy of The Land Institute; **p. 516** left to right: Vasilius/Shutterstock, bergamont/Shutterstock, natu/Shutterstock, Madlen/Shutterstock; **p. 518** left to right: Stephen Ausmus/USDA ARS, Anthony Mercieca/Science Source, courtesy of The Land Institute; **p. 519:** courtesy of The Land Institute.

Index

Note: Page numbers in *italic* refer to figures and tables.

E

N

Q

R

T

U

V

W